MECHANISM AND THEORY IN
ORGANIC CHEMISTRY

MECHANISM and THEORY in ORGANIC CHEMISTRY

Second Edition

Thomas H. Lowry
Smith College

Kathleen Schueller Richardson
Capital University

HARPER & ROW, PUBLISHERS, New York
Cambridge, Hagerstown, Philadelphia, San Francisco,
London, Mexico City, São Paulo, Sydney

1817

To Nancy and Frank

Sponsoring Editor: Malvina Wasserman
Project Editor: Eleanor Castellano
Designer: T. R. Funderburk
Senior Production Manager: Kewal K. Sharma
Compositor: York Graphic Services, Inc.
Printer and Binder: R. R. Donnelley & Sons, Inc.
Art Studio: Vantage Art, Inc.

MECHANISM AND THEORY IN ORGANIC CHEMISTRY, Second Edition

Copyright © 1981 by Thomas H. Lowry and Kathleen Schueller Richardson

Library of Congress Cataloging in Publication Data

Lowry, Thomas H
   Mechanism and theory in organic chemistry.

   Includes bibliographical references and index.
   1. Chemistry, Physical organic.   I. Richardson,
Kathleen Schueller, joint author.   II. Title.
QD476.L68   1981   547.1′3   80-25474
ISBN 0-06-044083-X

CONTENTS

PREFACE

This book is intended as a text for undergraduate and first-year graduate students who have completed a one-year course in organic chemistry. We hope that it provides a structure that will help the student organize and interrelate the factual information obtained in the earlier course and that it serves as a basis for study in greater depth of individual organic reactions and of methods by which chemists obtain information about chemical processes.

The primary focus of the book is on reaction mechanisms, not only because knowledge of mechanism is essential to understanding chemical processes but also because theories about reaction mechanisms can explain diverse chemical phenomena in terms of a relatively small number of general principles. It is this latter capability of mechanistic theory that makes it important as an organizing device for the subject of organic chemistry as a whole.

In treating mechanisms of the important classes of organic reactions, we have tried to emphasize the experimental evidence upon which mechanistic ideas are built and to point out areas of uncertainty and controversy where more work still needs to be done. In this way we hope to avoid giving the impression that all organic mechanisms are well understood and completely agreed upon but instead to convey the idea that the field is a dynamic one, still very much alive and filled with surprises, excitement, and knotty problems.

The organization of the book is traditional. We have, however, been selective in our choice of topics in order to be able to devote a significant portion of the book to the pericyclic reaction theory and its applications and to include a chapter on photochemistry.

The advent and subsequent development of the pericyclic theory have made it necessary for the mechanistic organic chemist to understand covalent bonding theory in greater detail than was required earlier. The discussion of frontier orbitals and of applications of symmetry that appeared in Chapter 10 of the first edition have therefore been incorporated in Chapter 1. The treatment remains qualitative, but the Appendix to Chapter 1 covers the Hückel molecular orbital method in detail.

Other significant changes in treatment from the first edition include an expanded review of stereochemistry in Chapter 2, a unified treatment of aliphatic nucleophilic substitution reactions in Chapter 4, an expanded discussion of carbanion chemistry in Chapter 6, a reorganized treatment of pericyclic reactions in Chapter 10, and discussion of potential energy surface crossings in connection with photochemical reactions in Chapter 12. Other material has been updated.

Problems of varying difficulty are included at the ends of the chapters. Some problems illustrate points discussed in the text, but others are meant to extend the text by leading the student to investigate reactions, or even whole categories of reactions, which we have had to omit because of limitations of space. References to review articles and to original literature are given for most problems. Problems that represent significant extensions of the text are included in the index.

The book is extensively footnoted. It is neither possible nor desirable in a book of this kind to present exhaustive reviews of the topics taken up, and we have made no effort to give complete references. We have tried to include references to review articles and monographs wherever recent ones are available, to provide key references to the original literature for the ideas discussed, and to give sources for all factual information presented. The text also contains numerous cross references.

The amount of material included is sufficient for a full-year course. For a one-semester course, after review of the first two chapters, material may be chosen to emphasize heterolytic reactions (Chapters 3–8), to cover a broader range including radicals and photochemistry (selections from Chapters 3–8 plus 9 and 12), or to focus primarily on pericyclic reactions and photochemistry (Chapters 10–12). In selecting material for a one-semester course, the following sections should be considered for possible omission: 1.9, 3.5, 5.3, 6.2, 6.3, 7.3, 7.5, 8.3, 9.6, 10.5, 10.6.

We would like to thank the following people who reviewed parts of the manuscript for the first edition: Professors D. E. Applequist, C. W. Beck, J. C. Gilbert, R. W. Holder, W. P. Jencks, J. R. Keeffe, C. Levin, F. B. Mallory, D. R. McKelvey, N. A. Porter, P. v. R. Schleyer, and T. T. Tidwell. For this second edition we would like to thank, for reviewing parts of the manuscript and for providing us with very helpful suggestions, formulations, and discussion, Professors B. K. Carpenter, W. P. Jencks, M. S. Platz, T. T. Tidwell, J. Swenton, and P. G. Gassman. We are grateful to Professor James M. Henle for making available the computer program used in preparing Figures 2.15, 2.17, 2.18, 4.3, 4.4, 7.5, 8.3, 8.5, and 11.4. We owe special thanks to the editors, Carol Dempster and Malvina Wasserman, and the Harper & Row staff for their help and encouragement.

THOMAS H. LOWRY
KATHLEEN SCHUELLER RICHARDSON

MECHANISM AND THEORY IN
ORGANIC CHEMISTRY

Chapter 1
THE COVALENT
BOND

Because the covalent bond is of central importance to organic chemistry, we begin with a review of bonding theory and of aromaticity. Starting with Section 1.6, ideas about covalent bonding are expanded to include symmetry and the perturbation theory of chemical interactions.

1.1 THE ELECTRON-PAIR BOND—
LEWIS STRUCTURES

The familiar Lewis structure is the simplest bonding model in common use in organic chemistry. It is based on the idea that, at the simplest level, the ionic bonding force arises from the electrostatic attraction between ions of opposite charge, and the covalent bonding force arises from sharing of electron pairs between atoms.

The starting point for the Lewis structure is a notation for an atom and its valence electrons. The element symbol represents the core, that is, the nucleus and all the inner-shell electrons. The core carries a number of positive charges equal to the number of valence electrons. This positive charge is called the *core charge*. Valence electrons are shown explicitly. For elements in the third and later rows of the periodic table, the *d* electrons in atoms of Main Groups III, IV, V, VI, and VII are counted as part of the core. Thus

$$:\overset{..}{\underset{.}{Br}}: \qquad :\overset{..}{Se}: \qquad :\overset{..}{\underset{.}{I}}:$$

1

Ions are obtained by adding or removing electrons. The charge on an ion is given by

$$\text{charge} = (\text{core charge}) - (\text{number of electrons shown explicitly})$$

An ionic compound is indicated by writing the Lewis structures for the two ions.

A covalent bond model is constructed by allowing atoms to share pairs of electrons. Ordinarily, a shared pair is designated by a line:

$$\text{H—H}$$

All valence electrons of all atoms in the structure must be shown explicitly. Those electrons not in shared covalent bonds are indicated as dots, for example:

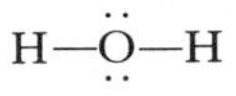

$$\text{H—}\overset{..}{\underset{..}{\text{O}}}\text{—H}$$

If an ion contains two or more atoms covalently bonded to each other, the total charge on the ion must equal the total core charge less the total number of electrons, shared and unshared:

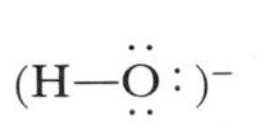

$$(\text{H—}\overset{..}{\underset{..}{\text{O}}}:)^-$$

$$
\begin{aligned}
\text{H core} &= +1 \\
\text{O core} &= \underline{+6} \\
\text{total core} &= +7 \\
\text{number of electrons} &= \underline{-8} \\
\text{total charge} &= -1
\end{aligned}
$$

In order to write correct Lewis structures, two more concepts are needed. The first is *valence-shell occupancy,* which is the total number of noncore electrons in the immediate neighborhood of each atom. To find valence-shell occupancy, all unshared electrons around the atom and all electrons in bonds leading to the atom must be counted. The valence-shell occupancy must not exceed 2 for hydrogen and must not exceed 8 for atoms of the first row (Li–F) of the periodic table. For elements of the second and later rows, the valence-shell occupancy may exceed 8. The structures

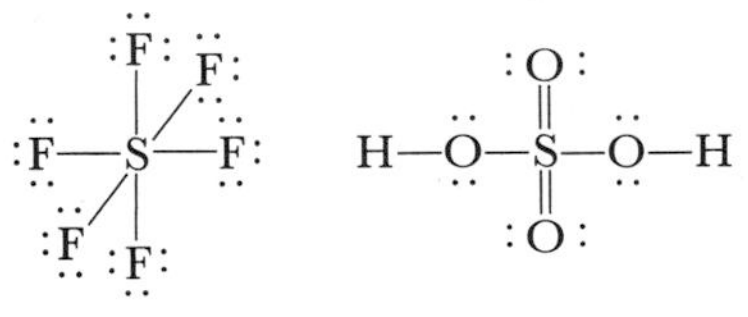

are acceptable.

The second idea is that of formal charge. For purposes of determining formal charge, partition all the electrons into groups as follows. Assign to each atom all of its unshared electrons and half of all electrons in bonds leading to it. Call the number of electrons assigned to the atom by this process its *electron ownership.* The formal charge of each atom is then given by

$$\text{formal charge} = (\text{core charge}) - (\text{electron ownership})$$

To illustrate formal charge, consider the hydroxide ion, H—$\overset{..}{\underset{..}{O}}$: $^-$. The electron ownership of H is 1, its core charge is $+1$, and its formal charge is therefore zero. The electron ownership of oxygen is 7, and the core charge is $+6$; therefore the formal charge is -1. All nonzero formal charges must be shown explicitly in the structure. The reader should verify the formal charges shown in the following examples:

$$H-\overset{\displaystyle H}{\underset{\displaystyle H}{\overset{|}{\underset{|}{N}}}}\overset{+}{-}\overset{\displaystyle H}{\underset{\displaystyle H}{\overset{|}{\underset{|}{B}}}}\overset{-}{-}H \qquad H-\overset{..}{N}=\overset{+}{N}=\overset{..}{N} : ^- \qquad \overset{\displaystyle \overset{..}{N}}{\underset{..}{O}} \underset{..}{O} : ^-$$

The algebraic sum of all formal charges in a structure is equal to the total charge.

Formal charge is primarily useful as a bookkeeping device for electrons, but it also gives a rough guide to the charge distribution within a molecule.

The following rules summarize the procedure for writing Lewis structures.

1. Count the total number of valence electrons contributed by the electrically neutral atoms. If the species being considered is an ion, add one electron to the total for each negative charge; subtract one for each positive charge.

2. Write the core symbols for the atoms and fill in the number of electrons determined in Step 1. The electrons should be added so as to make the valence-shell occupancy of hydrogen 2 and the valence-shell occupancy of other atoms not less than 8 wherever possible.

3. Valence-shell occupancy must not exceed 2 for hydrogen and 8 for a first-row atom; for a second-row atom it may be 10 or 12.

4. Maximize the number of bonds, and minimize the number of unpaired electrons, always taking care not to violate Rule 3.

5. Find the formal charge on each atom.

We shall illustrate the procedure with two examples.

Example 1. NO_2

STEP 1   17 valence electrons, 0 charge = 17 electrons.

STEP 2 $\overset{..}{\underset{..}{O}}=\overset{.}{N}-\overset{..}{\underset{..}{O}}$:

(Formation of another bond, $\overset{..}{\underset{..}{O}}=\overset{.}{N}=\overset{..}{\underset{..}{O}}$, would give nitrogen valence-shell occupancy 9.)

STEP 3   Formal charge:

Left O	Ownership 6	0 charge
Right O	Ownership 7	-1 charge
N	Ownership 4	$+1$ charge

Correct Lewis Structure:

$$\overset{..}{\underset{..}{O}}=\overset{+}{\underset{\pm}{N}}\overset{..}{\underset{..}{O}} : ^- \qquad (\text{or} \quad ^-: \overset{..}{\underset{..}{O}}-\overset{\pm}{N}\overset{..}{\underset{..}{O}}, \quad \text{see p. 6})$$

Example 2. $CO_3{}^{2-}$ Ion

STEP 1   22 valence electrons, $+2$ electrons for charge, $=24$ electrons.

STEP 2

$$:\overset{\cdot\cdot}{\underset{}{O}}:$$
$$:\overset{\cdot\cdot}{\underset{\cdot\cdot}{O}}-\overset{|}{C}=\overset{\cdot\cdot}{O}$$

(More bonds to C would exceed its valence-shell limit.)

STEP 3   Formal charge:

$:\overset{\cdot\cdot}{\underset{\cdot\cdot}{O}}{-}$	Ownership 7	-1 charge
$:\overset{\cdot\cdot}{\underset{\cdot\cdot}{O}}{-}$	Ownership 7	-1 charge
$=\overset{\cdot\cdot}{\underset{\cdot\cdot}{O}}$	Ownership 6	0 charge
C	Ownership 4	0 charge

Correct Lewis Structure:

$$^{-}:\overset{:\overset{\cdot\cdot}{O}:^{-}}{\underset{}{O}}-C=\overset{\cdot\cdot}{\underset{\cdot\cdot}{O}} \quad \text{(or} \quad ^{-}:\overset{:\overset{\cdot\cdot}{O}:}{\underset{\cdot\cdot}{O}}-\overset{\|}{C}-\overset{\cdot\cdot}{\underset{\cdot\cdot}{O}}:^{-} \quad \text{or} \quad \overset{\cdot\cdot}{\underset{\cdot\cdot}{O}}=C-\overset{:\overset{\cdot\cdot}{O}:^{-}}{\underset{\cdot\cdot}{O}}:^{-}, \quad \text{see p. 6)}$$

Resonance

The Lewis structure notation is useful because it conveys the essential qualitative information about properties of chemical compounds. The main features of the chemical properties of the groups that make up organic molecules,

$$H-\overset{|}{\underset{|}{C}}-H \qquad -\overset{\overset{\textstyle H}{|}}{\underset{\textstyle H}{C}}-H \qquad \overset{\diagdown}{\diagup}C=\overset{\cdot\cdot}{\underset{\cdot\cdot}{O}} \qquad -\overset{|}{\underset{|}{C}}-\overset{\cdot\cdot}{\underset{\cdot\cdot}{O}}-H$$

and so forth, are to a first approximation constant from molecule to molecule, and one can therefore tell immediately from the Lewis structure of a substance that one has never encountered before roughly what the chemical properties will be.

There is a class of structures, however, for which the properties are not those expected from the Lewis structure. A familiar example is benzene, for which the heat of hydrogenation (Equation 1.1) is less exothermic by about 36 kcal mole^{-1} than one would have expected from Lewis structure **1** on the basis of the measured

$$\text{(benzene)} + 3H_2 \longrightarrow \text{(cyclohexane)} \tag{1.1}$$

1

heat of hydrogenation of cyclohexene. The thermochemical properties of various types of bonds are in most instances transferable with good accuracy from molecule to molecule; a discrepancy of this magnitude therefore requires a fundamental modification of the bonding model.

The difficulty with model **1** for benzene is that there is another Lewis structure (**2**) that is identical to **1** except for the placement of the double bonds.

2

Whenever there are two alternative Lewis structures, one alone will be an inaccurate representation of the molecular structure. A more accurate picture will be obtained by the superposition of the two structures into a new model, which for benzene is indicated by **3**. The superposition of two or more Lewis structures into a composite picture is called *resonance*.

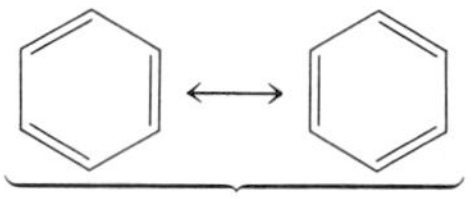

3

This terminology is well established but unfortunate, because the term resonance when applied to a pair of pictures tends to convey the idea of a changing back and forth with time. It is therefore difficult to avoid the pitfall of thinking of the benzene molecule as a structure with three conventional double bonds, of the ethylene type, jumping rapidly back and forth from one location to another. This idea is incorrect. The electrons in the molecule move in a field of force created by the six carbon and six hydrogen nuclei arranged around a regular hexagon (**4**). Each of the six sides of the hexagon is entirely equivalent to each other side; there is no reason why electrons should, even momentarily, seek out three sides and make them different from the other three, as the two alternative pictures **3** seem to imply that they do.

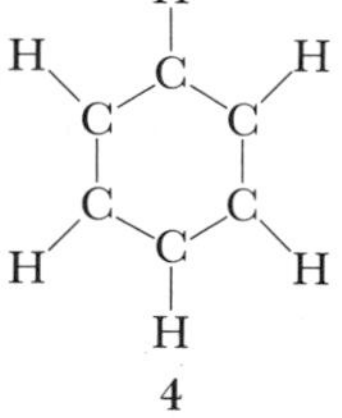

4

The symmetry of the ring of nuclei (**4**) is called a sixfold symmetry because rotating the picture by one-sixth of a circle will give the identical picture again. This sixfold symmetry must be reflected in the electron distribution. An alternative picture would be **5**, in which the circle in the middle of the ring is meant to imply a distribution of the six double-bond electrons of the same symmetry as the arrangement of nuclei. In this book we shall use the notation **2**, or the abbreviation ϕ—, for the phenyl ring. The resonance **3** will always be understood. Circles within rings will indicate delocalized ions.

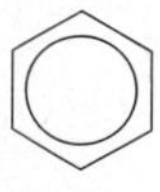

5

The most important features of substances for which the resonance concept is needed are two. First, the molecule has a different energy content from what we would have expected by looking at one of the individual structures. (The actual energy content is nearly always lower than predicted; that is, the molecule is more stable than a single structure indicates. The exceptions to this generalization are the antiaromatic compounds, discussed in Section 1.5.) Second, the actual distribution of electrons in the molecule is different from what we would expect on the basis of one of the structures. Since the composite picture shows that certain electrons are free to move over a larger area of the molecule than a single one of the structures implies, resonance is often referred to as *delocalization*.

While the benzene ring is perhaps the most familiar example of the necessity for modifying the Lewis structure language by the addition of the resonance concept, there are many others. The carboxylic acids, for example, are much stronger acids than the alcohols; this difference must be due largely to greater stability of the carboxylate ion (**6**) over the alkoxide ion (**7**). It is the possibility of writing two equivalent Lewis structures for the carboxylate ion that alerts us to this difference.

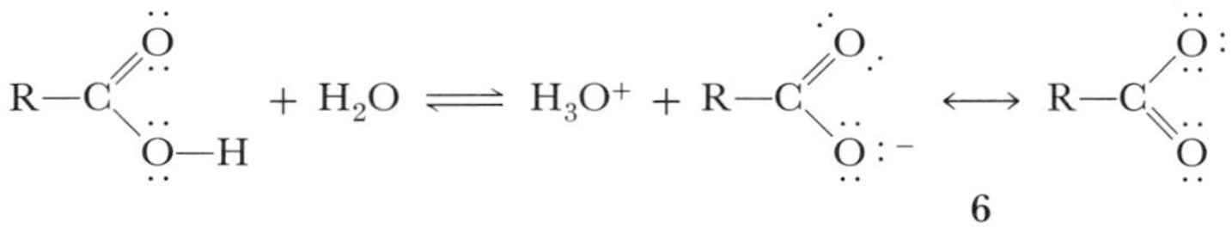

$$R-C\overset{\displaystyle \ddot{O}}{\underset{\displaystyle \ddot{O}-H}{\diagup}} + H_2O \rightleftharpoons H_3O^+ + R-C\overset{\displaystyle \ddot{O}.}{\underset{\displaystyle \ddot{O}:^-}{\diagup}} \longleftrightarrow R-C\overset{\displaystyle \ddot{O}:^-}{\underset{\displaystyle \ddot{O}}{\diagup}} \tag{1.2}$$

6

$$R-CH_2-OH + H_2O \rightleftharpoons H_3O^+ + R-CH_2-\ddot{O}:^- \tag{1.3}$$

7

Another example is the allylic system. The allyl cation (**8**), anion (**9**), and

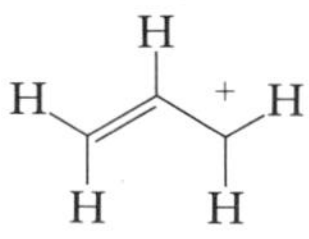

$$
\begin{array}{ccc}
\text{8} & \text{9} & \text{10}
\end{array}
$$

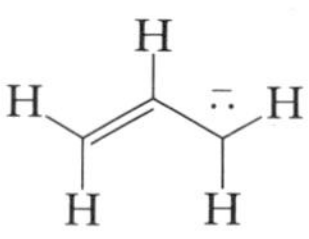

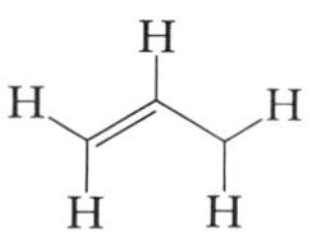

radical (**10**) are all more stable than their saturated counterparts. There is for each an alternative structure:

8   9   10

The alternative structures will not always be equivalent, as the following examples illustrate:

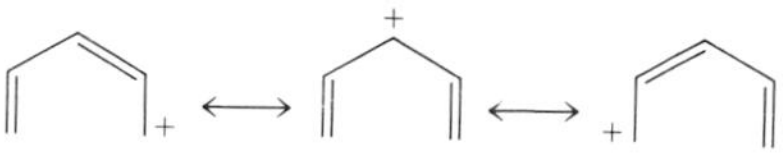

$$H-\overset{..}{\underset{..}{N}}-\overset{+}{N}\equiv N: \longleftrightarrow H-\overset{..}{N}=\overset{+}{N}=\overset{..}{N}:^-$$

Whenever there are nonequivalent structures, each will contribute to the composite picture to a different extent. The structure that would represent the thermodynamically most stable (lowest-energy) molecule, were such a molecule actually to exist, contributes the most to the composite, and others contribute successively less as they represent higher-energy molecules.

It is because the lowest-energy structures are most important that we specified in the rules for writing Lewis structures that the number of bonds should be maximum and the valence-shell occupancy not less than 8 whenever possible. Structures that violate these stipulations, such as 11 and 12, represent high-energy forms and hence do not contribute significantly to the structural pictures of ethene and 1,3-butadiene, which are quite adequately represented by 13 and 14.

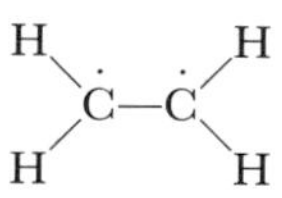

11 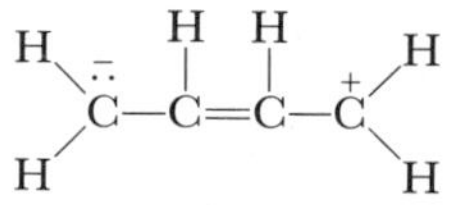   12

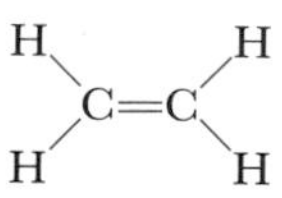

13 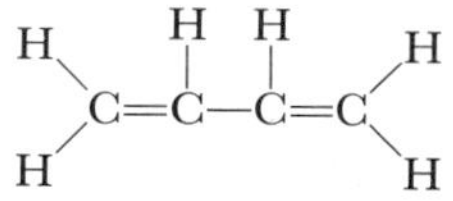   14

The following rules are useful in using resonance notation:

1. All nuclei must be in the same location in every structure. Structures with nuclei in different locations, for example, 15 and 16, are chemically distinct substances, and interconversions between them are actual chemical changes, always designated by ⇌.

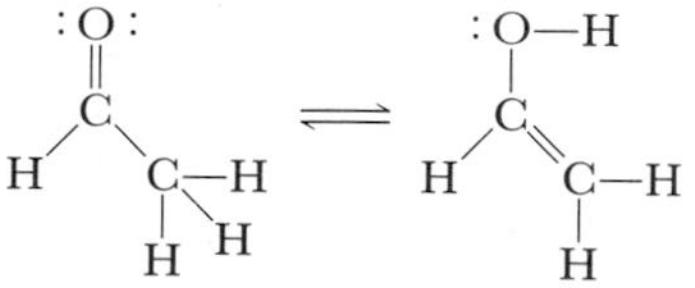

15   16

    2. Structures with fewer bonds or with greater separation of formal charge are of higher energy than those with more bonds or less charge separation. Thus **11** and **12** are of higher energy, respectively, than **13** and **14**.

    3. When two structures with formal charge have the same number of bonds and approximately the same degree of charge separation, the structure with negative charge on the more electronegative atom (or positive charge on the more electropositive atom) will be the more important contributor to the composite, but the difference will ordinarily be small enough that both structures must be included. Thus in **17a** ⟷ **17b**, **17a** should have a lower energy, but the chemistry of the ion can be understood only if it is described by the superposition of both structures.

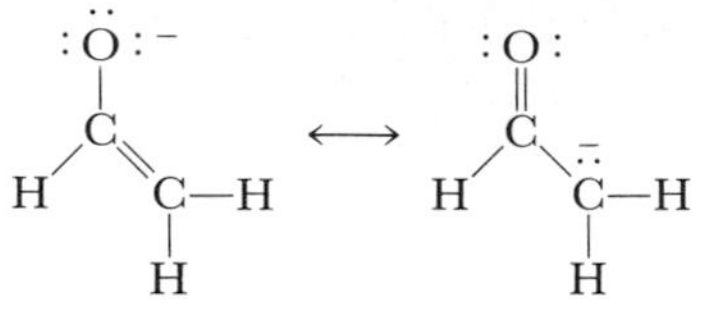

                    17a   **17b**

    4. All four groups attached to a pair of carbon atoms joined by a double bond in any structure must lie in or very nearly in the same plane. For example, Structure **18b** cannot contribute significantly because the bridged ring prevents carbons 6 and 7 from lying in the same plane as carbon 3 and the hydrogen on carbon 2. The prohibition against double bonds at bridgeheads of small rings is known as Bredt's rule. Double bonds can occur at a bridgehead if the rings are sufficiently large.[1]

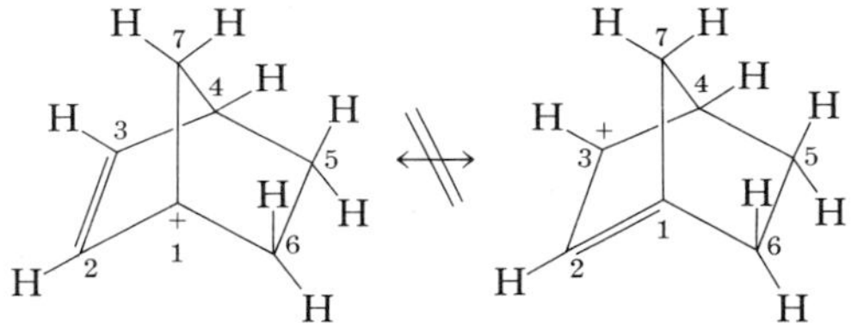

                    18a   **18b**

Molecular Geometry

Lewis structures provide a simple method of estimating molecular shapes. The geometry about any atom covalently bonded to two or more other atoms is found by counting the number of electron groups around the atom. Each unshared pair

[1](a) F. S. Fawcett, *Chem. Rev.*, **47**, 219 (1950); (b) J. R. Wiseman and W. A. Pletcher, *J. Am. Chem. Soc.*, **92**, 956 (1970); (c) C. B. Quinn and J. R. Wiseman, *J. Am. Chem. Soc.*, **95**, 6120 (1973); (d) C. B. Quinn, J. R. Wiseman, and J. C. Calabrese, *J. Am. Chem. Soc.*, **95**, 6121 (1973). Highly strained molecules with small-ring, bridgehead double bonds can exist but are likely to be thermodynamically unstable and reactive. See for example (e) J. Harnisch, O. Baumgärtel, G. Szeimies, M. Van Meerssche, G. Germain, and J. Declercq, *J. Am. Chem. Soc.*, **101**, 3370 (1979).

counts as one group, and each bond, whether single or multiple, counts as one group. The number of electron groups around an atom is therefore equal to the sum of the number of electron pairs on the atom and the number of other atoms bonded to it. The geometry is linear if the number of electron groups is two, trigonal if the number is three, and tetrahedral if the number is four.

The rule is based on the *valence-shell, electron-pair repulsion* model (VSEPR), which postulates that because electron pairs repel each other, they will try to stay as far apart as possible. In trigonal and tetrahedral geometries, the shape will be exactly trigonal (120° bond angles), or exactly tetrahedral (109.5° bond angles) if the electron groups are all equivalent, as, for example, in BH_3 or CH_3^+ (trigonal), or in CH_4 or NH_4^+ (tetrahedral).

If the groups are not all equivalent, the angles will deviate from the ideal values. Thus in NH_3 (four electron groups, three in N—H bonds, one an unshared pair), the unshared pair, being attracted only by the nitrogen nucleus, will be closer to the nitrogen on the average than will the bonding pairs, which are also attracted by a hydrogen nucleus. Therefore the repulsion between the unshared pair and a bonding pair is greater than between two bonding pairs, and the bonding pairs will be pushed closer to each other. The H—N—H angle should therefore be less than 109.5°. It is found experimentally to be 107°. Similarly, in H_2O (four electron groups: two unshared pairs and two O—H bonds), the H—O—H angle is 104.5°.[2]

Ambiguity may arise when more than one structure contributes. Then unshared pairs in one structure may become multiple bonds in another, so that the number of electron groups around a given atom is not the same in both structures. An example is methyl azide (**19**). The central nitrogen is clearly linear (two electron groups), but the nitrogen bonded to CH_3 has three electron groups in **19a** and

$$H_3C—\overset{..}{N}=\overset{+}{N}=\overset{..}{N}:^- \longleftrightarrow H_3C—\overset{\overset{\_}{..}}{N}—\overset{+}{N}\equiv N:$$

$$\text{19a} \qquad\qquad\qquad \text{19b}$$

four in **19b**. In such a situation the number of electron groups is determined from the structure with the *larger number of bonds*. Thus the nitrogen in question is trigonal, with a C—N—N angle of 120° ± 5°, as expected from Structure **19a**.[3]

Conventions for Structural Formulas

This book contains large numbers of Lewis structural formulas. Frequently we shall not write out the full Lewis structure; unshared pairs of electrons not shown explicitly are implied. When there are two or more contributing structures, we shall show them all only if that is essential to the point being illustrated; again, it will be assumed that the reader will understand that the missing structures are implied.

[2]The theoretical basis of the VSEPR rules is not entirely clear: (a) J. L. Bills and R. L. Snow, *J. Am. Chem. Soc.*, **97**, 6340 (1975); (b) M. B. Hall, *J. Am. Chem. Soc.*, **100**, 6333 (1978).
[3]*Tables of Interatomic Distances and Configuration in Molecules and Ions*, Special Publication No. 11, The Chemical Society, London, 1958, p. M113.

1.2 MOLECULAR ORBITALS

Lewis structures serve admirably for many aspects of mechanistic organic chemistry. Frequently, however, we need a more accurate bonding model.

Models Based on the Quantum Theory

The description of chemical bonding must ultimately be based on an understanding of the motions of electrons. In order to improve our model, we need to appeal to the quantum theory, which summarizes the current understanding of the behavior of particles of atomic and subatomic size.

The quantum theory provides the mathematical framework for describing the motions of electrons in molecules. When several electrons are present, all interacting strongly with each other through their mutual electrostatic repulsion, the complexity is so great that exact solutions cannot be found. Therefore approximate methods must be used even for simple molecules. These methods take various forms, ranging from complex *ab initio* calculations, which begin from first principles and have no parameters adjusted to fit experimental data, to highly approximate methods such as the Hückel theory, which is discussed further in Appendix 1 at the end of this chapter. The more sophisticated of these methods now can give results of quite good accuracy for small molecules, but they require extensive use of computing equipment.[4] Such methods are hardly suited to day-to-day qualitative chemical thinking. Furthermore, the most generally applicable and therefore most powerful methods are frequently simple and qualitative.

Our ambitions in looking at bonding from the point of view of the quantum theory are therefore modest. We want to make simple qualitative arguments that will provide a practical bonding model.

Atomic Orbitals

The quantum theory specifies the mathematical machinery required to obtain a complete description of the hydrogen atom. There are a large number of functions that are solutions to the appropriate equation; they are functions of the x, y, and z

---

[4] A number of texts cover methods for obtaining complete orbital descriptions of molecules. Examples are (a) A. Liberles, *Introduction to Molecular-Orbital Calculations,* Holt, Rinehart and Winston, New York, 1966; (b) J. D. Roberts, *Notes on Molecular Orbital Theory,* W. A. Benjamin, New York, 1962; (c) K. B. Wiberg, *Physical Organic Chemistry,* Wiley, New York, 1964; (d) A. Streitwieser, Jr., *Molecular Orbital Theory for Organic Chemists,* Wiley, New York, 1961; (e) M. J. S. Dewar, *The Molecular Orbital Theory of Organic Chemistry,* McGraw-Hill, New York, 1969; (f) M. J. S. Dewar and R. C. Dougherty, *The PMO Theory of Organic Chemistry,* Plenum, New York, 1975; (g) H. E. Zimmernam, *Quantum Mechanics for Organic Chemists,* Academic Press, New York, 1975; (h) E. Heilbronner and H. Bock, *The HMO Model and Its Application,* Vols. 1, 2, 3, Wiley, London, 1976; (i) J. N. Murrell and A. J. Harget, *Semi-empirical Self-consistent-field Molecular Orbital Theory of Molecules,* Wiley, London, 1972; (j) P. O'D. Offenhartz, *Atomic and Molecular Orbital Theory,* McGraw-Hill, New York, 1970; (k) S. P. McGlynn, L. G. Vanquickenborne, M. Kinoshita, and D. G. Carroll, *Introduction to Applied Quantum Chemistry,* Holt, Rinehart and Winston, New York, 1972; (l) W. G. Richards and J. A. Horsley, *Ab Initio Molecular Orbital Calculations for Chemists,* Oxford University Press (Clarendon Press), New York, 1970; (m) J. A. Pople and D. L. Beveridge, *Approximate Molecular Orbital Theory,* McGraw-Hill, New York, 1970; (n) M. Karplus and R. N. Porter, *Atoms and Molecules,* W. A. Benjamin, Menlo Park, Calif. 1970; (o) J. N. Murrell, S. F. A. Kettle, and J. M. Tedder, *The Chemical Bond,* Wiley, New York, 1978.

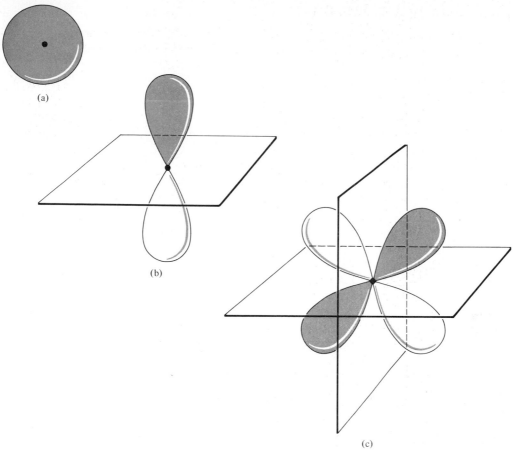

Figure 1.1 Hydrogen atomic orbital functions. (a) 1s; (b) 2p; (c) 3d. The surfaces drawn are meant to include some large fraction of the function within their boundaries and hence to show the function's shape. They are artificial, because orbital functions actually have no edges but merely decrease in magnitude as distance from the nucleus increases. The important features of the orbitals are the nodal planes indicated and the algebraic signs of the orbital functions, positive in the shaded regions and negative in the unshaded regions.

coordinates of a coordinate system centered at the nucleus.[5] Each of these functions describes a possible condition, or state, of the electron in the atom, and each has associated with it an energy, which is the total energy (kinetic plus potential) of the electron when it is in the state described by the function in question.

    The functions we are talking about are the familiar 1s, 2s, 2p, 3s, . . . atomic orbitals, which are illustrated in textbooks by diagrams like those in Figure 1.1.

[5] Actually, the origin is at the center of mass, which, because the nucleus is much more massive than the electron, is very close to the nucleus.

Each orbital function (or *wave function*) is a solution to the quantum mechanical equation for the hydrogen atom called the Schrödinger equation. The functions are ordinarily designated by a symbol such as φ, χ, ψ, and so on. We shall call atomic orbitals φ or χ, and designate by a subscript the orbital meant, as, for example, φ_{1s}, φ_{2s}, and so on. Later we may abbreviate the notation by simply using the symbols $1s$, $2s$, . . . , to indicate the corresponding orbital functions. Each function has a certain numerical value at every point in space; the value at any point can be calculated once the orbital function is known. We shall never need to know these values and shall therefore not give the formulas; they can be found in other sources.[6] The important things for our purposes are, first, that the numerical values are positive in certain regions of space and negative in other regions, and second, that the value of each function approaches zero as one moves farther from the nucleus. In Figure 1.1, and in other orbital diagrams used throughout this book, regions of positive value are shaded and regions of negative value are unshaded.

Imagine walking around inside an orbital, and suppose that there is some way of sensing the value—positive, negative, or zero—of the orbital function as you walk from point to point. On moving from a positive region to a negative region, you must pass through some point where the value is zero. The collections of all adjacent points at which a function is zero are called *nodes;* they are surfaces in three-dimensional space, and most of the important ones for our purposes are planes, like those shown in Figure 1.1 for the p and d orbitals illustrated. (Nodes can also be spherical, and of other shapes, but these are of less concern to us.)

The Physical Significance of Atomic Orbital Functions

The fact that an orbital function φ is of different algebraic sign in different regions has no particular physical significance for the behavior of an electron that finds itself in the state defined by the orbital. (We shall see shortly that the significance of the signs comes from the way in which orbitals can be combined with each other.) The quantity that has physical meaning is the value at each point of the function φ^2, which is positive everywhere, since the square of a negative number is positive. The squared function, φ^2, gives the probability of finding the electron at various points in space. Diagrams like that in Figure 1.2, with shading of varying density showing the relative probability of finding the electron in various regions or, more succinctly, the *electron distribution* or *electron density,* are actually pictures of φ^2, not of φ itself. The general shape of φ^2 will be similar to the shape of φ. The orbitals and their squares have no edges, even though definite outlines are usually drawn in diagrams; the values merely approach closer and closer to zero as one goes farther and farther from the nucleus.

Extension to Other Atoms

The hydrogen atomic orbitals would not do us a great deal of good if orbitals of other atoms were radically different, since in that case different pictures would be

---

[6]See, for example, K. B. Wiberg, *Physical Organic Chemistry,* Wiley, New York, 1964, pp. 17, 19, and 25.

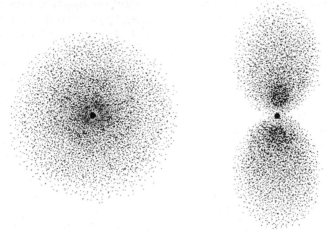

Figure 1.2 Cross sections of electron density, φ^2, for $1s$ and $2p$ atomic orbitals. The density of shading is roughly proportional to φ^2.

required for each atom. But the feature of the hydrogen atom problem that determines the most important characteristics of the hydrogen atom orbitals is the spherical symmetry of the force field provided by the positive nuclear charge. Because all the atoms have this same spherically symmetric force field, the atomic orbitals of all atoms are similar, the main difference being in their radial dependence, that is, in how rapidly they approach zero as one moves away from the nucleus. Because the radial dependence is of minimal importance in qualitative applications, one may simply use orbitals of the shapes found for hydrogen to describe behavior of electrons in all the atoms.

Ground and Excited States

We know that an electron in a hydrogen atom in a stationary state will be described by one of the atomic orbital functions φ_{1s}, φ_{2s}, φ_{2p_x}, and so forth.[7] We can make this statement in a more abbreviated form by saying that the electron is in one of the orbitals φ_{1s}, φ_{2s}, φ_{2p_x}, . . . , and we shall use this more economical kind of statement henceforth.

The orbital that has associated with it the lowest energy is φ_{1s}; if the electron is in this orbital, it has the lowest total energy possible, and we say the hydrogen atom is in its *electronic ground state*. If we were to give the electron more energy, say enough to put it in the φ_{2p_x} orbital, the atom would be in an *electronic excited state*. For any atom or molecule, the state in which all electrons are in the lowest possible energy orbitals (remembering always that the Pauli exclusion principle

---

[7]We assume from here on that the reader is familiar with the number and shape of each type of atomic orbital function. This information may be found in standard introductory college chemistry texts.

prevents more than two electrons from occupying the same orbital) is the electronic ground state. Any other state is of higher energy and is an electronic excited state.

An Orbital Model for the Covalent Bond

Suppose that we bring together two ground-state hydrogen atoms. Initially the two electrons are in φ_{1s} orbitals centered on their respective nuclei. We shall call one atom A and the other B, so that the orbitals are φ_{1sA} and φ_{1sB}. When the atoms are very close, say within 1 Å $(= 10^{-8}$ cm) of each other, each electron will feel strongly the attractive force of the other nucleus as well as of its own. Clearly, then, the spherical φ_{1s} orbitals will no longer be appropriate to the description of the electron motions. We need to find new orbital functions appropriate to the new situation, but we would prefer to do so in the simplest way possible, since going back to first principles and calculating the correct new orbital functions is likely to prove an arduous task.

We therefore make a guess that a possible description for a new orbital function will be obtained by finding at each point in space the value of φ_{1sA} and of

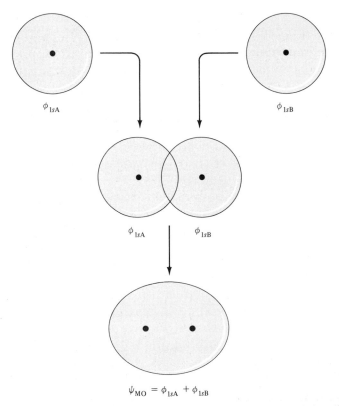

Figure 1.3 The linear combination of $1s$ orbital functions on hydrogen atoms A and B to yield a new orbital function, $\psi_{MO} = \varphi_{1sA} + \varphi_{1sB}$.

φ_{1sB} and adding the two numbers together. This process will give us a new orbital function, which, since φ_{1sA} and φ_{1sB} are both positive everywhere, will also be positive everywhere. Figure 1.3 illustrates the procedure. Mathematically the statement of what we have done is Equation 1.4:[8]

$$\psi_{MO} = \varphi_{1sA} + \varphi_{1sB} \qquad (1.4)$$

The symbol MO means that the new function is a *molecular orbital*; a molecular orbital is any orbital function that extends over more than one atom. Because the technical term for a sum of functions of the type 1.4 is a *linear combination*, the procedure of adding up atomic orbital functions is called *linear combination of atomic orbitals*, or *LCAO*.

   This simple procedure turns out to fit quite naturally into the framework of the quantum theory, which with little effort provides a method for finding the energy associated with the new orbital ψ_{MO}. This energy is lower than the energy of either of the original orbitals φ_{1sA}, φ_{1sB}.

   Instead of adding φ_{1sA} and φ_{1sB}, we might have subtracted them. We would then have obtained Equation 1.5:

$$\psi_{MO}^* = \varphi_{1sA} - \varphi_{1sB} \qquad (1.5)$$

Figure 1.4 illustrates the formation of this combination. Note that there is a node in this molecular orbital, because at any point equidistant from the two nuclei the value of φ_{1sA} is numerically equal to the value of φ_{1sB}, so that $\varphi_{1sA} - \varphi_{1sB}$ is zero.

   The procedures of the quantum theory require that the negative combination be made as well as the positive, and they show also that the energy associated with ψ_{MO}^* will be higher than that of φ_{1sA} and φ_{1sB}.

Energies of Molecular Orbitals

We can summarize the process of constructing our bonding model in an energy-level diagram. Figure 1.5 introduces the conventions we shall use for showing the formation of new orbitals by combining others. On either side we place the starting orbitals, and at the center the orbitals resulting from the combination process. In Figure 1.5 we have also shown orbital occupancies. Before the interaction, we have one electron in φ_{1sA} and one in φ_{1sB}; afterward we can place both electrons in ψ_{MO} to obtain a model for the ground state of the H_2 molecule, which will be of lower energy than the separated atoms by an amount $2\,\Delta E$ (two electrons each decrease in energy by ΔE).[9]

---

[8] Normalizing factors are omitted from Equations 1.4 and 1.5.

[9] The energy change on formation of the molecule, known from experiment to be 104 kcal mole^{-1}, will not actually be equal to $2\,\Delta E$, because the quantum mechanical procedures count the mutual repulsion of the electrons twice and neglect the mutual repulsion of the nuclei. The two corrections to $2\,\Delta E$ are opposite in sign and roughly cancel, but they are both large numbers (on the order of 400–450 kcal mole^{-1} for H_2), and their difference (about 35 kcal mole^{-1}) is significant. The actual energy lowering is less than $2\,\Delta E$ by this amount; in other words, for hydrogen the actual experimental dissociation energy is 104 kcal mole^{-1}, but $2\,\Delta E$ calculated from theory is about 139 kcal mole^{-1} and ΔE is about 69 kcal mole^{-1}. See C. A. Coulson, *Valence,* 2d ed., Oxford University Press, London, 1963, p. 90.

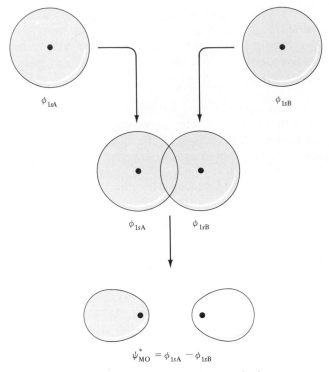

Figure 1.4 The linear combination of $1s$ orbital functions on hydrogen atoms A and B to yield orbital function $\psi_{MO}^* = \varphi_{1sA} - \varphi_{1sB}$.

The process of forming ground-state H_2 would be described in our LCAO model by saying that H_A, with its electron in φ_{1sA}, and H_B, with its electron in φ_{1sB}, will come together to give H_2 with a pair of electrons in ψ_{MO}, and will in the process give off energy $2\,\Delta E$ to the surroundings. We can also obtain models for a singly excited state and for a doubly excited state of H_2 by adding energy $2\,\Delta E$ or $4\,\Delta E$ to the ground-state molecule and placing either one or both electrons in ψ_{MO}^*.

Electrons in ψ_{MO} are stabilizing for the molecule, and electrons in ψ_{MO}^* are destabilizing. Therefore we call ψ_{MO} a *bonding* orbital and ψ_{MO}^* an *antibonding* orbital. In antibonding orbitals there is always a node between the nuclei, so that electrons are excluded from that region, whereas bonding orbitals have no such node and concentrate electrons between the nuclei.

Interaction of Orbitals

It is convenient to summarize the process of molecular orbital formation in Figure 1.5 as occurring through an interaction of φ_{1sA} and φ_{1sB} with each other to produce the two new orbitals ψ_{MO} and ψ_{MO}^*. The interaction has associated with it an energy change ΔE; measuring from the energy of the orbitals φ_{1sA} and φ_{1sB} before the

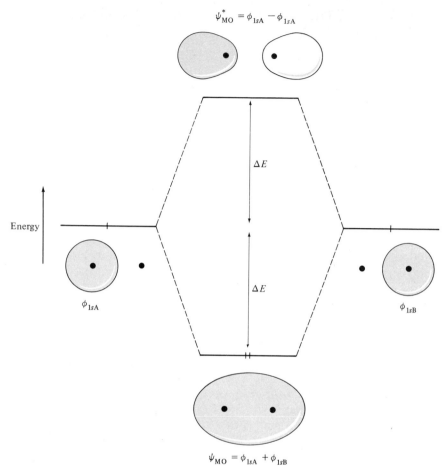

Figure 1.5 The energy-level diagram for the interaction of φ_{1sA} with φ_{1sB}. On either side are the atomic orbitals before interaction; at the center are the two molecular orbitals. Orbital occupancies are indicated for the two separate hydrogen atoms and for the molecule.

interaction occurs, ψ_{MO} moves down by interaction energy ΔE and ψ_{MO}^* moves up by interaction energy ΔE.[10]

The quantity ΔE can be calculated, but for our purposes we need to know only qualitatively how its magnitude depends on the nature of the interaction. The interaction energy is greater the more strongly the two orbitals overlap; overlap is large when both orbitals have large values of the same algebraic sign in the same

---

[10] A somewhat more careful treatment shows that ψ_{MO}^* will actually have moved up above the φ_{1sA} level by somewhat more than ψ_{MO} moved down. This fact will be important in certain applications later, but need not concern us now.

region of space. The overlap of two orbital functions φ_1 and φ_2 is obtained by multiplying the values of the two functions at each point and summing the products over all points, in other words, by integrating over all the spatial coordinates the quantity $\varphi_1\varphi_2$.

The second factor affecting the magnitude of ΔE is whether or not the two interacting orbitals are of the same or different energy. The interaction is maximum when the energies of the interacting pair are the same, and becomes smaller the farther apart in energy they are. We shall return to consider the overlap and the energy differences between the initial orbitals in more detail in Section 1.6.

Basis Functions

The initial functions taken for the starting point in the model-building process are called *basis functions*. We shall use this terminology henceforth. The reason for introducing a new term instead of just continuing to call the starting functions atomic orbitals is that molecular orbitals can themselves serve as starting functions in an interaction model.

The H_2 model has illustrated an important point about orbital interactions which must be remembered: *Whenever basis orbital functions interact to form new orbital functions, the number of new functions obtained is equal to the number of basis functions used.*

σ Bonds and π Bonds

In Figure 1.6 are shown the three-dimensional shapes of the electron distributions ψ_{MO}^2 and ψ_{MO}^{*2} corresponding to the H_2 molecular orbitals. Suppose that we were to

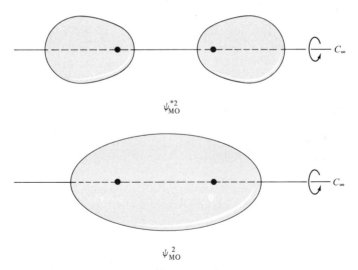

$$\psi_{MO}^{*2}$$

$$\psi_{MO}^2$$

Figure 1.6 The three-dimensional shapes of ψ_{MO}^2 and ψ_{MO}^{*2}. Each has infinite-fold rotational symmetry, because one can rotate each picture around the internuclear axis in an infinite number of steps and have at every step an identical picture.

rotate one of these pictures around an axis coinciding with the line joining the nuclei. We can rotate around this axis by any angle at all, and we shall get an identical picture. If you were to close your eyes while the rotation was done and then to open them, you would have no way of telling that any change had been made. To state this idea another way, we can say that we could divide one full rotation around the axis into an infinite number of steps, and have after each step an identical picture.

The rotation described above is called a *symmetry operation*. A symmetry operation is defined as any operation performed on an object that results in a new orientation of the object which is indistinguishable from the original. The most commonly encountered symmetry operations are rotation about an axis and reflection in a plane or through a point. The point, plane, or axis associated with a symmetry operation is called a *symmetry element*. A symmetry axis is designated by the letter C. Because the axis illustrated in Figure 1.6 is an infinite-fold rotation axis, it is called a C_∞ axis. (Symmetry is discussed further in Section 1.7.)

Any molecular orbital that has the symmetry property shown in Figure 1.6 is called a σ orbital. Both ψ_{MO} and ψ_{MO}^* of our hydrogen molecule model are σ orbitals.

Suppose that we make a molecular orbital by combining p orbitals on two atoms. We can do this in two different ways. If we choose the p orbitals that are oriented toward each other, we get MO's with the same C_∞ symmetry we had before. Figure 1.7 shows this process for the $2p_z$ orbitals. But if we choose the $2p_x$ orbitals, oriented as shown in Figure 1.8, we get a new type of molecular orbital.

Figure 1.9(a) shows the three-dimensional shape of the electron distributions $\psi_{\text{MO}_{2px}}^2$ and $\psi_{\text{MO}_{2px}}^{*2}$. Now the symmetry is different: One full rotation about the internuclear line must be divided into two equal steps if an identical picture is to be obtained after each step. This symmetry is a twofold rotation, and the symmetry element is called a C_2 axis. A second symmetry element, a mirror plane that reflects the top half of the orbital into the bottom half, illustrated in Figure 1.9(b), is also present. An orbital with this kind of symmetry is called a π orbital. Atomic orbitals of the s type can form only σ molecular orbitals; atomic orbitals of the p type can form either σ or π orbitals, depending on their orientation relative to each other. Because the overlap of p orbitals interacting in the π manner is smaller than overlap in the σ manner, the ΔE for interaction in π MO formation is usually less than in σ MO formation.

1.3 HYBRID ORBITALS

Suppose that we wish to construct an LCAO bonding model for methane. We set up the problem by defining an x, y, z coordinate system and placing the carbon at the origin. The molecule is tetrahedral, as determined from the electron-pair repulsion model. The orientation of the molecule is arbitrary; we choose to arrange it as

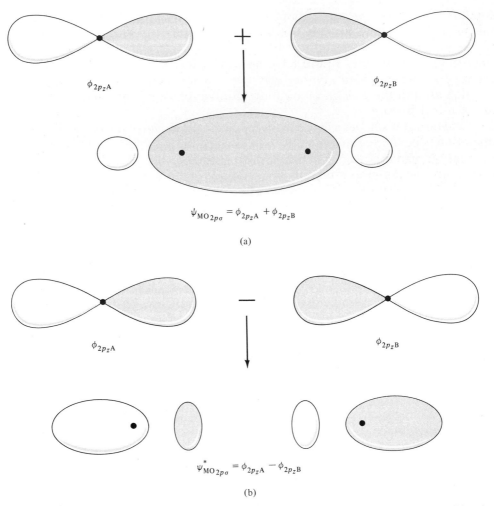

$$\psi_{MO_{2p\sigma}} = \phi_{2p_zA} + \phi_{2p_zB}$$

(a)

$$\psi^*_{MO_{2p\sigma}} = \phi_{2p_zA} - \phi_{2p_zB}$$

(b)

Figure 1.7 Combination of two p orbitals to give σ molecular orbitals. (a) Bonding combination. (b) Antibonding combination.

shown in **20**, with the hydrogen atoms in the $+x$, $+y$, $+z$ quadrant, the $-x$, $-y$, $+z$ quadrant, the $+x$, $-y$, $-z$ quadrant, and the $-x$, $+y$, $-z$ quadrant.

20

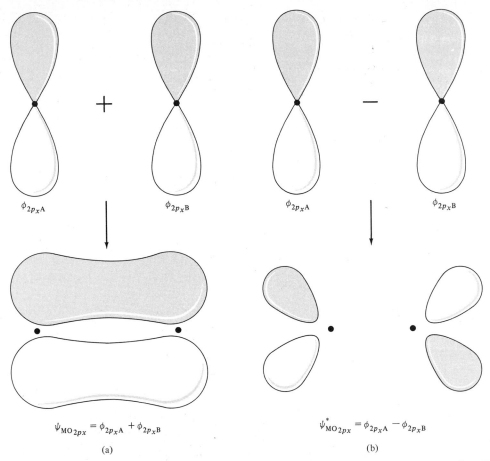

$$\psi_{\mathrm{MO}_{2px}} = \phi_{2p_xA} + \phi_{2p_xB}$$

(a)

$$\psi_{\mathrm{MO}_{2px}}^* = \phi_{2p_xA} - \phi_{2p_xB}$$

(b)

Figure 1.8 Combination of p orbitals to give π molecular orbitals $\psi_{\mathrm{MO}_{2px}}$ and $\psi_{\mathrm{MO}_{2px}}^*$. (a) Bonding combination. (b) Antibonding combination.

We have on each hydrogen a 1s orbital, $\varphi_{\mathrm{H}1s}$, and on the carbon a 2s, 2p_x, 2p_y, and 2p_z (Figure 1.10).

The Need for Hybrid Orbitals

We could simply proceed to inspect these orbitals to see which overlap with each other, and then begin to make molecular orbitals in the way described in the previous section. Unfortunately, the situation is now quite complicated. The φ_{1s1} orbital of hydrogen number 1 interacts with *all four* of the carbon valence orbitals.

The quantum theory gives procedures for dealing with this situation; for calculations done with the aid of a computer, there is no disadvantage in using the orbitals in Figure 1.10 directly. But the algebraic manipulations required are cumbersome. We are looking for a simpler model that will allow us to see quickly and clearly what the final outcome of this complex set of interactions will be.

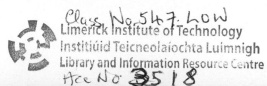

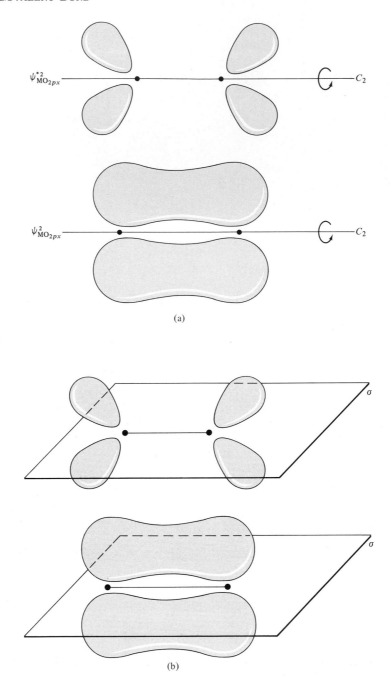

(a)

(b)

Figure 1.9 (a) The symmetry of the electron distributions $\psi^{*2}_{MO_{2px}}$ and $\psi^{2}_{MO_{2px}}$. Rotational symmetry is C_2. (b) These orbitals are also characterized by a mirror plane, σ, which reflects the top half of the orbital into the bottom half, and so leaves it unchanged.

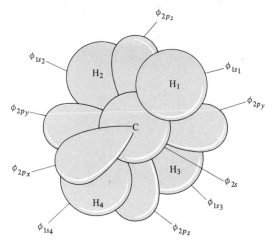

Figure 1.10 The valence atomic orbitals of the carbon and four hydrogens in methane.

Constructing Hybrids

The strategy we adopt is to look first at the atomic orbitals of the central atom, and to decide on the basis of the geometry which orbitals are going to interact with an orbital on a given ligand atom. For methane set up as in **20**, all four carbon orbitals will be involved in bonding to H_1. We then simply add together the four carbon orbitals to obtain a new orbital, χ_1, which will have the shape shown in Figure 1.11. The new function is called a *hybrid orbital* and is designated in this instance as sp^3 because it is formed from an s and three p orbitals.

The process of forming hybrids is not the same as the orbital interaction process that occurs on bringing two atoms together. There is no molecular orbital formation involved, because we are still talking about only one atom, and there is no energy lowering. The energy of a hybrid orbital is *between* the energies of the orbitals from which it is made, rather than being higher or lower.

The readers should convince themselves that the following four ways of adding together the s and p orbitals of the carbon will give four hybrid orbitals, each identical in shape to χ_{sp^31}, shown in Figure 1.11, but each oriented toward a different one of the four hydrogen atoms:[11]

$$\begin{aligned}
\chi_{sp^31} &= \varphi_{2s} + \varphi_{2p_x} + \varphi_{2p_y} + \varphi_{2p_z} \\
\chi_{sp^32} &= \varphi_{2s} - \varphi_{2p_x} - \varphi_{2p_y} + \varphi_{2p_z} \\
\chi_{sp^33} &= \varphi_{2s} + \varphi_{2p_x} - \varphi_{2p_y} - \varphi_{2p_z} \\
\chi_{sp^34} &= \varphi_{2s} - \varphi_{2p_x} + \varphi_{2p_y} - \varphi_{2p_z}
\end{aligned} \tag{1.6}$$

The actual correct mathematical forms are not exactly as indicated in Equations 1.6. The sign of each term, which is the important quantity for our

---

[11]Coefficients omitted.

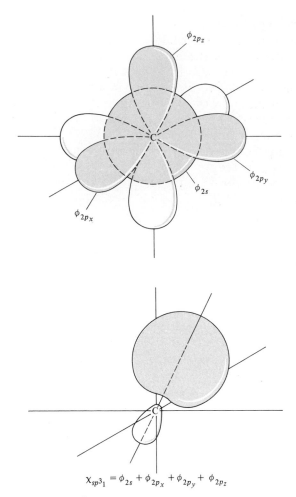

$$\chi_{sp^3{}_1} = \phi_{2s} + \phi_{2p_x} + \phi_{2p_y} + \phi_{2p_z}$$

Figure 1.11 The formation of an sp^3 hybrid by adding together the four valence atomic orbitals. Orbital shapes and locations of the nodes are approximate in these diagrams. For a more accurate description, see K. B. Wiberg, *Physical Organic Chemistry,* John Wiley & Sons, New York, 1964, pp. 29–33.

present purpose, is correctly represented there, but each orbital function must be multiplied by a coefficient that specifies the correct relative contribution of each orbital to the hybrid.

The advantage we gain by making hybrid orbitals is that we now have four new atomic orbitals on carbon, each one oriented directly toward one of the hydrogen atoms. Each hybrid will have a large overlap and therefore a large interaction with one, but only one, hydrogen. Our complicated original problem, in which each hydrogen $1s$ orbital had to interact with all four carbon atomic orbitals, is now replaced by four separate but simple problems.

MO's from Hybrid Orbitals

We can now proceed to make molecular orbitals in the same way we did for H_2. Figure 1.12 shows the form of the bonding molecular orbital obtained from φ_{1s1} and χ_{sp^31}; there will also be an antibonding combination that has a node between the atoms. The energy changes (Figure 1.13) follow the pattern we found in H_2. The only difference is that now the two interacting atomic orbitals are not the same and have different energies. The energy difference in this instance is not large and makes no fundamental change in our model. We shall return to this point in Section 1.6. Note that our new molecular orbitals have infinite-fold symmetry about the C—H axis, and so are σ orbitals.

The reader should now complete the bonding model for CH_4 by constructing a bonding–antibonding pair for each of the other three interacting pairs of atomic orbitals.

Other Types of Hybridization

A hybridization scheme can be constructed for each of the various possible geometries about the central atom. The sp^3 hybridization discussed above gives hybrids oriented at 109.5° angles to each other and is appropriate to tetrahedral atoms. For trigonal atoms the two p orbitals lying in the plane containing the nuclei are combined with the s to yield three sp^2 hybrids, as shown in Figure 1.14. For a linear geometry the appropriate hybridization is sp (Figure 1.15).

Hybrid orbitals are often characterized by the proportion of p character or s character they contain. An sp^3 orbital has 75 percent p character and 25 percent s character; the sp^2 is $66\frac{2}{3}$ percent p and $33\frac{1}{3}$ percent s; the sp orbital is 50 percent p and 50 percent s. It is possible to construct orbitals containing other proportions of s and p; the angles between the orbitals will vary accordingly. The more p charac-

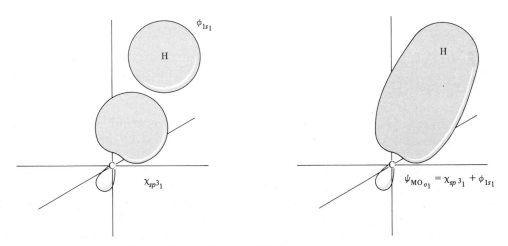

Figure 1.12 Formation of a bonding molecular orbital from φ_{1s1} and χ_{sp^31}.

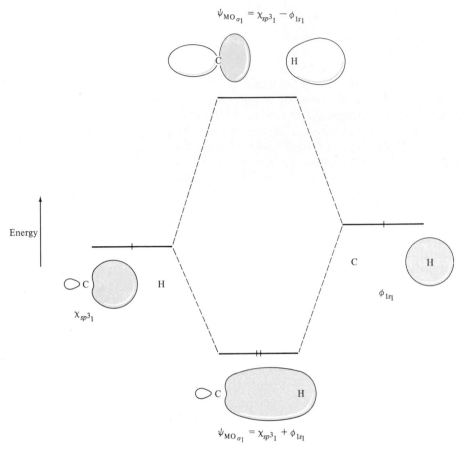

Figure 1.13 The energy relationships in MO formation from $\chi_{sp^3{}_1}$ and $\varphi_{1s{}_1}$.

ter, the more acute is the angle. For example, if one establishes two equivalent hybrids with 70 percent p character, the angle α between them will be 115.4° and the orbitals would be designated $sp^{2.33}$. Once such a pair has been constructed, the two other orbitals (if also required to be equivalent to each other) are determined by the mathematics of the hybridization procedure; they would be $sp^{4.0}$ orbitals with 80 percent p character, 20 percent s character, and an angle β of 104.5° between them. The geometry of this set is illustrated in Figure 1.16.

Because s electrons can penetrate closer to the nucleus than p electrons, which have a node at the nucleus, s electrons are held more tightly. Therefore an atom is effectively more electronegative in bonds that use a larger proportion of s orbital. Nuclear magnetic resonance coupling constants also reflect the orbital hybridization. Because nuclear spin information is conveyed from one nucleus to another by the electrons in the bonds, s electrons, with their greater penetration, carry the information more effectively. As a result the empirical relationships ex-

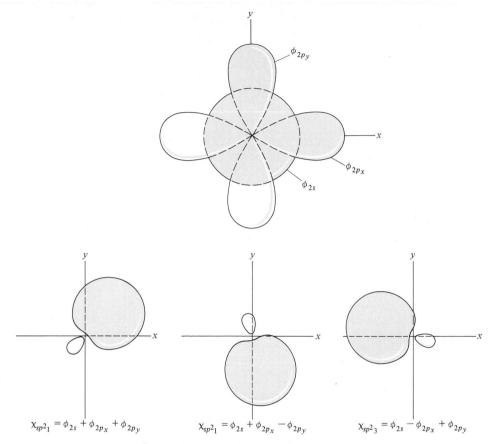

$$\chi_{sp^2{}_1} = \phi_{2s} + \phi_{2p_x} + \phi_{2p_y} \qquad \chi_{sp^2{}_1} = \phi_{2s} + \phi_{2p_x} - \phi_{2p_y} \qquad \chi_{sp^2{}_3} = \phi_{2s} - \phi_{2p_x} + \phi_{2p_y}$$

Figure 1.14 Formation of sp^2 hybrids from an s and two p orbitals.

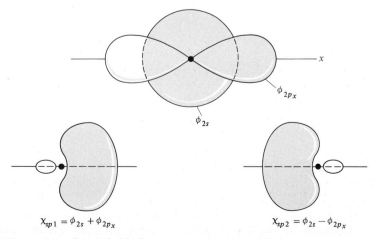

$$\chi_{sp1} = \phi_{2s} + \phi_{2p_x} \qquad\qquad \chi_{sp2} = \phi_{2s} - \phi_{2p_x}$$

Figure 1.15 Formation of sp hybrids from an s and one p orbital.

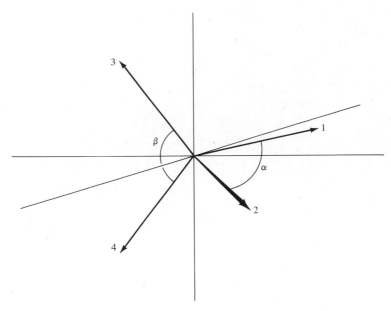

Figure 1.16 Geometrical relationship for the set of hybrid orbitals described in the text in which hybrids 1 and 2 have 70 percent p character and 3 and 4 have 80 percent p character. Hybrids point along the directions of the arrows, with angles $\alpha = 115.4°$ and $\beta = 104.5°$.

pressed by Equations 1.7 and 1.8 exist between s character of hybrids and one-bond carbon–hydrogen and carbon–carbon coupling constants.[12]

$$J_{^{13}C_i - H} = 500s(i) \qquad (1.7)$$

$$J_{^{13}C_i - ^{13}C_j} = 550s(i)s(j) \qquad (1.8)$$

In these equations $J_{^{13}C_i - H}$ is a one-bond coupling constant between a carbon and its attached hydrogen; $J_{^{13}C_i - ^{13}C_j}$ is a one-bond coupling constant between two directly bonded carbons; and $s(i)$ is the fractional s character of the carbon atom i in the bond.

σ and π Bonding. Ethylene

The ethylene molecule will illustrate construction of a model containing both σ and π bonding. The Lewis structure (**21**) shows that each carbon should be approx-

$$\begin{array}{c} H \qquad\qquad H \\ \diagdown \qquad\qquad \diagup \\ C{=}C \\ \diagup \qquad\qquad \diagdown \\ H \qquad\qquad H \end{array}$$

21

[12] (a) K. Frei and H. J. Bernstein, *J. Chem. Phys.*, **38**, 1216 (1963); (b) N. Muller and D. E. Pritchard, *J. Chem. Phys.*, **31**, 768 (1959); (c) H. Günther and W. Herrig, *J. Am. Chem. Soc.*, **97**, 5594 (1975).

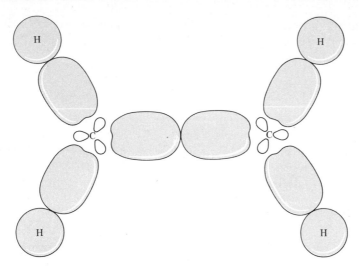

Figure 1.17 The basis orbitals for the σ MO's of ethylene.

imately trigonal. Therefore we need sp^2 hybrids on each carbon. Figure 1.17 shows these hybrids, together with the hydrogen $1s$ orbitals. The orbitals are allowed to interact in pairs, each pair yielding a bonding and an antibonding σ MO. There remain two p orbitals, one on each carbon, which were not used in the hybridization. These can overlap to form a π bonding–antibonding pair; this process is the same as illustrated earlier in Figure 1.8. Now we have obtained five bonding and five antibonding σ MO's and one bonding and one antibonding π MO. These can all be put on an approximate energy-level diagram as in Figure 1.18, which also shows how the 12 valence electrons are assigned to the molecular orbitals to repre-

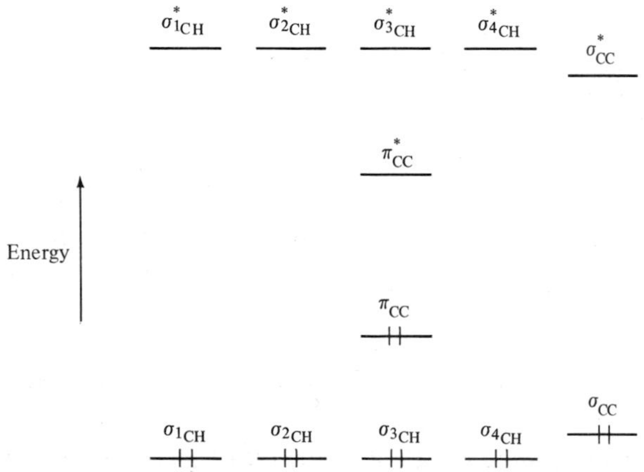

σ_{1CH}^{*} $\qquad$ σ_{2CH}^{*} $\qquad$ σ_{3CH}^{*} $\qquad$ σ_{4CH}^{*} $\qquad$ σ_{CC}^{*}

π_{CC}^{*}

Energy

π_{CC}

σ_{CC}

σ_{1CH} $\qquad$ σ_{2CH} $\qquad$ σ_{3CH} $\qquad$ σ_{4CH}

Figure 1.18 Energy-level diagram for the localized bonding model of ethylene.

sent the electronic ground state.[13] Note that the highest-energy bonding MO and the lowest-energy antibonding MO in ethylene will be the π and $\pi*$ levels, with the σ's lower than the π and the $\sigma*$'s higher than the $\pi*$.

1.4 DELOCALIZED π BONDING

In the allyl system (**22**) each carbon is trigonal, and each uses sp^2 hybrids to make bonds to its neighbors. The procedure outlined in the previous section is therefore adequate for constructing the σ MO's. The system of σ orbitals obtained is called the σ *framework*. After constructing the σ framework, a p orbital remains on each carbon. These p orbitals are the basis orbitals for the π *system* of molecular orbitals.

$$\left(\begin{array}{c} \text{H} \\ \text{H} \diagup\!\!\diagdown \text{H} \\ \text{H} \qquad\qquad \text{H} \end{array}\right)^{+,\ -,\ or\ \cdot}$$

22

Formation of π Systems

In allyl the three basis p orbitals can be symbolized as shown in **23**. Now there is no way to avoid the problem of the central p orbital interacting with more than one other orbital. One approach is to go to the quantum theory rules and work through

23

the prescribed procedures to find how the three orbitals will combine. The method, at an approximate level, is the *Hückel theory*. It is described in detail in the references cited earlier (see footnote 4, p. 10), and a brief derivation and examples are given in Appendix 1. Here we illustrate the results for some simple systems; in Section 1.6 we shall develop a method of obtaining the same results qualitatively in a simple way.

The first rule to remember in making π system orbitals is that the number of MO's is going to be the same as the number of basis p orbitals used. Thus for allyl we shall get three π MO's. The lowest-energy one will be the combination

$$\psi_{\text{MO}_1} = p_1 + p_2 + p_3 \tag{1.9}$$

[13] In a delocalized model, which would be needed if spectroscopic properties were under consideration, the σ orbitals would have different forms as dictated by the molecular symmetry. Then the C—H orbitals would not all have the same energy. See Section 1.7 for further discussion.

It will have the shape shown in **24**, and it will be bonding.

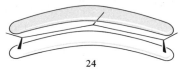

24

The next-higher energy MO is

$$\psi_{MO_2} = p_1 - p_3 \qquad (1.10)$$

It looks like **25**:

25

ψ_{MO_2} has a node cutting across it and passing through the central carbon; basis orbital p_2 does not contribute to this MO. The orbital ψ_{MO_3} is nonbonding. Its energy is the same as that of the basis orbitals themselves, so that electrons in it do not contribute to bonding. The third MO is

$$\psi_{MO_3} = p_1 - p_2 + p_3 \qquad (1.11)$$

and looks like **26**:

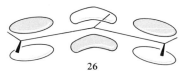

26

It has a node between each bonded pair of carbons and is antibonding. Figure 1.19 shows these π MO's in an energy-level diagram.

In the allyl cation, with two π electrons, ψ_{MO_1} is occupied; in the radical, with three π electrons, the additional electron is in the nonbonding ψ_{MO_2}, and in the anion, with four π electrons, ψ_{MO_1} and ψ_{MO_2} are both doubly occupied. Note that the nonbonding ψ_{MO_2} is concentrated at the ends of the chain; the molecular orbital pictures for these species thus correspond closely to the resonance pictures (see **8**, **9**, **10**, p. 6), which show the charge or unpaired electron to be concentrated at the ends.

Figures 1.20 and 1.21 show the π molecular orbitals for butadiene and pentadienyl. In each case the lowest-energy orbital has no vertical nodes, and each higher-energy orbital has one more vertical node than the orbital below it had, with the highest-energy orbital always having a node between every adjacent pair of atoms. Chains with an odd number of atoms have a nonbonding orbital, in which there is no contribution from alternate p orbitals.

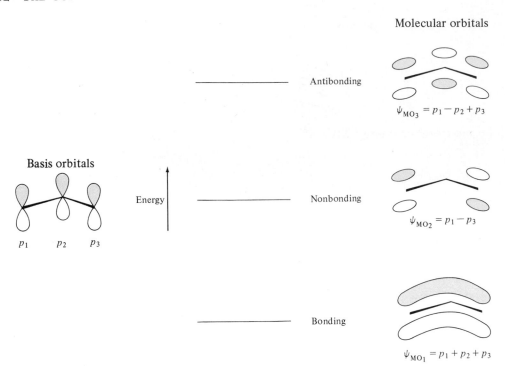

Molecular orbitals

— Antibonding

$\psi_{MO_3} = p_1 - p_2 + p_3$

Basis orbitals

$p_1 \quad p_2 \quad p_3$

Energy

— Nonbonding

$\psi_{MO_2} = p_1 - p_3$

— Bonding

$\psi_{MO_1} = p_1 + p_2 + p_3$

Figure 1.19 The π molecular orbitals of the allyl system. At the left are shown the three basis p orbitals that are to be used to construct the π molecular orbitals, drawn as though there were no interaction between them. On the right are the three molecular orbitals, arranged in order of increasing energy from bottom to top.

The π molecular orbitals in these systems extend over several atoms, rather than encompassing only two, as have the MO's we considered earlier. Orbitals that extend over more than two atoms are said to be *delocalized*.

1.5 AROMATICITY

The concept of aromaticity has been extremely fruitful for both theoretical and experimental organic chemists. Aromatic compounds are cyclic unsaturated molecules characterized by certain magnetic effects and by substantially lower chemical reactivity and greater thermodynamic stability than would be expected from a localized bond model. Antiaromatic compounds are cyclic unsaturated compounds that are thermodynamically less stable than would be expected from a localized bonding model.

Resonance and Aromaticity

The familiar properties of benzene illustrate the characteristics of aromatic compounds. Benzene is much less reactive toward electrophiles, such as molecular

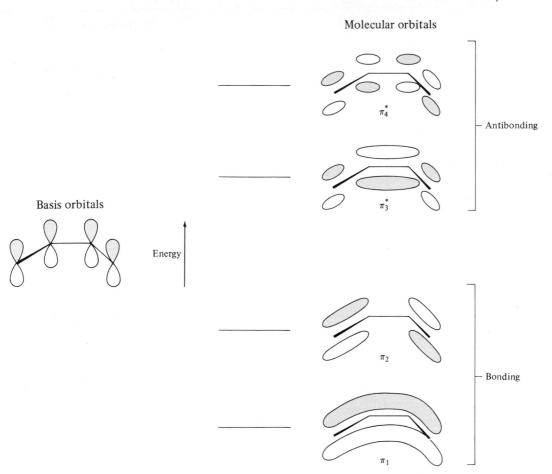

Figure 1.20 The π MO's for butadiene.

halogens, than are simple alkenes, and when it reacts it gives substitution rather than addition; and the heat evolved on hydrogenation is less by 37 kcal mole^{-1} than predicted for a cyclic C_6H_6 with three localized ethylene-type double bonds. (A more sophisticated estimate, discussed in the Appendix, gives 21 kcal mole^{-1}.) Furthermore, the nuclear magnetic resonance spectrum of benzene and its derivatives shows the protons bonded to the ring to be experiencing a stronger effective magnetic field than do protons attached to simple alkenes.[14]

Not all cyclic conjugated molecules for which such equivalent structures may be written share with benzene these special properties. For example, cyclobu-

[14]The term *alkene* is now used in most introductory organic chemistry texts for hydrocarbons containing nonconjugated C=C double bonds, but the term *olefin* is still used almost exclusively in the chemical literature. In this book we shall use the two terms interchangeably.

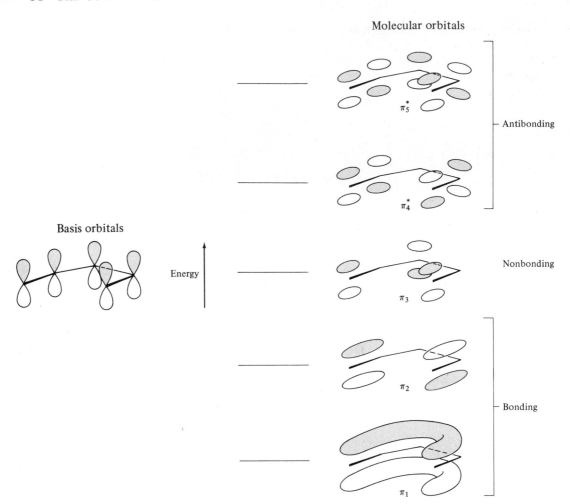

Figure 1.21 The π MO's for pentadienyl.

tadiene (**27**) eluded synthesis for many years; when finally prepared, it and its simple derivatives proved to be extremely reactive and capable of existence only when immobilized by freezing in an inert matrix at very low temperature.[15]

27

[15](a) L. Watts, J. D. Fitzpatrick, and R. Pettit, *J. Am. Chem. Soc.,* **87,** 3253 (1965); (b) O. L. Chapman, C. L. McIntosh, and J. Pacansky, *J. Am. Chem. Soc.,* **95,** 614 (1973); (c) G. Maier and M. Schneider, *Angew. Chem. Int. Ed. Engl.,* **10,** 809 (1971); (d) G. Maier, *Angew. Chem. Int. Ed. Engl.,* **13,** 425 (1974); (e) G. Maier, H.-G. Hartan, and T. Sayrac, *Angew. Chem. Int. Ed. Engl.,* **15,** 226 (1976); (f) S. Masamune, M. Suda, H. Ona, and L. M. Leichter, *J. Chem. Soc., Chem. Commun.,* 1268 (1972). For a summary of attempts to prepare cyclobutadiene, see (g) M. P. Cava and M. J. Mitchell, *Cyclobutadiene and Related Compounds,* Academic Press, New York, 1967.

Cyclooctatetraene, though a stable compound, does not have a planar ring; whatever stabilization we might have expected it to gain from the delocalization 28 ↔ 29 is evidently not sufficient to cause the molecule to abandon its tub-shaped conformation (30) for the planar structure that would allow cyclic conjugation.

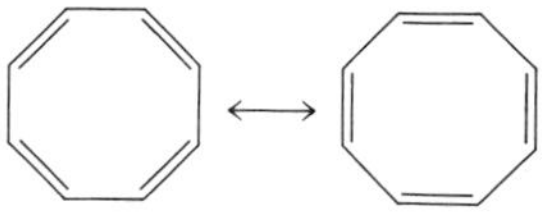

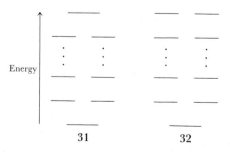

        28   29

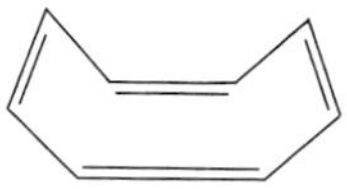

                30

The simple resonance theory fails to explain the singular lack of effectiveness of delocalization in cyclobutadiene and cyclooctatetraene, but we may turn to molecular orbitals for the solution.

π Electron Theory and the Hückel 4n + 2 Rule

In order to construct a bonding model for a planar conjugated ring, we follow the procedure outlined for the allyl system in the previous section and make the choice of sp^2 hybridization on each carbon. The σ framework is then constructed from these sp^2 hybrids and the hydrogen $1s$ orbitals, leaving a p orbital on each carbon. We next concentrate on the interactions among these p orbitals.

The interactions among the p orbitals are examined according to the rules of quantum mechanics, with the aid of a number of simplifying approximations.

Erich Hückel used this approach in developing a molecular orbital theory of cyclic conjugated molecules in the 1930s. (See the Appendix to this chapter for a description of the Hückel π electron theory.) Hückel discovered that the energies of the π molecular orbitals for any regular plane polygon with an even number of atoms will fall in the pattern 31.[16] A polygon with an odd number of atoms gives the pattern 32. These patterns, a single lowest level with higher levels in pairs of the

Energy

        31   32

[16](a) E. Hückel, *Z. Physik,* **70,** 204 (1931); (b) E. Hückel, *Z. Physik,* **76,** 628 (1932); (c) E. Hückel, *Z. Electrochem.,* **43,** 752 (1937).

same energy (*degenerate* pairs), are actually a consequence of the *n*-fold symmetry of the C_nH_n rings.

Hückel noted that if electron pairs are filled into the energy-level pattern **31** or **32**, a closed-shell structure (all electrons paired) will result only when the total number of pairs is odd (total number of electrons $= 4n + 2, n = 0, 1, 2, \ldots$); if the number of pairs is even (total number of electrons $= 4n, n = 1, 2, \ldots$), the last pair will be the only occupants of a doubly degenerate level and so one electron will go into each orbital with spins parallel. Diagrams **33** and **34** show the level filling for $4n + 2$ and $4n$ electrons, respectively. Because open-shell molecules are ordinarily highly reactive, the $4n$ electron rings should be chemically unstable.

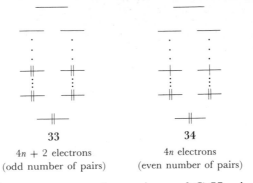

33	**34**
$4n + 2$ electrons	$4n$ electrons
(odd number of pairs)	(even number of pairs)

When the π electron energy of a conjugated C_nH_n ring is calculated from the energy-level diagram, it is found to be different from the π energy calculated for an open-chain C_nH_{n+2} conjugated polyene. The difference is termed the *resonance energy*, or *delocalization energy*. When the comparison of open-chain and cyclic π electron energies is made in the proper way (see the Appendix to this chapter for further discussion), it is found that the $4n + 2$ rings are stabilized compared with the open-chain analog, whereas the $4n$ systems are destabilized.[17] Theory thus explains the classification of cyclic conjugated systems as aromatic or antiaromatic according to the energy criterion, which we may define as follows: A molecule is aromatic if it is thermodynamically more stable than expected for the open-chain analog, and it is antiaromatic if it is thermodynamically less stable. A compound showing neither stabilization nor destabilization would be classed as nonaromatic.

A second criterion of aromaticity comes from the influence of a magnetic field on the π electrons of cyclic conjugated systems that have $4n + 2$ π electrons. Theory suggests that a magnetic field perpendicular to the ring plane will cause the electrons to behave as though they were circulating around the ring and generating their own small magnetic field, which will be superimposed on the applied field.[18] This circulation is called the *ring current*. Whereas ring currents occur in $4n + 2$ π systems, other local motions of electrons within bonding regions occur in all mole-

[17] See, for example, M. J. S. Dewar and G. J. Gleicher, *J. Am. Chem. Soc.*, **87**, 685 (1965).
[18] J. A. Pople and K. G. Untch, *J. Am. Chem. Soc.*, **88**, 4811 (1966).

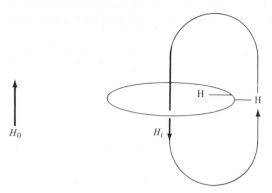

Figure 1.22 Induced fields caused by ring current in cyclic conjugated molecules. In a $4n + 2$ ring, the induced field H_i opposes the applied field H_0 for protons inside the ring and adds to it for protons outside.

cules and generate *local anisotropic* magnetic fields. (The term *anisotropic* means that the field strength varies with direction from the bond.) The sum at the position of a magnetic nucleus of the local anisotropic magnetic fields together with any fields generated by ring current is responsible for the chemical shift of that nucleus.

Figure 1.22 illustrates the direction of the fields generated by ring current in a $4n + 2$ π system. Inside the ring the induced field arising from ring current, H_i, opposes the applied field H_0. For resonance to occur for protons inside the ring, a larger H_0 must be applied than would be needed in the absence of ring current, and the protons are shifted upfield. Outside the ring H_i reinforces H_0 and outside protons are shifted downfield.

Conjugated rings of $4n$ electrons could in theory support ring currents as well, but their bonds alternate between single and double instead of being of uniform length. This bond alternation interrupts the ring current, leaving chemical shifts determined by local anisotropies. These anisotropic fields also differ for protons inside and outside the ring, but because of greater crowding inside, in the sense opposite to that shown in Figure 1.22.[19]

Homoaromaticity

Theory of aromaticity is not restricted to the simple planar conjugated rings. Any system that has extra stability by virtue of being cyclic would be classed as aromatic. The *homoaromatics* form one such category; they are systems in which the π system is interrupted at one or more points by a saturated center but in which geometry still permits significant overlap of the p orbitals across the insulating gap. Examples of homoaromatic stabilization have been found primarily in ions such as the trishomocyclopropenium ion (**36**), which forms as a short-lived intermediate

[19] M. Barfield, D. M. Grant, and D. Ikenberry, *J. Am. Chem. Soc.*, **97**, 6956 (1975). According to theory, ring currents in $4n$ rings would circulate in the sense opposite to those in $4n + 2$ rings. See note 18.

from the toluenesulfonate **35** in acetic acid. Products from coupling of **36** with nucleophiles show that all three CH_2 groups are equivalent in the intermediate.[20] Attempts to construct neutral homoaromatic systems have failed.[21]

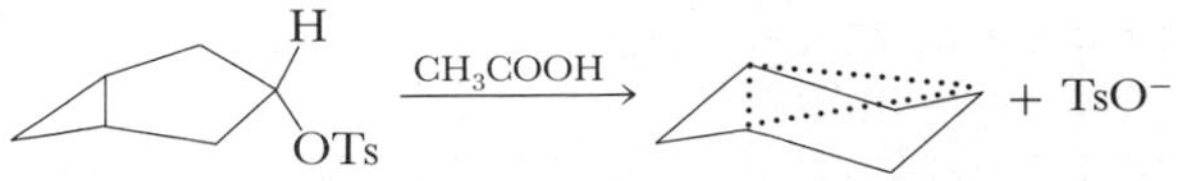

$$-OTs = -OSO_2-\langle\text{benzene ring}\rangle-CH_3$$

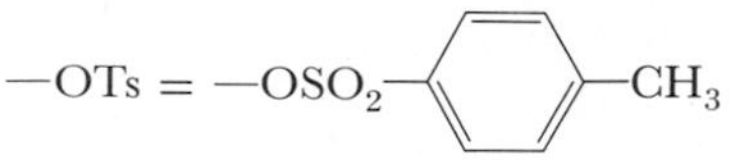

Some Examples of Aromatic and Antiaromatic Systems: Neutral Even-Membered Rings

The macrocyclic conjugated rings C_nH_n are known as *annulenes*. The nomenclature convention is to designate the ring size by a number in square brackets; thus [n]annulene is the ring C_nH_n. For benzene, [6]annulene, with six π electrons ($4n + 2$, $n = 1$), the theory clearly meets the tests both of stability and of ring current. As we have pointed out, there is a substantial stabilization of about 20 to 36 kcal mole^{-1} compared with a hypothetical localized model. The familiar chemical properties also point to a strong tendency for maintenance of the six π electron unsaturated system. The proton magnetic resonance spectrum of benzene and its derivatives shows the proton resonances in the range of $\delta = +7$ to $+8$ ppm (downfield from tetramethylsilane), 1–2 ppm lower than protons attached to nonbenzenoid double bonds.

Cyclobutadiene, [4]annulene, has proved to be a particularly elusive compound. The great difficulty chemists have experienced in its preparation suggests the conclusion that it lacks aromatic stabilization. The compound is stable as a ligand of low-valent transition metals. Oxidation of such a complex, cyclobutadieneiron tricarbonyl (**37**), produces free cyclobutadiene. Equation 1.13 illustrates one of several other methods of generation. This compound is so reactive that it

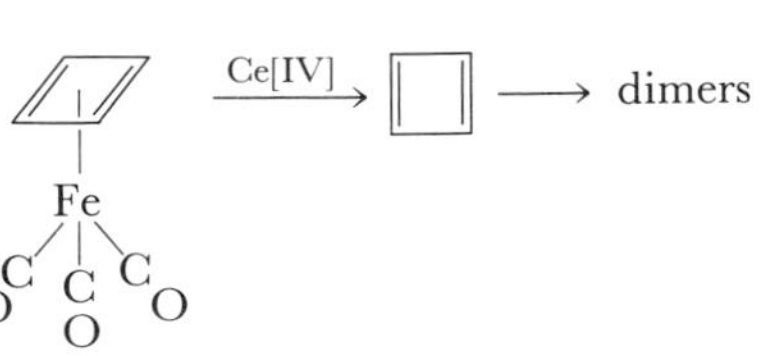

$$(1.12)$$

[20] S. Winstein, *Q. Rev. Chem. Soc.*, **23**, 141 (1969).
[21] (a) G. C. Christoph, J. L. Muthard, L. A. Paquette, M. C. Böhm, and R. Gleiter, *J. Am. Chem. Soc.*, **100**, 7882 (1978); (b) K. N. Houk, R. W. Gandour, R. W. Strozier, N. G. Rondan, and L. A. Paquette, *J. Am. Chem. Soc.*, **101**, 6797 (1979). See also (c) L. A. Paquette, C. C. Liao, R. Burson, R. E. Wingard, Jr., C. N. Shih, J. Fayos, and J. Clardy, *J. Am. Chem. Soc.*, **99**, 6935 (1977); (d) G. C. Christoph and M. A. Beno, *J. Am. Chem. Soc.*, **100**, 3156 (1978); (e) C. Santiago, K. N. Houk, G. J. DeCicco, and T. L. Scott, *J. Am. Chem. Soc.*, **100**, 692 (1978).

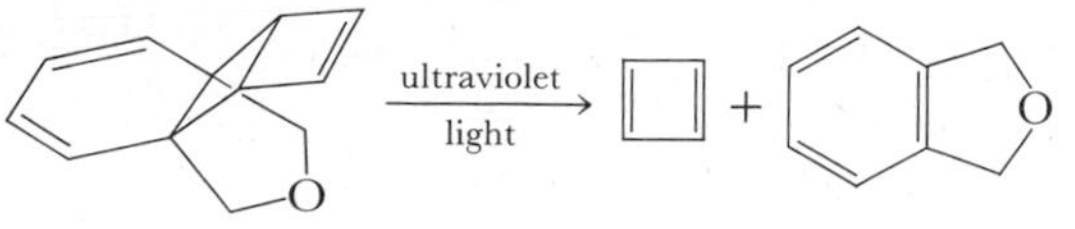

$$\text{(structure)} \xrightarrow[\text{light}]{\text{ultraviolet}} \square + \text{(structure)} \qquad (1.13)^{23}$$

dimerizes rapidly unless maintained frozen in a dilute matrix of argon or of an inert organic solvent at low temperature.[22]

Although the simple Hückel molecular theory, based on a square geometry, predicts that cyclobutadiene should have the energy level pattern shown in **38**, with

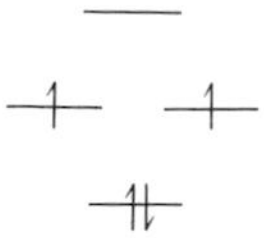

38

each of the degenerate orbitals containing one unpaired electron, the infrared spectrum of the matrix-isolated material at 7 K shows that the molecule is not square but instead has alternating single and double bonds.[23] Furthermore, it does not have unpaired electrons.[24] In order to avoid the more severe destabilization it would have as a fully antiaromatic $4n$ system, the molecule distorts from a square geometry to one with alternating single and double bonds, thereby reducing conjugation. It nevertheless retains sufficient antiaromatic character to be very reactive.

Ronald Breslow and his collaborators have given some attention to the problem of estimating the degree of destabilization of cyclobutadiene with respect to nonconjugated models. They have concluded from electrochemical measurements of oxidation–reduction potentials of the system **39a** ⇌ **39b**, of which only the quinone (**39b**) has the cyclobutadiene fragment, that the cyclobutadiene ring is destabilized by some 12–16 kcal mole^{-1} and so is definitely antiaromatic.[25]

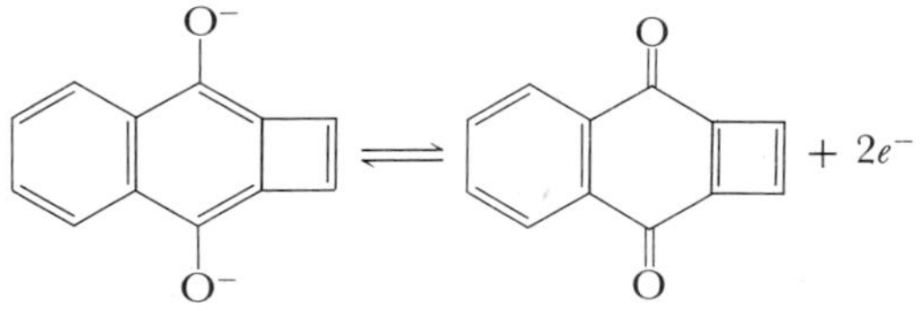

39a   **39b**

Cyclooctatetraene, as we have noted earlier, avoids the difficulty that we would predict it would encounter if it were planar. It takes up a nonplanar confor-

[22] See note 15.

[23] (a) S. Masamune, F. A. Souto-Bachiller, T. Machiguchi, and J. E. Bertie, *J. Am. Chem. Soc.,* **100,** 4889 (1978). In solution cyclobutadiene also has been shown to be rectangular, with the rate of isomerization through the square structure comparable to its rate of reaction with dienophiles. See (b) D. W. Whitman and B. K. Carpenter, *J. Am. Chem. Soc.,* **102,** 4272 (1980).

[24] S. Masamune, N. Nakamura, M. Suda, and H. Ona, *J. Am. Chem. Soc.,* **95,** 8481 (1973).

[25] R. Breslow, D. R. Murayama, S. Murahashi, and R. Grubbs, *J. Am. Chem. Soc.,* **95,** 6688 (1973).

mation in which the double bonds are effectively isolated from each other by twisting; in this way the p orbitals of one do not interact appreciably with those of the next. The molecule has conventional single and double bonds and behaves chemically like a typical olefin.[14] One might argue that this evidence is only suggestive, because the angle strain which would be introduced were the ring to become planar could be the cause of its preferred shape. But as we shall see, the angle strain is not sufficient to overcome the tendency toward planarity for an eight-membered ring with $4n + 2 \pi$ electrons.

A number of larger cyclic conjugated systems have been prepared, many of them by Sondheimer and co-workers.[26] [10]Annulene and [12]annulene are subject to considerable steric difficulties because the angle strain is too great to permit an all-cis ring with all hydrogens outside, while the rings are too small to accommodate a structure like **40** with trans double bonds and hydrogens inside. One way to avoid these problems is to bridge the ring, as, for example, in 1,6-methano[10]annulene (**41**). The bridge bends the ring out of planarity to some extent, but aromatic character expected for a $4n + 2$ system is retained, as indicated by the nmr

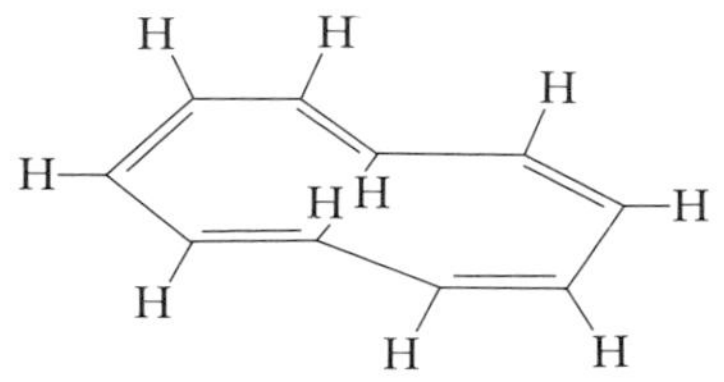

40

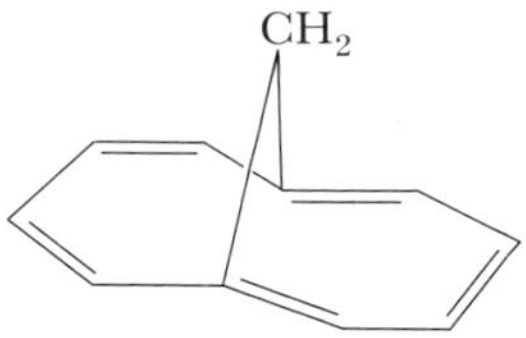

41

spectrum.[27] The larger rings are big enough to accommodate hydrogen atoms inside the rings and so can have trans double bonds and still be planar or nearly so. Likely conformations for some of these compounds are shown in Structures **42, 43,** and **44.** Annulenes as large as the 30-membered ring have been prepared, and many dehydroannulenes, which contain one or more triple bonds, are also known.

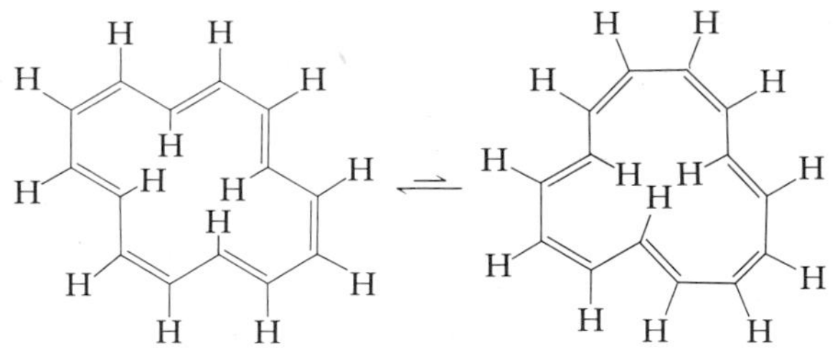

[14]annulene

42

[26] F. Sondheimer, *Acc. Chem. Res.,* **5,** 81 (1972).
[27] (a) E. Vogel and H. D. Roth, *Angew. Chem. Int. Ed. Engl.,* **3,** 228 (1964). See also (b) S. Masamune, D. W. Brooks, K. Morio, and R. L. Sobczak, *J. Am. Chem. Soc.,* **98,** 8277 (1976).

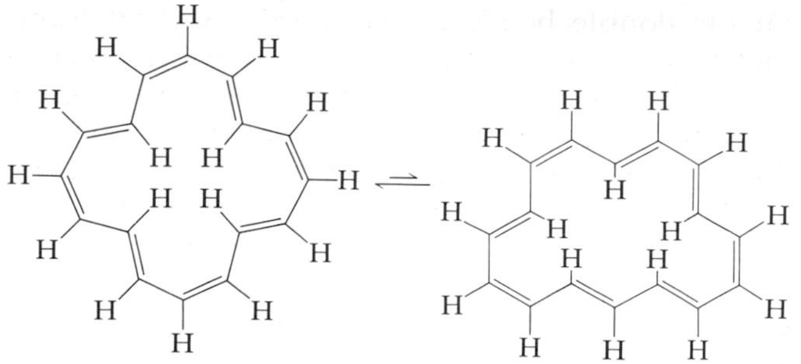

[16]annulene

43

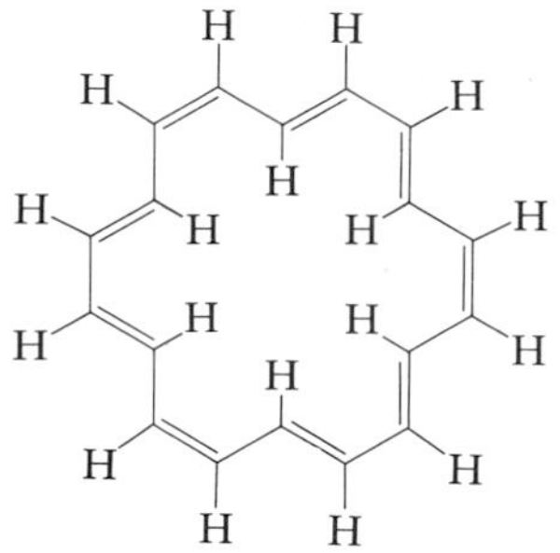

[18]annulene

44

These large rings, even the $4n + 2$ ones, do not show the kind of chemical stability that benzene has, although [18]annulene does undergo electrophilic substitutions. Ring currents provide the most useful criterion for testing their aromaticity. The molecules have protons both inside and outside the ring. Conformational equilibria such as those indicated in **42** and **43** exchange the inner and outer protons rapidly at room temperature, but at lower temperatures the rates are sufficiently slow so that the two types of proton can be observed. The spectra provide a dramatic confirmation of theory. The [14], [18], and [22]annulenes, $4n + 2$ systems, have outside proton resonances between about $\delta = +7.8$ and $\delta = +9.6$ ppm, shifts somewhat larger than those in benzene, whereas the inside protons appear between $\delta = -0.4$ and $\delta = -3$ ppm. (Positive δ values are downfield from tetramethylsilane (TMS); negative δ values are upfield.) The $4n$ rings [16] and [24]annulene have outside protons at $\delta = +4.7$ to $\delta = +5.3$ ppm and inside protons at much lower field, $\delta = +10$ to $\delta = +12$ ppm.[26]

Even-Membered Rings: Cations and Anions

Addition of two electrons to, or removal of two electrons from, a $4n$ antiaromatic ring converts it to a $4n + 2$ system, which should be aromatic. Several examples of such ions are known.

Tetramethylcyclobutadiene dication (**46**), has been prepared by Olah and co-workers by dissolving the dichloride **45** in a mixture of antimony pentafluoride and sulfur dioxide at low temperature.[28] It was identified by its proton magnetic

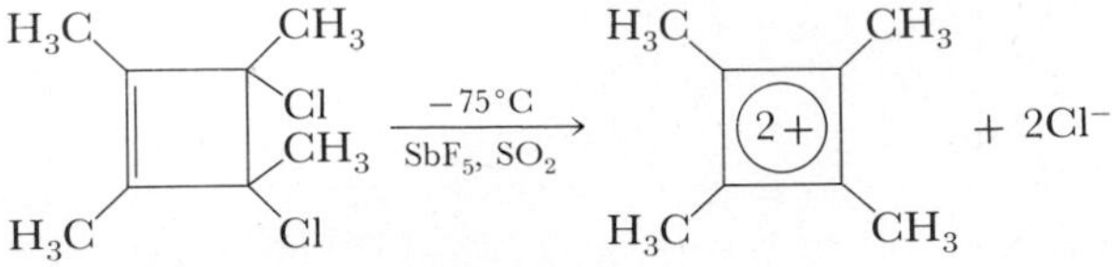

resonance spectrum, a single peak at $\delta = +3.7$ ppm. The tetraphenyl dication has also been observed.[29] A report of the dianion **47**, a six π electron system, has appeared.[30]

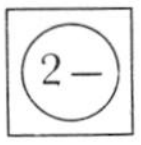

Addition of two electrons to cyclooctatetraene yields the dianion **48**, which shows a single peak in the proton magnetic resonance spectrum.[31] The conclusion

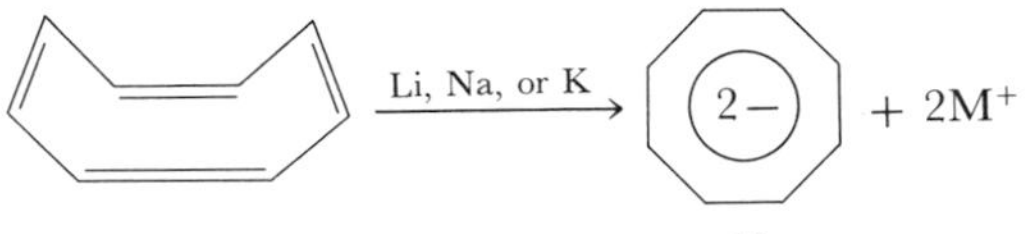

that the ion is a planar regular octagon is confirmed by the X-ray crystallographic structure determination of the 1,3,5,7-tetramethylcyclooctatetraenyl dianion, which shows the eight-membered ring in a planar conformation with equal bond lengths.[32] Note that the energy gain associated with establishment of the conjugated $4n + 2\,\pi$ electron aromatic system is sufficient to overcome the angle strain, which tends to oppose a planar structure. Several substituted cyclooctatetraenyl dications have also been prepared. Equation 1.14 shows an example. The proton magnetic resonance spectrum of tetramethylcyclooctatetraenyl dication **49** at

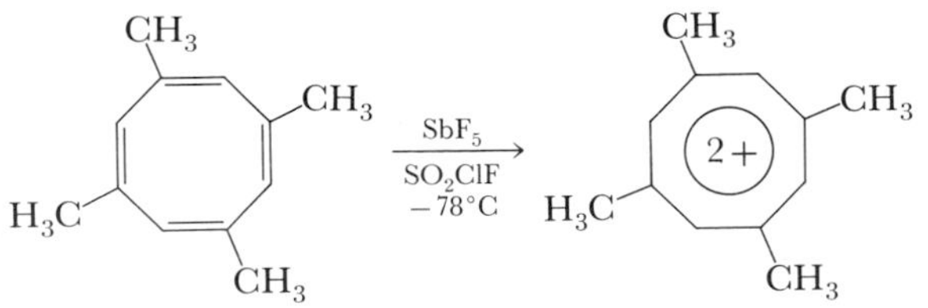

(1.14)

[28]G. A. Olah, J. M. Bollinger, and A. M. White, *J. Am. Chem. Soc.*, **91**, 3667 (1969).
[29]G. A. Olah and G. D. Mateescu, *J. Am. Chem. Soc.*, **92**, 1430 (1970).
[30]J. S. McKennis, L. Brener, J. R. Schweiger, and R. Pettit, *J. Chem. Soc., Chem. Comm.*, 365 (1972).
[31]T. J. Katz, *J. Am. Chem. Soc.*, **82**, 3784 (1960).
[32]S. Z. Goldberg, K. N. Raymond, C. A. Harmon, and D. H. Templeton, *J. Am. Chem. Soc.*, **96**, 1348 (1974).

$-50°C$ shows the methyl protons at $\delta = +3.60$ ppm and the ring protons at $\delta = +10.13$ ppm.[33]

Dianions of several of the large ring annulenes have also been prepared. The $4n + 2$ system [18]annulene, which has outer protons at $\delta = +9.3$ ppm and inner protons at $\delta = -3$ ppm, is converted by potassium to the dianion, a $4n$ system with outer protons at $\delta = -1$ ppm and inner protons at $\delta = +29$ ppm, the lowest field resonance known for a proton bound to carbon. (The largest known upfield shift, $\delta = -9$ ppm, occurs for the inner protons of an $18\,\pi$ electron $(4n + 2)$ mono-anion.)[34]

Another intriguing ion, hexachlorobenzene dication (**50**), a four π electron system, has been observed by Wasserman and his collaborators. As predicted by the simple molecular orbital energy-level pattern, the ion has two unpaired electrons. In this instance the distortion to alternating single and double bonds observed for cyclobutadiene does not occur.[35]

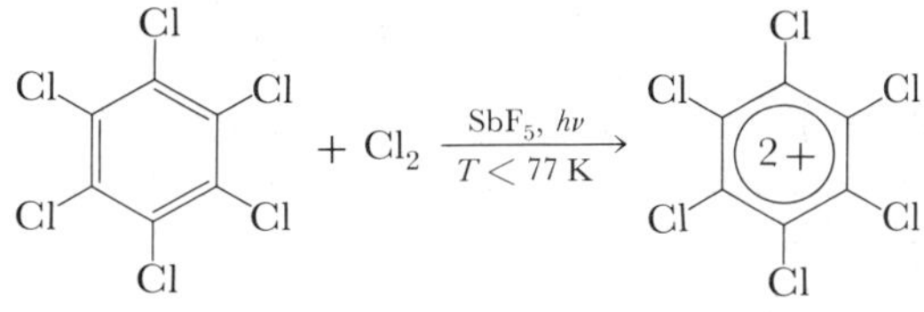

$$(1.15)$$

50

Odd-Membered Rings

Neutral rings composed of an odd number of C—H groups have an odd number of electrons and hence cannot be closed-shell molecules. In study of odd-membered rings, attention has focused on ions with even numbers of electrons, obtained by processes like those indicated in Equations 1.16 and 1.17. Rings containing a hetero atom that contributes two electrons to the π system are isoelectronic with the C—H ring anion of the same size. Rings containing one carbonyl group resemble the

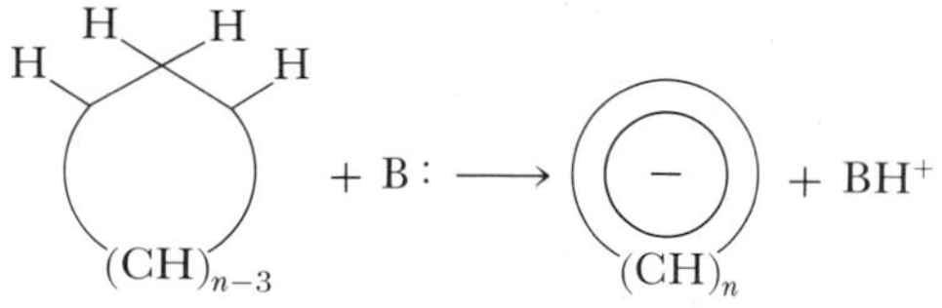

$$(1.16)$$

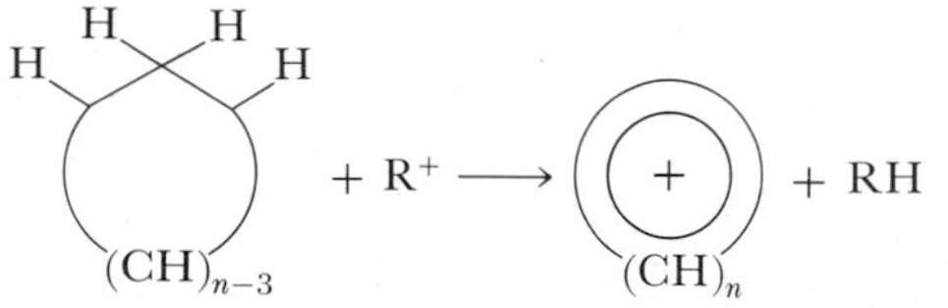

$$(1.17)$$

[33](a) G. A. Olah, J. S. Staral, and L. A. Paquette, *J. Am. Chem. Soc.*, **98**, 1267 (1976); (b) G. A. Olah, J. S. Staral, G. Liang, L. A. Paquette, W. P. Melega, and M. J. Carmody, *J. Am. Chem. Soc.*, **99**, 3349 (1977).
[34] See note 26.
[35] E. Wasserman, R. S. Hutton, V. J. Kuck, and E. A. Chandross, *J. Am. Chem. Soc.*, **96**, 1965 (1974).

C—H ring cation of the same size because the electron-withdrawing carbonyl oxygen leaves the carbonyl carbon electron-deficient. According to theory, rings of 3, 7, 11, . . . members should yield aromatic cations and antiaromatic anions, whereas rings of 5, 9, 13, . . . members should give aromatic anions and antiaromatic cations.

The best-known examples in this series are the cyclopentadienide anion (**51**) and the cycloheptatrienyl cation (**52**), both six π electron systems and both remarkably stable.[36]

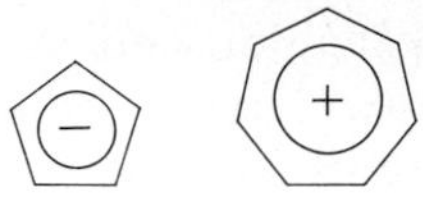

51   **52**

Attempts to prepare cyclopentadienone (**53**), an analog of $C_5H_5^+$, yield only the dimer (Equation 1.18).[37] Cycloheptatrienone (tropone, **54**), on the other hand, is stable and readily prepared by a number of methods.[38] Cycloheptatriene, however, resists conversion to an eight π electron anion. The acidity of cycloheptatriene

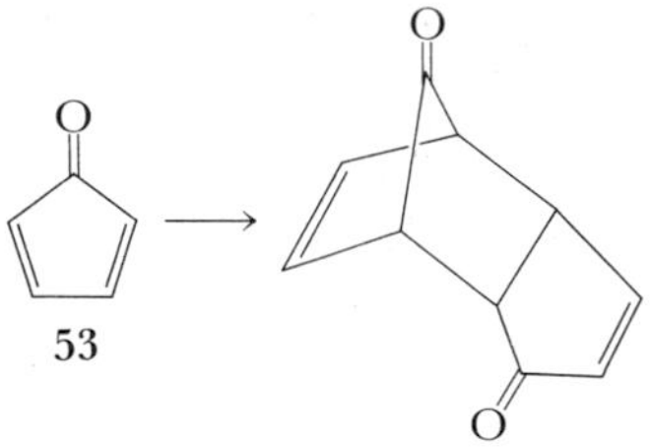

(1.18)

53

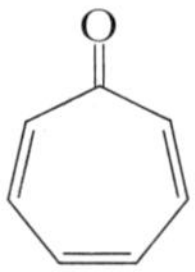

54

(Equation 1.19) is less than that of cyclopentadiene (Equation 1.20) by roughly 23 powers of ten,[39] a difference in reaction free energies of some 31 kcal mole^{-1}. Al-

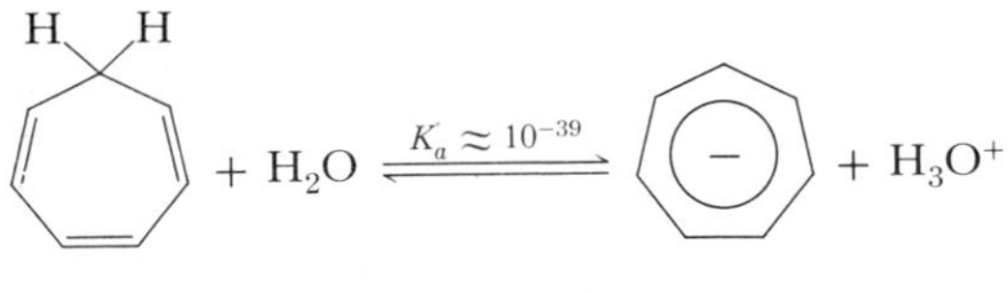

(1.19)

[36](a) Cyclopentadienide anion: P. L. Pauson, in *Non-Benzenoid Aromatic Compounds*, D. Ginsberg, Ed., Wiley, New York, 1959; (b) cycloheptatrienyl cation: F. Pietra, *Chem. Rev.*, **73**, 293 (1973).
[37]M. A. Ogliaruso, M. G. Romanelli, and E. I. Becker, *Chem. Rev.*, **65**, 261 (1965).
[38]See note 36 (b).
[39]See Table 3.2, pp. 260–261.

$$\text{H} \quad \text{H} \qquad + \text{H}_2\text{O} \xrightleftharpoons{K_a \approx 10^{-16}} \quad \ominus \quad + \text{H}_3\text{O}^+ \qquad (1.20)$$

though the eight π electron cycloheptatrienide anion forms reluctantly, the trianion, a ten π electron system, can be prepared by the method shown in Equation 1.21[40]

$$\xrightarrow[\text{H}_2\text{N(CH}_2)_4\text{NH}_2]{n\text{-C}_4\text{H}_9\text{—Li}} \quad \boxed{3-} \qquad (1.21)$$

Three-membered ring systems have also provided examples of aromatic and antiaromatic behavior. Despite the very substantial angle strain, Breslow and his collaborators have succeeded in preparing a number of cyclopropenyl cations (55).[41] Cyclopropenone (56) has been isolated and is stable in pure form below its melting point of -28 to $-29°\text{C}$ and in solution at room temperature,[42] even though the saturated analog cyclopropanone (57), which should be less strained, polymerizes rapidly in solution at room temperature by a self-addition to the carbonyl group which relieves some of the strain.[43] Analogs of 56, with C=S, S=O, and SO_2 groups in place of the C=O also have aromatic properties.[44]

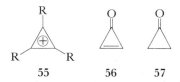

$$55 \qquad\qquad 56 \qquad\qquad 57$$

The contrasting reluctance of the three-membered ring π system to take on four electrons is illustrated by the very low acidity of triphenyl cyclopropene (58), estimated to be roughly 18 powers of ten less than that of triphenylmethane.[45] A number of ionic and hetero-atom large ring systems are also known.[46]

$$\phi \quad \text{H} \qquad + \text{H}_2\text{O} \xrightleftharpoons{K_a \approx 10^{-50}} \qquad + \text{H}_3\text{O}^+ \qquad (1.22)$$

$$58$$

[40] J. J. Bahl, R. B. Bates, W. A. Beavers, and C. R. Launer, *J. Am. Chem. Soc.*, **99**, 6126 (1977).
[41] See, for example, R. Breslow, H. Höver, and H. W. Chang, *J. Am. Chem. Soc.*, **84**, 3168 (1962).
[42] R. Breslow and M. Oda, *J. Am. Chem. Soc.*, **94**, 4787 (1972).
[43] N. J. Turro and W. B. Hammond, *J. Am. Chem. Soc.*, **88**, 3672 (1966).
[44] (a) H.-L. Hase, C. Müller, and A. Schweig, *Tetrahedron*, **34**, 2983 (1978); (b) C. Müller, A. Schweig, and H. Vermeer, *J. Am. Chem. Soc.*, **100**, 8056 (1978).
[45] R. Breslow and K. Balasubramanian, *J. Am. Chem. Soc.*, **91**, 5182 (1969).
[46] See note 26.

1.6 PERTURBATION THEORY

The process of forming molecular orbitals from atomic orbitals described in Section 1.2 may be thought of as the consequence of a *perturbation*. A perturbation is a disturbance (in this case the interaction between the atomic basis orbitals) applied to a starting state (the noninteracting basis atomic orbitals) that leads to a new, altered state (the molecular orbitals). The qualitative and quantitative assessment of the consequences of perturbations is known as *perturbation theory*.

Interactions Between Molecular Orbitals

In Section 1.2 we considered only interactions among atomic orbitals. It would be useful, however, to be able also to deal with interactions between molecular orbitals. For example, suppose that we have two molecules of ethylene, which are approaching each other as shown in **59** (wedge bonds in front of the plane of the page,

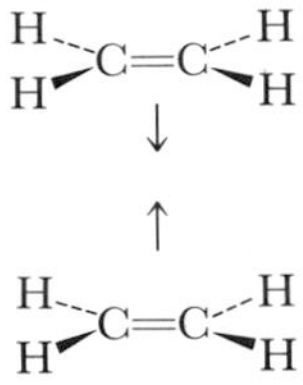

$$
\begin{array}{c}
\mathrm{H} \diagdown \quad \diagup \mathrm{H} \\
\qquad \mathrm{C}{=}\mathrm{C} \\
\mathrm{H} \diagup \quad \diagdown \mathrm{H} \\
\downarrow \\
\\
\uparrow \\
\mathrm{H} \diagdown \quad \diagup \mathrm{H} \\
\qquad \mathrm{C}{=}\mathrm{C} \\
\mathrm{H} \diagup \quad \diagdown \mathrm{H} \\
\mathbf{59}
\end{array}
$$

dashed bonds behind), so that the π orbitals of the two molecules come in closer and closer contact. We might ask whether the following reaction will occur:

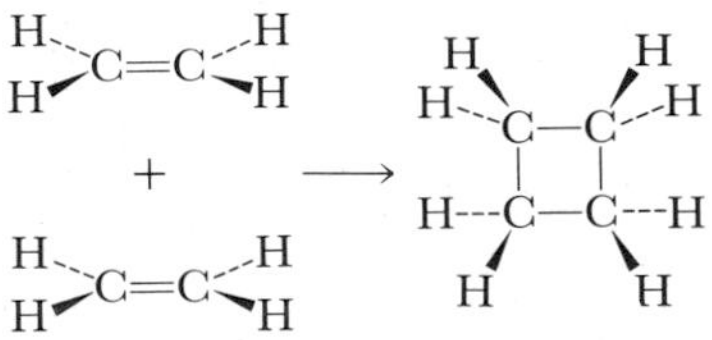

$$
\begin{array}{c}
\mathrm{H}{\diagdown}\mathrm{C}{=}\mathrm{C}{\diagup}\mathrm{H} \\
\mathrm{H}{\diagup}\quad\diagdown\mathrm{H} \\
+ \\
\mathrm{H}{\diagdown}\mathrm{C}{=}\mathrm{C}{\diagup}\mathrm{H} \\
\mathrm{H}{\diagup}\quad\diagdown\mathrm{H}
\end{array}
\longrightarrow
\begin{array}{c}
\mathrm{H}\quad\mathrm{H} \\
\mathrm{H}{-}\mathrm{C}{-}\mathrm{C}{-}\mathrm{H} \\
\ \ |\quad| \\
\mathrm{H}{-}\mathrm{C}{-}\mathrm{C}{-}\mathrm{H} \\
\mathrm{H}\quad\mathrm{H}
\end{array}
\tag{1.23}
$$

In order to answer this question, we need to be able to tell how the orbital energies of each molecule will be altered by the presence nearby of the orbitals of the other molecule. The situation is thus just the same as in bringing together two hydrogen atoms, except that here we are talking about molecular orbitals instead of atomic orbitals.

The same rules apply to interaction between two molecular orbitals as apply to interaction between two atomic orbitals. We shall be able to make a new model to cover the interacting situation by adding and subtracting the molecular orbital functions that were correct for the separate molecules before the interaction occurred.

From here on, then, everything we say will apply to orbitals in general, be they atomic or molecular.

Interaction of Basis Orbitals of the Same Energy

If the two interacting orbitals are of the same energy, the situation is just as described in Section 1.2 for the interaction of the two hydrogen $1s$ orbitals to give σ
bonding and antibonding MO's. We continue to use the term *basis* for the starting
orbitals, which we now identify as the orbitals appropriate to the unperturbed
situation. The perturbation (interaction) causes two new orbitals to form, one of
lower energy than the original unperturbed orbitals and one of higher energy.

The forms of the two new orbitals (the *perturbed* orbitals) are determined by
the following rule:

> **Rule 1.1.** In the *lower-energy orbital* the two basis orbitals are added
> together with the choice of relative signs that makes the result *bonding*
> in the region where the interaction is taking place. In the *higher-energy
> orbital* the two basis orbitals are added together with the choice of
> relative signs that makes the result *antibonding* in the region where the
> interaction is taking place.

The readers should apply this rule to the examples in Section 1.2 to convince
themselves that it is essentially equivalent to the methods used there.

When the unperturbed basis orbitals are of equal energy, each makes an
equal contribution to the perturbed orbital. In other words, at each point the values of
the two basis orbital functions are added (or subtracted, according to Rule 1.1)
with equal weights.

These ideas are best expressed by writing the orbital functions that result
from the perturbation in terms of the basis, Equations 1.24 and 1.25, where ψ_+ is
the lower-energy combination, ψ_- the higher-energy combination, and φ_1 and φ_2
the unperturbed basis orbital functions.[47]

$$\psi_+ = \varphi_1 + \varphi_2 \qquad (1.24)$$

$$\psi_- = \varphi_1 - \varphi_2 \qquad (1.25)$$

The magnitude of the energy lowering for the bonding combination, or
raising for the antibonding combination, is greater the greater the overlap between
the interacting orbitals. In the simplest theory the bonding combination is lowered
by the same amount as the antibonding is raised. More exact theory shows that the
antibonding combination is actually raised by somewhat more than the bonding
combination is lowered.

In summary, the case of basis orbitals of the same energy is exactly the one
we considered in Section 1.2.

---

[47]Normalizing factors are omitted from Equations 1.24, 1.25, 1.26, and 1.27.

Basis Orbitals of Different Energies

When the unperturbed basis orbitals are initially of unequal energies, the situation is altered only slightly. Again the perturbed orbitals are found by making linear combinations of the basis orbitals. But consider an electron initially in the lower-energy basis orbital. It is already in a relatively favorable situation; not much is to be gained by providing interaction with a second orbital that represents a higher energy state. So the orbital that was initially of lower energy will be altered relatively little. It will be perturbed and pushed to lower energy by adding in a contribution from the other orbital in a bonding way, but the change will be small. The basis orbital initially of higher energy, conversely, is going to be pushed still higher by an antibonding contribution from the first orbital, but again the contribution will be small.

These conclusions are summarized in Equations 1.26 and 1.27. Here φ_1 is the basis orbital of lower energy and φ_2 the basis orbital of higher energy, and λ is a number between zero and unity.[47] The farther apart φ_1 and φ_2 are in energy, the closer λ is to zero; the closer they are, the closer λ is to unity, until when the energies are the same, Equations 1.26 and 1.27 reduce to Equations 1.24 and 1.25. The changes in the orbitals brought about by adding in to each some portion of the other is sometimes called *mixing*. Figure 1.23 summarizes in an energy-level diagram the interaction between orbitals of different energies.

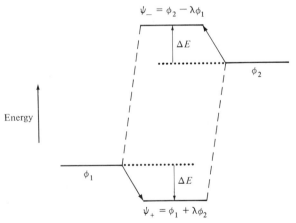

Figure 1.23 Energy changes of two interacting orbitals, φ_1 and φ_2 according to perturbation theory. φ_1 moves down by ΔE and φ_2 moves up by ΔE. The perturbed orbitals, ψ_+ and ψ_-, are linear combinations of φ_1 and φ_2 in which the lower-energy orbital, φ_1, has been modified by mixing in some of the higher ($\lambda \varphi_2$) in a bonding way and the higher-energy orbital, φ_2, has been modified by mixing in some of the lower ($-\lambda \varphi_1$) in an antibonding way. The magnitudes of energy change ΔE and of mixing parameter λ are larger the larger the overlap and are smaller the larger the initial energy difference. The solid arrows correlate each orbital with its perturbed counterpart, and the dashed lines indicate the sources of mixing that modify the form of each. A more exact treatment would show φ_2 raised by more than φ_1 is lowered.

$$\psi_+ = \varphi_1 + \lambda\varphi_2 \tag{1.26}$$

$$\psi_- = \varphi_2 - \lambda\varphi_1 \tag{1.27}$$

This discussion leads to the following rules for interactions between orbitals of different energies:[48]

Rule 1.2. When the unperturbed orbitals are initially of different energies, the perturbation causes the one initially lower in energy to be lowered further, and the one initially higher to be raised further. The energy changes are larger the larger the overlap, and smaller the larger the initial energy difference.

Rule 1.3. The orbital initially of lower energy is altered by adding in (mixing) a portion of the higher one in a bonding way. The orbital initially of higher energy is altered by adding in (mixing) a portion of the lower one in an antibonding way. The mixing in each case is larger the larger the overlap, and smaller the larger the initial energy difference.

1.7 SYMMETRY[49]

Symmetry plays a fundamental role in orbital theory. As has already been pointed out in Section 1.1, the nuclei provide a field of positive charge in which the electrons move. If the nuclei are arrayed in some symmetrical way, the field of positive charge will have that same symmetry, and the electron distribution, being controlled by the positive charge field, will also have the same symmetry.

It may be useful to illustrate this idea with one or two examples. The H_2 molecule has cylindrical symmetry. An electron that finds itself at a particular point off the internuclear axis experiences exactly the same forces as it would at another point obtained from the first by a rotation through any angle about the axis. The internuclear axis is therefore called an axis of symmetry; we have seen in Section 1.2 that such an axis is called an infinite-fold rotation axis, C_∞. Figure 1.24 illustrates the C_∞ symmetry and also some of the other symmetries. Perpendicular to the C_∞ axis and midway between the nuclei is a twofold rotation axis, called C_2; the C_2 operation consists of rotation by half a revolution, or $180°$. A C_2 axis can be

---

[48] For more quantitative statements of these rules, see (a) R. Hoffmann, *Acc. Chem. Res.*, **4**, 1 (1971); (b) L. Libit and R. Hoffman, *J. Am. Chem. Soc.*, **96**, 1370 (1974).

[49] The application of symmetry to chemical problems is treated in a number of texts on quantum chemistry. An introduction at an elementary level may be found in (a) H. H. Jaffé and M. Orchin, *Symmetry in Chemistry*, Wiley, New York, 1965; a more advanced treatment, including proofs of the important theorems of group theory is (b) M. Tinkham, *Group Theory and Quantum Mechanics*, McGraw-Hill, New York, 1964. See also (c) F. A. Cotton, *Chemical Applications of Group Theory*, 2d ed., Wiley, New York, 1971. An excellent insight into the nature of the quantum theory and its relationship with symmetry may be obtained from (d) R. P. Feynman, R. B. Leighton, and M. Sands, *The Feynman Lectures on Physics*, Vol. III, Addison-Wesley, Reading, Mass., 1965.

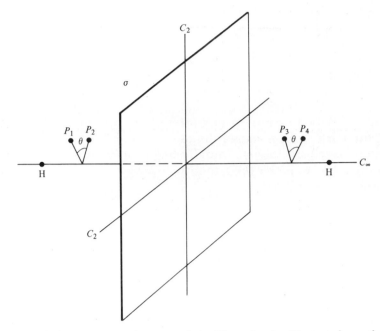

Figure 1.24 Some of the symmetry elements of the H_2 molecule. Illustrated are the infinite-fold rotation axis, the mirror plane perpendicular to it, and two of the twofold rotation axes. There are in addition an infinite number of C_2 axes in the same plane as those shown, an infinite number of mirror planes perpendicular to the one shown, and a point of inversion. P_1, P_2, P_3, and P_4 are symmetry equivalent points.

drawn at any angle around the internuclear line; there are actually infinitely many of them, although only two are shown in the figure. The H_2 molecule also has a mirror plane, abbreviated σ, a plane perpendicular to the C_∞ axis and halfway between the nuclei. To visualize the operation of reflection in this plane, imagine an infinitely thin two-sided mirror at the location of the plane; the reflection operation consists of replacing each point by its image point as seen in the mirror. The molecule also has infinitely many mirror planes perpendicular to the one illustrated, all containing the internuclear axis but lying at various angles around it. In addition there is a point of inversion on the C_∞ axis halfway between the nuclei. The operation of reflection through a point of inversion is accomplished by drawing a line from the point to be reflected to the point of inversion and extending the line an equal distance beyond. The resulting location is the image of the original point.

As an example of a simple molecule with different symmetry, consider HCl. The cylindrical symmetry, characterized by the C_∞ axis, is still present, as are the reflection planes that pass through the nuclei and contain the axis (Figure 1.25). But now the two ends of the molecule are not the same. The plane σ_h of Figure 1.24, the C_2 axes, and the point of inversion are no longer present. An electron still finds

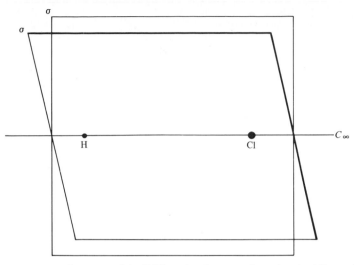

Figure 1.25 The symmetry elements of the HCl molecule are a C_∞ axis coinciding with the internuclear line and all the planes σ that contain the C_∞ axis.

the same forces at different points around the axis, but the situation near the
chlorine is quite different from that near the hydrogen.

As we have seen in Section 1.2, each mirror plane, axis, or point of inversion
is called a *symmetry element,* and the operations associated with these elements (reflection, rotation through a given angle) that leave the molecule exactly as before
are called *symmetry operations.* All the symmetry operations that leave a particular
object unchanged constitute the *symmetry group* of the object; there are only a limited number of these groups, and their properties have been thoroughly studied.[50]
Table 1.1 lists some of the commonly encountered molecular symmetry groups and
the symmetry elements they contain. Figure 1.26(a) illustrates the symmetry elements of the H_2O molecule, symmetry point group C_{2v}, and Figures 1.26(b)–
1.26(d) show the symmetry elements of the benzene molecule, symmetry group
D_{6h}. The H_2 molecule belongs to the symmetry group $D_{\infty h}$, and HCl belongs to $C_{\infty v}$.

We may summarize the significance of symmetry for electronic structure of
molecules by the following argument. A symmetry operation generates from any
arbitrary point in the molecule a new point. If the operation leaves the nuclei
unchanged, or interchanges equivalent nuclei so that the physical situation before
and after the operation is exactly the same, then the two points are equivalent
points, an electron will experience exactly the same forces at the two points, and the
probability of finding the electron in the neighborhoods of the two points must be
the same. In terms of orbital theory, this conclusion means that the function ψ^2
describing the electron density must be the same at the two points, or, put another

---

[50] Refer to the sources cited in note 49 (a), (b), and (c).

Table 1.1 SOME COMMON SYMMETRY POINT GROUPS AND THE
SYMMETRY ELEMENTS BELONGING TO THEM[a,b,c]

Group	Symmetry Elements	Example
C_1	None	
C_s	σ	
C_i	i	
C_2	C_2	
$C_{2v}{}^d$	$C_2\ 2\sigma_v{}^d$	
C_{3v}	$C_3\ 3\sigma_v$	
C_{4v}	$C_4\ C_2'\ 2\sigma_v\ 2\sigma_d{}^d$	
$C_{2h}{}^d$	$C_2\ \sigma_h{}^d\ i$	
D_{2h}	$3C_2{}^d\ 3\sigma\ i$	

way, ψ^2 must have the full symmetry of the molecule. If, then, we are constructing the molecular orbitals of a molecule with several elements of symmetry, the electron distribution, ψ^2, of each molecular orbital must remain unchanged under all the symmetry operations appropriate to the molecule itself.

We may carry the argument one step further. If ψ^2 is to remain unchanged under a given operation, the orbital function ψ may either remain unchanged or may change sign everywhere, but must not change in absolute value. Another way of expressing this idea is to say that the transformation property of ψ must be either

Table 1.1 (*Continued*)

Group	Symmetry Elements	Example
D_{4h}	$C_4 \ C_2'^{d} \ 4C_2 \ \sigma_h \ 2\sigma_v \ 2\sigma_d \ S_4^{d} \ i$	
D_{6h}	$C_6 \ C_3' \ C_2' \ 6C_2 \ \sigma_h$ $3\sigma_v \ 3\sigma_d \ S_6 \ S_3 \ i$	
T_d	$4C_3 \ 3S_4 \ 3C_2 \ 6\sigma_d$	$\begin{array}{c} H \\ \diagdown \\ H{-}C{-}H \\ \diagup \\ H \end{array}$
$C_{\infty v}$	$C_\infty \ \infty\sigma_v$	H—Cl
$D_{\infty h}$	$C_\infty \ \infty C_2 \ S_\infty \ \infty\sigma_v \ \sigma_h \ i$	H—H

[a] The term *point group* refers to the fact that all the operations leave the object at the same point in space. Other operations, such as translation, are included in the *space groups*, which describe, for example, the repeated patterns of wallpaper or the arrangement of ions or molecules in a crystal.

[b] This table lists all symmetry *elements* of each group but does not include the full systematic lists of symmetry *operations* that make up the mathematically correct descriptions of the groups. For example, a C_3 axis has associated with it two distinct operations: clockwise rotation by 120° and counterclockwise rotation by 120°.

[c] For more information, illustrations, and tables of all the point groups, see H. H. Jaffe and M. Orchin, *Symmetry in Chemistry*, Wiley, New York, 1965.

[d] Subscript v means the mirror plane contains the axis of highest order (principal axis). A subscript h means the mirror plane is perpendicular to the principal axis. A subscript d indicates a mirror plane bisecting the angle made by a pair of C_2 or C_3 axes. A number in front of an element symbol specifies the number of elements of that type present. A prime indicates an axis coincident with the principal axis. S_n elements, called improper axes, are elements whose operation consists of n-fold rotation followed by reflection in a plane perpendicular to the axis. See note 2, Chapter 2, for further discussion of improper axes.

symmetric or antisymmetric under each symmetry operation.[51] An orbital that is symmetric with respect to a given symmetry operation will look exactly the same after that symmetry operation has been carried out as it did before. An orbital that is antisymmetric will have its sign reversed at every point by the operation. An orbital that satisfies one or the other of these criteria is said to be *symmetry correct*

---

[51] It might at first seem inconsistent to allow ψ to change sign under a symmetry operation; it must be remembered, however, that ψ is not itself a physically observable quantity. All that can be required is that the physically meaningful quantity ψ^2 obey the symmetry; this restriction is satisfied equally well for ψ symmetric or antisymmetric.

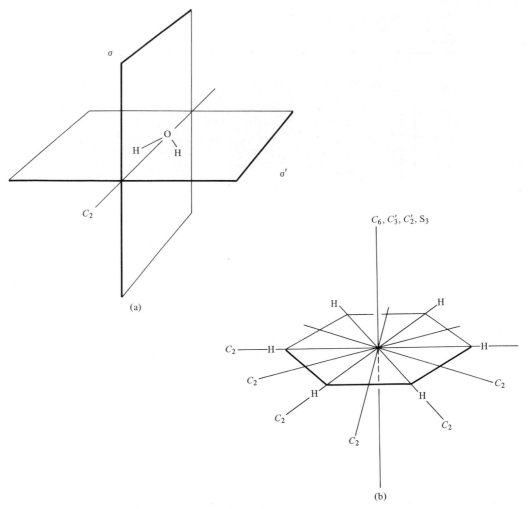

Figure 1.26 Symmetry elements. (a) Of the H_2O molecule. (b) Of the D_{6h} point group illustrated with the benzene molecule. The rotation axes C_6, C_3', C_2', and S_3, all of which coincide on a single line perpendicular to the plane of the ring and passing through its center.

with respect to the operation in question. To be completely symmetry correct, an orbital must meet the test for each of the symmetry operations that leaves the molecule unchanged.

The Application of Symmetry to Orbital Interaction

To illustrate the use of symmetry, let us examine a simple example. Consider the possibility of an interaction between the vacant p orbital in a methyl cation and the $1s$ orbital of one of the hydrogens. Figure 1.27 shows that a mirror plane is a

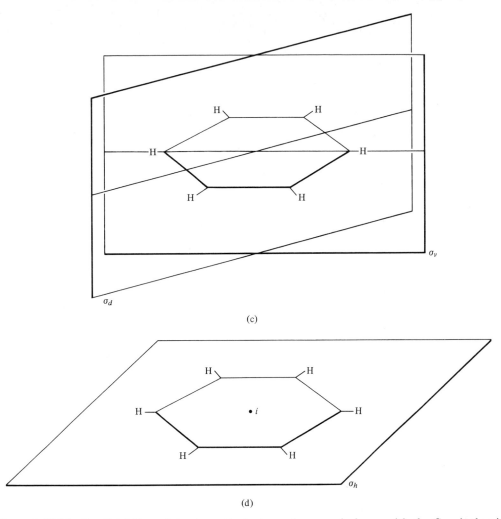

Figure 1.26 (*Continued*) (c) Two of the six vertical mirror planes, each shown with the C_2 axis that it contains (the other four mirror planes are omitted for clarity). (d) The horizontal mirror plane and the point of inversion.

symmetry element of the molecule. The p orbital is antisymmetric with respect to the mirror plane, since it changes sign on reflection, whereas the s orbital is symmetric because it is unchanged on reflection. Interaction between the s and the p orbitals will be possible only if their overlap is nonzero. Recall from Section 1.2 that the overlap is found by evaluating the product of the two orbital functions at each point and summing over all points. Inspection of Figure 1.27 will show that every contribution to overlap above the mirror plane will be canceled by a contribution of the same magnitude but opposite sign below the plane. There is therefore

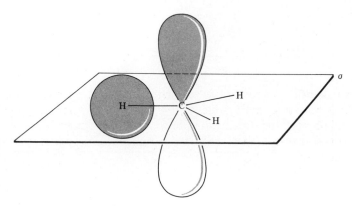

Figure 1.27 Mirror symmetry in a methyl cation. The p orbital is antisymmetric with respect to σ, and the s orbital of hydrogen is symmetric. The two have different symmetries and cannot interact.

no overlap and no interaction. This situation arises because one of the orbitals changes sign on reflection and the other does not.

The conclusion is general. Whenever two orbitals have opposite symmetry behavior with respect to a symmetry element, their overlap and therefore their interaction will be zero. If there are several symmetry elements, a symmetry mismatch with respect to any one of them will prevent interaction. We can summarize the argument as follows:

> **Only orbitals of the same symmetry can interact.**

Formation of Symmetry Correct Orbitals

In order to take advantage of the symmetry criterion for orbital interaction, it is necessary to have orbitals that are symmetry correct. The molecular orbitals obtained from atomic orbitals by the methods described in Sections 1.2 and 1.6 will sometimes be symmetry correct and sometimes not.

As an example, let us look at a bonding model for the H_2O molecule. The Lewis structure (**60**) leads to the choice of sp^3 hybridization on the oxygen. We then

$$:O:$$
$$H \qquad H$$

60

use two of the hybrids to make σ bonding and antibonding pairs by combination with the hydrogens. Let us focus on the two bonding σ orbitals. One of these, ψ_1, is shown in **61**, the other, ψ_2, in **62**. Next we identify the symmetry elements of the molecule. These are two mirror planes and a C_2 rotation axis (Figure 1.26(a)). Inspection of the orbital ψ_1 will show that it is symmetry correct with respect to the reflection σ′, since it is transformed into itself (i.e., is completely unchanged) by this

61   62

reflection. But it is not symmetry correct with respect to σ or C_2. These operations each transform ψ_1 into another orbital of the same shape but in a different location, as illustrated in **63** for the reflection. (The operation C_2 yields the same result.) Since ψ_1 is changed into neither itself nor its negative by these two operations, it is not symmetry correct.

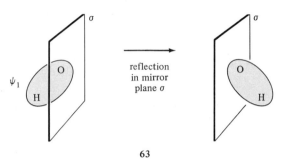

63

It is evident from a comparison of **63** and **62** that the reflection transforms ψ_1 into ψ_2. The reader should verify that the σ reflection and the C_2 rotation also transform ψ_2 into ψ_1, and that ψ_2 is therefore also not symmetry correct. The result we have found will hold when molecular orbitals constructed by the LCAO method from equivalent hybrid atomic orbitals are subjected to symmetry operations. Each of those orbitals in the set of MO's that is not already symmetry correct will be transformed by a symmetry operation into another orbital of the set.

How can we obtain symmetry correct orbitals to use in place of ψ_1 and ψ_2? The procedure is simple and follows the method we always use in dealing with changes in forms of orbitals. We combine ψ_1 and ψ_2 into a new set of two by adding and subtracting them as follows:[52]

$$\Psi_S = \psi_1 + \psi_2 \tag{1.28}$$
$$\Psi_A = \psi_1 - \psi_2 \tag{1.29}$$

These new orbitals are shown in **64** and **65**. The reader should verify that they are symmetry correct.

64   65

[52]Normalization omitted.

The symmetry correct orbitals can now be classified as to symmetry type. The positive combination (**64**) is unchanged by reflection in mirror plane σ or by rotation C_2; we therefore call it *symmetric* (*S*). The negative combination (**65**) changes sign on reflection or on rotation, and is *antisymmetric* (*A*). It happens in this instance that each of our new orbitals behaves the same way under reflection as it does under rotation; note, however, that Ψ_S and Ψ_A are both symmetric on reflection in mirror plane σ'. In general, a given orbital may be symmetric under some symmetry operations and antisymmetric under others.

The energy changes that accompany the construction of the symmetry correct orbitals are easily obtained qualitatively from our perturbation rules. The combinations represent interactions of ψ_1 and ψ_2, but the overlap between these two is small. (Remember that the reason we decided to use hybrid orbitals in the first place was precisely so that the overlaps would be confined to a localized bonding region.) The energy of interaction will therefore be small, even though the two interacting orbitals ψ_1 and ψ_2 are of the same energy. Figure 1.28 shows the changes. Note that the conclusion that both of the new orbitals will still be bonding is confirmed by the observation that both orbitals are bonding in the regions between the oxygen and the hydrogens.

Let us summarize the steps to be followed in obtaining symmetry correct orbitals.

1. Construct a bonding model of molecular orbitals from a basis set of atomic orbitals.

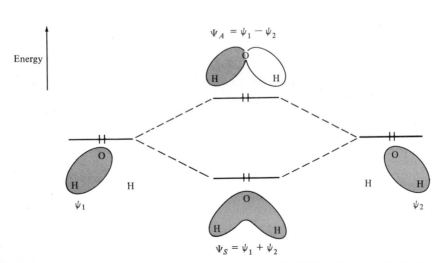

Figure 1.28 Energy changes in the interaction of two O—H bonding σ orbitals ψ_1 and ψ_2 of the water molecule to form the symmetry correct orbitals Ψ_S and Ψ_A. Energy changes are small because overlap is small; both orbitals remain bonding.

2. Identify the symmetry elements of the molecule.

3. Examine each molecular orbital to see whether it is already symmetry correct.

4. For orbitals that are not already symmetry correct, group together orbitals that are transformed into each other by the symmetry operations.

5. Form sums and differences within these symmetry related groups to obtain symmetry correct orbitals. Remember that the total number of orbitals obtained at the end of the process must always equal the total number at the start.

6. It may sometimes happen that not all symmetry elements are needed. Orbitals can be made symmetry correct with respect to selected symmetries, as long as subsequent conclusions are based on these symmetries only.

These procedures will be successful as long as there is no rotational symmetry axis of order higher than two. If there is a threefold or higher axis, some further complications arise. Because of Step 6, it will often be possible to make symmetry arguments on the basis of other symmetry elements in cases of this kind, and we shall not need to consider this problem further.

Localized and Delocalized Models

An obvious question is, when does one need symmetry correct orbitals? The H_2O example illustrates that the symmetry correct orbitals will usually extend over a larger region of the molecule than did the symmetry incorrect orbitals from which they were made. The symmetry correct model corresponds to a more highly delocalized picture of electron distribution. We believe that electrons are actually able to move over the whole molecule, and in this sense the delocalized symmetry correct pictures are probably more accurate than their localized counterparts.[53] Nevertheless, for most purposes we are able to use the more easily obtained localized model. The reason the localized model works is illustrated in Figure 1.28. The interaction that produced the delocalized symmetry correct orbitals made one electron pair go down in energy and another go up by an approximately equal amount. Thus the *total* energy of all the electrons in the molecule is predicted to be about the same by the localized and by the more correct delocalized model.

The key point here is that both of the orbitals were filled to start with. Delocalization of orbitals already filled will not affect the total energy in the first approximation, whereas delocalization involving orbitals not completely filled, as in the π systems considered in Section 1.4, will lead to significant lowering of the total energy and so must be considered. If we are interested in energies of electrons in *particular* individual orbitals, or if we want to take advantage of the help symmetry can give in evaluating interactions between orbitals within molecules or be-

---

[53] Computer-generated diagrams of delocalized orbitals of small molecules are available. See (a) W. L. Jorgensen and L. Salem, *The Organic Chemist's Book of Orbitals,* Academic Press, New York, 1973; (b) A. Streitwieser, Jr., and P. H. Owens, *Orbital and Electron Density Diagrams,* Macmillan, New York, 1973; (c) J. R. Van Wazer and I. Absar, *Electron Densities in Molecules and Molecular Orbitals,* Academic Press, New York, 1975; (d) I. Absar and J. R. Van Wazer, *Angew. Chem. Int. Ed. Engl.,* **17, 80** (1978).

tween orbitals on different molecules, we shall need symmetry correct orbitals. Otherwise, because total energies are reasonably accurately represented by a model in which completely filled levels are localized, we can use the localized model for these filled orbitals even though some of them may not be symmetry correct.

An Application of Perturbation Theory. Interaction of a Methyl Group with an Adjacent Cationic Center

As an example of the use of perturbation theory and symmetry, consider the stabilization that results when a hydrogen of the methyl cation is replaced by a methyl group to form the ethyl cation.

In CH_3^+ itself (**66**), we have already seen that the different symmetries of the C—H bonds and the vacant p orbital prevent any interaction. Therefore no electrons from the C—H bonds can be delocalized into the p region to help disperse positive charge toward the hydrogens.

$$\left[\begin{array}{c} H - C \overset{H}{\underset{H}{\diagdown}} \end{array} \right]^+$$

66

Now substitute a CH_3 group for one hydrogen. Our starting point will be a model in which we suppose that there is no interaction between the new group and the vacant p orbital. We will then use the perturbation theory to see how allowing such an interaction modifies the model. We must first decide which conformation we want to consider, and we will choose **67**, in which H_c lies in the molecular plane, and H_a and H_b are above and below. Next, we need a bonding model for the CH_3 group. An obvious choice is the set of σ MO's obtained from sp^3 hybrids on C_2 and $1s$ orbitals on H_a, H_b, and H_c. The bonding members of this set are shown in **68**.

67

68

(For this particular problem we shall not need the antibonding orbitals; their importance in other circumstances will become clear later.) We have already concluded that the C_1—H bonds cannot interact with the vacant p orbital, and we need not consider them further. In Figure 1.29 the methyl orbitals are shown individually in order to make clear their symmetry properties with respect to the mirror plane that is the only symmetry element of the methyl-substituted ion. Orbital ψ_c is symmetry correct; it is symmetric and therefore of different symmetry from the vacant p orbital on C_1. The C_2—H_c bond, like the C_1—H bonds, cannot interact with the p orbital. We therefore need not consider H_c or ψ_c further.

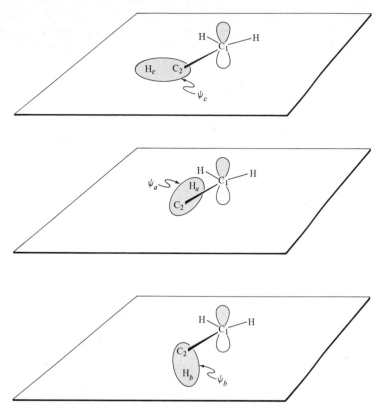

Figure 1.29 The three bonding C—H orbitals of the methyl group in $CH_3CH_2^+$ illustrated in the conformation

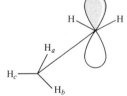

ψ_c is symmetry correct with respect to mirror plane σ, but ψ_a and ψ_b are not.

The orbitals ψ_a and ψ_b are not symmetry correct. They are, however, transformed into each other on reflection, so we can take combinations according to Equations 1.30 and 1.31.[54] We expect Ψ_S to be slightly lower in energy than Ψ_A. Figure 1.30 shows these new symmetry correct orbitals.

$$\Psi_S = \psi_a + \psi_b \tag{1.30}$$

$$\Psi_A = \psi_a + \psi_b \tag{1.31}$$

---

[54] Normalization omitted.

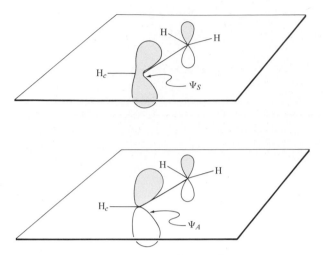

Figure 1.30 The symmetry correct C_2—H orbitals Ψ_S and Ψ_A.

Now we are ready to assess the interactions of the electrons in the C—H_a and C—H_b bonds with the vacant p orbital on C_1. First we notice that of our two new symmetry correct orbitals, only Ψ_A has the same symmetry as the vacant p orbital. Since Ψ_S has different symmetry, it cannot interact and need not be considered further. With the help of the symmetry we have reduced what looked initially like a moderately complicated problem to a simple one. We have only to consider the interaction of two orbitals. In Figure 1.31 we have placed on the left at the nonbonding energy the vacant p orbital. On the right, lower in energy because it is a bonding orbital, is Ψ_A. The two interact according to the perturbation theory rules, as shown at the center. Orbital Ψ_A moves down, carrying with it its electron pair, while the p moves up. Since the p is unoccupied, there is no change in the total energy contributed by its moving up. We conclude that the interaction is stabilizing to the ion as a whole, since one electron pair moves to lower energy.

If we look at the form of the perturbed Ψ_A, we see that it has been altered by mixing in a small contribution of the p orbital of C_1. This change means that the electron pair that in the unperturbed model was confined to the CH_3 group is actually free, according to the perturbed model, to wander into the electron-deficient p orbital. The perturbation has therefore introduced an electron donation by CH_3 to the C^+.

This delocalization, which would be represented in a Lewis structure model by adding a contribution from resonance structure **69**, is commonly known as *hyperconjugation*. It should be clear from this example that the hyperconjugation

$$^+C—C{\overset{H}{\underset{H}{-}}}H \longleftrightarrow C{=}C{\overset{H^+}{\underset{H}{-}}}H$$

69

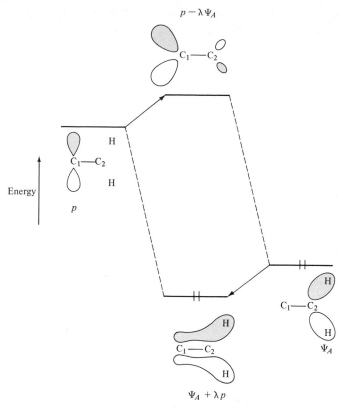

Figure 1.31 Interaction of the methyl group orbital Ψ_A with the p orbital of the cationic carbon. Ψ_A moves down, stabilizing the structure. Its perturbed form shows delocalization of the electron pair into the cation p orbital. Orbital p moves up, but it is unoccupied and so does not alter the energy.

concept arises solely from our particular model-building procedures. When we ask whether hyperconjugation is important in a given situation, we are asking only whether the localized model is adequate for that situation at the particular level of precision we wish to use, or whether the model must be corrected by including some delocalization in order to get a good enough description. In the case of the carbocation problem we know experimentally that the ion is significantly stabilized by methyl substitution; we therefore say that some delocalization is needed in the model. Note that the interaction is not restricted to C—H bonds, as has sometimes been assumed, but should occur with C—C bonds also, a conclusion that is in agreement with experimental results.[55] The arguments are easily extended to show that with a CH_2X substituent the preferred conformation should be determined by

[55](a) R. C. Bingham and P. v. R. Schleyer, *J. Am. Chem. Soc.*, **93**, 3189 (1971); (b) K. Eberl, and L. Mayring, *Angew. Chem. Int. Ed. Engl.*, **15**, 554 (1976).

the electronegativity of X compared with H.[56] We leave it as an exercise to the reader (Problems 10 and 11) to analyze the interaction between an anion center and an adjacent methyl group, and between a cation center and an adjacent unshared pair.[57]

1.8 INTERACTIONS BETWEEN MOLECULES

One of the most useful applications of the symmetry and perturbation theories is to interactions between molecules.

HOMO–LUMO Interactions

As long as the molecules whose interaction we want to consider are far apart, each has its own set of molecular orbitals undisturbed by the other. These MO's form the unperturbed basis from which the interaction is to be evaluated. As the molecules approach sufficiently closely that overlap between their orbitals becomes significant, the new interaction constitutes a perturbation that will mix orbitals of each molecule into those of the other. The strongest interactions will be between those orbitals that are close to each other in energy, but interaction between two filled levels will cause little change in the total energy because one orbital moves down nearly as much as the other moves up. The significant interactions are therefore between filled orbitals of one molecule and empty orbitals of the other; furthermore, since the interaction is strongest for orbital pairs that lie closest in energy, the most important interactions are between the highest occupied molecular orbital (HOMO) of one molecule and the lowest unoccupied molecular orbital (LUMO) of the other. These orbitals are sometimes referred to as the *frontier orbitals*. Figure 1.32 illustrates this conclusion for the approach of two molecules that have their HOMO and LUMO levels at comparable energies. If HOMO–LUMO interaction cannot occur, for example, because the orbitals are of different symmetry types, this stabilizing interaction is absent, the small energy increase arising from the filled level interactions will dominate, and no reaction will occur.

It is instructive also to look at an example in which the HOMO levels of the two molecules are of different energies. In Figure 1.33 the HOMO and LUMO levels are indicated for molecules D and A, D having its highest filled level substantially higher than that of A. This is a donor–acceptor situation, with D the donor and A the acceptor. Note that the HOMO of D is much closer in energy to the LUMO of A than in the previous example, but the A HOMO is much farther from the D LUMO. Hence the A HOMO will be relatively little affected, and most of the stabilization will occur by lowering the D HOMO. As it is lowered, it will mix in substantial amounts of the A LUMO; charge is thereby transferred from D to A.

---

[56] R. Hoffmann, L. Radom, J. A. Pople, P. v. R. Schleyer, W. J. Hehre, and L. Salem, *J. Am. Chem. Soc.*, **94**, 6221 (1972).
[57] For further discussion of hyperconjugation, see (a) M. J. S. Dewar, *Hyperconjugation*, Ronald Press, New York, 1962; (b) D. Holtz, *Prog. Phys. Org. Chem.*, **8**, 1 (1971).

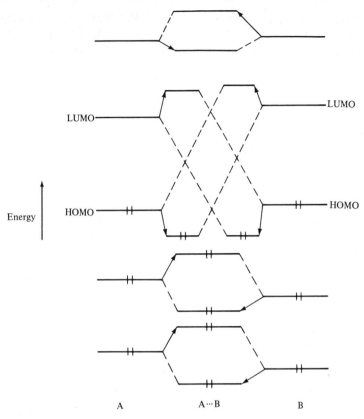

Figure 1.32 Interaction between molecular orbitals of A and B as they approach each other. Perturbations between filled levels have little effect on the total energy. Interaction of HOMO of B with LUMO of A and of HOMO of A with LUMO of B lowers the energy significantly because filled levels move to lower energy while unfilled levels move to higher energy.

Note that charge transfer occurs primarily to the lowest antibonding acceptor orbital.

Applications of HOMO–LUMO Theory

Qualitative arguments based on the HOMO–LUMO theory are applicable to a variety of chemical processes.[58] Consider as an example Reaction 1.32. Will this reaction occur in a single step through a cyclic transition state (Equation 1.33)? Since fluorine is more electronegative than hydrogen, F_2 is the acceptor and H_2 is

$$H_2 + F_2 \longrightarrow 2HF \tag{1.32}$$

[58] See, for example: (a) R. G. Pearson, *Acc. Chem. Res.*, **4**, 152 (1971); (b) R. G. Pearson, *J. Am. Chem. Soc.*, **94**, 8287 (1972); (c) G. Klopman, in *Chemical Reactivity and Reaction Paths*, G. Klopman, Ed., Wiley, New York, 1974, p. 55; (d) I. Fleming, *Frontier Orbitals and Organic Chemical Reactions*, Wiley, London, 1976.

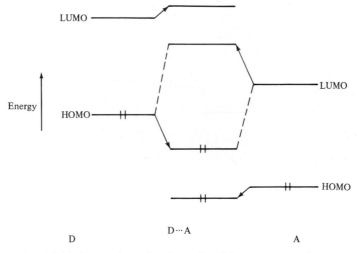

Figure 1.33 HOMO–LUMO interaction of a donor D with an acceptor A.

$$
\begin{array}{ccc}
\text{H—H} \\
& \longrightarrow & \begin{bmatrix} \text{H}\cdots\text{H} \\ \vdots \quad \vdots \\ \text{F}\cdots\text{F} \end{bmatrix} & \longrightarrow & \begin{array}{cc} \text{H} & \text{H} \\ | & | \\ \text{F} & \text{F} \end{array}
\end{array} \tag{1.33}
$$

the donor. Figure 1.34 shows the frontier orbitals of F_2 and of H_2 as the molecules are brought together. If these orbitals are put together in the geometry of the hypothetical transition state (**70**), it becomes clear that their different symmetries prevent interaction. Hence no stabilization can occur, and we would expect this reaction to have a high energy barrier. HOMO–HOMO and LUMO–LUMO interactions can occur (Figure 1.34), but these do not lower the energy.

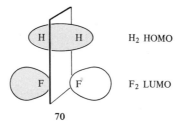

70

If we try instead a mechanism in which a hydrogen atom attacks the F_2 molecule at an end (Figure 1.35), there is no symmetry property that prevents HOMO–LUMO interaction. The perturbed orbital at the center in Figure 1.35 shows that the electron from hydrogen is being transferred to the fluorine.[59]

---

[59] Pearson [see note 58 (a) and (b)] has pointed out that the essential feature of a bond-breaking reaction that will occur with a low activation energy is transfer of electron density out of orbitals that are bonding or into orbitals that are antibonding with respect to the bond being broken. Bond formation will be favored by transfer of electron density into orbitals that are bonding or out of orbitals that are antibonding with respect to the bond being formed.

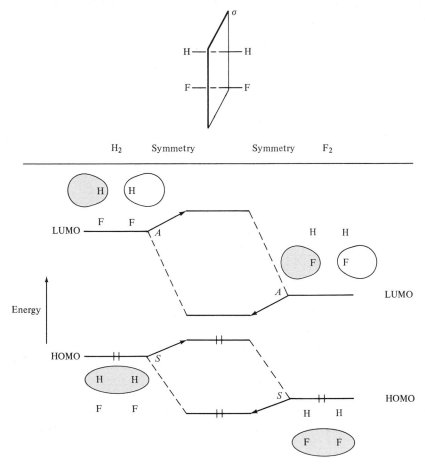

Figure 1.34 In a four-center approach of H_2 and F_2, symmetry prevents HOMO–LUMO interaction. Only HOMO–HOMO and LUMO–LUMO interactions can occur, and these do not lower the energy. Symmetries of the orbitals are indicated with respect to the mirror plane shown.

Bimolecular Aliphatic Substitution

Another application is to bimolecular S_N2 and S_E2 substitutions. As we shall see in more detail in Chapter 4, the nucleophilic reaction prefers backside attack by nucleophile on substrate, whereas the electrophilic reaction (Chapter 6) often prefers frontside attack. Figure 1.36 shows the appropriate frontier orbitals for frontside attack by a nucleophile. The nucleophile, symbolized by N, is the donor, and the C—X bond is the acceptor. In this example, because C and X are different, there is not strictly speaking any symmetry element present that will be of use in classifying the orbitals. However, we observe that the main feature of the C—X antibonding orbital is a node between the C and X nuclei. This node is not actually

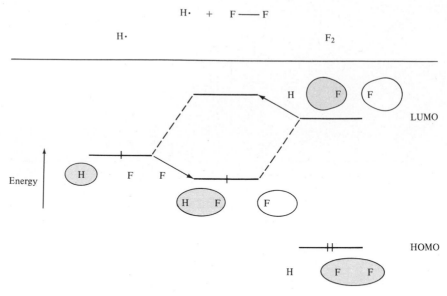

Figure 1.35 In the reaction $H \cdot + F_2 \longrightarrow HF + F \cdot$, the hydrogen atom HOMO can interact with the F_2 LUMO and stabilization occurs. (There will also be a smaller interaction of the $H \cdot$ orbital with the F_2 HOMO.)

symmetrically placed exactly between the nuclei as it would be if the two ends of the bond were identical, but for purposes of qualitative argument we can say that the node is approximately symmetrically placed. The mirror plane shown in Structures **71** and **72**, and in Figures 1.36 and 1.38, is thus an approximate symmetry element of the system. With respect to this mirror plane, the symmetries of the nucleophile HOMO and the C—X LUMO do not match (**71**); therefore only the filled–filled HOMO–HOMO interaction is possible and no stabilization can occur.[60] If, on the other hand, the nucleophile approaches at the back (Figure 1.37), stabilization is possible and the reaction can occur.

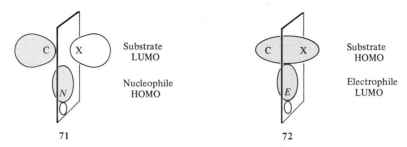

71   **72**

[60] N. Trong Anh and C. Minot, *J. Am. Chem. Soc.*, **102**, 103 (1980).

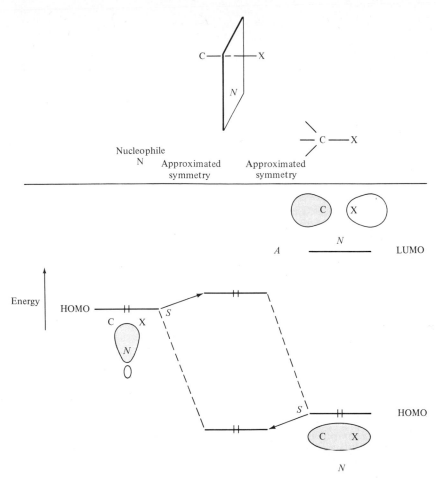

Figure 1.36 Frontside attack of a nucleophile, symbolized by *N*, on a C—X bond. Symmetry prevents HOMO–LUMO interaction; the only interaction is between filled levels. The reaction will not take this path. Because C and X are different, the plane indicated at top center is only an approximate symmetry element for the system.

In electrophilic substitution the electrophile is the acceptor and the C—X bond the donor. Now frontside attack gives donor HOMO and acceptor LUMO of the same approximated symmetry (**72**), and stabilization can occur (Figure 1.38).

It is possible to be misled by HOMO–LUMO arguments, because they are greatly oversimplified. Unwary acceptance of Figures 1.37 and 1.38, for example, might lead one to suppose that the reactions in question should have no activation energy, since only stabilization occurs. There are obviously other factors that are not shown in diagrams of this kind, in particular, energy increases caused by

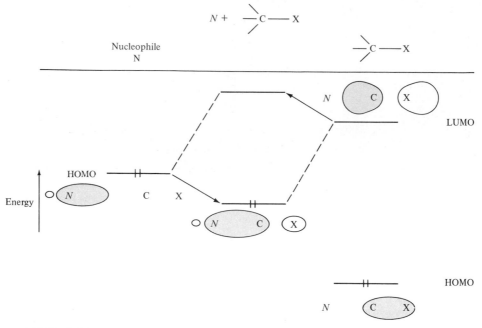

Figure 1.37 Backside attack by a nucleophile. Now HOMO–LUMO interaction is possible, and the transition state is stabilized. (As in the H · + F_2 process (Figure 1.35), there will also be a smaller HOMO–HOMO interaction here. Because it is a filled–filled interaction, it will not alter the energy.)

stretching bonds that will break. Perturbation and symmetry arguments give only a rough guide to the gross features of a reaction, but since they are often successful at picking out the dominant energy changes, they are particularly useful for qualitative understanding.[61]

1.9 APPLICATION OF PERTURBATION THEORY AND SYMMETRY TO π SYSTEMS

The perturbation and symmetry theories can be used to find molecular orbitals of conjugated chains and to explain the aromaticity $4n + 2$ rule.

[61] For other examples see (a) R. Hoffmann, *Acc. Chem. Res.*, **4**, 1 (1971); (b) W. Stohrer and R. Hoffmann, *J. Am. Chem. Soc.*, **94**, 779 (1972); (c) R. Hoffmann, R. W. Alder, and C. F. Wilcox, Jr., *J. Am. Chem. Soc.*, **92**, 4992 (1970); (d) W. Stohrer and R. Hoffmann, *J. Am. Chem. Soc.*, **94**, 1661 (1972); (e) R. Gleiter, W. Stohrer, and R. Hoffmann, *Helv. Chim. Acta*, **55**, 893 (1972); (f) R. Hoffmann and R. B. Davidson, *J. Am. Chem. Soc.*, **93**, 5699 (1971); (g) R. Hoffmann, L. Radom, J. A. Pople, P. v. R. Schleyer, W. J. Hehre, and L. Salem, *J. Am. Chem. Soc.*, **94**, 6221 (1972); (h) D. B. Boyd and R. Hoffmann, *J. Am. Chem. Soc.*, **93**, 1064 (1971); (i) S. David, O. Eisenstein, W. J. Hehre, L. Salem, and R. Hoffmann, *J. Am. Chem. Soc.*, **95**, 3806 (1973); (j) R. Hoffmann, C. C. Levin, and R. A. Moss, *J. Am. Chem. Soc.*, **95**, 629 (1973).

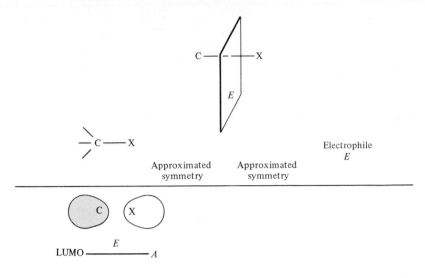

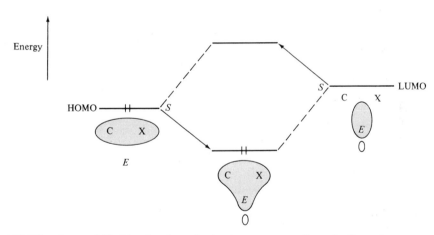

Figure 1.38 The electrophilic bimolecular substitution can occur from the front, because symmetry permits HOMO–LUMO interaction and stabilization.

Molecular Orbitals of Linear π Systems

Ethylene, the first member of the series, is already familiar. Two adjacent p orbitals interacting will yield a bonding orbital, symmetric with respect to reflection in the mirror plane lying midway between the two carbons and perpendicular to the C—C axis, and an antibonding orbital antisymmetric with respect to that mirror plane. Figure 1.39 illustrates the interaction and the resulting orbitals. In this figure a new convention is introduced for illustrating orbitals. Molecular orbitals, instead

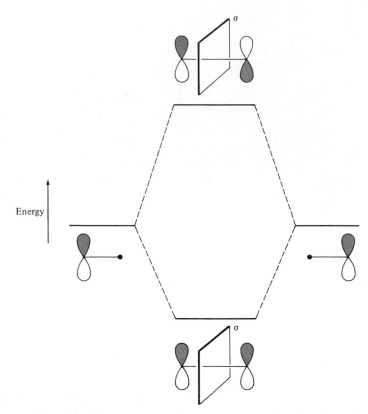

Figure 1.39 Interaction of *p* orbitals on two adjacent centers yields a bonding orbital, symmetric with respect to the mirror plane σ, and an antibonding orbital, antisymmetric with respect to the mirror plane.

of being drawn out according to their actual shapes, are depicted schematically in terms of the basis atomic orbitals of which they are composed, with shading to show the phase with which each enters into the combination. Confusion can arise from this widely used convention because an illustration of a set of basis orbitals from which molecular orbitals are going to be constructed is easily confused with an illustration of a molecular orbital. Ordinarily any diagram in which more than one atomic orbital appears is meant to be a molecular orbital; henceforth we shall follow this convention and shall make explicit specification if a set of basis orbitals is intended.

To find the orbitals for a three-carbon chain, we bring up a third *p* orbital to the end of the ethylene and use perturbation theory in Figure 1.40 to assess the results of the interaction. The ethylene bonding orbital mixes into itself the new orbital in a bonding way, moving down in energy, while the ethylene antibonding orbital mixes into itself the new orbital in an antibonding way and moves up. The

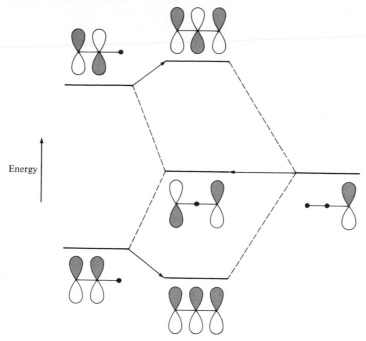

Figure 1.40 Interaction of a p orbital at the end of the ethylene system.

added p orbital mixes into itself both the ethylene bonding and antibonding orbitals.

If we follow the perturbation rules, we conclude that this double mixing must occur with the phases arranged as shown in **73**, the lower-energy ethylene bonding orbital mixing in an antibonding way to push the single p orbital up, but

73

the ethylene antibonding orbital mixing in a bonding way and pushing it down. Note that there is a reinforcement at the left-hand end of the chain but a canceling out at the center. If the perturbation is carried to its conclusion, a mirror plane again appears in the three-atom chain, this time passing directly through the central atom. Since we know that each of the orbitals we obtain in the end must be either symmetric or antisymmetric with respect to this plane, we can see at once that their forms must be those shown at the center in Figure 1.40. The lowest level is clearly bonding and the highest antibonding; the central level, antisymmetric with respect to the reflection, must necessarily have zero contribution from the central p orbital and is therefore nonbonding.

It is a straightforward task to carry the procedure one step further. In order to find the butadiene orbitals, it is easiest to allow two ethylene units to come together end to end. Figure 1.41 illustrates the results. The important interactions are between orbitals of the same energy; we simply treat the ethylene orbitals as the basis and combine the bonding pair in the two possible ways and the antibonding pair in the two possible ways.

From these three examples we can already see the pattern that emerges. We summarize it in a simple set of rules on page 76, which will allow construction directly by inspection of the π molecular orbitals for any length chain.

Figure 1.42 illustrates schematically the forms of the molecular orbitals for chains up to seven atoms in length.

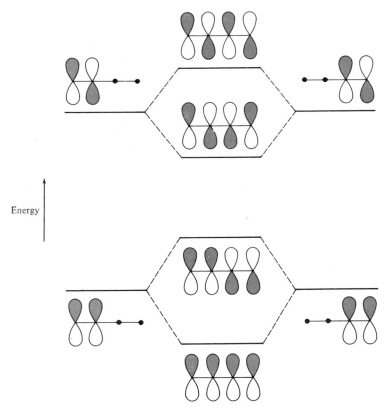

Figure 1.41 Construction of the π orbitals of butadiene by end-to-end interaction of two ethylenes.

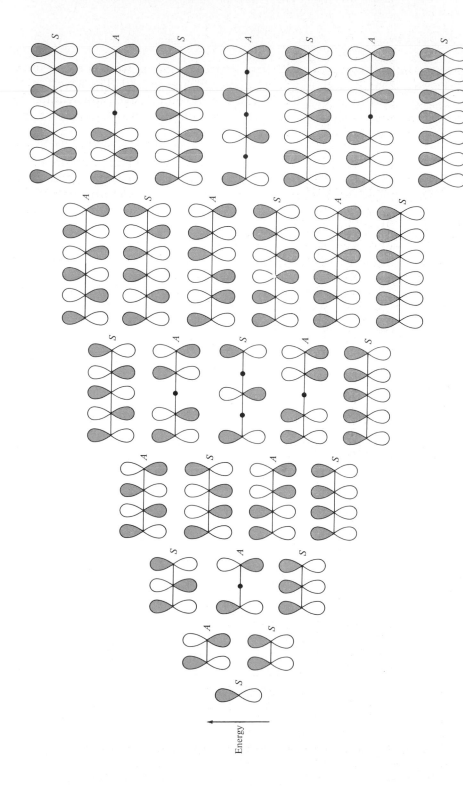

Figure 1.42 Forms of the π molecular orbitals for chains of from one to seven atoms. Symmetry labels (S, symmetric; A, antisymmetric) refer to symmetry with respect to a perpendicular mirror plane cutting the chain at its midpoint. Reprinted with permission from M. J. Goldstein and R. Hoffmann, *J. Am. Chem. Soc.*, **93**, 6193 (1971). Copyright 1971 American Chemical Society.

Rule 1.4. The orbitals alternate in symmetry with respect to the mirror plane passing through the midpoint of the chain, with the lowest-energy orbital always being symmetric.

Rule 1.5. The lowest-energy-level orbital has no nodes; the number of nodes increases by one in going from one level to the next higher level. The highest level has a node between each adjacent pair. Nodes must always be symmetrically located with respect to the central mirror plane.

Rule 1.6. In chains with an odd number of atoms, where the central atom lies in the mirror plane, antisymmetric orbitals must have zero contribution from the central atom. There is a nonbonding level to which alternate p orbitals make zero contribution.

Aromaticity

In order to find out about aromaticity, we want to see how the energy of an open-chain π system will change when it is made into a ring. If the energy goes down, the ring is aromatic; if the energy goes up, it is antiaromatic. Following the treatment of Goldstein and Hoffmann,[62] we note first that in order to make rings by joining two ends of a chain or by bringing two chains together, we shall have to distort the chains so that the central mirror plane may no longer strictly speaking be a symmetry element. But what is really important is the pattern of interaction of the p orbitals; this pattern is established by the linear sequence of p orbitals down the chain and is not disturbed by secondary distortions. Therefore, even if the mirror plane is not a proper symmetry element for the molecule as a whole, it is in effect a local symmetry for the p orbital sequence.[63] We may bend or otherwise distort linear conjugated chains and still feel confident in the use of the orbital patterns in Figure 1.42 as long as we maintain intact the sequence of interactions among the p orbitals.

Having established that the mirror plane is a proper symmetry element even if the chains are distorted, we look at HOMO–LUMO interactions. If we restrict our attention to chains with even numbers of electrons and focus on the HOMO and the LUMO, we can see that there are only two kinds of chains: those in which the HOMO is symmetric and the LUMO antisymmetric, and those in which the HOMO is antisymmetric and the LUMO symmetric. Goldstein and Hoffman have named these types respectively Mode 2 and Mode 0. Table 1.2 shows a few examples. Note that anions and cations are covered as well as neutral molecules.

---

[62] M. J. Goldstein and R. Hoffmann, *J. Am. Chem. Soc.*, **93**, 6193 (1971).
[63] Goldstein and Hoffmann (see note 62) refer to the mirror as a *pseudo symmetry* of the π system.

Table 1.2 Some Examples of Mode 2 and Mode 0 Chains

Mode 2 LUMO Antisymmetric HOMO Symmetric	Mode 0 LUMO Symmetric HOMO Antisymmetric

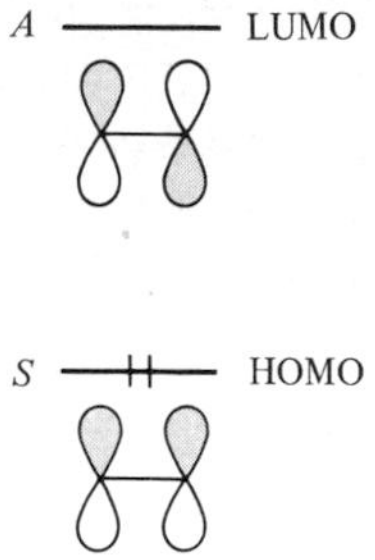

Ethylene

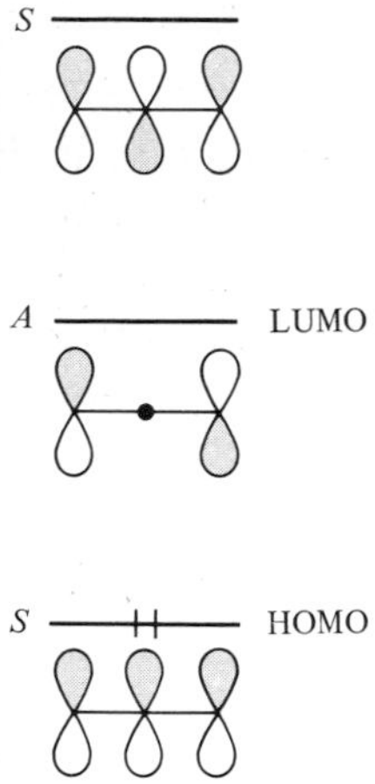

Allyl cation

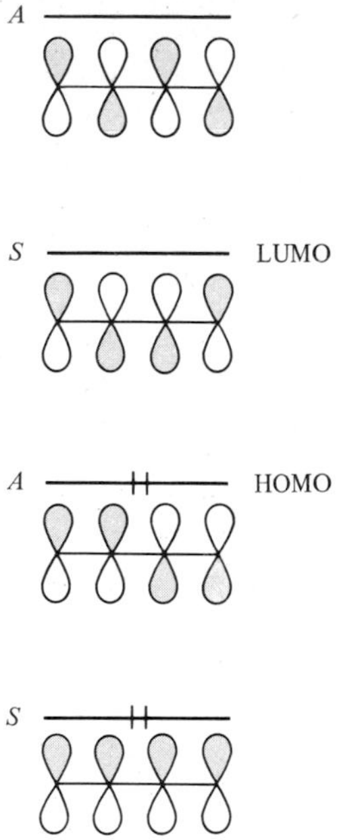

Butadiene

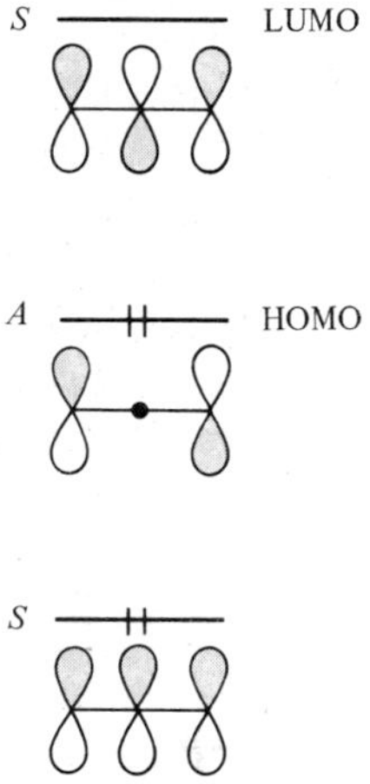

Allyl anion

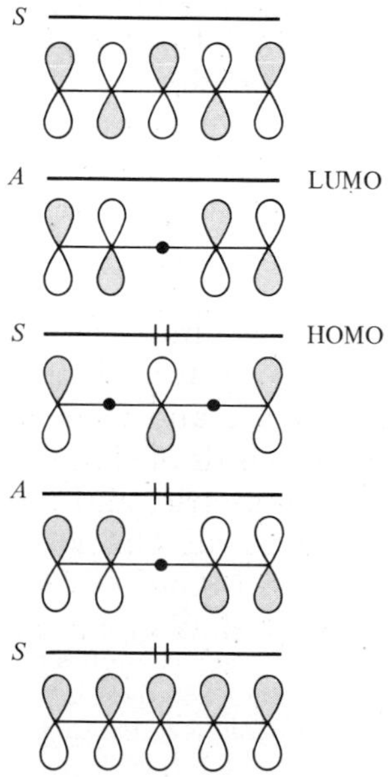

Pentadienyl anion

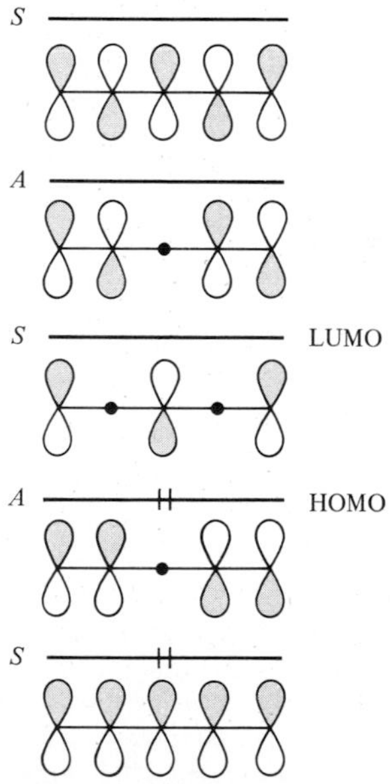

Pentadienyl cation

77

Now consider making a cyclic conjugated ring by joining two fragments at their ends. Stabilization will occur if the HOMO of one chain is of the proper symmetry to interact with the LUMO of the other. Figure 1.43 shows that stabilization occurs when a Mode 2 and a Mode 0 chain interact; we leave it as an exercise (Problem 15) for the reader to show that HOMO–LUMO stabilization will occur neither when two chains of Mode 2 are joined together nor when two chains of Mode 0 are joined together.

Since the lower levels are all filled, there will be to a first approximation no energy change resulting from their interactions. Therefore in the Mode 2 plus Mode 0 union, where HOMO–LUMO interactions are strongly stabilizing, there is an overall stabilization. But with the other two possibilities, Mode 2 plus Mode 2 or Mode 0 plus Mode 0, where HOMO–LUMO interactions are prevented by the symmetry mismatch, we should consider the interactions of the lower, filled levels more carefully. Although we have made the approximation that the higher level moves up by the same amount that the lower one moves down, a more exact theory shows that the higher level always moves up more than the lower one moves down, so that interaction of two filled levels is actually somewhat destabilizing. Hence we conclude that union of a Mode 2 chain with a Mode 2, or of a Mode 0 with a Mode 0, will be destabilizing, and the resulting ring will be antiaromatic, whereas union of a Mode 2 with a Mode 0 chain will produce a stabilized, aromatic ring.

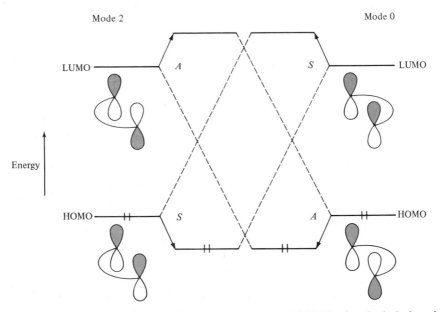

Figure 1.43 Interaction of a Mode 2 with a Mode 0 chain. The HOMO of each chain has the same symmetry as the LUMO of the other, and stabilization occurs. Adapted in part with permission from M. J. Goldstein and R. Hoffmann, *J. Am. Chem. Soc.*, **93**, 6193 (1971). Copyright 1971 American Chemical Society.

The connection can easily be made with number of electrons. The Mode 2 chains have an odd number of pairs ($4n + 2$ electrons), whereas the Mode 0 chains have an even number of pairs ($4n$ electrons). Therefore joining Mode 2 with Mode 0 will give an odd number of pairs, $4n + 2$ total electrons; but joining Mode 2 with Mode 2, or joining Mode 0 with Mode 0, will give an even number of pairs, $4n$ electrons. Therefore the $4n + 2$ rings should be aromatic, the $4n$ rings antiaromatic.

Goldstein and Hoffmann have extended their arguments to a number of possible geometries of interaction in addition to the simple single ring. Dewar has also presented a perturbation approach to aromaticity that is historically antecedent to that of Goldstein and Hoffmann. We shall return to the topic of aromaticity, and describe Dewar's method, when we discuss pericyclic reactions in Chapter 10.

PROBLEMS

1. (a) Write Lewis structures for each of the following molecules or ions. Show all significant contributing structures whenever there are more than one. Specify geometry where appropriate.
(b) In each molecule that has a delocalized bonding system, identify the orbitals that interact to form the delocalized molecular orbitals.

K_2CO_3	Sodium nitrate
Allene	$(H_3C)_3OBF_4$
Butadiene	Benzyl cation
Sodium phenoxide	Phenyl anion
Nitrobenzene	3,5-di-*tert*-butyl-4-nitrophenoxide
N—N—O	dimethylcarbodiimide, H_3C—N—C—N—CH_3

2. What kinds of symmetry are possible for interactions of d orbitals?

3. In the pentadienyl radical, predict the distribution of the unpaired electron (a) from the resonance model and (b) from the molecular orbital model.

4. Construct a complete qualitative orbital model for HN_3, showing both σ and π molecular orbitals, and giving an approximate energy-level diagram showing electron occupancy. Compare the orbital model with the resonance model.

5. Construct a qualitative orbital model for twisted ethylene, in which the two CH_2 groups lie in mutually perpendicular planes. Why does the molecule prefer coplanarity?

6. Explain why dehydroannulenes, which have some of the double bonds of the annulene replaced by triple bonds, can be considered in aromaticity theory as equivalent to the parent annulene. What advantages might dehydroannulenes have over annulenes in the study of aromaticity?

Problem 7 requires the material in Appendix 1.

7. Find the Hückel energy levels, molecular orbitals, and Hückel delocalization energy for cyclobutadiene.

8. Predict approximate proton nmr chemical shifts for the following compound:

9. Find symmetry correct orbitals in each of the following:

(a) H_2CO
(b) H_2CCH_2
(c) *cis-* and *trans-*1,2-dichloroethylene
(d) allyl cation

10. Analyze the interaction between a filled p orbital of a carbanion and an adjacent methyl group.

11. Analyze the interaction between the vacant p orbital of a carbocation and an unshared electron pair on an adjacent nitrogen. Assume sp^3 hybridization for nitrogen, and the conformation in which the unshared pair has maximum overlap with the carbon p orbital.

12. Consider the model for the transition state for a 1,2 rearrangement of an alkyl group R as being constructed from π orbitals on C_1, C_2, and a hybrid orbital on R. Analyze the HOMO–LUMO interactions (a) for rearrangement in a carbocation and (b) for rearrangement in a carbanion.

$$\overset{\displaystyle R}{C_1\text{---}C_2}$$

13. Apply Pearson's criteria (see note 59, p. 66) to the examples discussed on pp. 67–70.

14. Draw schematically the π molecular orbitals for unbranched eight- and nine-membered chains.

15. Analyze by perturbation theory the energy changes on forming a ring by joining these:

(a) a Mode 0 and a Mode 0 chain
(b) a Mode 2 and a Mode 2 chain

16. Consider 1,2-diaminoethane in the conformation shown in **1**.

(a) Construct a localized bonding model for the central C—C bond and for the nitrogen unshared pairs, using the basis orbitals shown in **2**. (Ignore the C—N, C—H, and N—H bonds.)

(b) Check to see if the orbitals found in part (a) are symmetry correct. Construct symmetry correct orbitals where necessary, and place the four molecular orbitals so obtained on an energy-level diagram. Of the two orbitals containing the nitrogen unshared pairs, which is of lower energy?

(c) Find the symmetry allowed interactions of unshared pair orbitals with C—C σ orbitals. Could the perturbation affect the conclusion reached in (b) about relative energies of the unshared pair orbitals?

REFERENCES FOR PROBLEMS

6. F. Sondheimer, *Acc. Chem. Res.*, **5,** 81 (1972).

7. A. Liberles, *Introduction to Molecular Orbital Theory*, Holt, Rinehart and Winston, New York, 1966; J. D. Roberts, *Notes on Molecular Orbital Calculations*, W. A. Benjamin, Menlo Park, Calif., 1962.

8. T. Otsubo, R. Gray, and V. Boekelheide, *J. Am. Chem. Soc.*, **100,** 2449 (1978).

12. (a) See Chapter 5. (b) N. F. Phelan, H. H. Jaffe, and M. Orchin, *J. Chem. Educ.*, **44,** 626 (1967); H. E. Zimmerman and A. Zweig, *J. Am. Chem. Soc.*, **83,** 1196 (1961); E. Grovenstein, Jr., and R. E. Williamson, *J. Am. Chem. Soc.*, **97,** 646 (1975).

15. M. J. Goldstein and R. Hoffmann, *J. Am. Chem. Soc.*, **93,** 6193 (1971).

16. (a) R. Gleiter, *Angew Chem. Int. Ed. Engl.*, **13,** 696 (1974); (b) L. A. Paquette, D. R. James, and G. Klein, *J. Org. Chem.*, **43,** 1287 (1978).

Appendix 1

MOLECULAR ORBITAL

THEORY[a]

The theory of quantum mechanics postulates that observable quantities of nature may be calculated by solving equations of the form

$$\mathcal{Q}\Phi = q\Phi \tag{A1.1}$$

where $\mathcal{Q}$ is an operator, Φ is a function of the coordinates of the system, and q is the value of the quantity being calculated. The term *operator* in this context means a set of mathematical operations to be carried out on the function written immediately to the right of the operator. Examples of simple operators are:

$\mathcal{P}$ = differentiate the function with respect to x

$\mathcal{R}$ = differentiate the function with respect to y, multiply the result by 5, and add 3 times the function

Operator $\mathcal{P}$ would be written

$$\mathcal{P} = \frac{d}{dx} \tag{A1.2}$$

Operator $\mathcal{R}$ would be

$$\mathcal{R} = 5\frac{d}{dy} + 3 \tag{A1.3}$$

---

[a] For further information consult the sources cited in note 4. For analysis of the relative accuracy of various molecular orbital methods, see (a) T. A. Halgren, D. A. Kleier, J. H. Hall, Jr., L. D. Brown, and W. N. Lipscomb, *J. Am. Chem. Soc.*, **100**, 6595 (1978); (b) M. J. S. Dewar and G. P. Ford, *J. Am. Chem. Soc.*, **101**, 5558 (1979).

Operator $\Re$ applied to the function $f(y) = y^2 + 4y$ would be written

$$\Re f(x) = \left(5\frac{d}{dy} + 3\right)f(x) \tag{A1.4}$$

$$= \left(5\frac{d}{dy} + 3\right)(y^2 + 4y) \tag{A1.5}$$

$$= 10y + 20 + 3y^2 + 12y \tag{A1.6}$$

An equation of the type A1.1 is called an *eigenvalue* equation; Φ is the *eigenfunction* and q is the *eigenvalue*. Note that eigenvalue equations have the form

(operator)(eigenfunction) = (eigenvalue)(eigenfunction)

The eigenfunctions of quantum mechanics are *wave functions,* because they resemble mathematically the functions describing wave motion.

Let us look at the eigenvalue equation A1.1 for the simple operator $\mathcal{P} = d/dx$. Equation A1.1 is then

$$\frac{d}{dx}\Phi = p\Phi \tag{A1.7}$$

This is a differential equation that asks us to find a function Φ (a function of the coordinate x in this example) that, when subjected to the operation "differentiate with respect to x," yields the same function Φ multiplied by some number p. For this example the solution is easily found: the function e^{bx}, where b is any number, has the desired property, because

$$\frac{d}{dx}e^{bx} = be^{bx} \tag{A1.8}$$

Therefore the function e^{bx} is the eigenfunction and b is the eigenvalue. Note that there are many possible eigenfunctions and eigenvalues corresponding to different values of b; this feature appears also in quantum mechanics, but it usually happens that *boundary conditions* (that is, physical restraints placed on the system under consideration, such as the requirement that electron density decrease smoothly toward zero at large distances from a molecule) dictate that only some subset of all the possible solutions are actually acceptable for the system.

THE ENERGY OF A MOLECULAR SYSTEM

In quantum mechanics, for each system one considers there are a number of operators corresponding to various possible physical quantities that one might want to observe. The most important operator for the problems we want to consider here is the Hamiltonian operator, abbreviated $\mathcal{H}$, which is the operator corresponding to the observable quantity energy. The energy of a molecule may be divided into contributions from the electron motions, the internal vibrations, the rotation of the molecule as a whole, and the translation of the molecule as a whole through space. A Hamiltonian operator can be written for each of these separate kinds of motion;

we are interested here in the behavior of the electrons, so we assume the nuclei are fixed in space and only the electrons move. The Hamiltonian operator for the energy of the electrons, $\mathcal{H}_e$, is a sum of terms of two types: first, terms for the kinetic energy that specify differentiation with respect to the coordinates (x_i, y_i, z_i) of the electrons (one term for each coordinate of each electron); and second, terms for the potential energy specifying division by the distances separating the electrons from the nuclei (R) and the electrons from each other (r). Each of these terms also specifies multiplication by certain constants, including the electron mass m and Planck's constant h for the kinetic energy terms, and electron charge e and nuclear charge Z for the potential energy terms.

For a very simple molecule consisting of two electrons and two nuclei, the electronic Hamiltonian $\mathcal{H}_e$ is only moderately complex:

$$\mathcal{H}_e = -\frac{\hbar^2}{2m}\left(\frac{\partial^2}{\partial x_1{}^2} + \frac{\partial^2}{\partial y_1{}^2} + \frac{\partial^2}{\partial z_1{}^2} + \frac{\partial^2}{\partial x_2{}^2} + \frac{\partial^2}{\partial y_2{}^2} + \frac{\partial^2}{\partial z_2{}^2}\right) -$$
$$Z_1 e^2\left(\frac{1}{R_{11}} + \frac{1}{R_{12}}\right) - Z_2 e^2\left(\frac{1}{R_{21}} + \frac{1}{R_{22}}\right) + \frac{e^2}{r_{12}} \qquad \text{(A1.9)}$$

Here x_i, y_i, z_i are the coordinates of electron i; $\hbar \equiv h/2\pi$; m is the mass of the electron; $\partial^2/\partial x_i{}^2$ specifies the second partial derivative with respect to the coordinate of electron i; the Z_i are charges of the nuclei; e is the charge of the electron; R_{ki} is the distance from nucleus k to electron i; and r_{ij} is the distance from electron i to electron j. Hamiltonians are usually written in condensed form with the help of the following abbreviations:

$$\nabla_i{}^2 \equiv \frac{\partial^2}{\partial x_i{}^2} + \frac{\partial^2}{\partial y_i{}^2} + \frac{\partial^2}{\partial z_i{}^2} \qquad \text{(A1.10)}$$

$$\sum_{i=1}^{n} t_i \equiv t_1 + t_2 + t_3 + \cdots + t_n \qquad \text{(A1.11)}$$

where t_i represents any general term that varies in a regular way with index i. We can now write Equation A1.8 more compactly:

$$\mathcal{H}_e = -\frac{\hbar^2}{2m}\sum_{i=1}^{2}\nabla_i{}^2 - \sum_{i=1}^{2}\frac{Z_1 e^2}{R_{1i}} - \sum_{i=1}^{2}\frac{Z_2 e^2}{R_{2i}} + \frac{e^2}{r_{12}} \qquad \text{(A1.12)}$$

With the help of the abbreviation for a double sum,

$$\sum_{i=1}^{n}\sum_{j=1}^{n} s_i t_j \equiv s_1 t_1 + s_1 t_2 + \cdots + s_1 t_n + s_2 t_1 + s_2 t_2 + \cdots$$
$$+ s_2 t_n + \cdots + s_n t_1 + s_n t_2 + \cdots + s_n t_n \qquad \text{(A1.13)}$$

we can even hope to write the electronic Hamiltonian for a molecule containing n electrons and N nuclei:

$$\mathcal{H}_e = -\frac{\hbar^2}{2m}\sum_{i=1}^{n}\nabla_i{}^2 - \sum_{i=1}^{n}\sum_{k=1}^{N}\frac{Z_k e^2}{R_{ik}} + \sum_{i=1}^{n-1}\sum_{\substack{j=i+1 \\ i<j}}^{n}\frac{e^2}{r_{ij}} \qquad \text{(A1.14)}$$

(The complicated summation in the last term assures that each electron–electron repulsion is counted only once.) Molecules differ in the number of electrons and in number, location, and charges of nuclei.

Knowing how to write Hamiltonian operators for molecules, we can write the operator equation that will allow us to find the electron energy:

$$\mathcal{H}_e \Psi_e = E_e \Psi_e \tag{A1.15}$$

This equation is called the Schrödinger equation. It says that there is some function Ψ_e, called the electronic eigenfunction or electronic wave function, a function of all the $3n$ coordinates x_i, y_i, z_i, which, when subjected to the differentiations and multiplications specified by the electronic Hamiltonian for the molecule of interest, will reproduce the same function Ψ_e multiplied by some number E_e, just as in the simple example of Equation A1.7 the solution function e^{bx} yielded e^{bx} multiplied by b. The quantum theory specifies that the number E_e will be the electronic energy, and the function Ψ_e will contain all the information that can be known about the motion of the electrons; in particular, its square, Ψ_e^2, will specify the spatial distribution of the electrons. For a given molecule we would expect to find a number of functions Ψ_e and energies E_e that are eigenfunctions and eigenvalues of the Hamiltonian operator. These eigenfunctions would represent the various possible electronic energy states of the molecule, and the eigenvalues would be the electronic energies of the respective states.

The Variation Principle

The difficulty with this prescription for finding wave functions and energies is that there is no way known to discover what the function Ψ_e is for any system with more than one electron. However, a number of approximate methods have been developed, some quite good but difficult to apply, others not so good but easier to apply. Most of these methods are based on the *variation principle,* which states that any approximate wave function Φ_e of the coordinates of the system, when operated on by the Hamiltonian, will yield an energy greater than the true lowest energy of the system.

In order to justify the variation principle, and to explain the way it is applied in approximate molecular quantum theory, we must introduce the following properties of eigenfunctions. First, eigenfunctions ordinarily exist as sets of functions Ψ_n, which are *complete* in the sense that any arbitrary function in the coordinate space of the problem can be expressed as a linear combination of the members of the set:

$$\xi = c_1 \Psi_1 + c_2 \Psi_2 + \cdots \tag{A1.16}$$

or

$$\xi = \sum_i c_i \Psi_i \tag{A1.17}$$

(The sum starts at $i = 0$; no limit is specified here and in what follows; the number of functions depends on the problem and may be infinite.) Second, the eigenfunctions of the set are *orthogonal*[b] to each other, which means

$$\int \Psi_i \Psi_j \, d\tau = 0 \quad \text{for} \quad i \neq j \tag{A1.18}$$

and are *normalized,*[b] which means

$$\int \Psi_i \Psi_j \, d\tau = 1 \quad \text{for} \quad i = j \tag{A1.19}$$

In these equations the symbol $d\tau$ means that the integral is a multiple one that is to be carried out over all the coordinates of the functions. These two statements may be combined into one by saying that the eigenfunctions constitute an ortho-normal set,

$$\int \Psi_i \Psi_j \, d\tau = \delta_{ij} \tag{A1.20}$$

where

$$\begin{aligned} \delta_{ij} &= 0 \quad \text{if} \quad i \neq j \\ \delta_{ij} &= 1 \quad \text{if} \quad i = j \end{aligned} \tag{A1.21}$$

Third, using the eigenvalue equation A1.15 for the Hamiltonian operator and the orthonormality property A1.20, we find:[c]

$$\int \Psi_i \mathcal{H} \Psi_j \, d\tau = E_i \delta_{ij} \tag{A1.22}$$

To see that this result is true, multiply the eigenvalue equation

$$\mathcal{H} \Psi_j = E_j \Psi_j \tag{A1.23}$$

by Ψ_i and integrate:

$$\int \Psi_i \mathcal{H} \Psi_j \, d\tau = \int \Psi_i E_j \Psi_j \, d\tau \tag{A1.24}$$

Because E_j is a number, it can be removed from the integral:

$$\int \Psi_i \mathcal{H} \Psi_j \, d\tau = E_j \int \Psi_i \Psi_j \, d\tau \tag{A1.25}$$

The integral on the right in Equation A1.25 is zero if $j \neq i$ and unity if $j = i$, which is the result expressed by Equation A1.22.

The variation principle can now be demonstrated as follows. (You may skip the derivation by going directly to Equation A1.39.) Suppose ϕ is some function of

---

[b] A more precise statement would be that the set of eigenfunctions can be made orthogonal and normalized without changing any fundamental property of the functions. We have assumed that the functions being considered are real, that is, that they do not contain $\sqrt{-1}$. If they are complex functions, the correct integrals are

$$\int \Psi_i^* \Psi_j \, d\tau$$

[c] From here on we shall omit the subscript e. All Hamiltonians, wave functions, and energies are understood to refer to the electrons.

the coordinates of the problem but not an eigenfunction. We can assume that ϕ is normalized in the sense specified above, that is,

$$\int \phi\phi \, d\tau = 1 \tag{A1.26}$$

Write ϕ in terms of the true eigenfunctions as follows:

$$\phi = \sum_i c_i \Psi_i \tag{A1.27}$$

(We can do this symbolically even though we do not know what the true eigenfunctions actually are.) Now operate with the Hamiltonian:

$$\mathcal{H}\phi = \mathcal{H} \sum_i c_i \Psi_i \tag{A1.28}$$

Because the c_i are numbers, they can be brought out in front:

$$\mathcal{H}\phi = \sum_i c_i \mathcal{H} \Psi_i \tag{A1.29}$$

Next multiply by ϕ and integrate:

$$\int \phi\mathcal{H}\phi \, d\tau = \int \left(\sum_j c_j \Psi_j\right)\left(\sum_i c_i \mathcal{H}\Psi_i\right) d\tau \tag{A1.30}$$

or

$$\int \phi\mathcal{H}\phi \, d\tau = \sum_j \sum_i c_j c_i \int \Psi_j \mathcal{H}\Psi_i \, d\tau \tag{A1.31}$$

(You may convince yourself that A1.30 and A1.31 are equivalent by writing out a few representative terms and by noting that the c's are constant numbers and may be taken outside the integral.) The result A1.22 may now be used to simplify the right-hand side:

$$\int \phi\mathcal{H}\phi \, d\tau = \sum_i c_i^2 E_i \tag{A1.32}$$

(The double sum has collapsed to a single sum because all terms in which $i \neq j$ are zero.) Now let E_0 be the true lowest energy of the system, corresponding to the eigenfunction Ψ_0. The following relations are true:

$$E_0 \int \phi\phi \, d\tau = E_0 \int \sum_i c_i \Psi_i \sum_j c_j \Psi_j \, d\tau \tag{A1.33}$$

$$= E_0 \sum_i \sum_j c_i c_j \int \Psi_i \Psi_j \, d\tau \tag{A1.34}$$

$$= E_0 \sum_i c_i^2 \tag{A1.35}$$

We now subtract the quantity $E_0 \int \phi\phi \, d\tau$ from the left side of A1.32, and its equivalent $E_0 \sum_i c_i^2$ from the right:

$$\int \phi \mathcal{H} \phi \, d\tau - E_0 \int \phi \phi \, d\tau = \sum_i c_i{}^2 (E_i - E_0) \qquad (A1.36)$$

On the right each term is greater than or equal to zero because $c_i{}^2$ is necessarily positive, and by the definition of E_0 as the lowest energy, all E_i are greater than or equal to E_0. Therefore

$$\int \phi \mathcal{H} \phi \, d\tau - E_0 \int \phi \phi \, d\tau \geqslant 0 \qquad (A1.37)$$

and

$$\int \phi \mathcal{H} \phi \, d\tau \geqslant E_0 \int \phi \phi \, d\tau \qquad (A1.38)$$

$$\frac{\int \phi \mathcal{H} \phi \, d\tau}{\int \phi \phi \, d\tau} \geqslant E_0 \qquad (A1.39)$$

Equation A1.39 expresses the variation principle. Its interpretation in terms of the molecular wave function problem is as follows. Suppose we have somehow guessed a function ϕ_e that we have reason to think may be a good approximation to the true lowest energy eigenfunction Ψ_0 for some molecule. We may put this approximate function into the Schrödinger equation:

$$\mathcal{H} \phi_e = E \phi_e \qquad (A1.40)$$

However, there is no guarantee that ϕ_e will be a solution (i.e., an eigenfunction); in fact, it probably will not be. But if we forge ahead anyway, multiplying both sides by ϕ_e, integrating, and dividing, we arrive at the result A1.42.

$$\int \phi_e \mathcal{H} \phi_e \, d\tau = \int \phi_e E \phi_e \, d\tau \qquad (A1.41)$$

$$\frac{\int \phi_e \mathcal{H} \phi_e \, d\tau}{\int \phi_e \phi_e \, d\tau} = E \qquad (A1.42)$$

We have obtained some number E, probably not one of the true eigenvalues, but guaranteed by the variation principle (Equation A1.39) to be greater than or equal to the true lowest eigenvalue E_0.

This result may not seem to be much help, but it does provide a strategy for constructing an approximate theory. We shall try to make an intelligent guess of an appropriate form for an approximate electronic function ϕ_e and include in it some variable parameters. Then if we adjust the parameters so as to give the lowest possible value for E in Equation A1.42, the variation principle guarantees that we have done as well as we can do with that particular form. Furthermore, if we try again with a different form of ϕ_e and get a lower energy, we shall know we are doing better; if we get a higher energy, we are doing less well.

APPROXIMATE LCAO MOLECULAR ORBITAL THEORY[d]

We choose as our approximate function, which we shall call ψ hereafter, a *linear combination of atomic orbitals,* LCAO:

$$\psi = \sum_{i=1}^{N} a_i \varphi_i \tag{A1.43}$$

The symbols φ_i stand for the basis atomic orbitals, N in number, and the a_i are parameters that specify the contribution of each atomic orbital to the trial function. The approximate function ψ will turn out to be a molecular orbital.

Application of the Variation Principle

We now wish to find the values of parameters a_i that will minimize the energy calculated according to Equation A1.42. (To skip the derivation, go directly to Equation A1.64.) First substitute A1.43 into A1.42:

$$E = \frac{\int \left(\sum_i a_i \varphi_i \right) \mathcal{H} \left(\sum_j a_j \varphi_j \right) d\tau}{\int \sum_i a_i \varphi_i \sum_j a_j \varphi_j \, d\tau} \tag{A1.44}$$

or

$$E = \frac{\sum_i \sum_j a_i a_j \int \varphi_i \mathcal{H} \varphi_j \, d\tau}{\sum_i \sum_j a_i a_j \int \varphi_i \varphi_j \, d\tau} \tag{A1.45}$$

(In these equations and in those that follow, sums run from 1 to N, where N is the number of atomic orbitals.) We now make the following definitions, to give a more manageable notation:

$$\int \varphi_i \mathcal{H} \varphi_j \, d\tau \equiv H_{ij} \tag{A1.46}$$

$$\int \varphi_i \varphi_j \, d\tau \equiv S_{ij} \tag{A1.47}$$

Our expression for E then takes the form:

$$E = \frac{\sum_i \sum_j a_i a_j H_{ij}}{\sum_i \sum_j a_i a_j S_{ij}} \tag{A1.48}$$

[d] (a) J. D. Roberts, *Notes on Molecular Orbital Calculations,* W. A. Benjamin, Menlo Park, Calif., 1962; (b) A. Liberles, *Introduction to Molecular Orbital Theory,* Holt, Rinehart and Winston, New York, 1966; (c) A. Streitwieser, Jr., *Molecular Orbital Theory for Organic Chemists,* Wiley, New York, 1961; (d) M. J. S. Dewar, *The Molecular Orbital Theory of Organic Chemistry,* McGraw-Hill, New York, 1969; (e) M. J. S. Dewar and R. C. Dougherty, *The PMO Theory of Organic Chemistry,* Plenum, New York, 1975; (f) H. E. Zimmerman, *Quantum Mechanics for Organic Chemists,* Academic Press, New York, 1975; (g) E. Heilbronner and H. Bock, *The HMO Model and its Application,* Vols. 1–3, Wiley, London, 1976.

The next step is to minimize E with respect to each of the a's. This task is accomplished by differentiating Equation A1.48 with respect to each of the a's in turn and setting the result of each differentiation equal to zero, so as to find the minimum of E with respect to all the a parameters simultaneously. We shall therefore have a total of N equations, the kth of which is

$$\frac{\partial E}{\partial a_k} = 0 = \frac{\partial}{\partial a_k} \left\{ \frac{\sum_i \sum_j a_i a_j H_{ij}}{\sum_i \sum_j a_i a_j S_{ij}} \right\} \tag{A1.49}$$

The quantity in braces is of the form AB, where

$$A = \sum_i \sum_j a_i a_j H_{ij} \tag{A1.50}$$

$$B = \left(\sum_i \sum_j a_i a_j S_{ij} \right)^{-1} \tag{A1.51}$$

We must carry out the differentiation:

$$\frac{\partial}{\partial a_k} AB = A \frac{\partial B}{\partial a_k} + B \frac{\partial A}{\partial a_k} \tag{A1.52}$$

The first of the indicated derivatives is

$$\frac{\partial B}{\delta a_k} = (-1) \left(\sum_i \sum_j a_i a_j S_{ij} \right)^{-2} \left(\frac{\partial}{\partial a_k} \sum_i \sum_j a_i a_j S_{ij} \right) \tag{A1.53}$$

In order to carry out the partial differentiation, we must locate all terms in the double sum that contain a_k.[e] To see which terms these are, we write out the double sum in an array of rows and columns:

$$\sum_{i=1}^{N} \sum_{j=1}^{N} a_i a_j S_{ij} = a_1 a_1 S_{11} + a_1 a_2 S_{12} + \cdots + a_1 a_k S_{1k} + \cdots + a_1 a_N S_{1N}$$
$$+ a_2 a_1 S_{21} + a_2 a_2 S_{22} + \cdots + a_2 a_k S_{2k} + \cdots + a_2 a_N S_{2N}$$
$$\longrightarrow + a_k a_1 S_{k1} + a_k a_2 S_{k2} + \cdots + a_k a_k S_{kk} + \cdots + a_k a_N S_{kN}$$
$$+ a_N a_1 S_{N1} + a_N a_2 S_{N2} + \cdots + a_N a_k S_{Nk} + \cdots + a_N a_N S_{NN} \tag{A1.54}$$

In this array terms containing a_k occur in the kth row and in the kth column (arrows). There is one unique term, $a_k{}^2 S_{kk}$. Its derivative with respect to a_k is $2a_k S_{kk}$. In addition, we have (in the kth column) terms $a_i a_k S_{ik}$ for which $i \neq k$, which have derivatives $a_i S_{ik}$. We also have (in the kth row) terms $a_k a_j S_{kj}$, which have derivatives $a_j S_{kj}$. However, the S integrals have the property

---

[e] Partial differentiation of a function of several variables is accomplished by pretending that all the variables except the one with respect to which differentiation is being carried out are constants, and then proceeding in the usual way.

$$S_{rs} = S_{sr} \tag{A1.55}$$

and so each term $a_i S_{ik}$ is equal to one of the terms $a_j S_{kj}$. The other terms in the double sum A1.54 do not contain a_k and contribute zero to the partial derivative. The required derivative is

$$\frac{\partial}{\partial a_k} \sum_i \sum_j a_i a_j S_{ij} = 2a_k S_{kk} + 2 \sum_{i \neq k} a_i S_{ki} \tag{A1.56}$$

On the right-hand side the first term is just the one missing from the sum, so

$$\frac{\partial}{\partial a_k} \sum_i \sum_j a_i a_j S_{ij} = 2 \sum_i a_i S_{ki} \tag{A1.57}$$

and

$$\frac{\partial B}{\partial a_k} = -\left(\sum_i \sum_j a_i a_j S_{ij} \right)^{-2} \left(2 \sum_i a_i S_{ki} \right) \tag{A1.58}$$

Returning to Equation A1.52, we also require the derivative

$$\frac{\partial A}{\partial a_k} = \frac{\partial}{\partial a_k} \left(\sum_i \sum_j a_i a_j H_{ij} \right) \tag{A1.59}$$

By the same argument we used above, this derivative is

$$\frac{\partial A}{\partial a_k} = 2 \sum_i a_i H_{ki} \tag{A1.60}$$

Combining Equations A1.50–A1.52, A1.58, and A1.60, we obtain:

$$\frac{\partial E}{\partial a_k} = 0 = -\left(\sum_i \sum_j a_i a_j H_{ij} \right) \left(\sum_i \sum_j a_i a_j S_{ij} \right)^{-2} \left(2 \sum_i a_i S_{ki} \right)$$
$$+ \left(\sum_i \sum_j a_i a_j S_{ij} \right)^{-1} \left(2 \sum_i a_i H_{ki} \right) \tag{A1.61}$$

Now multiply by $\frac{1}{2}(\Sigma_i \Sigma_j a_i a_j S_{ij})$:

$$\frac{\partial E}{\partial a_k} = 0 = -\left\{ \frac{\displaystyle\sum_i \sum_j a_i a_j H_{ij}}{\displaystyle\sum_i \sum_j a_i a_j S_{ij}} \right\} \sum_i a_i S_{ki} + \sum_i a_i H_{ki} \tag{A1.62}$$

Comparing this result with Equation A1.48, we find that the quantity in braces is just E, and we have therefore

$$\sum_i (H_{ki} - ES_{ki}) a_i = 0 \tag{A1.63}$$

Recall that differentiation is to be carried out for each of the a's; there are therefore N equations A1.63, one for each value of k from $k = 1$ to N:

$$\sum_{i=1}^{N} (H_{1i} - ES_{1i})a_i = 0$$

$$\sum_{i=1}^{N} (H_{2i} - ES_{2i})a_i = 0 \qquad\qquad\text{(A1.64)}$$

$$\vdots$$

$$\sum_{i=1}^{N} (H_{Ni} - ES_{Ni})a_i = 0$$

Equations A1.64 constitute a set of linear homogeneous equations of the form

$$\begin{aligned}
b_{11}x_1 + b_{12}x_2 + \cdots + b_{1n}x_n &= 0 \\
b_{21}x_1 + b_{22}x_2 + \cdots + b_{2n}x_n &= 0 \\
&\vdots \\
b_{n1}x_1 + b_{n2}x_2 + \cdots + b_{nn}x_n &= 0
\end{aligned} \qquad\qquad\text{(A1.65)}$$

Such a set of equations has a (nontrivial) solution for the unknowns x_i only if the determinant of the coefficients equals zero:

$$\begin{vmatrix}
b_{11} & b_{12} & \cdots & b_{1n} \\
b_{21} & b_{22} & \cdots & b_{2n} \\
\vdots & \vdots & & \vdots \\
b_{n1} & b_{n2} & \cdots & b_{nn}
\end{vmatrix} = 0 \qquad\qquad\text{(A1.66)}$$

In our case the unknowns are the parameters a_i (corresponding to the x's in A1.65) and the coefficients are the quantities $(H_{ki} - ES_{ki})$; there will be a solution (other than the trivial one of all a's equal to zero) only if the determinantal equation A1.67 holds:

$$\begin{vmatrix}
H_{11} - ES_{11} & H_{12} - ES_{12} & \cdots & H_{1N} - ES_{1N} \\
H_{21} - ES_{21} & H_{22} - ES_{22} & \cdots & H_{2N} - ES_{2N} \\
\vdots & \vdots & & \vdots \\
H_{N1} - ES_{N1} & H_{N2} - ES_{N2} & \cdots & H_{NN} - ES_{NN}
\end{vmatrix} = 0 \qquad\qquad\text{(A1.67)}$$

This result is called the *secular equation.*

Solving the Secular Equation

Solution of the secular equation is accomplished by expanding the determinant to yield an equation of order N in the unknown energy E. The quantities H_{ij}, called the *Hamiltonian matrix elements,* and S_{ij}, called the *overlap integrals,* are treated as known (see further discussion later in this Appendix). The equation has N roots corresponding to N possible values of energy, $E_J, J = 1, 2, \ldots, N$. Each of these energies E_J corresponds to a molecular orbital ψ_J of the form

$$\psi_J = \sum_{i=1}^{N} a_{Ji}\varphi_i \qquad\qquad\text{(A1.68)}$$

where φ_i are the basis atomic orbitals and a_{Ji} is the coefficient that gives the contribution of the ith atomic orbital to the Jth molecular orbital. To find the coefficients a_{Ji}, one substitutes E_J back into the set of Equations A1.64. One obtains a set of N equations in the N unknown coefficients a_{Ji}; only $N-1$ of these equations are independent, however, and so in order to solve for the a_{Ji}'s, one also requires that the orbital ψ_J be normalized. This condition provides the Nth independent equation needed. Following this procedure for each root in turn leads to the complete set of molecular orbitals expressed in terms of the basis atomic orbitals. This process is illustrated for the Hückel π electron method in the examples discussed below. Once energies and orbitals have been obtained, a model for the electronic ground state of the molecule is constructed by assigning electrons in pairs to the orbitals, starting with the lowest.

Equation A1.67 may be used for a model that includes all valence electrons. The basis orbitals φ_i are then all the valence orbitals of all the atoms. The overlap integrals are calculated by using atomic orbitals expressed explicitly as functions of the coordinates, and the Hamiltonian matrix elements H_{ij} are calculated with the help of various approximations. The extended Hückel method, described at the end of the Appendix, is of this type.

THE HÜCKEL π ELECTRON METHOD

The Hückel π electron theory is based on Equation A1.67 and A1.68, but it makes several additional simplifying assumptions. It is the only one of the quantitative molecular orbital methods that is practical for hand computation. Hückel energies are unreliable, but the results are nevertheless useful for semiquantitative arguments and for comparisons among similar systems.

The Hückel method has the following characteristics:

1. Only π electrons are treated.

2. The basis consists of N p orbitals, φ_{p_i}, one on each of the N carbon atoms of the π system.

3. All overlap integrals S_{ij}, $i \neq j$, are assumed to be zero; overlaps S_{ii} are unity because the basis orbitals are normalized.

4. Hamiltonian integral H_{ij}, $i \neq j$, which represents the interaction energy of an electron in basis orbital i with one in basis orbital j, is assumed to be zero except for pairs of basis orbitals i and j that are on carbons directly bonded to each other. All H_{ij} for bonded pairs are assumed to have the same value, which is not calculated but is simply called β. (β represents an energy lowering and is therefore negative.)

5. H_{ii}, which represents the energy of an electron in the basis orbital i in the absence of any interaction with its neighbors, is the same for all i (because all basis orbitals are the same). It is called α.

The secular equation then takes the form of the N-by-N determinantal equation A1.69. The asterisk in row i, column j is zero if atom i is not bonded to atom j, and it is β if i is bonded to j. A further simplification of the algebraic manipulations

$$
\begin{vmatrix}
\alpha - E & * & * & \cdots \\
* & \alpha - E & * & \cdots \\
* & * & \alpha - E & \cdots \\
\vdots & \vdots & \vdots &
\end{vmatrix} = 0
\tag{A1.69}
$$

required is obtained by dividing Equation A1.69 by β and substituting

$$
x = \frac{\alpha - E}{\beta}; \qquad E = \alpha - x\beta
\tag{A1.70}
$$

Then we have Equation A1.71, where $*'_{ij}$ is zero if i and j are not bonded and unity if they are.

$$
\begin{vmatrix}
x & *' & *' & \cdots \\
*' & x & *' & \cdots \\
*' & *' & x & \cdots \\
\vdots & \vdots & \vdots &
\end{vmatrix} = 0
\tag{A1.71}
$$

Expansion of Determinants

In order to solve secular equations by hand, the determinants must be expanded. This task is accomplished by using cofactors to reduce the determinant size to two by two. A cofactor of element ij of a determinant of order N is a determinant of order $N - 1$ found from the original determinant by striking out row i and column j. The algebraic sign of a cofactor is positive if $i + j$ is even, negative if $i + j$ is odd. A determinant is expanded in cofactors by summing the elements of any one row or column, each multiplied by its signed cofactor. We illustrate this rule by expanding a 4×4 determinant:

$$
\begin{vmatrix}
b_{11} & b_{12} & b_{13} & b_{14} \\
b_{21} & b_{22} & b_{23} & b_{24} \\
b_{31} & b_{32} & b_{33} & b_{34} \\
b_{41} & b_{42} & b_{43} & b_{44}
\end{vmatrix}
= + b_{11}
\begin{vmatrix}
b_{22} & b_{23} & b_{24} \\
b_{32} & b_{33} & b_{34} \\
b_{42} & b_{43} & b_{44}
\end{vmatrix}
- b_{21}
\begin{vmatrix}
b_{12} & b_{13} & b_{14} \\
b_{32} & b_{33} & b_{34} \\
b_{42} & b_{43} & b_{44}
\end{vmatrix}
$$
$$
+ b_{31}
\begin{vmatrix}
b_{12} & b_{13} & b_{14} \\
b_{22} & b_{23} & b_{24} \\
b_{42} & b_{43} & b_{44}
\end{vmatrix}
- b_{41}
\begin{vmatrix}
b_{12} & b_{13} & b_{14} \\
b_{22} & b_{23} & b_{24} \\
b_{32} & b_{33} & b_{34}
\end{vmatrix}
\tag{A1.72}
$$

The expansion may be carried out by using any row or column; the amount of work involved is minimized if the row or column containing the maximum number of zeros is chosen. The expansion process is continued until determinants of order two are obtained; these are evaluated as follows:

$$
\begin{vmatrix}
b_{11} & b_{12} \\
b_{21} & b_{22}
\end{vmatrix} = b_{11}b_{22} - b_{21}b_{12}
\tag{A1.73}
$$

The Allyl System in the Hückel Approximation

As an example of the Hückel method we will examine the allyl system. There are three basis orbitals, numbered as shown in **1**. Atoms 1 and 2 are bonded to each other, as are 2 and 3; 1 and 3 are not bonded. The secular equation is A1.74.

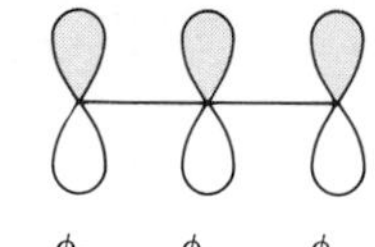

$$\phi_{p_1} \qquad \phi_{p_2} \qquad \phi_{p_3}$$
$$1$$

$$\begin{vmatrix} x & 1 & 0 \\ 1 & x & 1 \\ 0 & 1 & x \end{vmatrix} = 0 \qquad\qquad (A1.74)$$

Expansion by cofactors of the first column gives

$$x \begin{vmatrix} x & 1 \\ 1 & x \end{vmatrix} - \begin{vmatrix} 1 & 0 \\ 1 & x \end{vmatrix} = 0 \qquad\qquad (A1.75)$$

$$x(x^2 - 1) - x = 0 \qquad\qquad (A1.76)$$
$$x^3 - 2x = 0 \qquad\qquad (A1.77)$$
$$x(x^2 - 2) = 0 \qquad\qquad (A1.78)$$

The roots are

$$x_3 = + \sqrt{2}$$
$$x_2 = 0 \qquad\qquad (A1.79)$$
$$x_1 = - \sqrt{2}$$

Recalling that $E = \alpha - x\beta$, the energies E_J are therefore $E_3 = \alpha - \sqrt{2}\beta$, $E_2 = \alpha$, $E_1 = \alpha + \sqrt{2}\beta$; because β is a negative energy, the first of these is the highest energy and the third is the lowest.

To find the orbitals, we substitute each x in turn into the set of equations from which the determinant was originally taken:

$$a_{J1}x_J + a_{J2} \cdot 1 + \partial_{J3} \cdot 0 = 0$$
$$a_{J1} \cdot 1 + a_{J2}x_J + a_{J3} \cdot 1 = 0 \qquad\qquad (A1.80)$$
$$a_{J1} \cdot 0 + a_{J2} \cdot 1 + a_{J3}x_J = 0$$

Choosing the lowest-energy orbital first, we substitute $x_1 = - \sqrt{2}$ to obtain the relationships A1.81:

$$a_{12} = \sqrt{2}\, a_{11}$$
$$a_{13} = a_{11} \qquad\qquad (A1.81)$$

The third equation needed for a solution is obtained from the requirement that the molecular orbital be normalized. Recall that normalization of an orbital means

$$\int \psi\psi \, d\tau = 1 \qquad\qquad (A1.82)$$

For the molecular orbital ψ_1 that we are seeking,

$$\int \psi_1\psi_1 \, d\tau = \int \sum_{i=1}^{3} a_{1i}\varphi_{p_i} \sum_{j=1}^{3} a_{1j}\varphi_{p_j} \, d\tau \qquad\qquad (A1.83)$$

$$= \sum_{i=1}^{3} \sum_{j=1}^{3} a_{1i} a_{1j} \int \varphi_{p_i} \varphi_{p_j} \, d\tau \tag{A1.84}$$

The integral in A1.84 is what we have called S_{ij}; in the Hückel approximation it is unity if $i = j$ and zero if $i \neq j$. Therefore,

$$\int \psi_1 \psi_1 \, d\tau = \sum_{k=1}^{3} a_{1k}{}^2 \tag{A1.85}$$

For the orbital ψ_1 to be normalized, this quantity must be unity; the normalization condition is therefore

$$\int \psi_1 \psi_1 \, d\tau = 1 = a_{11}{}^2 + a_{12}{}^2 + a_{13}{}^2 \tag{A1.86}$$

Combining Equations A1.81 and A1.86, we obtain the coefficients for the lowest-energy molecular orbital:

$$a_{11} = \frac{1}{2}$$
$$a_{12} = \frac{\sqrt{2}}{2} = \frac{1}{\sqrt{2}} \tag{A1.87}$$
$$a_{13} = \frac{1}{2}$$

The coefficients for the other orbitals are obtained in the same way starting with $x_2 = 0$ and $x_3 = +\sqrt{2}$. (An alternative method of finding coefficients, using cofactors, is illustrated in the next example.) The orbitals are, in order of decreasing energy,

$$E_3 = \alpha - \sqrt{2}\beta \qquad \psi_3 = \frac{1}{2}\varphi_{p_1} - \frac{1}{\sqrt{2}}\varphi_{p_2} + \frac{1}{2}\varphi_{p_3}$$

$$E_2 = \alpha \qquad \psi_2 = \frac{1}{\sqrt{2}}\varphi_{p_1} - \frac{1}{\sqrt{2}}\varphi_{p_3} \tag{A1.88}$$

$$E_1 = \alpha + \sqrt{2}\beta \qquad \psi_1 = \frac{1}{2}\varphi_{p_1} + \frac{1}{\sqrt{2}}\varphi_{p_2} + \frac{1}{2}\varphi_{p_3}$$

These orbitals are illustrated schematically in Figure 1.19 and are shown in more detail in Figure A1.1. Note that the sum of the displacements of the orbital energies from the nonbonding level (the sum of the coefficients of β) is zero; this feature is general for Hückel orbital energies.

π Electron Energy and Delocalization Energy

The Hückel π electron energy is the sum of the respective orbital energies multiplied by the number of electrons in each orbital,

$$E_\pi = \sum_i n_i E_i \tag{A1.89}$$

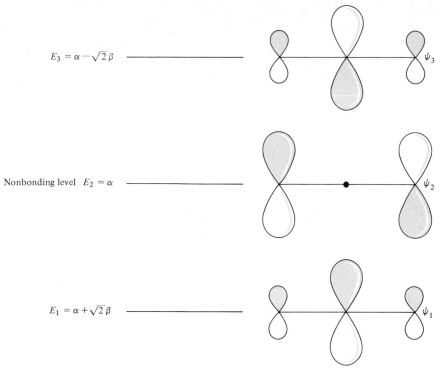

$$E_3 = \alpha - \sqrt{2}\,\beta \quad\underline{\hspace{6em}} \qquad \psi_3$$

Nonbonding level $E_2 = \alpha \quad\underline{\hspace{6em}} \qquad \psi_2$

$$E_1 = \alpha + \sqrt{2}\,\beta \quad\underline{\hspace{6em}} \qquad \psi_1$$

Figure A1.1 The Hückel π energy levels and molecular orbitals of the allyl system. The orbitals are depicted by drawing the basis orbitals of which they are composed, shaded in regions of positive algebraic sign, with relative sizes approximately corresponding to the relative contributions to the molecular orbital, as determined by the square of the orbital coefficient.

For allyl cation,

$$E_\pi = 2(\alpha + \sqrt{2}\beta) = 2\alpha + 2\sqrt{2}\beta \tag{A1.90}$$

For allyl radical,

$$E_\pi = 2(\alpha + \sqrt{2}\beta) + 1(\alpha) = 3\alpha + 2\sqrt{2}\beta \tag{A1.91}$$

For allyl anion,

$$E_\pi = 2(\alpha + \sqrt{2}\beta) + 2(\alpha) = 4\alpha + 2\sqrt{2}\beta \tag{A1.92}$$

The Hückel delocalization energy, HDE, is the difference between E_π and the Hückel localized energy, $\mathrm{HE_{loc}}$. $\mathrm{HE_{loc}}$ is the π electron energy calculated by the Hückel method for a single localized Lewis structure of the system. The bonding orbital of a localized double bond (ethylene) has energy $\alpha + \beta$, and a nonbonding electron in a p orbital has energy α, so

$$\mathrm{HE_{loc}} = 2B_d(\alpha + \beta) + n_p\alpha \tag{A1.93}$$

where B_d is the number of conventional double bonds in the Lewis structure and n_p is the number of nonbonding electrons localized in p orbitals. Thus for allyl radical,

$$\text{HE}_{\text{loc}} = 2(\alpha + \beta) + 1\alpha = 3\alpha + 2\beta \tag{A1.94}$$

and

$$\text{HDE} = 3\alpha + 2\sqrt{2}\beta - (3\alpha + 2\beta) = 0.83\beta \tag{A1.95}$$

The Hückel delocalization energy is not a particularly good guide to the importance of conjugation in a molecule. We shall discuss later a more reliable method that gives estimates of delocalization energy of cyclic conjugated molecules.

Butadiene

The secular determinant for butadiene is

$$\begin{vmatrix} x & 1 & 0 & 0 \\ 1 & x & 1 & 0 \\ 0 & 1 & x & 1 \\ 0 & 0 & 1 & x \end{vmatrix} = 0 \tag{A1.96}$$

Expanding by cofactors in the first row,

$$x \begin{vmatrix} x & 1 & 0 \\ 1 & x & 1 \\ 0 & 1 & x \end{vmatrix} - 1 \begin{vmatrix} 1 & 1 & 0 \\ 0 & x & 1 \\ 0 & 1 & x \end{vmatrix} = 0 \tag{A1.97}$$

Expanding again,

$$x \left\{ x \begin{vmatrix} x & 1 \\ 1 & x \end{vmatrix} - 1 \begin{vmatrix} 1 & 1 \\ 0 & x \end{vmatrix} \right\} - \left\{ 1 \begin{vmatrix} x & 1 \\ 1 & x \end{vmatrix} \right\} = 0 \tag{A1.98}$$

$$x\{x(x^2 - 1) - x\} - (x^2 - 1) = 0 \tag{A1.99}$$

$$x^4 - 3x^2 + 1 = 0 \tag{A1.100}$$

We can use the quadratic formula to find x:

$$x^2 = \frac{3 \pm \sqrt{9 - 4}}{2} = 0.382; \ 2.618 \tag{A1.101}$$

$$x = \pm 0.618, \ \pm 1.618 \tag{A1.102}$$

The energies are

$$\begin{aligned} x_4 &= +1.618; & E_4 &= \alpha - 1.618\beta \\ x_3 &= +0.618; & E_3 &= \alpha - 0.618\beta \\ x_2 &= -0.618; & E_2 &= \alpha + 0.618\beta \\ x_1 &= -1.618; & E_1 &= \alpha + 1.618\beta \end{aligned} \tag{A1.103}$$

These energy levels are illustrated in Figure A1.2. The Hückel ground-state model of butadiene has a pair of electrons in each of the two lower-energy orbitals; the π electron energy is

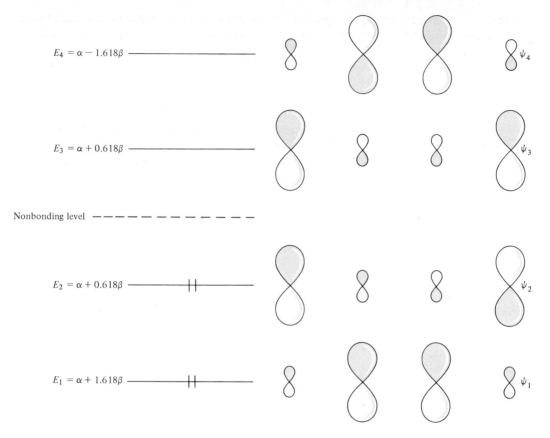

$E_4 = \alpha - 1.618\beta$ ——————————

$E_3 = \alpha + 0.618\beta$ ——————————

Nonbonding level —— — — — — — — — —

$E_2 = \alpha + 0.618\beta$ —————⫫—————

$E_1 = \alpha + 1.618\beta$ —————⫫—————

Figure A1.2 The energy levels and molecular orbitals of butadiene. Energy levels and orbital contributions are on the same scale as in Figure A1.1. Orbital occupancy represents the Hückel model approximation for the ground electronic state.

$$E_\pi = 2(\alpha + 1.62\beta) + 2(\alpha + 0.62\beta)$$
$$= 4\alpha + 4.48\beta \tag{A1.104}$$

and Hückel delocalization energy is

$$\text{HDE} = E_\pi - 4(\alpha + \beta) = 0.48\beta \tag{A1.105}$$

The orbital coefficients for the butadiene π orbitals can be found by the method illustrated above for allyl; another procedure, using cofactors, which is easier to apply for larger systems, is illustrated as follows.[f]

The ratio (coefficient a_{Jk})/(coefficient a_{J1}) is given by the quotient: (cofactor of element 1,k)/(cofactor of element 1,1). The cofactors are derived from the

[f]J. D. Roberts, *Notes on Molecular Orbital Calculations,* W. A. Benjamin, Menlo Park, Calif., 1962, p. 48.

secular determinant (A1.96 in our case), and proper signs must be included. For butadiene the ratios of coefficients are as follows:

$$\frac{a_{J1}}{a_{J1}} = 1.000$$

$$\frac{a_{J2}}{a_{J1}} = -\frac{\begin{vmatrix} 1 & 1 & 0 \\ 0 & x_J & 1 \\ 0 & 1 & x_J \end{vmatrix}}{\begin{vmatrix} x_J & 1 & 0 \\ 1 & x_J & 1 \\ 0 & 1 & x_J \end{vmatrix}} = -\frac{x_J^2 - 1}{x_J^3 - 2x_J}$$

$$\frac{a_{J3}}{a_{J1}} = +\frac{\begin{vmatrix} 1 & x_J & 0 \\ 0 & 1 & 1 \\ 0 & 0 & x_J \end{vmatrix}}{\begin{vmatrix} x_J & 1 & 0 \\ 1 & x_J & 1 \\ 0 & 1 & x_J \end{vmatrix}} = \frac{x_J}{x_J^3 - 2x_J} = \frac{1}{x_J^2 - 2}$$

(A1.106)

$$\frac{a_{J4}}{a_{J1}} = -\frac{\begin{vmatrix} 1 & x_J & 1 \\ 0 & 1 & x_J \\ 0 & 0 & 1 \end{vmatrix}}{\begin{vmatrix} x_J & 1 & 0 \\ 1 & x_J & 1 \\ 0 & 1 & x_J \end{vmatrix}} = -\frac{1}{x_J^3 - 2x_J}$$

A fourth relation comes from normalization:

$$\sum_i a_{Ji}^2 = 1 \tag{A1.107}$$

or

$$\frac{\sum_i a_{Ji}^2}{a_{J1}^2} = \frac{1}{a_{J1}^2} \tag{A1.108}$$

or

$$a_{J1}^2 = \frac{1}{\sum_i (a_{Ji}/a_{J1})^2} \tag{A1.109}$$

Now each root x_J in turn is substituted into A1.106. The results are illustrated for the lowest-energy orbital, $x_1 = -1.618$:

$$\frac{a_{11}}{a_{11}} = 1.000$$

$$\frac{a_{12}}{a_{11}} = 1.618$$

(A1.110)

$$\frac{a_{13}}{a_{11}} = 1.618$$

$$\frac{a_{14}}{a_{11}} = 1.000$$

We now apply Equation A1.109. From A1.110, squaring each result and summing,

$$\left(\frac{a_{11}}{a_{11}}\right)^2 = 1.000$$

$$\left(\frac{a_{12}}{a_{11}}\right)^2 = 2.618$$

$$\left(\frac{a_{13}}{a_{11}}\right)^2 = 2.618 \tag{A1.111}$$

$$\left(\frac{a_{14}}{a_{11}}\right)^2 = 1.000$$

$$\sum_i \left(\frac{a_{1i}}{a_{11}}\right)^2 = 7.236 \tag{A1.112}$$

Therefore

$$a_{11}{}^2 = \frac{1}{\sum_i (a_{1i}/a_{11})^2} = 0.138 \tag{A1.113}$$

and

$$a_{11} = 0.372 \tag{A1.114}$$

Then combining A1.110 and A1.114,

$$
\begin{aligned}
a_{11} &= 0.372 \\
a_{12} &= 0.602 \\
a_{13} &= 0.602 \\
a_{14} &= 0.372
\end{aligned} \tag{A1.115}
$$

The other orbitals are found in the same way by substituting $x_2 = -0.618$, $x_3 = +0.618$, and $x_4 = +1.618$ successively into the Equations A1.106. The resulting orbitals are

$$
\begin{aligned}
E_4 &= \alpha - 1.618\beta; & \psi_4 &= 0.372\varphi_{p_1} - 0.602\varphi_{p_2} + 0.602\varphi_{p_3} - 0.372\varphi_{p_4} \\
E_3 &= \alpha - 0.618\beta; & \psi_3 &= 0.602\varphi_{p_1} - 0.372\varphi_{p_2} - 0.372\varphi_{p_3} + 0.602\varphi_{p_4} \\
E_2 &= \alpha + 0.618\beta; & \psi_2 &= 0.602\varphi_{p_1} + 0.372\varphi_{p_2} - 0.372\varphi_{p_3} - 0.602\varphi_{p_4} \\
E_1 &= \alpha + 1.618\beta; & \psi_1 &= 0.372\varphi_{p_1} + 0.602\varphi_{p_2} + 0.602\varphi_{p_3} + 0.372\varphi_{p_4}
\end{aligned} \tag{A1.116}
$$

These orbitals are illustrated in Figure A1.2.

Degenerate Systems in the Hückel Approximation

In systems of high symmetry, degeneracies will occur; that is, there will be two equal roots of the secular equation. In these instances energies can be found in the usual way, but the procedures illustrated earlier will not give the orbital coefficients unambiguously. Degeneracy always implies a certain degree of arbitrariness in the choice of orbital coefficients. As an example consider the cyclopropenyl system. The reader may verify that the secular equation is A1.117, the solutions of which are A1.118:

$$\begin{vmatrix} x & 1 & 1 \\ 1 & x & 1 \\ 1 & 1 & x \end{vmatrix} = 0 \tag{A1.117}$$

$$\begin{aligned} x_3 &= +1; & E_3 &= \alpha - \beta \\ x_2 &= +1; & E_2 &= \alpha - \beta \\ x_1 &= -2; & E_1 &= \alpha + 2\beta \end{aligned} \tag{A1.118}$$

Substitution of $x_1 = -2$ into the set of equations from which the determinant was derived leads to A1.119:

$$\begin{aligned} -2a_{11} + a_{12} + a_{13} &= 0 \\ a_{11} - 2a_{12} + a_{13} &= 0 \\ a_{11} + a_{12} - 2a_{13} &= 0 \end{aligned} \tag{A1.119}$$

Combining the first two equations of this set yields A1.120:

$$-3a_{11} + 3a_{12} = 0 \tag{A1.120}$$

or

$$a_{11} = a_{12} \tag{A1.121}$$

Substituting A1.121 into the third of the set A1.119 gives

$$a_{12} = a_{13} \tag{A1.122}$$

Normalization will then require

$$a_{11} = a_{12} = a_{13} = \frac{1}{\sqrt{3}} \tag{A1.123}$$

The lowest-energy orbital is thus unambiguously defined:

$$E_1 = \alpha + 2\beta; \quad \psi_1 = \frac{1}{\sqrt{3}} \varphi_{p_1} + \frac{1}{\sqrt{3}} \varphi_{p_2} + \frac{1}{\sqrt{3}} \varphi_{p_3} \tag{A1.124}$$

The cofactor method will give this same solution for ψ_1.

For the degenerate orbital $x_2 = +1$, all three equations are identical:

$$\begin{aligned} a_{21} + a_{22} + a_{23} &= 0 \\ a_{21} + a_{22} + a_{23} &= 0 \\ a_{21} + a_{22} + a_{23} &= 0 \end{aligned} \tag{A1.125}$$

Similarly, $x_3 = +1$ yields only

$$a_{31} + a_{32} + a_{33} = 0 \tag{A1.126}$$

There is no unique solution even when normalization is imposed. The cofactor method is no help; the ratios have the indeterminate value of $0/0$.

We can get out of the difficulty by choosing arbitrarily $a_{21} = 0$; this choice will lead to one of many possible alternatives for the coefficients of the degenerate orbitals. With this choice,

$$a_{22} = -a_{23} \tag{A1.127}$$

and normalization requires

$$a_{22} = \frac{1}{\sqrt{2}} \tag{A1.128}$$

$$a_{23} = \frac{-1}{\sqrt{2}} \tag{A1.129}$$

The first degenerate orbital is therefore characterized by

$$E_2 = \alpha - \beta; \quad \psi_2 = \frac{1}{\sqrt{2}} \varphi_{p_2} - \frac{1}{\sqrt{2}} \varphi_{p_3} \tag{A1.130}$$

The other degenerate orbital, $x_3 = +1$, is determined by the requirements that it be normalized, orthogonal to the other two orbitals, and also obey A1.125. Two different orbitals

$$\psi_s = \sum_{i=1}^{N} a_{si} \varphi_i \tag{A1.131}$$

$$\psi_t = \sum_{i=1}^{N} a_{ti} \varphi_i \tag{A1.132}$$

are orthogonal if

$$\int \psi_s \psi_t \, d\tau = \int \sum_{i=1}^{N} a_{si} \varphi_i \sum_{j=1}^{N} a_{tj} \varphi_j \, d\tau = 0 \tag{A1.133}$$

or

$$\int \psi_s \psi_t \, d\tau = \sum_{i=1}^{N} \sum_{j=1}^{N} a_{si} a_{tj} \int \varphi_i \varphi_j \, d\tau = 0 \tag{A1.134}$$

Recalling the derivation of the normalization, we note that the integral on the right side of A1.134 is S_{ij}; in the Hückel approximation it is zero if $i \neq j$ and unity if $i = j$. Therefore orbitals ψ_s and ψ_t are orthogonal if

$$\int \psi_s \psi_t \, d\tau = \sum_{k=1}^{N} a_{sk} a_{tk} = 0 \tag{A1.135}$$

To apply this result to our problem, we have from A1.130 the coefficients of ψ_2: $a_{21} = 0$; $a_{22} = 1/\sqrt{2}$; $a_{23} = -1/\sqrt{2}$. Therefore the requirement that ψ_3 be orthogonal to ψ_2 means that the coefficients of ψ_3 must satisfy

$$a_{31} \cdot 0 + a_{32} \cdot \frac{1}{\sqrt{2}} - a_{33} \cdot \frac{1}{\sqrt{2}} = 0 \qquad \text{(A1.136)}$$

or

$$a_{32} = a_{33} \qquad \text{(A1.137)}$$

Then from A1.126 and A1.137,

$$a_{31} + 2a_{32} = 0 \qquad \text{(A1.138)}$$

or

$$a_{31} = -2a_{32} \qquad \text{(A1.139)}$$

Substituting A1.139 and A1.137 into the normalization condition then leads to

$$4a_{32}{}^2 + a_{32}{}^2 + a_{32}{}^2 = 1 \qquad \text{(A1.140)}$$

or

$$a_{32} = \frac{1}{\sqrt{6}} \qquad \text{(A1.141)}$$

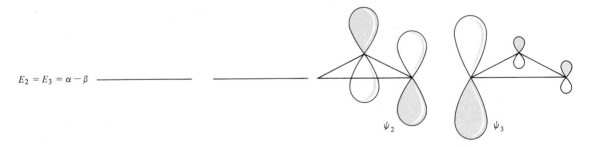

$E_2 = E_3 = \alpha - \beta$

ψ_2 ψ_3

Nonbonding level

$E_1 = \alpha + 2\beta$

ψ_1

Figure A1.3 The π energy levels and molecular orbitals of cyclopropenyl, on the same scale as Figures A1.1 and A1.2.

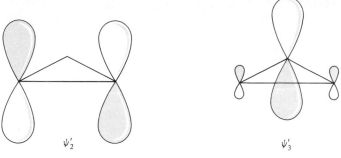

Figure A1.4 A different, equivalent set of degenerate orbitals ψ'_2 and ψ'_3 for cyclopropenyl.

and the third orbital is characterized by

$$E_3 = \alpha - \beta; \qquad \psi_3 = \frac{-2}{\sqrt{6}}\varphi_{p_1} + \frac{1}{\sqrt{6}}\varphi_{p_2} + \frac{1}{\sqrt{6}}\varphi_{p_3} \qquad (A1.142)$$

To summarize, this particular choice of cyclopropenyl orbitals is then

$$E_3 = \alpha - \beta; \qquad \psi_3 = \frac{-2}{\sqrt{6}}\varphi_{p_1} + \frac{1}{\sqrt{6}}\varphi_{p_2} + \frac{1}{\sqrt{6}}\varphi_{p_3}$$

$$E_2 = \alpha - \beta; \qquad \psi_2 = 0\,\varphi_{p_1} \quad + \frac{1}{\sqrt{2}}\varphi_{p_2} - \frac{1}{\sqrt{2}}\varphi_{p_3} \qquad (A1.143)$$

$$E_1 = \alpha + 2\beta; \qquad \psi_1 = \frac{1}{\sqrt{3}}\varphi_{p_1} + \frac{1}{\sqrt{3}}\varphi_{p_2} + \frac{1}{\sqrt{3}}\varphi_{p_3}$$

These energy levels and orbitals are illustrated in Figure A1.3.

  In solving for the a coefficients of the degenerate orbitals, we could have made some other arbitrary choice; for example, assigning $a_{22} = 0$ would have led to a degenerate orbital pair ψ'_2, ψ'_3, of identical shape to the ones found above but rotated, as illustrated in Figure A1.4. Other initial choices would lead to orbitals of different shapes; all the alternatives are alike, however, in having one node.

REGULARITIES OF HÜCKEL ORBITALS. CYCLIC POLYENES, DELOCALIZATION ENERGY, AND AROMATICITY

The Hückel π orbitals and energies have a number of useful regular properties.[g] We have already noted that the "center of gravity" of the energy levels is the nonbonding level α, so that the sum of the coefficients of β over all the orbitals is zero. We have also observed in Section 1.9 the regularities of symmetry properties and numbers of nodes for the π orbitals of linear conjugated chains.

[g]M. J. S. Dewar and R. C. Dougherty, *The PMO Theory of Organic Chemistry*, Plenum, New York, 1975, Chap. 3.

Linear conjugated chains and cyclic conjugated molecules containing only rings with even numbers of carbons belong to a type of molecule called *alternant*. An alternant system is one in which the atoms can be divided into two classes such that atoms of one class are bonded only to atoms of the other class. To determine if a hydrocarbon skeleton is alternant, put a star at an arbitrary position, then alternate unstarred, starred, until all atoms have been marked. If starred atoms are adjacent only to unstarred and unstarred are adjacent only to starred, the system is alternant. Some alternant and nonalternant hydrocarbons are shown below.

Alternant:

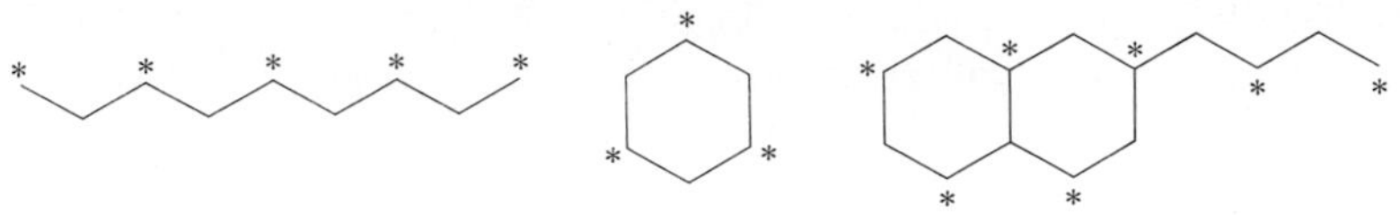

Nonalternant:

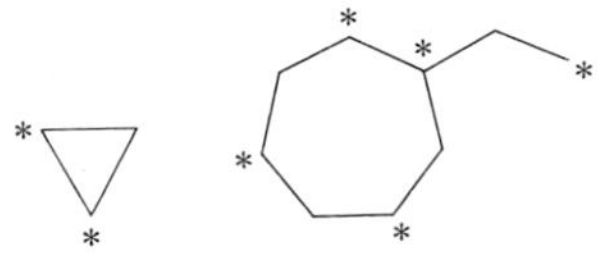

The energies of π orbitals of alternant hydrocarbons (with the exception of possible nonbonding orbitals) are in pairs, symmetrically disposed about the nonbonding level. This property is illustrated by the energies of allyl and of butadiene. *Even-alternant hydrocarbons* (butadiene, for example) have an even number of orbitals and (with the exception of certain cyclic systems, see below) have no nonbonding orbital; *odd-alternant hydrocarbons* (allyl, for example) have an odd number of orbitals and hence have a nonbonding level. Nonalternant hydrocarbons (cyclopropenyl, for example) do not obey this rule.

The Hückel π orbital energies of unbranched, linear conjugated polyenes can be written in closed form. They are given by

$$E_J = \alpha + 2\beta \cos\left(\frac{\pi}{N+1}J\right), \qquad J = 1, 2, \ldots, N \qquad (A1.44)$$

where J is the orbital number, counting upward from the lowest-energy orbital $J = 1$, and N is the number of carbon atoms (also the number of basis orbitals) in the chain. (The units of $\pi J/N + 1$ are radians.)

If a pair of orbitals of an alternant hydrocarbon of energies $\alpha + c\beta$ and $\alpha - c\beta$ are compared, the coefficient of each basis p orbital will be found to be numerically the same in the two. The coefficients in the two paired orbitals are related by inverting the signs of all the coefficients of either the starred set or the unstarred set of atoms. (It does not matter which set is inverted, because the negative of an orbital is equivalent to the orbital itself.)

An electrically neutral, even-alternant hydrocarbon has the right number of electrons to just fill all the bonding molecular orbitals, leaving all the antibonding ones empty.

When the atoms of an odd-alternant hydrocarbon are classified into starred and unstarred sets, one set has one more atom than the other. By convention, we always make the more numerous set the starred set. With this assignment the nonbonding molecular orbital is confined to the starred set of atoms; the coefficients of all the unstarred atoms in the nonbonding orbital are zero. (See Figure 1.42 for an illustration of this rule.)

π electron density is defined for each atom in a conjugated system as the sum over all occupied π molecular orbitals of the number n_J of electrons in the molecular orbital times the square of the coefficient of the atom's p orbital in that molecular orbital:

$$q_i = \sum_{J \text{ occ}} n_J a_{Ji} \qquad (A1.145)$$

To illustrate, for carbon number 1 of butadiene, a_{11} in orbital ψ_1 is $+0.372$ and a_{21} in ψ_2 is $+0.602$. Each of these orbitals has two electrons, and the other orbitals are unoccupied. Therefore

$$q_1 = (2)(0.372)^2 + (2)(0.602)^2 = 1.00 \qquad (A1.146)$$

The π electron density in a neutral ground-state alternant hydrocarbon (for the even molecules, all bonding orbitals doubly occupied; for the odd molecules, all bonding orbitals doubly occupied and the nonbonding orbital singly occupied) is unity at each atom. (A π electron density of unity is just balanced by the unit positive charge of the carbon atom skeleton, consisting of the carbon atom, its inner-shell electrons, and its associated sigma electrons.)

An odd-alternant anion differs from the neutral molecule in having one electron more in the nonbonding orbital; a cation differs in having one electron less. Because the nonbonding orbital is confined to the starred atoms, the excess charge in these ions appears only at the starred positions.

The coefficients of a nonbonding molecular orbital of an odd-alternant system may be found without solving the secular determinant by using the following rule: The sum of the nonbonding molecular orbital coefficients of the starred atoms adjacent to any unstarred atom is zero. We illustrate the application of this rule for the nonbonding orbital of pentadienyl:

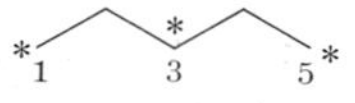

Recall that the coefficients at unstarred atoms 2 and 4 are zero. Let the coefficient a_1 at atom 1 equal $+c$. By the rule, $a_3 = -c$. If $a_3 = -c$, $a_5 = +c$. Normalization requires

$$a_1^2 + a_3^2 + a_5^2 = 1 \qquad (A1.147)$$

or

$$3c^2 = 1 \qquad (A1.148)$$

$$c = \frac{1}{\sqrt{3}} \tag{A1.149}$$

$$\psi_{\text{NBMO}} = \frac{1}{\sqrt{3}} \, \varphi_{p_1} - \frac{1}{\sqrt{3}} \, \varphi_{p_3} + \frac{1}{\sqrt{3}} \, \varphi_{p_5} \tag{A1.150}$$

Cyclic Systems

Energies for monocyclic conjugated π systems can be found with the aid of the following device. Draw the ring (a regular polygon) with one vertex down. Draw a circle passing through all the vertices. Call the radius of the circle 2β. Draw a horizontal line at the level of each vertex. The resulting diagram gives the energy levels, determined geometrically with energy α at the level of the center of the circle. The procedure is illustrated in Figure A1.5. The general formula for the π orbital energies for monocyclic conjugated rings is

$$E_J = \alpha + 2\beta \cos\left(\frac{2\pi}{N}J\right), \qquad J = 0, 1, 2, \ldots, N-1 \tag{A1.151}$$

where J is the orbital number (counting cyclically as explained in the caption to Figure A1.5) and N is the ring size. (The units of $2\pi J/N$ are radians.) The same results could be obtained, with a great deal more work, by solving the secular determinants as we illustrated for the cyclopropenyl system. Note that the energy-level patterns that give rise to the Hückel $4n + 2$ rule are generated directly by the projection procedure. Note also the pair of degenerate, nonbonding molecular orbitals that appears in each $4n$ ring. Coefficients for these nonbonding orbitals may be found with the nonbonding orbital rule discussed above, applied successively to the two alternative assignments of starred and unstarred atoms. (The coefficients found in this way are not unique.)

From the energy levels for the monocyclic conjugated rings we can calculate delocalization energies that reveal something about aromaticity and antiaromaticity. The Hückel delocalization energy HDE is not very useful for this purpose because it is based on an oversimplified model for the localized reference state. A reasonable criterion of aromaticity would be stabilization occurring solely because of the cyclic character of the conjugation; this criterion may be applied by comparing the π electron energy of the ring with that of a hypothetical open-chain compound that has the same number and kinds of bonds.

For acyclic molecules Hess and Schaad found the π energy to be given to good accuracy by adding together the contributions listed in Table A1.1 for the different types of bonds in the molecule.[h] A good estimate of a localized π electron energy E_{loc} can be obtained by writing a single localized Lewis structure for the system and adding up the contribution of each bond as found in the table. For benzene there are three bonds of type HC=CH and three of type HC—CH in the localized Lewis structure. From Table A1.1,

$$E_{\text{loc}} = 3(\alpha + 2.0699\beta) + 3(\alpha + 0.4660\beta) \tag{A1.152}$$

---

[h] B. A. Hess, Jr., and L. J. Schaad, *J. Am. Chem. Soc.*, **93**, 305 (1971).

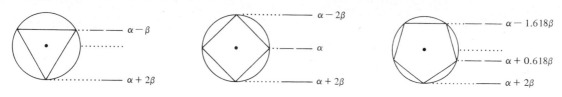

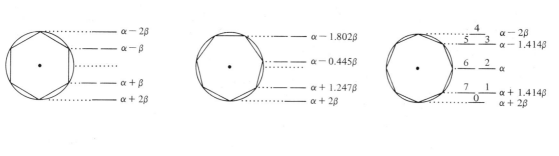

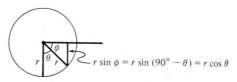

$$r \sin \phi = r \sin (90° - \theta) = r \cos \theta$$

Figure A1.5 The projection method for finding π orbital energies of monocyclic conjugated rings. The ring is inscribed point downward in a circle of radius 2β. Projection horizontally of each vertex generates an energy-level diagram with energy α at the level of the center of the circle. The circle at the bottom of the figure demonstrates that the vertical distance below the horizontal of a point on the circle of radius r is $r \cos \theta$, where θ is the angle measured from the vertical line. The polygon vertices (numbered by the index J, counting counterclockwise starting at the bottom with $J = 0$) are at angles $2\pi J/N$, where N is the number of vertices. The energy at the level of the center of the circles is α and the radii are 2β; therefore the energies are given by the formula

$$E_J = \alpha + 2\beta \cos\left(\frac{2\pi}{N}J\right)$$

Note that the orbital numbering starts at $J = 0$ and is cyclic around the ring, as illustrated for the cyclooctatetraene model. The method originates with A. A. Frost and B. Musulin, *J. Chem. Phys.*, **21**, 572 (1953).

$$E_{loc} = 6\alpha + 7.61\beta \tag{A1.153}$$

The Hückel π energies for benzene give

$$\begin{aligned} E_\pi &= 2(\alpha + 2\beta) + 2(\alpha + \beta) + 2(\alpha + \beta) \\ &= 6\alpha + 8\beta \end{aligned} \tag{A1.154}$$

Therefore delocalization energy

$$DE_\pi = E_\pi - E_{loc} = +0.39\beta \tag{A1.155}$$

Table A1.1 CALCULATED HÜCKEL π BOND
ENERGIES OF CARBON–CARBON DOUBLE
BONDS AND CARBON–CARBON SINGLE
BONDS OF ACYCLIC POLYENES[a]

Designation[b]	Type of Bond	Calculated π bond energy
23	$H_2C{=}CH$	$\alpha + 2.0000\beta^c$
22	$HC{=}CH$	$\alpha + 2.0699\beta$
22′	$H_2C{=}C$	$\alpha + 2.0000\beta^c$
21	$HC{=}C$	$\alpha + 2.1083\beta$
20	$C{=}C$	$\alpha + 2.1716\beta$
12	$HC{-}CH$	$\alpha + 0.4660\beta$
11	$HC{-}C$	$\alpha + 0.4362\beta$
10	$C{-}C$	$\alpha + 0.4358\beta$

[a] Adapted with permission from B. A. Hess, Jr., and L. J. Schaad, *J. Am. Chem. Soc.*, **93**, 305 (1971). Copyright 1971 American Chemical Society.
[b] The first index gives the bond order, the second the number of attached hydrogens.
[c] Arbitrarily assigned.

A crude experimental delocalization energy is the difference between heat of hydrogenation of benzene and that of a hypothetical "cyclohexatriene" containing three cyclohexene double bonds. The cycle shown in Scheme 1 leads to an estimate of 36 kcal mole^{-1} for benzene delocalization energy. A number of corrections need to be made to this figure, however. For example, a "cyclohexatriene"

SCHEME 1

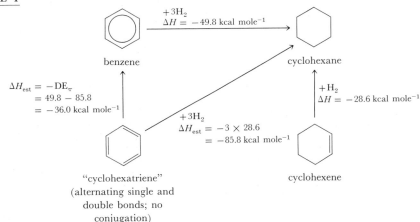

$+3H_2$
$\Delta H = -49.8$ kcal mole^{-1}

benzene

cyclohexane

$\Delta H_{est} = -DE_\pi$
$= 49.8 - 85.8$
$= -36.0$ kcal mole^{-1}

$+H_2$
$\Delta H = -28.6$ kcal mole^{-1}

$+3H_2$
$\Delta H_{est} = -3 \times 28.6$
$= -85.8$ kcal mole^{-1}

"cyclohexatriene"
(alternating single and
double bonds; no
conjugation)

cyclohexene

made of cyclohexene double bonds has alternating long and short bonds, whereas benzene has C—C bonds of equal length; furthermore, the bond type changes that occur on hydrogenating cyclohexene are different from those that occur on hydrogenating benzene. Dewar found that the different types of bond occurring in noncyclic conjugated systems make additive contributions to the heat of formation of the molecule; this result is analogous to the bond additivity property of Hückel energies found in Table A1.1. Hence a reliable delocalization energy, called the Dewar resonance energy, DRE, can be found by comparing experimental heat of formation of benzene with that calculated from the bond contributions. The DRE found in this way for benzene is 21.2 kcal mole^{-1}.[i]

The close analogy between the Hess-Schaad procedure for Hückel DE_π and the DRE approach allows us to estimate a value for β:

$$0.39\beta = -21.2 \text{ kcal mole}^{-1} \tag{A1.156}$$

$$\beta = -54.4 \text{ kcal mole}^{-1} \tag{A1.157}$$

In view of the fact that published estimates of β range from 10 to 60 kcal mole^{-1}, we should not place too much confidence in this value. Estimates based on different thermochemical models, or on properties other than thermochemistry (for example, electronic excitation energies measured by ultraviolet spectroscopy), differ substantially from this one.[j]

Despite uncertainty about the correct value of β, if we employ a self-consistent system, we can use the delocalization energy idea to investigate, semiquantitatively, aromaticity and antiaromaticity. From Figure A1.5, cyclobutadiene has

$$E_\pi = 4\alpha + 4\beta \tag{A1.158}$$

(one electron in each of the nonbonding orbitals and two in the bonding orbital). E_{loc} from Table A1.1 is

$$\begin{aligned} E_{loc} &= 2(\alpha + 2.0699\beta) + 2(\alpha + 0.4660\beta) \\ &= 4\alpha + 5.07\beta \end{aligned} \tag{A1.159}$$

and

$$DE_\pi = E_\pi - E_{loc} = -1.07\beta \tag{A1.160}$$

This calculation predicts a substantial antiaromatic destabilization for this $4n$ π electron molecule. The hypothetical planar cyclooctatetraene is also calculated to be antiaromatic, but less strongly so:

$$E_\pi = 8\alpha + 9.66\beta \tag{A1.161}$$

$$E_{loc} = 4(\alpha + 2.0699\beta) + 4(\alpha + 0.4660\beta) \tag{A1.162}$$

$$= 8\alpha + 10.14\beta \tag{A1.163}$$

$$DE_\pi = E_\pi - E_{loc} = -0.48\beta \tag{A1.164}$$

---

[i] N. C. Baird, *J. Chem. Educ.*, **48**, 509 (1971).
[j] (a) A. Streitwieser, reference *d*(c); (b) E. Heilbronner and H. Bock, reference *d*(g), Vol. 1.

THE EXTENDED HÜCKEL METHOD

The extended Hückel method, developed by Hoffmann,[k] has the following characteristics:

1. All valence orbitals are included in the basis. The method is not restricted to π systems.
2. All overlaps S_{ij} are calculated and included.
3. Energies H_{ii} of an electron in each of the basis orbitals are estimated empirically from spectroscopic data.
4. Interactions H_{ij} are approximated as a simple function of S_{ij}, H_{ii}, and H_{ij}.

A computer is required. This method provides a practical way of carrying out calculations on moderately large organic molecules with minimal expenditure of computer time. Although energies are still not particularly reliable, energy differences within a series of similar compounds are revealing.

Hückel and extended Hückel methods are termed *semiempirical* because they rely on experimental data for the quantification of parameters. There are other semiempirical methods, such as CNDO, MINDO, INDO, in which experimental data are still used, but more care is taken in evaluating the H_{ij}. These methods are discussed in various works on molecular orbital theory.[l]

AB INITIO METHODS

Ab initio methods calculate all quantities needed from first principles and use no experimentally determined parameters. The computations require more computer time and are therefore more expensive than semiempirical ones. Good energies can now be obtained for small molecules by these techniques. The reader is referred to specialized treatments for further information.[m]

---

[k] R. Hoffmann, *J. Chem. Phys.*, **39**, 1397 (1963).
[l] See note 4(i), (m).
[m] See note 4(k), (l), (m).

Chapter 2
SOME FUNDAMENTALS
OF PHYSICAL
ORGANIC CHEMISTRY

In this chapter we review several aspects of the physical chemistry of organic compounds that are particularly useful in the investigation of reaction mechanisms.

2.1 STEREOCHEMISTRY

Because excellent treatments of stereochemistry are available in other sources, we shall limit our discussion to a brief review of the more important definitions and principles.[1]

Isomers are compounds with the same molecular formula but different molecular structures. There are two broad categories of isomerism: *constitutional isomerism* and *stereoisomerism*. Constitutional isomers have different connectivity; that is, they differ in the order in which the atoms are connected by bonds. Stereoisomers

---

[1]For the best discussions in introductory textbooks, see (a) R. T. Morrison and R. N. Boyd, *Organic Chemistry,* 3d ed., Allyn & Bacon, Boston, 1973, pp. 73–80, 115–140, 225–246; (b) J. B. Hendrickson, D. J. Cram, and G. S. Hammond, *Organic Chemistry,* 3d ed., McGraw-Hill, New York, 1970, pp. 175–230; (c) N. L. Allinger, M. P. Cava, D. C. DeJong, C. R. Johnson, N. A. Lebel, and C. L. Stevens, *Organic Chemistry,* Worth, New York, 1971, pp. 105–126. More comprehensive introductory treatments are available: (d) G. Natta and M. Farina, *Stereochemistry,* Longmans, London, 1972; (e) K. Mislow, *Introduction to Stereochemistry,* W. A. Benjamin, Menlo Park, Calif., 1966; (f) E. L. Eliel and F. Basolo, *Elements of Stereochemistry,* Wiley, New York, 1969; (g) J. A. Hirsch, *Concepts in Theoretical Organic Chemistry,* Allyn & Bacon, Boston, 1974, Chaps. 10–12. For advanced treatments see (h) M. S. Newman, *Steric Effects in Organic Chemistry,* Wiley, New York, 1956; (i) E. L. Eliel, *Stereochemistry of Carbon Compounds,* McGraw-Hill, New York, 1962; (j) E. L. Eliel, N. L. Allinger, S. J. Angyal, and G. A. Morrison, *Conformational Analysis,* Wiley, New York, 1965.

have the same connectivity but differ in the way atoms or groups of atoms are oriented in space. Our discussion will be limited to stereoisomerism. Structures **1** through **16** illustrate pairs of stereoisomers. In these examples, as in structures throughout this book, bonds projecting behind the plane of the page are indicated by a dashed line, and bonds projecting in front of the page are indicated by a wedge. Dotted lines will be used in structures to designate partial bonds.

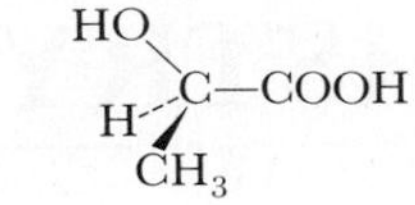

1

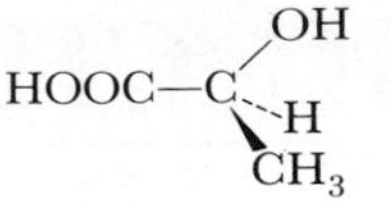

2

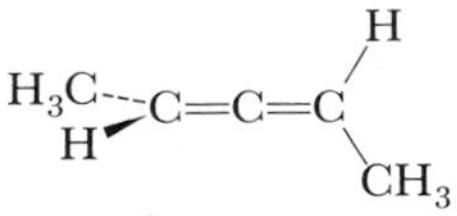

3

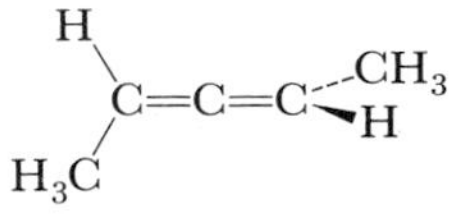

4

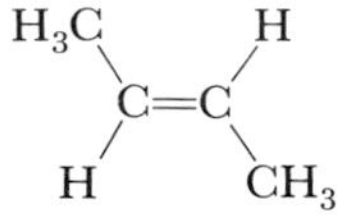

5

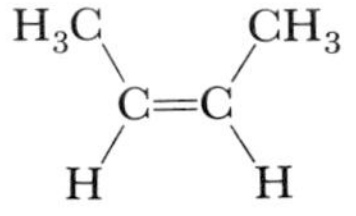

6

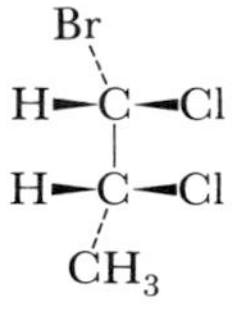

7

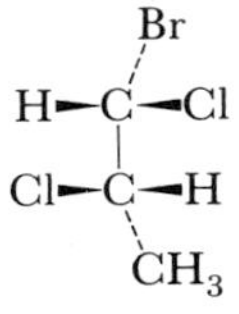

8

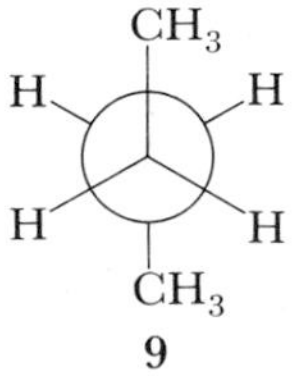

9

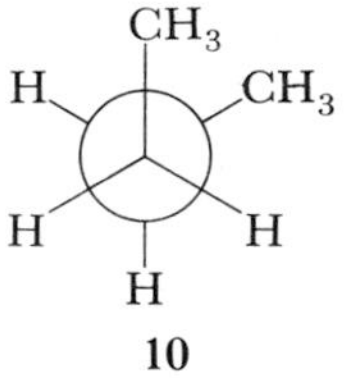

10

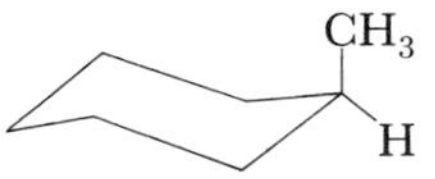

11

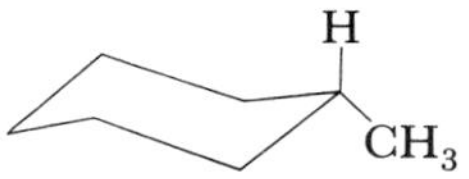

12

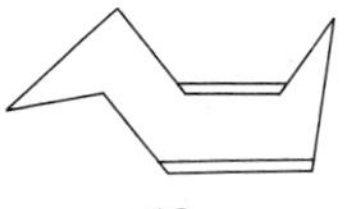

13

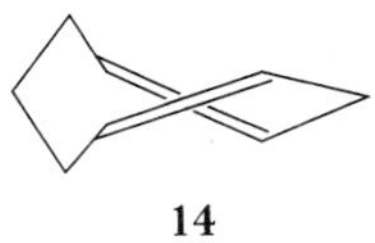

14

15   16

These examples illustrate several kinds of stereoisomerism. *Configurational stereoisomers* cannot be made superimposable by any rotations about single bonds; making them identical requires one or more single bonds to be broken, or double bonds to be reduced to single bonds. Pairs **1**–**2**, **3**–**4**, **5**–**6**, and **7**–**8** are configurational stereoisomers. *Conformational stereoisomers* (or *conformers*) can be made superimposable by rotations about single bonds. Pairs **9**–**10**, **11**–**12**, **13**–**14**, and **15**–**16** are conformational stereoisomers. The precise specification of the location in space of the groups in a configurational isomer (i.e., **1** as opposed to **2** or **3** as opposed to **4**) is called the *configuration;* a similar specification for a conformational isomer (**9** as opposed to **10** or **11** as opposed to **12**) is the *conformation.*

From an operational point of view, configurational isomers, having energy barriers between them of the order of bond dissociation energies (50 to 100 kcal mole^{-1}), can ordinarily be separated and stored at room temperature without interconverting. Conformational isomers, having energy barriers between them on the order of the rotational energy barriers of single bonds (about 3 to 20 kcal mole^{-1}), usually cannot be physically separated. Because single-bond rotational barriers can be dramatically increased by the presence of bulky groups, there are many exceptions now known to the latter generalization; we shall return to this point later.

Any pair of stereoisomers related as an object and its mirror image are *enantiomers;* any pair of stereoisomers not so related are *diastereomers.* The pairs **1**–**2**, **3**–**4**, and **15**–**16** are enantiomeric pairs; the other pairs illustrated above are diastereomeric. Note that enantiomerism may be based either on configurational isomerism (**1**–**2**; **3**–**4**) or on conformational isomerism (**15**–**16**).

Any object not identical with its mirror image, and therefore capable of exhibiting enantiomerism, is said to be *chiral.* In terms of the symmetry properties discussed in Chapter 1, the necessary and sufficient condition for chirality is that the object not possess any point of inversion, any mirror plane, or any alternating axis of symmetry.[2] An object may possess an axis of rotation and still be chiral; *trans*-cyclooctene, **15**–**16**, for example, has a C_2 axis. Most chiral molecules, how-

---

[2]An alternating axis, or reflection–rotation axis, is an element of symmetry for which the operation consists of rotation about the axis followed by reflection in a plane perpendicular to the axis. The molecule spiro[4.4]nonane has a fourfold alternating axis (designated S_4):

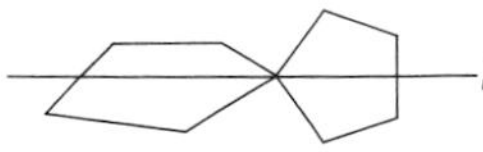

$$\text{——}\diagup\!\!\diagdown\!\!\diagdown\!\!\diagup\text{——}S_4$$

Actually, a mirror plane is equivalent to an S_1 axis and a point of inversion is equivalent to an S_2 axis; therefore the most concise specification of the necessary and sufficient condition for chirality is the absence of an alternating axis of any order. For further discussion and examples see (a) H. H. Jaffé and M. Orchin, *Symmetry in Chemistry*, Wiley, New York, 1965; (b) K. Mislow, *Introduction to Stereochemistry*, W. A. Benjamin, Menlo Park, Calif., 1966.

ever, lack any symmetry element and may properly be called *asymmetric*. Lactic acid (**1-2**) is an example. The less restrictive term *dissymmetric* includes objects with no symmetry together with those that have rotation axes but no point of inversion, no mirror plane, and no alternating axis. Hence asymmetry is sufficient but not necessary for chirality, while dissymmetry is both necessary and sufficient.

Compounds with pairs of identical chiral centers will have, in addition to chiral isomers, isomers in which each chiral center is the mirror image of another. This type of isomer is achiral and is designated *meso*.

The distinguishing physical property of substances whose molecules are chiral is optical activity, the rotation of the plane of polarization of light. A solution containing an excess of one enantiomer over the other will be optically active;[3] a solution containing exactly equal amounts of the two enantiomers will give no rotation. A material consisting of equal amounts of two enantiomers is a *racemic modification*. In the solid state a racemic modification may consist of a mixture of crystals each of which contains only one enantiomer (*racemic mixture*), crystals containing molecules of the two enantiomers arranged in a regular alternating pattern (*racemic compound* or *racemate*), or crystals containing both enantiomers but in random order (*racemic solid solution*), depending on the relative affinity of the molecules for their own kind as opposed to their affinity for the mirror image molecules.[4] The distinctions among these various categories of racemic modifications disappear in dilute solution and in the gas phase.

Any chemical process that converts an enantiomer into its mirror image will cause loss of optical activity and is called *racemization*. Note that conversion of half the molecules in a sample of a pure enantiomer into the mirror image causes complete racemization; the optical activity will have fallen to zero and will not change further if all other components of the system are achiral or racemic. Any process that generates an achiral intermediate will also cause loss of optical activity; any product derived by way of an achiral intermediate (assuming all subsequent participants in the reaction to be achiral or racemic) will be either achiral or racemic.

Specification of Configuration

The most common origin of molecular dissymmetry in organic compounds is the presence of one or more *chiral centers* (or asymmetric centers): saturated carbon atoms bearing four different substituents. The central carbon in lactic acid (**1–2**) is a chiral center; the halogen-bearing carbons in **7** and **8** are chiral centers. Compounds of the allene type (**3–4**) have no chiral center but show *axial dissymmetry*.

The configuration of a chiral center is specified by the Cahn–Ingold–Prelog

---

[3] The magnitude of optical rotation varies with the wavelength of the light and may even change sign. Hence it is possible that a pure enantiomer fails to rotate the plane of polarization at some particular wavelength, but it will do so at other wavelengths.

[4] For further discussion see E. L. Eliel, *Stereochemistry of Carbon Compounds,* McGraw-Hill, New York, 1962, p. 43.

system.[5] The ligands of the chiral center are assigned an order of priority: the higher the atomic number of the directly bonded atom, the higher is the priority. If two atoms are isotopes, the one with higher atomic mass has higher priority. An unshared pair of electrons is treated as a ligand atom of atomic number zero. If two attached atoms are the same, the atoms attached to them are compared, and so on, until priority can be established. Multiple bonds are replaced by single bonds, with both atoms considered to be duplicated or triplicated, as illustrated below:

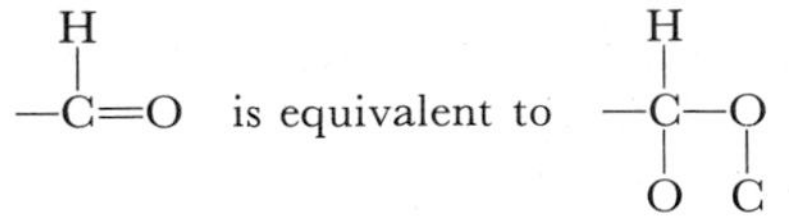

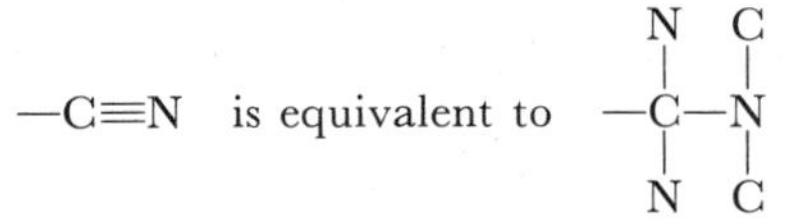

$$
\begin{array}{ccc}
\overset{\displaystyle H}{\underset{}{|}} & & \overset{\displaystyle H}{\underset{}{|}} \\
-\text{C}=\text{O} & \text{is equivalent to} & -\overset{|}{\underset{\displaystyle O}{C}}-\overset{}{\underset{\displaystyle C}{O}} \\
\end{array}
$$

$$
-\text{C}\equiv\text{N} \quad \text{is equivalent to} \quad -\overset{\displaystyle N \quad C}{\underset{\displaystyle N \quad C}{C}}-
$$

Having established the priority order of the groups, one views the molecule along the bond from the chiral center to the lowest-priority group, with the chiral center in front of the low-priority group. If the remaining three groups, from highest priority to lowest, are arranged clockwise, the center is R; if counterclockwise, S. The procedure is illustrated for lactic acid (**1–2**). Of the atoms attached to the chiral center, O has the highest atomic number and H the lowest. The other two atoms are both carbon; the carboxyl group,

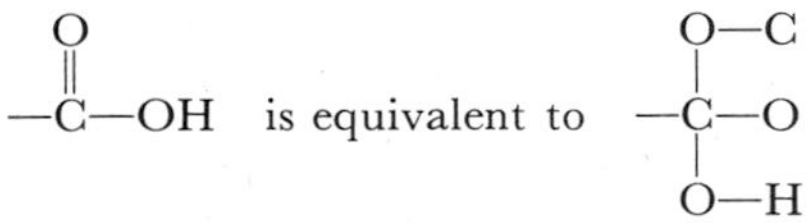

$$
\overset{\displaystyle O}{\underset{}{\parallel}} \\
-\text{C}-\text{OH} \quad \text{is equivalent to} \quad -\overset{\displaystyle O-C}{\underset{\displaystyle O-H}{C}}-\text{O}
$$

and so takes precedence over
$$
-\overset{\displaystyle H}{\underset{\displaystyle H}{C}}-\text{H}.
$$

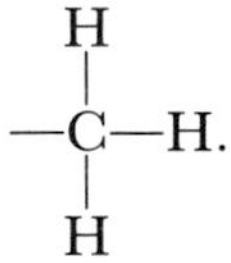

Structure **1** is viewed as shown below; it has the R configuration:

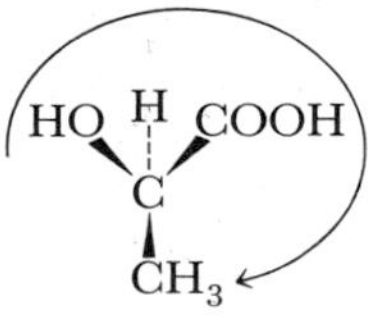

HO H COOH
C
CH$_3$

R

[5](a) R. S. Cahn, C. Ingold, and V. Prelog, *Angew. Chem. Int. Ed. Engl.*, **5**, 385 (1966); (b) R. S. Cahn and C. K. Ingold, *Experientia*, **12**, 81 (1956); (c) R. S. Cahn and C. K. Ingold, *J. Chem. Soc.*, 612 (1951); (d) E. L. Eliel, *Stereochemistry of Carbon Compounds*, McGraw-Hill, New York, 1962, pp. 92, 166, 311; (e) R. T. Morrison and R. N. Boyd, *Organic Chemistry*, 3d ed., Allyn & Bacon, Boston, 1973, p. 130.

Structure **2** is S:

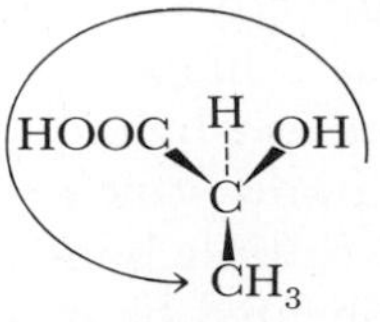

S

Molecules with axial symmetry, for example, **3–4**, are viewed along the axis so that the attached groups appear on mutually perpendicular planes, as shown in **17**, where the solid line represents the plane containing the substituents nearer the viewer. An additional priority rule is now needed: All near groups take precedence

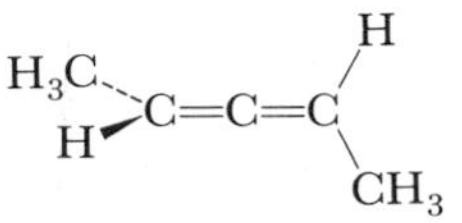

      3   **17**

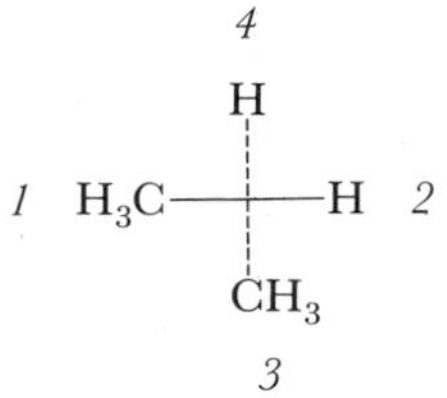

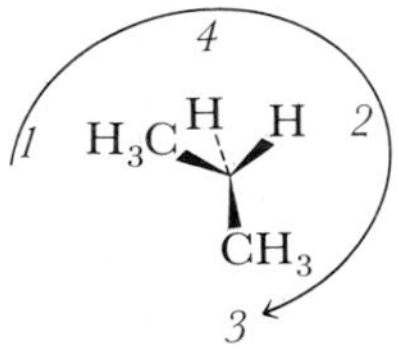

R

18

over all far groups. The priorities are thus as indicated in **17**. A small change in orientation will then bring the priority *3* group forward (**18**) and leave the priority *4* group remote so that the chirality designation can be determined as before.

Stereoisomers of the type **5–6**, sometimes called *geometrical isomers,* can be specified by the familiar *cis–trans* nomenclature system in simple cases. Thus **5** is unambiguously described as *trans*-2-butene and **6** is *cis*-2-butene. A systematic method of nomenclature is needed for more complex substitution patterns. The four substituents of the double bond are assigned priorities according to the Cahn–Ingold–Prelog system. If the higher-priority group at one end of the double bond is on the same side of the plane perpendicular to the plane of the nuclei and passing through the two double-bonded carbons as the higher-priority group at the other end, the configuration is Z (from German *zusammen,* together); if on the opposite side, the configuration is E (from German *entgegen,* opposite). Thus **19** is Z-1-bromo-1-chloro-1-propene; **20** is E-4-phenyl-3-decene.

$$\begin{array}{c} \text{Br} \quad\quad \text{CH}_3 \\ \diagdown \ \ \diagup \\ \text{C}{=}\text{C} \\ \diagup \ \ \diagdown \\ \text{Cl} \quad\quad \text{H} \\ \textbf{19} \end{array} \qquad \begin{array}{c} \text{H}_3\text{C}{-}\text{CH}_2 \quad\quad (\text{CH}_2)_5{-}\text{CH}_3 \\ \diagdown \ \ \diagup \\ \text{C}{=}\text{C} \\ \diagup \ \ \diagdown \\ \text{H} \quad\quad \phi \\ \textbf{20} \end{array}$$

Prochirality[6]

If replacement of one ligand of an achiral center by a new ligand generates a chiral center, the original molecule is said to be *prochiral*.[7] For example, replacement of one methylene proton of chloroethane by bromine will generate a chiral center, as illustrated by Structures **21** and **22**. Hence chloroethane is prochiral. The methyl-

$$\begin{array}{c} \text{H} \quad\ \text{H} \\ \diagdown\text{C}\diagup \\ \diagup\ \ \diagdown \\ \text{H}_3\text{C}\ \ \text{Cl} \\ \textbf{21} \\ \textit{prochiral} \end{array} \xrightarrow{\ \text{replace H by Br}\ } \begin{array}{c} \text{H} \quad\ \text{Br} \\ \diagdown\text{C}\diagup \\ \diagup\ \ \diagdown \\ \text{H}_3\text{C}\ \ \text{Cl} \\ \textbf{22} \\ \textit{chiral} \end{array}$$

ene protons are said to be *stereoheterotopic*. Dichloromethane, on the other hand, is not prochiral, as illustrated by **23** and **24**. Its protons are *homotopic* or *equivalent*.

$$\begin{array}{c} \text{H} \quad\ \text{H} \\ \diagdown\text{C}\diagup \\ \diagup\ \ \diagdown \\ \text{Cl}\ \ \text{Cl} \\ \textbf{23} \\ \textit{achiral} \end{array} \xrightarrow{\ \text{replace H by Br}\ } \begin{array}{c} \text{H} \quad\ \text{Br} \\ \diagdown\text{C}\diagup \\ \diagup\ \ \diagdown \\ \text{Cl}\ \ \text{Cl} \\ \textbf{24} \\ \textit{achiral} \end{array}$$

If the separate replacement of two stereoheterotopic ligands produces enantiomers, the ligands are enantiotopic; if diastereomers are produced, they are diastereotopic.

The apparently identical ligands of a prochiral center are actually not equivalent; they are distinguishable by chiral reagents in the enantiotopic case and by achiral reagents in the diastereotopic case. A ligand is named *pro-R* or *pro-S* by determining whether enhancement of its priority in the sequence rules over that of the other ligand of the same kind will generate configuration *R* or *S*. (The priority enhancement must not change the priority relative to any of the other ligands.) In Structure **25** the prochiral hydrogens are labeled according to this rule.

$$\begin{array}{c} \textit{pro-S}\ \ \text{H} \quad\ \text{H}\ \ \textit{pro-R} \\ \diagdown\text{C}\diagup \\ \diagup\ \ \diagdown \\ \text{H}_3\text{C}\ \ \text{Cl} \\ \textbf{25} \end{array}$$

---

[6](a) K. R. Hanson, *J. Am. Chem. Soc.*, **88**, 2731 (1966); (b) E. L. Eliel, *J. Chem. Educ.*, **48**, 163 (1971); (c) W. B. Jennings, *Chem. Rev.*, **75**, 307 (1975).

[7]In its more general form, the definition includes axially prochiral systems as well. See reference 6(a) for further details.

Trigonal centers are prochiral if addition of a new ligand to generate a tetrahedral center produces chirality. Thus 2-butanone is prochiral but acetone is not:

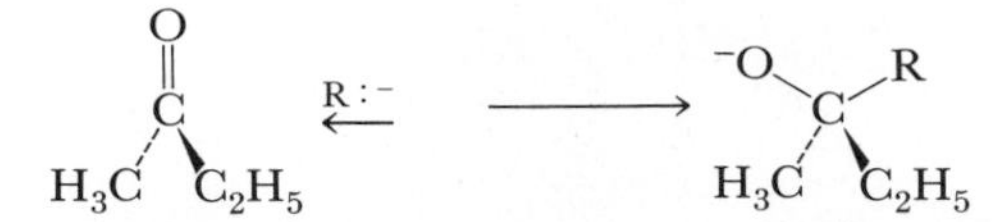

$$
\underset{\substack{\textbf{26} \\ \text{prochiral}}}{\underset{H_3C \quad C_2H_5}{\overset{\displaystyle O}{\overset{\|}{C}}}} \quad \xleftarrow{R:^-} \quad \longrightarrow \quad \underset{\substack{\textbf{27} \\ \text{chiral}}}{\underset{H_3C \quad C_2H_5}{\overset{\displaystyle {}^-O \quad R}{C}}}
$$

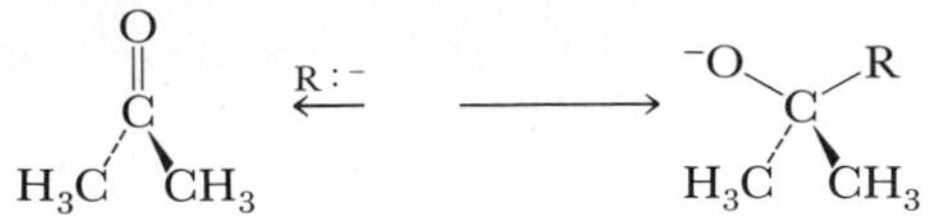

$$
\underset{\substack{\textbf{28}}}{\underset{H_3C \quad CH_3}{\overset{\displaystyle O}{\overset{\|}{C}}}} \quad \xleftarrow{R:^-} \quad \longrightarrow \quad \underset{\substack{\textbf{29} \\ \text{achiral}}}{\underset{H_3C \quad CH_3}{\overset{\displaystyle {}^-O \quad R}{C}}}
$$

A face of a prochiral trigonal system is labeled *re* if viewing the molecule from that face reveals a clockwise arrangement of ligands in order of highest priority to lowest; a face is *si* if the arrangement is counterclockwise. This rule is illustrated in **30**.

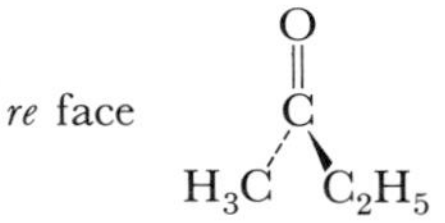

$$
re \text{ face} \quad \underset{\substack{H_3C \quad C_2H_5 \\ \textbf{30}}}{\overset{\displaystyle O}{\overset{\|}{C}}} \quad si \text{ face}
$$

Stereoselectivity and Stereospecificity[8]

The terms *stereoselective* and *stereospecific* properly refer only to reactions in which diastereomerically different materials may be formed or destroyed during the course of the reaction. *Stereoselective reactions* are all those in which one diastereomer (or one enantiomeric pair of diastereomers) is formed or destroyed in considerable preference to others that might have been formed or destroyed. Thus, for example, bromination of *trans*-2-butene might give either racemic or *meso*-2,3-dibromobutane. However, the meso compound is produced stereoselectively, as shown in Equation 2.1. Similarly, Equation 2.2 shows that *cis*-2-butene on bromination gives

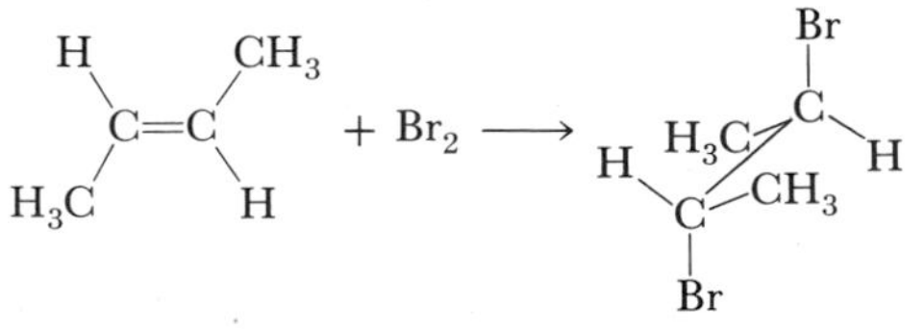

$$
\underset{H_3C \qquad H}{\overset{H \qquad CH_3}{C=C}} \quad + \ Br_2 \ \longrightarrow \quad \underset{\substack{Br \\ \text{meso}}}{\overset{Br}{\underset{H_3C}{\overset{\displaystyle C}{\underset{\displaystyle C}{}}}}} \text{CH}_3 \qquad \text{only} \tag{2.1}
$$

[8]H. E. Zimmerman, L. Singer, and B. S. Thyagarajan, *J. Am. Chem. Soc.*, **81**, 108 (1959).

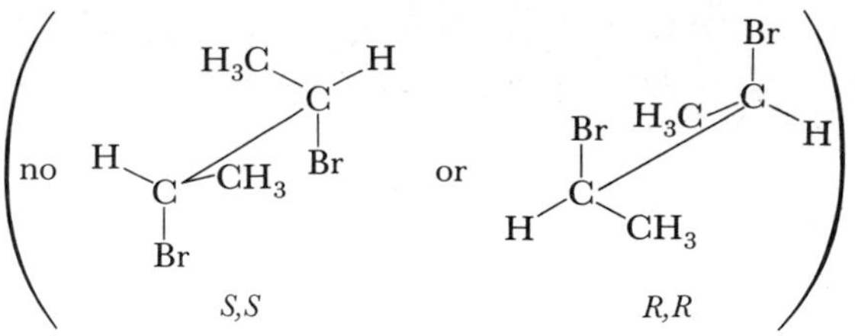

only racemic 2,3-dibromobutane. Another example of a stereoselective reaction is

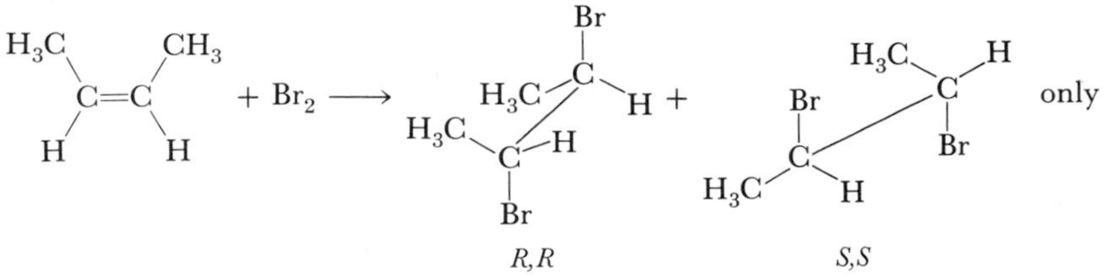

$$\text{(2.2)}$$

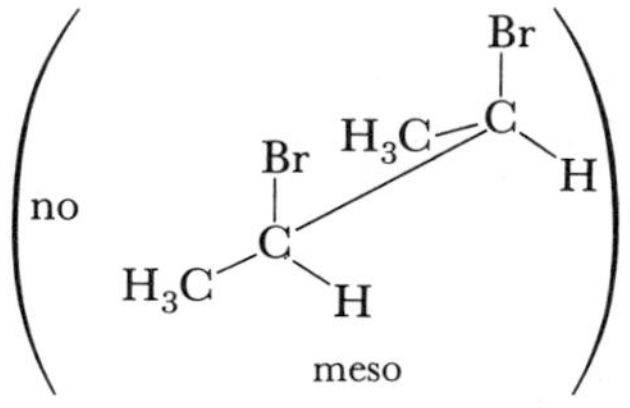

loss of *p*-toluenesulfonic acid

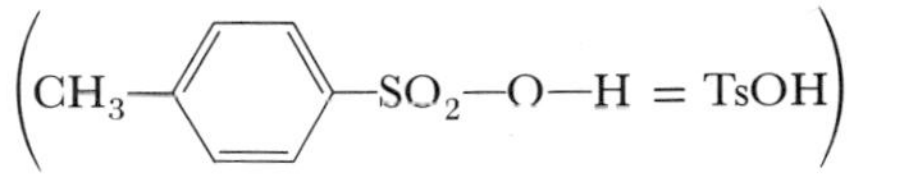

from *trans*-2-phenylcyclohexyl tosylate (Equation 2.3). Only *cis*-1-phenylcyclohexene is produced. Likewise, as Equation 2.3 shows, *cis*-2-phenylcyclohexyl tosylate also loses TsOH stereoselectively to form *cis*-1-phenylcyclohexene.

In a *stereospecific* reaction diastereomerically different starting materials give diastereomerically different products. Thus the bromination of the 2-butenes (Equations 2.1 and 2.2) is stereospecific, since one diastereomer gives one product and the other isomer a diastereomerically different product. Elimination of TsOH from the two 2-phenylcyclohexyl tosylates, however, is not stereospecific. As Equation 2.3 shows, both compounds give the same product. All stereospecific reactions must be stereoselective, but the converse is not true.[9]

The terms *regioselective* and *regiospecific* refer to reactions in which bonds can be made or broken in two or more different orientations. If one orientation is significantly favored, the reaction is regioselective; if one orientation occurs to the

---

[9] Reactions may be "partially", "90 percent," "60 percent," etc., stereoselective or stereospecific.

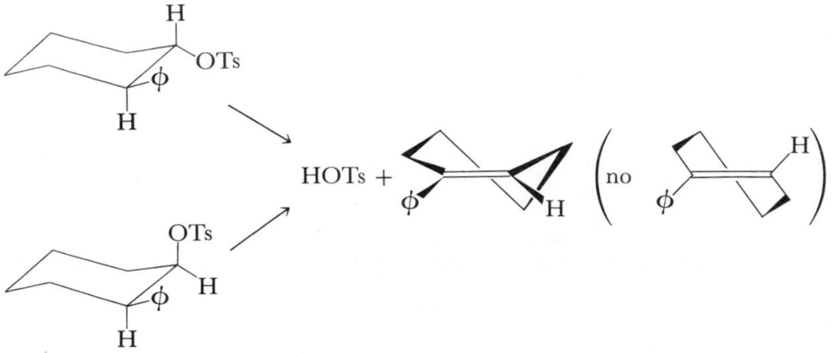

$$\text{HOTs} + \quad \left(\text{no} \quad \right) \tag{2.3}$$

exclusion of the others, the reaction is regiospecific. For example, hydration of styrene (Equation 2.4) is regiospecific.[10]

$$\phi\text{CH}=\text{CH}_2 + \text{H}_2\text{O} \xrightarrow{\text{H}^+} \phi\text{CHOH}-\text{CH}_3 \quad \text{only; no} \quad \phi\text{CH}_2\text{OH} \tag{2.4}$$

The term *stereoelectronic* refers to the effect of orbital overlap requirements on the steric course of a reaction. Thus because of stereoelectronic effects, the S_N2 substitution gives inversion (see Section 1.8) and E2 elimination proceeds most readily when the angle between the leaving groups is 0° or 180° (see Chapter 7). Stereoelectronic effects also play an important role in pericyclic reactions, which are the subject of Chapters 10 and 11.

Conformational Stereoisomerism[11]

Rotation about carbon–carbon single bonds is impeded by torsional energy barriers that vary with the size of the groups attached. In ethane the staggered conformation (**31**) is lower in energy than the eclipsed conformation (**32**) by about 2.9 kcal mole^{-1}.[12] Figure 2.1 shows the energy as a function of torsional angle.

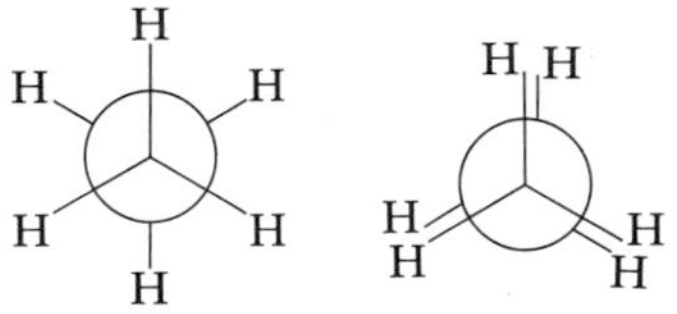

Substitution of bulkier groups for one or more of the hydrogens raises the barrier; for example, chloroethane has a barrier of 3.7 kcal mole^{-1}.[12] When one

---

[10] A. Hassner, *J. Org. Chem.*, **33**, 2684 (1968). The term *regiospecific* with qualifying adjectives *high* or *low* is often used in place of *regioselective*.

[11] (a) E. L. Eliel, N. L. Allinger, S. J. Angyal, and G. A. Morrison, *Conformational Analysis*, Wiley, New York, 1965; (b) M. Hanack, *Conformation Theory*, Academic Press, New York, 1965; theoretical: (c) A. Golebiewski and A. Parczewski, *Chem. Rev.*, **74**, 519 (1974).

[12] J. P. Lowe, *Prog. Phys. Org. Chem.*, **6**, 1 (1968).

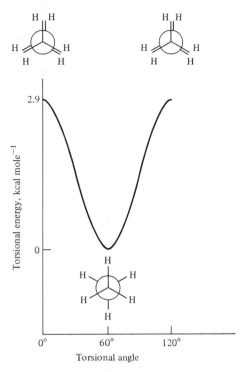

Figure 2.1 Conformational energy profile for ethane. The torsional angle is the angle of rotation about the carbon–carbon bond.

hydrogen on each carbon is replaced, there are two distinct staggered conformations, the *anti* (**33**) and *gauche* (**34**). The anti conformer, with the bulkier groups as far as possible from each other, is usually lower in energy. Figure 2.2 illustrates an approximate torsional energy profile of 1,2-dichloroethane.

Bulky substituents dramatically increase the rotational barriers about single bonds. Biphenyls substituted at the four ortho positions, as indicated in Structure **35,** have high rotational barriers. Compounds of the type **36** belong to the category of *atropisomers* (isomers separable because of hindered rotation about a single bond). They have axial chirality of the same type as shown by the allenes (**3–4**). In most biphenyls substituted at the 2,2′,6,6′ positions, the rotational barrier is high enough that the enantiomers can be resolved, assuming that the substitution pattern

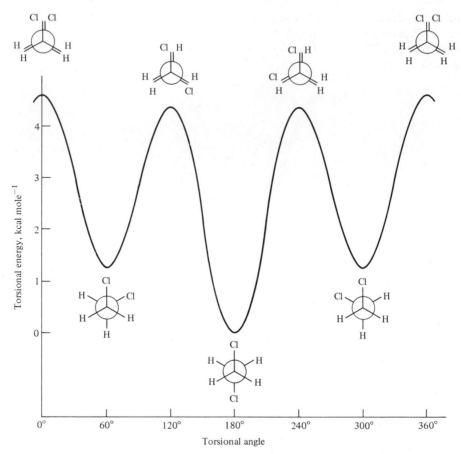

Figure 2.2 Approximate conformational energy profile for 1,2-dichloroethane. Energy data from J. P. Lowe, *Prog. Phys. Org. Chem.*, **6**, 1 (1968).

confers chirality. But if one ortho substituent is hydrogen, the enantiomers interconvert readily on heating.[13] Optically active 2,2′-dimethylbiphenyl (**37**) has been

35   36

[13] E. L. Eliel, *Stereochemistry of Carbon Compounds*, McGraw-Hill, New York, 1962, p. 156ff.

obtained by deamination of optically stable 2,2′-dimethyl-6,6′-diaminobiphenyl at low temperature. Compound **37** racemizes with an activation energy $E_a = 15.1$ kcal mole^{-1}. The half-reaction time is 7 minutes at $-32°C$ and an estimated 1.1 seconds at 25°C.[14]

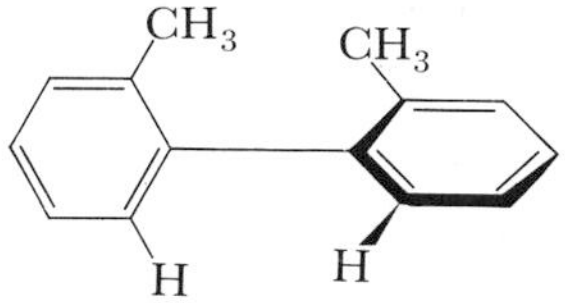

Rotational barriers for single bonds between saturated centers can also be raised by appropriate substitution; 2,2′-dimethyl-9,9′-bitriptycyl (**38**) has a rotational barrier about the central carbon–carbon bond in excess of 54 kcal mole^{-1}.[15]

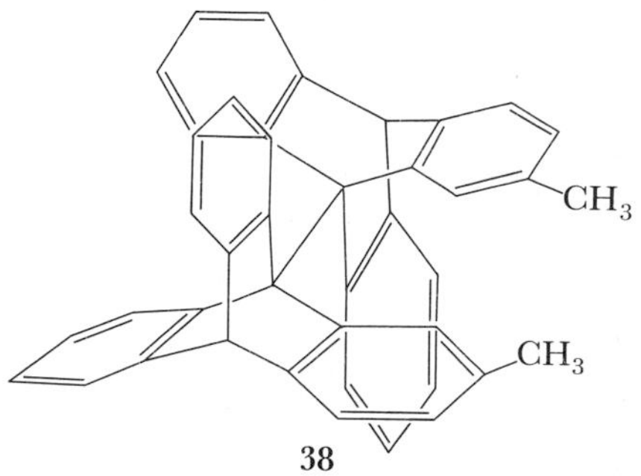

Conformations of Rings

The lowest-energy conformation of cyclohexane is the familiar chair (**39**) in which all the carbon–carbon bonds take up a perfectly staggered conformation.

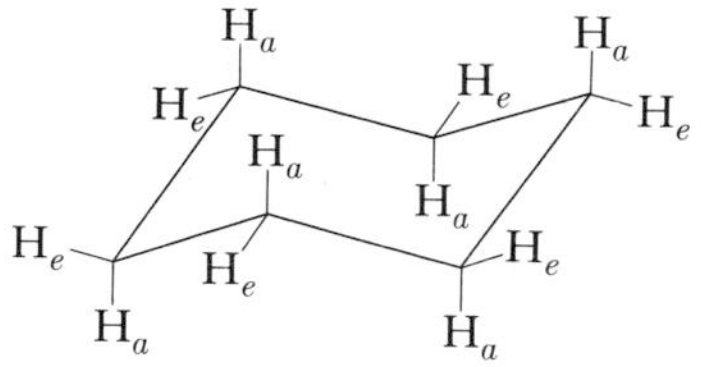

The substituents of the chair cyclohexane are divided into an axial set and an equatorial set, as indicated in **39**; axial and equatorial groups are interchanged by the chair–chair interconversion indicated for methylcyclohexane in Equation 2.5.

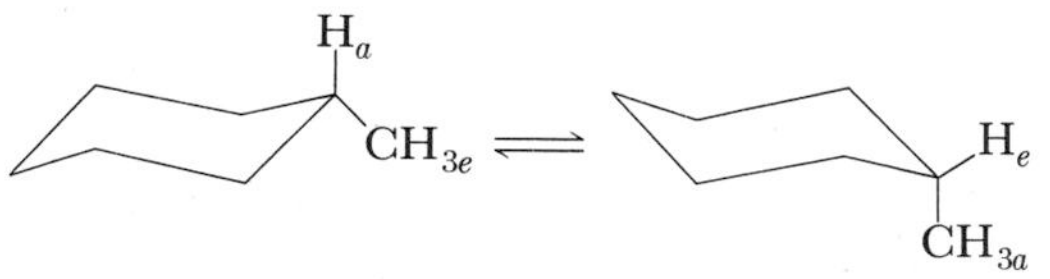

$$\text{(2.5)}$$

Conformations with subtituents in the equatorial position are usually of lower energy than those with substituents axial. Table 2.1 lists free-energy differences (A

[14]W. Theilacker and H. Böhm, *Angew. Chem. Int. Ed. Engl.,* **6**, 251 (1967).
[15](a) L. H. Schwartz, C. Koukotas, and C-S. Yu, *J. Am. Chem. Soc.,* **99**, 7710 (1977). See also (b) M. Oki, *Angew. Chem. Int. Ed. Engl.,* **15**, 87 (1976).

Table 2.1 Conformational Free-Energy
Differences (A values) Between Axial
and Equatorial Conformations of
Monosubstituted Cyclohexanes

Substituent	$\Delta G°$ (kcal mole^{-1})	Reference
CH_3	1.70	a
C_2H_5	1.75	a
$HC(CH_3)_3$	2.15	a
$C(CH_3)_3$	~5	b
F	0.28	c
Cl	0.53	c
Br	0.48	c
I	0.47	c
OH	0.5^d; 0.9^e	a
SH	1.17	c
NH_2	1.2^d; 1.6^e	a
$COOCH_3$	1.31	c
CN	0.24	c
NO_2	1.05	c

[a] J. A. Hirsch, *Top. Stereochem.*, **1**, 199 (1967).
[b] J. A. Hirsch, *Concepts in Theoretical Organic Chemistry,* Allyn &
Bacon, Boston, 1974, p. 254.
[c] F. R. Jensen, C. H. Bushweller, and B. H. Beck, *J. Am. Chem.
Soc.*, **91**, 344 (1969).
[d] Nonpolar solvents.
[e] Polar solvents.

values) between axial and equatorial conformers for various monosubstituted
cyclohexanes. The bulky *t*-butyl group, $(CH_3)_3C^-$, has a strong tendency to be
equatorial and will keep the ring predominantly in the conformation in which it
remains equatorial, even if a smaller substituent is thereby forced into the axial
position.

The chair conformations interconvert by way of the boat, of which the most
stable form is the twist boat (**40**). The twist boat is about 5.5 kcal mole^{-1} above the
chair in enthalpy.[16] The activation enthalpy from chair to twist boat is 10.8 kcal
mole^{-1}, by way of the half-chair conformation.

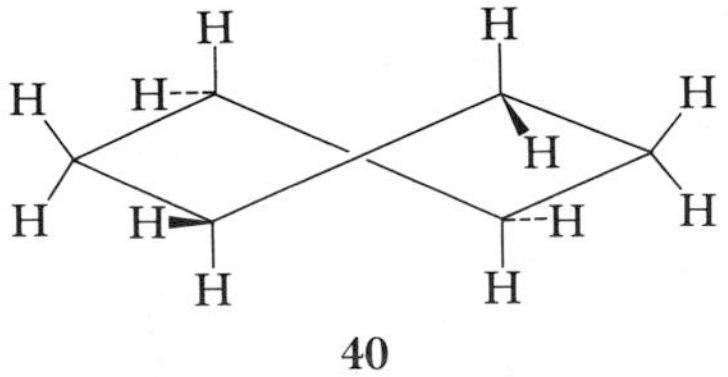

40

The twist boat passes by means of a low-activation-energy flexing motion called
pseudorotation through the boat (**41**) to a new twist boat equivalent to the first but

[16] M. Squillacote, R. S. Sheridan, O. L. Chapman, and F. A. L. Anet, *J. Am. Chem. Soc.,* **97**, 3244 (1975).

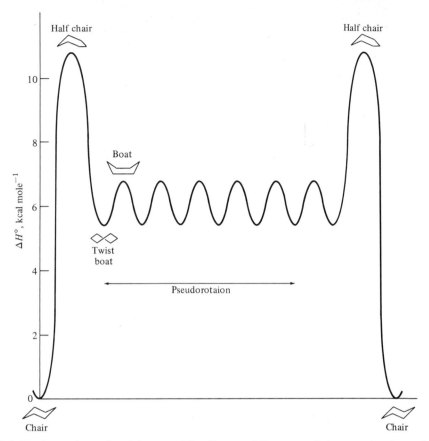

Figure 2.3 Conformations of cyclohexane. The diagram follows a cyclohexane ring from chair conformation over the half-chair barrier to twist boat, through one complete cycle of pseudorotation that passes through six identical but rotated twist boats and boats, and back to chair. Energy relationships are from (a) M. Squillacote, R. S. Sheriden, O. L. Chapman, and F. A. L. Anet, *J. Am. Chem. Soc.,* **97,** 3244 (1975) and (b) J. A. Hirsch, *Concepts in Theoretical Organic Chemistry,* Allyn & Bacon, Boston, 1974, p. 249–252.

rotated 60° around the vertical axis. Figure 2.3 illustrates the energy relationships of these conformational interconversions.

Although simple cyclohexane derivatives prefer the chair conformation to the boat, certain substitution patterns may lead to boat forms that are comparable to or lower than the chair in energy. Examples are cyclohexanes with two very bulky groups arranged so that only one can be equatorial (e.g., *trans*-1,3 or *cis*-1,4) or with two or more carbonyl groups or other trigonal centers in the ring.[17]

Cyclobutane Cyclobutane in a planar conformation would have all carbon–carbon and carbon–hydrogen bonds eclipsed; it therefore bends away from planarity to a conformation (**42**) in which the planes determined by C_1—C_2—C_3 and by C_1—C_4—C_3 are at an angle of about 35° to each other.

This bending generates axial and equatorial positions (**43**), and these interconvert as do those in cyclohexane. The cyclobutane conformational energy barriers are very low, a barrier of 1.48 kcal mole^{-1} having been found for the unsubstituted molecule.[18]

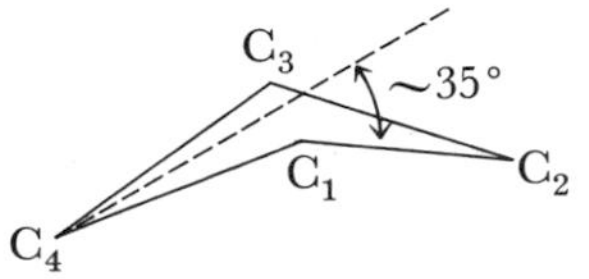

42 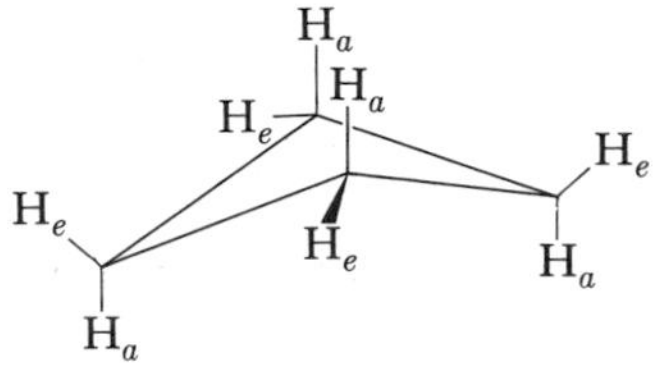   **43**

Cyclopentane Cyclopentane also prefers nonplanar conformations. There are two conformations: a half chair (**44**) and an envelope **45**. Energy differences are small, and axial and equatorial positions for substituents are not well defined.[19]

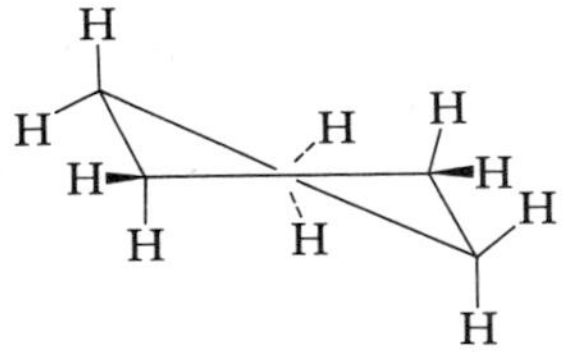

44 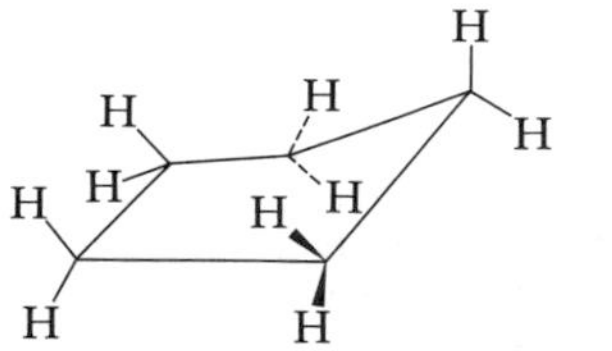   **45**

Larger rings Conformational analysis becomes increasingly complex as rings become larger, but a considerable amount of information has been obtained.[20] The most fruitful technique for investigation of these complex conforma-

[17] G. M. Kellie and F. G. Riddell, *Top. Stereochem.*, **8**, 225 (1974).
[18] R. M. Moriarty, *Top. Stereochem.*, **8**, 271 (1974).
[19] B. Fuchs, *Top. Stereochem.*, **10**, 1 (1978).
[20] For examples see (a) (cyclooctane) F. A. L. Anet and M. St. Jacques, *J. Am. Chem. Soc.*, **88**, 2585 (1966); (b) (cyclooctane) F. A. L. Anet and V. J. Basus, *J. Am. Chem. Soc.*, **95**, 4424 (1973); (c) (1,5-cyclooctadiene) F. A. L. Anet and L. Kozerski, *J. Am. Chem. Soc.*, **95**, 3407 (1973); (d) (1,4-cyclooctadiene) F. A. L. Anet and I. Yavari, *J. Am. Chem. Soc.*, **99**, 6986 (1977); (e) (1,3-cyclooctadiene) F. A. L. Anet and I. Yavari, *J. Am. Chem. Soc.*, **100**, 7814 (1978); (f) (cyclododecane) F. A. L. Anet and T. N. Rawdah, *J. Am. Chem. Soc.*, **100**, 7166 (1978); (g) (larger rings) F. A. L. Anet and T. N. Rawdah, *J. Am. Chem. Soc.*, **100**, 7810 (1978).

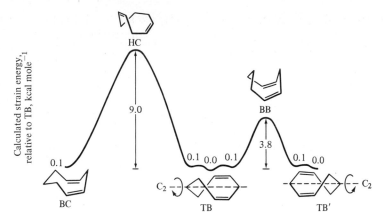

Figure 2.4 Conformational strain energy of *cis-cis*-1,4-cyclooctadiene as calculated by the force-field method. The twist boat (TB) is the lowest-energy conformation, taken as the zero of energy on this scale. The calculations show shallow minima, with structures close to TB, that may be artifacts. TB and TB′ are enantiomeric; the boat chair (BC) conformation is achiral. Reprinted with permission from F. A. L. Anet and I. Yavari, *J. Am. Chem. Soc.*, **99**, 6986 (1977). Copyright 1977 American Chemical Society.

tional problems is proton and ^{13}C nuclear magnetic resonance spectroscopy interpreted with the aid of force-field (molecular mechanics) calculations. These calculations are based on a classical mechanical model that includes appropriate potential energy functions for the torsional, bending, and nonbonded repulsion forces of the molecule.[21] Figure 2.4 shows an example of results obtained by force-field calculations of conformations of *cis-cis*-1,4-cyclooctadiene. There are two principal stable conformations, the boat chair (BC) and a chiral twist boat (TB). The BC and TB conformations are very close in energy and are interconverted by way of the half chair (HC), which lies 9 kcal mole^{-1} higher. The two enantiomeric twist boats are interconverted by way of a 3.8-kcal-mole^{-1} barrier through the boat–boat (BB). Experimentally, the ^{13}C nmr spectrum reveals two conformations at $-105\,°C$ separated by a free-energy difference of 0.58 kcal mole^{-1} and with a barrier of 8.0 kcal mole^{-1} between them. The spectra of these conformations are consistent with their assignment as the BC and TB forms, and the calculated and experimental energy differences agree quite well.[22]

[21]N. L. Allinger, *Adv. Phys. Org. Chem.*, **13**, 1 (1976).
[22]F. A. L. Anet and I. Yavari, *J. Am. Chem. Soc.*, **99**, 6986 (1977). Calculated energies are strain energies and not strictly comparable to the $\Delta G°$ and $\Delta G^{‡}$ obtained from the spectra. The difference between the two measures should be small, however, because conformational entropy and zero-point energy differences are expected to be small.

2.2 LINEAR FREE-ENERGY RELATIONSHIPS[23]

A problem that has challenged chemists for years is the determination of the electronic influence that substituents exert on the rate and course of a reaction. One of the difficulties involved in determining electronic substituent effects is that if the substituent is located close to the reaction center, it may affect the reaction by purely steric processes, so that electronic effects are masked; if placed far away in order to avoid steric problems, the electronic effects will be severely attenuated.

σ_{meta} and σ_{para} Constants

In 1937 Hammett recognized[24] that the electronic influence of a substituent, X, might be assessed by studying reactions in a side chain at Y in benzene derivatives (**46** and **47**).

The substituent X is separated physically from the reaction site, but its electron-donating or -withdrawing influence is transmitted through the relatively polarizable π electron system. The Hammett approach is to take as a standard reaction for evaluation of substituent effects the dissociation of substituted benzoic acids (**46**) and **47**, Y = COOH) at 25°C in H_2O. Substitution of an electron-withdrawing group (such as nitro) in the para position of benzoic acid causes an increase in strength of the acid, while an electron-donating group (for example, amino) decreases the strength. The same substituents in the meta position have slightly different effects. Ortho substituents are not included, because their proximity to the reaction site introduces interactions not present if the substituent is at the meta or para position.[25]

   If the free-energy change on dissociation of unsubstituted benzoic acid (X = H) is designated as ΔG_0°, the free energy on dissociation of a substituted

[23] For reviews see (a) R. D. Topsom, *Prog. Phys. Org. Chem.*, **12**, 1 (1976); (b) S. H. Unger and C. Hansch, *Prog. Phys. Org. Chem.*, **12**, 91 (1976); (c) L. S. Levitt and H. F. Widing, *Prog. Phys. Org. Chem.*, **12**, 119 (1976); (d) J. Hine, *Structural Effects on Equilibria in Organic Chemistry*, Wiley, New York, 1975; (e) S. Ehrenson, R. T. C. Brownlee, and R. W. Taft, *Prog. Phys. Org. Chem.*, **10**, 1 (1973); (f) M. Charton, *Prog. Phys. Org. Chem.*, **10**, 81 (1973); (g) K. F. Johnson, *The Hammett Equation*, Cambridge University Press, New York, 1973; (h) M. Godfrey, *Tetrahedron Lett.*, 753 (1972); (i) O. Exner, in *Advances in Linear Free Energy Relationships*, N. B. Chapman and J. Shorter, Eds., Plenum, London, 1972, p. 72; (j) J. Shorter, *Q. Rev. Chem. Soc.*, **24**, 433 (1970); (h) P. R. Wells, *Linear Free Energy Relationships*, Academic Press, New York, 1968; (l) L. P. Hammett, *Physical Organic Chemistry*, 2d ed., McGraw-Hill, New York, 1970, pp. 347ff; (m) C. D. Ritchie and W. F. Sager, *Prog. Phys. Org. Chem.*, **2**, 323 (1964); (n) S. Ehrenson, *Prog. Phys. Org. Chem.*, **2**, 195 (1964); (o) J. E. Leffler and E. Grunwald, *Rates and Equilibria of Organic Reactions*, Wiley, New York, 1963; (p) H. H. Jaffé, *Chem. Rev.*, **53**, 191 (1953).
[24] L. P. Hammett, *J. Am. Chem. Soc.*, **59**, 96 (1937).
[25] The *ortho effect* is due to steric effects, proximity electric effects, hydrogen bonding, and other intramolecular interactions. For discussions of the *ortho effect* see (a) T. Fujita and T. Nishioka, *Prog. Phys. Org. Chem.*, **12**, 49 (1976); (b) J. Shorter, in *Advances in Linear Free Energy Relationships*, N. B. Chapman and J. Shorter, Eds., Plenum, London, 1972, p. 71; (c) M. Charton, *Prog. Phys. Org. Chem.*, **8**, 235 (1971).

benzoic acid ($\Delta G°$) can be considered to be $\Delta G_0°$ plus an increment, $\Delta\Delta G°$, contributed by the substituent (Equation 2.6). Because the substituent X will make differ-

$$\Delta G° = \Delta G_0° + \Delta\Delta G° \tag{2.6}$$

ent contributions at the meta and para positions, it is always necessary to designate the position of substitution.

In order to bring the relationship 2.6 into more convenient form, a parameter σ is defined for each substituent according to Equation 2.7, so that Equation 2.6 becomes Equation 2.8. (Each substituent actually has two σ's—a σ_{meta} and a σ_{para}.)

$$\Delta\Delta G° = -2.303RT\sigma \tag{2.7}$$

$$-\Delta G° = -\Delta G_0° + 2.303RT\sigma \tag{2.8}$$

By using the Relationship 2.9 between free energy and equilibrium, Equation 2.8 can be rewritten as Equation 2.10, which in turn simplifies to Equation 2.11. Table 2.2 lists the σ_{meta} and σ_{para} constants for some of the common substituents.[26]

$$-\Delta G° = 2.303RT\log_{10} K \tag{2.9}$$

$$2.303RT\log_{10} K = 2.303RT\log_{10} K_0 + 2.303RT\sigma \tag{2.10}$$

$$\log_{10}\frac{K}{K_0} = pK_0 - pK = \sigma \tag{2.11}$$

Table 2.2 σ Values of Common Substituents

Substituent	σ_{meta}[a,c]	σ_{para}[a,c]	σ^{+}[a,d]	σ^{-}[a,d]	$\mathscr{F}$[b]	$\mathscr{R}$[b]
NH$_2$	−0.16	−0.66	−1.3	—	0.037	−0.68
CH$_3$	−0.07	−0.17	−0.31	—	−0.052	−0.14
C$_6$H$_5$	0.06	−0.01	−0.17	—	0.14	−0.088
OH	0.12	−0.37	−0.92	—	0.49	−0.64
OCH$_3$	0.12	−0.27	−0.78	−0.2	0.41	−0.50
F	0.34	0.06	−0.07	−0.02	0.71	−0.34
I	0.35	0.18	0.13	—	0.67	−0.20
CO$_2$H	0.37	0.45	0.42	—	0.55	0.14
Cl	0.37	0.23	0.11	—	0.69	−0.16
COCH$_3$	0.38	0.50	—	0.87	0.53	0.20
Br	0.39	0.23	0.15	—	0.73	−0.18
CO$_2$R	0.37	0.45	0.48	0.68	0.55	0.14
CF$_3$	0.43	0.54	—	—	0.63	0.19
CN	0.56	0.66	0.66	0.90	0.85	0.18
NO$_2$	0.71	0.78	0.79	1.24	1.11	0.16

[a] Values are those given by C. D. Ritchie and W. F. Sager, *Prog. Phys. Org. Chem.*, **2**, 323 (1964).
[b] Values are those given by C. G. Swain and E. C. Lupton, Jr., *J. Am. Chem. Soc.*, **90**, 4328 (1968).
[c] Values are those given by C. Hansch, A. Leo, S. Unger, K. H. Kim, D. Nikaitani, and E. J. Liem, *J. Med. Chem.*, **16**, 1207 (1973).
[d] σ^+ and σ^- values are given for para substituents only. σ^+ values for some meta substituents have been measured, but they do not differ appreciably from the σ_{meta} values.

---

[26] For much extended lists, see (a) C. Hansch, A. Leo, S. Unger, K. H. Kim, D. Nikaitani, and E. J. Lien, *J. Med. Chem.*, **16**, 1207 (1973); (b) C. Hansch and A. J. Leo, *Substituent Constants for Correlation Analysis in Chemistry and Biology*, Wiley, New York, 1979.

If we now examine the effect of substituents on another reaction, for example, acid dissociation of phenylacetic acids (Equation 2.12), we can anticipate that

$$XC_6H_4CH_2COOH + H_2O \rightleftharpoons XC_6H_4CH_2COO^- + H_3O^+ \tag{2.12}$$

the various substituents will exert the same kind of effects on these equilibrium constants as they did on the benzoic acid equilibrium constants; but the greater separation between substitution site and reaction site in the phenylacetic acids makes the reaction less sensitive to the substituent effects. The change in free energy due to the substituent, $-2.303RT\sigma$, which was appropriate for benzoic acid dissociations, must now be multiplied by a factor ρ, which characterizes the sensitivity of the new reaction to electron donation and withdrawal. We can therefore write Equation 2.13, where $\Delta G^{\circ\prime}$ is the free-energy change for the new reaction with substituent X and $\Delta G_0^{\circ\prime}$ is the free-energy change for the new reaction with no

$$-\Delta G^{\circ\prime} = -\Delta G_0^{\circ\prime} + 2.303RT\rho\sigma \tag{2.13}$$

substituents. Rewriting Equation 2.13 as before, we have the relationship in the more useful form of Equation 2.14:

$$\log_{10} \frac{K'}{K_0'} = \rho\sigma \tag{2.14}$$

It is important to keep in mind the equality between σ and $\log K/K_0$ for benzoic acid dissociations (Equation 2.11). Equation 2.14 then states that $\log K'/K_0'$ for the new reaction is directly proportional to $\log K/K_0$ for benzoic acid dissociations with proportionality constant ρ. A closely related way at looking at this proportionality is to solve Equation 2.8 for $2.303RT\sigma$ and substitute that into Equation 2.13. One obtains

$$\Delta G_0^{\circ\prime} - \Delta G^{\circ\prime} = \rho(\Delta G_0^{\circ} - \Delta G^{\circ}) \tag{2.15}$$

Thus in the new reaction series the difference between the free-energy changes in the unsubstituted and substituted acid dissociations is directly proportional to the analogous difference in benzoic acid dissociations. Since this proportionality of free energies is directly implied by Equation 2.14, Equation 2.14 is an example of a *linear free-energy relationship*. Although it was derived above for the dissociation of phenylacetic acids, it is applicable to a variety of equilibria of substituted aromatic compounds.

Equation 2.16, correlating rates of substituted aromatic compounds with the equilibrium constants of substituted benzoic acids, can also be easily derived if one uses free energies of activation rather than changes in ground-state free energies. Equation 2.16 is then, of course, also a linear free-energy relationship.

$$\log \frac{k}{k_0} = \sigma\rho \tag{2.16}$$

Either Equation 2.15 or Equation 2.16 is known as the Hammett equation. We shall see that it is a very useful tool for obtaining information about reaction mechanisms.

Usually the most convenient way to use the Hammett equation is to plot $\log K/K_0$, or just $\log K$ (for equilibria), or $\log k/k_0$, or just $\log k$ (for rates), of the reaction of interest on the vertical axis and σ values for the substituents on the horizontal axis. (Again, K and k are constants for the substituted compounds and K_0 and k_0 are constants for the unsubstituted compounds.) A straight line indicates that the free-energy relationship of Equation 2.13 is valid. The slope of the line is ρ for the reaction. A positive value of ρ means that the reaction responds to substituents in the same sense as does benzoic acid ionization; that is, the equilibrium constant (or reaction rate) is increased by electron-withdrawing groups. If $\rho > 1$, then the reaction is more sensitive to the effect of the substituent than benzoic acid dissociation; if $0 < \rho < 1$, then electron-withdrawing groups still increase the rate or equilibrium constant, but less than in benzoic acid dissociation. A negative ρ shows that electron-donating groups increase the reaction constant. A small ρ often means that the mechanism of the reaction involves radical intermediates or a cyclic transition state with little charge separation (see also, however, the discussion of isokinetic temperatures at the end of Section 2.2). Sometimes the slope of the plot of $\log k$ vs. σ changes more or less abruptly as substituents are varied, so that two straight lines are obtained. The reason for this behavior usually is that the mechanism is changing in response to the varying electron demand of the substituents.

Figure 2.5 shows the results of application of this method to ionization of ring-substituted phenylacetic and 3-phenylpropionic acids. The ρ values are positive for both equilibria but decrease in magnitude as the substituted ring is placed

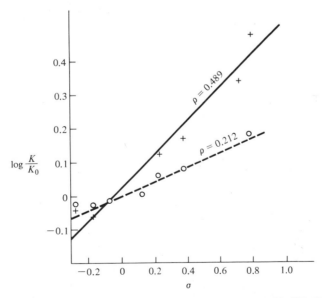

Figure 2.5 Plot of K/K_0 vs. σ constants for dissociation of X—$C_6H_4CH_2COOH$ (×) and of X—$C_6H_4CH_2CH_2COOH$ (○). The data are from J. F. J. Dippy and J. E. Page, *J. Chem. Soc.*, 357 (1938), and the ρ values from a least-squares analysis of the data by H. H. Jaffé, *J. Chem. Phys.*, **21**, 415 (1953).

farther from the ionization site. Tables 2.3 and 2.4 list several additional ρ values for the correlation of equilibrium constants and reaction rate constants, respectively. Note, for example, in Table 2.3 that ρ for the dissociation of substituted benzoic acids is much higher in ethanol than it is in water. This is because acid strength in the less-ionizing solvent, ethanol, is more dependent on any help it can get from substituents than it is in water.

σ^+ and σ^- Constants

When the reaction site comes into direct resonance with the substituent, the σ constants of the substituents do not succeed in correlating equilibrium or rate constants. For example, a p-nitro group increases the ionization constant of phenol much more than would be predicted from the $\sigma_{p\text{-}NO_2}$ constant obtained from the ionization of p-nitrobenzoic acid. The reason is readily understood when one realizes that the p-nitrophenoxide ion has a resonance structure (**48**) in which the nitro group participates in *through-resonance* with the O^-. The extra stabilization of the anion provided by this structure is not included in the $\sigma_{p\text{-}NO_2}$ constant because the

48

COO^- group in the benzoate anion cannot come into direct resonance interaction with any ring substituent. Similarly, a p-methoxy group is much more effective at

Table 2.3 ρ Values for Acid Dissociations[a]

Acid	Solvent	Temperature (°C)	ρ
X—⟨ring⟩—COOH	H_2O	25	1.00
X—⟨ring⟩—COOH	C_2H_5OH	25	1.957
X—⟨ring⟩—COOH (NO₂)	H_2O	25	0.905
X—⟨ring⟩—OH	H_2O	25	2.113

[a] Reprinted with permission from H. H. Jaffe, *Chem. Rev.*, **53**, 191 (1953). Copyright by the American Chemical Society. Refer to this source for more complete data.

Table 2.4 ρ Values Derived from Rates of Heterolytic Reactions[a]

Reaction	Solvent	Temperature (°C)	ρ
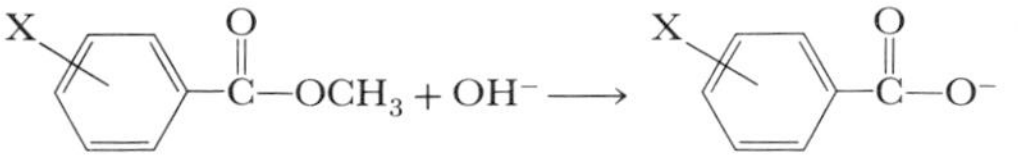	60% acetone	25	2.229
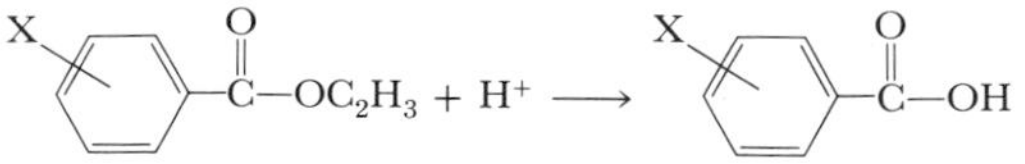	60% acetone	100	0.106
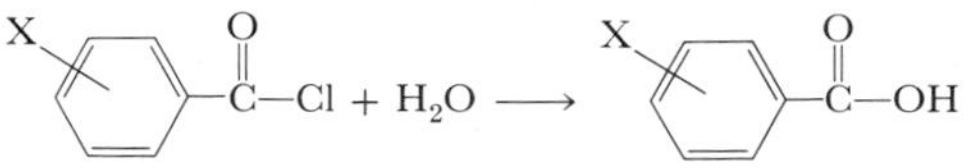	50% acetone	0	0.797
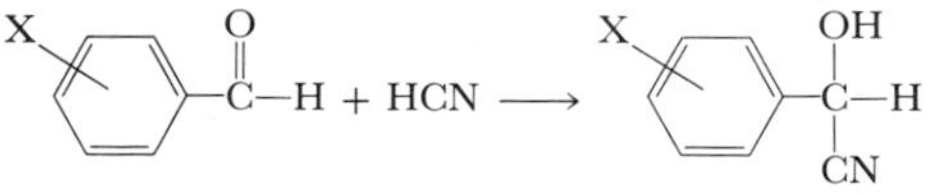	95% ethanol	20	2.329
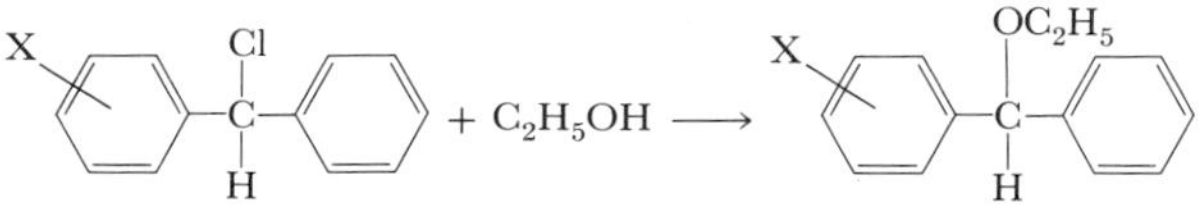	ethanol	25	−5.090

[a] Reprinted with permission from H. H. Jaffé, *Chem. Rev.*, **53**, 191 (1953). Copyright by the American Chemical Society. Refer to this source for more complete data.

increasing the rate of ionization of triphenylmethyl chloride than would be predicted from the $\sigma_{p\text{-OCH}_3}$ constant (see Equation 2.17).

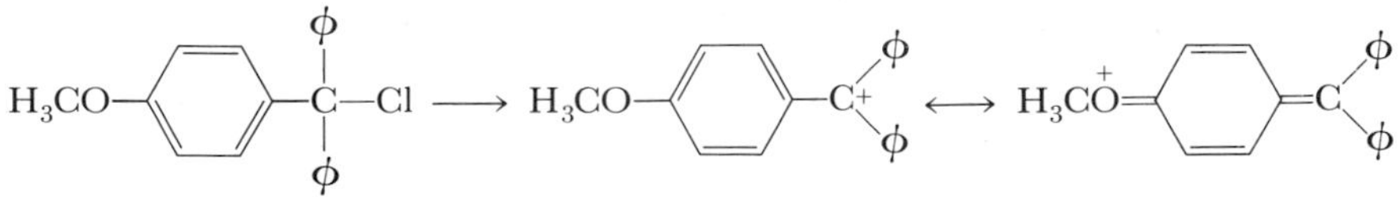

$$(2.17)$$

Several investigators found[27] that rate and equilibrium constants can be better correlated by the Hammett equation if two new types of σ constants are introduced. When there is through-resonance between a reaction site that becomes electron-rich and a substituent electron-withdrawing by resonance, the σ^- constant should be used. The standard reactions for the evaluation of σ^- constants are the ionizations of para-substituted phenols and of para-substituted anilinium ions.[28] The σ^+ constant should be used whenever a substituent electron-donating by resonance is para to a reaction site that becomes electron-deficient and when through-resonance is possible between the two groups. The standard reaction for the evaluation of σ^+ is the solvolysis of para-substituted *t*-cumyl chlorides in 90 percent

[27] For reviews see (a) note 23(m); (b) L. M. Stock and H. C. Brown, *Adv. Phys. Org. Chem.*, **1**, 35 (1963).
[28] See note 23(p).

aqueous acetone (Equation 2.18).[29] Table 2.2 lists a number of σ^+ and σ^- constants.

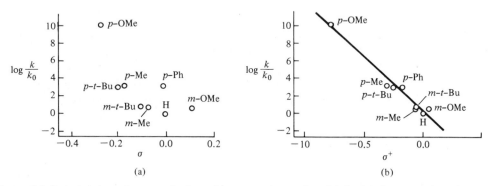

$$X\!\!-\!\!\underset{CH_3}{\overset{CH_3}{C}}\!\!-\!\!Cl \xrightarrow[\text{acetone 90\%}]{H_2O\ 10\%} X\!\!-\!\!\underset{CH_3}{\overset{CH_3}{C^+}} \longleftrightarrow X\!\!-\!\!\underset{CH_3}{\overset{CH_3}{C}}\!\!-\!\!OH \quad (2.18)$$

Figure 2.6, in which σ constants are plotted against log k/k_0 for the bromination of monosubstituted benzenes, shows an example of the usefulness of these new parameters. As can be seen from Structures **49** and **50**—which are representations of the intermediates in the ortho and para bromination of anisole—substituents electron-donating by resonance ortho or para to the entering bromine can stabilize the positive charge in the intermediate and therefore also in the transition state by through-resonance.

49   **50**

In Figure 2.6(a), in which Hammett σ constants are plotted, there is a scatter of points; but in Figure 2.6(b) σ^+ parameters are used and a straight line is obtained.

Figure 2.6 Bromination of monosubstituted benzenes in acetic acid. In (a) data are plotted against σ; in (b) against σ^+. Reprinted with permission from H. C. Brown and Y. Okamoto, *J. Am. Chem. Soc.*, **80**, 4979 (1958). Copyright 1958 American Chemical Society.

[29] H. C. Brown and Y. Okamoto, *J. Am. Chem. Soc.*, **80**, 4979 (1958).

Polar and Steric Substituent Constants in Aliphatic Systems

In the 1950s Taft proposed a method of extending linear free energy relationships to aliphatic systems in which steric factors are important.[30] He suggested that since the electronic nature of substituents has little effect on the rate of acid-catalyzed hydrolysis of meta- or para-substituted benzoates (ρ values are near 0; see Table 2.4), the electronic nature of substituents will also have little effect on acid-catalyzed hydrolysis of aliphatic esters. All rate changes due to substituents in the latter reactions are, therefore, probably due to steric factors.[31] Taft defined E_s, a steric substituent constant, by Equation 2.19

$$E_s = \log\left(\frac{k}{k_0}\right)_A \tag{2.19}$$

in which k and k_0 are the rate constants for hydrolysis of XCOOR and CH_3COOR, respectively, and in which the subscript A denotes acid-catalyzed hydrolysis. Table 2.5 gives a number of E_s values.[32] The rates of other reactions in which the polar effect of substituents is small can sometimes be correlated by E_s.

Taft suggested further that once the steric parameter has been evaluated, the polar[33] parameter should be available. Taft noted that the transition state structures for acid- and base-catalyzed hydrolysis of esters (**51** and **52**, respectively)

$$\left[\begin{array}{c} OH \\ | \\ X-C\cdots OR \\ | \\ OH_2 \end{array}\right]^+ \qquad \left[\begin{array}{c} O \\ | \\ X-C\cdots OR \\ | \\ OH \end{array}\right]^-$$

$$\quad\quad \textbf{51} \quad\quad\quad\quad\quad \textbf{52}$$

differ from each other by only two tiny protons and argued that therefore the steric effect of X will be approximately the same in the two transition states. But in base-catalyzed hydrolysis the electronic influence of a substituent cannot be neglected, as can be seen from the large values of ρ for base-catalyzed hydrolysis of *m*- or para-substituted benzoates (Table 2.4). The polar substituent constant, σ^*, was

---

[30] Taft followed a suggestion of Ingold [C. K. Ingold, *J. Chem. Soc.*, 1032 (1930)]. R. W. Taft, Jr., *J. Am. Chem. Soc.*, **74**, 2729, 3120 (1952); **75**, 4231 (1953).

[31] Swain and Lupton's work (see note 38) seems to confirm that this is a valid assumption. See, however, note 23(j).

[32] Better correlations are usually obtained by using modified steric parameters (E_s^c) that include a contribution to E_s from the hyperconjugative effect of σ hydrogens. Unger and Hansch have compiled an extensive list of these as well as of other steric parameters. See note 23(b). In their lists hydrogen always has an E_s value of unity so that steric constants will be analogous to other substituent constants. See also M. Charton, *J. Org. Chem.*, **43**, 3995 (1978).

[33] The term *polar effect* refers to the influence, other than steric, that nonconjugated substituents exert on reaction rates. It does not define whether the mechanism for its transmission is through bonds (inductive effect) or through space (field effect). Recent evidence, however, suggests that the field effect is most important. See note 23(a).

Table 2.5 STERIC AND POLAR PARAMETERS
FOR ALIPHATIC SYSTEMS[a]

X	E_s	σ^*
H	+1.24	+0.49
CH_3	0.00	0.00
CH_3CH_2	−0.07	−0.10
i-C_3H_7	−0.47	−0.19
t-C_4H_9	−1.54	−0.30
n-C_3H_7	−0.36	−0.115
n-C_4H_9	−0.39	−0.13
i-C_4H_9	−0.93	−0.125
neo-C_5H_{11}	−1.74	−0.165
$ClCH_2$	−0.24	+1.05
ICH_2	−0.37	+0.85
Cl_2CH	−1.54	+1.94
Cl_3C	−2.06	+2.65
CH_3OCH_2	−0.19	+0.52
$C_6H_5CH_2$	−0.38	+0.215
$C_6H_5CH_2CH_2$	−0.38	+0.08
$CH_3CH{=}CH$	−1.63	+0.36
C_6H_5	−2.55	+0.60

[a] J. Shorter, *Quart. Rev.* (London), **24**, 433 (1970), using data of R. W. Taft, in *Steric Effects in Organic Chemistry,* M. S. Newman, Ed., Wiley, New York, 1956, Chap. 13. Reproduced by permission of the Chemical Society, Wiley-Interscience, and J. Shorter.

therefore defined as

$$\sigma^* = \frac{\log\left(\frac{k}{k_0}\right)_B - \log\left(\frac{k}{k_0}\right)_A}{2.48} \tag{2.20}$$

The subscript B denotes base-catalyzed hydrolysis, and the factor 2.48 is present in order that the σ and σ^* constants of a substituent will have approximately the same value. Table 2.5 lists a number of σ^* constants.

The general Taft equation, then, for structure-rate correlations in aliphatic systems where steric factors are important is Equation 2.21.

$$\log\frac{k}{k_0} = \sigma^*\rho^* + \delta E_s \tag{2.21}$$

The proportionality constant δ represents the susceptibility of the reaction to steric factors.

In recent years the Taft σ^* values have been shown to only sometimes give good rate correlations and thus have been the subject of a good deal of controversy.

Some authors feel that $\sigma*$ is probably a measure of a steric effect and others defend its status as a polar parameter.[34] Another important polar substituent constant, σ_I, was originally defined[35,36] as

$$\sigma_{I(X)} = 0.450\,\sigma^*_{(XCH_2)} \qquad (2.22)$$

However, since the $\sigma*$ scale has become controversial, the σ_I scale has been slightly changed[37] to make hydrogen rather than methyl the reference substituent. Furthermore, new values for the alkyl groups have been proposed that give better agreement when the alkyl groups are attached directly to a saturated or aromatic carbon (as opposed to $\sigma*$ where they are attached to a carbonyl carbon). Table 2.6 lists a number of σ_I parameters. Note that all alkyl groups have small negative σ_I values. It has been argued that these values, which often do not give good rate and equilib-

Table 2.6 VALUES OF $\sigma_I{}^a$

Substituent	σ_I	Substituent	σ_I	Substituent	σ_I
$NMe_3{}^+$	0.92	COF	0.42	CH_2Cl	0.17
SO_2Cl	0.86	I	0.39	$NHNH_2$	0.15
SO_2F	0.86	OAc	0.39	NH_2	0.12
NO_2	0.65	OPh	0.38	CH_2OH	0.10
$SOCF_3$	0.64	$CH{=}CHNO_2$	0.38	Ph	0.10
SO_3Et	0.63	NO	0.37	NMe_2	0.06
$NH_3{}^+$	0.60	CHO	0.31	$CH{=}CH_2$	0.05
SO_2Me	0.59	CO_2Et	0.30	CH_2NH_2	0.00
SF_5	0.57	Ac	0.28	H	0.00
CN	0.56	OMe	0.27	Me	-0.04
COCN	0.55	NHAc	0.26	Et	-0.05
OCF_3	0.55	$CH_2NH_3{}^+$	0.25	i-Pr	-0.06
SOMe	0.50	OH	0.25	t-Bu	-0.07
F	0.50	SH	0.25	$B(OH)_2$	-0.08
SO_2NH_2	0.46	$SO_3{}^-$	0.25	$SiMe_3$	-0.10
Cl	0.46	$N{=}NPh$	0.25	CH_2SiMe_3	-0.11
CF_3	0.45	SMe	0.23	O^-	-0.12
Br	0.44	CH_2CN	0.23	$CO_2{}^-$	-0.35
SCF_3	0.42	$CONH_2$	0.21	$B(OH)_3{}^-$	-0.36

[a] J. Hine, *Structural Effects on Equilibria in Organic Chemistry*, Wiley, New York, 1975. Reprinted by permission of John Wiley & Sons, Inc.

[34] See for example (a) A.-J. MacPhee and J.-E. Dubois, *Tetrahedron Lett.*, 2471 (1976); (b) M. Charton, *J. Am. Chem. Soc.*, **99**, 5687 (1977); (c) M. Charton, *J. Org. Chem.*, **44**, 903 (1979) and references therein; (d) F. G. Bordwell, J. E. Bartmess, and J. A. Hautala, *J. Org. Chem.*, **43**, 3095 (1978).
[35] R. W. Taft, Jr., and I. C. Lewis, *J. Am. Chem. Soc.*, **80**, 2436 (1958).
[36] This definition was originally proposed so that σ_{para} for any substituent would be approximately equal to the sum of the polar and resonance substituent constants, i.e., $\sigma_{para} = \sigma_I + \sigma_R$.
[37] See note 23(e).

rium constant correlations, should properly be zero for all alkyl groups. For reviews on both sides of this question see notes 1(c), 1(j), and 34.

Dual-Parameter Substituent Constants

Although σ_{meta}, σ_{para}, σ^+, σ^-, and σ^* constants have been widely and successfully used in the correlation of rates and equilibria, over the years it has become apparent that these sets of substituent constants are not sufficient for the correlation of all rates and equilibria even when steric effects are absent. The relative effects of, for example, a nitro and a chloro group are sometimes not equal to their relative effects either on the ionization of benzoic acid (σ_{meta} and σ_{para}) or on the ionization of para-substituted phenols (σ^-) or on the ionization of para-substituted t-cumyl chlorides (σ^+) or on the hydrolysis of aliphatic esters (σ^*). This failure apparently arises because the overall substituent effect (σ) has both polar and resonance parts, and the relative contribution of the two depends on the nature of the reaction. Thus a plethora of sets of σ constants has grown up,[38] some for aromatic and some for aliphatic systems and each useful for a different type of reaction. As a result it is often not clear which set of σ constants should be chosen to correlate the rates of a particular reaction.

Swain and Lupton[39] have suggested a solution to this problem based on the assumption that the σ constant of any substituent in any reaction can be expressed by

$$\sigma = f\mathcal{F} + r\mathcal{R} \tag{2.23}$$

where $\mathcal{F}$ and $\mathcal{R}$ are polar and resonance constants, respectively, different for each substituent but constant for an individual substituent over all reactions. The factors f and r are empirical sensitivities or weighting factors independent of substituent but dependent on, and different for, each reaction (or closely related set of reactions). The linear free-energy relationship then becomes

$$\log \frac{k}{k_0} \quad \text{or} \quad \log \frac{K}{K_0} = \rho\sigma = \rho f\mathcal{F} + \rho r\mathcal{R}. \tag{2.24}$$

$\mathcal{F}$ and $\mathcal{R}$ can be determined for each substituent as described below, and f and r for a particular reaction series can then be calculated by a computer program that provides a "best-fit" correlation for Equation 2.24 for a number of different substituents. For example, to find the weighting factors f and r that would apply to the rate of Reaction 2.25, the σ constants of a number of substituents for this reaction would be found experimentally by using the Hammett approach described below. Then using those σ's and the $\mathcal{F}$ and $\mathcal{R}$ values of the substituents, the iterative computer program would provide an f and an r for Reaction 2.25 that best fit Equation 2.23 for all the substituents studied. Once the f and r values are known,

---

[38] For a list of many of these see C. G. Swain and E. C. Lupton, Jr., *J. Am. Chem. Soc.,* **90,** 4328 (1968).
[39] See note 38.

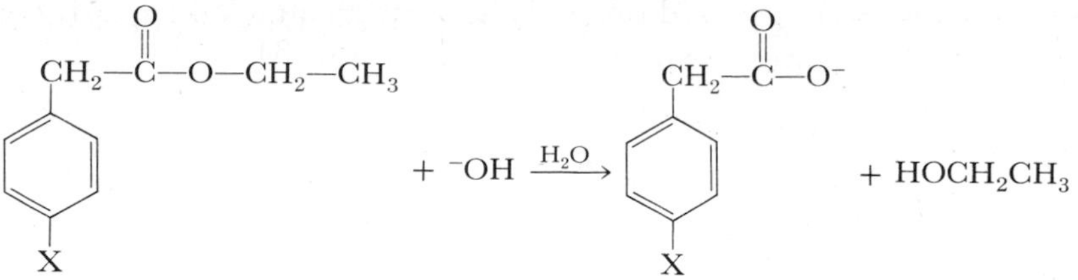

$$+ \ ^-OH \xrightarrow{H_2O} \qquad\qquad + \ HOCH_2CH_3 \qquad (2.25)$$

they can be used to calculate the σ constants of substituents not previously studied in that reaction.

To find the $\mathscr{F}$ value for a substituent, Swain and Lupton focused on the effect of that substituent on the dissociation of 4-substituted bicyclo[2.2.2]octane-1-carboxylic acids (**53**) in 50 percent ethanol and 50 percent water. Substituent

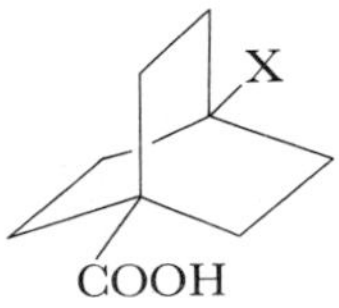

53

constants in this series have been determined and are known as the σ' series. Swain and Lupton (1) assumed that in this reaction the substituent effect is entirely a polar effect (i.e. $r = 0$), a reasonable assumption since there is no conjugation, and (2) assigned f for this reaction the value of unity. Thus when $\sigma = \sigma'$, Equation 2.23 becomes

$$\sigma' = \mathscr{F} \qquad (2.26)$$

and determination of σ' for a substituent directly gives that substituent's $\mathscr{F}$ value. Some of these are listed in Table 2.2.

The $\mathscr{R}$ values of the substituents are obtained by (1) assuming that $\mathscr{R} = 0.0$ for the trimethylammonium substituent, $(CH_3)_3\overset{+}{N}$— (again, reasonable since the nitrogen has no unshared electron pair and, being a first row element, cannot expand its valence shell to accommodate more than eight electrons), and (2) assigning r the value unity in the dissociation of para-substituted benzoic acids. Thus Equation 2.23 becomes

$$\sigma_{para} = f_{para}\mathscr{F} + \mathscr{R} \qquad (2.27)$$

For trimethylammonium, then,

$$\sigma_{para}^{\overset{+}{N}(CH_3)_3} = f\mathscr{F}^{\overset{+}{N}(CH_3)_3} + \mathscr{R}^{\overset{+}{N}(CH_3)_3} = f\mathscr{F}^{\overset{+}{N}(CH_3)_3} \qquad (2.28)$$

From the known $\mathscr{F}^{\overset{+}{N}(CH_3)}$ (determined as above) and the observed $\sigma_{para}^{\overset{+}{N}(CH_3)_3}$, f_{para} can be calculated to be 0.56. With this constant and known $\mathscr{F}$ values, the $\mathscr{R}$ value of each substituent that has a known σ_{para} can be obtained from Equation 2.27. Some of these are listed in Table 2.2.

An interesting by-product of the Swain–Lupton approach is that the relative contributions from resonance and polar effects to each set of substituent constants can be calculated from the f and r values for that set of substituent constants. The results are sometimes surprising. For example, σ_{meta} has a 22 percent resonance component (as compared to 53 percent for σ_{para}).[40]

The Swain-Lupton analysis has been used by a number of investigators. Others, however, have criticized it on the grounds that one resonance parameter does not seem adequate[41] and have suggested the dual-substituent-parameter approach in Equation 2.29.[42,43] σ_I is the aliphatic, polar substituent constant dis-

$$\log \frac{k}{k_0} \quad \text{or} \quad \log \frac{K}{K_0} = \rho_I \sigma_I + \rho_R \sigma_R \tag{2.29}$$

cussed earlier and σ_R is one of four established sets of resonance substituent constants. These sets are σ_R°, $\sigma_{R(BA)}$, σ_R^+, and σ_R^-. Representative values are shown in Table 2.7. The first is used when the system under investigation is a relatively unperturbed system, the second when it is a substituted benzoic acid, and the third and fourth when it is a particularly electron-deficient or electron-rich benzene ring, respectively. Again, the investigator must choose between sets of constants but now the choice is more limited. The σ_R scale that works for a particular system seems always to be consistent with expectation, and Equation 2.29 gives excellent correlations of a wide variety of kinetic and thermodynamic phenomena.[44] The analysis of Equation 2.29 has also been extended to nonaromatic unsaturated systems.[45]

The Isokinetic or Isoequilibrium Temperature[46]

Careful consideration of the Hammett equation leads one to understand pitfalls in its interpretation that have not yet been mentioned.

Equation 2.15 can be rewritten as

$$\Delta \Delta G' = \rho(\Delta \Delta G^\circ) \tag{2.30}$$

where $\Delta \Delta G'$ is identically equal to $\Delta G^{\circ\prime} - \Delta G_0^{\circ\prime}$, for example, the difference between the free-energy changes in the dissociations of unsubstituted and substituted

[40] See note 38.

[41] A. R. Katritzky and R. D. Topsom, *J. Chem. Educ.*, **48**, 427 (1971).

[42] (a) R. W. Taft and I. C. Lewis, *J. Am. Chem. Soc.*, **80**, 2436 (1958); (b) S. Ehrenson, *Prog. Phys. Org. Chem.*, **2**, 195 (1964); (c) P. R. Wells, S. Ehrenson, and R. W. Taft, *Prog. Phys. Org. Chem.*, **6**, 147 (1968); (d) see notes 23(a) and 23(e); (e) J. Bromilow and R. T. C. Brownlee, *J. Org. Chem.*, **44**, 1261 (1979).

[43] For other related dual-parameter equations see, for example, P. R. Young and W. P. Jencks, *J. Am. Chem. Soc.*, **101**, 3288 (1979) and references therein.

[44] See notes 23(a) and 23(e).

[45] See note 23(f).

[46] J. E. Leffler, *J. Org. Chem.*, **20**, 1202 (1955).

Table 2.7 σ_R VALUES[a]

Substituent	σ_R°	σ_R	σ_R^-	σ_R^+
—N(CH₃)₂	−0.52	−0.83	−0.34	−1.75
—NH₂	−0.48	−0.82	−0.48	−1.61
—NHCOCH₃	−0.25	−0.36	—	−0.86
—OCH₃	−0.45	−0.61	−0.45	−1.02
—OC₆H₅	−0.34	−0.58	—	−0.87
—SCH₃	−0.20	−0.32	−0.14	
—CH₃	−0.11	−0.11	−0.11	−0.25
—C₆H₅	−0.11	−0.11	0.04	−0.30
—F	−0.34	−0.45	−0.45	−0.57
—Cl	−0.23	−0.23	−0.23	−0.36
—Br	−0.19	−0.19	−0.19	−0.30
—I	−0.16	−0.16	−0.11	−0.25
—H	0.00	0.00	0.00	0.00
—SOCH₃	0.00	0.00	—	0.00
—SCF₃	0.04	0.04	0.14	0.04
—Si(CH₃)₃	0.06	0.06	0.14	0.06
—SF₅	0.06	0.06	0.20	0.06
—SOCF₃	0.08	0.08	—	0.08
—CF₃	0.08	0.08	0.17	0.08
—SO₂CH₃	0.12	0.12	0.38	0.12
—CN	0.13	0.13	0.33	0.13
—CO₂R	0.14	0.14	0.34	0.14
—NO₂	0.15	0.15	0.46	0.15
—COCH₃	0.16	0.16	0.47	0.16

[a] Data from S. Ehrenson, R. T. C. Brownlee, and R. W. Taft, *Prog. Phys. Org. Chem.*, **10**, 1 (1973). Reprinted by permission of John Wiley & Sons, Inc.

phenylacetic acids and $\Delta\Delta G^\circ$ is the corresponding difference in the benzoic acid series.

Substituting

$$\Delta\Delta G' = \Delta\Delta H' - T\,\Delta\Delta S' \tag{2.31}$$

into Equation 2.30, we obtain

$$\Delta\Delta H' - T\,\Delta\Delta S' = \rho(\Delta\Delta G^\circ) \tag{2.32}$$

In order for this equation, which is just one way of writing the Hammett equation, to be true, either (a) $\Delta\Delta H'$ must equal zero—that is, substitution on the substrate causes no change in the ΔH for the equilibrium—or (b) $\Delta\Delta S'$ must equal zero—substitution on the substrate causes no change in the ΔS for the equilibrium—or (c) $\Delta\Delta H'$ must be proportional to $\Delta\Delta S'$—that is, substitution on the substrate causes a change in the ΔH for the equilibrium, which is always proportional to the change in the ΔS for the equilibrium on the same substitution. If none of these possibilities are true, Equation 2.32 is an equation in two independent variables and there can be no linear relationship between $\Delta\Delta G^\circ$ and $\Delta\Delta H'$ —

$T \Delta\Delta S'$ (and therefore, of course, also none between $\Delta\Delta G°$ and $\Delta\Delta G'$, Equation 2.30). In fact, in sets of closely related equilibria or rates, one finds that $\Delta\Delta H$ and $\Delta\Delta S$ (or $\Delta\Delta H^{\ddagger}$ and $\Delta\Delta S^{\ddagger}$) most often *are* directly proportional to one another. It is only these equilibria and rates, then, that can be correlated by the Hammett equation. This restriction should be noted.

The required proportionality between $\Delta\Delta S'$ and $\Delta\Delta H'$ causes another hazard in the interpretation of the Hammett equation. If β, which has the units of absolute temperature, is defined as the proportionality constant between $\Delta\Delta H'$ and $\Delta\Delta S'$, then one can write

$$\Delta\Delta H' = \Delta\Delta H'_0 + \beta \, \Delta\Delta S' \qquad (2.33)$$

where $\Delta\Delta H'_0$ is the hypothetical $\Delta\Delta H'$ when $\Delta\Delta S' = 0$. Substituting Equation 2.33 into Equation 2.31, one obtains

$$\Delta\Delta G' = \Delta\Delta H'_0 + \beta\Delta\Delta S' - T \, \Delta\Delta S' \qquad (2.34)$$

Rearranging, one obtains

$$\Delta\Delta G' = \Delta\Delta H'_0 - (T - \beta) \, \Delta\Delta S' \qquad (2.35)$$

When $T = \beta$, the free-energy change of all equilibria in this reaction series will be the same, irrespective of substitution. Thus β is called the *isoequilibrium temperature*. When the Hammett equation correlates rates rather than equilibria, Equation 2.35 becomes

$$\Delta\Delta G^{\ddagger} = \Delta\Delta H_0^{\ddagger} - (T - \beta) \, \Delta\Delta S^{\ddagger} \qquad (2.36)$$

and β is called the *isokinetic temperature*.

What then happens to ρ at the isoequilibrium or isokinetic temperature? Solving Equation 2.33 for $\Delta\Delta S'$ gives

$$\Delta\Delta S' = \frac{\Delta\Delta H' - \Delta\Delta H'_0}{\beta} \qquad (2.37)$$

Substituting this into Equation 2.32 gives

$$\Delta\Delta H' \left(1 - \frac{T}{\beta}\right) - \Delta\Delta H'_0 = \rho \, \Delta\Delta G° \qquad (2.38)$$

If one assumes that $\Delta\Delta H'_0$ is very small (that is, where $\Delta\Delta S' = 0$, $\Delta\Delta H'$ is also near 0), then from Equation 2.38 one can see that near the isoequilibrium temperature, ρ becomes equal to 0. Furthermore, Equation 2.38 indicates that at temperatures below the isoequilibrium temperature, the sign of ρ is opposite to that at temperatures above the isoequilibrium temperature. Thus if rates or equilibria have been measured at only one temperature or over only a small temperature range, a small ρ near zero may mean only that the reaction is being followed near the isokinetic or isoequilibrium temperature.[47] Thus a ρ value close to zero may have an ambiguous

---

[47] See, for example, H. Kwart and W. E. Barnette, *J. Am. Chem. Soc.*, **99**, 614 (1977), and note 23(i).

meaning, but a ρ value greater than one or less than minus one is not ambiguous and does indicate charge separation. Most of the ρ values that we shall encounter fall in the latter category.

2.3 THERMOCHEMISTRY[48]

Of importance to the problem of relating structure and reactivity is the thermochemistry of the reaction—that is, the net enthalpy and entropy changes that occur upon the making of new bonds and the breaking of old ones. If we consider the reaction in Equation 2.39, for example, a large positive standard free-energy change for the reaction, $\Delta G°$, means that the equilibrium constant defined in Equation 2.40 is much less than unity and the equilibrium will lie predominantly

$$A + B \rightleftharpoons C + D \qquad (2.39)$$

$$K_{eq} = e^{-\Delta G°/RT} \qquad (2.40)$$

to the left. On the other hand, if $\Delta G°$ is large and negative, the equilibrium constant is much greater than unity, and the equilibrium will lie predominantly to the right.[49] $\Delta G°$ in turn is a function of $\Delta H°$ and $\Delta S°$, the standard enthalpy and entropy of reaction, respectively (Equation 2.41). $\Delta H°$ is a function of the heats of

$$\Delta G° = \Delta H° - T\Delta S° \qquad (2.41)$$

formation of the molecules being formed or destroyed, and $\Delta S°$ is a function of the entropies of the molecules being formed or destroyed.[50] Thus for the reaction in Equation 2.39,

$$\Delta H° = \Delta H_f° (C) + \Delta H_f° (D) - \Delta H_f° (A) - \Delta H_f° (B) \qquad (2.42)$$

where $\Delta H_f° (X)$ is the standard heat of formation of X. Similarly,

$$\Delta S° = S° (C) + S° (D) - S° (A) - S° (B) \qquad (2.43)$$

where $S°(X)$ is the standard entropy of X.[51]

A simple but relatively crude way to estimate energy changes for reactions is to note the bonding changes that occur and add together the energy contributions from the various bonds broken and formed. These bond energies are listed in Table 2.8. The values given in the first part of the table are bond dissociation

---

[48](a) S. W. Benson, F. R. Cruickshank, D. M. Golden, G. R. Haugen, H. E. O'Neal, A. S. Rodgers, R. Shaw, and R. Walsh, *Chem. Rev.,* **69,** 279 (1969); (b) S. W. Benson, *J. Chem. Educ.,* **42,** 502 (1965).

[49] Even if ΔG is a large negative quantity the reaction toward the right is, of course, not necessarily fast. The rate depends on the activation barrier that the reactants must overcome to reach the transition state. However, in the absence of special effects there is usually a qualitative correlation between a reaction's net free-energy change and its energy of activation. This point is discussed further in Section 2.6.

[50] The enthalpy change involved in the formation of one mole of a substance from the elements in their standard states is called the heat of formation of the substance. The standard heat of formation is the heat of formation when all substances in the reaction are in their standard states.

[51] The standard entropy of a substance is its entropy in the state specified based on $S° = 0$ at 0 K.

Table 2.8 BOND DISSOCIATION ENERGIES AND AVERAGE BOND ENERGIES FOR VARIOUS TYPES OF BONDS (kcal mole^{-1})

Bond Dissociation Energies[a]—Single Bonds: Diatomic Molecules

Bond	Energy	Bond	Energy	Bond	Energy
H—H	104.2	F—Cl	61	H—F	135.8[b]
D—D	106.0	F—Br	60	H—Cl	103.0[b]
F—F	38	F—I	58	H—Br	87.5[b]
Cl—Cl	58	Cl—Br	52	H—I	71.3[b]
Br—Br	46.0	Cl—I	50		
I—I	36.1				

Polyatomic Molecules

Bond	Energy	Bond	Energy	Bond	Energy
H—CH$_3$	104	Cl—CCl$_3$	73	H—OH	119
H—CH$_2$CH$_3$	98	Br—CCl$_3$	54	H—O$_2$H	90
H—CHCH$_2$	103	F—COCH$_3$	119	H—SH	90
H—C$_6$H$_5$	103	Cl—COCH$_3$	83.5	H—OCH$_3$	102
H—CCH	~125	I—COCH$_3$	52.5	H—OC$_6$H$_5$	85
H—CH$_2$C$_6$H$_5$	85			H—O$_2$CCH$_3$	112
H—CH$_2$CHCH$_2$	85	CH$_3$—CH$_3$	88		
H—CH$_2$OH	93	CH$_3$—CH$_2$CH$_3$	85	HO—CH$_3$	91.5
H—CF$_3$	104	CH$_3$—CH$_2$OH	83	HO—CH$_2$CH$_3$	91.5
H—CCl$_3$	96	CH$_3$—CF$_3$	100	HO—C$_6$H$_5$	103
H—COCH$_3$	87.5	CH$_3$—CHCH$_2$	92	HO—COCH$_3$	109
H—CN	130	CH$_3$—C$_6$H$_5$	93		
		CH$_3$—CCH	117	CH$_3$O—CH$_3$	80
F—CH$_3$	108	CH$_3$—CH$_2$C$_6$H$_5$	72	CH$_3$O—CH$_2$CH$_3$	80
Cl—CH$_3$	83.5	CH$_3$—CH$_2$CHCH$_2$	72	CH$_3$O—CHCH$_2$	87
Br—CH$_3$	70	CH$_3$CH$_2$—CHCH$_2$	89	CH$_3$O—C$_6$H$_5$	91
I—CH$_3$	56	CH$_3$CH$_2$—C$_6$H$_5$	90	CH$_3$O—COCH$_3$	97
F—CH$_2$CH$_3$	106	CH$_2$CH—CHCH$_2$	100		
Cl—CH$_2$CH$_3$	81.5	HCC—CCH	150	HO—OH	51
Br—CH$_2$CH$_3$	69	C$_6$H$_5$—C$_6$H$_5$	100	HO—Br	57
I—CH$_2$CH$_3$	53.5	CH$_2$CH—C$_6$H$_5$	99	CH$_3$O—OCH$_3$	36
Cl—CHCH$_2$	84				
F—C$_6$H$_5$	116	CH$_3$—COCH$_3$	82	H$_2$N—H	103
Br—C$_6$H$_5$	72	CH$_3$CH$_2$—COCH$_3$	79	H$_2$N—CH$_3$	79
I—C$_6$H$_5$	65	CH$_3$—CN	122	H$_2$N—CH$_2$CH$_3$	78
F—CF$_3$	129	CH$_2$CH—COCH$_3$	89	H$_2$N—C$_6$H$_5$	91
Cl—CF$_3$	85	CH$_2$CH—CN	128	H$_2$N—COCH$_3$	~96
Br—CF$_3$	70	CH$_3$CO—COCH$_3$	83		
I—CF$_3$	54	NC—CN	144	O$_2$N—NO$_2$	13.6
F—CCl$_3$	106	CF$_3$—CF$_3$	97	O$_2$N—COCH$_3$	97

Multiple Bonds

Bond	Energy	Bond	Energy
O=O	119	CF$_2$=CF$_2$	76.3
O=CO	128	CH$_2$=NH	~154
O=CH$_2$	175	C≡O	257
O=NH	115	N≡N	226
HN=NH	~109	N≡CH	224
CH$_2$=CH$_2$	163	HC≡CH	230

Table 2.8 (*Continued*)

Representative Average Bond Energies[c]—Single Bonds

	C	N	O	F	Cl	Br	I	Si
H	100	93	110	136	103	88	71	72
C	81	69	84	105	79	67	57	69
N		38	43	65	48			
O			33	50	50	53	57	103
F				60	60	67	141
Cl					53	50	96
Br						43	69
I							50
Si							45

Multiple Bonds

Elements	Single bond	Double bond	Triple bond
O—O	33	96	
N—N	38	100	226
C—C	81	148	194
C—O	84	172	
C—N	69	148	213

[a] From A. J. Gordon and R. A. Ford, *The Chemists Companion*, Wiley, New York, 1972. Reprinted by permission of John Wiley & Sons, Inc.
[b] S. W. Benson, *J. Chem. Educ.*, **42**, 502 (1965). Reprinted by permission of the Division of Chemical Education.
[c] From J. Waser, K. N. Trueblood, and C. M. Knobler, *Chem One*, McGraw-Hill, New York, 1976. Adapted by permission of McGraw-Hill.

energies; that is, the enthalpy change of the reaction:

$$X{-}Y \longrightarrow X\cdot + Y\cdot \qquad \Delta H = D(X{-}Y) \qquad (2.44)$$

Bond dissociation energies are useful for simple reactions, particularly of the free-radical type, in which one or two single bonds are formed or broken. Energy changes in more complex reactions may be approximated using the averaged bond energies given at the end of the table. The sum of the averaged bond energies of a molecule represents the enthalpy of its formation from its constituent atoms. The averaging is necessary because successive breaking of the bonds of a given type in a molecule requires different amounts of energy for each bond.

More accurate estimates of reaction thermochemistry may be obtained by the use of Benson's additivity rules, which allow calculation of ΔH_f° and S° for molecules in the gas phase. When an accurate experimental value is known, the calculated value is almost always within a few tenths of a kilocalorie of it, and usually the agreement is even better.

Benson's approach is to determine the ΔH_f° of a molecule by adding together the ΔH_f°'s of the various groups in the molecule. A group is defined as an atom and its ligands. For example, CH_3CH_3 is made up of two identical groups.

The central atom in the group is carbon, and the ligands are carbon and three hydrogens. By Benson's notation this group is designated $C\!-\!(H)_3(C)$: the central atom in the group is given first and then the ligand atoms in parentheses. ΔH_f° of each group is calculated from experimentally determined ΔH_f°'s of compounds that contain that group. Then ΔH_f° for a new molecule in the gas phase is obtained by simply adding together the contributions from each group. ΔH_f° for the $C\!-\!(H)_3(C)$ group is -10.08 kcal mole^{-1}. Thus ethane is calculated to have a ΔH_f° of -20.16 kcal mole^{-1}. Propane also has two $C\!-\!(H)_3(C)$ groups and a $C\!-\!(H)_2(C)_2$ group ($\Delta H_f^\circ = -4.95$ kcal mole^{-1}). Therefore ΔH_f° ($CH_3CH_2CH_3$) $= -20.16 - 4.95 = -25.11$ kcal mole^{-1}. The experimental ΔH_f°'s for ethane and propane are -20.24 and -24.82 kcal mole^{-1}, respectively. Benson's additivity rules do not apply to condensed-phase compounds because of the contribution of solvation and of lattice and hydrogen bond energies to $\Delta H_f^\circ(X)$ in the liquid and solid phases. These contributions are not additive.

Tables 2.9 through 2.13 list ΔH_f° values for a large number of groups (see Section 9.1 for additivity data for radicals). In these tables C_d refers to a carbon that is forming a carbon–carbon double bond. The notation $C_d\!-\!(H)_2(C_d)$ is shortened to $C_d\!-\!(H_2)$, since all carbon–carbon double bonds are between two sp^2 carbons. Similarly, $C_t\!-\!(X)$ refers to a carbon triply bonded to another sp carbon and to an X ligand; $C_B\!-\!(X)$ refers to an aromatic ring carbon bonded to two other ring carbons and to a substituent X; and C_a refers to the central carbon of the allenic group $C\!=\!C\!=\!C$. Other group abbreviations are noted at the end of the appropriate table.

Guide to the Use of the Group Tables (Tables 2.9–2.13)

1. ΔH_f° and S° are the heat of formation and entropy, respectively, of a group when that group is in a molecule in its standard state of hypothetical ideal gas at 1 atm pressure and 25°C. All values of ΔH_f° are in kilocalories per mole, and all values of S° are in calories per mole per degree (K). For a simple method of converting S° and ΔH_f° to other temperatures, see Benson et al., *Chem. Rev.*, **69** (1969), p. 313.

2. In order to assign values to all groups, some groups have had to be assigned arbitrary values. Groups in brackets in Tables 2.9–2.13 are those groups. Estimated values obtained from a single compound are in parentheses.

In simply adding together the ΔH_f°'s of all the groups in a molecule to obtain the ΔH_f° of the molecule, we make the assumption that only the nearest neighbors of a bond affect that bond. This is not always true, and we shall now discuss the more important corrections that must be applied if the group additivity scheme for molecular enthalpies is to be used successfully.

Alkanes   In an alkane, gauche interactions may raise the enthalpy content of the molecule. The correction is made as follows. Arrange the alkane in its most stable conformation, sight along each of the nonterminal chain carbon–carbon bonds, and count the number of gauche interactions. Then add $+0.80$ kcal mole^{-1}

to the calculated ΔH_f° of the compound for each gauche interaction. Thus, for example, in its most stable conformation n-butane (**54**) (and all unbranched open-chain alkanes) has no gauche interactions, and no gauche corrections should be applied. The most stable conformer of 2,3-dimethylbutane (**55**) has two gauche

54   **55**

interactions. Thus to obtain the ΔH° for the molecule, we add together the group ΔH_f° contributions and $+1.60$ for two gauche corrections:

$$
\begin{array}{rl}
4\ \text{C—(H)}_3\text{(C)} = & -40.32 \\
2\ \text{C—(H)(C)}_3 = & -3.80 \\
\hline
2\ \text{gauche corrections} = & +1.60 \\
\hline
\Delta H_f^\circ\ \text{(2,3-dimethylbutane)} = & -42.52\ \text{kcal mole}^{-1}
\end{array}
$$

The experimental value is -42.49 kcal mole^{-1}.

Alkenes   There are two types of corrections that are sometimes necessary in calculating ΔH_f° for alkenes. For a compound that contains a cis double bond, a correction of $+1.00$ must be added. (If one or both of the cis substituents is t-butyl, the correction is larger: see Table 2.9, footnote b.) For example, the ΔH_f° (cis-2-butene) is calculated as follows:

$$
\begin{array}{rl}
2\ \text{C—(H}_3\text{)(C)} = & -20.16 \\
2\ \text{C}_d\text{—(H)(C)} = & +17.18 \\
\hline
1\ \text{cis correction} = & +1.00 \\
\hline
\Delta H_f^\circ\ (cis\text{-2-butene}) = & -1.98\ \text{kcal mole}^{-1}
\end{array}
$$

The experimental value is -1.67 kcal mole^{-1}. Note that there is no ΔH_f° for C—(H)$_3$(C$_d$) in Table 2.9; a methyl bonded to an sp^2 carbon has the same ΔH_f° group value as a methyl bonded to an sp^3 carbon. This assumption was made in the original determination of ΔH_f° group values.

If one side of the double bond is substituted as in **56** or **57**, in which R stands for an alkyl group, an alkene gauche correction of 0.50 kcal mole^{-1} must be added. Thus the calculated ΔH_f° for 2,3-dimethylbut-1-ene (**58**) is as follows on page 156:

$$
\begin{array}{rl}
3\ \text{C—(H)}_3\text{(C)} = & -30.24 \\
1\ \text{C}_d\text{—(H)}_2 = & +6.26 \\
1\ \text{C}_d\text{—(C)}_2 = & +10.34 \\
1\ \text{C—(C}_d\text{)(C}_2\text{)(H)} = & -1.48 \\
\hline
1\ \text{alkene gauche correction} = & +0.50 \\
\hline
\Delta H_f^\circ\ \text{(2,3-dimethylbut-1-ene)} = & -14.62\ \text{kcal mole}^{-1}
\end{array}
$$

Table 2.9 HYDROCARBON GROUPS[a]

Group	ΔH°_{f298}	S°_{298}
C—(H)$_3$(C)	−10.08	30.41
C—(H)$_2$(C)$_2$	−4.95	9.42
C—(H)(C)$_3$	−1.90	−12.07
C—(C)$_4$	0.50	−35.10
C$_d$—(H)$_2$	6.26	27.61
C$_d$—(H)(C)	8.59	7.97
C$_d$—(C)$_2$	10.34	−12.7
C$_d$—(C$_d$)(H)	6.78	6.38
C$_d$—(C$_d$)(C)	8.88	−14.6
[C$_d$—(C$_B$)(H)]	6.78	6.4
C$_d$—(C$_B$)(C)	8.64	(−14.6)
[C$_d$—(C$_t$)(H)]	6.78	6.4
C—(C$_d$)(C)(H)$_2$	−4.76	9.8
C—(C$_d$)$_2$(H)$_2$	−4.29	(10.2)
C—(C$_d$)(C$_B$)(H)$_2$	−4.29	(10.2)
C—(C$_t$)(C)(H)$_2$	−4.73	10.3
C—(C$_B$)(C)(H)$_2$	−4.86	9.3
C—(C$_d$)(C)$_2$(H)	−1.48	(−11.7)
C—(C$_t$)(C)$_2$(H)	−1.72	(−11.2)
C—(C$_B$)(C)$_2$(H)	−0.98	−(12.2)
C—(C$_d$)(C)$_3$	1.68	(−34.72)
C—(C$_B$)(C)$_3$	2.81	(−35.18)
C$_t$—(H)	26.93	24.7
C$_t$—(C)	27.55	6.35
C$_t$—(C$_d$)	29.20	(6.43)
C$_t$—(C$_B$)	(29.20)	6.43
C$_B$—(H)	3.30	11.53
C$_B$—(C)	5.51	−7.69
C$_B$—(C$_d$)	5.68	−7.80
[C$_B$—(C$_t$)]	5.7	−7.80
C$_B$—(C$_B$)	4.96	−8.64
C$_a$	34.20	6.0

Next-Nearest Neighbor Corrections

Alkane gauche correction	0.80	
Alkene gauche correction	0.50	
cis Correction	1.00[b]	[c]
ortho Correction	0.57	−1.61

Corrections to be Applied
to Ring Compound Estimates

Ring (σ)	ΔH°_{f298}	S°_{298}
Cyclopropane (6)	27.6	32.1
Methylenecyclopropene	40.9	
Cyclopropene (2)	53.7	33.6
Cyclobutane (8)	26.2	29.8
Cyclobutene (2)	29.8	29.0
Cyclopentane (10)	6.3	27.3
Cyclopentene (2)	5.9	25.8
Cyclopentadiene	6.0	

Table 2.9 (*Continued*)

Ring (σ)	ΔH°_{f298}	S°_{298}
Cyclohexane (6)	0	18.8
Cyclohexene (2)	1.4	21.5
Cyclohexadiene-1,3	4.8	
Cyclohexadiene-1,4	0.5	
Cycloheptane (1)	6.4	15.9
Cycloheptene	5.4	
Cycloheptadiene-1,3	6.6	
Cycloheptatriene-1,3,5 (1)	4.7	23.7
Cyclooctane (8)	9.9	16.5
cis-Cyclooctene	6.0	
trans-Cyclooctene	15.3	
Cyclooctatriene-1,3,5	8.9	
Cyclooctatetraene	17.1	
Cyclononane	12.8	
cis-Cyclononene	9.9	
trans-Cyclononene	12.8	
Spiropentane (4)	63.5	67.6
Bicyclo[1.1.0]butane (2)	67.0	69.2
Bicyclo[2.1.0]pentane	55.3	
Bicyclo[3.1.0]hexane	32.7	
Bicyclo[4.1.0]heptane	28.9	
Bicyclo[5.1.0]octane	29.6	
Bicyclo[6.1.0]nonane	31.1	

[a] Reprinted with permission from S. W. Benson, F. R. Cruick-shank, D. M. Golden, G. R. Haugen, H. E. O'Neal, A. S. Rodgers, R. Shaw, and R. Walsh, *Chem. Rev.*, **69**, 279 (1969). Copyright by the American Chemical Society.
[b] When one of the groups is *t*-butyl, cis correction = 4.00; when both are *t*-butyl, cis correction = ~10.00; and when there are two corrections around one double bond, the total correction is 3.00.
[c] +1.2 for but-2-ene, 0 for all other 2-enes, and −0.6 for 3-enes.

Table 2.10 OXYGEN-CONTAINING GROUPS[a]

Group	ΔH°_f	S°
CO—(CO)(C)	−29.2	
CO—(O)(C_d)	−33.5	
CO—(O)(C_B)	−46.0	
CO—(O)(C)	−33.4	14.78
[CO—(O)(H)]	−29.5	34.93
CO—(C_d)(H)	−31.7	
CO—(C_B)$_2$	−39.1	
CO—(C_B)(C)	−37.6	
[CO—(C_B)(H)]	−31.7	
CO—(C)$_2$	−31.5	15.01
CO—(C)(H)	−29.6	34.93
CO—(H)$_2$	−27.7	53.67
O—(CO)$_2$	−50.9	

Table 2.10 (*Continued*)

Group	ΔH_f°	S°
O—(CO)(O)	−19.0	
[O—(CO)(C$_d$)]	−41.3	
O—(CO)(C)	−41.3	8.39
O—(CO)(H)	−60.3	24.52
O—(O)(O)	(19.0)	(9.4)
O—(O)(C)	(−4.5)	(9.4)
O—(O)(H)	−16.27	.27.85
O—(C$_d$)$_2$	−32.8	
O—(C$_d$)(C)	−31.3	
O—(C$_B$)$_2$	−19.3	
O—(C$_B$)(C)	−22.6	
[O—(C$_B$)(H)]	−37.9	29.1
O—(C)$_2$	−23.7	8.68
O—(C)(H)	−37.88	29.07
C$_d$—(CO)(O)	6.3	
C$_d$—(CO)(C)	9.4	
[C$_d$—(CO)(H)]	7.7	
[C$_d$—(O)(C$_d$)]	8.9	
[C$_d$—(O)(C)]	10.3	
[C$_d$—(O)(H)]	8.6	
C$_B$—(CO)	9.7	
C$_B$—(O)	−1.8	−10.2
C—(CO)$_2$(H)$_2$	−7.2	
C—(CO)(C)$_3$	1.58	
C—(CO)(C)$_2$(H)	−1.83	−12.0
C—(CO)(C)(H)$_2$	−5.0	9.6
[C—(CO)(H)$_3$]	−10.08	30.41
C—(O)$_2$(C)$_2$	−16.8	
C—(O)$_2$(C)(H)	−17.2	
C—(O)$_2$(H)$_2$	−17.7	
C—(O)(C$_B$)(H)$_2$	−6.6	9.7
C—(O)(C$_d$)(H)$_2$	−6.9	
C—(O)(C)$_3$	−6.60	−33.56
C—(O)(C)$_2$(H)	−7.00	−11.00
C—(O)(C)(H)$_2$	−8.5	10.3
[C—(O)(H)$_3$]	−10.08	30.41

Strain corrections		
Ether oxygen gauche	0.3	
Ditertiary ethers	8.4	
	27.6	31.4
	26.4	27.7
	6.7	

Table 2.10 (*Continued*)

Group	ΔH_f°	S°
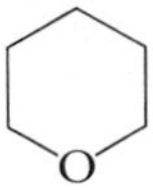	2.2	
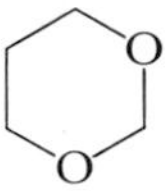	3.5	
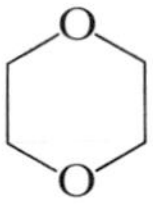	5.4	
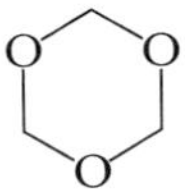	3.4	
	−6.2	
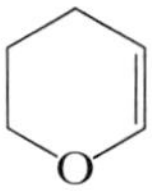	2.5	
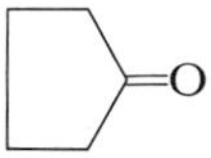	6.0	
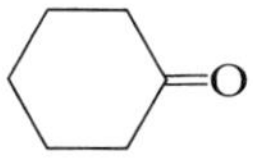	3.4	
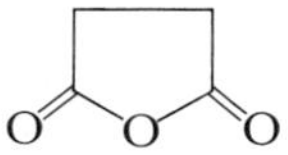	1.1	
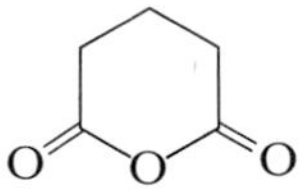	1.4	
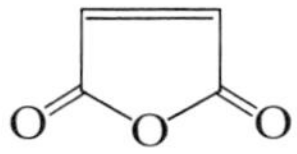	4.6	

[a]Reprinted with permission from S. W. Benson, F. R. Cruickshank, D. M. Golden, G. R. Haugen, H. E. O'Neal, A. S. Rodgers, R. Shaw, and R. Walsh, *Chem. Rev.*, **69,** 279 (1969). Copyright by the American Chemical Society.

153

Table 2.11 Nitrogen-containing Groups[a-e]

Group	ΔH_f°	S°
[C—(N)(H)$_3$]	−10.08	30.41
C—(N)(C)(H)$_2$	−6.6	(9.8)[f]
C—(N)(C)$_2$(H)	−5.2	(−11.7)[f]
C—(N)(C)$_3$	−3.2	(−34.1)[f]
C—(N$_A$)(C)(H)$_2$	(−5.5)	(9.8)
C—(N$_A$)(C)$_2$(H)	(−3.3)	.(−11.7)
C—(N$_A$)(C)$_3$	(−1.9)	(−34.7)
N—(C)(H)$_2$	4.8	29.71
N—(C)$_2$(H)	15.4	8.94
N—(C)$_3$	24.4	−13.46
N—(N)(H)$_2$	11.4	29.13
N—(N)(C)(H)	20.9	9.61
N—(N)(C)$_2$	29.2	−13.80
N—(N)(C$_B$)(H)	22.1	
N$_I$—(H)		
N$_I$—(C)	21.3	
N$_I$—(C$_B$)[e]	16.7	
N$_A$—(H)	25.1	26.8
N$_A$—(C)	(32.5)	(8.0)
N—(C$_B$)(H)$_2$	4.8	29.71
N—(C$_B$)(C)(H)	14.9	
N—(C$_B$)(C)$_2$	26.2	
N—(C$_B$)$_2$(H)	16.3	
C$_B$—(N)	−0.5	−9.69
N$_A$—(N)	23.0	
CO—(N)(H)	−29.6	34.93
CO—(N)(C)	−32.8	16.2
N—(CO)(H)$_2$	−14.9	24.69
N—(CO)(C)(H)	−4.4	(3.9)[f]
N—(CO)(C)$_2$		
N—(CO)(C$_B$)(H)	+0.4	
N—(CO)$_2$(H)	−18.5	
N—(CO)$_2$(C)	−5.9	
N—(CO)$_2$(C$_B$)	−0.5	
C—(CN)(C)(H)$_2$	22.5	40.20
C—(CN)(C)$_2$(H)	25.8	19.80
C—(CN)(C)$_2$		−2.80
C—(CN)$_2$(C)$_2$		28.40
C$_d$—(CN)(H)	37.4	36.58
C$_d$—(CN)$_2$	84.1	
C$_d$—(NO$_2$)(H)		44.4
C$_B$—(CN)	35.8	20.50
C$_t$—(CN)	63.8	35.40
C—(NO$_2$)(C)(H)$_2$	−15.1	(48.4)[f]
C—(NO$_2$)(C)$_2$(H)	−15.8	(26.9)[f]
C—(NO$_2$)$_2$(C)$_2$		(3.9)[f]
C—(NO$_2$)$_2$(C)(H)	−14.9	
O—(NO)(C)	−5.9	41.9
O—(NO)$_2$(C)	−19.4	48.50

Table 2.11 (*Continued*)

Group	ΔH_f°	S°

Corrections to be Applied to Ring
Compound Estimates

Ethylene imine

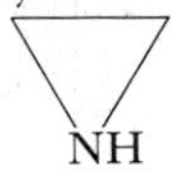

| | 27.7 | $(31.6)^f$ |

Azetidine

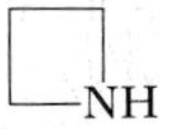

| | $(26.2)^f$ | $(29.3)^f$ |

Pyrrolidine

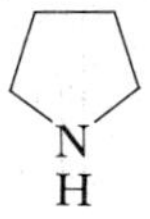

| | 6.8 | 26.7 |

Piperidine

| | 1.0 | |

| 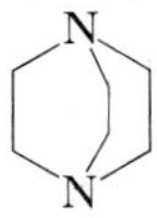 | 3.4 | |

| 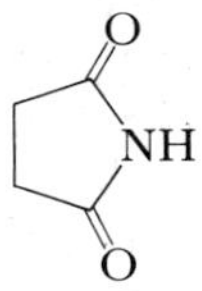 | 8.5 | |

[a] Reprinted with permission from S. W. Benson, F. R. Cruickshank, D. M. Golden, G. R. Haugen, H. E. O'Neal, A. S. Rodgers, R. Shaw, and R. Walsh, *Chem. Rev.*, **69**, 279 (1969). Copyright by the American Chemical Society.

[b] N_I = double-bonded nitrogen in imines; N_I—(C_B) = pyridine N. N_A = double-bonded nitrogen in azo compounds.

[c] No cis corrections applied to imines or azo compounds.

[d] gauche corrections of $+0.8$ kcal mole^{-1} to ΔH_f° applied just as for hydrocarbons.

[e] For ortho or para substitution in pyridine add -1.5 kcal mole^{-1} per group.

[f] Estimates

Table 2.12 Halogen-containing Groups[a]

Group	ΔH_f°	S°	Group	ΔH_f°	S°
C_d—$(F)_2$	−77.5	37.3	C—$(Cl)_3$(C)	−20.7	50.4
C_d—$(Cl)_2$	−1.8	42.1	C—$(Cl)_2$(H)(C)	(−18.9)	43.7
C_d—$(Br)_2$		47.6	C—(Cl)(H)$_2$(C)	−15.6	37.8
C_d—(F)(Cl)		39.8	C—$(Cl)_2(C)_2$	(−19.5)	
C_d—(F)(Br)		42.5	C—(Cl)(H)$(C)_2$	−12.8	17.6
C_d—(Cl)(Br)		45.1	C—(Cl)$(C)_3$	−12.8	−5.4
C_d—(F)(H)	−37.6	32.8	C—$(Br)_3$(C)		55.7
C_d—(Cl)(H)	2.1	35.4	C—(Br)(H)$_2$(C)	−5.4	40.8
C_d—(Br)(H)	12.7	38.3	C—(Br)(H)$(C)_2$	−3.4	
C_d—(I)(H)	24.5	40.5	C—(Br)$(C)_3$	−0.4	−2.0
C_t—(Cl)		33.4	C—(I)(H)$_2$(C)	7.95	42.5
C_t—(Br)		36.1	C—(I)(H)$(C)_2$	10.7	22.2
C_t—(I)		37.9	C—(I)$(C)_3$	13.0	(0.0)
C_B—(F)	−42.8	16.1	C—(Cl)(Br)(H)(C)		45.7
C_B—(Cl)	−3.8	18.9	N—$(F)_2$(C)	−7.8	
C_B—(Br)	10.7	21.6	C—(Cl)(C)(O)(H)	−22.3	15.4
C_B—(I)	24.0	23.7			
C—$(C_B)(F)_3$	−162.7	42.8	Corrections for Next-Nearest Neighbors		
C—(C_B)(Br)(H)$_2$	−6.9		ortho (F)(F)	5.0	0
C—(C_B)(I)(H)$_2$	8.4		ortho (Cl)(Cl)	2.2	
C—$(F)_3$(C)	−158.4	42.5	ortho (alk)(halogen)	0.6	
C—$(F)_2$(H)(C)	(−109.3)	39.1	cis (halogen) (halogen)	(0)	
C—(F)(H)$_2$(C)	−51.8	35.4	cis (halogen)(alk)	(0)	
C—$(F)_2(C)_2$	−97.0	17.8	gauche (halogen)(alk)	0.0	0
C—(F)(H)$(C)_2$	−48.4	(14.0)	gauche (halogen)		
C—(F)$(C)_3$	−43.9		(exclusive of F)	1.0	
C—$(F)_2$(Cl)(C)	−106.3	40.5	gauche (F) (halogen)	0.0	

[a] Reprinted with permission from S. W. Benson, F. R. Cruickshank, D. M. Golden, G. R. Haugen, H. E. O'Neal, A. S. Rodgers, R. Shaw, and R. Walsh, *Chem. Rev.*, **69**, 279 (1969). Copyright by the American Chemical Society.

The experimental value is −15.85 kcal mole^{-1}.

$$\begin{array}{ccc}
\overset{\displaystyle R}{\underset{\displaystyle H}{R-C-C=}} & \overset{\displaystyle R}{\underset{\displaystyle R}{R-C-C=}} & \overset{\displaystyle CH_3}{\underset{\displaystyle CH_3}{CH_3-C-C=CH_2}} \\
\text{56} & \text{57} & \text{58}
\end{array}$$

Aromatic Compounds   In alkylated benzenes a correction of 0.57 must be added if two substituents are ortho to each other. For example, ΔH_f° of 1,2-dimethylbenzene is calculated as follows:

$$
\begin{array}{rl}
4\ C_B\text{—(H)} = & +13.20 \\
2\ C_B\text{—(C)} = & +11.02 \\
2\ C(H)_3(C) = & -20.16 \\
\hline
1\ \text{ortho correction} = & +0.57 \\
\hline
\Delta H_f^\circ\ \text{(1,2-dimethylbenzene)} = & +4.63\ \text{kcal mole}^{-1}
\end{array}
$$

Table 2.13 SULFUR-CONTAINING GROUPS[a]

Group	ΔH_f°	S°
[C—(H)$_3$(S)]	−10.08	30.41
C—(C)(H)$_2$(S)	−5.65	9.88
C—(C)$_2$(H)(S)	−2.64	−11.32
C—(C)$_3$(S)	−0.55	−34.41
C—(C$_B$)(H)$_2$(S)	−4.73	
C—(C$_d$)(H)$_2$(S)	−6.45	
[C$_B$—(S)]	−1.8	10.20
[C$_d$—(H)(S)]	8.56	8.0
C$_d$—(C)(S)	10.93	−12.41
S—(C)(H)	4.62	32.73
S—(C$_B$)(H)	11.96	12.66
S—(C)$_2$	11.51	13.15
S—(C)(C$_d$)	9.97	
S—(C$_d$)$_2$	−4.54	16.48
S—(C$_B$)(C)	19.16	
S—(C$_B$)$_2$	25.90	
S—(S)(C)	7.05	12.37
S—(S)(C$_B$)	14.5	
S—(S)$_2$	3.04	13.36
[C—(SO)(H)$_3$]	−10.08	30.41
C—(C)(SO)(H)$_2$	−7.72	
C—(C)$_3$(SO)	−3.05	
C—(C$_d$)(SO)(H)$_2$	−7.35	
[C$_B$—(SO)]	2.3	
SO—(C)$_2$	−14.41	18.10
SO—(C$_B$)$_2$	−12.0	
[C—(SO$_2$)(H)$_3$]	−10.08	30.41
C—(C)(SO$_2$)(H)$_2$	−7.68	
C—(C)$_2$(SO$_2$)(H)	−2.62	
C—(C)$_3$(SO$_2$)	−0.61	
C—(C$_d$)(SO$_2$)(H)$_2$	−7.14	
C—(C$_B$)(SO$_2$)(H)$_2$	−5.54	
[C$_B$—(SO$_2$)]	2.3	
C$_d$—(H)(SO$_2$)	12.53	
C$_d$—(C)(SO$_2$)	14.47	
SO$_2$—(C)$_2$	−69.74	20.90
SO$_2$—(C)(C$_B$)	−72.29	
SO$_2$—(C$_B$)$_2$	−68.58	
[SO$_2$—(C$_d$)(C$_B$)]	−68.58	
SO$_2$—(C$_d$)$_2$	−73.58	
SO$_2$—(SO$_2$)(C$_B$)	−76.25	
[CO—(S)(C)]	−31.56	15.43
S—(H)(CO)	−1.41	31.20
C—(S)(F)$_3$		38.9
S—(C)(CN)	37.18	41.06
[CS—(N)$_2$]	−31.56	15.43
N—(CS)(H)$_2$	12.78	29.19
[S—(S)(N)]	−4.90	

Table 2.13 (*Continued*)

Group	ΔH_f°	S°
N—(S)(C)$_2$	29.9	
[SO—(N)$_2$]	−31.56	
N—(SO)(C)$_2$	16.0	
[SO$_2$—(N)$_2$]	−31.56	
N—(SO$_2$)(C)$_2$	−20.4	

Organosulfur Compounds Ring Corrections

Ring (σ)	ΔH_f°	S°
thiirane ring, S (2)	17.7	29.47
 thietane ring, S (2)	19.37	27.18
 thiolane ring, S (2)	1.73	23.56
 thiane ring, S (2)	0	17.46
 thiepane ring, S (2)	3.89	
 dihydrothiophene ring, S (2)	5.07	
 dihydrothiophene ring, S (1)	5.07	
 sulfolene ring, SO$_2$ (2)	5.74	
 thiophene ring, S (2)	1.73	

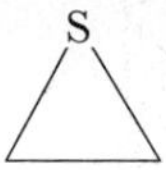

[a] Reprinted with permission from S. W. Benson, F. R. Cruickshank, D. M. Golden, G. R. Haugen, H. E. O'Neal, A. S. Rodgers, R. Shaw, and R. Walsh, *Chem. Rev.*, **69**, 279 (1969). Copyright by the American Chemical Society.

The experimental value is $+4.54$ kcal mole^{-1}.

Note again, C—(H)$_3$(C$_B$) is assumed to be equal to C—(H)$_3$(C).

Cycloalkanes Corrections that must be added for ring strain are given in Table 2.9. Contributions to ΔH_f° due to gauche interactions between substituents and between substituent and ring must also be taken into account. (Contributions from ring–ring gauche interactions are included in the ring strain.) Thus, for example, in the most stable conformation of *trans*-1,4-dimethylcyclohexane (**59**) both CH$_3$'s are equatorial and no gauche interactions exist between the methyl and the ring carbons. In *cis*-1,4-dimethylcyclohexane (**60**), one of the methyls must be axial. Sighting along the C$_2$—C$_1$ bond, as in **61**, we see that there is one methyl–ring

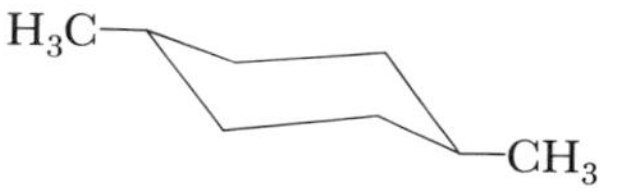

59 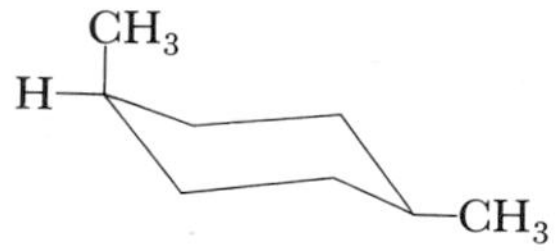   **60**

gauche interaction. Likewise, sighting along the C$_1$—C$_6$ bond as in **62**, we see another methyl–ring gauche interaction. Thus ΔH_f° (*cis*-1,4-dimethylcyclohexane)

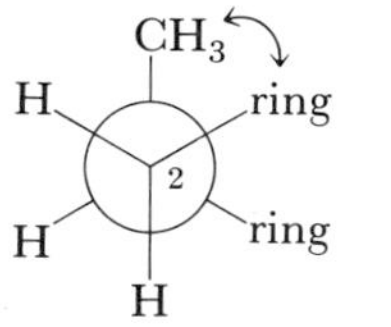

61 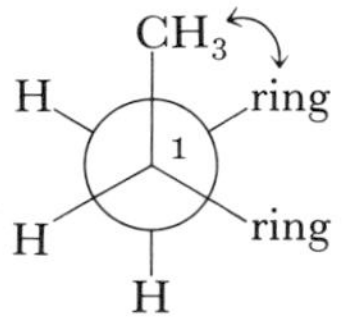   **62**

is calculated as follows:

$$
\begin{array}{rl}
2\ \text{C—(H)}_3\text{(C)} = & -20.16 \\
4\ \text{C—(H)}_2\text{(C)}_2 = & -19.80 \\
2\ \text{C—(H)(C)}_3 = & -3.80 \\
\text{cyclohexane ring strain} = & 0.00 \\
2\ \text{alkane gauche corrections} = & +1.60 \\
\hline
\Delta H_f^\circ\ (\textit{cis}\text{-1,4-dimethylcyclohexane}) = & -42.16\ \text{kcal mole}^{-1}
\end{array}
$$

The experimental value is -42.22 kcal mole^{-1}. (The calculated value for the trans isomer is $(-42.16 - 1.60) = -43.76$ kcal mole^{-1}, and the experimental value is -44.12 kcal mole^{-1}.)

Compounds Containing Heteroatoms The examples we have discussed have all been hydrocarbons, and all the group enthalpy values have been obtained from Table 2.9. However, by using Tables 2.10–2.13, ΔH_f° for a wide variety of compounds containing N, O, S, and the halogens can be calculated. The procedure is just the same as for the hydrocarbons; all necessary corrections are given in the tables. In the reference from which Tables 2.9–2.13 are taken, there are tables giving group enthalpy and entropy values for still other types of heteroatoms.

There the reader can also find a very large number of compounds listed with their estimated and observed enthalpies and entropies of formation.

Entropy

In this section we have so far emphasized only the $\Delta H°$ and $\Delta H_f°$ of the reaction components. This is because the entropy change in many reactions is small and can often be neglected in comparison to the enthalpy change. When $S°$'s are of interest, they too can be estimated by Benson's additivity rules. In order to calculate $S°$ for a molecule, the group $S°$ contributions are added together just as they are for $\Delta H_f°$, but now a correction for the overall rotational symmetry of the molecule must be added. The correction is $-R \ln \sigma$, where σ is the symmetry number, defined as the total number of independent permutations of identical atoms or groups in a molecule that can be arrived at by simple rotations of the entire molecule, or by rotations about single bonds within the molecule.[52] In practice, the symmetry number is found by multiplying the symmetries of the independent axes. These axes may be of two types, external or internal. External axes generate an identical arrangement of atoms by rigid rotation of the molecule as a whole; internal axes do so by rotations around bonds within the molecule. The H_2O molecule has a single external twofold axis and has $\sigma = 2$. CH_4 has four external threefold axes (one along each C—H bond) and $\sigma = 3 \times 4 = 12$. Ethane has an external twofold axis, an external threefold axis, $\sigma_{ext} = 2 \times 3 = 6$, and an internal threefold axis of rotation about the C—C bond, $\sigma_{int} = 3$, as illustrated by **63**. The overall symmetry number

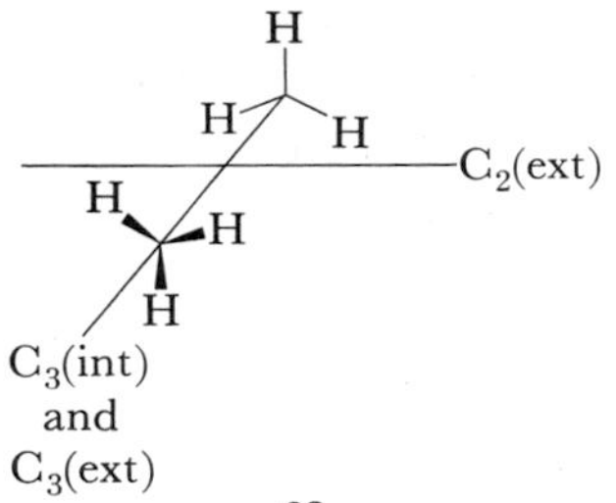

63

$\sigma = \sigma_{ext} \times \sigma_{int} = 18$. Acetone has an external twofold axis and two internal threefold axes (**64**), $\sigma = 2 \times 3 \times 3 = 18$. In a highly symmetrical molecule like cyclopropane or benzene, one must count only independent axes. Thus cyclopropane is

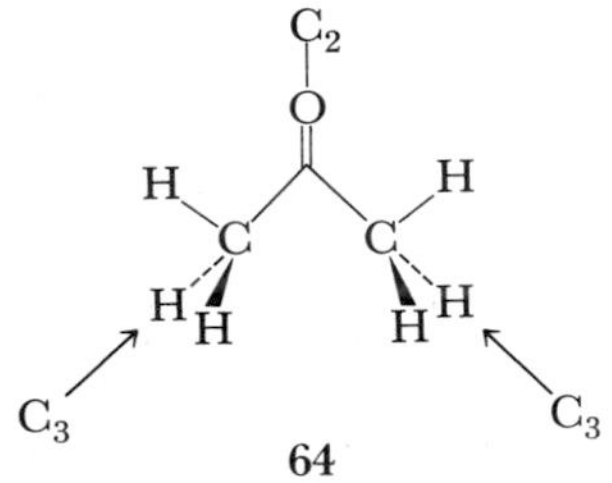

64

[52] S. W. Benson, *Thermochemical Kinetics,* 2d ed., Wiley, New York, 1976, p. 48.

considered to have one threefold and one twofold (**65**) but not a threefold and three twofold axes, because two of the twofold axes are generated from the first by the threefold rotation.

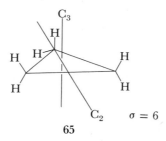

$\sigma = 6$

65

Table 2.14 Symmetry Numbers for Various Point Groups[a]

Point Group	σ	Point Group	σ	Point Group	σ
C_1, C_i, C_s	1	$D_2, D_{2d}, D_{2h} \equiv V$	4	$C_{\infty v}$	1
C_2, C_{2v}, C_{2h}	2	D_3, D_{3d}, D_{3h}	6	$D_{\infty h}$	2
C_3, C_{3v}, C_{2h}	3	D_4, D_{4h}, D_{4h}	8	T, T_d	12
C_4, C_{4v}, C_{4h}	4	D_6, D_{6d}, D_{6h}	12	O_h	24
C_6, C_{6v}, C_{6h}	6	S_6	3		

[a] From S. W. Benson, *Thermochemical Kinetics*, 2d ed., Wiley, New York, 1976. Reprinted by permission of John Wiley & Sons, Inc.

Table 2.14 lists the symmetry numbers associated with the various point groups.
   The following example illustrates the calculation of the entropy $S°$ for acetone:

$$2 \ C\text{—}(H)_3(C) = +60.82$$
$$1 \ CO\text{—}(C)_2 = +15.01$$
$$\underline{-R \ln \sigma = -R \ln 18 = \quad -5.74}$$
$$S° \text{ (acetone)} = +70.09 \text{ cal mol}^{-1} \text{ K}^{-1}$$

The experimental value is 70.5 cal mole^{-1} K^{-1}.
   If a molecule is chiral, $R \ln n$ must be added to its entropy estimate, where n is the total number of stereoisomers of equal energy present in the material.

2.4 SOLUTIONS[53]

The thermochemical additivity scheme outlined in the previous section is based on gas-phase data. Since most organic reactions are carried out in solution, it would be

[53] The following discussions of solvents and solvent effects will provide further information: (a) T. C. Waddington, *Non-Aqueous Solvents*, Thomas Nelson, London, 1969; (b) E. M. Kosower, *An Introduction to Physical Organic Chemistry*, Wiley, New York, 1968, p. 259; (c) T. C. Waddington, Ed., *Non-Aqueous Solvent Systems*, Academic Press, London, 1965; (d) E. S. Amis and J. F. Hinton, *Solvent Effects on Chemical Phenomena*, Academic Press, New York, 1973; (e) J. F. Coetzee and C. D. Ritchie, Eds., *Solute–Solvent Interactions*, Dekker, New York, Vol. 1, 1969; Vol. 2, 1976; (f) A. J. Parker, *Chem. Rev.*, **69**, 1 (1969); (g) F. Franks, Ed., *Water, a Comprehensive Treatise*, Plenum, New York, Vols. 1–5, 1972–1975.

most useful to be able to understand and predict the thermochemical changes that molecules, ions, and transition states undergo when dissolved in various solvents. Although our knowledge of the structure of liquids and of their interactions with solutes on the molecular level is still incomplete, progress is being made in this area and useful quantitative data are becoming available. We shall proceed by first discussing briefly the relation between solvent properties and the most important parameters characterizing liquids.

Dielectric Constant

The dielectric constant of a substance, ϵ, measures the reduction of the strength of the electric field surrounding a charged particle immersed in the substance, compared to the field strength around the same particle in a vacuum. The dielectric constant is a macroscopic property; that is, its definition and measurement assume that the substance of interest is continuous, with no microscopic structure. Electrostatic attractions and repulsions between ions are smaller the higher the dielectric constant of the medium, and ions of opposite charge therefore have a greater tendency to dissociate when the dielectric constant is larger. Table 2.15 lists dielectric constants for some common solvents.

The dielectric constant gives only a rough guide to solvent properties and does not correlate well with measured effects of solvents on reaction rates. It is nevertheless useful for making a division of solvents into two broad categories: polar and nonpolar. In nonpolar solvents, $\epsilon < {\sim}15$, ionic substances will be highly associated. Indeed, they will be very sparingly soluble in most of these solvents except as, for example, in the case of acetic acid, when hydrogen bonding is available, and even then solubility is low. Ionic substances are more soluble in solvents of higher dielectric constant, and the ions are dissociated if the dielectric constant is sufficiently high.

Dipole Moment and Polarizability

In order to gain a better understanding of solution phenomena, it is necessary to evaluate solvent properties on the molecular level. Here the most important properties are the dipole moment, μ, and the molecular polarizability. Values are listed in Table 2.15.

The dipole moment measures the internal charge separation of the molecules and is important in evaluating how the solvent molecules will cluster around a solute particle that itself has a dipole moment or a net charge. The solvent dipoles will tend to orient themselves around the solute in the manner indicated in Figure 2.7. The first solvent layer will be the most highly ordered, with randomness increasing as the influence of the solute particle decreases with increasing distance. A smaller solute ion will generate a more intense electric field than will a large one, and so will have a stronger and more far-ranging capacity to orient solvent dipoles around it. In the solvent molecule itself, one end of the dipole may be exposed while the other end is buried inside the bulk of the molecule. In dimethylsulfoxide,

Table 2.15 Properties of Selected Solvents[a]

Solvent	ϵ[b]	μ (debyes)	Polarizability[c] ($cm^3 \times 10^{24}$)	δ[d]	E_T[e]	π^*[f]	DN[g]	AN[g]
Nonpolar Protic[h]								
$CH_3-C(=O)-OH$	6.15[i]	1.68	5.16	8.9	51.9	0.664		52.9
$(CH_3)_3C-OH$	12.47	1.66	8.82	10.5	43.9			
$CH_3-(CH_2)_5-OH$	13.3	1.55	12.46					
cyclohexanol (OH, H)	15.0	1.86	11.33					
ϕOH	9.78[j]	1.45	11.11[k]					
Nonpolar Aprotic								
CCl_4	2.24[i]	0	10.49	8.6	32.5	0.294		8.6
pyridine	12.4[l]	2.37	9.55		40.2	0.867	33.1	14.2
$n\text{-}C_6H_{14}$	1.88	0.085	11.87	7.3	30.9	−0.081		(0)[m]
$cyclo\text{-}C_6H_{12}$	2.02[i]	—	10.99	8.2	31.2[d]	0.000		
$\phi-H$	2.28	0	10.39	9.2	34.5	0.588	0.1	8.2
$(C_2H_5)_2O$	4.34[i]	1.15	8.92	7.8	34.6	0.273	19.2	3.9
dioxane	2.21	0.45	8.60	9.8	36.0	0.553	14.8	10.8
$CH_3OCH_2CH_2OCH_3$	7.20	1.71	9.56		38.2		~24.	10.2
$CH_3O(CH_2CH_2O)_2CH_3$	—	1.97	13.79		38.6[n]			9.9
tetrahydrofuran	7.58	1.75	7.92[i]	9.3	37.4	0.576	20.0	8.0

Table 2.15 (*Continued*)

Solvent	ϵ^b	μ (debyes)	Polarizability^c (cm³ × 10²⁴)	δ^d	E_T^e	π^{*f}	DN^g	AN^g
Dipolar Protic								
H₂O^o	78.5	1.84	1.48	23.4	63.1	1.090	18.0^p	54.8
HCOOH	58.5^q	1.82	3.39					41.3
CH₃OH	32.70	2.87	3.26	14.3	55.5	0.586		37.1
C₂H₅OH	24.55	1.66	5.13	12.7	51.9	0.540		
HCNH₂ (O=)	111.0^i	3.37	4.22			1.118		39.8
Dipolar Aprotic								
CH₃—C(=O)—CH₃	20.70	2.69	6.41	9.6	42.2	0.683	17.0	12.5
H—C(=O)—N(CH₃)₂	36.71	3.86	7.90	11.8	43.8	0.875	26.6	16.0
CH₃—S(=O)—CH₃	46.68	3.9	7.99	13.0	45.0	1.000	29.8	19.3
CH₃—CN	37.5^i	3.44	4.41	11.7	46.0	0.713	14.1	18.9
CH₃NO₂	35.87^r	3.56	4.95	12.6	46.3	0.848	2.7	20.5
(CH₃)₂N—P(=O)—N(CH₃)₂ with N(CH₃)₂ (hexamethylphosphoramide)	30^i	5.54	18.90^i	8.6	40.9^d	0.871	38.8	10.6

[a] Dielectric constant and dipole moment data are from J. A. Riddick and W. B. Bunger, *Organic Solvents*, 3d ed., Vol. II of A. Weissberger, Ed., *Techniques of Chemistry*, Wiley-Interscience, New York, 1970. Other physical constants may also be found in this source.

[b] $T = 25°C$ except where noted.

[c] Calculated from the refractive index n according to the formula

$$\text{polarizability} = \frac{n^2 - 1}{n^2 + 2} \frac{M}{d} \frac{3}{4\pi N_0}$$

where M = molecular weight; d = density; and N_0 = Avagadro's number. See E. A. Moelwyn-Hughes, *Physical Chemistry*, 2d ed., Pergamon Press, Elmsford, N.Y., 1961, p. 382. $T = 25°C$ except where noted.

[d] M. H. Abraham, *Prog. Phys. Org. Chem.*, **11**, 1 (1974).

[e] C. Reichardt, *Angew. Chem. Int. Ed.*, **4**, 29 (1965).

[f] M. J. Kamlet, J. L. Abboud, and R. W. Taft, *J. Am. Chem. Soc.*, **99**, 6027 (1977).

[g] V. Gutmann, *Coord. Chem. Rev.*, **18**, 225 (1976).

[h] The dividing line between nonpolar and dipolar solvents is arbitrarily set at $\epsilon = 15$. See A. J. Parker, *Chem. Rev.*, **69**, 1 (1969).

[i] $T = 20°C$.

[j] $T = 60°C$.

[k] $T = 45°C$.

[l] $T = 21°C$.

[m] n-Hexane is the reference solvent for the AN scale.

[n] C. Reichardt, *Justus Liebigs Ann. Chem.*, **752**, 64 (1971).

[o] Data from A. J. Parker, *Chem. Rev.*, **69**, 1 (1969).

[p] Value for donicity of water in the associated liquid solvent is 33.0[g].

[q] $T = 16°C$.

[r] $T = 30°C$.

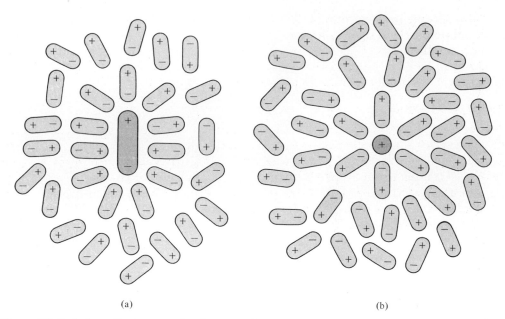

(a)   (b)

Figure 2.7 Ordering of solvent molecules around (a) a dipolar solute molecule and (b) a solute
positive ion. The orientation will be most pronounced in the innermost shell of solvent
molecules and will become increasingly random as distance from the solute particle
increases. The strength of the interaction will depend on the molecular sizes and shapes
and on the magnitudes of the dipole moments of both solute and solvent.

for instance, the negative oxygen end of the dipole is exposed (**66**), whereas the
positive end is not; this solvent interacts much more strongly with the cations of an

$$
\begin{array}{c}
O^{\delta-} \\
\parallel \\
H{\diagdown} \quad S^{\delta+} \quad {\diagup}H \\
C \qquad C \\
H{\diagup}\;{\uparrow}\qquad{\uparrow}\;{\diagdown}H \\
H \qquad\ H
\end{array}
$$

66

ionic solute than with the anions. Formation of solvation layers around the solute
particles will be accompanied by heat evolution (negative ΔH of solution) and an
increase in order (negative ΔS of solution).

Polarizability is a measure of the ease with which the electrons of a molecule
are distorted. It is the basis for evaluating the nonspecific attraction forces (London
dispersion forces) that arise when two molecules approach each other. Each molecule distorts the electron cloud of the other and thereby induces an instantaneous
dipole. The induced dipoles then attract each other. Dispersion forces are weak and
are most important for the nonpolar solvents where other solvation forces are absent. They do, nevertheless, become stronger the larger the electron cloud, and they

may also become important for some of the higher-molecular-weight polar solvents. Large solute particles such as iodide ion interact by this mechanism more strongly than do small ones such as fluoride ion. Furthermore, solvent polarizability may influence rates of certain types of reactions because transition states may be of different polarizability from reactants and so be differently solvated.

Hydrogen Bonding

Hydrogen bonding probably has a greater influence on solvent–solute interactions than any other single solvent property. Solvents that have O—H or N—H bonds are hydrogen bond donors, whereas most hydrogens bound to carbon are too weakly acidic to form hydrogen bonds. Any site with unshared electrons is a potential hydrogen bond acceptor, although the more strongly basic and the less polarizable the acceptor site the stronger will be the hydrogen bond.[54]

We class as *protic* those solvents that are good hydrogen bond donors and as *aprotic* those that are not.[55] Since the protic solvents have hydrogen bound to oxygen or nitrogen, they are also good hydrogen bond acceptors; the aprotic solvents may or may not be hydrogen bond acceptors.

Because negative ions have extra electrons, they are hydrogen bond acceptors and can be expected to be strongly solvated by protic solvents. Many neutral molecules also contain basic sites that will act as acceptors. Aprotic solvents, on the other hand, will be less able to solvate negative ions and basic molecules. Positive ions will ordinarily be solvated by dipolar interactions with the polar solvents, whether protic or not. Protic solutes will ordinarily interact by hydrogen bonding with protic solvents.

Protic and Dipolar Aprotic Solvents

It is useful to classify the more polar solvents ($\epsilon > \sim 15$) into two categories, depending on whether they are protic or aprotic. It is found that reactions involving bases, as for example $S_N 2$ substitutions (Chapter 4), E2 eliminations (Chapter 7), and substitutions at carbonyl groups (Chapter 8), proceed much faster in dipolar aprotic than in protic solvents, typically by factors of three to four powers of ten and sometimes by as much as six powers of ten.[56]

The phenomenon can be explained by considering the various aspects of solvent–solute interactions that we have discussed. The reactions typically take place through the attack of an anionic reagent on a neutral molecule. The protic solvents solvate the anions strongly by hydrogen bonding, whereas the aprotic solvents cannot. Furthermore, although the dipolar aprotic solvents ordinarily have large dipole moments, they are relatively ineffective at solvating the negative ions by dipole interactions because the positive ends of the dipoles are usually buried in the middle of the molecule. The dipolar aprotic solvents are, however, effective at

---

[54] For further discussion of these concepts see Section 3.5.
[55] Note that aprotic solvents may nevertheless contain hydrogen.
[56] A. J. Parker, *Chem. Rev.,* **69,** 1 (1969).

solvating the positive counter ion because the negative end of the dipole is ordinarily an exposed oxygen or nitrogen atom. The result of these differences is that the anions are more free of encumbrance by solvation in the dipolar aprotic solvents, and less energy is required to clear solvent molecules out of the way so that reaction can occur.

Solution Thermodynamics

Considerable insight into solution phenomena can be gained from thermodynamic measurements of the transfer of molecules and ions from the gas phase to solution, or from one solvent to another. The data for ionic compounds are perhaps the most interesting, but thermodynamic properties of individual ions cannot be easily determined. A convention is therefore adopted, based on the assumption that for certain ions, in which the charged center is buried inside a large shell of organic material, there will be no difference in behavior between a positive and a negative ion. Specifically, it is assumed that for any given solution process, such as transfer from one solvent to another, each of the quantities ΔG_{tr}, ΔH_{tr}, and ΔS_{tr} will be the same for the ϕ_4As^+ ion as it is for the ϕ_4B^- ion. This assumption can then be used as a basis for partitioning the thermodynamic data for electrolytes into contributions from the positive and negative ions. Because the partitioning assumption is not thermodynamically rigorous, it is well to use single-ion data cautiously.[57]

With this cautionary note in mind, we can look at free energies of transfer of positive and negative ions from a protic solvent (methanol) to a dipolar aprotic solvent (dimethylformamide).[58] For the chloride ion, the value is $\Delta G_{tr}^{CH_3OH \to DMF} = +8.9$ kcal mole^{-1}; this negative ion is better solvated in the protic solvent methanol. For the sodium ion, $\Delta G_{tr}^{CH_3OH \to DMF} = -5.3$ kcal mole^{-1}; here the ion is better solvated by the dipolar aprotic solvent. These trends are in agreement with the qualitative arguments presented above. We shall examine more data of this kind in Section 4.4 in considering solvent effects on the rates of nucleophilic substitution reactions.

Another example of a way in which thermodynamic data of solutions help to elucidate chemistry is in revealing the importance of hydrogen bonding. Thus the behavior of amines in strongly acidic aqueous media is influenced by the number of protons on the nitrogen in the protonated ion. An ion of the type $Ar\overset{+}{N}H_3$, for example, makes greater demands for hydrogen bonding with surrounding solvent molecules than does $Ar\overset{+}{N}H_2R$; hence the former ions suffer a greater

---

[57] C. V. Krishnan and H. L. Friedman, in *Solute–Solvent Interactions*, J. F. Coetzee and C. D. Ritchie, Eds., Vol. 2, Dekker, New York, 1976, p. 1.

[58] A. J. Parker, *Chem. Rev.*, **69**, 1 (1969). The data are given as transfer activity coefficients, $^\circ\gamma^S$, where superscript $^\circ$ stands for the reference solvent and superscript S for the solvent to which the transfer is made. The conversion to ΔG_{tr} is as follows:

$$\Delta G_{tr}^{\circ \to S} = 2.303 \, RT \log_{10} {}^\circ\gamma^S$$

Data for ΔH_{tr} are also available. See Krishnan and Friedman, note 57.

increase in free energy as the medium becomes more acidic (and free water becomes less available) than do the latter. These points are considered further in Section 3.2.

Water is a unique solvent in many ways, and the reasons for its behavior are only beginning to be understood. The prevailing theory of liquid water structure (not yet definitely established as correct) postulates a three-dimensional, hydrogen-bonded network, with each water molecule tetrahedrally coordinated, similar to the structure in the solid. This structure persists in the liquid over moderate ranges, though it is dynamic in the sense that the boundaries and extent of the structured regions are continually changing. Some of the molecules break away from this structure, the hydrogen bonds re-form to close the gap, and the free molecules remain in the interstices of the lattice. It is this interstitial nature of the structure that accounts for the unique increase of density on melting.[59]

An ion placed in the water structure disrupts it, immobilizing a strongly coordinated shell of water molecules in the solvation sphere. Outside this sphere another layer of water molecules is balanced between the forces tending to align them with the inner solvation sphere and the forces tending to join them to the hydrogen-bonded lattice of the bulk solvent. This intermediate layer is less ordered than either the inner solvation sphere or the bulk solvent. This hypothesis accounts for the curious observation that while the small highly charged ions (Li^+, F^-) cause a net increase in order (negative ΔS_{tr}) on transfer to aqueous solution from the gas phase, as one would expect for formation of a highly oriented solvation sphere, most ions cause the solvent to become less ordered, a change attributed to dominance of the increased randomness in the intermediate layer.[60] It is also interesting to note that nonpolar solutes, such as hydrocarbons, owe their very low water solubility not to energetically unfavorable interactions (the enthalpies of solution are actually exothermic[57]) but to a negative ΔS of solution caused by an increase in the structure of the water around the solute molecule. This entropy effect is responsible for the *hydrophobic interaction* that causes the hydrocarbon parts of large molecules in aqueous solution to cluster together. The hydrophobic interaction (often called hydrophobic bonding) is of great importance in determining the shape that proteins assume in an aqueous environment.[61]

Ion Pairing

In solvents of low to moderate polarity, ions are extensively associated into pairs.[62] Because the reactivity of an ion can be significantly affected by the presence of its counter-ion very nearby, the ion-pairing phenomenon plays an important role in the chemistry of ionic species in these solvents. Examples are the influence of ion

[59] H. S. Frank, in *Water, a Comprehensive Treatise*, F. Franks, Ed., Vol. 1, Plenum, New York, 1972, p. 536.
[60] (a) H. S. Frank and W.-Y. Wen, *Disc. Faraday Soc.*, **24**, 133 (1957); (b) E. A. Arnett and D. R. McKelvey, in *Solute–Solvent Interactions*, J. F. Coetzee and C. D. Ritchie, Eds., Vol. 1, Dekker, New York, 1969, p. 343.
[61] F. Franks, Ed., *Water, a Comprehensive Treatise*, Vol. 4, Plenum, New York, 1975, p. 1.
[62] M. Szwarc, Ed., *Ions and Ion Pairs in Organic Reactions*, Wiley, New York, Vol. 1, 1972; Vol. 2, 1974.

pairing in the behavior of carbocation intermediates in aliphatic nucleophilic substitution and on the chemistry of carbanions, topics that are treated in Chapters 4 and 6, respectively.

The existence of two distinct types of ion pairs has been established: the contact ion pair, M^+X^-, in which cation and anion have no solvent between them, and the solvent-separated ion pair, $M^+||X^-$, in which a molecule of solvent is between the ions. The two types of ion pairs can be detected spectroscopically and the equilibrium between them determined. In the case of conjugated carbanions, visible-ultraviolet spectrophotometry reveals the contact and solvent-separated ion pairs by their different wavelengths of absorption maxima.[63] Nuclear magnetic resonance spectroscopy, both of ^{13}C in organic anions, and of alkali metal cations (particularly $^7Li^+$, $^{23}Na^+$, $^{39}K^+$, $^{87}Rb^+$, $^{133}Cs^+$), is also fruitful in revealing details of the ion-pairing process.[64] An important aspect of ion-pairing behavior is the coordination of cations with electron-donor sites of the solvent. The oxygen atoms in ethers, for example, coordinate to metal ions and tend to surround them, at the same time removing them from contact with the anion by forming a solvent-separated ion pair. The smaller the cation, the higher is its electric field and the stronger is its coordination to a donor solvent; comparative studies of formation

Table 2.16 Equilibrium Constant for the Contact to Solvent-Separated Ion Pair Process[a]

$$F^-M^+ + \text{Solv.} \xrightleftharpoons{K_{SSIP}} F^- \cdot \text{Solv} \cdot M^+$$

Solvent	K_{SSIP}[b]	
	Li^+	Na^+
$H_3CO + (CH_2CH_2O)_2CH_3$	3.1	1.2
$H_3CO + (CH_2CH_2O)_3CH_3$	130	9.0
$H_3CO + (CH_2CH_2O)_4CH_3$	240	170

[a] F^- = fluorenyl anion,

[b] Determined by adding the solvent indicated to a solution of F^-M^+ in dioxane or tetrahydrofuran. From J. Smid, in *Ions and Ion Pairs in Organic Reactions*, Vol. 1, M. Szwarc, Ed., Wiley, New York, 1972, p. 117. Reprinted by permission of John Wiley & Sons, Inc.

[63] J. Smid, Vol. 1, p. 85, of reference 62.
[64] (a) A. I. Popov, in *Solute–Solvent Interactions*, J. F. Coetzee and C. D. Ritchie, Eds., Vol. 2, Dekker, New York, 1976, p. 271; (b) H. W. Vos, C. MacLean, and N. H. Velthorst, *J. Chem. Soc., Faraday Trans. 2*, **73**, 327 (1977); (c) D. H. O'Brien, C. R. Russell, and A. J. Hart, *J. Am. Chem. Soc.*, **101**, 633 (1979).

of solvent-separated ion pairs for lithium and sodium salts reveal this trend. Table 2.16 illustrates this point for lithium and sodium ion pairs of the fluorenyl anion (**67**).

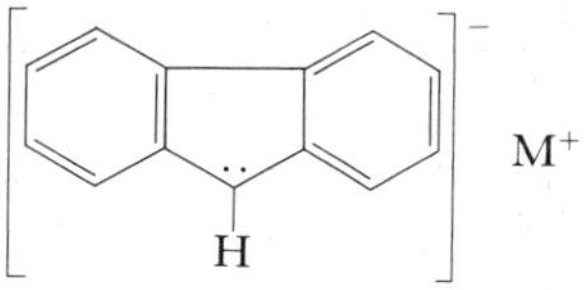

67

The tendency of electron-pair donors to coordinate with cations is exploited in constructing cation complexing agents, particularly the macrocyclic crown ethers of which **68** and **69** are examples, and the macrobicyclic cryptands of which **70** and **71** are examples.[65]

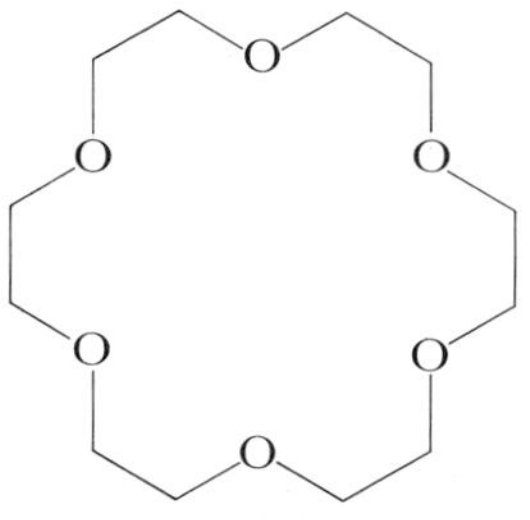

68

18-crown-6

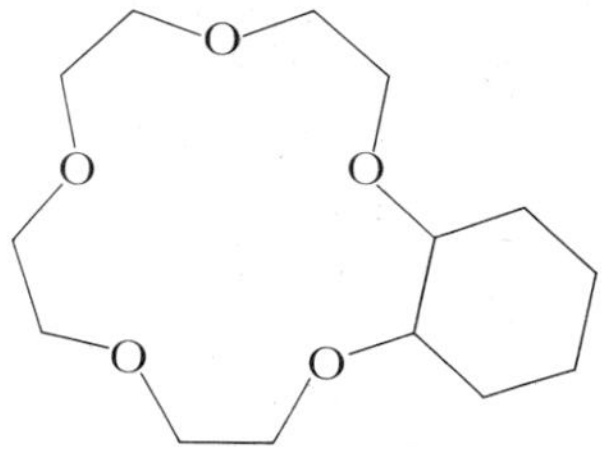

69

cyclohexyl-15-crown-5

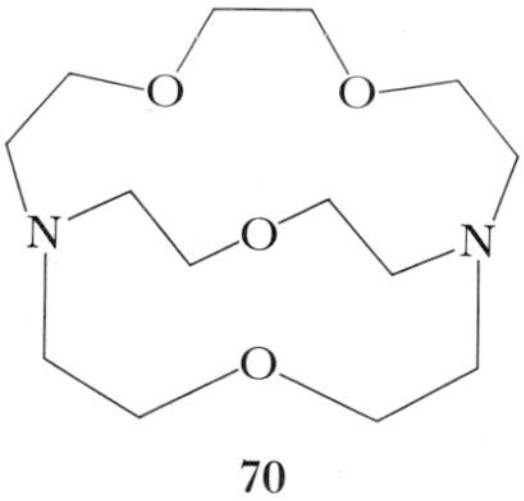

70

cryptand-2.1.1

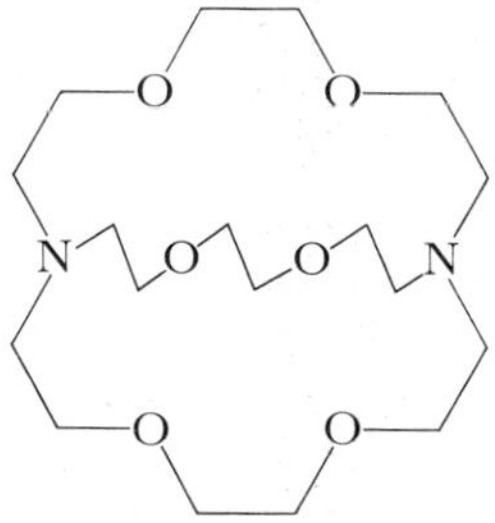

71

cryptand-2.2.2

The nomenclature systems illustrated are in common use for these compounds. The crown ethers are named according to the size of the ring and the number of oxygens, with any fused rings indicated by the prefix; the cryptands are designated according to the number of oxygen atoms in the three bridges. These molecules

[65](a) C. J. Pedersen and H. K. Frensdorff, *Angew. Chem. Int. Ed. Engl.,* **11**, 16 (1972); (b) J.-M. Lehn, *Structure and Bonding,* **16**, 2 (1973); (c) J.-M. Lehn, *Acc. Chem. Res.,* **11**, 49 (1978); (d) J. J. Christensen, D. J. Eatough, and R. M. Izatt, *Chem. Rev.,* **74**, 351 (1974).

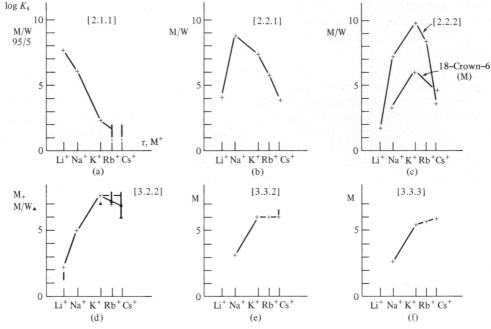

Figure 2.8 Stability constants, log K_s, for cryptate complexes with alkali metal ions. The horizontal axis is linear in ionic radius. The numbers in brackets give the numbers of oxygen atoms in the bridging chains; thus [2.1.1] is **70** and [2.2.2] is **71**. Data for the crown ether 18-crown-6 (**68**) is included in graph (c) for comparison. The solvent is methanol (M) or 95:5 methanol–water (M/W) at 25°C. Adapted with permission from J.-M. Lehn, *Acc. Chem. Res.,* **11**, 49 (1978). Copyright 1978 American Chemical Society. Data for 18-crown-6 are from J.-M. Lehn, *Structure and Bonding,* **16**, 1 (1973).

have a strong affinity for cations of the proper size to fit the cavity. The cryptands, which take the cation into the space bounded by the three bridges to form complexes called cryptates, are particularly effective. They show a high selectivity among the alkali metal cations according to size. Anion cryptates are also known.[66] Figure 2.8 shows the stability constants for several cryptands with alkali metal ions and compares the cryptands with a crown ether.

Observations of the influence of solvation and ion pairing on the behavior of ions has led to the development of a particularly effective method, known as phase transfer catalysis, of carrying out reactions rapidly and in high yield between organic molecules and ionic reagents.[67] The organic substrate is dissolved in a nonpolar organic solvent, and the reagent, an ionic substance of which the anion is to

[66] J.-M. Lehn, E. Sonveaux, and A. K. Willard, *J. Am. Chem. Soc.,* **100**, 4914 (1978).

[67] (a) W. P. Weber and G. W. Gokel, *Phase Transfer Catalysis in Organic Synthesis,* Springer-Verlag, Berlin, 1977; (b) R. A. Jones, *Aldrichimica Acta,* **9** (3), 35 (1976); (c) C. M. Starks and C. Liotta, *Phase Transfer Catalysis: Principles and Techniques,* Academic Press, New York, 1978.

act as a nucleophile, base, reducing or oxidizing agent, is dissolved in an aqueous phase. A phase transfer catalyst, a substance with a cation containing several hydrocarbon groups (such as a tetraalkyl ammonium or a trialkylphosphonium halide) is added. The phase transfer cations form ion pairs with the reagent anion and the pairs move into the organic layer because of the large hydrocarbon substituents on the cation. In the organic phase the anion is solvated only very weakly by the nonpolar solvent molecules and is paired with a large cation of low electric field intensity. It is consequently highly reactive. The phase transfer cations can continue to migrate back and forth between the phases, bringing more anionic reagent to the reacting system. Crown ethers can also function as phase transfer catalysts by complexing a cation and then bringing the complexed–cation-anion pair into the nonpolar phase.

Solvent Polarity Scales

A number of scales have been established in efforts to quantify the influence of solvents on chemical properties. Some of these are restricted to characterizing solvent effects in particular reaction types. Perhaps the best known of these is the Grunwald–Winstein $m\mathbf{Y}$ scale, which applies to processes in which an intermediate with positive charge is being generated. We shall consider the uses of this scale and its extensions in Chapter 4. Of a more general character are scales that attempt to quantify solvent behavior in a wider variety of circumstances. We shall consider three of these scales briefly here. The data are included in Table 2.15.

A very simple scale, proposed by Hildebrand, is the δ scale. The parameter δ for a substance (whether solute or solvent) is defined by Equation 2.45, where ΔH_v is the latent heat of vaporization and $\overline{V}$ is the molar volume. The activity coefficient, γ_i, of a solute in a solution is then given by Equation 2.46, where $\overline{V}_i$ is the molar volume of solute and δ_i and δ are parameters for solute and solvent, respectively. This treatment applies to nonpolar solutes in nonpolar solvents.[68]

$$\delta = \left(\frac{H_v - RT}{\overline{V}}\right)^{1/2} \tag{2.45}$$

$$RT \ln \gamma_i = \overline{V}_i(\delta_i - \delta)^2 \tag{2.46}$$

Another approach is to use solvent effects on ultraviolet absorption maxima of appropriately chosen salts to characterize a solvent. Kosower's $\mathbf{Z}$ scale,[69] and the Reichardt–Dimroth E_t scale[70] are of this type, based on electronic transition energies of the pyridinium iodide **72** and of the pyridinium zwitterion **73**, respectively. The absorption maxima of these compounds are sensitive to solvent because the absorption is a charge-transfer transition, in which an electron is transferred from the anion (or the negatively charged portion of the molecule in **73**) to the cation.

[68] M. H. Abraham, *Prog. Phys. Org. Chem.*, **11**, 1 (1974).
[69] E. M. Kosower, *An Introduction to Physical Organic Chemistry*, Wiley, New York, 1968, p. 293.
[70] (a) K. Dimroth, C. Reichardt, T. Siepmann, and E. Bohlmann, *Justus Liebigs Ann. Chem.*, **661**, 1 (1963); (b) C. Reichardt, *Angew. Chem. Int. Ed. Engl.*, **4**, 29 (1965); (c) C. Reichardt, *Justus Liebigs Ann. Chem.*, **752**, 64 (1971).

The ground state is highly polar, and the excited state is nonpolar. The ground-state energy is therefore much more affected by the solvent than is the excited state, and solvent polarity differences show up as shifts in the positions of bands in the spectra. Taft and co-workers have applied this approach in a somewhat more general way by analyzing absorption data from a number of different absorbing species to produce the π^* scale, so named because it is based on the solvent effect on the energy required to promote an electron to an antibonding π^* orbital.[71] Values are included in Table 2.15.

A different approach to the quantification of solvent properties is to measure the ability of the solvent to interact with specific substances. Gutmann has developed the scales of donor number (DN) and acceptor number (AN) to characterize a solvent's ability to donate electron density to electron-deficient centers and to accept electron density from electron-rich centers.[72] (These interactions are of the Lewis acid–base type. See Section 3.5 for further discussion.) Donor number is defined as the negative of the enthalpy of reaction of the substance in question with the strong Lewis acid (electron-pair acceptor) $SbCl_5$, both compounds being in dilute solution in 1,2-dichloroethane. The acceptor number AN is defined as the ^{31}P nmr chemical shift observed for the reference base $(C_2H_5)_3PO$ when dissolved in the solvent in question, relative to the chemical shift observed in the solvent hexane. Values of DN and AN are also included in Table 2.15.

2.5 KINETICS AND MECHANISM[73]

A *reaction mechanism* is a specification, by means of a sequence of elementary chemical steps, of the detailed process by which a chemical change occurs. An *elementary reaction step* is a process in which the reacting molecule or molecules pass through a single transition state to products by means of a simple set of motions of nuclei occurring simultaneously without the intervention of any intermediate. An *intermediate* is any chemical structure occurring during a reaction (necessarily having two

[71](a) M. J. Kamlet, J. L. Abboud, and R. W. Taft, *J. Am. Chem. Soc.*, **99**, 6027 (1977); (b) J. L. Abboud, M. J. Kamlet, and R. W. Taft, *J. Am. Chem. Soc.*, **99**, 8325 (1977).
[72](a) V. Gutmann, *Coord. Chem. Rev.*, **18**, 225 (1976); (b) W. B. Jensen, *Chem. Rev.*, **78**, 1 (1978).
[73](a) A. A. Frost and R. G. Pearson, *Kinetics and Mechanism*, 2d ed., Wiley, New York, 1961; (b) K. J. Laidler, *Chemical Kinetics*, 2d ed., McGraw-Hill, New York, 1965; (c) C. Capellos and B. H. J. Bielski, *Kinetic Systems*, Wiley, New York, 1972; (d) J. F. Bunnett, in *Investigation of Rates and Mechanisms of Reactions*, 3d ed., E. S. Lewis, Ed., Vol. VI, Part I, of *Techniques of Chemistry*, A. Weissberger, Ed., Wiley, New York, 1974, Chaps. IV and VIII.

or more elementary steps) that exists at a minimum of energy (i.e., for which any distortion raises the energy) and has a lifetime longer than the period of typical molecular vibrations (on the order of 10^{-13} to 10^{-14} sec). Each elementary reaction step is characterized by forward and reverse *microscopic rate constants*. A reaction mechanism may include detailed proposals about such things as structure, stereochemistry, and energetics at various stages of the process.

A reaction mechanism is a hypothesis, that is, a theoretical construct or model proposed to explain a chemical change. From the hypothesis certain consequences may be deduced and these consequences may be checked by experiment. Experimental results in agreement with the consequences are evidence in support of the proposed mechanism, but a mechanism may never be said to be proved. A mechanism becomes established through the accumulation of experimental evidence consistent with it, but there is never any guarantee against some future experiment revealing an inconsistency that will require the mechanism to be discarded or modified. Of course, mechanistic chemistry is seldom as clear-cut as these statements imply because experimental results must often be interpreted through the use of other theoretical concepts. It is therefore not unusual for disagreements to arise as to whether a given experimental result is or is not consistent with a particular mechanism.

There are a number of kinds of experimental information that may be used in testing mechanisms.[74] Among these are careful identification of product structure (including search for minor products); tests of reaction stereochemistry; testing of suspected intermediates by independent synthesis and subjection to reaction conditions; isolation, trapping, or spectroscopic detection of intermediates; use of substituents to block or modify suspected reactive sites; and isotopic labeling. We shall encounter examples of these techniques in subsequent chapters.

The most important single tool for mechanistic investigation is probably kinetics. Measurements of reaction rates are made for the purpose of verifying the consistency of the proposed pattern of reaction steps through dependence of rates on concentration, and also in order to derive activation parameters, to test the effect of changes of substituents or solvent, to determine kinetic isotope effects, and to evaluate differences between related structures. In the following paragraphs we shall discuss some of the basic principles of reaction kinetics.

Molecularity

For convenience in discussing various types of reactions, the terms *unimolecular* and *bimolecular* (also *termolecular,* though this type is uncommon) are used to characterize elementary reaction steps of a mechanism. A unimolecular step is one that involves, on the reactant side, only one chemical species. A bimolecular step involves the interaction of two reactants. These terms may be illustrated by the following examples.

---

[74] A discussion of techniques used in mechanistic investigations may be found in E. S. Gould, *Mechanism and Structure in Organic Chemistry,* Holt, Rinehart and Winston, New York, 1959, Chaps. 5 and 6.

trans-3,4-Dimethylcyclobutene (**74**) reacts when heated at 175°C to produce *trans,trans*-2,4-hexadiene (**75**). This reaction is stereospecific; **74** produces only **75**,

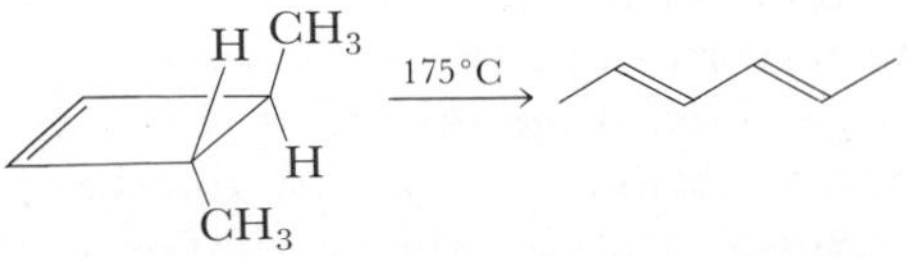

$$\text{(structure 74)} \xrightarrow{175°C} \text{(structure 75)} \qquad (2.47)$$

74   **75**

and the diastereomer, *cis*-3,4-dimethylcyclobutene (**76**) yields only *cis,trans*-2,4-hexadiene (**77**).[75]

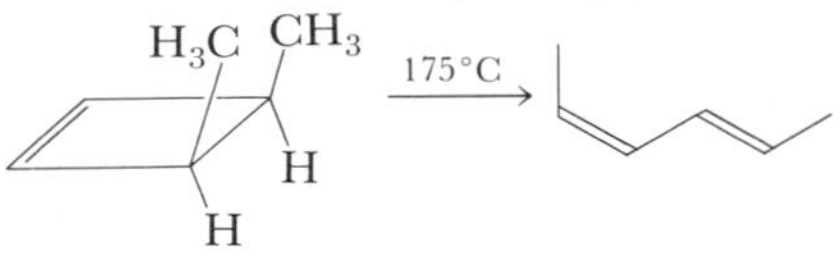

$$\text{(structure 76)} \xrightarrow{175°C} \text{(structure 77)} \qquad (2.48)$$

76   **77**

A possible mechanism for Reaction 2.47 (and the one currently accepted—see Chapter 10 for further discussion) consists of a single unimolecular step (Equation 2.49), in which molecules of **74** spontaneously isomerize without the intervention of any other reactant.

$$74 \xrightarrow{k_a} 75 \qquad (2.49)$$

Methyl iodide in the presence of sodium ethoxide in ethanol yields methyl ethyl ether; here a possible mechanism (again the one currently accepted—see Chapter 4) consists of a single bimolecular step (Equation 2.50) in which an encounter between a methyl iodide molecule and an ethoxide ion must occur. The microscopic rate constants are symbolized by k_a and k_b respectively, in Equations 2.49 and 2.50.

$$CH_3I + C_2H_5O^- \xrightarrow{k_b} CH_3OC_2H_5 + I^- \qquad (2.50)$$

As these examples illustrate, molecularity is a concept related to mechanism but applying only to single elementary reaction steps. A mechanism consisting of a sequence of steps could not properly be termed unimolecular, bimolecular, or termolecular, although this terminology might correctly be applied to individual steps of such a mechanism.

The Rate Equation

Once a sequence of elementary reaction steps constituting a mechanistic hypothesis has been proposed, rate equations can be obtained.[76] A differential rate equation relates the rate of change of some concentration to concentrations and to the micro-

---

[75] R. E. K. Winter, *Tetrahedron Lett.*, 1207 (1965).
[76] At least in principle. The algebraic manipulations required can become formidable.

scopic rate constants. Rate equations may be written in terms of rate of decrease of a concentration (i.e., $-d[X]/dt$) or in terms of rate of increase of a concentration (i.e., $+d[X]/dt$). A rate equation may be written for any of the components of the reaction; the object is to obtain a rate equation characterizing the reaction that contains only concentrations that can be determined experimentally. Each elementary reaction step in which a component appears will contribute a term to the rate equation for that component. Each term in a rate equation corresponds to an elementary reaction step and consists of the product of the microscopic rate constant of that step and the concentration of each of the species that come together to form the transition state of that step. For Reaction 2.49 the following rate equation may be written:

$$-\frac{d[74]}{dt} = k_a[74] \tag{2.51}$$

For Reaction 2.50

$$-\frac{d[CH_3I]}{dt} = k_b[CH_3I][C_2H_5O^-] \tag{2.52}$$

For a more complex scheme

$$A \underset{k_{-1}}{\overset{k_1}{\rightleftharpoons}} X + Y \tag{2.53}$$

$$Y \xrightarrow{k_2} C + D \tag{2.54}$$

one could write

$$-\frac{d[A]}{dt} = k_1[A] - k_{-1}[X][Y] \tag{2.55}$$

$$\frac{d[Y]}{dt} = k_1[A] - k_{-1}[X][Y] - k_2[Y] \tag{2.56}$$

$$\frac{d[C]}{dt} = k_2[Y] \tag{2.57}$$

Some, but not all, rate equations may be characterized by their *kinetic order*. The order n of a rate equation consisting of a single term on the right-hand side is the sum of the exponents of all the concentrations appearing in that term. Thus Equations 2.51 and 2.57 are first-order and Equation 2.52 is second-order; but order is undefined for Equations 2.55 and 2.56. One may also speak of the order with respect to a given component, which is the exponent of the concentration of that component. Thus Equation 2.52 is second-order overall, first-order in CH_3I, and first-order in $C_2H_5O^-$. The hypothetical Equation 2.58 is third-order overall, second-order in A, and first-order in B. One may also speak of the order of individ-

$$-\frac{d[A]}{dt} = k_1[A]^2[B] \tag{2.58}$$

ual terms in a rate equation. If an equation has only first-order terms, it is first-order; if only second-order terms, it is second-order. However, if terms of different orders appear, or if the rate expression contains terms that are not simple products of concentrations, the overall order is undefined.

Note that a unimolecular reaction step leads to a first-order equation, and a bimolecular reaction step leads to a second-order equation. The converse is not necessarily true, however; it is quite possible, for example, to propose a mechanism of more than one step involving several species that will nevertheless yield a first-order rate equation. It is therefore important to distinguish between molecularity, which is a concept of mechanism, and kinetic order, which is a property of rate equations and their solutions and an experimentally determinable quantity.

Experimental kinetics consists of finding the form of rate equations and determining rate constants by measuring concentrations as a function of time. Reaction 2.47, for example, could be investigated by measuring rate of change of concentration of the dimethylcyclobutene **74** for various concentrations of **74**; Equation 2.51 would be verified by finding that the rate is always proportional to the first power of concentration. In examining experimental data it is usually more convenient to have the equation in integrated form. The solution to the differential equation 2.51 is

$$[\textbf{74}] = [\textbf{74}]_0 e^{-k_a t} \tag{2.59}$$

or

$$\ln \frac{[\textbf{74}]}{[\textbf{74}]_0} = -k_a t \tag{2.60}$$

where $[\textbf{74}]_0$ is the initial concentration of **74**. Hence according to the proposed mechanism, a plot of the logarithm of the concentration ratio $\ln [\textbf{74}]/[\textbf{74}]_0$ vs. time should be a straight line with slope $-k_a$. The integrated form of Equation 2.52 (assuming unequal initial concentrations of the two reactants) is

$$\frac{1}{[CH_3I]_0 - [C_2H_5O^-]_0} \ln \frac{[C_2H_5O^-]_0[CH_3I]}{[CH_3I]_0[C_2H_5O^-]} = k_b t \tag{2.61}$$

which may be verified by plotting the left-hand side against t. Integrated forms of other types of kinetic equations may be found in the literature.[77]

Reversible Reactions

The mechanisms proposed in Equations 2.49 and 2.50 show reactions proceeding to the right only. Every elementary reaction must be reversible; this statement is known as the *principle of microscopic reversibility,* and to violate it is to violate the laws of thermodynamics. However, if one heats pure *trans,trans*-1,4-hexadiene to 175°C, no detectable amount of dimethylcyclobutene is formed; similarly, methyl ethyl ether with iodide in ethanol does not produce any observable amount of methyl iodide. These reverse reactions, although microscopic reversibility tells us that they

---

[77] See notes 73(a) and (c).

must occur, are so much slower than the forward reactions that as a practical matter we may ignore them. The omission of a reverse arrow in a reaction scheme is part of the mechanistic hypothesis; it means that for that step we are proposing that the reverse reaction is sufficiently slow that it does not occur to any measurable extent during the time the system will be under observation. Note that because the equilibrium constant for an elementary reaction step and the rate constants are related by Equation 2.62, the neglect of k_{-1} compared with k_1 is equivalent to postulating $K_{eq} \gg 1$, or that the products on the right have a much lower free energy than the reactants on the left.

$$K_{eq} = \frac{k_1}{k_{-1}} \tag{2.62}$$

In many cases neglect of reverse reactions is not justified. A simple illustration is Equation 2.63. Cope found that heating either 1,3,5-cyclooctatriene (**78**) or

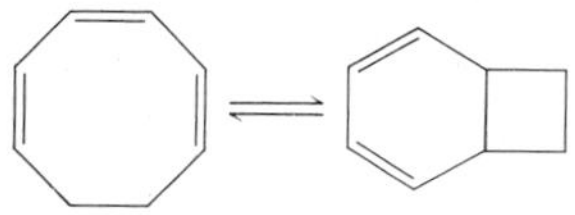

$$\tag{2.63}$$

78   79

bicyclo[4.2.0]octa-2,4-diene (**79**) at 100°C for one hour produced the same mixture containing 85 percent of **78** and 15 percent of **79**.[78] Clearly a mechanistic hypothesis for this reaction must include the reverse reaction pathway. If formulated as a single-step process, the mechanism must be

$$\textbf{78} \underset{k_{-1}}{\overset{k_1}{\rightleftharpoons}} \textbf{79} \tag{2.64}$$

For this mechanism one can write rate equations as follows:

$$\frac{d[\textbf{78}]}{dt} = -k_1[\textbf{78}] + k_{-1}[\textbf{79}] \tag{2.65}$$

$$\frac{d[\textbf{79}]}{dt} = k_1[\textbf{78}] - k_{-1}[\textbf{79}] \tag{2.66}$$

If the concentration of each species is expressed relative to the equilibrium concentration of that species, the solutions to Equations 2.65 and 2.66 take the form

$$\ln \frac{\Delta[\textbf{78}]}{\Delta[\textbf{78}]_0} = -(k_1 + k_{-1})t \tag{2.67}$$

$$\ln \frac{\Delta[\textbf{79}]}{\Delta[\textbf{79}]_0} = -(k_1 + k_{-1})t \tag{2.68}$$

Here the symbol $\Delta[\textbf{78}]$ means the difference between the observed concentration of **78** at time t and its equilibrium concentration, and $\Delta[\textbf{78}]_0$ is the difference between

[78] A. C. Cope, A. C. Haven, Jr., F. L. Ramo, and E. R. Trumbull, *J. Am. Chem. Soc.*, **74**, 4867 (1952).

initial and equilibrium concentrations of **78**.[79] Note that with concentrations measured relative to the equilibrium concentration, Equations 2.67 and 2.68 have exactly the same form as Equation 2.60 for the first-order reaction with reverse reaction omitted. If $\ln \Delta[\mathbf{78}]/\Delta[\mathbf{78}]_0$ is plotted vs. t for the reversible reaction, however, the negative of the slope is not identifiable as one of the microscopic rate constants. We refer to the empirically determined rate constant, found as the slope of a graph of an appropriate function of concentrations vs. time, as the *macroscopic* or *observed* rate constant, k_{obs}. Only for particularly simple mechanisms will k_{obs} be identifiable directly with a microscopic rate constant for an elementary chemical step.

Pseudo First-Order Kinetics

If one of the components of a second-order reaction remains effectively constant throughout a kinetic run, the observed behavior of the system will be first order. For example, suppose that the reaction of methyl iodide with ethoxide were set up so that initial concentration $[CH_3I]_0 = 0.01M$ and initial concentration $[C_2H_5O^-]_0 = 1.00M$. If we follow the reaction until the concentration of CH_3I has decreased to $0.0001M$, a change of a factor of 100, the $C_2H_5O^-$ concentration will have decreased to $0.99M$, a change of a factor of 1.01. Clearly the $C_2H_5O^-$ concentration is effectively constant throughout the experiment because $C_2H_5O^-$ is present in large excess. This situation will also arise when one of the reactants is a catalyst that is continuously regenerated; or if there is buffering action, as when hydrogen ion concentration is kept constant by an acid–base buffer, or if one of the reactants is the solvent. The second-order rate equation for the methyl iodide–ethoxide reaction,

$$\frac{d[CH_3I]}{dt} = k_b[C_2H_5O^-][CH_3I] \tag{2.69}$$

can be rewritten as follows:

$$\frac{d[CH_3I]}{dt} = k_{obs}[CH_3I] \tag{2.70}$$

where

$$k_{obs} = k_b[C_2H_5O^-] \tag{2.71}$$

Experimentally, one will find first-order kinetics with rate constant k_{obs}. From this kinetic run alone one would never know that $C_2H_5O^-$ was involved in the reaction at all; but if one tries another kinetic run with a different (but still large) concentration of $C_2H_5O^-$, say $1.5M$, one will find a different value of k_{obs}. Such a system is said to follow *pseudo first-order* kinetics. It is also possible to have pseudo second-order kinetics, which would occur if a third- (or higher-) order reaction showed second-order behavior because of constant concentration of one (or more) reactant. Investigators frequently arrange conditions deliberately so that reactions will follow pseudo first-order kinetics, because first-order data is easier to treat than data from

[79] G. M. Fleck, *Chemical Reaction Mechanisms,* Holt, Rinehart and Winston, New York, 1971.

higher-order systems. The higher-order rate constants are then found by doing several experiments at different constant concentrations of the reagents that are in excess or buffered. For example, in the CH_3I–$C_2H_5O^-$ reaction, k_b would be obtained by measuring k_{obs} at a number of different constant concentrations of $C_2H_5O^-$ and then plotting k_{obs} vs. $[C_2H_5O^-]$.

Kinetic and Thermodynamic Control

When two or more products are formed in a reaction, it is frequently of interest to determine their relative rates of formation. For example, 2-bromobutane reacting with a mixture of water and ethanol will yield a mixture of 2-butanol and 2-ethoxybutane (Equation 2.72).

$$\text{(structure, Br)} \xrightarrow{\text{C}_2\text{H}_5\text{OH, H}_2\text{O}} \text{(structure, OH)} + \text{(structure, OC}_2\text{H}_5) + \text{HBr} \tag{2.72}$$

One might propose competing bimolecular single-step reactions of 2-bromobutane with water and ethanol, respectively (Equations 2.73 and 2.74) and interpret the ratio of alcohol to ether in the product to be the ratio k_1/k_2. This procedure implic-

$$\text{(structure, Br)} + \text{H}_2\text{O} \xrightarrow{k_1} \text{(structure, OH)} + \text{HBr} \tag{2.73}$$

$$\text{(structure, Br)} + \text{C}_2\text{H}_5\text{OH} \xrightarrow{k_2} \text{(structure, OC}_2\text{H}_5) + \text{HBr} \tag{2.74}$$

itly assumes that the reaction is subject to *kinetic control,* that is, that the relative amounts of products formed are determined by the relative rates of formation of those products. Suppose, however, that the mechanism is actually as shown in Equations 2.75 and 2.76, and the product molecules, once formed, have time to

$$\text{(structure, Br)} + \text{H}_2\text{O} \underset{k_{-1}}{\overset{k_1}{\rightleftharpoons}} \text{(structure, OH)} + \text{HBr} \tag{2.75}$$

$$\text{(structure, Br)} + \text{C}_2\text{H}_5\text{OH} \underset{k_{-2}}{\overset{k_2}{\rightleftharpoons}} \text{(structure, OC}_2\text{H}_5) + \text{HBr} \tag{2.76}$$

revert to the bromide while the system is under observation. Under these circumstances equilibrium will be established among the components of the mixture, and the product ratio will reflect the free-energy difference between the alcohol and the ether, not the ratio k_1/k_2. The reaction would then be subject to *thermodynamic control,* in which the product ratio is determined by the thermodynamics, that is, by the relative free energies. In the example cited kinetic control might pertain under one set of conditions and thermodynamic control under another, because the ether and the alcohol would be expected to be quite unreactive toward bromide ion in neutral solution but considerably more reactive in strongly acidic solution. The wise experimentalist therefore checks that samples of the individual products remain unchanged under the reaction conditions before drawing conclusions based on the assumption of kinetic control.

Multistep Reactions

Reaction mechanisms of two or more steps are often analyzed with the aid of simplifying assumptions.

The stationary-state assumption One might imagine the following mechanism for the reaction of methyl iodide with water (probably not the correct one—see Chapter 4):

$$CH_3I \underset{k_{-1}}{\overset{k_1}{\rightleftharpoons}} CH_3^+ + I^- \tag{2.77}$$

$$CH_3^+ + H_2O \underset{k_{-2}}{\overset{k_2}{\rightleftharpoons}} CH_3OH + H^+ \tag{2.78}$$

One might then argue as follows. The cation CH_3^+ will probably react much faster with H_2O than with I^-, and the product CH_3OH is unlikely under the conditions of the reaction to revert to CH_3^+ to any significant extent during the time the reaction is under observation. We would then be justified in omitting both the k_{-1} and the k_{-2} arrows, to give 2.79 and 2.80:

$$CH_3I \overset{k_1}{\longrightarrow} CH_3^+ + I^- \tag{2.79}$$

$$CH_3^+ + H_2O \overset{k_2}{\longrightarrow} CH_3OH + H^+ \tag{2.80}$$

One might then assume further that the ion CH_3^+ will be very reactive, so that whenever a CH_3^+ ion is formed it reacts extremely rapidly with H_2O. If this assumption is correct, the CH_3^+ concentration will always be very small, and a concentration that is always very small cannot be changing very much. Hence we can conclude

$$\frac{d[CH_3^+]}{dt} \approx 0 \tag{2.81}$$

All the assumptions we have made about the reaction serve to simplify the kinetic analysis, but they are also part of the mechanism being proposed. Assumptions that simplify kinetics must find their ultimate justification in the known chemical properties of the substances involved.

By proposing that the second step of the reaction is much faster than the first, we have introduced the idea of a *rate-determining step*. The first step (2.79) acts as the bottleneck for the process. In any reaction that has a true rate-determining step the kinetics will reflect events up to and including the rate-determining step but will be unaffected by events that follow the rate-determining step.

The assumption of constant concentration $[CH_3^+]$ (Equation 2.81) is known as the *stationary-state approximation* (sometimes called the steady-state approximation). We use it as follows, by writing the rate equation for $[CH_3^+]$:

$$0 \approx \frac{d[CH_3^+]}{dt} = k_1[CH_3I] - k_2[H_2O][CH_3^+] \tag{2.82}$$

Solving for $[CH_3^+]$:

$$[CH_3^+] = \frac{k_1[CH_3I]}{k_2[H_2O]} \tag{2.83}$$

Now look at the rate equation for formation of the product CH_3OH:

$$\frac{d[CH_3OH]}{dt} = k_2[H_2O][CH_3^+] \tag{2.84}$$

This equation is not practical for use directly because checking its validity would require measurement of $[CH_3^+]$, a very difficult experimental task. But by combining Equations 2.83 and 2.84 we obtain 2.85, which can be checked experimentally.

$$\frac{d[CH_3OH]}{dt} = k_1[CH_3I] \tag{2.85}$$

Let us relax one of the restrictions on the proposed mechanism by allowing the possibility of reaction between CH_3^+ and I^-, but reactions of CH_3^+ are still assumed to be fast compared with its formation. The mechanism is now

$$CH_3I \underset{k_{-1}}{\overset{k_1}{\rightleftharpoons}} CH_3^+ + I^- \tag{2.86}$$

$$CH_3^+ + H_2O \overset{k_2}{\longrightarrow} CH_3OH + H^+ \tag{2.87}$$

Because the rates of the k_{-1} and the k_2 steps are comparable in this mechanism, there is no true rate-determining step; as we shall see, the kinetics will reflect events taking place in both steps.

The stationary-state approximation can still be used in analyzing mechanism 2.86–2.87. This approximation is appropriate whenever the reactions by which an intermediate disappears are (collectively) very fast compared with the reactions (collectively) by which it is formed. The expression for rate of change of $[CH_3^+]$ is now

$$0 \approx \frac{d[CH_3^+]}{dt} = k_1[CH_3I] - k_{-1}[I^-][CH_3^+] - k_2[H_2O][CH_3^+] \tag{2.88}$$

or

$$[CH_3^+] = \frac{k_1[CH_3I]}{k_{-1}[I^-] + k_2[H_2O]} \tag{2.89}$$

and

$$\frac{d[CH_3OH]}{dt} = k_2[H_2O][CH_3^+] \tag{2.90}$$

$$= \frac{k_1[CH_3I]}{1 + \dfrac{k_{-1}[I^-]}{k_2[H_2O]}} \tag{2.91}$$

As these examples illustrate, the stationary-state approximation is used to eliminate from a rate equation a concentration of a highly reactive intermediate. It is applied by writing the rate equation for the intermediate, equating its rate of change to zero, and solving for the intermediate concentration in terms of concentrations of the other species.

Preliminary Equilibrium

Let us look once again at the methyl iodide–water reaction, and this time assume that both steps k_1 and k_{-1} are fast compared with step k_2. This statement is equivalent to saying $k_1[CH_3I] \approx k_{-1}[CH_3^+][I^-] \gg k_2[CH_3^+][H_2O]$.

$$CH_3I \underset{k_{-2}}{\overset{k_1}{\rightleftharpoons}} CH_3^+ + I^- \tag{2.92}$$

$$CH_3^+ + H_2O \overset{k_2}{\longrightarrow} CH_3OH + H^+ \tag{2.93}$$

The second step is rate determining in this process. If this proposal is correct, we may say that CH_3I is in equilibrium with CH_3^+ and I^-, because when CH_3^+ and I^- form, they almost always go back to CH_3I, and only very rarely to CH_3OH. We therefore write an equilibrium constant for Reaction 2.92:

$$K_{eq} = \frac{[CH_3^+][I^-]}{[CH_3I]} \tag{2.94}$$

or

$$[CH_3^+] = K_{eq}\frac{[CH_3I]}{[I^-]} \tag{2.95}$$

Then for the rate of product formation,

$$\frac{d[CH_3OH]}{dt} = k_2[H_2O][CH_3^+] \tag{2.96}$$

and

$$\frac{d[CH_3OH]}{dt} = K_{eq}k_2[CH_3I]\frac{[H_2O]}{[I^-]} \tag{2.97}$$

A mechanism of this type is said to have a preliminary equilibrium. Again, the relationship introduced by the assumption is used to eliminate a concentration from the rate equation. Note that the rate equation includes both the events of the rate-determining step (the second) and of the step preceding it.

It is, of course, not reasonable to expect that all the mechanisms we have discussed would be observable for reactions of methyl iodide under the limited range of conditions we have mentioned. It is nevertheless interesting that all these mechanisms (or mechanisms closely related to them) have either been established or been under serious consideration for substitution reactions not very different from the ones considered here. These reactions are discussed further in Chapter 4.

More Complex Systems

We shall illustrate the application of the methods discussed above in an example from the work of Cordes and Jencks.[80] The reaction is hydrolysis of an imine, or Schiff's base (Equation 2.98). Although the overall reaction looks simple enough, a

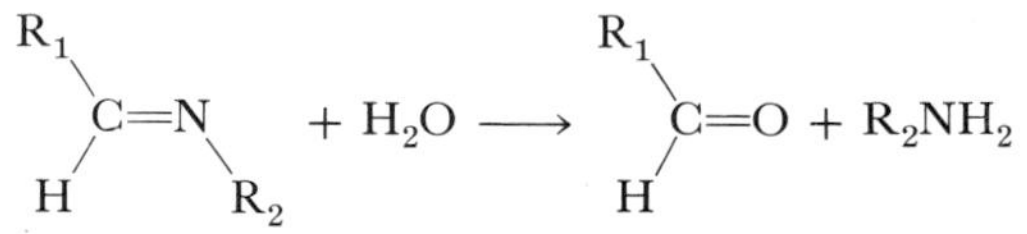

$$
\begin{array}{c}
R_1 \\
\diagdown \\
C{=}N \\
\diagup \quad \diagdown \\
H \qquad R_2
\end{array}
+ H_2O \longrightarrow
\begin{array}{c}
R_1 \\
\diagdown \\
C{=}O \\
\diagup \\
H
\end{array}
+ R_2NH_2
\qquad (2.98)
$$

rather complex dependence of rate on pH suggests that H^+ and OH^- ions are somehow involved. The mechanism proposed is shown in Scheme 1. (We shall discuss reactions of this general type in Chapter 8.) In Scheme 1 we have introduced labels for the various components to make writing kinetic equations less

SCHEME 1

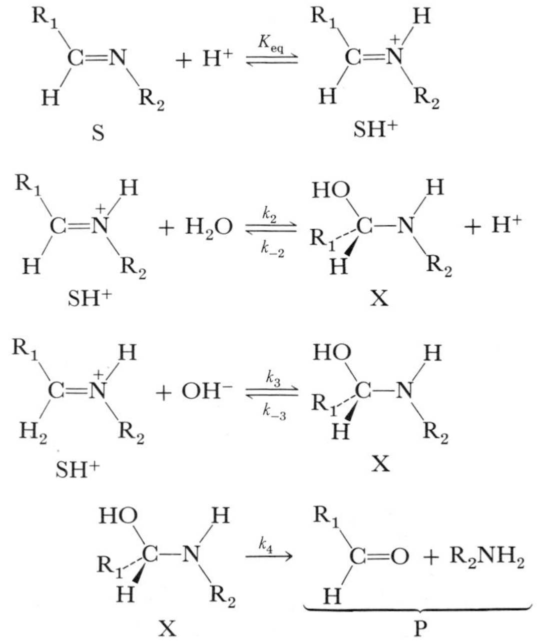

(1)

(2)

(3)

(4)

[80] E. H. Cordes and W. P. Jencks, *J. Am. Chem. Soc.*, **85**, 2843 (1963).

cumbersome. Note that the two products, an aldehyde and an amine, are necessarily formed at the same rate; the reaction may be followed by monitoring concentration of either one, and so for purposes of kinetic analysis we may simply refer to product as P, which can mean either of these substances.

The rate of product formation is easily expressed in terms of intermediate X (called a carbinolamine):

$$\frac{d[P]}{dt} = k_4[X] \tag{2.99}$$

Because carbinolamines are very reactive, it is reasonable to assume that X will be in the stationary state, and we may therefore write

$$0 \approx \frac{d[X]}{dt} = k_2[H_2O][SH^+] - k_{-2}[H^+][X] + k_3[OH^-][SH^+] - k_{-3}[X] - k_4[X] \tag{2.100}$$

Solving for [X],

$$[X] = \frac{[SH^+]\{k_2[H_2O] + k_3[OH^-]\}}{k_{-2}[H^+] + k_{-3} + k_4} \tag{2.101}$$

and substituting into 2.99;

$$\frac{d[P]}{dt} = \frac{k_4[SH^+]\{k_2[H_2O] + k_3[OH^-]\}}{k_{-2}[H^+] + k_{-3} + k_4} \tag{2.102}$$

We want the rate equation expressed in terms of the substrate S itself, rather than in terms of the protonated form, and to do this we take advantage of the preliminary equilibrium of Step (1). This is an acid-base reaction, and K_{eq} is the reciprocal of the acid dissociation constant of the acid SH^+:

$$K_{eq} = \frac{[SH^+]}{[S][H^+]} = \frac{1}{K_{a_{SH^+}}} \tag{2.103}$$

Because it is customary to discuss such reactions in terms of the K_a, we shall use the equilibrium equation in the form of Equation 2.104:

$$K_{a_{SH^+}} = \frac{[S][H^+]}{[SH^+]} \tag{2.104}$$

If the acid SH^+ were very strong, so that nearly all the substrate was in its unprotonated form S, it would suffice to solve Equation 2.104 for $[SH^+]$ and substitute into 2.102. However, in this instance a significant proportion of the substrate may be protonated. The rate equation should be expressed in terms of *total* substrate present in all forms, $[S_T]$. Noting that

$$[S_T] = [S] + [SH^+] \tag{2.105}$$

we find

$$K_{a_{SH^+}} = \frac{\{[S_T] - [SH^+]\}[H^+]}{[SH^+]} \tag{2.106}$$

Solving for [SH$^+$],

$$[SH^+] = \frac{[S_T][H^+]}{K_{a_{SH^+}} + [H^+]}$$

(2.107)

Equation 2.107 can now be substituted into Equation 2.102 to yield 2.108:

$$\frac{d[P]}{dt} = \frac{k_4[H^+]\{k_2[H_2O] + k_3[OH^-]\}}{\{K_{a_{SH^+}} + [H^+]\}\{k_{-2}[H^+] + k_{-3} + k_4\}} S_T$$

(2.108)

or

$$\frac{d[P]}{dt} = \frac{k_4\{k_2[H_2O][H^+] + k_3[OH^-][H^+]\}}{\{K_{a_{SH^+}} + [H^+]\}\{k_{-2}[H^+] + k_{-3} + k_4\}} S_T$$

(2.109)

Noting that [OH$^-$][H$^+$] is the ion product of water, equal to 10^{-14} mole^2 l^{-2} (see Chapter 3), we find:

$$\frac{d[P]}{dt} = \frac{k_2k_4[H_2O][H^+] + 10^{-14}k_3k_4}{\{K_{a_{SH^+}} + [H^+]\}\{k_2[H^+] + k_{-3} + k_4\}} S_T$$

(2.110)

We now observe that water is the solvent and therefore [H$_2$O] is constant, and if the reaction is carried out with an acid–base buffer present to keep [H$^+$] constant, the reaction becomes pseudo first-order, with the particularly simple form of Equation 2.111:

$$\frac{d[P]}{dt} = k_{obs}[S_T]$$

(2.111)

where

$$k_{obs} = \frac{k_2k_4[H_2O][H^+] + 10^{-14}k_3k_4}{\{K_{a_{SH^+}} + [H^+]\}\{k_{-2}[H^+] + k_{-3} + k_4\}}$$

(2.112)[81]

Experimental verification of the predicted kinetics consists of determining the first-order rate constant k_{obs} at a number of different pH values and seeing whether the behavior of k_{obs} as a function of pH follows the functional form predicted by Equation 2.112.

First-Order Methods

We have seen that such strategies as buffering or supplying some reactants in large excess can cause complex systems to follow first-order kinetics. If an entire reaction scheme can be made to contain only first-order and pseudo first-order terms, it is possible to analyze mechanisms having many steps without making any simplifying assumptions. Experimentally such systems show a series of exponential changes of concentration with time, each characterized by a different observed rate con-

---

[81] This expression differs from that of Cordes and Jencks in the numbering of the k's and in the presence of [H$_2$O] in the numerator. Those authors have included [H$_2$O] in their constant k_3, a customary procedure in analyzing reactions in aqueous solution.

stant. This set of rate constants constitutes the *relaxation spectrum*. The observed rate constants are related to microscopic rate constants for the assumed mechanism, and in favorable cases microscopic k's may be obtained from the data. First-order methods have been used extensively in analyzing the results of fast reactions but have found relatively little application in organic reaction mechanism investigations.[82] The reader is referred to specialized treatments for further information.[83]

2.6 INTERPRETATION OF RATE CONSTANTS

The utility of rate constants for understanding reaction mechanisms depends largely on interpreting them in terms of energies. Energy information is ordinarily obtained from rate data by either of two methods, one empirical and the other more theoretical.

The Arrhenius Equation

The temperature dependence of observed rate constants follows the Arrhenius equation (2.113) with good accuracy for most reactions. A and E_a are parameters determined experimentally, R is the gas constant, 1.986 cal K^{-1} mole^{-1}, and T is the Kelvin temperature. The units of A, called the *preexponential factor*, are the same as

$$k_{\text{obs}} = A \exp\left(-\frac{E_a}{RT}\right) \tag{2.113}$$

those of k_{obs}: for a first-order rate constant, time^{-1}; for a second-order rate constant, l mole^{-1} time^{-1}. We use the notation k_{obs} to emphasize that the equation applies to the observed rate constant, which may or may not be simply related to the microscopic k's characterizing the individual steps of a reaction sequence.

If we write Equation 2.113 in the form of Equation 2.114, we see at once a resemblance to the familiar relation 2.115 between the equilibrium constant of a reaction and its free-energy change. Hence it is natural to interpret E_a as an energy.

$$-E_a = RT \ln\left(\frac{k_{\text{obs}}}{A}\right) \tag{2.114}$$

$$-\Delta G° = RT \ln K \tag{2.115}$$

This energy is called the *Arrhenius activation energy*, or simply *activation energy*. The activation energy of a reaction consisting of one elementary step may be thought of as corresponding roughly to a threshold energy for the reaction. This approximation is best for reactions with high E_a and is rather poor for those with low E_a.[84]

---

[82] For an example see J. I. Seeman and W. A. Farone, *J. Org. Chem.*, **43**, 1854 (1978).

[83] (a) G. M. Fleck, *Chemical Reaction Mechanisms,* Holt, Rinehart and Winston, New York, 1971; (b) C. F. Bernasconi, *Relaxation Kinetics,* Academic Press, New York, 1976; (c) M. Eigen and L. De Maeyer, in *Investigation of Rates and Mechanisms of Reactions,* G. G. Hammes, Ed., Vol IV, Part II, of *Techniques of Chemistry,* A. Weissberger, Ed., Wiley, New York, 1974, Chapter III.

[84] For an elementary, one-step, gas-phase, bimolecular reaction at equilibrium, E_a = {average total energy, translational plus internal, of all reacting pairs of molecules} − {average total energy, translational plus internal, of all pairs of molecules}. See (a) D. G. Truhlar, *J. Chem. Educ.*, **55**, 309 (1978); (b) M. Menzinger and R. Wolfgang, *Angew. Chem. Int. Ed. Engl.*, **8**, 438 (1969).

A more extensive theoretical framework for understanding the relation between rate constants and molecular structure is provided by the transition state theory.

Transition State Theory[85]

The transition state theory is confined to consideration of single elementary reaction steps and is meaningful only when applied to a single microscopic rate constant. The theory postulates that when two molecules come together in a collision that leads to products (or when a single molecule in a unimolecular step follows the motions that cause the chemical change), they pass through a configuration of maximum potential energy called the *transition state*. In order to understand this concept fully, we must first digress to consider some ideas about potential energy surfaces.

Potential energy surfaces   Because each of the N atoms in a molecule can move in three mutually perpendicular and therefore independent directions, a molecule has a total of $3N$ degrees of freedom. But since we think of a molecule as a unit, it is useful to divide these degrees of freedom into three categories. If the atoms were fixed relative to each other, the position of the rigid molecule in space would be defined by specifying six quantities: the three Cartesian coordinates of its center of mass and three rotational angles to indicate its orientation in space. Hence there remain $3N - 6$ degrees of freedom which are internal vibrational motions of the atoms with respect to each other. (A linear molecule has only two rotational coordinates, hence $3N - 5$ vibrational degrees of freedom. We shall continue to say $3N - 6$, with the understanding that $3N - 5$ is to be substituted if the molecule is linear.)

The total molecular vibration is complex, but to a good approximation the vibration may be divided into $3N - 6$ independent *normal modes,* with the entire vibration being a superposition of these.[86] Each normal mode will in general involve many atoms and may include bond stretching or bending or both, but as all motions are in phase with each other, just one parameter suffices to follow the vibration of a single mode, and each mode can be thought of as being essentially equivalent to the stretching vibration of a diatomic molecule. The appropriate model for vibration of a diatomic is two masses joined by a spring, with restoring force proportional to the displacement from the equilibrium separation.

The potential energy of such an oscillator can be plotted as a function of the separation r, or, for a normal mode in a polyatomic molecule, as a function of a

[85]Transition state theory is discussed in standard texts on physical chemistry, kinetics, and physical organic chemistry. See, for example, (a) W. J. Moore, *Physical Chemistry,* 4th ed., Prentice-Hall, Englewood Cliffs, N.J., 1972, p. 381; (b) S. W. Benson, *Thermochemical Kinetics,* 2d ed., Wiley, New York, 1976; (c) K. J. Laidler, *Chemical Kinetics,* 2d ed., McGraw-Hill, New York, 1965, (d) K. B. Wiberg, *Physical Organic Chemistry,* Wiley, New York, 1964; (e) L. P. Hammett, *Physical Organic Chemistry,* 2d ed., McGraw-Hill, New York, 1970. For alternative approaches to chemical dynamics, see (f) D. L. Bunker, *Acc. Chem. Res.,* **7,** 195 (1974); (g) F. K. Fong, *Acc. Chem. Res.,* **9,** 433 (1976).

[86]The vibrations are separable if they follow simple harmonic motion. Molecular vibrations are not quite harmonic but are nearly so. Everything that follows will assume harmonic vibration.

parameter characterizing the phase of the oscillation. For a simple harmonic oscillator the potential energy function is parabolic, but for a molecule its shape is that indicated in Figure 2.9. The true curve is close to a parabola at the bottom, and it is for this reason that the assumption of simple harmonic motion is justified for vibrations of low amplitude.

For a polyatomic molecule there will be a potential energy curve like that of Figure 2.9 for each of the $3N - 6$ vibrational modes. The potential energy is therefore characterized by a surface in $3N - 6 + 1$-dimensional space. To plot such a surface is clearly impossible; we must be content with slices through it along the coordinates of the various normal modes, each of which will resemble Figure 2.9. (Figure 2.9 represents a stretching mode; bending mode potential curves will not level off on one side, but will resemble more closely a symmetrical parabola.)

The reaction coordinate When two molecules come together and react, it is the potential energy surface for the whole process that is of interest. Let us imagine a reaction in which A and B come together and the constituent atoms move over the potential energy surface of the combined system to produce C and D. We shall suppose that we have identified the particular set of atomic motions that has to occur to accomplish this change, and we make a slice through the surface along the dimension of this particular motion. We shall find that the shape of the surface along the line of the slice is something like that shown in Figure 2.10. The motion of atoms characterizing the change is called the *reaction coordinate*. It is convenient to define a parameter x that characterizes progress of the system along

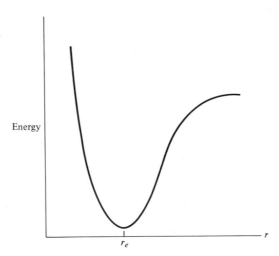

Figure 2.9 Potential energy of a diatomic molecule as a function of internuclear separation r. The equilibrium separation is r_e. A normal mode in a polyatomic molecule would have a similar potential curve, with a parameter characterizing the phase of the motion replacing r.

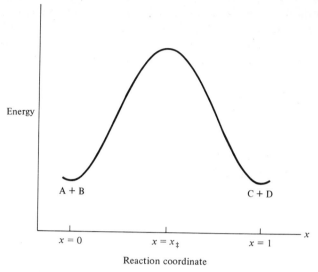

Figure 2.10 The reaction coordinate diagram for the reaction A + B $\rightleftharpoons$ C + D.

the reaction coordinate. The graph of potential energy as a function of x is the *reaction coordinate diagram.*[87]

When A and B are separate and not yet interacting, we are at the left-hand side of the diagram, $x = 0$; as they come together in a reactive collision, the potential energy rises as the atoms begin to execute the motion that will carry over to products. At some configuration the potential passes through a maximum, and then falls as we proceed to the right, finally reaching a minimum again with separated products C and D, $x = 1$. An entirely similar process can be imagined for a unimolecular reaction. The configuration of atoms corresponding to the maximum in the reaction coordinate diagram is the transition state, symbolized by ‡. It occurs at $x = x_{\ddagger}$.

[87] We shall have occasion to use reaction coordinate diagrams frequently throughout this book. While we shall sometimes, as here, be plotting potential energy as a function of reaction coordinate, we shall often want to use a more comprehensive quantity such as enthalpy or free energy. The internal energy E of a system is the sum of the potential and kinetic energies of its constituent parts:

$$E = ke + pe \tag{1}$$

Its enthalpy is a function of its internal energy, pressure, and volume:

$$H = E + PV \tag{2}$$

Its free energy depends on both its enthalpy and its entropy:

$$G = H - TS \tag{3}$$

The free energy includes all the other energy terms, and any changes in the individual terms will be reflected in it. Thus free energy can always be used as the y coordinate in a reaction coordinate diagram. When the changes being considered chiefly affect one of the less comprehensive terms, it may be more meaningful to plot that energy term against the reaction coordinate.

A related idea that helps in thinking about pathways of reactions is the *principle of least motion*. It states that *those elementary reactions will be favored that involve the least change in atomic position and electronic configuration.*[88] Applied to electronic configuration, this statement finds expression in the theory of pericyclic reactions (Chapter 10); the atomic position aspect, known as the *principle of least nuclear motion,* postulates an economy of movement in chemical changes. Indeed, although profound structural change is a frequent occurrence in chemical reactions, close investigation usually reveals that the overall change consists of several simpler steps. In considering reaction coordinates, we are therefore usually able to concentrate on a relatively simple set of motions.

There are two perhaps obvious but easily overlooked points about the reaction coordinate diagram that must be stressed. First, it is only a one-dimensional slice of a $3N - 6 + 1$-dimensional surface. (N is the total number of atoms in A and B.) We can imagine, at each point of the line, motions off the line corresponding to vibrations other than the single one that is carrying the molecules over to products. These motions are all ordinary vibrations, having nothing to do (in a first approximation at least) with the reaction and proceeding quite independently of it. If we assume that the reaction coordinate corresponds to a normal mode of the reacting system,[89] the reaction coordinate is "perpendicular" (in $3N - 6$-dimensional space) to each of these other normal modes. Our curve passes along the equilibrium position of each of the other vibrations, so that if we were to leave the reaction coordinate line and follow the potential energy surface in the direction of some other mode, the energy would always go up.

The situation can be visualized if only one vibrational degree of freedom besides the reaction coordinate is included. Then we have the three-dimensional potential energy surface of Figure 2.11, two valleys meeting over a mountain pass, or saddle point. If we climb along the reaction coordinate out of one valley over the pass into the other, we go over an energy maximum along the reaction coordinate, but the surface rises in the perpendicular direction and we are therefore following a potential energy minimum with respect to the motion perpendicular to the reaction coordinate.

The second point about the surface is that it shows only the potential energy. The total energy of the molecular system is the sum of its kinetic and potential energies. The molecules exchange kinetic energy by collisions and are distributed over a range of total energies, with many at low energies and fewer at higher energies. It is tempting to think of a reaction as following the path of a pack horse up out of one valley and over the pass into the other one. This model is quite inappropriate; a much better way to think of the situation is to imagine many birds flying in the valleys at various levels, the levels representing the various possible

---

[88](a) F. O. Rice and E. Teller, *J. Chem. Phys.*, **6**, 489 (1938); (b) J. Hine, *Adv. Phys. Org. Chem.*, **15**, 1 (1977). The principle of least motion has been developed into a quantitative method for predicting pathways of reactions such as eliminations and rearrangements. (c) O. S. Tee, *J. Am. Chem. Soc.*, **91**, 7144 (1969); (d) J. A. Altmann, O. S. Tee, and K. Yates, *J. Am. Chem. Soc.*, **98**, 7132 (1976).

[89]This assumption is implicit in the transition state theory, although it may not be entirely correct.

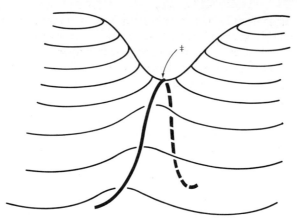

Figure 2.11 Contour diagram of the saddle point on a hypothetical three-dimensional energy surface. The heavy line traces the near end of the reaction coordinate, which continues
down the back (dashed line) into a similar valley on the far side. The point at the top of
the saddle is the transition state. From J. E. Leffler and E. Grunwald, *Rates and Equilibria
of Organic Reactions,* John Wiley & Sons, New York, 1963, p. 65. Adapted by permission of
John Wiley & Sons, Inc.

total energies. The individual birds can go up or down by receiving or giving up
some energy to their surroundings, but the vertical distribution of birds is in equilibrium and remains unchanged. The birds are flying around at random, and those
that are high enough may in their wanderings happen to sail over the pass and join
the population in the other valley. The rate of passage of the birds from one side to
the other depends on the height of the pass and on the vertical distribution of the
birds. In the molecular system the vertical distribution is determined by the temperature.

Thermodynamics of the transition state   In developing the transition state
theory, we shall take advantage of the fact that most of the motions in a reacting
molecular system are ordinary vibrations, rotations, and translations. Only the one
normal mode corresponding to the reaction coordinate is doing something peculiar
by breaking and forming bonds to yield new molecules. We shall postulate therefore that the molecules going over the barrier are in equilibrium with all the other
reactant molecules, just as in our bird analogy we said that the birds that can get
over the pass are just those that happen to be high enough up and headed in the
right direction.

We assume that in Reaction 2.116 there are at any instant some molecules at
the transition state going in each direction over the barrier, and we shall concen-

$$A \underset{k_{-1}}{\overset{k_1}{\rightleftharpoons}} B \tag{2.116}$$

trate on one direction only, A $\rightarrow$ B. We are therefore dealing with rate constant k_1;

exactly the same arguments will apply to the A ← B reaction and k_{-1}. We suppose that the transition state molecules moving from left to right, $A^{\ddagger}$, are at equilibrium with the bulk of the A molecules in the restricted sense specified above. The concentration $[A^{\ddagger}]$ can therefore be written in terms of an equilibrium constant, $K_{\ddagger}$ (Equation 2.117). The rate of the reaction from left to right is $k^{\ddagger}[A^{\ddagger}]$, the concentra-

$$[A^{\ddagger}] = K_{\ddagger}[A] \tag{2.117}$$

tion of molecules at the transition state multiplied by a rate constant characterizing their rate of passage over the barrier. Then since $k_1[A]$ is the conventional reaction rate,

$$k_1[A] = k^{\ddagger}[A^{\ddagger}] \tag{2.118}$$

and, substituting for $[A^{\ddagger}]$ from Equation 2.117, and canceling $[A]$ from both sides, we find Equation 2.119 relating the first-order rate constant to properties of the transition state.

$$k_1 = k^{\ddagger}K_{\ddagger} \tag{2.119}$$

The equilibrium constant $K_{\ddagger}$ is then analyzed by the methods of statistical thermodynamics to separate out the contribution of the reaction coordinate from other contributions. The rate constant $k^{\ddagger}$ is also calculated by statistical thermodynamic methods. These calculations are given in Appendix 1 to this chapter. The results of the analysis are expressed by Equation 2.120, where k is the Boltzmann

$$k_1 = \frac{kT}{h}K^{\ddagger} \tag{2.120}$$

constant, h is Planck's constant, T is the Kelvin temperature, and $K^{\ddagger}$ is a new equilibrium constant that excludes the contributions from the reaction coordinate. The new equilibrium constant $K^{\ddagger}$ can be written in terms of a free energy of activation, $\Delta G^{\ddagger}$ (Equation 2.121), and $\Delta G^{\ddagger}$ can in turn be divided into contributions from enthalpy of activation, $\Delta H^{\ddagger}$, and entropy of activation, $\Delta S^{\ddagger}$ (Equation 2.122). Equations 2.123 and 2.124 then follow. Equation 2.123 is called the Eyring equation, after Henry Eyring, who was instrumental in the development of the transition state theory.

$$\Delta G^{\ddagger} = -RT \ln K^{\ddagger} \tag{2.121}$$

$$\Delta G^{\ddagger} = \Delta H^{\ddagger} - T\,\Delta S^{\ddagger} \tag{2.122}$$

$$k_1 = \frac{kT}{h}\exp\left(-\frac{\Delta G^{\ddagger}}{RT}\right) \tag{2.123}$$

$$k_1 = \frac{kT}{h}\exp\left(-\frac{\Delta H^{\ddagger}}{RT}\right)\exp\left(\frac{\Delta S^{\ddagger}}{R}\right) \tag{2.124}$$

Comparison between the transition state expression (2.124) and the Arrhenius equation (2.113) may be made if both are applied to the microscopic rate constant for a single reaction step.[90] The correspondence is as follows:

$$A = \frac{ekT}{h} \exp\left(\frac{\Delta S^{\ddagger}}{R}\right) \qquad (2.125)$$

$$E_a = \Delta H^{\ddagger} + RT \qquad (2.126)$$

The term RT is small at ordinary temperatures; in the neighborhood of 298 K the difference between E_a and $\Delta H^{\ddagger}$ is only about 0.6 kcal mole^{-1}.

Weaknesses of transition state theory The transition state theory has been extensively used for interpretation of rate constants, but doubts remain as to its validity. In its usual form, it is not quantum mechanically correct. The main quantum effect is to permit a small extent of reaction by molecules that have less energy than is required to surmount the barrier according to a classical picture. This phenomenon is called tunneling. Tunneling is more important the smaller the mass of the atom or atoms involved in the reaction coordinate motion; hence it is most important for hydrogen atom or hydrogen ion transfers, and probably does not have a significant effect in other processes.[91] A second weakness of the transition state theory is the assumption that the reaction coordinate can be separated from the other motions. Detailed calculations for very simple reactions (for example, $H \cdot + H_2 \rightarrow H_2 + H \cdot$) indicate that the error introduced by the separability assumption into a rate constant calculated from a known potential energy surface is significant at temperatures up to about 1000 K, and again the error is most important for reactions involving motion of light atoms.[92]

Magnitudes of Kinetic Quantities
The factor kT/h is equal to $10^{12.8}$ sec^{-1} at 298 K, and ekT/h is $10^{13.2}$ sec^{-1} at this temperature. These figures should therefore represent roughly the rate to be expected for a unimolecular gas-phase reaction step of zero enthalpy and entropy of activation.

In solution, bimolecular reaction rates are limited by the rate at which the partners can diffuse among the solvent molecules; the *diffusion-controlled rate* determined by this process varies with solvent viscosity and temperature and is on the order of $10^{10} M^{-1}$ sec^{-1} for common solvents at ambient temperature.

In order to give an idea of the relation between activation parameters and rate constants, Table 2.17 lists values of $kT/h\{\exp(\Delta H^{\ddagger}/RT)\}$ and of $\exp(\Delta S^{\ddagger}/R)$.

[90] S. W. Benson, *Thermochemical Kinetics*, Wiley, New York, 1968.
[91] (a) R. P. Bell, *Chem. Soc. Rev.*, **3**, 513 (1974). It is also important to remember that quantum behavior restricts the knowledge we can obtain about events occurring near the top of the energy barrier. If we tried to follow in detail the motions of the atoms in a molecule crossing the barrier with just enough total energy to get over, we would come up against the uncertainty principle just as we do in trying to follow electron motions, and we would be unable to say just how the atoms got from one place to the other. For further discussion see (b) W. F. Sheehan, *J. Chem. Educ.*, **47**, 254 (1970).
[92] W. H. Miller, *Acc. Chem. Res.*, **9**, 306 (1976).

Table 2.17 VALUES OF $kT/h\{\exp(-\Delta H^{\ddagger}/RT)\}$ AND OF $\exp(\Delta S^{\ddagger}/R)$
FOR VARIOUS VALUES OF $\Delta H^{\ddagger}$ AND $\Delta S^{\ddagger}$ AT 298 K[a]

$\Delta H^{\ddagger}$ (kcal mole^{-1})	$kT/h\{\exp(-\Delta H^{\ddagger}/RT)\}$ (sec^{-1})	$\Delta S^{\ddagger}$ (cal mole^{-1} K^{-1})	$\exp(\Delta S^{\ddagger}/R)$
1	1.15×10^{12}	-30	2.75×10^{-7}
2	2.12×10^{11}	-25	3.41×10^{-6}
5	1.33×10^{9}	-20	4.23×10^{-5}
10	2.85×10^{5}	-15	5.25×10^{-4}
15	6.10×10^{1}	-10	6.50×10^{-3}
20	1.31×10^{-2}	-5	8.07×10^{-2}
25	2.80×10^{-6}	0	$1.$
30	6.00×10^{-10}	5	2.14×10^{1}
35	1.29×10^{-13}	10	1.54×10^{2}
40	2.76×10^{-17}	15	1.91×10^{3}
		20	2.36×10^{4}
		25	2.93×10^{5}
		30	3.63×10^{6}

[a] $k/h = 2.0837 \times 10^{10}$ sec^{-1}K^{-1}.

Rates of different reactions are often compared. Table 2.18 shows the relationship between relative rate and difference of activation parameters between two reactions. Note particularly the relatively small differences in activation parameters that correspond to rather large ratios of rates. The following relation may sometimes be useful:

$$\Delta E_a \approx 1.37 \log \frac{k_a}{k_b} \tag{2.127}$$

Table 2.18 DIFFERENCES IN $\Delta H^{\ddagger}$ OR E_a AND IN $\Delta S^{\ddagger}$ CORRESPONDING
TO VARIOUS RATE CONSTANT RATIOS FOR TWO
ELEMENTARY REACTION STEPS, a AND b

k_a/k_b	$\Delta H^{\ddagger}_b - \Delta H^{\ddagger}_a$, or $E_{a_b} - E_{a_a}$ (kcal mole^{-1}, 298 K, constant $\Delta S^{\ddagger}$ or A)
2	0.41
10	1.36
10^{2}	2.73
10^{4}	5.45
10^{6}	8.18

k_a/k_b	$\Delta S^{\ddagger}_a - \Delta S^{\ddagger}_b$ (entropy units, e.u., cal mole^{-1} K^{-1}, constant $\Delta H^{\ddagger}$)
2	1.38
10	4.57
10^{2}	9.15
10^{4}	18.29
10^{6}	27.44

Table 2.19 PERCENT COMPOSITION OF THE EQUILIBRIUM MIXTURE
FOR THE REACTION A $\xrightleftharpoons{K_{eq}}$ B
FOR VARIOUS VALUES OF $\Delta G°$ AND K_{eq}; $T = 298\ K$

$\Delta G°$(kcal mole^{-1})	K_{eq}	Percent A at Equilibrium
+6.0	3.95×10^{-5}	99.996
+5.0	2.14×10^{-4}	99.98
+4.0	1.16×10^{-3}	99.88
+3.0	6.29×10^{-3}	99.38
+2.0	3.41×10^{-2}	96.71
+1.0	1.85×10^{-1}	84.42
+0.5	4.30×10^{-1}	69.95
+0.1	8.45×10^{-1}	54.21
−0.1	1.18	45.78
−0.5	2.33	30.05
−1.0	5.42	15.58
−2.0	2.94×10^{1}	3.29
−3.0	1.59×10^{2}	0.625
−4.0	8.62×10^{2}	0.116
−5.0	4.67×10^{3}	0.0214
−6.0	2.53×10^{4}	0.00395

Table 2.19 shows the relation between an equilibrium $\Delta G°$ and the composition of a two-component equilibrium mixture,

$$A \xrightleftharpoons{K_{eq}} B \tag{2.128}$$

The Hammond Postulate

Consider a reaction in which starting materials and products lie at significantly different energies. We have no *a priori* way of predicting, short of carrying out time-consuming and expensive calculations, where along the reaction coordinate the transition state will occur. But it seems intuitively reasonable that if the starting materials are of high energy (exothermic reaction), relatively little change of geometry will be required to reach the transition state; whereas if the reaction is endothermic, the reorganization required will be considerable and the transition state will not be reached until the geometry already closely resembles the high-energy products. This idea, illustrated in Figure 2.12, is known as the *Hammond postulate*.[93]

A reaction that is highly exothermic is expected on the basis of the Hammond postulate to have a small activation energy and therefore a high rate. Chemists therefore sometimes speak of a feature of a structure that makes a large exothermic contribution to the equilibrium free-energy change as a *driving force* for the reaction. The formation of a particularly strong bond, or relief of an unfavorable

[93](a) G. S. Hammond, *J. Am. Chem. Soc.*, **77**, 334 (1955). Analytical functions that describe the reaction coordinate and reproduce the Hammond behavior have been developed. See (b) W. J. Le Noble, A. R. Miller, and S. D. Hamann, *J. Org. Chem.*, **42**, 338 (1977); (c) A. R. Miller, *J. Am. Chem. Soc.*, **100**, 1984 (1978).

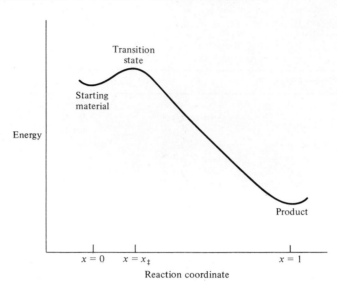

(a)

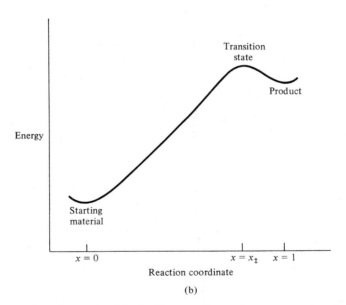

(b)

Figure 2.12 In an exothermic reaction (a), the Hammond postulate assumes that the transition state should resemble the starting material, whereas in an endothermic process (b), it should resemble the product.

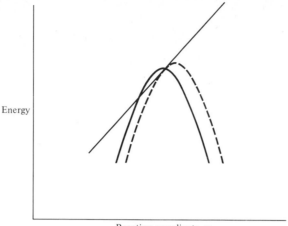

Energy

Reaction coordinate, x

Figure 2.13 Addition to the reaction coordinate potential (solid curve) of a perturbation of positive slope makes the reaction toward the right more difficult and shifts the transition state to the right (dashed curve). Reprinted with permission from E. R. Thornton, *J. Am. Chem. Soc.*, **89**, 2915 (1967). Copyright 1967 American Chemical Society.

steric interaction, might constitute a driving force. It is well to remember that there is no direct connection between equilibrium thermodynamics and rate; the driving force idea is therefore only a rough qualitative one and must be used cautiously.

Transition State Structure and Three-dimensional Reaction Coordinate Diagrams

It is often useful to have the Hammond postulate stated in the context of a small change in structure or *perturbation*, brought about, for example, by changing a substituent. Thornton has given an analysis that we follow here.[94] We approximate our potential energy curve in the region of the transition state by a parabola, opening downward as shown in Figure 2.13. We then suppose that we make some small change in structure that makes it more difficult to proceed to the right. This change is equivalent to raising the energy at the right-hand side of the reaction coordinate curve more than the left-hand side, and can be accomplished by adding to the energy at each point along the curve an increment $\delta \Delta E°$ that increases to the right. Here the symbol δ signifies the effect on the quantity $\Delta E°$ of the structural change.[95] The simplest approach is to make the increment increase linearly with x, that is,

$$\delta \Delta E° = mx \qquad (2.129)$$

[94] E. R. Thornton, *J. Am. Chem. Soc.*, **89**, 2915 (1967).
[95] δ is known as a *Leffler–Grunwald operator* and is used to designate the change in any quantity resulting from a structural change. See J. E. Leffler and E. Grunwald, *Rates and Equilibria of Organic Reactions*, Wiley, New York, 1963, p. 26.

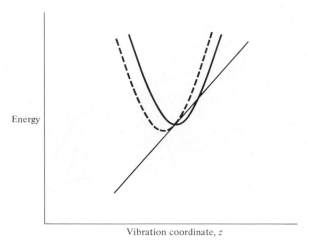

Energy

Vibration coordinate, z

Figure 2.14 Addition to a vibration potential (solid curve) of a perturbation of positive slope makes bond stretching more difficult and decreases the equilibrium separation (dashed curve). Reprinted with permission from E. R. Thornton, *J. Am. Chem. Soc.*, **89**, 2915 (1967). Copyright 1967 American Chemical Society.

where x is the reaction parameter defined earlier (Figure 2.10), and m is the slope, positive in the present example. In Figure 2.13 the straight line superimposed on the reaction coordinate potential curve represents the perturbation $\delta \Delta E^{\circ}$. If we place the origin at the vertex of the parabola, it is easy to verify by inspection that the result of adding the perturbation to the potential energy curve will be to shift its maximum, and thus the transition state, to the right (dashed curve). A perturbation with a negative slope, that is, a structural change making motion from left to right easier, will shift the curve to the left.

It will also be of interest to know how structural changes affect the position of the transition state on the potential energy surface with respect to degrees of freedom other than the reaction coordinate. Recall that these other degrees of freedom correspond to ordinary vibrations. They cut across the surface perpendicular to the reaction coordinate and are valleys rather than hills. Suppose that we make a change in structure that will make a certain bond, not corresponding to the one breaking, more difficult to stretch. We show in Figure 2.14 the potential surface cut along the stretching degree of freedom, with a perturbation

$$\delta \Delta E^{\circ} = mz \tag{2.130}$$

where m is positive. Now the perturbed potential (dashed curve) is shifted to the left. Making the bond more difficult to stretch has changed the structure of the transition state so that the equilibrium bond distance is shorter.

We may summarize the effects of energy changes on transition state structure as follows:[96]

---

[96]These rules, originally known as the reacting bond rules, were formulated by Thornton, note 94.

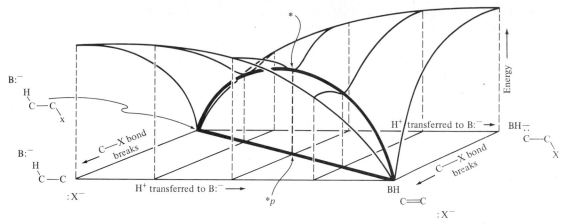

Figure 2.15 Energy surface and reaction coordinate for a hypothetical E2 elimination reaction in which H^+ transfer from carbon to base $B:^-$ and cleavage of carbon–halogen bond occur together, with the transition state at the midpoint of the process. The curved heavy line is the reaction coordinate and the straight heavy line is its projection on the horizontal plane. The transition state is marked by *, and its projection by $*_p$. Vertical dashed lines connect points on the surface with the projections of those points on the horizontal plane.

    1. *Along the reaction coordinate: Transition state structure shifts toward the end that is raised, away from the end that is lowered.*

    2. *Perpendicular to the reaction coordinate: Transition state structure shifts toward the side that is lowered, away from the side that is raised.*

    The utility of these rules is apparent in the analysis of reactions in which two processes must take place. An example is the 1,2 elimination reaction (discussed in Chapter 7), where a base removes a proton from one carbon and a leaving group departs from the adjacent carbon. These processes may occur together (E2 elimination); proton removal may come first (E1cB elimination); or leaving group may depart first (E1 elimination). Figure 2.15 represents schematically the energy surface for the E2 elimination reaction (Equation 2.131).[97] Each point on the

$$B:^- + \quad \underset{\beta}{\overset{H}{C}}{-}\underset{\alpha}{\overset{}{C}}\underset{X}{\overset{}{\diagdown}} \quad \longrightarrow \quad BH + \; {>}C{=}C{<} \; + \; :X^- \tag{2.131}$$

horizontal plane represents a particular combination of values of the coordinates for C—X bond breaking (back-to-front coordinate) and of H^+ transfer from carbon to the base $B:^-$ (left-to-right coordinate), and hence a particular structure for the ensemble $B:^-\cdots H\cdots C{\mathrel{\underline{\cdots}}}C\cdots X$. The vertical height of the surface above each point

[97] Three-dimensional reaction coordinate diagrams of this type were proposed by More O'Ferrall and developed by More O'Ferrall, Thornton, Jencks, and others. See, for example, (a) E. R. Thornton, *J. Am. Chem. Soc.*, **89**, 2915 (1967); (b) R. A. More O'Ferrall, *J. Chem. Soc. B*, 274 (1970); (c) W. P. Jencks, *Chem. Rev.*, **72**, 705 (1972).

in the plane represents the energy corresponding to that particular structure. The E2 reaction starts at the back left corner with base $B:^-$ and an alkyl halide, represented by the partial structure H—C—C—X. At the front right corner are the products: BH; an alkene represented by partial structure C=C; and the halide ion $:X^-$. The other two corners represent energy maxima for the E2 reaction: at the front left, the carbocation H—C—C$^+$; and at the back right the carbanion $^-:C$—C—X. Hence moving from back to front in the diagram corresponds to breaking the C—X bond, and moving from left to right corresponds to transferring the proton from carbon to the base $B:^-$. The lowest-energy pathway from reactants to products in the hypothetical example depicted in Figure 2.15 is a combination of both H$^+$ transfer and C—X bond cleavage occurring together; therefore the reaction coordinate follows the heavy curved line diagonally across the surface. The transition state comes at the highest point of this curve, marked by * in the figure.

Figure 2.15 also shows, as the straight heavy line, a projection of the reaction coordinate on the horizontal plane. If the surface were viewed from directly above, the reaction coordinate would be seen superimposed on this projection line. The projection of the transition state is designated by $*_p$. Figure 2.16 shows only the projection on the horizontal plane, viewed from directly above. In this projection the information about the energy at different points has not been reproduced explicitly. It is understood that there are energy minima at corners P and R, that the energy rises to maxima at corners Q and S, and that the reaction coordinate shown follows the lowest-energy path from reactants to products. In this hypothetical example the dashed lines show that the transition state (marked $*_p$) has a structure in which the C—X bond is half-broken and the proton is half-transferred.

Figure 2.17 shows the effect of raising the energy of the reactant corner of the diagram while leaving the energies of the other corners unchanged. The transition state is moved to an earlier position along the reaction coordinate. In the projection (Figure 2.16) this change corresponds to raising the energy of corner R and moving the transition state in the direction of arrow 1 (Rule 1.1, p. 201). Also by Rule 1, raising energy of corner P would move the transition state in the direction of arrow 3. By Rule 1, raising energy of corner Q (or lowering that of S) would shift the transition state structure in the direction 2, and raising energy of S (or lowering that of Q) would shift the transition state structure in the direction 4.

It is common to find a structural change that affects the energy of one edge. For example, making X a better leaving group raises the energy along the back edge of Figure 2.15, or along edge RQ in Figure 2.16. Figure 2.18 shows the effect on the surface of this change. The transition state structure is moved toward the front and toward the left, and the reaction coordinate projection onto the horizontal plane has become a curved line. In Figure 2.16, raising the energy of the edge RQ moves the transition state structure in direction 1 along the reaction coordinate and in direction 2 perpendicular to it; the sum of these motions is a net shift in the direction 5. Figure 2.19 shows, in projection on the horizontal plane, the new reaction coordinate and transition state location that result from this change. Note the interesting conclusion that the principal effect on transition state structure of

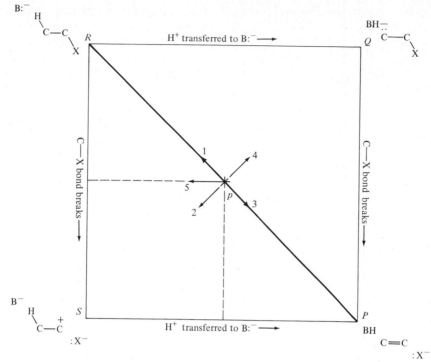

Figure 2.16 The projection on the horizontal plane of the reaction coordinate shown in Figure 2.15, viewed from directly above. The transition state projection, $*_p$, is at the center of the diagram in this hypothetical example and has coordinates (shown by the dashed lines) that specify a structure in which the C—X bond is half-broken and the H^+ is half-transferred. Points R (reactants) and P (products) correspond to energy minima; Q and S correspond to maxima. The arrows show the direction in which the transition state coordinates move when the energy is changed at various points of the diagrams as follows:

Raising energy at corner R
(or lowering energy at corner P):   1
Raising energy at corner P
(or lowering energy at corner R):   3
Raising energy at corner Q
(or lowering energy at corner S):   2
Raising energy at corner S
(or lowering energy at corner Q):   4
Raising energy along edge RQ
(or lowering energy along edge SP): $1 + 2 = 5$

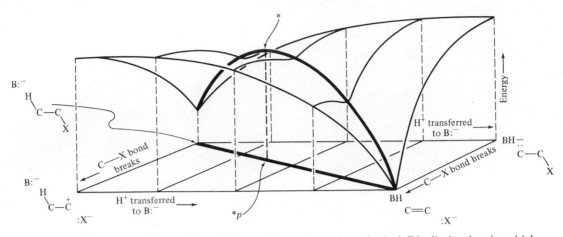

Figure 2.17 Energy surface and reaction coordinate for a hypothetical E2 elimination in which energy of the reactants is higher than in Figure 2.15 but in which the energies of the other corners are unchanged. The shape of the surface has changed so as to bring the transition state closer to the reactants in structure. The change from Figure 2.15 corresponds to arrow 1 in Figure 2.16.

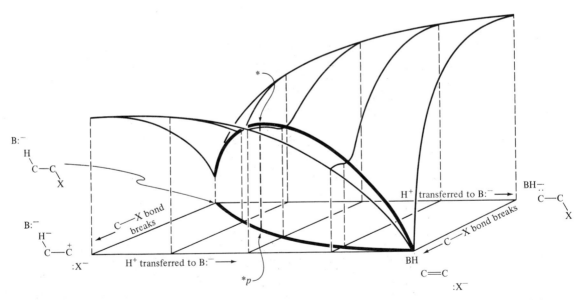

Figure 2.18 Energy surface and reaction coordinate for a hypothetical E2 elimination in which energy has been raised along the back edge to simulate the effect of changing to a better leaving group X. The shape of the surface has changed so as to bring the transition state nearer to the left edge (H^+ transferred to a smaller extent). The projection of the reaction coordinate on the horizontal plane is now a curved line. The change from Figure 2.15 corresponds to arrow 5 in Figure 2.16.

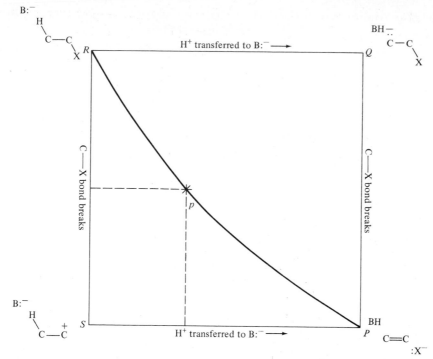

Figure 2.19 Projection on the horizontal plane of the reaction coordinate for the energy surface shown in Figure 2.18. The new transition state coordinates (dashed lines) correspond to a structure in which the C—X bond is still about half-broken but in which H^+ is less than half-transferred to the base B:$^-$.

making X a better leaving group is not to change the degree of C—X bond breaking but instead to reduce the extent of H^+ transfer. We shall examine reaction coordinates of this kind in more detail in Chapters 4, 7, and 8.

2.7 ISOTOPE EFFECTS

The kinetic isotope effect, a change of rate that occurs upon isotopic substitution, is a widely used tool for elucidating reaction mechanism.[98] The most common isotopic substitution is D for H, although isotope effects for heavier atoms are also frequently measured. Our discussion will be in terms of hydrogen isotope effects; the same principles apply to other atoms.

To a good approximation, substitution of one isotope for another does not

[98] For general treatments of the isotope effect, see (a) K. B. Wiberg, *Physical Organic Chemistry,* Wiley, New York, 1964, pp. 273 and 351; (b) L. Melander and W. H. Saunders, Jr., *Reaction Rates of Isotopic Molecules,* Wiley, New York, 1980; (c) F. H. Westheimer, *Chem. Rev.,* **61,** 265 (1961); (d) J. Bigeleisen and M. Wolfsberg, *Adv. Chem. Phys.,* **1,** 15 (1958), (e) C. J. Collins and N. S. Bowman, Eds., *Isotope Effects in Chemical Reactions,* ACS Monograph 167, Van Nostrand Reinhold, New York, 1970.

alter the potential energy surface. The electronic structure, and thus all binding forces, remains the same. All differences are attributable solely to the change in mass, which manifests itself primarily in the frequencies of vibrational modes. For a hypothetical model of a small mass m attached to a much larger mass by a spring of force constant k, the classical vibrational frequency is given by[99]

$$\nu = \frac{1}{2\pi}\sqrt{\frac{k}{m}} \qquad (2.132)$$

The quantum mechanical treatment of the same model leads to energy levels

$$\epsilon_n = (n + \tfrac{1}{2})h\nu \qquad n = 0, 1, 2, \ldots \qquad (2.133)$$

and thus to energy-level separations $\Delta\epsilon = h\nu$, where ν is the classical frequency given by Equation 2.132. Energies are measured from the lowest point on the potential energy curve.

An important feature of the vibrational energy levels is that the energy of the lowest possible level lies $\tfrac{1}{2}h\nu$ above the minimum of the potential curve. From Equation 2.132 this zero-point energy is inversely proportional to the square root of the mass.

Primary Isotope Effects

Figure 2.20 illustrates the zero-point energy level for a C—H stretching vibration and compares it with the zero-point energy of the same stretch for a C—D bond. In a reaction in which the C—H (C—D) bond breaks, there will be a *primary isotope effect*. The stretching vibration of the reactants is converted to the translational motion over the barrier, and the zero-point energy disappears for that particular degree of freedom. Since the C—H molecule starts out at a higher energy, its activation energy is lower, and k_H/k_D will be greater than 1.

We can easily calculate the isotope effect to be expected were this loss of zero-point energy the sole contributor. The C—D frequency should be smaller than the C—H frequency by a factor of roughly $1/\sqrt{2} = 1/1.41$ according to Equation 2.132; the observed ratio is closer to $1/1.35$.[100] The activation energy difference is therefore[101]

$$\epsilon_H^{\ddagger} - \epsilon_D^{\ddagger} = -\tfrac{1}{2}hc(\nu_H - \nu_D) = -\tfrac{1}{2}hc\left(1 - \frac{1}{1.35}\right)\nu_H \qquad (2.134)$$

---

[99] If the two masses joined by the spring are comparable, m in Equation 2.132 must be replaced by the reduced mass,

$$\mu = \frac{m_1 m_2}{m_1 + m_2}$$

When one of the masses is much larger than the other, as would be the case for a hydrogen attached to a large molecule, μ is approximately equal to the smaller mass.

[100] A. Streitwieser, Jr., R. H. Jagow, R. C. Fahey, and S. Suzuki, *J. Am. Chem. Soc.,* **80**, 2326 (1958).

[101] Multiplication by the speed of light, c, converts frequency expressed in cm^{-1} to sec^{-1}.

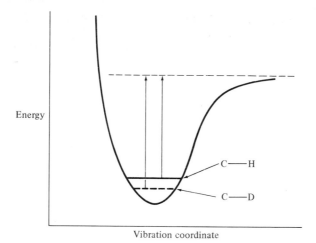

Figure 2.20 The zero-point energy is proportional to ν and thus to $\sqrt{1/m}$; the C—D bond therefore has a lower zero-point energy than the C—H bond, and greater activation energy is required for cleavage.

If we take this energy difference as being approximately equal to the difference in enthalpy of activation, we may use Equation 2.124 (p. 194) to find the rate constant ratio. We may assume $\Delta S^{\ddagger}$ will be equal for the two molecules that differ only by substitution of H for D. Equation 2.135, where the Boltzmann constant k has been

$$\frac{k_H}{k_D} = \exp\left(-\frac{\epsilon_H^{\ddagger} - \epsilon_D^{\ddagger}}{kT}\right) \tag{2.135}$$

used in place of R in order to express energy in units of erg molecule^{-1} to match the units of $hc\nu$, will then express the desired ratio of rate constants. Substituting Equation 2.134 into 2.135, we obtain Equation 2.136

$$\frac{k_H}{k_D} \approx \exp\left[\frac{hc}{2kT}\left(1 - \frac{1}{1.35}\right)\nu_H\right] = \exp\left(\frac{0.1865}{T}\nu_H\right) \tag{2.136}$$

The C—H stretching vibration appears in the infrared spectrum around 3000 cm^{-1}; the isotope effect at $T = 298$ K would therefore be

$$\frac{k_H}{k_D} \approx \exp\left[\frac{0.1865}{298}(3000)\right] = 6.5 \tag{2.137}$$

Note that the isotope effect is larger the lower the temperature.

   This simple model is too crude to account for the observed range of isotope effects. We have neglected bending vibrations and other changes that a more careful treatment must take into account. Appendix 2 to this chapter gives a derivation that shows that the isotope effect is more closely approximated by Equation 2.138. The Π symbols signify a product of terms. The first is a product over normal modes

$$\frac{k_{\mathrm{H}}}{k_{\mathrm{D}}} \approx \prod_i^{\ddagger} \exp[-\tfrac{1}{2}(u_{i\mathrm{H}} - u_{i\mathrm{D}})_{\ddagger}] \prod_i^{r} \exp[+\tfrac{1}{2}(u_{i\mathrm{H}} - u_{i\mathrm{D}})_r] \tag{2.138}$$

of vibration of the transition state, and the second over normal modes of the reactants. The quantity u_i is defined as $h\nu_i/kT$, where ν_i is the frequency of normal mode i; each of the exponential terms thus contains a difference in vibrational frequency between the hydrogen and the deuterium compound. The products are over bound vibrations only. In other words, the reaction coordinate itself, which is a vibration in the reactants but not in the transition state, contributes only to the reactant part of Equation 2.138. It is necessary to include in Equation 2.138 only those vibrations that involve changes of force constants at isotopically substituted positions. An expression for the isotope effect on an equilibrium is given in Appendix 2.

The following qualitative statement of the direction of an isotope effect is sometimes useful. The heavy isotope will concentrate at that site where it is bound more strongly, that is, has the larger force constant and frequency. For a kinetic effect this statement means that deuterium will prefer the reactant, where the force constant is higher, and the hydrogen will prefer the transition state, where the force constant is lower; the hydrogen compound will react faster. For an equilibrium,

$$AH + BD \xrightleftharpoons{K_{\mathrm{H/D}}} AD + BH \tag{2.139}$$

if the force constant is higher in AH(D) than in BH(D), the deuterium will prefer to be in A and the hydrogen will prefer to be in B; $K_{\mathrm{H/D}}$ will be greater than 1.

Isotope effects in linear transition states   In order to see how the isotope expression Equation 2.138 is applied to a reaction, we shall analyze the kinetic isotope effect in a simple system, a transfer of hydrogen from AH to B through a linear transition state (Equation 2.140).[102] We assume that A and B are polyatomic fragments. This process provides a somewhat more realistic model for a chemical

$$AH + B \longrightarrow A\cdots H\cdots B \to A + HB \tag{2.140}$$

reaction than does the simple dissociation considered above. Because the hydrogen is being transferred from one molecule to another, there will be vibrations in addition to the C—H or C—D stretch that will have to be considered. In the reactants there are both A—H stretching and A—H bending modes. In the transition state the A—H stretch has become the reaction coordinate (**80**),

$$\overset{\longleftarrow}{\underset{A}{\bigcirc}} \quad \overset{\longrightarrow}{\underset{H}{\bigcirc}} \quad \overset{\longleftarrow}{\underset{B}{\bigcirc}}$$

80

and contributes nothing to the transition state term in Equation 2.138, leaving

[102](a) L. Melander and W. H. Saunders, Jr., *Reaction Rates of Isotopic Molecules,* Wiley, New York, 1980; (b) F. H. Westheimer, *Chem. Rev.,* **61,** 265 (1961).

$\exp[+\frac{1}{2}(u_{iH} - u_{iD})]$ for this mode to contribute to the reactant term. It is this factor that we evaluated earlier as being about 6.5. But there are also in the transition state other vibrations to be considered. There will be two degenerate bends, **81** and **82**, which are identical but occur in mutually perpendicular planes. These

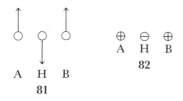

motions are roughly comparable to the reactant A—H bending; and since bending frequencies are lower than stretching and therefore contribute less to the isotope effect in any event, bending frequencies are usually considered to cancel approximately between reactant and transition state when a primary isotope effect is being evaluated.[103]

We are then left with one final transition state vibration, a symmetric stretch (**83**), which has no counterpart in the reactants. If the transition state is

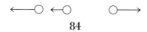

highly symmetric, so that the A···H and the H···B force constants are equal, this stretch will involve only A and B moving in and out together, with no motion of the H (or D). The frequency will then be the same for H and D, and its contribution to the transition state term in Equation 2.138 will cancel. We shall then be left with only the reaction coordinate mode, and an isotope effect around 6.5. If the transition state is not symmetric, the H (D) will be closer to A or to B; then the H (D) will move in the symmetric stretch and since $\nu_H > \nu_D$, $\exp[-\frac{1}{2}(u_H - u_D)_\ddagger]$ for this mode will be less than unity. Part of the contribution from the reactant zero-point energy of the reaction coordinate mode will be canceled and the isotope effect will be lowered. In the limit that the transition state is nearly the same as reactant, the symmetric stretch (**84**) will involve nearly as much motion of H or D as the reactant stretch, and its zero-point difference will largely cancel the contribution

$$\longleftarrow\!\!\circ\!\!-\!\longleftarrow\!\!\circ\qquad\circ\!\!-\!\!\longrightarrow$$
84

from the reactant stretch zero-point energy. The isotope effect, in this simple model at least, thus becomes a rough measure of the position of the transition state along the reaction coordinate. The isotope effect is expected to be largest for the most

[103] See K. B. Wiberg, *Physical Organic Chemistry*, Wiley, New York, 1970, pp. 332–361, for calculations that roughly justify this assumption for a specific example.

symmetrical location of the transition state, and smaller the closer the transition state is to either reactant or product.[104]

Primary isotope effects in nonlinear transition states If the transition state is nonlinear, the vibration corresponding to the symmetric stretch looks like **85**.

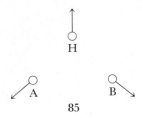

85

Now even for the symmetrical case the H (D) moves with relatively high frequency, and this mode cancels most of the zero-point contribution from the reaction coordinate mode of the reactants. Hence a bent transition state should show a small isotope effect. This mode is furthermore little affected by dissymmetry, and so the isotope effect for a nonlinear transition state will not be a sensitive indicator of position of the transition state along the reaction coordinate.[105]

Secondary Isotope Effects[106]

A *secondary isotope effect* is one that results from isotopic substitution at a bond not being broken in the reaction. The reaction cordinate, not being affected by the substitution, does not make any contribution to the isotope effect. Secondary effects must arise solely from changes of zero-point energies of ordinary vibrations. Thus if an isotopically substituted C—H bond experiences a change of force constant on going from reactant to transition state, the effect is approximately

$$\frac{k_H}{k_D} = \exp\left\{-\tfrac{1}{2}[u_{\ddagger H} - u_{\ddagger D} - (u_{rH} - u_{rD})]\right\} \tag{2.141}$$

or, using the approximation that $\nu_D = \nu_H/1.35$,

$$\frac{k_H}{k_D} = \exp\left[-\frac{0.1865}{T}(\nu_{\ddagger H} - \nu_{rH})\right] \tag{2.142}$$

Any vibration for which the frequency decreases on going to the transition state contributes a factor greater than unity to k_H/k_D, and any vibration for which the

[104] Various series of reactions have been studied in which k_H/k_D has a maximum for a presumably symmetrical transition state. See (a) R. P. Bell and B. G. Cox, *J. Chem. Soc. B*, 783 (1971); (b) R. A. More O'Ferrall, in *Proton Transfer Reactions*, E. Caldin and V. Gold, Eds., Chapman and Hall, London, 1975, p. 201. Bell has found evidence for the importance of tunneling in proton transfer reactions and has suggested that a major part of the isotope effect variation attributed by the simple theory described here to variation of the transition state may actually be caused by tunneling. (c) R. P. Bell, *Chem. Soc. Rev.*, **3**, 513 (1974).

[105] R. A. More O'Ferrall, *J. Chem. Soc. B*, 785 (1970).

[106] For a review of secondary isotope effects, see E. A. Halevi, *Prog. Phys. Org. Chem.*, **1**, 109 (1963).

frequency increases contributes a factor less than unity. A commonly observed secondary isotope effect occurs when deuterium substitution is made at a carbon that changes hybridization, as in Equations 2.143 and 2.144.

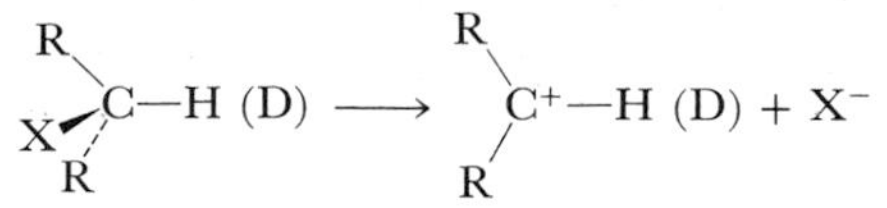

R
X
C—H (D)
R
R
C+—H (D) + X−
R

(2.143)

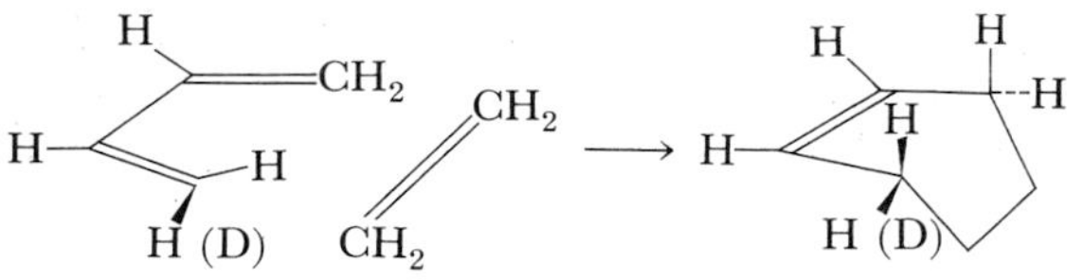

H
CH2
H—
H
H (D)
CH2
CH2
H—
H
H
H
H
H (D)

(2.144)

Streitwieser and collaborators have analyzed this process and concluded that, in going from sp^3 to sp^2 (Equation 2.143) the three C—H vibrations, one stretch and two bends, change as indicated below:[107]

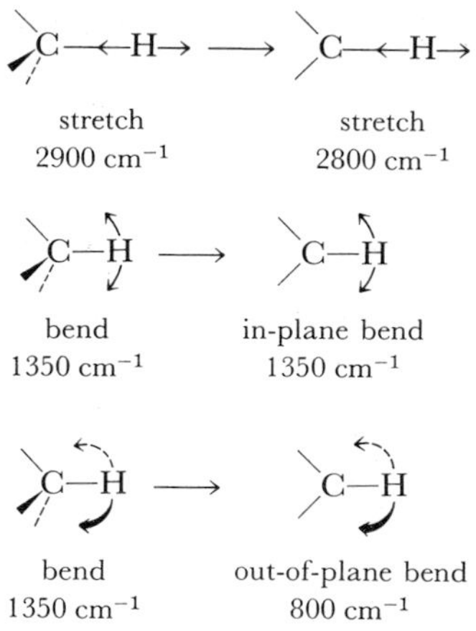

stretch
2900 cm−1
stretch
2800 cm−1
bend
1350 cm−1
in-plane bend
1350 cm−1
bend
1350 cm−1
out-of-plane bend
800 cm−1

The first change is small and the second nearly zero; the last one is significant and would contribute a factor of approximately

$$\frac{k_H}{k_D} = \exp\left[-\frac{0.1865}{T}(800 - 1350)\right] = 1.41$$

at 298 K if the transition state were very close to sp^2 hybridized product. The isotope effect will be smaller if the transition state comes earlier; it is typically around 1.15 to 1.25 for reactions of the type of Equation 2.143. (See Section 4.4 for further discussion.) For a reaction in which hybridization changes from sp^2 to sp^3, as in Equation 2.144, the effect will be inverse, k_H/k_D less than 1, with a minimum

[107] A. Streitwieser, Jr., R. H. Jagow, R. C. Fahey, and S. Suzuki, *J. Am. Chem. Soc.*, **80**, 2326 (1958).

of roughly $1/1.41 = 0.71$ for a transition state closely resembling sp^3 hybridized product, but typical values being between 0.8 and 0.9.[108]

    Substitution of deuterium at the β position leads to the β deuterium isotope effect in reactions like 2.145:

$$\text{(D) H}\diagdown\text{C–C}\diagdown\diagup \longrightarrow \text{(D) H}\diagdown\text{C–}\overset{+}{\text{C}}\diagdown \tag{2.145}$$

Here the C—H bond is weakened and the frequencies lowered by hyperconjugation. A bending mode is probably again the most important one;[107] k_H/k_D is greater than 1, values ranging up to about 1.4 for favorably situated hydrogens,[109] but more typically on the order of 1.1.

Solvent Isotope Effects

Isotope effects are frequently observed when reactions are carried out in solvents with O—H (O—D) groups. The reader is referred to the literature for further information.[110]

PROBLEMS

    1. Identify all the symmetry elements for allene, $H_2C{=}C{=}CH_2$, and for the dimethylallene shown in structures **3** and **4,** p. 114.

    2. Specify the configuration according to the R-S system of the biphenyls shown in structures **36** and **37**.

    3. The ρ value of hydroxide-catalyzed hydrolysis of substituted methyl benzoates is 2.4. The corresponding value when the excellent S_N2 nucleophile, ^-CN, is the catalyst is 1.74. Suggest a reason for the lower ρ value when ^-CN is the catalyst.

    4. Protonation of 1,1,3,3-tetramethyl-2-X-guanidine (**1**) has been shown to occur on the imino nitrogen when X is electron donating. A controversy about which nitrogen is protonated when X is —NO_2 was solved by a Hammett plot of the basicity of the guanidine vs. the σ constants of the substituents—including nitro. The plot was linear, with a negative ρ, for all substituents. What do you conclude about the protonation site when X = NO_2? What would you expect the plot to look like if an amine nitrogen were protonated when X = NO_2?

[108] Streitwieser's treatment as applied to addition to double bonds has been criticized as being oversimplified. However, a calculation including an important isotope-sensitive vibration of the transition state not present in reactants (a CH_2 twist) and also including the effect of moment-of-inertia changes on going to the transition state gives $k_H/k_D = 1/1.38 = 0.72$, only a small change from Streitwieser's value. See O. P. Strausz, I. Safarik, W. B. O'Callaghan, and H. E. Gunning, *J. Am. Chem. Soc.,* **94**, 1828 (1972).

[109] (a) V. J. Shiner, Jr., and J. G. Jewett, *J. Am. Chem. Soc.,* **86**, 945 (1964). Isotope effects can be observed from more remote positions. See (b) D. E. Sunko, S. Hiršl-Starčević, S. K. Pollack, and W. J. Hehre, *J. Am. Chem. Soc.,* **101**, 6163 (1979).

[110] (a) R. L. Schowen, *Prog. Phys. Org. Chem.,* **9**, 275 (1972); see also (b) P. M. Laughton and R. E. Robertson, in *Solute–Solvent Interactions,* Vol. 1, J. F. Coetzee and C. D. Ritchie, Eds., Dekker, New York, 1969, p. 399.

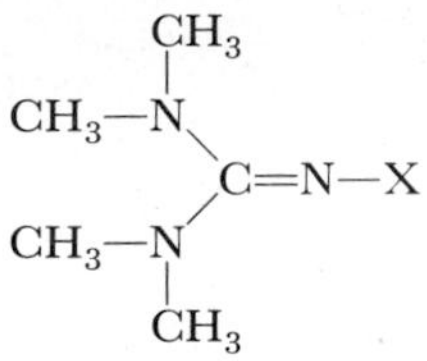

$$
\begin{array}{c}
\overset{\displaystyle CH_3}{\underset{\displaystyle |}{}} \\
CH_3-N \\
\end{array}
$$

1

5. 1-Substituents, Y, have approximately the same effect on the acidity of a carboxylic acid group in the 4 position of [2,2,2]bicyclooctane (**2**) as 1-substituents have on an analogously placed carboxylic acid group in the cubane system (**3**). What does this mean about the relative importance of field and inductive effects in saturated systems?

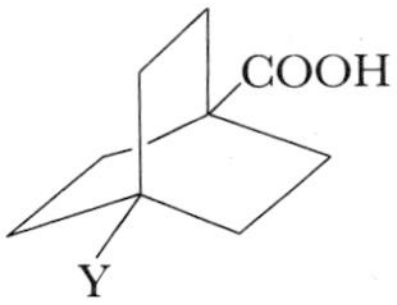

2 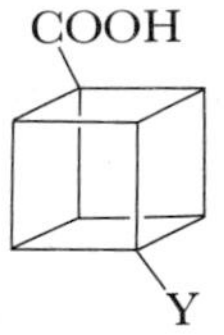   3

6. Benzaldehyde cyanohydrin formation, shown below, may involve rate-determining attack of either H^+ or ^-CN. From the ρ value for the rate of formation of cyanohydrins from substituted benzaldehydes (Table 2.4), which step do you think is rate determining?

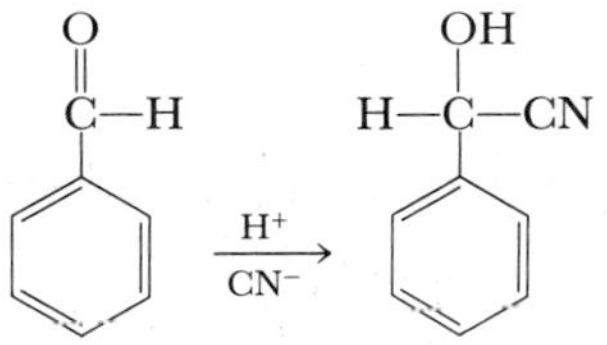

7. Derive a rate equation for formation of C in the following mechanism, assuming the stationary state for B and constant concentrations of D and E.

$$A \underset{k_{-1}}{\overset{k_1}{\rightleftharpoons}} B + D$$

$$E + B \xrightarrow{k_2} C$$

8. Derive the rate equation for rate of formation of E in terms of concentrations of reactants A and B in the following mechanism, assuming that the rates of steps k_1 and k_{-1} are both fast compared with the rate of step k_2. What is the kinetic order?

$$2A + B \underset{k_{-1}}{\overset{k_1}{\rightleftharpoons}} C + D$$

$$C + D \xrightarrow{k_2} E$$

9. Derive the rate equation for formation of F in terms of concentrations of A, B, and D in the following mechanism, assuming that A, B, and C are in equilibrium and E is a highly reactive intermediate.

$$A + B \underset{k_{-1}}{\overset{k_1}{\rightleftharpoons}} C$$

$$C + D \xrightarrow{k_2} E$$

$$2E \xrightarrow{k_3} F$$

10. Estimate the enthalpy of hydrogenation of benzene and of a hypothetical benzene with three fixed double bonds, each reacting with three moles of H_2 to yield cyclohexane, (a) from the data in Table 2.8 (use the average bond energies for C—C, C=C, and C—H) and (b) from the data in Table 2.9.

11. How do you know, from the information given in the text, that the Reaction 2.47, *trans*-2,3-dimethylcyclobutene → *trans,trans*-2,4-hexadiene, is subject to kinetic control?

12. In the reaction coordinate diagram of Figure 2.16, find the effect on transition state structure of (a) raising the energy of edge *SP* and (b) lowering the energy of edge *PQ*. To what changes in the nature of the reactants do these energy changes correspond?

13. Verify by reference to the equilibrium isotope effect equation (A2.18 in Appendix 2) the statement that the heavy isotope will concentrate, relative to the light, at that site where it is more strongly bound.

14. Rationalize the observation that D_3O^+ is a stronger acid than H_3O^+.

15. Verify that a decrease in H—C (D—C) vibrational frequency on dissociation will cause the observed secondary equilibrium isotope effect $K_{a_H}/K_{a_D} > 1$ for dissociation of HCOOH (DCOOH).

16. Estimate (a) ΔH_f° for triethylamine and (b) S° for 2-iodobutane.

17. Find the symmetry number for the *t*-butyl radical, $(CH_3)_3C \cdot$.

REFERENCES FOR PROBLEMS

1. K. Mislow, *Introduction to Stereochemistry*, W. A. Benjamin, Menlo Park, Calif., 1966, pp. 27, 32.
3. P. Müller and B. Siegfried, *Helv. Chim. Acta*, **57**, 987 (1974).
4. M. Liler, *Adv. Phys. Org. Chem.*, **11**, 267 (1975), p. 306.
5. T. W. Cole, Jr., C. J. Mayers, and L. M. Stock, *J. Am. Chem. Soc.*, **96**, 4555 (1974); note 101 in Chapter 3.
16. S. W. Benson, F. R. Cruickshank, D. M. Golden, G. R. Haugen, H. E. O'Neal, A. S. Rodgers, R. Shaw, and R. Walsh, *Chem. Rev.*, **69**, 279 (1969).

Appendix 1

DERIVATION OF
THE TRANSITION STATE
THEORY EXPRESSION
FOR A RATE CONSTANT[a]

In order to analyze the transition state equilibrium, we need to know how a collection of molecules divides up the available energy.

THE BOLTZMANN DISTRIBUTION

Molecules distribute their total energy among translational, rotational, vibrational, and electronic motions. These motions are all quantized, with energy-level separations very small for translation, larger for rotation, still larger for vibration, and very large for electronic motion. There are therefore many discrete energy states available. At very low temperature, approaching absolute zero, nearly all the molecules are in their lowest energy state; but as the temperature is raised, the molecules acquire more energy and begin to populate higher states. The ratio of numbers of molecules in any two states depends on the energy difference between the states and on the temperature, and is given by the Boltzmann distribution law,

$$\frac{n_2}{n_1} = \exp\left[\frac{-(\epsilon_2 - \epsilon_1)}{kT}\right] \tag{A1.1}$$

Here n_1 is the number of molecules in state 1, energy ϵ_1, n_2 is the number of molecules in state 2, energy ϵ_2, k is the Boltzmann constant, 1.3806×10^{-16} erg K^{-1}, and T is the Kelvin temperature.[b]

---

[a] Derivations may be found in the sources cited in note 85 of Chapter 2.
[b] For a derivation see K. B. Wiberg, *Physical Organic Chemistry,* Wiley, New York, 1964, p. 211, or W. J. Moore, *Physical Chemistry,* 4th ed., Prentice-Hall, Englewood Cliffs, N.J., 1972, p. 180.

Let us assume that we have two isomeric substances A and B in equilibrium at very low temperature, so we can say that for practical purposes all molecules are in their lowest energy states. Then we could regard A and B as two energy states of a system. Since there are only two populated states, the equilibrium constant K, the ratio of the number of B molecules to the number of A molecules, would be given by Equation A1.2:

$$K = \frac{N_B}{N_A} = \exp\left[\frac{-(\epsilon_{0B} - \epsilon_{0A})}{kT}\right] \tag{A1.2}$$

The quantity ϵ_{0B} is the energy of the lowest state of B, and ϵ_{0A} is the energy of the lowest state of A. As it is more convenient to have energies on a molar basis, we multiply the energies and the Boltzmann constant by Avagadro's number, N_0, and since kN_0 is equal to the gas constant R, 1.986 cal K mole^{-1}, we obtain Equation A1.3:

$$K = \exp\left[\frac{-(E_{0B}^\circ - E_{0A}^\circ)}{RT}\right] \tag{A1.3}$$

E_0° is the standard state energy at 0 K. Equation A1.3 may be written in the form A1.4, and since at absolute zero $\Delta E_0^\circ = \Delta H_0^\circ = \Delta G_0^\circ$, this equation is indeed the familiar thermodynamic expression for the equilibrium constant.

$$-\Delta E_0^\circ = RT \ln K \tag{A1.4}$$

The Partition Function

The example we have used is, of course, unrealistic; we are interested in what goes on at temperatures above absolute zero, where many energy levels of the molecules are populated. All we need do to correct our equilibrium constant (A1.3) is to find out how many molecules of A and of B are in each energy level at the temperature of interest. The total number of molecules of A, N_A, is the sum of numbers of molecules in each energy state,

$$N_A = n_{0A} + n_{1A} + n_{2A} + \cdots \tag{A1.5}$$

So far we have assumed that each state has a different energy, but this will not always be true. We have to allow for *degeneracies*, that is, groups of more than one state at the same energy. When two states have the same energy, their populations must be identical; we therefore modify Equation A1.5 to A1.6, where g_{iA} is the

$$N_A = g_{0A}n_{0A} + g_{1A}n_{1A} + g_{2A}n_{2A} + \cdots \tag{A1.6}$$

number of states that have energy ϵ_{iA}, and n_{iA} is the number of molecules occupying a single state of energy ϵ_{iA}. Since the Boltzmann distribution deals with ratios of numbers in various states, we divide both sides of Equation A1.6 by n_{0A} and obtain Equation A1.7, which gives the ratio of the total number of molecules to the

$$\frac{N_A}{n_{0A}} = g_{0A} + g_{1A}\frac{n_{1A}}{n_{0A}} + g_{2A}\frac{n_{2A}}{n_{0A}} + \cdots \tag{A1.7}$$

number in the lowest energy state. Then using Equation A1.1, we obtain

$$\frac{N_A}{n_{0A}} = g_{0A} + g_{1A} \exp\left[\frac{-(\epsilon_{1A} - \epsilon_{0A})}{kT}\right] + g_{2A} \exp\left[\frac{-(\epsilon_{2A} - \epsilon_{0A})}{kT}\right] + \cdots \qquad \text{(A1.8)}$$

or

$$\frac{N_A}{n_{0A}} = \sum_i g_{iA} \exp\left[\frac{-(\epsilon_{iA} - \epsilon_{0A})}{kT}\right] \qquad i = 0, 1, 2, \ldots \qquad \text{(A1.9)}$$

The ratio N_A/n_{0A} is defined as the partition function for A, Q_A,

$$Q_A = \frac{N_A}{n_{0A}} = \sum_i g_{iA} \exp\left[\frac{-(\epsilon_{iA} - \epsilon_{0A})}{kT}\right] \qquad \text{(A1.10)}$$

If we want to know the equilibrium constant for the isomerization $A \rightleftharpoons B$, we need again $K = N_B/N_A$. But we have N_A and N_B summed up over all the energy states in the partition functions,

$$N_A = n_{0A}Q_A \qquad \text{(A1.11)}$$

$$N_B = n_{0B}Q_B \qquad \text{(A1.12)}$$

and the equilibrium constant is given by Equation A1.13:

$$K = \frac{N_B}{N_A} = \frac{n_{0B}Q_B}{n_{0A}Q_A} \qquad \text{(A1.13)}$$

The ratio n_{0B}/n_{0A}, the numbers in the lowest states, is, as we have already seen, just $\exp\left[-(E_{0B}^\circ - E_{0A}^\circ)/RT\right]$, so our equilibrium constant is

$$K = \frac{Q_B}{Q_A} \exp\left[\frac{-(E_{0B}^\circ - E_{0A}^\circ)}{RT}\right] \qquad \text{(A1.14)}$$

or

$$RT \ln K = -\Delta E_0^\circ + RT \ln \left(\frac{Q_B}{Q_A}\right) \qquad \text{(A1.15)}$$

For an isomerization, in which there is no change in the number of molecules, the expression A1.15 is equal to $-\Delta G_T^\circ$. When the number of molecules changes, Equation A1.15 must be modified to Equation A1.16.[c]

$$RT \ln K = -\Delta G_T^\circ = -\Delta E_0^\circ + \Delta(PV) + RT \ln \left(\frac{Q_B}{Q_A}\right) \qquad \text{(A1.16)}$$

The Components of the Partition Function

For thinking about transition states, it is useful to divide the energy levels into categories and to associate a fraction of the partition function with each category.

[c] The other thermodynamic functions are readily derived from Equation A1.16. See, for example, K. B. Wiberg, *Physical Organic Chemistry*, Wiley, New York, 1964, p. 216.

Each energy level has contributions from translational, rotational, vibrational, and electronic substates,

$$\epsilon_i = \epsilon_{it} + \epsilon_{ir} + \epsilon_{iv} + \epsilon_{ie} \tag{A1.17}$$

If the state multiplicity g_i is the product of the multiplicities of the substates, each term in the partition function sum can be written in terms of these energies as in Equation A1.18,

$$g_i \exp\left[\frac{-(\epsilon_i - \epsilon_0)}{kT}\right] = g_{it} \exp\left[\frac{-(\epsilon_{it} - \epsilon_{0t})}{kT}\right] g_{ir} \exp\left[\frac{-(\epsilon_{ir} - \epsilon_{0r})}{kT}\right] g_{iv} \exp\left[\frac{-(\epsilon_{iv} - \epsilon_{0v})}{kT}\right]$$

$$\times\, g_{ie} \exp\left[\frac{-(\epsilon_{ie} - \epsilon_{0e})}{kT}\right] \tag{A1.18}$$

There is a term like Equation A1.18 for each energy level, and there is an energy level for every combination of each ϵ_{it}, ϵ_{ir}, ϵ_{iv}, ϵ_{ie} with every other, so the whole partition function is a multiple sum over all combinations,

$$Q = \sum_{\epsilon_{it}} g_{it} \exp\left[\frac{-(\epsilon_{it} - \epsilon_{0t})}{kT}\right] \sum_{\epsilon_{ir}} g_{ir} \exp\left[\frac{-(\epsilon_{ir} - \epsilon_{0r})}{kT}\right] \sum_{\epsilon_{iv}} g_{iv} \exp\left[\frac{-(\epsilon_{iv} - \epsilon_{0v})}{kT}\right]$$

$$\times \sum_{\epsilon_{ie}} g_{ie} \exp\left[\frac{-(\epsilon_{ie} - \epsilon_{0e})}{kT}\right] \tag{A1.19}$$

or

$$Q = f_t f_r f_v f_e \tag{A1.20}$$

where the f's are separated partition functions for the different kinds of motion.

Of these separated partition functions, we shall need to evaluate only that for vibration. The translational and rotational functions can be obtained from the quantum mechanics of a freely moving particle and of a rotating object, respectively, but we shall not need them in explicit form here.[d] The electronic partition function is unity for most molecules at ordinary temperatures, because electronic energy levels are so widely separated that for practical purposes only the lowest is populated.

The vibrational partition function is found by summing over the vibrational energy levels for each vibrational mode and multiplying together the results for all the modes. Assuming simple harmonic motion, the lowest energy level for a normal mode, the zero-point level, has energy $\epsilon_0 = \frac{1}{2}h\nu$, measured from the minimum of the potential energy curve. Here ν is the excitation frequency for the vibration (equal to the frequency observed for that mode in the infrared or Raman spectrum). The other levels are spaced upwards from this one at intervals of $h\nu$. The

---

[d]W. J. Moore, *Physical Chemistry*, 4th ed., Prentice-Hall, Englewood Cliffs, N.J., 1972, p. 196. The symmetry number, σ (Section 2.3) is introduced into the partition functions through the rotational part and appears as a divisor of the partition function for the species to which it applies.

levels thus fall at integral multiples of $h\nu$ above the lowest and, since each normal mode is nondegenerate, the vibrational partition function for each normal mode is

$$f_v^{\text{mode }i} = \sum_{n=0}^{\infty} \exp\left(-\frac{nh\nu_i}{kT}\right) \tag{A1.21}$$

Since an infinite sum of terms of the form e^{-ax} converges to $1/(1 - e^{-ax})$, the partition function A1.21 is more simply written

$$f_v^{\text{mode }i} = [1 - \exp(-u_i)]^{-1} \tag{A1.22}$$

where $u_i = h\nu_i/kT$. The total vibrational partition function is then a product of terms for the $3N - 6$ modes,

$$f_v = \prod_{i=1}^{3N-6} [1 - \exp(-u_i)]^{-1} \tag{A1.23}$$

THE TRANSITION STATE EQUILIBRIUM

Now consider Reaction A1.24 in the k_1 direction.

$$\text{A} \underset{k_{-1}}{\overset{k_1}{\rightleftharpoons}} \text{B} \tag{A1.24}$$

We have from Section 2.6 the following relations:

$$[\text{A}^{\ddagger}] = K_{\ddagger}[\text{A}] \tag{A1.25}$$

$$k_1[\text{A}] = k^{\ddagger}[\text{A}^{\ddagger}] \tag{A1.26}$$

$$k_1 = k^{\ddagger}K_{\ddagger} \tag{A1.27}$$

We express the equilibrium constant $K_{\ddagger}$ in terms of the partition function ratio $Q_{\ddagger}/Q_{\text{A}}$ to yield Equation A1.28, where $\Delta E_0^{\ddagger}$ is the difference between the lowest energy level of A and the lowest energy level of the transition state.

$$k_1 = k^{\ddagger}\frac{Q_{\ddagger}}{Q_{\text{A}}}\exp\left(-\frac{\Delta E_0^{\ddagger}}{RT}\right) \tag{A1.28}$$

Now we must analyze $Q_{\ddagger}$. It contains the usual translational and rotational functions, $f_t^{\ddagger}$ and $f_r^{\ddagger}$; the electronic contribution $f_e^{\ddagger}$ is unity. It is in the vibrational part that the difference from the ordinary stable molecule appears. Since one degree of freedom (that corresponding to the reaction coordinate) is no longer a vibration, there are only $3N - 7$ vibrations in $Q_{\ddagger}$, and these contribute in the usual way according to Equation A1.29.

$$f_v^{\ddagger} = \prod_{i=1}^{3N\ddagger-7} [1 - \exp(-u_{i\ddagger})]^{-1} \tag{A1.29}$$

Rewriting the contribution from the reaction coordinate motion as[e]

$$f_{RC} = \left[1 - \exp\left(-\frac{h\nu_{RC}}{kT}\right)\right]^{-1} \tag{A1.30}$$

we can then express k_1 in terms of a reduced partition function, $Q^{\ddagger}$, which contains only the translational, rotational, and vibrational contributions an ordinary molecule would have, and the special contribution f_{RC}:

$$k_1 = k^{\ddagger}\left[1 - \exp\left(-\frac{h\nu_{RC}}{kT}\right)\right]^{-1} \frac{Q^{\ddagger}}{Q_A} \exp\left(-\frac{\Delta E_0^{\ddagger}}{RT}\right) \tag{A1.31}$$

Now we make the following observations about ν_{RC}. It represents the "frequency" of the motion along the reaction coordinate. It is much smaller than ordinary vibration frequencies, and the quantity $\exp(-h\nu_{RC}/kT)$ can be approximated by using the first two terms of the series expansion:

$$\exp\left(-\frac{h\nu_{RC}}{kT}\right) = 1 - \frac{h\nu_{RC}}{kT} + \frac{1}{2!}\left(\frac{h\nu_{RC}}{kT}\right)^2 - \cdots \tag{A1.32}$$

$$\approx 1 - \frac{h\nu_{RC}}{kT} \tag{A1.33}$$

Furthermore, ν_{RC} corresponds to the frequency with which the transition state passes over to product and so may be set equal to $k^{\ddagger}$. Therefore we may write

$$k_1 = k^{\ddagger}\left[1 - \left(1 - \frac{h\nu_{RC}}{kT}\right)\right]^{-1} \frac{Q^{\ddagger}}{Q_A} \exp\left(-\frac{\Delta E_0^{\ddagger}}{RT}\right) \tag{A1.34}$$

or

$$k_1 = \frac{kT}{h} \frac{Q^{\ddagger}}{Q_A} \exp\left(-\frac{\Delta E_0^{\ddagger}}{RT}\right) \tag{A1.35}$$

Another way to express this result is to define a new equilibrium constant, $K^{\ddagger}$, which includes all features of the transition state except the reaction coordinate, and write

$$k_1 = \frac{kT}{h} K^{\ddagger} \tag{A1.36}$$

We then define free energy of activation as the free energy of the transition state excluding the reaction coordinate mode, so that Equations A1.37 and A1.38 hold.

$$\Delta G^{\ddagger} = -RT \ln K^{\ddagger} \tag{A1.37}$$

$$K^{\ddagger} = \exp\left(\frac{-\Delta G^{\ddagger}}{RT}\right) \tag{A1.38}$$

---

[e]This treatment follows that of K. J. Laidler, *Theories of Chemical Reaction Rates,* McGraw-Hill, New York, 1969, p. 46. Laidler also gives alternative, more rigorous derivations.

A factor κ, called the *transmission coefficient,* is sometimes included in the expression for k_1 to allow for the possibility that some transition states may be reflected back at the barrier, or that some may tunnel through it even though classically they do not have the requisite energy. These corrections are usually considered to be small, although the tunneling correction is probably significant for proton transfer reactions. We leave it as an exercise to the reader to extend the transition state treatment to bimolecular reactions.

Appendix 2

THE TRANSITION
STATE THEORY OF
ISOTOPE EFFECTS

We begin with the transition state theory result for the rate constant, Equation A1.35. We shall need to consider bimolecular processes, $A + B \rightarrow C$, for which the proper adaptation of Equation A1.35 is Equation A2.1. The quantity $\Delta E_0^{\ddagger}$ is the

$$k = \frac{kT}{h} \frac{Q^{\ddagger}}{Q_A Q_B} \exp\left(-\frac{\Delta E_0^{\ddagger}}{RT}\right) \tag{A2.1}$$

energy difference from the lowest level of the reactants up to the lowest energy level of the transition state. But each of these lowest levels is above the potential energy curve by the sum of the zero-point vibrational energies of all the modes. It is in these zero-point energies that the differences between the H and D compounds lie; we must therefore measure energies instead from the potential energy surface, which is the same for both. The quantity $\Delta E_0^{\ddagger}/RT$ is given by

$$\frac{\Delta E_0^{\ddagger}}{RT} = \frac{(\epsilon_{0\ddagger} - \epsilon_{0r})}{kT} \tag{A2.2}$$

where $\epsilon_{0\ddagger}$ is the zero-point energy of the transition state and ϵ_{0r} is the zero-point energy of reactants. Also,

$$\frac{\epsilon_{0\ddagger}}{kT} = \sum_i^{3N\ddagger-7} \frac{\frac{1}{2}h\nu_{i\ddagger}}{kT} + \frac{E_{\ddagger}}{RT} \tag{A2.3}$$

and

$$\frac{\epsilon_{0r}}{kT} = \sum_i^{3N_r-6} \frac{\frac{1}{2}h\nu_{ir}}{kT} + \frac{E_r}{RT} \tag{A2.4}$$

222

The transition state sum omits the reaction coordinate degree of freedom since it is not a bound vibration and does not contribute to the zero-point energy in the transition state. $E_{\ddagger}$ and E_r are respectively the energy of the potential energy surface at transition state and reactants. Then,

$$\frac{\Delta E_0^{\ddagger}}{RT} = \frac{1}{kT}\left[\sum_i^{3N\ddagger-7}\frac{1}{2}h\nu_{i\ddagger} - \sum_i^{3N_r-6}\frac{1}{2}h\nu_{ir}\right] + \frac{\Delta E}{RT} \tag{A2.5}$$

and

$$\exp\left(-\frac{\Delta E_0^{\ddagger}}{RT}\right) = \exp\left[-\frac{1}{kT}\left(\sum_i^{3N\ddagger-7}\frac{1}{2}h\nu_{i\ddagger} - \sum_i^{3N_r-6}\frac{1}{2}h\nu_{ir}\right) - \frac{\Delta E}{RT}\right] \tag{A2.6}$$

or

$$\exp\left(-\frac{\Delta E_0^{\ddagger}}{RT}\right) = \prod_i^{3N\ddagger-7}\exp\left(-\frac{1}{2}u_{i\ddagger}\right)\prod_i^{3N_r-6}\exp\left(\frac{1}{2}u_{ir}\right)\exp\left(-\frac{\Delta E}{RT}\right) \tag{A2.7}$$

where $u_i = h\nu_i/kT$, and ΔE is the energy difference along the potential surface from reactants to transition state. The expression for k is now given by Equation A2.8:

$$k = \frac{kT}{h}\frac{Q^{\ddagger}}{Q_A Q_B}\prod_i^{3N\ddagger-7}\exp\left(-\frac{1}{2}u_{i\ddagger}\right)\prod_i^{3N_r-6}\exp\left(\frac{1}{2}u_{ir}\right)\exp\left(-\frac{\Delta E}{RT}\right) \tag{A2.8}$$

The isotope effect is now found by taking the ratio of rate constants for the two isotopic systems (Equation A2.9):

$$\frac{k_H}{k_D} = \frac{Q_{AD}Q_{\ddagger H}}{Q_{AH}Q_{\ddagger D}}\prod_i^{3N\ddagger-7}\exp\left[-\frac{1}{2}(u_{iH} - u_{iD})\right]\prod_i^{3N_r-6}\exp\left[+\frac{1}{2}(u_{iH} - u_{iD})\right] \tag{A2.9}$$

The energy difference ΔE is independent of isotopic substitution and cancels. We have assumed that the isotopic substitution is in A, so Q_B cancels also.

The partition functions must now be written in terms of their component parts, which have the following values:

$$f_t = \left(\frac{2\pi MkT}{h^2}\right)^{3/2}V \tag{A2.10}$$

where M is the mass and V is the volume, and

$$f_r = \left(\frac{8\pi^2 kT}{h^2}\right)^{3/2}\frac{(\pi I_x I_y I_z)^{1/2}}{\sigma} \tag{A2.11}$$

where I_x, I_y, I_z are moments of inertia about three mutually perpendicular axes, and σ is the symmetry number. Of the quantities appearing in the expressions for these components, only the molecular mass M, the moments of inertia I, the vibrational frequencies u_i, and the symmetry numbers σ are different for the isotopic molecules; all other factors cancel, leaving Equation A2.12.

$$\frac{k_H}{k_D} = \frac{\sigma_H}{\sigma_D} \frac{\sigma_D^{\ddagger}}{\sigma_H^{\ddagger}} \left(\frac{M_D}{M_H}\right)_r^{3/2} \left(\frac{I_{xD}I_{yD}I_{zD}}{I_{xH}I_{yH}I_{zH}}\right)_r^{1/2} \left(\frac{M_H}{M_D}\right)_{\ddagger}^{3/2} \left(\frac{I_{xH}I_{yH}I_{zH}}{I_{xD}I_{yD}I_{zD}}\right)_{\ddagger}^{1/2} \prod_i^{3N\ddagger-7} \exp\left[-\frac{1}{2}(u_{iH}-u_{iD})_{\ddagger}\right]$$

$$\times \frac{1-\exp(-u_{iD_{\ddagger}})}{1-\exp(-u_{iH_{\ddagger}})} \prod_i^{3N_r-6} \exp\left[+\frac{1}{2}(u_{iH}-u_{iD})_r\right] \frac{1-\exp(-u_{iH_r})}{1-\exp(-u_{iD_r})} \qquad (A2.12)$$

This expression can fortunately be simplified by use of a theorem known as the Teller–Redlich rule, which expresses the molecular mass and moment of inertia ratios in terms of a ratio of a product of all the atomic masses m_j and the vibrational frequencies:[a]

$$\left(\frac{M_H}{M_D}\right)^{3/2} \left(\frac{I_{xH}I_{yH}I_{zH}}{I_{xD}I_{yD}I_{zD}}\right)^{1/2} = \prod_j^N \left(\frac{m_{jH}}{m_{jD}}\right)^{3/2} \prod_i^{3N-6} \left(\frac{\nu_{iH}}{\nu_{iD}}\right) \qquad (A2.13)$$

For the transition state, of course, one of the $3N - 6$ vibrations is really a translation; for the moment we single it out and write for its frequency ratio $\nu_{LH}^{\ddagger}/\nu_{LD}^{\ddagger}$. When Equation A2.13 is substituted into Equation A2.12, the products of atomic masses will cancel, leaving Equation A2.14:

$$\frac{k_H}{k_D} = \frac{\sigma_H}{\sigma_D} \frac{\sigma_D^{\ddagger}}{\sigma_H^{\ddagger}} \frac{\nu_{LH}^{\ddagger}}{\nu_{LD}^{\ddagger}} \prod_i^{3N\ddagger-7} \frac{\nu_{iH}}{\nu_{iD}} \exp\left[-\frac{1}{2}(u_{iH}-u_{iD})_{\ddagger}\right] \frac{1-\exp(-u_{iD_{\ddagger}})}{1-\exp(-u_{iH_{\ddagger}})} \prod_i^{3N_r-6} \frac{\nu_{iD}}{\nu_{iH}}$$

$$\times \exp\left[+\frac{1}{2}(u_{iH}-u_{iD})_r\right] \frac{1-\exp(-u_{iH_r})}{1-\exp(-u_{iD_r})} \qquad (A2.14)$$

This expression gives the isotope effect in terms of vibrational frequencies only; if the molecules are simple enough, a complete vibrational analysis and direct calculation of the isotope effect will be possible. But for most purposes we want an expression that will be easier to apply. Some simplification can be achieved by noting that for all those vibrational modes that involve no substantial motion at the isotopically substituted position, $\nu_{iH} = \nu_{iD}$ (and therefore also $u_{iH} = u_{iD}$) in both reactant and transition state. These modes will therefore cancel and need not be considered further. Moreover, any mode that does involve motion at the isotopically substituted position but that has the same force constant in reactant and transition state will have ν_H in the reactant equal to ν_H in the transition state and likewise for ν_D, and will also cancel. We therefore need consider only those modes for which force constants of vibrations involving the isotopically substituted position change on going from reactant to transition state. For vibrations involving hydrogen, most of which have frequencies above 1000 cm^{-1}, the factor $1 - e^{-u}$ is approximately unity. Furthermore, since all the ratios ν_H/ν_D should be about $\sqrt{2}$, they will approximately cancel.[b] If we ignore for the moment the symmetry number ratio, which can always be put in later if needed, we then have

---

[a] (a) K. B. Wiberg, *Physical Organic Chemistry*, Wiley, New York, 1964, p. 275; (b) J. Bigeleisen and M. Wolfsberg, *Adv. Chem. Phys.*, **1**, 15 (1958).
[b] See (a) L. Melander, *Isotope Effects on Reaction Rates*, Ronald Press, New York, 1960, p. 38; (b) J. Bigeleisen, *Pure Appl. Chem.*, **8**, 217 (1964), for further discussion.

$$\frac{k_{\mathrm{H}}}{k_{\mathrm{D}}} \approx \prod_i^{\ddagger} \exp\left[-\frac{1}{2}(u_{i\mathrm{H}} - u_{i\mathrm{D}})_{\ddagger}\right] \prod_i^r \exp\left[+\frac{1}{2}(u_{i\mathrm{H}} - u_{i\mathrm{D}})_r\right] \tag{A2.15}$$

where the products are over only those vibrations that involve force constant changes at isotopically substituted positions.

It is frequently also necessary to assess isotope effects on equilibria. For an equilibrium

$$\mathrm{AH} + \mathrm{BD} \xrightleftharpoons{K_{\mathrm{H/D}}} \mathrm{AD} + \mathrm{BH} \tag{A2.16}$$

the appropriate expression is[c]

$$K_{\mathrm{H/D}} = \frac{\sigma_{\mathrm{AH}}\sigma_{\mathrm{BD}}}{\sigma_{\mathrm{AD}}\sigma_{\mathrm{BH}}} \prod_i^{3N_A-6} \frac{\nu_{i\mathrm{AD}}}{\nu_{i\mathrm{AH}}} \exp\left[+\frac{1}{2}(u_{i\mathrm{AH}} - u_{i\mathrm{AD}})\right] \frac{1 - \exp(-u_{i\mathrm{AH}})}{1 - \exp(-u_{i\mathrm{AD}})} \prod_i^{3N_B-6} \frac{\nu_{i\mathrm{BH}}}{\nu_{i\mathrm{BD}}}$$

$$\times \exp\left[-\frac{1}{2}(u_{i\mathrm{BH}} - u_{i\mathrm{BD}})\right] \frac{1 - \exp(-u_{i\mathrm{BD}})}{1 - \exp(-u_{i\mathrm{BH}})} \tag{A2.17}$$

Again the terms $1 - e^{-u}$ will all be close to unity and the ratios $\nu_{\mathrm{H}}/\nu_{\mathrm{D}}$ should all be close to $\sqrt{2}$. If symmetry numbers are omitted, we have

$$K_{\mathrm{H/D}} \approx \prod_i^{3N_A-6} \exp\left[+\tfrac{1}{2}(u_{i\mathrm{AH}} - u_{i\mathrm{AD}})\right] \prod_i^{3N_B-6} \exp\left[-\tfrac{1}{2}(u_{i\mathrm{BH}} - u_{i\mathrm{BD}})\right] \tag{A2.18}$$

[c] K. B. Wiberg, *Physical Organic Chemistry*, Wiley, New York, 1964, p. 275.

Chapter 3
ACIDS
AND BASES

3.1 BRØNSTED ACIDS AND BASES

Of the concepts that chemists use to make sense of chemical transformations, ideas about acids and bases are among the most fruitful. Nearly all of the heterolytic reactions that we shall be considering can be thought of as acid–base processes; it is therefore appropriate to begin our discussion of the chemical properties of organic compounds with a review of these ideas and of their applications in organic chemistry.

Definition of Brønsted Acids and Bases

Acids and bases have been known for centuries, but the definitions in common use today are of comparatively recent origin. In 1923 J. N. Brønsted proposed the following definitions:[1]

> **An acid is a proton donor.**
> **A base is a proton acceptor.**

An acid HA is thus any substance that reacts according to Equation 3.1, and a base B is any substance that reacts according to Equation 3.2:

$$HA \longrightarrow H^+ + A^- \tag{3.1}$$
$$B + H^+ \longrightarrow BH^+ \tag{3.2}$$

[1] J. N. Brønsted, *Rec. Trav. Chim. Pays-Bas,* **42,** 718 (1923).

Equations 3.1 and 3.2 are oversimplified as descriptions of acid–base chemistry in the liquid phase. The proton, H^+, does not exist free to any appreciable extent in solution but is always solvated by one or more molecules of some other species. Bonding to one solvent molecule would produce in water H_3O^+; in ammonia NH_4^+; in alcohols, ROH_2^+.[2] In general, in solvent S, coordination of a proton to solvent yields SH^+, referred to as the lyonium ion of the solvent. Even these formulations are oversimplifications, because the lyonium ion, once formed, will itself be solvated by close association with more solvent molecules, the number and strength of binding depending on the solvent.

It has proved possible to study by mass spectrometry the binding in the gas phase of protons to water molecules and to molecules of various other basic species.[3] The binding of the first water molecule to yield H_3O^+ is about 165 kcal mol^{-1} exothermic;[3a] coordination of the second water molecule[3b] releases 36 kcal mol^{-1}. Addition of subsequent water molecules up to at least the eighth releases decreasing but still significant amounts of energy. Kebarle and co-workers are of the opinion that exothermicity on adding more water molecules to the cluster indicates that the proton is bound equally to several water molecules, as would be indicated by the formula $H^+(H_2O)_n$, rather than being coordinated more tightly to one, as would be indicated by $H_3O^+(H_2O)_m$.[3b] Gas-phase binding of the proton to CH_3OH, $(CH_3)_2O$,[4] H_2S, HCN, and CH_3CN[5] have also been investigated. The small size and consequent large electrostatic field of the proton makes it seem very likely that in solution association of H^+ with a base is a general phenomenon.[6]

Although it is clear that the proton in aqueous solution is solvated by a number of water molecules as yet not precisely determined but certainly greater than one, the shorthand designations H_3O^+ or even more simply H^+, where the remaining solvation is understood, are commonly used. The H_3O^+ formulation emphasizes an important symmetry of Brønsted acid–base reactions, illustrated in Equation 3.3 for the dissociation of acetic acid in water. Here the acid donates a proton and the base accepts it; this chemical change constitutes an acid–base reaction in the Brønsted sense.

In this reaction, note that the reverse process is an acid–base reaction just as

[2] See (a) R. P. Bell, *The Proton in Chemistry*, 2d ed., Cornell University Press, Ithaca, N.Y., 1973, p. 13; (b) M. Eigen, *Agnew. Chem. Int. Ed. Engl.*, **3**, 1 (1964). The H_3O^+ ion has been detected by ^{17}O nmr spectroscopy in a solution containing $1.5M$ H_2O and a slight excess of $HF-SbF_5$ in SO_2 as solvent. See C. D. Mateescu and G. M. Benedikt, *J. Am. Chem. Soc.*, **101**, 3959 (1979).

[3] (a) M. S. B. Munson, *J. Am. Chem. Soc.*, **87**, 2332 (1965); (b) P. Kebarle, S. K. Searles, A. Zolla, J. Scarborough, and M. Arshadi, *J. Am. Chem. Soc.*, **89**, 6393 (1967).

[4] E. P. Grimsrud and P. Kebarle, *J. Am. Chem. Soc.*, **95**, 7939 (1973). The first $(CH_3)_2O$ molecule interacts more strongly than subsequent ones, probably for steric reasons.

[5] (a) M. Meot-Ner and F. H. Field, *J. Am. Chem. Soc.*, **99**, 998 (1977); (b) M. Meot-Ner, *J. Am. Chem. Soc.*, **100**, 4694 (1978).

[6] G. A. Olah, A. M. White, and D. H. O'Brien, *Chem. Rev.*, **70**, 561 (1970), review evidence for coordination of protons with a large number of substances.

the forward process is. The acetate ion is a base that can accept a proton from the acid H_3O^+. This reciprocal relationship is emphasized by the terminology. Acetate ion is called the *conjugate base* of the acid CH_3COOH, and H_3O^+ is called the *conjugate acid* of the base H_2O.

$$H_3C-\overset{\overset{\textstyle O}{\|}}{C}-OH + H_2O \rightleftharpoons H_3O^+ + H_3C-\overset{\overset{\textstyle O}{\|}}{C}-O^- \qquad (3.3)$$

$$\text{acid} \qquad \text{base} \qquad \text{conjugate} \qquad \text{conjugate}$$
$$\text{acid} \qquad \text{base}$$

In considering an acid–base reaction, it is important to realize that the choice of which acid is to be called the conjugate acid is completely arbitrary. In Equation 3.3 we could just as well have decided to call H_2O the conjugate base of the acid H_3O^+ and CH_3COOH the conjugate acid of the base CH_3COO^-.

Note that acid-base reactions can be of a variety of charge types; a general representation is

$$HA^{m+} + B^{n+} \rightleftharpoons A^{(m-1)+} + HB^{(n+1)+} \qquad (3.4)$$

where m and n can each be a positive or negative integer or zero.

We must also recognize that many molecules that we ordinarily think of as exhibiting neither acidic nor basic behavior are in fact acids or bases, or, frequently, both. For example, acetone, which is neutral in water solution, reacts as a base in sulfuric acid according to the equilibrium in 3.5; and in dimethylsulfoxide containing sodium methoxide, acetone is an acid (Equation 3.6).

$$H_3C-\overset{\overset{\textstyle :O:}{\|}}{C}-CH_3 + H_2SO_4 \rightleftharpoons H_3C-\overset{\overset{\textstyle :\overset{+}{O}H}{\|}}{C}-CH_3 + HSO_4^- \qquad (3.5)$$

$$H_3C-\overset{\overset{\textstyle :O:}{\|}}{C}-CH_3 + CH_3-\overset{\cdot\cdot}{\underset{\cdot\cdot}{O}}:^- \rightleftharpoons H_3C-\overset{\overset{\textstyle :O:}{\|}}{C}-CH_2^- + CH_3OH \qquad (3.6)$$

Logical extension of these ideas leads to the conclusion that acetic acid is a base as well as an acid, and that aniline, a substance ordinarily considered as a base, can also act as an acid (Equations 3.7 and 3.8).

$$H_3C-\overset{\overset{\textstyle :O:}{\|}}{C}-\overset{\cdot\cdot}{\underset{\cdot\cdot}{O}}H + H_2SO_4 \rightleftharpoons H_3C-\overset{\overset{\textstyle :\overset{+}{O}H}{\|}}{C}-\overset{\cdot\cdot}{\underset{\cdot\cdot}{O}}H + HSO_4^- \qquad (3.7)$$

$$C_6H_5-\overset{\cdot\cdot}{N}H_2 + \overset{\cdot\cdot}{N}H_2^- \rightleftharpoons C_6H_5\overset{\cdot\cdot}{N}H^- + \overset{\cdot\cdot}{N}H_3 \qquad (3.8)$$

Indeed, one may conclude that any molecule containing hydrogen is a potential Brønsted acid, whereas any molecule at all is a potential Brønsted base.

Acid and Base Strength

For acids that can be studied in aqueous solution, we measure the strength by the magnitude of the equilibrium constant for dissociation, K_a. This quantity is defined by first writing the equilibrium constant K'_a for Reaction 3.9, using activities,[7] and then converting to the form given in Equation 3.11 by incorporating the water activity, which is essentially constant in dilute solution when water is the solvent, into the equilibrium constant.

$$HA^{m+} + pH_2O \rightleftharpoons A^{(m-1)+} + H^+(H_2O)_p \tag{3.9}$$

$$K_a' = \frac{a_{A^{(m-1)+}} a_{H^+(H_2O)_p}}{a_{HA^{m+}} (a_{H_2O})^p} \tag{3.10}$$

$$K_a = \frac{a_{A^{(m-1)+}} a_{H^+(H_2O)_p}}{a_{HA^{m+}}} \tag{3.11}$$

Because we will be considering acid dissociation in various solvents, and in recognition of the fact that the degree of solvation of H^+ is imprecisely known and varies from solvent to solvent, we will simplify the nomenclature by referring to the proton henceforth simply as H^+; it is to be understood that this symbol represents the proton in a state of solvation as it exists in the medium under consideration. With this simplification, Equation 3.11 becomes 3.12, which can also be written in terms of activity coefficients (Equation 3.13).[8]

$$K_a = \frac{a_{A^{(m-1)+}} a_{H^+}}{a_{HA^{m+}}} \tag{3.12}$$

$$K_a = \frac{[A^{(m-1)+}][H^+]\gamma_{A^{(m-1)+}} \gamma_{H^+}}{[HA^{m+}]\gamma_{HA^{m+}}} \tag{3.13}$$

The standard state is defined as the hypothetical state that would exist if the solute were at a concentration of $1M$, but with the molecules experiencing the environment of an extremely dilute solution; with this standard state, activity coefficients approach unity with increasing dilution. For electrolytes in dilute solution in water, the departure of the coefficients from unity can be calculated from Debye–Hückel limiting law.[9]

It is possible to define another equilibrium constant, K_c (Equation 3.14), which does not include the activity coefficients and hence will not be a true con-

$$K_c = \frac{[A^{(m-1)+}][H^+]}{[HA^{m+}]} \tag{3.14}$$

stant except in very dilute solutions, where it approaches the thermodynamic K_a that we have been considering so far. The constant K_c is often used for convenience, but it is not satisfactory for careful work, nor where comparisons between different solvents must be made.

---

[7] W. J. Moore, *Physical Chemistry,* 4th ed., Prentice-Hall, Englewood Cliffs, N.J., 1972, p. 233.
[8] The activity coefficient γ is defined so that $a = \gamma c$, where a is activity and c is concentration.
[9] See W. J. Moore, *Physical Chemistry,* 4th ed., Prentice-Hall, Englewood Cliffs, N.J., 1972, pp. 449–457.

Base strengths can be defined similarly by the equilibrium constant for Reaction 3.15:

$$B^{m+} + pH_2O \rightleftharpoons BH^{(m+1)+} + OH^-(H_2O)_p \tag{3.15}$$

$$K_b' = \frac{a_{BH^{(m+1)+}} \, a_{OH^-(H_2O)_p}}{a_{B^{m+}} (a_{H_2O})^p} \tag{3.16}$$

Or, adopting the same conventions as before,

$$K_b = \frac{a_{BH^{(m+1)+}} \, a_{OH^-}}{a_{B^{m+}}} \tag{3.17}$$

However, it is more convenient to consider instead of Reaction 3.15 the acid dissociation of the acid $BH^{(m+1)+}$:

$$BH^{(m+1)+} + pH_2O \rightleftharpoons B^{m+} + H^+(H_2O)_p \tag{3.18}$$

$$K_a = \frac{a_{B^{m+}} \, a_{H^+}}{a_{BH^{(m+1)+}}} \tag{3.19}$$

If K_a for equilibrium 3.18 is known, K_b, as defined by Equation 3.17, can easily be found by use of the constant K_w, the ionization constant of pure water. K_w is defined by Equations 3.20 and 3.21 and has been carefully measured at various

$$H_2O \rightleftharpoons H^+ + OH^- \tag{3.20}$$

$$K_w = a_{H^+} a_{OH^-} \tag{3.21}$$

temperatures.[10] It has the value $10^{-14.00}$ at $25°C$.[11] From Equations 3.17, 3.19, and 3.21, it is easy to verify that the relation between K_b of a substance and K_a of its conjugate acid is Equation 3.22:

$$K_a K_b = K_w \tag{3.22}$$

In order to avoid proliferation of tables, it is customary to report only one constant for each conjugate acid–conjugate base pair. The reader may easily verify that if acid A is a stronger acid than acid B, the conjugate base of A will be a weaker base than the conjugate base of B.

The Leveling Effect

We are now in a position to consider the experimental problems involved in measuring equilibrium constants for acids of differing strengths. One may use any of a number of methods of determining the concentrations of the various species involved in the reaction; the most common procedure for aqueous solutions is to use the glass electrode, which allows a convenient and accurate determination of hy-

---

[10]Values of K_w at various temperatures are given by R. G. Bates, *Determination of pH, Theory and Practice*, 2d ed, Wiley, New York, 1973, p. 448.
[11]H. S. Harned and R. A. Robinson, *Trans. Faraday Soc.*, **36**, 973 (1940).

drogen ion activity over a wide range.[12] Other possibilities include spectrophotometric determinations of acid and conjugate base, and conductimetric measurement of ion concentrations.[13]

It generally happens that the range of acidity that can be determined in a given solvent is limited by the acid–base reactions of the solvent itself. Consider, for example, the hypothetical situation of two acids, HA_1 and HA_2, with dissociation constants of 10^{+2} and 10^{+3} ($pK_a = -2$ and -3).[14] If we add enough of each of these acids to water to give solutions $0.1M$ in total acid, the solutions will be respectively $0.09990M$ and $0.09999M$ in hydrogen ion, a difference of only 0.0004 pH unit. This difference is too small to measure. Qualitatively, both the substances, being stronger acids than the hydrated proton, are able to transfer their protons essentially completely to the water, and so their strengths cannot be distinguished. Note that, if two acids are again separated by one pK unit, but this time have dissociation constants of 10^{-4} and 10^{-5} ($pK_a = +4$ and $+5$), the pH of the two solutions will differ by an easily measurable 0.5 unit. Similar difficulties arise with very weak acids; in this case the amount of H^+ produced by dissociation of the acid is less than the amount present by virtue of the ionization of water itself (Equation 3.20) and so cannot be determined. As a rough rule we can state that in water solution it is possible to measure strengths only of those acids that are stronger than water and weaker than the hydrated proton; by the same token, bases can be studied in water only if they are stronger bases than water and weaker than OH^-.

The phenomenon described above for water also applies to other amphoteric solvents. It is termed the *leveling effect* and may be summarized by the following statements:

1. No acid stronger than the conjugate acid of a solvent can exist in appreciable concentration in that solvent.

2. No base stronger than the conjugate base of a solvent can exist in appreciable concentration in that solvent.

Useful corallaries of these statements are the following:

1. Relative strengths of acids stronger than the conjugate acid of a solvent cannot be determined in that solvent.

---

[12](a) R. G. Bates (note 10) gives a comprehensive discussion of pH measurements. See also (b) C. C. Westcott, *pH Measurements*, Academic Press, New York, 1978.
[13]For discussions of methods for determining K_a, see (a) E. J. King, *Acid–Base Equilibrium*, Pergamon Press, Elmsford, N.Y., 1965; (b) A. Albert and E. P. Serjeant, *The Determination of Ionization Constants*, Chapman and Hall, London, 1971; (c) R. F. Cookson, *Chem. Rev.*, **74**, 5 (1974).
[14]pK_a is defined by the equation

$$pK_a = -\log K_a$$

A pK difference of one unit thus corresponds to a factor of ten difference in K_a.

2. Relative strengths of bases stronger than the conjugate base of the solvent cannot be determined in that solvent.

The acids in which we are interested span a range of roughly 60 pK units, from the strongest acids (HI, $HClO_4$) to the weakest (methane, cyclohexane), and since there is no single solvent that is suitable for the entire range, it is necessary to use several different solvents and to try to make connections among the results obtained.

Water is taken as the standard solvent for setting up a scale of acidity. It has the advantage, in addition to convenience, of having a high dielectric constant and being effective in solvating ions. As we noted in Section 2.4, the result of these properties is that positive and negative ions separate, and complications that result from association of ions in pairs or in larger aggregates are avoided. For acids too strong to be investigated in water solution, more acidic media such as acetic acid or mixtures of water with sulfuric or perchloric acid are commonly used; for very weak acids solvents such as liquid ammonia, dimethylsulfoxide, and cyclohexylamine have been employed.

Ideally, one might hope to be able to relate results in different solvent systems to each other and thus to construct a universal scale of acidity. As we shall see in more detail in Sections 3.2 and 3.3, it has not been possible to establish such a scale, primarily because solvent–solute interactions affect acid–base equilibria differently in different solvents and in ways as yet not thoroughly understood. Despite this difficulty a great deal of useful information about structures of molecules and ions, about reaction mechanisms, and about the solvent–solute interactions themselves can be obtained from acid–base measurements.

3.2 STRENGTHS OF WEAK BRØNSTED BASES; ACIDITY FUNCTIONS

A variety of organic reactions, including dehydration of alcohols, cleavage of ethers, many additions to alkenes, a number of nucleophilic substitutions, and various rearrangements, are catalyzed by acids. Since the substrates in these processes are bases, it is reasonable to postulate that the reactions involve acid–base interactions. In order to obtain further information about the detailed course of these types of reactions, it is often desirable to be able to make quantitative measurements of the acid–base properties of the substances involved.

Acidity Functions

One solution to the problem of achieving appreciable concentrations of the protonated form of very weak bases is to use as a solvent a mixture of water with some strong mineral acid. It can be demonstrated by measurement of freezing-point depressions[15] that many organic compounds that contain basic atoms such as N, O, or S, but that are too weakly basic to be protonated to a significant extent in water,

---

[15] For a discussion of freezing-point depression, see W. J. Moore, *Physical Chemistry*, 4th ed., Prentice-Hall, Englewood Cliffs, N.J., 1972, p. 247.

are essentially completely converted to their conjugate acids in concentrated sulfuric acid.[16] In appropriately chosen mixtures of water and sulfuric acid, appreciable concentrations of both base and conjugate acid may be expected to be present.

Hammett and Deyrup, in 1932, proposed a method of determining quantitatively acid–base behavior in water–strong acid mixtures.[17-19] In order to understand their contribution, we begin with the general expression for the equilibrium constant for the dissociation of an acid of the charge type Hammett and Deyrup used for setting up their scale:

$$BH^+ \rightleftharpoons H^+ + B \tag{3.23}$$

$$K_a = \frac{a_{H^+} a_B}{a_{BH^+}} \tag{3.24}$$

or

$$K_a = \frac{a_{H^+}[B]\gamma_B}{[BH^+]\gamma_{BH^+}} \tag{3.25}$$

BH^+ represents the protonated form of a weak base, typically an organic molecule protonated at an atom with unshared pairs of electrons or at a double bond. Structures **1–4** are examples.

1	**2**	**3**	**4**

In Equations 3.24 and 3.25 we have taken the solvation of the proton as understood and have written its abbreviated form. The solvent molecules that enter Reaction 3.23 to accept the proton from BH^+ are taken into K_a and are not

---

[16] For example, the observation that the freezing point of a 1 molal solution of acetone in sulfuric acid is depressed by twice the molal freezing-point depression constant of sulfuric acid is interpreted in terms of the reaction

$$H_3C-\overset{O}{\underset{\parallel}{C}}-CH_3 + H_2SO_4 \rightleftharpoons H_3C-\overset{+OH}{\underset{\parallel}{C}}-CH_3 + HSO_4^-$$

This equilibrium lies far to the right; 2 moles of ions are produced for each mole of acetone added. Similar results are obtained with many other compounds that are neutral in water but contain unshared electron pairs.

[17] L. P. Hammett and A. J. Deyrup, *J. Am. Chem. Soc.*, **54**, 2721 (1932).
[18] L. P. Hammett, *Physical Organic Chemistry,* 2d ed., McGraw-Hill, New York, 1970, Chap. 9.
[19] For reviews, see (a) M. A. Paul and F. A. Long, *Chem. Rev.*, **57**, 1 (1957); (b) R. A. Boyd, in *Solute–Solvent Interactions*, Vol. 1, J. F. Coetzee and C. D. Ritchie, Eds., Dekker, New York, 1969, p. 97; (c) E. M. Arnett, *Prog. Phys. Org. Chem.*, **1**, 223 (1963); (d) E. M. Arnett and G. Scorrano, *Adv. Phys. Org. Chem.*, **13**, 83 (1976); (e) C. H. Rochester, *Acidity Functions*, Academic Press, London, 1970; (f) M. Liler, *Reaction Mechanisms in Sulphuric Acid*, Academic Press, London, 1971; (g) R. A. Cox and K. Yates, *J. Am. Chem. Soc.*, **100**, 3861 (1978) (pK_a values given in supplementary material available from the American Chemical Society).

shown. Note that the nature of the solvent S and of the protonated solvent SH^+ are not well defined, because in mixed solvents each consists of more than one species; however, the proton-donating ability of SH^+ and the proton-accepting ability of S, whatever they may be, together determine the effectiveness of the particular solvent mixture in protonating the base B. It is this "protonation effectiveness" that Hammett and Deyrup set out to measure.

The first step is to choose a series of bases, $B_1, B_2, B_3, \ldots, B_n, \ldots$, each weaker than the previous one. We also require that these substances absorb light in the visible or ultraviolet region, and that the absorption spectra of the free bases differ from the spectra of their respective conjugate acids. The reason for this requirement is that we must have some means of determining the concentrations [B] and $[BH^+]$ for each of the base–conjugate acid pairs; the visible-ultraviolet spectrophotometric method is convenient and is the one that has been employed most frequently.[20] Other methods, notably nmr spectroscopy, have come into more frequent use in recent years.[21] Hammett and Deyrup picked as their series of bases various substituted anilines with increasing numbers of electron-withdrawing substituents to provide successively weaker bases. The first base, B_1, is sufficiently strong that the acid dissociation constant of its conjugate acid can be determined in pure water. We next go to water containing a small amount of sulfuric acid, for example, 10 percent of H_2SO_4, in which the base B_1 will still give appreciable concentrations of both the conjugate acid and conjugate base forms, and that will also allow measurements to be made on the weaker base B_2, which is too weak to give measurable amounts of B_2H^+ in pure water.

We may now write two equations of the type 3.25 describing the behavior of our two bases in the new solvent:

$$K_{a_{B_1H^+}} = \frac{a_{H^+}[B_1]\gamma_{B_1}}{[B_1H^+]\gamma_{B_1H^+}} \tag{3.26}$$

$$K_{a_{B_2H^+}} = \frac{a_{H^+}[B_2]\gamma_{B_2}}{[B_2H^+]\gamma_{B_2H^+}} \tag{3.27}$$

Note that $K_{a_{B_1H^+}}$ is known from the measurements in dilute water solution; we have defined the quantities in the equations in such a way that the K_a's are truly constants (at constant temperature and pressure), and all nonideal behavior resulting from changing the solvent is incorporated into the activities. Furthermore, the concentrations $[B_1]$, $[B_1H^+]$, $[B_2]$, $[B_2H^+]$ are directly measurable spectrophotometrically. If we divide Equation 3.26 by Equation 3.27, we obtain Equation 3.28:

[20] The spectrophotometric measurements often present experimental difficulties. See (a) E. M. Arnett, R. P. Quirk, and J. J. Burke, *J. Am. Chem. Soc.*, **92**, 1260 (1970); (b) E. M. Arnett, R. P. Quirk, and J. W. Larsen, *J. Am. Chem. Soc.*, **92**, 3977 (1970). A method for overcoming some of these problems has been proposed by (c) J. T. Edward and S. C. Wong, *J. Am. Chem. Soc.*, **99**, 4229 (1977).
[21] E. M. Arnett and G. Scorrano, *Adv. Phys. Org. Chem.*, **13**, 83 (1976).

$$\frac{K_{a_{B_1H^+}}}{K_{a_{B_2H^+}}} = \frac{[B_1][B_2H^+]\gamma_{B_1}\gamma_{B_2H^+}}{[B_2][B_1H^+[\gamma_{B_2}\gamma_{B_1H^+}}\tag{3.28}$$

In Equation 3.28 all quantities are known or measurable except $K_{a_{B_2H^+}}$ and the ratio involving activity coefficients.

If it were possible to obtain the activity coefficients, Equation 3.28 would provide a way of obtaining $K_{a_{B_2H^+}}$. In dilute aqueous solution the Debye–Hückel theory, which is based on calculation of interionic forces in a medium containing dissociated ions, provides a method for estimating activity coefficients of ions. However, even for ionic strengths as low as 0.01 there are significant deviations from the theory.[22] In the strong acid–water mixtures under consideration here, the concentration of ionic species ($H^+(H_2O)_n$, HSO_4^-) is high; thus even if the concentrations of the acids and bases under study are kept small, the Debye–Hückel theory is of no help. It is possible, however, to make the following qualitative argument. The departure of the activity coefficients from unity is the result of some nonideal behavior of the species involved. Departures from ideality therefore depend on the structure, and probably particularly on the charges, of the components. If B_1 and B_2 (and thus also B_1H^+ and B_2H^+) are sufficiently close in structure, one might guess that in a given solvent the ratio $\gamma_{B_1}/\gamma_{B_1H^+}$ would be approximately the same as $\gamma_{B_2}/\gamma_{B_2H^+}$. If this were the case, the ratio of activity coefficients in Equation 3.28 would equal unity and the equation would become

$$\frac{K_{a_{B_1H^+}}}{K_{a_{B_2H^+}}} = \frac{[B_1][B_2H^+]}{[B_2][B_1H^+]}\tag{3.29}$$

An experimental check on this assumption about activity coefficients is possible over a limited range of solvent acidity. If the composition of water–sulfuric acid mixtures is varied over the range in which all four species, B_1, B_2, B_1H^+, and B_2H^+, are present in appreciable concentration, then since $K_{a_{B_1H^+}}/K_{a_{B_2H^+}}$ is (by definition) constant, a constant ratio $[B_1][B_2H^+]/[B_2][B_1H^+]$ implies that the assumption of the ratio of γ's being constant is correct in this range of solvents. Experimentally, for bases that are substituted anilines this test is fairly successful. The question of how similar two compounds must be to be "sufficiently close in structure" will be considered later.

Proceeding with our analysis, we find that if we can assume that Equation 3.29 is valid, we know all quantities necessary to obtain $K_{a_{B_2H^+}}$, the equilibrium constant for the second base. This base is now used in conjunction with a third base, B_3, in a solvent system containing a larger proportion of strong acid, and the procedure is continued until equilibrium constants are established for the whole range of bases.

Having found equilibrium constants for the series of bases, we may now use them to characterize the proton-donating ability of any mixture of sulfuric acid

---

[22] Hammett, *Physical Organic Chemistry*, 2d ed., McGraw-Hill, New York, 1970, p. 192.

and water. Rearranging Equation 3.25, we have

$$K_a \frac{[BH^+]}{[B]} = \frac{a_{H^+}\gamma_B}{\gamma_{BH^+}} \tag{3.30}$$

The quantity on the right side of Equation 3.30 is defined as h_0; it gives the desired information about proton-donating ability.

$$h_0 \equiv \frac{a_{H^+}\gamma_B}{\gamma_{BH^+}} = K_a \frac{[BH^+]}{[B]} \tag{3.31}$$

Because of the magnitudes of the numbers involved, it is more convenient to use a logarithmic scale. A new quantity, H_0, is therefore defined by Equation 3.32:

$$H_0 \equiv -\log h_0 = -\log \left\{ K_a \frac{[BH^+]}{[B]} \right\} = -\log \frac{a_{H^+}\gamma_B}{\gamma_{BH^+}} \tag{3.32}$$

or

$$\log \frac{[B]}{[BH^+]} = H_0 - pK_a \tag{3.33}$$

H_0 is known as the Hammett acidity function, and the series of substituted anilines used to establish the scale are called Hammett indicators. Note that as the proportion of sulfuric acid in the water–acid mixture approaches zero, γ_B and γ_{BH^+} approach the pure water value of unity and H_0 merges into pH.

The procedure outlined above serves to define H_0 for mixtures of water and various strong acids. Ideally, once H_0 has been found for a number of different mixtures, one could obtain, by use of Equation 3.33, the pK_a for bases other than those used in setting up the scale. One must be able to measure the ratio $[B]/[BH^+]$, and one must assume that the ratio of activity coefficients of the new base and its conjugate acid is the same as that of the indicator bases. Ranges of pK_a values that have been found for various types of compounds will be given in Section 3.4. Values for particular compounds may be found in reviews.[23]

Figure 3.1 shows the H_0 function in aqueous sulfuric acid. It is sometimes useful to have H_0 broken down into its various components, γ_B, γ_{BH^+}, and a concentration and activity coefficient of H^+. Because the actual concentration of hydrogen ions, $[H^+]$, cannot be determined in water–strong acid mixtures, an activity coefficient γ_{H^+} is defined so that

$$a_{H^+} = C_{H^+}\gamma_{H^+} \tag{3.34}$$

where C_{H^+} is the formal molar concentration of acid in the mixture.[24]

---

[23] See notes 19(a), (c), (d), and (g).
[24] For example, a water–acid mixture prepared by mixing 5 moles of H_2SO_4 with enough water to make 1 l of solution has $C_{H^+} = 5$. (K_a of HSO_4^- is 1.2×10^{-2}; therefore very little SO_4^{2-} forms under these conditions, and H_2SO_4 is for practical purposes a monoprotic acid.)

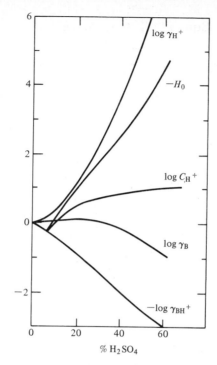

Figure 3.1 The behavior of the acidity function H_0 in sulfuric acid–water mixtures. The contributions to H_0 are broken down according to the equation:

$$-H_0 = \log C_{H^+} + \log \gamma_{H^+} + \log \gamma_B - \log \gamma_{BH^+}$$

(See text for definition of C_{H^+}.) Vertical scale in logarithmic units. From K. Yates and R. A. McClelland, in *Progress in Physical Organic Chemistry*, Vol 11, A. Streitwieser, Jr., and R. W. Taft, Eds., Wiley, New York, 1974, p. 372. Reprinted by permission of John Wiley & Sons, Inc.

Other Acidity Scales

For the H_0 treatment to be successful, the substituted anilines used to determine the H_0 scale, and also those bases that are to be investigated using the scale, must be of sufficiently similar structure that the activity coefficient ratio of Equation 3.28 will be unity. It has become increasingly evident as data have accumulated that this requirement is more restricting than one might have hoped. Arnett and Mach,[25] using a set of *N,N*-dialkylnitroanilines and *N*-alkylnitroanilines as indicator bases, found an acidity scale, designated H''', which is different from H_0. A group of cyclic amines, indoles of general structure **5**, were investigated by Hinman

[25] E. M. Arnett and G. W. Mach, *J. Am. Chem. Soc.*, **86**, 2671 (1964).

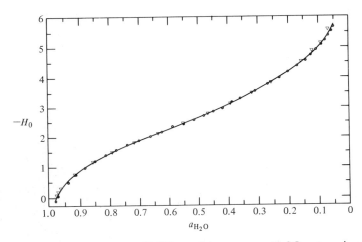

5

and Lang;[26] these indicators gave still another acidity scale, denoted H_I, which differed slightly from the H''' scale. Another scale, H_A, was established by Yates, Stevens, and Katritzky[27] with a series of amides as indicators. Still another function, H_R, is based on the behavior of triarylcarbinols. These substances, studied by Deno, Jaruzelski, and Schriesheim,[28] typically react according to Equation 3.35 to form water (which is converted partly to oxonium ion) and a carbocation. The H_R function thus includes the activity of water in addition to the quantities of Equation 3.33.

$$(C_6H_5)_3COH + H^+ \rightleftharpoons (C_6H_5)_3C^+ + H_2O \qquad (3.35)$$

A function H'_R, derived from H_R by subtracting the logarithm of the water activity (Figure 3.2), is appropriate to protonation of diaryl alkenes.[29] The function

Figure 3.2 Activity of water in aqueous $HClO_4$ and in aqueous H_2SO_4 at various values of H_0. Reprinted with permission from K. Yates and H. Wai, *J. Am. Chem. Soc.*, **86**, 5408 (1964). Copyright 1964 American Chemical Society.

[26] R. L. Hinman and J. Lang, *J. Am. Chem. Soc.*, **86**, 3796 (1964).
[27] K. Yates, J. B. Stevens, and A. R. Katritzky, *Can. J. Chem.*, **42**, 1957 (1964).
[28] (a) N. C. Deno, J. J. Jaruzelski, and A. Schriesheim, *J. Am. Chem. Soc.*, **77**, 3044 (1955); (b) N. C. Deno, H. E. Berkheimer, W. L. Evans, and H. J. Peterson, *J. Am. Chem. Soc.*, **81**, 2344 (1959).
[29] N. C. Deno, P. T. Groves, and G. Saines, *J. Am. Chem. Soc.*, **81**, 5790 (1959).

$$H'_R = H_R - \log a_{H_2O} \tag{3.36}$$

H_C is a more recent substitute for H'_R, derived directly from measurements on alkenes, and a modification of it, $H_C^{\ddagger}$, correlates acid-catalyzed proton–deuterium exchange rates in carbon bases.[30] Figure 3.3 shows the behavior of several acidity functions in sulfuric acid–water mixtures, and Figure 3.4 illustrates the H_0 function for mixtures of water with various strong acids.[31]

The proliferation of acidity scales demonstrates the need to take careful account of solvation. The difficulty is that the ratio γ_B/γ_{BH^+}, instead of being constant for all organic bases of the same charge type, varies significantly with the

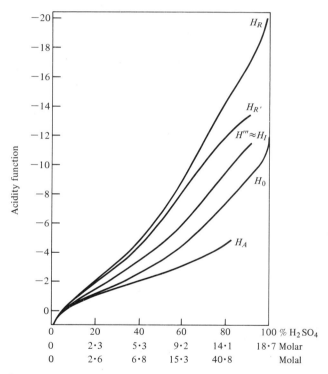

Figure 3.3 Values of various acidity functions in mixtures of water and sulfuric acid. With permission from C. H. Rochester, *Acidity Functions,* Academic Press, London, 1970 p. 90. Copyright by Academic Press Inc. (London) Ltd.

[30](a) M. T. Reagan, *J. Am. Chem. Soc.,* **91,** 5506 (1969); (b) C. C. Greig, C. D. Johnson, S. Rose, and P. G. Taylor, *J. Org. Chem.,* **44,** 745 (1979).
[31]For a more complete discussion of the various acidity functions, see (a) L. P. Hammett, *Physical Organic Chemistry,* 2d ed., McGraw-Hill, New York, 1970, Chap. 9; (b) C. H. Rochester, *Acidity Functions,* Academic Press, London, 1970. Acidity function measurements in ethanol–water–sulfuric acid mixtures have been reported: (c) C. Capobianco, F. Magno, and G. Scorrano, *J. Org. Chem.,* **44,** 1654 (1979).

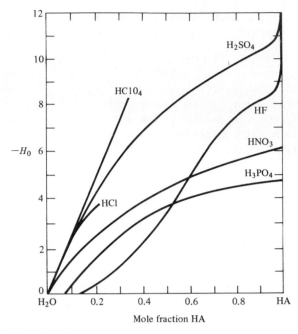

Figure 3.4 The H_0 acidity function for mixtures of water with various acids. From G. A. Olah, *Friedal–Crafts Chemistry,* Wiley, New York, 1973, p. 368. Reprinted by permission of John Wiley & Sons, Inc.

functional group being protonated and, to a lesser degree, with changing substituents on a given functional group. Hence a universally applicable treatment of weak bases will require a second parameter, in addition to H_0, that will characterize the response of the ratio γ_B/γ_{BH^+} for the particular compound under consideration to the particular solvation properties of the water–acid mixture. Two attempts to establish such a two-parameter acidity scale will be considered briefly.

Bunnett and Olsen derived a scheme that allows just one acidity function, H_0, to be used for all bases, but with a correction factor characteristic of each base.[32] Referring to Equation 3.32 and writing $C_{H^+}\gamma_{H^+}$ for a_{H^+}, we obtain Equation 3.37,[33]

$$H_0 = -\log C_{H^+} - \log \frac{\gamma_{H^+}\gamma_B}{\gamma_{BH^+}} \qquad (3.37)$$

where B represents the substituted aniline indicator bases used to establish the H_0 scale. Bunnett and Olsen replaced Hammett's assumption that the activity coeffi-

---

[32] J. F. Bunnett and F. P. Olsen, *Can. J. Chem.,* **44**, 1899, 1917 (1966).
[33] C_{H^+} is the formal concentration of acid in the mixture. See note 24.

cient ratio would be the same for other bases by the assumption that the ratios would be linearly related:

$$\log \frac{\gamma_{H^+}\gamma_Z}{\gamma_{ZH^+}} = (1 - \phi) \log \frac{\gamma_{H^+}\gamma_B}{\gamma_{BH^+}} \tag{3.38}$$

where Z is any other base. Rearranging 3.37 and substituting it into 3.38 yields Equation 3.39:

$$\log \frac{\gamma_{H^+}\gamma_Z}{\gamma_{ZH^+}} = -(1 - \phi)(H_0 + \log C_{H^+}) \tag{3.39}$$

Using the definition of pK_a, Equation 3.40, we obtain Equation 3.41:

$$pK_{a_{ZH^+}} = -\log K_{a_{ZH^+}} = -\log \frac{C_{H^+}[Z]}{[ZH^+]} - \log \frac{\gamma_{H^+}\gamma_Z}{\gamma_{ZH^+}} \tag{3.40}$$

$$pK_{a_{ZH^+}} = -\log \frac{C_{H^+}[Z]}{[ZH^+]} + (1 - \phi)(H_0 + \log C_{H^+}) \tag{3.41}$$

Equation 3.41 can be rewritten as Equation 3.42, which, when compared to Equation 3.33, can be seen to specify the quantity $H_0 - \phi(H_0 + \log C_{H^+})$ as a new

$$\log \frac{[Z]}{[ZH^+]} = H_0 - \phi(H_0 + \log C_{H^+}) - pK_a \tag{3.42}$$

corrected acidity function. H_0 and $\log C_{H^+}$ depend only on the medium; ϕ is characteristic of the base Z and is the parameter that relates all the various acidity functions mentioned above to H_0. For a particular base Z, pK_a and ϕ are found by plotting the quantity $\log ([ZH^+]/[Z]) + H_0$ against $H_0 + \log C_{H^+}$.[34]

A second approach to the acidity function problem has been proposed by Marziano and Passerini and developed by Cox and Yates.[35] The Bunnett–Olsen approach still requires the H_0 scale and so depends on the constancy of the ratio γ_B/γ_{BH^+} for the Hammett indicators. Cox and Yates define the excess acidity **X** by Equation 3.43, where B* is a hypothetical reference base. They then postulate that

$$\mathbf{X} = \log \frac{\gamma_{H^+}\gamma_{B^*}}{\gamma_{B^*H^+}} \tag{3.43}$$

the logarithm of the activity coefficient ratio for the base Z will be linearly related to **X**:

$$\log \frac{\gamma_{H^+}\gamma_Z}{\gamma_{ZH^+}} = m^*\mathbf{X} \tag{3.44}$$

[34] For applications of the Bunnett–Olsen treatment, see, for example, (a) E. M. Arnett and G. Scorrano, *Adv. Phys. Org. Chem.*, **13**, 83 (1976); (b) J. T. Edward and S. C. Wong, *J. Am. Chem. Soc.*, **99**, 4229 (1977); (c) G. Scorrano, *Acc. Chem. Res.*, **6**, 132 (1973); (d) P. Bonvicini, A. Levi, V. Lucchini, G. Modena, and G. Scorrano, *J. Am. Chem. Soc.*, **95**, 5960 (1973); (e) A. Levi, G. Modena, and G. Scorrano, *J. Am. Chem. Soc.*, **96**, 6585 (1974).
[35] (a) N. C. Marziano, P. G. Traverso, A. Tomasin, and R. C. Passerini, *J. Chem. Soc., Perkin Trans.* 2, 309 (1977); (b) R. A. Cox and K. Yates, *J. Am. Chem. Soc.*, **100**, 3861 (1978).

Substituting this expression into Equation 3.40 and rearranging yields Equation 3.45, which shows that in the Cox–Yates treatment the quantity $-\log C_{H^+} - m^*\mathbf{X}$

$$\log \frac{[Z]}{[ZH^+]} = -\log C_{H^+} - m^*\mathbf{X} - pK_a \tag{3.45}$$

serves as the acidity function. The $\mathbf{X}$ parameter cannot be derived directly from experiment; the procedure is an iterative analysis of data for a number of different bases. The ratio $[Z]/[ZH^+]$ is known from experiment for each base in each water–acid mixture, and $\log C_{H^+}$ is known from the formal acid concentration. The quantities m^* and pK_a are unknowns characteristic of the particular base, and $\mathbf{X}$ is an unknown characteristic of the solvent but the same for all bases. An iterative analysis of data for a number of bases over the range of pure water to pure sulfuric acid, or some other strong mineral acid, yields values for the unknowns.

Applications of Acidity Functions in Mechanism Studies

In order to investigate the many organic reactions catalyzed by strong acids, it is necessary to be able to determine the concentration of the protonated form of the substrate. The acidity function approach allows determination of the pK_a for weak bases and also the extent of protonation in various acidic media, through Equation 3.33, 3.42, or 3.45, provided that the acidity function (or its substitute in the more recent treatments) is known for the medium in question and provided that the concentration ratio $[Z]/[ZH^+]$ can be determined.

A question that often arises in the study of acid-catalyzed reactions is what role water plays in the transition state for the rate-determining step. For example, the acid-catalyzed hydrolysis of methyl benzoate might proceed by either Scheme 1 or Scheme 2.[36] In Scheme 1, termed A-1, the slow step is unimolecular and the

SCHEME 1

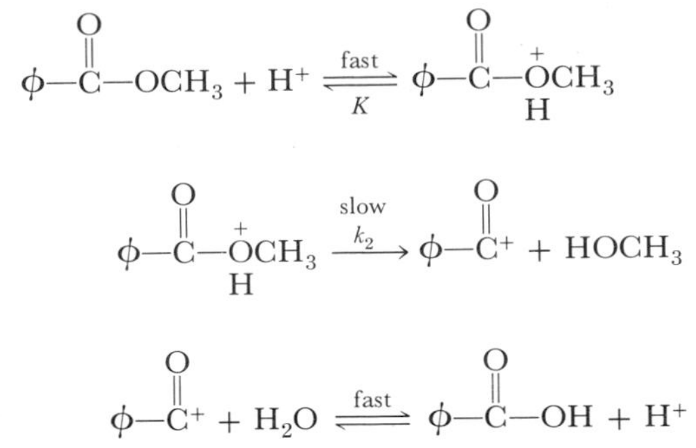

---

[36]These schemes are somewhat oversimplified as to details of possible intermediates. See Chapter 8 for further discussion.

SCHEME 2

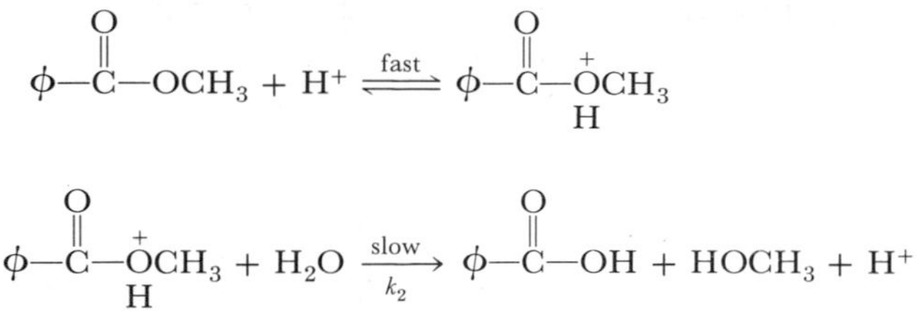

$$\phi-\overset{\overset{\textstyle O}{\|}}{C}-OCH_3 + H^+ \underset{}{\overset{fast}{\rightleftharpoons}} \phi-\overset{\overset{\textstyle O}{\|}}{C}-\overset{+}{\underset{H}{O}}CH_3$$

$$\phi-\overset{\overset{\textstyle O}{\|}}{\underset{H}{C}}-\overset{+}{O}CH_3 + H_2O \xrightarrow[k_2]{slow} \phi-\overset{\overset{\textstyle O}{\|}}{C}-OH + HOCH_3 + H^+$$

transition state contains only protonated substrate; in Scheme 2, called A-2, the slow step is bimolecular with a water molecule in the transition state. The kinetics are followed by monitoring the total amount of substrate, $[S] + [SH^+]$, where we have written S for the ester and SH^+ for its conjugate acid. Scheme 1 predicts

$$-\frac{d([S] + [SH^+])}{dt} = k_2[SH^+]\frac{\gamma_{SH^+}}{\gamma_{\ddagger}} \tag{3.46}$$

where the activity coefficients of SH^+ and of the transition state for the second step are included to account for nonideal behavior in the strong acid medium. Using Equation 3.47 for the acid dissociation constant of SH^+ and Equation 3.31 relating a_{H^+} and h_0, we obtain Equation 3.48 for the rate of disappearance of substrate.

$$K_a = \frac{a_{H^+}a_S}{a_{SH^+}} = \frac{a_{H^+}[S]}{[SH^+]}\frac{\gamma_S}{\gamma_{SH^+}} \tag{3.47}$$

$$\frac{d([S] + [SH^+])}{dt} = \frac{k_2}{K_a}[S]h_0\frac{\gamma_S\gamma_{BH^+}}{\gamma_{\ddagger}\gamma_B} \tag{3.48}$$

(In Equation 3.48 B and BH^+ still refer to the Hammett indicator bases used to establish the H_0 scale.) A kinetic run at fixed acidity will appear as a pseudo first-order reaction, with observed rate constant

$$k_{obs} = -\frac{1}{[S] + [SH^+]}\frac{d([S] + [SH^+])}{dt} = \frac{[S]}{[S] + [SH^+]}\frac{k_2}{K_a}h_0\frac{\gamma_S\gamma_{BH^+}}{\gamma_{\ddagger}\gamma_B} \tag{3.49}$$

If most of the substrate is in the unprotonated form, that is, if $[S] \gg [SH^+]$, then Equation 3.49 reduces to Equation 3.50. An early treatment of acid-catalyzed

$$k_{obs} = \frac{k_2}{K_a}h_0\frac{\gamma_S\gamma_{BH^+}}{\gamma_{\ddagger}\gamma_B} \tag{3.50}$$

kinetics, by Zucker and Hammett,[37] assumed that the activity coefficient canceling assumption that goes into the H_0 function definition would apply to Equation 3.50,

---

[37] L. Zucker and L. P. Hammett, *J. Am. Chem. Soc.*, **61**, 2791 (1939).

yielding 3.51. Equation 3.52 then predicts that a plot of log k_{obs} vs. $-H_0$ should be a straight line with a slope of unity.

$$k_{obs} \approx \frac{k_2}{K_a} h_0 \tag{3.51}$$

$$\log k_{obs} = -H_0 + \log \frac{k_2}{K_a} \tag{3.52}$$

For Scheme 2, a similar derivation leads to Equation 3.53,

$$k_{obs} = \frac{k_2}{K_a} C_{H^+} \frac{a_{H_2O} \gamma_{H^+} \gamma_S}{\gamma_\ddagger} \tag{3.53}$$

again assuming $[S] \gg [SH^+]$. Zucker and Hammett proposed that in this case $a_{H_2O} \gamma_{H^+} \gamma_S / \gamma_\ddagger$ would be approximately unity, and hence that a plot of log k_{obs} vs. log C_{H^+} would be linear with unit slope (Equation 3.54).[38]

$$\log k_{obs} \approx \log C_{H^+} + \log \frac{k_2}{K_a} \tag{3.54}$$

Although the correlations found according to the Zucker–Hammett criterion frequently failed to have the predicted unit slope, they did indicate that for the hydrolysis of many esters the correlation was better with log C_{H^+} than with $-H_0$, a result consistent with the A-2 mechanism of Scheme 2. For hydrolysis of acetals, on the other hand, log k_{obs} is linear when plotted against $-H_0$; these reactions presumably proceed by Scheme 1.[39]

As might be expected from the earlier discussion, the assumptions of Zucker and Hammett about activity coefficient ratios have turned out to fail frequently, as indicated by nonunit slopes of the correlations described above. There have been several other approaches to the problem. One that has been widely used is that of Bunnett.[40] Equation 3.53 can be written as 3.55; then making the same substitution for a_{H^+} as was done in deriving Equation 3.48, we obtain Equations 3.56–3.58.

$$k_{obs} = \frac{k_2}{K_a} a_{H^+} a_{H_2O} \frac{\gamma_S}{\gamma_\ddagger} \tag{3.55}$$

$$k_{obs} = \frac{k_2}{K_a} h_0 a_{H_2O} \frac{\gamma_S \gamma_{BH^+}}{\gamma_\ddagger \gamma_B} \tag{3.56}$$

$$k_{obs} \approx \frac{k_2}{K_a} h_0 a_{H_2O} \tag{3.57}$$

$$\log k_{obs} + H_0 = \log a_{H_2O} + \log \frac{k_2}{K_a} \tag{3.58}$$

[38] C. H. Rochester, *Acidity Functions*, Academic Press, London, 1970, p. 113.
[39] For a summary, see Chapter 5 of reference 38.
[40] J. F. Bunnett, *J. Am. Chem. Soc.*, **83**, 4956 (1961), and following papers.

Bunnett introduced a parameter w to characterize the extent of participation of water, as indicated in Equation 3.59, where the final term in Equation 3.58 has

$$\log k_{\text{obs}} + H_0 = w \log a_{\text{H}_2\text{O}} + Const \qquad (3.59)$$

been replaced by the constant *Const*. Comparison of values found for w in a number of reactions indicated that for substrates protonated on nitrogen or oxygen, negative values of w meant no involvement of water in the rate-determining step (Scheme 1), and values above $\sim +1.2$ meant water is involved (Scheme 2).

An alternative treatment, based on the Bunnett–Olsen acidity function, is to plot the equation

$$\log k_{\text{obs}} + H_0 = \phi(H_0 + \log C_{\text{H}^+}) + Const \qquad (3.60)$$

Negative values of ϕ indicate that water is not involved in the rate-determining step; values of ϕ above $+0.22$ indicate that water is involved.[41] Table 3.1 shows results of application of these methods to a few of the many acid-catalyzed reactions that have been investigated. The Bunnett criteria indicate that the ester hydrolyses involve a water molecule, while the acetal hydrolyses do not. In these cases the older Zucker–Hammett criterion led to the same conclusions, although slopes were not unity in all cases. Epoxide hydrolysis should involve no water molecule according to the Zucker–Hammett criterion, although the slope is not

Table 3.1 APPLICATION OF ZUCKER–HAMMETT AND OF BUNNETT
CRITERIA TO SOME ACID-CATALYZED HYDROLYSES

Reaction	$-H_0$	Slope $\log k_{\text{obs}}$ vs. C_{H^+}	w^a	ϕ^b
$CH_3COOCH_3 + H_2O \xrightarrow{\text{aq HCl}} CH_3COOH + CH_3OH$		1.0^c	$+5.83$	$+0.80$
$\phi COOCH_3 + H_2O \xrightarrow{\text{aq HClO}_4} \phi COOH + CH_3OH$		1.02^d	$+7.02$	$+0.93$
$H_2C(OCH_3)_2 + H_2O \xrightarrow{\text{aq HCl}} H_2CO + 2CH_3OH$	1.15^e		-5.26	-0.70
Sucrose hydrolysis, aq HCl	0.96^f		$+0.43$	-0.45
$H_2C\overset{O}{\diagup \diagdown}CHCH_2Cl \xrightarrow{\text{aq HClO}_4} H_2\overset{HO}{\underset{\vert}{C}}-\overset{OH}{\underset{\vert}{C}}HCH_2Cl$	0.87^g		$+2$ to $+3^h$	

[a] J. F. Bunnett, *J. Am. Chem. Soc.*, **83**, 4956 (1961).
[b] J. F. Bunnett and F. P. Olsen, *Can. J. Chem.*, **44**, 1917 (1966).
[c] M. Duboux and A. de Sousa, *Helv. Chim. Acta*, **23**, 1381 (1940).
[d] C. T. Chmiel and F. A. Long, *J. Am. Chem. Soc.*, **78**, 3326 (1956).
[e] D. McIntyre and F. A. Long, *J. Am. Chem. Soc.*, **76**, 3240 (1954).
[f] P. M. Leininger and M. Kilpatrick, *J. Am. Chem. Soc.*, **60**, 2891 (1938).
[g] J. G. Pritchard and F. A. Long, *J. Am. Chem. Soc.*, **78**, 2667 (1956).
[h] J. F. Bunnett, *J. Am. Chem. Soc.*, **83**, 4978 (1961).

[41] Bunnett has correlated various roles the water molecule can play with the value of w in the range above $+1.2$ and of ϕ above $+0.22$. Up to $w = 3.3$ ($\phi = +0.56$), the water molecule acts as a nucleophile; above $w = +3.3$ ($\phi = +0.58$) as a proton transfer agent. Further examples will be considered in Chapter 8.

unity; the Bunnett criterion, which is probably more reliable, indicates that a water molecule does enter the rate-determining step.

Superacids

There has been a great deal of interest in recent years in properties and uses of extremely acidic media, which have been termed *superacids*. The most commonly used principal constituent of these mixtures is fluorosulfuric acid, HSO_3F, to which is ordinarily added SbF_5. The mixture $HF–SbF_5$ has also been used. These mixtures can protonate extremely weak bases,[42] and because of their very low nucleophilicity, they can provide stable solutions of carbocations from alcohols or halides. Applications to the study of carbocations will be discussed in Chapters 4 and 5. Figure 3.5 illustrates the acidity of $HSO_3F–SbF_5$ and of $HF–SbF_5$ mixtures.

Measurements in Nonaqueous Solvents

An alternative to the acidity function method for making measurements with weak base–strong acid conjugate pairs is to choose a pure solvent that is more acidic and less basic than water. The convenient glass electrode and pH meter can often be

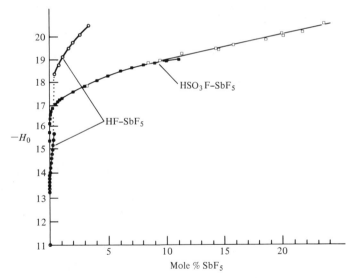

Figure 3.5 The H_0 acidity function for mixtures of SbF_5 and HSO_3F and for $SbF_5–HF$. Data up to 8 percent SbF_5 in $SbF_5–HSO_3F$ and up to 0.5 percent SbF_5 in $SbF_5–HF$ are from R. S. Gillespie and T. E. Peel, *J. Am. Chem. Soc.*, **95**, 5173 (1973). Adapted with permission from J. Sommer, S. Schwartz, P. Rimmelin, and P. Canivet, *J. Am. Chem. Soc.*, **100**, 2576 (1978). Copyright 1978 American Chemical Society.

[42] G. A. Olah and J. Shen, *J. Am. Chem. Soc.*, **95**, 3582 (1973).

used successfully in nonaqueous media as long as the reference solution used for standardization of the meter employs the same solvent.[43] The pH values determined, however, will be characteristic of the particular solvent system and will not be directly transferable to the water scale. The use of a single solvent avoids the difficulties inherent in making comparisons between different solvent systems, as is done in work with acid–water mixtures, but at the same time the range of acidities that can be considered is more limited. Furthermore, if the dielectric constant is low, constants measured represent ion-pair acidities.

Acetic acid has been used, for example, to determine that $HClO_4$, HBr, and HCl, which all behave as strong acids in water, differ significantly in acidity.[44] Other useful nonaqueous solvents are acetonitrile, methanol, ethanol, dimethylformamide, and dimethylsulfoxide.[45] When acids whose strengths can also be measured in water are studied in these solvents, the constants obtained are generally quite different. However, if the relative acidities of two compounds in water and another solvent are compared, the difference in pK_a between the two acids is usually approximately independent of solvent (within about one pK unit) *as long as the acids being compared are of the same charge type and are very similar in structure.*[46] For example, two particular substituted carboxylic acids may be expected to differ in acidity by roughly the same amount in dimethylformamide as they do in water, even though the values of pK_a found in the two solvents will be quite different; but no such correlation would be expected if the comparison were between a carboxylic acid and an anilinium ion (different charge type) or between a carboxylic acid and a phenol (same charge type but different structural type). It should be noted, however, that there appear to be exceptions even to this rough rule of thumb.

Other methods of making quantitative measurements on weak bases, less commonly used than those described here, have been reviewed by Arnett.[47] The reader is referred to that article for further information.

Heats of Protonation

Arnett has summarized the difficulties inherent in the currently available methods of dealing with weak bases in solution. He notes, for example, that the pK_a values

[43](a) L. P. Hammett, *Physical Organic Chemistry,* 2d ed., McGraw-Hill, New York, 1970, p. 265; (b) J. F. Coetzee, *Prog. Phys. Org. Chem.,* **4,** 64 (1967); (c) I. M. Kolthoff and T. B. Reddy, *Inorg. Chem.,* **1,** 189 (1962); (d) C. D. Ritchie and R. E. Uschold, *J. Am. Chem. Soc.,* **89,** 1721, 2752 (1967); (e) R. F. Cookson, *Chem. Rev.,* **74,** 5 (1974); (f) C. C. Westcott, *pH Measurements,* Academic Press, New York, 1978.

[44]R. P. Bell, *The Proton in Chemistry,* 2d ed., Cornell University Press, Ithaca, N.Y., 1973, p. 46.

[45](a) M. M. Davis, *Acid–Base Behavior in Aprotic Organic Solvents,* Nat. Bur. Stds. Monograph 105, 1968; (b) I. M. Kolthoff, M. K. Chantooni, Jr., and S. Bhowmik, *J. Am. Chem. Soc.,* **90,** 23 (1968); (c) J. F. Coetzee and G. R. Padmanabhan, *J. Am. Chem. Soc.,* **87,** 5005 (1965); (d) B. W. Clare, D. Cook, E. C. F. Ko, Y. C. Mac, and A. J. Parker, *J. Am. Chem. Soc.,* **88,** 1911 (1966); (e) C. D. Ritchie and G. H. Megerle, *J. Am. Chem. Soc.,* **89,** 1447, 1452 (1967).

[46]R. P. Bell, *The Proton in Chemistry,* 2d ed., Cornell University Press, Ithaca, N.Y., 1973, p. 56; see also footnotes 45 (c) and 45 (d).

[47]E. M. Arnett, *Prog. Phys. Org. Chem.,* **1,** 223 (1963).

given in the literature for ketones, a very important class of compounds that undergo a variety of acid-catalyzed reactions, vary over an unacceptably wide range. The variations arise not only from the activity coefficient problems mentioned above, but also from such practical problems as the effect of differing media on position of the absorption peaks in the ultraviolet spectrum. Although statistical methods that will be of considerable help in avoiding the problems of medium effects on ultraviolet absorption,[48] and nmr methods that are less severely affected by medium changes, are coming into more widespread use, an alternative approach to the problem of defining pK_a for weak bases proposed by Arnett and co-workers promises to be useful.[49] They have measured the heats of protonation of a number of weak bases in HSO_3F, in which most of the bases of interest are known from freezing-point depression, electrical conductivity, ultraviolet spectroscopy, and nuclear magnetic resonance measurements to be completely protonated. They found a good correlation of these heats of protonation with recorded pK_a values for series like the original Hammett nitroaniline indicators that are well behaved in acidity function experiments. The heat of protonation method has the advantage over the acidity function procedure that all measurements are made in the same solvent.

Activity Coefficients and Solvent–Solute Interactions

Interest in acidity function behavior, together with developing knowledge of solution phenomena in general, has stimulated a search for the underlying factors that lead to nonideal acidity behavior. Two approaches have been taken. One method seeks understanding of solution phenomena by looking at the activity coefficients of individual species involved in the equilibria. In contrast to the acidity function method, which looks for uniform behavior of large categories of compounds and so tries to minimize the effect of activity coefficient variation from compound to compound, this approach studies these variations explicitly to uncover their underlying causes. Yates and co-workers have collected activity coefficient data for molecules and ions in acidic media.[50]

A second approach is to measure thermodynamics of solution of molecules and ions from the gas phase. A thermodynamic cycle such as that shown in Scheme 3 will allow evaluation of the free energy (or enthalpy or entropy) of solution of the conjugate acid BH^+ relative to a standard conjugate acid B_0H^+ from the free energy of proton transfer from the standard B_0H^+ to the base B in solution and in the gas phase, and from the free energies of solution of the bases. (The symbol δ_R in the scheme means that the reaction measures the magnitude of the indicated thermodynamic property for substance B relative to the standard B_0.)

[48] J. T. Edward and C. S. Wong, *J. Am. Chem. Soc.*, **99**, 4229 (1977).

[49] (a) E. M. Arnett, R. P. Quirk, and J. J. Burke, *J. Am. Chem. Soc.*, **92**, 1260 (1970); (b) E. M. Arnett, R. P. Quirk, and J. W. Larsen, *J. Am. Chem. Soc.*, **92**, 3977 (1970); (c) E. M. Arnett and J. F. Wolf, *J. Am. Chem. Soc.*, **97**, 3262 (1975); (d) E. M. Arnett and G. Scorrano, *Adv. Phys. Org. Chem.*, **13**, 83 (1976).

[50] K. Yates and R. A. McClelland, *Prog. Phys. Org. Chem.*, **11**, 323 (1974).

SCHEME 3

$$B_0H^+(g) \ + \ B(g) \ \xrightarrow{\ \delta_R \Delta G_i(g)\ } \ B_0(g) \ + \ BH^+(g)$$

$$\Big\downarrow \Delta G_{sB_0H^+} \qquad \Big\downarrow \Delta G_{sB} \qquad\qquad\qquad \Delta G_{sB_0}\Big\downarrow \qquad \Delta G_{sBH^+}\Big\downarrow$$

$$B_0H^+(soln) + B(soln) \ \xrightarrow{\ \delta_R \Delta G_i(soln)\ } \ B_0(soln) + BH^+(soln)$$

$$\Delta G_{sBH^+} - \Delta G_{sB_0H^+} = \delta_R \Delta G_i(soln) - \delta_R \Delta G_i(g) + \Delta G_{sB} - \Delta G_{sB_0}$$

The solution proton transfer parameters are evaluated for alkyl amines by direct measurements in water and for weaker bases by acidity function methods. The gas-phase data come from the ion cyclotron resonance (ICR) technique or from high-pressure mass spectrometry, which allow direct observation and quantitative measurement of proton transfers in the gas phase.[51]

From the activity coefficient and the thermodynamic approaches, the following points have emerged as important factors in determining how a molecule or ion interacts with solvent:[52] hydrogen bonding; creation of a cavity in the solvent; electrostatic and electrical dispersion (van der Waals) forces between solvent and solute molecules or ions; and increase or decrease of solvent–solvent structure on introducing the solute. As might be expected, the variations in solvation on changing from one solvent mixture to another are greater for ions than for neutral species. One result of this fact is that variations in the activity coefficients for the ions dominate the activity coefficient ratios that appear in acidity function formulations.

Both the activity coefficient approach and the thermodynamic measurements show the importance of hydrogen bonding, particularly for ions, in cases where the number of acidic protons is varied while keeping number of carbons, charge, and basic atom constant. Thus, for example, in the series RNH_3^+, $R_2NH_2^+$, R_3NH^+, R_4N^+, one finds decreasing solvation because of successively fewer hydrogen-bonding sites. The consequence in strong acid mixtures is that as water becomes less and less available in mixtures richer in strong mineral acid, the ions that have the larger number of acidic protons, and hence have larger solvation demands, show the larger increases in activity coefficient. Thus, for example, activity coefficients of $ArNH_3^+$ ions in water–sulfuric acid mixtures increase faster than those of $ArNHR_2^+$ ions with increasing sulfuric acid content. Because the effect is stronger on the ionic conjugate acid than on the neutral free base, the $ArNH_2$ bases (indicators for the H_0 scale) are harder to protonate than $ArNR_2$ bases (indicators for the H''' scale). Hence for a given water–sulfuric acid mixture H_0 will show a

[51] R. W. Taft, in *Proton Transfer Reactions*, E. Caldin and V. Gold, Eds., Chapman and Hall, London, 1975, Chap. 2.
[52] (a) R. W. Taft, J. F. Wolf, J. L. Beauchamp, G. Scorrano, and E. M. Arnettt, *J. Am. Chem. Soc.*, **100**, 1240 (1978); (b) K. Yates and R. A. McClelland, *Prog. Phys. Org. Chem.*, **11**, 323 (1974).

lower effective acidity than will H'''. The greater solvation of oxonium ions as compared with ammonium ions and the increase of solvation of an ion caused by a more electronegative substituent are also attributable largely to hydrogen bonding. Hydrogen bonding can also be affected by steric bulk and by charge delocalization, although here effects on the size of the solvent cavity, polarizability, and solvent structure terms will complicate the situation. Clearly, more experimental and theoretical work will be needed for a complete understanding of the phenomena.

3.3 STRENGTHS OF WEAK BRØNSTED ACIDS[53,54]

Conant and Wheland[55] and McEwen[56] initiated the quantitative exploration of acidity of very weak acids in the 1930s. They equilibrated an organosodium or organopotassium compound, R_1M, with a weak acid, R_2H, according to Equation 3.61, and used the equilibrium constant as a measure of the relative acidities of

$$R_1^-M^+ + R_2H \rightleftharpoons R_1H + R_2^-M^+ \tag{3.61}$$

R_1H and R_2H.[57] The solvent was ether or benzene; in these low dielectric solvents, ions are strongly associated into ion pairs.[58] Because acid dissociation constants are defined in terms of dissociated ions (Equation 3.62), the equilibria measured do not

$$R_1^- + R_2H \rightleftharpoons R_1H + R_2^- \tag{3.62}$$

reflect true pK_a differences unless the ion-pair dissociation constants of $R_1^-M^+$ and $R_2^-M^+$ are equal. Conant and Wheland assumed that they would be. This expectation is at least approximately justified as long as R_1 and R_2 are similar in structure,[59] but it would not be expected to hold for all the compounds included in the investigations.

[53] For reviews see (a) A. Streitwieser, Jr., and J. H. Hammons, *Prog. Phys. Org. Chem.*, **3**, 41 (1965); (b) H. Fischer and D. Rewicki, *Prog. in Org. Chem.* (Cook and Carruthers, Eds.), **7**, 116 (1968); (c) J. R. Jones, *The Ionisation of Carbon Acids*, Academic Press, London, 1973; (d) E. Buncel, *Carbanions: Mechanistic and Isotopic Aspects*, Elsevier, Amsterdam, 1975, Chap. 1.

[54] We denote acids in which the acidic proton is attached to carbon as carbon acids, those with the proton attached to oxygen as oxygen acids, and so forth for acids of other types. Many of the very weak acids of interest in organic chemistry are carbon acids, so that their conjugate bases are carbanions. We will discuss chemistry of carbanions in more detail in Chapter 6.

[55] J. B. Conant and G. W. Wheland, *J. Am. Chem. Soc.*, **54**, 1212 (1932).

[56] W. K. McEwen, *J. Am. Chem. Soc.*, **58**, 1124 (1936).

[57] See D. J. Cram, *Fundamentals of Carbanion Chemistry*, Academic Press, New York, 1965, for discussion of these results.

[58] M. Szwarc, in *Ions and Ion Pairs in Organic Reactions*, Vol. 1, M. Szwarc, Ed., Wiley, New York, 1972, Chap. 1.

[59] M. Szwarc, A. Streitwieser, and P. C. Mowery, in *Ions and Ion Pairs in Organic Reactions*, Vol. 2, M. Szwarc, Ed., Wiley, New York, 1974, Chap. 2.

Acidities in Cyclohexylamine and in Dimethylsulfoxide

The general approach pioneered by Conant, Wheland, and McEwen has been developed extensively by Streitwieser and co-workers.[60] They used cyclohexylamine (CHA) as solvent and cesium cyclohexylamide (Cs^+CHA^-) as the base to form the conjugate base of a hydrocarbon substrate R_1H, as indicated in Equation 3.63.

$$R_1H + Cs^+CHA^- \longrightarrow R_1^-Cs^+ + CHA \qquad (3.63)$$

A second hydrocarbon, R_2H, was then added, and relative acidity of R_1H and R_2H was evaluated from the equilibrium constant for Reaction 3.64, which was deter-

$$R_1^-Cs^+ + R_2H \rightleftharpoons R_1H + R_2^-Cs^+ \qquad (3.64)$$

mined spectrophotometrically. Because the pK_a of cyclohexylamine itself is about 41.6, the limit of pK_a that can be determined by this technique is about 38.

In recognition of the negligibly small dissociation of ion pairs in cyclohexylamine,[61] Streitwieser used the term *ion-pair acidity* for results of this method. Because the conjugate bases of most of the substances investigated have charge delocalized over extended π systems, the ion-pair dissociation constants are expected to be quite similar; indeed there is a high degree of consistency between Streitwieser's ion-pair acidities and other results (discussed below) determined under conditions of ion-pair dissociation. It has even been possible to make a direct link of the ion-pair acidity scale for hydrocarbons in cyclohexylamine with the water pK_a scale through cyclopentadiene, which is sufficiently acidic, and sufficiently soluble, for pK_a determination (via an acidity function method) in water containing 9 molar NaOH. It is important to realize, however, that ion-pair acidities for substances whose conjugate bases have localized charge (phenylacetylene; alcohols) are not comparable to those of substances that yield delocalized anions. Ion-pair acidities for a number of hydrocarbons, placed on a scale relative to 9-phenylfluorene, $pK_a = 18.5$, are included in the table at the end of this section.

Bordwell and co-workers have developed a scale of acidity in dimethylsulfoxide (DMSO).[62] This solvent has the advantage of a relatively high dielectric constant ($\epsilon = 47$); ions are therefore dissociated so that problems of differential ion

[60] See, for example, (a) A. Streitwieser, Jr., J. I. Brauman, J. H. Hammons, and A. H. Pudjaatmaka, *J. Am. Chem. Soc.*, **87**, 384 (1965); (b) A. Streitwieser, Jr., J. H. Hammons, E. Ciuffarin, and J. I. Brauman, *J. Am. Chem. Soc.*, **89**, 59 (1967); (c) A. Streitwieser, Jr., E. Ciuffarin, and J. H. Hammons, *J. Am. Chem. Soc.*, **89**, 63 (1967); (d) A. Streitwieser, Jr., and D. M. E. Reuben, *J. Am. Chem. Soc.*, **93**, 1794 (1971); (e) A. Streitwieser, Jr., J. R. Murdoch, G. Häfelinger, and C. J. Chang, *J. Am. Chem. Soc.*, **95**, 4248 (1973); (f) A. Streitwieser, Jr., and F. Guibé, *J. Am. Chem. Soc.*, **100**, 4532 (1978).

[61] A. Streitwieser, Jr., W. M. Padgett II, and I. Schwager, *J. Phys. Chem.*, **68**, 2922 (1964).

[62] (a) W. S. Matthews, J. E. Bares, J. E. Bartmess, F. G. Bordwell, F. J. Cornforth, G. E. Drucker, Z. Margolin, R. J. McCallum, G. J. McCollum, and N. R. Vanier, *J. Am. Chem. Soc.*, **97**, 7006 (1975); (b) F. G. Bordwell, J. E. Bares, J. E. Bartmess, G. J. McCollum, M. Van Der Puy, N. R. Vanier, and W. S. Matthews, *J. Org. Chem.*, **42**, 321 (1977); (c) F. G. Bordwell, J. E. Bartmess, and J. A. Hautala, *J. Org. Chem.*, **43**, 3095 (1978), and following papers.

pairing are reduced. DMSO has a pK_a of 35.1, and the limit of pK_a accessible to measurement is about 31, a more restrictive limit than applies to the cyclohexylamine system. Although the results are referred to a standard state in DMSO instead of in water, a link with the aqueous pK_a scale has been made. For compounds whose acidities are measurable both in water and in DMSO, acids whose conjugate bases have their charge localized are stronger acids in water (which is better at solvating anions); acids whose conjugate bases have their charge delocalized over a large area are stronger in DMSO (which is more polarizable and hence better at solvating large molecules).

Acidity Function Methods

The acidity function approach discussed in the previous section in connection with mixtures of water and strong acids can be extended to basic solutions.[63] The solvent system is water–dimethylsulfoxide containing a small amount of tetramethyl-

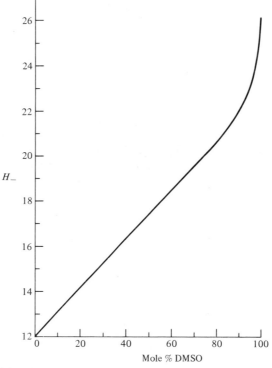

Figure 3.6 The H_- acidity function for mixtures of water and dimethylsulfoxide containing 0.011 molar tetramethylammonium hydroxide. Data from R. A. Cox and R. Stewart, *J. Am. Chem. Soc.*, **98**, 488 (1976).

[63] For reviews, see (a) K. Bowden, *Chem. Rev.*, **66**, 119 (1966); (b) J. R. Jones, *Prog. Phys. Org. Chem.*, **9**, 241 (1972); (c) E. Buncel and H. Wilson, *Adv. Phys. Org. Chem.*, **14**, 133 (1977); (d) C. H. Rochester, *Acidity Functions*, Academic Press, London, 1970.

ammonium hydroxide (or methanol–DMSO containing sodium methoxide). Increasing basicity of the medium is obtained by increasing the proportion of DMSO. The H_- function is defined by Equation 3.65.[64] (See Problem 14.) Figure 3.6 shows the H_- function in water–DMSO mixtures.

$$H_- = -\log K_w - \log a_{H_2O} - \log \frac{\gamma_{A^-}}{a_{OH^-}\gamma_{HA}} = pK_a + \log \frac{[A^-]}{[HA]} \tag{3.65}$$

Like H_0, the H_- function is not unique. Defined with substituted aniline indicators of the same type as used for H_0 (but now acting as acids rather than as bases), the H_- scale is not quite correct for acids of other types because of variations in the ratio γ_{A^-}/γ_{HA} from one type of compound to another. It is possible, however, to avoid this difficulty to some extent by the same tactics as were discussed in the previous section for weak bases. Both the Bunnett–Olsen and the Marziano–Passerini methods have been applied by Cox and Stewart.[64] These authors list pK_a's derived for a number of weak carbon and nitrogen acids and also give the quantity $-\log \gamma_{A^*}/\gamma_{HA^*}\gamma_{OH^-}$ that corresponds to the Cox and Yates **X** parameter for acid media.

Compared to Bordwell's measurements in pure DMSO, the H_- method has the advantage of an aqueous standard state but the disadvantage of extrapolation across varying solvent media.

Enthalpies of Protonation

An alternative to the equilibrium constant measurements is the study of thermodynamics of proton transfer. In solution, Arnett and co-workers have extended their proton transfer measurements to include enthalpies of deprotonation in dimethylsulfoxide with the dimethylsulfoxide conjugate base, $H_3CSOCH_2^-$, as proton acceptor.[65] Compounds must be more acidic than DMSO to be measured. There is a linear correlation between enthalpies of deprotonation, ΔH_D, and pK_a measured in pure DMSO. This method is applicable also to carboxylic acids and even to HCl, HBr, and HI. It thus provides a measure of acidity over a very wide range in a single solvent.

Gas-Phase Acidity

Relative acidities in the gas phase have been determined by high-pressure mass spectrometry[66] and by ion cyclotron resonance.[67] Gas-phase acidities for a given

[64] R. A. Cox and R. Stewart, *J. Am. Chem. Soc.*, **98**, 488 (1976).

[65] (a) E. M. Arnett, T. C. Moriarity, L. E. Small, J. P. Rudolph, and R. P. Quirk, *J. Am. Chem. Soc.*, **95**, 1492 (1973); (b) E. M. Arnett and L. E. Small, *J. Am. Chem. Soc.*, **99**, 808 (1977); (c) E. M. Arnett, D. E. Johnston, L. E. Small, and D. Oancea, *Faraday Symp. Chem. Soc.*, **10**, 20 (1975).

[66] (a) T. B. McMahon and P. Kebarle, *J. Am. Chem. Soc.*, **96**, 5940 (1974); (b) J. B. Cumming and P. Kebarle, *J. Am. Chem. Soc.*, **100**, 1835 (1978).

[67] (a) J. I. Brauman and L. K. Blair, *J. Am. Chem. Soc.*, **92**, 5986 (1970); (b) R. T. McIver, Jr., and J. S. Miller, *J. Am. Chem. Soc.*, **96**, 4323 (1974); (c) J. E. Bartmess, W. J. Hehre, R. T. McIver, Jr., and L. E. Overman, *J. Am. Chem. Soc.*, **99**, 1976 (1977); for a summary, see (d) R. W. Taft, in *Proton Transfer Reactions*, E. Caldin and V. Gold, Eds., Chapman and Hall, London, 1975, Chap. 2.

type of compound parallel the acidity in dimethylsulfoxide and the enthalpy of deprotonation in dimethylsulfoxide. For acids yielding relatively localized anions, for example substituted benzoic acids, acidity in the gas phase is much more sensitive to substituent effects than in DMSO; but for hydrocarbon acids yielding highly delocalized anions substituent effects are nearly the same in DMSO and in the gas phase.[68] We shall consider these substituent effects in more detail in Section 3.4.

Electrochemical Techniques

Breslow and co-workers have developed an electrochemical method that allows measurement of pK_a of extremely weak acids. The analysis is carried out with the aid of the thermodynamic cycle shown in Scheme 4. The desired quantity is the

SCHEME 4

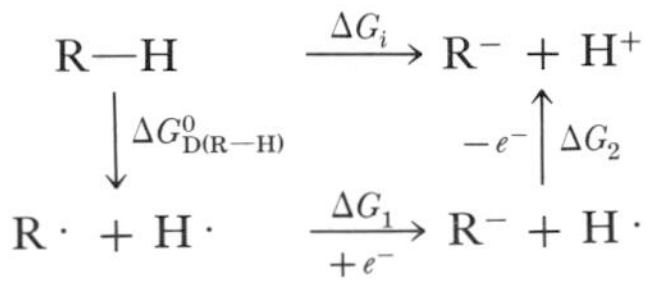

$$
\begin{array}{ccc}
\text{R—H} & \xrightarrow{\ \Delta G_i\ } & \text{R}^- + \text{H}^+ \\[4pt]
\Big\downarrow {\scriptstyle \Delta G^0_{D(R-H)}} & \quad {\scriptstyle -e^-}\Big\uparrow {\scriptstyle \Delta G_2} & \\[4pt]
\text{R} \cdot + \text{H} \cdot & \xrightarrow[{\ +e^-\ }]{\ \Delta G_1\ } & \text{R}^- + \text{H} \cdot
\end{array}
$$

$$ \Delta G_i = \Delta G_{D(R-H)} + \Delta G_1 + \Delta G_2 $$

free energy of ionization of the hydrocarbon, ΔG_i. The R—H bond dissociation energy, $\Delta G^0_{D(R-H)}$, is estimated from known thermodynamic data (see Table 9.5). The reduction step, labeled ΔG_1, the one actually involved in the experiment, is carried out by electrochemical reduction of RBr or RI to R $\cdot$, which is immediately reduced again at the electrode to R $\colon^-$. The free-energy change for this second reduction is ΔG_1. Then instead of measuring ΔG_2 directly, the reduction is carried out on the radical derived from triphenylmethane, the pK_a of which is taken as 31.5. This pK_a is converted into ΔG_i for triphenylmethane; because the step ΔG_2 is common to all hydrocarbons, ΔG_i (and therefore pK_a) can be found for any hydrocarbon relative to that of triphenylmethane by comparing its $\Delta G_{D(R-H)}$ and ΔG_1 to those of triphenylmethane.[69]

The Brønsted Catalysis Law

The solution techniques for determining dissociation constants of weak acids considered so far are limited, with the exception of Breslow's electrochemical method, to pK_a less than about 38. Most compounds in this range owe their acidity to some

[68] W. S. Matthews, J. E. Bares, J. E. Bartmess, F. G. Bordwell, F. J. Cornforth, G. E. Drucker, Z. Margolin, R. J. McCallum, G. J. McCollum, and N. R. Vanier, *J. Am. Chem. Soc.*, **97**, 7006 (1975).
[69] (a) R. Breslow and W. Chu, *J. Am. Chem. Soc.*, **95**, 411 (1973); (b) R. Breslow, *Pure Appl. Chem.*, **40**, 493 (1974); (c) M. R. Wasielewski and R. Breslow, *J. Am. Chem. Soc.*, **98**, 4222 (1976); (d) R. Breslow and J. L. Grant, *J. Am. Chem. Soc.*, **99**, 7745 (1977). (In the earlier papers a more complex scheme of less general applicability involving ROH was used.)

structural feature that allows the negative charge of the conjugate base to be delocalized. We turn now to a brief discussion of another method by which measurements can be extended, at least in a semiquantitative way, into the region of still weaker acids.

In the acid–base reaction (3.66) it would seem reasonable that if the rate (k_1) at which a proton is removed by a particular base B^- were compared for various

$$HA + B^- \underset{k_{-1}}{\overset{k_1}{\rightleftharpoons}} A^- + BH \tag{3.66}$$

acids HA, the base might remove the proton more rapidly from the stronger acids. Relationships between rate of an acid–base reaction and an equilibrium have been observed in many cases and are frequently found to obey an equation known as the Brønsted catalysis law:

$$k = CK_a{}^\alpha \tag{3.67}$$

or

$$\log k = \alpha \log K_a + \log C \tag{3.68}$$

where k is the rate constant for the reaction, K_a is the acid dissociation constant, and C is a constant of proportionality. If such a relationship could be shown to hold between acid strength and rate of transfer of the proton to some particular base, a means would be available to find equilibrium acidities through kinetic measurements.

An appreciation for the form of the catalysis law may be gained by consideration of the energy relationships involved. In Figure 3.7 is plotted schematically the free energy (ΔG) vs. reaction coordinate for proton transfer reactions between a series of acids, HA_n and a single base, B^-. The differing pK_a values of the acids are reflected in the different free-energy changes in going from reactants to products, ΔG_1°, ΔG_2°, ..., ΔG_n°,..., and are caused by structural differences among the acids HA_n and among the conjugate bases $A_n{}^-$. If one assumes that the factors that cause these free-energy differences also cause the differences in the transition state free energies, it is reasonable to suppose as a first approximation that the activation free energy for proton transfer, $\Delta G_n^\ddagger$, might be related to the ΔG_n° in a linear fashion. This relationship is expressed in Equation 3.69, where we have arbitrarily chosen the first acid, HA_1 as a reference compound for the series.

$$\Delta G_1^\ddagger - \Delta G_n^\ddagger = \alpha(\Delta G_1^\circ - \Delta G_n^\circ) \tag{3.69}$$

We have from equilibrium thermodynamics Relation 3.70 between standard free-energy change, ΔG°, and equilibrium constant, K, and from transition state theory Equation 3.71 (Section 2.6), where $\Delta G^\ddagger$ is the free energy of activation, k is the rate constant, k is the Boltzmann constant, and h is Planck's constant.

$$-\Delta G^\circ = 2.303RT \log K \tag{3.70}$$
$$-\Delta G^\ddagger = 2.303(RT \log k - RT \log kT/h) \tag{3.71}$$

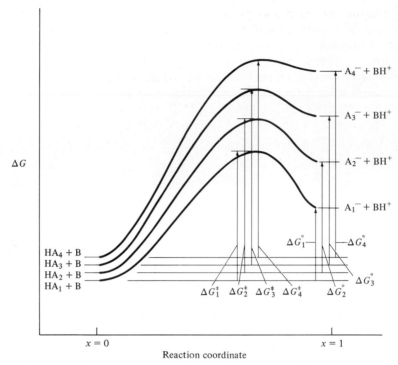

Figure 3.7 Hypothetical free energy vs. reaction coordinate curves for proton transfer from four different acids, HA_1, HA_2, HA_3, HA_4, to base B^-. The Brønsted catalysis law presumes that the effects of structural change on the transition state free energies will be some constant fraction of their effects on the overall free-energy changes.

By substituting Equations 3.70 and 3.71 in Equation 3.69, we obtain

$$\log k_n - \log k_1 = \alpha(\log K_n - \log K_1) \tag{3.72}$$

The acid HA_1 serves as our standard for comparison of all the others, so that $\log k_1$ and $\log K_1$ are constants for a series of measurements; therefore we can write Equation 3.73 (where C is a constant), which is equivalent to Equation 3.68 and, when written in exponential form, to Equation 3.67.[70]

$$\log k_n = \alpha \log K_n + \log C \tag{3.73}$$

[70] An alternative way of expressing the argument presented here is to assume that $\Delta G^{\ddagger}$ is some unspecified function of $\Delta G°$,

$$\Delta G^{\ddagger} = f(\Delta G°)$$

and to expand that function in a Taylor series about the reference point $\Delta G_1°$:

$$\Delta G^{\ddagger} = \text{constant} + \alpha(\Delta G° - \Delta G_1°) + \alpha'(\Delta G° - \Delta G_1°)^2 + \alpha''(\Delta G° - \Delta G_1°)^3 + \cdots$$

The approximation involved in stating the catalysis law is equivalent to dropping terms of order higher than the first in the power series expansion:

$$\Delta G^{\ddagger} \approx \text{constant} + \alpha(\Delta G° - \Delta G_1°)$$

This expression leads to Equation 3.74.

The Brønsted law is a linear free-energy relationship, similar in form to the Hammett and Taft correlations discussed in Section 2.2. We emphasize that the connection between rate and equilibrium expressed by Equation 3.68 is in no sense predicted by or derived from the laws of equilibrium thermodynamics. The relationship is an empirical one that must be verified experimentally in each particular case, and that is subject to severe limitations. We have assumed in drawing Figure 3.7 and in making the arguments we have presented rationalizing the catalysis law that the position of the transition state along the reaction coordinate will not change as the acid strengths change. We have seen in Section 2.6, where we considered the Hammond postulate, that this assumption is unlikely to be true if we make more than a rather small change in the reactant-to-product free-energy difference. As a result, we can expect that over a wide range of acidities α will not be a constant. It should be close to unity for a very endothermic process of type 3.66 (the transition state closely resembles $A^- + BH$ and the entire $\Delta G°$ differences show up in $\Delta G^{\ddagger}$), and close to zero for a very exothermic process (the transition state closely resembles $HA + B^-$ and none of the $\Delta G°$ differences show up in $\Delta G^{\ddagger}$). For carbon acids of similar structural type, α changes relatively slowly with changing equilibrium constants,[71] but its value for acids yielding localized anions is quite different from that for acids yielding resonance-stabilized delocalized ions. We must therefore proceed cautiously if we wish to use the catalysis law to assist us in estimating equilibrium acidities, and we expect difficulties if the range of equilibrium constants is large. We shall return to consider these points in more detail in Section 8.1.

Kinetic Acidity

The Brønsted catalysis law can be applied to the problem of determination of acidity of very weak acids in the following way. First, a suitable base is chosen; the base must be sufficiently strong to remove protons from the carbon acids in question at a measurable rate. The acids to be investigated are then prepared with deuterium or tritium substituted for hydrogen, and the rate of exchange of the isotopic label out of the carbon acid in the presence of the base is measured. Experiments of this type have been carried out with weak acids by various workers.[72-75]

[71] M. Eigen, *Angew. Chem. Int. Ed. Engl.*, **3**, 1 (1964).

[72] R. G. Pearson and R. L. Dillon, *J. Am. Chem. Soc.*, **75**, 2439 (1953).

[73] A. I. Shatenshtein, *Adv. Phys. Org. Chem.*, **1**, 155 (1963).

[74] See, for example, (a) A. Streitwieser, Jr., R. A. Caldwell, R. G. Lawler, and G. R. Ziegler, *J. Am. Chem. Soc.*, **87**, 5399 (1965); (b) A. Streitwieser, Jr., W. B. Hollyhead, G. Sonnichsen, A. H. Pudjaatmaka, C. J. Chang, and T. L. Kruger, *J. Am. Chem. Soc.*, **93**, 5096 (1971); (c) A. Streitwieser, Jr., and W. C. Langworthy, *J. Am. Chem. Soc.*, **85**, 1757, (1963); (d) A. Streitwieser, Jr., R. A. Caldwell, and M. R. Granger, *J. Am. Chem. Soc.*, **86**, 3578 (1964); (e) A. Streitwieser, Jr., and D. Holtz, *J. Am. Chem. Soc.*, **89**, 692 (1967); (f) A. Streitwieser, Jr., A. P. Marchand, and A. H. Pudjaatmaka, *J. Am. Chem. Soc.*, **89**, 693 (1967); (g) A. Streitwieser, Jr., and F. Mares, *J. Am. Chem. Soc.*, **90**, 644, 2444 (1968); (h) A. Streitwieser, Jr., and D. W. Boerth, *J. Am. Chem. Soc.*, **100**, 755 (1978). See also references cited in Table 3.2.

[75] R. E. Dessy, Y. Okuzumi, and A. Chen, *J. Am. Chem. Soc.*, **84**, 2899 (1962).

Two points must be established before the exchange data can be used to measure acidity. First, it is necessary to show that the measured exchange rate does indeed reflect the rate of proton removal from the substrate, and, second, the Brønsted slope α relating proton transfer to pK_a must be established. The isotope exchange consists of at least two steps, illustrated in Equations 3.74 and 3.75, where

$$RD + Sol^-M^+ \longrightarrow R^-M^+ + SolD \tag{3.74}$$
$$R^-M^+ + SolH \longrightarrow RH + Sol^-M^+ \tag{3.75}$$

we assume that the base, Sol^-, is the conjugate base of the solvent, and we have indicated ion pairs because these experiments are usually carried out in solvents of relatively low dielectric constant. The ion pairs are solvated, and it is reasonable to suppose that immediately after its formation in Equation 3.74, the ion pair R^-M^+ will have in its solvent shell the molecule SolD. R^-M^+ will now rapidly be reprotonated; in order for the act of ionization that formed it to be detected, it must be reprotonated by a SolH molecule. But because the SolD is close, there is a chance, if the reprotonation follows very rapidly, that the deuterium will go directly back on from SolD. This process is called *internal return* and is one that we will meet again in considering ionization leading to nucleophilic substitution.

With the possibility of internal return, Equations 3.74 and 3.75 may be reformulated as follows:

$$RD + Sol^-M^+ \underset{k_{-1}}{\overset{k_1}{\rightleftharpoons}} R^-M^+ \cdots SolD \tag{3.76}$$
$$R^-M^+ \cdots SolD + SH \xrightarrow{k_2} R^-M^+ \cdots SolH + SD \tag{3.77}$$
$$R^-M^+ \cdots SolH \longrightarrow RH + Sol^-M^+ \tag{3.78}$$

(Note that if the molar amount of RD is small compared with the amount of SolH, there will be no significant exchange of D back into RH once Reaction 3.77 has occurred; hence the reverse of 3.77 and 3.78 need not be included.) Assuming stationary-state kinetics for R^-M^+,

$$k_{obs} = \frac{k_1 k_2}{k_{-1} + k_2} \tag{3.79}$$

If $k_2 \gg k_{-1}$, $k_{obs} \approx k_1$ and the deuterium removal is rate determining. In this case there should be a primary isotope effect. On the other hand, if $k_{-1} \gg k_2$, $k_{obs} \approx k_1 k_2 / k_{-1}$. Isotope effects on k_1 and k_{-1} should be approximately equal and that on k_2 should be small. Hence a large isotope effect indicates no internal return, and a small one implies internal return. If there is significant internal return, exchange rates will not accurately reflect deprotonation rates, unless the ratio k_2 / k_{-1} is the same for all compounds being compared.[76]

[76] A. Streitwieser, Jr., D. Holtz, G. R. Ziegler, J. D. Stoffer, M. L. Brokaw, and F. Guibé, *J. Am. Chem. Soc.*, **98**, 5229 (1976).

Streitwieser and co-workers have developed a method of using measured k_T/k_H and k_D/k_H ratios to determine the amount of internal return.[77] In cyclohexylamine, internal return is found to be relatively unimportant for most substrates. It becomes significant for some compounds in methanol.

Once the importance of internal return has been established, the logarithm of the exchange rate, corrected for internal return if necessary, is plotted against pK_a for a series of compounds for which acidities are known by equilibrium measurements, and the Brønsted slope α is determined. Streitwieser and co-workers have established such a correlation for hydrocarbons with benzylic hydrogens including diphenylmethane and triphenylmethane; the slope α is 0.31.[78] Use of this slope, together with exchange rates measured for toluene, cumene (isopropyl benzene), and others of the same type allows assignment of approximate pK_a's. For carbon acids yielding localized anions, α is close to unity.[79] The reason presumably is that these compounds have a transition state for proton transfer that comes later along the reaction coordinate than in the case of compounds forming delocalized ions.

The Scale of Carbon Acidity

In 1965 Cram established a scale of acidity for weak carbon acids using data available at the time.[80] Known as the MSAD scale after the chemists W. K. McEwen, A. Streitwieser, D. E. Applequist, and R. E. Dessy, it was based on a pK_a of 18.5 for 9-phenylfluorene, established by the acidity function technique and therefore referred to the aqueous pK scale.[81] The MSAD scale, modified by incorporating new data, is given in Table 3.2. The value of 18.5 for pK_a of 9-phenylfluorene, to which most other measurements are referred, has been retained, although there is evidence from the newer acidity function methods as applied by Cox and Stewart that this value should be raised to 20.2.[82] Above pK_a 40, the exchange rate method with $\alpha = 0.31$ for conjugated anions and $\alpha = 0.9$ for localized anions, or the electrochemical technique of Breslow, is used.

It is important to remember when using this scale of acidities that data are determined by a number of different methods in different solvents. Specific solvent effects, and particularly ion pairing, can cause variations of several pK units. Results from measurements in cyclohexylamine are ion-pair acidities only roughly

[77] A. Streitwieser, Jr., W. B. Hollyhead, G. Sonnichsen, A. H. Pudjaatmaka, C. J. Chang, and T. L. Kruger, *J. Am. Chem. Soc.*, **93**, 5096 (1971).

[78] (a) A. Streitwieser, Jr., M. R. Granger, F. Mares, and R. A. Wolf, *J. Am. Chem. Soc.*, **95**, 4257 (1973); (b) A. Streitwieser, Jr., and F. Guibé, *J. Am. Chem. Soc.*, **100**, 4532 (1978).

[79] (a) A. Streitwieser, Jr., P. J. Scannon, and H. M. Niemeyer, *J. Am. Chem. Soc.*, **94**, 7936 (1972); (b) A. Streitwieser, Jr., D. Holtz, G. R. Ziegler, J. D. Stoffer, M. L. Brokaw, and F. Guibé, *J. Am. Chem. Soc.*, **98**, 5229 (1976).

[80] D. J. Cram, *Fundamentals of Carbanion Chemistry*, Academic Press, New York, 1965, p. 19.

[81] C. H. Langford and R. L. Burwell, Jr., *J. Am. Chem. Soc.*, **82**, 1503 (1960).

[82] R. A. Cox and R. Stewart, *J. Am. Chem., Soc.*, **98**, 488 (1976).

Table 3.2 APPROXIMATE pK_a VALUES OF WEAK HYDROCARBON ACIDS[a]

Compound (Acidic H Indicated)		pK_a[b]	Method[c]	References[d]
Cyclopentadiene	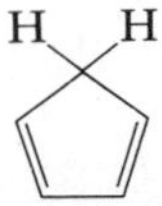	16.0	H_-, eq	1
9-Phenylfluorene	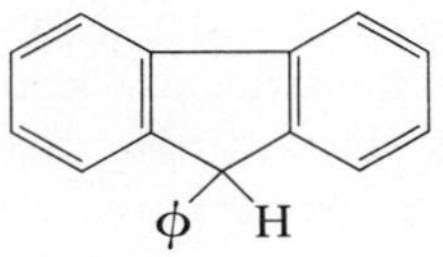	18.5 17.9	H_-, eq DMSO	2, 3, 4, 5, 6 7
Indene	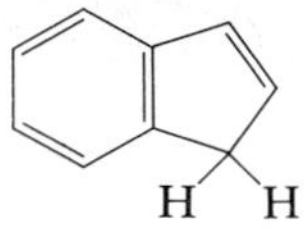	19.9	eq, H_-	4, 5, 6
Phenylacetylene	ϕ—C≡C—H	23.2[b] 28.8	eq DMSO	7 8
Fluorene	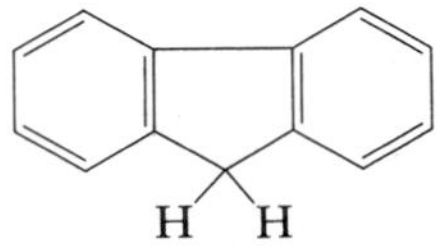	22.6	eq, H_-, DMSO	4, 5, 6 8
Acetylene	H—C≡C—H	24	eq	9
Triphenylmethane	ϕ_3CH	31.5	eq, ex, H_-	4, 5, 10
Diphenylmethane	ϕ_2CH$_2$	33.1	eq, ex, H_-	6, 10, 11
Cycloheptatriene	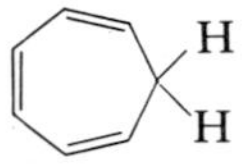	38.8	el	12
Toluene	ϕ—CH$_3$	41.2	ex	13
Benzene	ϕ—H	43	ex	14
Ethylene	H$_2$C=CH$_2$	44	ex	15
Triptycene	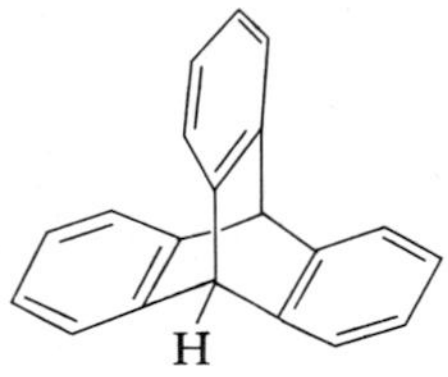	44	ex	10, 17
Cyclohexene	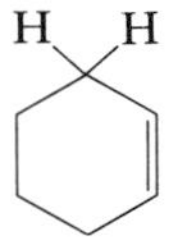	46	ex	16
Cyclopropane	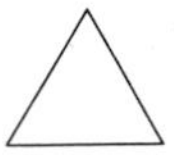	46	ex	18

260

Table 3.2 Approximate pK_a Values of Weak Hydrocarbon Acids[a]

Compound (Acidic H Indicated)		pK_a[b]	Method[c]	References[d]
Methane	CH$_4$	48	ex	19
		68–70	el	20
Triphenylcyclopropene	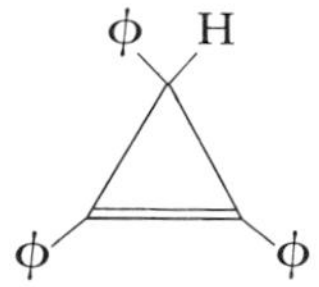	50	el	12
Propene	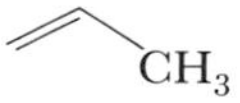	53	el	20
Cyclopropene	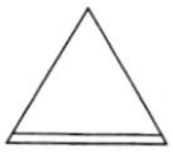	61	el	12
Isobutane	(CH$_3$)$_3$CH	71	el	21
Cyclobutane		50	ex	18
Cyclopentane	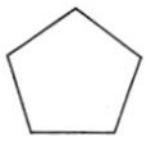	51	ex	18
Cyclohexane	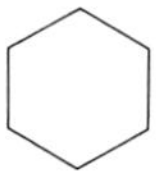	52	ex	22

[a] Values up to pK_a 33 (diphenylmethane) are those reported for equilibrium methods and were measured either directly using the H_- acidity function or by comparing acidity with 9-phenylfluorene. Exchange measurements above pK_a 43 use $\alpha = 0.9$. The scale is based on the Langford and Burwell value of 18.5 for 9-phenylfluorene
[b] Rounded to three significant figures.
[c] Method code:

DMSO = equilibrium in dimethylsulfoxide
  el = electrochemical
  eq = equilibrium methods related to the water pK scale by direct or indirect comparison with 9-phenylfluorene
  ex = H–D or H–T exchange rate
  H_- = equilibrium measurement using H_- acidity function

[d] References:
1. A. Streitwieser, Jr., and L. L. Nebenzahl, *J. Am. Chem. Soc.*, **98**, 2188 (1976).
2. C. H. Langford and R. L. Burwell, Jr., *J. Am. Chem. Soc.*, **82**, 1503 (1960).
3. R. A. Cox and R. Stewart, *J. Am. Chem. Soc.*, **98**, 488 (1976).
4. A. Streitwieser, Jr., E. Ciuffarin, and J. H. Hammons, *J. Am. Chem. Soc.*, **89**, 63 (1967).
5. C. D. Ritchie and R. E. Uschold, *J. Am. Chem. Soc.*, **89**, 2752 (1967).
6. E. C. Steiner and J. D. Starkey, *J. Am. Chem. Soc.*, **89**, 2751 (1967); E. C. Steiner and J. M. Gilbert, *J. Am. Chem. Soc.*, **87**, 382 (1965).
7. A. Streitwieser, Jr., and D. M. E. Reuben, *J. Am. Chem. Soc.*, **93**, 1794 (1971).
8. W. S. Matthews, J. E. Bares, J. E. Bartmess, F. G. Bordwell, F. J. Cornforth, G. E. Drucker, Z. Margolin, R. J. McCallum, G. J. McCollum, and N. R. Vanier, *J. Am. Chem. Soc.*, **97**, 7006 (1975).
9. N. S. Wooding and W. E. C. Higginson, *J. Chem. Soc.*, 774 (1952). In liquid NH$_3$, corrected to triphenylmethane = 31.
10. A. Streitwieser, Jr., R. A. Caldwell, and M. R. Granger, *J. Am. Chem. Soc.*, **86**, 3578 (1964).

11. A. Streitwieser, Jr., W. B. Hollyhead, G. Sonnichsen, A. H. Pudjaatmaka, C. J. Chang, and T. L. Kruger, *J. Am. Chem. Soc.,* **93**, 5096 (1971).
12. M. R. Wasielewski and R. Breslow, *J. Am. Chem. Soc.,* **98**, 4222 (1976).
13. A. Streitwieser, Jr., and F. Guibé, *J. Am. Chem. Soc.,* **100**, 4532 (1978).
14. A. Streitwieser, Jr., P. J. Scannon, and H. M. Niemeyer, *J. Am. Chem. Soc.,* **94**, 7936 (1972).
15. M. J. Maskornick and A. Streitwieser, Jr., *Tetrahedron Lett.,* 1625 (1972).
16. A. Streitwieser, Jr., and D. W. Boerth, *J. Am. Chem. Soc.,* **100**, 755 (1978).
17. A. Streitwieser, Jr., and G. R. Ziegler, *J. Am. Chem. Soc.,* **91**, 5081 (1969).
18. A. Streitwieser, Jr., R. A. Caldwell, and W. R. Young, *J. Am. Chem. Soc.,* **91**, 529 (1969).
19. A. Streitwieser, Jr., and D. R. Taylor, *J. Chem. Soc. D,* 1248 (1970).
20. R. Breslow and J. L. Grant, *J. Am. Chem. Soc.,* **99**, 7745 (1977).
21. R. Breslow and R. Goodin, *J. Am. Chem. Soc.,* **98**, 6076 (1976).
22. A. Streitwieser, Jr., W. R. Young, and R. A. Caldwell, *J. Am. Chem. Soc.,* **91**, 527 (1969).

comparable to results from dimethylsulfoxide, where ion pairing is much less important. Furthermore, the connection to the aqueous standard state is tenuous, and particularly so for the higher pK's. Comparison of results obtained by the same method should be valid, but comparisons between methods are likely to show discrepancies, sometimes substantial ones. Note particularly the large difference between exchange and electrochemical estimates for methane.

3.4 SUBSTITUENT EFFECTS ON STRENGTHS OF BRØNSTED ACIDS AND BASES

Acid–base reactions have long served as a starting point for consideration of the effects of changes in a structure on the course of chemical reactions.[83] Table 3.3 summarizes solution data for a variety of Brønsted acids and bases; because of the problems of measurement, any such table necessarily contains a fair amount of uncertainty. The pK_a values that fall between 2 and 10 may be used with considerable confidence, since they are based on accurate measurements in dilute aqueous solutions; the values outside this range must be regarded with a certain amount of skepticism. As we have noted in the two previous sections, uncertainties become severe for very strong and very weak acids; the extreme values have only qualitative significance.

In Table 3.3 we follow the convention of giving the strength of a base in terms of the pK_a of the conjugate acid. It is useful to keep in mind that the weaker the base, the stronger is its conjugate acid. Hence the weakest bases and strongest acids (those with negative pK_a) appear at the beginning of the table, whereas the strongest bases and weakest acids (large positive pK_a) are at the end. Many compounds are both acids and bases; it is worthwhile to note that although there is a definite relationship between strength of a base and strength of the *conjugate acid of that base,* there is not any quantitative correlation between the strength of a given

[83] J. Hine, *Structural Effects on Equilibria in Organic Chemistry,* Wiley, New York, 1975.

substance as a base and the strength *of the same substance* as an acid, although it is often true that a strongly acidic molecule will be weakly basic and vice versa.[84]

Solution and Gas-Phase Acidity

As we have seen in the previous two sections, the solvent plays a large part in determining acidities in solution. In the gas phase, intrinsic acidities uninfluenced by solvent can be determined. One must decide which quantity is appropriate for a given application. Clearly, if the purpose is to assess the influence of an acidic or basic site on the course of a reaction in solution, the solution acidity is appropriate, preferably determined in the solvent being used for the reaction or one as similar to it as possible. On the other hand, acid–base equilibrium data are frequently used in assessing influence of substituents and for making arguments about inductive, resonance, and steric effects. For this purpose an acidity determined largely by specific interactions with a particular solvent can be misleading, and gas-phase data are better. It is true, of course, that solution data have been used in this way for many years, and the availability of a large amount of solution data provides a strong incentive to continue to do so. The existence of even a relatively limited amount of gas-phase information, however, serves as a warning of the risks involved and also as an aid in identifying the circumstances that are likely to lead to difficulties in the uncritical use of solution data.

Tables 3.4 and 3.5 illustrate this point in the case of water and the simple aliphatic alcohols. The first feature to note about these data is the large discrepancy between estimates of acidity in water (which are uncertain at best) and in dimethylsulfoxide. Because of the solvating properties of DMSO, acids whose conjugate bases have highly localized charge are much less acidic in DMSO than in water. The order of acidity is, however, the same in the two solvents; the alcohols become weaker acids as more substituent hydrocarbon groups are added. From a practical point of view, this order will mean, for example, that in solution the *t*-butoxide ion, $(CH_3)_3CO^-$, will be a stronger base than hydroxide. On the basis of the solution data alone, one would conclude that substitution by successively more bulky groups causes a steady lowering of acidity, although the relative positions of water and methanol are somewhat uncertain. Before the advent of the gas-phase measurements, these data were the only ones available and were generally interpreted in terms of the inductive effects of the alkyl groups. It is well known, for example, that increasing alkyl substitution stabilizes carbocations (see Section 4.4), and so it was presumed that an alkyl group, being evidently electron donating, should destabilize a negative charge. Hence it was reasonable that the alcohols with more or larger groups should have less tendency to form a negative ion by loss of a proton and hence should be less acidic. This interpretation was apparently supported by

[84](a) E. M. Arnett, *Prog. Phys. Org. Chem.*, **1**, 223 (1963); (b) R. J. Gillespie, in *Friedel–Crafts and Related Reactions*, Vol. 1, G. A. Olah, Ed., Wiley, New York, 1963, p. 181.

Table 3.3 SOLUTION DISSOCIATION CONSTANTS OF ACIDS AND BASES[a]

Conjugate Acid	Conjugate Base	pK_a	References[b]
RNO_2H^+	RNO_2	-12	1, 2
$RC{\equiv}NH^+$	$RC{\equiv}N$	-12	3
PH_4^+	PH_3	-12	24
Ar—C(=$\overset{+}{O}H$)—Cl	Ar—C(=O)—Cl	-11	4
HI	I^-	-9	5
HBr	Br^-	-8	5
HCl	Cl^-	-7	5
RC(=$\overset{+}{O}H$)G	RC(=O)G	-2 to -8	1, 4, 6, 22
	(G = H, R, Ar, OR, OH)		
$ArOH_2^+$	ArOH	-7	1, 7
RSH_2^+	RSH	-7	1
$Ar\overset{+}{O}R$ with H	ArOR	-6 to -8	1, 6, 7, 8
ArC(=$\overset{+}{O}H$)Ar	ArC(=O)Ar	-3 to -5	6
$R\overset{+}{O}R$ with H	ROR	-2 to -5	1, 8, 9, 10, 22
ROH_2^+	ROH	-2 to -5	1, 9, 10
H_3O^+	H_2O	-1.74	23
$CH_3\overset{+}{S}(=\overset{+}{O}H)CH_3$	$CH_3S(=O)CH_3$	-1.5	22
R—C(=$\overset{+}{O}H$)—NH_2	R—C(=O)—NH_2	0 to -4	1, 6
$ArNH_3^+$	$ArNH_2$	-10 to $+5$	1, 6, 10, 11
HF	F^-	3.17	5
RCOOH	$RCOO^-$	4 to 5	11
H_2S	HS^-	7.0	5
ArSH	ArS^-	8	7
HCN	CN^-	9.22	5
NH_4^+	NH_3	9.24	5
RNH_3^+	RNH_2	10 to 12	1, 11
ArOH	ArO^-	9 to 11	11
R—C(=O)—CH_2—C(=O)—R	R—C(=O)—CH—C(=O)—R	9	12
RCH_2NO_2	$RCHNO_2^-$	10 (17, DMSO)	12, 13
RSH	RS^-	12	11
H_2O	OH^-	15.7	14
ROH	RO^-	17 to 20	14
cyclopentadiene	cyclopentadienyl anion	16	15

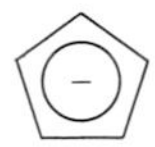

Table 3.3 (*Continued*)

Conjugate Acid	Conjugate Base	pK_a	References[b]
$ArNH_2$	$ArNH^-$	18 to ~28	16, 17, 18
$R-\overset{\displaystyle O}{\overset{\|}{C}}-CH_2-R'$	$R-\overset{\displaystyle O}{\overset{\|}{C}}-\overset{-}{C}H-R'$	19 to 20 (24 to 27, DMSO)	12 19
$R-O-\overset{\displaystyle O}{\overset{\|}{C}}-CH_2-R'$	$R-O-\overset{\displaystyle O}{\overset{\|}{C}}-\overset{-}{C}H-R'$	24	12
PH_3	PH_2^-	27	24
$CH_3SO_2CH_3$	$CH_3SO_2CH_2^-$	31	19
CH_3CN	$^-CH_2CN$	31	19
ϕ_3CH	ϕ_3C^-	31	15
NH_3	NH_2^-	33	5
CH_3SOCH_3	$CH_3SOCH_2^-$	35	19
H_2	H^-	35	20
ϕCH_3	ϕCH_2^-	41	15
cyclohexyl–NH_2	cyclohexyl–NH^-	42	21
ϕH	ϕ^-	43	15
CH_4	CH_3^-	~50	15

[a] Values less than ~3 and greater than ~10 are approximate, and values at the extremes of the scale probably have only qualitative significance. Compilations of pK data may be found in D. D. Perrin, *Dissociation Constants of Organic Bases in Aqueous Solution*, Butterworths, London, 1965; E. P. Serjeant and B. Dempsey, *Ionisation Constants of Organic Acids in Aqueous Solution*, Pergamon Press, Oxford, 1979; and in references 1, 5, 11, and 12.

[b] References.

1. E. M. Arnett, *Prog. Phys. Org. Chem.*, **1**, 223 (1963).
2. N. C. Deno, R. W. Gaugler, and T. J. Schulze, *J. Org. Chem.*, **31**, 1968 (1966).
3. N. C. Deno, R. W. Gaugler, and M. J. Wisotsky, *J. Org. Chem.*, **31**, 1967 (1966).
4. E. M. Arnett, R. P. Quirk, and J. W. Larsen, *J. Am. Chem. Soc.*, **92**, 3977 (1970).
5. D. D. Perrin, *Dissociation Constants of Inorganic Acids and Bases in Aqueous Solution*, Butterworths, London, 1969.
6. R. A. Cox and K. Yates, *J. Am. Chem. Soc.*, **100**, 3861 (1978).
7. E. M. Arnett and C. Y. Wu, *J. Am. Chem. Soc.*, **82**, 5660 (1960).
8. E. M. Arnett and C. Y. Wu, *J. Am. Chem. Soc.*, **82**, 4999 (1960).
9. N. C. Deno and J. O. Turner, *J. Org. Chem.*, **31**, 1969 (1966).
10. E. M. Arnett, R. P. Quirk, and J. J. Burke, *J. Am. Chem. Soc.*, **92**, 1260 (1970).
11. H. C. Brown, D. H. McDaniel, and O. Häflinger, in *Determination of Organic Structures by Physical Methods*, E. A. Braude and F. C. Nachod, Eds., Academic Press, New York, 1955, p. 567.
12. R. G. Pearson and R. G. Dillon, *J. Am. Chem. Soc.*, **75**, 2439 (1953).
13. F. G. Bordwell, J. E. Bartmess, and J. A. Hautala, *J. Org. Chem.*, **43**, 3095 (1978).
14. See Table 3.4.
15. See Table 3.2.
16. K. Bowden, *Chem. Rev.*, **66**, 119 (1966).
17. D. Dolman and R. Stewart, *Can. J. Chem.*, **45**, 911 (1967); T. Birchall and W. L. Jolly, *J. Am. Chem. Soc.*, **88**, 5439 (1966).
18. E. M. Arnett, T. C. Moriarity, L. E. Small, J. P. Rudolph, and R. P. Quirk, *J. Am. Chem. Soc.*, **95**, 1492 (1973).
19. W. S. Matthews, J. E. Bares, J. E. Bartmess, F. G. Bordwell, F. J. Cornforth, G. E. Drucker, Z. Margolin, R. J. McCallum, G. J. McCollum, and N. R. Vanier, *J. Am. Chem. Soc.*, **97**, 7006 (1975).
20. E. Buncel and B. Menon, *J. Am. Chem. Soc.*, **99**, 4457 (1977).
21. A. Streitwieser, Jr., and F. Guibé, *J. Am. Chem. Soc.*, **100**, 4532 (1978).
22. G. Perdoncin and G. Scorrano, *J. Am. Chem. Soc.*, **99**, 6983 (1977).
23. See Table 3.17.
24. C. A. Streuli, *Anal. Chem.*, **32**, 985 (1960).

Table 3.4 SOLUTION ACIDITIES OF WATER
AND THE SIMPLE ALCOHOLS

Compound	pK_a in water	$pK_e{}^a$	pK_a in DMSO
H_2O	15.7^b	-0.1	27.5^g
CH_3OH	16^c 15.5^e	-0.6	27.9^g 29.1^h
CH_3CH_2OH	18^d 15.9^e $(16)^f$	0.0	28.2^g 29.5^h
$(CH_3)_2CHOH$	18^d	$(+1.1)$	29.3^g
$(CH_3)_3COH$	19^d	—	29.4^g 31.3^h

[a] J. Hine and M. Hine, *J. Am. Chem. Soc.*, **74**, 5266 (1952). In isopropyl alcohol,

$$HA + i\text{-prO}^- \rightleftharpoons A^- + i\text{-prOH}$$

$$K_e = \frac{[A^-]}{[HA][i\text{-prO}^-]}$$

The value for isopropyl alcohol is determined by the definition of K_e.
[b] Calculated for

$$K_a = \frac{[H_3O^+][OH^-]}{[H_2O]}$$

using $K_w = 10^{-14}$ and $[H_2O] = 55.5\ M$.
[c] A. Unmack, *Z. Phys. Chem.*, **129**, 349 (1927); **131**, 371 (1928); **133**, 45 (1928).
[d] W. K. McEwen, *J. Am. Chem. Soc.*, **58**, 1124 (1936). Measured in benzene using Unmack's value for CH_3OH as standard.
[e] P. Ballinger and F. A. Long, *J. Am. Chem. Soc.*, **82**, 795 (1960). Measured by conductivity in water.
[f] Reference d, by extrapolation of a correlation with Taft's σ^* parameters.
[g] Derived from enthalpy of deprotonation. E. M. Arnett and L. E. Small, *J. Am. Chem. Soc.*, **99**, 808 (1977).
[h] From equilibrium measurements in DMSO. C. D. Ritchie, in *Solute–Solvent Interactions*, Vol. 2, J. F. Coetzee and C. D. Ritchie, Eds., Dekker, New York, 1976, p. 233.

the establishment of a correlation between alcohol acidity and the Taft σ^* inductive parameter (see Section 2.2).[85]

The gas-phase results (Table 3.5) show that, in the absence of solvent, water has the most endothermic heat of ionization and is therefore the weakest acid, while *tert*-butyl alcohol is the strongest acid. (It is generally assumed that $T\Delta S$ for gas-phase ionization will be about the same for different compounds; ΔH_i and ΔG_i are therefore used interchangeably.[86]) The gas-phase order should reflect intrinsic molecular properties; solvation is entirely responsible for the observed order in solution. The reason is presumably that the bulky $(CH_3)_3CO^-$ ion is much less well solvated than the OH^- ion. The intrinsic basicity of OH^- is thus reduced by

[85] (a) P. Ballinger and F. A. Long, *J. Am. Chem. Soc.*, **82**, 795 (1960); (b) E. M. Arnett and L. E. Small, *J. Am. Chem. Soc.*, **99**, 808 (1977).
[86] R. T. McIver, Jr., and J. S. Miller, *J. Am. Chem. Soc.*, **96**, 4323 (1974).

Table 3.5 GAS-PHASE ACIDITIES (ΔH_i) OF WATER AND SIMPLE ALCOHOLS[a]

Compound	ΔH_i(gas) (kcal mole^{-1})[b]
Most Acidic	
$(CH_3)_3COH$	373.3
$(CH_3)_2CHOH$	374.1
CH_3CH_2OH	376.1
CH_3OH	379.2
H_2O	390.8
Least Acidic	

[a] Reprinted in part with permission from J. E. Bartmess, J. A. Scott, and R. T. McIver, Jr., *J. Am. Chem. Soc.*, **101**, 6046 (1979). Copyright 1979 American Chemical Society.
[b] ΔH_i refers to the reaction
$$ROH \longrightarrow RO^- + H^+.$$

solvation more than that of $(CH_3)_3CO^-$, and the difference is large enough to reverse the intrinsic tendency. We shall return later in this section to a discussion of the nature of the alkyl substituent effect on gas-phase acidities.[87]

Table 3.6 lists comparative gas-phase acidities for a variety of compounds. Comparison of the relative gas-phase acidities with the solution pK_a values given in Table 3.3 reveals a number of changes in order. The most striking difference is the position of water, which, in comparison with other compounds, is a very much weaker acid in the gas phase than in the liquid phase. One may conclude that the strong propensity for water to solvate ions and polar molecules, particularly through hydrogen bonding, influences its acid–base properties so strongly as to overshadow other effects arising from the internal structure and bonding.

The contrast between solution and gas-phase data warns that in interpreting solution acidity in terms of molecular structure one must be careful to minimize solvation differences.

Acidities of Amines

Acidities of aliphatic amines are less well known than those of alcohols. Cyclohexyl-amine has about the same acidity as toluene (Table 3.3) and is apparently less

---

[87] For further discussion of gas-phase acidities, see (a) E. M. Arnett, D. E. Johnston, L. E. Small, and D. Oancea, *Faraday Symp. Chem. Soc.*, **10**, 20 (1975); (b) C. D. Ritchie, in *Solute–Solvent Interactions*, Vol. 2, J. F. Coetzee and C. D. Ritchie, Eds., Dekker, New York, 1976, Chap. 12; (c) R. W. Taft, in *Proton Transfer Reactions*, E. Caldin and V. Gold, Eds., Chapman and Hall, London, 1975, Chap. 2. For quantitative data on a variety of carbon and nitrogen acids, see (d) T. B. McMahon and P. Kebarle, *J. Am. Chem. Soc.*, **98**, 3399 (1976).

Table 3.6 Selected Gas-Phase Acidities $(\Delta H_i)^a$

Compound	$\Delta H_i(\text{gas})$ (kcal mole^{-1})^b
Most Acidic	
ϕOH	351.4
H_2S	353.4
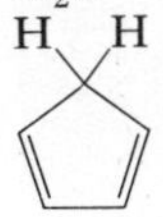 cyclopentadiene (H H)	356.1
CH_3NO_2	358.7
CH_3SH	359.0
ϕNH_2	367.1
$CH_3-\overset{\overset{\displaystyle O}{\|}}{C}-CH_3$	368.8
HF	371.5
CH_3CN	372.2
$CH_3-\overset{\overset{\displaystyle O}{\|}}{S}-CH_3$	372.7
$H-C{\equiv}C-H$	375.4
CH_3CH_2OH	376.1
ϕCH_3	379.0
CH_3OH	379.2
H_2O	390.8
NH_3	399.6
H_2	400.4
CH_4	416.6
Least Acidic	

^{a} Reprinted in part with permission from J. E. Bartmess, J. A. Scott, and R. T. McIver, Jr., *J. Am. Chem. Soc.*, **101**, 6046 (1979). Copyright 1979 American Chemical Society.
$^b \Delta H_i$ refers to the reaction $HA \longrightarrow H^+ + A^-$.

acidic in solution than ammonia. In the gas phase, Brauman and Blair found the order (most acidic to least acidic)

$$(C_2H_5)_2NH > (CH_3)_3CCH_2NH_2 \geq (CH_3)_3CNH_2 \geq (CH_3)_2NH \geq$$
$$(CH_3)_2CHNH_2 > CH_3CH_2CH_2NH_2 > C_2H_5NH_2 > CH_3NH_2 > NH_3.[88]$$

Water falls between diethylamine and ammonia. Note that the gas-phase order for the amines corresponds to that for the alcohols: more and larger alkyl groups provide greater stabilization to the ion.

There is a substantial amount of information on acidity of substituted anilines and some diphenylamines, as these compounds have been used as indicators in constructing the H_- function. Table 3.7 lists pK_a's for a few of these com-

[88] J. I. Brauman and L. K. Blair, *J. Am. Chem. Soc.*, **91**, 2126 (1969).

Table 3.7 SOLUTION ACIDITIES OF
SUBSTITUTED ANILINES[a]

Substituent	pK_a
4-CN	23.64
3-CN	24.80
3-CF$_3$	26.70
3-Cl	26.77
4-NO$_2$	18.35
2-NO$_2$	17.94
3,4-di-Cl	25.07
3,5-di-Cl	24.05
4-Cl-2-NO$_2$	16.60
4-NO$_2$-2-CH$_3$	18.29
4-NO$_2$-3,5-di-CH$_3$	22.23

[a] Reprinted in part with permission from R. A. Cox and R. Stewart, *J. Am. Chem. Soc.*, **98**, 488 (1976). Copyright 1976 American Chemical Society.

pounds. Here it is probably reasonable to use solution values as measures of relative intrinsic acidity, because structural changes are remote from the NH$_2$ group and should have relatively little effect on solvation. The substituent effects of the electron-withdrawing groups can be seen in the enhanced acidity of the nitro compound.

Acidities of Carbon Acids

Another class of acids of interest in organic chemistry is the group of carbon acids. Here we may discern three kinds of effects on acidity. The first of these is illustrated by the acidity of methane compared with that of cyclohexane (p$K_a \approx 48$ and 52 respectively as measured by the exchange method, Table 3.2). It would appear that the trend is in the direction of decreasing acid strength with substitution of hydrogen by alkyl. Note that the tendency here is in the direction opposite to the effect in alcohols if we take the gas-phase results to be the more accurate indication of intrinsic acid strength. The hydrocarbon data are from solution measurements subject to considerable uncertainty, and the differences are small. It seems risky to interpret the results in terms of intrinsic molecular properties.

A second effect of structure on acidity is evident from the data in Table 3.8. Here the differences are considered to be due primarily to the change in hybridization of the orbital that bears the negative charge in the conjugate base. The large contribution of the *s* orbital in the *sp*-hybridized carbon of acetylene results in greater electronegativity than is found in hybrids with high *p* orbital contributions, because the *s* orbital function puts the electrons on the average nearer to the nucleus. A regular trend toward weaker acids is evident from the data as the hybridization changes from *sp* (C$_6$H$_5$C≡CH) to *sp*^{2} (H$_2$C=CH$_2$ and C$_6$H$_6$) to *sp*^{3}

Table 3.8 Acidities of Selected Hydrocarbons

Compound	Approximate pK_a[a]
$C_6H_5C{\equiv}CH$	23
C_6H_6	43
$H_2C{=}CH_2$	44
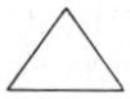 △	46
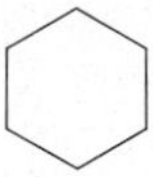 ⬡	52

[a] See Table 3.2.

(cyclohexane). Although we are still dealing with solution values, the interpretation in terms of molecular structure may be considered to be more reliable in this case than for the alcohols or saturated hydrocarbons, as the differences observed are larger. Streitwieser and co-workers have correlated kinetic acidity with s character through nmr ^{13}C—H coupling constants.[89]

The data in Table 3.2 show clearly the effect of conjugative stabilization of negative charge by a phenyl group attached to the acidic center. Note that triptycene (**6**) is about 8 powers of 10 more acidic than cyclohexane and about 12

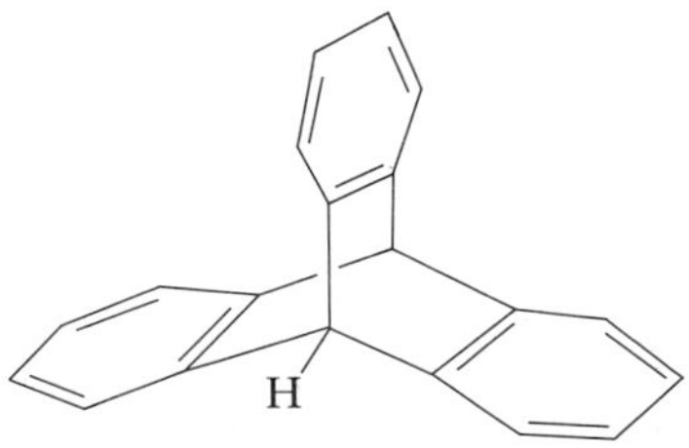

6

powers of 10 less acidic than triphenylmethane. The electron-withdrawing inductive effect of the aryl rings, (5 to 6 powers of 10) together with greater s character in the triptycene C—H bond (2 to 3 powers of 10), accounts for the enhancement of acidity over cyclohexane.[90] The difference between triptycene and triphenylmethane is caused by the geometry of the bridged ring, which forces the acidic C—H bond to lie in the nodal plane of the π orbitals and prevents conjugation. Following Streitwieser's argument, one can estimate the roughly 20-power-of-10 acidifying effect of three phenyl groups on an adjacent C—H bond to be due very approxi-

[89] A. Streitwieser, Jr., R. A. Caldwell, and W. R. Young, *J. Am. Chem. Soc.*, **91**, 529 (1969).
[90] (a) A. Streitwieser, Jr., R. A. Caldwell, and M. R. Granger, *J. Am. Chem. Soc.*, **86**, 3578 (1964); (b) A. Streitwieser, Jr., and G. R. Ziegler, *J. Am. Chem. Soc.*, **91**, 5081 (1969).

mately one-fourth to one-third (5 to 6 powers of 10) to inductive effect and the remaining two-thirds to three-fourths to conjugation.

The series cyclohexane–toluene–diphenylmethane–triphenylmethane (approximate pK_a 52, 41, 33, 31) illustrates a commonly observed saturation effect. The first phenyl group enhances acidity by 11 powers of 10, the second by 8, the third by only 2. The alkyl anion, with localized charge, gains more in stability by addition of a phenyl group than does an anion in which charge is already partly delocalized. There is another effect hidden in this sequence, however. The phenyl groups bonded to the anionic carbon in the conjugate base must be coplanar for maximum conjugation, but the ortho hydrogens of adjacent rings interfere sterically and prevent perfect coplanarity. The effect of this distortion can be seen by comparing diphenylmethane (p$K_a \approx 33$) with fluorene (p$K_a \approx 23$); the latter has the two rings tied together by elimination of the interfering ortho hydrogens, an arrangement that ensures coplanarity and efficient conjugation. Streitwieser and Nebenzahl have synthesized an all-coplanar 9-phenylfluorene analog (**7**) that has ion-pair pK_a of 15.4. They estimated from a ρ^*–σ^* correlation that a hypothetical 90°-twisted 9-phenylfluorene (**10**) would have a pK_a of about 19.5 to 20. Hence in **7** about 40 percent of the acidifying effect of the coplanar phenyl group is due to polar effects and about 60 percent to conjugation.[91]

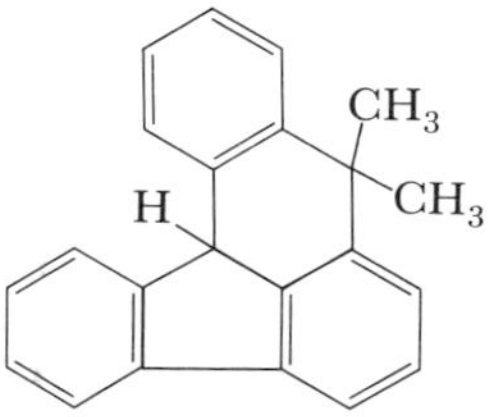

7

pK_a 15.4

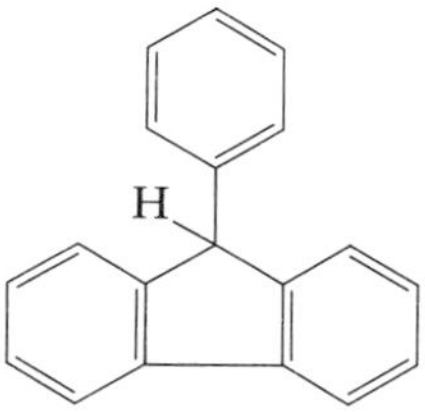

8

pK_a 18.5

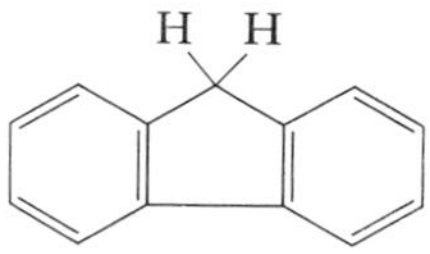

9

pK_a 22.6

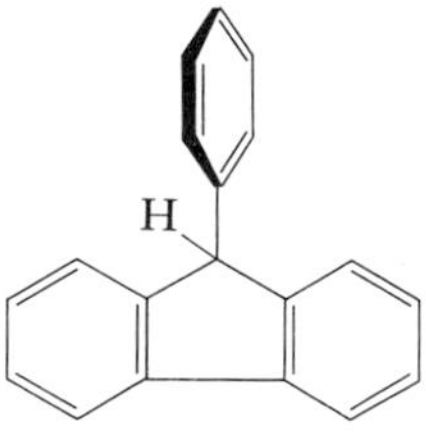

10

pK_a 19.5 to 20
(estimate)

[91]A. Streitwieser, Jr., and L. L. Nebenzahl, *J. Org. Chem.*, **43**, 598 (1978). Inductive and rotational effects of the bridge in **7** were neglected because they were estimated to be opposite in direction and of comparable small magnitude. By considering also a compound similar to **7** but with a two-carbon bridge, Streitwieser and Nebenzahl estimated that the angle of twist out of the fluorene ring plane of the phenyl group in 9-phenylfluorene is about 50°.

The final effect to be noted in the carbon acids, and a most important one from the point of view of organic reactions in general, is illustrated by the data in Table 3.9. It is a well-known feature of organic molecules that certain electron-withdrawing groups increase the acidity of neighboring carbon–hydrogen bonds. A few of these groups are represented in Table 3.9, which also indicates the cumulative effects observed when more than one such group is bonded to the same carbon. The acidifying groups shown have unsaturated structures containing nitrogen or oxygen or both, and the acid-strengthening effect is attributable primarily to the stabilization of the conjugate base by delocalization of the negative charge onto an electronegative center, as illustrated in the alternative formulations **11** and **12**.

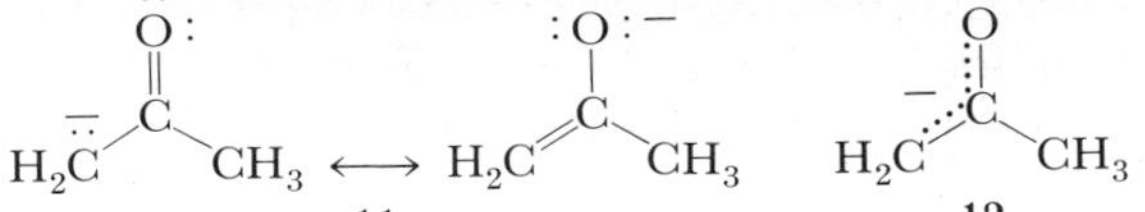

Table 3.9 Acidities of Carbon Acids Containing Electron-Withdrawing Groups[a]

Compound	pK_a
CH_3NO_2	11
$CH_2(NO_2)_2$	4
$CH(NO_2)_3$	0
$CH_3-\overset{O}{\overset{\|}{C}}-CH_3$	20
$CH_3-\overset{O}{\overset{\|}{C}}-CH_2-\overset{O}{\overset{\|}{C}}-CH_3$	9
$CH(\overset{O}{\overset{\|}{C}}-CH_3)_3$	6
$CH_3-\overset{O}{\overset{\|}{C}}-O-C_2H_5$	24.5
$CH_2(\overset{O}{\overset{\|}{C}}-O-C_2H_5)_2$	13.3
$CH_3SO_2CH_3$	31[b]
$CH_2(SO_2C_2H_5)_2$	12[c]
$CH(SO_2CH_3)_3$	0
$CH_3C{\equiv}N$	31[b]
$CH_2(C{\equiv}N)_2$	12
$CH(C{\equiv}N)_3$	0

[a] Reprinted with permission from R. G. Pearson and R. L. Dillon, *J. Am. Chem. Soc.*, **75**, 2439 (1953). Copyright by the American Chemical Society.
[b] See Table 3.3.
[c] F. Hibbert, *J. Chem. Soc. Perkin II*, 1171 (1978).

Again, solution acidities are being interpreted in terms of intrinsic properties; however, the differences are large enough (CH_4 to acetone over 20 pK units) that we may feel fairly confident of our theory in this case.

The cumulative effects of adding more anion-stabilizing groups (Table 3.9) indicate to some extent the saturation mentioned above for the phenylmethanes, although the data were not all obtained under the same conditions and hence are not strictly comparable. Bordwell and co-workers have made a careful analysis of the saturation effects in adding phenyl groups by comparing acidity in dimethylsulfoxide solution for CH_3—EWG, ϕCH_2—EWG, $\phi_2 CH$—EWG, where EWG is an electron-withdrawing group (CN—, $CH_3 SO_2$—, $CH_3 CO$—, etc.). They observed a definite resonance saturation effect that decreased the acidifying influence of the added phenyl group as the negative charge in the anion became smaller with more effective electron withdrawal by the substituent EWG group. They were also able to separate, through steric inhibition of resonance, the conjugation and polar effects of phenyl. The ratio of conjugation to polar contributions in the phenyl substituent effect were estimated at 4:1 to 6.6:1, depending on the nature of EWG.[92]

Carboxylic Acids

Another important class of organic acids are the carboxylic acids. Since the pK_a's of these substances usually fall in the range 4–5, their acidities can be determined with precision and compared with considerable confidence, despite the fact that the differences are small.[93] The pK_a's of a large number of carboxylic acids have been determined; we list in Table 3.10 only a few representative values.[94] The data illustrate qualitatively the acid-strengthening effect of electron-withdrawing substituents. Quantitative correlations have been made for many series of carboxylic acids, both through the Hammett σ substituent constants for acids bearing a phenyl substituent and through the Taft σ^* or σ_I constants for aliphatic systems. Table 3.11 lists a few of the results obtained by these methods and includes also some correlations by the two-parameter method for separation of resonance and polar effects.[95]

---

[92] F. G. Bordwell, J. E. Bares, J. E. Bartmess, G. J. McCollum, M. Van Der Puy, N. R. Vanier, and W. S. Matthews, *J. Org. Chem.*, **42**, 321 (1977).

[93] The greatly increased acidity of the carboxylic acids over water and the alcohols is accounted for by delocalization of charge in the conjugate base, as indicated by Structures **a**.

[94] References to compilations are given in Table 3.3, note *a*.

[95] R. D Topsom, *Prog. Phys. Org. Chem.*, **12**, 1 (1976).

Table 3.10 Acid Dissociation Constants of Some Representative Carboxylic acids

Compound	$pK_a{}^a$
HCOOH	3.77
CH_3COOH	4.76
CH_3CH_2COOH	4.88
$CH_3CH_2CH_2COOH$	4.82
$CH_3CH_2CH_2CH_2COOH$	4.86
$H_3N^+CH_2COOH$	2.31
O_2NCH_2COOH	1.68
$ClCH_2COOH$	2.86
$Cl_2CHCOOH$	1.29
Cl_3CCOOH	0.65
$^-OOCCH_2COOH$	5.69
$\phi COOH$	4.20
$p\text{-}(CH_3)_3\overset{+}{N}\phi COOH$	3.43
$p\text{-}\overset{-}{O}OC\phi COOH$	4.82

[a] H. C. Brown, D. H. McDaniel, and O. Häflinger, in *Determination of Organic Structures by Physical Methods,* Vol. I, E. A. Braude and F. C. Nachod, Eds., Academic Press, New York, 1955, p. 567. Reprinted by permission of Academic Press and H. C. Brown.

Nitrogen and Phosphorus Bases

The effect in the liquid phase of substituting hydrogen by alkyl groups on the nitrogen and phosphorus bases is illustrated by the solution data presented in Table 3.12. The phosphorus basicities are much more strongly affected than are the nitrogen. The tertiary amine $(CH_3)_3N$ is in an anomalous position with respect to the other amines. We suspect immediately that solvation is the culprit. In the gas phase the amine order is (most basic to least) tertiary $>$ secondary $>$ primary $>$ ammonia (Table 3.13).[96] The phosphines have the same basicity order in the gas phase as in solution, tertiary $\gg$ phosphine.[97]

[96](a) E. M. Arnett, F. M. Jones III, M. Taagepera, W. G. Henderson, J. L. Beauchamp, D. Holtz, and R. W. Taft, *J. Am. Chem. Soc.,* **94,** 4724 (1972); (b) D. H. Aue, H. M. Webb, and M. T. Bowers, *J. Am. Chem. Soc.,* **94,** 4726 (1972); **98,** 311 (1976); (c) W. G. Henderson, M. Taagepera, D. Holtz, R. T. McIver, Jr., J. L. Beauchamp, and R. W. Taft, *J. Am. Chem. Soc.,* **94,** 4728 (1972).

[97] E. M. Arnett, *Acc. Chem. Res.,* **6,** 404 (1973). Gas-phase data are reported for $(CH_3)_3P$ and PH_3 only, but heat of protonation results indicate that secondary and primary phosphines will fall in that order between these two.

Table 3.11 CORRELATION OF SUBSTITUENT EFFECTS ON ACIDITY BY THE HAMMETT AND TAFT RELATIONS[a]

Series	ρ	References	$\rho_I^{\text{meta}b}$	$\rho_R^{\text{meta}b}$	$\rho_I^{\text{para}b}$	$\rho_R^{\text{para}b}$
ArCOOH	$(1.00)^c$		1.00	0.28	1.00	1.00
ArCH$_2$COOH	0.49	1			0.48	0.43
ArCH$_2$CH$_2$COOH	0.21	1				
trans-ArCH=CHCOOH	0.47	1				
ArOCH$_2$COOH	0.30	2				
ArOH	2.23	3				
ArNH$_3^+$	2.89	3			3.09	3.48
ArCH$_2$NH$_3^+$	1.06	4				
	ρ^*					
RCOOH	1.72	5				
RNH$_3^+$	3.14	6				

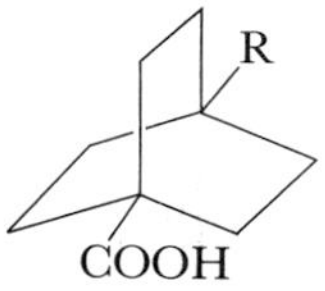

1.63[d]   7

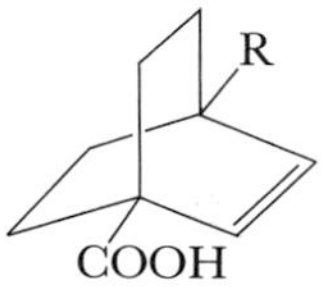

1.75[d]   7

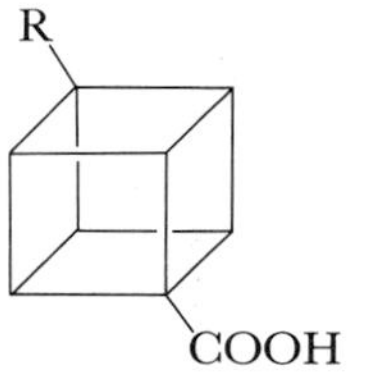

1.52[d]   8

[a] In water at 25°C. For other correlations, see J. Hine, *Structural Effects on Equilibria in Organic Chemistry,* Wiley, New York, 1975, p. 162.
[b] R. D. Topsom, *Prog. Phys. Org. Chem.,* **12,** 1 (1976).
[c] ρ defined as 1.00 for this series.
[d] ρ_I.

References:
1. H. H. Jaffé, *Chem. Rev.,* **53,** 191 (1953).
2. L. D. Pettit, A. Royston, C. Sherrington, and R. J. Whewell, *J. Chem. Soc. B,* 588 (1968).
3. A. I. Biggs and R. A. Robinson, *J. Chem. Soc.,* 388 (1961).
4. L. F. Blackwell, A. Fischer, I. J. Miller, R. D. Topsom, and J. Vaughan, *J. Chem. Soc.,* 3588 (1964).
5. R. W. Taft, Jr., in *Steric Effects in Organic Chemistry,* M. S. Newman, Ed., Wiley, New York, 1956, p. 607.
6. H. K. Hall, Jr., *J. Am. Chem. Soc.,* **79,** 5441 (1957).
7. F. W. Baker, R. C. Parish, and L. M. Stock, *J. Am. Chem. Soc.,* **89,** 5677 (1967).
8. T. W. Cole, Jr., C. J. Mayers, and L. M. Stock, *J. Am. Chem. Soc.,* **96,** 4555 (1974).

Table 3.12 Solution pK_a Values of Conjugate Acids
of Some Nitrogen and Phosphorus Bases

Base	pK_a (BH$^+$)	Base	pK_a (BH$^+$)
NH_3	9.24^a	PH_3	$\sim -12^b$
CH_3NH_2	10.6^c	$n\text{-}C_4H_9PH_2$	0^d
$(CH_3)_2NH$	10.7^c	$(n\text{-}C_4H_9)_2PH$	4.5^d
$(CH_3)_3N$	9.8^c	$(n\text{-}C_4H_9)_3P$	8.4^d

^{a} See Table 3.3.
^{b} R. E. Weston and J. Bigeleisen, *J. Am. Chem. Soc.*, **76**, 3074 (1954).
^{c} D. D. Perrin, *Dissociation Constants of Organic Bases in Aqueous Solution*, Butterworths, London, 1965. Values at 25°C.
^{d} C. A. Streuli, *Anal. Chem.*, **32**, 985 (1960). Determined by titration in nitromethane and corrected to water solution.

Arnett has presented an analysis of the solvation thermochemistry of the amines and their conjugate acids.[98] Table 3.14 gives data for the four processes of Equations 3.80 through 3.83. The subscript w refers to water solution, g to gas

$$BH_w^+ \longrightarrow B_w + H_w^+ \qquad \Delta G_{iw} \qquad (3.80)$$
$$BH_g^+ \longrightarrow B_g + H_g^+ \qquad \Delta G_{ig} \qquad (3.81)$$
$$B_g \longrightarrow B_w \qquad \Delta G_s(B) \qquad (3.82)$$
$$BH_g^+ \longrightarrow BH_w^+ \qquad \Delta G_s(BH^+) \qquad (3.83)$$

phase, i to the ionization process, and s to the transfer from gas to solution. Values reported in Table 3.14 are $\delta \Delta G$, free energies measured relative to the value for NH_3. The more positive $\delta \Delta G_i$, the smaller the tendency for BH$^+$ to ionize and hence the stronger the base. The more positive $\delta \Delta G_s$, the more reluctant is that species to enter solution from the gas phase.

The first column of Table 3.14 reflects the gas-phase order, $(CH_3)_3N$ most basic. The second column reflects the solution basicities (compare Table 3.12). The third column reveals that the differences in free energies of solution among the free bases are small. In the fourth column we find large differences among solvation free energies of the ions BH$^+$. This result agrees with our earlier discussion in Section 3.2 of activity coefficients, where the major variation of γ_B/γ_{BH^+} on changing solvents came from γ_{BH^+}. The more substituted BH$^+$, the less favorable is its transfer to solution. Recall that this effect arises primarily from the less effective hydrogen bonding when there are fewer acidic hydrogens. In solution the more substituted amines will be reduced in basicity compared with their gas-phase behavior, because the solvation of BH$^+$ becomes poorer the more highly substituted it is. Note that the solvation free-energy differences very nearly cancel the intrinsic basicity differences revealed by the gas-phase ionization free-energy differences in the first

[98](a) E. M. Arnett and G. Scorrano, *Adv. Phys. Org. Chem.*, **13**, 83 (1976); (b) R. W. Taft, J. F. Wolf, J. L. Beauchamp, G. Scorrano, and E. M. Arnett, *J. Am. Chem. Soc.*, **100**, 1240 (1978).

Table 3.13 Gas-phase Basicities of Selected Compounds

Compound	Proton affinity (relative to NH_3)[a] (kcal mole^{-1})	References[b]
Least basic		
CH_4	122	1
$H_2C{=}CH_2$	154	1
CH_3Cl	155	1
H_2O	170	2
H_2S	174	2
$CH_3{-}CH{=}CH_2$	177	1
HCOOH	180	2
CH_3SH	180	3
PH_3	180, 183	3, 4
CH_3OH	182	2
CH_3COOH	183	1
CH_3CH_2OH	188	1
$(CH_3)_2CHOH$	188	1
$CH_3{-}O{-}CH_3$	191.3	2
$(CH_3)_3COH$	193	1
CH_3COCH_3	195	2
$CH_3CH_2{-}O{-}CH_2CH_3$	198.6	2
$CH_3{-}S{-}CH_3$	198.8	2
CH_3PH_2	202	4
NH_3	202.3	5
ϕNH_2	211	3
$\phi_2 NH$	—	6
CH_3NH_2	213	7
$\phi_3 N$	—	6
$(CH_3)_2PH$	214	4
$(CH_3)_2NH$	220	7
$(CH_3)_3P$	223	4
$(CH_3)_3N$	224	7
Most basic		

[a] All values referred to $NH_3 = 202.3$. Some data indicate a slightly higher value. See F. A. Houle and J. L. Beauchamp, *J. Am. Chem. Soc.*, **101**, 4067 (1979).

[b] References:

1. J. Long and B. Munson, *J. Am. Chem. Soc.*, **95**, 2427 (1973); data referred to $NH_3 = 202$.
2. J. F. Wolf, R. H. Staley, I. Koppel, M. Taagepera, R. T. McIver, Jr., J. L. Beauchamp, and R. W. Taft, *J. Am. Chem. Soc.*, **99**, 5417 (1977).
3. E. M. Arnett, *Acc. Chem. Res.*, **6**, 404 (1973).
4. R. H. Staley and J. L. Beauchamp, *J. Am. Chem. Soc.*, **96**, 6252 (1974).
5. R. H. Staley, M. Taagepera, W. G. Henderson, I. Koppel, J. L. Beauchamp, and R. W. Taft, *J. Am. Chem. Soc.*, **99**, 326 (1977).
6. I. Dzidic, *J. Am. Chem. Soc.*, **94**, 8333 (1972). Relative basicities only.
7. D. H. Aue, H. M. Webb, and M. T. Bowers, *J. Am. Chem. Soc.*, **98**, 311 (1976).

Table 3.14 Free Energies of Ionization and Solvation of Amines and Ammonium Ions in the Gas Phase and in Water[a]

Amine	$\delta \Delta G_{ig}$	$\delta \Delta G_{iw}$	$\delta \Delta G_s(B)$	$\delta \Delta G_s(BH^+)$
NH_3	0	0	0	0
CH_3NH_2	9.2	1.92	−0.3	7.0
$(CH_3)_2NH$	16.0	2.09	0.00	13.9
$(CH_3)_3N$	20.6	0.75	+1.1	20.9

[a] Reprinted in part with permission from R. W. Taft, J. F. Wolf, J. L. Beauchamp, G. Scorrano, and E. M. Arnett, *J. Am. Chem. Soc.*, **100**, 1240 (1978). Copyright 1978 American Chemical Society.

column. The observed solution order (second column) results from the small free-energy variations remaining after combining two large, opposing terms [$\delta \Delta G_{ig}$ and $\delta \Delta G_s(BH^+)$] and one small term [$\delta \Delta G_s(B)$].

Table 3.15 dissects the important $\delta \Delta G_s(BH^+)$ term into enthalpy and entropy contributions. Increasing substitution on BH^+ makes both of these quantities less favorable in the gas → solution direction. For the bases themselves, on the other hand, the solution enthalpies and entropies are in opposition (Table 3.16), enthalpies being more favorable to the solution process the more substituents, but entropies becoming less favorable with more substituents. The cancelation of these opposing effects leaves the small $\delta \Delta G_s(B)$ values shown in Table 3.14.

Oxygen and Sulfur Bases

Table 3.17 lists pK_a values for conjugate acids of some oxygen and sulfur bases obtained with the Bunnett–Olsen H_0–ϕ acidity function. As alkyl groups are substituted for hydrogen on an oxygen, the compound becomes a weaker base. The gas phase trend is the reverse (Table 3.13). As with the amines, decreased effectiveness of hydrogen bonding to the conjugate acid as hydrogens are replaced by alkyl groups is largely responsible for reversing the intrinsic gas phase basicities when the ions are put into water. The effect is stronger for the oxygen bases than for

Table 3.15 Enthalpy and Entropy Contributions to Relative Free Energies of Solution of Ammonium Ions in Water[a]

Amine	$\delta \Delta H_s(BH^+)$	$-\delta T\Delta S_s(BH^+)$
NH_3	0	0
CH_3NH_2	5.7	1.3
$(CH_3)_2NH$	11.4	2.5
$(CH_3)_3N$	18.8	2.1

[a] Reprinted in part with permission from R. W. Taft, J. F. Wolf, J. L. Beauchamp, G. Scorrano, and E. M. Arnett, *J. Am. Chem. Soc.*, **100**, 1240 (1978). Copyright 1978 American Chemical Society.

Table 3.16 Enthalpy and Entropy Contributions
to Relative Free Energies of Solution
of Amines in Water[a]

Amine	$\delta \Delta H_s(B)$	$-\delta T\Delta S_s(B)$
NH_3	0	0
CH_3NH_2	-2.6	2.3
$(CH_3)_2NH$	-4.7	4.7
$(CH_3)_3N$	-4.7	5.8

[a] Reprinted in part with permission from R. W. Taft, J. F. Wolf, J. L. Beauchamp, G. Scorrano, and E. M. Arnett, *J. Am. Chem. Soc.*, **100**, 1240 (1978). Copyright 1978 American Chemical Society.

the amines; free energy of transfer of oxonium ions are more negative than those for the corresponding ammonium ions by 19 kcal mole^{-1} ($H_3O^+ - NH_4^+$) and 12 kcal mole^{-1} ($CH_3OH_2^+ - CH_3NH_3^+$).[98b]

The solution results indicate that dimethylsulfide is a considerably weaker base than dimethyl ether; in the gas phase it is slightly stronger. The greater solvation of the oxonium ion compared with the sulfonium ion is responsible. The free energy of transfer, gas phase to aqueous solution, is 14 kcal mole^{-1} more negative for $(CH_3)_2OH^+$ than for $(CH_3)_2SH^+$; free energies of transfer of $(CH_3)_2O$ and $(CH_3)_2S$ differ by only 0.4 kcal mole^{-1}.

Alkyl Substituent Effects

It is striking that in gas-phase equilibria the accumulation of more or larger alkyl groups on N, O, P, or S favors the ion, whether the ion be negatively charged (Table 3.5) or positively charged (Table 3.13). It should be clear that when the alkyl substituent influences interaction of an ionic center with solvent, as in the primary–secondary–tertiary amine comparison or in the methyl–ethyl–isopropyl–*tert*-butyl alcohol comparison, solvation will be the prime determinant of chemical behavior. But what is the intrinsic polar property of an alkyl group? Taft σ^* parameters for alkyl groups are all electron donating with respect to hydrogen. The σ^* scale, however, has been criticized as an inaccurate measure of inductive effects for the alkyl groups and as reflecting mainly steric rather than electronic factors.[99] The σ_I scale (Table 2.5), in which the alkyl groups differ very little from hydrogen, is probably closer to a reliable measure of a pure inductive effect.

Analysis of the gas-phase data gives a possible explanation for the puzzle of why alkyl groups stabilize both positive and negative charge in the gas phase.[100]

[99] (a) F. G. Bordwell, J. E. Bartmess, and J. A. Hautala, *J. Org. Chem.*, **43**, 3095 (1978); (b) F. G. Bordwell and H. E. Fried, *Tetrahedron Lett.*, 1121 (1977); (c) M. Charton, *J. Org. Chem.*, **44**, 903 (1979).

[100] (a) R. W. Taft, M. Taagepera, J. L. M. Abboud, J. F. Wolf, D. J. DeFrees, W. J. Hehre, J. E. Bartmess, and R. T. McIver, Jr., *J. Am. Chem. Soc.*, **100**, 7765 (1978); (b) S. R. Smith and T. D. Thomas, *J. Am. Chem. Soc.*, **100**, 5459 (1978); (c) D. H. Aue, H. M. Webb, and M. T. Bowers, *J. Am. Chem. Soc.*, **98**, 311 (1976).

Table 3.17 SOLUTION BASICITIES OF SOME
OXYGEN AND SULFUR BASES

Compound	pK_{aBH^+}	ϕ
H_2O	$-1.11;^a$ -1.74^b	—
CH_3OH	-2.05^c	0.87^c
CH_3CH_2OH	-1.94^c	0.86^c
CH_3OCH_3	-2.48^c	0.82^c
$C_2H_5OC_2H_5$	-2.39^c	0.78^c
CH_3SCH_3	-6.99^c	-0.27^c
$CH_3\overset{\overset{\displaystyle O}{\|}}{C}CH_3$	-2.85^c	0.75^c
$CH_3\overset{\overset{\displaystyle O}{\|}}{S}CH_3$	-1.54^c	0.58^c
$\phi\overset{\overset{\displaystyle O}{\|}}{C}CH_3$	-4.02^d	0.44^d

^{a} Calculated from data quoted by R. W. Taft, J. F. Wolf,
J. L. Beauchamp, G. Scorrano, and E. M. Arnett, *J. Am.
Chem. Soc.*, **100**, 1240 (1978).
^{b} Formal value, calculated for the equilibrium

$$H_3O^+ \rightleftharpoons H^+ + H_2O$$

$$K_a = \frac{[H^+][H_2O]}{[H_3O^+]} = [H_2O] = 55.5$$

^{c} From Bunnett and Olsen's H_0–ϕ acidity function. Re-
printed in part from G. Perdoncin and G. Scorrano, *J. Am.
Chem. Soc.*, **99**, 6983 (1977). Copyright 1977 American
Chemical Society.
^{d} From Bunnett and Olsen's H_0–ϕ acidity function. A.
Levi, G. Modena, and G. Scorrano, *J. Am. Chem. Soc.*, **96**,
6585 (1974).

Stabilization of charge takes place by two mechanisms. The inductive effect originates in the electric dipole of the substituent, depends on the first power of the
charge, and falls off as the inverse square of the distance between the group and the
charge. The dispersion effect operates by a redistribution of electron density in the
substituent group induced by the charge itself; it depends on the polarizability of
the group and on the square of the charge and falls off as the inverse fourth power
of the distance. Taft and co-workers separated the two effects by taking advantage
of the fact that the polarizability (P) stabilizes both positive and negative charge;
the inductive effect (I) stabilizes one and destabilizes the other. They assumed that
the magnitude of each effect is about the same on a positive or on a negative
charge; the polarizability contribution could then be canceled out by subtracting
Equation 3.85 from Equation 3.84, leaving Equation 3.86 as a measure of the pure

$$ROH(g) + CH_3OH_2{}^+(g) \longrightarrow ROH_2{}^+(g) + CH_3OH(g); \quad -\Delta G_1^\circ = I + P \quad (3.84)$$
$$ROH(g) + CH_3O^-(g) \longrightarrow RO^-(g) + CH_3OH(g); \quad -\Delta G_2^\circ = -I + P \quad (3.85)$$
$$RO^-(g) + CH_3OH_2{}^+(g) \longrightarrow ROH_2{}^+(g) + CH_3O^-(g); \quad (3.86)$$
$$\Delta G_3^\circ = -\Delta G_1^\circ - (-\Delta G_2^\circ) = 2I$$

inductive effect. The inductive effects found in this way correlate well with σ_I for the groups R. Alkyl groups stabilize both positive and negative charge in the gas phase because the dispersion mechanism dominates. In solution, however, charge is spread out among solvent molecules. Because the dispersion mechanism has higher power charge and distance dependences, it is attenuated more than is the inductive mechanism, leaving in solution a slight dominance of the electron-donating inductive effect of the alkyl groups.[101]

3.5 LEWIS ACIDS AND BASES[102]

In 1923 G. N. Lewis proposed a definition of acids and bases somewhat different from that of Brønsted:[103]

An acid is an electron-pair acceptor.
A base is an electron-pair donor.

Lewis acids are thus electron-deficient molecules or ions such as BF_3 or carbocations, whereas Lewis bases are molecules or ions containing available electrons, such as amines, ethers, alkoxide ions, and so forth. A Lewis acid–base reaction in its

---

[101] The substituent effect commonly termed *inductive* is more properly a *field* effect because it is transmitted directly through space rather than along bonds. The field effect mechanism is established by varying number and type of bonding pathways between substituent and acidic group while keeping separation and orientation constant, as in these systems:

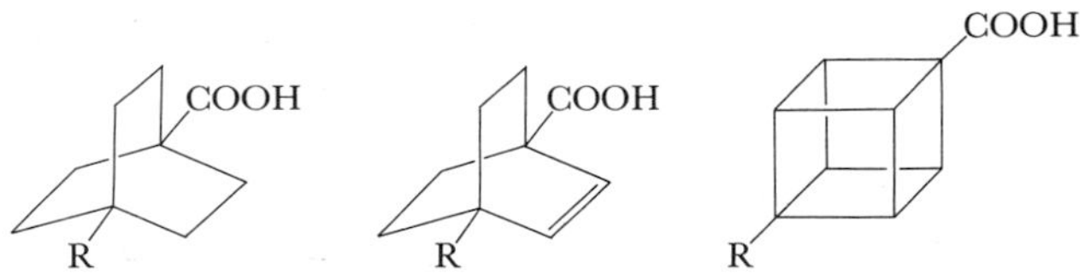

An electrostatic model due to Kirkwood and Westheimer accounts well for relative acidity in these systems. See (a) F. W. Baker, R. C. Parish, and L. M. Stock, *J. Am. Chem. Soc.*, **89**, 5677 (1967); (b) T. W. Cole, Jr., C. J. Mayers, and L. M. Stock, *J. Am. Chem. Soc.*, **96**, 4555 (1974).
[102] Reviews: (a) R. J. Gillespie, in *Friedel–Crafts and Related Reactions*, Vol. 1, G. A. Olah, Ed., Wiley, New York, 1963, p. 169; (b) W. B. Jensen, *Chem. Rev.*, **78**, 1 (1978).
[103] G. N. Lewis, *Valence and Structure of Atoms and Molecules*, American Chemical Society Monograph, The Chemical Catalog Co., New York, 1923. Lewis also gave a definition equivalent to that of Brønsted.

broadest sense is any process in which Lewis acid and base molecules or ions associate, dissociate, or exchange partners. Reactions can thus take any of the forms indicated in Equations 3.87–3.90. A few examples are shown in Table 3.18. Most of

$$A + B \rightleftharpoons AB \tag{3.87}$$
$$A + A'B \rightleftharpoons AB + A' \tag{3.88}$$
$$B + AB' \rightleftharpoons AB + B' \tag{3.89}$$
$$AB + A'B' \rightleftharpoons AB' + A'B \tag{3.90}$$

the intermediates and reactive species encountered in heterolytic chemistry are Lewis acids or bases, and most heterolytic reactions can be classified within the Lewis group.

The proton is a Lewis acid, and Brønsted acid–base reactions fall in the Lewis category also, where they fit into type 3.89. The Lewis classification is thus a much broader one than the Brønsted. We will continue to follow the usual custom of reserving the terms *acid* and *base* to the Brønsted definitions and will specify Lewis acid or base when the broader type is meant.

Strengths of Lewis Acids and Bases

Because the Lewis acid–base concept is an extremely useful one in chemistry, quantitative relationships of the types discussed in the previous sections for Brønsted acids and bases would be helpful. The task of classifying the Lewis acids and bases according to some criterion of strength has nevertheless proved to be a difficult one, and methods being developed still yield largely qualitative results. Brønsted acid–base reactions always involve transfer of a proton; this common feature allows meaningful quantitative comparisons of strengths to be made. Dif-

Table 3.18 EXAMPLES OF LEWIS
ACID–BASE REACTIONS

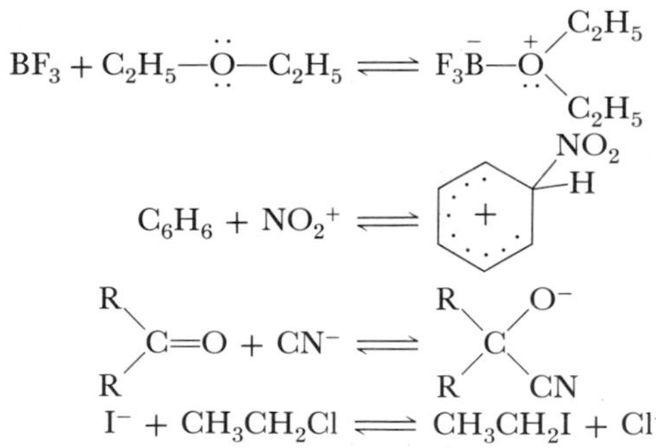

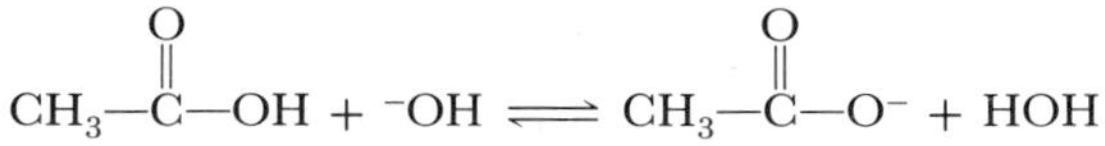

ferent Lewis acid–base reactions, on the other hand, do not necessarily have any feature in common, and the result is that the term *strength* does not have a well-defined meaning.

The problem may be illustrated by a simple example.[104] Suppose that we wish to compare the "coordinating power" of two Lewis bases, say an amine, NR_3, and a phosphine, PR_3. We might do this by comparing the equilibrium constants for Reactions 3.91 and 3.92 of the two bases, B and B′, with the same Lewis acid, for

$$A + B \xrightleftharpoons{K} AB \tag{3.91}$$

$$A + B' \xrightleftharpoons{K'} AB' \tag{3.92}$$

example, BF_3. Quantitative data are not always available, but it is often possible to make qualitative decisions about orders of reactivity. The information we have shows that the nitrogen base should be judged to have the greater coordinating power, since the equilibrium constant is greater for the formation of nitrogen base complex.[105] A similar qualitative result is found with H^+ as the reference acid.[106] If, on the other hand, the equilibrium constants for a nitrogen and a phosphorus base with Ag^+ are measured, it is found that with respect to this Lewis acid the phosphine has much greater coordinating power than does the amine. A similar situation arises with another set of bases, the halide ions. If H^+ is taken as the reference acid, fluoride is the most effective base in solution, followed by chloride, bromide, and iodide. With silver ion, however, the order is exactly reversed; iodide forms the most stable complex and fluoride the least stable.[104]

Hard and Soft Lewis Acids and Bases

Despite the apparent chaos of the picture presented by these results, it is possible to find some qualitative relationships that are useful. Schwarzenbach,[107] and also Ahrland, Chatt, and Davies,[104] classified Lewis acids into two categories, Class *a* and Class *b*. Class *a* acceptors are those that form their most stable complexes with donors of the first row of the periodic table: N, O, and F. Class *b* acids complex best with donors of the second or subsequent row: P, S, Cl, Br, I.[108]

Attempts have been made to quantify complexing tendencies of Lewis acids and bases, notably by Gutmann, who established a series of donor numbers, DN, and acceptor numbers, AN, for various solvents,[109] and by Drago and Wayland,

[104]S. Ahrland, J. Chatt, and N. R. Davies, *Q. Rev. Chem. Soc.*, **12**, 265 (1958).

[105]W. A. G. Graham and F. G. A. Stone, *J. Inorg. Nuc. Chem.*, **3**, 164 (1956).

[106]See Tables 3.12 and 3.13.

[107](a) G. Schwarzenbach, *Experientia*, Suppl., **5**, 162 (1956); (b) G. Schwarzenbach, *Adv. Inorg. Chem. Radiochem.*, **3**, 257 (1961); (c) G. Schwarzenbach and M. Schellenberg, *Helv. Chim. Acta*, **48**, 28 (1965).

[108]See also J. O. Edwards and R. G. Pearson, *J. Am. Chem. Soc.*, **84**, 16 (1962).

[109](a) V. Gutmann, *Coord. Chem. Rev.*, **18**, 225 (1976); (b) A. J. Parker, U. Mayer, R. Schmid, and V. Gutmann, *J. Org. Chem.*, **43**, 1843 (1978); (c) W. B. Jensen, *Chem. Rev.*, **78**, 1 (1978). See Section 2.4.

who assigned a parameter E, measuring electrostatic bonding potential, and a parameter C, measuring covalent bonding potential, to each of a series of Lewis acids and bases.[110]

R. G. Pearson has proposed a qualitative scheme in which each Lewis acid and base is characterized by two parameters, one of which is referred to as *strength* and the other of which is called *softness*.[111] Thus the equilibrium constant for a simple acid–base reaction would be a function of four parameters, two for each partner.

The next step in Pearson's argument is to classify acids and bases as *hard* or *soft* according to their properties. Hard acids correspond roughly in their behavior to the Class *a* acids of Schwarzenbach and of Ahrland, Chatt, and Davies. They are characterized by small acceptor atoms that have outer electrons not easily excited and that bear considerable positive charge. Soft acids have acceptor atoms of lower positive charge, large size, and with easily excited outer electrons. Hard and soft bases are defined analogously. Hard bases contain highly electronegative donor atoms of low polarizability, are typically difficult to oxidize, and have no empty low-energy orbitals available; soft bases are polarizable, have less electronegative donor atoms, and have empty orbitals of low energy and electrons that are more easily removed by oxidizing agents. Table 3.19 gives Pearson's classification of acids and bases into the hard and soft categories.

Having defined the terminology, we may now state Pearson's *principle of hard and soft acids and bases* (commonly abbreviated HSAB principle): Hard acids prefer to bind to hard bases and soft acids prefer to bind to soft bases.

The HSAB principle has proved useful in rationalizing and classifying a large amount of chemical data,[112] but it is intended only as a qualitative theory, and is not subject to direct quantitative test.

Applications of the HSAB principle   A few brief examples of applications of the HSAB principle follow.

The proton is a hard Lewis acid; it binds more strongly to a hard base (nitrogen) than to a softer one (phosphorus). Thus ammonia is a stronger base toward protons than is phosphine, both in solution and in the gas phase. Small, hard cations (Li^+) are more strongly solvated by hard solvents (oxygen donors) than are larger, softer cations (Cs^+). In hydroxylic solvents, I^- is a better nucleophile toward carbon than is F^-, because the F^- (hard) is strongly solvated by the hard

[110](a) R. S. Drago and B. B. Wayland, *J. Am. Chem. Soc.*, **87**, 3571 (1965); (b) A. P. Marks and R. S. Drago, *J. Am. Chem. Soc.*, **97**, 3324 (1975); (c) W. B. Jensen, *Chem. Rev.*, **78**, 1 (1978).

[111](a) R. G. Pearson, *J. Am. Chem. Soc.*, **85**, 3533 (1963); (b) R. G. Pearson and J. Songstad, *J. Am. Chem. Soc.*, **89**, 1827 (1967); (c) R. G. Pearson, *Survey of Progress in Chemistry*, **5**, 1 (1969); (d) R. G. Pearson, Ed., *Hard and Soft Acids and Bases*, Dowden, Hutchinson, and Ross, Stroudsberg, Pa., 1973.

[112]T-L. Ho, *Chem. Rev.*, **75**, 1 (1975).

Table 3.19 PEARSON'S CLASSIFICATION OF LEWIS ACIDS AND BASES[a]

Hard	Borderline	Soft
Acids		
H^+, Li^+, Na^+, K^+	Fe^{2+}, Co^{2+}, Ni^{2+}	Cu^+, Ag^+, Hg^+
Be^{2+}, Mg^{2+}, Ca^{2+}, Sr^{2+}, Mn^{2+}	Cu^{2+}, Zn^{2+}	Hg^{2+}
Al^{3+}, Cr^{3+}, Co^{3+}, Fe^{3+}	Pb^{2+}, Sn^{2+}	BH_3, RS^+, I^+
BF_3, $B(OR)_3$	$B(CH_3)_3$, SO_2	Br^+, HO^+, RO^+
$Al(CH_3)_3$, $AlCl_3$, AlH_3	NO^+, R_3C^+	I_2, Br_2
RPO_2^+, $ROPO_2^+$	$C_6H_5^+$	Trinitrobenzene, etc.
RSO_2^+, $ROSO_2^+$, SO_3		Chloranil, quinones, etc.
RCO^+, CO_2, NC^+		Tetracyanoethylene, etc.
HX (hydrogen-bonding molecules)		CH_2, carbenes
Bases		
H_2O, OH^-, F^-	$C_6H_5NH_2$, C_5H_5N	R_2S, RSH, RS^-
CH_3COO^-, PO_4^{3-}, SO_4^{2-}	N_3^-, Br^-, NO_2^-	I^-, SCN^-, $S_2O_3^-$
Cl^-, CO_3^{2-}, ClO_4^-, NO_3^-	SO_3^{2-}	R_3P, $(RO)_3P$
ROH, RO^-, R_2O	N_2	CN^-, RNC, CO
NH_3, RNH_2		C_2H_4, C_6H_6
		H^-, R^-

[a] From R. G. Pearson, *Survey of Progress in Chemistry*, **5**, 1 (1969). Reproduced by permission of Academic Press and R. G. Pearson.

acid ROH; the order is reversed in aprotic solvents (DMSO), whose softer acidic center does not solvate F^- well.[113]

Pearson has proposed that the accumulation of several soft or of several hard centers together leads to stabilization; he has called this effect symbiosis. Thus an ambident nucleophile like an enolate ion (**13**), with a hard oxygen center and a

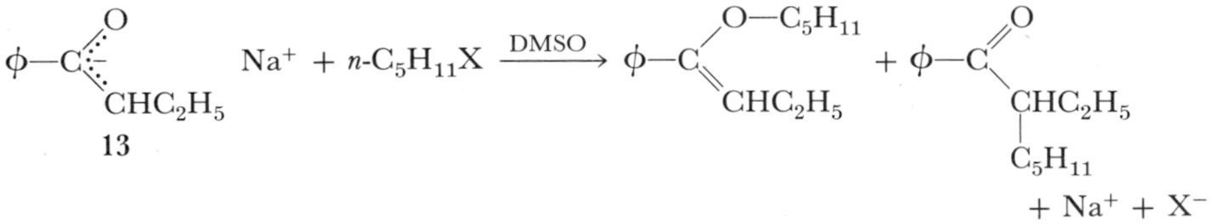

$$X = Cl, \quad 1.2:1$$
$$Br, \quad 0.64:1$$
$$I, \quad 0.23:1$$

softer carbon center, will substitute for a leaving group X in RX preferentially through oxygen if X is hard, but preferentially through carbon if X is soft. Behavior of ambident nucleophiles can be modified by choice of solvent; a hydroxylic solvent coordinates to the oxygen and so favors C alkylation. Small cations have a similar effect.[115]

---

[113] For further discussion see Section 4.4.
[114] H. D. Zook and J. A. Miller, *J. Org. Chem.*, **36**, 1112 (1971).
[115] H. O. House, *Modern Synthetic Reactions*, 2d ed., W. A. Benjamin, Menlo Park, Calif., 1972, p. 521.

Theory of Lewis acid–base interactions Lewis acid–base interactions, and the basis of the hard–soft approach, can be understood in molecular orbital terms through frontier orbital arguments, applied to this problem by Klopman.[116] The diagram in Scheme 5 shows some possible arrangements of frontier orbitals for

SCHEME 5

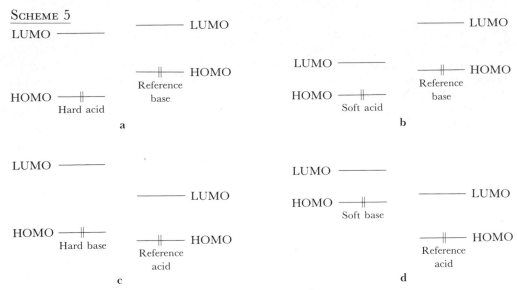

Lewis acids and bases relative to orbitals of a reference compound. A hard acid has a large HOMO–LUMO separation (Scheme 5, **a**) coupled with a high concentration of positive charge at the acidic center. Because of the high-energy LUMO, the reference base cannot transfer much charge from its HOMO to the acid LUMO; there will be relatively little covalent bonding. If the base also has a high charge concentration, there will be an effective electrostatic interaction, and a relatively stable hard–hard complex will result. But if the base is soft, with low charge, there will be little interaction, and the complex will be less stable. A soft acid, on the other hand, has a low-energy LUMO. The energies of acid LUMO and base HOMO will be similar (Scheme 5, **b**). A strong covalent interaction will result, with transfer of electron density from base to acid. High charge densities are not needed, and the soft–soft complex will owe its stability primarily to covalent bonding. Scheme 5 **c** and **d** illustrate the analogous situation for hard and soft bases interacting with a reference acid.

Klopman has derived an expression for the energy of Lewis acid–base interactions by combining an electrostatic term with the expression for orbital interaction energy derived from perturbation theory. The formula is given in Equation

[116](a) G. Klopman, *J. Am. Chem. Soc.,* **90,** 223 (1968); (b) G. Klopman, in *Chemical Reactivity and Reaction Paths,* G. Klopman, Ed., Wiley, New York, 1974, Chap. 4; (c) I. Fleming, *Frontier Orbitals and Organic Chemical Reactions,* Wiley, London, 1976.

3.94, which includes all interactions of occupied orbitals of the donor (base, B) with unoccupied orbitals of the acceptor (acid, A). For the frontier orbital approxima-

$$E = \frac{-q_a q_b}{R_{ab}\epsilon} + 2 \sum_m \sum_n \frac{(c_a{}^m c_b{}^n \beta_{ab})^2}{E_m - E_n} \qquad (3.94)$$

$$\underset{\substack{\text{unocc.} \quad \text{occ.}\\ \text{of A} \quad \text{of B}}}{}$$

tion, the second term would cover only the base HOMO and the acid LUMO. In Equation 3.94, A is the acid, with orbitals m, acceptor center a, and charge q_a; B is the base with orbitals n, basic center b, and charge q_b; the c's are the coefficients of the molecular orbitals at the acidic and basic centers; β_{ab} is the interaction integral; R_{ab} is the separation of the acidic and basic centers; ϵ is the dielectric constant of the solvent. The first term is the electrostatic attraction of the charges of acid and base; it corresponds roughly to the contribution from hard–hard interactions. The second term is the perturbation energy; it represents the covalent (soft–soft) inter-action and depends on the orbital overlap and energy separation.

The perturbation molecular orbital approach is strictly valid only for the initial stages of an interaction, when forces between molecules are still small. Mak-ing inferences from it about overall energy changes in going to a transition state or to a Lewis acid–base complex requires the assumption that the initial bonding tendencies manifest in the first interactions will be maintained throughout the process. This requirement is equivalent to assuming a linear free-energy relation-ship between initial interaction energy and ultimate equilibrium or activation energy change.

PROBLEMS

1. With the aid of Figures 3.3 and 3.6, and of Tables 3.2, 3.7, 3.9, and 3.17, find the conditions required for (a) 40 percent deprotonation of acetone in aqueous DMSO; (b) half deprotonation of cyclopentadiene in aqueous DMSO; (c) 80 percent deprotonation of *p*-nitroaniline in aqueous DMSO. (d) What is the extent of protonation of dimethyl sulfide in 80 percent aqueous H_2SO_4?

2. What would be the pK_a of a base that was 25 percent protonated in HSO_3F containing 10 percent by weight SbF_5?

3. Draw orbital structures for $\phi CH_2{}^-$ and for ϕ^-, and explain on the basis of the structures why benzene would be expected to be a weaker acid than toluene, and why α might be different for the two compounds.

4. Data are given below for rates of hydrolysis of esters **1** and **2** in aqueous perchloric acid. Plot the data according to the Bunnett w method, and propose mechanisms for the reactions.

1   2

$HClO_4$ (mole 1^{-1})	$h_0{}^a$	$k_{obs} \times 10^5$, Compound 1[b]	$k_{obs} \times 10^5$, Compound 2[b]
0.98	1.62	18.7	
1.00	1.66		0.073
1.93	5.62	36.6	0.42
3.82	42.7	74.5	
3.90	46.8		5.6
4.80	135		26

[a] Calculated from data of M. A. Paul and F. A. Long, *Chem. Rev.*, **57**, 1 (1957).
[b] Adapted with permission from C. T. Chmiel and F. A. Long, *J. Am. Chem. Soc.*, **78**, 3326 (1956). Copyright 1956 American Chemical Society.

5. Values for the Bunnett–Olsen parameter ϕ are given below. Recalling that $\log \gamma_{H^+}\gamma_Z/\gamma_{ZH^+} = (1 - \phi) \log \gamma_{H^+}\gamma_B/\gamma_{BH^+}$, where B represents substituted aniline bases used to define the H_0 scale and Z is the base under investigation, explain the trends observed in ϕ.

Compound	ϕ
$C_2H_5OC_2H_5$	0.78[a]
CH_3COCH_3	0.75[a]
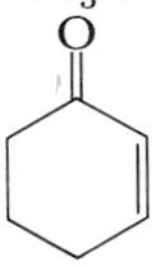	0.52[b]
ϕCHO	0.50[b]
$\phi COCH_3$	0.40[b]
$\phi CO\phi$	0.30[b]
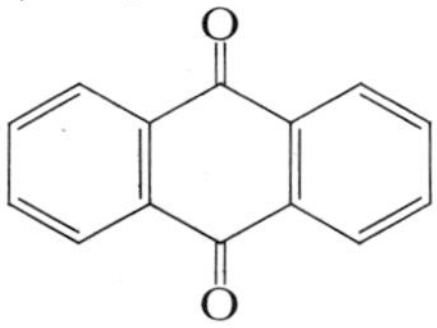	0.06[b]
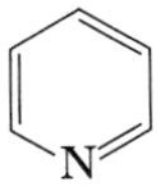	-0.40[a]
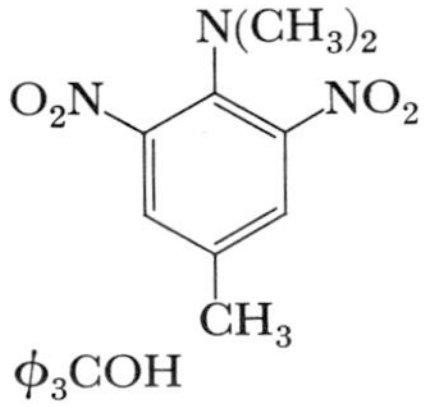	-0.58[c]
$\phi_3 COH$	-1.02[c]

[a] E. M. Arnett and G. Scorrano, *Adv. Phys. Org. Chem.*, **13**, 83 (1976), p. 144.
[b] J. T. Edward and S. C. Wong, *J. Am. Chem. Soc.*, **99**, 4229 (1977).
[c] J. F. Bunnett and F. P. Olsen, *Can. J. Chem.*, **44**, 1899 (1966).

6. In the reaction below, predict which product will predominate when X = F and when X = I.

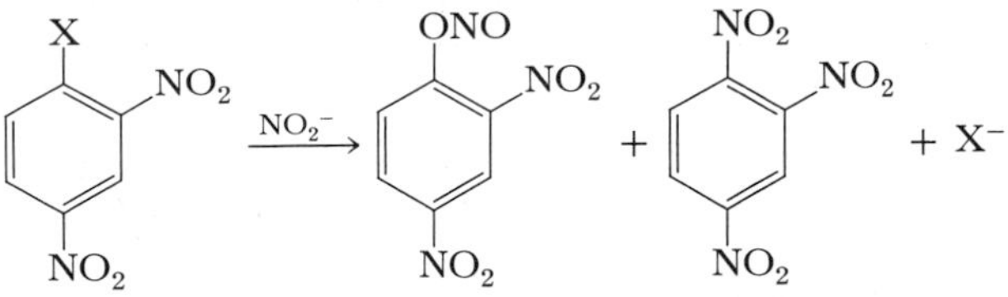

7. Explain the following observations:

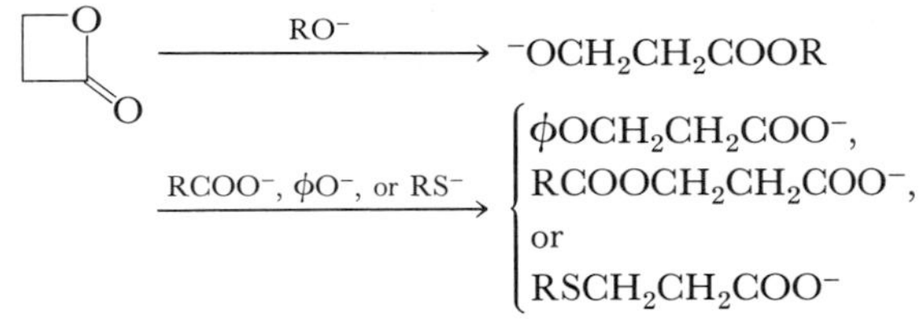

8. Predict the equilibrium site of protonation of the following:

$$H_2N-C=NH$$
$$| $$
$$CH_3$$

9. Explain the following observation:

$$K_T = 1.4 \text{ in } 6.6\% \ H_2SO_4$$
$$K_T = 2.8 \text{ in } 24.2\% \ H_2SO_4$$

10. Construct a substituent constant correlation for the aniline acidity data of Table 3.7.

11. Explain the following differences in pK_a (DMSO):

$$\phi CH_2CN \text{ vs. } \phi_2CHCN, \qquad \Delta pK = 4.7$$
$$\phi CH_2SO_2\phi \text{ vs. } \phi_2CHSO_2\phi, \qquad \Delta pK = 1.4$$

12. For ArCOOH in ethanol, ρ for the acid dissociation is 1.96. Explain.

13. Show why the acidity function $H'_R = H_R - \log a_{H_2O}$ should be appropriate to describe protonation of aryl-substituted alkenes.

14. Show that the expression for H_- (Equation 3.65) is equivalent to that for H_0 (Equation 3.32). Note that a_{H_2O} is not constant and must be included explicitly in the expression for K_w.

15. Explain the trends observed in pK_a for the following compounds:

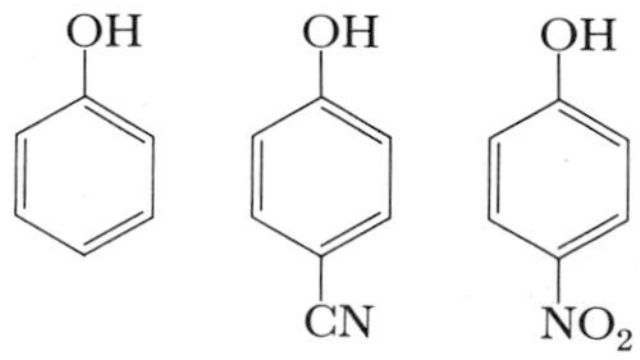

pK_a: 9.99	7.95	7.16

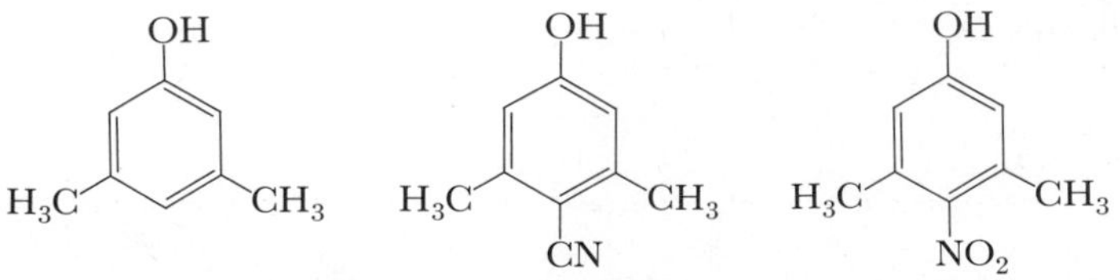

pK_a: 10.15	8.21	8.25

16. What can be concluded from the following pK_a data about the effect on basicity of interaction of a nitrogen unshared pair with an attached aryl group?

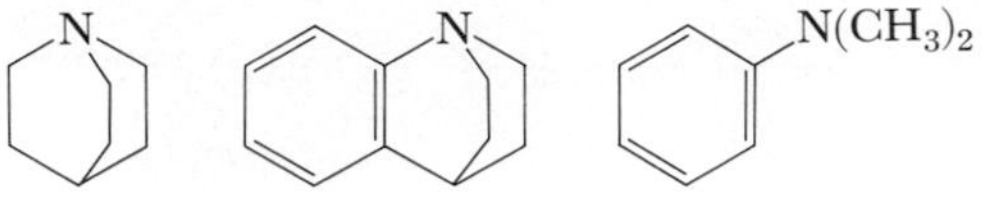

$pK_{a_{BH^+}}$ 10.58   7.79   5.06

REFERENCES FOR PROBLEMS

4. J. F. Bunnett, *J. Am. Chem. Soc.*, **83**, 4968, 4978 (1961).
5. J. T. Edward and S. C. Wong, *J. Am. Chem. Soc.*, **99**, 4229 (1977).
6. D. H. Rosenblatt, W. H. Dennis, Jr., and R. D. Goodin, *J. Am. Chem. Soc.*, **95**, 2133 (1973); T-L. Ho, *Chem. Rev.*, **75**, 1 (1975).
7. T. L. Gresham, J. E. Jansen, F. W. Shaver, and J. T. Gregory, *J. Am. Chem. Soc.*, **70**, 999 (1948), and following papers; T. L. Gresham, J. E. Jansen, F. W. Shaver, R. A. Bankert, W. L. Beears, and M. G. Pendergast, *J. Am. Chem. Soc.*, **71**, 661 (1949); T-L. Ho, *Chem. Rev.*, **75**, 1 (1975).
8. M. Liler, *Adv. Phys. Org. Chem.*, **11**, 267 (1975), p. 301.
9. S-J. Yeh and H. H. Jaffe, *J. Am. Chem. Soc.*, **81**, 3283 (1959); M. Liler, *Adv. Phys. Org. Chem.*, **11**, 267 (1975), p. 310.
11. F. G. Bordwell, J. E. Bares, J. E. Bartmess, G. J. McCollum, M. Van Der Puy, N. R. Vanier, and W. S. Matthews, *J. Org. Chem.*, **42**, 321 (1977).
12. H. H. Jaffe, *Chem. Rev.*, **53**, 191 (1953); J. Hine, *Structural Effects on Equilibria in Organic Chemistry*, Wiley, New York, 1975, p. 161.
13. C. H. Rochester, *Acidity Functions*, Academic Press, London, 1970, pp. 73, 87.
15. J. Hine, *Structural Effects on Equilibria in Organic Chemistry*, Wiley, New York, 1975, p. 170.
16. B. M. Wepster, *Rec. Trav. Chim. Pays-Bas*, **71**, 1159, 1171 (1952); M. Liler, *Adv. Phys. Org. Chem.*, **11**, 267 (1975), p. 280.

Chapter 4
Aliphatic
Nucleophilic
Substitution

A large number of organic reactions, differing often in mechanism and in the nature of the attacking reagent, are overall substitution reactions on carbon in which Y replaces X. Equation 4.1, which, in order to be as general as possible, ignores charges and substituent groups, is a schematized representation of these displacements.

$$Y + C—X \longrightarrow C—Y + X \tag{4.1}$$

Nucleophilic aliphatic substitution is the displacement from saturated carbon of a group with its bonding electrons by a group with an extra pair of electrons (Equation 4.2).

$$Y: + R—\overset{\displaystyle R}{\underset{\displaystyle R}{C}}—X \longrightarrow Y—\overset{\displaystyle R}{\underset{\displaystyle R}{C}}—R + X: \tag{4.2}$$

$$\quad B_1 \qquad AB_2 \qquad\qquad AB_1 \qquad B_2$$

Since both the entering group (or *nucleophile*) and the leaving group are Lewis bases, Equation 4.2 is an example of a Lewis acid–base reaction in which one base replaces another in the Lewis acid–base adduct.

Reactions corresponding to Equation 4.2 fall mainly into four charge types, illustrated in Table 4.1.[1] In these processes the Lewis bases are either uncharged or carry a single negative charge.[2] The nucleophile is frequently also the solvent; in

---

[1] E. D. Hughes and C. K. Ingold, *J. Chem. Soc.*, 244 (1935).
[2] A few nucleophiles such as $S_2O_3{}^{2-}$ carry a double negative charge.

Table 4.1 Charge Types for Nucleophilic Substitution

Type	Y	X	Example
1	Negative	Negative	$I^- + RCl \longrightarrow R{-}I + Cl^-$
2	Neutral	Negative	$H_2O + R{-}I \longrightarrow R{-}\overset{+}{O}H_2 + I^-$
3	Negative	Neutral	$Cl^- + R\overset{+}{N}H_3 \longrightarrow R{-}Cl + NH_3$
4	Neutral	Neutral	$NH_3 + R{-}\overset{+}{S}R_2 \longrightarrow R\overset{+}{N}H_3 + SR_2$

that case the reaction is called a *solvolysis*. The illustration for charge type 2 in Table 4.1 is a typical solvolysis.

4.1 GENERAL FEATURES OF ALIPHATIC NUCLEOPHILIC SUBSTITUTION[3]

In 1933 two mechanisms for nucleophilic substitution reactions were proposed by Hughes, Ingold, and Patel.[4] They found that decomposition of quartenary ammonium salts, $R_4N^+Y^-$, to give R_3N and RY exhibited two different kinds of kinetic behavior depending on the ammonium salt used. For example, when methyl alcohol was formed from trimethyl-*n*-decylammonium hydroxide (Equation 4.3), the rate of formation of methyl alcohol was found to be second-order, first-order each in trimethyl-*n*-decylammonium cation and in hydroxide ion, as in Equation 4.4. On

$$\text{HO}^{-}\text{ CH}_3{-}\underset{\underset{CH_3}{|}}{\overset{\overset{CH_3}{|}}{\overset{+}{N}}}{-}\text{CH}_2(\text{CH}_2)_8\text{CH}_3 \longrightarrow \text{HOCH}_3 + \ :\underset{\underset{CH_3}{|}}{\overset{\overset{CH_3}{|}}{N}}{-}\text{CH}_2(\text{CH}_2)_8\text{CH}_3 \quad (4.3)$$

$$\text{rate} = k[(\text{CH}_3)_3\text{NC}_{10}\text{H}_{21}][\text{OH}^-] \tag{4.4}$$

the other hand, the rate of formation of diphenylmethanol (benzhydrol) from (diphenylmethyl)trimethylammonium hydroxide was found to be overall first-order—dependent only on the ammonium ion concentration, as in Equation 4.6.

$$ \quad (4.5)$$

$$\text{rate} = k_1[\phi_2\text{CH}\overset{+}{\text{N}}(\text{CH}_3)_3] \tag{4.6}$$

[3] Reviews: (a) C. A. Bunton, *Nucleophilic Substitution at a Saturated Carbon Atom,* Elsevier, Amsterdam, 1963; (b) A. Streitwieser, Jr., *Solvolytic Displacement Reactions,* McGraw-Hill, New York, 1962; (c) E. R. Thornton, *Solvolysis Mechanisms,* Ronald Press, New York, 1964; (d) C. K. Ingold, *Structure and Mechanism in Organic Chemistry,* 2d ed., Cornell University Press, Ithaca, N.Y., 1969; (e) D. J. Raber, J. M. Harris, and P. v. R. Schleyer, in *Ions and Ion Pairs in Organic Reactions.*, Vol. 2, M. Szwarc, Ed., Wiley, New York, 1974; (f) J. M. Harris, *Prog. Phys. Org. Chem.,* **11,** 89 (1974); (g) T. W. Bentley and P. v. R. Schleyer, *Adv. Phys. Org. Chem.,* **14,** 1 (1977).
[4] E. D. Hughes, C. K. Ingold, and C. S. Patel, *J. Chem. Soc.,* 526 (1933).

Added hydroxide ion did not change the rate. This and related evidence led the authors to postulate that these reactions, so closely related in starting materials and products, nevertheless proceed by two different mechanisms.

In the decomposition of trimethyl-*n*-decylammonium hydroxide (Equation 4.3), they suggested, OH⁻ attacks one of the methyl groups, forcing the amine to depart. Both ammonium and hydroxide ions are part of the transition state of this single-step reaction, and therefore both enter into the rate equation. This mechanism is known as S_N2 (substitution nucleophilic bimolecular).

The decomposition of (diphenylmethyl)trimethylammonium hydroxide (Equation 4.5), on the other hand, according to Ingold, proceeds by initial rate-determining formation of the relatively stable diphenylmethyl cation and subsequent fast attack on the cation by hydroxide. Because hydroxide is not involved until after the rate-determining step of this reaction, it does not enter into the rate equation. This mechanism is known as S_N1 (substitution nucleophilic unimolecular).

Hughes and Ingold, in 1935, postulated that these mechanisms, or a combination of them in which the nucleophile plays an intermediate role in the departure of the leaving group, are general for all aliphatic nucleophilic substitutions.[5]

According to the Hughes–Ingold picture, the S_N2 mechanism would be expected to operate if the substitution site is sterically unhindered (for example, a methyl or primary carbon). The nucleophile approaches and by donation of its electron pair forms a partial bond to carbon while the leaving-group–carbon bond begins to break, as illustrated for charge type 1 in Equation 4.7. At the transition

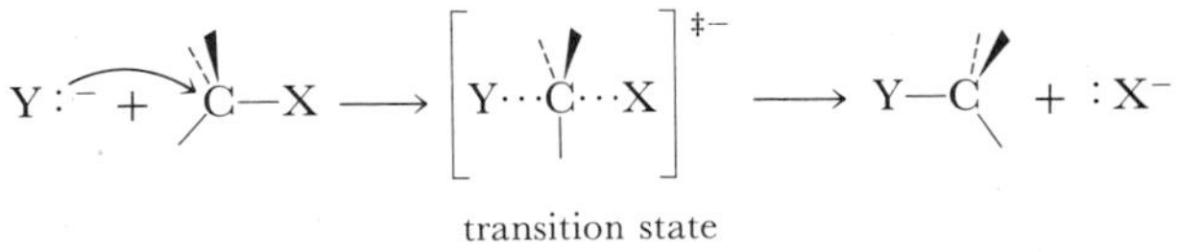

$$ Y:^- + \overset{}{\underset{}{C}}{-}X \longrightarrow \left[Y{\cdots}\overset{}{\underset{}{C}}{\cdots}X \right]^{\ddagger-} \longrightarrow Y{-}\overset{}{\underset{}{C}} + :X^- \qquad (4.7) $$

transition state

state there are partial bonds to both entering and leaving groups, although bond making and bond breaking need not have occurred to the same extent.[6] The reaction coordinate diagram for this proposed mechanism is shown in Figure 4.1.

---

[5]J. L. Gleave, E. D. Hughes, and C. K. Ingold, *J. Chem. Soc.*, 236 (1935).

[6]A bistrifluoromethanesulfonate salt with the proposed structure shown below has been observed in solution by nmr spectroscopy. The carbon bonded to the two sulfur atoms has the geometry of the S_N2 transition state. If the structure is confirmed, this compound will be the first example of a stable molecule containing a pentacoordinate carbon. T. R. Forbus, Jr., and J. C. Martin, *J. Am. Chem. Soc.*, **101**, 5057 (1979).

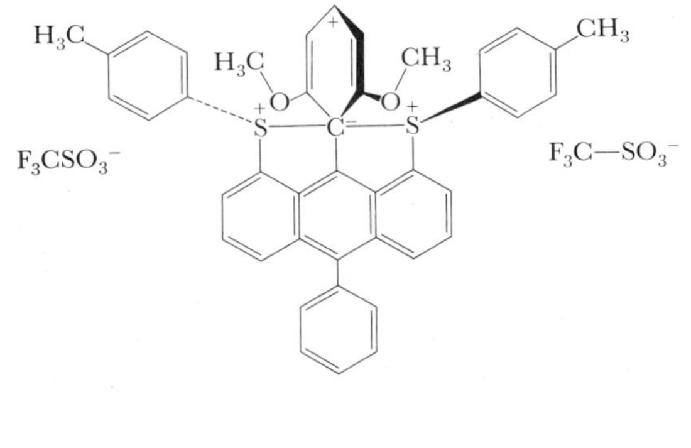

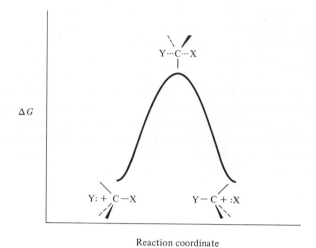

Reaction coordinate

Figure 4.1 Reaction coordinate diagram for the S_N2 reaction.

The S_N1 mechanism would be expected to operate if the substitution site bears electron-donating substituents, for example, three alkyl groups or two or three aryl groups, that will provide strong stabilization of positive charge. In this case the bond to the leaving group cleaves heterolytically and a carbocation is formed. Then in a second step the nucleophile attacks this highly reactive intermediate, as shown in Equation 4.5. Figure 4.2 shows the reaction coordinate diagram for the S_N1 mechanism. The carbocation is an intermediate and thus lies at an energy minimum; energy maxima occur both when the C—X bond is stretched and when the C—Y bond is formed.

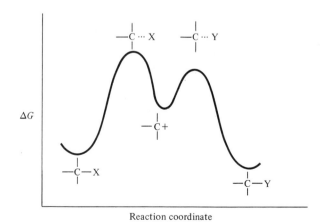

Reaction coordinate

Figure 4.2 Reaction coordinate diagram for the S_N1 reaction.

A number of the experimentally observed phenomena associated with aliphatic nucleophilic substitutions are readily understood in terms of the Hughes–Ingold S_N1–S_N2 classifications, but others are more difficult to rationalize.

Kinetics

The S_N2 mechanism, consisting of a single elementary reaction step in which the nucleophile and substrate come together to form the final product, predicts reaction kinetics first-order both in substrate and in added nucleophile. This behavior is often observed and is good evidence for the operation of the bimolecular mechanism. (As we shall see later, however, there is another possible explanation for second-order kinetics.) Failure to observe overall second-order kinetics, on the other hand, is not necessarily proof that the bimolecular mechanism is not operating because in solvolysis, a common category of nucleophilic substitution, the nucleophile is the solvent. Because solvent is always present in large excess, its concentration will not change during the reaction and kinetic order with respect to it cannot be determined. In a solvolysis, the kinetics alone, therefore, does not distinguish the S_N2 type of mechanism from the S_N1 type.

Methyl iodide in methanol solvent is a system that would show characteristics of a typical S_N2 reaction. Various nucelophiles that might be added (for example, CH_3O^-, Br^-, RS^-, etc.) would each react at a characteristic rate, generally with a rate constant several powers of ten larger than the rate constant for reaction with the solvent.

The Hughes–Ingold S_N1 mechanism postulates a rate-determining dissociation to a carbocation followed by rapid capture of the ion by a nucleophile.[7] Equations 4.8 and 4.9 delineate the S_N1 route for the case of ion capture by solvent.

$$RCl \underset{k_{-1}}{\overset{k_1}{\rightleftharpoons}} R^+ + Cl^- \tag{4.8}$$

$$R^+ + SOH \xrightarrow{k_2} ROS + H^+ \tag{4.9}$$

The reverse of the second step, although it should be included to be rigorously correct, is frequently omitted because the final product in many cases is sufficiently unreactive that no experimentally significant amount will return to carbocation during the time the reaction is under observation.

This simple two-step mechanism, when combined with the stationary-state assumption for the presumably highly reactive positive ion (see Section 2.5) leads to the prediction given in Equation 4.10 for the rate of product formation. (See Problem 4.) The term in parentheses in Equation 4.10 will reduce to unity whenever $k_2[SOH] \gg k_{-1}[Cl^-]$; in that case a simple first-order kinetic behavior is predicted. First-order kinetics is indeed the result usually observed for reactions of this type. A

[7]The mechanism was proposed by S. C. J. Olivier and G. Berger, *Recl. Trav. Chim.*, **45**, 712 (1926); A. M. Ward, *J. Chem. Soc.*, 2285 (1927); and C. K. Ingold, *Annu. Rep. Chem. Soc.*, **24**, 156 (1927). It was set out in detail by E. D. Hughes, C. K. Ingold, and C. S. Patel, *J. Chem. Soc.*, 526 (1933); the S_N1 terminology was introduced by J. L. Gleave, E. D. Hughes, and C. K. Ingold, *J. Chem. Soc.*, 236 (1935).

sufficiently large added concentration of the leaving group (in this case Cl⁻) might lead to a rate depression, called a *common ion effect*.

$$\frac{d[\text{ROS}]}{dt} = k_1[\text{RCl}]\left(\frac{k_2[\text{SOH}]}{k_{-1}[\text{Cl}^-] + k_2[\text{SOH}]}\right) \tag{4.10}$$

Diphenylchloromethane (benzhydryl chloride, **1**) in solvents such as aqueous ethanol, aqueous acetone, acetic acid, or formic acid, behaves as anticipated for an S_N1 process. The observed kinetics of disappearance of substrate is first-order in

$$\phi-\underset{\underset{\text{H}}{|}}{\overset{\overset{\phi}{|}}{\text{C}}}-\text{Cl} + \text{SOH} \longrightarrow \phi-\underset{\underset{\text{H}}{|}}{\overset{\overset{\phi}{|}}{\text{C}}}-\text{OS} + \text{HCl} \tag{4.11}$$

1

benzhydryl chloride. This fact alone tells little about the mechanism because the nucleophile, being the solvent, is always present in large excess. However, added nucleophiles have relatively small effects on the rate, and rates of reaction with various nucleophilic salts differ only slightly from each other. Most added salts accelerate the process, but common ion salts (chlorides in the case of benzhydryl chloride) make it slower. Addition of sodium azide causes formation of some benzhydryl azide but affects the rate in the same way as other noncommon ion salts.[8] The small rate accelerations are caused by salt effects (changes of ionizing power of the medium on adding electrolytes). Added nucleophiles, like azide ion, that can form stable products compete with solvent for the R⁺ ion, but not being involved in the rate-determining formation of R⁺, do not affect the overall rate aside from the salt effect.

Stereochemistry

The Ingold S_N2 reaction, with a transition state formulated as in Equation 4.7, predicts inversion of configuration at the substitution center. A number of experiments have demonstrated that inversion of configuration occurs in certain nucleophilic substitution reactions.

In the 1890s, many years before the mechanism of direct substitution was proposed by Hughes and Ingold, Walden had observed that some reactions of optically active compounds give products of opposite absolute configuration from the starting materials.[9] Walden, however, was not able to discover what conditions brought about this inversion of configuration. His task was complicated by the fact that two compounds of the same absolute configuration may nevertheless have opposite signs of optical rotation. In the following 40 years a great deal of work and

[8](a) L. C. Bateman, E. D. Hughes, and C. K. Ingold, *J. Chem. Soc.*, 974 (1940); (b) L. C. Bateman, M. G. Church, E. D. Hughes, C. K. Ingold, and N. A. Taher, *J. Chem. Soc.*, 979 (1940), and references cited therein; (c) C. G. Swain, C. B. Scott, and K. H. Lohmann, *J. Am. Chem. Soc.*, **75**, 136 (1953); (d) D. Kovačević, Z. Majerski, S. Borčić, and D. E. Sunko, *Tetrahedron*, **28**, 2469 (1972).
[9]P. Walden, *Chem. Ber.*, **26**, 210 (1893); **29**, 133 (1896); **32**, 1855 (1899).

thought was given to the problem of the relation of Walden inversion to mechanism.[10] Then, in 1935, Hughes and co-workers in ingenious experiments clearly showed that Walden inversion occurs in nucleophilic substitution.[11]

These workers studied the exchange reaction of optically active 2-iodooctane with radioactive iodide ion in acetone (Equation 4.12) and found that (1) the kinetics is second-order, first-order each in 2-iodooctane and in iodide ion, and thus

$$*I^- + CH_3(CH_2)_5\underset{\underset{I}{|}}{C}HCH_3 \longrightarrow I^- + CH_3(CH_2)_5\underset{\underset{*I}{|}}{C}HCH_3 \qquad (4.12)$$

the mechanism is bimolecular; and (2) the rate of racemization is twice the rate of incorporation of labeled iodide ion into the organic molecule. The rate of racemization must be twice the rate of inversion. (If an optically active compound begins to racemize, each molecule that undergoes inversion is one of a racemic pair of molecules; for example, pure levorotatory starting material is 100 percent racemized when only 50 percent of it has been converted to the dextrorotatory isomer.) So if the rate of racemization is twice the rate of incorporation of radioactive iodide, then *each attacking iodide ion inverts the molecule it enters.*

This one-to-one correlation of inversion with displacement must mean that the incoming iodide enters the molecule from the side of the substitution site opposite to the departing iodide every single time. It initially attacks the back lobe of the sp^3 orbital used for bonding with the iodide. In the transition state proposed by Hughes and co-workers (**2**), carbon has rehybridized and is using three sp^2

$$
\begin{array}{cc}
\underset{2}{\overset{\displaystyle Y\cdots\overset{\overset{\textstyle R\;\;R}{\diagdown\,\diagup}}{\underset{|}{C}}\cdots X}{\underset{R}{}}} &
\underset{3}{\overset{\displaystyle \overset{\textstyle R\,R}{R\diagdown\!\diagup}C\diagdown\cdots Y}{\;\;\;R\;\;\;X}}
\end{array}
$$

orbitals for bonding with the nonreacting ligands and the remaining p orbital for forming partial bonds with X and Y. The geometry is that of a trigonal bipyramid with the entering and leaving groups in the apical positions.

One might inquire at this point why the entering nucleophile chooses a backside displacement, rather than coming from the front as indicated in **3**. Inversion of configuration in the displacement of iodide by radioactive iodide and in all reactions of charge type 1 might be explained on the grounds that the bipyramidal transition state shown in **2** allows the entering and leaving groups, both of which carry a partial negative charge, to be as far away from each other as possible, thus minimizing electrostatic repulsion.[12]

[10] For a comprehensive summary of this work see C. K. Ingold, *Structure and Mechanism in Organic Chemistry*, 2d ed., Cornell University Press, Ithaca, N.Y., 1969, pp. 509*ff.*
[11] (a) E. D. Hughes, F. Juliusburger, S. Masterman, B. Topley, and J. Weiss, *J. Chem. Soc.*, 1525 (1935); (b) E. D. Hughes, F. Juliusburger, A. D. Scott, B. Topley, and J. Weiss, *J. Chem. Soc.*, 1173 (1936); (c) W. A. Cowdrey, E. D. Hughes, T. P. Nevell, and C. L. Wilson, *J. Chem. Soc.*, 209 (1938).
[12] N. Meer and M. Polanyi, *Z. Phys. Chem.*, **B19**, 164 (1932).

SCHEME 1

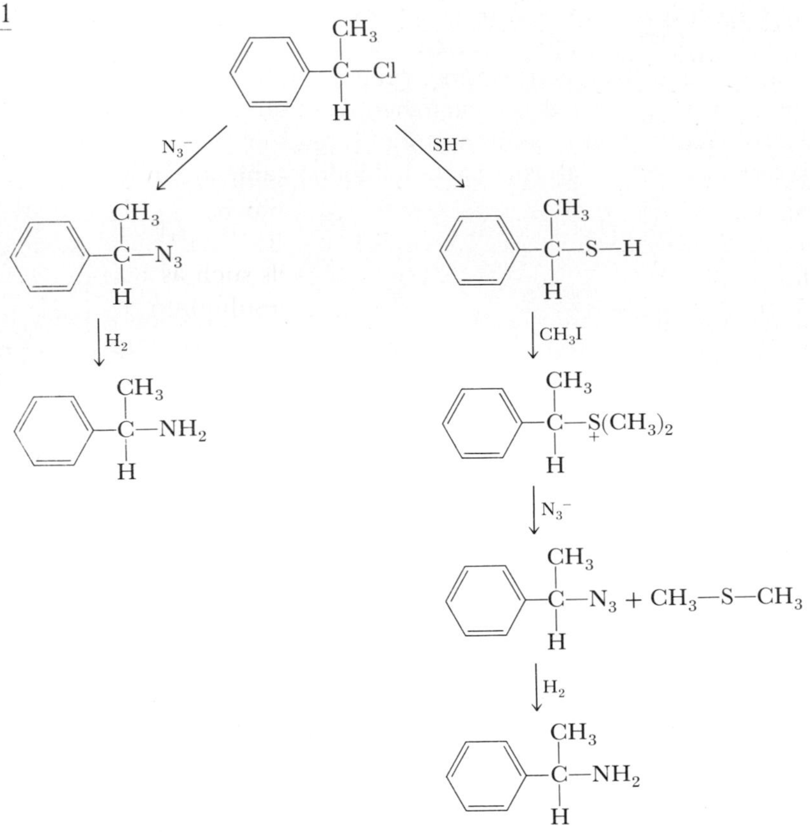

In order to demonstrate that this explanation is not correct, Hughes, Ingold, and co-workers carried out the reaction sequences shown in Scheme 1, which demonstrate that even reactions of charge type 3, with entering and leaving groups bearing opposite charges, proceed with inversion.[13] A sample of optically active 1-chloro-1-phenylethane was converted to the corresponding azide with sodium azide while another was converted to the thiol with sodium hydrogen sulfide. Both of these second-order reactions are of charge type 1, processes already shown to proceed with inversion. Thus both the thiol and the azide have the configuration opposite to that of the starting chloride. The azide was then reduced with hydrogen over platinum to the corresponding amine, and the thiol was converted to the dimethylsulfonium salt. Neither of these processes disturbs the chiral center, and therefore both of these compounds have the opposite configuration to that of the starting material. Then in another second-order substitution reaction the sulfonium salt was converted to the azide and the azide reduced to the amine. This

[13] S. H. Harvey, P. A. T. Hoye, E. D. Hughes, and C. K. Ingold, *J. Chem. Soc.*, 800 (1960).

amine had the *opposite* configuration of the amine produced by the first route, and therefore the substitution (of charge type 3) of azide ion on the sulfonium salt must occur with inversion of configuration.

There is now a great deal of evidence that all S_N2 reactions of all charge types proceed with inversion.[14]

Proof that a site incapable of undergoing inversion is also incapable of undergoing a second-order substitution reaction has been obtained from bicyclic compounds. The bridgehead carbon of rigid bicyclic systems cannot invert without fragmenting the molecule, and indeed, compounds such as 1-bromotriptycene (**4**) and 7,7-dimethyl-[2.2.1]-bicycloheptyl-1-*p*-toluenesulfonate (**5**) are completely inert when treated with a nucleophile under S_N2 conditions.[15]

Some insight concerning the preference for displacement from the back can be gained from frontier orbital theory. Recall from the examples discussed in Chapter 1 (p. 67) that nucleophilic displacement must consist of transfer of electron density from the nucleophile HOMO (generally an unshared pair in an atomic orbital on the nucleophilic center) into the substrate LUMO, which will be an antibonding σ orbital between the substitution center and the leaving group. Attack from the front would bring the nucleophile HOMO onto the node of the antibonding LUMO and the interaction would be minimal; attack from the back, however, presents no such problems and so is the favored route.

The stereochemistry of the S_N1 process is more complex. Carbocations prefer a geometry in which the cationic carbon and the three atoms attached to it are coplanar. The intermediate R^+ of Equation 4.8 should take up this structure. If an optically active substrate that has a chiral center at the site of substitution follows the mechanism of Equations 4.8 and 4.9, the cation R^+, when it takes this preferred structure, has a plane of symmetry and so cannot be chiral. Attack on the ion by a nucleophile from the two sides of the plane, yielding the two enantiomers of the product (Equation 4.13), must occur at equal rates. (In Equation 4.13 and in subsequent examples we use *Nuc* to symbolize a generalized nucleophile.) The benz-

---

[14] See (a) note 11; (b) note 13; (c) H. M. R. Hoffmann and E. D. Hughes, *J. Chem. Soc.*, 1252, 1259 (1964).
[15] (a) P. D. Bartlett and L. H. Knox, *J. Am. Chem. Soc.*, **61**, 3184 (1939); (b) P. D. Bartlett and E. S. Lewis, *J. Am. Chem. Soc.*, **72**, 1005 (1950).

$$
\text{Nuc}:^- \ \cdots\cdots\rightarrow \quad
\underset{R_1 \ R_2}{\overset{R_3}{C^+}} \quad
\leftarrow\cdots\cdots \ :\text{Nuc}^-
\quad
\begin{array}{c}
\nearrow \quad \text{Nuc}-\underset{R_2}{\overset{R_3}{C}}{\diagdown}^{R_1} \\
\\
\searrow \quad \underset{R_1 \ R_2}{\overset{R_3}{C}}-\text{Nuc}
\end{array}
\tag{4.13}
$$

hydryl system apparently fits this prediction: the solvolysis products of optically active phenyl-*p*-chlorophenylchloromethane (*p*-chlorobenzhydryl chloride) are almost completely racemic.[16]

Although totally racemic products from optically active starting materials are sometimes found, it is much more common to observe partial racemization. Ingold and co-workers recognized that many of these reactions could not be clearly categorized into S_N1 or S_N2.[17]

Another prediction of the simple S_N1 mechanism is that in cases where a common ion effect is observed in reaction of a chiral starting material, the unreacted substrate should racemize during the course of the reaction. In such instances the ion forms (Equation 4.8) but returns to starting material at a rate comparable to the rate of product formation. Because R^+ is achiral, the starting material regenerated in this process should be racemic. Starting material racemization is observed, but it is also observed in cases where there is no common ion effect. We have here another difficulty with the simple Hughes–Ingold scheme.

Rearrangements and Internal Nucleophiles

Frequently observed phenomena consistent with the Hughes–Ingold scheme are rearrangement and attack by internal nucleophiles. After an ion forms, an alkyl group or a hydrogen may migrate from an adjacent center, as illustrated in Equation 4.14. This process occurs most commonly when the ion formed is more stable than the original one, although rearrangement pathways proceeding through less stable ions are also known.

$$
\underset{H_3C}{\overset{R}{\diagdown}}\underset{CH_3}{\overset{}{C}}-\underset{X}{\overset{H}{C}}{\diagup}^{H}
\ \longrightarrow \ X^- + \
\underset{H_3C}{\overset{R}{\diagdown}}\underset{CH_3}{\overset{}{C}}-\overset{+}{C}\underset{H}{\overset{H}{\diagdown}}
\ \longrightarrow \
\underset{H_3C}{\overset{H_3C}{\diagdown}}\overset{+}{C}-\underset{H}{\overset{R}{C}}{\diagup}^{}
\ \overset{Nuc}{\longrightarrow} \ \text{products}
\tag{4.14}
$$

The migration may occur, as indicated in Equation 4.14, after the ion has formed, or it may take place simultaneously with departure of the leaving group (Equation 4.15), in which case the rearrangement assists the bond breaking by providing backside nucleophilic push. Such a reaction is said to be subject to

---

[16]This system will be discussed further in Section 4.3.
[17]C. K. Ingold, *Structure and Mechanism in Organic Chemistry,* 2d ed., Cornell University Press, Ithaca, N.Y., 1969, pp. 427–457.

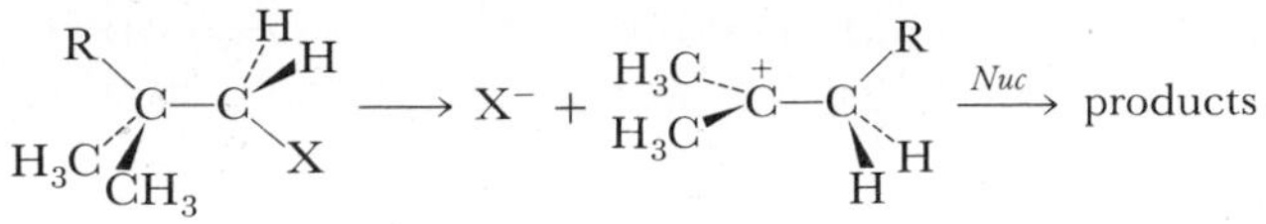

$$\underset{\substack{H_3C \\ \quad CH_3}}{\overset{R}{\underset{}{}}}C-C\underset{X}{\overset{H\;H}{}} \longrightarrow X^- + \underset{H_3C}{\overset{H_3C}{}}\overset{+}{C}-C\underset{H}{\overset{R}{}} \xrightarrow{Nuc} \text{products} \qquad (4.15)$$

anchimeric assistance. A variant of this process occurs when the half-migrated structure is an intermediate; the bridged ion (**6**) has positive charge distributed over more than one center, each of which has coordination number greater than

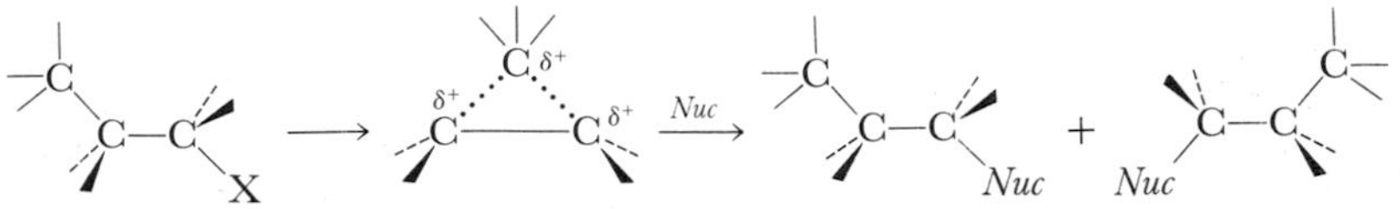

$$\qquad (4.16)$$

6

three. Such species, sometimes referred to as *nonclassical ions,* are more properly called *carbonium ions,* while the ordinary three-coordinate carbocations are called *carbenium ions.* Structures **7** and **8** are examples of ions thought to have the carbonium structure and to be formed with anchimeric assistance. It is, of course, also

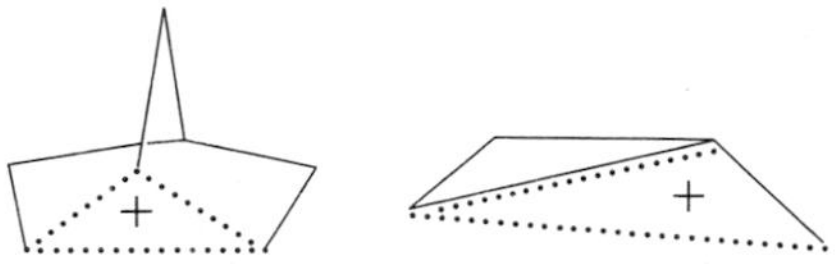

7   **8**

possible for a more commonplace nucleophile to act intramolecularly, as illustrated in Equation 4.17. The rates of such reactions are significantly enhanced over those of model systems with similar nucleophiles not part of the same molecule. Car-

$$\qquad (4.17)$$

bon–carbon double bonds can also serve as intramolecular nucleophiles. Note that substitutions involving internal nucleophiles are unimolecular and will follow first-order kinetics, although in terms of the bonding changes occurring they are more closely related to S_N2 than to S_N1 reactions. Rearrangements and carbonium ions are considered in more detail in Chapter 5.

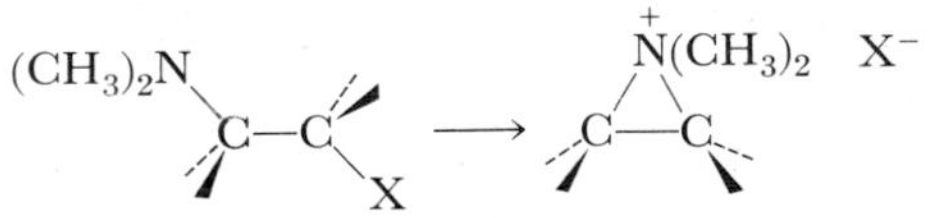

Substitution and Elimination

Nucleophilic substitution and β-elimination are closely related. If the substrate has one or more hydrogens bonded to a carbon (the β carbon) attached to the carbon bearing the leaving group, that hydrogen can be lost as a proton, generating a multiple bond. In close analogy to substitution, there are two general types of mechanism. In a bimolecular elimination, E2 (Equation 4.18), a base removes the proton simultaneously with bond breaking to the leaving group. The unimolecular elimination, E1 (Equation 4.19), has the same first step as the S_N1; instead of

$$\text{B}:^- + \quad \overset{H}{\underset{X}{\text{C}-\text{C}}} \quad \longrightarrow \text{BH} + \text{C}=\text{C} + \text{X}^- \tag{4.18}$$

$$\overset{H}{\underset{X}{\text{C}-\text{C}}} \quad \longrightarrow \text{X}^- + \quad \overset{H}{\text{C}-\overset{+}{\text{C}}} \quad \longrightarrow \text{C}=\text{C} + \text{H}^+ \tag{4.19}$$

combining with nucleophile, the ion loses a proton to generate the alkene. (In a third type of β elimination, not so closely related to nucleophilic substitution as these two, the proton is removed first.) Eliminations are covered in more detail in Chapter 7.

Generally speaking, the conditions that favor S_N2 types of substitutions (strong nucleophiles, solvents of low ionizing power, primary carbon at the substitution center) tend to favor E2 eliminations, and conditions that favor S_N1 type of substitutions (weak nucleophiles, solvents of high ionizing power, secondary or tertiary carbon at the substitution center) tend to favor E1 eliminations. Hence it is common to find competing S_N2- and E2-type reactions, or competing S_N1- and E1-type reactions. In fact, in studying the unimolecular reactions, where the point of interest is frequently the initial ionization step, investigators often look at the composite of substitution and elimination, with little regard to the partitioning of the ion between the pathways open to it.

A Unified Theory of Aliphatic Nucleophilic Substitution

The Hughes–Ingold S_N1–S_N2 scheme provides an appropriate overall framework for classification of aliphatic nucleophilic substitution reaction mechanisms. It focuses attention on what is perhaps the most fundamental aspect of the process: the presence or absence of bonding between a nucleophile and the substitution center during breaking of the bond between substitution center and leaving group. However, some confusion can arise because a substitution by internal nucleophile is unimolecular, therefore S_N1, even though it may be closely related in terms of bonding changes to an S_N2 reaction. We will use the term *nucleophile-assisted* (or simply *assisted*) to refer to any process in which specific covalent bonding to any nucleophilic group, internal or external, takes place during departure of the leaving group. We will also continue to use the term S_N2 in its traditional sense of a nucleophile-assisted reaction in which the nucleophile is external. Substitutions are *unassisted* only when the leaving group departs without any nucleophilic assistance. The term S_N1 (in the sense we use here, sometimes called *limiting* S_N1) can refer either to an unassisted reaction or to an internally assisted reaction. Commonly used notation for rate constants of these various processes is as follows:

$$k_c = \text{unassisted ionization}$$

k_s = assisted by solvent as nucleophile

k_Δ = assisted by an internal nucleophile

k_N = assisted by an external (nonsolvent) nucleophile

The elaboration of the S_N1–S_N2 framework to explain many phenomena uncovered since the original Hughes–Ingold hypothesis relies primarily on ideas about ion pairs, which we shall summarize here and consider in more detail in Section 4.3. The theory currently in use was introduced by Winstein, who also provided much of the experimental evidence on which it is based.[18] Winstein identified two types of ion pairs important in nucleophilic substitutions. The contact ion pair (or intimate ion pair), R^+X^-, has the positive and negative ions close together surrounded by a single solvation shell. In the solvent-separated ion pair, $R^+\|X^-$, the positive and negative ions are still specifically associated but are separated by a solvent molecule. The final stage of dissociation leads to the independent free ions, $R^+ + X^-$, each with its own solvation shell. Each stage may return to the previous stage, may advance to the next stage, or may react with solvent or with some other nucleophile to yield products. It is also possible for a nucleophile to be involved in the initial ionization either to give direct substitution on the substrate (Hughes–Ingold S_N2) or to assist departure of the leaving group by providing nucleophilic solvation of the incipient ion. In this latter situation a contact ion pair with the nucleophile as part of its solvation shell would be formed; this ion pair could (1) react with the nucleophile to form product; (2) return to starting material; (3) proceed on to a nucleophilically solvated, solvent-separated ion pair and perhaps eventually to free solvated ions, or (4) cross over to a nonnucleophilically solvated ion pair. Scheme 2 is an outline of these possibilities, showing some of the

SCHEME 2

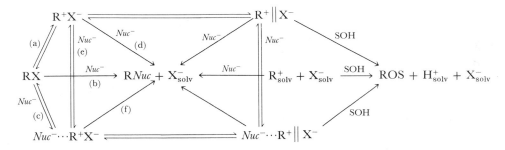

pathways a nucleophilic substitution might be expected to follow.[19] (Scheme 2 assumes a negatively charged nucleophile Nuc^-. The nucleophile could, of course,

[18] For a summary and complete references, see P. D. Bartlett, "The Scientific Work of Saul Winstein," *J. Am. Chem. Soc.*, **94**, 2161 (1972).

[19] (a) S. Winstein and G. C. Robinson, *J. Am. Chem. Soc.*, **80**, 169 (1958); (b) J. M. Harris, *Prog. Phys. Org. Chem.*, **11**, 89 (1974).

be uncharged.) The scheme does not explicitly include variations that may occur in systems that rearrange or in which ionization is assisted by an internal nucleophile; in these instances there may be other intermediate ions, each of which may have associated with it the various stages of ion pairing and the various reaction pathways outlined in Scheme 2.

With so many possibilities open, it is a challenging task to sort out which pathways are followed in a particular instance. There are few if any systems known in which more than a few of the possibilities have been demonstrated,[20] but the entire scheme does provide a reasonable outline for viewing the range of behavior observed with a variety of compounds and conditions.

The following sections will consider methods for discovering the degree of involvement of an external nucleophile in the initial stage of the substitution, methods of obtaining information about involvement of ion pairs, and the effect of solvent, nucleophile, leaving group, and substrate structure on the rate of reaction and on the pathway followed. Consideration of internal nucleophiles and rearrangements will be taken up in Chapter 5.

4.2 DETERMINING THE ROLE OF EXTERNAL NUCLEOPHILES

The classical S_N2 process provides the simplest, but not the only possible, explanation for substitutions by relatively strong nucleophiles on methyl and primary substrates and in many cases on secondary substrates also.

The S_N2 mechanism is supported by the high degree of inversion of configuration already discussed. Replacement of hydrogen by alkyl groups at the substitution site and at the carbon attached to the substitution site, which we will consider in more detail in Section 4.4, slows these reactions dramatically, as would be expected from steric hindrance to approach of a nucleophile. Furthermore, the rates are not very sensitive to electron donation and withdrawal where steric factors are constant, a finding consistent with a low degree of charge accumulation at the substitution center.

Replacement of a hydrogen at the substitution site by deuterium leads to the α deuterium isotope effect, and at the β carbon to the β deuterium isotope effect; both of these isotope effects are typically close to unity for the substitutions we are considering. The α-d isotope effect arises from changes of C—H(D) bending force constants from reactant to transition state. The change to sp^2 hybridization tends to allow easier bending and hence to lower these constants, but the presence of both the nucleophile and the leaving group in the apical positions of the trigonal bipyramid (**2**, p. 297) has the opposite effect. The overall result is that there is little net change in force constants, and hence the isotope effect is close to unity. The α-d isotope effect per deuterium in hydrolysis of methyl tosylate at 25°C is 0.984, and of ethyl tosylate 1.020. In contrast, the α-d isotope effect for solvolysis of isopropyl

---

[20]T. W. Bentley and P. v. R. Schleyer, *Adv. Phys. Org. Chem.*, **14**, 1 (1977).

tosylate in 100 percent trifluoroacetic acid, a very weakly nucleophilic solvent that favors unassisted reactions, is 1.22.[21] The β-d isotope effect is considered to arise primarily from hyperconjugative electron transfer from the β—C—H bond to a center of developing positive charge. Again, the values are close to unity for S_N2-type reactions (1.006 per deuterium for ethyl tosylate in water) in contrast to larger values in S_N1-type reactions where positive charge is developing (1.100 per deuterium for t-butyl chloride in 60 percent aqueous ethanol.)[21]

Despite the apparently consistent picture presented by these results, evidence has been presented that at least some systems that might be expected to react by the nucleophilically assisted S_N2 process actually are following the unassisted route shown in Equation 4.20. If the rate of return from ion pair to substrate,

$$\text{RX} \underset{k_{-1}}{\overset{k_1}{\rightleftharpoons}} \text{R}^+\text{X}^- \xrightarrow{Nuc^-}_{k_2} \text{R}Nuc + \text{X}^- \tag{4.20}$$

$k_{-1}[\text{R}^+\text{X}^-]$, is large compared to rate of attack of nucleophile on ion pairs, the rate expression is

$$-\frac{d[\text{RX}]}{dt} = k_2[\text{R}^+\text{X}^-][Nuc^-] \tag{4.21}$$

$$\approx \frac{k_1 k_2}{k_{-1}}[\text{RX}][Nuc^-] \tag{4.22}$$

a result indistinguishable from that for a conventional S_N2 reaction of Nuc^- with RX. In principle, it is possible to distinguish between the two mechanisms by careful measurements of rates and products formed when two nucleophiles are present. For example, Sneen and Larsen found that 2-octyl methanesulfonate reacts in aqueous dioxane containing added azide ion to yield a mixture of 2-octyl alcohol and 2-octylazide.[22] The two nucleophiles present, water and azide ion, compete for the substrate. But the ratio of the rate of disappearance of the methanesulfonate in the presence of azide to the rate in the absence of azide was found to be that expected if there were an intermediate that could react in any one of three ways: return to substrate, combine with water, or combine with azide. Sneen and Larsen proposed that the intermediate is an ion pair. This finding, in a system that would have been expected to react by an S_N2-like process, led them to propose that all nucleophilic substitutions, S_N2 and S_N1 alike, react through ion pairs.

Earlier, Swain and Kreevoy had suggested the possibility of rate-determining attack by methanol on ion pairs from triphenylmethyl chloride in benzene

[21] V. J. Shiner, Jr., in *Isotope Effects in Chemical Reactions*, C. J. Collins and N. S. Bowman, Eds., Van Nostrand Reinhold, New York, 1970, pp. 120, 123, 144, 145. See also (b) M. Wolfsberg, *Acc. Chem. Res.*, **5**, 225 (1972).
[22] (a) R. A. Sneen and J. W. Larsen, *J. Am. Chem. Soc.*, **91**, 362, 6031 (1969); (b) R. A. Sneen, *Acc. Chem. Res.*, **6**, 46 (1973).

solvent,[23] and Shiner and co-workers have also found evidence for rate-determining attack on ion pairs,[24] but the Sneen–Larsen proposal would eliminate from consideration the Hughes–Ingold S_N2 pathway and leave only variants of the unassisted reaction.

In analyzing their data, Sneen and Larsen had to correct for salt effects, since they were comparing rate with azide present to rate without. Schleyer and co-workers criticized Sneen's conclusions by pointing out the uncertainties involved in such corrections,[25] and Sneen replied, justifying his earlier conclusions and presenting similar evidence for α-phenylethyl systems,[26] and for an allylic system.[27]

The Sneen proposal remains a subject of controversy.[28] Friedberger and Thornton have pointed out that it is inconsistent with α-d isotope effects, as well as with sulfur primary isotope effects in benzyl sulfonium salt solvolysis; they suggested that the Sneen mechanism is not general for the S_N2-like reactions.[29] McLennan has analyzed data for a number of reactions that show relationships of rates and product distributions consistent with the traditional S_N2 process. He has also pointed out that other features of S_N2-like reactions, such as lack of rearrangement, are better explained by the traditional mechanism, and that rapid formation of poorly stabilized ion pairs without any nucleophilic assistance is chemically unreasonable.[30] It appears on the basis of these arguments that although there undoubtedly are instances of formation of ion-pair intermediates with nucleophilic assistance by solvent or some other nucleophile, the traditional Hughes–Ingold S_N2 mechanism, with no intermediate, is still valid as a description of many reactions customarily placed in this category.

Solvent-Assisted Ionization

If we admit the viability of pathways (b) and (c) in Scheme 2, the next question is under what circumstances pathway (a) is followed. That is, when does the leaving group depart without any covalent bonding to a particular nucleophile, and when does a nucleophile assist that process, going either directly to product, path (b), or to a nucleophilically solvated ion pair, path (c)? This question arises particularly in solvolysis reactions, because nucleophilic participation by the solvent cannot be detected kinetically.

Investigations reported by Schleyer and a number of co-workers have contributed significantly to clarification of some of the points in question. A study of

[23] C. G. Swain and M. M. Kreevoy, *J. Am. Chem. Soc.*, **77**, 1122 (1955).

[24] (a) V. J. Shiner, Jr., R. D. Fisher, and W. Dowd, *J. Am. Chem. Soc.*, **91**, 7748 (1969); (b) V. J. Shiner, Jr., S. R. Hartshorn, and P. C. Vogel, *J. Org. Chem.*, **38**, 3604 (1973).

[25] (a) D. J. Raber, J. M. Harris, R. E. Hall, and P. v. R. Schleyer, *J. Am. Chem. Soc.*, **93**, 4821 (1971); (b) T. W. Bentley and P. v. R. Schleyer, *Adv. Phys. Org. Chem.*, **14**, 1 (1977).

[26] R. A. Sneen and H. M. Robbins, *J. Am. Chem. Soc.*, **94**, 7868 (1972).

[27] R. A. Sneen and W. A. Bradley, *J. Am. Chem. Soc.*, **94**, 6975 (1972).

[28] See, for example, (a) T. W. Bentley, S. H. Liggero, M. A. Imhoff, and P. v. R. Schleyer, *J. Am. Chem. Soc.*, **96**, 1970 (1974); (b) F. G. Bordwell and G. A. Pagani, *J. Am. Chem. Soc.*, **97**, 118 (1975), and following papers.

[29] M. P. Friedberger and E. R. Thornton, *J. Am. Chem. Soc.*, **98**, 2861 (1976).

[30] D. J. McLennan, *Acc. Chem. Res.*, **9**, 281 (1976).

rearrangements occurring during solvolysis of tosylates bearing substituted phenyl groups on the β carbon (**9**) led them to suspect that nucleophilic assistance to

9

ionization by solvent is more general than had previously been thought.[31] (These rearrangement experiments are discussed in Section 5.1.)

Their approach in looking into the problem further was to find structures in which specific covalent bonding to the back side of the carbon undergoing substitution is difficult or impossible. As models for reactions at tertiary carbon they chose bridgehead substitutions. We shall see in Section 4.4 that rates in these systems are usually retarded, in some cases by many powers of ten, because of the increase in strain upon ionization. But the important point in the present context is that it is impossible for a solvent molecule to approach from the back side of a bridgehead carbon; the only possibilities are frontside attack, known to be strongly disfavored, or ionization by path (a), Scheme 2.

An important tool used in comparing the bridgehead systems with others that might follow either paths (a) or path (b) and (c) is the influence of solvent on reaction rate. Grunwald and Winstein developed a linear free-energy relation (Equation 4.23) for solvent ionizing power, defined in terms of the solvolysis rate of *t*-butyl chloride, a system assumed to react without nucleophilic assistance.[32] In

$$\log \frac{k_S}{k_{80\% \text{ EtOH}}} = m\mathbf{Y} \tag{4.23}$$

Equation 4.23, k_S is the solvolysis rate in the solvent S; $k_{80\% \text{ EtOH}}$ is the rate in the standard solvent, 80 percent aqueous ethanol; m measures the sensitivity of the particular system to solvent change; and $\mathbf{Y}$ is the ionizing power of solvent S, determined from *t*-butyl chloride solvolysis rates by defining $m = 1$ for this substrate. In addition to allowing comparison among experiments carried out in different solvents, the m–$\mathbf{Y}$ system serves as an important tool for study of mechanism. Sensitivity to ionizing power, measured by m, is an index of the degree of charge separation at the transition state. If *t*-butyl chloride does indeed solvolyze by path (a), without nucleophilic assistance, the $\mathbf{Y}$ scale will not contain any contribution from solvent nucleophilicity. Two different solvent systems can be set up, for exam-

[31] D. J. Raber, J. M. Harris, and P. v. R. Schleyer, *J. Am. Chem. Soc.*, **93**, 4829 (1971), and references cited therein.
[32] E. Grunwald and S. Winstein, *J. Am. Chem. Soc.*, **70**, 846 (1948); a compilation of Y values may be found in A. H. Fainberg and S. Winstein, *J. Am. Chem. Soc.*, **78**, 2770 (1956).

ple, ethanol and acetic acid, each mixed with the correct amount of water to achieve the same value of **Y**. (See Table 4.7, p. 329, for data.) *t*-Butyl chloride will solvolyze at the same rate in both mixtures (by the definition of **Y**), but another substrate that requires nucleophilic assistance will react faster in the ethanol–water mixture than in the less nucleophilic acetic acid–water mixture.

In order to use this criterion, it is necessary first to check that *t*-butyl chloride solvolysis is not nucleophilically assisted; if it were, the **Y** scale would reflect the fact, and a substrate requiring less nucleophilic assistance would show $k_{\text{EtOH}}/k_{\text{HOAc}}$ less than unity. The 1-adamantyl system (**10**) was chosen as the refer-

ence to check this point. Schleyer and co-workers found that in most solvents the sensitivity of rate to changes of solvent was the same for 1-adamantyl bromide as for *t*-butyl chloride.[33] The correlation between **Y** and rate of 1-adamantyl solvolysis is good in a variety of solvents of varying nucleophilicity; however, there are deviations for the very weakly nucleophilic solvents formic acid, acetic acid, trifluoroacetic acid, and hexafluoro-2-propanol.[34] The deviations imply some nucleophilic assistance to *t*-butyl solvolysis in alcohol solvents. The adamantyl systems cannot solvolyze with nucleophilic assistance; Schleyer has therefore proposed that 1-adamantyl (or 2-adamantyl, see below) tosylate be the standard for the **Y** scale.[35] This revised **Y** scale measures solvent ionizing power only and does not include any contribution from solvent nucleophilicity.

In order to extend this line of argument to secondary systems, Schleyer and his collaborators chose the 2-adamantyl structure (**11**). They reasoned that the axial hydrogens in this rigid molecule would block backside approach of a nucleophile. Indeed, they found that 2-adamantyl tosylate solvolysis rates correlate with those of 1-adamantyl, showing the same lack of sensitivity to solvent nucleophilicity. Open-chain secondary tosylates, for example, isopropyl, proved to be markedly sensitive to nucleophilicity.[36] These compounds react at different rates in solvents of the same **Y** but different nucleophilicity; therefore the solvent must be assisting the departure of the leaving group by nucleophilic attack.

---

[33] D. J. Raber, R. C. Bingham, J. M. Harris, J. L. Fry, and P. v. R. Schleyer, *J. Am. Chem. Soc.*, **92**, 5977 (1970).
[34] T. W. Bentley, C. T. Bowen, W. Parker, and C. I. F. Watt, *J. Am. Chem. Soc.*, **101**, 2486 (1979).
[35] (a) J. M. Harris, D. J. Raber, W. C. Neal, Jr., and M. D. Dukes, *Tetrahedron Lett.*, 2331 (1974); (b) F. L. Schadt, P. v. R. Schleyer, and T. W. Bentley, *Tetrahedron Lett.*, 2335 (1974).
[36] (a) J. L. Fry, C. J. Lancelot, L. K. M. Lam, J. M. Harris, R. C. Bingham, D. J. Raber, R. E. Hall, and P. v. R. Schleyer, *J. Am. Chem. Soc.*, **92**, 2538 (1970); (b) P. v. R. Schleyer, J. L. Fry, L. K. M. Lam, and C. J. Lancelot, *J. Am. Chem. Soc.*, **92**, 2542 (1970).

A variation of the solvent response method is to compare solvolysis rate of the substrate of interest with that of 1-adamantyl bromide in ethanol and in a very nonnucleophilic solvent such as 2,2,2-trifluoroethanol, trifluoroacetic acid, or hexafluoro-2-propanol.[37] This technique has shown that in ethanol the hindered cyclooctyl and 3,3-dimethyl-2-butyl tosylates (**12** and **13**) solvolyze by the k_c process, without nucleophilic assistance, as does 2-adamantyl tosylate; in 3-methyl-2-butyl tosylate (**14**) nucleophilic assistance by solvent (k_s process) and assistance internally by a migrating hydrogen (k_Δ process) compete, while benzyl chloride solvolysis (**15**) is solvent-assisted (k_s process).[38]

Use of other methods has contributed further to the emerging picture of solvolysis of most secondary systems as being solvent-assisted. For example, the solvolysis rate acceleration on substituting α hydrogen by CH_3 in 2-adamantyl bromide is $10^{7.5}$, much larger than that found for other secondary–tertiary pairs such as isopropyl–t-butyl. In molecules less hindered than 2-adamantyl, the secondary substrate is accelerated by nucleophilic attack of solvent; there is less positive charge at the substitution center in the transition state and consequently less demand for stabilization and a smaller effect of substituting hydrogen by methyl. Rate accelerations and product distributions found on adding azide ion to solvolysis mixtures (Problem 8) also provide confirmatory evidence for these conclusions.[39] The α deuterium isotope effect provides another criterion: values close to unity indicate strong nucleophilic assistance, while high values (near 1.2 for sulfonate leaving groups) are characteristic of unassisted ionization.[40]

[37] Secondary systems solvolyze with little or no nucleophilic assistance in these solvents. See (a) T. W. Bentley and P. v. R. Schleyer, *Adv. Phys. Org. Chem.*, **14**, 1 (1977); (b) T. W. Bentley, C. T. Bowen, W. Parker, and C. I. F. Watt, *J. Am. Chem. Soc.*, **101**, 2486 (1979).

[38] (a) D. J. Raber, W. C. Neal, Jr., M. D. Dukes, J. M. Harris, and D. L. Mount, *J. Am. Chem. Soc.*, **100**, 8137 (1978); (b) J. M. Harris, D. L. Mount, M. R. Smith, W. C. Neal, Jr., M. D. Dukes, and D. J. Raber, *J. Am. Chem. Soc.*, **100**, 8147 (1978). For further evidence on the benzyl system, see (c) M. P. Friedberger and E. R. Thornton, *J. Am. Chem. Soc.*, **98**, 2861 (1976). The k_Δ process is identified in separate experiments by isotope effect measurements.

[39] (a) J. L. Fry, J. M. Harris, R. C. Bingham, and P. v. R. Schleyer, *J. Am. Chem. Soc.*, **92**, 2540 (1970); (b) J. M. Harris, D. J. Raber, R. E. Hall, and P. v. R. Schleyer, *J. Am. Chem. Soc.*, **92**, 5729 (1970).

[40] Shiner has proposed that the higher isotope effects indicate rate-determining conversion of contact ion pair to solvent-separated ion pair, while intermediate values (1.06 to 1.15) are characteristic of the Sneen mechanism, rate-determining attack of nucleophile on a contact ion pair, or of assistance to ionization by an internal nucleophile, and values of 1.15 to 1.16 indicate rate-determining ionization to contact ion pair without nucleophilic assistance. Schleyer has criticized these proposals and considers values as high as 1.15 consistent with nucleophilic assistance to ionization, while values of 1.2 are characteristic of unassisted rate-determining formation of contact ion pair. See (a) V. J. Shiner, Jr., in *Isotope Effects in Chemical Reactions*, C. J. Collins and N. S. Bowman, Eds., Van Nostrand Reinhold, New York, 1970, p. 90; (b) T. W. Bentley and P. v. R. Schleyer, *Adv. Phys. Org. Chem.*, **14**, 1 (1977); (c) G. A. Gregoriou and F. S. Varveri, *Tetrahedron Lett.*, 287, 291 (1978).

In conclusion, it seems evident that tertiary and sterically hindered secondary substrates solvolyze by path (a), Scheme 2, while most secondary systems and probably nearly all primary systems solvolyze by paths (b) or (c), except possibly in the extremely nonnucleophilic fluorinated solvents.[41]

The methods we have considered here for the most part detect assistance by external nucleophile. As we have pointed out earlier, assistance by an internal nucleophile is also possible, and assistance by internal and external nucleophiles may compete.[42]

Reaction Coordinate Diagrams

Figure 4.3 shows schematically the potential energy surface and reaction coordinate for an S_N2-like reaction in a diagram of the type discussed in Section 2.6.[43] Reactants designated as C—X and Nuc^-, where C signifies the carbon at the substitution center of RX, are at the energy minimum at the back left corner. The coordinate running from back to front is the distance from the substitution center C

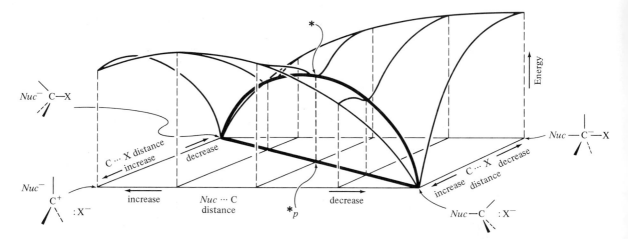

Figure 4.3 Three-dimensional energy surface and reaction coordinate for the S_N2 mechanism. Reactants are at the back left corner and products are at the front right corner. The reaction coordinate is the heavy curved line joining the reactants and products; its projection on the horizontal plane is the heavy straight line. Dotted lines join points on the surface and their projections on the horizontal plane. The transition state is marked * and its projection $*_p$.

[41] For a summary see (a) T. W. Bentley and P. v. R. Schleyer, *J. Am. Chem. Soc.*, **98**, 7658 (1976); (b) F. L. Schadt, T. W. Bentley, and P. v. R. Schleyer, *J. Am. Chem. Soc.*, **98**, 7667 (1976); (c) note 40(b).

[42] (a) F. L. Schadt and P. v. R. Schleyer, *J. Am. Chem. Soc.*, **95**, 7860 (1973); (b) G. A. Dafforn and A. Streitwieser, Jr., *Tetrahedron Lett.*, 3159 (1970); (c) P. C. Myhre and E. Evans, *J. Am. Chem. Soc.*, **91**, 5641 (1969).

[43] (a) E. R. Thornton, *J. Am. Chem. Soc.*, **89**, 2915 (1967); (b) R. A. More O'Ferrall, *J. Chem. Soc. B.*, 274 (1970); (c) W. P. Jencks, *Chem. Rev.*, **72**, 705 (1972).

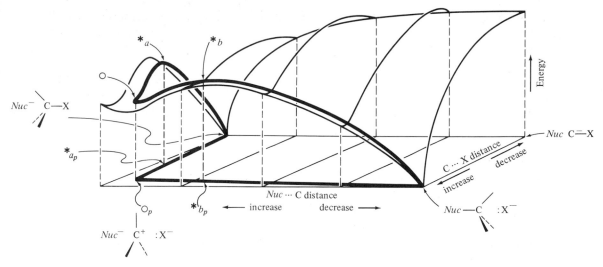

Figure 4.4 Three-dimensional energy surface and reaction coordinate for the S_N1 mechanism. The
reaction has two steps. The first follows the heavy line along the left edge of the surface
over transition state $*_a$ to intermediate ion pair o. The second follows the heavy line along
the front edge of the surface over transition state $*_b$ to products.

to the leaving group X, increasing toward the front. At the front left corner is ion
pair C^+X^-. The coordinate running from left to right is the distance from the
substitution center C to the nucleophile Nuc^-, decreasing toward the right. At the
back right corner is the hypothetical intermediate Nuc—C—X, which has a
pentacovalent carbon. The products C—Nuc and $:X^-$ are at the front right corner.
In the S_N2 process C—X bond breaking and Nuc—C bond making occur together;
therefore the reaction coordinate (heavy curved line) runs diagonally from the
reactant corner to the product corner, with the transition state assumed for purposes of this illustration to be at the midpoint. Figure 4.3 also shows the projection
of the reaction coordinate on the horizontal plane (heavy straight line).

   Figure 4.4 shows the energy surface and the two-step reaction coordinate for
an S_N1-like reaction. The coordinates are the same as for Figure 4.3. In this instance
the lowest-energy path goes along the left edge to an intermediate ion pair, marked
by o on the diagram, in which the C—X bond has broken but the Nuc—C bond has
not yet begun to form. There are two transition states, $*_a$ and $*_b$, one for each of the
two elementary reaction steps. Transition state $*_a$ is at a higher energy than $*_b$, and
so the first step is rate determining.

   The processes described by the diagrams are paths (a) through (f) of
Scheme 2; processes involving solvent-separated ion pairs and free ions are not
included in the diagrams as formulated. Hence the discussion here is confined to
questions concerning the degree of involvement of nucleophile in the initial dissoci-

SCHEME 2 (Repeated from p. 303.)

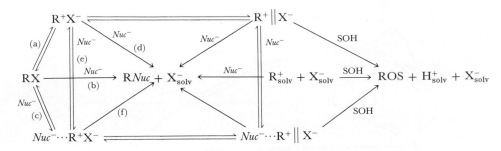

ation process. The pathway across the surface shown in Figure 4.3 (S_N2-like) corresponds to the reaction (b) of Scheme 2. The pathway shown in Figure 4.4 corresponds to the sequence (a), (d). Which of these routes is lower in energy, and whether there will be an alternate route intermediate between the two [for example, (c), (f) in Scheme 2], will depend on the details of the particular reaction. Among the factors that determine the choice are the strength of the nucleophile, the propensity of X^- to depart, the degree of stabilization of positive charge provided by substituents on C, and the ionizing power of the solvent.

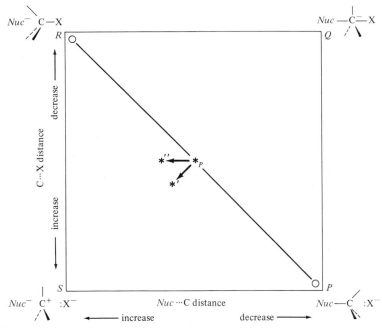

Figure 4.5 Projection onto the horizontal plane of the S_N2 reaction coordinate shown in Figure 4.3. The transition state $*_p$ is shifted to $*'$ by lowering the energy of corner S (making the carbocation more stable) and to $*''$ by lowering the energy of edge SP (making X a better leaving group).

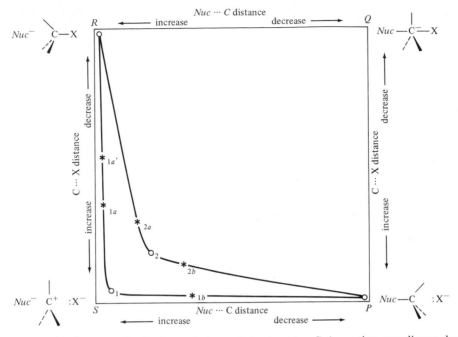

Figure 4.6 Projection onto the horizontal plane of the two-step S_N1 reaction coordinate shown in Figure 4.4. The transition states are at $*_{1a}$ and $*_{1b}$, and the intermediate ion pair is at $\circ_1$. Lowering the energy along edge *SP* (making X a better leaving group) moves the transition state $*_{1a}$ to position $*_{1a'}$. Raising the energy of corner *S* (making the carbocation less stabilized) shifts the reaction path to $*_{2a}$–$\circ_2$–$*_{2b}$ and thereby introduces nucleophilic assistance.

Figure 4.5 shows the projection on the horizontal plane of the reaction coordinate for the S_N2-like reaction of Figure 4.3. The reactant and product corners *R* and *P* correspond to energy minima (open circles); the corners *Q* and *S* are higher energy. The transition state is at the energy maximum encountered along the reaction path, marked $*_p$. Figure 4.6 shows the projection onto the horizontal plane of the reaction coordinate for the two-step, S_N1-like reaction of Figure 4.4. This projected reaction path differs from that for the S_N2-like reaction in having an energy minimum, the carbocation, near corner *S* at the position marked $\circ_1$. The projections of the two transition states are at $*_{1a}$ and $*_{1b}$.

The effects of various parameters on the locations of the transition states and the relation between the S_N2-like and the S_N1-like reactions can be understood from these diagrams, using the rules given in Section 2.6, p. 201. For example, changing substituents so as to supply electrons at the substitution center C will lower the energy of the ion pair at corner S of the figures. For the S_N2 reaction (Figure 4.5) this change results in a shift of transition state structure perpendicular to the reaction coordinate to $*'$. The new transition state is looser than the first one;

the C—X bond breaking has proceeded farther and *Nuc*—C bond making has proceeded less far than before. This change is reflected experimentally in a larger α deuterium isotope effect for the looser transition state. For example, the k_H/k_D ratios for p—Y—ϕ—CH$_2$—OSO$_2\phi$—Br—p in 80 percent aqueous ethanol are Y = NO$_2$, 1.005; Y = CF$_3$, 1.016; Y = H, 1.074.[44]

Replacing the leaving group by a better one is equivalent to lowering the energy along the edge *SP* of the diagrams. For the S$_N$1-like reaction (Figure 4.6) the first transition state $*_{1a}$, which corresponds to the rate-determining step, moves toward corner *R* to position $*_{1a'}$. The change makes the transition state come earlier, and it therefore has a tighter structure. Experimentally the ratio of α deuterium isotope effects for X=Cl/X=Br is 1.021 in solvolysis of ϕ—CH(CH$_3$)X.[45] The transition state is tighter (smaller α-d isotope effect) for the better leaving group Br.

For the hypothetical symmetrical S$_N$2 process depicted in Figure 4.5, changing to a better leaving group (lowering energy along edge *SP*) should move the transition state structure from $*_p$ along the reaction coordinate toward *R* and perpendicular to the reaction coordinate toward *S*. If both of these motions occur to an equal extent, the net result expected is a new transition state structure at $*''$. Compared with $*_p$, the structure would be tighter with respect to *Nuc*—C distance and approximately unchanged with respect to C—X distance. The α deuterium isotope effect ratio for X=Cl/X=Br should be less than unity in such a case. For CH$_3$—X solvolysis the observed ratio is 1.008;[46] the trend compared with the more S$_N$1-like ϕ—CH(CH$_3$)X solvolysis is in the direction expected, although the ratio is still greater than unity. The discrepancy may be partly accounted for if the methyl chloride reaction does not follow exactly the idealized diagonal reaction path depicted in Figure 4.5.

The nature of the path (c), (f) of Scheme 2 is illustrated in Figure 4.6. Starting from a reaction with no nucleophilic assistance, $*_{1a} \rightarrow \circ_1 \rightarrow *_{1b}$, making the ion less stabilized raises the energy at corner *S* and changes the path to one like $*_{2a} \rightarrow \circ_2 \rightarrow *_{2b}$. The new intermediate, $\circ_2$, has the nucleophile and leaving group closer to C. The less stable ion pair now requires some nucleophilic assistance to form, but it still exists as a discrete intermediate although its energy well is shallower. Continued changes in the same direction will push the ion pair farther toward the upper right, and the two transition states closer to it, until the energy minimum disappears and the path has been converted into S$_N$2.

It should in principle be possible to quantify coordinates on these reaction coordinate diagrams by applying experimental parameters such as the heavy atom primary isotope effects for nucleophiles and leaving groups, the secondary α and β deuterium isotope effects, and Hammett σ-ρ substituent effect correlations in the

[44]V. J. Shiner, Jr., in *Isotope Effects in Chemical Reactions,* C. J. Collins and N. S. Bowman, Eds., Van Nostrand Reinhold, New York, 1970, p. 116.
[45]See p. 114 of the reference given in note 44.
[46]See p. 120 of the reference given in note 44.

manner outlined by Jencks.[47] Work is proceeding along these lines, and increasingly quantitative applications are likely as understanding of substitution reactions and of their isotope effects progresses.[48]

4.3 ION PAIRS AS INTERMEDIATES IN NUCLEOPHILIC SUBSTITUTION[49]

If two processes occur through the same intermediate, the products should be identical. This principle can be used to test for the presence of free carbocations in nucleophilic substitution reactions by measuring the ratio of elimination to substitution as a function of the leaving group. Scheme 3 formulates the solvolysis pathway for a *t*-butyl derivative according to the Hughes–Ingold S_N1 mechanism. If the *t*-butyl cation were present as a free ion, completely dissociated from the leaving group X^-, the nature of X^- should have no influence on subsequent reactions. The two products should form in proportions determined by the medium but independent of X^-.

SCHEME 3

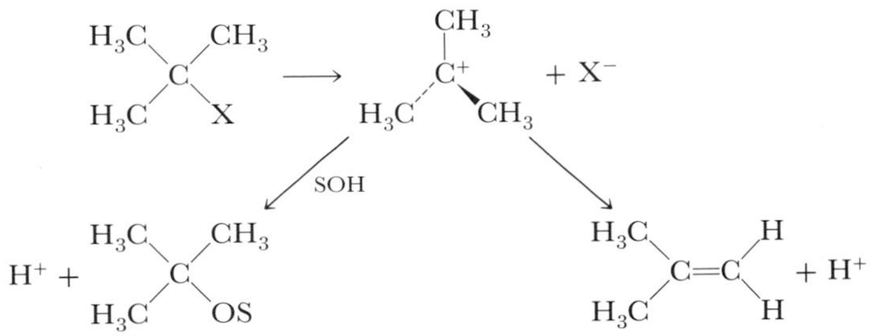

Table 4.2 gives product distributions for solvolysis of a series of *t*-butyl derivatives in three different solvent systems. In water, the most strongly ionizing of the solvents, the products are indeed independent of leaving group. But in less strongly ionizing solvents, the marked dependence of product distribution on leaving group shows that the intermediate is an ion pair in which X^- is still closely

[47]Jencks has proposed quantification for acid-catalyzed carbonyl addition reactions using the Brønsted catalysis law for parameterization. D. A. Jencks and W. P. Jencks, *J. Am. Chem. Soc.*, **99**, 7948 (1977).

[48](a) J. C. Harris and J. L. Kurz, *J. Am. Chem. Soc.*, **92**, 349 (1970); (b) V. J. Shiner, Jr., and R. C. Seib, *J. Am. Chem. Soc.*, **98**, 862 (1976); (c) M. P. Friedberger and E. R. Thornton, *J. Am. Chem. Soc.*, **98**, 2861 (1976); (d) R. L. Julian and J. W. Taylor, *J. Am. Chem. Soc.*, **98**, 5238 (1976). See also (e) R. T. Hargreaves, A. M. Katz, and W. H. Saunders, Jr., *J. Am. Chem. Soc.*, **98**, 2614 (1976); (f) K. C. Westaway, *Can. J. Chem.*, **56**, 2691 (1978); (g) J. M. Harris, S. C. Shafer, J. M. Moffatt, and A. R. Becker, *J. Am. Chem. Soc.*, **101**, 3295 (1979); (h) P. R. Young and W. P. Jencks, *J. Am. Chem. Soc.*, **101**, 3288 (1979).

[49]Reviews: (a) D. J. Raber, J. M. Harris, and P. v. R. Schleyer, in *Ions and Ion Pairs in Organic Reactions*, Vol. 2, M. Szwarc, Ed., Wiley, New York, 1974, p. 247; (b) J. M. Harris, *Prog. Phys. Org. Chem.*, **11**, 89 (1974); (c) L. P. Hammett, *Physical Organic Chemistry*, McGraw-Hill, New York, 1st ed., 1940, pp. 171–173; 2d ed., 1970, pp. 157–158.

Table 4.2 PARTITION BETWEEN ELIMINATION AND
SUBSTITUTION IN t-BUTYL-X SOLVOLYSIS[a]

	Mole percent alkene		
X in t-bu-X	H_2O 75°C	C_2H_5OH 75°C	CH_3COOH 75°C
Cl	7.6	44.2	73
Br	6.6	36.0	69.5
I	6.0	32.3	—
$S^+(CH_3)_2$ ClO_4^-	6.5	17.8	11.7

[a] Reprinted with permission from M. Cocivera and S. Winstein, *J. Am. Chem. Soc.*, **85**, 1702 (1963). Copyright by the American Chemical Society.

associated with the cation and so is able to influence the course of subsequent events.

A similar conclusion comes from study of allylic systems. Because the resulting cationic center is stabilized by interaction with the π electrons, allylic halides ionize readily to produce the delocalized allylic ion (**16**). The free-ion theory predicts that isomeric allylic halides that give the same intermediate upon ionization should yield a product distribution independent of the isomeric origin of the ion. Scheme 4 illustrates the argument. The prediction is sometimes, but not always, verified in practice; there is a marked tendency in many systems to favor the unrearranged product.[50]

SCHEME 4

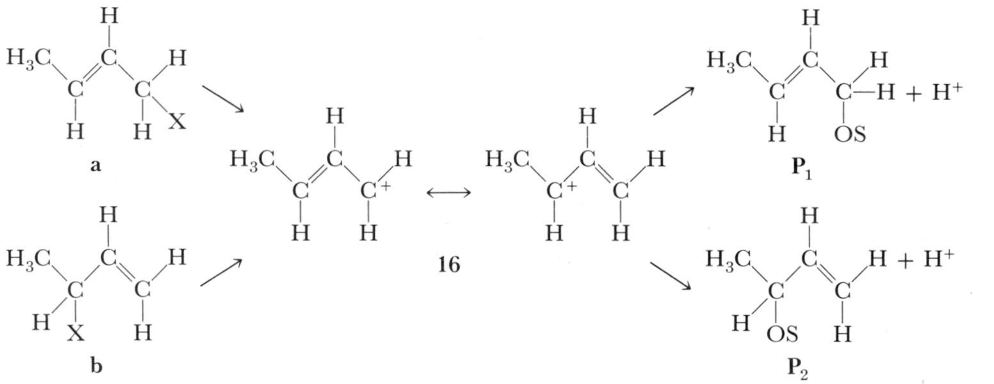

For a free ion, P_1/P_2 is the same starting from **a** or **b**.

Again, the result indicates an intermediate in which the leaving group is still closely associated with the ion (in this instance specifically with the carbon of the allylic system to which it was originally attached) and hence able to influence the point of attack of the incoming nucleophile.

[50] R. H. DeWolfe and W. G. Young, *Chem. Rev.*, **56**, 753 (1956), give an extensive table (pp. 794–796).

A closely related phenomenon is the rearrangement during acetolysis of α,α-dimethylallyl chloride to γ,γ-dimethylallyl chloride (Equation 4.24). Young,

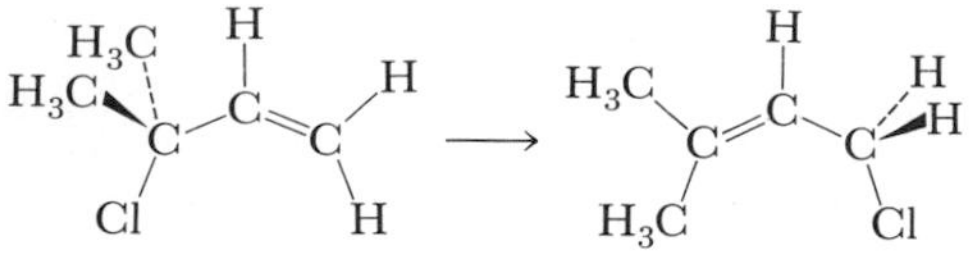

$$(4.24)$$

Winstein, and Goering showed that the reaction is unaffected by added chloride ion. The same chloride that leaves the substituted end of the allylic system therefore returns to the other end, and the process must occur through an ion pair.[51] Goering's work has subsequently supplied a number of details about allylic ion pair structure.[52]

Evidence for Ion Pairs from Stereochemistry and Isotope Exchange

Another kind of evidence for ion pairs comes from product stereochemistry. Although some solvolyses of chiral derivatives, such as the benzhydryl compounds discussed earlier, lead to complete racemization of products, other cases are known in which tertiary systems that in all probability ionize without nucleophilic assistance nevertheless give only partly racemic products. One example is the tertiary phthalate (**17**), which yields solvolysis product in methanol that is 46 percent racemic and 54 percent of inverted configuration.[53] The highly hindered 1-phenylneopentyl tosylate (**18**) in acetic acid yields acetate that is 90 percent racemic and 10

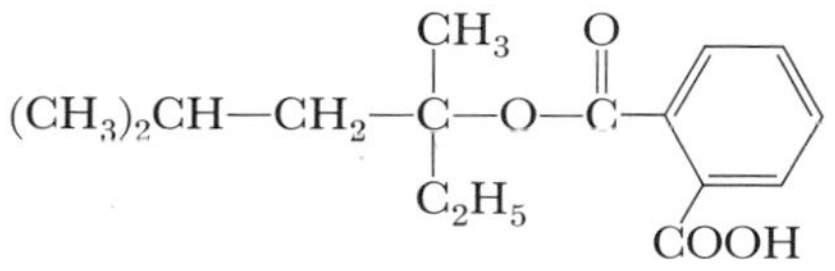

17

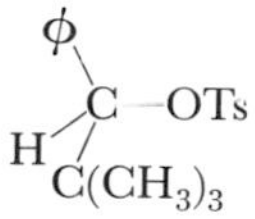

18

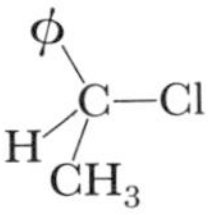

19

percent inverted.[54] 1-Phenylethyl chloride (**19**) in acetic acid yields 1-phenylethyl acetate partly racemic and partly of inverted configuration, even though addition

[51]W. G. Young, S. Winstein, and H. L. Goering, *J. Am. Chem. Soc.*, **73**, 1958 (1951).

[52](a) H. L. Goering and E. C. Linsay, *J. Am. Chem. Soc.*, **91**, 7435 (1969); (b) H. L. Goering, G. S. Koermer, and E. C. Linsay, *J. Am. Chem. Soc.*, **93**, 1230 (1971); (c) H. L. Goering, M. M. Pombo, and K. D. McMichael, *J. Am. Chem. Soc.*, **85**, 965 (1963); (d) H. L. Goering and R. P. Anderson, *J. Am. Chem. Soc.*, **100**, 6469 (1978).

[53]W. v. E. Doering and H. H. Zeiss, *J. Am. Chem. Soc.*, **75**, 4733 (1953).

[54]S. Winstein and B. K. Morse, *J. Am. Chem. Soc.*, **74**, 1133 (1952).

of acetate ion does not increase the rate.[55] These results can be explained if the ionization leads to a contact ion pair. The front side of the carbon is then protected by the counter-ion, and attack of solvent occurs preferentially from the back. The partial racemization can occur either by reorientation of the cation within the contact ion pair or by progression to the next stage of dissociation, the solvent-separated ion pair, which can lead to product by attack of solvent from either side.

A number of investigators, notably Winstein and co-workers, have contributed kinetic evidence for the presence of ion pairs in solvolysis reactions. Table 4.3 gives the notation used to designate various rate constants that can be determined. The rate constant for disappearance of reactant (or appearance of total product) is frequently determined by titration of the acid formed, and is referred to as the titrimetric rate constant, k_t. Two kinds of rate constant connected with loss of chirality can be obtained: k_α, the overall rate of loss of optical activity, and k_{rac}, the rate of racemization of the starting material that remains unsolvolyzed. Finally, there are two rate constants for isotopic exchange: k_{ex}, which is an exchange in unsolvolyzed substrate of leaving group with isotopically labeled ions in the medium, and k_{eq}, observed with carboxylate leaving groups when a C—O bond breaks and forms again to the other oxygen of the same carboxylate ion.

Winstein's group found that for substrates that have the leaving group bound to a chiral center, k_α frequently exceeds k_t by substantial factors.[56] For example, with *p*-chlorobenzhydryl chloride in acetic acid, the ratio k_α/k_t is between 30 and 70, and is about 5 in aqueous acetone.[57] The excess of rate of loss of optical activity, k_α, over rate of product formation, k_t, means that some process racemizes

Table 4.3 Rate Processes in Solvolysis Reactions

Reaction[a]	Rate constant notation
RX + SOH $\longrightarrow$ ROS + HX	k_t
d-RX *or l*-RX $\longrightarrow$ *optically inactive products*	k_α
d-RX *or l*-RX $\longrightarrow$ *dl*-RX	k_{rac}
RX + *X⁻ $\rightleftharpoons$ R—*X + X⁻	k_{ex}
R—O—$\overset{\text{*O}}{\overset{\|}{\text{C}}}$—Ar $\rightleftharpoons$ R—*O—$\overset{\text{O}}{\overset{\|}{\text{C}}}$—Ar	k_{eq}

[a] * denotes isotopically labeled atom; *d*- or *l*- indicates a chiral and *dl*- a racemic substance.

[55] (a) J. Steigman and L. P. Hammett, *J. Am. Chem. Soc.*, **59**, 2536 (1937). See also (b) A. Streitwieser, Jr., *Solvolytic Displacement Reactions,* McGraw-Hill, New York, 1962, p. 59.
[56] (a) S. Winstein and D. Trifan, *J. Am. Chem. Soc.*, **74**, 1154 (1952); (b) S. Winstein and K. C. Schreiber, *J. Am. Chem. Soc.*, **74**, 2165 (1952).
[57] S. Winstein, J. S. Gall, M. Hojo, and S. Smith, *J. Am. Chem. Soc.*, **82**, 1010 (1960).

the substrate more rapidly than the substrate can form products. The observation that in these same systems k_α is also larger than k_{ex} rules out the possibility that racemization occurs through free ions. If free ions were forming, the X^- anions would become equivalent to any other X^- ions present; if isotopically labeled $*X^-$ ions are added, the process shown in Equation 4.25 would have to occur. Since the

$$R^+ + {}^*X^- \longrightarrow R—{}^*X \qquad (4.25)$$

experimental results show that racemization is faster than this process, there must be present some intermediate or intermediates, presumably ion pairs, in which the C—X bond is broken, but in which the X^- ion is still closely associated with the particular C^+ ion to which it was bonded and to which it can return. It is important to note that formation of ion pairs and their return to the covalently bonded state does not give rise to the common ion rate depression.[58] As long as the X^- ion remains associated with a particular carbocation in the ion pair, it is part of a species chemically distinct from the free X^- ions.

Excess of rate of racemization over rate of product formation supports the idea of ion pairs that can return to substrate; nevertheless, as it is possible that some, or even most, of the ion pairs return without racemizing (Scheme 5), considerable doubt remains about k_i, the rate of formation of the ion pairs. The difficulty is that there is no way to detect the event represented by k_i if it is followed immediately by k_{-i}. We shall know that something has happened only when k_i is followed by k_r or by k_p. If k_{-i} competes with these processes, some ionization will go undetected.

SCHEME 5

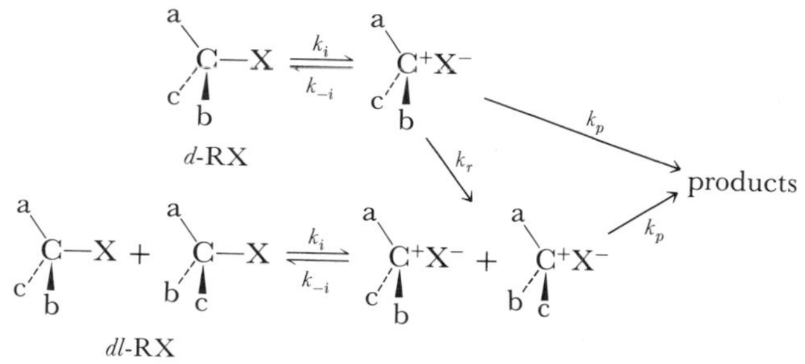

Goering and his collaborators developed a method for finding a better approximation to the ionization rate.[59] Their technique uses as leaving group an aryl ester, usually a p-nitrobenzoate. Equation 4.26 illustrates that if one oxygen of the carboxyl group in the substrate is labeled, ionization and return to the covalent state may bring about equilibration of the label between the two oxygens. The

[58] S. Winstein, E. Clippinger, A. H. Fainberg, R. Heck, and G. C. Robinson, *J. Am. Chem. Soc.*, **78**, 328 (1956).
[59] (a) H. L. Goering, R. G. Briody, and J. F. Levy, *J. Am. Chem. Soc.*, **85**, 3059 (1963); (b) H. L. Goering and H. Hopf, *J. Am. Chem. Soc.*, **93**, 1224 (1971).

$$R-O-\overset{\overset{\displaystyle *O}{\|}}{C}-Ar \longrightarrow R^+ \quad \overset{\overset{\displaystyle *O\cdots}{}}{\underset{\displaystyle O}{-\ C-Ar}} \longrightarrow R-*O-\overset{\overset{\displaystyle O}{\|}}{C}-Ar + R-O-\overset{\overset{\displaystyle *O}{\|}}{C}-Ar \qquad (4.26)$$

structural change necessary to make the two oxygens equivalent in the ion pair is undoubtedly less than that required for racemization (Scheme 5), where the cation must turn over. Indeed, in various substituted benzhydryl systems, racemization of unreacted substrate is slower than equilibration of the oxygens. The α-phenyl-ethyl systems yield similar results.[60] Oxygen equilibration thus provides a more sensitive test for ionization followed by return than does loss of chirality. It is nevertheless possible that, after the ionization but before the two oxygens become equivalent, there might be time for the oxygen originally bonded to carbon to return. Goering has shown that return without equilibration does occur in α,γ-dimethylallyl p-nitrobenzoate,[61] although equilibration is complete or nearly so in α-phenyl-γ-methylallyl p-nitrobenzoate.[62]

Contact and Solvent-Separated Ion Pairs

Winstein's work demonstrated that, in some systems at least, there is more than one kind of ion pair on the solvolysis pathway. The original evidence comes from the effect of added salts on solvolysis rates. Nearby ions affect the free energy of an ion in solution; hence a change in the concentration of dissolved salt will alter the rate of any elementary step in which ions form or are destroyed. For S_N1 solvolyses, the rate increases with addition of noncommon ion salt. In the usual solvolysis solvents, for example, acetic acid, aqueous acetone, and ethanol, the increase follows the linear equation 4.27, where k_{salt} is the rate constant with added salt and k_0 is the rate constant in the absence of salt.[63]

$$k_{salt} = k_0(1 + b[salt]) \qquad (4.27)$$

Certain systems depart from this behavior. Addition of a low concentration of a noncommon ion salt such as lithium perchlorate causes a large increase in k_t, but as more salt is added the increase levels off and finally parallels the expected linear relation. This *special salt effect*[64] is illustrated in Figure 4.7 for solvolysis of the rearranging system **20**. Note that k_α exhibits only the normal linear salt effect. A

[60] H. L. Goering, R. G. Briody, and G. Sandrock, *J. Am. Chem. Soc.*, **92**, 7401 (1970).

[61] (a) H. L. Goering, M. M. Pombo, and K. D. McMichael, *J. Am. Chem. Soc.*, **85**, 965 (1963); (b) H. L. Goering and M. M. Pombo, *J. Am. Chem. Soc.*, **82**, 2515 (1960).

[62] H. L. Goering and E. C. Linsay, *J. Am. Chem. Soc.*, **91**, 7435 (1969).

[63] (a) A. H. Fainberg and S. Winstein, *J. Am. Chem. Soc.*, **78**, 2763, 2780 (1956). Salt effects in nonpolar solvents such as ether are of dramatic magnitude and follow a more complex relationship. See S. Winstein, E. C. Friedrich, and S. Smith, *J. Am. Chem. Soc.*, **86**, 305 (1964). In water the relationship is logarithmic. (b) C. L. Perrin and J. Pressing, *J. Am. Chem. Soc.*, **93**, 5705 (1971) discuss the mechanism of the linear salt effect.

[64] S. Winstein and G. C. Robinson, *J. Am. Chem. Soc.*, **80**, 169 (1958).

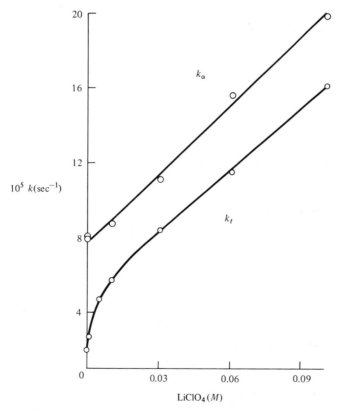

20

related phenomenon is the *induced common ion effect*. Addition of X^- ions to RX solvolyzing in the presence of lithium perchlorate may partly cancel the special salt effect rate acceleration even though X^- ions do not depress the rate in the absence of the perchlorate.[65] These results require the presence of both the contact ion pair

Figure 4.7 Effect of added $LiClO_4$ on k_a and k_t in solvolysis of *threo*-3-*p*-anisyl-2-butyl-*p*-bromobenzenesulfonate (**20**) in acetic acid. Reprinted with permission from S. Winstein and G. C. Robinson, *J. Am. Chem. Soc.*, **80**, 169 (1958). Copyright 1958 American Chemical Society.

[65] S. Winstein, P. E. Klinedinst, Jr., and G. C. Robinson, *J. Am. Chem. Soc.*, **83**, 885 (1961).

and the solvent-separated ion pair. Return from ion pair to substrate can occur from the contact ion pair (*ion-pair return* or *internal return*), from the solvent-separated ion pair (*external ion-pair return*), or from the free ions (*external ion return*). The term *external return* refers to the sum of external ion-pair return and external ion return. The special salt effect operates by diversion of the solvent-separated ion pair, probably through the mechanism shown in Equation 4.28, so that it can no longer

$$R^+ \parallel X^- + Y^- \rightleftharpoons R^+ \parallel Y^- + X^- \tag{4.28}$$

return to RX. No stable compound can form by return from an ion pair whose anion is perchlorate; the net rate of disappearance of RX therefore increases more than it would if the new salt were exerting only the normal linear salt effect. Racemization and oxygen equilibration occur through the contact ion pair and are not subject to the special salt effect.[66] The induced common ion effect arises because added X^- reduces the diversion of ion pairs to $R^+ \parallel Y^-$. When fewer ion pairs are diverted, more can return to RX, and part of the acceleration caused by Y^- is canceled. Benzhydryl systems show special salt effects with added azide through Equation 4.29; in this case the $R^+ \parallel N_3^-$ ion pair collapses to RN_3, which is stable and accumulates as one of the products.[67]

$$R^+ \parallel X^- + N_3^- \rightleftharpoons R^+ \parallel N_3^- + X^- \tag{4.29}$$

Other more recent results have confirmed the importance of both contact and solvent-separated ion pairs. For example, solvolysis of chiral 2-adamantyl derivatives yields a slight excess of retention of configuration,[68] a result consistent with product formation from a solvent-separated ion pair but not from a contact ion pair. Furthermore, the ion pairs in 2-adamantyl tosylate solvolysis in ethanol–water mixtures react faster with water than with ethanol, even though ethanol is more nucleophilic and usually reacts faster than water. The interpretation of this result is that with tosylate counter-ion, the solvent-separated ion pair with water between the ions is slightly favored over that with ethanol because water provides more hydrogen-bonding sites for stabilization of the oxygens of the sulfonate.[69]

In solvolysis of deuteriohomoadamantyl tosylate (**21**), two types of rearrangements can take place in the ion **22**, one a hydride shift to **23** and the other an alkyl shift to **24**. Each yields a homoadamantyl ion with a characteristic location of

[66] In certain favorable cases of rearranging systems, the occurrence of internal return distinct from external ion-pair return can be demonstrated without recourse to optical rotation or isotopic labeling experiments. See S. Winstein and A. H. Fainberg, *J. Am. Chem. Soc.*, **80**, 459 (1958); S. Winstein, P. E. Klinedinst, Jr., and E. Clippinger, *J. Am. Chem. Soc.*, **83**, 4986 (1961).

[67] See note 56(b).

[68] J. A. Bone and M. C. Whiting, *Chem. Commun.*, 115 (1970).

[69] (a) J. M. Harris, A. Becker, J. F. Fagan, and F. A. Walden, *J. Am. Chem. Soc.*, **96**, 4484 (1974). See also (b) F. G. Bordwell and T. G. Mecca, *J. Am. Chem. Soc.*, **97**, 123, 127 (1975); (c) F. G. Bordwell, P. F. Wiley, and T. G. Mecca, *J. Am. Chem. Soc.*, **97**, 132 (1975); (d) H. Aronovitch and A. Pross, *Tetrahedron Lett.*, 2729 (1977); (e) T. Ando and S. Tsukamoto, *Tetrahedron Lett.*, 2775 (1977).

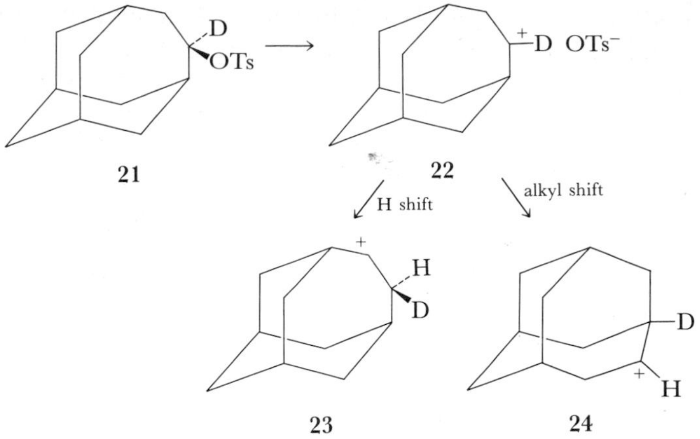

the deuterium label. Elimination occurs from an intermediate (probably a contact ion pair) in which only the faster alkyl rearrangement has occurred; substitution products arise from another intermediate (probably a solvent-separated ion pair) in which the slower hydrogen migration has also occurred.[70]

Shiner has interpreted α deuterium isotope effects in the high part of their range (~ 1.2 for sulfonate leaving groups) as indicating rate-determining conversion from contact to solvent-separated ion pair.[71] This interpretation has, however, been questioned.[72]

4.4 INFLUENCE OF SOLVENT, NUCLEOPHILE, LEAVING GROUP, AND SUBSTRATE STRUCTURE

The effects of changes in reaction parameters vary considerably over the range of mechanisms possible for aliphatic nucleophilic substitution. Our discussion will focus primarily on the two extremes of that range, the traditional S_N2 reaction with direct substitution by an external nucleophile and S_N1 ionization to an ion pair with no nucleophilic assistance.

The Solvent

The solvent plays two types of roles in nucleophilic substitutions. It is the medium for the reaction, and affects it by solvating the starting compounds and transition state, and it also frequently itself acts as nucleophile. We will concentrate mainly

[70] J. E. Nordlander, J. B. Hamilton, Jr., F. Y.-H. Wu, S. P. Jindal, and R. R. Gruetzmacher, *J. Am. Chem. Soc.*, **98**, 6658 (1976).

[71] (a) R. C. Seib, V. J. Shiner, Jr., V. Sendijarević, and K. Humski, *J. Am. Chem. Soc.*, **100**, 8133 (1978); (b) V. J. Shiner, Jr., in *Isotope Effects in Chemical Reactions*, C. J. Collins and N. S. Bowman, Eds., Van Nostrand Reinhold, 1970.

[72] T. W. Bentley and P. v. R. Schleyer, *Adv. Phys. Org. Chem.*, **14**, 1 (1977).

on the former role here and consider the latter further when we take up nucleophiles.

The effect of solvent polarity on the reaction rate depends in part on the relative charge densities in the starting material and in the transition state. If the starting materials have a high charge density but in the transition state the charge is already dispersed, a more polar solvent should lower the energy of the starting material more than the energy of the transition state. The result would be an increase in activation energy and a decrease in rate. This situation is shown in Figure 4.8(a). On the other hand, if the transition state has a higher charge density than the starting material, increasing solvent polarity should lower the activation energy and increase the rate [Figure 4.8(b)].

S_N2 reactions S_N2 reactions of charge types 1, 3, and 4 (Table 4.1) all have more highly dispersed charge in the transition state than in the ground state.[73] The effect on them of changing solvent polarity is therefore described by Figure 4.8(a); an increase in solvent polarity should be accompanied by a decrease in reaction rate.

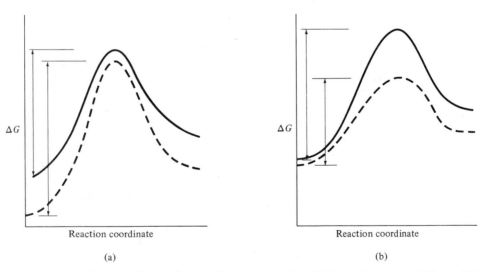

(a) (b)

Figure 4.8 (a) Reaction coordinate diagram for a reaction in which starting material has a higher charge density than transition state. (b) Reaction coordinate diagram for a reaction in which the transition state has a higher charge density than the starting materials. Solid line indicates less polar solvent; dashed line indicates more polar solvent.

[73] For reviews of solvent effects on S_N2 reactions, see (a) A. J. Parker, *Adv. Phys. Org. Chem.*, **5**, 173 (1967); (b) A. J. Parker, *Chem. Rev.*, **69**, 1 (1969).

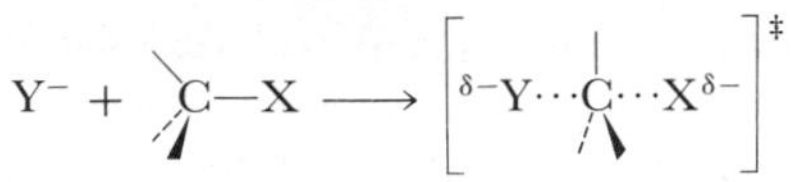

$$Y^- + \underset{\diagup}{\overset{\diagdown}{C}}-X \longrightarrow \left[\delta^- Y\cdots\underset{\diagup|\diagdown}{C}\cdots X^{\delta-} \right]^{\ddagger} \qquad \text{Charge type 1}$$

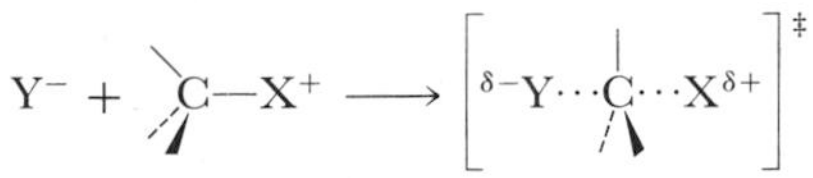

$$Y^- + \underset{\diagup}{\overset{\diagdown}{C}}-X^+ \longrightarrow \left[\delta^- Y\cdots\underset{\diagup|\diagdown}{C}\cdots X^{\delta+} \right]^{\ddagger} \qquad \text{Charge type 3}$$

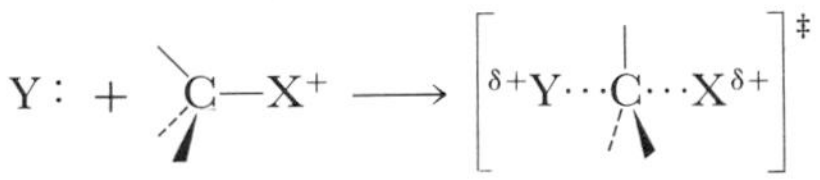

$$Y: + \underset{\diagup}{\overset{\diagdown}{C}}-X^+ \longrightarrow \left[\delta^+ Y\cdots\underset{\diagup|\diagdown}{C}\cdots X^{\delta+} \right]^{\ddagger} \qquad \text{Charge type 4}$$

The only S_N2 reactions in which the transition states have a higher charge density than the starting materials are those of charge type 2 in which two neutral starting materials produce a dipolar transition state.

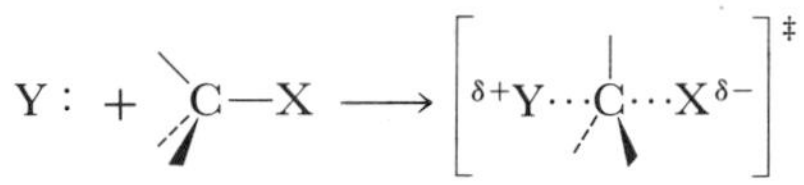

$$Y: + \underset{\diagup}{\overset{\diagdown}{C}}-X \longrightarrow \left[\delta^+ Y\cdots\underset{\diagup|\diagdown}{C}\cdots X^{\delta-} \right]^{\ddagger} \qquad \text{Charge type 2}$$

Only these, then, would be expected to show a rate increase when run in a more polar solvent.

These predictions of effect of solvent polarity on reaction rates were first made by Hughes and Ingold in 1935. They searched the literature of direct displacement reactions and found that for charge types 1–3 the experimental facts agreed with their predictions. For example, the reaction of ethyl iodide with triethylamine (Equation 4.30) an S_N2 displacement of type 2, does proceed more rapidly

$$(CH_3CH_2)_3N: + CH_3CH_2I \longrightarrow (CH_3CH_2)_4N^+ + I^- \tag{4.30}$$

in alcohols than in hydrocarbons;[74] and on the other hand, both the rate of bromine exchange between radioactive bromide ion and n-butyl bromide in acetone (Equation 4.31), a substitution of charge type 1, and the rate of alkaline hydrolysis

$$*Br^- + BrCH_2CH_2CH_2CH_3 \longrightarrow Br^*CH_2CH_2CH_2CH_3 + Br^- \tag{4.31}$$

of trimethylsulfonium ion (Equation 4.32) in water–methanol (charge type 3), are slower if water is added to the solvent.[75]

$$OH^- + (CH_3)_3S^+ \longrightarrow CH_3OH + (CH_3)_2S \tag{4.32}$$

[74] N. Menshutkin, *Z. Physik. Chem.*, **5**, 589 (1890).
[75] (a) L. J. le Roux and S. Sugden, *J. Chem. Soc.*, 1279 (1939); (b) L. J. le Roux, C. S. Lu, S. Sugden, and R. H. K. Thomson, *J. Chem. Soc.*, 586 (1945); (c) Y. Pocker and A. J. Parker, *J. Org. Chem.*, **31**, 1526 (1966).

More recently, the theory has been shown to hold for reactions of charge type 4 also. The reaction of trimethylamine with trimethylsulfonium ion (Equation 4.33) proceeds more rapidly in nonpolar than in polar solvents.[76]

$$(CH_3)_3N + (CH_3)_3S^+ \longrightarrow (CH_3)_4N^+ + (CH_3)_2S \tag{4.33}$$

Most of the early work on solvent effects was done with protic solvents. Dipolar aprotic solvents often accelerate S_N2 reactions dramatically, and it is of interest to analyze these effects.[77] Thermodynamics of the solvent effects are analyzed in terms of free energies of transfer of substrate, nucleophile, and transition state from a reference solvent to the solvent of interest.[78] (Free energy of transfer of the transition state is determined from the free energies of activation in the two solvents and the free energies of transfer of the reactants.) Small ions with localized charge, such as OH^-, F^-, Cl^-, which are good hydrogen bond acceptors, are much better solvated by protic than by aprotic solvents; they have large positive ΔG_{tr} in the direction protic → aprotic (see, for example, Cl^- in Table 4.4). For large polarizable anions, and large polarizable neutrals, ΔG_{tr} is closer to zero or even negative. In an S_N2 reaction of charge type 1, a concentrated charge on Y^- is dispersed in the transition state. Table 4.5 shows that $\Delta G_{tr}(Y^-)$ is large and positive, $\Delta G_{tr}(RX)$ is slightly negative, and $\Delta G_{tr}(\ddagger)$, the free energy of transfer of the transition state, is slightly positive. Therefore $\Delta G^{\ddagger}$ for the reaction will be smaller in the dipolar aprotic solvent, largely as a result of the poor solvation of the nucleophile, as

Table 4.4 FREE ENERGIES OF TRANSFER OF SOME IONS FROM METHANOL TO DIPOLAR APROTIC SOLVENTS[a]

Ion	$\Delta G_{tr}^{CH_3OH \rightarrow solvent}$ (Y^-)			
	$HCON(CH_3)_2$	H_3CSOCH_3	H_3CCN	$((H_3C)_2N)_3PO$
Cl^-	8.9	7.5	8.6	12.0
Br^-	6.7	4.9	5.7	9.7
I^-	3.5	1.8	3.3	—
N_3^-	6.7	4.8	6.4	9.8
CN^-	8.5	—	—	—
SCN^-	3.7	1.9	3.5	4.6
Picrate$^-$	−0.55	—	1.5	—
I_3^-	−2.0	−4.9	−0.55	—
Ag^+	−7.0	−11.1	−8.6	−13.6
Na^+	−5.3	−4.9	1.9	−8.6
$(H_3C)_3S^+$	−4.2	—	−2.2	—

[a] ΔG_{tr} in kcal mole^{-1}, based on the assumption $\Delta G_{tr}(\phi_4As^+) = \Delta G_{tr}(\phi_4B^-)$. Adapted with permission from A. J. Parker, *Chem. Rev.*, **69**, 1 (1969). Copyright 1969 American Chemical Society.

[76] E. D. Hughes and D. J. Whittingham, *J. Chem. Soc.*, 806 (1960). For other examples see (b) note 72a; (c) C. K. Ingold, *Structure and Mechanism in Organic Chemistry*, 2d ed., Cornell University Press, Ithaca, N.Y., 1969, pp. 457–463.
[77] Reviews: (a) A. J. Parker, *Chem. Rev.*, **69**, 1 (1969); (b) M. H. Abraham, *Prog. Phys. Org. Chem.*, **11**, 1 (1974).
[78] Conventions used in determining thermodynamics of transfer of ions are discussed in Section 2.4, p. 168. The data are sometimes expressed in terms of an activity coefficient of transfer. See note 58 in Chapter 2.

Table 4.5 Free Energies of Transfer of Substrate, Nucleophile, and Transition State from Methanol to Dipolar Aprotic Solvents[a]

Reaction (n')[b]	Solvent[c]	$\log k_S/k_{CH_3OH}$	$\Delta G_{tr}^{CH_3OH \to S}$ (RX)	$\Delta G_{tr}^{CH_3OH \to S}$ (Y⁻)	$\Delta G_{tr}^{CH_3OH \to S}$ (‡)
$CH_3I + Cl^-$	DMF	5.9	−0.68	8.9	0.14
$(n' = 1.30)$	ACN	4.6	−0.55	8.6	1.8
$CH_3I + Br^-$	DMF	4.2	−0.68	6.7	0.27
$(n' = 1.0)$					
$CH_3I + SCN^-$	DMF	2.2	−0.68	3.7	0.0
$(n' = 0.45)$	ACN	1.4	−0.55	3.5	1.1
$CH_3I + N_3^-$	DMF	4.6	−0.68	6.7	−0.27
$CH_3I + CH_3O^-$	DMSO	5.4	−0.68	7.2	−0.82
$CH_3I + CN^-$	DMF	5.7	−0.68	8.5	0.0

[a] ΔG_{tr} in kcal mole⁻¹. For ions, based on the assumption $\Delta G_{tr}(\phi_4As^+) = \Delta G_{tr}(\phi_4B^-)$. Adapted with permission from A. J. Parker, *Chem. Rev.*, **69**, 1 (1969). Copyright 1969 American Chemical Society.
[b] n' defined by

$$\Delta G_{tr}^{\ddagger}(S_N2) = n' \Delta(AN) - \Delta G_{tr}(RX)$$

A. J. Parker, U. Mayer, R. Schmid, and V. Gutmann, *J. Org. Chem.*, **43**, 1843 (1978).
[c] DMF = $HCON(CH_3)_2$; ACN = H_3CCN; DMSO = H_3CSOCH_3.

reflected in $\Delta G_{tr}(Y^-)$. Similar analysis for the other charge types shows lower sensitivity to transfer from protic to dipolar aprotic solvent. For example, in reactions of charge type 3 (Y⁻ + RX⁺) the positive $\Delta G_{tr}(Y^-)$ is partly offset by a negative $\Delta G_{tr}(RX^+)$, because large polarizable cations tend to be better solvated in dipolar aprotic than in protic solvents.[79]

Parker, Gutmann, and co-workers have analyzed the relationship between $\Delta G_{tr}^{\ddagger}$ for S_N2 reactions of charge type 1 and solvent donor numbers (DN) and acceptor numbers (AN).[80] The changes in free energies of activation of S_N2 reactions were found to obey Equation 4.34. $\Delta(AN)$ is the change of acceptor number.

$$\Delta G_{tr}^{\ddagger}(S_N2) = n' \Delta(AN) - \Delta G_{tr}(RX) \tag{4.34}$$

The quantity $\Delta(AN)$ is itself linearly related to $\Delta G_{tr}(Cl^-)$; an alternative formulation is therefore Equation 4.35. Values of n' are indicated for three of the reactions

$$\Delta G_{tr}^{\ddagger}(S_N2) = -n \Delta G_{tr}(Cl^-) - \Delta G_{tr}(RX) \tag{4.35}$$

in Table 4.5. Note that the smaller the anion, the larger is the effect on reaction rate of transfer to an aprotic solvent. An approximate relation that is satisfactory for semiquantative predictions is Equation 4.36. This remarkably simple equation

[79] See note 77(a).
[80] A. J. Parker, U. Mayer, R. Schmid, and V. Gutmann, *J. Org. Chem.*, **43**, 1843 (1978). The DN and AN parameters are defined in Section 2.4, p. 174.

dramatizes the importance of anion solvation as a controlling factor in reactions of this type.

$$\Delta G_{tr}^{\ddagger}(S_N2) \approx +n' \Delta(AN) \approx -n \Delta G_{tr}(Cl^-) \tag{4.36}$$
$$\text{(charge type 1)}$$

This kind of analysis provides a probe for transition state structure. The looser the transition state in an S_N2 reaction, the more negative charge is localized on Y and X. (Recall that the limit of a loose transition state would be $Y^- + R^+ + X^-$). The looser the transition state, the less well it is solvated in a dipolar aprotic solvent compared with a protic solvent; $\Delta G_{tr}(\ddagger)$, protic $\rightarrow$ aprotic, should become more positive with looser transition states. Table 4.6 gives an example.

S_N1 reactions As we have noted earlier, the Grunwald–Winstein m–Y correlation has been the relation used most commonly for analysis of solvent effects in reactions of the S_N1 type.[81] Recall that the standard reaction is solvolysis of t-butyl chloride and the standard solvent is 80 percent aqueous ethanol; for that combination $m = 1$ and $Y = 1$. According to Equation 4.37, one establishes the Y parame-

$$\log \frac{k_S}{k_{80\% \text{ EtOH}}} = mY \tag{4.37}$$

ters characteristic of various solvent mixtures as the logarithm of the ratio of t-butyl chloride solvolysis rate in the solvent S to rate in 80 percent ethanol; then a plot of this rate ratio for some other substrate vs. Y will give the slope m for that substrate. As we have seen earlier, m can be used as a criterion of mechanism, because a substrate solvolyzing with nucleophilic assistance of the solvent will have reduced separation of positive and negative charge at the transition state and so will require less assistance from ionizing power of solvent and hence will have m less than unity. Note, however, that unusually early transition states, as might occur if ionization were accompanied by relief of steric strain, or if there were a particularly good leaving group, could have the same effect in the absence of any nucleophilic assistance.[82] Our earlier discussion indicated that there may be some nucleophilic assistance even in t-butyl halide solvolysis in alcohol solvents; therefore, a solvent-ionizing power scale based on 2-adamantyl tosylate instead of on t-butyl chloride has advantages. This parameter is designated Y_{OTs}. Table 4.7 lists values of Y and Y_{OTs} for several pure solvents and solvent mixtures.

Other solvent parameters based on the influence of solvent on electronic excitation energies have been developed by Kosower,[83] Smith, Fainberg, and Winstein;[84] and Dimroth and co-workers.[85]

[81](a) E. Grunwald and S. Winstein, *J. Am. Chem. Soc.*, **70**, 846 (1948); (b) A. H. Fainberg and S. Winstein, *J. Am. Chem. Soc.*, **78**, 2770 (1956).

[82](a) J. S. Lomas and J.-E. Dubois, *J. Org. Chem.*, **40**, 3303 (1975); (b) X. Creary, *J. Am. Chem. Soc.*, **98**, 6608 (1976).

[83](a) E. M. Kosower, *J. Am. Chem. Soc.*, **80**, 3253, 3261, 3267 (1958); (b) E. M. Kosower, *An Introduction to Physical Organic Chemistry*, Wiley, New York, 1968, p. 295; solvation is discussed in detail beginning on p. 260.

[84]S. G. Smith, A. H. Fainberg, and S. Winstein, *J. Am. Chem. Soc.*, **83**, 618 (1961).

[85]K. Dimroth, C. Reichardt, T. Siepmann, and F. Bohlmann, *Justus Liebigs Ann. Chem.*, **661**, 1 (1963).

Table 4.6 FREE ENERGY OF TRANSFER OF THE S_N2 TRANSITION STATE FROM ETHANOL TO DIMETHYLFORMAMIDE[a]

Reaction	Z	$\Delta G_{tr}^{EtOH \to DMF}$ ($\ddagger$)
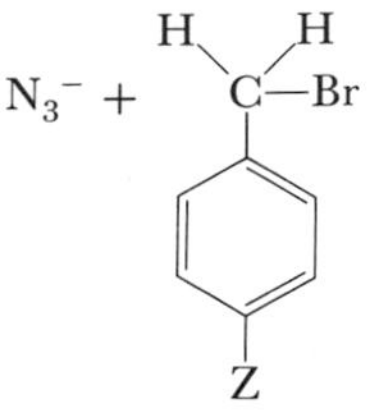	NO$_2$	0.27
N$_3^-$ + (see structure)	H	3.0
	OCH$_3$	4.1

Structure: N_3^- + a carbon bearing two H's and bonded to Br, attached to a benzene ring with substituent Z at the para position.

[a] In kcal mole^{-1}. Adapted with permission from A. J. Parker, *Chem. Rev.*, **69**, 1 (1969). Copyright 1969 American Chemical Society.

Table 4.7 VALUES OF SOLVENT IONIZING POWER PARAMETERS Y AND Y_{OTs} AND NUCLEOPHILICITY PARAMETERS N AND N_{OTs}

Solvent	Y^a	N^b	$Y_{OTs}^{\ b}$	$N_{OTs}^{\ b}$
F$_3$CCOOH	1.84[b]	−4.74	4.57	−5.56
H$_2$O	3.49	−0.26	4.0[c]	−0.41[c]
F$_3$CCHOHCF$_3$–H$_2$O				
97:3	2.46[b]	−3.93	3.61	−4.27
F$_3$CCH$_2$OH	1.04[b]	−2.78	1.80	−3.0
HCOOH	2.05	−2.05	3.04	−2.35
HCOOH–H$_2$O				
80:20	2.32			
50:50	2.64			
25:75	3.10			
H$_3$CCOOH	−1.64	−2.05	−0.61	−2.35
H$_3$CCOOH–H$_2$O				
50:50	1.94			
25:75	2.84			
C$_2$H$_5$OH	−2.03	0.09	−1.75	0.00
C$_2$H$_5$OH–H$_2$O				
80:20	0.00	0.00	0.00	0.00
50:50	1.66	−0.20	1.29	−0.09
20:80	3.05			
H$_3$CCOCH$_3$–H$_2$O				
90:10	−1.86	−0.43		
50:50	1.40	−0.44		
20:80	2.91			
Dioxane–H$_2$O				
90:10	−2.03	−0.65		
50:50	1.36	−0.39		
20:80	2.88			

[a] Reprinted in part with permission from A. H. Fainberg and S. Winstein, *J. Am. Chem. Soc.*, **78**, 2770 (1956). Copyright 1956 American Chemical Society.
[b] Reprinted in part with permission from F. L. Schadt, T. W. Bentley, and P. v. R. Schleyer, *J. Am. Chem. Soc.*, **98**, 7667 (1976). Copyright 1976 American Chemical Society.
[c] T. W. Bentley and P. v. R. Schleyer, *Adv. Phys. Org. Chem.*, **14**, 1 (1977).

The Grunwald–Winstein **Y** scale is purely empirical and gives relatively little insight into the reasons for the observed solvent effects. Abraham, by looking at free energies of transfer of substrates and transition states from methanol to methanol–water to water, and comparing the results to those for ion pairs, has demonstrated the polar character of the S_N1 transition state for solvolysis of *t*-butyl chloride in contrast to the much less polar transition state for reaction of trimethylamine with methyl iodide. This work also shows that the rate increase observed for typical S_N1 reactions on increasing the proportion of water in the solvent is due partly to improved solvation of the polar transition state and partly to increase of free energy of the nonpolar substrate.[86]

Parker and Gutmann's relation shown in Equation 4.38 allows donor properties of the solvent to be taken into account.[80] For solvolysis of *t*-butyl chloride,

$$\Delta G^{\ddagger}_{tr}(S_N1) = -p'\,\Delta(DN) - n'\,\Delta(AN) - \Delta G_{tr}(RX) \tag{4.38}$$

or

$$\Delta G^{\ddagger}_{tr}(S_N1) = p\,\Delta G_{tr}(K^+) + N\,\Delta G_{tr}(Cl^-) - \Delta G_{tr}(RX) \tag{4.39}$$

$p' = 0.28$ and $n' = 0.65$. Solvation of the anionic leaving group is the most important contributor. The solvent owes its ionizing power more to its ability to provide electrophilic solvation of the leaving group than to its ability to solvate the incipient cation. In solvolysis of 2-methyl-2-(*p*-methoxyphenyl)propyl tosylate (*p*-methoxyneophyl tosylate) (**25**), a rearranging system that reacts by a k_Δ process

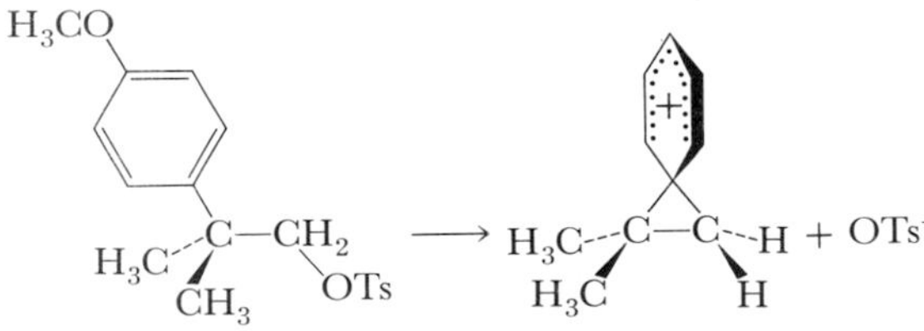

$$\tag{4.40}$$

and hence needs no assistance by external nucleophile even though it is primary, there is no dependence on DN, and rate is correlated by the same simple relation that works for S_N2 reactions, but with opposite sign: Equation 4.41 with $n' = 0.52$.

$$\Delta G^{\ddagger}_{tr} = -n'\,\Delta(AN) \tag{4.41}$$

The cation **26** has charge delocalized over such a large volume, and hence has so low an electric field, that its solvation becomes unimportant compared with that of the sulfonate leaving group with its negative charge relatively concentrated on the oxygens.

[86](a) M. H. Abraham and G. F. Johnston, *J. Chem. Soc. A*, 1610 (1971); (b) M. H. Abraham, *Prog. Phys. Org. Chem.*, **11**, 1 (1974).

The Nucleophile in S_N2 Reactions[87]

In an S_N2 reaction the role of the entering Lewis base is to use its unshared pair of electrons to push away the leaving Lewis base with its bonding pair. Thus a good nucleophile is one that readily donates its unshared pair to the substrate, allowing rapid reaction. If S_N2 reactions on carbon only are considered, a reagent that is a good nucleophile for one substrate is usually a good nucleophile for all substrates *in the same type of solvent*. Swain and Scott proposed that the nucleophilicity of a reagent can be represented by a constant value, n, which holds for carbon S_N2 reactions in protic solvents in general. The rate of an S_N2 reaction in a protic solvent can be predicted quantitatively if the n value of the attacking reagent and a second parameter, s, which represents the sensitivity of the substrate to the reagent's nucleophilicity, are known.[88] The quantitative relationship between these two parameters and the rate is the linear free-energy relationship:

$$\log \frac{k}{k_0} = ns \tag{4.42}$$

in which k is the rate of an S_N2 reaction in which the nucleophile has nucleophilicity n and the substrate has sensitivity s. The constant k_0 is the rate constant for the reaction of methyl bromide with water at 25°C. The substrate methyl bromide is arbitrarily assigned an s value of 1. The parameter n for a given nucleophile, Y, then is defined by the following equation:

$$n = \log \frac{k_{CH_3Br+Y}}{k_{CH_3Br+H_2O}} \tag{4.43}$$

To determine n directly for a specific nucleophile, the rate constant for its reaction with methyl bromide is measured at 25°C; the logarithm of the ratio of this rate constant to the rate constant for water with methyl bromide gives n. If Y is a better nucleophile than water, n will be positive; if Y is a worse nucleophile, n will be negative. Water itself has an n value of zero. In Table 4.8 are listed the majority of the n values that have been determined. Since so few are known, they are of limited usefulness. Table 4.9, however, gives some of the more plentiful, analogous n_{CH_3I} values,[89] which are defined in just the same way as n values except that the arbi-

---

[87] Reviews: (a) R. G. Pearson, H. Sobel, and J. Songstad, *J. Am. Chem. Soc.*, **90**, 319 (1968); (b) K. M. Ibne-Rasa, *J. Chem. Educ.*, **44**, 89 (1967); (c) A. J. Parker, *Chem. Rev.*, **69**, 1 (1969).

[88] (a) C. G. Swain and C. B. Scott, *J. Am. Chem. Soc.*, **75**, 141 (1953). Edwards has proposed a two-parameter equation for S_N2 rates;

$$\log \frac{k}{k_0} = \alpha E_n + \beta H$$

where $H = pK_{a_{HY}} + 1.74$ and E_n is a measure of polarizability. See (b) J. O. Edwards, *J. Am. Chem. Soc.*, **76**, 1540 (1954); (c) J. O. Edwards, *J. Am. Chem. Soc.*, **78**, 1819 (1956). For another theoretical treatment of nucleophilic reactivities, see (d) R. F. Hudson, *Chimia*, **16**, 173 (1962).

[89] R. G. Pearson, H. Sobel, and J. Songstad, *J. Am. Chem. Soc.*, **90**, 319 (1968).

Table 4.8 NUCLEOPHILIC CONSTANTS (n VALUES)[a]

Nucleophile	n	Nucleophile	n
H_2O	0.00	Br^-	3.89
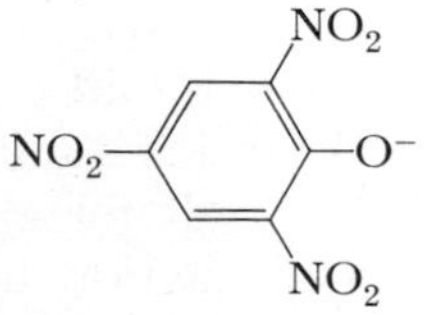		N_3^-	4.0
(2,4,6-trinitrophenolate) $-O^-$	1.9	$NH_2-\overset{\displaystyle S}{\underset{\displaystyle \parallel}{C}}-NH_2$	4.1
		HO^-	4.20
$CH_3-C\overset{\displaystyle O}{\underset{\displaystyle O^-}{}}$	2.72	(phenyl)$-NH_2$	4.49
Cl^-	3.04	I^-	5.04
		HS^-	5.1
(pyridine) N	3.6	$S_2O_3^{2-}$	6.36

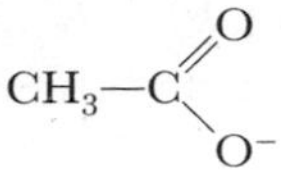

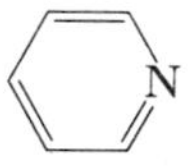

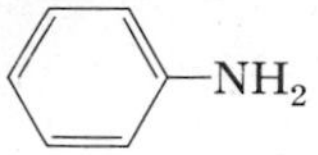

[a] Reprinted with permission from C. G. Swain and C. B. Scott, *J. Am. Chem. Soc.*, **75,** 141 (1953). Copyright by the American Chemical Society.

tarily chosen standard reaction is the displacement on CH_3I by methanol in methanol solvent.[90] Thus

$$n_{CH_3I} = \log \frac{k_{CH_3I+Y}}{k_{CH_3I+CH_3OH}} \tag{4.44}$$

The s parameter for a particular substrate RX is determined by measuring the rate of a number of S_N2 reactions of RX with nucleophiles of known n. Log k/k_0 is determined for each reaction, and these values are plotted against the corresponding n values. The best straight line that can be drawn through those points has slope s. A substrate that is more dependent than methyl bromide on the nucleophilicity of the attacking group will have an s value greater than 1, and one that is less dependent will have a smaller value. Table 4.10 lists s values for a few substrates.

The first four compounds have lower s values than methyl bromide because each is quite reactive in itself and therefore is not very dependent on help from the nucleophile: p-toluenesulfonate is a good leaving group and does not need much assistance to begin to depart; ring strain in propiolactone and in the mustard cation make a ring-opening S_N2 reaction very favorable; and the transition state of benzyl chloride is stabilized by resonance and therefore is easily reached.

We have seen earlier that the solvent plays an important nucleophilic role in many reactions. An early attempt to establish nucleophilicities for solvents was made by Swain, Moseley, and Bown.[91] They proposed the correlation given in

[90] There is a linear relationship between Swain's n values and Pearson's n_{CH_3I} values: $n_{CH_3I} = 1.4n$.
[91] C. G. Swain, R. B. Mosely, and D. E. Bown, *J. Am. Chem. Soc.*, **77,** 3731 (1955).

Table 4.9 Nucleophilic Constants (n_{CH_3I} Values)[a]

Nucleophile	n_{CH_3I}	n_{Pt}^b	pK$_a$ of Conjugate acid in methanol
CH_3OH	0.0	0.0	-1.7
NO_3^-	(1.5)	—	-1.3
CO	<2.0	<2.0	—
$C_6H_{11}NC$	<2.0	6.34	—
ϕ_3Sb	<2.0	6.79	—
F^-	~2.7	<2.2	3.45
SO_4^{2-}	3.5	—	2.0
$SnCl_3^-$	~3.84	5.44	—
$CH_3\overset{O}{\overset{\|}{C}}{-}O^-$	4.3	<2.0	4.75
Cl^-	4.37	3.04	-5.7
$\phi{-}\overset{O}{\overset{\|}{C}}{-}O^-$	4.5	—	4.19
ϕ_3As	4.77	6.89	—
$(\phi CH_2)_2S$	4.84	3.43	—
Imidazole	4.97	3.44	7.10
ϕSO_2NH^-	5.1	—	8.5
$(CH_3O)_3P$	~5.2	7.23	—
$(\phi CH_2)_2Se$	5.23	5.53	—
Pyridine	5.23	3.19	5.23
$(CH_3CH_2)_2S$	5.34	4.52	—
NO_2^-	5.35	3.22	3.37
 (thiane, S-ring)	5.42	5.02	—
NH_3	5.50	3.07	9.25
$(CH_3)_2S$	5.54	4.87	-5.3
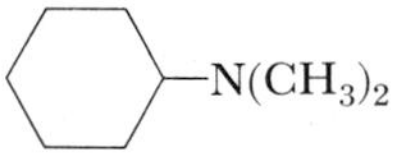 (cyclohexyl)$-N(CH_3)_2$	5.64	—	5.06
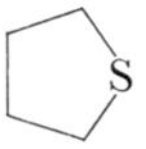 (tetrahydrothiophene, S-ring)	5.66	5.14	-4.8
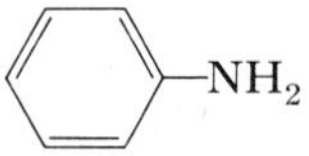 (phenyl)$-NH_2$	5.70	3.16	4.58
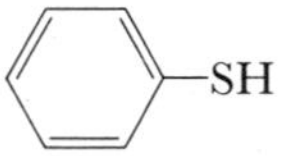 (phenyl)$-SH$	5.70	4.15	—
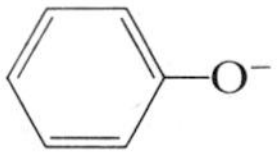 (phenyl)$-O^-$	5.75	—	9.89
N_3^-	5.78	3.58	4.74
Br^-	5.79	4.18	-7.77
CH_3O^-	6.29	<2.4	15.7
$(CH_3)_2Se$	6.32	5.70	—

Table 4.9 (*Continued*)

Nucleophile	n_{CH_3I}	$n_{Pt}{}^b$	pK_a of Conjugate acid in methanol
NH_2OH	6.60	3.85	5.82
NH_2NH_2	6.61	3.86	7.93
$(CH_3CH_2)_3N$	6.66	—	10.7
SCN^-	6.70	5.75	−0.7
CN^-	6.70	7.14	9.3
ϕSO_2NCl^-	6.8	—	3.0
$(C_2H_5)_3As$	6.90	7.68	2.6
ϕ_3P	7.00	8.93	2.73
$(C_2H_5)_2NH$	~7.0	—	11.0
$(CH_3O)_2PO^-$	7.00	5.01	—
pyrollidene	7.23	—	11.27
$SC(NH_2)_2$	7.27	7.17	−0.96
Piperidine	7.30	3.13	11.21
I^-	7.42	5.46	−10.7
$SeCN^-$	7.85	7.11	—
(HS^-)	(8.)	—	7.8
$SO_3{}^{2-}$	8.53	5.79	7.26
$[(C_2H_5)_2N]_3P$	8.54	4.54	—
$(C_4H_9)_3P$	8.69	8.96	8.43
$(C_2H_5)_3P$	8.72	8.99	8.69
$S_2O_3{}^{2-}$	8.95	7.34	1.9
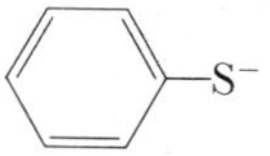 ⬡—S^-	9.92	7.17	6.52
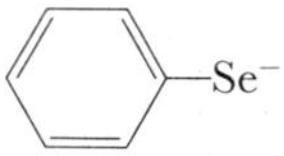 ⬡—Se^-	~10.7	—	—

^{a} Reprinted with permission from R. G. Pearson, H. Sobel and J. Songstad, *J. Am. Chem. Soc.*, **90**, 319 (1968). Copyright by the American Chemical Society.

$$n_{Pt} = \log \frac{k_{Pt(Py)_2Cl_2 + Y}}{k_{Pt(Py)_2Cl_2 + CH_3OH}}$$

Equation 4.45, where c_1, c_2 are parameters characteristic of the substrate and d_1, d_2

$$\log \frac{k}{k_0} = c_1 d_1 + c_2 d_2 \qquad (4.45)$$

are parameters characteristic of the solvent. The Swain–Moseley–Bown equation was not extensively used because the choice of standards left each parameter with contributions from both ionizing power and nucleophilicity and so the values obtained had no direct mechanistic significance. Several investigators have subsequently proposed relationships of the form of Equation 4.45 but parameterized so

Table 4.10 SUBSTRATE PARAMETERS FOR
NUCLEOPHILIC ATTACK[a]

Substrate		s
CH$_3$CH$_2$OTs	(Ethyl toluenesulfonate)	0.66
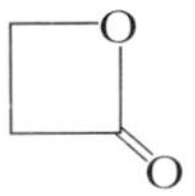	(β-propionolactone)	0.77
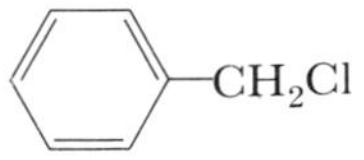 —CH$_2$Cl	(Benzyl chloride)	0.87
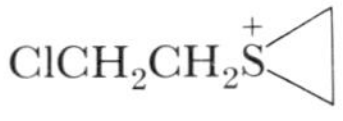 ClCH$_2$CH$_2$S	(Mustard cation)	0.95
CH$_3$Br	(Methyl bromide)	1.00

[a] Reprinted with permission from C. G. Swain and C. B. Scott, *J. Am. Chem. Soc.*, **75**, 141 (1953). Copyright by the American Chemical Society.

as to separate solvent nucleophilicity from ionizing power, and so as to be consistent with the Grunwald–Winstein equation.[92] These correlations use Equation 4.46,

$$\log \frac{k}{k_0} = l\mathbf{N} + m\mathbf{Y} \tag{4.46}$$

where **N** represents the nucleophilicity of solvent, l is sensitivity of substrate to solvent nucleophilicity, and m and **Y** have the same significance as in the Grunwald–Winstein equation. In the formulation of Schleyer and co-workers, the parameter l is unity for methyl tosylate; **N** values are obtained for various solvents through Equation 4.47, where the term 0.3**Y** corrects for the sensitivity of methyl

$$\mathbf{N} = \log \left(\frac{k}{k_0}\right)_{\text{CH}_3\text{OTs}} - 0.3\mathbf{Y} \tag{4.47}$$

tosylate solvolysis rate to solvent ionizing power.[93] A scale based on 2-adamantyl tosylate solvolysis uses **Y**$_{\text{OTs}}$. Table 4.7 (p. 329) includes values of **N** and **N**$_{\text{OTs}}$ for various solvents. Values of l and m from Equation 4.46 are given for some typical solvolyses in Table 4.11. Peterson and co-workers have derived the relation of the Swain c and d parameters to those of Equation 4.46.[94]

---

[92](a) P. E. Peterson and F. J. Waller, *J. Am. Chem. Soc.*, **94**, 991 (1972); (b) T. W. Bentley, F. L. Schadt, and P. v. R. Schleyer, *J. Am. Chem. Soc.*, **94**, 992 (1972).

[93](a) F. L. Schadt, T. W. Bentley, and P. v. R. Schleyer, *J. Am. Chem. Soc.*, **98**, 7667 (1976). The 0.3Y correction has been criticized: (b) D. A. Kevill and G. M. L. Lin, *J. Am. Chem. Soc.*, **101**, 3916 (1979).

[94]P. E. Peterson, D. W. Vidrine, F. J. Waller, P. M. Henrichs, S. Magaha, and B. Stevens, *J. Am. Chem. Soc.*, **99**, 7968 (1977).

Table 4.11 VALUES OF SOLVENT SENSITIVITY
PARAMETERS ℓ AND m FOR
SOLVOLYSIS OF TOSYLATES[a]

R	ℓ	m
2-adamantyl	$(0.0)^b$	$(1.00)^b$
Bicyclo[2.2.2]octyl	0.02	1.05
Cyclohexyl	0.23	0.75
Cyclopentyl	0.26	0.71
3-Pentyl	0.26	0.72
2-Propyl	0.40	0.58
Ethyl	0.83	0.41
Methyl	$(1.0)^b$	$(0.3)^b$
Benzyl	0.75	0.64

[a] With permission from T. W. Bentley and P. v. R.
Schleyer, *Adv. Phys. Org. Chem.*, **14**, 1 (1977), p. 53.
Copyright by Academic Press Inc. (London) Ltd.
[b] By definition.

Nucleophilicities considered so far apply to protic solvents. We have already seen how the change to a dipolar aprotic solvent can have a profound effect on reaction rates for S_N2 substitutions, particularly for charge type 1. In addition to solvent, another agent that modifies the reactivity of anionic nucleophiles is the counter-ion. A positive ion tightly ion-paired to a negatively charged nucleophile in a solvent of moderate to low dielectric constant will lower the nucleophile's reactivity. Cation complexing agents, such as the crown ethers (18-crown-6, **27**, is

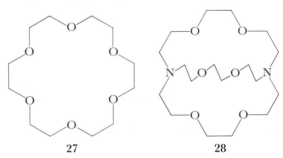

27   28

an example), will therefore enhance reactivity of the anionic nucleophiles.[95] Even more effective cation complexing agents are the macrobicyclic cryptands, of which **28** is an example.[96] These complexing agents act effectively as phase transfer catalysts for nucleophilic substitutions by solubilizing an ionic reagent in a nonpolar solvent such as benzene. Under these circumstances the anions, separated from

[95] (a) S. A. DiBiase and G. W. Gokel, *J. Org. Chem.*, **43**, 447 (1978); (b) J. J. Christensen, D. J. Eatough, and R. M. Izatt, *Chem. Rev.*, **74**, 351 (1974); (c) L. F. Lindoy, *Chem. Soc. Rev.*, **4**, 421 (1975). Crown ethers and cryptands are discussed in Section 2.4.

[96] J.-M. Lehn, *Acc. Chem. Res.*, **11**, 49 (1978).

their counter-ions by the complexing molecule and unable to obtain solvation from the solvent, are particularly reactive.[97]

In view of the profound influence of solvent on nucleophilicity, it is risky to interpret relative reactivity of various nucleophiles in terms of intrinsic properties. For example, one might be tempted to argue that the nucleophilicity order for the halides, $I^- > Br^- > Cl^- > F^-$, could be explained in terms of the hard–soft acid–base theory by saying that the softer ions are better nucleophiles toward the soft carbon center. This nucleophilicity order, however, holds only in protic solvents, which, as we have seen, solvate anions strongly. In dipolar aprotic solvents the order is reversed.[98] Perhaps the hard–soft theory will still work, if we argue that the hard-acid protic solvents are coordinating most strongly with, and thereby deactivating most strongly, the smaller, harder anions. This last explanation at least fits with the thermodynamics of solvent transfer as shown in Table 4.4.

As with acidity, gas-phase data should provide a measure of intrinsic nucleophilicity. Brauman has measured rates of nucleophilic substitution on methyl halides in the gas phase by ion cyclotron resonance; the order of nucleophilicity of halide ions is $F^- > Cl^- > Br^- > I^-$, the same as in the dipolar aprotic solvents.[99]

The Nucleophile in S_N1 Reactions

The rate of an S_N1 reaction proceeding exclusively by the k_c (or k_Δ) process should be unaffected by presence of a nucleophile, except for salt effects and possibly the common ion effect. The carbocation formed, however, will react with a nucleophile, and one can ask what its selectivity will be if given a choice of more than one nucleophile. On the basis of the Hammond postulate, we anticipate that a highly reactive intermediate, facing low activation barriers, will find small differences between various paths and will be relatively nondiscriminating in its choice of reaction partner; whereas a less reactive one, confronted with higher barriers, will encounter larger differences and be more selective. When coupled with the expectation that a less reactive ion is formed more rapidly, this principle predicts a correlation between solvolysis rate and selectivity for S_N1 processes. Figure 4.9 illustrates the reasoning and Figure 4.10 presents the experimental evidence.[100] There is a good deal of scatter in the correlation, but there does seem to be a trend of the expected kind.

Ritchie has reported extensive measurements of rates of reactions of nucleophiles with more stable carbocations that show no such selectivity.[101] These reac-

[97](a) D. Bethell, K. McDonald, and K. S. Rao, *Tetrahedron Lett.*, 1447 (1977); (b) W. P. Weber and G. W. Gokel, *Phase Transfer Catalysis in Organic Synthesis,* Springer-Verlag, Berlin, 1977; (c) A. Brändström, *Adv. Phys. Org. Chem.*, **15**, 267 (1977).
[98](a) A. J. Parker, *J. Chem. Soc. A,* 220, (1966); (b) S. Winstein, L. G. Savedoff, S. Smith, I. D. R. Stevens, and J. S. Gall, *Tetrahedron Lett.*, No. 9, 24 (1960).
[99] W. N. Olmstead and J. I. Brauman, *J. Am. Chem. Soc.*, **99**, 4219 (1977).
[100] D. J. Raber, J. M. Harris, R. E. Hall, and P. v. R. Schleyer, *J. Am. Chem. Soc.*, **93**, 4821 (1971).
[101](a) C. D. Ritchie, *Acc. Chem. Res.*, **5**, 348 (1972); (b) C. D. Ritchie, R. J. Minasz, A. A. Kamego, and M. Sawada, *J. Am. Chem. Soc.*, **99**, 3747 (1977); (c) C. D. Ritchie and M. Sawada, *J. Am. Chem. Soc.*, **99**, 3754 (1977).

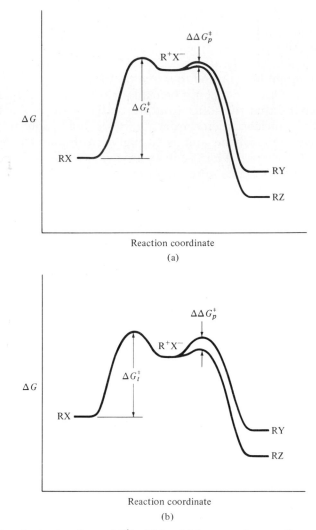

Figure 4.9 (a) A slow ionization (large $\Delta G_t^{\ddagger}$) yields a high-energy intermediate which is expected to be relatively unselective (small $\Delta\Delta G_p^{\ddagger}$) and to react with Y and Z at nearly equal rates. (b) Rapid ionization (small $\Delta G_t^{\ddagger}$) produces a more stable intermediate which is expected to be more discriminating (large $\Delta\Delta G_p^{\ddagger}$) and to favor combination with Z strongly over Y.

tions follow the simple relation 4.48, where N_+ is a parameter characteristic of the

$$\log \frac{k}{k_0} = N_+ \qquad (4.48)$$

nucleophile. This equation means that the relative rates of reaction of nucleophiles are always the same, no matter what the electrophile. This surprising result is clearly inconsistent with the conclusion drawn from Figure 4.10. The reason for the discrepancy is not altogether clear, but Ritchie has presented evidence that in at

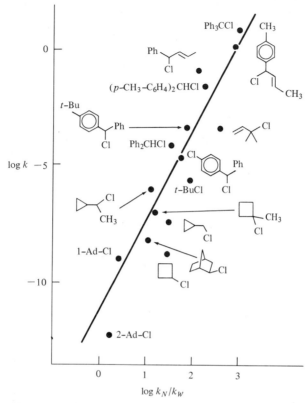

Figure 4.10 Correlation between stability, measured by solvolysis rate in 80 percent aqueous acetone, and selectivity, determined by relative rate of reaction with azide ion (k_N) and water (k_W), for carbocations derived from alkyl chlorides. Reprinted with permission from D. J. Raber, J. M. Harris, R. E. Hall, and P. v. R. Schleyer, *J. Am. Chem. Soc.*, **93**, 4821 (1971). Copyright 1971 American Chemical Society.

least one case, solvolysis of a benzhydryl derivative, the apparent selectivity is an artifact arising from trapping by the nucleophile of two different intermediates, the solvent-separated ion pair and the free ion.[102]

The Leaving Group[103]

Because the leaving group is becoming more negatively charged in the transition state of a nucleophilic substitution reaction, a good leaving group must readily take on negative charge. The acidity of the conjugate acid might be expected to provide a rough guide to leaving-group ability. As Table 4.12 shows, this expectation is fulfilled for comparisons within a series of closely related groups, but the analogy of

[102](a) C. D. Ritchie, *J. Am. Chem. Soc.*, **93**, 7324 (1971). See also (b) A. Pross, *Adv. Phys. Org. Chem.*, **14**, 69 (1977).
[103]Stirling has measured relative leaving-group ability (nucleofugality) for a number of groups in elimination reactions. The halides are not included. C. J. M. Stirling, *Acc. Chem. Res.*, **12**, 198 (1979).

Table 4.12 RELATIVE REACTION RATES FOR SOME COMMON LEAVING GROUPS IN S_N1 REACTIONS[a]

Group	Approximate k_X/k_{Br}	Approximate pK_{aXH}
—OSO$_2$CF$_3$ (—OTf, triflate)	$5 \times 10^{8\,b,c}$	0.3^c
—OSO$_2$—⟨benzene⟩—Br (—OBs, brosylate)	$1.5 \times 10^{4\,d}$	
—OSO$_2$—⟨benzene⟩—CH$_3$ (—OTs, tosylate)	$5 \times 10^{3\,b,d}$	1^e
Br	1^b	-8^f
Cl	$2.5 \times 10^{-2\,b,g}$	-7^f
—O—C(=O)—⟨benzene⟩—NO$_2$ (—OPNB)	$2 \times 10^{-6\,h}$	3.4^i

[a] Relative rates are approximate because they are not independent of the structure of the rest of the solvolyzing molecule. All comparisons except —OBs and —OPNB are for bridgehead solvolyses. p-Bromobenzenesulfonate is compared with —OTs for 3-anisyl-2-butyl; —OPNB is compared with —Cl for benzhydryl (data for chloride obtained by extrapolation).
[b] R. C. Bingham and P. v. R. Schleyer, *J. Am. Chem. Soc.*, **93**, 3189 (1971); R. D. Howells and J. D. McCown, *Chem. Rev.*, **77**, 69 (1977).
[c] J. B. Hendrickson, D. D. Sternbach, and K. W. Bair, *Acc. Chem. Res.*, **10**, 306 (1977).
[d] S. Winstein and G. C. Robinson, *J. Am. Chem. Soc.*, **80**, 169 (1958).
[e] *Handbook of Chemistry and Physics*, 57th ed., CRC Press, Cleveland, Ohio, 1976, p. D-150, reports 0.7 for pK$_a$ of benzenesulfonic acid.
[f] See Table 3.3.
[g] L. C. Bateman, M. G. Church, E. D. Hughes, C. K. Ingold, and N. A. Taher, *J. Chem. Soc.*, 979 (1940).
[h] H. L. Goering and H. Hopf, *J. Am. Chem. Soc.*, **93**, 1224 (1971).
[i] G. Kortum, W. Vogel, and K. Andrussow, *Dissociation Constants of Organic Acids in Aqueous Solution*, Butterworths, London, 1961.

basicity toward proton and basicity toward carbon fails, as we would expect, for leaving groups of different types. As indicated in Table 4.12, which gives approximate relative leaving-group reactivities for S_N1 reactions, the sulfonates are much better leaving groups relative to halogen than would be expected from their basicity toward protons.

In S_N2 reactions sulfonate and halide have comparable leaving-group ability, and the relative reactivities of the leaving groups are dependent on nucleophile as well as on solvent. Table 4.13 gives leaving-group reactivities for some S_N2 reactions. In ethanol, with the strong nucleophile p-toluenethiolate, tosylate reactivity is between that of bromide and of chloride, but with the weaker ethoxide as nucleophile, and hence presumably with a looser transition state, tosylate becomes more reactive than iodide. Hoffmann proposed that the ratio k_{OTs}/k_{Br} is a measure of the amount of substrate–leaving-group bond breaking at the transition state.[104]

[104] H. M. R. Hoffmann, *J. Chem. Soc.*, 6753 (1965), and references therein.

Table 4.13 Dependence of Leaving-Group Reactivity on the Nucleophile

Leaving Group, X	Reaction number	Substrate and nucleophile[a]	Temp. (°C)	k_X/k_{Br}	Ref.
I	1	$nC_3H_7X + p\text{-}CH_3C_6H_4S^-$	25	3.5	b
I	2	$C_2H_5X + C_2H_5O^-$	25	1.9	c
Br				1.0	
Cl	3	$nC_3H_7X + p\text{-}CH_3C_6H_4S^-$	25	0.0074	b
Cl	4	$C_2H_5X + C_2H_5O^-$	40	0.0024	c
$-OSO_2-\!\!\!\bigcirc\!\!\!-CH_3$	5	$nC_3H_7X + p\text{-}CH_3C_6H_4S^-$	25	0.44	b
$-OSO_2-\!\!\!\bigcirc\!\!\!-CH_3$	6	$C_2H_5X + C_2H_5O^-$	25	3.6	d

[a] All reactions were run in ethanol solvent.
[b] H. M. R. Hoffmann, *J. Chem. Soc.*, 6753 (1965), and references therein.
[c] A. Streitwieser, *Solvolytic Displacement Reactions*, McGraw-Hill, New York, 1962, pp. 30–31 and references therein.
[d] The relative reactivity here was estimated by using the $k_{OSO_2-\bigcirc}/k_{Br}$ value of 5.8 calculated by

Streitwieser (reference c above, p. 30) and multiplying that by 0.63, the relative reactivity of tosylate to benzene sulfonate [M. S. Morgan and L. H. Cretcher, *J. Am. Chem. Soc.*, **70**, 375 (1948)]. Since the latter value is for reaction at 35°C, the real k_{OTs}/k_{Br} value might be a little smaller.

This proposal accounts for behavior of S_N2 reactions, but for S_N1 reactions Hoffmann's hypothesis that faster solvolysis rates and larger k_{OTs}/k_{Br} ratios indicate greater charge separation is in conflict with the Hammond postulate prediction that the more reactive substrates should have earlier transition states and less charge separation. Hoffmann's group of supposed S_N1 reactions actually probably contained many reactions proceeding by the k_s process; Bingham and Schleyer, examining dependence of reaction rate on solvent, were unable to detect any significant variation in transition state charge separation in a series of bridgehead derivatives of varying reactivity.[105]

Table 4.14 shows the effect of dipolar aprotic solvent on leaving-group reactivity. With azide as nucleophile, chloride leaving group is accelerated the least by transfer from methanol to dimethylformamide, and iodide the most. The effect is due mainly to solvation of the transition state. The transition state with the smaller Cl^- leaving group and hence larger charge density is relatively better solvated by methanol, and so suffers an increase of free energy on transfer to DMF; the transition state with the larger, more polarizable iodide leaving group has slightly lower free energy in DMF. The same trend occurs with the more polarizable nucleophile SCN^-, but the differences are smaller.

Substituents that form strong bases, such as —OH, —OR, —NH$_2$, —SR, are poor leaving groups; converting them into a form with a positive charge

[105] (a) R. C. Bingham and P. v. R. Schleyer, *J. Am. Chem. Soc.*, **93**, 3189 (1971); (b) J. Slutsky, R. C. Bingham, P. v. R. Schleyer, W. C. Dickason, and H. C. Brown, *J. Am. Chem. Soc.*, **96**, 1969 (1974).

Table 4.14 EFFECT OF SOLVENT ON LEAVING-GROUP REACTIVITY IN S_N2 REACTIONS[a]

Reaction	$\log k_{DMF}{}^{b}/k_{CH_3OH}$	$\Delta G_{tr}^{CH_3OH \rightarrow DMF}$ (RX)	$\Delta G_{tr}^{CH_3OH \rightarrow DMF}$ (Y$^-$)	$\Delta G_{tr}^{CH_3OH \rightarrow DMF}$ ($\ddagger$)
$N_3{}^- + CH_3Cl$	3.3	-0.5	6.7	1.6
$N_3{}^- + CH_3Br$	3.9	-0.4	6.7	1.0
$N_3{}^- + CH_3I$	4.6	-0.7	6.7	-0.3
$SCN^- + CH_3Cl$	1.4	-0.5	3.7	1.2
$SCN^- + CH_3Br$	1.7	-0.4	3.7	1.0
$SCN^- + CH_3I$	2.2	-0.7	3.7	0.0

[a] In kcal mole^{-1}. Adapted with permission from A. J. Parker, *Chem. Rev.*, **69**, 1 (1969). Copyright 1969 American Chemical Society.
[b] DMF = dimethylformamide.

($-OH_2{}^+$, $-N_2{}^+$, $-SR_2{}^+$) will make them much more reactive. Protonation is a feasible way to do this in many instances.

The leaving group $-N_2{}^+$, formed by diazotization of a primary amine, departs as the very stable N_2 molecule and is one of the most reactive leaving groups known.[106] Products formed on decomposition of diazonium ions are typical of carbocation reactions, but product distributions and stereochemistry often differ from those obtained with less reactive leaving groups. The diazonium reactions show a greater tendency for rearrangement and less selectivity in reaction with competing nucleophiles.[107] One explanation of this observation was that the loss of N_2 leaves a vibrationally excited carbocation that has different properties from one formed in solvolysis.[108] This theory is no longer considered tenable; the differences in behavior can probably be attributed to differences in degree of neighboring group and solvent participation and in ion pairing arising from the reactive leaving group.[109] A number of examples of diazotization reactions are considered in Chapter 5.

Effect of Substrate Structure in S_N2 Reactions

Because the nucleophile must approach the center at which substitution occurs in the rate-determining step, more and larger substituents in the substrate decrease the rate. The effect of electron supply or withdrawal, however, is not so easy to predict because the substitution center could have either excess positive or excess negative charge or neither, depending on the location of the transition state in the reaction coordinate space.

[106](a) C. J. Collins, *Acc. Chem. Res.*, **4**, 315 (1971); (b) R. A. Moss, *Chem Eng. News*, **49** (48), 28 (1971); (c) L. Friedman, in *Carbonium Ions*, Vol. II, G. A. Olah and P. v. R. Schleyer, Eds., Wiley, New York, 1970, p. 655; (d) W. Kirmse, *Angew. Chem. Int. Ed.*, **15**, 251 (1976).
[107] For further discussion see Section 5.1.
[108] D. Semenow, C.-H. Shih, and W. G. Young, *J. Am. Chem. Soc.*, **80**, 5472 (1958).
[109](a) W. Kirmse, *Angew. Chem. Int. Ed.*, **15**, 251 (1976); see also (b) H. Zollinger, *Angew. Chem. Int. Ed.*, **17**, 141 (1978).

Table 4.15 AVERAGE RELATIVE S_N2 RATES OF ALKYL SYSTEMS[a]

R in R—X	Relative rate
CH_3	1
CH_3CH_2	3.3×10^{-2}
$CH_3CH_2CH_2$	1.3×10^{-2}
$(CH_3)_2CH$	8.3×10^{-4}
$(CH_3)_3C$	5.5×10^{-5}[b]
$(CH_3)_3CCH_2$	3.3×10^{-7}
$H_2C=CHCH_2$	1.3
ϕCH_2	4.0

[a] Data from A. Streitwieser, Jr., *Solvolytic Displacement Reactions*, McGraw-Hill, New York, 1962, p. 13. Reproduced by permission of McGraw-Hill.

[b] This value is not from Streitwieser but is the reactivity of *t*-butyl bromide to S_N2 substitution by free Cl^- in DMF at 25°C relative to the reactivity of CH_3Br under the same conditions [D. Cook and A. J. Parker, *J. Chem. Soc. B*, 142 (1968)]. This value is corrected for the substitution that actually proceeds by an elimination addition mechanism.

Substrate alkylation It is well established that S_N2 reactions occur less readily in molecules where the α or β carbons bear alkyl substituents. For example, the relative reactivities shown in Table 4.15 are derived from studies of 15 different S_N2 reactions in various solvents. Although it was at one time postulated that electron-donating polar effects of the substituents are responsible for the reduction in reactivity,[110] it is now clear that alkyl group polar effects are much too small to account for the large retardation observed with increasing substitution.[111] A striking example is the neopentyl system, $(CH_3)_3CCH_2$—X, where the branching is one carbon atom removed from the reaction site but the rate of substitution is about 10^{-5} times slower than in the ethyl system.

The currently accepted explanation is that the rate decrease with increasing substitution is of steric origin, arising from nonbonded interactions in the transition state between the substituents and the entering and leaving groups.[112] On going from ground state to transition state in an S_N2 reaction, the angles H—C—H in CH_3X increase from close to 109° to near 120°; at the same time the H—C—X angles decrease to about 90° and a fifth ligand, Y, is introduced with Y—C—H angles also near 90°. Ingold and co-workers calculated that in CH_3X, because the

[110] (a) C. N. Hinshelwood, K. J. Laidler, and E. W. Timm, *J. Chem. Soc.*, 848 (1938); (b) P. B. D. de la Mare, L. Fowden, E. D. Hughes, C. K. Ingold, and J. D. H. Mackie, *J. Chem. Soc.*, 3200 (1955).

[111] (a) H. D. Holtz and L. M. Stock, *J. Am. Chem. Soc.*, **87**, 2404 (1965); (b) H. J. Hinze, M. A. Whitehead, and H. H. Jaffé, *J. Am. Chem. Soc.*, **85**, 148 (1963).

[112] (a) I. Dostrovsky, E. D. Hughes, and C. K. Ingold, *J. Chem. Soc.*, 173 (1946); (b) P. B. D. de la Mare, L. Fowden, E. D. Hughes, C. K. Ingold, and J. D. H. Mackie, *J. Chem. Soc.*, 3169 (1955) and following papers; (c) C. K. Ingold, *Quart. Rev. (London)*, **11**, 1 (1957); (d) C. K. Ingold, *Structure and Mechanism in Organic Chemistry*, 2d ed., Cornell University Press, Ithaca, New York, 1969, pp. 544ff. See also (e) D. Cook and A. J. Parker, *J. Chem. Soc. B*, 142 (1968); (f) M. Charton, *J. Am. Chem. Soc.*, **97**, 3694 (1975).

protons are small, there is little, if any, increase of nonbonding interaction between X and Y and the protons in going from the ground state to the transition state. However, when one of the H's is replaced by the much larger CH_3 group (i.e., when CH_3CH_2X is the substrate), the interference between X and Y and the methyl group does increase as the angle between them decreases. This leads to compressions of the covalent bonds to lengths shorter than normal and a corresponding increase in potential energy. Thus the transition state of the ethyl system has a higher potential energy relative to the starting material than that of the methyl system. Accompanying the increase in potential energy is a decrease in entropy as motions of the atoms are restricted.[113] The tighter the transition state, the greater are the energy increases and entropy decreases arising from the steric effects. A looser transition state suffers less from steric crowding but loses bonding energy because groups X and Y are farther away from the substitution center.[114] Each system must adopt the transition state geometry that achieves the optimum balance of these opposing trends.

It is interesting to note in Table 4.15 that the primary neopentyl systems, in which all the substituents are in the β position, are substituted more slowly than t-butyl systems in which the substituents are directly on the reaction site. Apparently, steric hindrance is less important when the larger substituents are α, because in the transition state they all lie in a plane perpendicular to X and Y and thus are fairly well out of the way; only one β substituent can lie in this plane, as shown in Figure 4.11. If there are only one or two β substituents, it is still possible to rotate them out of the way of entering and leaving groups. However, when there are three, as in the neopentyl group, it is not possible, and substantial steric hindrance occurs in the transition state.[115]

Substituent polar effects The reaction coordinate diagram for the S_N2 reaction (Figure 4.5) suggests that the carbon undergoing substitution might be either more positive, more negative, or of the same polarity in the transition state as in the ground state. Experimentally, electron-withdrawing substituents are found sometimes to accelerate and sometimes to decelerate these reactions.[116] Table 4.16 shows an example of substituent effects in reactions of substituted benzyl chlorides with the anionic nucleophile $S_2O_3{}^{2-}$ and in solvolysis with the neutral nucleophiles ethanol and water. In the former case all the para-substituted compounds react faster than the unsubstituted, whether the substituent is electron donating or withdrawing. The solvolysis example shows more regular behavior in which electron-donating substituents accelerate the reaction and electron-withdrawing substituents retard it.

[113]N. Ivanoff and M. Magat, *J. Chim. Phys.*, **47**, 914 (1950); E. Bauer and M. Magat, *J. Chim. Phys.*, **47**, 922 (1950).
[114]A. Streitwieser, Jr., *Solvolytic Displacement Reactions,* McGraw-Hill, New York, 1962, p. 23.
[115]C. K. Ingold, *Structure and Mechanism in Organic Chemistry,* 2d ed., Cornell University Press, Ithaca, N.Y., 1969, pp. 547*ff.*
[116]For a summary see Streitwieser, *Solvolytic Displacement Reactions,* pp. 16–20.

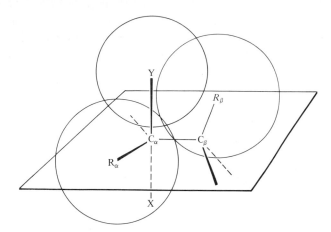

Figure 4.11 Transition state for the S_N2 reaction, showing steric interactions of nucleophile with substituent groups. R_α and C_β are 1.54 Å from C_α, and R_β is 1.54 Å from C_β with angle $C_\alpha-C_\beta-R_\beta = 109.5°$. Distance $C_\alpha-Y$ is 2.2 Å. Circles are drawn with radii of 2.0 Å for R_α and R_β (the van der Waals radius of the methyl group) and radius 1.8 Å (Cl^- ion) for Y. Distance from center of Y to center of R_α is 2.69 Å; from center of Y to center of R_β is 2.19 Å.

Young and Jencks have analyzed these substituent effects in terms of a two-parameter Hammett $\sigma-\rho$ correlation according to Equation 4.49. The substitu-

$$\log \frac{k}{k_0} = \rho\sigma^n - \rho^r(\sigma^+ - \sigma^n) \tag{4.49}$$

ent constant σ^n measures polar effects only and excludes resonance; the difference $(\sigma^+ - \sigma^n)$ measures only resonance effects. For reactions of substituted benzyl chlorides with thiophenoxide ion, ϕS^-, the analysis shows $\rho = +1.06$ and $\rho^r = -1.3$. This result means that the reaction is accelerated by inductive electron withdrawal and by resonance electron donation. For substitution with aniline as nucleophile, $\rho = -0.5$ and $\rho^r = -1.8$; in this reaction, which is of a different charge type, both polar and resonance effects act in the same direction.[117] The cause of the anomalous behavior for the $S_2O_3{}^{2-}$ reaction in Table 4.16 (and for other anionic nucleophiles) is primarily the opposition of the two types of substituent effects. In solvolysis, and in other reactions with neutral nucleophiles, the two effects act in the same direction and the behavior with change of substituent is normal.

Response of transition state structure to substituent effects and to solvent effects has been assessed with the help of reaction coordinate diagrams, as we have outlined in Section 4.2. For example, in the reaction of substituted benzyl chlorides with aniline as nucleophile, ρ changes from -0.8 to -0.2 as the substituent on the aniline is changed from m-Cl to p-OCH$_3$. This change corresponds to

[117]P. R. Young and W. P. Jencks, *J. Am. Chem. Soc.*, **101**, 3288 (1979).

Table 4.16 Relative Rates of Reaction of X—⟨benzene ring⟩—CH₂Cl with Nucleophiles $S_2O_3^{2-}$ and Ethanol–Water

X	p—X, $S_2O_3^{2-b}$	m—X, $S_2O_3^{2-b}$	p—X, EtOH/H$_2$O^c	m—X, EtOH/H$_2$O^c
(CH$_3$)$_3$C—	1.29	—	—	—
H$_3$C—	1.56	0.920	9.04	1.45
(CH$_3$)$_2$CH—	1.21	—	5.70	—
ϕ—	1.47	—	—	—
H—	1.00	1.00	1.00	1.00
F—	1.29	0.801	1.48	0.224
Cl—	1.43	0.810	0.560	0.189
NO$_2$—	2.55	1.40	0.0656	0.0765

[a] Calculated from data of R. Fuchs and D. M. Carleton, *J. Am. Chem. Soc.*, **85**, 104 (1963).
[b] In 60 percent aqueous acetone at 30°C.
[c] In 50 percent aqueous ethanol at 60°C.

making the nucleophile more basic; the transition state becomes tighter with respect to the carbon–chlorine distance (see Figure 4.5 and Problem 14) and consequently has less positive charge on carbon. The observed change of ρ in the positive direction is the expected response to this change.[118]

Because S$_N$2 reactions are so sensitive to steric effects, investigation of other influences on rates is difficult. In order to study pure polar effects, Holtz and Stock have carried out rate studies of displacements by thiophenoxide ion on 4-Z-bicyclo-[2.2.2]-octylmethyl toluenesulfonate (Equation 4.50).[119]

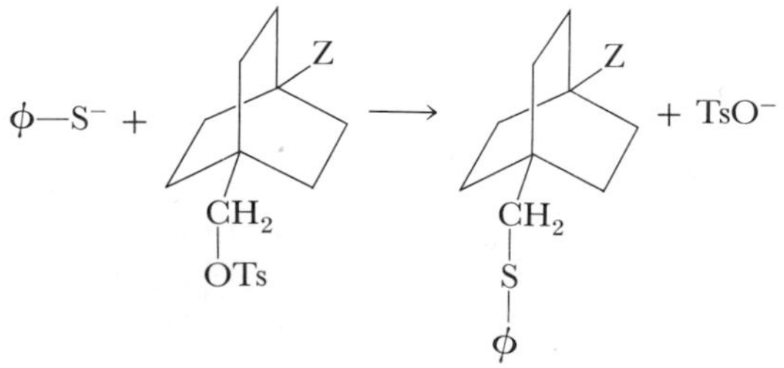

$$\phi\text{—S}^- + \text{[bicyclo[2.2.2]octyl structure with Z and CH}_2\text{—OTs]} \longrightarrow \text{[bicyclo[2.2.2]octyl structure with Z and CH}_2\text{—S—}\phi\text{]} + \text{TsO}^- \qquad (4.50)$$

This system has the virtues of being completely rigid so that changing Z does not change the steric environment of the transition state and of having the substituent so far removed from the reaction site through saturated bonds that conjugation is impossible. Holtz and Stock found that although alkyl substituents had little effect on the reaction rate, electron-withdrawing groups in general did increase the rate (see Table 4.17). In these displacements the anionic nucleophile thiophenoxide is

[118] (a) Note 117; (b) J. M. Harris, S. G. Shafer, J. R. Moffatt, and A. R. Becker, *J. Am. Chem. Soc.*, **101**, 3295 (1979).
[119] H. D. Holtz and L. M. Stock, *J. Am. Chem. Soc.*, **87**, 2404 (1965).

Table 4.17 RELATIVE RATES OF
4-Z-BICYCLO[2.2.2]OCTYLMETHYL
TOLUENESULFONATE WITH
THIOPHENOXIDE ION[a]

Z	Relative rate
H	1.00
CH$_3$	1.07
CH$_2$OH	1.17
CO$_2$Et	1.52
Cl	1.84
Br	2.01

[a] Reprinted with permission from H. D. Holtz and L. M. Stock, *J. Am. Chem. Soc.*, **87**, 2404 (1965). Copyright by the American Chemical Society.

always the attacking reagent, and therefore the results are not surprising within the context of the discussion above.

Substitution α to a carbonyl or cyano group Compounds that have an α-carbonyl or α-cyano group are usually particularly reactive in S$_N$2 reactions. For example, when treated with potassium iodide in acetone, α-chloroacetone (ClCH$_2$COCH$_3$) reacts 35,000, and α-chloroacetonitrile (ClCH$_2$C≡N) 3000 times faster than *n*-butyl chloride.[120] The probable reason for this increased reactivity is that there is partial bonding between the incoming nucleophile and the electrophilic carbon of the carbonyl or cyano group in the transition state. Figure 4.12

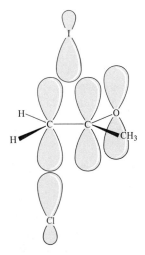

Figure 4.12 Transition state for S$_N$2 displacement of iodide ion on α-chloroacetone.

[120] J. B. Conant, W. R. Kirner, and R. E. Hussey, *J. Am. Chem. Soc.*, **47**, 488 (1925).

shows an orbital representation of the activated complex for the displacement of chloride by iodide in α-chloroacetone.

Bartlett and Trachtenberg have studied the kinetics of Reactions 4.51 and 4.52, and their results provide strong support for this hypothesis.[121] When 5,7-dinitro-3-coumaranone (**29**) reacts with potassium iodide in acetone the enthalpy of activation is 20 kcal mole^{-1} higher than when the ketone **30** undergoes the analogous reaction.

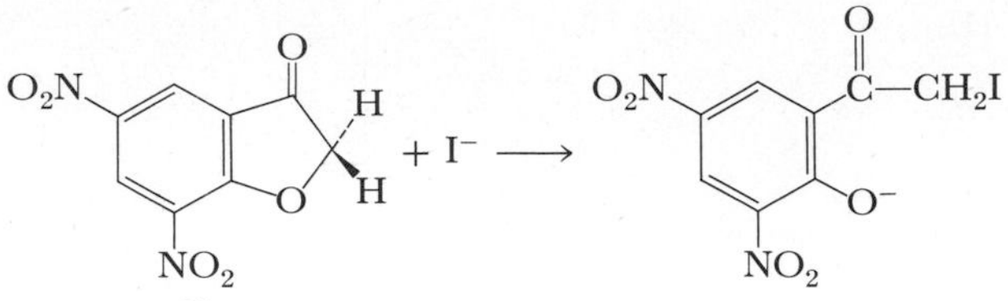

$$\Delta H^{\ddagger} = 31.3 \text{ kcal mole}^{-1} \quad (4.51)$$

29

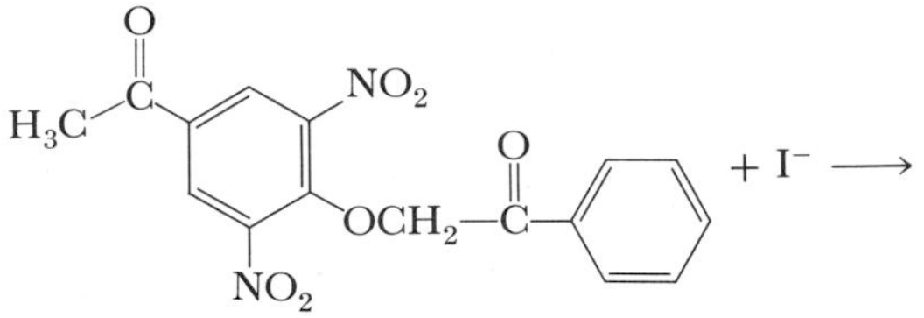

30

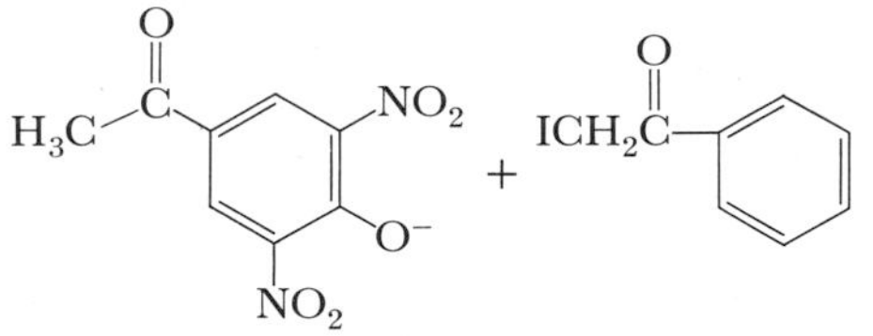

$$\Delta H^{\ddagger} = 10.2 \text{ kcal mole}^{-1} \quad (4.52)$$

Figure 4.13 shows that in order to carry out a backside displacement on **29**, the iodide must attack the reacting carbon in the plane of the ring. The carbonyl π bond is, however, perpendicular to the plane of the ring and because of the rigid ring structure of **29**, cannot rotate to overlap with the incoming iodide. On the other hand, the activated complex of iodide substitution on **30** is probably very similar to that illustrated in Figure 4.12 for substitution on α-chloroacetone.

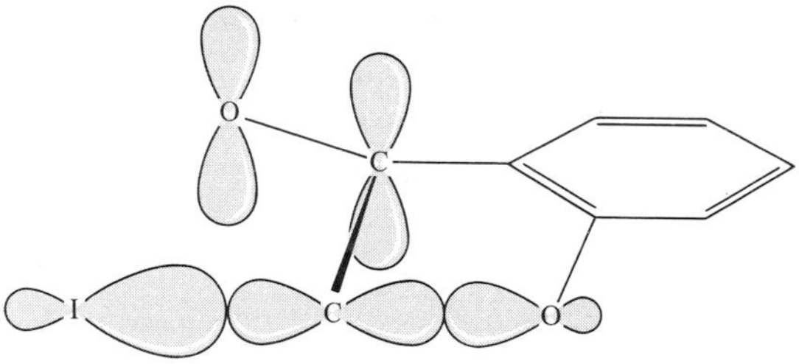

Figure 4.13 Transition state for S$_N$2 displacement of iodide ion on 5,7-dinitro-3-coumaranone.

[121]P. D. Bartlett and E. N. Trachtenberg, *J. Am. Chem. Soc.*, **80**, 5808 (1958).

The S_N2' reaction Allylic systems, in addition to undergoing ordinary substitutions without rearrangement, can react with rearrangement of the double bond, as indicated in Equation 4.53.[122] This transformation is called the S_N2'

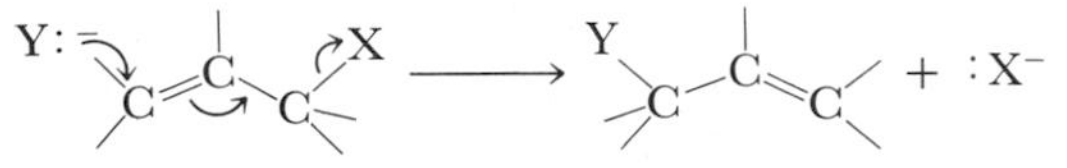

$$\text{Y:} \quad \searrow_{C=C}^{C} \diagup^{X} \longrightarrow \quad \searrow_{C-C}^{Y} =C\diagup + \text{:X}^- \tag{4.53}$$

$$\underset{D}{\overset{H}{\searrow}}C=C\underset{CH_3}{\overset{Cl}{\diagdown}}_H \xrightarrow{Et_2NH} \underset{H}{\overset{Et_2N}{\diagdown}}C=C\underset{H}{\overset{H}{\diagup}}\underset{CH_3}{\overset{}{\diagdown}} + \underset{Et_2N}{\overset{D}{\diagdown}}C-C\underset{CH_3}{\overset{H}{\diagup}}\underset{H}{\overset{H}{=}}C\diagdown_H \tag{4.54}$$

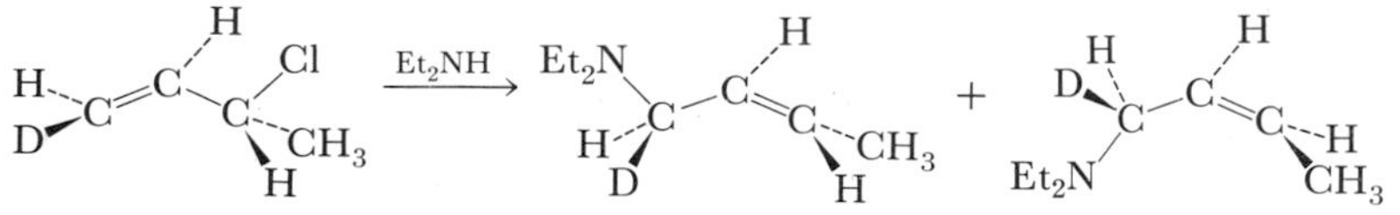

reaction. The stereochemistry of the S_N2' substitution is variable, but in an acyclic system free of potential complications by ring conformations, the reaction is stereo-specific, with the nucleophile entering syn to the leaving group. An experiment demonstrating this stereochemistry is illustrated in Equation 4.54. Note that the two products arise from different rotational conformations of the substrate, in which the chlorine is respectively above and below the plane of the double bond; the position of the methyl group in the product shows that in each case the nucleo-phile and leaving group were on the same side of the molecule.[123]

Figure 4.14 shows a frontier orbital analysis of the syn and antistereochem-istry for the S_N2' reaction. The transition state is modeled as a three-carbon allylic cation, the orbitals of which are depicted at the center in the figure, and two hybrid orbitals, one on nucleophile Y and the other on leaving group X. (X and Y are considered to be equivalent.) When X and Y are syn, there is some overlap between them; they form a weakly bonding and a weakly antibonding combination, as indicated on the left in the figure. Both of these orbitals are filled; hence the HOMO is the upper one, which has correct symmetry to interact with the allyl LUMO. When X and Y are anti, there is practically no overlap between them; the orbitals remain at their original energy, as shown at the right-hand side in Figure 4.14. There is a larger HOMO–LUMO stabilization for the syn configuration because of the smaller energy separation for that arrangement.[124]

Effect of Substrate Structure in S_N1 Reactions

In the S_N1 reaction the substrate ionizes to the high-energy ion pair, and the Hammond postulate predicts that the transition state should resemble the ion pair. Hence any structural change that lowers the carbocation energy should lower tran-sition state energy and increase reaction rate. Because of this close relationship

[122](a) F. G. Bordwell, *Acc. Chem. Res.*, **3**, 281 (1970); (b) R. H. DeWolfe and W. G. Young, *Chem. Rev.*, **56**, 753 (1956).
[123](a) R. M. Magid and O. S. Fruchey, *J. Am. Chem. Soc.*, **101**, 2107 (1979). For other examples see (b) W. Kirmse, F. Scheidt, and H.-J. Vater, *J. Am. Chem. Soc.*, **100**, 3945 (1978); (c) G. Stork and A. F. Kreft III, *J. Am. Chem. Soc.*, **99**, 3850, 3851 (1977).
[124]R. L. Yates, N. D. Epiotis, and F. Bernardi, *J. Am. Chem. Soc.*, **97**, 6615 (1975).

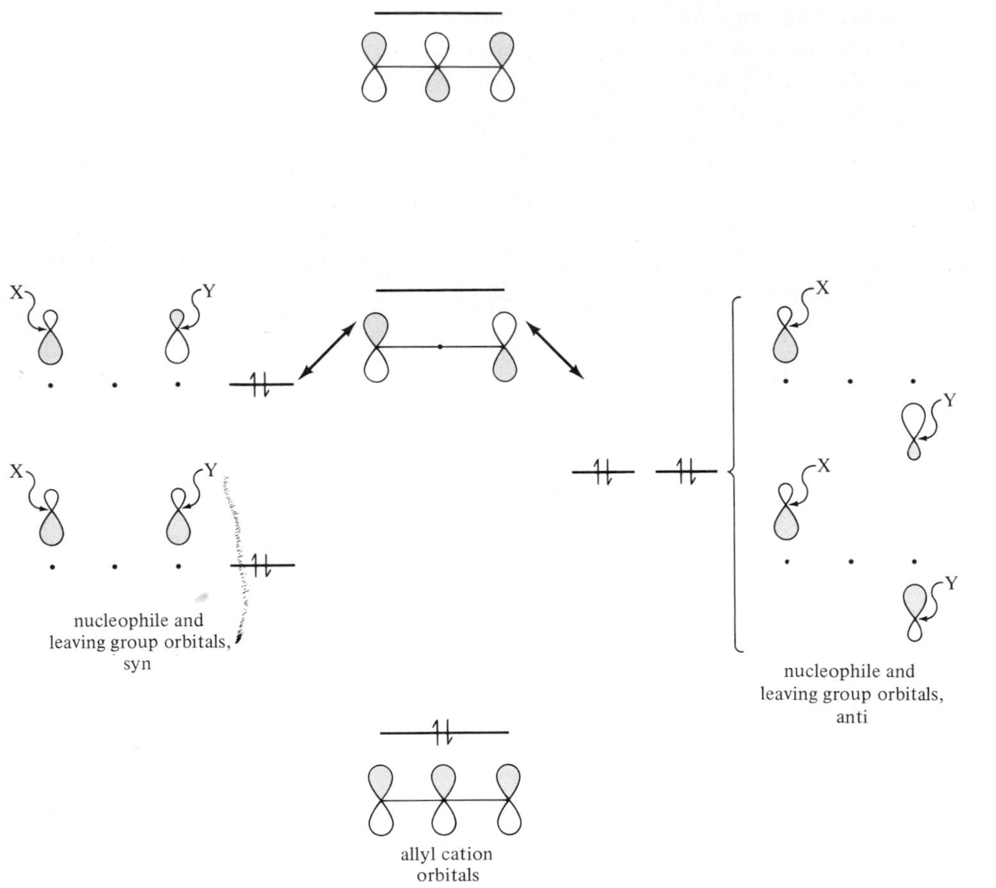

Figure 4.14 Frontier orbital interactions for the S_N2' reaction. At the center the three-carbon allylic system is represented by an allyl cation, with three molecular orbitals and one pair of electrons. The two nucleophiles X and Y are each represented by a hybrid orbital. On the left, the syn configuration, the X and Y orbitals interact and split into a symmetric and an antisymmetric pair. At the right, the anti configuration, X and Y are too far separated to interact. HOMO–LUMO stabilization is greater for the syn configuration because of the smaller energy gap. Adapted with permission from R. L. Yates, N. D. Epiotis, and F. Bernardi, *J. Am. Chem. Soc.*, **97**, 6615 (1975). Copyright 1975 American Chemical Society.

between S_N1 reaction rates and carbocation stability, we will consider the two topics together.[125]

    The nomenclature of positive carbon ions used here is that suggested by Olah.[126] The term *carbocation* is a generic name referring to ions with positive charge

[125] Reviews: (a) G. A. Olah and P .v. R. Schleyer, Eds., *Carbonium Ions,* Wiley, New York, Vol. I, 1968; Vol. II, 1970; Vol. III, 1972; Vol. IV, 1973; Vol. V, 1976. Other reviews: (b) D. Bethell and V. Gold, *Carbonium Ions,* Academic Press, London, 1967; (c) N. C. Deno, *Prog. Phys. Org. Chem.,* **2**, 129 (1964); (d) P. J. Stang, *Prog. Phys. Org. Chem.,* **10**, 205 (1973); (e) V. Buss, P. v. R. Schleyer, and L. C. Allen, *Top. Stereochem.,* **7**, 253 (1973).
[126] G. A. Olah, *J. Am. Chem. Soc.,* **94**, 808 (1972).

on carbon. The two main types of carbocation are the *carbenium* ions, in which the positive carbon has coordination number 3, as in trimethylcarbenium ion (*t*-butyl cation) (**31**) and the *carbonium* ions, in which the positive carbon or carbons has coordination number 4 or 5, as in the bridged structure **32**. The former structures,

for many years designated *classical* ions, have ordinary two-electron bonds; the latter, known earlier as *nonclassical* ions, have a three-center, two-electron bond. Carbenium ions derived from alkenes by protonation may also be called *alkenium* ions. A third type, the carbynium ions, or vinyl cations (**33**), have been studied more recently. This nomenclature has not yet been universally adopted; in the referring to all types of carbocations, and ions of type **32** are called nonclassical ions. Our discussion here will be restricted mainly to ions of types **31** and **33**; ions of type **32** are discussed in Chapter 5.

Effects of alkyl and aryl substituents   As we have seen in Chapter 3, the polar effects of alkyl groups when bonded to saturated centers are of small magnitude and ambiguous direction, being apparently either electron donating or withdrawing in the gas phase, as the center to be stabilized is positively or negatively charged.[127] When bonded to a cationic sp^2 center, however, the effect of alkyl compared to hydrogen is definitely electron donating. Schleyer and co-workers[128] estimated that for limiting S_N1 reactions, with no assistance to ionization by nucleophiles, the substitution of H by CH_3 on the reacting carbon in 2-adamantyl solvolysis accelerates the rate by a factor of 10^8, a difference in activation energy of about 11 kcal mole^{-1}. Other data of a similar sort, but subject to uncertainty because of uncertain mechanisms, have been given by Brown and Rei.[129] Some of this information is reproduced in Table 4.18. The differences are not as large as that reported by Schleyer, because reactions of secondary substrates are being accelerated by nucleophilic attack of solvent; nevertheless, the trends are clear.

Data obtained by methods independent of rates of substitution reactions are available on relative stabilities of carbocations. Saunders and Hagen observed, by proton magnetic resonance spectroscopy of carbenium ions in $SO_2ClF–SbF_5$ solution, rearrangements of ions from primary to secondary,[130] and from secondary to

[127] See the discussion in Section 3.4.

[128] J. L. Fry, E. M. Engler, and P. v. R. Schleyer, *J. Am. Chem. Soc.*, **94**, 4628 (1972).

[129] (a) H. C. Brown and M. Rei, *J. Am. Chem. Soc.*, **86**, 5008 (1964); (b) See also G. A. Olah, P. W. Westerman, and J. Nishimura, *J. Am. Chem. Soc.*, **96**, 3548 (1974).

[130] M. Saunders and E. L. Hagen, *J. Am. Chem. Soc.*, **90**, 6881 (1968).

Table 4.18 SOLVOLYSIS RATES OF ALKYL CHLORIDES[a]

Compound	R = H	R = CH$_3$	R = ϕ	k_{CH_3}/k_H	k_ϕ/k_{CH_3}
CH$_3$—C(CH$_3$)(R)—Cl	1.57×10^{-6}	0.086	394	55,000	4580
CH$_3$—C(φ)(R)—Cl	0.216	394	19,900	1,800	50
φ—C(φ)(R)—Cl	575	19,900	578,000	34.6	29

[a] First-order rate constants, corrected to 25°C in ethanol, $10^6 \times$ sec^{-1}. Reprinted with permission from H. C. Brown and M. Rei, *J. Am. Chem. Soc.,* **86**, 5008 (1964). Copyright by the American Chemical Society.

tertiary[131] structures, Equations 4.55 and 4.56. They concluded that the energy

$$CH_3-\overset{+}{C}H-CH_3 \rightleftharpoons CH_3-CH_2-\overset{+}{C}H_2 \tag{4.55}$$

$$CH_3-CH_2-\overset{+}{C}\overset{CH_3}{\underset{CH_3}{\diagdown}} \rightleftharpoons CH_3-\overset{+}{C}H-\overset{CH_3}{\underset{CH_3}{\overset{|}{C}}}-H \tag{4.56}$$

differences between primary and secondary and between secondary and tertiary are both on the order of 11 to 15 kcal mole^{-1}. Arnett has reported a secondary-to-tertiary enthalpy difference of 12.2 kcal mole^{-1} in SO$_2$ClF, using the heat of ionization method.[132] In the gas phase there is a definite trend to smaller stabilization by methyl the more stabilizing groups are already present. The magnitudes of differences found between secondary and tertiary cations in the gas phase are comparable to differences found by Schleyer and by Saunders in solution. Table 4.19 presents some of the gas-phase data. Note the large stabilizations by hydroxyl and by phenyl, and also the small size of the stabilization on adding a methyl group when a phenyl group is already present. The qualitative agreement among results obtained in solution and in the gas phase suggests that solvation does not change dramatically with cation structure.[133]

In aryl-stabilized and in allylic cations, favorable overlap of the p orbitals of the π system should require a coplanar arrangement; evidence that such a structure is indeed preferred comes, for example, from proton magnetic resonance observations that demonstrate barriers to bond rotation in the isomeric dimethylallyl ions

---

[131] M. Saunders and E. L. Hagen, *J. Am. Chem. Soc.,* **90**, 2436 (1968).
[132] E. M. Arnett and N. J. Pienta, *J. Am. Chem. Soc.,* **102**, 3329 (1980).
[133] W. L. Jorgensen, *J. Am. Chem. Soc.,* **99**, 280 (1977).

Table 4.19 RELATIVE STABILITIES OF CARBOCATIONS IN THE GAS PHASE, AS MEASURED BY DISSOCIATION ENERGY OF R—H AND OF R—Br[a]

R	$D_{R^+H^-}$[b]	$\delta_{CH_3}D_{R^+H^-}$	$D_{R^+Br^-}$[c]	$\delta_{CH_3}D_{R^+Br^-}$
CH_3^+	313.		217.7	
$CH_3CH_2^+$	272.5	41.	181.9	35.8
$(CH_3)_2CH^+$	249.7	22.8	162.9	19.0
$(CH_3)_3C^+$	234.8	14.9	148.7	14.2
ϕCH_2^+	239.			
$\phi\overset{+}{C}HCH_3$	229.2	10.		
	$\delta_{OH}D_{R^+H^-}$			
CH_3^+	313.			
$HOCH_2^+$	247.8	65.		
	$\delta_\phi D_{R^+H^-}$			
CH_3^+	313.			
ϕCH_2^+	239.	74.		

[a] In kcal mole^{-1}.
[b] Reprinted in part with permission from J. F. Wolf, R. H. Staley, I. Koppel, M. Taagepera, R. T. McIver, Jr., J. L. Beauchamp, and R. W. Taft, *J. Am. Chem. Soc.*, **99**, 5417 (1977). Copyright 1977 American Chemical Society.
[c] R. D. Wieting, R. H. Staley, and J. L. Beauchamp, *J. Am. Chem. Soc.*, **96**, 7552 (1974).

(**34, 35,** and **36**). These ions form stereospecifically from the three dimethylcyclopropyl chlorides (Section 11.2), and barriers to rotation about the partial double

| 34 | 35 | 36 |
| cis,cis | cis,trans | trans,trans |

bonds are sufficiently high to prevent interconversion at low temperature. At $-10°C$, **34**, the least stable isomer, changes with a half-life of about 10 min to the more stable cis,trans isomer (**35**), and this in turn at $+35°C$ converts to **36**.[134]

Similar considerations might be expected to apply to the triarylmethyl ions. The most favorable charge delocalization would be obtained if the rings were all coplanar. But inspection of a model shows that a completely planar triphenylmethyl ion can be made only at the expense of unacceptable crowding of the ortho hydrogens.[135] Triphenylmethyl ions are sufficiently stable to be isolated as salts in the crystalline state. In the solid perchlorate, the actual structure, although coplanar about the central cationic carbon, has the rings twisted out of this plane by an angle of about 32°.[136] In solution, nuclear magnetic resonance evidence, obtained with ring-fluorinated derivatives, suggests a similar structure, with a barrier to

[134] P. v. R. Schleyer, T. M. Su, M. Saunders, and J. C. Rosenfeld, *J. Am. Chem. Soc.*, **91**, 5174 (1969).
[135] G. N. Lewis and M. Calvin, *Chem. Rev.*, **25**, 273 (1939).
[136] A. H. Gomes de Mesquita, C. H. MacGillavry, and K. Eriks, *Acta Crystallogr.*, **18**, 437 (1965).

rotation of all three rings to the enantiomeric conformation (**37** $\rightleftharpoons$ **38**) of about 9 kcal mole^{-1}.[137]

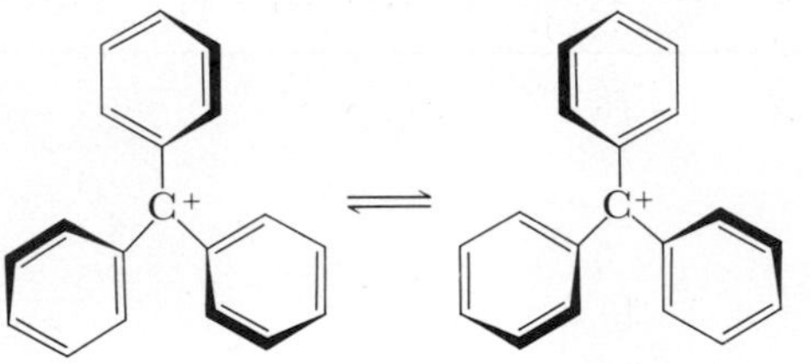

$$\Delta H^{\ddagger} \approx 9 \text{ kcal mole}^{-1}$$

      37   **38**

    The Hammett σ–ρ linear free-energy relationship is useful for evaluating substituent effects in systems **39**. Table 4.20 shows some ρ^+ values, obtained with

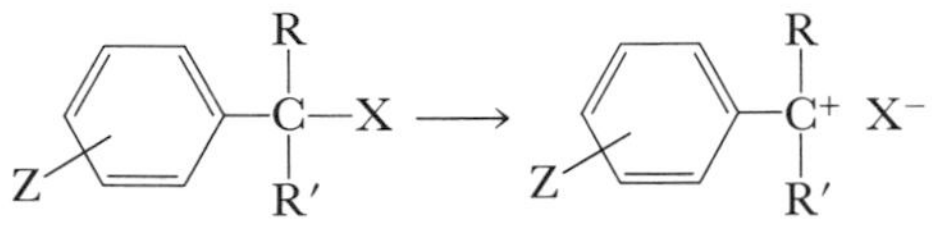

          (4.57)

      39

the σ^+ substituent parameters. One expects that the less stabilized the incipient ion, the later the transition state will occur along the reaction coordinate. The later the transition state, the larger will be the positive charge at the substitution center, and a larger positive charge will mean a larger demand for electron density from the substituents and hence a more negative ρ^+. The triphenylmethyl system, number 4 in Table 4.20, which is well stabilized, has $\rho^+ = -2.7$. The less-stabilized benzhydryl system, numbers 2 and 3, has ρ^+ of about -4.0. If the incipient cation is part of a small ring (numbers 6 and 7), angle strain increases as the transition state is approached because the cation prefers a bond angle of 120° as opposed to 109° preferred by the sp^3 substrate.[138] This increased angle strain causes a later transition state and a more negative ρ^+.[139] The cyclohexyl system (number 9) also resists

---

[137] I. I. Schuster, A. K. Colter, and R. J. Kurland, *J. Am. Chem. Soc.*, **90**, 4679 (1968).

[138] H. C. Brown, M. Ravindranathan, E. N. Peters, C. G. Rao, and M. M. Rho, *J. Am. Chem. Soc.*, **99**, 5373 (1977).

[139] Cyclopropyl cations are difficult to study because they open to allyl cations:

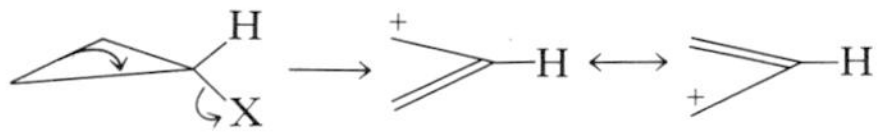

(see Problem 6 and Section 11.2). They can be formed if the cyclopropyl group is incorporated into a bicyclic ring, for example,

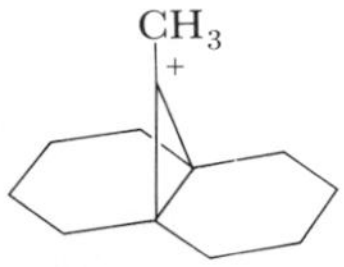

See (a) G. A. Olah, G. Liang, D. B. Ledlie, and M. G. Costopoulos, *J. Am. Chem. Soc.*, **99**, 4196 (1977); (b) D. B. Ledlie, W. Barber, and F. Switzer, *Tetrahedron Lett.*, 607 (1977); (c) X. Creary, *J. Org. Chem.*, **41**, 3734, 3740 (1976).

Table 4.20 Values of Hammett Reaction Constant for Solvolysis Reactions

Substrate	Conditions	ρ^+	Reference
1. 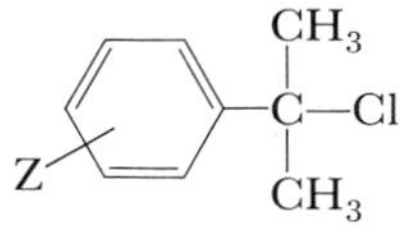 Z—C₆H₄—C(CH₃)₂—Cl	90% aqueous acetone, 25°C	-4.54	a
2. 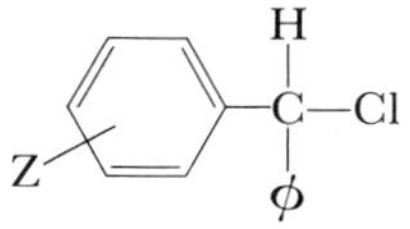 Z—C₆H₄—CH(φ)—Cl	2-Propanol, 25°C	-4.06	b
3. 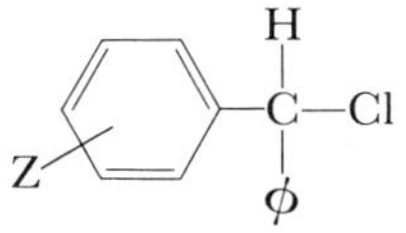 Z—C₆H₄—CH(φ)—Cl	Ethanol, 25°C	-4.05	b
4. 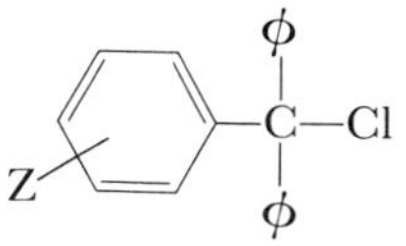 Z—C₆H₄—C(φ)₂—Cl	40% ethanol–60% diethylether, 0°C	-2.68	b
5. 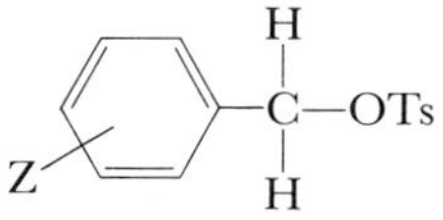 Z—C₆H₄—CH₂—OTs	Acetic acid, 40°C	$\begin{cases} -5.71^1 \\ -2.33^2 \end{cases}$	c
6. 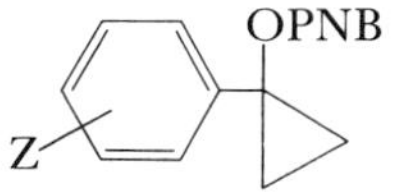 Z—C₆H₄—(cyclopropyl)—OPNB	80% aqueous acetone, 25°C	-5.15	d
7. 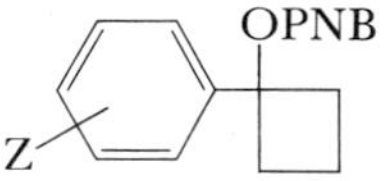 Z—C₆H₄—(cyclobutyl)—OPNB	80% aqueous acetone, 25°C	-4.91	d
8. 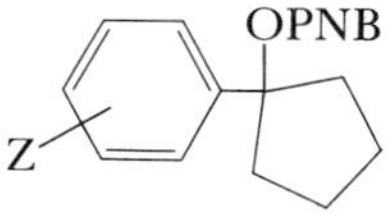 Z—C₆H₄—(cyclopentyl)—OPNB	80% aqueous acetone, 25°C	-3.82	d
9. 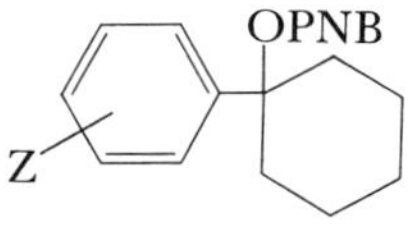 Z—C₆H₄—(cyclohexyl)—OPNB	80% aqueous acetone, 25°C	-4.60	d
10. 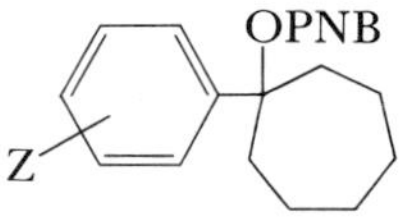 Z—C₆H₄—(cycloheptyl)—OPNB	80% aqueous acetone, 25°C	-3.87	d
11. 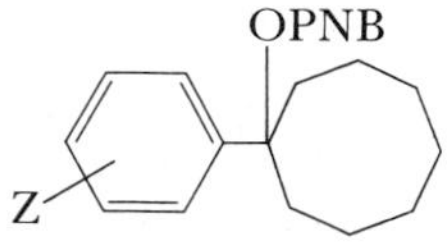 Z—C₆H₄—(cyclooctyl)—OPNB	80% aqueous acetone, 25°C	-3.83	d

Table 4.20 (*Continued*)

Substrate	Conditions	ρ^+	Reference
12. 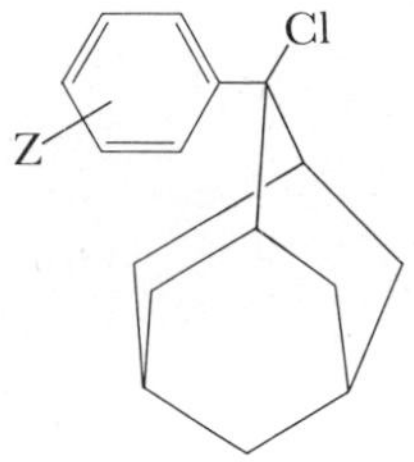	90% aqueous acetone, 25°C	−4.83	e
13. 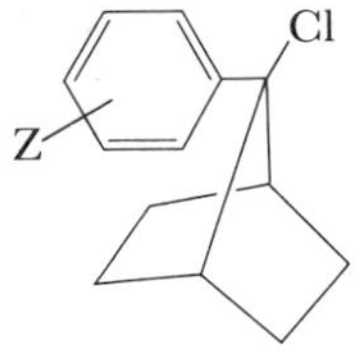	90% aqueous acetone, 25°C	−5.64	e
14. 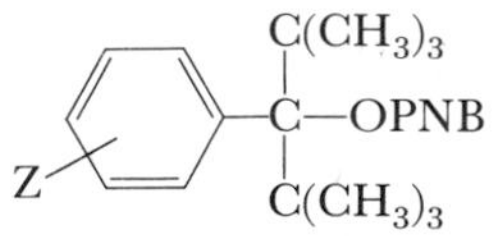	70% aqueous acetone, 100°C	−1.30	f

[1] Substituents *p*—CH$_3$, *m*—CH$_3$, *p*—F, H, *p*—Cl.
[2] Substituents *m*—Cl, *m*—CF$_3$, *p*—CF$_3$.
[a] H. C. Brown and Y. Okamoto, *J. Am. Chem. Soc.*, **80**, 4979 (1958).
[b] Y. Okamoto and H. C. Brown, *J. Org. Chem.*, **22**, 485 (1957).
[c] A. Streitwieser, Jr., H. A. Hammond, R. H. Jagow, R. M. Williams, R. G. Jesaitis, C. J. Chang, and R. Wolf, *J. Am. Chem. Soc.*, **92**, 5141 (1970).
[d] H. C. Brown, M. Ravindranathan, E. N. Peters, C. G. Rao, and M. M. Rho, *J. Am. Chem. Soc.*, **99**, 5373 (1977).
[e] H. Tanida and T. Tsushima, *J. Am. Chem. Soc.*, **92**, 3397 (1970).
[f] H. Tanida and H. Matsumura, *J. Am. Chem. Soc.*, **95**, 1586 (1973).

introduction of a cationic center, presumably partly because of increased eclipsing as the transition state is approached. A sterically crowded reactant (number 14) can cause an early transition state and a small ρ^+. There is another factor involved in this instance: the bulky substituents prevent coplanarity of the aryl group, and conjugation is decreased.[140]

Heteroatom substituents When atoms with unshared pairs of electrons are bonded to the reaction center in an S$_N$1 reaction, two effects must be considered: the inductive effect, which is usually electron withdrawing, and the electron-donating conjugative effect (**40**). For the more basic atoms, O, N, S, conjugation is

[140] (a) S. P. McManus and J. M. Harris, *J. Org. Chem.*, **42**, 1422 (1977); (b) J. S. Lomas and J.-E., Dubois, *J. Org. Chem.*, **40**, 3303 (1975).

$$\overset{+}{\underset{/}{\overset{\diagdown}{C}}}-\overset{..}{E}: \longleftrightarrow \overset{\diagdown}{\underset{/}{C}}=\overset{+}{E}$$

40

dominant, as the relative solvolysis rates of **41** and **42** show. Insulation from the

$$C_2H_5-O-CH_2-Cl \qquad CH_3CH_2CH_2CH_2-Cl \qquad C_2H_5-O-CH_2CH_2-Cl$$

41   **42**   **43**

Relative rate: 10^9 1   0.2

conjugative influence by one CH_2 group (**43**) leaves the inductive effect to cause a rate depression.[141] These comparisons have only qualitative significance, because the compounds probably solvolyze with different degrees of nucleophilic assistance.

With α halogens the two effects are more nearly balanced. Fluorine is highly electronegative and decreases the rate slightly despite its unshared electrons; chlorine and bromine increase rates, but much less than does oxygen ($k_{halogen}/k_H$ ratios 10 to 500). The reason could be that there is less effective overlap with carbon by the larger $3p$ or $4p$ orbitals that contain the unshared pairs on Cl and Br than for a substituent with unpaired electrons in a $2p$ orbital. In the gas phase, sulfur appears to be better at stabilizing positive charge than oxygen,[142] probably because of greater polarizability. Chemical shifts in ^{13}C nuclear magnetic resonance spectra independently demonstrate the decreasing effectiveness of donation of electron density by conjugation to an adjacent positive carbon in the order $F > Cl > Br$.[143]

Bridgehead systems Another important consequence of structural change, first observed and explained by Bartlett and Knox for the apocamphyl system (**44**), is the resistance to cation formation at a strained bridgehead position.[144] Table 4.21 lists approximate rates for some bridgehead systems relative to t-butyl.

44

By analogy with the isoelectronic $(CH_3)_3B$, in which the boron and three carbons are coplanar,[145] $(CH_3)_3C^+$ should prefer a geometry with all carbons co-

[141] A. Streitwieser, Jr., *Solvolytic Displacement Reactions,* McGraw-Hill, New York, 1962, pp. 102–103.
[142] J. K. Pau, M. B. Ruggera, J. K. Kim, and M. C. Caserio, *J. Am. Chem. Soc.,* **100,** 4242 (1978).
[143] G. A. Olah, Y. K. Mo, and Y. Halpern, *J. Am. Chem. Soc.,* **94,** 3551 (1972).
[144] P. D. Bartlett and L. H. Knox, *J. Am. Chem. Soc.,* **61,** 3184 (1939).
[145] H. A. Lévi and L. O. Brockway, *J. Am. Chem. Soc.,* **59,** 2085 (1937).

planar. Many theoretical calculations,[146] and spectroscopic observations[147] are consistent with coplanar structures in unconstrained carbenium ions. In small-ring bridgehead ions, coplanarity entails a large increase in ring strain. The systems with the greatest strain increase upon passing from ground state to transition state react slowest. In larger rings constraints of the ring systems can cause rate increases, a point illustrated by the first example in Table 4.21. In this bicyclo[3.3.3] system, which consists of fused eight-membered rings, the ground-state molecule is strained by nonbonded interactions between the bridging CH_2 groups; a planar cation at the bridgehead relieves some of this strain. Reaction rates and calculated strain energies correlate well for the bridgehead systems.[148]

The bridgehead ions, even though they usually cannot achieve a coplanar geometry, are stabilized by hyperconjugative electron donation from attached groups. In bridgehead systems the conformation is fixed by the rings, and the transition state for S_N1 solvolysis is usually **45**. The numbering in **45** is keyed to the

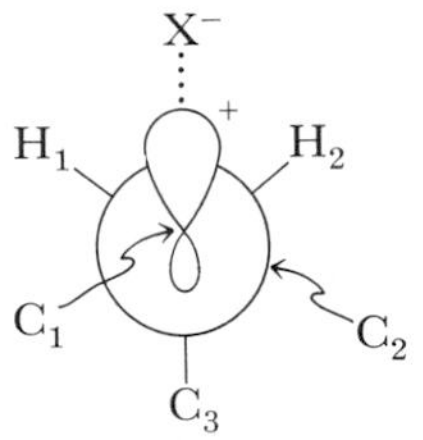

45

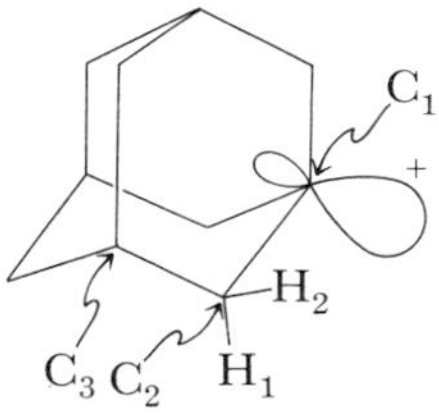

46 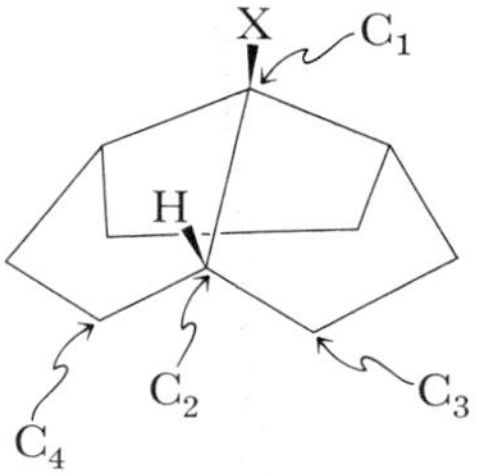   **47** 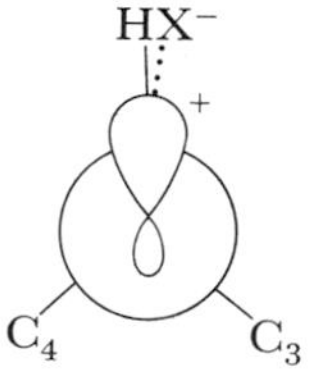   **48**

1-adamantyl cation (**46**), a typical bridgehead ion. The carbon C_3 is in the trans periplanar position with respect to the leaving group. The system **47**, which solvolyzes at roughly 3×10^{-5} times the rate of 1-adamantyl (**46**), has the conformation **48**, with a cis periplanar hydrogen and no trans periplanar group. It thus

[146] See, for example, (a) J. E. Williams, Jr., R. Sustmann, L. C. Allen, and P. v. R. Schleyer, *J. Am. Chem. Soc.*, **91**, 1037 (1969); (b) L. Radom, J. A. Pople, V. Buss, and P. v. R. Schleyer, *J. Am. Chem. Soc.*, **94**, 311 (1972); (c) L. Radom, P. C. Hariharan, J. A. Pople, and P. v. R. Schleyer, *J. Am. Chem. Soc.*, **95**, 6531 (1973); for a review see (d) V. Buss, P. v. R. Schleyer, and L. C. Allen, *Top. Stereochem.*, **7**, 253 (1973).
[147] G. A. Olah, J. R. DeMember, A. Commeyras, and J. L. Bribes, *J. Am. Chem. Soc.*, **93**, 459 (1971).
[148] (a) R. C. Bingham and P. v. R. Schleyer, *J. Am. Chem. Soc.*, **93**, 3189 (1971); (b) W. Parker, R. L. Tranter, C. I. F. Watt, L. W. K. Chang, and P. v. R. Schleyer, *J. Am. Chem. Soc.*, **96**, 7121 (1974); (c) M. R. Smith and J. M. Harris, *J. Org. Chem.*, **43**, 3588 (1978).

Table 4.21 APPROXIMATE SOLVOLYSIS RATES OF BRIDGEHEAD SYSTEMS RELATIVE TO *t*-BUTYL

Structure	Relative rate
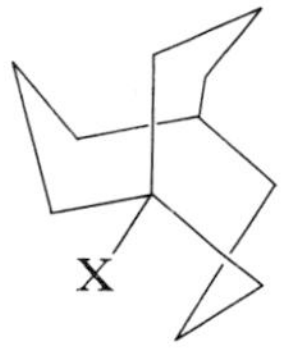	10^{4a}
$(CH_3)_3C—X$	1^{b}
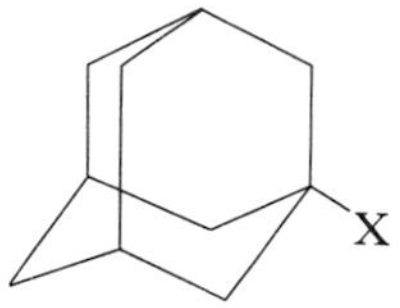	10^{-3b}
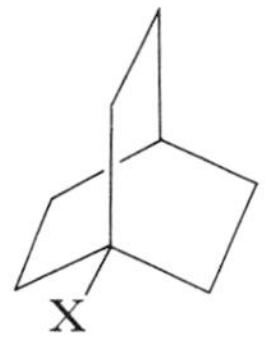	10^{-7b}
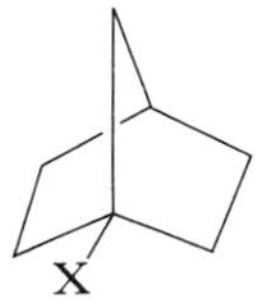	10^{-13b}

[a] W. Parker, R. L. Tranter, C. I. F. Watt, L. W. K. Chang, and P. v. R. Schleyer, *J. Am. Chem. Soc.,* **96,** 7121 (1974).
[b] Calculated for tosylates in acetic acid at 70°C using data of R. C. Bingham and P. v. R. Schleyer, *J. Am. Chem. Soc.,* **93,** 3189 (1971) and of E. Grunwald and S. Winstein, *J. Am. Chem. Soc.,* **70,** 846 (1948).

appears that hyperconjugative electron donation occurs more readily at the back of an incipient electron-deficient center than at the front; just as in S_N2 substitutions, attack by nucleophile from the side opposite the departing group is more favorable than attack from the same side.[149]

Studies of substituent effects in the bridgehead bicyclo[2.2.2]octyl (**49**) and adamantyl (**50**) series have been carried out by Schleyer and Woodworth.[150] Corre-

[149] See note 148(a), (b).
[150] P. v. R. Schleyer and C. W. Woodworth, *J. Am. Chem. Soc.,* **90,** 6528 (1968).

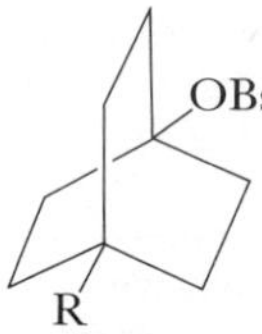

49 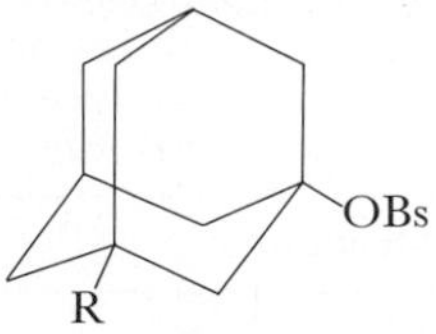   50

lations with the Taft inductive $\sigma^*_{CH_2}$ parameter had the negative slope expected for a reaction accelerated by electron-donating groups. The electron-donating inductive effect of the alkyl groups increases along the series methyl < ethyl < isopropyl < *t*-butyl. The rate differences on which this order is based are small, a factor of 2 between methyl and *t*-butyl in **49** and of 3 in **50**. The position of hydrogen in the series is different in **49** and **50**; this fact, and the small rate differences, make it unclear whether alkyl groups are electron donating or withdrawing compared to hydrogen in these compounds.

Vinyl Cations

The inertness of vinyl halides to substitution led to the expectation that vinyl cations (alkynium ions, **51**) would be very high in energy and difficult to observe. It has been found, however, that given a sufficiently reactive leaving group, like trifluoromethanesulfonate (triflate, OTf) the process indicated in Equation 4.58

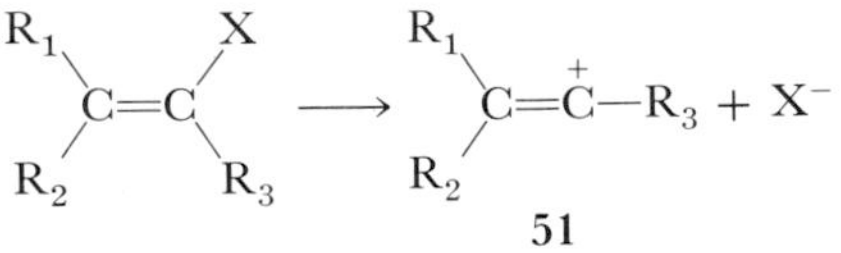

$$\tag{4.58}$$

51

can be achieved.[151] Detailed molecular orbital calculations predict that the vinyl cations should be linear (if $R_1 = R_2$),[152] as does simple VSEPR theory; experimental confirmation comes from observation of decreasing rate of solvolysis with decreasing ring size in cyclic triflates of Structure **52**. The smaller the ring, the more strained is a linear structure for the ion **53**.[153]

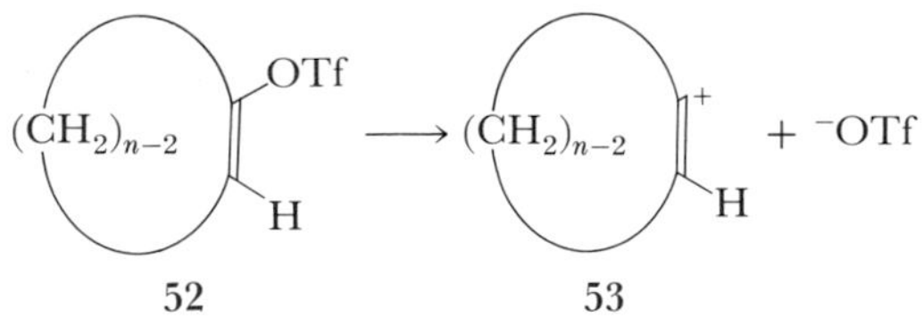

$$\tag{4.59}$$

52   53

[151] Reviews: (a) G. Modena and U. Tonellato, *Adv. Phys. Org. Chem.*, **9**, 185 (1971); (b) M. Hanack, *Acc. Chem. Res.*, **3**, 209 (1970); **9**, 364 (1976); (c) P. J. Stang, *Prog. Phys. Org. Chem.*, **10**, 205 (1973); (d) P. J. Stang, *Acc. Chem. Res.*, **11**, 107 (1978).
[152] (a) W. A. Latham, W. J. Hehre, and J. A. Pople, *J. Am. Chem. Soc.*, **93**, 808 (1971). See also (b) R. H. Summerville, and P. v. R. Schleyer, *J. Am. Chem. Soc.*, **96**, 1110 (1974).
[153] W. D. Pfeifer, C. A. Bahn, P. v. R. Schleyer, S. Bocher, C. E. Harding, K. Hummel, M. Hanack, and P. J. Stang, *J. Am. Chem. Soc.*, **93**, 1513 (1971).

The solvolysis of vinyl triflates **54** show $\rho^+ = -4.1$, indicative of a typical S_N1 transition state. The β hydrogen isotope effect k_H/k_D of 1.20 per deuterium is

$$H_2C=C \begin{smallmatrix} \phi-Z \\ \\ OTf \end{smallmatrix} \qquad \begin{smallmatrix} Ar \\ \end{smallmatrix} C=C \begin{smallmatrix} Ar \\ \\ Br \end{smallmatrix}$$

54   55

unusually large; the reasons are the favorable 0° and 180° dihedral angles between the C—H (D) bonds and the leaving group, which leads to particularly effective hyperconjugation, and the presence of only one other stabilizing group on the cationic carbon to help delocalize charge.[154] When there are three charge-stabilizing aryl groups, solvolysis is possible with bromide—a less reactive leaving group, as in Structure **55**.[155] Vinyl cations undergo rearrangements,[156] and can be formed by other methods, notably protonation of alkynes.[157]

The phenyl cation is apparently significantly less stable than the vinyl cations. A number of attempts to generate it by solvolysis have failed,[158] although it does appear to be an intermediate in decomposition of phenyl diazonium ion, ϕN_2^+,[159] and in the cyclization that occurs during solvolysis of the trifluoromethanesulfonate **56**.[160]

$$(4.60)$$

56   9% of product

Isotope Effects[161]

Isotope effects measured for isotopic substitution in the substrate are usually for H—D substitution, either at the reaction site (α-d effects) or at the β carbon (β-d effects). Both of these substitutions lead to secondary isotope effects (Section 2.7), with k_H/k_D less than 1.5.

[154] P. J. Stang, R. J. Hargrove, and T. E. Dueber, *J. Chem. Soc., Perkin Trans. 2*, 1486 (1977).

[155] Z. Rappoport and Y. Apeloig, *J. Am. Chem. Soc.*, **97**, 821, 826 (1975).

[156] P. J. Stang and T. E. Dueber, *Tetrahedron Lett.*, 563, (1977).

[157] (a) G. A. Olah and H. Mayr, *J. Am. Chem. Soc.*, **98**, 7333 (1976); (b) M. J. Chandy and M. Hanack, *Tetrahedron Lett.*, 4377 (1977).

[158] (a) L. R. Subramanian, M. Hanack, L. W. K. Chang, M. A. Imhoff, P. v. R. Schleyer, F. Effenberger, W. Kurtz, P. J. Stang, and T. E. Dueber, *J. Org. Chem.*, **41**, 4099 (1976); (b) A. Streitwieser, Jr., and A. Dafforn, *Tetrahedron Lett.*, 1435 (1976).

[159] (a) R. G. Bergstrom, R. G. M. Landells, G. H. Wahl, Jr., and H. Zollinger, *J. Am. Chem. Soc.*, **98**, 3301 (1976); (b) C. G. Swain, J. F. Sheats, and K. G. Harbison, *J. Am. Chem. Soc.*, **97**, 783 (1975).

[160] M. Hanack and U. Michel, *Angew. Chem. Int. Ed. Engl.*, **18**, 870 (1979).

[161] (a) V. J. Shiner, Jr., in *Isotope Effects in Chemical Reactions*, C. J. Collins and N. S. Bowman, Eds., Van Nostrand Reinhold, New York, 1970, Chap., 2; (b) H. Simon and D. Palm, *Angew. Chem. Int. Ed. Engl.*, **5**, 920 (1966); (c) A. Streitwieser, Jr., *Solvolytic Displacement Reactions*, McGraw-Hill, New York, 1962, pp. 98–101; 172–174.

For each leaving group there is a maximum for the α-d effect, typically 1.22 to 1.25 for sulfonate leaving groups, about 1.15 for Cl, and about 1.13 for Br.[162] These maxima occur for substrate structures and solvents that we have associated with S_N1-type reactions, while lower ratios, close to 1.00 but perhaps extending up to 1.10 or even higher, are found under conditions favoring S_N2 reactions. The precise interpretation of these effects is controversial, with some investigators considering effects near the maximum for the particular leaving group to indicate rate-determining transformation of contact to solvent-separated ion pair, intermediate values to indicate rate-determining attack of nucleophile on contact ion pair, and low values to indicate nucleophilic attack on substrate.[163] Others consider that nucleophilic attack on substrate encompasses low to medium values, with values near the maxima indicating rate-determining contact-ion-pair formation without assistance of external nucleophile.[164]

In general terms, as we have seen in Section 2.7, the origin of the rate change is in the out-of-plane bending vibration, which decreases in frequency on going from the sp^3-hybridized starting material to the transition state, where hybridization is approaching sp^2. The presence of the leaving group and an entering nucleophile nearby stiffens the bond and makes the frequency change smaller; hence S_N2 reactions show little or no rate change on isotopic substitution. It is on the basis of this reasoning, as well as by comparison with other indicators of mechanism, that the α-d isotope effect is considered to be a measure of the degree of participation by nucleophile, and of the degree of ionization of the substrate, in the transition state.

The β-d effects are typically around $k_H/k_D = 1.10$ per deuterium, but they can be as high as 1.44 for situations in which the H (D) is held at a 180° dihedral angle with the axis of the carbon–leaving-group bond.[165] The β-d effects are thought to reflect delocalization of the developing positive charge to the β C—H (D) bonds.

Direct Observation of Carbocations[166]

Most carbocations are too reactive to be directly observable in ordinary solvents, and until the advent of superacids evidence was obtained indirectly, primarily

[162](a) J. M. Harris, R. E. Hall, and P. v. R. Schleyer, *J. Am. Chem. Soc.*, **93**, 2551 (1971); (b) V. J. Shiner, Jr., and R. D. Fisher, *J. Am. Chem. Soc.*, **93**, 2553 (1971); (c) T. W. Bentley, S. H. Liggero, M. A. Imhoff, and P. v. R. Schleyer, *J. Am. Chem. Soc.*, **96**, 1970 (1974); (d) E. A. Halevi, *Prog. Phys. Org. Chem.*, **1**, 109 (1963); (e) V. J. Shiner, Jr., W. E. Buddenbaum, B. L. Murr, and G. Lamaty, *J. Am. Chem. Soc.*, **90**, 418 (1968); (f) A. Streitwieser, Jr. and G. A. Dafforn, *Tetrahedron Lett.*, 1263 (1969); (g) G. A. Dafforn and A. Streitwieser, Jr., *Tetrahedron Lett.*, 3159 (1970).
[163] V. J. Shiner, Jr., in *Isotope Effects in Chemical Reactions*, C. J. Collins and N. S. Bowman, Eds., Van Nostrand Reinhold, New York, 1970, Chap. 2.
[164] T. W. Bentley and P. v. R. Schleyer, *Adv. Phys. Org. Chem.*, **14**, 1 (1977).
[165] V. J. Shiner, Jr., and J. G. Jewett, *J. Am. Chem. Soc.*, **86**, 945 (1964).
[166] We will not cover the extensive work that has been done on molecular orbital calculation of carbocation structure and properties. See (a) W. A. Latham, L. A. Curtiss, W. J. Hehre, J. B. Lisle, and J. A. Pople, *Prog. Phys. Org. Chem.*, **11**, 175 (1974); (b) W. J. Hehre, *Acc. Chem. Res.*, **8**, 369 (1975); (c) L. Radom, D. Poppinger, and R. C. Haddon, in *Carbonium Ions*, G. A. Olah and P. v. R. Schleyer, Eds., Wiley, New York, 1976, Vol. V, p. 2302; (d) M. J. S. Dewar and H. S. Rzepa, *J. Am. Chem. Soc.*, **99**, 7432 (1977); (e) M. J. S. Dewar and D. Landman, *J. Am. Chem. Soc.*, **99**, 7439, (1977); (f) Y. Apeloig, P. v. R. Schleyer, and J. A. Pople, *J. Am. Chem. Soc.*, **99**, 5901 (1977); (g) H.-J. Köhler and H. Lischka, *J. Am. Chem. Soc.*, **101**, 3479 (1979).

through the study of reaction kinetics and trapping processes. Nevertheless, a few types of compounds have long been known to produce observable concentrations of positive ions relatively easily. The triarylmethyl derivatives were the first of this type to be investigated; the halides ionize readily in nonnucleophilic solvents such as sulfur dioxide,[167] and the alcohols yield solutions of the ions in concentrated sulfuric acid. Early observations by the freezing-point depression technique established that each mole of triphenyl carbinol yields 4 moles of ions in sulfuric acid, the reaction presumably being by way of Equation 4.61.[168] Results in methanesulfonic acid are similar.[169]

$$\phi_3COH + 2H_2SO_4 \longrightarrow \phi_3C^+ + H_3O^+ + 2HSO_4^- \tag{4.61}$$

The cryoscopic method is also applicable to other triarylmethyl systems, to some diarymethyl and allylic ions, and, when ortho substituents are present, to aryl acylium ions (**57**) (Equation 4.62);[170] however, side reactions frustrate most attempts to generate carbocations in sulfuric acid.

$$\text{CH}_3\text{-(ring)-C(=O)-OH} + 2H_2SO_4 \longrightarrow \text{CH}_3\text{-(ring)-}\overset{+}{C}{=}O + H_3O^+ + 2HSO_4^- \tag{4.62}$$

57

Development of the superacid solvent systems has permitted the preparation at low temperature of stable solutions of carbocations of many structural types. The solvents ordinarily used consist of the strong Lewis acid antimony pentafluoride with or without an added protonic acid, usually hydrofluoric or fluorosulfuric acid. A substance of very low basicity such as SO_2, SO_2ClF, or SO_2F_2 serves as diluent when required. As we have seen in Section 3.2, these solvent systems are considerably more acidic than concentrated sulfuric acid as measured by the H_0 acidity function.[171] Olah and his co-workers have made extensive contributions to this field.[172] The ready availability of solutions of many types of carbocations has

[167] N. N. Lichtin, *Prog. Phys. Org. Chem.*, **1**, 75 (1963).

[168] (a) A. Hantzsch, *Z. Phys. Chem.*, **61**, 257 (1907); (b) L. P. Hammett and A. J. Deyrup, *J. Am. Chem. Soc.*, **55**, 1900 (1933).

[169] R. A. Craig, A. B. Garrett, and M. S. Newman, *J. Am. Chem. Soc.*, **72**, 163 (1950).

[170] (a) H. P. Treffers and L. P. Hammett, *J. Am. Chem. Soc.*, **59**, 1708 (1937); (b) M. S. Newman and N. C. Deno, *J. Am. Chem. Soc.*, **73**, 3644 (1951); (c) N. C. Deno, H. G. Richey, Jr., J. D. Hodge, and M. J. Wisotsky, *J. Am. Chem. Soc.*, **84**, 1498 (1962).

[171] R. J. Gillespie and T. E. Peel, *Adv. Phys. Org. Chem.*, **9**, 1 (1971). See Figure 3.5.

[172] For reviews see (a) G. A. Olah and J. A. Olah in *Carbonium Ions*, Vol. II, Olah and Schleyer, Eds., Wiley, New York, 1970, p. 715; (b) R. J. Gillespie, *Acc. Chem. Res.*, **1**, 202 (1968).

made possible spectroscopic observations of a greatly expanded variety of structures. Nuclear magnetic resonance, both proton and ^{13}C, has been fruitful and has yielded information not only about structure but also about rearrangement processes; other methods, particularly infrared and Raman spectroscopy, have proved informative as well. X-Ray photoelectron spectroscopy (electron spectroscopy for chemical analysis, ESCA), which measures binding energies of $1s$ electrons of the carbon atoms, yields information about delocalization of charge with the ion.[173]

Figures 4.15 through 4.17 show proton magnetic resonance spectra of some simple carbocations in superacid solutions. Note the large shifts to low field that

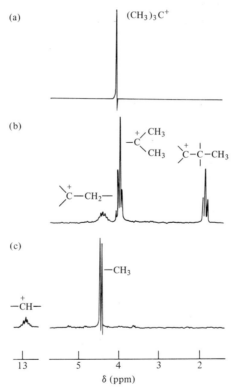

Figure 4.15 Proton magnetic resonance spectra of (a) $(CH_3)_3C^+$; (b) $(CH_3)_2\overset{+}{C}-C_2H_5$; (c) $(CH_3)_2\overset{+}{C}H$, recorded at 60 MHz in SbF_5-SO_2ClF at $-60°C$. G. A. Olah, *Angew. Chem. Int. Ed. Engl.*, **12**, 173 (1973). Reprinted by permission of Verlag Chemie Gmbh.

[173] See the following reviews in G. A. Olah and P. v. R. Schleyer, Eds., *Carbonium Ions,* Vol. I, Wiley, New York, 1968: (a) electronic spectra, G. A. Olah, C. U. Pittman, Jr., and M. C. R. Symons, p. 153; (b) vibrational spectra, J. C. Evans, p. 223; (c) NMR spectra, G. Fraenkel and D. G. Farnum, p. 237; a review of applications of all the spectroscopic techniques to carbocation structures and reactions is (d) G. A. Olah, *Angew. Chem. Int. Ed.*, **12**, 173 (1973); see also (e) D. G. Farnum, *Adv. Phys. Org. Chem.*, **11**, 123 (1975).

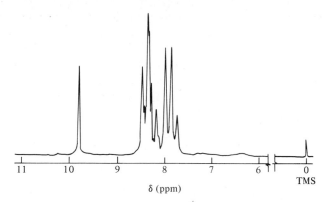

Figure 4.16 Proton magnetic resonance spectrum of $\phi_2\overset{+}{C}H$, recorded at 60 MHz in SO_2 at $-30°C$. The proton assignments are ortho, -7.92δ; meta, -8.49δ; para, -8.37δ, $\overset{+}{C}H$, -9.8δ. Reprinted with permission from G. A. Olah, *J. Am. Chem. Soc.*, **86**, 932 (1964). Copyright 1964 American Chemical Society.

accompany the introduction of positive charge. In the adamantyl cation (Figure 4.17), the γ protons are shifted farther downfield than are the δ protons, even though they are farther from the positive charge. The reason is overlap inside the molecular cage structure of the back lobes of the C—H bonding orbitals with the vacant orbital at the positive center, as shown in Structure **58**.

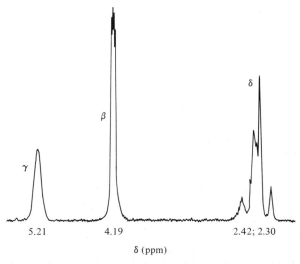

Figure 4.17 Proton magnetic resonance spectrum of 1-adamantyl cation, recorded at 100 MHz. G. A. Olah, *Angew. Chem. Int. Ed. Engl.*, **12**, 173 (1973). Reprinted by permission of Verlag Chemie Gmbh.

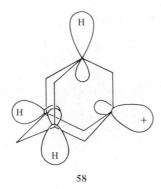

58

Recall the earlier discussion of Equation 4.63, which shows that all carbocation react with roughly the same selectivity toward all nucleophiles.[174] It has been

$$\log \frac{k}{k_0} = N_+ \qquad (4.63)$$

proposed that this simple relationship implies that the activation energies of nucleophile–cation combination reactions are controlled entirely by changes in solvation of nucleophile and cation. The more reactive cations are more strongly solvated, so that there is an overall cancellation of varying intrinsic reactivity (which should lead to varying selectivity) by desolvation energy varying in the opposite way.[175] This explanation has been criticized as requiring an unreasonably high degree of correlation of properties; however, it does appear that solvation must in some way control the combination reaction of free cations with nucleophiles.[176]

PROBLEMS

1. Predict which would be more reactive as a nucleophile in S_N2 substitution:

or $(CH_3CH_2CH_2)_3N$:

2. Suggest an explanation for the fact that the order of reactivity of the halides toward n-butyl brosylate in acetone is $Cl^- > Br^- > I^-$ when $(C_4H_9)_4N^+$ is the cation of the halide salt but $I^- > Br^- > Cl^-$ when Li^+ is the cation.

[174](a) C. D. Ritchie, *Acc. Chem. Res.*, **5**, 348 (1972); (b) C. D. Ritchie and P. O. I. Virtanen, *J. Am. Chem. Soc.*, **95**, 1882 (1973); (c) C. D. Ritchie, D. J. Wright, D.-S. Huang, and A. A. Kamego, *J. Am. Chem. Soc.*, **97**, 1163 (1975); (d) C. D. Ritchie, R. J. Minasz, A. A. Kamego, and M. Sawada, *J. Am. Chem. Soc.*, **99**, 3747 (1977).
[175] A. Pross, *Adv. Phys. Org. Chem.*, **14**, 69 (1977).
[176] C. D. Ritchie and M. Sawada, *J. Am. Chem. Soc.*, **99**, 3754 (1977).

3. Suggest a reason for the secondary isotope effect, $k_H/k_D = 1.10$ in the reactions

$$CH_3I + I^{*-} \longrightarrow CH_3I^* + I^-$$
$$CD_3I + I^{*-} \longrightarrow CD_3I^* + I^-$$

4. Show that the simple unassisted S_N1 mechanism predicts the kinetic behavior given in Equation 4.10, p. 296.

5. Explain the rate ratio for compounds **1** and **2** estimated for unassisted solvolysis.

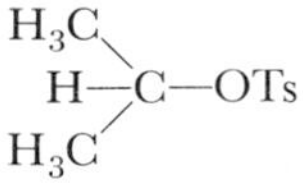

$$
\begin{array}{cc}
\underset{\text{H}_3\text{C}}{\overset{\text{H}_3\text{C}}{\diagdown}}\text{H}-\text{C}-\text{OTs} & \underset{(\text{CH}_3)_3\text{C}}{\overset{(\text{CH}_3)_3\text{C}}{\diagdown}}\text{CHOTs} \\
\mathbf{1} & \mathbf{2}
\end{array}
$$

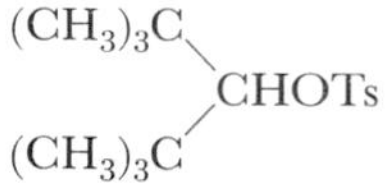

Relative rate
(estimate):   1 10^5

6. Explain the rate ratio for compounds **3** and **4** estimated for unassisted S_N1 solvolysis by correcting the cyclopropyl system for anchimeric assistance which occurs when the ring opens.

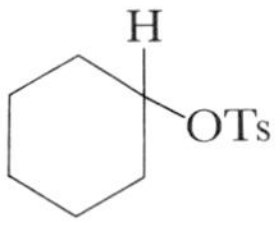

$$
\begin{array}{cc}
\mathbf{3} & \mathbf{4}
\end{array}
$$

Relative rate
(estimate):   1 10^{-10}

7. Nucleophiles such as NH_2OH and NH_2NH_2, which have an electronegative atom with one or more pairs of unshared electrons adjacent to the nucleophilic center, are more reactive than might be expected from the nature of the nucleophilic center, as can be seen from Table 4.9. Suggest a reason for this increased reactivity (this phenomenon is called the α effect).

8. Solvolysis rates of isopropyl tosylate and 2-adamantyl tosylate in 80 percent ethanol are measured with and without added azide. Define rate enhancement, R.E., as the ratio of rate with azide to rate without, and designate by f_{RN_3} the fraction of alkyl azide in the product. Explain the significance of the fact that the isopropyl results fit the equation

$$1 - \frac{1}{\text{R.E.}} = f_{RN_3}$$

while the 2-adamantyl results do not.

9. Silicon compounds of the type R_3SiOR undergo nucleophilic substitution. From the following experimental observations, decide whether the mostly likely mechanism is

(a) S_N1: $R_3SiOR' \longrightarrow R_3Si^+ \ ^-OR \xrightarrow{Y^-} R_3SiY$

(b) S_N2: $R_3SiOR' \xrightarrow{Y^-} Y-SiR_3 + \ ^-OR$

(c) Two-step: $R_3SiOR' \xrightarrow{Y^-(fast)} R_3\bar{S}iOR \xrightarrow{slow} R_3SiY + {}^-OR$
                                              $|$
                                              Y

(d) Two-step: $R_3SiOR' \xrightarrow[slow]{Y^-} R_3\bar{S}iOR \rightleftharpoons R_3Si-Y + {}^-OR$
                                              $|$
                                              Y

Experimental observations:

(a) Substitution almost always proceeds with retention or inversion—never with racemization.

(b) The reaction shown below gives only the product depicted:

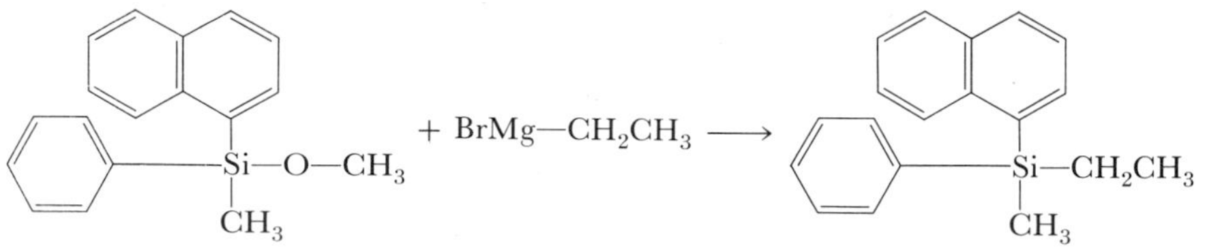

$+ BrMg-CH_2CH_3 \longrightarrow$

No

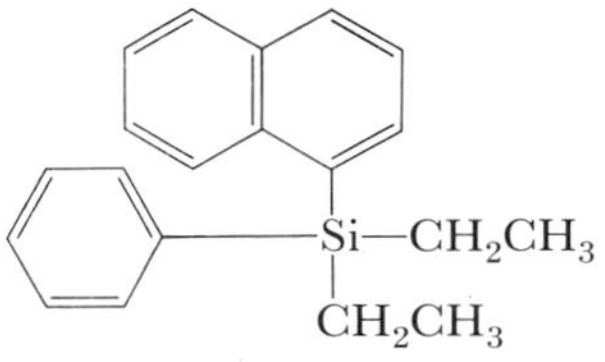

is formed.

(c) Acyloxysilanes (leaving group $R'-\overset{\overset{O}{\|}}{C}-O$) are considerably more reactive than alkoxysilanes.

10. The stereochemistry of the reaction shown below depends on the amount of butanol in the solvent. For example, when 2.3 percent by volume butanol is present in benzene, the reaction proceeds with 100 percent retention of configuration. When the solvent is 100 percent butanol, the reaction proceeds with 77 percent inversion.

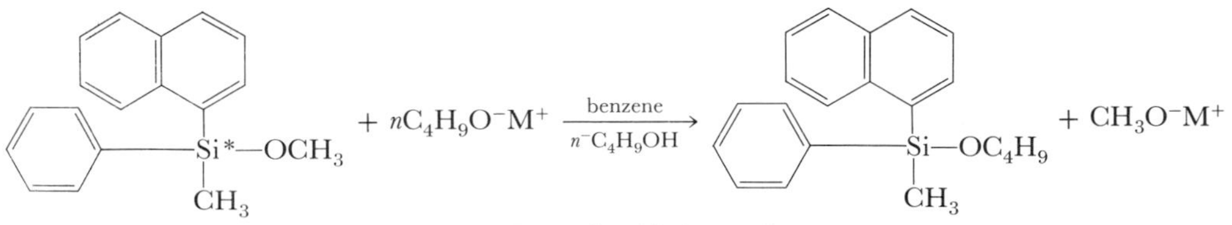

$+ nC_4H_9O^-M^+ \xrightarrow[n\text{-}C_4H_9OH]{benzene}$

$+ CH_3O^-M^+$

(* indicates the chiral center)

Suggest a reason for the change in stereochemistry and draw probable transition states.

11. Explain the relative rates of solvolysis of tosylates **5** and **6**.

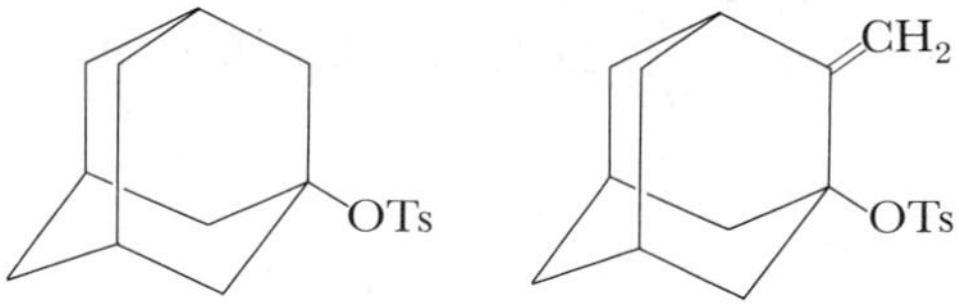

<table>
<tr><td></td><td>**5**</td><td>**6**</td></tr>
<tr><td>Relative rate:</td><td>1</td><td>6×10^{-5}</td></tr>
</table>

12. Nucleophiles that have more than one atom that may attack the substrate are called *ambident*. Using HSAB theory, predict with which atom the ambident nucleophile

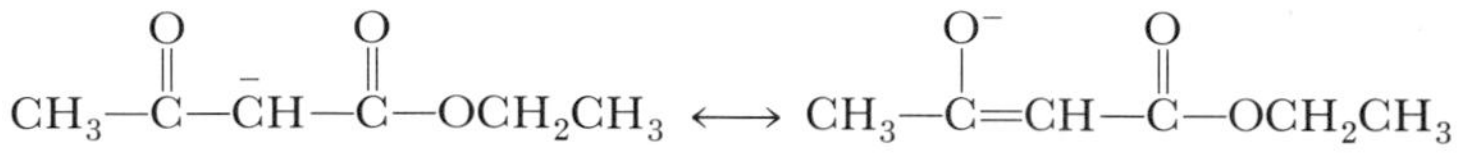

$$\underset{\text{O}}{\overset{\text{O}}{CH_3-\overset{\|}{C}-\overset{-}{C}H-\overset{\|}{C}-OCH_2CH_3}} \longleftrightarrow \underset{\text{O}}{\overset{\text{O}^-}{CH_3-\overset{|}{C}=CH-\overset{\|}{C}-OCH_2CH_3}}$$

will attack

(a) CH_3I (b) $\underset{Cl}{\overset{|}{C}H_2-OCH_3}$

13. Show how reaction paths with transition states in which *Nuc*···C bond making has progressed to a greater or lesser extent than C···X bond breaking are accommodated in a two-dimensional reaction coordinate diagram.

14. Using reaction coordinate diagrams, analyze the change in location of the S_N2 transition state expected when the nucleophile is replaced by a better one.

15. Explain why 1-adamantyl and 2-adamantyl derivatives cannot undergo elimination during solvolysis.

16. Alcohols react with thionyl chloride to yield the unstable chlorosulfites (**7**), which react further to the alkyl chloride and SO_2. (Rearrangement and elimination can also occur.) In dioxane, the product is formed with retention of configuration (**8**). If pyridinium hydrochloride is present, configuration is inverted (**9**). Explain.

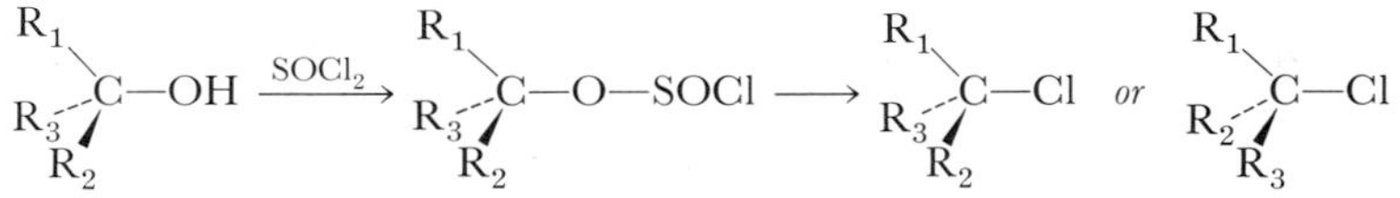

$$\underset{\text{R}_2}{\overset{\text{R}_1}{\underset{\diagup}{\overset{\diagdown}{C}}}}-OH \xrightarrow{SOCl_2} \underset{\text{R}_2}{\overset{\text{R}_1}{\underset{\diagup}{\overset{\diagdown}{C}}}}-O-SOCl \longrightarrow \underset{\text{R}_2}{\overset{\text{R}_1}{\underset{\diagup}{\overset{\diagdown}{C}}}}-Cl \; or \; \underset{\text{R}_3}{\overset{\text{R}_1}{\underset{\diagup}{\overset{\diagdown}{C}}}}-Cl$$

$$\qquad\qquad\qquad\qquad\quad **7** \qquad\qquad\qquad **8** \qquad\qquad **9**$$

17. Explain the relative magnitudes of the following pairs of β-deuterium isotope effects (per deuterium) for solvolysis:

(a)

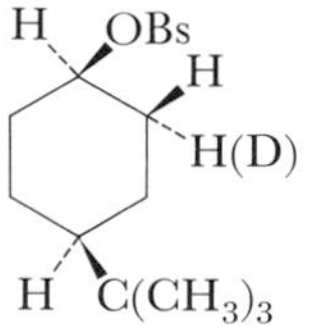

$\dfrac{k_H}{k_D} = 1.44$ $\dfrac{k_H}{k_D} = 1.10$

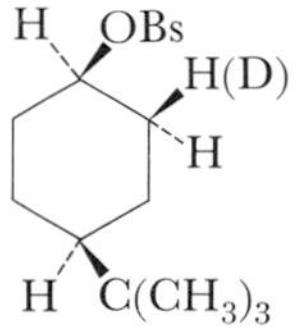

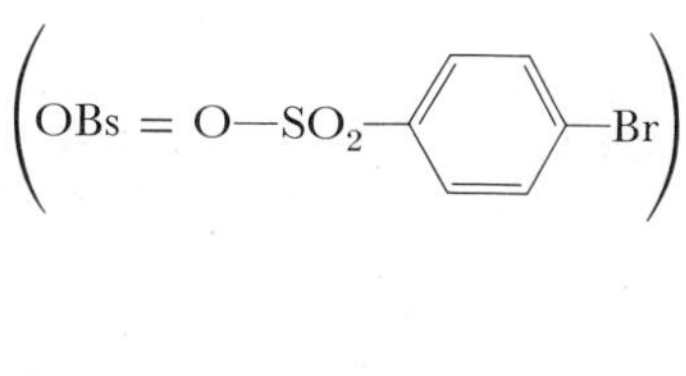

$$\left(OBs = O-SO_2-\!\!\left\langle\!\!\bigcirc\!\!\right\rangle\!\!-Br \right)$$

(b) $H_2(D_2)C=C$ with phenyl and OTf groups $\qquad \dfrac{k_H}{k_D} = 1.20$ $\qquad H_2(D_2)C=C$ with p-NO_2-phenyl and OTf groups $\qquad \dfrac{k_H}{k_D} = 1.31$

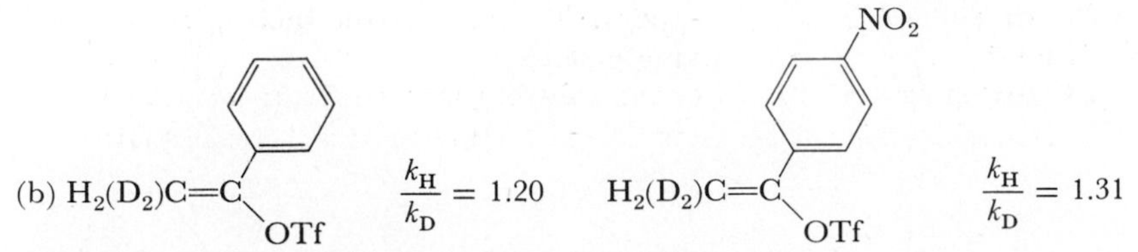

18. Predict products of solvolysis of the following bromides:

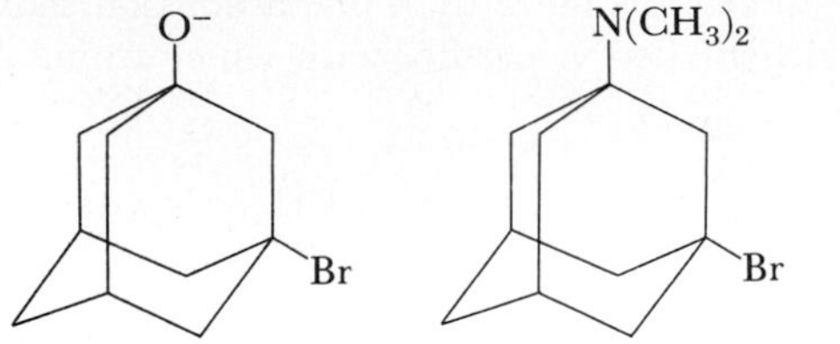

19. Explain the following observations:

$$3\% \qquad 97\%$$

$$23\% \qquad 77\%$$

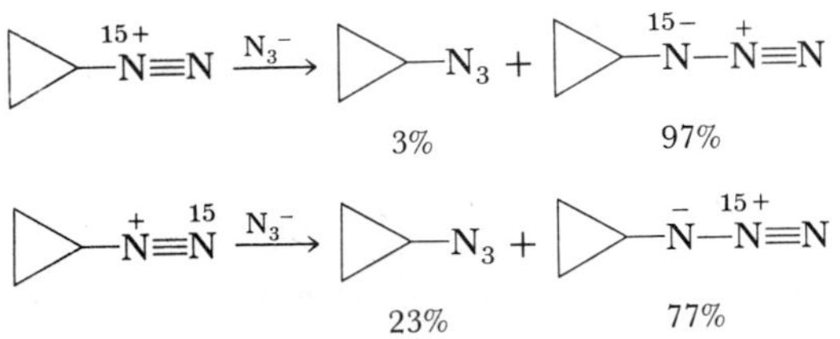

20. Explain the following observations:

$$-OCO\phi(NO_2)_2 \qquad \text{only}$$

$$OCO\phi(NO_2)_2 \qquad \text{only}$$

same mixture from either isomer

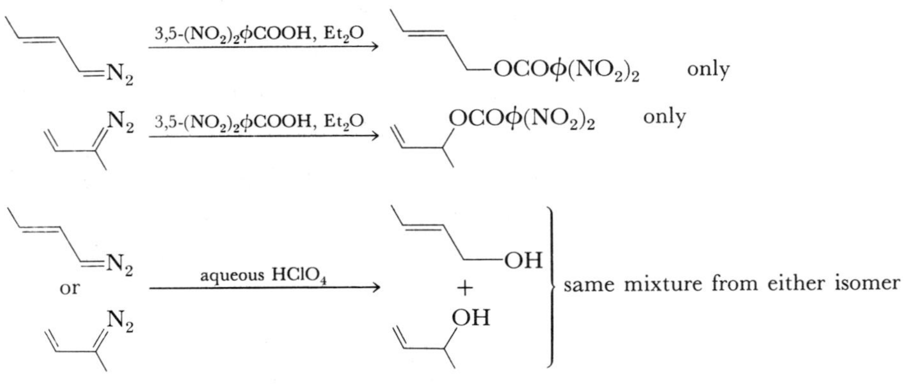

21. Explain the following observation:

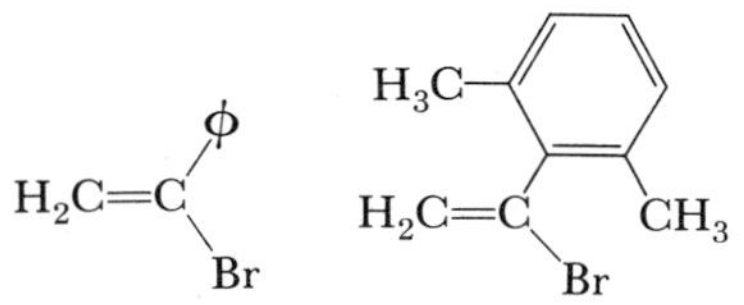

Relative solvolysis
rate, 80% percent ethanol: 1 10^5

22. In Table 4.2, p. 316, explain the order of the mole percent alkene in *t*-butyl solvolysis for the three halide leaving groups.

23. In Figure 4.17 interpret the splitting pattern for the δ protons.

24. Consider the solvolysis of the two isomeric chiral *p*-nitrobenzoates **10** and **11**.

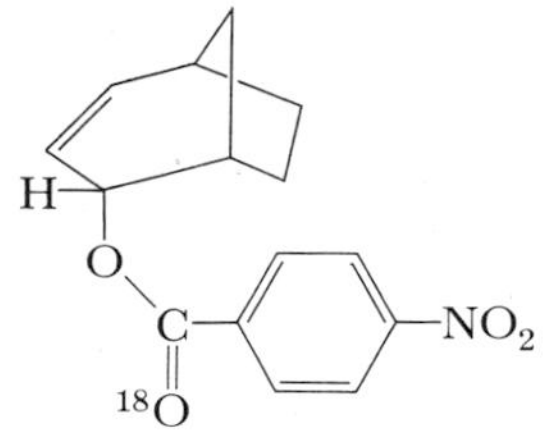

10 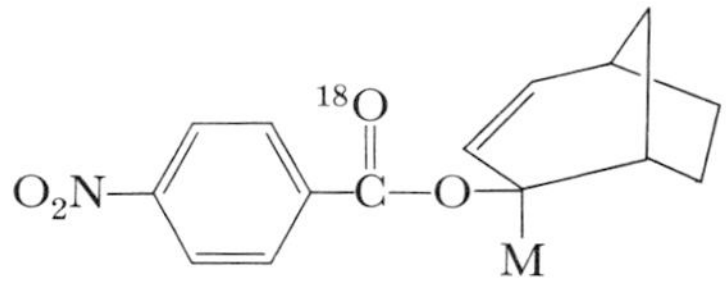   11

(a) Draw the structure of the free ion that would be formed by complete dissociation. What would be the stereochemistry of product formed from this ion?

(b) The compounds labeled with ^{18}O as indicated racemize during solvolysis; ^{18}O equilibration is also observed. For Isomer **10**, $k_{eq} = k_{rac}$. What does this result imply about the structure of the ion pair?

(c) For Isomer **11**, $k_{eq} > k_{rac}$. What does this result imply about the structure of the ion pair?

(d) Propose a reason for the difference of behavior of the two isomers.

25. McManus and Harris have defined a substituent constant γ^+ for the methyl group as a substituent directly attached to the reaction site by including the compound

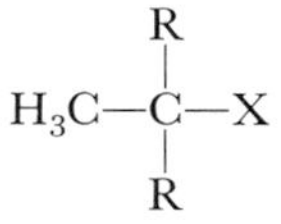

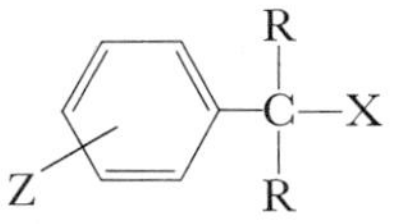

Hence γ^+ is on the same scale as σ^+ but refers to the methyl group as a member of the series

. $\gamma^+_{methyl} = 0.63$ in aqueous acetone and 0.79 in ethanol. Once γ^+ is measured, it is used to predict the rate for

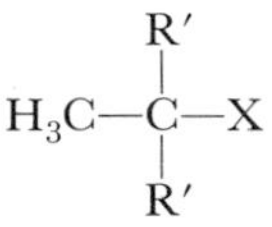

in a new series, using ρ^+ determined for

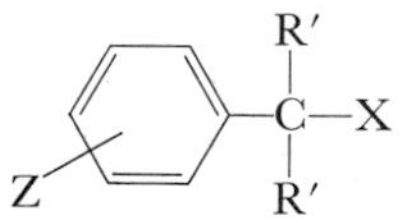

What is the significance of the observed deviation of measured from predicted rate in the following series

The methyl compound reacts faster than predicted.

REFERENCES FOR PROBLEMS

1. H. C. Brown and N. R. Eldred, *J. Am. Chem. Soc.*, **71**, 445 (1949).
2. S. Winstein, L. G. Savedoff, S. Smith, I. D. R. Stevens, and J. S. Gall, *Tetrahedron Lett.*, (9) 24 (1960).
3. M. Wolfsberg, *Acc. Chem. Res.*, **5**, 225 (1972).
5. S. H. Liggero, J. J. Harper, P. v. R. Schleyer, A. P. Krapcho, and D. E. Horn, *J. Am. Chem. Soc.*, **92**, 3789 (1970).
6. P. v. R. Schleyer, F. W. Sliwinski, G. W. Van Dine, U. Schöllkopf, J. Paust, and K. Fellenberger, *J. Am. Chem. Soc.*, **94**, 125 (1972).
7. R. F. Hudson, in *Chemical Reactivity and Reaction Rates*, G. Klopman, Ed., Wiley, New York, 1974, p. 203.
8. (a) J. M. Harris, D. J. Raber, R. E. Hall, and P. v. R. Schleyer, *J. Am. Chem. Soc.*, **92**, 5729 (1970); (b) R. A. Sneen and J. W. Larsen, *J. Am. Chem. Soc.*, **91**, 362 (1969).
9. L. H. Sommer, *Stereochemistry, Mechanism, and Silicon*, McGraw-Hill, New York, 1965, pp. 48*ff*; p. 67.
10. L. H. Sommer and H. Fujimoto, *J. Am. Chem. Soc.*, **90**, 982 (1968).
11. V. Buss, R. Gleiter, and P. v. R. Schleyer, *J. Am. Chem. Soc.*, **93**, 3927 (1971).
12. R. G. Pearson and J. Songstad, *J. Am. Chem. Soc.*, **89**, 1827 (1967).
16. (a) D. J. Cram, *J. Am. Chem. Soc.*, **75**, 332 (1953); (b) C. C. Lee and A. J. Finlayson, *Can. J. Chem.*, **39**, 260 (1961); (c) C. C. Lee, J. W. Clayton, D. G. Lee, and A. J. Finlayson, *Tetrahedron*, **18**, 1395 (1962); (d) C. E. Boozer and E. S. Lewis, *J. Am. Chem. Soc.*, **75**, 3182 (1953).
17. (a) V. J. Shiner, Jr., and J. G. Jewett, *J. Am. Chem. Soc.*, **86**, 945 (1964); (b) P. J. Stang, R. Hargrove, and T. E. Dueber, *J. Chem. Soc., Perkin Trans. 2*, 1486 (1977).
18. C. A. Grob, *Angew. Chem. Int. Ed. Engl.*, **15**, 569 (1976).
19. W. Kirmse, *Angew. Chem. Int. Ed. Engl.* **15**, 251 (1976), p. 258.
20. (a) R. A. More O'Ferrall, *Adv. Phys. Org. Chem.*, **5**, 331 (1967); (b) W. Kirmse, *Angew. Chem. Int. Ed. Engl.*, **15**, 251 (1976), p. 254.
21. (a) M. Hanack, *Acc. Chem. Res.*, **9**, 364 (1976); (b) K. Yates and J.-J. Périé, *J. Org. Chem.*, **39**, 1902 (1974).
22. M. Cocivera and S. Winstein, *J. Am. Chem. Soc.*, **85**, 1702 (1963).
24. H. L. Goering and R. P. Anderson, *J. Am. Chem. Soc.*, **100**, 6469 (1978).
25. S. P. McManus and J. M. Harris, *J. Org. Chem.*, **42**, 1422 (1977).

Chapter 5

INTRAMOLECULAR

CATIONIC

REARRANGEMENTS

In this chapter we shall discuss intramolecular rearrangements to electron-deficient carbon, nitrogen, and oxygen.

5.1 1,2-SHIFTS IN CARBENIUM IONS[1]

The intramolecular migration, shown in Equation 5.1, of a hydrogen, an alkyl, or an aryl group with its pair of electrons from a β carbon (*migration origin*) to the adjacent carbocationic center (*migration terminus*) is called a 1,2-shift (or in the case of migration of an alkyl or an aryl group, a Wagner–Meerwein shift). The new

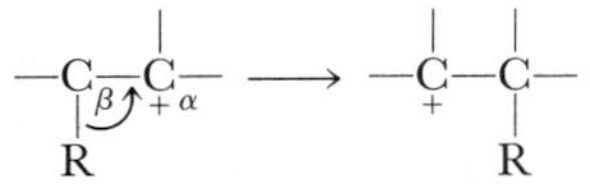

$$-\underset{\underset{R}{|}}{C}\underset{\beta}{\overset{}{\frown}}\underset{+\alpha}{\overset{|}{C}}- \longrightarrow -\underset{+}{\overset{|}{C}}-\underset{\underset{R}{|}}{\overset{|}{C}}- \tag{5.1}$$

carbocation thus formed can subsequently add to a Lewis base, lose a proton from an adjacent atom, or rearrange further.

The first such rearrangement to be studied was that of pinacol (**1**) to pinacolone (**2**) in acid solution (Equation 5.2).[2]

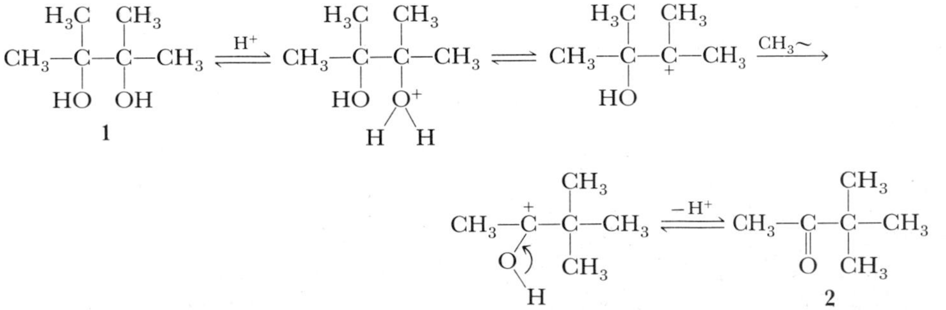

$$\underset{\substack{1}}{\underset{\substack{HO\ OH}}{\overset{\overset{\displaystyle H_3C\ \ CH_3}{|\quad\ \ |}}{CH_3-C-C-CH_3}}} \underset{}{\overset{H^+}{\rightleftharpoons}} \underset{\substack{HO\ \ O^+\\ \quad\ \ |\ \ |\\ \quad\ \ H\ H}}{\overset{\overset{\displaystyle H_3C\ \ CH_3}{|\quad\ \ |}}{CH_3-C-C-CH_3}} \rightleftharpoons \underset{\substack{HO}}{\overset{\overset{\displaystyle H_3C\ \ CH_3}{|\quad\ \ |}}{CH_3-C-C-CH_3}} \overset{CH_3\sim}{\longrightarrow}$$

$$\underset{\substack{O\\ \diagdown\\ H}}{\overset{\overset{\displaystyle CH_3}{|}}{CH_3-\overset{+}{C}-\overset{|}{C}-CH_3}} \underset{CH_3}{} \overset{-H^+}{\rightleftharpoons} \underset{\substack{O\ \ CH_3}}{\overset{\overset{\displaystyle CH_3}{|}}{CH_3-C-C-CH_3}} \tag{5.2}$$

$$\qquad\qquad\qquad\qquad\qquad\qquad\qquad\qquad\qquad\qquad\mathbf{2}$$

---

[1] For review articles, see (a) J. L. Fry and G. J. Karabatsos, in *Carbonium Ions*, Vol. II, G. A. Olah and P. v. R. Schleyer, Eds., Wiley, New York, 1970. p. 521; (b) C. J. Collins, *Quart. Rev.* (London), **14**, 357 (1960); (c) D. M. Brouwer and H. Hogeveen, *Prog. Phys. Org. Chem.*, **9**, 179 (1972); and (d) S. P. McManus, *Organic Reactive Intermediates*, Academic Press, New York, 1973.

[2] R. Fittig, *Justus Liebigs Ann. Chem.*, **114**, 54 (1860).

The name *pinacol rearrangement* is now given to the general type of rearrangement exemplified by Equation 5.2, in which the methyl groups on the 1,2-diol may be replaced by other alkyl, hydrogen, or aryl groups.

In the pinacol rearrangement the driving force to migration is the formation of a carbonyl group. The driving force to migration in solvolyses and similar reactions is usually the formation of a more stable carbocation. Since the energy differences between a tertiary and a secondary and between a secondary and a primary carbocation are about 11–15 kcal mole^{-1} each,[3] rearrangements converting a less to a more highly substituted carbocation are exothermic. Thus, for example, reaction of neopentyl iodide with aqueous silver nitrate gives entirely rearranged products (Equation 5.3).[4]

$$CH_3-\underset{\underset{CH_3}{|}}{\overset{\overset{CH_3}{|}}{C}}-CH_2-I \xrightarrow[H_2O]{AgNO_3} CH_3-\underset{\underset{CH_3}{|}}{\overset{\overset{CH_3}{|}}{C}}-CH_2^+ \rightleftharpoons CH_3-\underset{+}{\overset{\overset{CH_3}{|}}{C}}-CH_2CH_3 \rightleftharpoons$$

$$CH_3-\underset{\underset{CH_3}{|}}{C}=CH-CH_3 + CH_3-\underset{\underset{OH}{|}}{\overset{\overset{CH_3}{|}}{C}}-CH_2-CH_3 \quad (5.3)$$

Similarly, the dehydration of 1-butanol leads to 2-butenes (Equation 5.4).[5,6]

$$CH_3CH_2CH_2CH_2OH \xrightarrow[-H_2O]{H^+} CH_3CH_2CH_2CH_2^+ \overset{H\sim}{\rightleftharpoons}$$

$$CH_3CH_2\overset{+}{C}HCH_3 \xrightarrow{-H^+} CH_3CH=CHCH_3 \quad (5.4)$$

Vinyl cations are less stable than their aliphatic counterparts. Solvolysis of 1,1-diphenyl-2-propenyl triflate (trifluoromethylsulfonate) leads to the rearranged product 1,2-diphenyl-1-propanone.[7] Apparently the mechanism is that shown in Equation 5.5. A 1,2-shift converts one vinyl cation to another, but the rearranged one is stabilized by conjugation with a benzene ring.

---

[3](a) F. D. Lossing and G. P. Semeluk, *Can. J. Chem.,* **48,** 955 (1970); (b) L. Radom, J. A. Pople, and P. v. R. Schleyer, *J. Am. Chem. Soc.,* **94,** 5935 (1972); (c) E. M. Arnett and C. Petro, *J. Am. Chem. Soc.,* **100,** 5408 (1978); see also Section 4.3.

[4]F. C. Whitmore, E. L. Wittle, and A. H. Popkin, *J. Am. Chem. Soc.,* **61,** 1586 (1939).

[5]F. C. Whitmore, *J. Am. Chem. Soc.,* **54,** 3274 (1932).

[6]Although the reactions shown in Equations 5.3 and 5.4 and in some equations and schemes found later in this chapter depict a primary carbocation as an intermediate, these highly unstable species do not exist as free ions in solution. If they exist as intermediates at all, they are so encumbered by solvent and counter-ion as to give them properties very different from those expected for free ions. See (a) P. C. Hariharan, L. Radom, J. A. Pople, and P. v. R. Schleyer, *J. Am. Chem. Soc.,* **96,** 599 (1974); (b) P. v. R. Schleyer, in H. C. Brown, *The Nonclassical Ion Problem,* Plenum, New York, 1977, p. 67; (c) P. Ausloos, R. E. Rebbert, L. W. Sieck, and T. O. Tiernan, *J. Am. Chem. Soc.,* **94,** 8939 (1972), and references therein; (d) J. M. Harris, *Prog. Phys. Org. Chem.,* **11,** 89 (1974).

[7]M. A. Imhoff, R. H. Summerville, P. v. R. Schleyer, A. G. Martinez, M. Hanack, T. E. Dueber, and P. J. Stang, *J. Am. Chem. Soc.,* **92,** 3802 (1970).

$$\underset{\phi}{\overset{\phi}{}}C=C\overset{OTf}{\underset{CH_3}{}} \xrightarrow[\text{20\% H}_2\text{O}]{\text{80\% EtOH}} \underset{\phi}{\overset{\phi}{}}C=\overset{+}{C}\underset{CH_3}{} \xrightarrow{\phi\sim} \phi-C=\overset{+}{C}\underset{CH_3}{\overset{\phi}{}} \xrightarrow[-\text{H}^+]{\text{H}_2\text{O}}$$

$$\underset{HO}{\overset{\phi}{}}C=C\underset{CH_3}{\overset{\phi}{}} \rightleftharpoons \phi-\underset{O}{\overset{\phi}{C}}-CH\underset{CH_3}{\overset{\phi}{}} \qquad (5.5)$$

$(-O-Tf = -OSO_2CF_3)$

In superacids, conditions under which cations have long lives (Section 4.4), 1,2-shifts interconverting ions of like stability also occur and are very rapid. For example, at $-180°C$ there is only one peak for the five methyl groups of 2-(2,3,3-trimethylbutyl) cation in the nmr. This observation implies that the methyl shift in Equation 5.6 occurs at the rate of 75×10^3 sec^{-1} or faster, with an activation barrier of <5 kcal mole^{-1}.[8]

$$CH_3-\overset{CH_3}{\underset{H_3C\ \ CH_3}{\overset{+}{C}}-C}-CH_3 \underset{\xrightarrow{CH_3\sim}}{\rightleftharpoons} CH_3-\overset{H_3C}{\underset{H_3C\ \ CH_3}{C}}-\overset{+}{C}-CH_3 \qquad (5.6)$$

Occasionally rearrangements from more stable to less stable carbocations occur, but only if (1) the energy difference between them is not too large and (2) the carbocation that rearranges has no other possible rapid reactions open to it.[9] For example, in superacid medium, in the temperature range 0–40°C, the proton nmr spectrum of isopropyl cation indicates that the two types of protons are exchanging rapidly. The activation energy for the process was found to be 16 kcal mole^{-1}, approximately that required for conversion of a secondary to a primary carbocation.

In addition to other processes, the equilibrium shown in Equation 5.7 apparently occurs.[10] In the superacid medium, no Lewis base is available either to

$$\underset{H\ \ \ H}{\overset{H}{}}\overset{H}{\underset{+}{\wedge}}\overset{H}{\underset{H\ \ H}{}} \rightleftharpoons \underset{H\ \ \ H}{\overset{H\ \ H}{}}\overset{}{\wedge}\overset{H}{\underset{+\ H}{}} \qquad (5.7)$$

add to the carbocation or to accept a proton from it in an elimination reaction, and in the absence of such competing reactions there is ample time for the endothermic 1,2-hydride shift to take place. Also, tertiary carbocations can rearrange to secondary ones in superacid. By repeated alkyl and hydride shifts, the strained 1-propyl-cyclodecyl cation rearranges to 1-heptylcyclohexyl cation (40 percent) and 1-methyl-cyclododecyl cation (60 percent), as shown in Equation 5.8.[11]

---

[8]G. A. Olah and J. Lukas, *J. Am. Chem. Soc.*, **89**, 4739 (1967).

[9]For references to a number of such rearrangements, see (a) note 3(b) and (b) M. Saunders, P. Vogel, E. L. Hagen, and J. Rosenfeld, *Acc. Chem. Res.*, **6**, 53 (1973).

[10]See note 9(b).

[11]R. P. Kirchen, T. S. Sorensen, and K. E. Wagstaff, *J. Am. Chem. Soc.*, **100**, 5134 (1978).

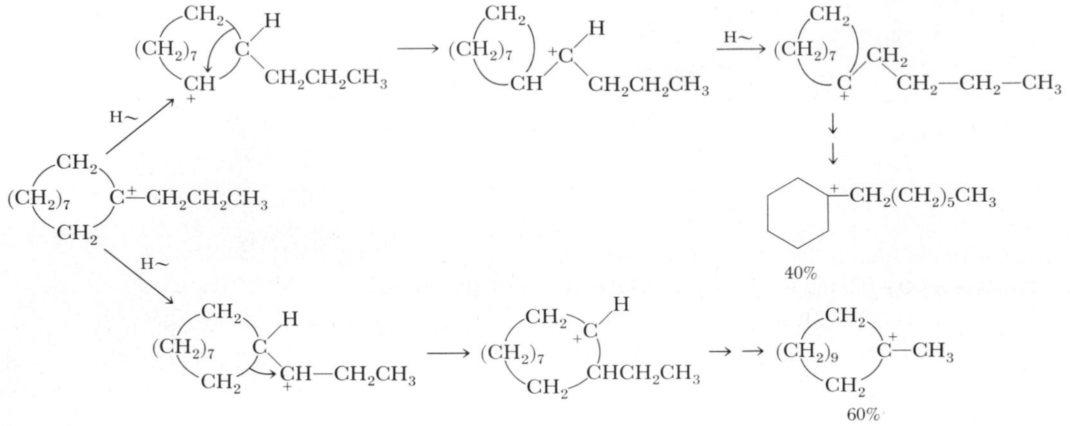

$$
\begin{array}{ccc}
\underset{\text{CH}}{\overset{\text{CH}_2}{(CH_2)_7}}\!\!\!\underset{+}{\overset{\text{H}}{C}}\text{CH}_2\text{CH}_2\text{CH}_3 & \longrightarrow & \underset{\text{CH}}{\overset{\text{CH}_2}{(CH_2)_7}}\,\underset{}{\overset{\text{H}}{+C}}\text{CH}_2\text{CH}_2\text{CH}_3
\end{array}
$$

(5.8)

40%

60%

1,2-Shifts have stereochemical as well as energetic requirements. In order for such rearrangements to occur, the C—Z (Z = migrating group) bond at the migration origin must lie in, or almost in, the plane described by the vacant p orbital on the adjacent carbon and the C_α—C_β bond as in Figure 5.1—that is, the dihedral angle between Z and the empty p orbital must be 0°. For example, the apparent 1,2-hydride shift (Equation 5.9) in the 2-adamantyl cation (3) was shown to be entirely quenched in highly dilute solution.[12] Thus it must be an *inter*-, not an *intra*molecular reaction. Further, the apparent 1,2-methyl shift in the 2-methyl-

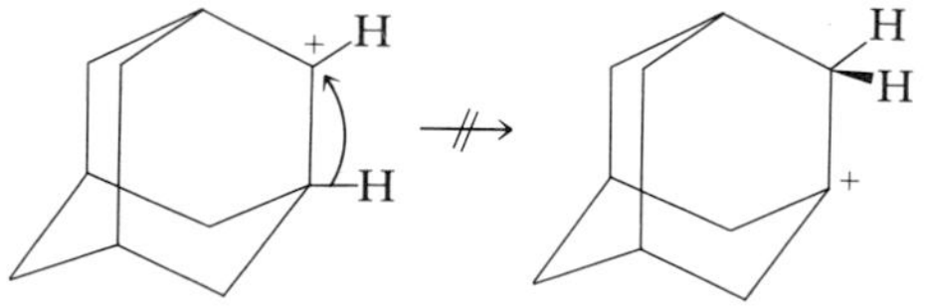

(5.9)

3

adamantyl cation (4, Equation 5.10) has been shown by isotope labeling to occur by a complicated skeletal rearrangement.[13] In both these cases the C—Z bond and the vacant p orbital, which in 3 and 4 is perpendicular to the plane of the page, form a dihedral angle of 90°.[14] In this worst of all possible stereochemical situations, the simple 1,2-shift cannot occur, and rearrangement must take another pathway.

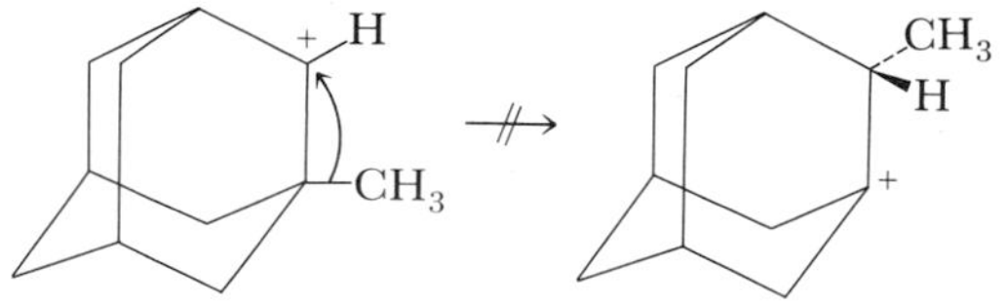

(5.10)

4

[12] P. v. R. Schleyer, L. K. M. Lam, D. J. Raber, J. L. Fry, M. A. McKervey, J. R. Alford, B. D. Cuddy, V. G. Keizer, H. W. Geluk, and J. L. M. A. Schlatman, *J. Am. Chem. Soc.,* **92,** 5246 (1970).
[13] Z. Majerski, P. v. R. Schleyer, and A. P. Wolf, *J. Am. Chem. Soc.,* **92,** 5731 (1970).
[14] (a) See notes 12 and 13; (b) D. M. Brouwer and H. Hogeveen, *Recl. Trav. Chim. Pays-Bas,* **89,** 211 (1970),

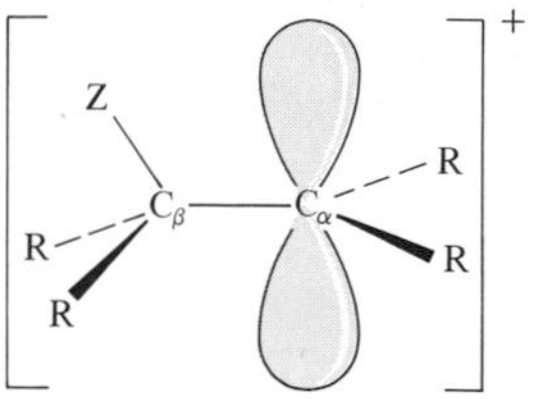

Figure 5.1 The ideal relationship of the empty p orbital to the migrating group (Z) for a 1,2-shift.

The Timing of the Migration in Acyclic Alkyl Systems

A common phenomenon in organic chemistry, illustrated further in Problems 1(a) and 1(b), is that a group adjacent to the leaving group acts as an intramolecular nucleophile. This *neighboring-group participation* can occur, for example, if the neighboring group has an unshared pair of electrons or a double bond, as shown, for example, in Equations 5.11 and 5.12. The cyclic structures **5** or **6** may sometimes be isolated but more often are attacked by a nucleophile: The ring is opened and some (or all) of the product may be rearranged. Because the effective concentration of

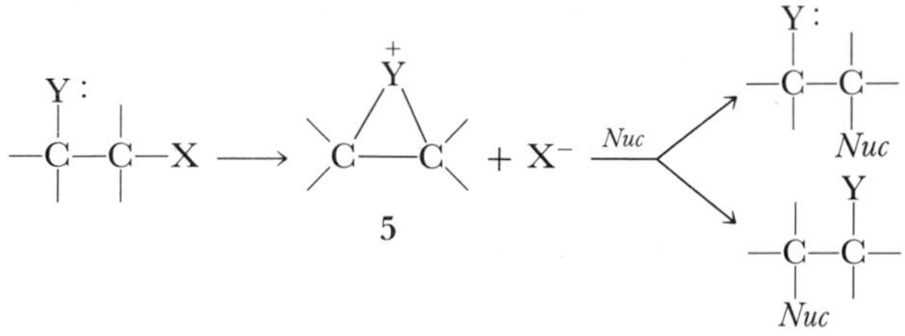

$$(5.11)$$

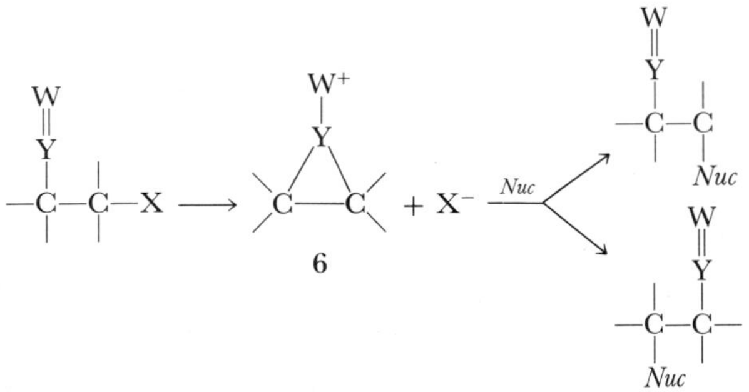

$$(5.12)$$

the neighboring group is very high (it is always in the immediate vicinity of the reaction site) and because of the relatively small degree of reorganization required to reach the transition state (and, therefore, small entropy change), reactions of this type are often faster than intermolecular or unimolecular substitutions. In fact, only if it is faster can it compete with other processes and be observed. If neighboring-group participation leads to an enhanced reaction rate, the group is said to provide *anchimeric assistance*.[15] Since one almost always goes hand in hand with the other, the terms are often used interchangeably.

[15] B. Capon and S. P. McManus, *Neighboring Group Participation*, Plenum, New York, 1976.

Analogy with intramolecular nucleophilic substitution reactions raises two fundamental problems in the study of rearrangements. The first is whether, when *a carbon or hydrogen migrates* in an electron-deficient structure, it does so only after the cationic center has fully formed in a previous step, or whether it migrates simultaneously with departure of the leaving group, thus providing neighboring-group participation. Such participation is conceivable even though carbon and hydrogen have no unshared pairs: The pair of electrons the migrating group takes with it from the β to the α carbon could be partially available to the α carbon at the transition state for the migration. Structure **7** and Figure 5.2 illustrate a *bridging*

$$\left[\begin{array}{c} \overset{\delta+}{Z} \\ \overset{\delta+}{C} \cdots \overset{}{C} \overset{\delta+}{} \\ \beta \quad \alpha \\ \overset{}{X}\delta- \end{array} \right]^{\ddagger}$$

7

transition state in which neighboring group Z (H or alkyl) participates in the departure of leaving group X and in which Z is symmetrically situated between the α and β carbons. There is, in fact, no requirement that Z be so situated—it could have hardly moved from the β carbon or it could already be almost entirely bonded to the α carbon. Winstein[16] was the first to suggest that bridging (Figure 5.2) and hyperconjugation (Figure 5.3) are different degrees of a single delocalization phenomenon. In both cases the C_{β}—Z bonding electrons are made partially available to the C_{α} carbon, but in bridging there is much nuclear motion of Z and in hyperconjugation there is little. Traylor[17] and Brown[18] have suggested that the term

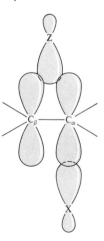

Figure 5.2 Orbital picture of the transition state for a 1,2-shift in which migration is concerted with ionization of the leaving group and in which the migrating group is symmetrically situated between C_{α} and C_{β} (a bridging transition state).

[16]S. Winstein, B. K. Morse, E. Grunwald, K. C. Schreiber, and J. Corse, *J. Am. Chem. Soc.*, **74**, 1113 (1952).
[17]T. G. Traylor, W. Hanstein, H. J. Berwin, N. A. Clinton, and R. S. Brown, *J. Am. Chem. Soc.*, **93**, 5715 (1971).
[18]H. C. Brown, *The Nonclassical Ion Problem*, Plenum, New York, 1977.

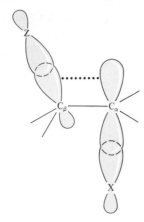

Figure 5.3 Orbital picture of hyperconjugative interaction of Z with C_α.

σ participation be used to imply some direct interaction through space of an incipient vacant *p* orbital with the electrons in a neighboring σ bond, as in Figure 5.4. When σ participation is so extensive that a fully delocalized, three-electron, two-center bond is formed, we have bridging. On the other hand, *σ conjugation* implies sideways interaction of orbitals, as in Figure 5.3.[19] As several authors have pointed out, σ participation and σ conjugation may act simultaneously.[20] We will use the terms *neighboring-group participation* and *anchimeric assistance* to cover the range of neighboring-group involvement and the other terms above when we wish to be more specific.

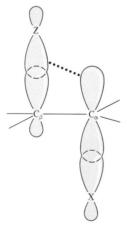

Figure 5.4 Orbital picture of a small amount of σ participation.

[19]The terms *σ conjugation* and *hyperconjugation* are usually synonymous, although σ conjugation may involve some nuclear motion of Z.

[20](a) J. M. Harris, *Prog. Phys. Org. Chem.*, **11**, 89 (1974); (b) P. v. R. Schleyer in H. C. Brown, *The Nonclassical Ion Problem*, Plenum, New York, 1977; (c) D. E. Eaton and T. G. Traylor, *J. Am. Chem. Soc.*, **96**, 1226 (1974), and references therein; (d) G. A. Olah and G. Liang, *J. Am. Chem. Soc.*, **95**, 3792 (1973).

A related question is, if there is neighboring-group participation by an alkyl group or hydrogen, is there an energy minimum on the reaction path after the departure of the leaving group in which the migrating group is bonded to both migration origin and terminus—that is, is there a bridged intermediate (8)?

$$\left[\begin{array}{c} Z \\ -C \cdots C- \\ \end{array} \right]^{+}$$

8

If the migrating group does provide neighboring-group participation, certain consequences should follow. One is kinetic: the rate should be faster than the rate of an exactly analogous, but unassisted, reaction. Another is stereochemical: neighboring-group participation is an intramolecular S_N2 displacement and therefore the migration terminus should be inverted by the rearrangement. Or looking at it another way, two closely related molecules may react by different paths if in one the neighboring group can attack the leaving group from the back side, but in the other it cannot adopt that position. Most experiments designed to determine whether anchimeric assistance occurs or not have centered on the kinetics and/or stereochemistry of the reaction under consideration. Rate acceleration is often difficult to ascertain because of problems in predicting the rate of the nonassisted reaction. Recently the kinetic *tool of increasing electron demand* (see Section 5.2) has been used to ascertain the presence or absence of neighboring-group participation. Inversion of configuration is, of course, experimentally observable only in chiral systems, but in systems that are otherwise achiral, the stereochemistry of the reaction can often be determined by isotope labeling.[21]

Experiments indicate that in open-chain and unstrained cyclic compounds, hydride and alkyl groups usually do not provide anchimeric assistance if the leaving group is on a secondary or tertiary carbon.[22] (For a discussion of participation in some primary and strained cyclic systems, see Section 5.2.) Early evidence against neighboring-group participation by alkyl groups came from oxygen-exchange studies in the pinacol rearrangement. When pinacol was allowed to rearrange in acidic ^{18}O-labeled water, recovered, unreacted pinacol was found to contain [18]O. This result is consistent with formation of a carbocation that either rearranges to pinacolone or adds water to return to pinacol as shown in Equation 5.2. The possibility that ionization and rearrangement occur in the same step, as shown in Equation

[21] For a detailed discussion of experimental methods used to study neighboring-group participation, see B. Capon and S. P. McManus, *Neighboring Group Participation*, Plenum, New York, 1976, Chap. 3.

[22] For cases where some participation by neighboring hydride or alkyl in the formation of secondary or tertiary carbocations has been suggested, see, for example, (a) V. J. Shiner, Jr., and J. G. Jewett, *J. Am. Chem. Soc.*, **87**, 1382 (1965); (b) S. Winstein and H. Marshall, *J. Am. Chem. Soc.*, **74**, 1120 (1952); (c) J. M. Harris, *Prog. Phys. Org. Chem.*, **11**, 89 (1974); (d) D. J. Raber, W. C. Neal, Jr., M. D. Dukes, J. M. Harris, and D. L. Mount, *J. Am. Chem. Soc.*, **100**, 8147 (1978), and references therein; (e) D. J. Raber, J. M. Harris, and P. v. R. Schleyer, *Ions and Ion Pairs in Organic Reactions*, Vol. 2, M. Szwarc, Ed., Wiley, New York, 1974, p. 247.

5.13, and that the ^{18}O is incorporated during a reverse rearrangement of pinacolone to pinacol is excluded by the observation that the addition of pinacolone to the reaction system does not affect the rate of rearrangement.[23]

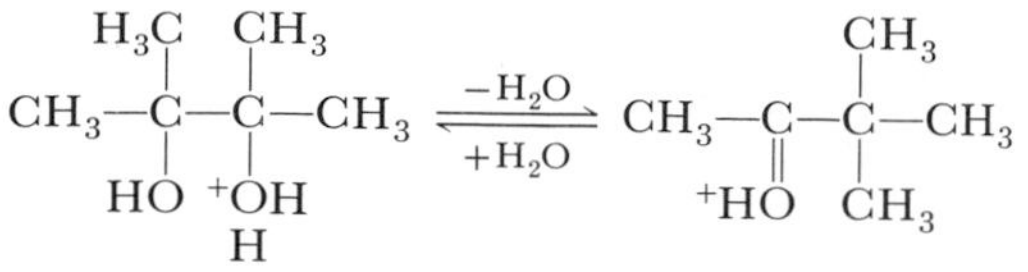
H3C CH3
CH3—C—C—CH3
HO +OH
H
−H2O
+H2O
CH3
CH3—C—C—CH3
+HO CH3

(5.13)

Scheme 1

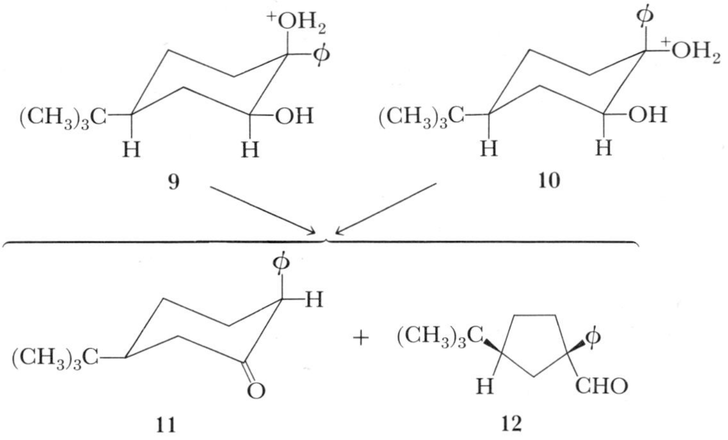
+OH2
—ϕ
(CH3)3C
OH
H
H
9
ϕ
+OH2
(CH3)3C
OH
H
H
10
ϕ
H
(CH3)3C
O
11
+
(CH3)3C
ϕ
H
CHO
12

Stereochemical evidence confirms that neither alkyl nor hydride provides neighboring-group participation in the pinacol rearrangement. Compounds **9** and **10** both give the same products, **11** and **12**, in the same ratio (9:1) when they undergo the pinacol rearrangement with BF_3–ether complex, as shown in Scheme 1.[24] (Note that the hydroxy group lost here, as usual, is the one that gives the most stable carbocation.[25] If the migrating group provided anchimeric assistance, it would have to come in from the back side of the departing $\overset{+}{O}H_2$. In Compound **9** the group that can come in from the back side is the hydride, and Compound **11** should be the principal product. Conversely, in Compound **10** it is the alkyl chain of the ring that can come in from the back side, and the chief product should be Compound **12**. The fact that the products are formed in a constant ratio indicates that a common intermediate—presumably the planar carbocation—must be

[23] C. A. Bunton, T. Hadwick, D. R. Llewellyn, and Y. Pocker, *Chem. Ind.* (*London*), 547 (1956).
[24] P. L. Barili, G. Berti, B. Macchia, F. Macchia, and L. Monti, *J. Chem. Soc. C*, 1168 (1970).
[25] (a) See note 1(b) for examples and exceptions; for another exception see (b) W. M. Schubert and P. H. LeFevre, *J. Am. Chem. Soc.*, **94**, 1639 (1972).

formed from both starting materials.[26] More direct stereochemical evidence has been provided by Kirmse and co-workers. Chiral (*S*)-2-methylbutan-1,2-diol (**13**) rearranges to racemic 2-methylbutanal (**14**), as shown in Equation 5.14.[27]

$$C_2H_5\!-\!\overset{\overset{\displaystyle CH_3}{|}}{\underset{\underset{\displaystyle OH}{|}}{C}}\!-\!CH_2OH \xrightarrow{\ H^+\ } C_2H_5\!-\!\overset{\overset{\displaystyle CH_3}{|}}{\underset{\underset{\displaystyle H}{|}}{C}}\!-\!\overset{\displaystyle O}{\underset{\displaystyle H}{C}} \tag{5.14}$$

S   R,S

13   14

Aryl Participation—The Phenonium Ion Controversy[28]

Can aryl groups provide anchimeric assistance? For 25 years that question provoked considerable controversy. In 1949 Cram solvolyzed the L-threo and L-erythro isomers of 3-phenyl-2-butyl tosylate in acetic acid. L-*Threo*-tosylate gave 96 percent racemic *threo*-acetate (plus olefins), whereas the L-erythro isomer gave 98 percent L-*erythro*-acetate.[29] To explain this stereospecificity, Cram postulated that neighboring phenyl begins a backside migration to C_α as the tosylate departs. At the first energy maximum both the tosylate and the phenyl groups are partially bonded to the α carbon. After heterolysis of the carbon–tosylate bond is complete, an intermediate *phenonium ion* is formed in which the phenyl is equally bonded to both the α and β carbons. The phenonium ion formed from the L-*threo*-tosylate (Equation 5.15) has a plane of symmetry perpendicular to and bisecting the C_α—C_β bond and therefore must yield racemic products. The phenonium ion from the L-*erythro*-tosylate is chiral (Equation 5.16) and therefore gives chiral products. Examination of the two possible paths of attack of acetic acid (it must come from the opposite side from the bulky phenyl ring) in each of the intermediates confirms that the products expected are those that are observed experimentally.

Cram provided further evidence for the existence of a phenonium ion intermediate by isolating starting tosylate after reaction had proceeded for 1.5 half-lives;

---

[26] Departure of the leaving group is apparently rate determining when the first-formed carbocation is not particularly stabilized. This is shown by the fact that the rate of rearrangement of alkyl glycols is dependent on the concentration of

$$R\!-\!\overset{\overset{\displaystyle R}{|}}{\underset{\underset{\displaystyle +OH_2}{|}}{C}}\!-\!\overset{\overset{\displaystyle R}{|}}{\underset{\underset{\displaystyle OH}{|}}{C}}\!-\!R$$

[J. F. Duncan and K. R. Lynn, *J. Chem. Soc.*, 3512, 3519 (1956); J. B. Ley and C. A. Vernon, *Chem. Ind.* (*London*), 146 (1956).] That the rate-determining step can be the migration when the first-formed carbocation is particularly stable has been shown by Schubert and LeFevre [note 25(b)]. These workers subjected 1,1-diphenyl-2-methyl-1,2-propanediol to the pinacol rearrangement and found that deuterium substitution in the migrating methyls caused the reaction to slow down.

[27] W. Kirmse, H. Arold, and B. Kornrumpf, *Chem. Ber.*, **104**, 1783 (1971).

[28] C. J. Lancelot, D. J. Cram, and P. v. R. Schleyer, in *Carbonium Ions*, Vol. III, G. A. Olah and P. v. R. Schleyer, Eds., Wiley, New York, 1972.

[29] D. J. Cram, *J. Am. Chem. Soc.*, **71**, 3863 (1949).

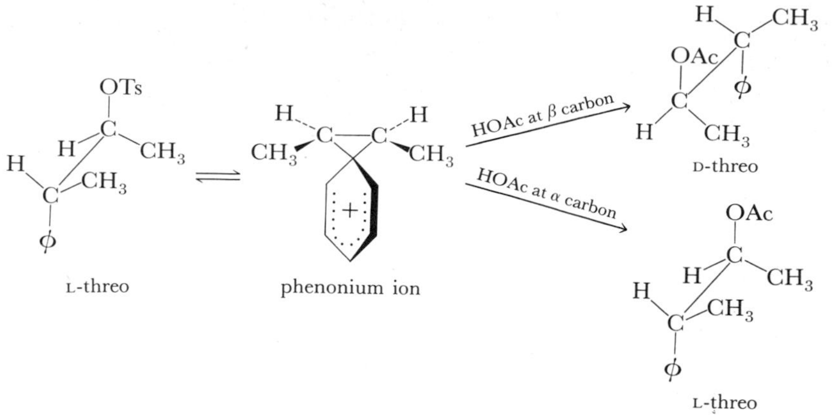

$$(5.15)$$

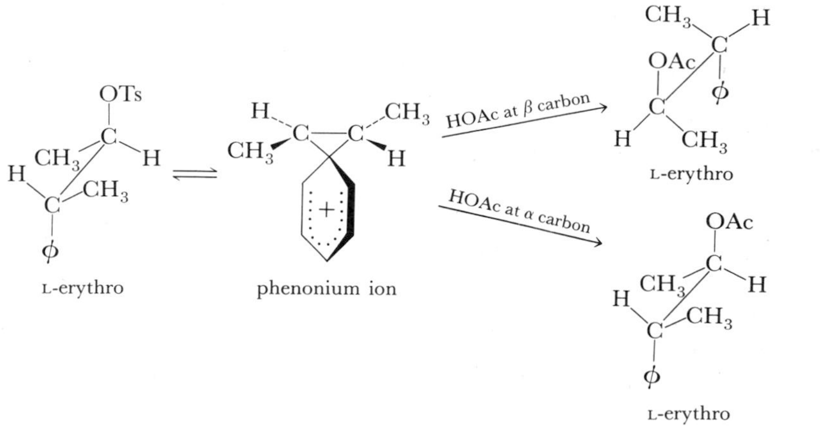

$$(5.16)$$

he found that the L-*threo*-tosylate was 94 percent racemized but the L-*erythro*-tosylate was still optically pure.[30] These results can be easily understood if the starting material first forms a phenonium–tosylate intimate ion pair, which can either revert to starting materials or go on to products. The achiral ion pair from the *threo*-tosylate will return to racemic starting material, whereas the chiral intermediate from the erythro isomer will return to optically active starting material.

Winstein strengthened the phenonium ion hypothesis in 1952. He followed the rate of acetolysis of *threo*-3-phenyl-2-butyl tosylate both titrimetrically, by titrating the toluenesulfonic acid formed, and polarimetrically, by watching the rate of loss of optical activity. He found that the polarimetric rate was five times the titrimetric rate and concluded that the intermediate phenonium ion is formed rapidly but reverts back to starting materials five times more often than it goes on

[30] D. J. Cram, *J. Am. Chem. Soc.*, **74**, 2129 (1952).

to products. That both solvolysis and racemization occur through a common intermediate seemed most probable because of the similar sensitivity of the rates of both reactions to solvent polarity.[31]

The concept of an intermediate phenonium ion was, at first, controversial, and its chief critic was H. C. Brown.[32] Although 3-phenyl-2-butyl tosylate showed the stereochemical behavior expected if an intermediate phenonium ion were formed, it did not, in his opinion, show the rate acceleration that should attend anchimeric assistance to ionization of the tosylate.[33] Brown pointed out that rapidly equilibrating open carbocations **15** would give the same stereochemistry. Ac-

$$H_3C \quad CH_3 \qquad\qquad H_3C \quad CH_3$$
$$H—\overset{|}{\underset{|}{C}}—\overset{|}{\underset{+}{C}}—H \;\rightleftharpoons\; H—\overset{|}{\underset{+}{C}}—\overset{|}{\underset{|}{C}}—H$$
$$\phi \qquad\qquad\qquad \phi$$

15

cording to his explanation, ionization of the tosylate occurs, for steric reasons, only when the phenyl and the tosylate are trans to each other. The phenyl then migrates rapidly back and forth, blocking solvent attack from the back side by the "windshield wiper effect." Rotation about the C_α—C_β bond does not occur because (1) the rapid phenyl transfer hinders it and (2) the large phenyl group must remain trans to the large, departing, tosylate group. Solvent attack from the back side, relative to the tosylate, is blocked by the phenyl group. Eventually, solvent attacks the α or β carbon from the front side, giving the stereochemical results obtained by Cram and by Winstein.

More recently, studies initiated by Schleyer[34] and completed by Brown and Schleyer[35] have convinced Brown of the existence of the phenonium ion. Schleyer, Brown, and their co-workers determined the rates and products of acetolysis of *threo*-3-aryl-2-butyl brosylate (**16**) with a number of different X substituents. Their goal was to calculate from both product and kinetic data the amount of aryl participation in ionization and then to compare the results. If both methods gave the same answer, that would be convincing evidence for the phenonium ion.[35]

They analyzed the product data by assuming that all the *threo*-acetate formed arises from a phenonium ion and all the *erythro*-acetate from backside assistance of the solvent to ionization. Then the percentage of aryl participation is

[31] S. Winstein and K. C. Schreiber, *J. Am. Chem. Soc.*, **74**, 2165 (1952).

[32] (a) H. C. Brown, *Chem. Soc. (London), Spec. Publ.*, **16**, 140 (1962); (b) H. C. Brown, K. J. Morgan, and F. J. Chloupek, *J. Am. Chem. Soc.*, **87**, 2137 (1965).

[33] The solvolysis of 3-phenyl-2-butyl tosylate is only half as fast as that of 2-butyl tosylate. However, Winstein suggested that the inductive effect of the phenyl group should retard the rate by a factor of ten and that neighboring-group participation therefore has given a fivefold rate enhancement (see note 16). See below for the actual rate enhancement.

[34] (a) C. J. Lancelot and P. v. R. Schleyer, *J. Am. Chem. Soc.*, **91**, 4291, 4296 (1969); (b) C. J. Lancelot, J. J. Harper, and P. v. R. Schleyer, *J. Am. Chem. Soc.*, **91**, 4294 (1969); (c) P. v. R. Schleyer and C. J. Lancelot, *J. Am. Chem. Soc.*, **91**, 4297 (1969).

[35] (a) H. C. Brown, C. J. Kim, C. J. Lancelot, and P. v. R. Schleyer, *J. Am. Chem. Soc.*, **92**, 5244 (1970); (b) H. C. Brown and C. J. Kim, *J. Am. Chem. Soc.*, **93**, 5765 (1971).

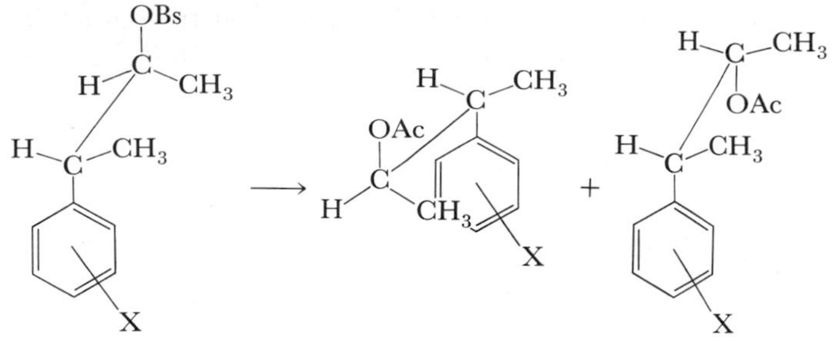

$$(5.17)$$

| 16 | 17a | 17b |
| | threo | erythro |

synonymous with the percentage of *threo*-acetate in the product. The product analyses for acetolysis at 75°C are in the sixth column of Table 5.1.

The rate data were analyzed by assuming Equation 5.18. In this equation k_t is the titrimetric rate constant for product formation, k_s is the solvent-assisted ionization constant, and Fk_Δ is the fraction of the aryl-assisted rate constant that gives rise to product (as opposed to the fraction that gives starting material through internal return). The constant k_t is determined experimentally, and k_s can be

$$k_t = k_s + Fk_\Delta \qquad (5.18)$$

calculated by use of a Hammett plot, as described below. Then Fk_Δ can be calculated by simply subtracting k_s from k_t. If Fk_Δ is the rate of formation of anchimerically assisted solvolysis product, it should lead to *threo*-acetate. The product arising from k_s should lead to *erythro*-acetate. Therefore the fraction of *threo*-acetate expected can be calculated by $Fk_\Delta/(Fk_\Delta + k_s)$ or Fk_Δ/k_t. Column five of Table 5.1 shows the calculated fraction of *threo*-acetate.

Table 5.1 Rates and Products of Acetolysis of Substituted
threo-3-Phenyl-2-butyl Brosylates (**16**) at 75.0°C[a]

16 X =	$k_t \times 10^5$	$k_s \times 10^5$	$Fk_\Delta \times 10^5$	$Fk_\Delta/k_t \times 100$	Percent *threo*-17-acetate
p-MeO	1060	14.9	1045	99	100
p-Me	81.4	10.7	70.7	87	88
m-Me	28.2	7.66	20.5	73	68
H	18.0	6.08	11.9	66	59
p-Cl	4.53	2.85	1.68	37	39
m-Cl	2.05	2.05			12
m-CF$_3$	1.38	1.38			6
p-CF$_3$	1.26	1.26			11
p-NO$_2$	0.495	0.495			1
m,m'-(CF$_3$)$_2$	0.330	0.330			1

In order to find k_s for each compound, the logarithms of the k_t's for acetolysis of **16** were determined and plotted against the Hammett σ constants of the X substituents (Figure 5.5). For electron-withdrawing X the plot is a straight line with a negative slope. This is just what would be expected from an aryl-nonassisted pathway in which a negative inductive effect from the phenyl ring decreases the rate of ionization (cf. the effect of substituents on the ionization of benzoic acid, Section 2.2). Thus for electron-withdrawing X, k_s equals k_t. The rates of solvolysis of **16** when X is electron donating, however, are faster than would be predicted from simple inductive effects. The deviations are ascribed by Brown and Schleyer to anchimeric assistance: The extrapolation of the line correlating the σ constants with the logarithms of the rates when X is electron withdrawing is taken as defining the k_s contribution when X is electron donating. For example, the *p*-methoxy substituent has a σ constant of -0.27. On the extrapolated line this corresponds to $\log k_s = -3.83$ or $k_s = 1.49 \times 10^{-4}$. Experimentally, k_t is found to be 106.0×10^{-4}, thus $Fk_\Delta = 104.5 \times 10^{-4}$. Data for the other substituents are found in Table 5.1.

The excellent correlation between the percentage of anchimeric assistance

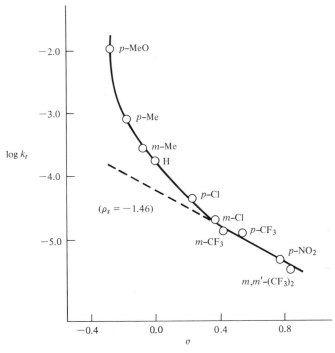

Figure 5.5 Logarithms of the rates of acetolysis of *threo*-3-aryl-2-butyl brosylates (**16**) at 75.0°C vs. the σ constants of the X substituent. Reprinted with permission from H. C. Brown, C. J. Kim, C. J. Lancelot, and P. v. R. Schleyer, *J. Am. Chem. Soc.*, **92**, 5244 (1970). Copyright 1970 American Chemical Society.

calculated from the rate (column 5) and product (column 6) data is compelling evidence that the analyses above are accurate and thus that separate, discrete pathways for solvolysis exist—one solvent-assisted and one aryl-assisted.[36]

Note that in acetic acid a very small rate enhancement corresponds to a large contribution of anchimerically assisted components. For example, the titrimetric rate constant for solvolysis of **16** (X = H) is only 18.0/6.08 times faster than the solvolytic rate constant—a rate enhancement of about three. This corresponds to 59 percent anchimeric assistance. The reason for the small rate enhancement with the large degree of anchimeric assistance is that the competition is not between an assisted and a nonassisted pathway but rather between an anchimerically assisted and a solvent assisted pathway.[36] In nonnucleophilic solvents a much larger rate enhancement is observed.[37]

Primary β-aryl tosylates have also been shown to undergo solvolysis by two distinct pathways—aryl- and solvent-assisted.[38] Tertiary β-aryl tosylates, however, ionize to a stable carbocation.[39]

The unsubstituted phenonium ion, as well as phenonium ions substituted with electron-donating groups, have been observed as stable ions in superacid medium.[40] That the structure is actually **18** and not an unsymmetrically bridged ion (**19**) nor a carbonium ion (**20**) (see Section 5.2) in which there are three-center bonds was shown by the nmr evidence. The ring carbon that is bonded to the aliphatic carbons was established by ^{13}C shifts to be tetrahedral in nature; and ^{13}C and proton chemical shifts in the ring were similar to those of cations shown to have Structure **21**.

18   **19**   **20**   **21**

Stereochemistry[41]

The central carbon of the migrating group and the carbons of the migration terminus and of the migration origin all undergo bonding changes during a Wagner–Meerwein shift, and the stereochemistry at each may change.

The orbital picture we have previously formulated (Figure 5.2) predicts that the stereochemistry of the migrating group will be retained during the migration since, in this picture, the migrating group uses the same lobe of the same orbital to

[36] For a discussion of solvent-assisted ionization, see Chapter 4.

[37] See note 34(c).

[38] F. L. Schadt and P. v. R. Schleyer, *J. Am. Chem. Soc.,* **95**, 7860 (1973).

[39] H. C. Brown and C. J. Kim, *J. Am. Chem. Soc.,* **90**, 2082 (1968).

[40] (a) G. A. Olah, R. J. Spear, and D. A. Forsythe, *J. Am. Chem. Soc.,* **98**, 6284 (1976), and references therein; (b) G. A. Olah, *Angew. Chem. Int. Ed. Engl.,* **12**, 173 (1973).

[41] D. J. Cram, in *Steric Effects in Organic Chemistry,* M. S. Newman, Ed., Wiley, New York, 1956.

bond to both migration origin and migration terminus. Predominant retention is in fact observed, but some racemization may occur. For example, in the semipinacol rearrangement[42] of (3S)-1-amino-2,3-dimethyl-2-pentanol (**22**), the product **23** that

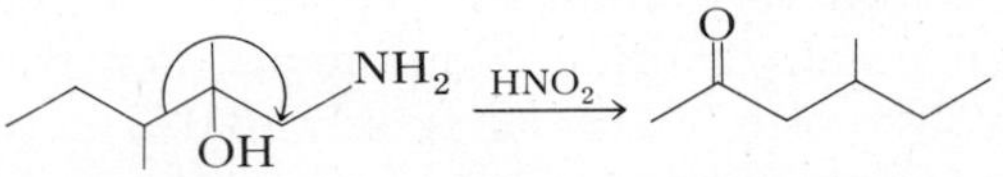

$$\text{(in 33\% yield)} \qquad (5.19)$$

3-*S*   86–88% retained

22   **23**

arises from migration of the *S*-butyl group accounts for 33 percent of the product. The chirality of the *S*-butyl group in **23** is only 86–88 percent retained.[43] Kirmse and co-workers have proposed that the pathway for racemization involves the cyclopropanol **24** as an intermediate. Their proposed mechanism is abbreviated in Equation 5.20.[44] The intermediacy of **24** is supported by the fact that deuterium is incorporated into **23** when the deamination is carried out in D_2O.

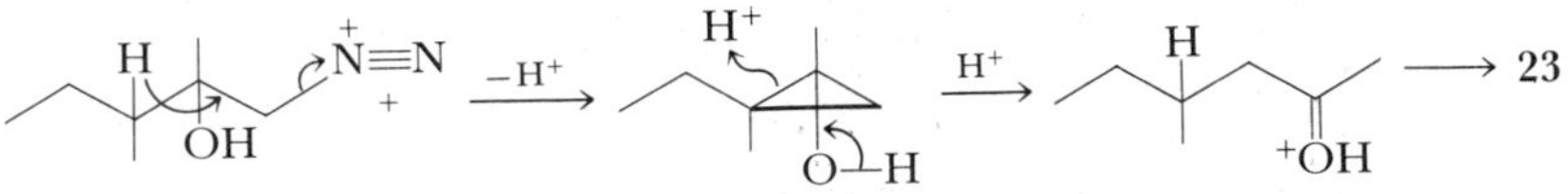

$$\longrightarrow \textbf{23} \qquad (5.20)$$

racemized

24

The stereochemistry at the migration terminus depends on the relative timing of the leaving group's departure and the 1,2-shift. In the

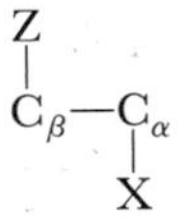

system, if Z begins to migrate before X has completely departed, the migration terminus, C_α, will be inverted. We have already seen in Cram's work that phenyl migration with neighboring-group participation leads to an inverted migration terminus. A novel example in which a carbonyl carbon provides the neighboring-group participation is shown in Equations 5.21 and 5.22.[45] In both reactions the migration terminus is inverted even though the ground-state conformer necessary

---

[42] The semipinacol rearrangement is the rearrangement that ensues when a β-amino alcohol is deaminated as in the following equation. See Problem 4 in Chapter 3.

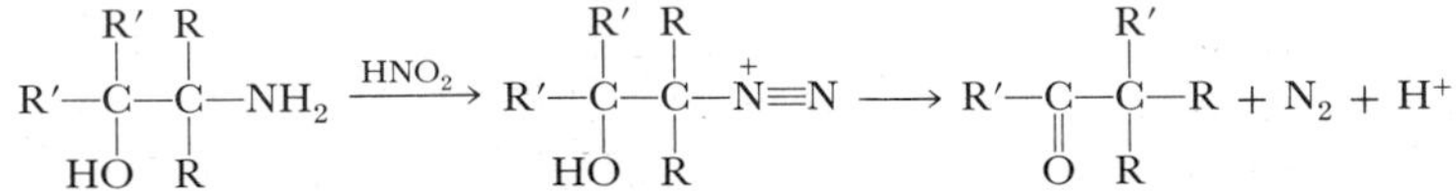

[43] W. Kirmse and W. Gruber, *Chem. Ber.*, **106**, 1365 (1973); W. Kirmse, W. Gruber, and J. Knist, *Chem. Ber.*, **106**, 1376 (1973).

[44] Kirmse and co-workers suggest that **24** is formed and destroyed via a protonated cyclopropane (see Section 5.2).

[45] J. M. Domagala and R. D. Bach, *J. Am. Chem. Soc.*, **101**, 3118 (1979).

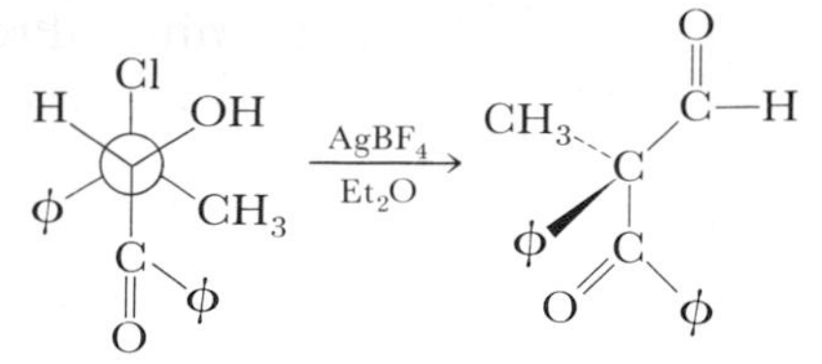

$$(5.21)$$

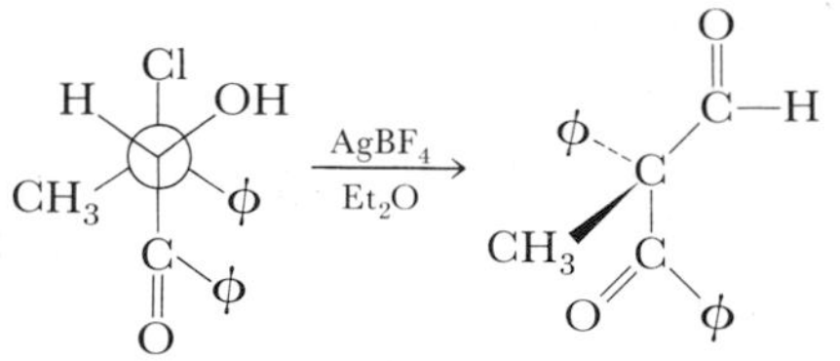

$$(5.22)$$

for simultaneous ionization and migration must be the least stable of the three conformers (phenyl is larger than methyl, which is larger than chloro).

If X departs before Z begins to move, either retention or inversion can occur at the migration terminus. If the lifetime of the carbocation is very short, retention will result if Z was gauche to the leaving group in the unreacted starting material

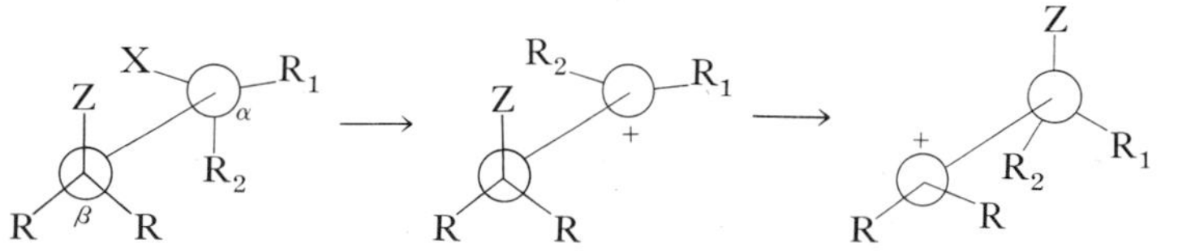

$$(5.23)$$

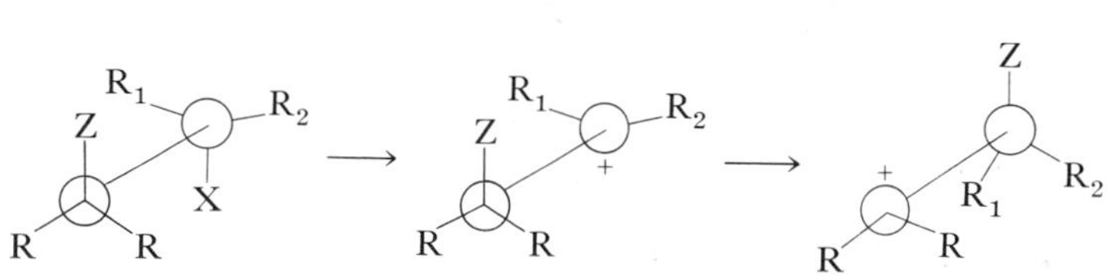

$$(5.24)$$

(Equation 5.23) and inversion will result if Z was anti (Equation 5.24). If the C_α carbocation has a long lifetime and there is free rotation about the C_α—C_β bond, the relative amounts of retention and inversion will depend on whether the transition state leading to retention or inversion is more stable. If they are of equal energy, racemization should result; we have already seen an example of this in Equation 5.14.

Deamination of amines often gives rise to short-lived carbocations. Deamination of (+)-1,1-diphenyl-2-amino-1-propanol specifically labeled with ^{14}C in one of the two phenyl groups (**25**) gives α-phenyl-propiophenone as product, 88 percent of it inverted and 12 percent retained. All the inverted ketone comes from migration of the ^{14}C-labeled phenyl and all the retained from migration of the unlabeled phenyl group (Equation 5.25).[46] This behavior can be understood if we look at the

[46] B. M. Benjamin, H. J. Schaeffer, and C. J. Collins, *J. Am. Chem. Soc.*, **79**, 6160 (1957).

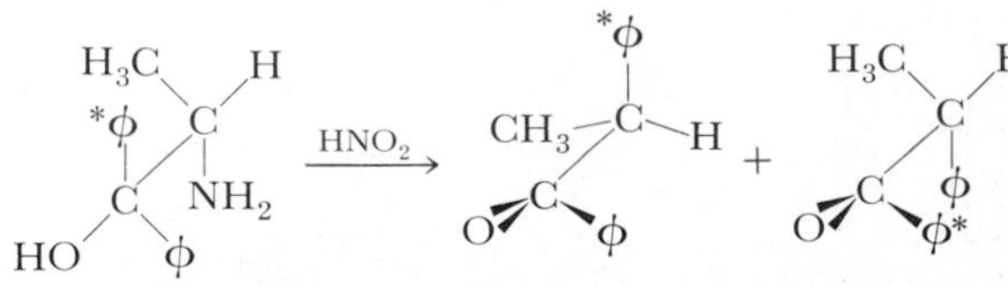

$$(5.25)$$

ground state of **25**. The most stable of the three staggered rotamers of **25** is **25a** (in this conformation each of the large phenyl groups has one small proton next to it), and therefore most of the amine molecules adopt this conformation. When the free carbocation is formed from **25a** there is not time for rotation about the C_β—C_α

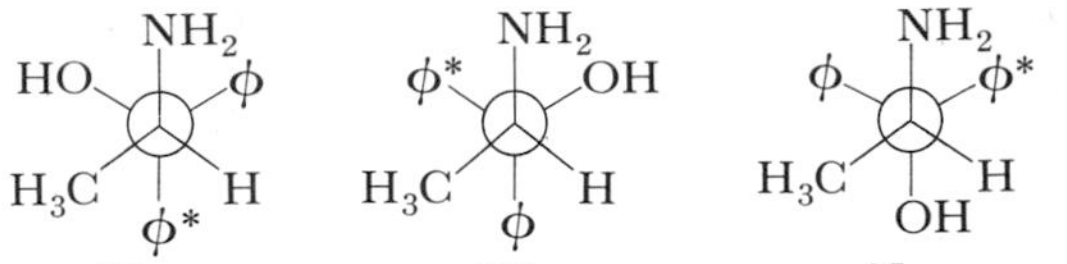

bond before a phenyl group migrates. The labeled phenyl is backside to the original amine group, and migration of it gives inversion. The unlabeled phenyl is frontside, and its migration gives retention.

The first-formed carbocation from the deamination of *threo*-1-amino-1-phenyl-2-*p*-tolyl-2-propanol (**26**) is stabilized by resonance and longer-lived than the carbocation formed from the deamination of **25**. In rotamers **26b** and **26c** the bulky phenyl and *p*-tolyl groups are next to each other, and thus again the ground-

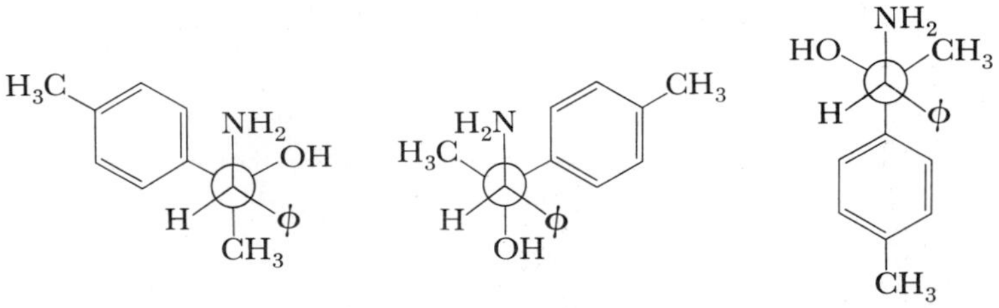

state amine will be almost entirely in the conformation represented by rotamer **26a**. The carbocation formed from **26a** presumably has time to rotate about the C_α—C_β bond before the migration because, although the product of tolyl migration is formed in >80 percent yield, 58 percent of it is retained and 42 percent inverted. Scheme 2 shows the probable reason for the predominance of the retained product. In the transition state for it (**27**) the two large groups (phenyl and methyl) are trans to each other but in the transition state for inversion (**28**) they are cis.[47] The absence of neighboring-group participation in deamination reactions seems to be a fairly general phenomenon, although when the neighboring group is as reactive as *p*-methoxyphenyl some participation may occur.[48]

[47] B. M. Benjamin and C. J. Collins, *J. Am. Chem. Soc.*, **83**, 3662 (1961).
[48] P. I. Pollak and D. Y. Curtin, *J. Am. Chem. Soc.*, **72**, 961 (1950).

SCHEME 2

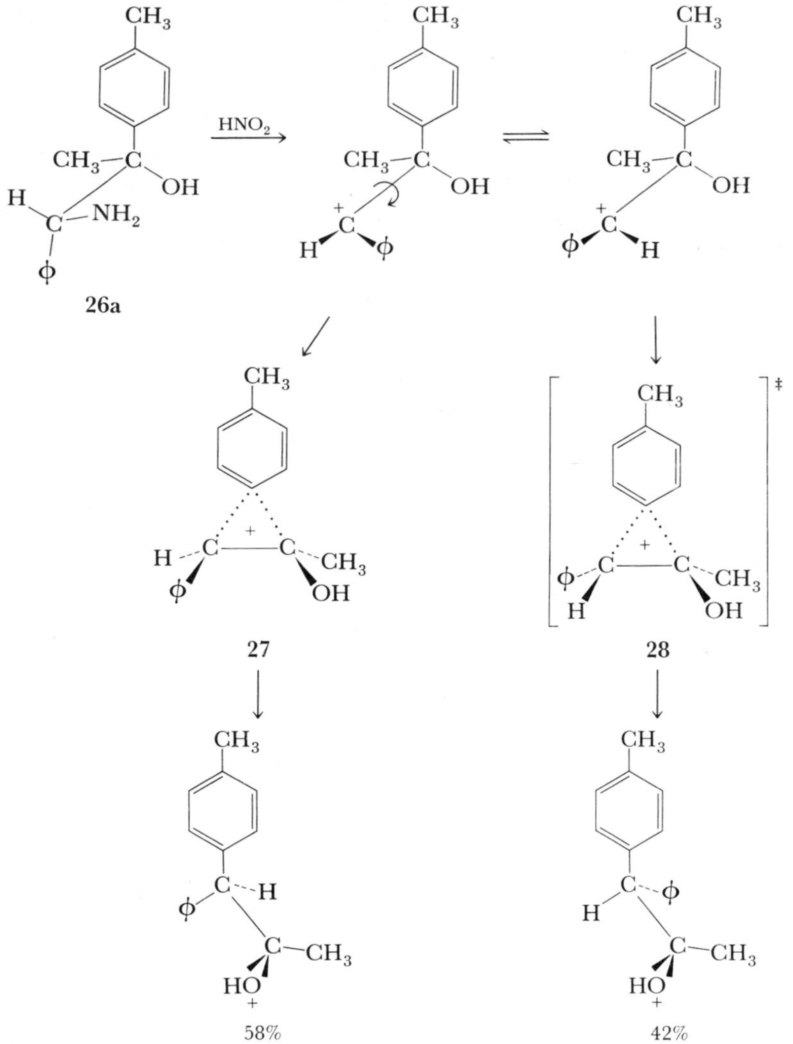

CH_3

CH_3

CH_3

$\xrightarrow{HNO_2}$

$\rightleftharpoons$

CH_3—C OH

CH_3—C OH

CH_3—C OH

H C—NH_2

H C^+ ϕ

ϕ C^+ H

ϕ

26a

CH_3

CH_3 ‡

H---C^+—C---CH_3

ϕ---C^+—C---CH_3

ϕ OH

H OH

27

28

CH_3

CH_3

ϕ C---H

H C---ϕ

C—CH_3

C—CH_3

HO $+$

HO $+$

58%

42%

    When the steric effects in the transition states for retention and inversion are of equal energy, attack of a migrating group on a carbocation formed by deamination occurs preferentially from the back side. For example, (R)-2-amino-2-methylbutanol-1 (**29**) is deaminated in aqueous $HClO_4$ to afford 2-methylbutanal (**14**) (16 percent) with 30 percent inversion of configuration at C_α.[49] (Comparison of Equations 5.26 and 5.14 shows how the stereochemistry at C_α may depend on the lifetime of the carbocation.)

[49] W. Kirmse, H. Arold, and B. Kornrumpf, *Chem. Ber.,* **104,** 1783 (1971).

$$CH_3\overset{C_2H_5}{\underset{NH_2}{\mid}}CH_2OH \xrightarrow[HClO_4]{HNO_2} CH_3-\overset{H}{\underset{C_2H_5}{\mid}}-C\overset{O}{\diagdown_H} \qquad (5.26)$$

30% inversion

29   14

The stereochemistry at the migration origin cannot always be studied because, as in the pinacol and semipinacol rearrangements and as in Wagner–Meerwein shifts where migration is followed by loss of a proton, the migration origin often becomes trigonal in the product. When it is tetrahedral in the product, its stereochemistry varies and is not yet fully understood. If the nucleophile attacks before the migrating group has fully departed, then it must come in from the back side and give inversion at the migration origin. We saw a dramatic example of this in Cram's work on the solvolysis of 3-phenyl-2-butyl tosylate, in which solvent attacks the phenonium ion directly and the migration origin, C_β, is almost entirely inverted. If the migrating group has completely departed from the migration origin before the nucleophile attacks *and* there is time for rotation about the $C_\alpha—C_\beta$ bond, then racemization should result. (If there is not time for free rotation, attack from one side might be less hindered than attack from the other.) Racemization at C_β occurs, for example, in the deamination of (S)-1-amino-2-cyclohexylbutane (**30**), which affords, among other products, racemic 2-cyclohexylbutan-2-ol (**31**).[50]

$$H\text{--}\overset{C_6H_{11}}{\underset{C_2H_5}{\mid}}CH_2NH_2 \xrightarrow[HNO_2]{HClO_4} HO-\overset{C_6H_{11}}{\underset{C_2H_5}{\mid}}CH_3 \qquad (5.27)$$

S R,S

30   31

Migratory Aptitudes

The relative ease with which alkyl and aryl groups migrate is called their *migratory aptitude*. Migratory aptitudes are not absolute quantities; values determined in one reaction under one set of conditions may differ enormously from values in another reaction or even in the same reaction under other conditions.

Both "intermolecular" and intramolecular migratory aptitudes have been studied in the pinacol rearrangement. For determination of the latter, a pinacol in which the β carbon is substituted with two different R groups is used, and the product is analyzed to see in what proportion the two groups have migrated. It is necessary to use symmetrical pinacols (**32**) and compare the migration of R_1 and R_2. If unsymmetrical pinacols (**33**) are used, the group with the higher migratory aptitude may not be able to migrate. For example, if 1,1-dimethyl-2,2-diphenylethylene glycol is treated with H_2SO_4 in acetic anhydride, only methyl migration

[50]W. Kirmse and W. Gruber, *Chem. Ber.,* **104,** 1789 (1971).

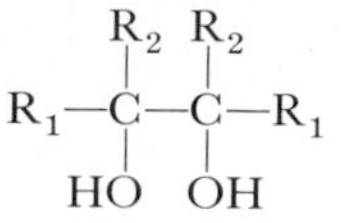

$$
\underset{\mathbf{32}}{\underset{\underset{\text{HO}\quad\text{OH}}{|}}{\overset{\overset{R_2\quad R_2}{|}}{R_1-C-C-R_1}}}
\qquad
\underset{\mathbf{33}}{\underset{\underset{\text{HO}\quad\text{OH}}{|}}{\overset{\overset{R_1\quad R_2}{|}}{R_1-C-C-R_2}}}
$$

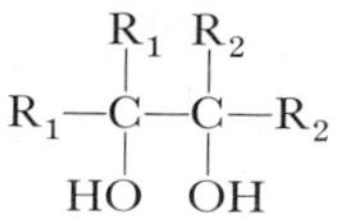

occurs.[51] This does not mean that methyl has the higher migratory aptitude, but simply that the more stable diphenyl-substituted cation (**34**) is formed in preference to the dimethyl-substituted cation (**35**), as shown in Equation 5.28. In **34** only methyl migration is possible.

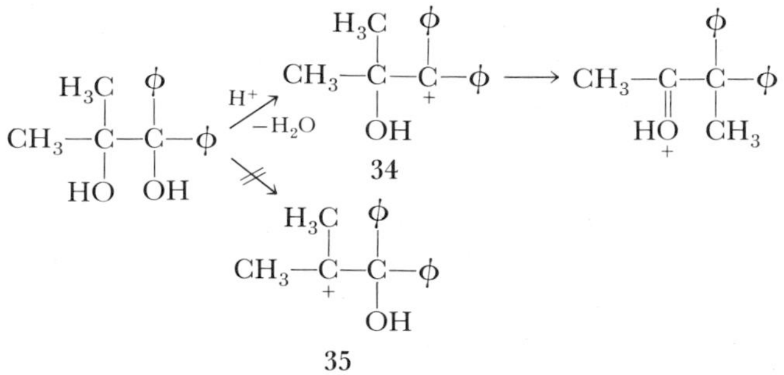

$$(5.28)$$

By analyzing the products of ten symmetrical glycols of type **32**, Bachmann determined the following migratory aptitudes relative to phenyl: *p*-ethoxyphenyl, 500; *p*-methoxyphenyl, 500; *p*-tolyl, 15.7; *p*-biphenyl, 11.5; *m*-tolyl, 1.95; *m*-methoxyphenyl, 1.6; phenyl, 1.0; *p*-chlorophenyl, 0.66; *m*-chlorophenyl, 0.[52]

Migratory aptitudes in the pinacol rearrangement have also been determined by comparing the rates of reaction of different pinacols, each with four identical substituents of the type **36**.

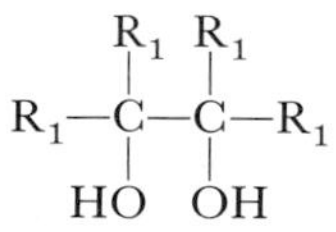

$$
\underset{\mathbf{36}}{\underset{\underset{\text{HO}\quad\text{OH}}{|}}{\overset{\overset{R_1\quad R_1}{|}}{R_1-C-C-R_1}}}
$$

The migratory aptitudes obtained in this way were *p*-methoxyphenyl, 880; *p*-tolyl, 40; phenyl, 1; *p*-chlorophenyl, 0.[53]

We have already mentioned that migratory aptitudes are dependent on the reaction and on the conditions under which the reaction is carried out. An example of the latter type of variation is that in the pinacol rearrangement of triphenylethylene glycol, the phenyl/hydrogen migration ratio may vary by a factor of 180 (from 7.33 to 0.41) when the catalyst is changed from concentrated sulfuric acid to

[51] Ramart-Lucas and M. F. Salmon-Legagneur, *Compt. Rend.*, **188**, 1301 (1929).
[52] W. E. Bachmann and J. W. Ferguson, *J. Am. Chem. Soc.*, **56**, 2081 (1934).
[53] P. Depovere and R. Devis, *Bull. Soc. Chim. Fr.*, 479 (1969).

HCl in water/dioxane.[54] Furthermore, groups may change in their apparent relative abilities to migrate as the temperature is raised.[55]

A striking example of the former type of variation can be seen in a comparison of the migratory aptitudes in the pinacol rearrangement with those in a deamination reaction. The semipinacol rearrangement of **37** gives the following migratory aptitudes: *p*-methoxyphenyl, 1.5; phenyl, 1; *p*-chlorophenyl, 0.9.[56] As was mentioned in Section 4.4, deamination reactions generally show a lower selectivity than pinacol and Wagner–Meerwein rearrangements.

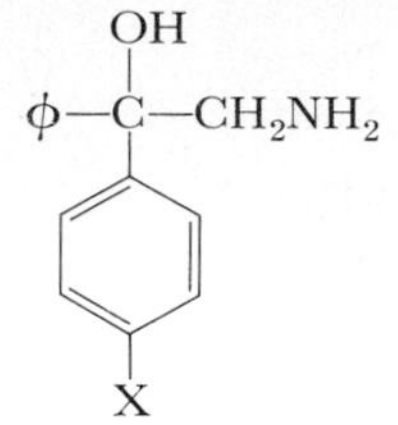

$$\phi-\underset{\underset{\displaystyle}{|}}{\overset{\overset{\displaystyle OH}{|}}{C}}-CH_2NH_2$$

37

A novel method of measuring migratory aptitudes has been published by Shubin and co-workers.[57] They studied the temperature at which the two methyls of **38** become equivalent in superacid solution in the nmr and found the following results for various substituents X: CH_3, $-100°$; F, $-70°$; H, $-70°$; Cl, $-55°$; CF^3, $0°C$.

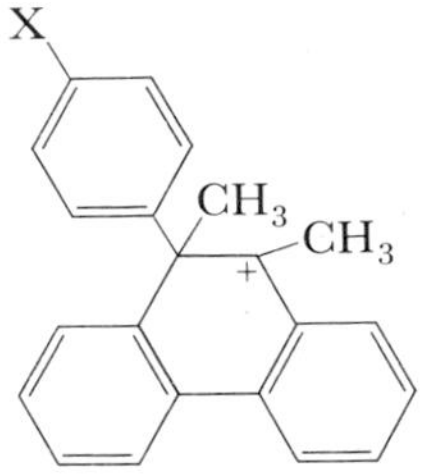

38

Memory Effects[58]

Heretofore we have been concerned mainly with single rearrangements. Multiple rearrangements also occur, in which the carbocation formed after the initial migration rearranges again (and again) before products are formed. Some of these consecutive rearrangements are remarkable in that presumably identical carbocations, which arise by rearrangement from different starting materials, retain a memory of their antecedent, and give different second rearrangements.

[54] C. J. Collins, *J. Am. Chem. Soc.*, **77**, 5517 (1955).
[55] J. Kagan, D. A. Agdeppa, Jr., S. P. Singh, D. A. Mayers, C. Boyajian, C. Poorker, and B. E. Firth, *J. Am. Chem. Soc.*, **98**, 4581 (1976).
[56] D. Y. Curtin and M. C. Crew, *J. Am. Chem. Soc.*, **76**, 3719 (1954).
[57] V. G. Shubin, D. V. Korchagina, G. I. Borodkin, B. G. Derendjaev, and V. A. Koptyug, *J. Chem. Soc. D*, 696 (1970).
[58] For a review, see J. A. Berson, *Angew. Chem. Int. Ed. Engl.*, **7**, 779 (1968).

For example, deamination of *syn*- and *anti*-2-norbornenyl-7-carbinyl amines (**39** and **40**) with sodium nitrite in aqueous acetic acid both give twice-rearranged products (Equations 5.29 and 5.30). The first rearrangement in both deaminations is a ring expansion to give **41**. If **41** is symmetrical, as would be expected if a flat

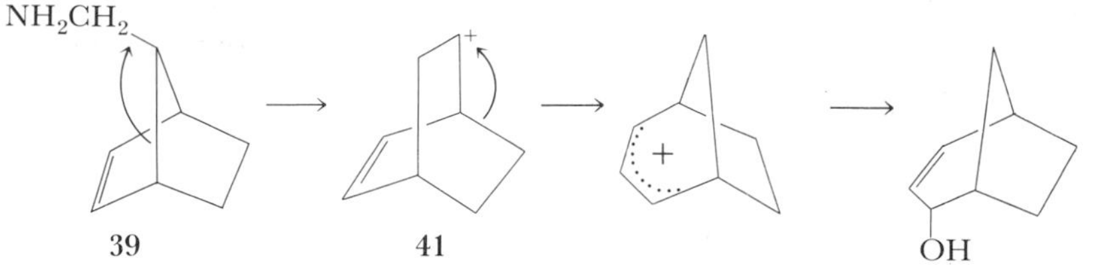

(5.29)

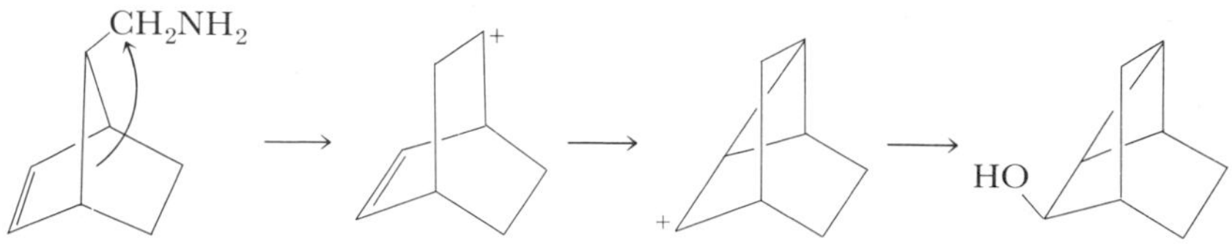

(5.30)

carbocation were formed, both reactions should go on to give the same products. But they do not.[59]

The cause of the memory effect is controversial. Berson has suggested that the symmetrical ion **41** is not the first-formed cation in both reactions, but that twisted cations that can rearrange further before they undergo the readjustments that convert them to **41** are formed first. In this view **39** would first form **42**, in which the σ bond is better able to migrate than the π bond, whereas **40** would first form **43**, in which the α bond is better situated for migration.[60,61]

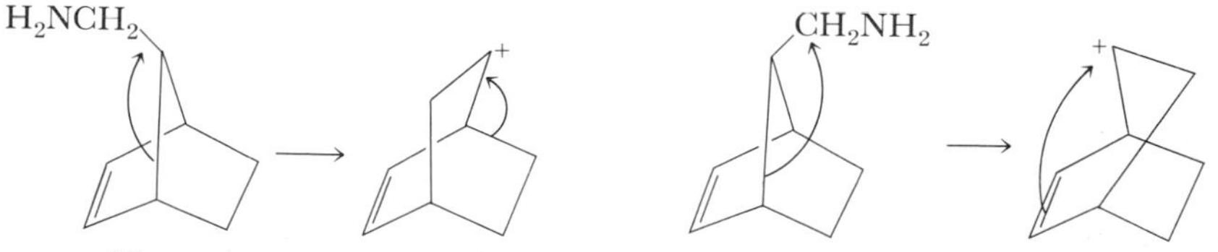

Collins, on the other hand, has proposed that the memory effect in deaminations can be explained by counter-ion control.[62] According to his explanation deamination in acetic acid occurs from a diazonium intermediate **44**. On decomposition of **44**, the acetate ion is left near the newly formed carbocation, as in **45**, and

[59] J. A. Berson, J. J. Gajewski, and D. S. Donald, *J. Am. Chem. Soc.*, **91**, 5550 (1969).
[60] See (a) note 58, (b) note 59; (c) J. A. Berson, J. M. McKenna, and H. Junge, *J. Am. Chem. Soc.*, **93**, 1296 (1971).
[61] For ^{13}C nmr evidence for a nonplanar carbocation, see R. P. Kirchen and T. S. Sorensen, *J. Am. Chem. Soc.*, **99**, 6687 (1977).
[62] C. J. Collins, *Q. Rev. Chem. Soc.*, **4**, 251 (1975), and references therein.

controls the stereochemistry of the product. Collins has, in fact, provided an example of a memory effect where the carbocation is too strained to be twisted.

$$R—N_2{}^+ \quad -\!\!:\!\!\overset{\displaystyle O}{\underset{\displaystyle O}{C}}\!\!-CH_3 \longrightarrow R^+ \quad -\!\!:\!\!\overset{\displaystyle O}{\underset{\displaystyle O}{C}}\!\!-CH_3 \tag{5.31}$$

44　　　　　　　　　　　**45**

Bridged ions can also control the stereochemistry of sequential rearrangements. On solvolysis in aqueous dioxane, (R)-3-methyl-2-phenyl-1-butyl tosylate (**46**) yields among other products (S)-3-methyl-4-phenyl-2-butanol (**47**) of 95 per-

46　　　　　　　　　　**48**　　　　　　　　　**47**

cent optical purity. Compound **47** is formed by consecutive 1,2-shifts of first phenyl and then methyl. The high optical purity of **47** indicates that the methyl shift must occur in the phenonium ion **48**.[63] This type of memory effect is, of course, closely related to inversion of the migration origin by solvent (see below).

5.2 CARBONIUM IONS[64-67]

In this section we will discuss carbocations in which the charge is delocalized in two-electron, three-center bonds. By Olah's terminology these are *carbonium ions* as opposed to the charge-localized *carbenium ions*.[68] (By older usage the former are *nonclassical carbonium ions* and the latter *classical carbonium ions*.) However, as was

[63] W. Kirmse and B. R. Gunther, *J. Am. Chem. Soc.*, **100**, 3619 (1978).

[64] For a review on the general subject, see P. D. Bartlett, *Nonclassical Ions*, W. A. Benjamin, Menlo Park, Calif., 1965.

[65] For reviews of homoallylic and small-ring participation, see (a) R. Breslow, in *Molecular Rearrangements*, Vol. 1, P. Mayo, Ed., Wiley, New York, 1963, p. 233; (b) M. Hanack and H.-J. Schneider, *Angew. Chem., Int. Ed. Engl.*, **6**, 666 (1967); (c) R. R. Story and B. C. Clark, in *Carbonium Ions*, Vol. III, G. A. Olah and P. v. R. Schleyer, Eds., Wiley, New York, 1972, p. 1007; (d) K. B. Wiberg, B. A. Hess, and A. J. Ashe, in *Carbonium Ions*, Vol. III, Olah and Schleyer, Eds., p. 1295; (e) H. G. Rickey, in *Carbonium Ions*, Vol. III, Olah and Schleyer, Eds., p. 1201.

[66] For reviews of bicyclic carbonium ions, see: (a) J. A. Berson, in *Molecular Rearrangements, Vol. I*, P. Mayo, Ed., New York, p. 111; (b) G. D. Sargent, *Quart. Rev.* (London), **20**, 301 (1966); (c) G. D. Sargent, in *Carbonium Ions*, Vol. III, Olah and Schleyer, Eds., p. 1099; (d) G. M. Kramer, *Adv. Phys. Org. Chem.*, **11**, 177 (1975); (e) G. A. Olah, *Acc. Chem. Res.*, **9**, 41 (1976); (f) S. E. Scheppelle, *Chem. Rev.*, **72**, 511 (1972).

[67] For an opposing view, see (a) H. C. Brown, *The Nonclassical Ion Problem*, Plenum, New York, 1977, and references therein; (b) H. C. Brown, *Tetrahedron*, **32**, 179 (1976), and references therein; (c) H. C. Brown, C. Gundu Rao, and D. L. Vander Jagt, *J. Am. Chem. Soc.*, **101**, 1780 (1979), and references therein.

[68] G. A. Olah, *J. Am. Chem. Soc.*, **94**, 808 (1972).

discussed in the previous section, there can be a continuum of electron delocalization in carbocations; thus these terms actually refer to limiting cases: Carbenium ions lie at one end of the spectrum and carbonium ions at the other end. There has been a good deal of controversy about how much electron delocalization there must be and of what type (both conjugation and participation or just participation) before a carbocation should be labeled a carbonium ion. Recently, the following careful definitions were worked out by Brown and Schleyer.[69]

A carbenium ion "can be represented by a single Lewis structure involving only two-electron, two-center bonds. Traditionally π-conjugated cations such as allyl are included in this category." A carbonium ion "cannot be represented adequately by a single Lewis structure. Such a cation contains one or more carbon or hydrogen bridges joining two electron-deficient centers. The bridging atoms have coordination numbers higher than usual, typically five or more for carbon and two or more for hydrogen. Such ions contain two-electron, three-center bonds." Structures **49** and **50** are the norbornyl carbonium ion and the 2-propylcarbenium ion, respectively.[70]

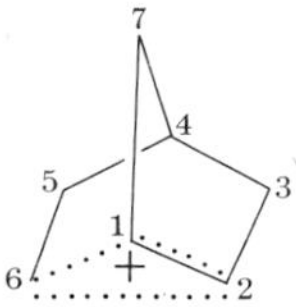

$$CH_3\overset{+}{C}HCH_3$$

49   50

Homoallylic Carbonium Ions[65]

In 1946 Shoppee noted that the reaction of 3-β-cholesteryl chloride with acetate ion proceeds entirely with retention of configuration (Equation 5.33). Substitutions on the analogous saturated compound proceed with the expected inversion. Shoppee

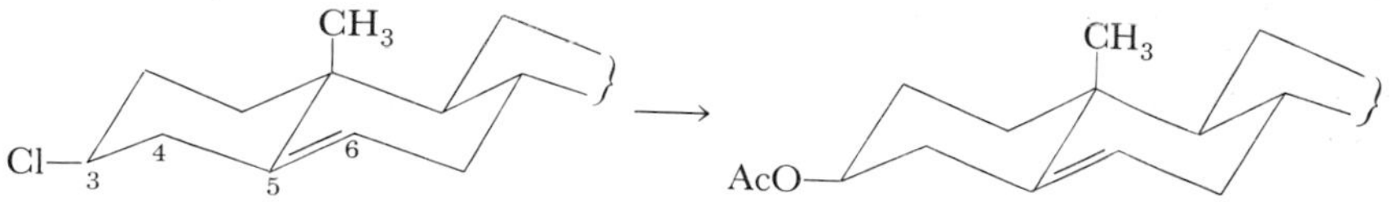

$$(5.33)$$

51   52

postulated some sort of assistance from the 5,6-double bond to explain these results.[71]

Winstein investigated the kinetics and products of 3-β-cholesteryl substitutions further. He found that under certain conditions 3-β-cholesteryl tosylate (or chloride) is acetolyzed to the cholesteryl i-acetate (Equation 5.34), and that this reaction is 100 times faster than the solvolysis of cyclohexyl tosylate. Moreover, if

[69] H. C. Brown, *The Nonclassical Ion Problem*, Plenum, New York, 1977, p. 49.
[70] Remember, by the convention that we are using, a dotted line means a partial bond. Thus in Structure **49** there is a partial bond between C_6 and C_1 and between C_6 and C_2 and a partial double bond between C_1 and C_2.
[71] C. W. Shoppee, *J. Chem. Soc.*, 1147 (1946).

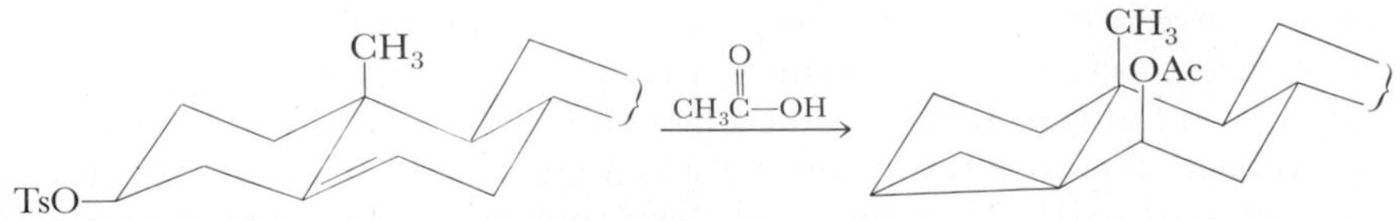

$$\text{(5.34)}$$

the conditions are slightly varied, the *i*-acetate undergoes rearrangement to form
3-β-cholesteryl acetate, also at an enhanced rate.[72] Rate enhancement for Reaction
5.34 might be explained if solvolysis of the tosylate leads immediately to the rearranged ion **53**, and if this ion is for some reason particularly stable. Its formation
would then be the driving force for the reaction. However, this explanation cannot
be correct. If the driving force for acceleration of Reaction 5.34 is the formation of
cation **53**, then the reverse ring opening of the *i*-acetate should not have a comparable driving force, but it does. Winstein suggested that a stabilized intermediate is
common to both reactions and is responsible for their accelerated rates.

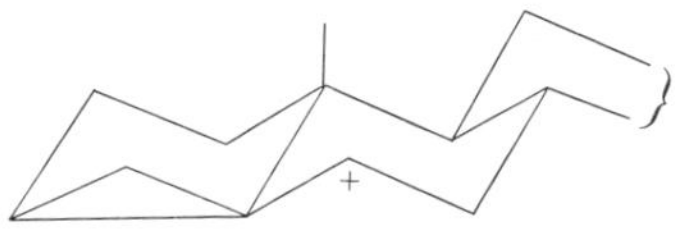

53

But what is the nature of the intermediate? A π bond between the empty *p*
orbital on C_3 and the *p* orbital on C_5 could not impart such stability, because α
overlap falls off rapidly with distance. Winstein suggested that the empty *p* orbital
on C_3 overlaps in an end-on, or σ, fashion with the *p* orbital on C_5, while at the
same time the 5,6-π bond is maintained, resulting in a two-electron, three-center

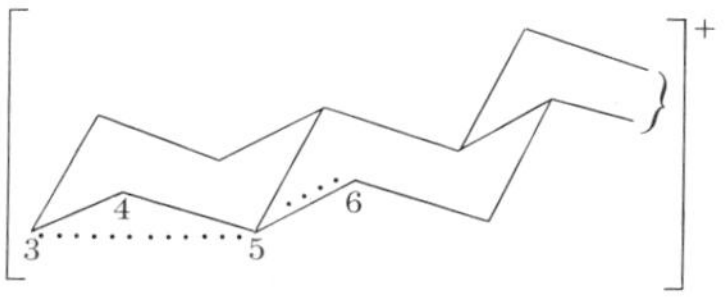

54

bond.[73] This structure is shown in an orbital representation in Figure 5.6 and in a
dotted-line representation in Structure **54**.

Intermediate **54** is responsible for the retention of configuration observed by
Shoppee. The 5,6-double bond that has displaced the leaving group from the back
side now shields this side from attack by the entering group, thus leaving the front
side as the only available route to substitution.

Because positive charge is delocalized by a *p* orbital one carbon atom fur-

[72] S. Winstein and R. Adams, *J. Am. Chem. Soc.*, **70**, 838 (1948).
[73] M. Simonetta and S. Winstein, *J. Am. Chem. Soc.*, **76**, 18 (1954).

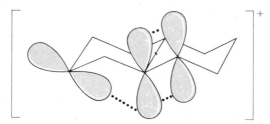

Figure 5.6. Orbital picture of homoallylic participation in the cholesteryl system.

ther removed than the allylic position, the kind of bonding shown in **54** is called *homoallylic* (the *homo* is for homologous) participation.[74]

The work on the cholesterol system stimulated investigation of other examples of homoallylic participation. Roberts found that *exo*-5- and *endo*-5-bicyclo[2.2.1]heptenyl (i.e., *exo*-5- and *endo*-5-norbornenyl) halides (**55** and **56**) both solvolyze in aqueous ethanol to give the same product (**57**); the exo compound (**55**) solvolyzes about ten times more rapidly than the endo compound (**56**). Roberts pointed out that backside homoallylic participation in ionization was possible in **55** but not in **56** (see Figure 5.7). Thus **55** forms a carbonium ion. Once **56** has ionized it can, in a second step, form either the rearranged carbenium ion (**58**) or the same carbonium ion formed in the solvolysis of **55**. By these pathways both compounds give the same product.[75]

A much more spectacular driving force was found in the acetolysis of *anti*-7-norbornenyl tosylate (**59**). This compound solvolyzes 10^{11} times faster than the

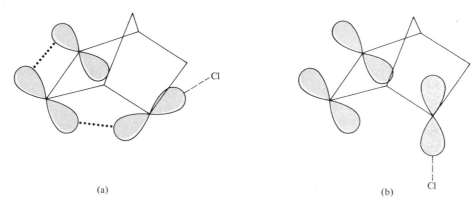

(a)   (b)

Figure 5.7 (a) Orbital picture of the transition state of solvolysis of *exo*-5-norbornenyl halides. (b) Orbital picture of the transition state of solvolysis of *endo*-5-norbornenyl halides.

[74] We have included homoallylic participation under carbonium ions because two-electron, three-center bonds *are* formed by end-on overlap through space. Because no carbon is coordinated to more than four other atoms, Brown calls these carbenium ions.

[75] J. D. Roberts, W. Bennett, and R. Armstrong, *J. Am. Chem. Soc.*, **72**, 3329 (1950).

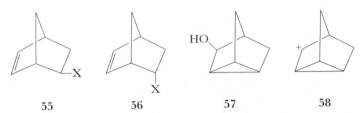

55   56   57   58

saturated analog and gives as the sole product the *anti*-acetate, **60**.[76] Winstein attributed the enormously accelerated rate to powerful anchimeric assistance of both p orbitals of the 2,3-double bond. Carbon-7 is located on a plane that bisects

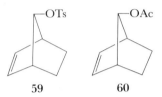

59   60

the 2,3-bond and is in fact homoallylic to both sides of the double bond. Therefore a developing p orbital on it is in a position to overlap equally with each of the p orbitals of the double bond as is shown in Figure 5.8. That both sides of the double bond do provide anchimeric assistance simultaneously is shown by the relative rates of solvolysis of the compounds shown below.[77]

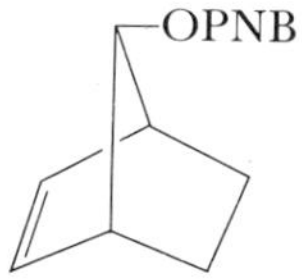

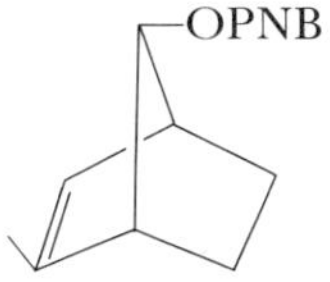

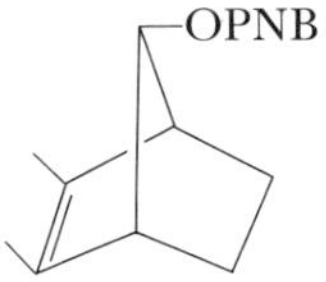

Relative rate: 1.0   13.3   148

Solvent does not attack at C_2 or C_3 to give a three-membered ring analogous to *i*-cholesteryl derivatives and to **57** because the resulting carbon skeleton is too strained. The fact that the product is 100 percent *anti*-acetate is a result of the back side of C_7 being hindered by the three-center bond that is fully formed in the intermediate carbonium ion.

There are strict geometrical requirements for homoallylic participation. For example, Bartlett and Rice found no indication of homoallylic participation on solvolysis of **61** in aqueous acids. Apparently the strain energy of bonding is greater than the stabilization so obtained.[78] The importance for homoallylic participation

[76](a) S. Winstein, M. Shatavsky, C. Norton, and R. B. Woodward, *J. Am. Chem. Soc.*, **77**, 4183 (1955); (b) S. Winstein and M. Shatavsky, *J. Am. Chem. Soc.*, **78**, 592 (1956).

[77]P. G. Gassman and D. S. Patton, *J. Am. Chem. Soc.*, **91**, 2160 (1969).

[78]P. D. Bartlett and M. R. Rice., *J. Org. Chem.*, **28**, 3351 (1963). Recently, however, π participation in the solvolysis of the corresponding tosylate has been observed [J. B. Lambert, R. B. Finzel, C. A. Belec, *J. Am. Chem. Soc.*, **102**, 3282 (1980)].

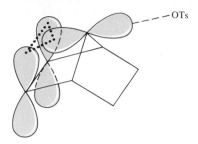

Figure 5.8 Transition state for solvolysis of *anti*-7-norbornenyl tosylate.

61

of the exact position of the *p* orbitals of the double bond in relation to the developing *p* orbital at the reaction site is also shown by the rate change attendant on puckering of the five-membered ring in the series **62**, **63**, and **64**[79] In the lower

	62	**63**	**64**
Relative rate:	1	5×10^2	2.5×10^6

homologs of these bicycloalkenes the five-membered ring is more puckered than in the higher homologs, the distance between the π bond and the developing positive charge is shortened, and backside participation is facilitated.

Electron demand at the incipient carbocation is also important in determining whether or not homoallylic participation takes place. Gassman and Fentiman have plotted the logarithms of the rates of solvolysis of both **65** and **66** in dioxane–water vs. the Hammett σ^+ constants of the X substituents (Figure 5.9). The straight

(OPNB = *p*-nitrobenzoate)

65   **66**

[79] B. A. Hess, Jr., *J. Am. Chem. Soc.*, **91**, 5657 (1969), and references therein.

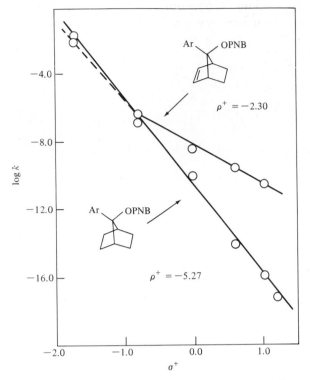

Figure 5.9 Logarithms of the rates of solvolysis of 7-arylnorbornyl and 7-aryl-*anti*-norbornenyl *p*-nitrobenzoates vs. the σ^+ constants of the aryl substituents. From H. C. Brown, *The Nonclassical Ion Problem*, Plenum Press, New York, 1977. Reprinted by permission of Plenum Press.

line for solvolysis of **65** is to be expected if the variations in rate are due to the electron-donating ability of X alone. The logarithms of the rates of **67** when X is a *p*-N,N-dimethyl or a *p*-methoxy group fall on the same straight line, signifying that the mechanism of ionization is the same as that of **65**. However, the logarithms of the rates of ionization of **66** when X is hydrogen, *p*-trifluoromethyl, or 3,5-bis-(trifluoromethyl) deviate from the line and are much larger than would be expected from the Hammett correlation.[80,81] Apparently, when X is electron donating, participation of the double bond is not required for ionization: The carbenium ion is more stable than the carbonium ion. However, when X is electron withdrawing, the carbenium ion is destabilized and the carbonium ion becomes the favored intermediate. This method of determining the onset of participation is called the "tool of increasing electron demand."

[80] P. G. Gassman and A. F. Fentiman, Jr., *J. Am. Chem. Soc.*, **92**, 2549 (1970).
[81] See also D. G. Farnum, R. E. Botto, W. T. Chambers, and B. Lam, *J. Am. Chem. Soc.*, **100**, 3847 (1978), for similar studies in superacid.

Several homoallylic carbonium ions have been observed by nmr in super-acid. For example, when 7-norbornenyl alcohol (**67**) is added to FSO_3H/SO_2CF at $-78°C$, the ^{13}C nmr spectrum indicates that C_7 of the resulting carbocation is very shielded relative to the cationic center in model carbenium ions, whereas C_2 and C_3 are deshielded relative to model double-bonded carbons. Thus the positive charge appears to be delocalized from C_7 to C_2 and C_3, as in **68**.[82]

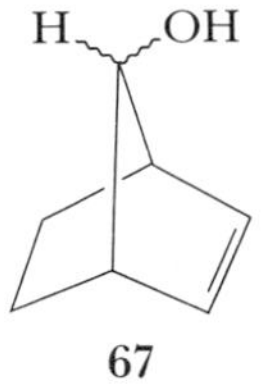

67 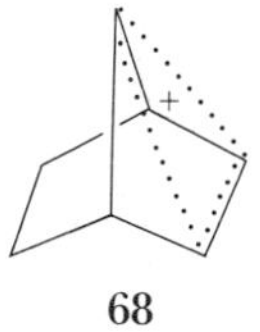   68

Double bonds further removed from the incipient carbocation than the homoallylic position can also assist in ionization if the geometry of the system allows it. For example, *trans*-5-cyclodecen-1-yl *p*-nitrobenzoate (**69**) solvolyzes 1500 times faster than the corresponding saturated compound in aqueous acetone. The product is *trans,trans*-1-decalol (**70**). If after 12 half-lives the product is isolated, a *p*-nitrobenzoate derivative is found which has the structure **71**. Furthermore, if the original *p*-nitrobenzoate has ^{18}O in the carbonyl carbon, it is found incompletely scrambled in **71**. These facts indicate that the rearrangement occurs in an intimate ion pair of which the ion **72** is the cation.[83]

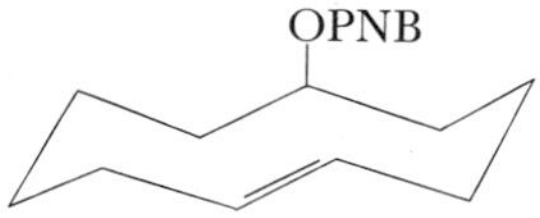

69 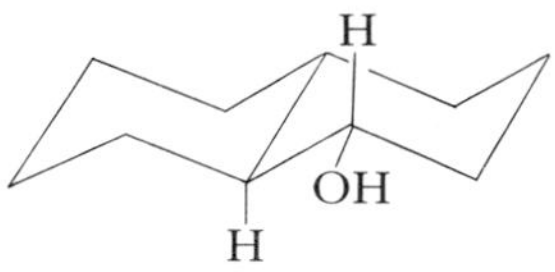   70 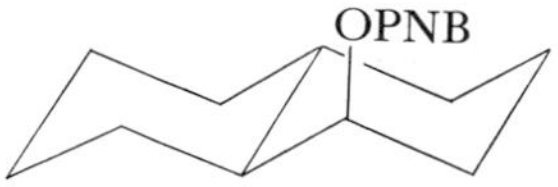   71

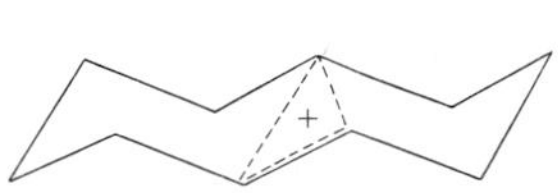

72

The Cyclopropylcarbinyl Cation

In 1951 Roberts observed that most cationic reactions of cyclopropylcarbinyl and of cyclobutyl derivatives give the same products in nearly the same ratio.[84] For example, cyclopropylcarbinyl and cyclobutyl amines (**73** and **74**) on deamination

[82] G. A. Olah and G. Liang, *J. Am. Chem. Soc.*, **97**, 6803 (1975). See also D. G. Farnum, *Adv. Phys. Org. Chem.*, **11**, 123 (1975), for some of the dangers in this kind of chemical shift–charge density analysis.
[83] H. L. Goering and W. D. Closson, *J. Am. Chem. Soc.*, **83**, 3511 (1961).
[84] J. D. Roberts and R. H. Mazur, *J. Am. Chem. Soc.*, **73**, 2509 (1951).

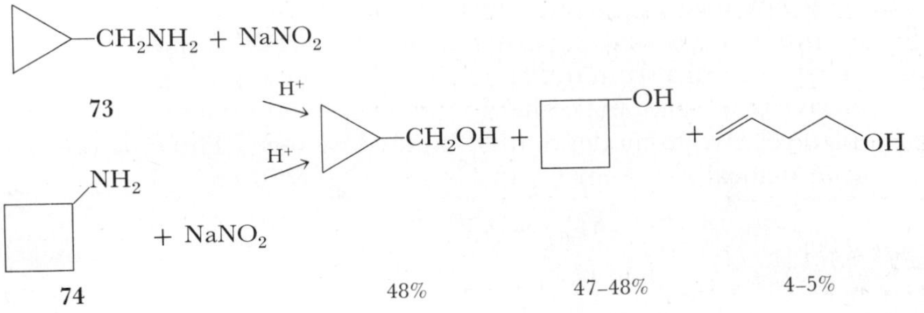

$$(5.35)$$

form the products shown in Equation 5.35.[85] Moreover, when allylcarbinyl tosylate (75) is solvolyzed in 98 percent formic acid, the products shown in Equation 5.35 are again formed, and their ratio is similar to that of Equation 5.35. If solvolysis of

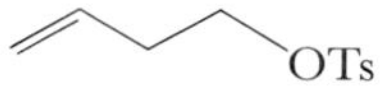

75

75 is carried out in more nucleophilic solvents (formic acid is strongly ionizing but weakly nucleophilic), predominant S_N2 reaction is observed.[86] When cyclopropylcarbinyl amine labeled with ^{14}C in the carbinyl position is deaminated, the label is found to be scrambled in the products so that the three methylene groups have almost—but not quite—achieved equivalence. The results are shown in Equation 5.36, in which the numbers at the carbons of the products refer to the percent ^{14}C found at that position in that product.[85] A set of rapidly

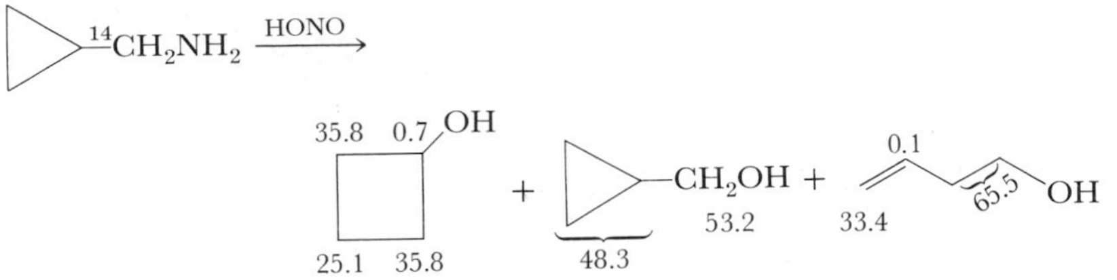

$$(5.36)$$

equilibrating carbenium ions might account for the rearrangements and the label scrambling; but this cannot be the correct explanation, for cyclopropylcarbinyl, cyclobutyl, and allylcarbinyl systems all solvolyze much more rapidly than would be expected from model compounds. Thus, for example, the rate of solvolysis of cyclopropylcarbinyl tosylate is 10^6 times that of the solvent-assisted solvolysis of isobutyl tosylate.[87] Cyclobutyl tosylate solvolyzes 11 times more rapidly than cyclo-

[85] R. H. Mazur, W. N. White, D. A. Semenow, C. C. Lee, M. S. Silver, and J. D. Roberts, *J. Am. Chem. Soc.*, **81**, 4390 (1959).
[86] K. L. Servis and J. D. Roberts, *J. Am. Chem. Soc.*, **86**, 3773 (1964).
[87] D. D. Roberts, *J. Org. Chem.*, **29**, 294 (1964); **30**, 23 (1965).

Relative rate:　　　　　10^6　　　　　1

hexyl tosylate.[88] And allylcarbinyl tosylate in 98 percent HCOOH solvolyzes 3.7

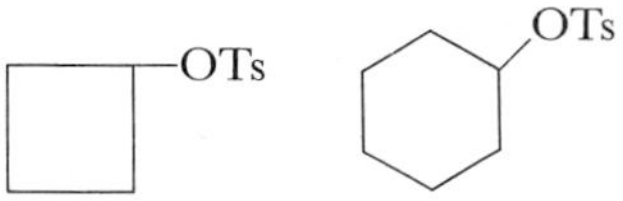

Relative rate:　　　11　　　　1

times faster than its saturated analog, in spite of the electron-withdrawing effect of the double bond.[86]

Relative rate:　　　3.7　　　　1

A set of rapidly equilibrating carbenium ions also cannot account for the ρ^+ constants obtained when the rates of solvolysis of Compounds **76** and **77** are plotted

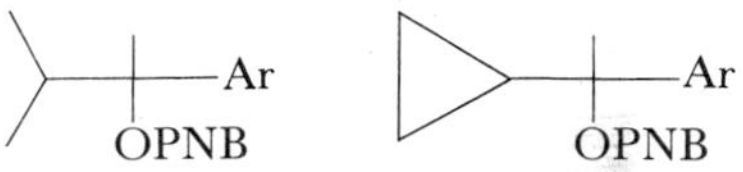

PNB = *para*-nitrobenzoate

ρ^+:　　　-4.76　　　　-2.76

against the σ^+ constants of the substituents in the aromatic ring. The much smaller need for electrons from the aromatic ring of the cation from **77** suggests that **77** must be getting its stabilization elsewhere—from neighboring-group participation by the cyclopropyl ring.[89]

Roberts suggested that a set of charge-delocalized, rapidly equilibrating carbonium ions, which he called *bicyclobutonium ions,* are the first-formed ions from **73, 74,** and **75.** In Scheme 3 are shown the bicyclobutonium ions formed from the deamination of ^{14}C-labeled cyclopropylcarbinyl amine (Equation 5.36). According to Roberts, there would be two equivalent first-formed carbonium ions: A three-center bond could be formed from the developing empty orbital on C_4 and either orbitals on C_1 and C_3 (path a) or orbitals on C_1 and C_2 (path b). Once these carbonium ions were formed, they could be converted to any of the other carbonium ions in Scheme 3. Equilibration arrows should be shown between all the structures, but their inclusion would further confuse this already conceptually difficult scheme. Figure 5.10 shows an orbital representation of bicyclobutonium ion (**78**).

[88] H. C. Brown and G. Ham, *J. Am. Chem. Soc.,* **78,** 2735 (1956).
[89] E. N. Peters and H. C. Brown, *J. Am. Chem. Soc.,* **95,** 2397 (1973).

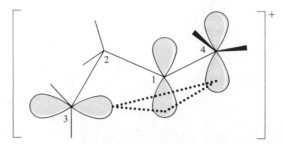

Figure 5.10 Orbital representation of the bicyclobutonium ion. Reprinted with permission from R. H. Mazur, W. N. White, D. A. Semonow, C. C. Lee, M. S. Silver, *J. Am. Chem. Soc.,* **81,** 4390 (1959). Copyright by the American Chemical Society.

Scheme 3

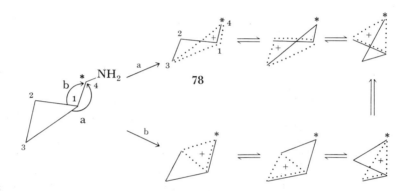

Positive charge in each of the bicyclobutonium ions would be distributed between the three carbon atoms of the three-center bond. For example, the positive charge on carbonium ion (**78**) would be delocalized over C_4, C_1, and C_3. If water attacked this ion at C_4, unrearranged cyclopropylcarbinol would be obtained; if water attacked at C_1, cyclobutanol would be the product; and if it attacked at C_3, allylcarbinol would be formed (Scheme 4). Addition of water to each of the bicyclobutonium ions would give the same three products, but the ^{14}C label would be scrambled differently in each.

Note that if the bicyclobutonium ion were formed directly upon ionization of an allylcarbinyl derivative, it would be a case of homoallylic participation in an acyclic system. In fact, the bicyclobutonium ion is similar to the carbonium ion proposed by Winstein for homoallylic participation in the 7-norbornenyl system—cf. Figures 5.8 and 5.10 and Structures **78** and **68**. The difference between them is that **68** is more symmetrical.

Since Roberts' work, a great deal of evidence, both experimental and theoretical, has accumulated that indicates that the bicyclobutonium ion is not the

SCHEME 4

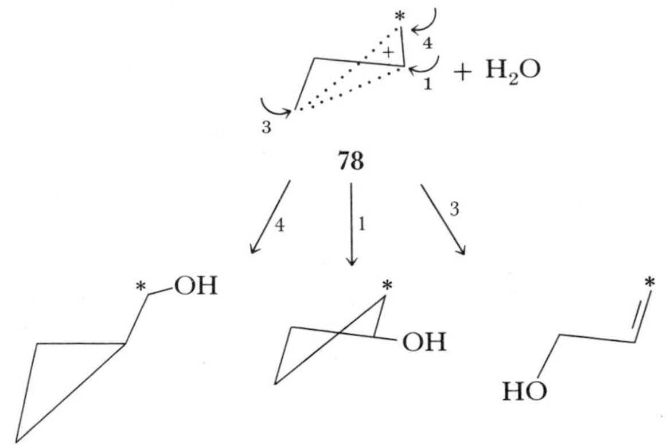

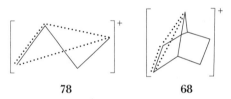

78

first-formed ion upon solvolysis of unstrained cyclopropylcarbinyl systems. Instead, the structure of the ion apparently is the bisected cyclopropylcarbinyl cation, as shown in an orbital diagram in Figure 5.11.

78   **68**

A comparison of Figures 5.10 and 5.11 shows that the bisected cyclopropyl-carbinyl cation differs from the bicyclobutonium ion in several ways. For example, in the bisected cyclopropylcarbinyl cation, the carbinyl carbon (C_4) and its substituents lie above or below the ring in a plane that is perpendicular to the plane of the ring and bisects C_1 and the C_2—C_3 bond. The vacant p orbital is *parallel* to the plane of the ring and to the C_2—C_3 bond. By π overlap with the C—C bonding orbitals of the cyclopropane ring, which, because of angle strain, have an abnormal

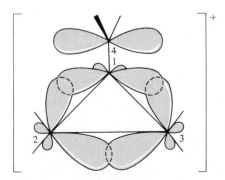

Figure 5.11. Orbital representation of the bisected cyclopropylcarbinyl cation.

amount of p character,[90] the positive charge at C_4 is delocalized to all three ring carbons. In the bicyclobutonium ion, C_4 is not equidistant from C_2 and from C_3; the vacant p orbital on C_4 is almost *perpendicular* to the C_2—C_3 bond. Furthermore, positive charge at C_4 in a single bicyclobutonium ion can be delocalized to C_1 and to either C_2 or C_3 but not to both.

Some of the evidence for the bisected cyclopropylcarbinyl cation follows. That the charge can be delocalized to both C_2 and C_3 simultaneously has been shown by the work of Schleyer and Van Dine.[91] These workers studied the solvolysis of cyclopropylcarbinyl 3,5-dinitrobenzoates and found that methyl substituents accelerated the rate by an amount dependent only on the number of such substituents and not on their position. Thus **79**, **80**, and **81** react at almost the same rate. If the transition state for ionization were similar to the bicyclobutonium ion, two methyl groups at C_2 should accelerate the rate more than one at C_2 and one at C_3. A symmetrical transition state for ionization similar to the bisected cyclopropylcarbinyl cation in which the charge is delocalized over all four carbon atoms best explains the results.

79	**80**	**81**

That maximum acceleration occurs when the vacant p orbital is parallel to the plane of the cyclopropyl ring can be seen from the solvolysis of spiro[cyclopropane-1,2′-adamantyl] chloride (**82**). The carbocation formed by departure of Cl⁻ is unable to adopt the geometry of the bisected cyclopropylcarbinyl cation, but can orient its empty p orbital properly to form the bicyclobutonium ion. This compound solvolyzes 10^3 times more slowly than 1-adamantyl chloride.[92] On the other hand, **83** solvolyzes 10^5 times faster than **84**. The cation from **83** *must* adopt the bisected cyclopropylcarbinyl conformation.[93]

There is presently controversy about whether the bisected cyclopropylcarbinyl cation should properly be called a carbonium or a carbenium ion.[94] Brown maintains that since the σ bonds of the ring are not involved in bridging

[90](a) D. Peters, *Tetrahedron* **19**, 1539 (1963); (b) M. Randi and A. Maksić, *Theor. Chim. Acta*, **3**, 59 (1965); (c) A. D. Walsh, *Trans. Faraday Soc.*, **45**, 179 (1949); (d) L. I. Ingraham, in *Steric Effects in Organic Chemistry*, M. S. Newman, Ed., Wiley, New York, 1956, Chap. 11; (e) C. A. Coulson and W. E. Moffitt, *Phil. Mag.*, **40**, 1 (1949); (f) R. Hoffmann and R. B. Davidson, *J. Am. Chem. Soc.*, **93**, 5699 (1971).

[91] P. v. R. Schleyer and G. W. Van Dine, *J. Am. Chem. Soc.*, **88**, 2321 (1966).

[92] B. R. Ree and J. C. Martin, *J. Am. Chem. Soc.*, **92**, 1660 (1970); V. Buss, R. Gleiter, P. v. R. Schleyer, *J. Am. Chem. Soc.*, **93**, 3927 (1971), and references therein.

[93] Y. E. Rhodes and V. G. Difate, *J. Am. Chem. Soc.*, **94**, 7582 (1972).

[94](a) H. C. Brown, E. N. Peters, and M. Ravindranathan, *J. Am. Chem. Soc.*, **99**, 505 (1977); (b) H. C. Brown, *The Nonclassical Ion Problem*, Plenum, New York, 1977, and comments by P. v. R. Schleyer therein; (c) H. C. Brown, *Tetrahedron*, **32**, 179 (1976).

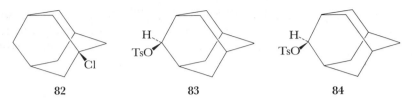

82   83   84

to, but only in conjugation with, the cationic carbinyl carbon, the bisected cyclo-propylcarbinyl cation should be termed a carbenium ion in analogy to the allyl cation. Schleyer, however, wants it in the carbonium ion category. He points out that it cannot be represented by a single Lewis structure but requires extensive charge delocalization into the ring by resonance to explain both its chemical and physical (e.g., bond lengths, charge distribution) properties.

Structures other than the bisected cyclopropylcarbinyl cation have also been suggested as intermediates in cyclopropylcarbinyl and cyclobutyl solvolyses.[95] Winstein has pointed out that the nature of the intermediate cation may differ with the geometrical requirements of the starting material.[96] Indeed, Figure 5.12 presents the results of CNDO calculations on the barrier to rotation of the carbinyl group.[97,98] It appears that a 30° rotation from the symmetrical structure ($\alpha = 0°$, see **85**) leads to only a small decrease in stabilization. This finding has been confirmed experimentally.[93]

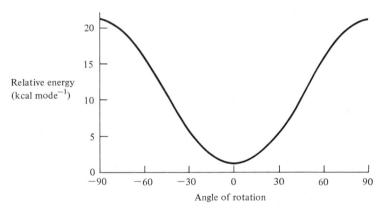

Figure 5.12 Change in energy of the cyclopropylcarbinyl cation as the cationic center is rotated. Reprinted with permission from K. B. Wiberg and J. G. Pfeiffer, *J. Am. Chem. Soc.,* **92,** 553 (1970). Copyright 1970 American Chemical Society.

[95] (a) Z. Majerski, S. Borčić, and D. E. Sunko, *J. Chem. Soc. D,* 1636 (1970); (b) C. D. Poulter, E. C. Friedrich, and S. Winstein, *J. Am. Chem. Soc.,* **92,** 4274 (1970); (c) C. D. Poulter and S. Winstein, *J. Am. Chem. Soc.,* **92,** 4282 (1970); (d) Z. Majerski and P. v. R. Schleyer, *J. Am. Chem. Soc.,* **93,** 665 (1971); (e) J. E. Baldwin and W. D. Foglesong, *J. Am. Chem. Soc.,* **90,** 4303, 4311 (1968).
[96] See note 95(b).
[97] K. B. Wiberg and J. G. Pfeiffer, *J. Am. Chem. Soc.,* **92,** 553 (1970).
[98] See also B. Andersen, O. Schallner, and A. de Meijere, *J. Am. Chem. Soc.,* **97,** 3521 (1975).

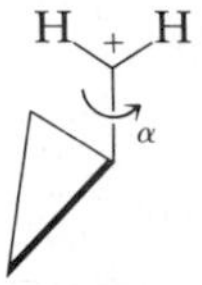

85

Is the cyclopropylcarbinyl system also the first-formed ion in solvolysis of cyclobutyl derivatives? The evidence is conflicting. Majerski, Borčić, and Sunko studied the reactions shown in Equations 5.37 and 5.38 and found that when the starting material is the cyclopropylcarbinyl methanesulfonate, the label scrambling is less complete in the cyclopropylcarbinol than in the cyclobutanol; similarly, cyclobutyl methanesulfonate gives less label scrambling in the cyclobutanol than in the cyclopropylcarbinol[99] (In Equations 5.37 and 5.38 the numbers indicate the percent of the methylene groups at that position that are CD_2.) This would make it appear that cyclopropyl and cyclobutyl derivatives each solvolyze to give ions that are similar in structure to the starting material. Solvent capture may occur at this stage. If it does not, the first-formed ion rearranges.

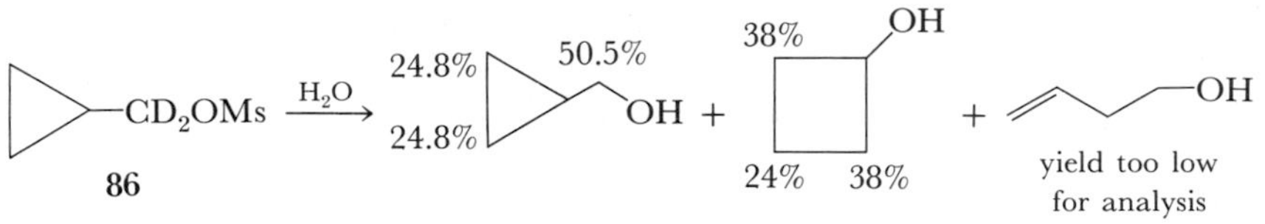

On the other hand, there is now a good deal of evidence that the solvolysis of most cyclobutyl derivatives does lead directly to the cyclopropylcarbinyl cation. For example, orbital symmetry considerations (Chapter 10) indicate that the conversion of cyclobutyl cations into cyclopropylcarbinyl cations should occur by disrotatory ring opening, as shown in Figure 5.13, and steric factors that hinder such a process are found to decelerate cyclobutyl solvolyses. Thus both **87** and **88** solvolzye in acetone–water to give 3-cyclopentenol (**89**), but **87** solvolyzes 10^7 times faster than **88**.[100] In **87** to overlap with the back side of the developing p orbital, the orbitals of the bond being broken must turn in such a way as to move the bridgehead hydrogens away from each other. In **88**, however, the same process would require that the bridgehead hydrogens move toward each other. This is energetically unfavorable.

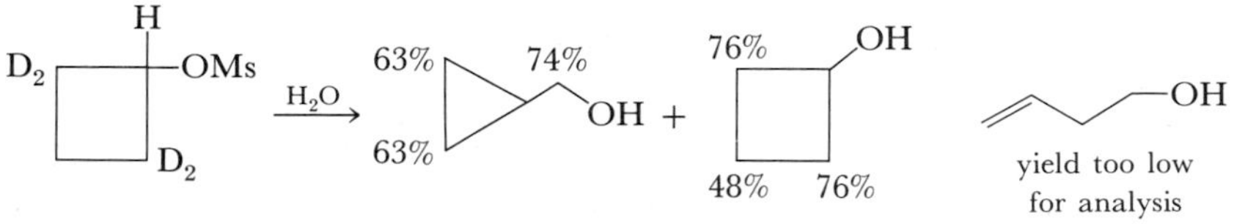

[99] See note 95(b).
[100] (a) K. B. Wiberg, V. Z. Williams, Jr., and L. E. Friedrich, *J. Am. Chem. Soc.*, **90**, 5338 (1968); (b) *J. Am. Chem. Soc.*, **92**, 564 (1970).

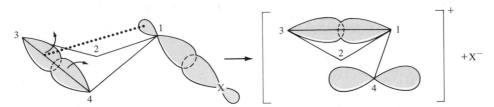

Figure 5.13 Orbital symmetry allowed (disrotatory) opening of a cyclobutyl cation. Note that the orbitals of the C—C bond being broken overlap with the back side of the orbital used for bonding to the departing group.

DNBO = dinitrobenzoate

87   88   89

Cyclobutyl cations certainly do exist if they are especially stabilized. For example, although 1-methylcyclobutyl cation rearranges to a methylcyclopropylcarbinyl structure in superacid,[101] the 1-phenylcyclobutyl cation does not rearrange.[102]

Ab initio calculations on the parent cyclopropylcarbinyl system indicate that the lowest energy minimum on the $C_4H_7^+$ potential energy surface is the bisected cyclopropylcarbinyl cation but, lying in a potential energy well only 0.5 kcal mole^{-1} higher, there is a structure very like the bicylobutonium ion. Interconversions between equivalent, bisected, cyclopropylcarbinyl cation structures occur via a puckered, cyclobutyl cation transition state. Figure 5.14(a) shows a cross section of the potential energy surface in the vicinity of the bisected cyclopropylcarbinyl cation. Calculations on the 4-methyl- and the 4,4-dimethylcyclopropylcarbinyl cations (**90** and **91**, respectively) also predict a bisected conformation, but now the

90   91

4-substituted ion is more stable than the others so that at least in **91** rearrangement is effectively blocked.[103] Figure 5.14(b) shows the proposed reaction coordinate diagram for the *gem*-dimethyl system, **91**.

[101]G. A. Olah, G. K. S. Prakash, D. J. Donovan, and I. Yavari, *J. Am. Chem. Soc.*, **100**, 7085 (1978).

[102]G. A. Olah, C. L. Jeuell, D. P. Kelly, and R. D. Porter, *J. Am. Chem. Soc.*, **94**, 146 (1972).

[103](a) See note 65(e); (b) W. J. Hehre and P. C. Hiberty, *J. Am. Chem. Soc.*, **96**, 302 (1974); (c) R. D. Bach and P. E. Blanchette, *J. Am. Chem. Soc.*, **101**, 46 (1979); (d) W. J. Hehre, *Acc. Chem. Res.*, **8**, 369 (1975); (e) B. A. Levi, E. S. Blurock, and W. J. Hehre, *J. Am. Chem. Soc.*, **101**, 5537 (1979).

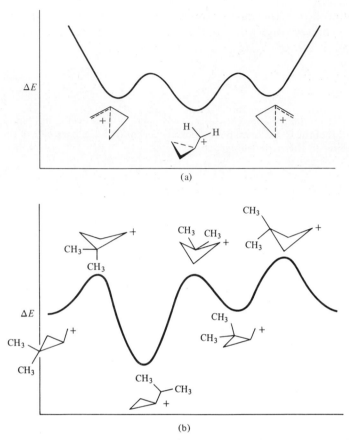

Figure 5.14 (a) Potential energy surface for $C_4H_7^+$ in the region of the bisected cyclopropylcarbinyl cation. Reprinted with permission from B. A. Levi, E. S. Blurock, and W. J. Hehre, *J. Am. Chem. Soc.*, **101**, 5537 (1979). Copyright 1979 American Chemical Society. (b) Potential energy surface for the dimethyl cyclopropylcarbinyl system. Reprinted with permission from W. J. Hehre and P. C. Hiberty, *J. Am. Chem. Soc.*, **96**, 302 (1974). Copyright 1974 American Chemical Society.

Nuclear magnetic resonance studies of the unsubstituted cyclopropylcarbinyl cation in superacid solution do not agree with theoretical predictions of its structure. The nmr results are most consistent with rapidly equilibrating cations with some bridging as in the bicyclobutonium ion, rather than with rapidly equilibrating, equivalent, bisected cyclopropylcarbinyl cations.[104] On the other hand, the nmr spectrum of **91** in superacid indicates that this ion *is* stable and nonequilibrat-

[104](a) G. A. Olah, C. L. Jeuell, D. P. Kelly, and R. D. Porter, *J. Am. Chem. Soc.*, **94**, 146 (1972); (b) G. A. Olah, R. J. Spear, P. C. Hiberty, and W. J. Hehre, *J. Am. Chem. Soc.*, **98**, 7470 (1976). For a different interpretation, see (c) D. P. Kelly and H. C. Brown, *J. Am. Chem. Soc.*, **97**, 3897 (1975).

ing, as predicted from the calculations. Two nonequivalent methyl groups are present—one syn and one anti to the ring; these peaks do not coalesce up to $-30°C$, at which temperature ring opening occurs.[105]

The Norbornyl Cation[106]

In discussing the cyclopropylcarbinyl cation before the norbornyl cation we have, chronologically, put the cart before the horse. The first example of anchimeric assistance by a C—C σ bond was published by Winstein and Trifan in 1949.[107] These workers studied the solvolysis of *exo*- and *endo*-2-norbornyl arenesulfonates (**92** and **93**, respectively) and found that the reactions had these interesting characteristics: (1) the exo compound solvolyzes 350 times more rapidly than the endo compound; (2) both exo and endo starting materials give exclusively (>99.9 percent) exo product, as shown in Equation 5.39; (3) chiral exo starting material gives entirely racemic product, but the product from endo starting material retains some chirality; and (4) chiral exo starting material, recovered before complete reaction, is partially racemized. The ratio of polarimetric and titrimetric rate constants in acetic acid is 4.6. Since recovered, unreacted endo starting material is not racemized, the rate of ionization of **92** relative to the rate of ionization of **93** in acetic acid is not 350 but 350×4.6 or 1550.

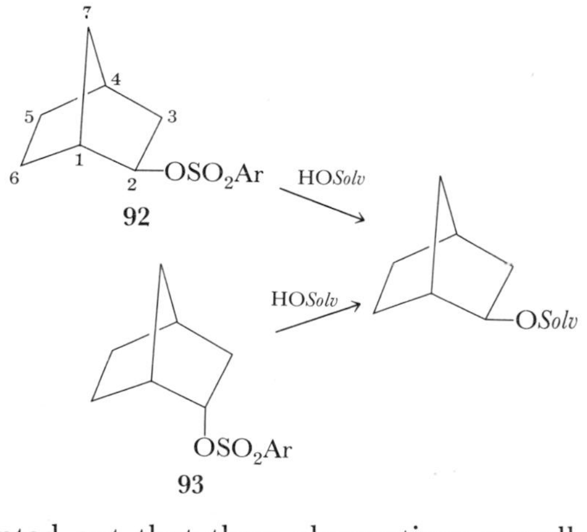

(5.39)

Winstein pointed out that these observations are all consistent if, in the solvolysis of **92**, the 1,6-bond assists in the ionization and the "norbornonium" ion, shown in **94** and Figure 5.15, is the first-formed intermediate. Below are shown the

[105](a) G. A. Olah, *Angew. Chem. Int. Ed. Engl.,* **12**, 173 (1973); (b) G. A. Olah and G. Liang, *J. Am. Chem. Soc.,* **97**, 1920 (1975).
[106] See note 66.
[107](a) S. Winstein and D. S. Trifan, *J. Am. Chem. Soc.,* **71**, 2953 (1949); (b) S. Winstein and D. S. Trifan, *J. Am. Chem. Soc.,* **74**, 1147, 1154 (1952); (c) S. Winstein, E. Clippinger, R. Howe, and E. Vogelfanger, *J. Am. Chem. Soc.,* **87**, 376 (1965).

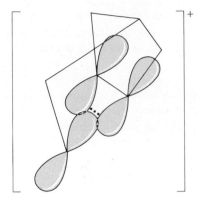

Figure 5.15 Orbital picture of the norbornyl cation.

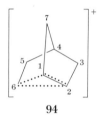

94

resonance structures implied by **94**. The driving force for the rearrangement would be the relief of skeletal strain.

The explanations, in terms of **94**, of the observed differences between the exo and endo arenesulfonates (**92** and **93**) follow.

Rate In the *exo*-norbornyl arenesulfonates, the C_1—C_6 bond is in the trans periplanar orientation to the leaving group and therefore in the optimum position to provide anchimeric assistance. The carbonium ion (**94**) is formed directly. In the *endo*-norbornyl arenesulfonate, the C_1—C_6 bond is not properly oriented for anchimeric assistance and although the C_1—C_7 bond is not badly oriented, the rearranged ion, **95**, resulting from participation of this bond is more highly strained than the starting material. Thus according to Winstein the *endo*-norbornyl derivative first ionizes to the charge-localized carbenium ion (**96**). Once formed, this ion can rearrange to the more stable carbonium ion (**94**). Figure 5.16 shows the pro-

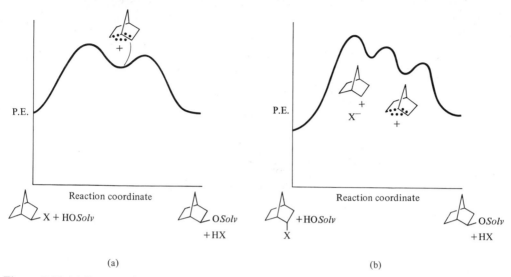

Figure 5.16 (a) Suggested reaction coordinate diagram for the solvolysis of *exo*-2-norbornyl derivatives. (b) Suggested reaction coordinate diagram for solvolysis of *endo*-2-norbornyl derivatives.

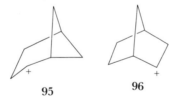

posed reaction coordinate diagrams for solvolysis of the *exo*- and *endo*-norbornyl
sulfonates.

Product The product from solvent attack on the carbonium ion (**94**) must
be from the exo direction because this mode of attack corresponds to backside
displacement of the three-center bond.

Stereochemistry The norbornonium ion has a plane of symmetry. This can
perhaps be seen most readily if **94** is rotated as shown in Equation 5.40. Thus chiral
starting material should give racemic product if the intermediate **94** lies on the
reaction path: Solvent attack at C_2 yields product of retained configuration, but
attack at the equivalent site, C_1, yields the enantiomer. If the chiral norbornenium
ion (**96**) were the intermediate, solvent attack could occur only at C_2 and retained
product would be obtained. As we have already seen, the product from chiral exo
starting material is entirely racemic. The *exo*-norbornyl derivatives obtained from
endo starting material are about 87 percent racemized in strongly nucleophilic

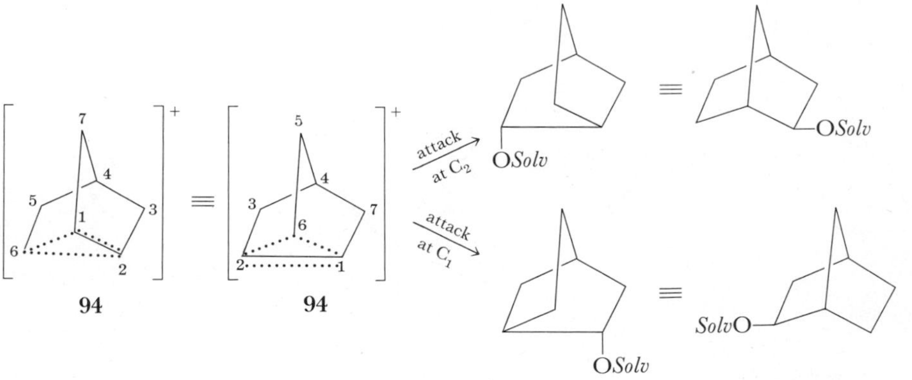

$$(5.40)$$

solvents. Thus most of the product is indeed formed by the reaction path shown in Figure 5.16, but some solvent must also attack C_2 before the carbenium ion has had time to rearrange. Less nucleophilic solvents give a greater extent of racemization.

Isomerization of starting material Partial racemization of recovered exo starting material is consistent with the hypothesis that a norbornonium ion is formed immediately on ionization if it is further postulated that an intimate ion pair is the first ionic species on the reaction path. Internal return from the achiral norbornonium–arenesulfonate intimate ion pair must give racemized starting material.

If the norbornonium ion is formed on solvolysis of *exo*-norbornyl derivatives, C_2 should become equivalent to C_1 and C_7 to C_3. In a most elegant tracer experiment, Roberts and Lee synthesized *exo*-2-norbornyl-[2,3-^{14}C] brosylate (**97**), solvolyzed it in acetic acid, and degraded the product. Equation 5.41 shows the

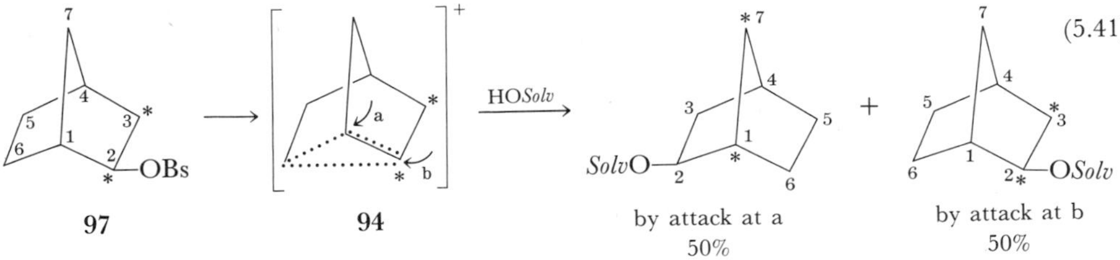

$$(5.41)$$

product distribution expected if the symmetrical carbonium ion (**94**), were formed. The label was found not only at C_1, C_2, C_3, and C_7, but also at C_5 and C_6. To account for this Roberts suggested that 6,2- or 6,1-hydride shifts occur in the carbonium ion simultaneously with the rearrangement of Equation 5.41 as shown in Equation 5.42.[108]

[108] J. D. Roberts and C. C. Lee, *J. Am. Chem. Soc.*, **73**, 5009 (1951); (b) J. D. Roberts, C. C. Lee, and W. H. Saunders, Jr., *J. Am. Chem. Soc.*, **76**, 4501 (1954).

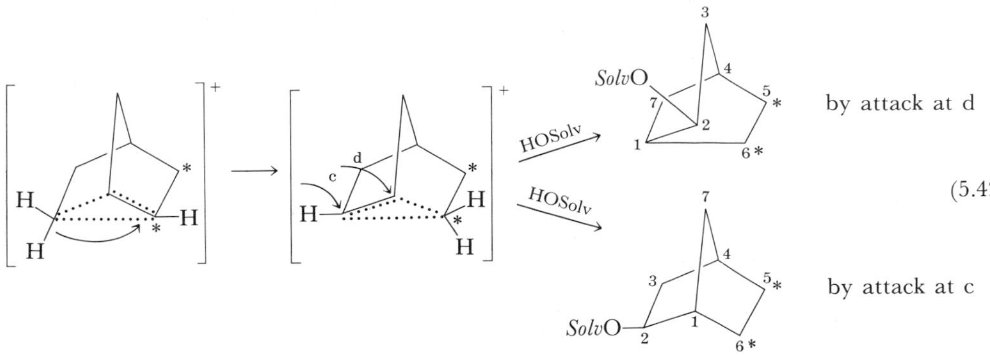

$$(5.42)$$

Although cationic reactions of 2-*exo*-norbornyl arenesulfonates have characteristics that would be associated with a charge-delocalized carbonium ion intermediate—driving force, stereospecific product formation, rearranged products, internal return to rearranged starting material, and special chiral characteristics—a storm of controversy has raged over its existence. Its opponents, of whom H. C. Brown is the chief, have maintained that the postulation of a bridged carbonium ion intermediate is not necessary to explain the characteristics of the norbornyl system. He has argued that the exo/endo rate ratio in solvolyses of norbornyl derivatives is large, not because k_{exo} is particularly large but because k_{endo} is particularly small. In his view a 2-endo substituent experiences steric hindrance to ionization by the three endo protons. Brown cites a number of systems in which the exo/endo rate ratios are large even when σ participation seems to be ruled out. For example, Brown reasons that since a *p*-methoxyphenyl group causes neighboring-group participation to vanish in the 7-norbornenyl system (see below), it should also cause any that exists to vanish in the 2-norbornyl system where the rate accelerations are smaller. However, in 80 percent aqueous acetone, *exo*-2-*p*-methoxyphenyl 2-norbornyl *p*-nitrobenzoate, **98** (X = *p*-OCH$_3$) solvolyzes 284 times faster than the endo isomer, **99** (X = *p*-OCH$_3$)—an exo/endo rate ratio similar to that in

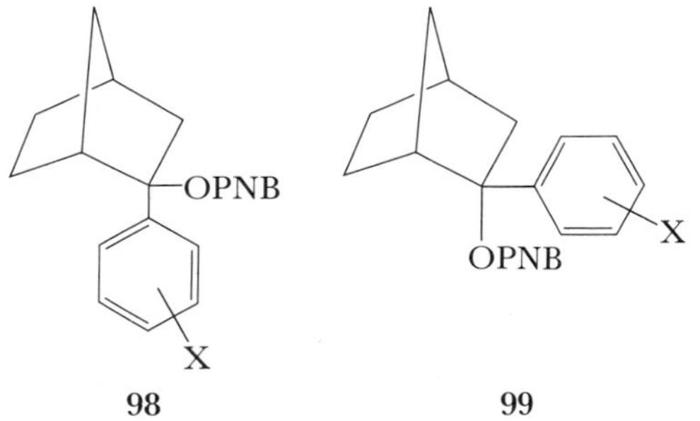

the unsubstituted 2-norbornyl tosylates. Furthermore, when the electron-demanding nature of X in **98** and **99** is varied widely and the logarithms of the resulting solvolysis rate constants are plotted against the σ$^+$ constants of the substituents, the plots are linear over all X's for both exo and endo isomers, as shown in Figure 5.17.

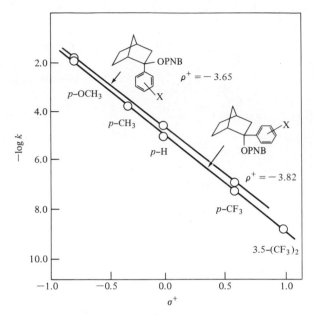

Figure 5.17 Logarithms of the rates of solvolysis of the tertiary 2-aryl-2-norbornyl *p*-nitrobenzoates in 80 percent aqueous acetone at 25°C vs. the σ^+ constants of the aryl substituents. Reprinted with permission from H. C. Brown, K. Takeuchi, and M. Ravindranathan, *J. Am. Chem. Soc.*, **99**, 2684 (1977). Copyright 1977 American Chemical Society.

The slopes of the lines, the ρ^+'s, are also very similar. Thus the tool of increasing electron demand reveals no σ participation in the solvolysis of these tertiary 2-norbornyl *p*-nitrobenzoates. Figure 5.18 compares the solvolysis of the 2-aryl-2-norbornyl *p*-nitrobenzoates with the 2-aryl-2-norbornenyl *p*-nitrobenzoates.[109]

Brown also does not believe that the exclusive formation of an exo product is evidence for a bridged intermediate. Rather, he suggests, the peculiar U-shaped structure of the C_6—C_1—C_2 segment hinders endo approach of a nucleophile to the classical 2-norbornyl cation and thus exo product is formed. Evidence that a reagent prefers to attack from the exo direction comes from reactions such as that shown in Equation 5.43. Although no free carbocation is formed in this reaction (see Section 7.1), 99.5 percent of the product is exo alcohol.[110] Brown points out

$$ \text{(5.43)} $$

99.5%

[109](a) H. C. Brown and K. Takeuchi, *J. Am. Chem. Soc.*, **99**, 2679 (1977); (b) H. C. Brown, K. Takeuchi, and M. Ravindranathan, *J. Am. Chem. Soc.*, **99**, 2684 (1977).
[110]H. C. Brown and J. H. Kawakami, *J. Am. Chem. Soc.*, **92**, 1900 (1970).

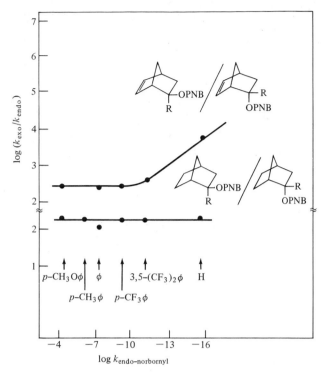

Figure 5.18 Effect of increasing electron demand on the exo/endo rate ratio in 2-norbornyl and 2-norbornenyl *p*-nitrobenzoates. Reprinted with permission from H. C. Brown, K. Takeuchi, and M. Ravindranathan, *J. Am. Chem. Soc.*, **99**, 2684 (1977). Copyright 1977 American Chemical Society.

that the energetics of the reaction, whatever their cause, preclude formation of endo product. The exo/endo rate ratio of 1550 for acetolysis of the norbornyl tosylates corresponds to a difference in the free energies for ionization of 4.5 kcal mole^{-1}. Schleyer has estimated the strain in the *endo*-norbornyl tosylate to be ~1.3 kcal mole^{-1} greater than in the exo isomer.[111] If one uses these data to construct the *Goering–Schewene* diagram for acetolysis of *exo* and *endo*-norbornyl tosylate, Figure 5.19(a), one sees that recombination of the intermediate carbocation to form *exo*-tosylate requires 5.8 kcal mole^{-1} less energy than recombination to form endo starting material.[112] Furthermore, if the analogous Goering–Schewene diagram is constructed, from the appropriate data, for capture of the common intermediate by acetic acid,[113] Figure 5.19(b), it is very similar to that of Figure 5.19(a). Thus the fate of the intermediate cation in both internal return and acetolysis is determined by the low activation energy leading to exo product. Whatever factor stabilizes the

[111] P. v. R. Schleyer, *J. Am. Chem. Soc.*, **86**, 1854, 1856 (1964).
[112] H. C. Brown, *Tetrahedron*, **32**, 179 (1976).
[113] H. L. Goering and C. B. Schewene, *J. Am. Chem. Soc.*, **87**, 3516 (1965).

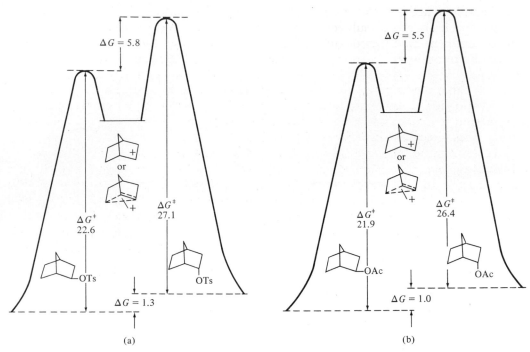

Figure 5.19 Free-energy diagram for the acetolysis of (a) *exo-* and *endo*-norbornyl tosylate and (b) *exo-* and *endo*-norbornyl acetate. Reprinted with permission from H. C. Brown, *Tetrahedron,* **32,** 179 (1976). Copyright 1976 Pergamon Press, Ltd.

transition state for exo relative to that for endo product—be it a bridged ion or less steric hindrance—causes the stereoselectivity. A unique interpretation of a bridged ion is not required.

Finally, Brown proposes that the skeletal rearrangements and loss of chirality are consistent with rapid 1,3-Wagner–Meerwein shifts, as shown in Equation 5.44.[114] Note that the two carbenium ions in Equation 5.44 are mirror images

$$(5.44)$$

of one another. The carbonium ion (**94**) would then be a transition state for Equation 5.44, not a stabilized intermediate.

The investigation of the 2-norbornyl cation has been intensive and detailed. Sargent said, it "may well be the most thoroughly investigated yet least thoroughly

[114](a) See note 67; (b) H. C. Brown and M. Ravindranathan, *J. Am. Chem. Soc.,* **100,** 1865 (1978).

understood reactive intermediate known to organic chemists. Seldom, if ever, has a single species been the subject of so many ingenious experiments conceived by so many eminent investigators utilizing such a variety of sophisticated methods. Despite the intensity of this effort, the structure of the 2-norbornyl cation remains an enigma."[115] We shall only touch on the controversy momentarily and give examples of experiments carried out to clarify one of its aspects—is the large exo/endo rate ratio due to a remarkably large (assisted) exo rate or a remarkably small (hindered) endo rate? For more detailed presentations, the reader is referred to the references cited in notes **66** and **67**.

A number of workers have criticized Brown's extrapolations of the effect of increasing electron demand in **98** and **99** to secondary systems. Schleyer has pointed out that all aryl groups—even $3,5\text{-}(CF_3)_2$-substituted ones—are electron donors, and thus the carbocations formed from **98** and **99** may well be carbenium ions while those from secondary systems may be carbonium ions.[116]

In fact, Grob has found that, in contrast to Brown's results with **98** and **99**, although the rates of solvolysis of both **100** and **101** in **70** percent aqueous dioxane

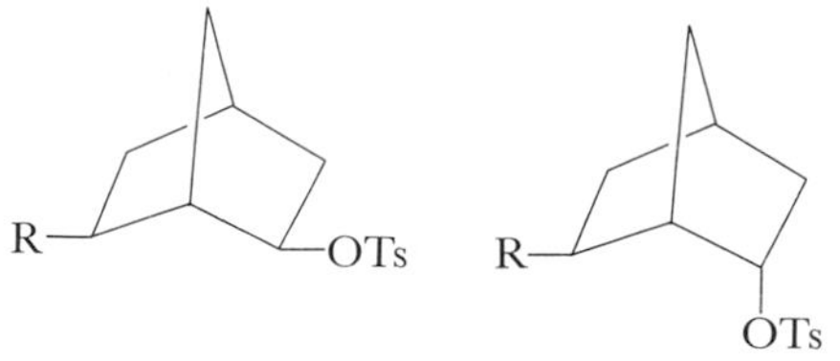

R———OTs   R———OTs

100   **101**

can be correlated with the σ_I constants of the R substituents, the rate of **100** is much more affected by the nature of R than is the rate of **101** (Figure 5.20). As is shown in Table 5.2, the exo/endo rate ratios vary from 425 to *less than one*.[117]

Brown's claim that the exo/endo rate ratio in 2-norbornyl tosylate solvolysis is due to steric hindrance to ionization in the endo isomer has been investigated. A number of cases of steric deceleration of solvolysis have been reported. For example, the nonbonded strain in **103** is approximately 1.9 kcal mole^{-1} greater than that

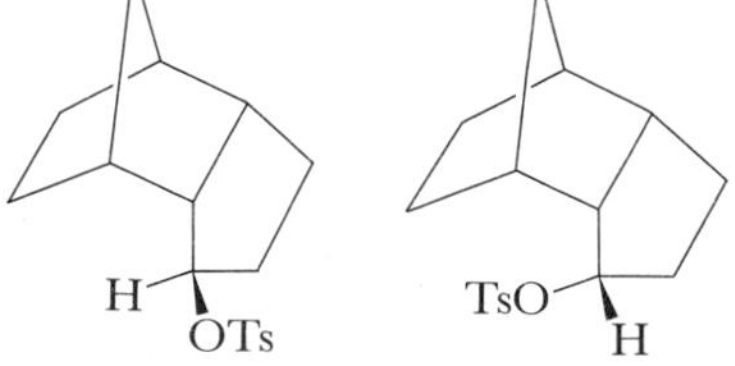

H———OTs   TsO———H

102   **103**

[115] See note 66(c).

[116] (a) P. v. R. Schleyer in H. C. Brown, *The Nonclassical Ion Problem*, Plenum, New York, 1977. See also (b) G. A. Olah, *Acc. Chem. Res.*, **9**, 41 (1976); (c) M. A. Battiste and R. A. Fiato, *Tetrahedron Lett.*, 1255 (1975).

[117] (a) W. Fischer, C. A. Grob, and G. von Sprecher, *Tetrahedron Lett.*, 473 (1979); (b) W. Fischer, C. A. Grob, G. von Sprecher, and A. Waldner, *Tetrahedron Lett.*, 1901, 1905 (1979).

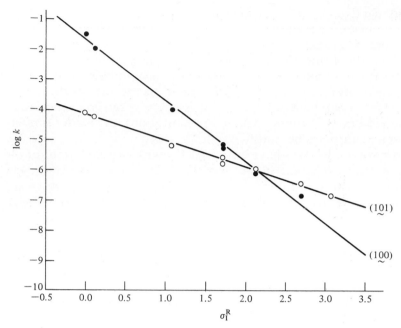

Figure 5.20 Logarithms of the rates of solvolysis of **100** and **101** vs. the σ_I constants of the R substituents. Reprinted with permission from W. Fischer, C. A. Grob, G. von Sprecher, and A. Waldner, *Tetrahedron Lett.*, 1905 (1979). Copyright 1979 Pergamon Press, Ltd.

in **102**. Assuming that the strain is fully relieved in the transition state for ionization, one would predict that the rate of solvolysis of the *endo*-tosylate (**103**) should exceed that of the *exo*-tosylate (**102**) by a factor of ~25. The exo compound actually solvolyzes 5.7 times faster than the endo compound. Since there is no obvious route for σ bond participation here, it appears that there must be an increase in nonbonded strain in the transition state of **103** of $RT \ln 5.7$ or ~1 kcal mole^{-1}.[118] As Sargent pointed out, experiments and examination of molecular models both indicate that this system should offer more extreme steric hindrance to endo ionization than the norbornyl system does. One kcal mole^{-1}, then, is an upper limit for the increase in nonbonded interaction experienced by the leaving group in going from the ground to the transition state in *endo*-2-norbornyl tosylate.[119] But, as stated above, the exo/endo rate ratio of 1550 reported by Winstein and Trifan, a ratio since corroborated by hundreds of other studies, requires a difference in the free energy of activation of 4.5 kcal mole^{-1}.

   When 2-norbornyl fluoride is dissolved in superacid, a solution of the 2-norbornyl cation is obtained, which has been examined by a number of physical methods. One method that has been employed is ESCA (electron spectroscopy for

[118] H. C. Brown, I. Rothberg, P. v. R. Schleyer, M. M. Donaldson, and J. J. Harper, *Proc. Natl. Acad. Sci. U.S.A.*, **56**, 1653 (1967).
[119] See note 66(c).

Table 5.2 Rates of Solvolysis of **100** and **101**[a]

R	$k(\text{sec}^{-1})$, 70.0°		$\dfrac{k_{exo}}{k_{endo}}$
	exo Series **100**	endo Series **101**	
H	3.58×10^{-2}	8.42×10^{-5}	425
CH$_3$	1.09×10^{-2}	6.02×10^{-5}	181
CH$_2$Br	1.06×10^{-4}	6.75×10^{-6}	16
COOH	5.97×10^{-6}	2.88×10^{-6}	2
COOCH$_3$	6.33×10^{-6}	1.73×10^{-6}	3.7
OCOCH$_3$	8.14×10^{-7}	1.21×10^{-6}	0.67
Br	1.51×10^{-7}	4.06×10^{-7}	0.37
CN	1.23×10^{-7}	1.40×10^{-7}	0.88

[a] Reprinted with permission from W. Fischer, C. A. Grob, G. von Sprecher, and A. Waldner, *Tetrahedron Let.*, 1905 (1979). Copyright (1979), Pergamon Press, Ltd.

chemical analysis). By ESCA one can determine the energy required to remove inner-shell electrons from around the nucleus.[120] A sample is exposed to high-energy X-rays of known wavelength, which cause electrons to be ejected from the molecule. The energy conservation expression for the photoemission process can be expressed by

$$E_{h\nu} = E_{\text{kin}} + E_b + E_\phi \tag{5.45}$$

where $E_{h\nu}$, E_{kin}, and E_b are the X-ray energy, the kinetic energy of the electron emitted, and the binding energy of the electron emitted, respectively. E_ϕ is a constant for a given system and can be determined. An electron multiplier detector counts the emitted electrons, and an electron energy analyzer determines the kinetic energies of the emitted electrons. Thus E_b can be determined from Equation 5.45.

The energy required to remove a $1s$ electron from a hydrocarbon is almost a constant. For example, by ESCA one cannot distinguish between benzene and neopentane. In classical, nonresonance-stabilized carbocations, the positive charge is usually centered on a single atom (the time required for ESCA ionization is approximately 10^{-16} sec—much faster than intramolecular interactions such as Wagner–Meerwein rearrangements or hydrogen shifts). Thus more energy must be applied to remove an electron from the positively charged carbon than from its uncharged neighbors. Figure 5.21 shows the carbon $1s$ electron spectrum for the t-butyl cation. The positive carbon is well separated from the carbons of the methyl groups. Figure 5.22 shows the carbon $1s$ electron spectrum for the 2-methylnorbornyl cation and the norbornyl cation. The former is a classical ion and has a spectrum similar to that of the t-butyl cation. By comparison, the spectrum of the 2-norbornyl cation shows no center of especially high binding energy. The peak that represents removal of a $1s$ electron from a positively charged carbon is only a poorly separated

[120] For a review of ESCA studies, see J. M. Hollander and W. L. Jolly, *Acc. Chem. Res.*, **3**, 193 (1970).

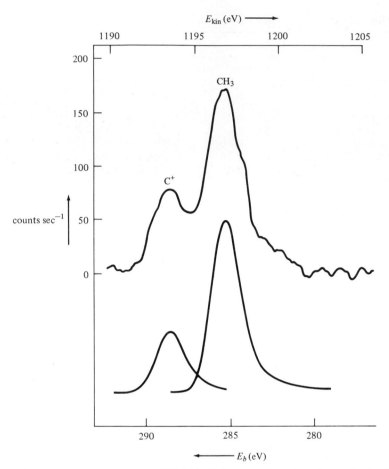

Figure 5.21 Carbon $1s$ electron spectrum (ESCA) of the t-butyl cation. From G. A. Olah, *Angew. Chem. Int. Ed. Engl.*, **12**, 173 (1973). Reproduced by permission of Verlag Chemie, GMBH.

shoulder on the overall $1s$ electron peak. Olah has suggested that the peak-area ratio of the shoulder to the rest is $2:5$—a result that would indicate that the charge is equally distributed between carbon-2 and carbon-6. There has, however, been controversy about the peak-area ratio. Kramer has suggested that it is actually closer to $1:6$, which would indicate charge localized on Carbon-2. Thus ESCA has not, as yet, given an unambiguous answer to the structure of the norbornyl cation.[121,122,123]

[121](a) G. A. Olah, *Angew. Chem. Int. Ed. Engl.*, **12**, 173 (1973). (b) D. T. Clark, B. M. Cromarty, and L. Colling, *J. Am. Chem. Soc.*, **99**, 8120 (1977), and references therein.
[122]G. A. Olah, *Acc. Chem. Res.*, **9**, 41 (1976), and references therein.
[123]For criticisms of Olah's ESCA spectrum, see G. M. Kramer, *Adv. Phys. Org. Chem.*, **11**, 177 (1975).

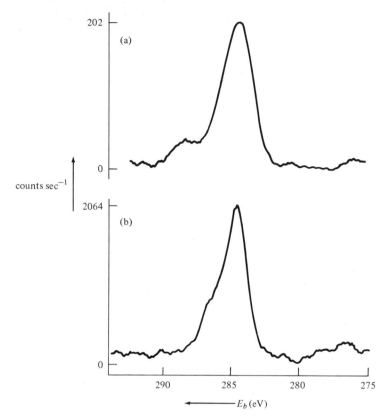

Figure 5.22 Carbon 1s electron spectrum (ESCA) of (a) the 2-methylnorbornyl cation and (b) the norbornyl cation. From G. A. Olah, *Angew. Chem. Int. Ed. Engl.*, **12**, 173 (1973). Reproduced by permission of Verlag Chemie, GMBH.

Proton and ^{13}C nmr studies and Raman spectral investigations in superacid are all consistent with, and have been interpreted as, evidence for the bridged structure **94**.[122] Kramer, however, has argued that all three tools are inadequate for deciding between the bridged structure and equilibrating classical ions.[124]

Another piece of evidence for a bridged norbornyl cation in superacid solution is that in SO_2ClF/SbF_5 at $-55°C$ the heat of isomerization of the secondary 4-methyl-2-norbornyl cation (**104**) to the tertiary 2-methyl-2-norbornyl cation (**105**) is only 6.57 kcal mole^{-1}. Rearrangement of *sec*-butyl cation (**106**) to *t*-butyl cation (**107**) under the same conditions releases 14.2 kcal mole^{-1}. This implies that the

[124](a) See the reference in note 123; see also (b) J. W. Larsen, *J. Am. Chem. Soc.*, **100**, 330 (1978) and (c) D. G. Farnum, *Adv. Phys. Org. Chem.*, **11**, 123 (1975), for some dangers in using ^{13}C nmr spectra to interpret the structure of carbocations.

norbornyl cation is stabilized by approximately 7.5 kcal mole^{-1} relative to normal saturated secondary ions.[125]

| 104 | 105 | 106 | 107 |

CH$_3$—CH—CH$_2$—CH$_3$ CH$_3$—C—CH$_3$

It appears that even 2-arylsubstituted-2-norbornyl cations (**108**) can exist as bridged cations in superacid solution. Farnum measured the ^{13}C chemical shifts of

| 108 | 109 |

the cationic carbon of a number of aryl-substituted carbocations and confirmed that the more electron-withdrawing the substituents are in the benzene ring, the more the cationic carbon is deshielded and therefore the more it is shifted to lower fields. Furthermore, the chemical shift in one carbenium ion is proportional to the chemical shift in another, as shown in Figure 5.23. However, when the chemical shifts of C$_2$ in **108** are plotted against the cationic carbon of a "normal" model compound, **109**, the correlation is not linear (Figure 5.24). The chemical shifts *are* proportional for electron-donating groups, but when the groups are very electron withdrawing, the norbornyl C$_2$ is not shifted as far downfield as expected. Apparently under these circumstances bridging sets in, delocalizing the positive charge to C$_1$ and C$_6$.[126]

There is always a question about what the relationship is between the structure of a carbocation in superacid solution and the structure of the transition state for solvolysis leading to the carbocation. Arnett and Schleyer have attempted to answer that question for the norbornyl cation. They found that there is a close correlation between the heats of ionization, ΔH_i, of most alkyl chlorides, including 2-norbornyl chloride, in superacid and the ethanolysis rates of the same compounds. This correlation is good evidence that for these compounds the transition state for solvolysis does resemble the carbocation. A major exception is 2-*endo*-

[125](a) E. M. Arnett, N. Pienta, and C. Petro, *J. Am. Chem. Soc.*, **102**, 398 (1980); see also (b) E. M. Arnett and C. Petro, *J. Am. Chem. Soc.*, **100**, 2563, 5408 (1978); (c) J. J. Solomon and F. H. Field, *J. Am. Chem. Soc.*, **98**, 1567 (1976); and (d) P. P. S. Saluja and P. Kebarle, *J. Am. Chem. Soc.*, **101**, 1084 (1979), for similar studies in the gas phase.

[126]D. G. Farnum, R. E. Botto, W. T. Chambers, and B. Lam, *J. Am. Chem. Soc.*, **100**, 3847 (1978); (b) G. A. Olah, G. K. S. Prakash, and G. Liang, *J. Am. Chem. Soc.*, **99**, 5683 (1977).

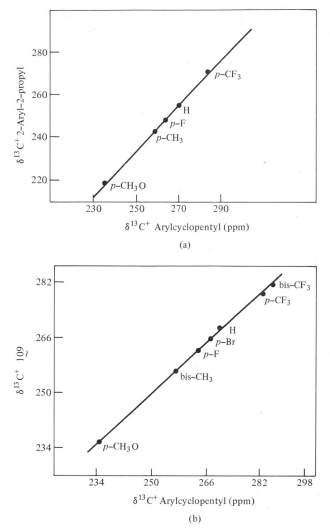

Figure 5.23 Correlation of the ^{13}C chemical shifts at the carbocationic center of (a) the 2-aryl-2-propyl cations and (b) the 6-aryl-6-bicylo[3.2.1]octyl cations (**109**) with those of the 1-arylcyclopentyl cations. Reprinted with permission from D. G. Farnum, R. E. Botto, W. T. Chambers, and B. Lam, *J. Am. Chem. Soc.*, **100**, 3847 (1978). Copyright 1978 American Chemical Society.

norbornyl chloride, of which the ΔH_i is much more exothermic than would be expected from the solvolysis data. Arnett and Schleyer point out that both *exo-* and *endo-*norbornyl chlorides ionize to the same ion. If one of the chlorides ionizes with rearrangement and the other without, then the correlation can succeed for only that chloride whose transition state for solvolysis is more closely related to the free

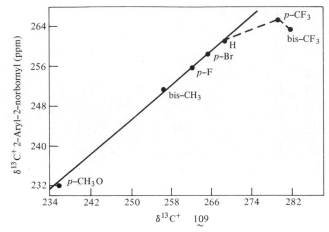

Figure 5.24 Correlation of the ^{13}C chemical shifts of the carbocationic center of the 2-aryl-2-norbornyl cations with those of **109**. Reprinted with permission from D. G. Farnum, R. E. Botto, W. T. Chambers, and B. Lam, *J. Am. Chem. Soc.*, **100**, 3847 (1978). Copyright 1978 American Chemical Society.

carbocation in superacid. Thus the transition state for solvolysis of *exo*-norbornyl chloride is closely related to the free norbornyl cation, but the transition state for solvolysis of the *endo*-2-norbornyl chloride is of much higher energy than the free carbocation.[127]

A number of theoretical calculations have been carried out to determine the most stable structure of the unsolvated norbornyl cation. Some have concluded that the carbenium ion **110** is slightly more stable[128] and others that the carbonium ion **94** is slightly more stable.[129] Dewar has, by using MINDO/3 calculations, concluded that **110** is more stable than **94** but that **111**, the ion formed by C_7 participation, is the most stable of all.[130]

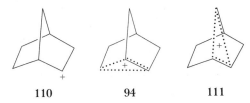

110	**94**	**111**

Although the 2-norbornyl cation question has not yet been completely resolved, **112** has been shown, even to H. C. Brown's satisfaction, to solvolyze with σ participation to the "Coates carbonium ion," **113**.[131]

[127] E. M. Arnett, C. Petro, and P. v. R. Schleyer, *J. Am. Chem. Soc.*, **101**, 522 (1979).
[128] F. K. Fong, *J. Am. Chem. Soc.*, **96**, 7638 (1974).
[129] D. W. Goetz, H. B. Schlegel, and L. C. Allen, *J. Am. Chem. Soc.*, **99**, 8118 (1977).
[130] M. J. S. Dewar, R. C. Haddon, A. Komornicki, and H. Rzepa, *J. Am. Chem. Soc.*, **99**, 377 (1977).
[131] (a) R. M. Coates and E. Robert Fretz, *J. Am. Chem. Soc.*, **99**, 297 (1977); (b) H. C. Brown and M. Ravindranathan, *J. Am. Chem. Soc.*, **99**, 299 (1977).

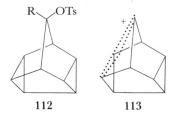

112   113

Protonated Cyclopropanes[132]

Deamination of 1-aminopropane-1-^{14}C gives unscrambled 2-propanol and partially scrambled 1-propanol, as shown in Equation 5.46. (The numbers at the

$$CH_3CH_2{}^{14}CH_2NH_2 \xrightarrow{HNO_2} CH_3\underset{\underset{OH}{|}}{CH}{}^{14}CH_3 + CH_3CH_2CH_2OH \qquad (5.46)$$
$$\phantom{CH_3CH_2{}^{14}CH_2NH_2 \xrightarrow{HNO_2} CH_3CH{}^{14}CH_3 + } 1.9 \quad 2.2 \quad 95.9$$

carbons of the 1-propanol indicate the percentage of ^{14}C found at each position.)[133] The 2-propanol arises from a 1,2-hydride shift. The label at C_3 of the primary alcohol could arise from a 1,3-hydride shift, but no simple hydride shift can bring about the label at C_2.

Aboderin and Baird suggested the mechanism of Scheme 5 for the label scrambling in Equation 5.46.[134] In each of the *edge-protonated cyclopropanes* (**114a–c**) two carbons are pentacoordinated; the third participant in the three-center bond is hydrogen. Figure 5.25 shows an orbital diagram of Structure **114**. In the product-determining step of Scheme 5, the nucleophilic oxygen of water attacks one of the carbons; this carbon then withdraws its orbital from the three-center bond and an ordinary σ bond is formed between the remaining carbon and the hydrogen.

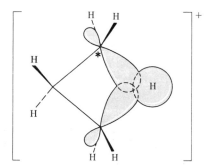

Figure 5.25 Orbital picture of edge-protonated cyclopropane.

[132]See note 1(a) and note 9(b).
[133](a) C. C. Lee and J. E. Kruger, *J. Am. Chem. Soc.*, **87**, 3986 (1965); (b) C. C. Lee and J. E. Kruger, *Tetrahedron*, **23**, 2539 (1967).
[134](a) See note 6; (b) A. A. Aboderin and R. L. Baird, *J. Am. Chem. Soc.*, **86**, 2300 (1964).

Scheme 5

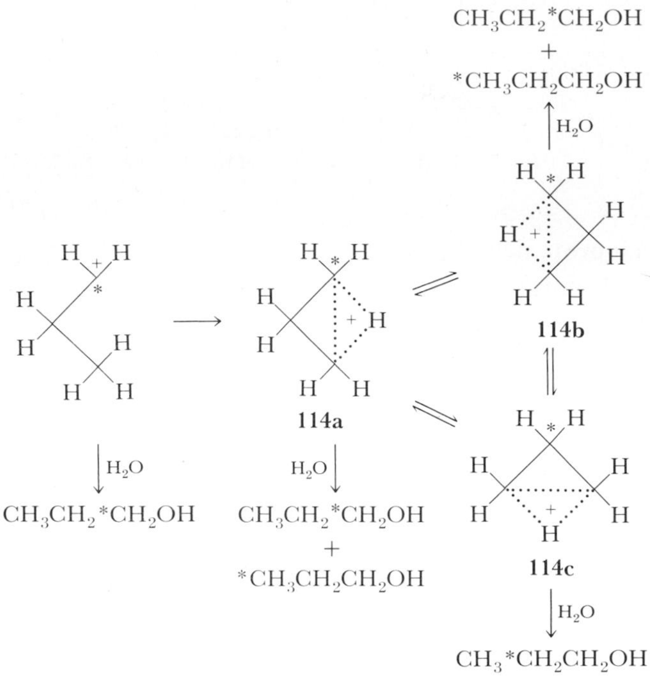

$$CH_3CH_2{}^*CH_2OH$$
$$+$$
$${}^*CH_3CH_2CH_2OH$$

↑ H_2O

114b

114a

114c

↓ H_2O

$$CH_3CH_2{}^*CH_2OH$$

H_2O ↓

$$CH_3CH_2{}^*CH_2OH$$
$$+$$
$${}^*CH_3CH_2CH_2OH$$

↓ H_2O

$$CH_3{}^*CH_2CH_2OH$$

Another mechanism, that shown in Scheme 6, has also been suggested for the label scrambling of Equation 5.46. In each of the *corner-protonated cyclopropanes* (**115a–c**), all three participants in the three-center bond are carbon. Structures **114a–c** would be the transition states for the interconversions of Structures **115a–c**.

Ab initio molecular orbital calculations carried out by Pople, Schleyer, and co-workers predict that the most stable form of $C_3H_7{}^+$ is the 2-propyl cation (**116**).

116

The second most stable structure they found to be the corner-protonated cyclopropane (**115**). The calculations suggest that **115** lies at an energy minimum and thus is an intermediate, not merely a transition state, but that **114** is probably not an intermediate. Figure 5.26 shows some of the structures and their calculated relative energies.[135]

A number of other reactions have been postulated to involve protonated cyclopropanes as intermediates.[136] For example, nmr studies of the *sec*-butyl cation in superacid show that from -100 to $-40°C$ a process with an activation energy of 7 kcal mole^{-1} occurs that scrambles all the protons. The activation energy is too low

[135] P. C. Hariharan, L. Radoni, J. A. Pople, P. v. R. Schleyer, *J. Am. Chem. Soc.,* **96**, 599 (1974).
[136] For a recent example, see C. H. DePuy, A. H. Andrist, and P. C. Fünfschilling, *J. Am. Chem. Soc.,* **96**, 948 (1974).

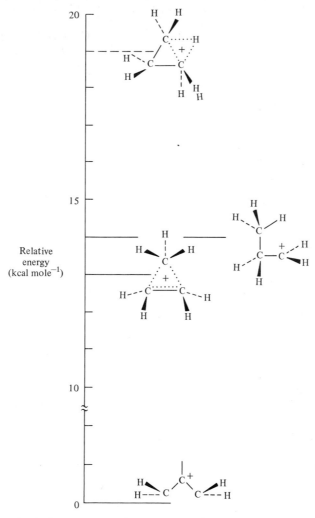

Figure 5.26 Calculated relative energies of various $C_3H_7^+$ ions. Reprinted with permission from P. C. Hariharan, L. Radom, J. A. Pople, and P. v. R. Schleyer, *J. Am. Chem. Soc.*, **96**, 599 (1974). Copyright 1974 American Chemical Society.

for the scrambling to occur by the hydride-shift mechanism shown in Equation 5.47. Saunders suggests that the most probable mechanism is that shown in Scheme 7.[137]

$$\text{CH}_3\text{CH}_2\text{CH}_2^+ \;\rightleftharpoons\; \text{CH}_3\text{CHCH}_3^+ \;\rightleftharpoons\; \text{CH}_3\text{CH}_2\text{CH}_2^+ \;\rightleftharpoons\; \text{etc.} \qquad (5.47)$$

1,3-Hydride shifts can take place directly, without the intervention of a carbonium ion intermediate, if the geometry of the system is favorable. For example, in

[137] See note 9(b).

SCHEME 6

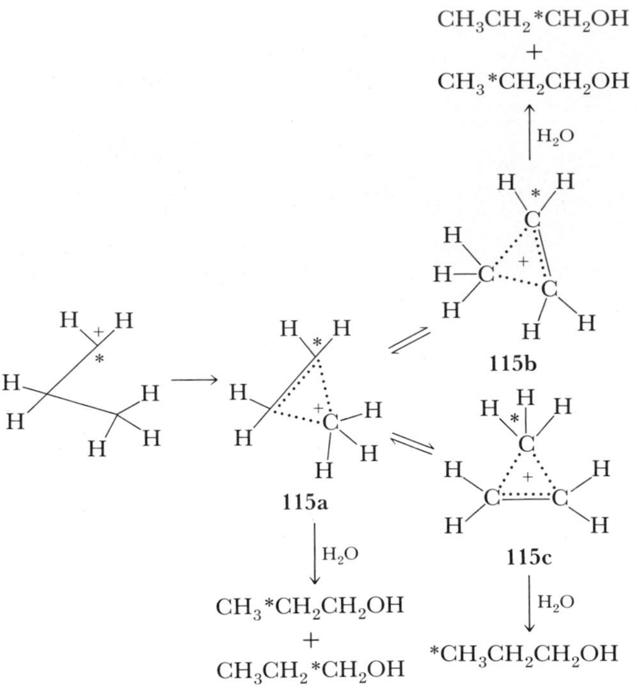

$$CH_3CH_2*CH_2OH$$
$$+$$
$$CH_3*CH_2CH_2OH$$

115a, 115b, 115c

$$CH_3*CH_2CH_2OH$$
$$+$$
$$CH_3CH_2*CH_2OH$$

$$*CH_3CH_2CH_2OH$$

the solvolysis of cyclohexyl-2,6-d$_2$ tosylate in 97 percent acetic acid, 1,3-hydride shifts have been reported to account for 33 percent of the product.[138] Apparently the reaction is made facile by the proximity of the 3-axial hydrogen to the empty p orbital.

Hydride shifts of higher order than 1,3 occur if (1) the migration origin and the migration terminus are close together and (2) the geometry of the system allows overlap of the orbitals of hydrogen and of the migration terminus. Because of the proximity requirement, such shifts are rare in acyclic systems[139] in nucleophilic solvents,[140] but in reactions of medium-ring (7–12 carbon atoms) carbocations, 1,3-, 1,4-, 1,5-, and 1,6-hydride shifts occur readily.[141] "Transannular shifts" were first

[138] Y. G. Bundel, V. A. Savin, A. A. Lubovich, and O. A. Reutov, *Proc. Acad. Sci. USSR Chem. Sect.* (in English), **165**, 1180 (1965).
[139] For a summary of these, see J. L. Fry and G. J. Karabatsos, in *Carbonium Ions*, Vol. II, G. A. Olah and P. v. R. Schleyer, Eds., Wiley, New York, 1972, pp. 555–557.
[140] Saunders and Stofko have observed 1,3- 1,4-, and 1,5-intramolecular shifts from tertiary to tertiary center in superacid and have calculated the activation barriers to be 8.5, 12–13, and 6–7 kcal mole^{-1}, respectively. [M. Saunders and J. J. Stofko, Jr., *J. Am. Chem. Soc.*, **95**, 252 (1973)].
[141] For reviews, see (a) V. Prelog and J. G. Traynham, in *Molecular Rearrangements*, Vol. I, P. Mayo, Ed., Wiley, New York, 1963, p. 593; (b) A. C. Cope, M. M. Martin, and M. A. McKervey, *Quart. Rev.* (London), **20**, 119 (1966).

SCHEME 7

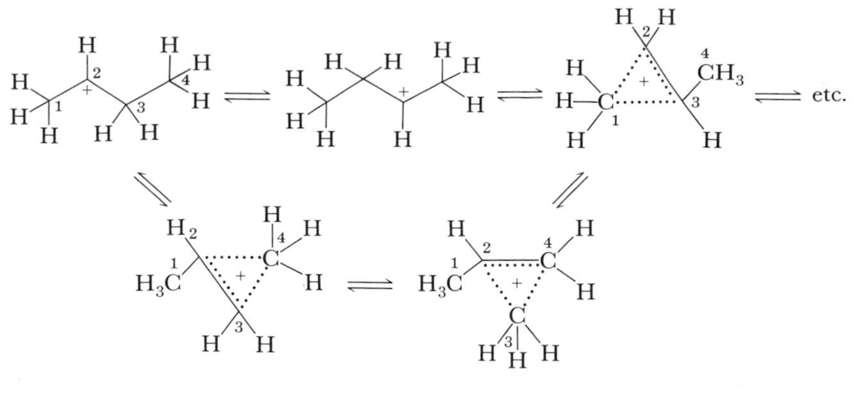

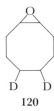

noted when unexpected products were found[142,143] in the peroxyformic acid oxidation of medium-ring alkenes. For example, cyclooctene gave, in addition to minor amounts of the expected *trans*-1,2-diol, cyclooctane-*cis*-1,4-diol and 3- and 4-cyclooctene-ols. Either a 1,3- or 1,5-hydride shift could bring about formation of the 1,4-diol and of the unsaturated alcohols (see Scheme 8). That both orders of hydride shift take place in this reaction was shown by Cope and co-workers, who treated 5,6-d$_2$-cyclooctene oxide (**120**) with 90 percent formic acid and, by deter-

120

mining the position of deuterium in the products, ascertained that 1,5-migration accounted for 94 percent of the formation of 3-cycloocten-1-ol and 61 percent of the 1,4-diol.[144]

The fact that only *trans*-1,2- and *cis*-1,4-diols are obtained implies that they cannot actually be formed by the simplified mechanism in Scheme 8. The carbenium ions **117–119** should give a mixture of cis and trans glycols. However, the reaction can be neither entirely concerted, as shown for a 1,5-hydride shift in Equation 5.48, nor involve initial formation of a carbonium ion, as shown in Equation 5.49. The k_H/k_D isotope effects are too small for C—H bond breaking to be involved in the rate-determining step.[145] The mechanism is probably similar to

[142](a) A. C. Cope, S. W. Fenton, and C. F. Spencer, *J. Am. Chem. Soc.*, **74**, 5884 (1952); (b) A. C. Cope, A. H. Keough, P. E. Peterson, H. E. Simmons, Jr., and G. W. Wood, *J. Am. Chem. Soc.*, **79**, 3900 (1957).
[143]V. Prelog and K. Schenker, *Helv. Chim. Acta*, **35**, 2044 (1952).
[144]A. C. Cope, G. A. Berchtold, P. E. Peterson, and S. H. Sharman, *J. Am. Chem. Soc.*, **82**, 6366 (1960).
[145]A. A. Roberts and C. B. Anderson, *Tetrahedron Lett.*, 3883 (1969).

SCHEME 8

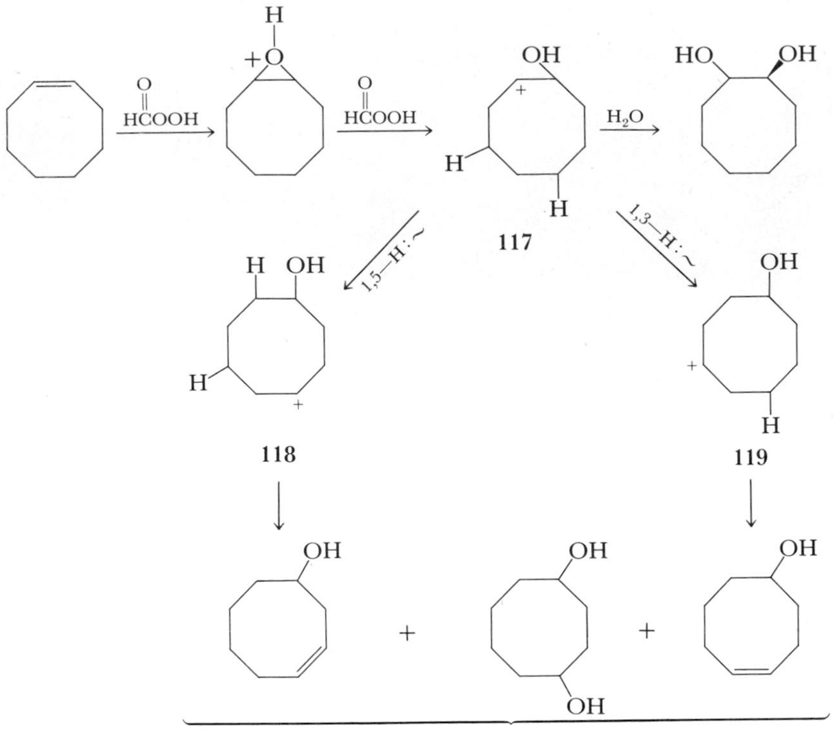

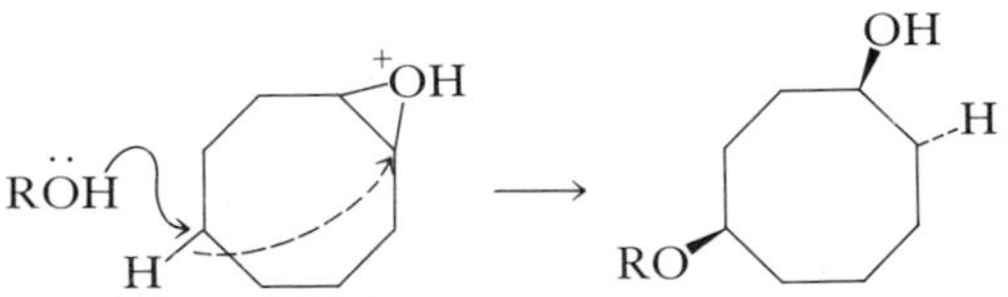

that shown in Equation 5.50, in which the slow step is breaking of the C—O bond (although some stereochemical-preserving attraction remains). Then, in a subsequent fast step, a carbonium ion is formed that is attacked from the back side by solvent.

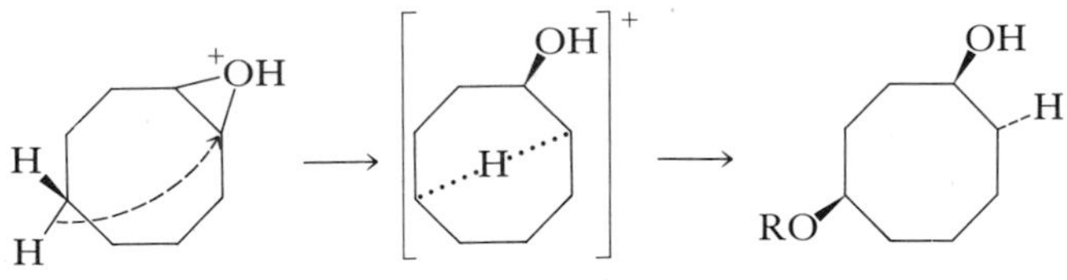

In some transannular hydride shifts, hydride participation in the rate-determining step does, however, seem to occur.[146]

---

[146](a) N. L. Allinger and W. Szkrybalo, *Tetrahedron*, **24,** 4699 (1968); (b) N. L. Allinger and S. Greenberg, *J. Am. Chem. Soc.*, **84,** 2394 (1962).

SCHEME 9

1. Homoallylic

cholesteryl

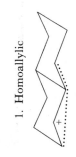

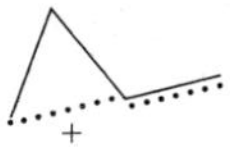

bicyclobutonium

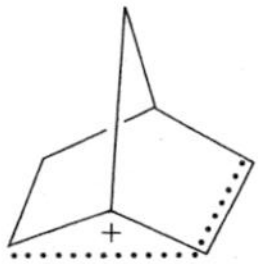

2-norbornenyl

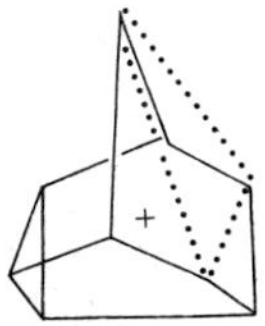

Coates cation

2. Corner-protonated cyclopropane

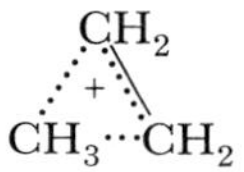

$$CH_2$$
$$CH_3 \cdots CH_2$$
$$+$$

corner-protonated cyclopropane

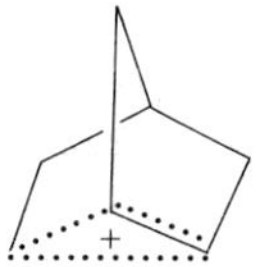

2-norbornyl

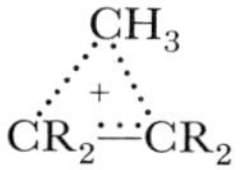

$$CH_3$$
$$CR_2 \cdots CR_2$$
$$+$$

transition state for neighboring-group participation by methyl

3. Edge-protonated cyclopropane

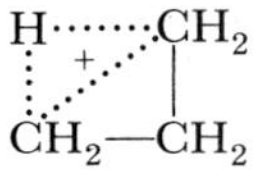

$$H \cdots CH_2$$
$$CH_2 — CH_2$$
$$+$$

edge-protonated cyclopropane

$$H$$
$$CR_2 \quad + \quad CR_2$$
$$(CR_2)_n$$

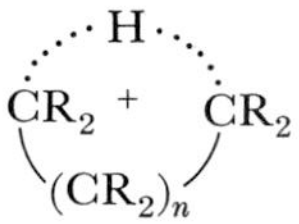

hydrogen bridging

4. Bisected cyclopropylcarbinyl cation

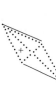

bisected cyclopropylcarbinylcation

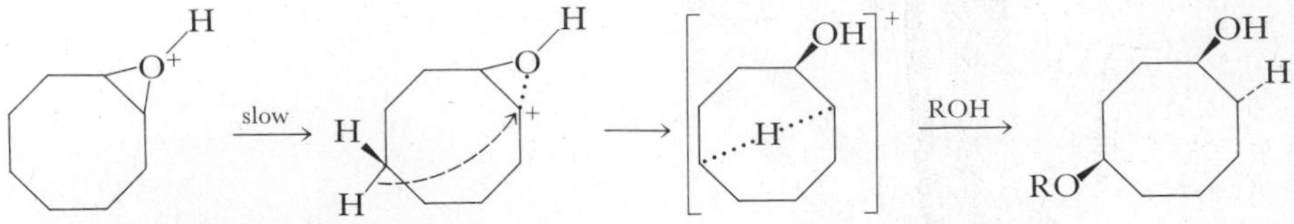

(5.50)

Higher-order shifts are facile in medium-sized rings. The geometry of the ring forces some of the transannular hydrogens to be within it, close to the lobe of the empty p orbital. In fact, in superacid solution hydride bridges have been observed in C_8–C_{11} rings.[147] For example, the cyclodecyl cation in superacid shows in its proton nmr spectrum one very shielded proton (-6.85δ). When the ^{13}C nmr spectrum is integrated, the lowest field peak has a relative area of two. The two carbons responsible for this peak are each coupled to two different protons. On the basis of these nmr spectra, the structure assigned to this stable cyclodecyl cation is **121**.

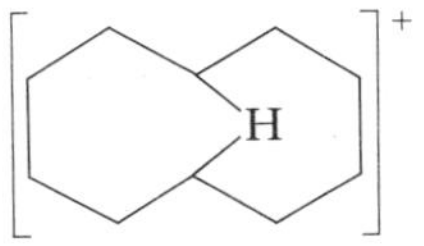

121

Before we leave the subject of carbonium ions, it might be useful to look at all the carbonium ion structures we have been considering in one place and to see how much some of them, not at first glance closely related, resemble each other. Thus in Scheme 9 the carbonium ions we have considered are grouped into four geometrical categories.

5.3 MIGRATIONS TO CARBONYL CARBON[148]

In the previous sections of this chapter we discussed migrations to electron-deficient carbon in which the electron deficiency was a result of departure of a leaving group with its pair of electrons. Although a carbonyl carbon is electron-deficient because of the electron-withdrawing ability of the oxygen, migrations to it in uncharged ground-state compounds do not occur. However, if (1) the carbonyl compound is converted to its conjugate acid (Equation 5.51) so that a full positive charge resides

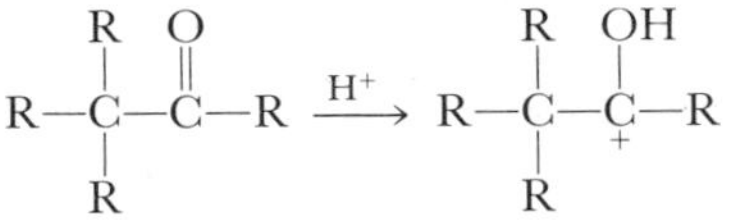

(5.51)

[147](a) R. P. Kirchen and T. S. Sorensen, *J. Am. Chem. Soc.,* **101**, 3240 (1979), and references therein; (b) R. P. Kirchen, T. S. Sorensen, and K. Wagstaff, *J. Am. Chem. Soc.,* **100**, 6761 (1978).
[148] For a general review, see C. J. Collins and J. F. Eastham, in *The Chemistry of the Carbonyl Group,* Vol. I, S. Patai, Ed., Wiley, New York, 1966, p. 761.

on it or (2) the migration origin is made especially electron-rich, increasing the tendency of a group to migrate with its pair of electrons, rearrangements do occur. The aldehyde–ketone rearrangement [Problem 2(f)] is an example of the first type, and the benzilic acid rearrangement is an example of the second type.

Benzilic Acid Rearrangement[149]

Liebig observed the first intramolecular rearrangement in 1838 when he found that benzil in basic solution forms a new compound.[150] In 1870 Jena correctly established the product of the reaction as benzilic acid, but proposed an incorrect structure for the starting material to avoid postulating a skeletal rearrangement.[151] In 1928 Ingold proposed the mechanism shown in Equation 5.52, which today is solidly supported by experimental evidence.[152]

$$\phi-\overset{\overset{O}{\|}}{C}-\overset{\overset{O}{\|}}{C}-\phi + {}^-OH \underset{\text{step 1}}{\rightleftharpoons} \phi-\overset{\overset{O}{\|}}{C}-\overset{\overset{O^-}{|}}{\underset{\underset{\phi}{|}}{C}}-OH \xrightarrow{\text{step 2}}$$

$$\textbf{122}$$

$$\phi-\overset{\overset{O^-}{|}}{\underset{\underset{\phi}{|}}{C}}-\overset{\overset{O}{\|}}{C}-OH \underset{\text{step 3}}{\overset{\text{step 3}}{\rightleftharpoons}} \phi-\overset{\overset{HO}{|}}{\underset{\underset{\phi}{|}}{C}}-\overset{\overset{O}{\|}}{C}-O^- \qquad (5.52)$$

The reaction is second-order overall, first-order each in benzil and in base.[153] This is consistent with any of the three steps being rate determining, since each depends on the concentrations of benzil and either of free base or of base that has already added to the benzil. Roberts and Urey carried out the rearrangement with ^{18}O-labeled base and found that the label was incorporated into unreacted benzil at a rate faster than that of the rearrangement.[154] Thus the first step must be rapid and reversible (although the first intermediate must exist long enough for the facile proton exchange,

$$\phi-\overset{\overset{O}{\|}}{C}-\overset{\overset{O^-}{|}}{\underset{\underset{OH}{|}}{C}}-\phi \rightleftharpoons \phi-\overset{\overset{O}{\|}}{C}-\overset{\overset{OH}{|}}{\underset{\underset{O^-}{|}}{C}}-\phi$$

to take place). That step 3 is not rate determining was shown by Hine, who used ${}^-OD$ as base and found no deuterium isotope effect.[155] By elimination, that leaves the migration, step 2, as the rate-determining process.

[149] For a review, see S. Selman and J. F. Eastham, *Quart. Rev.* (London), **14**, 221 (1960).
[150] J. Liebig, *Justug Liebigs Ann. Chem.,* **25**, 1 (1838).
[151] A. Jena, *Justug Liebigs Ann. Chem.,* **155**, 77 (1870).
[152] C. K. Ingold, *Ann. Rept. Chem. Soc.,* **25**, 124 (1928).
[153] F. H. Westheimer, *J. Am. Chem. Soc.,* **58**, 2209 (1936).
[154] I. Roberts and H. C. Urey, *J. Am. Chem. Soc.,* **60**, 880 (1938).
[155] J. Hine and H. W. Haworth, *J. Am. Chem. Soc.,* **80**, 2274 (1958).

An interesting aspect of this rearrangement is that the phenyl group with the lower-electron-donating ability usually migrates. For example, in **123** the sub-

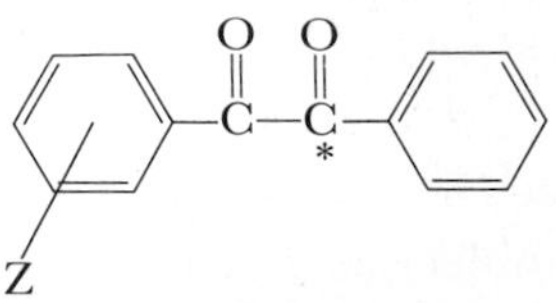

123

stituted phenyl migrates 81 percent of the time if Z is *m*-chloro, but only 31 percent of the time if Z is *p*-methoxy.[156] (Note that which group migrates can be determined only if one of the carbonyl carbons is labeled with ^{14}C.) Consideration of the mechanism in Equation 5.52 explains the anomaly. If the second step is rate determining, then the observed rate is given in Equation 5.53,

$$k_{obs} = k_2[\mathbf{122}] \tag{5.53}$$

where k_2 is the rate constant for step 2. The concentration of **122** is given by Equation 5.54:

$$[\mathbf{122}] = K_1[\text{benzil}]\,[\text{OH}^-] \tag{5.54}$$

in which K_1 is the equilibrium constant for step 1. Substituting Equation 5.54 into Equation 5.53, we obtain Equation 5.55. The observed rate is dependent on the equilibrium constant for the formation of **122** as well as on the rate of migration of the aryl group.

$$k_{obs} = k_2 K_1[\text{benzil}]\,[\text{OH}^-] \tag{5.55}$$

If the substituted phenyl is to migrate, then the intermediate (**122a**) must be formed; migration of the phenyl requires **122b**. Electron-withdrawing substituents will increase K_1 for the formation of **122a**; if K_1 is increased more than k_2 is decreased, more substituted phenyl will migrate than unsubstituted phenyl.

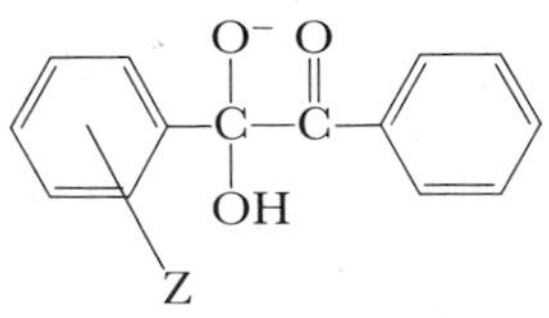

122a 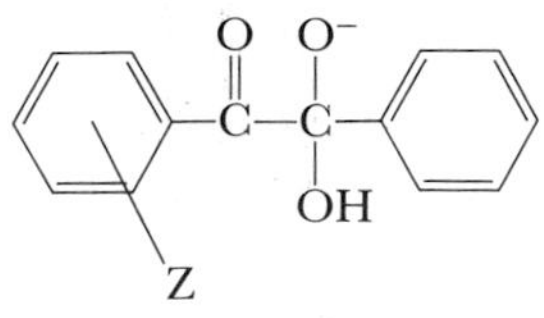   **122b**

[156] M. T. Clark, E. C. Hendley, and O. K. Neville, *J. Am. Chem. Soc.*, **77**, 3280 (1955).

5.4 REARRANGEMENTS TO ELECTRON-DEFICIENT NITROGEN AND OXYGEN[157]

Our consideration of rearrangements to electron-deficient hetero atoms must be brief. In discussing migrations to electron-deficient nitrogen, we first discuss three rearrangements that occur in carbonyl derivatives, the Beckmann, Hofmann, and Schmidt rearrangements, and then consider rearrangements of nitrenium ions.

Because of the high electronegativity of oxygen, an O—X bond will cleave heterolytically, producing a positive oxygen, only if X is an excellent leaving group. As a result, electron-deficient oxygen is formed most frequently in reactions of peresters and aromatic peroxides, R—O—O—R' (R' = aryl or acyl). In these compounds when $^-$OR' departs, the negative charge on the leaving group is stabilized by resonance. Even here heterolytic cleavages are not universal: The energy required for a heterolytic cleavage in the absence of anchimeric assistance is about 50 kcal mole^{-1},[158] whereas the energy for a homolytic cleavage to two alkoxy radicals is only about 30–40 kcal mole^{-1}.[159] Thus heterolytic cleavage usually takes place only with anchimeric assistance.

The Beckmann Rearrangement[160]

The acid-catalyzed conversion of ketoximes and aldoximes to amides or amines (the amide is often hydrolyzed to the corresponding amine under the reaction conditions) is known as the Beckmann rearrangement after its discoverer.[161] The reaction and its widely accepted mechanism are shown in Equation 5.56.

$$R-\underset{\underset{N}{\underset{|}{\parallel}}}{C}-R' \xrightarrow{HA} R-\underset{+}{N}\equiv C-R' \xrightarrow{H_2O} R-N=\underset{\underset{OH}{|}}{C}-R' \rightleftharpoons RNH-\underset{\underset{O}{\parallel}}{C}-R' \qquad (5.56)$$

124

The observation that picryl ethers of oximes (**125**) rearrange without a catalyst established that the role of the catalyst was to convert the hydroxyl into a better leaving group.[162] Some acids catalyze by simply protonating the oxime as in **126**. Other acids may esterify the oximes. For example, Schofield has suggested[163]

[157] For a general review, see (a) P. A. S. Smith, in *Molecular Rearrangements,* Vol. I, P. Mayo, Ed., Wiley, New York, 1963, p. 457. For reviews of rearrangements to electron-deficient nitrogen, see (b) D. V. Banthorpe, in *The Chemistry of the Amino Group,* S. Patai, Ed., Wiley, New York, 1968, p. 623. For a review of rearrangements to electron-deficient oxygen, see (c) J. B. Lee and B. C. Uff, *Quart. Rev.* (London), **21,** 429 (1967); (d) R. Curci and J. O. Edwards, in *Organic Peroxides,* Vol. I, D. Swern, Ed., Wiley, New York, 1970, p. 199.

[158] E. Hedaya and S. Winstein, *J. Am. Chem. Soc.,* **89,** 1661, 5314 (1967).

[159] S. W. Benson and R. Shaw, in *Organic Peroxides,* Vol. 1, D. Swern, Ed., Wiley, New York, 1970, p. 147.

[160] L. G. Donaruma and W. Z. Heldt, *Org. Reactions,* **11,** 1 (1960).

[161] E. Beckmann, *Ber. Deut. Chem. Ges.,* **20,** 1507 (1887).

[162] A. W. Chapman and F. A. Fidler, *J. Chem. Soc.,* 448 (1936).

[163] B. J. Gregory, R. B. Moodie, and K. Schofield, *J. Chem. Soc. B,* 338 (1970).

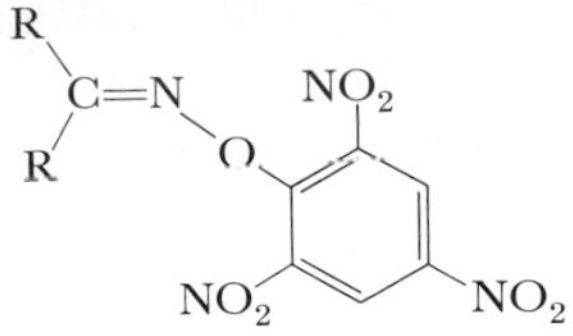

125

that catalysis of rearrangement of **127** by sulfuric acid is due to the preliminary conversion of **127** to **128**.[164]

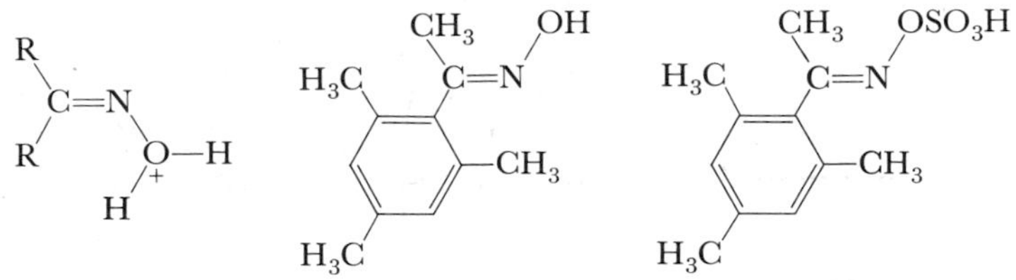

126   **127**   **128**

The oxime of an aldehyde or ketone can often be separated into two geometrical isomers, the syn and anti forms. When the Beckmann rearrangement is carried out under nonisomerizing conditions, it is always the groups anti to the —OH that migrate.[165] For example, Curtin and co-workers carried out Beckmann rearrangements on **129** and **130** in the solid phase by gently heating crystals of the

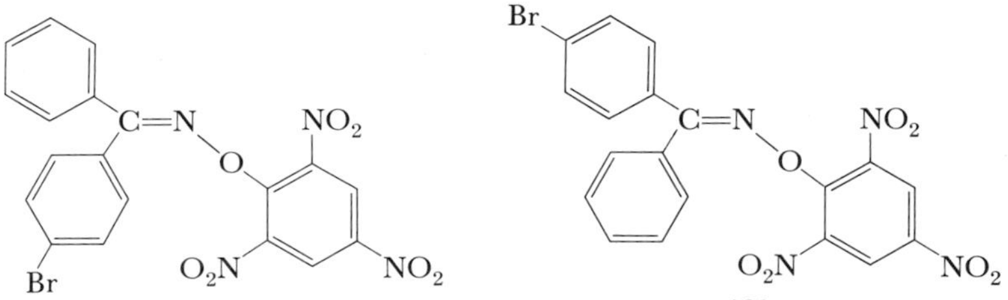

129   **130**

compounds. The conditions do not allow interconversion of **129** and **130**; in **129** only the phenyl group migrates, whereas in **130** it is the *p*-bromophenyl group that shifts.[166] When the catalyst is a Brønsted acid, migration is not stereospecific. Under these conditions, syn and anti forms are readily interconverted, presumably via the pathway shown in Equation 5.57.

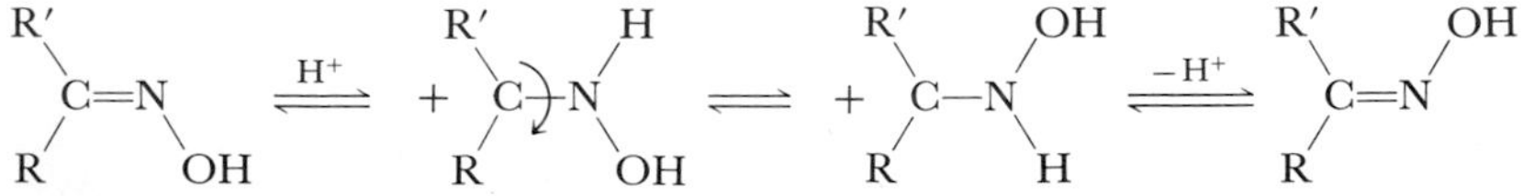

$$(5.57)$$

[164] This intermediate has been observed in the nmr. See Y. Yukawa and T. Ando, *J. Chem. Soc. D*, 1601 (1971).
[165] See note 157(a), (b) and note 160.
[166] J. D. McCullough, Jr., D. Y. Curtin, and I. C. Paul, *J. Am. Chem. Soc.*, **94**, 874 (1972).

The stereochemistry of the reaction indicates that rearrangement is concerted with departure of the leaving group, as is implied by step 1 of Equation 5.56. The question then remains whether this step or another is rate determining. An answer can be found in the effect of the nature of the migrating group on the rate of reaction. If the migration step is rapid, it should not matter to the overall rate whether an electron-rich or an electron-poor group is migrating. On the other hand, if migration is the slow step, electron-donating substituents in the migrating group should increase the rate, and electron-withdrawing substituents should decrease it. Kinetic studies of the rearrangements of meta- and para-substituted acetophenone oximes (**131**) in concentrated H_2SO_4 show that the rates do indeed vary with the electron-donating ability of the substituents and that a fairly good correlation exists between the rates of rearrangement and the Hammett σ^+ constants for the substituents as shown in Figure 5.27. This observation indicates that some participation by phenyl occurs in the rate-determining step and suggests **132**

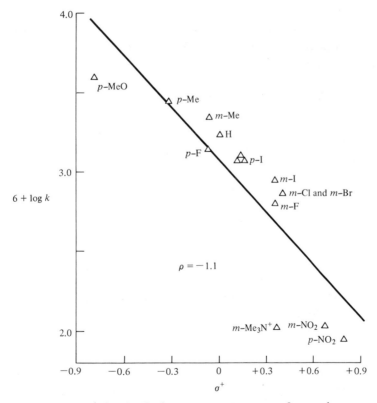

Figure 5.27 Plot of log k vs. σ^+ for the Beckmann rearrangement of acetophenone oximes in 98.2 percent sulfuric acid at 80°C. From B. J. Gregory, R. B. Moodie, and K. Schofield, *J. Chem. Soc.*, **13**, 338 (1970). Reprinted by permission of K. Schofield and The Chemical Society.

as the transition state.[163] Not all Beckmann rearrangements need necessarily, however, have the same rate-determining step.

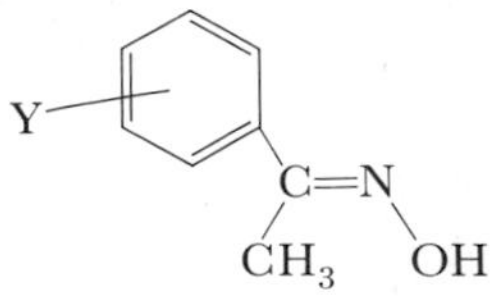

Structures **131** and **132**

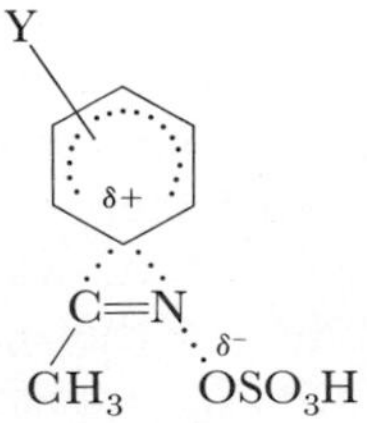

That intermediate **124** in Equation 5.56 is formed in the Beckmann rearrangement has been amply demonstrated both by diverting it to other products[167] and by direct observation by nmr.[168]

If the mechanism in Equation 5.56 is correct, then oxygen transfer should be an intermolecular process. And, in fact, when unlabeled acetophenone oximes (**131**) are allowed to rearrange in ^{18}O-enriched solvent, the product amide contains the same percentage of ^{18}O as the solvent.[163] Alkyl or aryl migration, on the other hand, must be intramolecular, since a chiral migrating group retains its chirality during the migration.[169]

The Hofmann Rearrangement[170]

In 1882 Hofmann discovered that when amides are treated with bromine in basic solution, they are converted to amines with one carbon less than the starting amide.[171] He also isolated the N-bromo amine (**133**) and the isocyanate (**134**) as intermediates on the reaction path. The mechanism in Equation 5.58 accounts for the products and the intermediates. This reaction (or the analogous rearrangement of the N-chloro amine) is now known as the Hofmann rearrangement or, because of its synthetic usefulness in eliminating a carbon atom, the Hofmann degradation.

$$\text{Br}^- + \text{O=C=N-R} \xrightarrow{\text{H}_2\text{O}} \text{HO-}\overset{\text{O}}{\underset{||}{\text{C}}}\text{-NHR} \longrightarrow \text{H}_2\text{NR} + \text{CO}_2 \qquad (5.58)$$
134

[167](a) A. Werner and A. Piguet, *Ber. Deut. Chem. Ges.*, **37**, 4295 (1904); (b) R. M. Palmere, R. T. Conley, and J. L. Rabinowitz, *J. Org. Chem.*, **37**, 4095 (1972).
[168]B. J. Gregory, R. B. Moodie, and K. Schofield, *J. Chem. Soc. D*, 645 (1969).
[169](a) A. Campbell and J. Kenyon, *J. Chem. Soc.*, 25 (1946); (b) J. Kenyon and D. P. Young, *J. Chem. Soc.*, 263 (1941).
[170]E. S. Wallis and J. F. Lane, *Org. Reactions*, **3**, 267 (1946).
[171]A. W. Hofmann, *Ber. Deut. Chem. Ges.*, **15**, 762 (1882).

In a most convincing demonstration of the intramolecularity of the migration step, Wallis and Moyer carried out the Hofmann degradation on chiral **135**. This compound can be prepared in optically active form because the groups in the

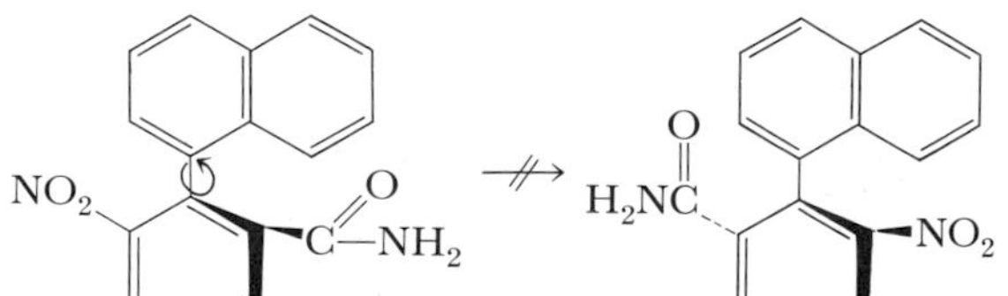

135 **136**

ortho position of the phenyl ring hinder the rotation that would convert **135** to its mirror image **136**. During rearrangement **135** would lose chirality if the migrating bond simply stretched enough to allow rotation about itself. However, loss of chirality is not observed: **135** rearranges with retention of configuration.[172]

There is a question whether Equation 5.58 shows all the intermediates on the reaction path. If instead of rearrangement being concerted with loss of halide ion as shown in Equation 5.58, the halide ion departed first, then a nitrene[173] would be formed as shown in Equation 5.59. To date no nitrene intermediate in the Hofmann reaction has been proved, but the possibility remains that it is at least sometimes formed.

$$\mathrm{R{-}\overset{\overset{\displaystyle O}{\|}}{C}{-}\underset{-}{\ddot{N}}{-}Br \longrightarrow R{-}\overset{\overset{\displaystyle O}{\|}}{C}{-}\ddot{N}\!: + Br^-} \qquad (5.59)$$

The Schmidt Rearrangements[174]

The group of rearrangements brought about by treatment of aldehydes, ketones, or carboxylic acids with hydrogen azide are known as Schmidt rearrangements. All are acid-catalyzed and all involve addition of HN_3 to the carbonyl group followed by dehydration. They are shown in Equations 5.60–5.62.

$$\mathrm{R{-}\overset{\overset{\displaystyle O}{\|}}{C}{-}H \xrightarrow{H^+} R{-}\overset{\overset{\displaystyle {}^+OH}{\|}}{C}{-}H \xrightarrow{HN_3} R{-}\overset{\overset{\displaystyle OH}{|}}{\underset{\underset{\displaystyle H}{|}}{C}}{-}\underset{\underset{\displaystyle H}{|}}{N}{-}\underset{+}{N}{\equiv}N \xrightarrow{-H_2O} R{-}\overset{\overset{\displaystyle H}{|}}{C}{=}N{-}\underset{+}{N}{\equiv}N \longrightarrow}$$

$$\mathrm{\underset{\mathbf{137a}}{H\underset{+}{C}{=}N{-}R} + N_2 \xrightarrow[-H^+]{H_2O} H{-}\overset{\overset{\displaystyle OH}{|}}{C}{=}N{-}R \longrightarrow H{-}\overset{\overset{\displaystyle O}{\|}}{C}{-}NHR} \qquad (5.60)$$

[172] E. S. Wallis and W. W. Moyer, *J. Am. Chem. Soc.,* **55**, 2598 (1933).
[173] A *nitrene* is a nitrogen-containing compound in which the nitrogen has only a sextet of electrons. Such a species is neutral but electron-deficient; cf. carbenes (Section 6.3).
[174] H. Wolff, *Org. Reactions,* **3**, 307 (1946).

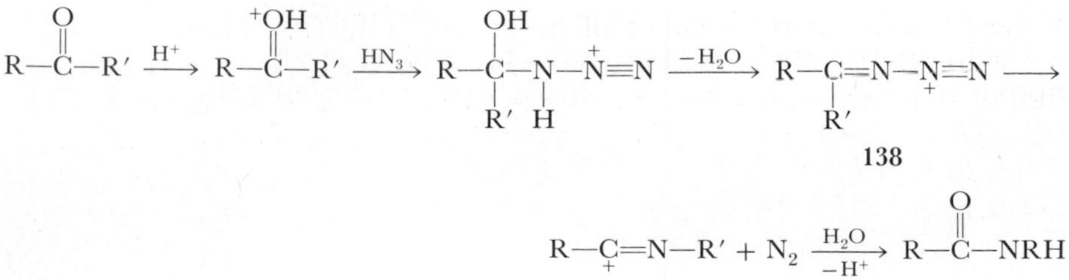

$$R-\overset{O}{\overset{\|}{C}}-R' \xrightarrow{H^+} R-\overset{+OH}{\overset{\|}{C}}-R' \xrightarrow{HN_3} R-\overset{OH}{\underset{\overset{|}{R'}\ \overset{|}{H}}{\overset{|}{C}}}-\overset{+}{N}-N\equiv N \xrightarrow{-H_2O} R-\overset{OH}{\underset{R'}{\overset{|}{C}}}=N-\overset{+}{N}=N \longrightarrow$$

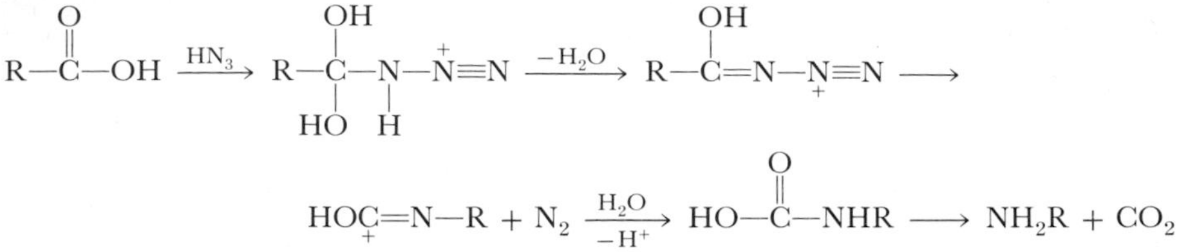

$$\text{\textbf{138}}$$

$$R-\overset{+}{C}=N-R' + N_2 \xrightarrow[-H^+]{H_2O} R-\overset{O}{\overset{\|}{C}}-NRH \qquad (5.61)$$

$$\text{\textbf{137b}}$$

$$R-\overset{O}{\overset{\|}{C}}-OH \xrightarrow{HN_3} R-\overset{OH}{\underset{\overset{|}{HO}\ \overset{|}{H}}{\overset{|}{C}}}-\overset{+}{N}-N\equiv N \xrightarrow{-H_2O} R-\overset{OH}{\overset{|}{C}}=N-\overset{+}{N}=N \longrightarrow$$

$$HO\overset{+}{C}=N-R + N_2 \xrightarrow[-H^+]{H_2O} HO-\overset{O}{\overset{\|}{C}}-NHR \longrightarrow NH_2R + CO_2 \qquad (5.62)$$

$$\text{\textbf{137c}}$$

Although mechanisms can be formulated that do not involve dehydration and subsequent formation of the intermediates **137a–137c**, there is strong evidence that these steps take place. For example, tetrazoles, which are formed from **137** as shown in Equation 5.63, have been isolated as side products. Further evidence for

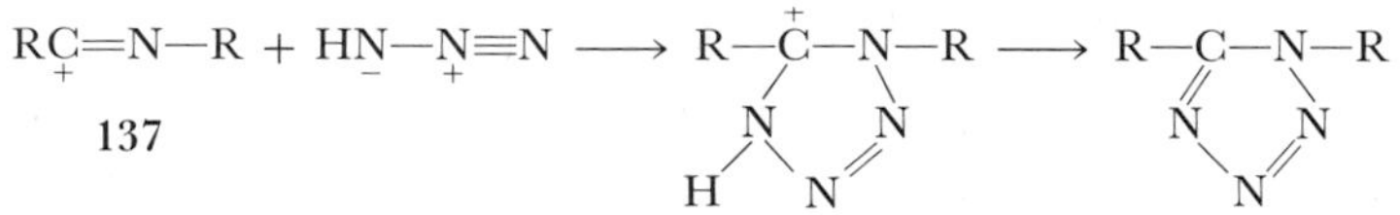

$$R\overset{+}{\underset{-}{C}}=N-R + H\overset{-}{N}-\overset{+}{N}\equiv N \longrightarrow R-\overset{+}{C}-N-R \longrightarrow R-C-N-R \qquad (5.63)$$

$$\text{\textbf{137}}$$

the dehydration step was obtained by Hassner and co-workers, who showed that acid-catalyzed rearrangement of vinyl azides gives the same products in the same ratio as the Schmidt rearrangement of the corresponding ketone under the same conditions. He postulated that the reaction paths of the two rearrangements converge at the common intermediate **138a** as shown in Scheme 10.[175]

SCHEME 10

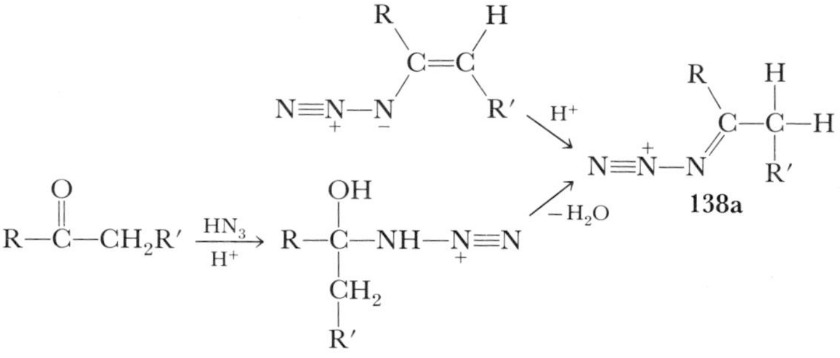

[175] A. Hassner, E. S. Ferdinandi, and R. J. Isbister, *J. Am. Chem. Soc.,* **92,** 1672 (1970).

In the Schmidt rearrangement of carboxylic acids the formation of the adduct is apparently usually not rate determining. The evidence for this comes from studies of the comparative rate of nitrogen evolution from HN_3 in the presence and in the absence of carboxylic acids: When *m*- or *p*-nitrobenzoic acid is added to HN_3 in H_2SO_4, the rate of nitrogen evolution decreases. Thus HN_3 must be rapidly converted to an adduct from which loss of nitrogen is slower than from HN_3 itself. Moreover, the adduct, to be formed at all, must be formed more rapidly than N_2 is lost from HN_3.[176]

The intramolecularity of the migration step in the Schmidt rearrangements has been convincingly demonstrated by showing the retention of chirality of the migrating group.[177]

In the Schmidt rearrangement of ketones the larger group, irrespective of its nature, tends to migrate. Apparently the intermediate **138** is formed so that the bulkier aryl or alkyl group is trans to the N_2 group. Then, as in the Beckmann rearrangement, the group trans to the leaving group prefers to migrate. The barriers to interconversion of the cis and trans forms are, however, lower in the Schmidt than in the Beckmann rearrangement.[175]

Nitrenium Ions[178]

The nitrenium ion (**139**) is isoelectronic with the carbene (Section 6.3). Until the middle 1960s it was unknown, but at that time Gassman began an intensive investi-

$$R-\overset{..}{\underset{+}{N}}-R$$
$$139$$

gation to determine whether or not it exists. Since nitrogen is more electronegative than carbon, it was to be expected that the nitrenium ion would be less stable than its carbon analog.

Gassman and Fox first synthesized and then solvolyzed N-chloroisoquinuclidine (**140**). In refluxing methanolic silver nitrate, **140** is converted in 60 percent yield to the rearranged product **141** as shown in Equation 5.64.[179] Since alkyl groups do not migrate to radical centers, this rearrangement clearly indicates that

$$140 \qquad\qquad 141$$

[176]L. H. Briggs and J. W. Lyttleton, *J. Chem. Soc.*, 421 (1943). But see also V. A. Ostrovskii, A. S. Enin, and G. I. Koldobski, *J. Org. Chem., U.S.S.R.*, **9**, 827 (1973).

[177]See note 157(a).

[178]P. G. Gassman, *Acc. Chem. Res.*, **3**, 26 (1970).

[179]P. G. Gassman and B. L. Fox, *J. Am. Chem. Soc.*, **89**, 338 (1967); P. G. Gassman, K. Uneyama, and J. L. Hahnfield, *J. Am. Chem. Soc.*, **99**, 647 (1977).

an electron-deficient nitrogen must have been formed. What it does not indicate is whether the reaction occurred via the nitrenium ion (**142**), as a discrete intermediate, or whether rearrangement is concerted with departure of the leaving group and **143** is the first-formed ion.

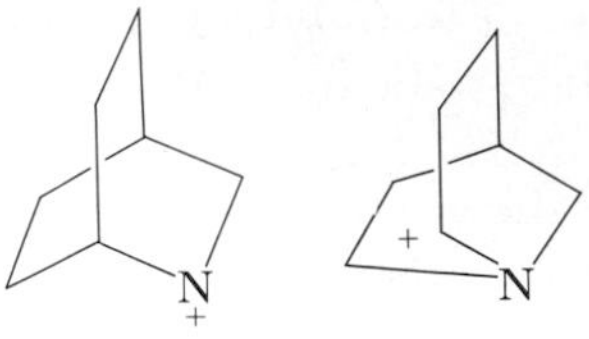

142   143

Nitrenium ions can also be generated by solvolyzing esters of N,N-dialkylhydroxylamines. 3,5-Dinitrobenzoates (**144**) were found to be the most useful hydroxylamine derivatives.[180]

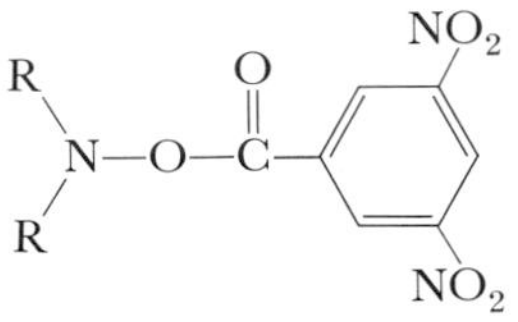

144

Heterolytic Peroxyester Decomposition: The Criegee Rearrangement

In 1944 Criegee observed that *trans*-9-decalyl peroxyesters rearrange on standing to 1,6-epoxycyclodecyl esters.[181] Further study of the reaction showed that it has the characteristics of an ionic rather than a radical pathway: The rate is proportional to the anionic stability of $^-O-\overset{\text{O}}{\overset{\|}{C}}-R$ and increases with the polarity of the solvent.[182] Because no products were obtained that were the result of solvent intervention or of exchange with added salts, the reaction was postulated to have only intimate ion-pair intermediates as shown in Equation 5.65.[183] Denney showed that

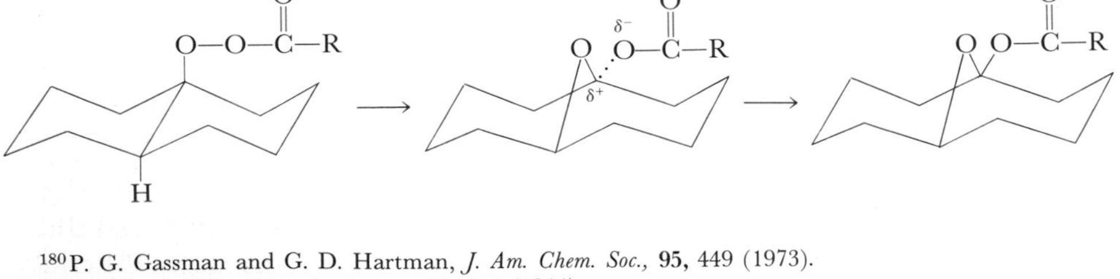

(5.65)

[180] P. G. Gassman and G. D. Hartman, *J. Am. Chem. Soc.*, **95**, 449 (1973).

[181] R. Criegee, *Ber. Deut. Chem. Ges.*, **77**, 722 (1944).

[182] R. Criegee and R. Kaspar, *Justus Liebigs Ann. Chem.*, **560**, 127 (1948).

[183] P. D. Bartlett and J. L. Kice, *J. Am. Chem. Soc.*, **75**, 5591 (1953); (b) H. L. Goering and A. C. Olson, *J. Am. Chem. Soc.*, **75**, 5853 (1953).

if the carbonyl oxygen in *trans*-9-decalyl peroxybenzoate is labeled with ^{18}O, almost all the label is found in the carbonyl oxygen of the product.[184] This experiment dramatically demonstrates the closeness with which the carboxylate anion must be connected to the decalyl cation during the rearrangement.

Winstein investigated the peroxybenzoate rearrangements further to see if there was neighboring-group participation in departure of the leaving group. He studied the products from rearrangement of 2-R-2-propyl-*p*-nitroperoxybenzoates and found that the more electron donating R is, the greater is its migratory aptitude. Equation 5.66 shows the principal products formed if R is more electron

$$CH_3\text{--}\underset{\underset{CH_3}{|}}{\overset{\overset{R}{|}}{C}}\text{--}O\text{--}O\text{--}\overset{\overset{O}{||}}{C}\text{--}\langle\ \rangle\text{--}NO_2 \xrightarrow{CH_3OH} CH_3\text{--}\underset{\underset{CH_3}{|}}{\overset{+}{C}}\text{--}OR\ +\ {}^-O\text{--}\overset{\overset{O}{||}}{C}\text{--}\langle\ \rangle\text{--}NO_2$$

$$\downarrow$$

$$CH_3\text{--}\underset{\underset{CH_3}{|}}{C}\text{=}O\ +\ ROH$$

$$+\ ROCH_3$$

$$+\ \text{olefin from } R^+ \qquad (5.66)$$

donating than methyl. Winstein also found that the rate is highly dependent on the electron-donating ability of R—much more so than the rates of other intramolecular rearrangements (see Table 5.3). He therefore postulated that the migrating alkyl group provides anchimeric assistance to the departure of the nitrobenzoate anion and that the transition state for the rearrangement can be represented by a bridged structure such as **145**.

$$\overset{R}{\underset{\diagup}{\overset{\diagdown}{C}}\overset{\delta^+}{\cdots\cdots}O}$$

$$\delta^-OPNB$$

145

Winstein estimated the rate acceleration due to anchimeric assistance in ionic perester decompositions as follows. By using Equation 5.67, in which the first term on the right-hand side is the difference in homolytic dissociation energies

$$\Delta\Delta E_H = (\Delta E_{oo} - \Delta E_{co}) + (I_o - I_c) \qquad (5.67)$$

between peroxide and carbon–oxygen bonds and the second term is the ionization potential difference between oxygen and carbon, he estimated that the differences in energies of heterolysis of peroxide and carbon–oxygen bonds ($\Delta\Delta E_H$) should be 22 kcal mole^{-1} (peroxide bond breaking being more costly). He then compared the observed enthalpies of activation ($\Delta H^{\ddagger}$) of acetolysis of neopentyl tosylate (**146**) and *t*-butyl pertosylate (**147**) and assumed that they are a measure of the energies of

[184]D. B. Denney and D. G. Denney, *J. Am. Chem. Soc.*, **79**, 4806 (1957).

Table 5.3 Relative Rates of Rearrangement Reactions with Different Migrating Groups[a]

R	Ionic Decomposition of 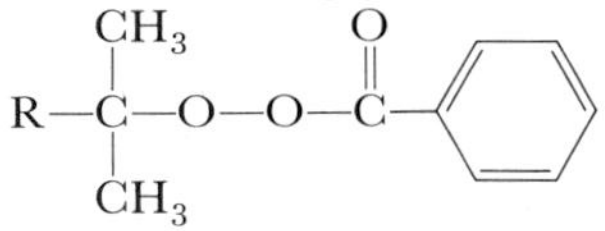 $R-C(CH_3)(CH_3)-O-O-C(=O)-\phi$	Pinacol Rearrangement of 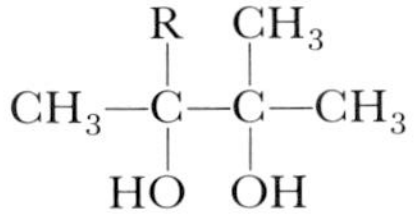 $CH_3-C(OH)(CH_3)-C(OH)(R)-CH_3$	Acetolysis of $(CH_3)_2CRCH_2OTs$	Lossen Rearrangement of 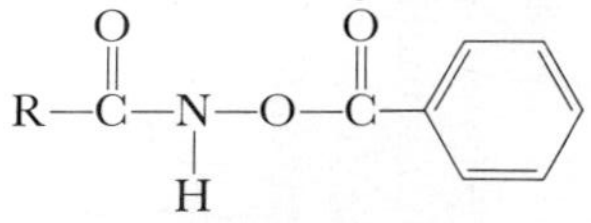 $R-C(=O)-N(H)-O-C(=O)-\phi$
Me	1	1	1	0.03
Et	45	17	—	1
i-Pr	2.9×10^3	—	5.3	14.9
t-Bu	2.3×10^5	4000	—	12.0
CH$_2$—ϕ	1.6×10^3	—	0.16	—
ϕ	1.1×10^5	—	335	12.2

[a]Reprinted with permission from A. E. Hedaya and S. Winstein, *J. Am. Chem. Soc.*, **89**, 1661 (1967). Copyright by the American Chemical Society.

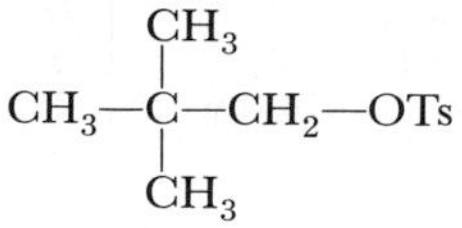

$$CH_3-\underset{\underset{CH_3}{|}}{\overset{\overset{CH_3}{|}}{C}}-CH_2-OTs \qquad CH_3-\underset{\underset{CH_3}{|}}{\overset{\overset{CH_3}{|}}{C}}-O-O-Ts$$

146 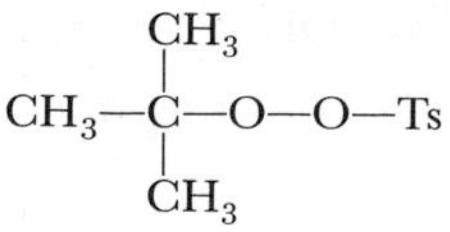   147

heterolysis. He found that $\Delta H^{\ddagger}$ for **146** was 10 kcal higher than for **147**. Thus methyl assistance is responsible for lowering the $\Delta H^{\ddagger}$ for heterolysis of **147** by about 32 kcal mole^{-1}, which corresponds to a rate acceleration of 10^{23}.[185]

Acid-catalyzed decompositions of hydroperoxides in which the leaving group is water also take place. An example is shown in Equation 5.68. That the

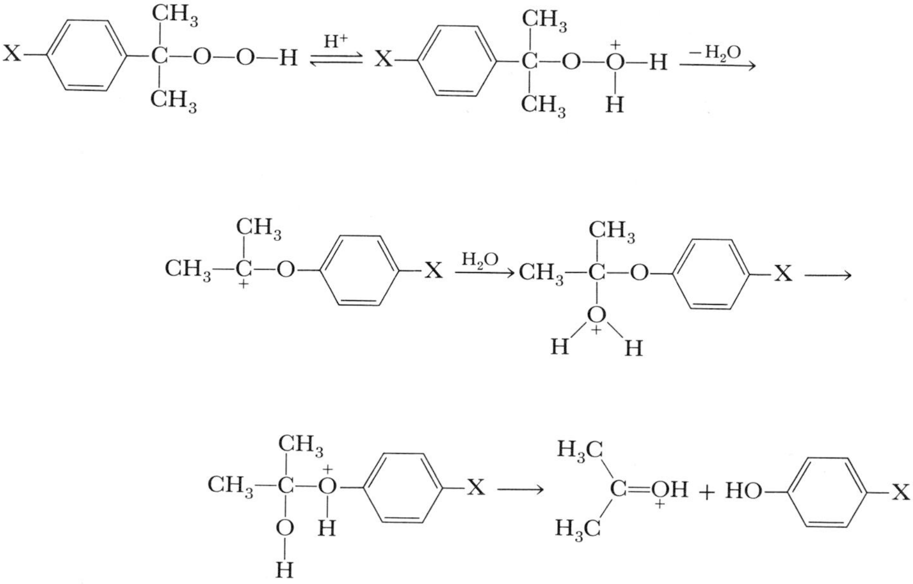

pathway is actually one of migration concerted with departure of water as shown, and does not include intervention of the high-energy species RO^{+}, has been demonstrated by the fact that the rates of rearrangement of Reaction 5.68 correlate with the σ^{+} constants of the X substituent.[186]

The Baeyer–Villiger Oxidation[187]

In 1899 Baeyer and Villiger observed that peroxy acids convert ketones to esters.[188] The reaction is first-order each in ketone and in peroxy acid, and it is general acid-catalyzed. Criegee first suggested the mechanism shown in Equation 5.69. The role of the acid catalyst is to protonate the leaving group, thereby facilitating its

[185] E. Hedaya and S. Winstein, *J. Am. Chem. Soc.,* **89,** 1661, 5314 (1967).
[186] A. W. de Ruyter van Steveninck and E. C. Kooyman, *Recl. Trav. Chim. Pays-Bas,* **79,** 413 (1960).
[187] C. H. Hassall, *Org. Reactions,* **9,** 73 (1957).
[188] A. Baeyer and V. Villiger, *Ber. Deut. Chem. Ges.,* **32,** 3625 (1899).

$$\begin{matrix} R \\ \diagdown \\ \quad C{=}O + HOOA \longrightarrow R{-}\underset{\underset{R}{|}}{\overset{\overset{OH}{|}}{C}}{-}O{-}OA \longrightarrow R{-}\overset{\overset{+OH}{\|}}{C}{-}OR + {}^-OA \longrightarrow \\ \diagup \\ R \end{matrix}$$

148

$$R{-}\overset{\overset{O}{\|}}{C}{-}OR + HOA \qquad (5.69)$$

departure.[189] The intermediate **148** has never been directly observed in a rearrangement reaction, but analogous structures are known for the addition of peroxyacetic acid to aldehydes.[190] Doering and Dorfman provided strong support for the mechanism of Equation 5.69 when they showed that oxidation of benzophenone labeled with ^{18}O gave phenyl benzoate in which all of the ^{18}O was retained in the carboxyl group (Equation 5.70).[191] This experiment ruled out symmetrical species such as **149** as intermediates on the reaction path.

$$\phi{-}\overset{\overset{18O}{\|}}{C}{-}\phi + HOOH \longrightarrow \phi{-}\overset{\overset{18O}{\|}}{C}{-}O\phi + HOA \qquad (5.70)$$

$$\begin{matrix} \phi \diagdown \quad \diagup O \\ \quad C \\ \phi \diagup \quad \diagdown O \end{matrix}$$

149

The question remains: Is the formation of **148** or its destruction rate determining? Experiment indicates that rearrangement is concerted and that in the oxidation of most ketones rearrangement is rate determining. For example, Palmer and Fry oxidized para-substituted acetophenones-1-^{14}C as shown in Equation 5.71

$$X{-}\langle\quad\rangle{-}\overset{*}{\underset{}{C}}\overset{\overset{O}{\|}}{}{-}CH_3 + \quad \text{(3-Cl-C}_6H_4){-}\overset{\overset{O}{\|}}{C}{-}O{-}OH \longrightarrow$$

$$X{-}\langle\quad\rangle{-}\overset{*}{}{-}O{-}\overset{\overset{O}{\|}}{C}{-}CH_3 + \quad \text{(3-Cl-C}_6H_4){-}\overset{\overset{O}{\|}}{C}{-}OH \qquad (5.71)$$

and compared these rates of oxidation with the rates of oxidation of the unlabeled ketones.[192] As shown in Table 5.4, for all substituents except p-OCH_3, there is a significant ^{14}C isotope effect. Thus for all the acetophenones other than the p-OCH_3-substituted one, the rate-determining step is rearrangement. Rate-deter-

[189] M. F. Hawthorne and W. D. Emmons, *J. Am. Chem. Soc.*, **80**, 6398 (1958).
[190] B. Phillips, F. C. Frostick, Jr., and P. S. Starcher, *J. Am. Chem. Soc.*, **79**, 5982 (1957).
[191] W. v. E. Doering and E. Dorfman, *J. Am. Chem. Soc.*, **75**, 5595 (1953).
[192] B. W. Palmer and A. Fry, *J. Am. Chem. Soc.*, **92**, 2580 (1970).

Table 5.4 Isotope Effects for the Oxidation of Para-Substituted Acetophenones-1-^{14}C with m-Chlorobenzoic Acid in Chloroform at 32°C (Equation 5.71)^a

X	k_{12}/k_{14}
CH$_3$O	0.998
CH$_3$	1.032
H	1.048
Cl	1.050
CN	1.084

^{a} Reprinted with permission from B. W. Palmer and A. Fry, *J. Am. Chem. Soc.*, **92**, 2580 (1970). Copyright by the American Chemical Society.

mining formation of **148** would not give an isotope effect, since this step does not involve significant bond alteration at the labeled position.

Further evidence for the mechanism of Equation 5.69 with the second step rate determining is provided by substituent effects on the rate. For example, Figure 5.28 shows a plot of the logarithms of the rates of oxidation of substituted acetophenones by trifluoroperoxyacetic acid vs. the σ values of the substituents. The ρ value is negative, indicating that electron-donating substituents in the migrating group increase the rate. Furthermore, the rate of oxidation of cyclohexanone with peroxyacetic acid is only 1/200th the rate with trifluoroperoxyacetic acid. The greater basicity of the unfluorinated acid should make it a better nucleophile toward the carbonyl group, and if formation of **148** were rate determining, it should be the better oxidizing agent. On the other hand, the electron-withdrawing ability of the CF$_3$ group should make trifluoroacetic acid the better leaving group, and thus if rearrangement concerted with O—O bond breaking is the rate-determining step, the trifluoroperoxyacetic acid should be the better oxidizing agent, as observed.[189]

Formation of the intermediate may become rate determining if the migrating group is especially reactive. For example when p-hydroxybenzaldehyde is oxidized by peroxybenzoic acid, the products are those shown in Equation 5.72. Over

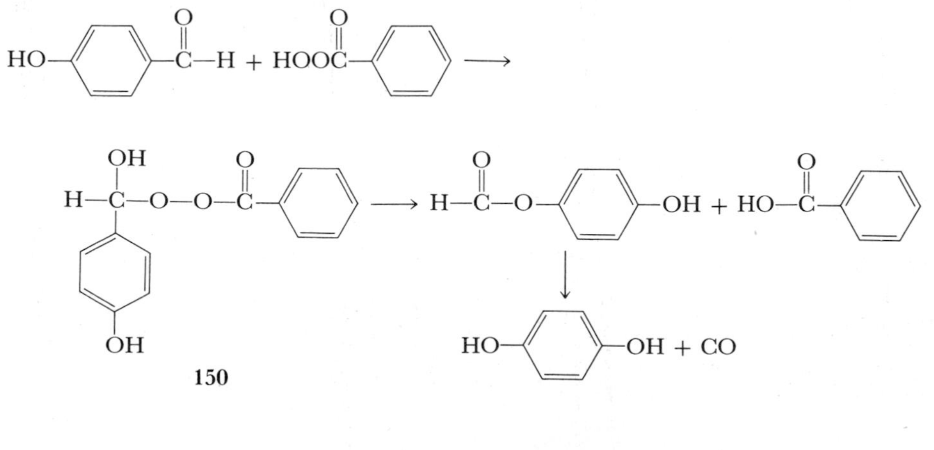

(5.72)

150

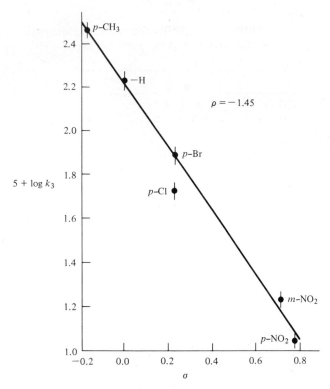

Figure 5.28 Plot of log k vs. σ for the Baeyer–Villiger oxidation of substituted acetophenones by
CF_3COOOH in acetonitrile at 29.8°C. Reprinted with permission from M. F. Haw-
thorne and W. D. Emmons, *J. Am. Chem. Soc.*, **80**, 6398 (1958). Copyright 1958 American Chemical Society.

the pH range 2–7, the rate of this reaction, instead of showing acid catalysis, is *faster*
at higher pH. Ogata and Sawaki suggest that their data are consistent with rate-
determining formation of **150**. This step would be accelerated in less acidic solution
because the peroxybenzoic acid would be more dissociated. They also suggest that
rate-determining formation of the intermediate adduct in Equation 5.71 when
X = OCH_3 is responsible for the unit ^{14}C isotope effect observed in that reaction.[193]

Theoretical calculations[194] and secondary deuterium isotope effects[195] are
also in agreement with the mechanism of Equation 5.69, as is the fact that the
chirality of a migrating group is retained.[196]

[193] Y. Ogata and Y. Sawaki, *J. Am. Chem. Soc.*, **94**, 4189 (1972).
[194] V. A. Stoute, M. A. Winnik, and I. G. Csizmadia, *J. Am. Chem. Soc.*, **96**, 6388 (1974).
[195] M. A. Winnik, V. Stoute, and P. Fitzgerald, *J. Am. Chem. Soc.*, **96**, 1977 (1974).
[196] See note 157(a), (c), (d).

The Aryloxenium Ion

When compound **151** is heated in benzene, compound **152** is formed, presumably via intermediate **153**. The analogous rearrangement is not observed in **154**.[197] Ap-

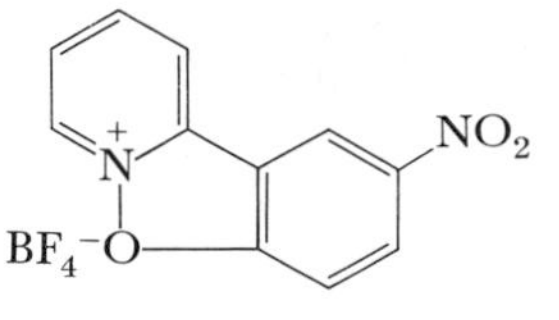

| **151** | **152** | **153** | **154** |

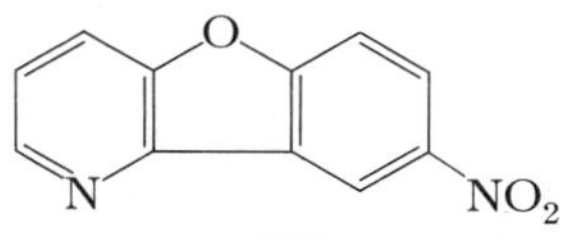

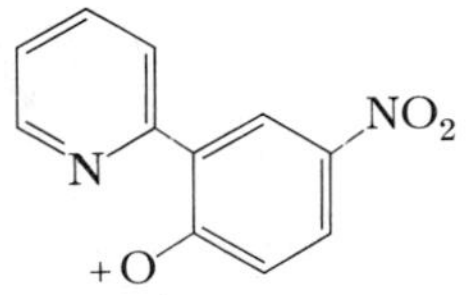

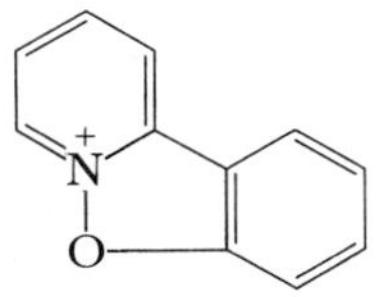

parently the intermediate oxenium ion **153** is either not nucleophilic enough to attack the 3 position of the pyridine ring or contributions from **153b** are more important than from **153a**.

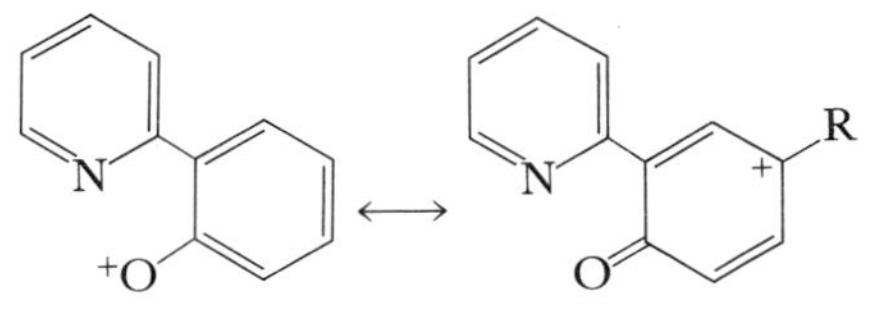

| **153a** | **153b** |

PROBLEMS

1. Explain each of the following observations.

(a) Mustard gas (**1**) owes its deadliness to the fact that it immediately gives off HCl when it mixes with atmospheric moisture. 1,5-Dichloropentane (**2**) hydrolyzes much more slowly.

$$CH_2-CH_2-S-CH_2-CH_2 \qquad CH_2-CH_2-CH_2-CH_2-CH_2$$
$$\ \ |\qquad\qquad\qquad\qquad |\qquad\qquad\ \ |\qquad\qquad\qquad\qquad\qquad\qquad |$$
$$Cl\qquad\qquad\qquad\qquad Cl\qquad\quad Cl\qquad\qquad\qquad\qquad\qquad\qquad Cl$$

| **1** | **2** |

(b) Acetolysis of **3** is a stereospecific reaction and gives only **4**.

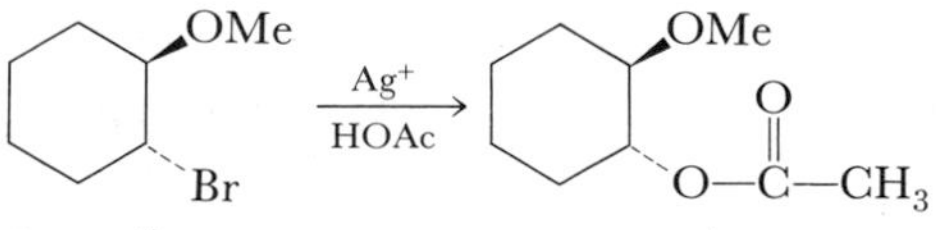

| **3** | **4** |

(c) The migratory aptitude of the *o*-methoxyphenyl group is only 0.3 in the pinacol rearrangement compared to ~500 for the *p*-Methoxyphenyl group.

(d) The titrimetric first-order rate constant for solvolysis of cyclopropylcarbinyl benzenesulfonate decreases as the reaction proceeds.

[197] R. A. Abramovitch and M. N. Inbasekaran, *J. Chem. Soc., Chem. Commun.,* 149 (1978).

(e) **5** solvolyzes 2×10^3 more rapidly in acetic acid than does **6**.

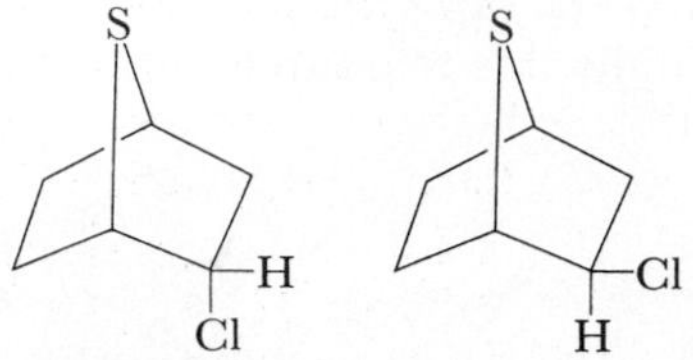

5   **6**

(f) Deamination of **7** proceeds with phenyl migration to give 97 percent inversion of configuration at the migration origin, while **8** undergoes the analogous reaction with only 54 percent inversion of configuration.

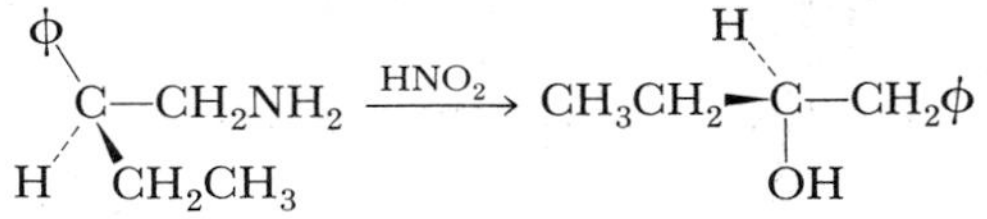

7

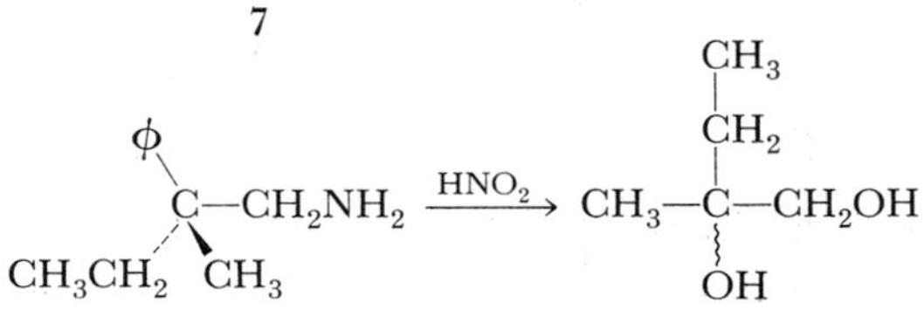

8

(g) The $\phi CH_2CH_2OTs/CH_3CH_2OTs$ solvolysis rate ratio changes from 0.2 in 50 percent aqueous ethanol to 1770 in trifluoroacetic acid.

(h) The relative rates of solvolysis of methylated p-cyclopropyl-t-cumyl chlorides are those shown below.

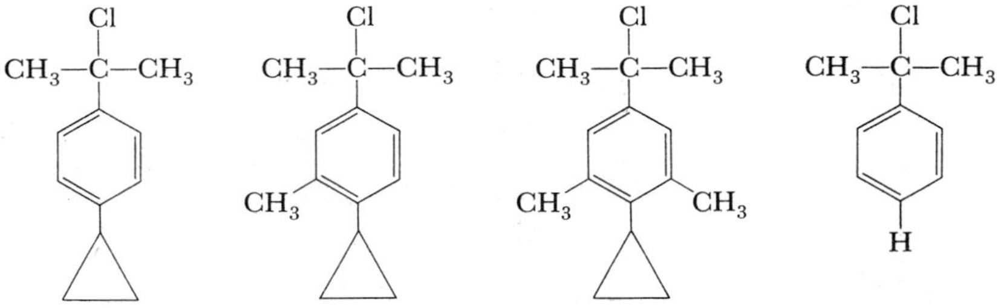

Relative rates: 157   86   9   1

(i) When **9** has chloride as leaving group X, the solvolysis product is entirely ring-opened

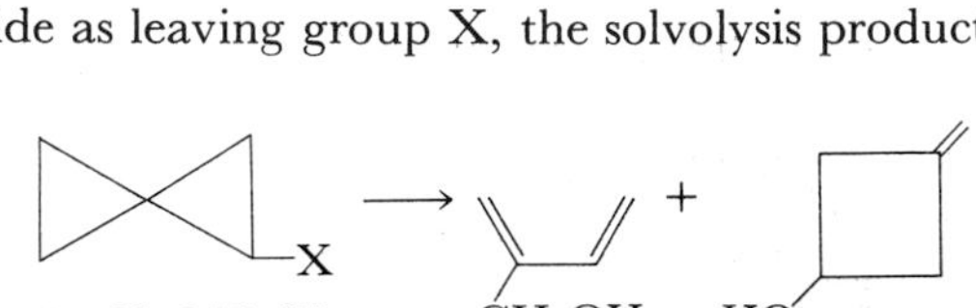

X=Cl, OSO$_2$CF$_3$ CH$_2$OH   HO

9   **10**   **11**

product **10**, but when X is the better leaving group, triflate, some **11** is obtained. (The product **11** is also obtained from the deamination of **9** when X is NH$_2$.)

(j) In general, 1-arylcylopropyl p-nitrobenzoates (**12**) solvolyze to give **13** and **14**. If the aryl group has strongly electron-donating substituents on it, plotting log k vs. the σ^+ constants of the substituents gives a good linear correlation ($\rho^+ = -.707$); no **14** is obtained. If the aryl group is less electron donating, a different correlation line exists between the logarithms of the rates and the σ^+ constants ($\rho^+ = -2.47$) and **14** as well as **13** is formed.

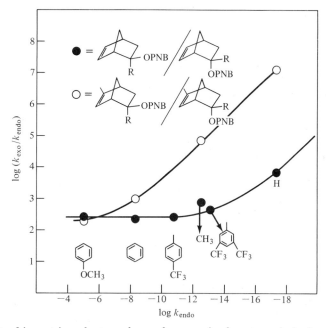

(k) Figure 5.29.

Figure 5.29 Effect of increasing electron demand on exo/endo rate ratio in 2-norbornenyl and 5-methyl-2-norbornenyl p-nitrobenzoates. Reprinted with permission from H. C. Brown and E. N. Peters, *J. Am. Chem. Soc.*, **97**, 7442 (1975); H. C. Brown, E. N. Peters, and M. Ravindranathan, *J. Am. Chem. Soc.*, **97**, 7449 (1975). Copyright 1975 American Chemical Society.

2. Propose a mechanism for each of the following reactions.

(a)

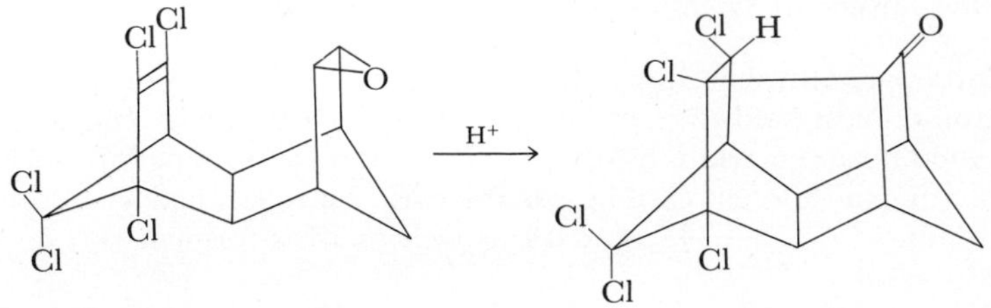

$\xrightarrow{H^+}$

(b)

$\xrightarrow{\text{HOAc}}$ AcO

+

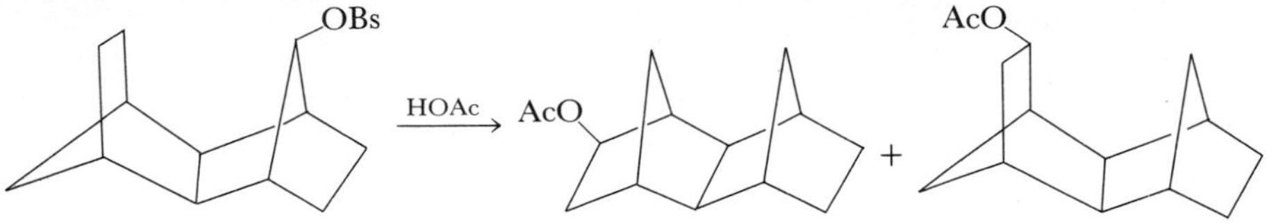

(c) Treatment of ketone (**15**), labeled with ^{14}C in the 9 position, with acid gives the phenol (**16**) with the label equally scrambled between the positions shown.

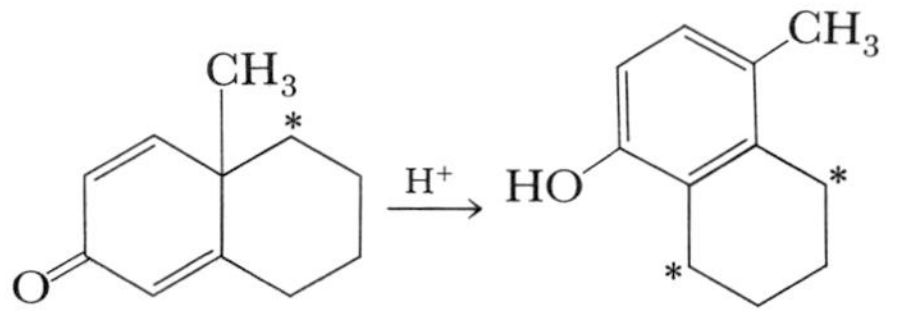

$\xrightarrow{H^+}$ HO

15   **16**

(d) $\phi-\overset{O}{\underset{\|}{C}}-\overset{O}{\underset{\|}{C}}-\phi$ +

$-MgBr \longrightarrow$

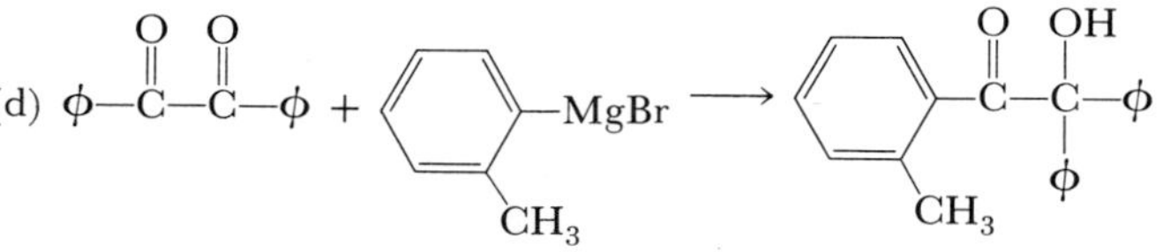

(e)

$\xrightarrow{N_3 \ H^+}$

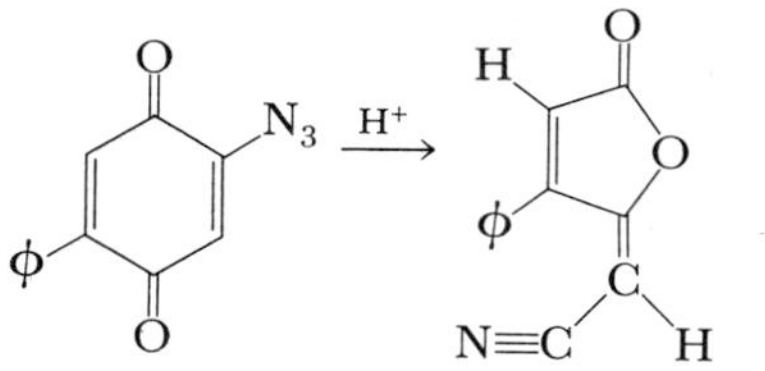

(f) An example of the *aldehyde–ketone* rearrangement:

$$CH_3-\overset{H_3C}{\underset{CH_3}{\overset{|}{C}}}-\overset{O}{\underset{}{\overset{14\|}{C}}}-H \xrightarrow{H^+} CH_3-\overset{H_3C}{\underset{H}{\overset{|}{C}}}-\overset{O}{\underset{}{\overset{14\|}{C}}}-CH_3$$

(g) An example of the *Lossen rearrangement:*

$$CH_3-\overset{O}{\underset{}{\overset{\|}{C}}}-NH-O-\overset{O}{\underset{}{\overset{\|}{C}}}-CH_2CH_3 \xrightarrow{^-OH} CH_3NH_2 + CO_2 + {}^-O-\overset{O}{\underset{}{\overset{\|}{C}}}-CH_2CH_3$$

(h) An example of the *Curtius rearrangement:*

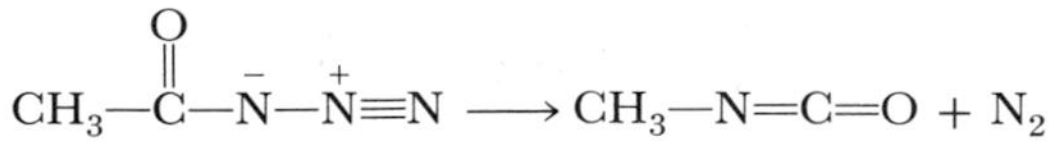

$$CH_3-\overset{\overset{O}{\parallel}}{C}-\overset{-}{N}-\overset{+}{N}\equiv N \longrightarrow CH_3-N=C=O + N_2$$

(i) 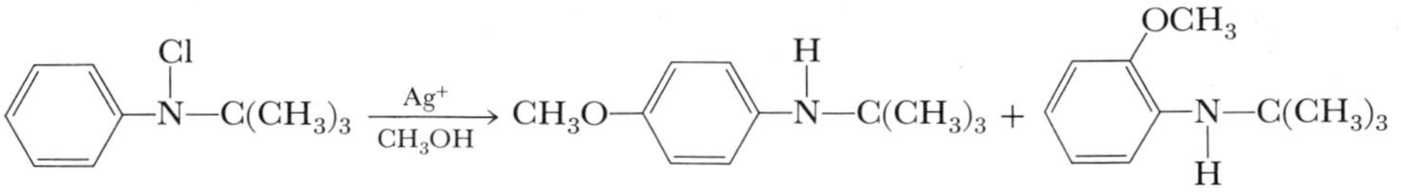 (phenyl)–N(Cl)–C(CH$_3$)$_3$ $\xrightarrow[\text{CH}_3\text{OH}]{\text{Ag}^+}$ CH$_3$O–(phenyl)–N(H)–C(CH$_3$)$_3$ + (o-OCH$_3$-phenyl)–N(H)–C(CH$_3$)$_3$

(j) 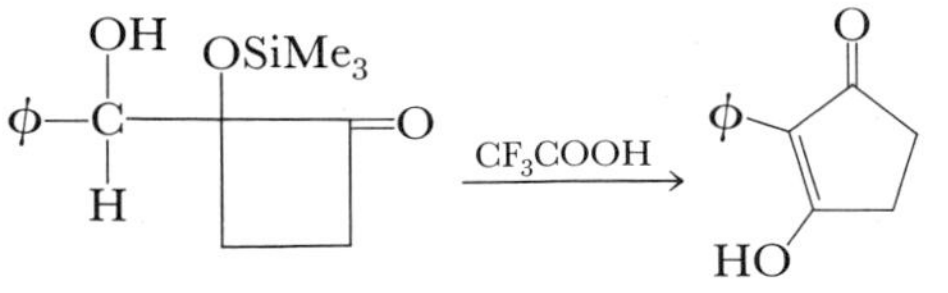 $\phi-\overset{\overset{OH}{|}}{\underset{H}{C}}$ —OSiMe$_3$ ... =O $\xrightarrow{\text{CF}_3\text{COOH}}$ (cyclopentenone with ϕ and HO substituents)

(k) 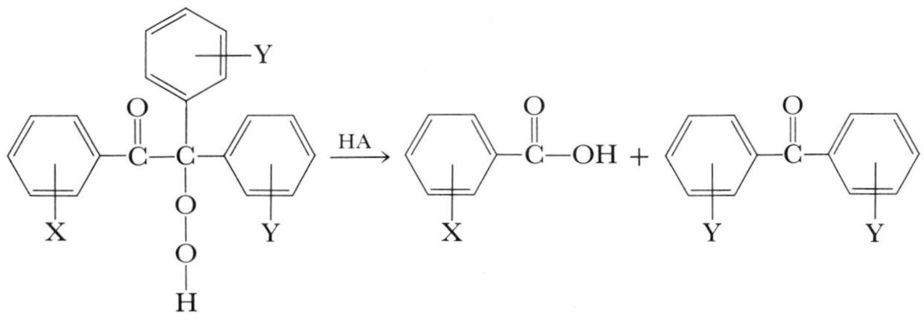 (substrate with X-phenyl, Y-phenyl, Y-phenyl groups, central C–C with O and O–O–H) $\xrightarrow{\text{HA}}$ (X-phenyl)–C(=O)–OH + (Y-phenyl)–C(=O)–(Y-phenyl)

$\rho = -2.33$ as X is changed (rates correlate with σ constants of X). $\rho^+ = -0.82$ as Y is changed (rates correlate with σ^+ constants of Y).

$$\text{rate} = k[\text{HA}][\text{substrate}]$$

(l) 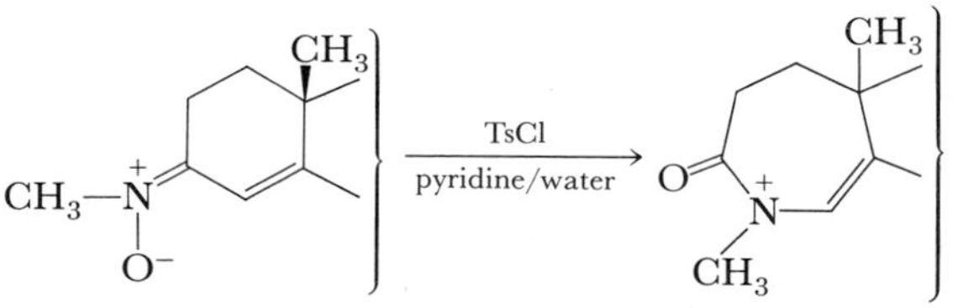 CH$_3-\overset{+}{N}(\text{O}^-)=$ (cyclohexene with CH$_3$) $\xrightarrow[\text{pyridine/water}]{\text{TsCl}}$ O= (seven-membered ring with $\overset{+}{N}$–CH$_3$ and CH$_3$)

(m) 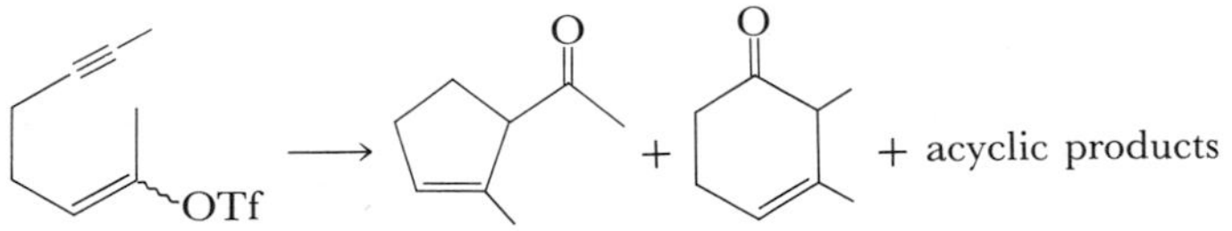 (enyne with OTf) $\longrightarrow$ (cyclopentene acetyl product) + (cyclohexenone product) + acyclic products

—OTf = triflate

The amount of ring-closed products is 1.5 percent in 50 percent aqueous ethanol and 35.5 percent in 99.5 percent 2,2,2-trifluoroethanol.

(n)

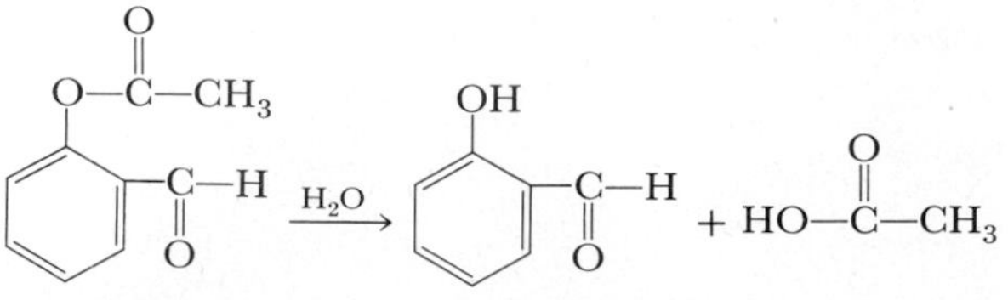

1. The rates of this reaction and of the hydration of the aldehyde **17** are the same.
2. If the reaction is carried out in $H_2{}^{18}O$, *half* the acetic acid has one ^{18}O atom; the other half has none.

(o)

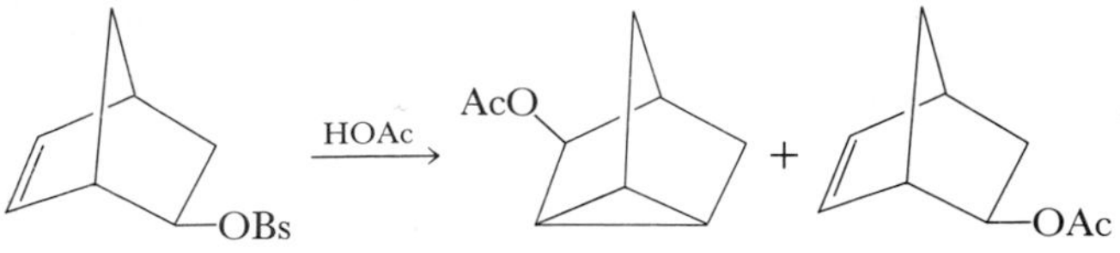

chiral   racemic   racemic

(p)

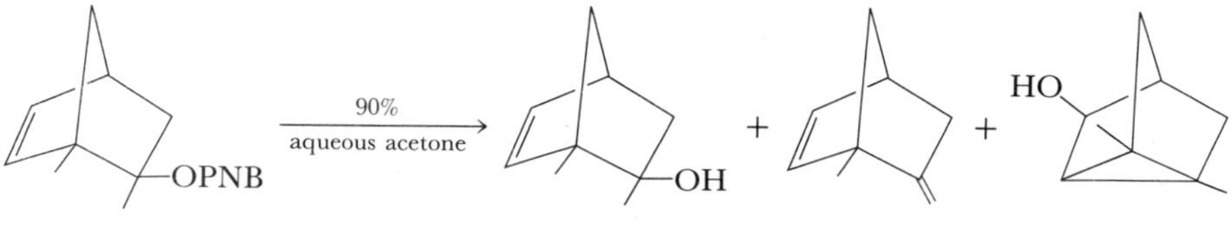

15% retention   62% retention   racemic

(q)

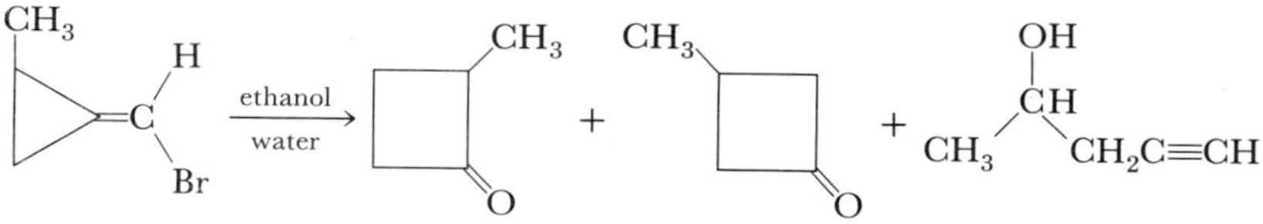

(r)

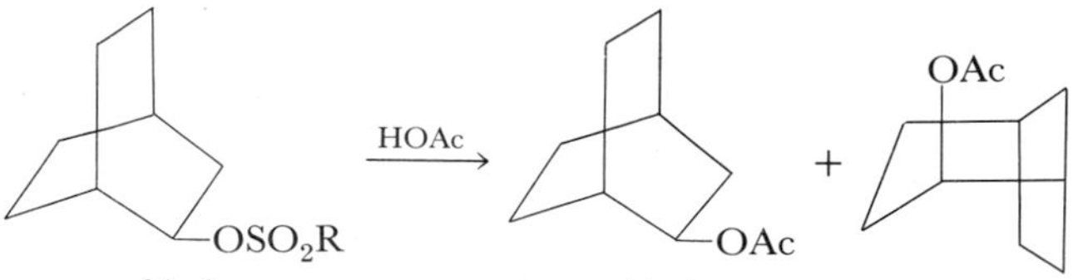

chiral   chiral

(s)

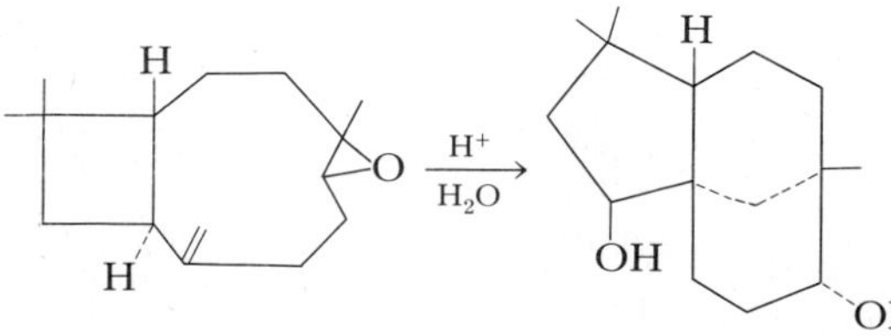

(t)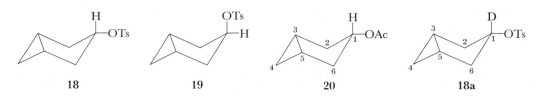

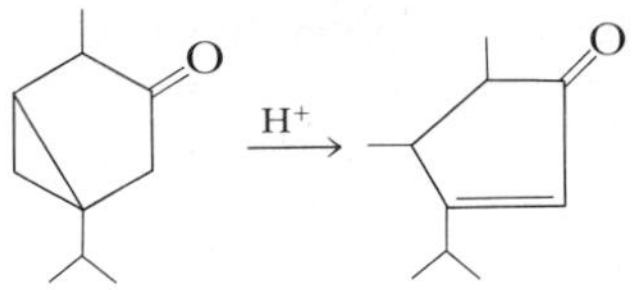

$\xrightarrow{H^+}$

3. All five of the following reactions give pinacol and pinacolone. The products are formed in a similar ratio from each reaction. Write a mechanism for each reaction. What does the constant ratio of products tell you about any intermediate that may be formed?

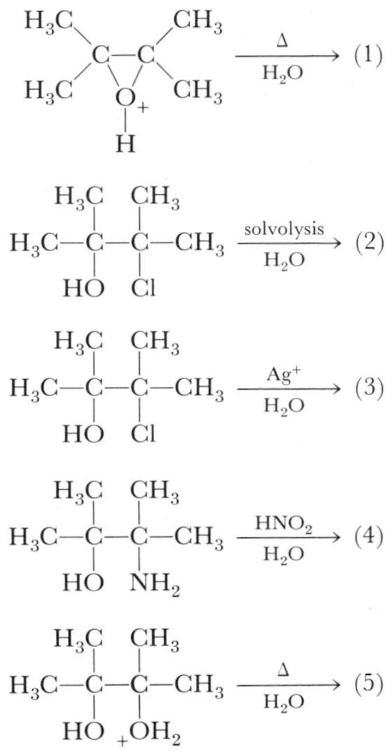

$$\underset{H_3C}{\overset{H_3C}{>}} \underset{\underset{\underset{H}{+}}{O}}{\overset{}{C-C}} \underset{CH_3}{\overset{CH_3}{<}} \xrightarrow[H_2O]{\Delta} \quad (1)$$

$$H_3C-\underset{\underset{HO}{|}}{\overset{\overset{H_3C}{|}}{C}}-\underset{\underset{Cl}{|}}{\overset{\overset{CH_3}{|}}{C}}-CH_3 \xrightarrow[H_2O]{solvolysis} \quad (2)$$

$$H_3C-\underset{\underset{HO}{|}}{\overset{\overset{H_3C}{|}}{C}}-\underset{\underset{Cl}{|}}{\overset{\overset{CH_3}{|}}{C}}-CH_3 \xrightarrow[H_2O]{Ag^+} \quad (3)$$

$$H_3C-\underset{\underset{HO}{|}}{\overset{\overset{H_3C}{|}}{C}}-\underset{\underset{NH_2}{|}}{\overset{\overset{CH_3}{|}}{C}}-CH_3 \xrightarrow[H_2O]{HNO_2} \quad (4)$$

$$H_3C-\underset{\underset{HO}{|}}{\overset{\overset{H_3C}{|}}{C}}-\underset{\underset{+OH_2}{|}}{\overset{\overset{CH_3}{|}}{C}}-CH_3 \xrightarrow[H_2O]{\Delta} \quad (5)$$

4. Predict the products expected from (a) S_N1 and (b) S_N2 substitutions of acetic acid on L-*threo*- and L-*erythro*-3-phenyl-2-butyl tosylate and compare with the results actually obtained in Section 5.1.

5. Explain the following facts: (a) Acetolysis of **18** and **19** gives only **20**. (b) The rate of acetolysis of **18** relative to **19** is about 44. (c) When **18** is solvolyzed, the product has the deuterium scrambled equally over carbons 1, 3, and 5.

| 18 | 19 | 20 | 18a |

6. The norbornonium ion is formed as an intermediate in the solvolysis of a mono-cyclic arenesulfonate. This pathway is called the π route to the norbornonium ion. What is the starting material for the π route to this ion?

7. Predict the products, and whether each will be chiral or not, from the acetolysis of (a) optically active **21** and (b) optically active **22**. (c) In the solvolysis of what monocyclic starting material would the same carbonium ion be formed as is formed in the solvolysis of **21**? (d) Repeat (c) for **22**.

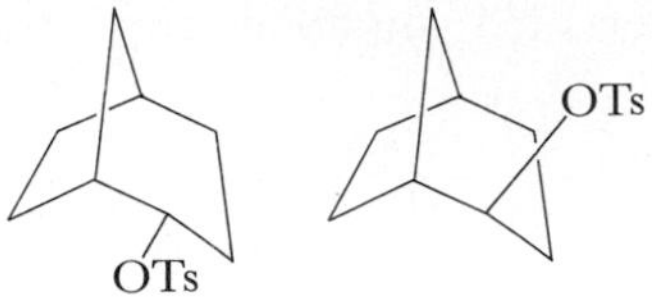

21   22

8. The rearrangement of α-diazoketones to carboxylic acids or esters shown in the equation is known as the Wolff rearrangement.

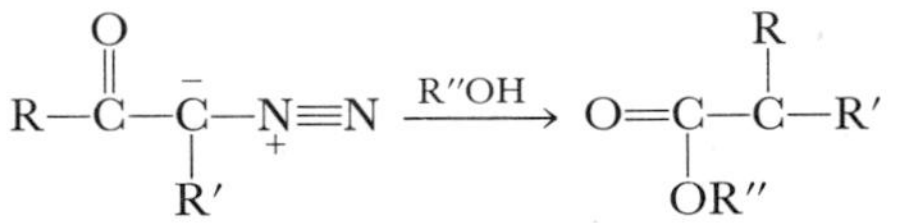

Taking into account the following facts, suggest a mechanism for the Wolff rearrangement. (a) The kinetics are clearly first-order in substrate and first-order overall. (b) In aprotic media ketenes can sometimes be isolated. (c) The rate of rearrangement of **23** is almost identical whether Z is $-OCH_3$ or $-NO_2$. (d) The rate of rearrangement of **24** is much faster if Z is $-OCH_3$ than if it is $-NO_2$.

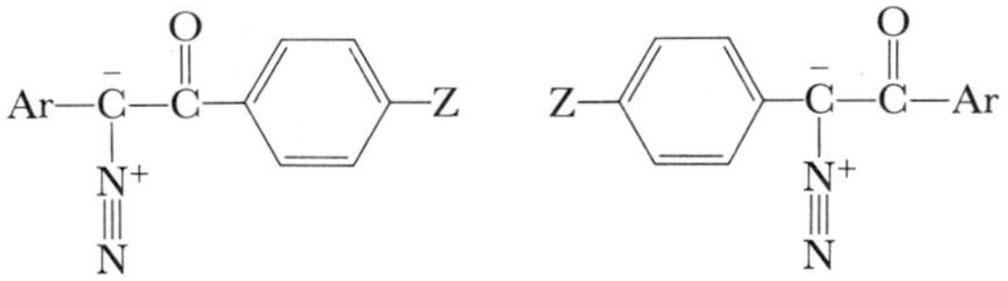

23   24

9. Of each of the following pairs of compounds, which would you expect to react more rapidly? Why?

(a) 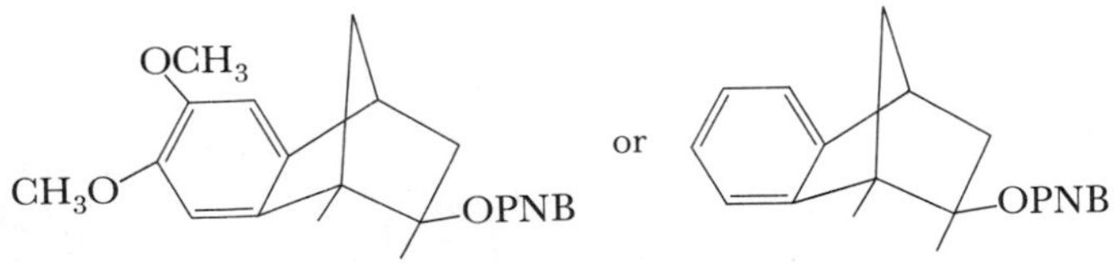   or

(b) 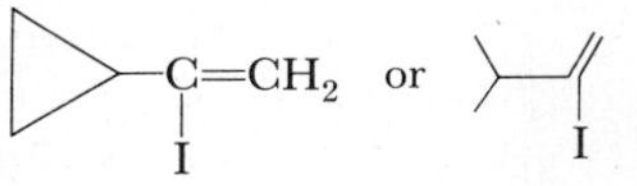 with $Ag^+/HOAc$.

(c) 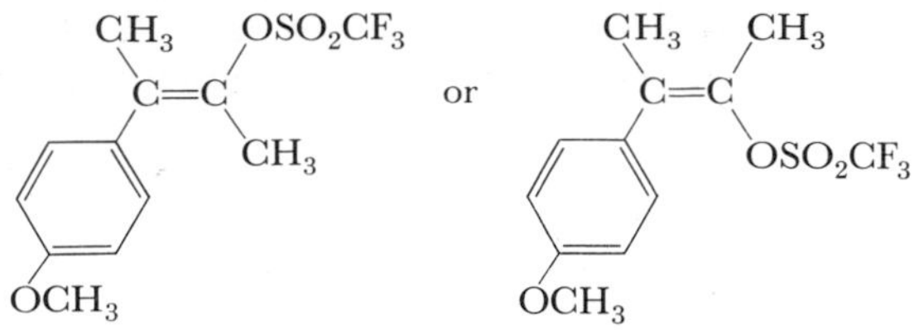 [structure: CH$_3$, OSO$_2$CF$_3$, C=C, CH$_3$, with benzene ring bearing OCH$_3$]   or   [structure: CH$_3$, CH$_3$, C=C, OSO$_2$CF$_3$, with benzene ring bearing OCH$_3$]

10. (a) Propose a mechanism for the reaction shown below. (b) Why does the carbethoxy group migrate in preference to the phenyl even though phenyl has a much greater migratory aptitude?

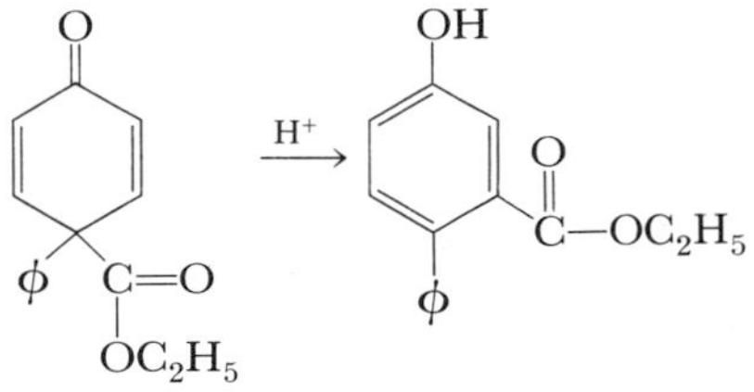

[reaction structure: cyclohexadienone with ϕ—C=O and OC$_2$H$_5$ substituent, $\xrightarrow{H^+}$, phenol product with OH and C—OC$_2$H$_5$ (C=O) and ϕ]

11. Winstein suggested that the exo/endo rate ratio of norbornyl tosylates is not greater because the ionization of the endo tosylate is also assisted—by solvent. When the rates of solvolysis of *endo*-norbornyl tosylate in a number of solvents varying widely in their nucleophilicity is plotted against the rates of solvolysis of 1-adamantyl bromide in the same solvents, a good straight line is obtained. Is this evidence for or against Winstein's argument?

REFERENCES FOR PROBLEMS

1. (a) P. D. Bartlett and C. G. Swain, *J. Am. Chem. Soc.*, **71**, 1406 (1949); (b) S. Winstein and R. B. Henderson, *J. Am. Chem. Soc.*, **65**, 2196 (1943); (c) W. E. Bachmann and J. W. Ferguson, *J. Am. Chem. Soc.*, **56**, 2081 (1934); (d) C. G. Bergstrom and S. Siegel, *J. Am. Chem. Soc.*, **74**, 145 (1952); (e) I. Tabushi, Y. Tamaru, Z. Yoshida, and T. Sugimoto, *J. Am. Chem. Soc.*, **97**, 2886 (1975); (f) W. Kirmse and P. Feyen, *Chem. Ber.*, **108**, 71 (1975); (g) F. L. Schadt and P. v. R. Schleyer, *J. Am. Chem. Soc.*, **95**, 7860 (1973); (h) H. C. Brown and J. D. Cleveland, *J. Am. Chem. Soc.*, **88**, 2051 (1966); (i) D. E. Applequist and G. W. Nickel, *J. Org. Chem.*, **44**, 321 (1979); (j) H. C. Brown, C. G. Rao, and M. Ravindranathan, *J. Am. Chem. Soc.*, **100**, 7946 (1978); (k) H. C. Brown and E. N. Peters, *J. Am. Chem. Soc.*, **97**, 7442 (1975); H. C. Brown, E. N. Peters, and M. Ravindranathan, *J. Am. Chem. Soc.*, **97**, 7449 (1975).
2. (a) R. C. Cookson and E. Crundell, *Chem. Ind. (London)*, 703 (1959); (b) S. Winstein and R. L. Hansen, *J. Am. Chem. Soc.*, **82**, 6206 (1960); (c) R. Futaki, *Tetrahedron Lett.*, 3059 (1964); (d) J. F. Eastham, J. E. Huffaker, V. F. Raaen, and C. J. Collins, *J. Am. Chem. Soc.*, **78**, 4323 (1956); (e) H. W. Moore, H. R. Shelden, D. W. Deters, and R. J. Wikholm, *J. Am. Chem. Soc.*, **92**, 1675 (1970); (f) M. Oka and A. Fry, *J. Org. Chem.*, **35**, 2801 (1970); (g) H. L. Yale, *Chem. Rev.*, **33**, 209 (1943); (h) A. A. Bothner-By and L. Friedman, *J. Am. Chem. Soc.*, **73**, 5391 (1951); (i) P. G. Gassman, G. A. Campbell, and R. C. Frederick, *J. Am. Chem. Soc.*, **94**, 3884 (1972); (j) E. Nakamura and I. Kuwajima, *J.*

Am. Chem. Soc., **99,** 961 (1977); (k) Y. Suwaki and Y. Ogata, *J. Am. Chem. Soc.*, **100,** 856 (1978); (l) D. H. R. Barton, M. J. Day, R. H. Hesse, and M. M. Pecket, *J. Chem. Soc., Perkin Trans. 1,* 1764 (1975); (m) M. J. Chandy and M. Hanack, *Tetrahedron Lett.*, 4515 (1975); (n) J. A. Walder, R. S. Johnson, and I. M. Klotz, *J. Am. Chem. Soc.*, **100,** 5156 (1978); (o) H. C. Goering and C.-S. Chang, *J. Am. Chem. Soc.*, **99,** 1547 (1977); (p) H. C. Goering and C.-S. Chang, *J. Am. Chem. Soc.*, **99,** 1547 (1977); (q) G. Hammen, T. Bassler, and M. Hanack, *Chem. Ber.*, **107** 1676 (1974); (r) H. M. Walborsky, M. E. Baum, and A. A. Youssef, *J. Am. Chem. Soc.*, **83,** 988 (1961); H. L. Goering and M. F. Sloan, *J. Am. Chem. Soc.*, **83,** 1397 (1961); (s) J. F. King and P. de Mayo, in *Molecular Rearrangements,* Vol. 2, P. de Mayo, Ed., Wiley, New York, 1964, p. 771; (t) R. H. Eastman and A. V. Winn, *J. Am. Chem. Soc.*, **82,** 5908 (1960).

3. Y. Pocker, *Chem. Ind. (London)*, 332 (1959).

5. S. Winstein and J. Sonnenberg, *J. Am. Chem. Soc.*, **83,** 3235, 3244 (1961).

6. R. G. Lawton, *J. Am. Chem. Soc.*, **83,** 2399 (1961); P. D. Bartlett and S. Bank, *J. Am. Chem. Soc.*, **83,** 2591 (1961).

7. H. L. Goering and G. N. Fickes, *J. Am. Chem. Soc.*, **90,** 2856, 2862 (1968).

8. A. Melzer and E. F. Jenny, *Tetrahedron Lett.*, 4503 (1968).

9. (a) H. L. Goering, C.-S. Chang, and D. Masilamani, *J. Am. Chem. Soc.*, **100,** 2506 (1978); (b) S. A. Sherrod and R. Bergman, *J. Am. Chem. Soc.*, **91,** 2115 (1969); M. Hanack and T. Bässler, *J. Am. Chem. Soc.*, **91,** 2117 (1969); (c) P. J. Stang and T. E. Dueber, *J. Am. Chem. Soc.*, **99,** 2602 (1977).

10. J. N. Marx, J. C. Argyle, and L. R. Norman, *J. Am. Chem. Soc.*, **96,** 2121 (1974).

11. J. M. Harris, D. L. Mount, and D. J. Raber, *J. Am. Chem. Soc.*, **100,** 3139 (1978).

Chapter 6

CARBANIONS, CARBENES, AND ELECTROPHILIC ALIPHATIC SUBSTITUTION

In this chapter we consider chemistry of carbanions and carbenes, which together with the carbocations (Chapters 4 and 5) and radicals (Chapter 9) comprise the important reactive intermediates of organic reactions. We shall also consider briefly electrophilic substitution reactions, which are related to carbanions in the same way that nucleophilic substitutions are related to carbocations.

6.1 CARBANIONS[1]

Carbanions may be encountered under two contrasting sets of circumstances. First, they occur as short-lived reactive intermediates in solution under basic conditions, as, for example, in base-catalyzed H—D exchange (Chapter 3), $E1_{CB}$ eliminations (Chapter 7), and α eliminations (Section 6.2). In these reactions some other substance in the solution (often the solvent) is a proton donor. The carbanion is in equilibrium with its conjugate acid and is likely to be present in low concentration and to have a short lifetime. Second, carbanions and the closely related organometallics can occur in solution at appreciable concentration and with long lifetime. The solvent must be substantially less acidic than the conjugate acid of the carbanion and not subject to nucleophilic attack. These conditions are typically achieved by use of an anhydrous solvent of the ether type.

The occurrence of carbanions in moderate to high concentration in aqueous solution, where they would be expected to be dissociated to free ions, is strictly limited because a C—H bond will be acidic enough to dissociate appreciably in aqueous solution only if substituted by two or three strongly electron-withdrawing groups. (See Table 3.9, p. 272 and Table 6.1.) The dipolar aprotic solvents, for example, dimethylsulfoxide, acetonitrile, or hexamethylphosphoramide, are sufficiently nonacidic to support significant concentrations of less strongly stabilized carbanions. (See Table 3.2, p. 260.) In these solvents the anions may sometimes exist as free ions, but more often they will be in the form of solvent-separated or

---

[1] Monographs: (a) E. Buncel, *Carbanions: Mechanistic and Isotopic Aspects,* Elsevier, Amsterdam, 1975; (b) D. J. Cram, *Fundamentals of Carbanion Chemistry,* Academic Press, New York, 1965.

Table 6.1 Effect of Substituent Groups on the Acidity of C—H Bonds[a]

Group G	Approximate pK_a of G—$\overset{\textstyle\mid}{\underset{\textstyle\mid}{C}}$—H	Approximate pK_a of G$_2$C—H	Approximate pK_a of G$_3$CH
R—	50–70		
φ—	41	33	32
R—$\overset{O}{\overset{\|}{C}}$—	20–30	9	6
R—O—$\overset{O}{\overset{\|}{C}}$—	24	13	—
N≡C—	31	12	0
R—$\overset{O}{\overset{\|}{S}}$—	35	—	—
R—SO$_2$—	31	12	0
CF$_3$—SO$_2$—	19[b]	—	—
O$_2$N—	11	4	0

[a] See Tables 3.2, 3.3, and 3.9.
[b] F. G. Bordwell, N. R. Vanier, W. S. Matthews, J. B. Hendrickson, and P. L. Skipper, *J. Am. Chem. Soc.*, **97**, 7160 (1975).

contact ion pairs. In the less acidic solvents (tetrahydrofuran, diethyl ether) frequently used for strongly basic systems, strong ion pairing of carbanion with its counter-ion will be the rule. When the anion is not stabilized by an electron-delocalizing group, it is likely to be covalently bonded to the cation, particularly if the latter has an electronegativity not too different from carbon.

When the carbanionic species is bonded to a metal, the compound is an organometallic. In organometallic compounds an organic group that provides a σ electron pair is said to be a σ donor ligand of the metal. Examples are the alkyl lithium and alkyl magnesium compounds, which will be considered briefly later in this section. Organic ligands can also bond to metals through π orbitals. Organometallic compounds containing π donor ligands are formed particularly readily by the transition metals, where donation of electrons back from filled d orbitals of the metal to antibonding π* orbitals of the ligand contributes to the stability of the compound. We shall not cover compounds of this type.[2]

Stability of Carbanions

Acidity of a C—H bond can be taken as a rough guide to stability of the conjugate base carbanion. As we have seen in Chapter 3, C—H acidities vary over a wide

[2] For detailed treatments of organometallic chemistry, see (a) K. F. Purcell and J. C. Kotz, *Inorganic Chemistry*, Saunders, Philadelphia, 1977; (b) G. E. Coates, M. L. H. Green, and K. Wade, *Organometallic Compounds*, 3d ed., Methuen, London, Vol. 1, *The Main Group Elements*, 1967; Vol. 2, *The Transition Elements*, 1968.

range, the acidity, and hence carbanion stability, being strongly enhanced by groups that will delocalize negative charge. Table 6.1 summarizes the more important groups of this type. The substituents shown in the table owe their anion-stabilizing ability to a greater or lesser degree to direct delocalization of the negative charge. There is also a category of stabilizing groups of general type **1**, in which

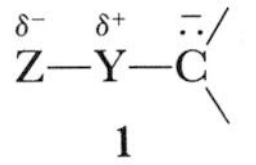

$$\overset{\delta-}{Z}{-}\overset{\delta+}{Y}{-}\overset{\ddots/}{C}$$

1

the positive end of a dipole helps to stabilize an adjacent negative charge.[3] Carbanions stabilized by groups of this sort typically gain extra stabilization through chelation incorporating the cation. Structures **2**, **3**, and **4** are examples.

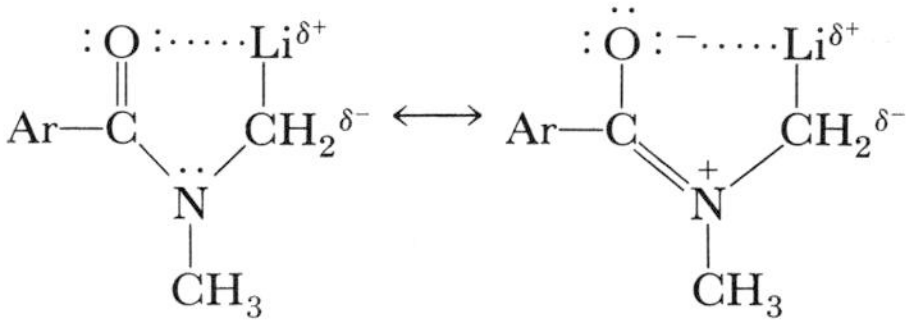

2

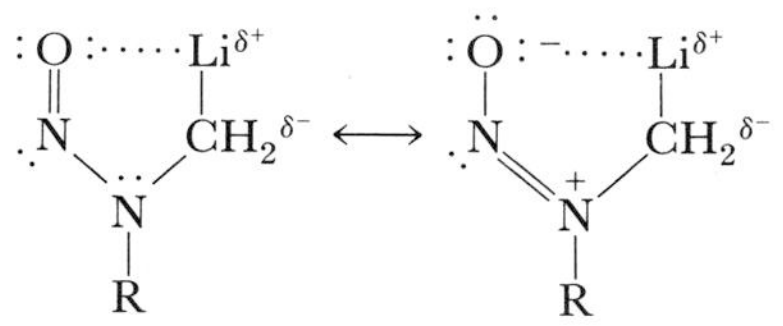

3

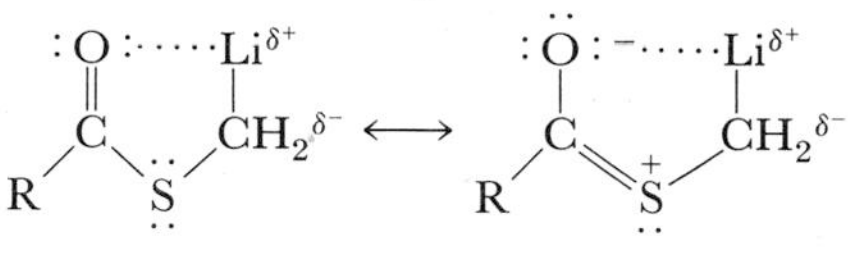

4

The logical extension of dipole stabilization leads to the ylides (**5** and **6** are examples) in which a full positive charge resides on an adjacent atom. Ylides are electrically neutral and hence not carbanions, but they do have similar reactivity and synthetic applications.

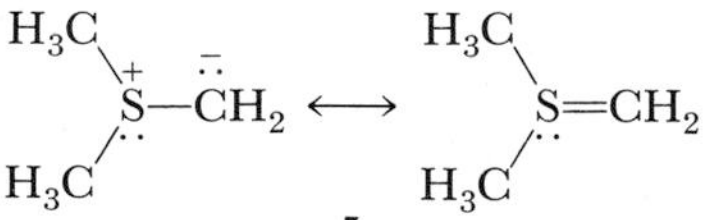

5

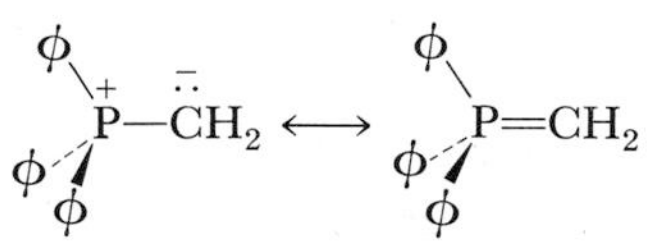

6

[3](a) A. Beak and D. B. Reitz, *Chem. Rev.* **78**, 275 (1978); (b) D. B. Reitz, P. Beak, R. F. Farney, and L. S. Helmick, *J. Am. Chem. Soc.,* **100**, 5428 (1978).

Stabilities of carbanions in the gas phase are indicated in Table 6.2 by the difference between R—H bond dissociation energy, $D(RH)$, and the electron affinity of the radical R, $EA(R)$, a quantity based on the thermochemical cycle shown in Scheme 1. Note that the term $IP(H)$, the ionization potential of hydrogen, which

SCHEME 1

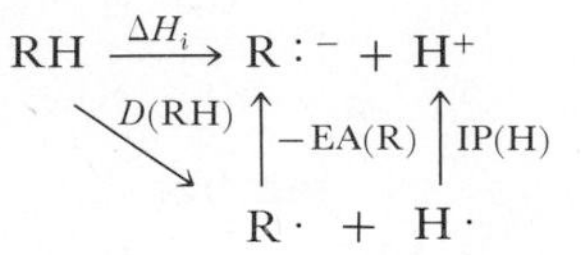

$$RH \xrightarrow{\Delta H_i} R:^- + H^+$$

$$D(RH) \nwarrow \quad \uparrow -EA(R) \quad \uparrow IP(H)$$

$$R \cdot \ + \ H \cdot$$

$$\Delta H_i = D(RH) - EA(R) + IP(H)$$

is 313.6 kcal mole^{-1}, is common to all the reactions and can be omitted for convenience without affecting the relative magnitudes of the results for different anions R^-. Figure 6.1 shows that except for cyclopentadienyl there is a rough correlation among compounds of a given structural type between gas phase and solution acidity. The large amount of scatter and the different slopes for different types of compounds show, as expected, that solvation affects ions of various structural types differently.[4]

Formation and Reactions of Carbanions

Methods of generation of carbanions vary depending on the degree of stabilization of the negative charge by substituents. For the more highly stabilized anions such as enolates, equilibrium concentrations suitable for synthetic reactions can be produced in aqueous or alcohol solutions with hydroxide or alkoxide as base, although a stronger base and an aprotic solvent may be required if it is necessary to attain a high concentration of the anion.[5] Less highly stabilized ions can sometimes be obtained by reaction with alkali metal hydrides, as, for example, in the standard procedure for preparing the conjugate base of dimethylsulfoxide (Equation 6.1).[6]

$$H_3C-\overset{\overset{O}{\|}}{S}-CH_3 + NaH \longrightarrow H_3C-\overset{\overset{O}{\|}}{S}-\overset{-}{C}H_2\overset{+}{N}a + H_2 \qquad (6.1)$$

The metal hydride method, however, is not widely applicable; the standard technique for preparation of the more basic carbanions is to use an organometallic compound in an aprotic solvent. The organometallic compounds in turn are prepared by reaction of lithium or magnesium with an alkyl halide.[7] n-Butyllithium is a common choice for generation of carbanions and other lithium organometallics.

Carbanions owe their great synthetic utility to their reactivity as nucleophiles, both in S_N2 substitutions and in additions to multiple bonds, particularly

[4]E. M. Arnett, D. E. Johnston, and L. E. Small, *J. Am. Chem. Soc.*, **97**, 5598 (1975).

[5]For examples see H. O. House, *Modern Synthetic Reactions*, 2d ed., W. A. Benjamin, Menlo Park, Calif., 1972, Chap. 9.

[6]E. J. Corey and M. Chaykovsky, *J. Am. Chem. Soc.*, **84**, 866 (1962); **87**, 1345 (1965).

[7]See note 2.

Table 6.2 RELATIVE GAS-PHASE ACIDITIES OF C—H BONDS[a]

Compound	$D(RH) - EA(R)^a$	$pK_a^{\ b}$
Fluorene	34.9	23
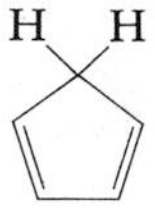	39.1	16
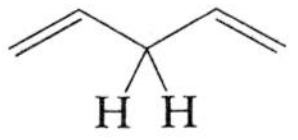	46.1	—
ϕ_2CH_2	47.0	33
$(CN)_2CH_2$	17.2	12
CH_3CN	47.9	31
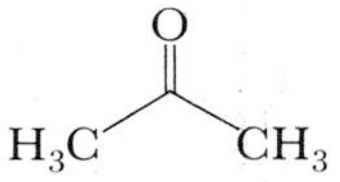	50.4	20
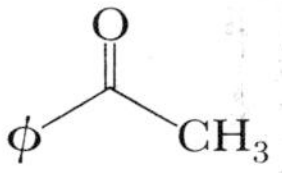	45.6	19[a]
F_3CCCH_3	33.5	—
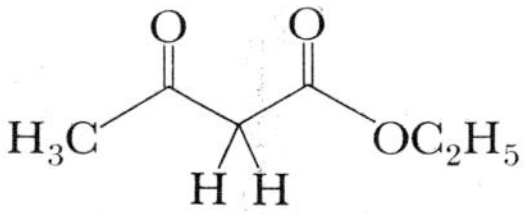	33.5	—
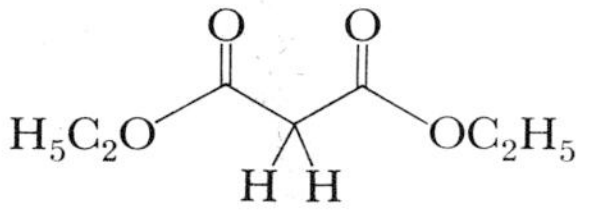	32.5	13
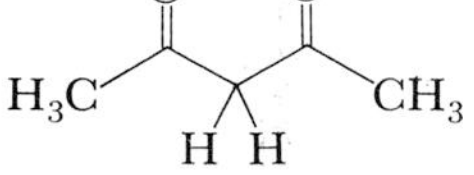	28.0	9
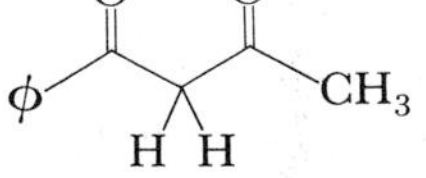	24.2	9.4[a]
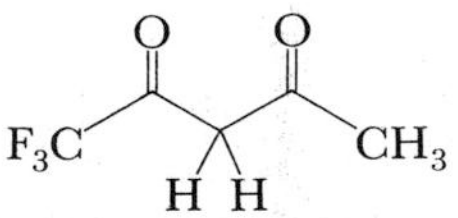	12.4	4.7[a]

[a] Reprinted in part with permission from T. B. McMahon and P. Kebarle, *J. Am. Chem. Soc.*, **98**, 3399 (1976). Copyright 1976 American Chemical Society.
[b] See Tables 3.2, 3.3, and 3.9.

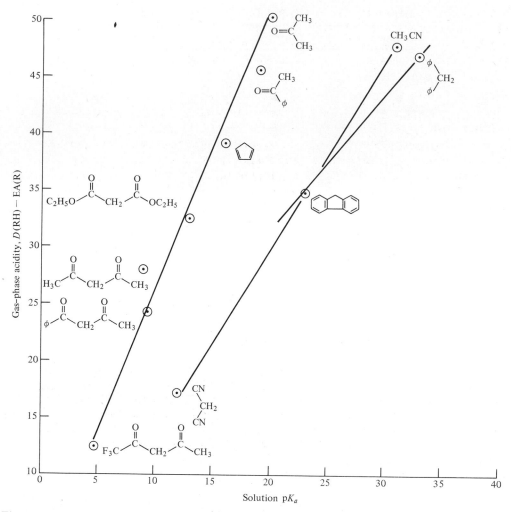

Figure 6.1 Comparison of gas phase and solution acidities of C—H bonds.

C=O and C=C. These reactions constitute some of the most useful methods of generating new carbon–carbon bonds. Enolate ions are particularly important in this connection. Because of the extensive charge delocalization, as indicated by resonance structures **7**, enolates are ambient ions that can react either at C or at O. In additions to carbonyl groups, the product forms by bonding to the carbon end, rather than to the oxygen end, but in S_N2 substitutions on alkyl halides significant amounts of O alkylation occur. The more acidic compounds, such as those with the β-dicarbonyl structure, yield enolates with the greater tendency toward O-alkylation. Protic solvents and small cations favor C alkylation, because

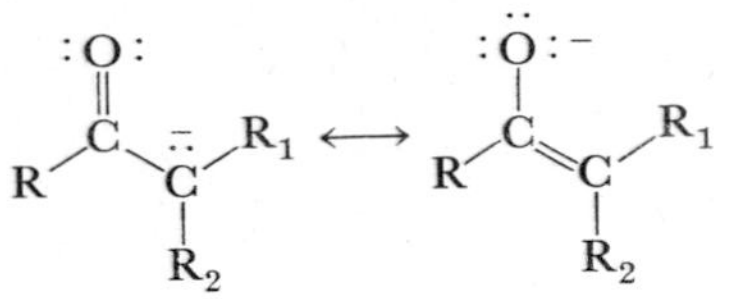

7

the harder oxygen base of the enolate coordinates more strongly than does the carbon with these hard Lewis acids.[8]

Unsymmetrical ketones can yield two different enolates, and in some cases the one that is the less stable thermodynamically is formed faster.[9] Scheme 2 illustrates the example of 2-methylcyclopentanone. When this ketone is added

SCHEME 2

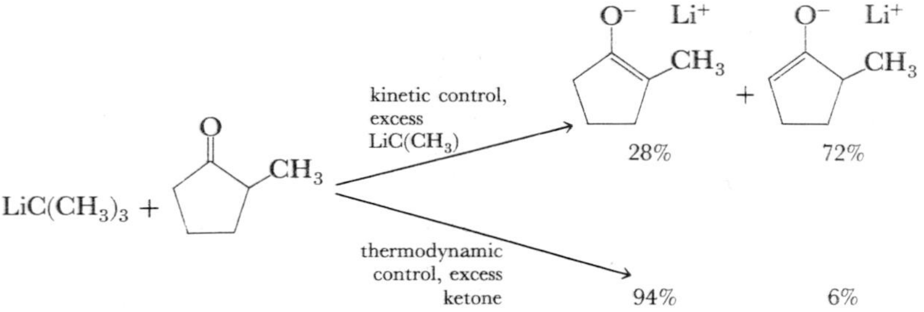

slowly to excess *t*-butyllithium, the proton is removed preferentially from the less substituted carbon. If excess ketone is added, it can serve as a proton donor to allow equilibrium to be established, and nearly all the enolate is then the more highly substituted one.[10] It may be possible in some cases to take advantage of such a selective formation of one of two possible enolates in synthesis. A more general procedure is to use a compound in which the position at which reaction is desired is activated by two electron-withdrawing groups, for example, keto ester (**8**) from which, with one equivalent of base, only one enolate will form. Excess base in such cases may yield a dienolate (**9**), which will react as a nucleophile preferentially at the position of the less stabilized anion (Equation 6.4). Anions stabilized by the other groups listed in Table 6.1 find similar synthetic use.

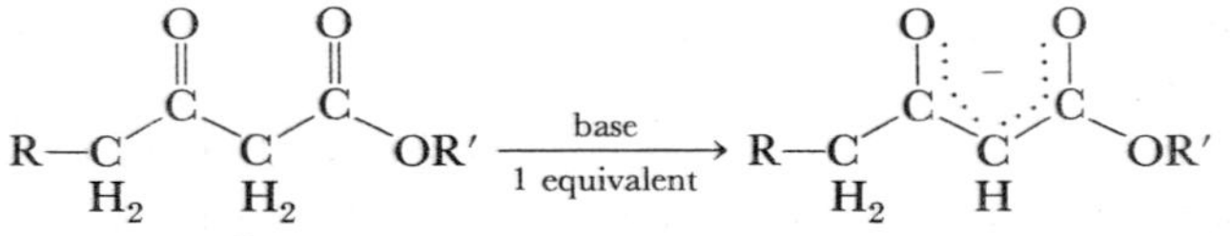

(6.2)

8

---

[8] H. O. House, *Modern Synthetic Reactions,* 2d ed., W. A. Benjamin, Menlo Park, Calif., 1972, p. 520.
[9] H. O. House, *Modern Synthetic Reactions,* 2d ed., W. A. Benjamin, Menlo Park, Calif., 1972, pp. 501–502.
[10] H. O. House and B. M. Trost, *J. Org. Chem.,* **30**, 1341, 4395 (1965).

$$
\underset{\substack{\text{H}_2 \quad \text{H}_2}}{R-C-C-C}\overset{\displaystyle O \quad\quad O}{}OR' \xrightarrow[\text{excess}]{\text{base}} \underset{\substack{\text{H} \quad \text{H}}}{R-C-C-C}\overset{\displaystyle O \quad\quad O}{}OR' \qquad (6.3)
$$

$$
\underset{\substack{\text{H} \quad \text{H}}}{R-C-C-C}OR' + R''X \longrightarrow \underset{\substack{\text{H}}}{R''-C-C-C}OR' \qquad (6.4)
$$

Ion Pairing

Ion pairing of stabilized carbanions with metal cations has been demonstrated by a number of spectroscopic techniques.[11] Methods used include ultraviolet-visible spectrophotometry, infrared and Raman spectroscopy, and nuclear and electron spin resonance spectroscopy, the latter technique being applicable to study of ion pairing in radical ions.

Fluorenyl anion (**9**) absorbs strongly in the ultraviolet between 345 and

9

375 nm, with the location of the maximum in the range determined by the nature of the ion pairing. In tetrahydrofuran the equilibrium between contact and solvent-separated ion pairs can be observed by monitoring their absorption spectra, which have maxima, respectively, at 355 and 373 nm when Na$^+$ is the counter-ion. The contact ion pairs predominate at room temperature, and the solvent-separated ion pairs at $-50°$C.[12] The absorption spectrum is affected also by the size of the cation, as indicated in Table 6.3. Note that the absorption maximum for the solvent-separated ion pair is very close to that of the free ion, indicating a much attenuated influence of the cation on the anion orbitals when separated by one solvent molecule. The large tetrabutylammonium ion even in the contact ion pair has a relatively small effect because of its low electric field intensity; there is a regular increase in the effect as the cation becomes smaller and its field becomes more intense.

The ability of a solvent to promote transformation of contact to solvent-separated ion pairs (Equation 6.5) depends primarily on its ability to solvate the cations. Hence the solvents with the more basic oxygens show larger values of the equilibrium constant K_{ip}.[13] The crown ethers and cryptands are particularly effec-

[11] M. Szwarc, Ed., *Ions and Ion Pairs in Organic Reactions*, Vol. 1, Wiley, New York, 1972.

[12] (a) T. E. Hogen-Esch and J. Smid, *J. Am. Chem. Soc.*, **87**, 669 (1965); **88**, 307, 318 (1966). See also (b) E. Buncel and B. Menon, *J. Org. Chem.*, **44**, 317 (1979).

[13] J. Smid, in *Ions and Ion Pairs in Organic Reactions*, Vol. 1, M. Szwarc, Ed., Wiley, New York, 1972, Chap. 3.

Table 6.3 Dependence of λ_{max} on the Radius of the Cation for 9-Fluorenyl Salts in Tetrahydrofuran at $25°$[a]

Cation	r_c (Å)	λ_{max} (nm)
Li^+	0.60	349
Na^+	0.96	356
K^+	1.33	362
Cs^+	1.66	364
N^+Bu_4	3.5	368
$\|M^+$	~4.5	373
Free ion		374

[a] Data from J. Smid, in *Ions and Ion Pairs in Organic Reactions,* Vol. 1, M. Szwarc, Ed., Wiley, New York, 1972. Reprinted by permission of John Wiley & Sons, Inc.

$$Fl^-Li^+ \rightleftharpoons Fl^- \| Li^+$$

$$K_{ip} = \frac{[Fl^- \| Li^+]}{[Fl^-Li^+]} \tag{6.5}$$

tive at coordinating with cations. The crown ethers can produce either crown-complexed contact pairs, in which the crown ether complexes the cation on the side away from the anion to form an anion–cation–crown ether sandwich, or crown-separated pairs, in which a crown ether molecule lies between the anion and the cation.[14]

Dorfman and co-workers generated carbanions in solution by pulse radiolysis. High-energy electrons in short pulses were beamed into a solution containing an organomercury compound; the electrons react to generate free anions according to Equation 6.6. It was then possible to observe the reactions of the anions, includ-

$$e^- + (\phi CH_2)_2Hg \longrightarrow \phi CH_2^- + \phi CH_2Hg \tag{6.6}$$

ing ion pairing to metal ions present in the solution (Equation 6.7) and proton

$$\phi CH_2^- + Na^+ \longrightarrow \phi CH_2^-Na^+ \tag{6.7}$$

transfer from water or alcohols. Benzyl anion forms pairs with sodium ions with a rate constant of $1.5 \times 10^{11} M^{-1}$ sec^{-1}.[15] Rates of reaction of the anions with proton donors were also determined; rate constants for those reactions are shown in Table

---

[14](a) U. Takaki, T. E. Hogen-Esch, and J. Smid, *J. Am. Chem. Soc.,* **93**, 6760 (1971); (b) J. M. Lehn, *Acc. Chem. Res.,* **11**, 49 (1978). See also (c) E. Kauffmann, J. L. Dye, J.-M. Lehn, and A. I. Popov, *J. Am. Chem. Soc.,* **102**, 2274 (1980).
[15] B. Bockrath and L. M. Dorfman, *J. Am. Chem. Soc.,* **96**, 5708 (1974).

Table 6.4 RATE CONSTANTS FOR THE PROTONATION OF FREE AND OF ION-PAIRED BENZYL CARBANIONS IN TETRAHYDROFURAN AT 24°C[a] (M^{-1} SEC^{-1})

Carbanion	CH_3OH	C_2H_5OH	$(CH_3)_3COH$	H_2O
$PhCH_2^-$	2.3×10^8	1.4×10^8	1.6×10^7	5.3×10^7
$PhCH_2^-,Na^+$	5.8×10^9	3.7×10^9	1.3×10^9	5.5×10^9
$PhCH_2^-,Li^+$	3.4×10^8		9.7×10^7	
$PhCH_2^-,N(C_4H_9)_4^+$	6.0×10^8		4.6×10^8	

[a] Reprinted with permission from L. M. Dorfman, R. J. Sujdak, and B. Bockrath, *Acc. Chem. Res.,* **9**, 352 (1976). Copyright 1976 American Chemical Society.

6.4. The ion pairs are more reactive toward proton donors than the free ions. The sodium ion pair reacts fastest, whereas the lithium pairs, which one might expect to be the tightest and therefore the most different from free ions, actually react at nearly the same rate as the free ions. The sodium ion pairs may be solvent-separated, while the lithium pairs are contact;[16] in the solvent-separated ion pairs the structure may facilitate transfer of the proton to the anion and transfer of the cation to a pairing location with the newly formed hydroxide or alkoxide ion. The contact ion pairs and free ions may be similar in not having the solvation shell ideally structured for proton transfer.[17]

   The dependence of carbanion reactivity on state of ion pairing is a generally observed phenomenon that applies to the nucleophilic reactions as well as to proton transfer. For example, in anionic polymerization of styrene the substituted benzylic-type anions **10**, ion-paired with sodium as the counter-ion in a $10^{-3}M$ solution in tetrahydrofuran, are less reactive by a factor of 810 than are the free ions in adding to the C=C double bond of the next styrene unit (Equation 6.8).[18] In

$$\text{(Equation 6.8)}$$

10

this instance the lithium salt reacts faster than the sodium; in tetrahydrofuran, which is relatively good at solvating small cations, the equilibrium contact $\rightleftharpoons$ solvent-separated ion pair may be shifted in favor of solvent-separated ion pairs when lithium is the counter-ion. In dioxane, which is poorer than tetrahydrofuran at solvating cations, the ion-pair reactivity increases with size of the cation, and the sodium salt is more reactive than the lithium.[19] This is the order one would expect if all the ion pairs are of the contact variety, in which case the larger the cation, the more closely the ion pair resembles the free ion.

[16] L. M. Dorfman, R. J. Sujdak, and B. Bockrath, *Acc. Chem. Res.,* **9**, 352 (1976).
[17] For further discussion of the effect of ion pairing on proton transfer rates, see M. Szwarc, A. Streitwieser, and P. C. Mowery, in *Ions and Ion Pairs in Organic Reactions*, Vol. 2, M. Szwarc, Ed., Wiley, New York, 1974, p. 180.
[18] D. N. Bhattacharyya, C. L. Lee, J. Smid, and M. Szwarc, *J. Phys. Chem.,* **69**, 612 (1965).
[19] T. E. Hogen-Esch, *Adv. Phys. Org. Chem.,* **15**, 153 (1977).

Nucleophilic reactivity of carbanions in substitution reactions is also affected by ion pairing.[20] De Palma and Anett made conductance measurements in dimethylsulfoxide to determine the degree of association of the β-ketoenolates **11** and **12** with alkali metal ions.[21] The dissociation constants at 25°C are between

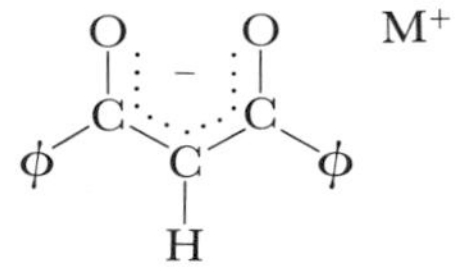

10^{-4} and 10^{-2}; the potassium salt of **11**, for example, has $K_{dis} \approx 2 \times 10^{-3}$, which means that in a $0.1M$ solution less than 1 percent is dissociated. Nevertheless the correlation of reaction rate with degree of dissociation demonstrated that the reaction of **11** with methyl iodide, a nucleophilic substitution leading to C methylation (Equation 6.9), occurs through the free ions.

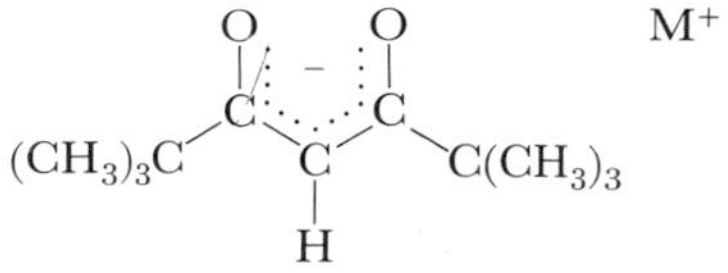

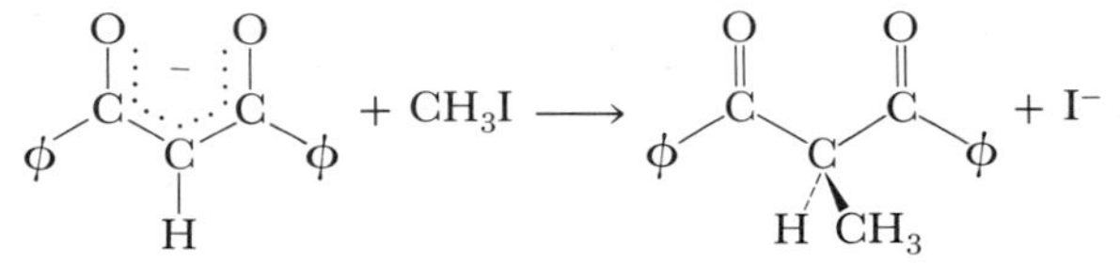

$$(6.9)$$

De Palma and Arnett also found that proton and ^{13}C magnetic resonance chemical shifts δ (downfield from TMS) vary inversely with the size of the cation, as do the frequencies of certain infrared absorption bands. These results indicate that the ion pairs are of the contact variety.

Addition of crown ether or cryptand cation-complexing agent increases rate of alkylation of β-ketoenolate **13** with ethyl tosylate by factors of 15 (crown ether) to 235 (cryptand) and also increases the proportion of O alkylation from 9 percent to 50 percent (crown ether) or 84 percent (cryptand).[22]

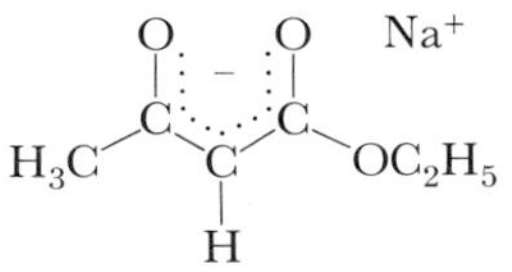

When carbanions are not stabilized by a conjugating group, they form polar covalent bonds to metals. The alkyllithium and alkylmagnesium organometallic compounds are commonly encountered examples. Organomagnesium compounds are usually prepared as Grignard reagents in ether solutions that also contain

[20](a) L. M. Jackman and B. C. Lange, *Tetrahedron*, **33**, 2737 (1977); (b) O. A. Reutov and A. L. Kurts, *Russ. Chem. Rev.*, **46**, 1040 (1977).

[21]V. M. DePalma and E. M. Arnett, *J. Am. Chem. Soc.*, **100**, 3514 (1978). For further examples see note 19.

[22]C. Cambillau, P. Sarthou, and G. Bram, *Tetrahedron Lett.*, 281 (1976).

magnesium halides.[23] The composition of these solutions is quite complex and depends on the solvent and on concentration. The species $RMgX$, R_2Mg, and MgX_2, where X is Cl, Br, or I, may all be present and in equilibrium according to Equation 6.10. In addition, these species may aggregate by bridge bonding, for

$$R_2Mg + MgX_2 \rightleftharpoons 2RMgX \qquad (6.10)$$

example, in structures like **14** and **15** or in longer-chain structures of the same type.

$$
\begin{array}{cc}
\text{R—Mg} \underset{X}{\overset{X}{<>}} \text{MgX} & \text{R—Mg} \underset{X}{\overset{R}{<>}} \text{MgX} \\
\textbf{14} & \textbf{15}
\end{array}
$$

In tetrahydrofuran this aggregation does not occur extensively, but the equilibrium constant K for Equation 6.10 is on the order of 1 to 10 so all three species are present in appreciable concentration. In diethyl ether K is 200 to 600; the monomeric reagent is therefore mainly $RMgX$, but aggregation is extensive.[24]

Organolithium compounds associate into tetramers or hexamers in solution; the solids retain the structure. X-Ray diffraction study shows that methyllithium has the structure **16**, a lithium tetrahedron with a methyl group centered over each face.[25]

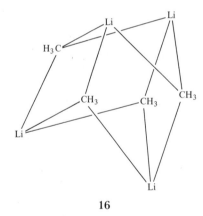

16

[23]Gringard reagents are probably generated by electron transfer or halogen atom abstraction at the magnesium surface. Free carbanion intermediates are unlikely. (a) H. R. Rogers, C. L. Hill, Y. Fujiwara, R. J. Rogers, H. L. Mitchell, and G. M. Whitesides, *J. Am. Chem. Soc.*, **102**, 217 (1980); (b) H. R. Rogers, J. Deutch, and G. M. Whitesides, *J. Am. Chem. Soc.*, **102**, 226 (1980); (c) H. R. Rogers, R. J. Rogers, H. L. Mitchell, and G. M. Whitesides, *J. Am. Chem. Soc.*, **102**, 231 (1980); (d) J. J. Barber and G. M. Whitesides, *J. Am. Chem. Soc.*, **102**, 239 (1980).
[24](a) E. C. Ashby, *Q. Rev., Chem. Soc.*, **21**, 259 (1967); (b) E. C. Ashby, J. Laemmle, and H. M. Neumann, *Acc. Chem. Res.*, **7**, 272 (1974).
[25](a) E. Weiss and E. A. C. Lucken, *J. Organomet. Chem.*, **2**, 197 (1964). For further information on alkyllithium compounds, see (b) B. J. Wakefield, *The Chemistry of Organolithium Compounds*, Pergamon Press, Elmsford, N.Y., (1974); (c) T. L. Brown, *Acc. Chem. Res.*, **1**, 23 (1968).

Stereochemistry

A carbanion that is not stabilized by an electron-withdrawing conjugating group should have a pyramidal geometry according to the VSEPR theory. The pyramidal structures would be expected to invert (Equation 6.11) as the isoelectronic

$$\ddot{C} \rightleftharpoons \dot{C} \tag{6.11}$$

ammonia and amines do.[26] The rates of inversions of this type, which have been studied most extensively with amines, are usually high but vary over a considerable range depending on the nature of the substituents and on the electronegativity of the central atom. Electronegative substituents and π donor substituents raise the inversion barrier. Because the halogens are more electronegative than hydrogen and also have unshared pairs of electrons, halogen-substituted carbanions would be expected to have inversion rates at the low end of the range. Equation 6.12 shows a

$$H_3C-\overset{\overset{\displaystyle O}{\|}}{C}-\overset{\overset{\displaystyle F}{|}}{\underset{\overset{|}{Br}}{C}}-Cl \xrightarrow{OH^-} H_3C-\overset{\overset{\displaystyle O}{\|}}{C}-OH + :\overset{\overset{\displaystyle -}{}}{\underset{\overset{|}{Br}}{C}}\overset{F}{<}-Cl \xrightarrow{H_2O} HCFClBr \tag{6.12}$$

$$[\alpha]_D = +0.39 \qquad\qquad\qquad\qquad\qquad\qquad\qquad [\alpha]_D = +0.25$$

reaction in which a chiral carbanion of this type is apparently trapped before it has time to racemize.[27] This experiment was carried out in 9 molar aqueous potassium hydroxide solution; under these conditions there are protons readily available and the carbanion lifetime is short. The product was formed under kinetic control; optical activity was lost if the product was allowed to remain in the basic solution where the reverse of the final protonation step could occur. The experiment thus demonstrated at best only very short-term configurational stability of the carbanion. It should also be noted that the relative configurations of reactant and product were not determined. It is therefore not clear whether the configuration was retained or inverted.

Placing the anionic center in a small ring gives inversion rate low enough for the process to be observable by dynamic nuclear magnetic resonance spectroscopy, because the increase in strain associated with the planar transition state for inversion (I strain[28]) raises the barrier. The carbanion **17** has an inversion barrier of 17 to 18 kcal mole^{-1};[29] the nitrogen analog **18** has a barrier of 10 kcal mole^{-1}.[30] In

[26]Reviews: (a) H. Kessler, *Angew. Chem. Int. Ed. Engl.*, **9**, 219 (1970); (b) A. Rauk, L. C. Allen, and K. Mislow, *Angew. Chem. Int. Ed. Engl.*, **9**, 400 (1970); (c) J. B. Lambert, *Top. Stereochem.*, **6**, 19 (1971).

[27]M. K. Hargreaves and B. Modarai, *J. Chem. Soc. C*, 1013 (1971).

[28](a) H. C. Brown and M. Gerstein, *J. Am. Chem. Soc.*, **72**, 2926 (1950); (b) H. C. Brown, R. S. Fletcher, and R. B. Johannesen, *J. Am. Chem. Soc.*, **73**, 212 (1951).

[29](a) A. Ratajczak, F. A. L. Anet, and D. J. Cram, *J. Am. Chem. Soc.*, **89**, 2072 (1967). See also (b) H. M. Walborsky and J. M. Motes, *J. Am. Chem. Soc.*, **92**, 2445 (1970); (c) J. M. Motes and H. M. Walborsky, *J. Am. Chem. Soc.*, **92**, 3697 (1970); (d) H. M. Walborsky and L. M. Turner, *J. Am. Chem. Soc.*, **94**, 2273 (1972).

[30]F. A. L. Anet, R. D. Trepka, and D. J. Cram, *J. Am. Chem. Soc.*, **89**, 357 (1967).

contrast, ammonia has a barrier of only 5.8 kcal mole^{-1} and methylamine a barrier of 4.8 kcal mole^{-1},[31] whereas NF_3 has a barrier of 50 kcal mole^{-1}.[32]

It is difficult to obtain stereochemical information on the simple unsubstituted carbanions because they exist primarily as organometallic compounds rather than as free ions. Theoretical calculations predict a barrier of only about 1.5 kcal mole^{-1} for free $H_3C:^-$;[33] a gas-phase observation consistent with a pyramidal geometry has been made, but detailed experimental structural information is not yet available.[34]

The more labile of the organometallic compounds, the alkali metal and magnesium derivatives, do not maintain chirality unless special structural features are present. For example, attempts to prepare optically active Grignard reagents have been unsuccessful, except in the case of a cyclopropyl derivative such as **19**.[35] The lithio- and even sodio-derivatives of the cyclopropyl system also show sufficient stereochemical stability to give optically active products.[36] The highly covalent organomercurials, on the other hand, are readily prepared in optically active form without any special structural requirements.

When carbanions are stabilized by π electron acceptor groups (phenyl, carbonyl, nitro, cyano), one expects the ions to be planar, or to have very low inversion barriers. This expectation does seem to be fulfilled in practice, although, as we shall see further below, the stereochemical outcome of formation and protonation of anions of this type is strongly influenced by the solvent and the cation. The 7-phenylnorbornyl anion (**20**) is planar with K^+ or Cs^+ as counter ion, but with Li^+ it is pyramidal. The inversion of the lithium salt has $\Delta H^{\ddagger} = 6.7$ kcal mole^{-1}, $\Delta S^{\ddagger} = -14$ cal mole^{-1} K^{-1}.[37]

[31] See note 26(b).

[32] See note 26(c).

[33] See, for example, C. E. Dykstra, M. Hereld, R. R. Lucchese, H. F. Schaefer III, and W. Meyer, *J. Chem. Phys.*, **67**, 4071 (1977). For other references see note 34.

[34] G. B. Ellison, P. C. Engelking, and W. C. Lineberger, *J. Am. Chem. Soc.*, **100**, 2556 (1978).

[35] H. M. Walborsky and A. E. Young, *J. Am. Chem. Soc.*, **86**, 3288 (1964).

[36] (a) H. M. Walborsky, F. J. Impastato, and A. E. Young, *J. Am. Chem. Soc.*, **86**, 3283 (1964); (b) J. B. Pierce and H. M. Walborsky, *J. Org. Chem.*, **33**, 1962 (1968).

[37] R. R. Peoples and J. B. Grutzner, *J. Am. Chem. Soc.*, **102**, 4709 (1980).

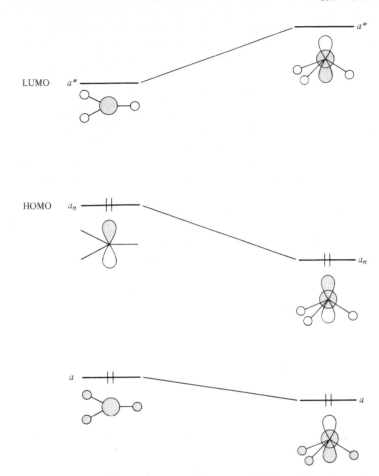

Figure 6.2 Orbitals and energy changes important in pyramidal inversion of a molecule XH_3. Four other orbitals of a different symmetry whose energy changes are important but not controlling are not shown. In NH_3 and CH_3^-, a_n is the HOMO and a^* is the LUMO. The planar structure is at the left. The dominant interaction that occurs on distorting the structure toward the pyramidal geometry at the right is that between a_n and a^*.

The effect of various types of substituents on the height of the barrier to pyramidal inversion in carbanions can be understood in molecular orbital terms, according to an analysis by Levin and by Epiotis.[38] Figure 6.2 illustrates three of the molecular orbitals appropriate to describing a molecule XH_3. These three orbitals, labeled a, a_n, and a^*, are the ones primarily responsible for the energy changes that occur as the pyramidal molecule is inverted through a planar structure. They are all of the same symmetry (designated symmetry type a) with respect to the threefold rotation axis; four other orbitals of different symmetries, whose energy changes are not controlling in the process are not shown. (The planar

[38](a) C. C. Levin, *J. Am. Chem. Soc.*, **97**, 5649 (1975); (b) W. Cherry and N. Epiotis, *J. Am. Chem. Soc.*, **98**, 1135 (1976).

structure has a symmetry element, the mirror plane that includes all four nuclei, that is not used in the analysis because it disappears as soon as the molecule begins to move toward the pyramidal structure.) For purposes of the analysis it is easiest to begin with the planar structure (at the left in Figure 6.2) and look at changes in the orbitals as the hydrogens are moved out of the plane. As this process occurs the three orbitals interact with each other. The strongest interaction is between the a_n and a^* orbitals, which are, respectively, the HOMO and the LUMO for NH_3, CH_3^-, and molecules isoelectronic with them. There is a smaller interaction between a and a_n. Note that the forms of the orbitals change according to the rules of the perturbation theory (Section 1.6, p. 49), with a_n gaining a bonding contribution from the central p orbital, a^* gaining an antibonding contribution from the central p orbital, and a_n gaining contributions from both a^* and a, of which that from a^* (bonding) dominates. Note also that the figure shows the molecular orbitals in terms of their constituent unhybridized atomic orbitals. The illustrations on the right could be expressed just as well in terms of hybrids, in which case the pyramidal a_n would consist primarily of an approximately sp^3 hybrid lone-pair orbital on the central atom, with small bonding contributions from the hydrogen orbitals, as illustrated in **21**. Because the greater stability of the pyramidal over the planar

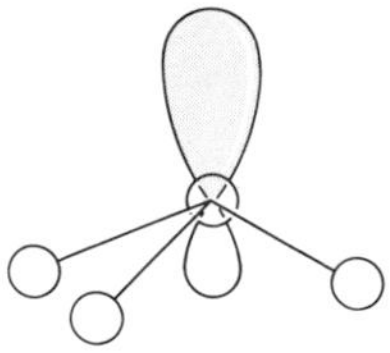

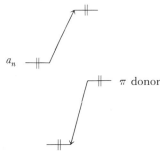

21

structure arises primarily from the energy decrease of a_n in going from left to right in Figure 6.2, and because this decrease in turn derives from interaction with a^*, any substituent change that tends to prevent energy decrease of a_n or to diminish interaction of a_n and a^* will favor the planar structure relative to the pyramidal and will lower the inversion barrier. Conversely, any change that lowers a_n further or that enhances interaction of a_n and a^* will favor the pyramidal geometry and raise the barrier.

A substituent that has π orbitals can interact through those π orbitals with a_n but, because of symmetry, not with a^*. If the substituent is a π donor, its effect on a_n in the planar structure is as follows:

A filled–filled interaction is somewhat destabilizing; therefore the π donor will raise the energy of the planar structure. Furthermore, the interaction of a_n and a^* is enhanced because the energy separation is now smaller. Change toward the pyramidal structure then will give a large stabilization from the a_n–a^* interaction and at the same time will diminish the unfavorable a_n–π donor interaction by reducing the p character and increasing the s character of a_n. Therefore a π donor should favor the pyramidal structure and increase the inversion barrier.

A π acceptor interacts with a_n as follows:

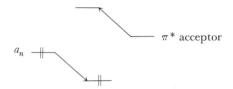

Now the interaction of a_n with a^* is diminished by the larger energy separation, and at the same time a strongly stabilizing filled–unfilled interaction is introduced, which is maximum for maximum p character of a_n. Hence π acceptor substituents strongly favor the planar geometry and lower the inversion barrier.[39]

The effect of electronegativity of the substituents can be understood in terms of the influence on the a^* orbital. Inductive electron withdrawal in the planar geometry will affect primarily the σ orbitals a and a^*, because the substituent σ orbitals interact with these and not with the π symmetry orbital a_n. Therefore withdrawal of electrons lowers the energy of a and a^* more than it lowers that of a_n. If a^* is lowered more than a_n, they come closer together in energy, their interaction is stronger, the change to a pyramidal structure is more strongly favored, and the inversion barrier is raised.[40] The converse should hold for inductive electron donation. These ideas are in accord with the observed high inversion barriers for halogenated amines, and with low barriers for carbanions substituted by π acceptor groups.

Stereochemistry of stabilized carbanions has been studied extensively by Cram and co-workers.[41] They observed the stereochemical result of base-catalyzed exchange of H for D at activated C—H bonds at a chiral center using the ratio of rate of exchange (k_e) to rate of loss of optical activity (k_α). The results were depen-

[39] The π acceptor argument puts into molecular orbital terms the familiar resonance argument for planarity in such systems, indicated by the structures

$$\overset{:\ddot{O}:}{\underset{}{\diagdown}}C-\overset{..}{C}\diagup\ \longleftrightarrow\ \overset{:\ddot{O}:^-}{\underset{}{\diagdown}}C=C\diagup$$

[40] Electronegativity of the central atom also affects the barrier, but in the opposite sense. Changing to a more electronegative central atom has the same effect on the orbitals as does changing to a less electronegative substituent. See note 38(a).

[41] Reviews: (a) D. J. Cram, *Fundamentals of Carbanion Chemistry*, Academic Press, New York, 1965, Chaps. III and IV; (b) E. Buncel, *Carbanions: Mechanistic and Isotopic Aspects*, Elsevier, Amsterdam, 1975, p. 54; (c) M. Szwarc, A. Streitwieser, and P. C. Mowery, in *Ions and Ion Pairs in Organic Reactions*, Vol. 2, M. Szwarc, Ed., Wiley, New York, 1974, p. 207; (d) T. E. Hogen-Esch, *Adv. Phys. Org. Chem.*, **15**, 153 (1977), p. 232.

dent on ion pairing, ranging all the way from high net retention of configuration ($k_e/k_\alpha > 1$) through racemization ($k_e/k_\alpha = 1$) and inversion ($k_e/k_\alpha = 0.5$) to a process Cram called *isoracemization,* in which racemization occurred without any exchange ($k_e/k_\alpha = 0$).

In the deuterated 9-methylfluorenyl system (**22**), Cram and co-workers found retention of configuration ($k_e/k_\alpha = 148$) in tetrahydrofuran with ammonia or a primary amine as base.[42] Streitwieser obtained similar results with the benzyl system in cyclohexylamine with cyclohexylamide as base.[43] Cram's proposed mechanism is shown in Scheme 3. Note that in these low-dielectric solvents the carbanion is closely paired with the ammonium cation, so that the reprotonation occurs from the same ammonium ion that was formed in the original deprotonation and from the same side.

SCHEME 3

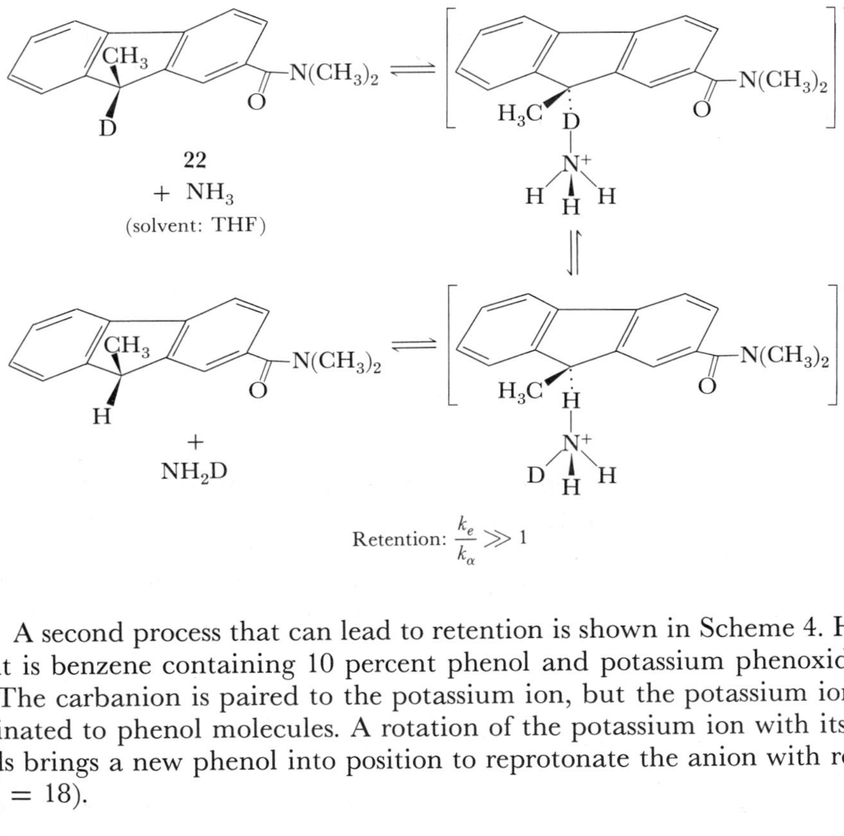

Retention: $\dfrac{k_e}{k_\alpha} \gg 1$

A second process that can lead to retention is shown in Scheme 4. Here the solvent is benzene containing 10 percent phenol and potassium phenoxide is the base. The carbanion is paired to the potassium ion, but the potassium ion is also coordinated to phenol molecules. A rotation of the potassium ion with its phenol ligands brings a new phenol into position to reprotonate the anion with retention ($k_e/k_\alpha = 18$).

[42](a) D. J. Cram and L. Gosser, *J. Am. Chem. Soc.,* **85**, 3890 (1963); (b) D. J. Cram and L. Gosser, *J. Am. Chem. Soc.,* **86**, 5445 (1964).
[43]A. Streitwieser, Jr., and J. H. Hammons, *Prog. Phys. Org. Chem.,* **3**, 41 (1965).

SCHEME 4

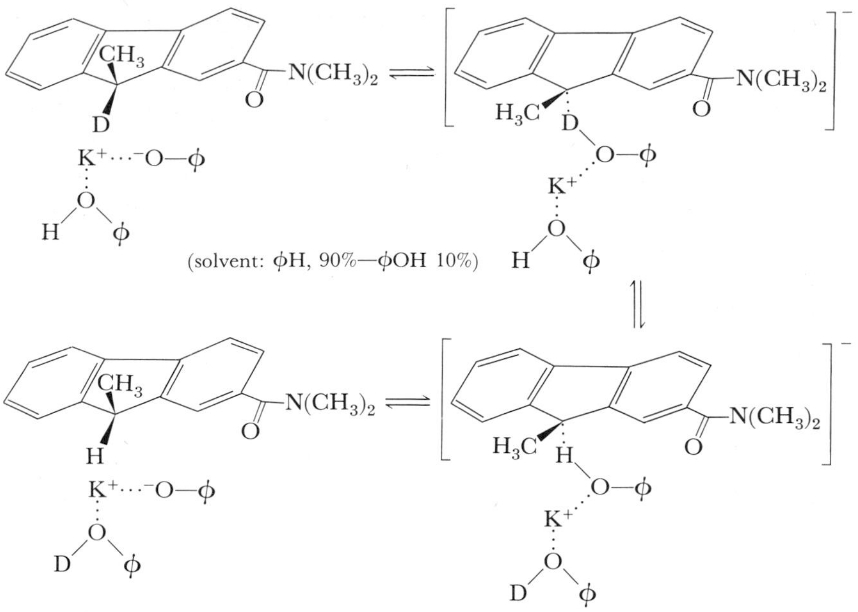

(solvent: φH, 90%—φOH 10%)

Retention: $\dfrac{k_e}{k_\alpha} \gg 1$

In dimethylsulfoxide exchange and racemization rates are equal. The more polar solvent presumably allows the ion pairs to dissociate; the intrinsically symmetric carbanion then enters a symmetric solvation environment and is protonated at equal rates from either side.[43]

In a protic solvent such as methanol, there are protons readily available from the medium. As the proton departs from one side, it can be replaced concertedly from the back. Predominant operation of this mechanism (Scheme 5) leads to an excess of inversion. For example, exchange of nitrile **23** in methanol with *tri-n*-propylamine as base leads to $k_e/k_\alpha = 0.84$.[44]

Finally, when the carbanion forms in a solvent of low polarity and low proton concentration, and with a base that has no exchangeable hydrogens, the initially formed ion pair collapses back to base and substrate without exchange. The anion will sometimes turn over in the ion pair to yield the unexchanged enantiomer upon reprotonation. This is the process called isoracemization (Scheme 6). It occurs, for example, with nitrile **23** in tetrahydrofuran containing $1.5M$ *t*-butyl alcohol and *tri-n*-propylamine as base ($k_e/k_\alpha = 0.05$).[44] Cram has also identified a second process that leads to $k_e/k_\alpha < 1$ in carbanions that have a substituent group capable of forming a tautomeric structure or having a high electron

[44] D. J. Cram and L. Gosser, *J. Am. Chem. Soc.*, **86**, 5457 (1964).

SCHEME 5

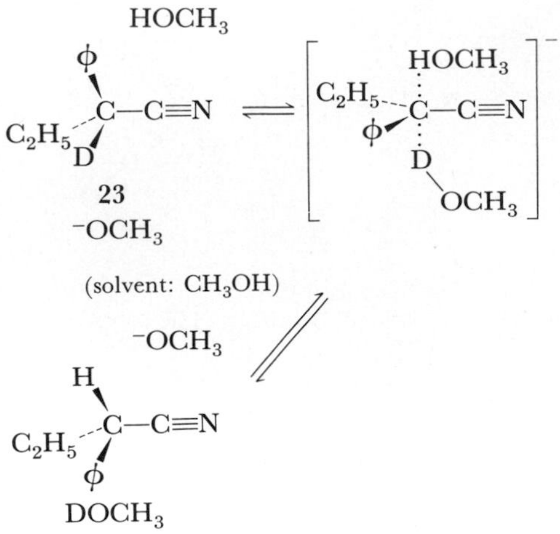

$$HOCH_3$$

$$\phi$$
$$C_2H_5 \overset{\displaystyle |}{\underset{\displaystyle D}{C}} -C \equiv N \quad \rightleftharpoons \quad \left[C_2H_5 \overset{HOCH_3}{\underset{\displaystyle D}{\underset{\displaystyle \phi}{C}}} -C \equiv N \atop OCH_3 \right]^{-}$$

23

$$^{-}OCH_3$$

(solvent: CH$_3$OH)

$$^{-}OCH_3$$

$$H$$
$$C_2H_5 \overset{\displaystyle |}{\underset{\displaystyle \phi}{C}} -C \equiv N$$

$$DOCH_3$$

Inversion: $\dfrac{k_e}{k_\alpha} \approx 0.5$

SCHEME 6

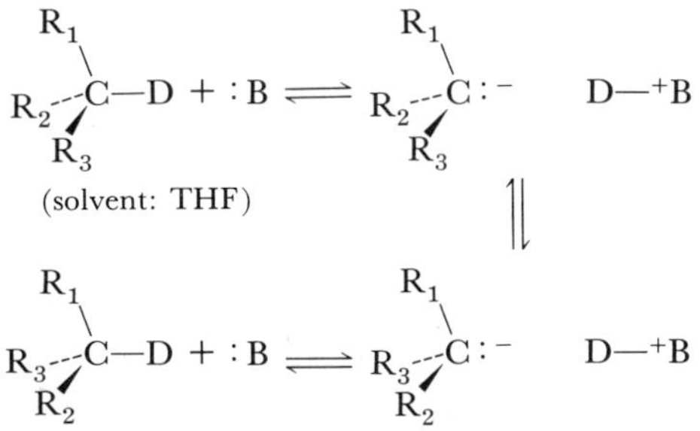

$$R_2 \overset{R_1}{\underset{R_3}{C}} -D + :B \rightleftharpoons R_2 \overset{R_1}{\underset{R_3}{C}}:^{-} \quad D-^{+}B$$

(solvent: THF)

$$R_3 \overset{R_1}{\underset{R_2}{C}} -D + :B \rightleftharpoons R_3 \overset{R_1}{\underset{R_2}{C}}:^{-} \quad D-^{+}B$$

Isoracemization: $\dfrac{k_e}{k_\alpha} \approx 0$

density. The initially formed dissymmetric ion pair **24**, for example, is easily transformed into the symmetric ion pair **25**. Protonation with equal probability through the original **24** or through the enantiomer **26** will lead to racemization without exchange.[45] This process is called the *conducted tour* mechanism. Compound **24**, for example, shows $k_e/k_\alpha = 0.10$ at 25°C in tetrahydrofuran—16 percent *t*-butyl alcohol with 10 mole percent *tri-n*-propylamine as base. These various mechanisms can, of course, operate in competition in any given system, leading to values of k_e/k_α intermediate between those expected for the various mechanisms operating in isolation.

[45](a) W. T. Ford and D. J. Cram, *J. Am. Chem. Soc.*, **90**, 2606 (1968); (b) K. C. Chu and D. J. Cram, *J. Am. Chem. Soc.*, **94**, 3521 (1972).

SCHEME 7

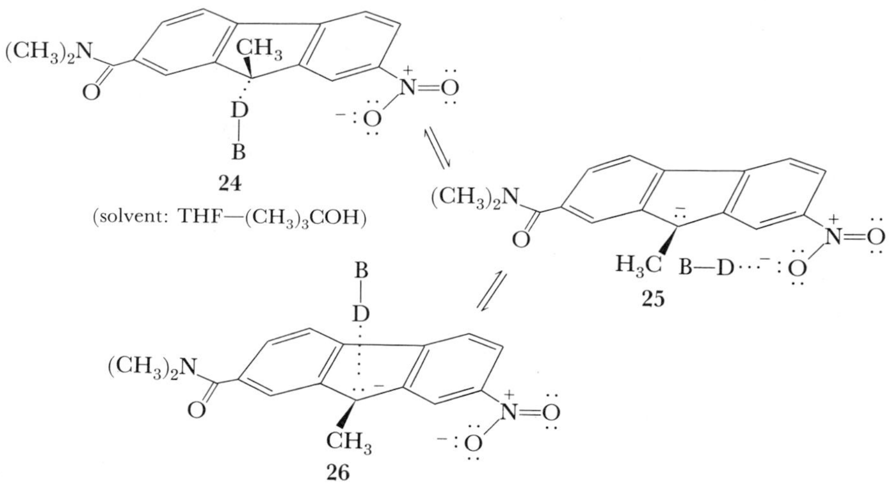

24

(solvent: THF—(CH₃)₃COH)

25

26

Conducted tour: $k_e/k_\alpha \approx 0$

Rearrangements of Carbanions

Carbanions are much less prone to rearrangement than are carbocations. The migration of an alkyl group to an adjacent electron-rich center leaving its bonding electrons behind does not occur at all readily; the explanation for this difference of behavior will be reserved for Chapter 11. Also covered in Chapter 11 are rearrangements involving extended π systems, which come under the category of pericyclic reactions.

Except for rearrangements of the pericyclic type, carbanion rearrangements usually occur stepwise either through intramolecular addition of a carbanion to a neighboring unsaturated group or by nucleophilic attack of the carbanion at a neighboring substitution site.[46]

Equation 6.13 shows an example of an anionic rearrangement in which a phenyl group migrates from an adjacent center, leaving a more stable anion.[47,48]

$$\phi_3C-CH_2Cl + Li \xrightarrow{-30°C} \phi_3C-CH_2-Li \xrightarrow{0°C} \overset{Li^+}{\phi_2\overset{..}{C}=CH_2\phi} \qquad (6.13)^{47}$$

The potassium derivative, being more ionic, rearranges on formation at $-66°C$:

[46]Reviews: (a) H. E. Zimmerman, in *Molecular Rearrangements,* Part 1, P. de Mayo, Ed., Wiley, New York, 1963, p. 345; (b) D. J. Cram, *Fundamentals of Carbanion Chemistry,* Academic Press, New York, 1965, Chap. VI; (c) E. Buncel, *Carbanions: Mechanistic and Isotopic Aspects,* Elsevier, Amsterdam, 1975.

[47]E. Grovenstein, Jr., and L. P. Williams, Jr., *J. Am. Chem. Soc.,* **83**, 412, 2537 (1961).

[48]H. E. Zimmerman and A. Zweig, *J. Am. Chem. Soc.,* **83**, 1196 (1961).

$$\phi_3C-CH_2Cl + K \xrightarrow{-66°C} [\phi_3C-\overset{..}{\overset{..}{C}}H_2K^+] \xrightarrow{-66°C} \phi_2\overset{\overset{K^+}{|}}{C}\overset{..}{=}CH_2\phi \qquad (6.14)^{47}$$

Zimmerman showed that when there is a choice between a phenyl and a tolyl group, the phenyl migrates in preference:

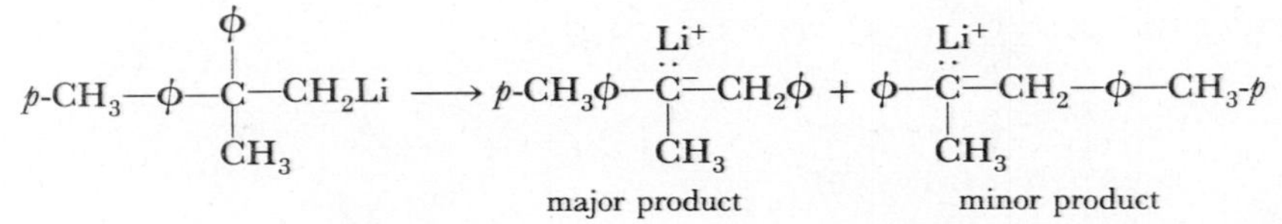

$$p-CH_3-\phi-\overset{\overset{\displaystyle\phi}{|}}{\underset{\underset{\displaystyle CH_3}{|}}{C}}-CH_2Li \longrightarrow p-CH_3\phi-\overset{\overset{\displaystyle Li^+}{\overset{..}{|}}}{\underset{\underset{\displaystyle CH_3}{|}}{C}}{\overset{..}{=}}CH_2\phi + \phi-\overset{\overset{\displaystyle Li^+}{\overset{..}{|}}}{\underset{\underset{\displaystyle CH_3}{|}}{C}}{\overset{..}{=}}CH_2-\phi-CH_3\text{-}p \qquad (6.15)^{48}$$

major product   minor product

This observation is consistent with a mechanism in which the migrating aryl group interacts with the carbanion center through its π system to give a transition state (or perhaps an intermediate) structure **27** or **28**. The phenyl migration **27** would be

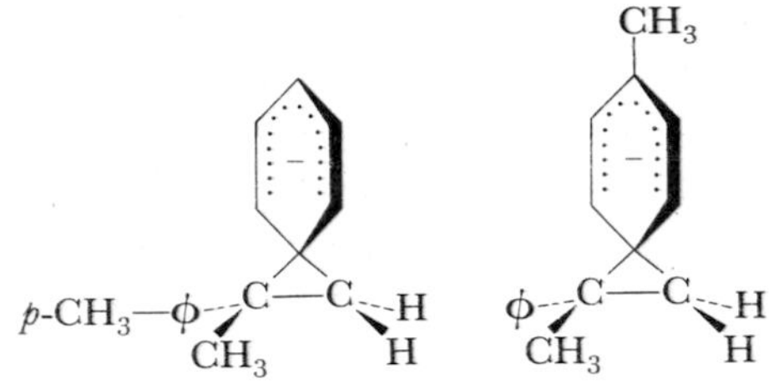

27   **28**

favored because in **28** the electron-donating methyl group would destabilize the system. (The transition state **28**, or the transition state leading to it if it is an intermediate, may also be less well solvated than **27**.) This mechanism, and the possibility that **27** and **28** might be intermediates, finds support in the observation of **29** in solution at low temperature.[49] This rearrangement finds close analogy in

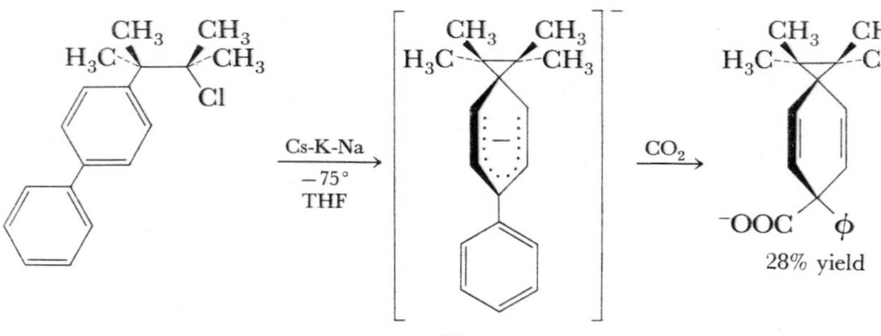

29

the phenonium ion mechanism for phenyl migration in carbocations (Section 5.1).

    Another carbanion rearrangement is the Favorskii rearrangement of α-haloketones.[50] A typical example is shown in Equation 6.16. If the ketone is not

[49] J. A. Bertrand, E. Grovenstein, Jr., P. Lu, and D. VanDerveer, *J. Am. Chem. Soc.*, **98**, 7835 (1976).
[50] Reviews: (a) A. S. Kende, *Org. Reactions*, **11**, 261 (1960); (b) A. A. Akhrem, T. K. Ustynyuk, and Yu. A. Titov, *Russ. Chem. Rev.*, **39**, 732 (1970); (c) F. G. Bordwell, *Acc. Chem. Res.*, **3**, 281 (1970).

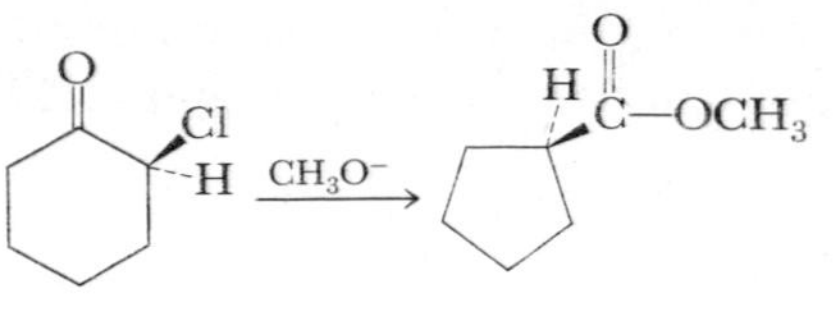

$$(6.16)$$

symmetrical, two products are obtained:

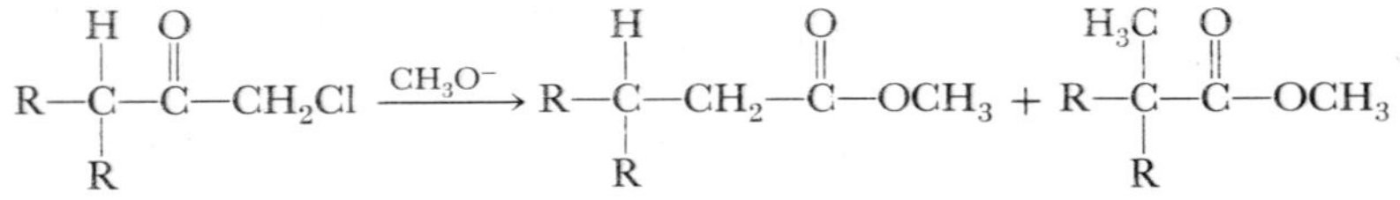

$$R-\underset{\underset{R}{|}}{\overset{\overset{H}{|}}{C}}-\overset{\overset{O}{||}}{C}-CH_2Cl \xrightarrow{CH_3O^-} R-\underset{\underset{R}{|}}{\overset{\overset{H}{|}}{C}}-CH_2-\overset{\overset{O}{||}}{C}-OCH_3 + R-\underset{\underset{R}{|}}{\overset{\overset{H_3C}{|}}{C}}-\overset{\overset{O}{||}}{C}-OCH_3 \qquad (6.17)$$

The mechanism of the rearrangement is shown in Scheme 8. It was established by Loftfield by means of the labeling experiment indicated, where the positions marked are labeled with ^{14}C.[51] The reaction is initiated by formation of an enolate ion, which carries out an intramolecular nucleophilic substitution to form a cyclopropanone. Cyclopropanones are extremely reactive, presumably because of the large strain associated with a trigonal carbon in a three-membered ring, and are very susceptible to nucleophilic attack which relieves some of the strain by returning the oxygenated carbon to sp^3 hybridization. The three-membered ring then opens to give a transient carbanion that is protonated by the solvent.

SCHEME 8

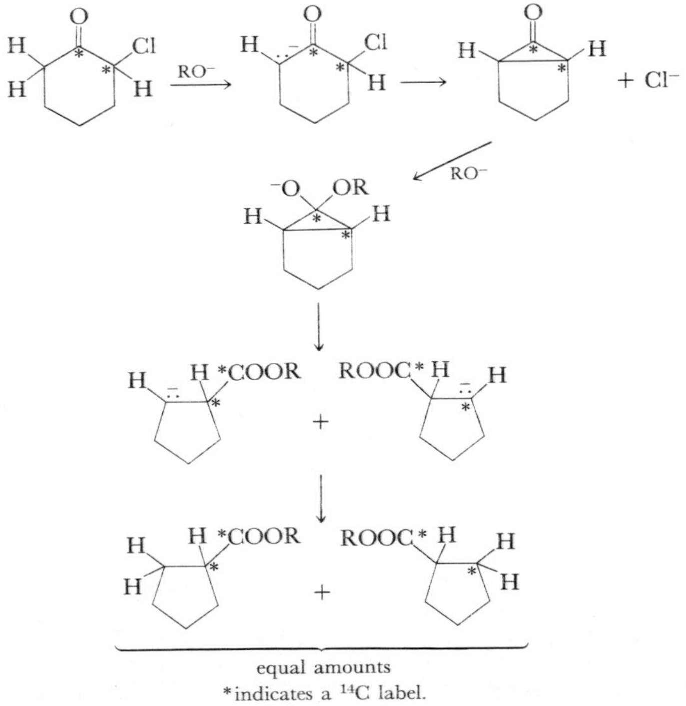

equal amounts
*indicates a ^{14}C label.

[51] R. B. Loftfield, *J. Am. Chem. Soc.*, **72**, 632 (1950); **73**, 4707 (1951).

Two rearrangements that involve migration of groups originally bonded to oxygen or nitrogen are shown in Equations 6.18 and 6.19.[52] There is some evidence that the Stevens rearrangement may in some cases actually be a radical rather than a carbanion reaction.[53]

$$\phi\text{CH}\underset{\underset{\text{Li}}{|}}{-}\text{O}-\text{CH}_3 \longrightarrow \phi\text{CH}\underset{\underset{\text{CH}_3}{|}}{-}\text{OLi} \xrightarrow{\text{H}_2\text{O}} \phi\text{CH}\underset{\underset{\text{CH}_3}{|}}{-}\text{OH} \tag{6.18}$$

Wittig rearrangement

$$\phi-\overset{\overset{\text{O}}{\|}}{\text{C}}-\text{CH}_2\underset{\underset{\text{CH}_3}{|}}{\overset{\overset{\text{CH}_2\phi}{|}}{\overset{+}{\text{N}}}}-\text{CH}_3 \xrightarrow{\text{OH}^-} \left[\phi-\overset{\overset{\text{O}}{\|}}{\text{C}}-\overset{..}{\text{CH}}\underset{\underset{\text{CH}_3}{|}}{\overset{\overset{\text{CH}_2\phi}{|}}{\overset{+}{\text{N}}}}-\text{CH}_3\right] \longrightarrow \phi-\overset{\overset{\text{O}}{\|}}{\text{C}}-\overset{\overset{\text{CH}_2\phi}{|}}{\text{CH}}-\text{N(CH}_3)_2 \tag{6.19}$$

Stevens rearrangement

6.2 ELECTROPHILIC ALIPHATIC SUBSTITUTION

Electrophilic aliphatic substitution reactions are formally related to carbanions in the same way that nucleophilic aliphatic substitution reactions are related to carbocations.[54] The substituting group is a Lewis acid or electrophile, therefore electron-deficient, and the leaving group is likewise a Lewis acid. The pair of electrons that is initially bonding the leaving group to the substrate carbon remains with the substrate and is used to form the new bond to the entering electrophile. Hence a unimolecular reaction, S_E1, would generate a carbanion as an intermediate (Equation 6.20), just as an S_N1 reaction generates a carbocation. A bimolecular reaction, S_E2, will involve direct displacement of leaving by entering group (Equation 6.21).

$$\text{R}-\text{E}_1 \xrightarrow{\text{slow}} \text{E}_1^+ + \text{R}:^- \xrightarrow[\text{fast}]{\text{E}_2^+} \text{R}-\text{E}_2 \tag{6.20}$$

$$\text{E}_2^+ + \text{R}-\text{E}_1 \longrightarrow \text{R}-\text{E}_2 + \text{E}_1^+ \tag{6.21}$$

Electrophilic substitutions can occur when the leaving group is more electropositive than carbon; hence the common leaving groups are the metals and hydrogen. Reactions involving free carbanions, such as the hydrogen–deuterium exchanges considered earlier, can be regarded as S_E1 substitutions. Other similar processes, such as decarboxylation (Equation 6.22) and anionic fragmentation (Equation 6.23), are also S_E1 substitutions.

[52] H. E. Zimmerman, in *Molecular Rearrangements,* Part 1, P. de Mayo, Ed., Wiley, New York, 1963, pp. 372*ff.*
[53] See Section 9.6.
[54] Reviews: (a) F. R. Jensen and B. Rickborn, *Electrophilic Substitution of Organomercurials,* McGraw-Hill, New York, 1968; (b) O. A. Reutov, *Pure Appl. Chem.,* **17,** 79 (1968); (c) D. S. Matteson, *Organomet. Chem. Rev. A.,* **4,** 263 (1969); (d) C. K. Ingold, *Structure and Mechanism in Organic Chemistry,* 2d ed., Cornell University Press, Ithaca, N.Y., 1969, pp. 563–584; (e) G. A. Olah, *Friedel–Crafts Chemistry,* Wiley, New York, 1973, pp. 526*ff*; (f) C. H. Bamford and C. F. H. Tipper, Eds., *Electrophilic Substitution at a Saturated Carbon Atom (Comprehensive Chemical Kinetics,* Vol. 12), Elsevier, Amsterdam, 1973.

$$G-CH_2-\overset{\overset{\displaystyle O}{\|}}{C}-O^- \longrightarrow CO_2 + G\overset{..}{C}H_2 \xrightarrow{ROH} GCH_3 + RO^- \tag{6.22}$$

$$G-CH_2-\overset{\overset{\displaystyle O^-}{|}}{\underset{\underset{\displaystyle R_1}{|}}{C}}-R_2 \longrightarrow R_1-\overset{\overset{\displaystyle O}{\|}}{C}-R_2 + G\overset{..}{C}H_2 \xrightarrow{ROH} GCH_3 + RO^- \tag{6.23}$$

$$(G = R-\overset{\overset{\displaystyle O}{\|}}{C}-; \ Ar-; \ O_2N-; \ N\equiv C-; \ X_3C-)$$

We shall be concerned here mainly with S_E2 substitutions. These reactions occur, for example, with organometallics that have a considerable degree of covalent character or with C—H bonds under conditions not sufficiently basic for proton removal. S_E2 reactions have been quite extensively studied for organomercury compounds. They have also been observed with a number of other metals as leaving group including lithium,[55] tin,[56] chromium,[57] iron,[58] and cobalt,[59] as well as hydrogen.[60]

Stereochemistry

Stereochemistry of S_E2 reactions was first investigated in organomercury compounds, which are easy to prepare in optically active form because they have essentially covalent carbon–mercury bonds. These compounds undergo exchange reactions in which both the leaving and entering groups are divalent mercury; Equations 6.24 through 6.28 summarize the types of reactions observed. In these equations X is a halide ligand of the mercury, and italics are used for one of the reactants and fragments derived from it as an aid to tracing the course of the exchange.

$$RHgR + XHgX \longrightarrow RHgX + XHgR \tag{6.24}$$
$$RHgR + RHgX \longrightarrow RHgR + XHgR \tag{6.25}$$
$$RHgX + XHgX \longrightarrow RHgX + XHgX \tag{6.26}$$
$$RHgX + RHgX \longrightarrow RHgR + XHgX \tag{6.27}$$
$$RHgR + RHgR \longrightarrow RHgR + RHgR \tag{6.28}$$

[55] (a) D. E. Applequist and G. N. Chmurny, *J. Am. Chem. Soc.,* **89,** 875 (1967); (b) W. H. Glaze, C. H. Selman, A. L. Ball, Jr., and L. E. Bray, *J. Org. Chem.,* **34,** 641 (1969).

[56] (a) F. R. Jensen and D. D. Davis, *J. Am. Chem. Soc.,* **93,** 4048 (1971); (b) A. Rahm and M. Pereyre, *J. Am. Chem. Soc.,* **99,** 1672 (1977); (c) N. S. Isaacs and K. Javaid, *Tetrahedron Lett.,* 3073 (1977).

[57] J. P. Leslie II and J. H. Espenson, *J. Am. Chem. Soc.,* **98,** 4839 (1976).

[58] (a) G. M. Whitesides and D. J. Boschetto, *J. Am. Chem. Soc.,* **93,** 1529 (1971); (b) P. L. Bock, D. J. Boschetto, J. R. Rasmussen, J. P. Demers, and G. M. Whitesides, *J. Am. Chem. Soc.,* **96,** 2814 (1974); (c) K. Stanley and M. C. Baird, *J. Am. Chem. Soc.,* **99,** 1808 (1977).

[59] (a) F. R. Jensen, V. Madan, and D. H. Buchanan, *J. Am. Chem. Soc.,* **93,** 5283 (1971); (b) H. L. Fritz, J. H. Espenson, D. A. Williams, and G. A. Molander, *J. Am. Chem. Soc.,* **96,** 2378 (1974).

[60] (a) G. A. Olah, Y. Halpern, J. Shen, and Y. K. Mo, *J. Am. Chem. Soc.,* **93,** 1251 (1971), and following papers; (b) D. H. R. Barton, R. H. Hesse, R. E. Markwell, M. M. Pechet, and H. T. Toh, *J. Am. Chem. Soc.,* **98,** 3034 (1976). See also note 54(e).

The stereochemistry of mercury-exchange reactions was first established in an experiment carried out as follows.[61] Di-2-butylmercury was prepared by reacting optically active 2-butylmercuric bromide with racemic 2-butylmagnesium bromide as shown in Equation 6.29 (the asterisk designates the chiral center responsible for the optical activity).

$$CH_3CH_2-\overset{\overset{\displaystyle CH_3}{|}}{\underset{\underset{\displaystyle H}{|}}{C}}*-HgBr + BrMg-\overset{\overset{\displaystyle CH_3}{|}}{\underset{\underset{\displaystyle H}{|}}{C}}-CH_2CH_3 \longrightarrow$$

$$CH_3CH_2-\overset{\overset{\displaystyle CH_3}{|}}{\underset{\underset{\displaystyle H}{|}}{C}}*-Hg-\overset{\overset{\displaystyle CH_3}{|}}{\underset{\underset{\displaystyle H}{|}}{C}}-CH_2CH_3 + MgBr_2 \quad (6.29)$$

Since the mercury–carbon bond is not disturbed in this reaction, one (and only one) of the 2-butyl groups in the di-2-butylmercury should have an optically active center. The mercury exchange reaction of Equation 6.30 was then carried out on

$$HgBr_2 + CH_3CH_2-\overset{\overset{\displaystyle CH_3}{|}}{\underset{\underset{\displaystyle H}{|}}{*C}}-Hg-\overset{\overset{\displaystyle CH_3}{|}}{\underset{\underset{\displaystyle H}{|}}{C}}-CH_2CH_3 \longrightarrow 2CH_3CH_2-\overset{\overset{\displaystyle CH_3}{|}}{\underset{\underset{\displaystyle H}{|}}{C}}-HgBr \quad (6.30)$$

the di-2-butylmercury thus formed, and the optical rotation of the product was compared to that of the 2-butylmercuric bromide used as starting material. If the substitution had proceeded with retention of configuration (Equation 6.31), then

$$CH_3CH_2-\overset{\overset{\displaystyle CH_3}{|}}{\underset{\underset{\displaystyle H}{|}}{*C}}-Hg-\overset{\overset{\displaystyle CH_3}{|}}{\underset{\underset{\displaystyle H}{|}}{C}}-CH_2CH_3 + HgBr_2 \longrightarrow CH_3CH_2-\overset{\overset{\displaystyle CH_3}{|}}{\underset{\underset{\displaystyle H}{|}}{*C}}-Hg-Br +$$

(retained)

$$BrHg-\overset{\overset{\displaystyle CH_3}{|}}{\underset{\underset{\displaystyle H}{|}}{C}}-CH_2CH_3 + CH_3CH_2-\overset{\overset{\displaystyle CH_3}{|}}{\underset{\underset{\displaystyle H}{|}}{*C}}-HgBr + BrHg-\overset{\overset{\displaystyle CH_3}{|}}{\underset{\underset{\displaystyle H}{|}}{C}}-CH_2CH_3 \quad (6.31)$$

(retained)

the specific rotation of the product would be half that of the original 2-butylmercuric bromide. If the substitution had proceeded with racemization (Equation 6.32), the specific rotation should be one-quarter that of the starting material. Finally, if the substitution had proceeded with inversion (Equation 6.33), the product should be racemic.

[61](a) H. B. Charman, E. D. Hughes, and C. Ingold, *J. Chem. Soc.*, 2530 (1959); (b) F. R. Jensen, *J. Am. Chem. Soc.*, **82**, 2469 (1960); (c) O. A. Reutov and E. V. Uglova, *Bull. Acad. Sci. USSR, Chem. Div. Sci.*, 1628 (1959).

$$CH_3CH_2-\overset{\overset{\displaystyle CH_3}{|}}{\underset{\underset{\displaystyle H}{|}}{*C}}-Hg-\overset{\overset{\displaystyle CH_3}{|}}{\underset{\underset{\displaystyle H}{|}}{C}}-CH_2CH_3 + HgBr_2 \longrightarrow CH_3CH_2-\overset{\overset{\displaystyle CH_3}{|}}{\underset{\underset{\displaystyle H}{|}}{*C}}-HgBr +$$

(retained)

$$CH_3CH_2-\overset{\overset{\displaystyle CH_3}{|}}{\underset{\underset{\displaystyle H}{|}}{C}}-HgBr + CH_3CH_2-\overset{\overset{\displaystyle CH_3}{|}}{\underset{\underset{\displaystyle H}{|}}{C}}-HgBr + BrHg-\overset{\overset{\displaystyle CH_3}{|}}{\underset{\underset{\displaystyle H}{|}}{C}}-CH_2CH_3 \qquad (6.32)$$

(racemized)

$$CH_3CH_2-\overset{\overset{\displaystyle CH_3}{|}}{\underset{\underset{\displaystyle H}{|}}{*C}}-Hg-\overset{\overset{\displaystyle CH_3}{|}}{\underset{\underset{\displaystyle H}{|}}{C}}-CH_2CH_3 + HgBr_2 \longrightarrow CH_3CH_2-\overset{\overset{\displaystyle CH_3}{|}}{\underset{\underset{\displaystyle H}{|}}{*C}}-HgBr +$$

(retained)

$$CH_3CH_2-\overset{\overset{\displaystyle CH_3}{|}}{\underset{\underset{\displaystyle H}{|}}{C}}-HgBr + CH_3CH_2-\overset{\overset{\displaystyle CH_3}{|}}{\underset{\underset{\displaystyle H}{|}}{*C}}-HgBr + CH_3CH_2-\overset{\overset{\displaystyle CH_3}{|}}{\underset{\underset{\displaystyle H}{|}}{C}}-HgBr \qquad (6.33)$$

(inverted)

The specific rotation of the initial 2-butylmercuric bromide used by Charman, Hughes, and Ingold was -15.2. The specific rotation of the product of the mercury-exchange reaction was exactly half that, -7.6. When mercuric acetate or mercuric nitrate was used as the cleaving salt, the products showed a specific rotation of -7.5 and -7.8, respectively. Thus this electrophilic substitution clearly proceeds with retention of configuration.

Stereochemical studies on other mercury exchange reactions have been carried out, and all point to retention as the predominant pathway.[62]

Electrophilic substitution of hydrogen by deuterium or by a carbocation with retention of configuration can occur in strongly acidic conditions, for example, in superacids such as HSO_3F-SbF_5 or $HF-SbF_5$.[60] Equation 6.34 gives an exam-

---

[62] (a) H. B. Charman, E. D. Hughes, C. Ingold, and F. G. Thorpe, *J. Chem. Soc.*, 1121 (1961); (b) E. D. Hughes, C. Ingold, F. G. Thorpe, and H. C. Volger, *J. Chem. Soc.*, 1133 (1961); (c) E. D. Hughes, C. K. Ingold, and R. M. G. Roberts, *J. Chem. Soc.*, 3900 (1964). The presence of the halide ligand of the mercuric ion adds some complication to the structure of the transition state for frontside displacement. The probable transition state structure for substitution of RHgX by HgX_2 is

R·····Hg
⋮ ⋮ X
Hg·····R
X

This pathway is referred to as the S_Ei reaction. See (d) F. R. Jensen and B. Rickborn, *Electrophilic Substitution of Organomercurials,* McGraw-Hill, New York, 1968, pp. 153*ff.*

$$\left[(H_3C)_3C\cdots\overset{H}{\underset{D}{\vdots}}\right]^+ \rightleftharpoons (H_3C)_3CD + H^+$$

$$\overset{CH_3}{\underset{CH_3}{H_3C-\overset{|}{\underset{|}{C}}-H}} \xrightarrow{DF-SbF_5} \Bigg\langle \quad \left[(H_3C)_2CHC\underset{H_2}{\cdots}\overset{H}{\underset{D}{\vdots}}\right]^+ \rightleftharpoons (H_3C)_2CHCH_2D + H^+ \qquad (6.34)$$

$$\left[(H_3C)_2C\cdots\overset{CH_3}{\underset{D}{\vdots}}\right]^+ \longrightarrow (H_3C)_2C^+H + H_3CD$$

ple. The conditions of these reactions are far removed from those under which carbanions could exist even as transient intermediates. Nevertheless, the reactions are legitimate S_E2 substitutions because the bonding pair of electrons remains at the substitution center and entering and leaving groups are both electron-deficient. Olah has formulated the transition state in terms of attack of the electrophile on the substrate's bonding electrons passing through a structure **30** with a three-center, two-electron bond. This structure is consistent with the picture based on frontier orbitals arrived at in Chapter 1.

$$\overset{\diagdown}{\underset{\diagup}{C}}-H + R^+ \longrightarrow \left[\overset{\diagdown}{\underset{\diagup}{\cdot\cdot C}}\cdots\overset{H}{\underset{R}{\vdots}}\right]^+ \longrightarrow \overset{\diagdown}{\underset{\diagup}{C}}-R + H^+ \qquad (6.35)$$

$$\mathbf{30}$$

The retention stereochemistry of mercury exchange and of electrophilic attack on C—H bonds contrasts with the inversion process universally observed in nucleophilic substitution (see Chapter 4). Recall from Chapter 1 that the frontier orbital theory permits frontside electrophilic displacement but does not require it. It is not surprising, therefore, that there are also examples of S_E2 substitutions that proceed with inversion of configuration. For example, organolithium compounds show inversion of configuration on reaction with bromine but predominant retention on reaction with chloroformate esters.[55] Cobalt is displaced with inversion,[59] and tin is displaced sometimes with nearly complete inversion and sometimes with nearly complete retention depending on the entering electrophile.[56] Theoretical calculations using *ab initio* molecular orbital methods also predict this dichotomy in stereochemistry: **31**, **32**, and **33** show the predicted lowest-energy structures for symmetrical substitution of hydrogen, lithium, and beryllium.[63]

$$\overset{H}{\underset{H}{\overset{\diagdown}{\underset{\diagup}{H}}}}\overset{\cdot\cdot H}{\underset{\cdot\cdot H}{C^+}} \qquad \underset{H\ H}{Li\cdots\overset{H}{\underset{|}{C}}\cdots Li} \qquad \underset{H\ H}{H-Be\cdots\overset{H}{\underset{|}{C}}\cdots Be-H}$$

$$\mathbf{31} \qquad\qquad \mathbf{32} \qquad\qquad \mathbf{33}$$

[63] E. D. Jemmis, J. Chandrasekhar, and P. v. R. Schleyer, *J. Am. Chem. Soc.*, **101**, 527 (1979).

6.3 CARBENES

Carbenes are intermediates that contain uncharged divalent carbon (34).[64] Car-

$$\overset{\displaystyle ..}{\underset{\displaystyle X \qquad Y}{C}}$$

34

benes are related to carbanions through the α elimination reaction, illustrated in Equation 6.36, a route that is frequently used in their preparation. They may also

$$\overset{X}{\underset{Z}{\overset{\diagdown}{\underset{\diagup}{C}}}}:^- \longrightarrow \overset{X}{\underset{Y}{\overset{\diagdown}{C}}}: + \ Z:^- \qquad (6.36)$$

be considered formally as the conjugate bases of carbocations (Equation 6.37), although this reaction is in most cases not a practical preparative pathway.[65]

$$\overset{X}{\underset{H\ Y}{\overset{|}{\underset{\diagup}{C^+}}}} \longrightarrow \overset{X}{\underset{Y}{\overset{\diagdown}{C}}}: + \ H^+ \qquad (6.37)$$

Carbenes are highly reactive, and undergo characteristic chemical changes, the most important of which are listed with examples in Table 6.5. Under some conditions carbenes are coordinated with other species and have modified reactivities, in which case they are called *carbenoid*. Free carbenes are formed thermally or photochemically in the gas phase and photochemically or by α elimination with appropriate conditions in solution.

Formation of Carbenes

Hine and his co-workers showed in the 1950s by kinetic and trapping experiments that dihalomethylene, $:CX_2$, is an intermediate in the reaction of haloforms with base in aqueous solution.[66] Scheme 9 depicts for chloroform the mechanism they

[64] The term *carbene* is used here as a generic designation; individual carbenes are named as substituted methylenes. For reviews of carbene chemistry, see (a) D. Bethell, *Adv. Phys. Org. Chem.*, **7**, 153 (1969); (b) G. L. Closs, *Top. Stereochem.*, **3**, 193 (1968); (c) W. Kirmse, *Carbene Chemistry*, 2d ed., Academic Press, New York, 1971; (d) J. Hine, *Divalent Carbon*, Ronald Press, New York, 1964; (e) M. Jones and R. A. Moss, Eds., *Carbenes*, Wiley, New York, Vol. 1, 1973; Vol. 2, 1975. The analogous nitrenes,

$$R\!-\!\overset{\displaystyle ..}{N}:$$

have chemistry that is similar in many ways to that of carbenes. See (f) W. Lwowski, Ed., *Nitrenes*, Wiley, New York, 1970; (g) E. Wasserman, *Prog. Phys. Org. Chem.*, **8**, 319 (1971); (h) H. Durr and H. Kober, *Top. Current Chem.*, **66**, 89 (1976). (i) For the origin of the term carbene, see footnote 9 of W. V. E. Doering and L. H. Knox, *J. Am. Chem. Soc.*, **78**, 4947 (1956).

[65] Certain types of carbocations can be deprotonated, with formation of typical carbene products. See R. A. Olofson, S. W. Walinsky, J. P. Marino, and J. L. Jernow, *J. Am. Chem. Soc.*, **90**, 6554 (1968).

[66] (a) J. Hine, *J. Am. Chem. Soc.*, **72**, 2438 (1950); (b) J. Hine and A. M. Dowell, Jr., *J. Am. Chem. Soc.*, **76**, 2688 (1954); (c) J. Hine, A. M. Dowell, Jr., and J. E. Singley, Jr., *J. Am. Chem. Soc.*, **78**, 479 (1956).

Table 6.5 CHARACTERISTIC REACTIONS OF CARBENES

Insertion	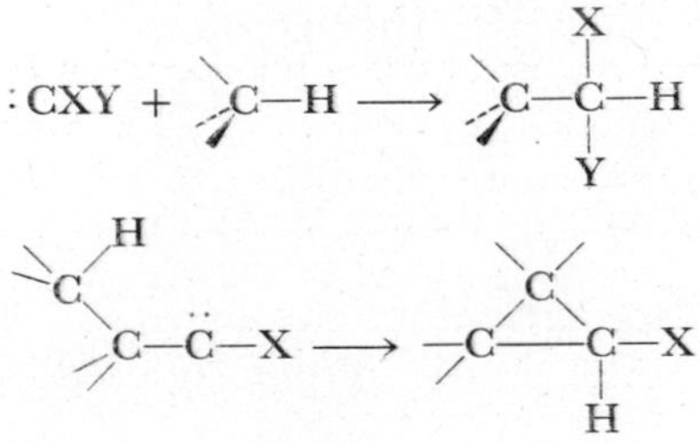
Addition	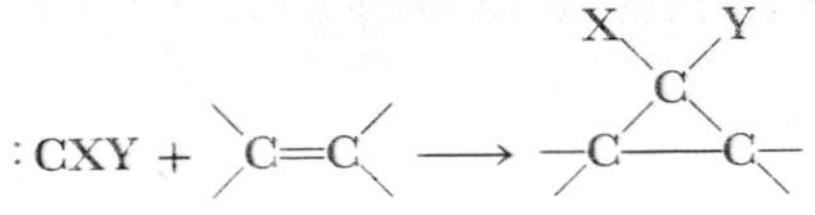
Dimerization	$2\ :CXY \longrightarrow XYC{=}CXY$
Rearrangement	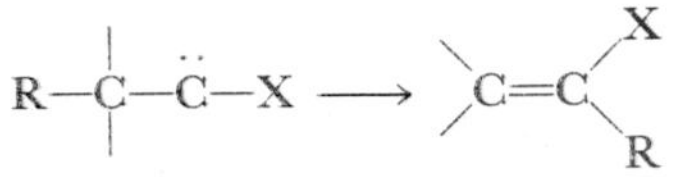

proposed. If the first step is a rapid equilibrium and k_2 is rate determining, the observed second-order kinetics are consistent with the mechanism, as are a number of other results.[67]

SCHEME 9

$$HCCl_3 + OH^- \underset{k_{-1}}{\overset{k_1}{\rightleftharpoons}} {}^-{:}CCl_3 + H_2O$$

$${}^-{:}CCl_3 \xrightarrow{k_2} {:}CCl_2 + Cl^-$$

$${:}CCl_2 \longrightarrow \xrightarrow[\text{fast}]{\substack{H_2O,\ OH^- \\ \text{several steps}}} \longrightarrow CO + HCOO^- + Cl^-$$

Phase transfer catalysis (Section 2.4) has proved particularly effective in generation of carbenes by α elimination. The halomethyl anion is transported as an ion pair with a tetraalkylammonium ion, or a crown ether-complexed alkali metal ion, from a strongly basic aqueous solution into an organic phase where the elimination takes place and the resulting carbene reacts with the substrate.[68] Comparison of selectivity of carbenes generated through α elimination toward various al-

[67] (a) J. Hine, N. W. Burske, M. Hine, and P. B. Langford, *J. Am. Chem. Soc.*, **79**, 1406 (1957), and references to earlier work cited therein; (b) E. D. Bergmann, D. Ginsberg, and D. Lavie, *J. Am. Chem. Soc.*, **72**, 5012 (1950); (c) R. Lombard and R. Boesch, *Bull. Soc. Chim. Fr.*, 733 (1953); (d) J. Hine and P. B. Langford, *J. Am. Chem. Soc.*, **79**, 5497 (1957).

[68] (a) M. Makosza and M. Ludwikow, *Angew. Chem. Int. Ed. Engl.*, **13**, 665 (1974); (b) M. Makosza, A. Kacprowicz, and M. Fedorynski, *Tetrahedron Lett.*, 2119 (1975); (c) S. Julia and A. Ginebreda, *Synthesis*, 682 (1977); (d) M. Fedorynski, *Synthesis*, 783 (1977); (e) T. Sasaki, S. Eguchi, M. Ohno, and F. Nakata, *J. Org. Chem.*, **41**, 2408 (1976).

kenes (see below) with selectivity of carbenes generated photochemically shows that the intermediates formed in α elimination are sometimes carbenoid. The carbenoid character presumably arises from complexing between the carbene and the cation associated with the base that initiated the elimination. Free carbenes can be obtained by carrying out the eliminations in the presence of crown ethers.[69]

The α elimination method is mainly applicable to the halomethylenes. A closely related method of obtaining intermediates of the carbene type is through organometallic derivatives of general structure **35**, where X is a halogen and M is a metal, usually Li, Zn, or Sn. These compounds, when heated in the presence of appropriate substrates, yield typical carbene products.[70]

$$R_1 \diagdown \underset{R_2 \diagup}{\overset{\diagup X}{C}} \diagdown M$$

35

A convenient and widely applicable method is the Simmons–Smith reaction, in which diiodomethane reacts with zinc–copper couple to yield an organozinc reagent that is a source of CH_2.[71] The intermediates in these reactions are definitely carbenoid and show different reactivity patterns from free carbenes.[72]

Other important routes to carbenes are thermal or photochemical decomposition of diazo compounds (**36**) and diazirines (**37**) and treatment of tosylhydrazones (**38**) with bases. These processes are illustrated in Equations 6.38–6.40.

$$\underset{R'}{\overset{R}{\diagdown}} C = \overset{+}{N} = \overset{..}{\overset{-}{N}} :$$

$$\updownarrow \qquad \xrightarrow[\Delta]{h\nu \text{ or}} \qquad \underset{R'}{\overset{R}{\diagdown}} C : + N_2 \qquad (6.38)[73]$$

$$\underset{R'}{\overset{R}{\diagdown}} \overset{\overset{..}{\overset{-}{}}}{C} - \overset{+}{N} \equiv N :$$

36

[69](a) R. A. Moss and F. G. Pilkiewicz, *J. Am. Chem. Soc.*, **96**, 5632 (1974); (b) R. A. Moss, M. A. Joyce, and F. G. Pilkiewicz, *Tetrahedron Lett.*, 2425 (1975).

[70] See note 64(a).

[71] H. E. Simmons, T. L. Cairns, S. A. Vladuchick, and C. M. Hoiness, *Org. Reactions*, **20**, 1 (1973).

[72] The trihalomethylmercury compounds are an exception. See D. Seyferth, J. Y. Mui, and J. M. Burlitch, *J. Am. Chem. Soc.*, **89**, 4953 (1967).

[73] Photochemical reactions, indicated in reaction schemes by the symbol $h\nu$, are considered in Chapter 12. For the present, it is sufficient to note that absorption of light transforms a molecule to an excited state, which, in the case of diazo compounds, has sufficient energy for rupture of the C—N bond. Generation of carbenes from diazo compounds is reviewed by W. J. Baron, M. R. DeCamp, M. E. Hendrick, M. Jones, Jr., R. H. Levin, and M. B. Sohn, in *Carbenes*, Vol. 1, M. Jones and R. A. Moss, Eds., Wiley, New York, 1973, p. 1. See also V. Dave and E. W. Warnhoff, *Org. Reactions*, **18**, 217 (1970).

$$\begin{array}{c}
\overset{R}{\underset{R'}{\diagdown}}C\!\!\begin{array}{c}N\\ \| \\ N\end{array} \xrightarrow[\Delta]{h\nu\ or} \overset{R}{\underset{R'}{\diagdown}}C: + N_2
\end{array} \qquad (6.39)^{74}$$

$$\mathbf{37}$$

$$\overset{R}{\underset{R'}{\diagdown}}C{=}\overset{..}{N}{-}\overset{H}{\underset{|}{N}}{-}Ts \xrightarrow[\Delta]{base} \overset{R}{\underset{R'}{\diagdown}}C{=}\overset{-}{N}{-}\overset{\bar{..}}{N}{-}Ts \longrightarrow Ts^- + \overset{R}{\underset{R'}{\diagdown}}C{=}\overset{+}{N}{=}\overset{\bar{..}}{N}: \longrightarrow \overset{R}{\underset{R'}{\diagdown}}C: + N_2 \quad (6.40)^{75}$$

$$\mathbf{38}$$

$$Ts = H_3C{-}\!\!\left\langle \rule{0pt}{1em}\right.\!\!\!\bigcirc\!\!\!\left.\rule{0pt}{1em}\right\rangle\!\!{-}SO_2$$

Structure of Carbenes

A carbene carbon uses two of its four valence orbitals for bonding to the attached groups. If the molecule is linear, the carbene carbon will be sp-hybridized. Two p orbitals will remain unhybridized, and the two unshared electrons should go one into each with spins parallel. If the molecule is bent, one of these orbitals remains p while the other acquires s character and so decreases in energy; if the energy separation becomes large enough, the electrons will prefer to pair and go into the lower-energy hybrid orbital, leaving the p-orbital empty. A structure with two unpaired electrons is said to be in a *triplet state,* a situation well known from spectroscopic observations of excited atoms and molecules but rare in ground-state molecular chemistry.[76] A structure with all electrons paired is in the *singlet state*. Structures **39** and **40** show the expected orbital occupancies for singlet and triplet carbenes, respectively.

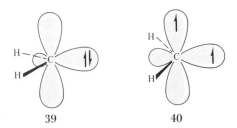

39   **40**

Herzberg provided the first definitive evidence on the geometry of $:CH_2$ through his observation of absorption spectra of both the lowest-energy triplet and the lowest-energy singlet.[77] The precise geometry of the triplet could not be deter-

[74] S. Braslavsky and J. Heicklen, *Chem. Rev.,* **77,** 473 (1977).
[75] R. H. Shapiro, *Org. Reactions,* **23,** 405 (1976).
[76] An important exception is O_2, which has a triplet ground state.
[77] (a) G. Herzberg and J. Shoosmith, *Nature,* **183,** 1801 (1959); (b) G. Herzberg, *Proc. Roy. Soc.,* **A262,** 291 (1961); (c) G. Herzberg and J. W. C. Johns, *Proc. Roy. Soc.,* **A295,** 107 (1966).

mined, but Herzberg originally concluded that it is linear or nearly so; the spectra did furnish an accurate measurement of the structure of the higher-energy singlet and showed the H—C—H angle to be 102.4°. Structural information is also available for a number of halomethylenes from absorption spectra.[78] Electron spin resonance spectroscopy (esr) is a second technique that has yielded information on carbene structures.[79] Similar in principle to nuclear magnetic resonance, esr detects energy changes accompanying changes in electron spin states in a magnetic field. Triplet spectra are characteristic and easily identified.[80] Wasserman and co-workers observed ground-state CH_2 by electron spin resonance and concluded, contrary to Herzberg's original determination, that the triplet is nonlinear with a H—C—H angle of about 136°.[81] Herzberg reinterpreted the ultraviolet data and found that a nonlinear structure is consistent with the spectrum.[82] Dichloromethylene, : CCl_2, has also been observed by ion cyclotron resonance spectrometry.[83]

   The picture that emerges from spectroscopic study of a variety of carbenes is that halomethylenes are ground-state singlets with bond angles in the range 100–110°, whereas methylene, arylmethylenes, and probably also alkylmethylenes are ground-state triplets with bond angles 130–180° and have excited singlet states with angles of 100–110°.[84] Accurate quantum mechanical calculations reproduce the experimental geometries well.[85] There is rough agreement between experimentalists and theoreticians that the energy separation between triplet and singlet methylene is about 8 to 11 kcal mole^{-1}.[86]

   The vinylidene carbenes (**41**) apparently have singlet ground states; their

$$\begin{array}{c} R_1 \\ \diagdown \\ \hspace{1em} C{=}C: \\ \diagup \\ R_2 \end{array}$$

41

[78] Summaries of structural information may be found in (a) D. Bethell, *Adv. Phys. Org. Chem.,* **7**, 153 (1969); (b) J. F. Harrison, *Acc. Chem. Res.,* **7**, 378 (1974).

[79] Esr spectroscopy is discussed in a number of sources; for a brief introduction see D. J. Pasto and C. R. Johnson, *Organic Structure Determination,* Prentice-Hall, Englewood Cliffs, N.J., 1969, Chap. 6.

[80] A. Carrington and A. D. McLachlan, *Introduction to Magnetic Resonance,* Harper & Row, New York, 1967, Chap. 8.

[81] E. Wasserman, V. J. Kuck, R. S. Hutton, and W. A. Yager, *J. Am. Chem. Soc.,* **92**, 7491 (1970).

[82] G. Herzberg and J. W. C. Johns, *J. Chem. Phys.,* **54**, 2276 (1971).

[83] B. A. Levi, R. W. Taft, and W. J. Hehre, *J. Am. Chem. Soc.,* **99**, 8454 (1977).

[84] (a) See note 81; (b) C. A. Hutchison, Jr., and B. E. Kohler, *J. Chem. Phys.,* **51**, 3327 (1969); (c) R. Hoffmann, G. D. Zeiss, and G. W. Van Dine, *J. Am. Chem. Soc.,* **90**, 1485 (1968). Ground-state multiplicity appears to correlate with electronegativity of substituents: (d) J. F. Harrison, R. C. Liedtke, and J. F. Liebman, *J. Am. Chem. Soc.,* **101**, 7162 (1979).

[85] (a) J. F. Harrison and L. C. Allen, *J. Am. Chem. Soc.,* **91**, 807 (1969); (b) C. F. Bender, H. F. Schaefer III, D. F. Franceschetti, and L. C. Allen, *J. Am. Chem. Soc.,* **94**, 6888 (1972); (c) M. J. S. Dewar, R. C. Haddon, and P. K. Weiner, *J. Am. Chem. Soc.,* **96**, 253 (1974); (d) J. F. Harrison, *J. Am. Chem. Soc.,* **93**, 4112 (1971); (e) C. W. Bauschlicher, Jr., H. F. Schaefer III, and P. S. Bagus, *J. Am. Chem. Soc.,* **99**, 7106 (1977); (f) N. C. Baird and K. F. Taylor, *J. Am. Chem. Soc.,* **100**, 1333 (1978).

[86] (a) R. K. Lengel and R. N. Zare, *J. Am. Chem. Soc.,* **100**, 7495 (1978); (b) R. R. Lucchese and H. F. Schaefer III, *J. Am. Chem. Soc.,* **99**, 6765 (1977); (c) B. O. Roos and P. M. Siegbahn, *J. Am. Chem. Soc.,* **99**, 7716 (1977); (d) C. W. Bauschlicher, Jr., and I. Shavitt, *J. Am. Chem. Soc.,* **100**, 739 (1978).

reactivity pattern is similar to that of carbenes discussed below, but they are some-what less electrophilic and do not insert as readily.[87]

Reactions of Carbenes

The first carbene reaction to be considered is insertion (Equation 6.41). It was first

$$R_2 \overset{R_1}{\underset{R_3}{\text{—}}} C\text{—}H + :CH_2 \longrightarrow R_2 \overset{R_1}{\underset{R_3}{\text{—}}} C\text{—}\overset{H}{\underset{H}{C}}\text{—}H \tag{6.41}$$

reported in 1942 by Meerwein and co-workers,[88] but its importance was not recog-nized until Doering investigated the reaction and noted that in the liquid phase $:CH_2$ generated by photolysis of diazomethane attacks the various types of C—H bonds of hydrocarbons with no discrimination.[89] More extensive results of Richardson, Simmons, and Dvoretsky have confirmed this finding.[90] In the gas phase the reaction is more selective, and when measures are taken to increase the lifetime of the $:CH_2$ intermediates by addition of an inert gas, so that more of the initially formed unselective singlet has time to decay to the somewhat less reac-tive triplet ground state, the insertion becomes more selective still.[91] Table 6.6 pre-sents representative experimental results. Insertion is common for methylene and carbon-substituted methylenes and can occur either inter- or intramolecularly.

When the carbene is in the triplet state, a hydrogen abstraction to yield a radical pair (Equation 6.42) seems a reasonable possibility for the insertion mecha-

$$:CH_2 + \overset{}{\underset{}{C}}\text{—}H \longrightarrow \overset{}{\underset{}{C}}\cdot + \cdot CH_3 \longrightarrow \overset{}{\underset{}{C}}\text{—}CH_3 \tag{6.42}$$

nism. The singlet-state carbenes, however, insert into the C—H bond with reten-tion of configuration, and a single-step process is likely.[92] The attack of the singlet carbene on the C—H bond could occur either through the occupied hybrid (**42**) or through the vacant *p* orbital (**43**); the latter possibility, which would be electro-philic S_E2 substitution, is in better accord with the strongly electrophilic character of carbenes and with the frontside attack required by the stereochemistry. Theoret-ical calculations of Hoffmann and co-workers suggest that attack is initially mainly

[87](a) P. J. Stang, *Acc. Chem. Res.*, **11**, 107 (1978); (b) P. J. Stang, *Chem. Rev.*, **78**, 383 (1978); (c) P. J. Stang, D. P. Fox, C. J. Collins, and C. R. Watson, Jr., *J. Org. Chem.*, **43**, 364 (1978).
[88] H. Meerwein, H. Rathjen, and H. Werner, *Chem. Ber.*, **75**, 1610 (1942).
[89] W. v. E. Doering, R. G. Buttery, R. G. Laughlin, and N. Chaudhuri, *J. Am. Chem. Soc.*, **78**, 3224 (1956).
[90] D. B. Richardson, M. C. Simmons, and I. Dvoretzky, *J. Am. Chem. Soc.*, **82**, 5001 (1960); **83**, 1934 (1961).
[91](a) H. M. Frey and G. B. Kistiakowsky, *J. Am. Chem. Soc.*, **79**, 6373 (1957); (b) H. M. Frey, *J. Am. Chem. Soc.*, **80**, 5005 (1958).
[92](a) Kirmse, *Carbene Chemistry*, 2d ed., Academic Press, New York, 1971, p. 220; (b) P. S. Skell and R. C. Woodworth, *J. Am. Chem. Soc.*, **78**, 4496 (1956) (Structures I and II in this paper are reversed. See p. 6427.); (c) W. v. E. Doering and H. Prinzbach, *Tetrahedron*, **6**, 24 (1959); (d) C. D. Gutsche, G. L. Bachman, W. Udell, and S. Bäuerlein, *J. Am. Chem. Soc.*, **93**, 5172 (1971).

Table 6.6 SELECTIVITY OF INSERTION OF CH_2 INTO C—H BONDS

Substrate	State	Method[a]	Product	Type of C—H bond reacting[b]	Number of equivalent bonds	Calculated yield (%)[c]	Observed yield (%)	Reference
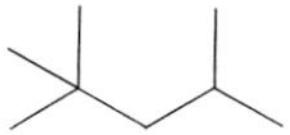	Liquid	A	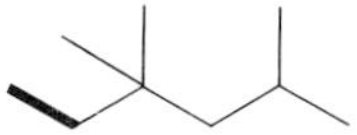	1	9	50.0	51	d
			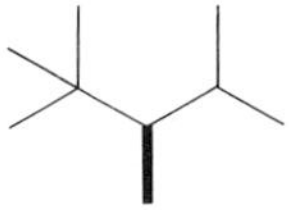	2	2	11.1	10	
			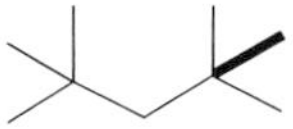	3	1	5.6	4	
			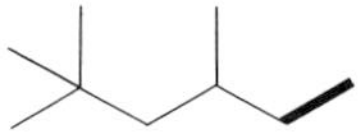	1	6	33.3	35	
	Gas	B		1	6	75	72	e
				2	2	25	28	
	Gas	B		1	9	90	86	e
				3	1	10	14	
	Gas	C		1	6	75	63	f
				2	2	25	37	
	Gas	D		1	6	75	57	f
				2	2	25	43	

[a] A: Liquid-phase photolysis of CH_2N_2.　B: Gas-phase photolysis of CH_2N_2.　C: Gas-phase photolysis of CH_2CO.　D: Gas-phase photolysis of CH_2CO with inert gas (CO_2) in eightfold excess over propane.

[b] 1 = primary; 2 = secondary; 3 = tertiary.　[c] Assuming random attack.

[d] D. B. Richardson, M. C. Simmons, and I. Dvoretzky, *J. Am. Chem. Soc.*, **83**, 1934 (1961). See this reference for further examples.

[e] H. M. Frey, *J. Am. Chem. Soc.*, **80**, 5005 (1958).　[f] H. M. Frey and G. B. Kistiakowsky, *J. Am. Chem. Soc.*, **79**, 6373 (1957).

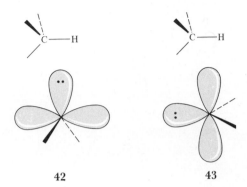

42   43

at the hydrogen end of the C—H bond (**44**), and that transfer of the hydrogen to the incoming CH_2 runs ahead of C—C bond making.[93] This proposal is similar to one by DeMore and Benson.[94] Dihalocarbenes do not insert as readily as does : CH_2;[95] carbenoids usually do not insert.[96]

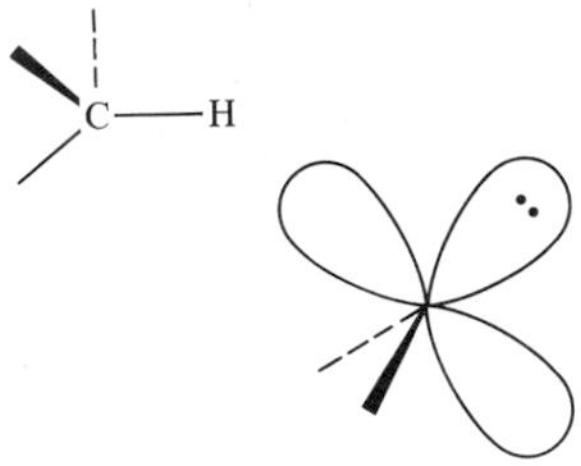

44

Intramolecular insertion reactions lead to rearranged structures and are therefore often classified in a separate category. The fundamental nature of the reaction is nevertheless the same as for intermolecular insertions. Insertion in a C—H bond on an adjacent carbon is equivalent to a 1,2-migration of hydrogen (Equation 6.43). These rearrangements take place through the singlet state, and the

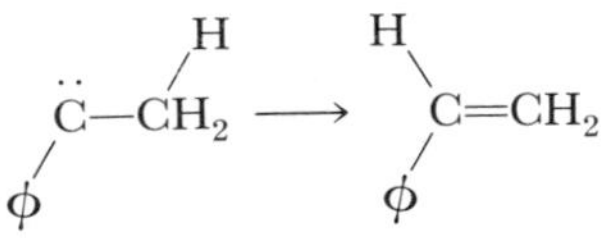

$$\ddot{C}-CH_2 \longrightarrow C=CH_2 \qquad (6.43)$$

preferred stereochemistry is that shown in **45** (carbene carbon in front).[97] Note the close analogy with 1,2-migration in a carbocation (**46**). Insertion in a C—H bond

[93] R. C. Dobson, D. M. Hayes, and R. Hoffmann, *J. Am. Chem. Soc.*, **93**, 6188 (1971). See also note 85(b).
[94] (a) W. B. DeMore and S. W. Benson, *Adv. Photochem.*, **2**, 219 (1964); for further discussion of the insertion pathway, see (b) E. A. Hill, *J. Org. Chem.*, **37**, 4008 (1972).
[95] V. Franzen and R. Edens, *Justus Liebigs Ann. Chem.*, **729**, 33 (1969).
[96] L. Friedman, R. J. Honour, and J. G. Berger, *J. Am. Chem. Soc.*, **92**, 4640 (1970), and references cited therein.
[97] (a) Y. Yamamoto, S.-I. Murahashi, and I. Moritani, *Tetrahedron*, **31**, 2663 (1975); (b) L. S. Press and H. Shechter, *J. Am. Chem. Soc.*, **101**, 509 (1979).

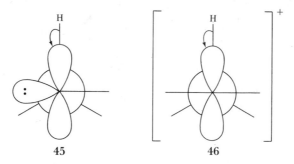

that is two carbons removed from the carbene center yields a cyclopropane (Equation 6.44). This reaction is primarily a singlet-state process.[98]

$$\underset{\underset{CH_3}{}{\overset{CH_3}{}{}}}{\phi}C-C\underset{H}{\overset{N_2}{}} \xrightarrow{-N_2} \underset{\underset{CH_3}{}{\overset{CH_2}{}{}}}{\phi}C-C\underset{H}{\overset{}{}}H \quad + \text{ other products} \qquad (6.44)^{98}$$

A second characteristic reaction of carbenes is addition to alkenes to yield cyclopropanes.[99] Singlet carbenes might react as either nucleophiles or electrophiles; triplets may be expected to behave like free radicals. The data in Table 6.7, showing the increase in rate of addition on substitution of electron-donating alkyl groups in the alkene, demonstrate that the carbenes and carbenoids are strong electrophiles.[100] Carbenoids are sterically more hindered and react relatively more slowly with the highly substituted alkenes than the free carbenes.[96]

A structure with unshared pairs of electrons adjacent to the carbene center (**47**) should be much less electrophilic. Carbenes of this type do show reduced

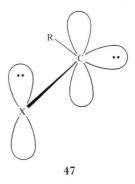

47

[98] J. J. Havel, *J. Org. Chem.*, **41**, 1464 (1976).

[99] Intramolecular additions also occur and are sometimes followed by formation of a new carbene through opening of a small ring, resulting in a rearrangement. See (a) C. Wentrup, *Top. Current Chem.*, **62**, 173 (1976); (b) C. Wentrup, *Tetrahedron*, **30**, 1301 (1974); (c) M. Jones, Jr., *Acc. Chem. Res.*, **7**, 415 (1974); (d) W. M. Jones, *Acc. Chem. Res.*, **10**, 353 (1977).

[100] Reactivity of a number of carbenes toward alkenes has been correlated with inductive and resonance substituent parameters. (a) R. A. Moss, C. B. Mallon, and C-T. Ho, *J. Am. Chem. Soc.*, **99**, 4105 (1977). See also (b) A. Pross, *Adv. Phys. Org. Chem.*, **14**, 69 (1977).

electrophilicity and can even be nucleophilic, as is the case with dimethoxy-carbene.[101]

A second question posed by the alkene additions is one of stereochemistry. A concerted ring formation of the type shown in Equation 6.45 implies stereospecific

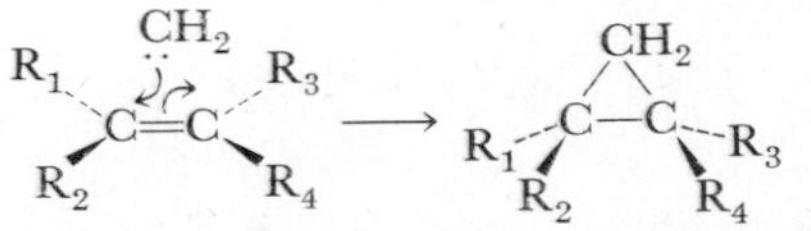

$$
\begin{array}{cc}
R_1 \quad CH_2 \quad R_3 \\
\diagdown C{=}C \diagup \\
R_2 \qquad R_4
\end{array}
\longrightarrow
\begin{array}{c}
CH_2 \\
R_1 \diagup C{-}C \diagdown R_3 \\
R_2 \quad R_4
\end{array}
\tag{6.45}
$$

cis addition, a suggestion first made in 1956 by Skell and Woodworth.[102] With a triplet carbene, however, the spin state of one of the electrons must change before bonding can be completed; if this process takes long enough for rotation to occur about bonds in the intermediate, a mixture of products should result (Scheme 10).[103] It should be pointed out that the singlet is not required to react stereospecifi-

SCHEME 10

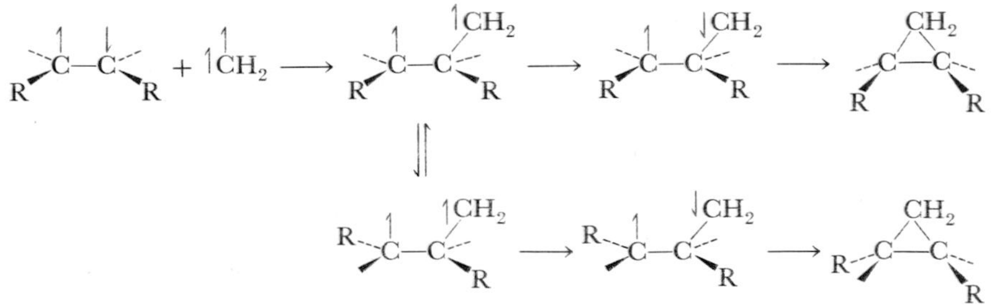

cally simply because it can, nor must the triplet necessarily add with loss of stereo-chemistry.[104] Table 6.8 presents the data. It will be noted that, with the exception of dicyanomethylene, loss of stereochemistry is not complete in triplet additions. The rates of spin inversion and bond rotation must be of comparable magnitudes. Stereochemistry of the concerted additions is considered further in Section 11.1. Carbenoids add stereospecifically,[105] but since free carbenes are not involved, the singlet–triplet considerations do not apply.

[101](a) U. Schöllkopf and E. Wiskott, *Angew. Chem. Int. Ed. Engl.*, **2**, 485 (1963); (b) D. Seebach, *Angew. Chem. Int. Ed. Engl.*, **6**, 443 (1967); see also (c) R. Gleiter and R. Hoffmann, *J. Am. Chem. Soc.*, **90**, 5457 (1968); (d) H.-J. Schönherr and H.-W. Wanzlick, *Chem. Ber.*, **103**, 1037 (1970); (e) R. W. Hoffmann and M. Reiffen, *Chem. Ber.*, **109**, 2565 (1976).

[102](a) See note 92(b); (b) R. Hoffmann, *J. Am. Chem. Soc.*, **90**, 1475 (1968), argues on the basis of orbital symmetry that the approach cannot be symmetrical, and that the $:CH_2$ must initially be closer to one end of the alkene than to the other; the one-step stereospecific nature of the addition is not affected by this argument. More refined *ab initio* calculations make a similar prediction: (c) B. Zurawski and W. Kutzelnigg, *J. Am. Chem. Soc.*, **100**, 2654 (1978).

[103]R. Hoffmann, note 102(b), gives a more rigorous discussion of the additions and reaches the same conclusions regarding stereochemistry.

[104]See note 64(b) and note 103.

[105]G. L. Closs and L. E. Closs, *Angew. Chem. Int. Ed. Engl.*, **1**, 334 (1962).

Table 6.7 RELATIVE REACTIVITIES OF CARBENES WITH ALKENES

Carbene	Alkene (structure 1)	Alkene (structure 2)	Alkene (cyclohexene)	Alkene (structure 4)	Alkene (structure 5)	Alkene (structure 6)	Alkene (structure 7)	Reference
:CF$_2$			1.0	12.8				a
:CCl$_2$	0.24	0.14	1.0	8.3	2.05	23	54	b
:CCl$_2$			1.0		3.53	21.5		c
:CBr$_2$		0.17	1.0	3.7		7.4	6.9	b
ICH$_2$ZnI	0.39		1.0		2.53	2.18	1.29	d

[a] From F$_2$C=N–N (structure), hv. R. A. Mitsch, *J. Am. Chem. Soc.*, **87**, 758 (1965).

[b] From HCCl$_3$ or HCBr$_3$, base. W. v. E. Doering and W. A. Henderson, Jr., *J. Am. Chem. Soc.*, **80**, 5274 (1958).

[c] From φHgCCl$_2$Br. D. Seyferth and J. M. Burlitch, *J. Am. Chem. Soc.*, **86**, 2730 (1964).

[d] E. P. Blanchard and H. E. Simmons, *J. Am. Chem. Soc.*, **86**, 1337 (1964).

Table 6.8 STEREOCHEMISTRY OF ADDITION OF CARBENES TO *cis*- AND *trans*-2-BUTENE

Carbene	Probable multiplicity, ground state	Multiplicity, reactive state[a]	Phase	Conditions	Products, percent cis addition — to  (cis)	Products, percent cis addition — to 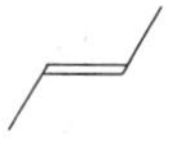 (trans)	References
:CH$_2$	Triplet	Singlet	Gas	500 mm pressure	83	95	b
:CH$_2$	Triplet	Triplet	Gas	>2000 mm; excess argon	51	—	c
:CH$_2$	Singlet	Singlet	Liquid		100	100	d
:CBr$_2$		(Singlet)	Liquid	α Elimination	100	100	e
:CCl$_2$		(Singlet)	Liquid	From φHgCCl$_2$Br	100	100	f
:CF$_2$	Singlet	Singlet	Liquid	From F$_2$C [structure] 	100	100	g
:C(CN)$_2$	Triplet[h]	(Singlet)	Liquid	Pure alkene as solvent	92	94	i
:C(CN)$_2$		(Triplet)	Liquid	Alkene diluted 100:1 with c-C$_6$H$_{12}$	30	70	i
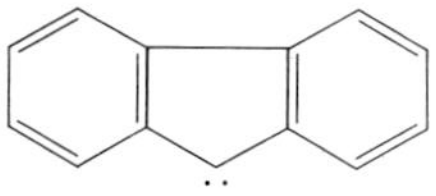 [fluorenylidene structure]	Triplet[j]	Singlet[k]	Liquid	Butadiene added	~98	—	l
		Triplet	Liquid	Diluted with C$_6$F$_6$	21	88	l, m

[a] Parentheses indicate that reactive state is deduced from stereochemistry of olefin addition.

[b] H. M. Frey, *Proc. Roy. Soc.*, **A251**, 575 (1959).

[c] H. M. Frey, *J. Am. Chem. Soc.*, **82**, 5947 (1960).

[d] P. S. Skell and R. C. Woodworth, *J. Am. Chem. Soc.*, **78**, 4496, 6427 (1956).

[e] P. S. Skell and A. Y. Garner, *J. Am. Chem. Soc.*, **78**, 3409, 5430 (1956); W. v. E. Doering and P. LaFlamme, *J. Am. Chem. Soc.*, **78**, 5447 (1956). It is possible that this α elimination does not involve a free carbene.

[f] D. Seyferth and J. M. Burlitch, *J. Am. Chem. Soc.*, **86**, 2730 (1964).

[g] R. A. Mitsch, *J. Am. Chem. Soc.*, **87**, 758 (1965).

[h] E. Wasserman, L. Barash, and W. A. Yager, *J. Am. Chem. Soc.*, **87**, 2075 (1965).

[i] E. Ciganek, *J. Am. Chem. Soc.*, **88**, 1979 (1966).

[j] R. W. Brandon, G. L. Closs, C. E. Davoust, C. A. Hutchison, Jr., B. E. Kohler, and R. Silbey, *J. Chem. Phys.*, **43**, 2006 (1965).

[k] Butadiene is an effective scavenger for triplets.

[l] M. Jones, Jr., and K. R. Rettig, *J. Am. Chem. Soc.*, **87**, 4015 (1965).

[m] M. Jones, Jr., and K. R. Rettig, *J. Am. Chem. Soc.*, **87**, 4013 (1965).

502

PROBLEMS

1. Propose mechanisms to account for the following rearrangements:

(a)

$$CH_3-\overset{O}{\overset{\|}{C}}-CH_2-\overset{\phi}{\underset{\phi}{\overset{|}{C}}}-\overset{O}{\overset{\|}{C}}-\phi \xrightarrow{CH_3O^-Na^+} CH_3-\overset{O}{\overset{\|}{C}}-\overset{\phi}{\underset{\phi}{\overset{|}{C}}}-CH_2-\overset{O}{\overset{\|}{C}}-\phi$$

(b)

$\xrightarrow{RO^-M^+}$

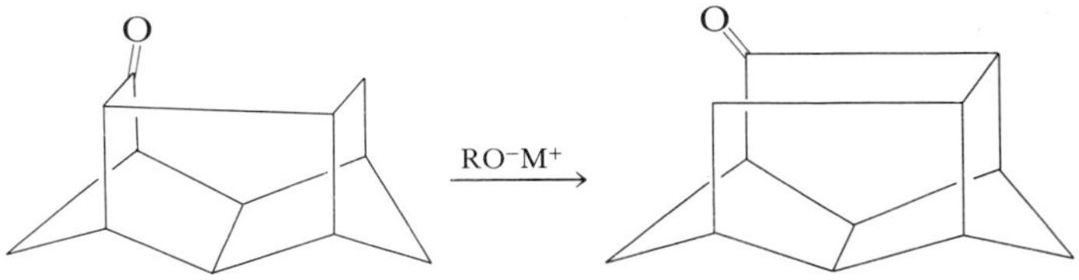

2. Explain why the β-keto acid **1** does not decarboxylate at 300°C, whereas **2** does so readily below 100°C.

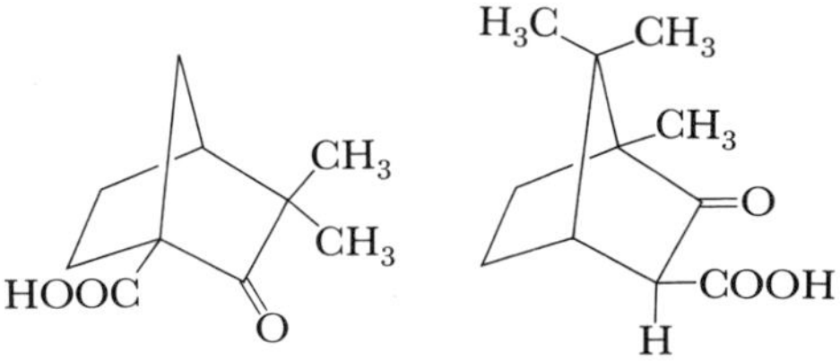

1   2

3. Explain the difference in product distributions in the following two reactions:

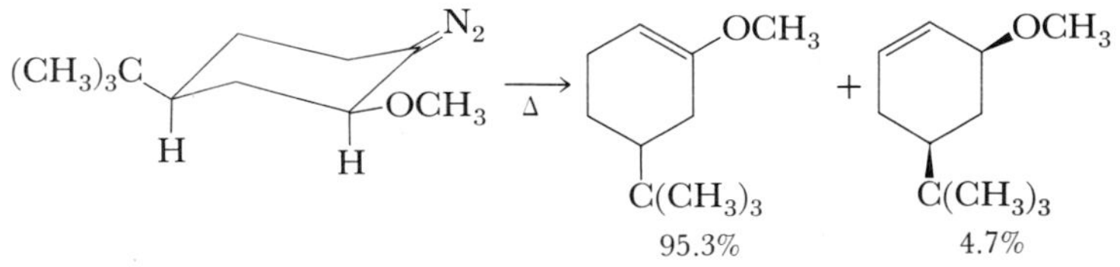

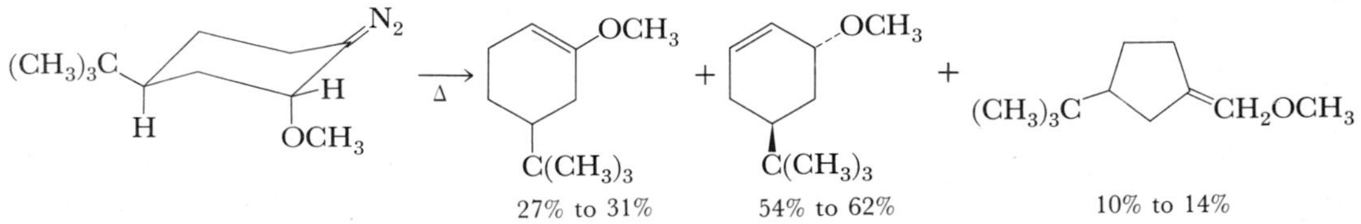

95.3%   4.7%

27% to 31%   54% to 62%   10% to 14%

4. Predict the products formed by reaction of ground-state carbon atoms with *cis*-2-butene and with *trans*-2-butene.

5. Propose a mechanism to account for the following result:

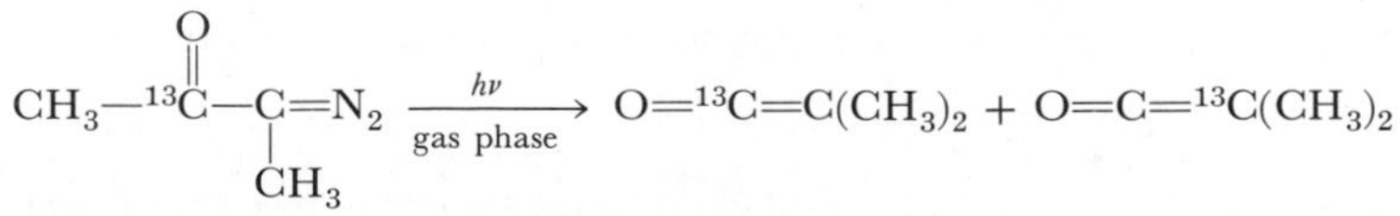

$$CH_3-{}^{13}\overset{\overset{\textstyle O}{\|}}{C}-\underset{\underset{\textstyle CH_3}{|}}{C}=N_2 \xrightarrow[\text{gas phase}]{h\nu} O={}^{13}C=C(CH_3)_2 + O=C={}^{13}C(CH_3)_2$$

6. Propose a mechanism to account for the following transformation:

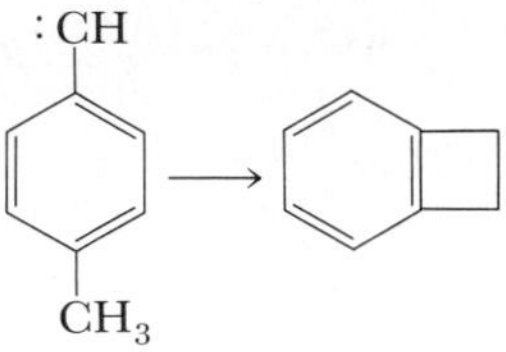

7. Explain why the carbene **3** does not react with cyclohexene in the manner of ordinary carbenes but does react with dimethyl fumarate (**4**) and maleate (**5**) to yield spiropentanes (**6**).

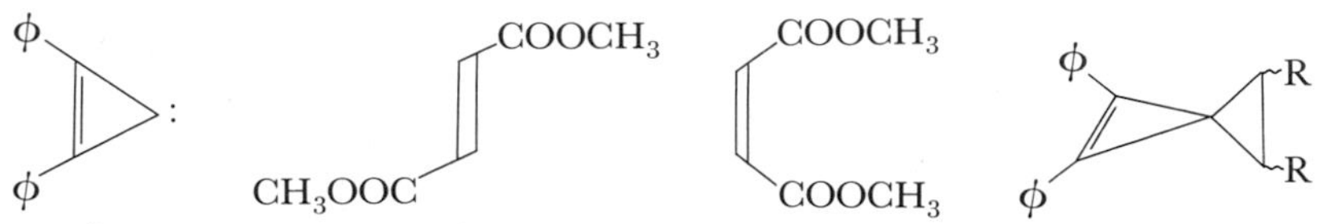

8. Predict whether, in the following reaction, the rate of reaction will be much faster, much slower, or about the same when R = —CH$_3$ than when

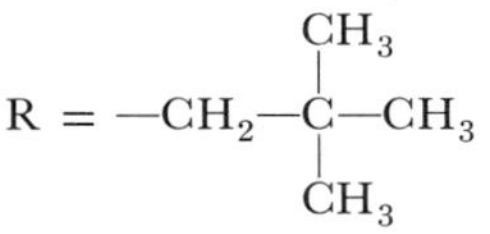

$$R = -CH_2-\underset{\underset{\textstyle CH_3}{|}}{\overset{\overset{\textstyle CH_3}{|}}{C}}-CH_3$$

Explain your reasoning.

$$RHgX + \overset{*}{HgX_2} \rightleftharpoons R\overset{*}{Hg}X + HgX_2$$

9. Explain the relative positions of fluorene and cyclopentadiene in Figure 6.1.

10. Explain why ethyl formate condenses with 2-methylcyclohexanone in the presence of base at the unsubstituted position:

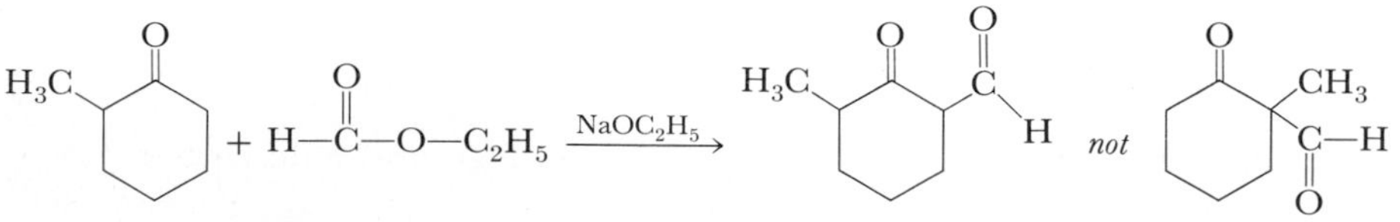

11. Propose a mechanism for the α-halosulfone rearrangement:

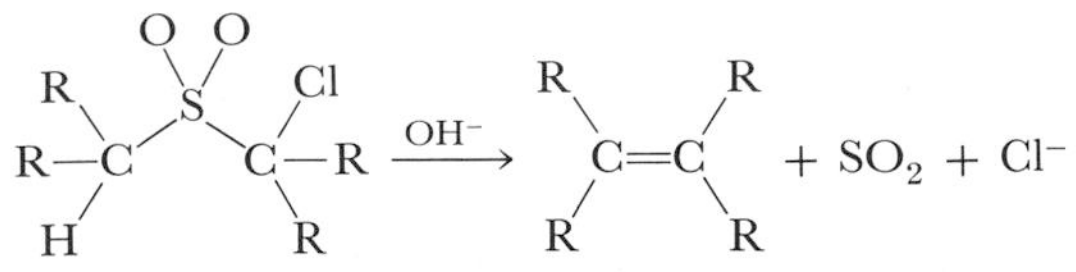

$$
\underset{H}{\overset{R}{R-C}}\overset{\overset{O\ O}{\underset{\|\ \|}{S}}}{\underset{R}{C-R}}\overset{Cl}{\underset{R}{C-R}} \xrightarrow{\ OH^-\ } \underset{R}{\overset{R}{C}}=\underset{R}{\overset{R}{C}} + SO_2 + Cl^-
$$

12. There are two possible products from the Favorskii rearrangement that occurs when **7** reacts with sodium methoxide. What are the two products, and which will form in larger amount?

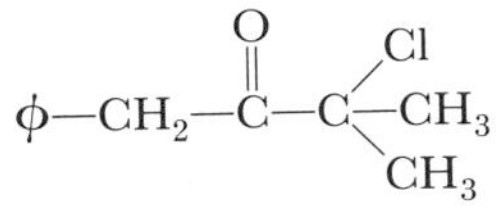

$$
\phi-CH_2-\overset{\overset{O}{\|}}{C}-\overset{\overset{Cl}{|}}{\underset{\underset{CH_3}{|}}{C}}-CH_3
$$

7

REFERENCES FOR PROBLEMS

1. (a) M. J. Betts and P. Yates, *J. Am. Chem. Soc.*, **92**, 6982 (1970); P. Yates and M. J. Betts, *J. Am. Chem. Soc.*, **94**, 1965 (1972); (b) R. Howe and S. Winstein, *J. Am. Chem. Soc.*, **87**, 915 (1965); T. Fukunaga, *J. Am. Chem. Soc.*, **87**, 915 (1965); T. Fukunaga, *J. Am. Chem. Soc.*, **87**, 916 (1965).
2. F. S. Fawcett, *Chem. Rev.*, **47**, 219 (1950).
3. L. S. Press and H. Shechter, *J. Am. Chem. Soc.*, **101**, 509 (1979).
4. P. S. Skell and R. R. Engel, *J. Am. Chem. Soc.*, **87**, 1135 (1965).
5. I. G. Csizmadia, J. Font, and O. P. Strausz, *J. Am. Chem. Soc.*, **90**, 7360 (1968); J. Fenwick, G. Frater, K. Ogi, and O. P. Strausz, *J. Am. Chem. Soc.*, **95**, 124 (1973); I. G. Csizmadia, H. E. Gunning, R. K. Gosavi, and O. P. Strausz, *J. Am. Chem. Soc.*, **95**, 133 (1973); O. P. Strausz, R. K. Gosavi, A. S. Denes, and I. G. Csizmadia, *J. Am. Chem. Soc.*, **98**, 4784 (1976).
6. M. Jones, Jr., *Acc. Chem. Res.*, **7**, 415 (1974); W. J. Baron, M. Jones, Jr., and P. W. Gaspar, *J. Am. Chem. Soc.*, **92**, 4739 (1970).
7. W. M. Jones, M. E. Stowe, E. E. Wells, Jr., and E. W. Lester, *J. Am. Chem. Soc.*, **90**, 1849 (1968).
8. E. D. Hughes and H. C. Volger, *J. Chem. Soc.*, 2359 (1961).
9. T. B. McMahon and P. Kebarle, *J. Am. Chem. Soc.*, **98**, 3399 (1976).
10. H. O. House, *Modern Synthetic Reactions*, 2d ed., W. A. Benjamin, Menlo Park, Calif., 1972, p. 750.
11. L. A. Paquette, *Acc. Chem. Res.*, **1**, 209 (1968).
12. A. S. Kende, *Org. Reactions*, **11**, 261 (1960), p. 267.

Chapter 7

ADDITION

AND ELIMINATION

REACTIONS

In this chapter we shall discuss the destruction and formation of carbon–carbon multiple bonds by addition and elimination reactions, respectively. The mechanism of aromatic substitution in which addition and elimination both occur in separate steps will also be discussed.

7.1 ELECTROPHILIC ADDITION TO DOUBLE AND TRIPLE BONDS[1]

Carbon–carbon π bonds are relatively weak ($\sim$65 kcal mole^{-1}). They are also, unless substituted by strong electron-withdrawing groups, electron-rich. For these reasons addition to them by electrophilic reagents usually occurs readily. The exact mechanism of addition depends on the reagent. To understand some of the mechanistic evidence, we must first consider general and specific acid and base catalysis.

General and Specific Acid and Base Catalysis

Suppose that acid catalyzes a reaction by forming the conjugate acid of the substrate, S, in a rapid equilibrium preceding a slower step, as indicated in Scheme 1.

[1](a) P. B. D. de la Mare and R. Bolton, *Electrophilic Additions to Unsaturated Systems,* Elsevier, Amsterdam, 1966; (b) R. C. Fahey, in *Topics in Stereochemistry,* Vol. 3, E. L. Eliel and N. L. Allinger, Eds., Wiley, New York, 1968, p. 237; (c) R. Bolton, in *Comprehensive Chemical Kinetics,* Vol. 9, C. H. Bamford and C. F. H. Tipper, Eds., Elsevier, New York, 1973, Chap. 1, (d) M. A. Wilson, *J. Chem. Educ.,* **52,** 495 (1975); (e) G. H. Schmid and D. G. Garratt, *The Chemistry of the Double-Bonded Functional Groups,* Part 2, Supplement A to *The Chemistry of the Functional Groups,* S. Patai, Ed., Wiley, New York, 1977, p. 725; (f) E. Winterfeldt, in *The Chemistry of Acetylenes,* H. G. Viehe, Ed., Dekker, New York, 1969; (g) G. H. Schmid in *The Chemistry of the Carbon–Carbon Triple Bond,* S. Patai, Ed., Wiley, New York, 1978, p. 275.

SCHEME 1

$$S + HA \xrightleftharpoons{fast} SH^+ + A^-$$

$$SH^+ \xrightarrow[k]{slow} products$$

The reaction rate is given by Equation 7.1. Concentration $[SH^+]$ is in turn determined by the preliminary equilibrium, for which we may write the equilibrium constant K (Equation 7.2). But concentrations $[HA]$ and $[A^-]$ are themselves related to $[H^+]$ through the K_a for the acid HA (Equation 7.3). Combination of Equations 7.2 and 7.3 shows that under these circumstances the concentration of reactive species, SH^+, is actually determined by the H^+ concentration and the K_a of SH^+ (Equations 7.4—7.6). The reaction rate (Equation 7.8) depends on the H^+ concentration; acids other than H^+ that may be present contribute to the reaction rate only through their contribution to determining pH. Thus the reaction will proceed at the same rate in two buffers of the same buffer ratio (and hence the same pH) but different concentrations. Such a reaction is said to exhibit *specific acid catalysis*. An entirely analogous argument can be made for a base-catalyzed reaction with a preliminary equilibrium to form the conjugate base of the substrate. Such a reaction shows *specific base catalysis. Specific catalysis does not involve any proton transfer in the rate-determining step.*

$$rate = k[SH^+] \tag{7.1}$$

$$K = \frac{[SH^+][A^-]}{[S][HA]} \tag{7.2}$$

$$K_{a_{HA}} = \frac{[H^+][A^-]}{[HA]} \tag{7.3}$$

$$\frac{K}{K_{a_{HA}}} = \frac{[SH^+][A^-]}{[S][HA]} \frac{[HA]}{[A^-][H^+]} \tag{7.4}$$

$$\frac{K}{K_{a_{HA}}} = \frac{[SH^+]}{[H^+][S]} = \frac{1}{K_{a_{SH^+}}} \tag{7.5}$$

$$[SH^+] = \frac{[H^+][S]}{K_{a_{SH^+}}} \tag{7.6}$$

$$rate = \frac{k}{K_{a_{SH^+}}} [H^+][S] \tag{7.7}$$

$$= k_{H^+}[H^+][S] = k_{obs}[S] \tag{7.8}$$

(where $k_{H^+} = k/K_{a_{SH^+}}$ and $k_{obs} = k_{H^+}[H^+]$).

Another possibility is that the proton transfer itself constitutes the rate-determining step, or that the rate-determining step consists of proton transfer occurring simultaneously with some other process. An example is the deprotonation of carbon acids which we discussed in Section 3.3 when we considered the Brønsted catalysis law relating the effectiveness of the catalyst to its equilibrium acidity.

Under these circumstances each individual acid (or base) present in the system can act as a proton donor (or acceptor) in the rate-determining step, and the rate of this step then depends on each of these acids (bases) individually, as indicated in Equation 7.9 for an acid-catalyzed process. A reaction that follows Equation 7.9 is said to be subject to *general acid catalysis*; the analogous situation with base catalysis is *general base catalysis*.

$$k_{obs} = k_0 + k_{H^+}[H^+] + k_1[HA_1] + k_2[HA_2] + \cdots \tag{7.9}$$

Acid catalysis: $\quad \log k_n = \alpha \log K_{a_n} + \log C \quad$ or $\quad k_n = CK_{a_n}^{\alpha} \tag{7.10}$

Base catalysis: $\quad \log k_n = -\beta \log K_{a_{BH^+}} + \log C' \quad$ or $\quad k_n = C'K_{a_{BH^+}}^{-\beta} \tag{7.11}$

The Brønsted catalysis law states that the individual catalytic constants, k_n, should be related to the equilibrium acidities by Equation 7.10, or, for a base-catalyzed process, by Equation 7.11. (See Section 3.3 for further discussion of these relations.) The Brønsted slope α (β for base catalysis) is a measure of the sensitivity of the reaction to the acid strengths of the various catalysts. If for a particular reaction α is near 1.0, most of the catalysis will be by the strongest acid (H^+ in aqueous solution); catalysis by weaker acids will then be difficult or impossible to detect, and the situation will be indistinguishable kinetically from specific catalysis. If α is near zero, all acids will be equally effective, but since the solvent is present in much higher concentration than any other acid, it will be the predominant catalyst and again the catalysis by other acids will be difficult to detect.

The usual means of finding general catalysis is to measure the reaction rate with various concentrations of the general acids or bases but a constant concentration of H^+. Since the pH depends only on the ratio of [HA] to [A$^-$] and not on the absolute concentrations, this requirement may be satisfied by the use of buffers.

Hydration

The rate of addition of water to a double bond depends strongly on the acidity of the medium and correlates with the appropriate acidity function.[2] Figure 7.1 shows the correlation for the hydration of ethylene with H_0. The steep slope (-1.54) indicates that the transition state for the rate-determining step contains a proton.

Taft originally suggested[3] that the mechanism for hydration involves the rapid, reversible formation of a π complex from a proton and the alkene, and that this complex then rearranges, in the rate-determining step, to the carbocation as shown in Equation 7.12. This mechanism has, however, been shown to be incorrect.

$$\tag{7.12}$$

[2](a) K. Yates and H. Wai, *J. Am. Chem. Soc.*, **86**, 5408 (1964); (b) N. C. Deno, P. T. Groves, and G. Saines, *J. Am. Chem. Soc.*, **81**, 5790 (1959); (c) W. K. Chwang, V. J. Nowlan, and T. T. Tidwell, *J. Am. Chem. Soc.*, **99**, 7233 (1977); (d) K. M. Koshy, D. Roy, and T. T. Tidwell, *J. Am. Chem. Soc.*, **101**, 357 (1979); (e) C. F. Bernasconi and W. J. Boyle, Jr., *J. Am. Chem. Soc.*, **96**, 6070 (1974).

[3] R. W. Taft, Jr., *J. Am. Chem. Soc.*, **74**, 5372 (1952).

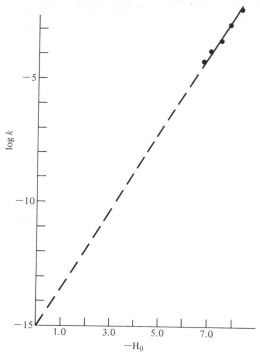

Figure 7.1 Dependence of rate of hydration of ethylene on acidity. Reprinted with permission from W. K. Chwang, V. J. Nowlan, and T. T. Tidwell, *J. Am. Chem. Soc.*, **99**, 7233 (1977). Copyright 1977 American Chemical Society.

Kresge and co-workers studied the hydration of *trans*-cyclooctene and of 2,3-dimethyl-2-butene in phosphoric acid–bisulfate buffer solutions in which the amount of undissociated acid varied and found that the reactions exhibit general acid catalysis.[4] This behavior parallels that observed earlier in the hydration of styrene and substituted styrenes[5] and many alkoxyethylenes[4] and more recently in the hydration of di-*t*-butylketene.[6] The process of Equation 7.12 should show only specific acid catalysis.

The currently accepted mechanism for hydration is the Ad_E2 (addition–electrophilic–bimolecular) mechanism in Equation 7.13. This mechanism not only takes into account the need, stated above, for a proton and for the conjugate base of the acid that is donating the proton in the transition state, but it also has a body of other evidence supporting it.

$$\text{C=C} + HA \xrightarrow[\text{step 1}]{\text{slow}} -\overset{|}{\underset{H}{C}}-\overset{|}{\underset{\oplus}{C}}- \xrightarrow[\text{step 2}]{H_2O} -\overset{|}{\underset{H}{C}}-\overset{|}{\underset{\overset{+}{O}H_2}{C}}- \underset{\text{step 3}}{\rightleftharpoons} -\overset{|}{\underset{H}{C}}-\overset{|}{\underset{OH}{C}}- \qquad (7.13)$$

[4] A. J. Kresge, Y. Chiang, P. H. Fitzgerald, R. S. McDonald, and G. H. Schmid, *J. Am. Chem. Soc.*, **93**, 4907 (1971).
[5] W. M. Schubert, B. Lamm, and J. R. Keeffe, *J. Am. Chem. Soc.*, **86**, 4727 (1964).
[6] S. H. Kabir, H. R. Seikaly, and T. T. Tidwell, *J. Am. Chem. Soc.*, **101**, 1059 (1979).

For example, the rate of hydration is greatly increased by electron-donating substituents (e.g., MeO, cyclopropyl, Me) and decreased by electron-withdrawing substituents (e.g., Cl, CF_3).[5,7,8] Figure 7.2 shows a plot of the second-order rate constants of a large number of 1,1-disubstituted alkenes (1) vs. the sum of the σ_p^+ constants of the R substituents. Note the good correlation even though the range of reactivity is 22 orders of magnitude! The steep slope of the line (-10.5) indicates that the substituted carbon has a high positive charge density in the transition state of the rate-determining step.

$$\begin{matrix} R \\ \diagdown \\ \diagup \\ R \end{matrix} C{=}CH_2$$

1

Furthermore, the accelerating effect of electron-donating substituents is cumulative only if the substituents are on the same side of the double bond[9] (see Table 7.1). This result is consistent only with an asymmetric transition state in which the positive charge is localized on one carbon of the original double bond. In accord with the substituent effects of Table 7.1, hydration always gives *regiospecific*[10] Markownikoff addition; that is, the proton adds to the less substituted side of the double bond. Also in agreement with a cationic intermediate but not with concerted addition is the fact that Wagner–Meerwein rearrangements sometimes occur during hydration.[11]

When hydrations are carried out in both deuterated and nondeuterated solvents, a primary isotope effect of 1–6 is observed depending on the alkene and the acid used.[6,12,13] Thus the proton must be undergoing a covalency change in the rate-determining transition state. This conclusion is in agreement with the mechanism in Equation 7.13 but not with a mechanism in which protonation is rapid and reversible and is followed by a slow attack by the oxygen atom of water.

The following observations also support the conclusion that the protonation is not reversible. Schubert found that when styrene-β,β-d_2 (2) is hydrated, recovered, unreacted 2 has not lost deuterons to the solvent. If a carbocation intermedi-

[7](a) V. J. Nowlan and T. T. Tidwell, *Acc. Chem. Res.*, **10**, 252 (1977), and references therein; (b) K. Oyama and T. T. Tidwell, *J. Am. Chem. Soc.*, **98**, 947 (1976); (c) W. M. Schubert and B. Lamm, *J. Am. Chem. Soc.*, **88**, 120 (1966); (d) W. M. Schubert and J. R. Keeffe, *J. Am. Chem. Soc.*, **94**, 559 (1972).

[8] See note 2(d).

[9](a) Unpublished results of R. W. Taft, Jr., cited in P. D. Bartlett and G. D. Sargent, *J. Am. Chem. Soc.*, **87**, 1297 (1965); (b) P. Knittel and T. T. Tidwell, *J. Am. Chem. Soc.*, **99**, 3408 (1977); (c) W. K. Chang and T. T. Tidwell, *J. Org. Chem.*, **43**, 1404 (1978).

[10] *Regiospecific* is a term introduced by A. Hassner, *J. Org. Chem.*, **33**, 2684 (1968). If bonds can be made or broken in two or more different orientations but only one of the possible isomers is formed, the reaction is regiospecific. If there is a significant preponderance of one isomer formed, Hassner calls that reaction regioselective. Most workers use only the former term with qualifying adjectives such as "high" or "low."

[11] See note 1(a).

[12] See note 2(c) and note 7(d).

[13](a) V. Gold and M. A. Kessick, *J. Chem. Soc.*, 6718 (1965); (b) A. J. Kresge and Y. Chiang, *J. Chem. Soc. B*, 58 (1967).

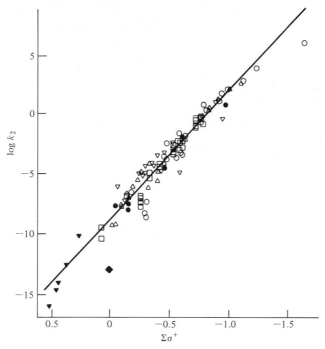

Figure 7.2 Correlation of acid-catalyzed hydration rates of alkenes vs. $\Sigma\,\sigma^+$ of the substituents on the double bond. The different types of points indicate very different types of substituents. Reprinted with permission from K. M. Koshy, D. Roy, and T. T. Tidwell, *J. Am. Chem. Soc.*, **101**, 357 (1979). Copyright 1979 American Chemical Society.

2

ate were formed by a rapid equilibrium prior to the rate-determining process, then deuterons as well as protons should be lost from the intermediate carbocation in the reverse of the first step of Equation 7.13.[5,14] Furthermore, when 2-methyl-1-butene is subjected to hydration conditions and recovered after 50 percent reaction, the starting material has not isomerized as would be expected if step 1 of Equation 7.13 were a rapid equilibrium (see Equation 7.14).[15]

$$(7.14)$$

[14] See note 7(c), (d).

[15] J. B. Levy, R. W. Taft, Jr., and L. P. Hammett, *J. Am. Chem. Soc.*, **75**, 1253 (1953).

Table 7.1 EFFECT OF METHYL SUBSTITUTION ON THE RATE OF HYDRATION OF CARBON–CARBON DOUBLE BONDS, H_2O, 25°C[a]

Alkene	k_2 (M^{-1} sec^{-1})
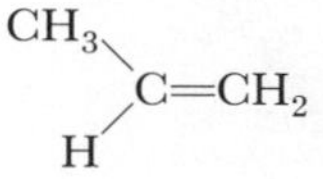 CH_3, H / C=CH$_2$	4.95×10^{-8}
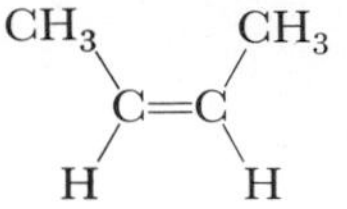 CH_3, H / C=C \ CH_3, H	8.32×10^{-8}
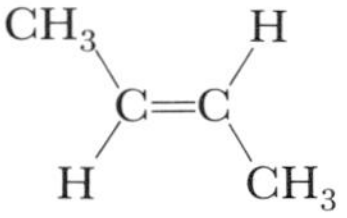 CH_3, H / C=C \ H, CH_3	3.51×10^{-8}
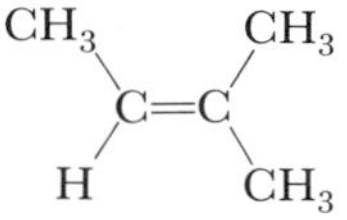 CH_3, H / C=C \ CH_3, CH_3	2.15×10^{-4}
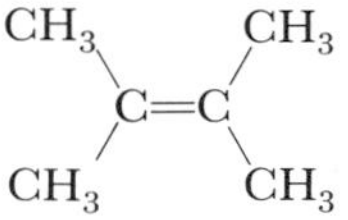 CH_3, CH_3 / C=C \ CH_3, CH_3	3.42×10^{-4}

[a] See P. Knittel and T. T. Tidwell, *J. Am. Chem. Soc.*, **99**, 3408 (1977), and references therein.

In the acid-catalyzed addition of methanol to **3**, the intermediate **4** is stable enough to build up to observable concentrations. Thus Bernasconi and Boyle were able to determine directly the rate constants for formation and destruction of **4**. Indeed, k_2 was found to be greater than k_{-1}; k_1 is the rate-determining step.[16]

Alkynes generally undergo acid-catalyzed hydration to form vinyl alcohols, which rapidly rearrange to ketones.[17] These hydrations exhibit general acid catalysis,[18] and unreacted acetylenes recovered after partial reaction have not exchanged deuterium with the solvent.[19] Noyce and Schiavelli have found that the rate of hydration of ring-substituted phenylacetylenes is very dependent on the nature of the substituent, giving a linear correlation with σ^+ ($\rho = -3.84$).[20] Thus all the

[16] See note 2(e).
[17] (a) G. Modena and U. Tonnellato, *Adv. Phys. Org. Chem.*, **9**, 185 (1971); (b) J. P. Stang, *Prog. Phys. Org. Chem.*, **10**, 205 (1973).
[18] (a) W. Drenth and H. Hogeveen, *Recl. Trav. Chim.*, **79**, 1002 (1960); (b) E. J. Stamhuis and W. Drenth, *Recl. Trav. Chim.*, **80**, 797 (1961); (c) E. J. Stamhuis and W. Drenth, *Recl. Trav. Chim.*, **82**, 394 (1963); (d) H. Hogeveen and W. Drenth, *Recl. Trav. Chim.*, **82**, 410 (1963); (e) D. S. Noyce and M. D. Schiavelli, *J. Am. Chem. Soc.*, **90**, 1020 (1968).
[19] See note 18(d).
[20] See note 18(e).

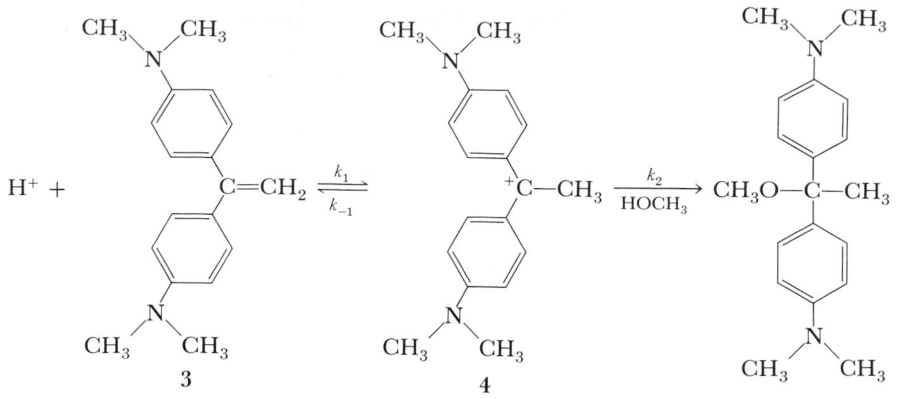

$$H^+ + \text{(structure 3)} \underset{k_{-1}}{\overset{k_1}{\rightleftharpoons}} \text{(structure 4)} \xrightarrow[\text{HOCH}_3]{k_2} CH_3O\text{—}\overset{|}{\underset{|}{C}}\text{—}CH_3 \quad (7.15)$$

3   **4**

evidence points to the slow formation of the transient unstable vinyl cation in a mechanism entirely analogous to that for hydration of alkenes as shown in Equation 7.16. As Table 7.2 shows, the rate of hydration of alkenes is usually greater than

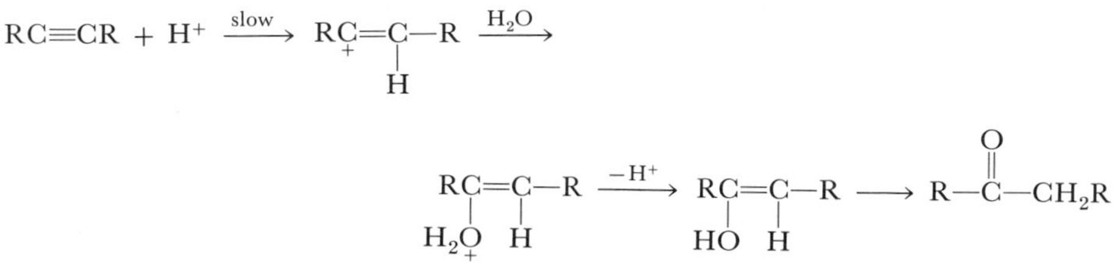

$$RC\equiv CR + H^+ \xrightarrow{\text{slow}} R\underset{+}{C}=\overset{|}{\underset{H}{C}}\text{—}R \xrightarrow{H_2O}$$

$$\underset{\overset{|}{\underset{H_2\overset{+}{O}}{}}}{R C}=\overset{|}{\underset{H}{C}}\text{—}R \xrightarrow{-H^+} R\underset{\overset{|}{HO}}{C}=\overset{|}{\underset{H}{C}}\text{—}R \longrightarrow R\text{—}\overset{\overset{O}{\|}}{C}\text{—}CH_2R \quad (7.16)$$

that of alkynes. It has been suggested that the greater reactivity of alkenes with electrophiles is due to the fact that the HOMO of alkenes is at a higher energy level than the HOMO of alkynes. Thus the HOMO of an alkene is closer in energy to the electrophiles LUMO than is that of the alkyne, and there is more charge stabilization of the transition state with alkenes than with alkynes.[21,22]

Addition of Hydrohalides and Acetic Acid

The addition of HX to double bonds in the dark and in the absence of free-radical initiators is closely related to hydration. The orientation of the elements of HX in the adduct always corresponds to Markownikoff addition; no deuterium exchange with solvent is found in unreacted alkenes recovered after partial reaction, nor is recovered starting material isomerized after partial reaction.[23] However, the addition of HX apparently can proceed by a number of different mechanisms depend-

[21](a) D. S. Noyce, M. A. Matesich, M. D. Schiavelli, and P. E. Peterson, *J. Am. Chem. Soc.*, **87**, 2295 (1965); (b) K. Yates, G. H. Schmid, T. W. Regulski, D. G. Garratt, H.-W. Leung, and R. McDonald, *J. Am. Chem. Soc.*, **95**, 160 (1973), and references therein.

[22] See also G. Modena, F. Rivetti, G. Scorrano and U. Tonnellato, *J. Am. Chem. Soc.*, **99**, 3392 (1977).

[23] Y. Pocker, K. D. Stevens, and J. J. Champoux, *J. Am. Chem. Soc.*, **91**, 4199 (1969).

Table 7.2 RELATIVE RATES OF
HYDRATION OF ALKENES
AND ALKYNES[a]

Alkene Alkyne	Relative rate
$\phi CH{=}CH_2$ $\phi C{\equiv}CH$	0.55–0.65
$\phi CH{=}CHCH_3$ $\phi C{\equiv}CCH_3$	1.5
$n BuCH{=}CH_2$ $n BuC{\equiv}CH$	3.6
$EtCH{=}CHEt$ $EtC{\equiv}CEt$	13–19

[a] Reprinted with permission from K. Yates, G. H. Schmid, T. W. Regulski, D. G. Garratt, Hei-Wun Leung, and R. McDonald, *J. Am. Chem. Soc.*, **95**, 160 (1973). Copyright by the American Chemical Society.

ing on the nature of the substrate and on the reaction conditions. Thus when HCl is added to *t*-butylethylene in acetic acid, the rate is first-order in each reactant and the products are those shown in Equation 7.17.[24] Since **5** and **6** were demonstrated to be stable to the reaction conditions, the rearranged product (**6**) can be formed

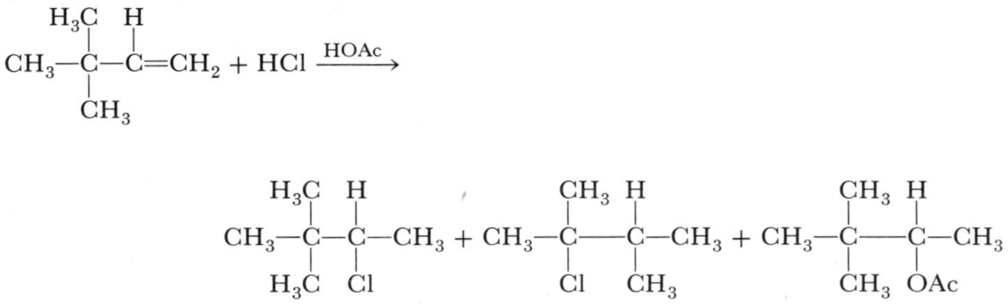

$$CH_3{-}\underset{\underset{\overset{|}{CH_3}}{\overset{|}{C}}}{\overset{\overset{H_3C}{|}\;\overset{H}{|}}{C}}{-}\overset{H}{\underset{|}{C}}{=}CH_2 + HCl \xrightarrow{\;HOAc\;}$$

$$CH_3{-}\underset{\underset{H_3C}{|}}{\overset{H_3C}{|}}C{-}\underset{\underset{Cl}{|}}{\overset{H}{|}}C{-}CH_3 + CH_3{-}\underset{\underset{Cl}{|}}{\overset{CH_3}{|}}C{-}\underset{\underset{CH_3}{|}}{\overset{H}{|}}C{-}CH_3 + CH_3{-}\underset{\underset{CH_3}{|}}{\overset{CH_3}{|}}C{-}\underset{\underset{OAc}{|}}{\overset{H}{|}}C{-}CH_3 \qquad (7.17)$$

Relative amounts:	2	2	1
	5	**6**	**7**

only if a carbocationic intermediate is formed during reaction. However, the carbocation exists almost solely in an intimate ion pair, and the rate of collapse of the ion pair to products must be faster than, or comparable to, the rate of diffusion of Cl⁻ away from the carbocation. This must be so because the ratio of chloride to acetate products is unaffected by the concentration of HCl or by added chloride ion in the form of tetramethylammonium chloride. If the nature of the products depended on

[24] (a) R. C. Fahey and C. A. McPherson, *J. Am. Chem. Soc.*, **91**, 3865 (1969); (b) rearranged acetate corresponding to **6** is not stable to the reaction conditions but reacts with Cl⁻ to form **6**.

the environment outside the ion pair, the ratio of chloride to acetate would increase with increasing chloride ion concentration in the solution. Such rapid collapse of the carbocation implies that the addition of the proton to the alkene is rate determining. This conclusion is supported by the fact that an isotope effect of $k_H/k_D = 1.15$ is found when the rate of addition of HCl in acetic acid is compared with the rate of addition of DCl in DOAc. Thus the mechanism can be classified as AD_E2 and is shown in Scheme 1.[25]

SCHEME 1

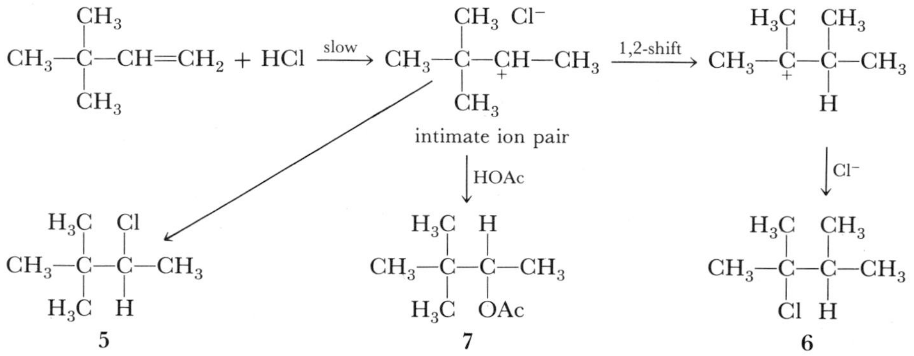

When HCl is added to cyclohexene under the same conditions as were employed in the addition to t-butylethylene, the nature of the products is similar (Equation 7.18), but some of the other characteristics of the reaction are quite

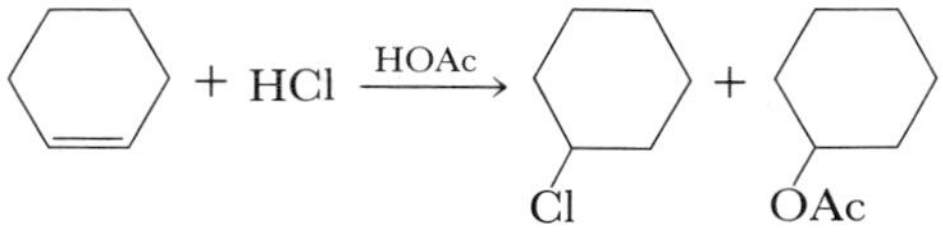

$$(7.18)$$

different. In this case the ratio of cyclohexyl chloride to cyclohexyl acetate is low (0.3) at low HCl concentrations and in the absence of added chloride ion, but it increases sharply to approximately 2 if the reaction mixture is made $0.226M$ in tetramethylammonium chloride (TMAC). Furthermore, if the stereochemistry of the reaction is studied by using cyclohexene-1,3,3-d$_3$ as substrate, the five products shown in Equation 7.19 are always found but their relative amounts depend strongly on the chloride ion concentration. Thus the ratio of cyclohexyl chloride derived from syn addition (**8**), relative to that formed by anti addition (**9**), decreases markedly with chloride ion concentration, whereas the ratio of the syn-formed chloride adduct (**8**) to the anti-formed acetate adduct (**10**) remains unchanged. (Note that **11** and **12** tell us nothing about the stereochemistry of addition but, since the amount of **11** formed by syn addition relative to that formed by anti

---

[25] See note 24(a).

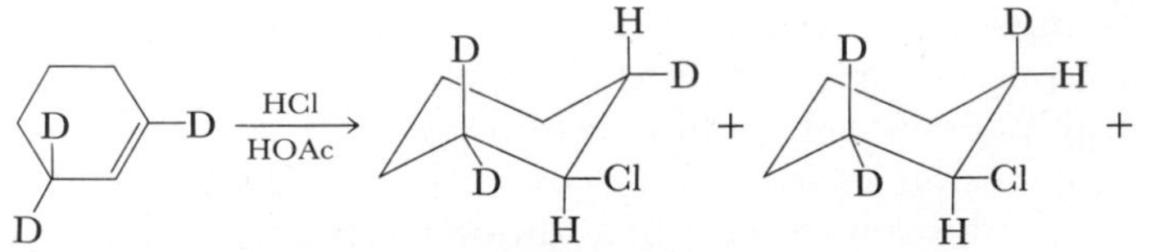

$$\text{rate} = k_1[\text{HCl}][\text{alkene}] + k_2[\text{HCl}][\text{alkene}][\text{Cl}^-] + k_3[\text{HCl}][\text{alkene}][\text{HOAc}] \qquad (7.20)$$

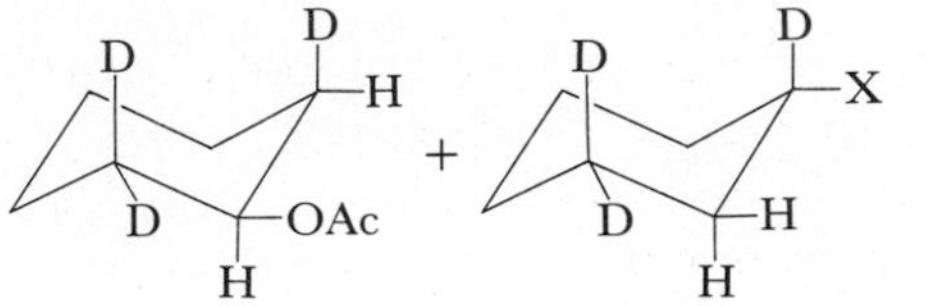

addition should be equal to the ratio of **8** : **9**, we can focus our discussion on **8**, **9**, and **10** alone.) No acetate formed by syn addition is found.[26]

Analysis of the rate and product data show that the rate equation is composed of three terms (Equation 7.20); each includes the concentrations of olefin and of acid, but one also includes the concentration of chloride ion and another the concentration of acetic acid.[27]

Fahey has suggested that all the facts are consistent with the products being formed by three competing reactions. The first, responsible for the second-order rate term, is an ion-pair mechanism like that found in the hydrochlorination of *t*-butylethylene. Such a pathway, which involves collapse of the ion pair (see above), accounts for the *syn*-HCl adduct and some of the *anti*-HCl adduct. The second and third reactions, responsible for the second and third terms in the rate equation and leading to *anti*-HCl and *anti*-HOAc adducts, respectively, are termolecular processes with transition states **13** and **14**. Such mechanisms are called Ad_E3 (addition–electrophilic–termolecular). Ad_E3 transition states analogous to **13** and **14** but leading to syn adducts are, in this case, precluded by the steric requirements of the addends.[26,28] Thus increased chloride ion concentration increases the

---

[26] R. C. Fahey, M. W. Monahan, and C. A. McPherson, *J. Am. Chem. Soc.*, **92**, 2810 (1970); R. C. Fahey and M. W. Monahan, *J. Am. Chem. Soc.*, **92**, 2816 (1970).

[27] This is actually a simplified form of the rate equation for Reaction 7.18. Fahey found that it is not the concentration of HCl, but its activity as represented by Satchell's acidity function, *A*[D. P. N. Satchell, *J. Chem. Soc.*, 1916 (1958)], that should be included in each term of the rate equation. See note 26.

[28] A syn Ad_E3 mechanism with a transition state analogous to **14** has been proposed for the addition of acetic acid to simple alkenes in the presence of only catalytic amounts of HCl[D. J. Pasto and J. F. Gadberry, *J. Am. Chem. Soc.*, **100**, 1469 (1978); D. J. Pasto, G. R. Meyer, and B. Lepesaka, *J. Am. Chem. Soc.*, **96**, 1858 (1974)].

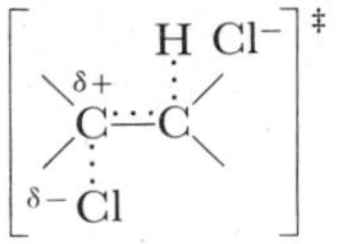

$$
\left[\begin{array}{c} H\ Cl^- \\ \overset{\delta+}{C}\cdots\overset{}{C} \\ \overset{\delta-}{Cl} \end{array} \right]^{\ddagger}
\qquad
\left[\begin{array}{c} H\ Cl^- \\ \overset{\delta+}{C}\cdots\overset{}{C} \\ O \\ CH_3-\overset{}{C}-OH \\ \overset{\delta+}{} \end{array} \right]^{\ddagger}
$$

13 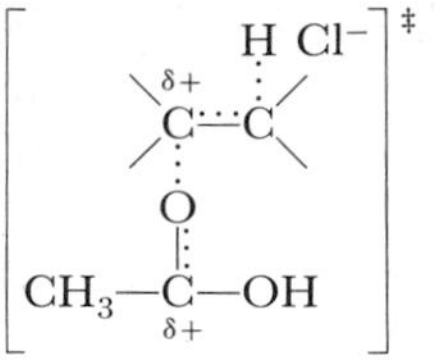   **14**

contribution of the second term of the rate equation relative to the other two, and *anti*-HCl adduct is formed more rapidly than *syn*-HCl or -HOAc adducts.

Why does hydrochlorination of *t*-butylethylene not also proceed in part by a termolecular mechanism? The apparent reason is shown in Table 7.3: The carbocation from protonation of *t*-butylethylene is formed more rapidly than that from protonation of cyclohexene (k_1 of Equation 7.20 is larger for *t*-butylethylene). Furthermore, *t*-butylethylene has a small k_2 because of steric interference of the bulky *t*-butyl group in a termolecular transition state. Table 7.3 gives the estimated rate constants, k_1, k_2, and k_3, of Equation 7.20 for four alkenes. The rate constant, k_1, increases with the ability of the substrate to stabilize a positive charge. The larger value of k_2 for 1,2-dimethylcyclohexene than for cyclohexene means that the β carbon in the transition state of the Ad_E3 mechanism has some cationic character as shown in **13**.[29] This conclusion is consistent with the universality of Markownikoff addition in hydrochlorinations.

With a variety of different mechanisms available, it is not surprising that the characteristics of hydrochlorination depend on the reaction conditions. Thus,

Table 7.3 ESTIMATED RATE CONSTANTS FOR ADDITION TO ALKENES IN HCl/HOAc SOLUTIONS AT 25°C[a]

Alkene	$10^8\,k_1$ $(M^{-1}\,sec^{-1})$	k_2 $(M^{-2}\,sec^{-1})$	$10^8\,k_3$ $(M^{-2}\,sec^{-1})$
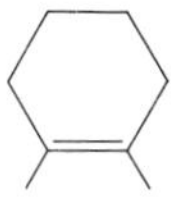 (1,2-dimethylcyclohexene)	22,000	1.8	700
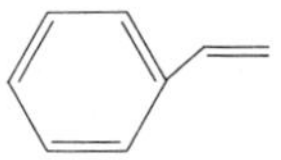 (allylbenzene/styrene)	2,300	<0.1	
 (t-butylethylene)	8	<10^{-5}	
 (cyclohexene)	2.4	1.0×10^{-3}	1.3

[a] Reprinted with permission from R. C. Fahey and C. A. McPherson, *J. Am. Chem. Soc.*, **93**, 2445 (1971). Copyright by the American Chemical Society.

[29] R. C. Fahey and C. A. McPherson, *J. Am. Chem. Soc.*, **93**, 2445 (1971).

in nitromethane, a medium that gives extensive Wagner–Meerwein shifts during hydrochlorination, alkenes react according to the third-order rate law,

$$k = [\text{alkene}][\text{HCl}]^2$$

The fact that rearrangements occur implicates a carbocation intermediate. When the reaction is carried out with DCl, alkene recovered after several half-lives contains no deuterium. Thus formation of the intermediate must be rate determining. Apparently the role of the second HCl molecule is to assist the first in ionizing, and the HCl_2^- anion is produced as shown in Equation 7.21. Predominant anti addition is observed—presumably because a third HCl attacks from the back side in a second fast step.[30] Note that this is also an Ad_E3 mechanism but the role of the second HCl in nitromethane is different from that in acetic acid.

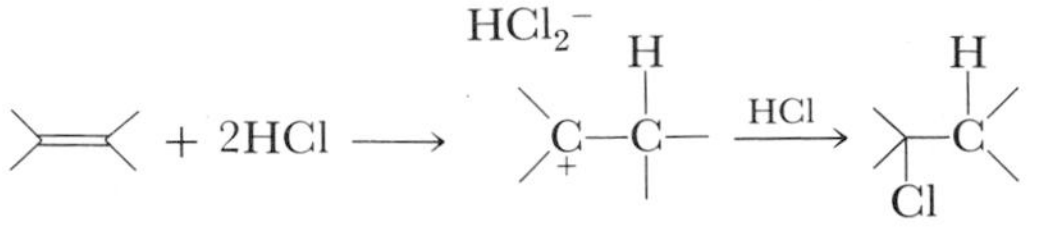

$$\text{(7.21)}$$

Another mechanism, third-order in HCl, with a six-center transition state as in **15**, has recently been proposed for the gas-phase addition of HCl to propene.[31]

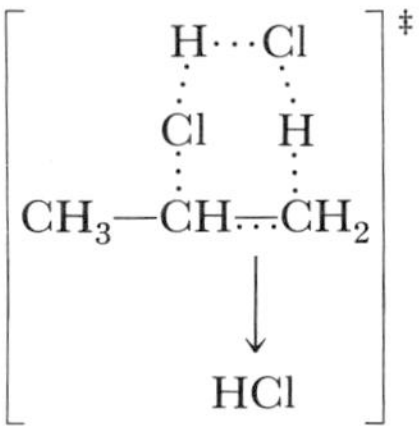

15

The characteristics of the addition of HBr to double bonds are similar to those of the addition of HCl. However, in acetic acid 1,2-dimethylcyclohexene gives more anti addition if HBr is the addend.[32] Also, as Figure 7.3 shows, when HX is added to a double bond in acetic acid, the ratio of alkyl halide to alkyl acetate increases sharply as the concentration of HBr is increased but that ratio is almost independent of the concentration of HCl. Fahey suggests that the much larger acid dissociation constant of HBr ($\Delta pK_a = 3$ to 4) is responsible for both of these facts. Hydrobromic acid acts as a better halide source, and the Ad_E3 addition with a transition state comparable to **13** is favored.[33]

The Ad_E3 mechanism for the addition of acetic acid to alkenes that we have

[30] Y. Pocker and K. D. Stevens, *J. Am. Chem. Soc.*, **91**, 4205 (1969).
[31] J. Haugh and D. R. Dalton, *J. Am. Chem. Soc.*, **97**, 5674 (1975).
[32] R. C. Fahey and R. A. Smith, *J. Am. Chem. Soc.*, **86**, 5035 (1964).
[33] R. C. Fahey, C. A. McPherson, and R. A. Smith, *J. Am. Chem. Soc.*, **96**, 4534 (1974); see also D. J. Pasto, G. R. Meyer, and B. Lepeska, *J. Am. Chem. Soc.*, **96**, 1858 (1974).

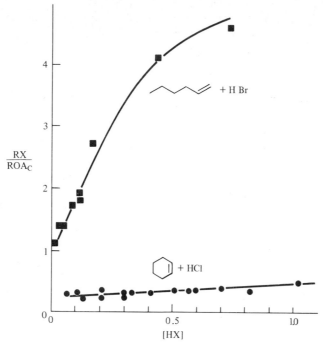

Figure 7.3 Variation in the alkyl halide to alkyl acetate product ratio with HX concentration for reaction in HOAc at 25°C. Reprinted with permission from R. C. Fahey, C. A. McPherson, and R. A. Smith, *J. Am. Chem. Soc.*, **96**, 4534 (1973). Copyright 1973 American Chemical Society.

discussed above apparently changes, when the acid catalyst is much stronger than HBr, to an Ad_E2 mechanism in which the cationic intermediate forms in a rapid, reversible first step. For example, when *cis-* or *trans-*2-butene is dissolved in acetic acid with small amounts of CF_3SO_3D present, acetic acid adds to the double bond. Starting material recovered after partial reaction contains deuterium and is partially isomerized.[34]

Electrophilic addition of HCl to triple bonds can apparently also go by bi- or termolecular mechanisms. Thus in acetic acid 3-hexyne (**16**) gives predominantly anti addition through an Ad_E3 pathway, but 1-phenylpropyne (**17**), which can form the resonance-stabilized vinyl cation (**18**), gives predominantly syn addition through an ion pair Ad_E2 mechanism.[35] In more polar solvents an Ad_E2 mechanism with a free carbocation intermediate may be involved.[36]

$$\phi-C\equiv C-CH_3 \qquad \phi-\overset{+}{C}=CHCH_3$$

    16   17   18

[34] See note 28.
[35] R. C. Fahey and D.-J. Lee, *J. Am. Chem. Soc.*, **88**, 5555 (1966); **89**, 2780 (1967); **90**, 2124 (1968).
[36] F. Marcuzzi, G. Melloni, and G. Modena, *Tetrahedron Lett.*, 413 (1974).

Addition of Halogens[37]

Bromination of double bonds is strongly accelerated by electron-releasing substituents and retarded by electron-withdrawing ones (see Tables 7.4 and 7.7) and is therefore clearly an electrophilic addition. The rate of reaction is always first-order in alkene, but for alkenes of widely different reactivity it can be either first- or second-order in molecular bromine depending on the reaction conditions. At low concentrations of bromine and in water and alcoholic solvents, the rate expression is second-order overall and first-order in bromine. Under these conditions, then, addition occurs by an Ad_E2 mechanism. However, in less polar solvents (e.g., acetic acid) or when the bromine concentration is high, a second molecule of bromine helps to polarize the first in the transition state as in **19**.[1] In the presence of added

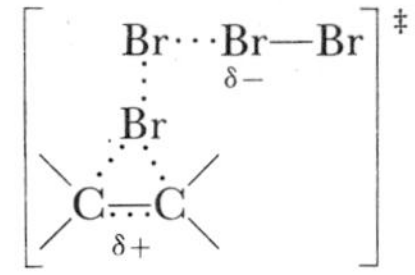

$$\left[\begin{array}{c} \text{Br} \cdots \text{Br}-\text{Br} \\[2pt] \vdots \quad\quad {}^{\delta-} \\ \text{Br} \\ \vdots \\ \underset{\delta+}{\overset{}{\text{C}\cdots\text{C}}} \end{array} \right]^{\ddagger}$$

19

nucleophiles or in hydroxylic solvents, a mixture of products is often obtained, as shown in Equation 7.22.

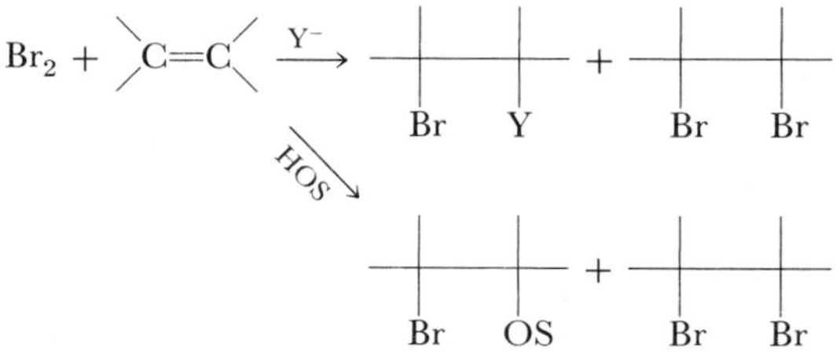

$$Br_2 + \;\;\diagdown\!\!\!C{=}C\!\!\!\diagup\;\; \xrightarrow{\;Y^-\;} \quad\underset{Br\quad Y}{\overset{}{\big|\big|}} \;+\; \underset{Br\quad Br}{\overset{}{\big|\big|}}$$

$$\downarrow{}^{HOS}$$

$$\underset{Br\quad OS}{\overset{}{\big|\big|}} \;+\; \underset{Br\quad Br}{\overset{}{\big|\big|}}$$

(7.22)

Bromine normally adds anti to a nonconjugated alkene. For example, *cis*-2-butene gives exclusively the D,L-2,3-dibromobutanes (Equation 7.23), whereas *trans*-2-butene gives only the corresponding meso compound (Equation 7.24).[38] Similarly, 4-*t*-butylcyclohexene gives only the trans dibromides (**20** and **21**).[39] In

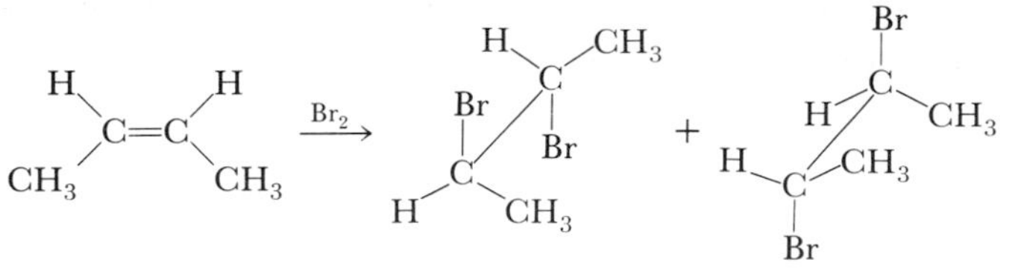

(7.23)

[37]F. Freeman, *Chem. Rev.*, **75**, 439 (1975).

[38](a) W. G. Young, R. T. Dillon, and H. J. Lucas, *J. Am. Chem. Soc.*, **51**, 2528 (1929); (b) R. T. Dillon, W. G. Young, and H. J. Lucas, *J. Am. Chem. Soc.*, **52**, 1953 (1930); (c) J. H. Rolston and Y. Yates, *J. Am. Chem. Soc.*, **91**, 1469 (1969).

[39]E. L. Eliel and R. G. Haber, *J. Org. Chem.*, **24**, 143 (1959); see also H. Weiss, *J. Am. Chem. Soc.*, **99**, 1670 (1977).

Table 7.4 RELATIVE RATES OF SECOND-ORDER
REACTIONS OF ALKENES WITH
BROMINE IN WATER AT $25°C^a$

R in RCH=CH$_2$	Relative rate
CH$_3$	11.4
CH$_2$OH	1.7
H	1.0
CH$_2$CN	1.1×10^{-3}
CO$_2$CH$_2$CH$_3$	3×10^{-7}

[a] Data from J. R. Atkinson and R. P. Bell, *J. Chem. Soc.*, 3260 (1963). Table reproduced from P. B. D. de la Mare and R. Bolton, *Electrophilic Additions to Unsaturated Systems*, Elsevier, Amsterdam, 1966, p. 115. Reprinted by permission of The Chemical Society, Elsevier, R. P. Bell, and P. B. D. de la Mare.

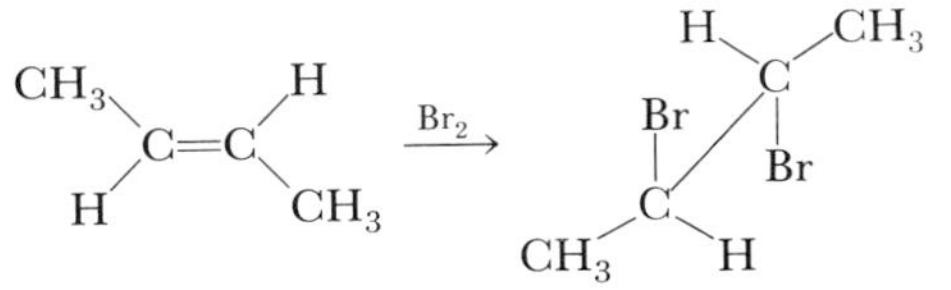

$$(7.24)$$

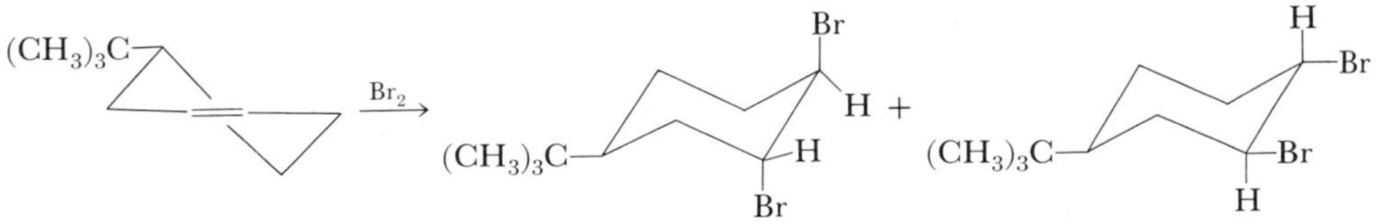

$$(7.25)$$

20

major

21

minor

1937 Roberts and Kimball pointed out that the observed stereochemistry is incompatible with the formation of an intermediate carbocation (**22**) (Equation 7.26) and suggested that an intermediate "bromonium ion" (**23**) is formed in which the entering bromine, using one of its unshared electron pairs, bonds to both carbons of the double bond (Equation 7.27). Rotation about the C_α—C_β bond is impossible in **23**, and Br$^-$ must attack back side from the Br$^+$ to give anti addition.[40]

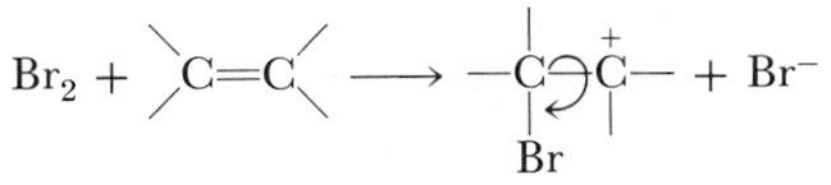

$$(7.26)$$

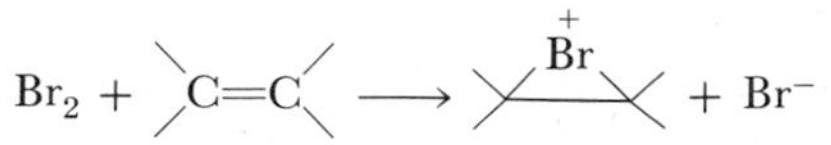

22

$$(7.27)$$

23

[40] I. Roberts and G. E. Kimball, *J. Am. Chem. Soc.*, **59**, 947 (1937).

Conjugated alkenes do not always give predominantly anti addition.[41,42] The stereochemistry of the bromination of methylated styrenes in acetic acid is summarized in Table 7.5. Furthermore, the stereoselectivity of these conjugated alkenes is solvent-dependent and decreases as the ionizing power of the medium increases. Table 7.6 gives the data for *cis*-phenylpropene.

Dubois[42] and Yates[43] suggest that competitive pathways exist for the bromination of 1-arylalkenes—a bromonium ion and an open carbenium ion mechanism. As the one carbon of the double bond becomes better able to stabilize a positive charge, bridging becomes less important and stereoselectivity decreases. Different interpretations have been given for the solvent effect. Yates[43] and Olah[44] believe that the more ionizing solvents better solvate the carbenium ion and thus more carbenium ion is formed. Dubois,[45] however, suggests that the amount of carbenium ion formed is not solvent-dependent but that the changing stereochemistry is due to the effect of the solvent on the competition between conformational equilibrium in and nucleophilic attack on the carbenium ion.

Stereochemistry alone is not conclusive evidence for a bromonium ion. For example, the stereochemistry could also be a result of competition between Ad_E3 and Ad_E2 mechanisms. Alkenes that cannot form stable carbocations might react via an Ad_E3 pathway to give anti addition, whereas conjugated alkenes might form carbocations via the Ad_E2 mechanism to give nonstereoselective addition.

There is, however, other evidence that speaks for the bromonium ion concept and against competition between Ad_E2 and Ad_E3 pathways.[46] We have already noted that, in polar solvents, addition of bromine to multiple bonds is first-order in bromine when bromine is present in low concentration. Moreover, as Table 7.7 shows, increasing the number of substituents on double bonds in aliphatic compounds cumulatively increases the rate of bromination of nonconjugated olefins in polar solvents irrespective of whether each new substituent is on the same or on the other olefinic carbon as the last.[47] Dubois has found that the bimolecular rate constants for addition of bromine to alkyl substituted ethylenes are correlated by Equation 7.28 where $\Sigma\sigma^*$ represents the sum of the Taft σ^* values

$$\log k_2 = -2.99\,\Sigma\sigma^* + 7.61 \tag{7.28}$$

for the four substituents on the double bond.[48] Thus in the transition state the positive charge is distributed over both carbons of the double bond—a very different situation from that obtaining in hydration or hydrochlorination of double bonds (see, for example, Table 7.1).

[41] See note 38(c).

[42] M. F. Ruasse, A. Argile, and J.-E. Dubois, *J. Am. Chem. Soc.*, **100**, 7645 (1978), and references therein.

[43] J. H. Rolston and K. Yates, *J. Am. Chem. Soc.*, **91**, 1469, 1477, 1483 (1969).

[44] G. A. Olah and T. R. Hockswender, Jr., *J. Am. Chem. Soc.*, **96**, 3574 (1974).

[45] M.-F. Ruasse and J.-E. Dubois, *J. Am. Chem. Soc.*, **97**, 1977 (1975).

[46] For a summary, see R. C. Fahey and H.-J. Schneider, *J. Am. Chem. Soc.*, **90**, 4429 (1968).

[47] J.-E. Dubois and G. Mouvier, *Bull. Soc. Chim. Fr.*, 1426 (1968).

[48] J.-E. Dubois and E. Goetz, *J. Chim. Phys.*, **63**, 780 (1966).

Table 7.5 STEREOCHEMISTRY OF DIBROMOADDUCTS FROM ALKENE AND BROMINE IN ACETIC ACID AT 25°C[a]

Alkene	Percent anti addition	Alkene	Percent anti addition
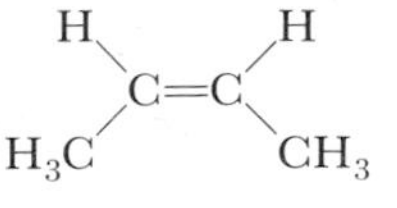	100	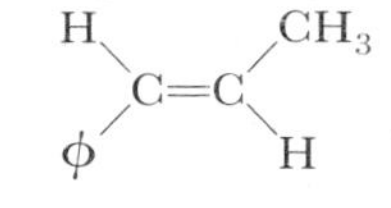	83
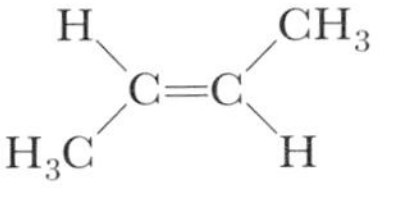	100	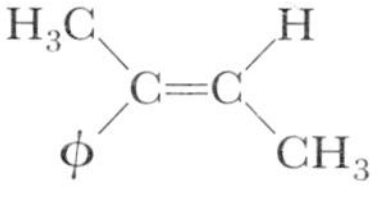	63
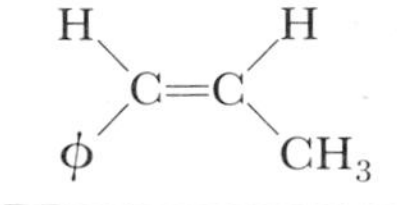	73	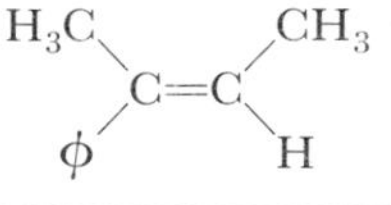	68

[a] Reprinted with permission from J. H. Rolston and K. Yates, *J. Am. Chem. Soc.*, **91**, 1469 (1969). Copyright by the American Chemical Society.

If substituted styrenes do react by competitive carbenium and bromonium ion mechanisms, it would be expected that the effect of electron-donating substituents on the double bond would not be cumulative but would depend on whether

$$\phi CH{=}CH_2 \qquad \phi CH_3 C{=}CH_2 \qquad \phi CH{=}CHCH_3$$

	24	**25**	**26**
Relative rates:	1	87	25

they were α or β to the phenyl group. Indeed, **24**, **25**, and **26**, when brominated under the conditions of Table 7.7, have the relative rates shown.[49]

Furthermore, a ρ–σ^+ plot for the bromination of stilbenes (**27**) shows curvature, indicating a change in mechanism as the electron-donating power of X changes.[42]

Table 7.6 STEREOCHEMISTRY OF DIBROMOADDUCT OF *cis*-PHENYLPROPENE IN VARIOUS SOLVENTS[a]

Solvent	Dielectric constant	Percent anti addition
Acetic acid	6.2	73
Tetrachloroethane	8.2	66
Methylene chloride	9.1	70
Acetic anhydride	21	49
Nitrobenzene	35	45

[a] Reprinted with permission from J. H. Rolston and K. Yates, *J. Am. Chem. Soc.*, **91**, 1477 (1969). Copyright by the American Chemical Society.

[49] J.-E. Dubois, J. Toullec, and G. Barbier, *Tetrahedron Lett.*, 4485 (1970).

Table 7.7 RELATIVE RATES OF BROMINATION OF ALKYL-SUBSTITUTED ALKENES WITH
MOLECULAR BROMINE IN METHANOLIC SODIUM BROMIDE AT 25°C[a]

Alkene	Relative rate	Alkene	Relative rate
$CH_2=CH_2$	1	$(CH_3)_2C=CH_2$	5,400
$CH_2=CH-CH_3$	61	$(CH_3)_2C=CHCH_3$	130,000
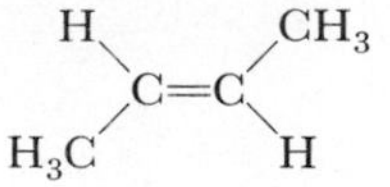 $\begin{array}{c} H \qquad CH_3 \\ C=C \\ H_3C \qquad H \end{array}$	1,700	$(CH_3)_2C=C(CH_3)_2$	1,800,000
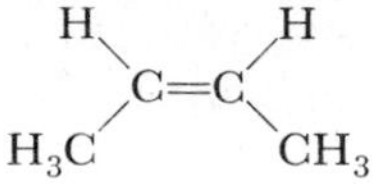 $\begin{array}{c} H \qquad H \\ C=C \\ H_3C \qquad CH_3 \end{array}$	2,600		

[a] Data from J.-E. Dubois and G. Mouvier, *Bull. Soc. Chim. Fr.,* 1426 (1968). Reprinted by permission of the Société Chemique de France.

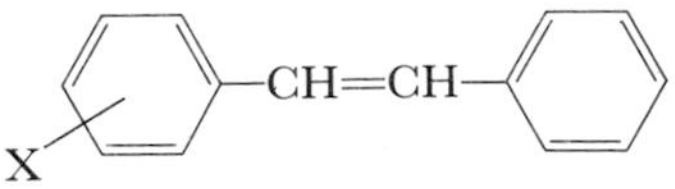

27

Another piece of evidence for the bromonium ion is that addition is less regiospecific when bromine is the electrophile than when H^+ is. With molecular bromine we cannot, of course, observe the site at which the original electrophilic bromine attacks, but with unsymmetrical reactants such as HOBr or BrCl we can. Thus, for example, the addition of BrCl to propene in aqueous HCl gives only 54 percent of the Markownikoff addition product (**28**) and 46 percent of the anti-Markownikoff product (**29**).[50] There seems, then, to be only a small difference between charge densities at the central and terminal carbons—a situation that might exist in the bromonium ion.

$$\begin{array}{cc} CH_3-CH-CH_2 & CH_3-CH-CH_2 \\ \quad | \quad\; | & \quad | \quad\; | \\ \;\; Cl \quad Br & \;\; Br \quad Cl \\ \textbf{28} & \textbf{29} \end{array}$$

Olah has observed the unsubstituted bromonium ion and several alkylated bromonium ions by nmr spectroscopy after dissolving α-bromohalides in SbF_5-SO_2 solution at low temperatures.[51] All four hydrogens of the unsubstituted ion were equivalent.

[50] P. B. D. de la Mare and S. Galandauer, *J. Chem. Soc.*, 36 (1958).

[51] (a) G. A. Olah, P. Schilling, P. W. Westerman, and H. C. Lin, *J. Am. Chem. Soc.*, **96**, 3581 (1974); (b) G. A. Olah, P. W. Westerman, E. G. Melby, and Y. K. Mo, *J. Am. Chem. Soc.*, **96**, 3565 (1974), and references therein; (c) G. A. Olah, *Halonium Ions,* Wiley, New York, 1975.

The bromonium ion (**30**) has actually been isolated as the tribromide salt. This ion is stable because it cannot be attacked from the back side.[52]

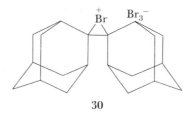

30

Molecular fluorine, because of its very low bond dissociation energy, usually reacts violently with organic compounds.[1] Merritt, however, has observed electrophilic addition of F_2 to *cis*- and *trans*-1-phenylpropenes at low temperatures. The mode of addition is predominantly syn. A fluoronium ion, in which the fluorine is positively charged, would be very unstable and apparently does not form.[53] Attempts to form a three-membered ring fluoronium ion in superacid medium have also failed.[51]

Electrophilic addition of Cl_2 and I_2 to alkenes is similar in mechanism to the electrophilic addition of Br_2.[54] The rate of chlorination in acetic acid is second-order, first-order each in alkene and in chlorine.[55] Predominantly anti addition to alkyl-substituted double bonds occurs, indicating that a chloronium ion is formed.[46,56] Further evidence for the chloronium ion is that addition of hypochlorous acid to double bonds is not entirely regiospecific. For example, addition to propene gives 91 percent of the Markownikoff product **31**, and 9 percent of the anti-Markownikoff product **32**. Phenyl-substituted alkenes give a mixture of syn and anti adducts with Cl_2, as they do with Br_2.[46,56]

$$CH_3CH-CH_2 \qquad CH_3CH-CH_2$$
$$\quad | \qquad | \qquad\qquad | \qquad |$$
$$\quad OH \quad Cl \qquad\quad Cl \quad OH$$

31   **32**

Iodination is usually second-[57] or third-[58] order in I_2. The role of the additional I_2 molecules, apparently, is to assist in breaking one I—I bond in the rate-determining step. Because iodine is less electronegative than bromine, the iodonium ion can compete with carbocation formation even when the bromonium ion

[52] J. Strating, J. H. Wieringa, and H. Wynberg, *J. Chem. Soc. D*, 907 (1969).

[53] R. F. Merritt, *J. Am. Chem. Soc.*, **89**, 609 (1967).

[54] See note 1. Chlorination of olefins in nonpolar media in the absence of radical inhibitors may proceed by a radical pathway. [M. L. Poutsma, *J. Am. Chem. Soc.*, **87**, 2161, 2172 (1965).]

[55] I. R. C. McDonald, R. M. Milburn, and P. W. Robertson, *J. Chem. Soc.*, 2836 (1950), and earlier papers in this series (by Robertson and co-workers).

[56] (a) R. C. Fahey and C. Schubert, *J. Am. Chem. Soc.*, **87**, 5172 (1965); R. C. Fahey, *J. Am. Chem. Soc.*, **88**, 4681 (1966); (b) M. L. Poutsma and J. L. Kartch, *J. Am. Chem. Soc.*, **89**, 6595 (1967).

[57] N. J. Bythell and P. W. Robertson, *J. Chem. Soc.*, 179 (1938).

[58] (a) J. Gróh and J. Szelestey, *Z. Anorg. Allgem. Chem.*, **162**, 333 (1927); (b) J. Gróh *Z. Anorg. Allgem. Chem.*, **162**, 287 (1927); (c) J. Gróh and E. Takács, *Z. Phys. Chem. (Leibzig)*, **149A**, 195 (1930).

cannot. Thus IN_3 with *cis-β*-deuterostyrene gives anti addition only, whereas BrN_3 with the same olefin gives a 1:1 mixture of syn and anti adducts.[59] Chloronium and iodonium ions have been observed in superacid medium.[51]

Electrophilic additions of the halogens to alkynes have not been much studied. In acetic acid a given alkene reacts with bromine 10^3 to 10^5 times more rapidly than the corresponding alkyne. The rate trends in chlorination parallel those for bromination.[60] The limited facts available indicate that the mechanism is similar to that of addition to olefins. Pincock and Yates have studied the addition of bromine to a number of alkyl- and arylacetylenes in acetic acid. At low bromine concentrations the reaction is second-order, first-order each in Br_2 and in acetylene. Alkylacetylenes give only anti addition, indicating that a bromonium ion lies on the reaction path. Ring-substituted phenylacetylenes, however, give both syn and anti addition; and the logarithms of the rates correlate linearly with the σ^+ constants of the substituents, giving a very large negative ρ value (-5.17). In these compounds, open vinyl cations are apparently formed as intermediates.[61] It has been suggested that the $k_{alkene}/k_{acetylene}$ ratios in bromination, so much greater than in hydration, are due to different stabilities and properties of the bromonium ions in each case.[62]

Hydroboration[63]

The addition of a boron hydride across a double or triple bond (Equation 7.29) is called hydroboration.

$$\text{\textbackslash B-H} + \text{C=C} \longrightarrow -\overset{|}{\underset{H}{C}}-\overset{|}{\underset{B}{C}}- \tag{7.29}$$

Hydroboration always gives syn addition of the elements of boron and hydrogen to a double bond. For example, nmr analysis shows that (E)- and (Z)-1-hexene-1,2-d_2 with dicyclohexylborane produce *threo*- and *erythro*-(1,2-dideuterohexyl)dicyclohexylboranes, respectively, as shown in Equations 7.30 and 7.31.[64]

In additions of diborane, the major product is formed by the attachment of boron to the less substituted carbon as shown in Table 7.8. For example, addition of diborane to 1-hexene (**33**) gives a product that has 94 percent of the boron attached to the terminal carbon. Similarly, diborane added to 2-methylbutene-2 (**34**) gives 98 percent of boron incorporation at C_3. Table 7.8 also shows that diborane does

[59] A. Hassner, F. P. Boerwinkle, and A. B. Levy, *J. Am. Chem. Soc.*, **92**, 4879 (1970).

[60] K. Yates, G. H. Schmid, T. W. Regulski, D. G. Garratt, H.-W. Leung, and R. McDonald, *J. Am. Chem. Soc.*, **95**, 160 (1973).

[61] (a) J. A. Pincock and K. Yates, *Can. J. Chem.*, **48**, 3332 (1970); (b) J. A. Pincock and K. Yates, *J. Am. Chem. Soc.*, **90**, 5643 (1968).

[62] G. Modena, F. Rivetti, G. Scorrano, and U. Tonellato, *J. Am. Chem. Soc.*, **99**, 3392 (1977).

[63] H. C. Brown, *Hydroboration*, W. A. Benjamin, Menlo Park, Calif., 1962; H. C. Brown, *Boranes in Organic Chemistry*, Cornell University Press, Ithaca, N.Y., 1972.

[64] G. W. Kabalka, R. J. Newton, Jr., and J. Jacobus, *J. Org. Chem.*, **43**, 1567 (1978).

$$\text{(7.30)}$$

$$\text{(7.31)}$$

not discriminate well between two carbons of very different steric requirements if they are equally substituted. Thus boron attacks the two doubly bonded carbons in 4-methyl-*cis*-2-pentene (**35**) about equally.[65]

If, instead of diborane, a boron hydride substituted with bulky alkyl groups is added to a double bond, the regiospecificity increases. Thus *bis*(3-methyl-2-butyl)borane (**36**)[66] and 9-borabicyclo[3.3.1]nonane (**37**)[67] react with 1-hexene to give 99 and 99.9 percent, respectively, terminal boron incorporation. Furthermore, these bulky boron hydrides are highly discriminate in choosing the less sterically hindered doubly bonded carbon, as can be seen in their mode of attack on 4-methyl-*cis*-2-pentene (Table 7.8).

Electronic effects as well as steric effects are important in determining the orientation of addition as is shown, for example, by the data in Table 7.9. The regioselectivity of diborane increases as 2-butene is substituted in the 1 position by increasingly strong electron-withdrawing groups.[68]

Rearrangements have not been observed in the addition of boron hydrides to double bonds under mild conditions.

All the evidence presented above would fit a concerted mechanism with a four-center transition state.[69] Orbital symmetry considerations (see Chapter 10), however, preclude the simple transition state shown in **38**. If the transition state is a four-center one, it must include the empty orbital on boron as in **38a**.[70] In this mechanism C_1 of the alkene would donate electrons to the empty orbital on boron and at the same time a hydrogen on boron would donate electrons back to C_2 on the alkene. Molecular orbital calculations carried out by Lipscomb indicate that the hydroboration mechanism *is* two-step and has as an intermediate a structure similar to **38a**.[71] On the other hand, Seyforth, Streitwieser, and Jones have sug-

[65] H. C. Brown and G. Zweifel, *J. Am. Chem. Soc.*, **82**, 4708 (1960).

[66] H. C. Brown and G. Zweifel, *J. Am. Chem. Soc.*, **83**, 1241 (1961).

[67] H. C. Brown, E. F. Knights, and C. G. Scouten, *J. Am. Chem. Soc.*, **96**, 7765 (1974).

[68] H. C. Brown and R. M. Gallivan, Jr., *J. Am. Chem. Soc.*, **90**, 2906 (1968); see also H. C. Brown, R. Liotta, and C. G. Scouten, *J. Am. Chem. Soc.*, **98**, 5297 (1976).

[69] H. C. Brown and G. Zweifel, *J. Am. Chem. Soc.*, **83**, 2544 (1961).

[70] [D. J. Pasto, B. Lapeska, and T.-C. Cheng, *J. Am. Chem. Soc.*, **94**, 6083 (1972). Both lobes of the empty orbital are involved in the orbital interaction and thus there is a phase inversion. See Chapter 10.]

[71] K. R. Sundberg, G. D. Graham, and W. N. Lipscomb, *J. Am. Chem. Soc.*, **101**, 2863 (1979).

Table 7.8 Regiospecificity of Boron Addition to Unsymmetrical Alkenes with Diborane, *bis*(3-Methyl-2-Butyl)Borane (36) and 9-Borabicyclo[3.3.1]Nonane (37)

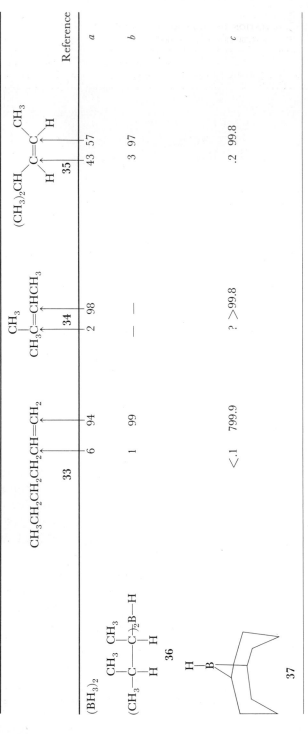

	CH₃CH₂CH₂CH₂CH=CH₂ **33**		CH₃C=CHCH₃ (CH₃) **34**		(CH₃)₂CH-C=C (H, CH₃, H) **35**		Reference
(BH₃)₂	6	94	2	98	43	57	a
36	1	99	—	—	3	97	b
37	<.1	>99.9	?	>99.8	.2	99.8	c

[a] H. C. Brown and G. Zweifel, *J. Am. Chem. Soc.*, **82**, 4708 (1960).
[b] H. C. Brown and G. Zweifel, *J. Am. Chem. Soc.*, **83**, 1241 (1960).
[c] H. C. Brown, E. F. Knights, and C. G. Scouten, *J. Am. Chem. Soc.*, **96**, 7765 (1974).

Table 7.9 ORIENTATION OF ADDITION OF DIBORANE
TO SUBSTITUTED 2-BUTENES[a]

X in $CH_3CH=CHCH_2X$	Percent of boron in position 2 of product	Percent of boron in position 3 of product
H	50	50
OEt	84	16
Oϕ	86	14
OH	90	10
OCH$_2\phi$	91	9
OAc	95	5
Cl	100	0

[a] Reproduced with permission from H. C. Brown and R. M. Gallivan, Jr., *J. Am. Chem. Soc.*, **90**, 2906 (1968). Copyright by the American Chemical Society.

gested that in a reversible first step a π complex (**39**) is formed between the boron hydride and the alkene and that **39** then rearranges to products.[72] Molecular or-

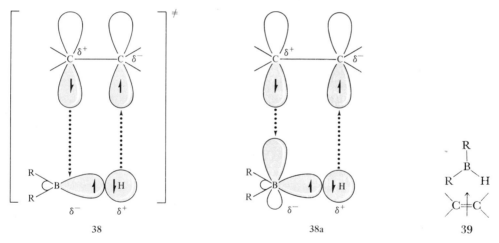

38   38a   39

bital calculations carried out by Schleyer and by Dewar agree with the idea that this mechanism would be the lowest energy pathway.[73] The controversy about the nature of the transition state remains unresolved.

The work of H. C. Brown has made hydroboration an enormously useful reaction. Oxidation of the adduct with alkaline hydrogen peroxide removes the boron smoothly without rearrangement and replaces it by a hydroxy group. The oxidation proceeds entirely with retention of configuration. For example, 1-methyl-cyclopentene is converted by hydroboration followed by oxidation to *trans*-2-

[72](a) D. Seyforth, *Prog. Inorg. Chem.*, **3**, 129 (1962); (b) A. Streitwieser, Jr., L. Verbit, and R. Bittman, *J. Org. Chem.*, **32**, 1530 (1967); (c) P. R. Jones, *J. Org. Chem.*, **37**, 1886 (1972); (c) see also F. Freeman, *Chem. Rev.*, **75**, 439 (1975).
[73](a) T. Clark and P. v. R. Schleyer, *J. Organomet. Chem.*, **156**, 191 (1978); M. J. S. Dewar and M. L. McKee, *Inorg. Chem.*, **17**, 1075 (1978).

methylcyclopentanol in high yields.[74] Since the overall reaction and the hydroboration proceed by retention, so must the oxidation step (Equation 7.32). Thus

$$\text{(7.32)}$$

hydroboration of a double bond followed by peroxide oxidation is a convenient procedure for converting an alkene into the alcohol corresponding to anti Markownikoff addition of water with retention of configuration about the original double bond.

Alkynes also react with boron hydrides bearing bulky organic groups to give attachment of boron to the less substituted position;[75] syn addition is again the rule. The adducts can be smoothly converted to the *cis*-alkenes by treatment with acetic acid at 0°C (Equation 7.33).[76]

$$\text{(7.33)}$$

7.2 1,2-ELIMINATION REACTIONS[77]

The opposite of an addition to a double bond is a 1,2-elimination reaction. In solution, where the reaction is promoted by solvent or by base, the most common eliminations (and those to which we shall limit our discussion) are those that involve loss of HX, although loss of X_2 from 1,2-dihalides and similar reactions are also well known. The mechanisms of eliminations of HX are of three main types: (1) The E1 (elimination, unimolecular), shown in Equation 7.34, which is the

$$\text{(7.34)}$$

reverse of the Ad_E2 reaction; (2) the E1cB (elimination, unimolecular, conjugate base) reaction of Equation 7.35, which involves initial abstraction of a proton followed by loss of X^-; and (3) the E2 (elimination, bimolecular) reaction shown in

[74] H. C. Brown and G. Zweifel, *J. Am. Chem. Soc.*, **83**, 2544 (1961).

[75] G. Zweifel, G. M. Clark, and N. L. Polston, *J. Am. Chem. Soc.*, **93**, 3395 (1971).

[76] H. C. Brown and G. Zweifel, *J. Am. Chem. Soc.*, **81**, 1512 (1959); H. C. Brown, C. G. Scouten, and R. Liotta, *J. Am. Chem. Soc.*, **101**, 96 (1979).

[77] For reviews, see (a) W. H. Saunders, in *The Chemistry of Alkenes*, S. Patai, Ed., Wiley, New York, 1964, p. 149; (b) D. V. Banthorpe, *Elimination Reactions*, Elsevier, Amsterdam, 1963; (c) J. F. Bunnett, *Survey Prog. Chem.*, **5**, 53 (1969); (d) W. H. Saunders, Jr., and A. F. Cockerill, *Mechanisms of Elimination Reactions*, Wiley, New York, 1973; (e) A. F. Cockerill, in *Comprehensive Chemical Kinetics*, Vol. 9, C. H. Banford and C. F. H. Tippett, Eds., Elsevier, New York, 1973, p. 163, (f) A. F. Cockerill and R. G. Harrison, in *The Chemistry of Double-Bonded Functional Groups*, Part 1, Supplement A to *The Chemistry of the Functional Groups*, S. Patai, Ed., Wiley, New York, 1977, p. 149.

$$-\underset{\underset{X}{|}}{\overset{\overset{|}{}}{C}}-\underset{\underset{H}{|}}{\overset{\overset{|}{}}{C}}- + B^- \longrightarrow -\underset{\underset{X}{|}}{\overset{\overset{|}{}}{C}}-\underset{\overset{|}{}}{\overset{\overset{|}{}}{C}}- + BH \longrightarrow \overset{}{\underset{}{>}}C=C\overset{}{\underset{}{<}} + X^- \qquad (7.35)$$

Equation 7.36, in which the base attacks the β proton at the same time as the C—X bond cleaves.[78]

$$-\underset{\underset{X}{|}}{\overset{\overset{H}{|}}{C}}-\underset{\overset{|}{}}{\overset{\overset{|}{}}{C}}- + B^- \longrightarrow \left[\begin{array}{c} B^{\delta-} \\ \vdots \\ H \\ \vdots \\ -\underset{\underset{\ddot{X}^{\delta-}}{|}}{\overset{}{C}}\cdots\overset{}{C}- \end{array}\right]^{\ddagger} \longrightarrow \overset{}{\underset{}{>}}C=C\overset{}{\underset{}{<}} + HB + X^- \qquad (7.36)$$

We shall discuss each of these mechanisms and also, very briefly, 1,2-eliminations that require assistance of neither solvent nor base.

The E1 Reaction

Solvolysis of alkyl derivatives often leads to a mixture of substitution and elimination products. As was noted in Section 4.1, although the rate of solvolysis changes with the leaving group, when the solvent is of high dielectric constant the ratio of substitution to elimination products is independent of the leaving group. For example, in 80 percent aqueous ethanol, t-butyl iodide solvolyzes over 100 times as rapidly as t-butyl chloride, but the ratio of elimination to substitution products is the same for the chloride and iodide.[79] It was evidence of this sort that made early investigators postulate that first-order elimination (E1) and first-order substitution (S_N1) share a preliminary, rate-determining step. Then, they suggested, in a second step, the fully solvated carbocation either adds solvent (S_N1 reaction) or gives up a proton to the solvent (E1 reaction).

Further investigation, however, showed that in solvents of low ionizing power the ratio of substitution to elimination depends on the nature of the leaving group. For example, as Table 4.2 shows, in glacial acetic acid, when the leaving group is Cl$^-$, elimination accounts for 73 percent of the product; but when it is CH_3—S—CH_3, only 12 percent alkene is formed. These facts are consistent with formation of intimate ion pairs in the less dissociating solvents in which the leaving group is the base that removes the proton from the β carbon. Shiner has concluded, from secondary deuterium isotope effects, that cyclopentyl brosylate (**40**) rapidly

---

[78] Sneen has suggested that most elimination reactions proceed by initial ionization of the leaving group to form an ion pair and that this first step (either fast or rate-determining) is followed by attack of base. Thus the S_N2, S_N1, E2, and E1 reactions all proceed by one "merged" mechanism. For a brief discussion of this view, see Section 4.2, and R. A. Sneen, *Acc. Chem. Res.*, **6**, 46 (1973). Bordwell, on the other hand, suggests that most eliminations proceed *either* by initial isomerization of the leaving group *or* by initial abstraction of the proton and that very few eliminations are concerted. See F. G. Bordwell, *Acc. Chem. Res.*, **3**, 281 (1970), and F. G. Bordwell, *Acc. Chem. Res.*, **5**, 374 (1972). For the purposes of the discussion in this chapter, we shall use the classification scheme just outlined, which is accepted by most workers in the field.

[79] K. A. Cooper, E. D. Hughes, and C. K. Ingold, *J. Chem. Soc.*, 1280 (1937).

and reversibly forms the intimate ion pair **41** and that rate-determining elimination then occurs from the ion pair.[80]

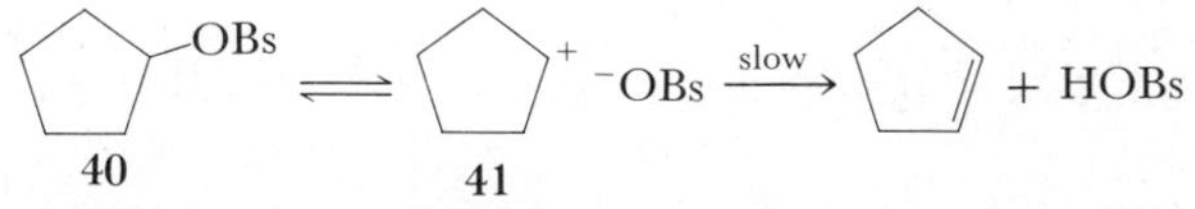

$$(7.37)$$

Whether a β proton is lost from the same or the opposite side of the molecule as the leaving group, that is, whether syn or anti elimination obtains in an E1 mechanism, depends on the reaction conditions. If a solvated, planar carbocation is formed, then the β proton lost to solvent in the second step should come with equal probability from either the same or the opposite side of the plane as the original leaving group. If, however, elimination occurs from an intimate ion pair and the leaving group rather than the solvent acts as the base, then syn elimination should result. These predictions have been borne out by experiment.[77,81] For example, Skell and Hall studied the elimination of *erythro*-3-d$_1$-2-butyl tosylate.[82] As shown in Scheme 2, syn elimination would give nondeuterated *cis*-2-butene and deuterated *trans*-2-butene, but anti elimination would yield deuterated *cis*- and nondeuterated *trans*-2-butene. In poorly ionizing nitromethane the product is almost entirely that of syn elimination; thus the tosylate pulls off the proton of deuteron from the same side of the molecule from which it departed. In aqueous ethanol, however, a mixture of syn and anti elimination products is obtained.[83]

SCHEME 2

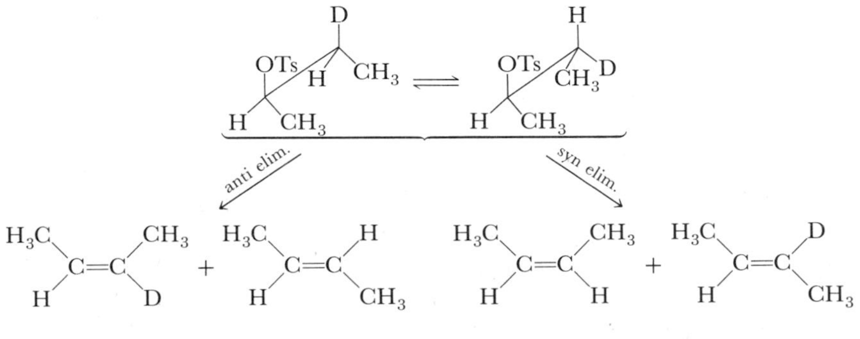

In elimination reactions in which a β hydrogen may be lost from one of two carbons, the question of which way the double bond will be oriented arises. *Saytzeff's rule* states that in E1 reactions the double bond will be oriented toward the more highly substituted carbon. (Thus elimination that gives the more highly

---

[80] R. C. Seib, V. J. Shiner, Jr., V. Sendijarević, and K. Humski, *J. Am. Chem. Soc.*, **100**, 8133 (1978), and references therein.
[81] (a) T. Cohen and A. R. Daniewski, *J. Am. Chem. Soc.*, **91**, 533 (1969); (b) P. S. Skell and W. L. Hall, *J. Am. Chem. Soc.*, **85**, 2851 (1963).
[82] See note 81(b).
[83] For a similar example, see K. Humski, V. Sendijarević, and U. J. Shiner, *J. Am. Chem. Soc.*, **96**, 6187 (1974).

substituted of two possible products is called *Saytzeff elimination*. Elimination that gives the less substituted product is called *Hofmann elimination*.) It is easy to see why E1 reaction usually leads to Saytzeff elimination. The transition state for the product-determining step has some double-bond character, and thus the lowest-energy transition state will be that leading to the most stable double bond. It is well substantiated that alkyl groups lower the energy of the double bond through hyperconjugation. Saytzeff's rule is, however, not necessarily obeyed when the carbocation remains part of an intimate ion pair. For example, elimination of HX from **42**

$$CH_3 - \overset{\overset{\displaystyle X}{|}}{\underset{\underset{\displaystyle \phi}{|}}{C}} - CH_2CH_3$$

42

in glacial acetic acid gives the products shown in Table 7.10. As the leaving group becomes more basic, Saytzeff elimination becomes less and less important and Hofmann product begins to predominate. The change in product composition can be explained by a consideration of Hammond's postulate. The more basic the counter-ion that pulls off the proton, the more the transition state for the product-determining step will look like carbocation and the less double-bond character it will have. Then the orientation of the double bond in the product depends more on the relative acidity of the two kinds of protons than on the relative stabilities of the possible double bonds. In solution the γ-methyl group renders the β-methylene hydrogens less acidic than the β-methyl hydrogens; thus the more basic the counter-ion, the more terminal olefin results.

The sensitivity of the orientation of the double bond to the leaving group when elimination occurs from an intimate ion pair can be seen from the ethanolysis of **43**. When X$^-$ is changed from Br$^-$ to ClO$_4^-$, the ratio of internal to terminal alkene changes from 16 to 7. Note that X$^-$ is not the leaving group but the counter-ion to the leaving group. Since CH$_3$SCH$_3$ is a weak base, the solvent acts as

Table 7.10

$$CH_3 - \overset{\overset{\displaystyle X}{|}}{\underset{\underset{\displaystyle \phi}{|}}{C}} - CH_2CH_3 \xrightarrow{\text{AcOH}} \underset{cis}{\overset{H_3C}{}\!\!\diagdown C = C \diagup\!\!\overset{CH_3}{\underset{H}{}}} + \underset{trans}{\overset{H_3C}{}\!\!\diagdown C = C \diagup\!\!\overset{H}{\underset{CH_3}{}}} + \underset{terminal}{CH_2 = \overset{}{\underset{\underset{\phi}{|}}{C}} - CH_2CH_3}{}^a$$

X	Percent cis	Percent trans	Percent terminal
Cl	68	9	23
OAc	53	2	45
NHNH$_2$	40	0	60

^{a} Reprinted with permission from D. J. Cram and M. R. V. Sahyun, *J. Am. Chem. Soc.,* **85**, 1257 (1963). Copyright by the American Chemical Society.

$$CH_3-CH-CH_2-CH_2-CH_3$$

$$\overset{+}{S}$$

$$CH_3 \diagdown \diagup CH_3$$

$$X^-$$

43

the base and abstracts the proton; even a change in the leaving group's counter-ion affects the solvent shell into which the ion is born.[84]

Carbanion Mechanisms[85]

If, instead of a good leaving group as is required for the E1 reaction, a compound has a poor leaving group but a highly acidic proton, elimination may take place through the consecutive reactions shown in Equations 7.38 and 7.39. These are

$$B^- + H-\overset{|}{\underset{|}{C}}-\overset{|}{\underset{|}{C}}-X \underset{k_{-1}}{\overset{k_1}{\rightleftharpoons}} BH + {}^-\overset{|}{\underset{|}{C}}-\overset{|}{\underset{|}{C}}-X \tag{7.38}$$

$$^-\overset{|}{\underset{|}{C}}-\overset{|}{\underset{|}{C}}-X \overset{}{\underset{k_2}{\longrightarrow}} \diagup C=C\diagdown + X^- \tag{7.39}$$

usually called E1cB reactions but, depending on the relative magnitudes of the rate constants and on the degree of separation between BH and the anion, Equations 7.38 and 7.39 actually describe four different mechanisms. Table 7.11 compares the predicted kinetic effects for all four with those for an E2 reaction whose transition state has considerable carbanionic character.

If k_1 is much greater than both k_{-1} and k_2 of Equations 7.38 and 7.39—that is, if the β hydrogen is very acidic but the leaving group is poor—then if sufficient base is present, formation of the anion will be complete before loss of X begins. An example of a reaction that goes by this [the $(E1)_{anion}$] mechanism is shown in Equations 7.40 and 7.41.[86] This reaction proceeds at the same rate with triethyl- or with tri-n-butylamine. Furthermore, if more than an equimolar amount of base is present, the rate is independent of the base concentration and is equal to $k_2[SH]$, where HS is the substrate.[87] Both these facts indicate that abstraction of the proton, which is rendered highly acidic by two electron-withdrawing groups, is not involved in the rate-determining step. The $(E1)_{anion}$ mechanism is rare because of the high acidity required of the β hydrogen.[88]

[84] I. N. Feit and D. G. Wright, *J. Chem. Soc., Chem. Commun.*, 776 (1975).

[85] D. J. McLennan, *Quart. Rev.*, **21**, 490 (1967).

[86] Z. Rappoport and E. Shohamy, *J. Chem. Soc. B*, 2060 (1971).

[87] Actually, in this case one does not need an equimolar quantity of base, because HCN is such a weak acid that free base is continually reformed by $R_3\overset{+}{N}H + CN^- \longrightarrow R_3N + HCN$.

[88] See also, however; (a) F. G. Bordwell, K. C. Yee, and A. C. Knipe, *J. Am. Chem. Soc.*, **92**, 5945 (1970); (b) F. G. Bordwell, M. M. Vestling, and K. C. Yee, *J. Am. Chem. Soc.*, **92**, 5950 (1970).

Table 7.11 KINETIC PREDICTIONS FOR BASE-INDUCED β ELIMINATIONS[a]

$$\bar{B} + (D)H{-}C_\beta{-}C_\alpha{-}X \longrightarrow BH + {>}C{=}C{<} + \bar{X}$$

Mechanism	Kinetic[b] order	β Protium exchange faster than elimination	General or specific base catalysis	k_H/k_D	Electron withdrawal at C_β[e]	Electron release at C_α[e]	Leaving-group isotope effect or element effect
$(E1)_{anion}$	1	Yes	General[d]	1.0	Rate decrease	Rate increase	Substantial
$(E1cB)_R$	2	Yes	Specific	1.0	Small rate increase	Small rate increase	Substantial
$(E1cB)_{ip}$	2	No	General[f]	$1.0 \rightarrow 1.2$	Small rate increase	Small rate increase	Substantial
$(E1cB)_{irr}$	2	No	General	$2 \rightarrow 8$	Rate increase	Little effect	Small to negligible
E2[c]	2	No	General	$2 \rightarrow 8$	Rate increase	Small rate increase	Small

[a] Data from A. F. Cockerill and R. G. Harrison, *The Chemistry of Double-Bonded Functional Groups,* Part 1, Supplement A to *The Chemistry of the Functional Groups,* S. Patai, Ed., Wiley, New York, 1977, p. 725. Copyright John Wiley and Sons, Ltd (1977). Reprinted by permission of John Wiley and Sons, Ltd.
[b] All mechanisms exhibit first-order kinetics in substrate.
[c] Only transition states with considerable carbanion character considered in this table.
[d] Specific base catalysis predicted if extent of substrate ionization reduced from almost complete.
[e] Effect on rate assuming no change in mechanism is caused; steric factors upon substitution at C_α and C_β have not been considered. The rate predictions are geared to substituent effects such as these giving rise to Hammett reaction constants on β- and α-aryl substitution.
[f] Depends on whether ion pair assists in removal of leaving group.

$$
\begin{array}{c}
\text{NC} \quad \text{CN} \\
| \quad\quad | \\
\text{Ar}-\text{C}-\text{C}-\text{H} \; + \; \text{R}_3\text{N}: \xrightarrow[\text{fast}]{k_1} \text{Ar}-\underset{|}{\overset{|}{\text{C}}}-\text{C(CN)}_2 \; + \; \text{R}_3\overset{+}{\text{N}}\text{H} \\
| \quad\quad | \\
\text{NC} \quad \text{CN}
\end{array}
\tag{7.40}
$$

$$
\text{Ar}-\underset{\text{CN}}{\overset{\text{CN}}{\underset{|}{\overset{|}{\text{C}}}}}-\text{C(CN)}_2 \xrightarrow{\text{slow}} \underset{\text{CN}}{\overset{\text{Ar}}{\diagdown}}\text{C}=\text{C}\underset{\diagdown \text{CN}}{\overset{\diagup \text{CN}}{}} \; + \; \text{CN}^-
\tag{7.41}
$$

In the other three variations of the carbanion mechanism, an equilibrium concentration of carbanion is formed, which then either returns to starting material or decomposes to products.

If the β proton is slightly less acidic than required for the $(\text{E1})_{\text{anion}}$ mechanism and k_{-1} is comparable to k_1 but k_2 is still small, the anion forms from the starting material in a rapid equilibrium and the leaving group departs in a subsequent slow step. This is called the $(\text{E1cB})_\text{R}$ ("R" for "reversible") mechanism. Because k_2 is much smaller than k_1 and k_{-1}, we can assume that k_2 does not affect the equilibrium concentration of the anion of the substrate, S^-; then the concentration of S^- can be found according to Equation 7.42 and the rate of elimination will be that of Equation 7.43.[89]

$$
[S^-] = \frac{k_1[\text{HS}][\text{B}]}{k_{-1}[\text{BH}^+]}
\tag{7.42}
$$

$$
\text{rate} = \frac{k_2 k_1[\text{HS}][\text{B}]}{k_{-1}[\text{BH}^+]}
\tag{7.43}
$$

Examination of Equation 7.43 shows that the rate of an $(\text{E1cB})_\text{R}$ reaction should be independent of the base concentration if the buffer ratio, B/BH^+, is kept constant—that is, the reaction should exhibit specific base catalysis (see Section 7.1). An example of such a reaction is elimination of methanol from **44**. Not only is

$$
\underset{\text{O}}{\overset{}{\text{CH}_3}}-\underset{\|}{\overset{}{\text{C}}}-\text{CH}_2-\underset{\text{OCH}_3}{\overset{}{\text{CH}_2}}
$$

44

specific base catalysis observed, but also, in agreement with rapid and reversible formation of carbanion, in deuterated solvent the rate of incorporation of deuterium into the substrate is 226 times faster than the rate of elimination.[90]

The $(\text{E1cB})_{\text{ip}}$ mechanism is a close cousin of the $(\text{E1cB})_\text{R}$ mechanism. The difference is that in the former the free anion is not formed but exists as an ion pair with the protonated base as counter-ion. An example of a reaction that goes by this

---

[89] Note that in this and the following E1cB mechanisms, the rates are not really independent of the base concentration and therefore the "1" part of the classification may be misleading.

[90] L. R. Fedor, *J. Am. Chem. Soc.*, **91**, 908 (1969). For other examples of the $(\text{E1cB})_\text{R}$ mechanism see J. Crosby and C. J. M. Stirling, *J. Chem. Soc. B*, 671, 679 (1970).

mechanism is the formation of bromoacetylene from cis-1,2-dibromoethylene and triethylamine (Equation 7.44).[91] If the rate of elimination from deuterated 1,2-

$$
\begin{array}{c}
\underset{Br}{\overset{H}{\diagdown}}C=C\underset{Br}{\overset{H}{\diagup}} \underset{k_{-1}}{\overset{k_1}{\rightleftharpoons}} \left[\underset{Br}{\overset{H}{\diagdown}}C=C\underset{Br}{\overset{\overset{+}{H}NEt_3}{\diagup}} \right] \xrightarrow{k_2} HC\equiv CBr + \overset{+}{H}NEt_3\ Br^-
\end{array}
\qquad (7.44)
$$

dibromoethylene is compared to the rate from nondeuterated material, $k_H/k_D \approx 1$. Therefore proton abstraction is not involved in the rate-determining step. Because added $Et_3\overset{+}{N}D\ X^-$ does not affect the rate and because deuterium exchange with solvent does not take place, the $(E1cB)_R$ mechanism cannot be involved. Apparently the intimate ion pair either goes back to starting material or loses Br^- in a slow step; a free carbanion is not formed.

   Finally, there is the $(E1cB)_{irr}$ ("irr" for "irreversible") mechanism, in which the leaving group is so good that proton abstraction becomes rate determining. For this case ($k_2 \gg k_1, k_{-1}$) the rate equation reduces to Equation 7.45. Reactions of

$$
\text{rate} = k_1[B][HS] \qquad (7.45)
$$

this sort, then, should be dependent on the base concentration—that is, they should be general-base catalyzed. Elimination of benzoic acid from **45** exhibits general-

$$
CH_3-\underset{\underset{O}{\|}}{C}-CH_2-CH_2-O-\underset{\underset{}{\|}}{\overset{\overset{O}{\|}}{C}}-\underset{Y}{\diagdown}
$$

45

base catalysis and the rate is independent of the nature of the substituents on the phenyl ring. The general-base catalysis excludes the $(E1cB)_R$ mechanism and the lack of leaving-group effect excludes all mechanisms in which leaving-group–carbon bond breaking occurs in the transition state. An $(E1cB)$ mechanism thus seems indicated.[92] Note that **45** differs from **44** only in that benzoate is a much better leaving group than methoxide. This is only one example of several in the literature that show how sensitive the various carbanion elimination mechanisms are to changes in the structure of the reactants and to the reaction condition.[93]

   Carbanion mechanisms may give either syn or anti elimination. For example, Hunter and Shearing studied the butoxide-catalyzed elimination of methanol from **46** and **47**. Since deuterium exchange with solvent is in close competition with elimination, the mechanism is probably $(E1cB)_R$. The ratio of syn/anti elimination varies by a factor of approximately 75, depending on the cation of the butoxide salt,

[91] W. K. Kwok, W. G. Lee, and S. I. Miller, *J. Am. Chem. Soc.*, **91**, 468 (1969).
[92] R. C. Cavestri and L. R. Fedor, *J. Am. Chem. Soc.*, **92**, 610 (1970).
[93] See also (a) L. R. Fedor and W. R. Glave, *J. Am. Chem. Soc.*, **93**, 985 (1971); (b) note 88(b); (c) R. A. More O'Ferrall and P. J. Warren, *J. Chem. Soc., Chem. Commun.*, 483 (1975).

and decreases in the order $Li^+ > K^+ > Cs^+ > (CH_3)_4N^+$. Hunter and Shearing attribute the changing stereochemistry to the tendency of the cation to coordinate with the methoxy group of the substrate. Li^+, which has the strongest coordinating ability, gives mostly syn elimination; $(CH_3)_4N^+$, which has the weakest, gives predominantly anti.[94]

In the carbanionic mechanisms for elimination, if the substrate has two proton-bearing β carbons, the more acidic protons will be removed. Thus in alkylated substrates the double bond will be oriented toward the less substituted carbon and Hofmann elimination is obtained in solution.

E2 Eliminations[95]

The rates of a large number of eliminations are (1) second-order, first-order each in base and in substrate; (2) decreased if β-deuterium is substituted for β-hydrogen; and (3) dependent on the nature of the leaving group. The mechanism of these reactions (shown in Equation 7.36), in which C—H and C—X bond breaking are concerted, is E2. Because several of the E1cB mechanisms are also second-order, differentiating between an E1cB and an E2 mechanism can be difficult. To do so, one can use the kinetic predictions in Table 7.11 or one can ascertain whether the elimination proceeds more rapidly than the hypothetical two-step reaction. For the concerted pathway to be utilized, it should be of lower energy than the possible stepwise paths. (The last criterion is, however, often difficult to apply because of the problem in calculating the hypothetical rates.) Brower has used volumes of activation to distinguish between E1cB and E2 mechanisms.[96] The former have positive and the latter negative $\Delta V^{\ddagger}$'s. Saunders has recently reviewed the experimental methods for distinguishing between concerted and nonconcerted eliminations.[97]

Substituent and isotope effects show that Equation 7.36 must actually describe a spectrum of transition states in which the relative extents of C—H and C—X bond breaking vary according to the specific substrate and to the reaction conditions. For example, a comparison of the rates of ethoxide-catalyzed elimination of HX from **48** and **49** in ethanol at 30°C shows that k_H/k_D varies from 3.0

[94]D. H. Hunter and D. J. Shearing, *J. Am. Chem. Soc.*, **93**, 2348 (1971); **95**, 8333 (1973).
[95](a) See note 77; (b) D. V. Banthorpe, in *Studies on Chemical Structure and Reactivity*, J. H. Ridd, Ed., Methuen, London, 1966, p. 33; (c) N. A. LeBel, in *Advances in Alicyclic Chem.*, Vol. 3, H. Hart and G. J. Karabatsos, Eds., Academic Press, New York, 1971, p. 196.
[96]K. R. Brower, M. Muhsin, and H. E. Brower, *J. Am. Chem. Soc.*, **98**, 779 (1976).
[97]W. H. Saunders, Jr., *Acc. Chem. Res.*, **9**, 19 (1976).

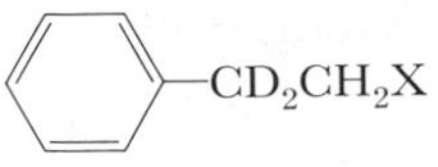

$$\text{48} \qquad\qquad\qquad \text{49}$$

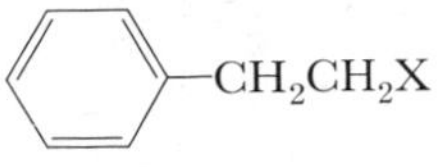

48 has structure phenyl—CD$_2$CH$_2$X, 49 has structure phenyl—CH$_2$CH$_2$X

when X = $^+$N(CH$_3$)$_3$ to 7.1 when X = Br.[98] Similarly, if X is kept constant, the logarithms of the rates of Reaction 7.46 correlate linearly with the σ values of the substituents but the slopes of the correlation lines depend on X and are given in Table 7.12. The extent of bond breaking in the transition state must, then, depend on X. Other factors that alter the nature of the transition state are base strength, solvent, and substrate structure.

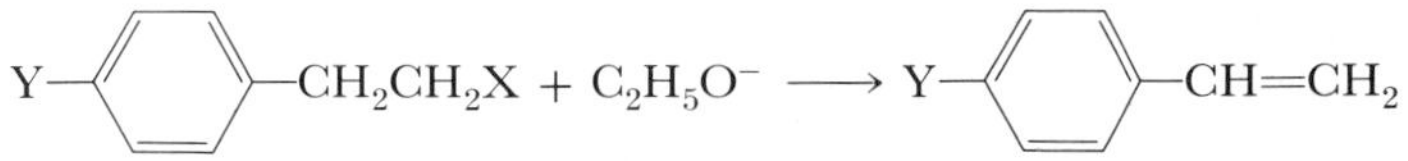

$$Y\!\!-\!\!\langle\bigcirc\rangle\!\!-\!\!CH_2CH_2X + C_2H_5O^- \longrightarrow Y\!\!-\!\!\langle\bigcirc\rangle\!\!-\!\!CH\!\!=\!\!CH_2 \qquad (7.46)$$

Bunnett[99] suggested that the spectrum of transition states ranges from one similar to that of E1cB elimination, in which C—H bond breaking has proceeded considerably further than C—X bond breaking (**50**), to one similar to that of the E1 reaction (**52**). Intermediate would be the symmetrical transition state **51**. This

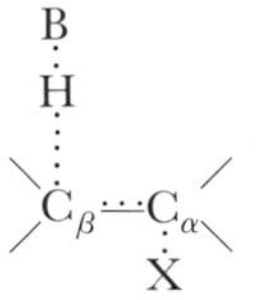

$$\text{50} \qquad\qquad \text{51} \qquad\qquad \text{52}$$

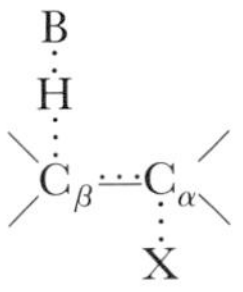

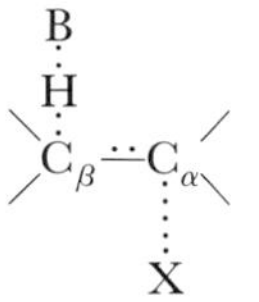

Table 7.12 HAMMETT ρ CONSTANTS
FOR REACTION 7.46[a]

X	ρ
I	+2.07
Br	2.14
OTs	2.27
Cl	2.61
S(CH$_3$)$_2$	2.75
F	3.12
N(CH$_3$)$_3$	3.77

[a] Data from W. H. Saunders, *The Chemistry of Alkenes*, S. Patai, Ed., Wiley, New York, 1964, p. 155, Table 1. Reprinted by permission of Wiley-Interscience.

[98] W. H. Saunders, Jr., and D. H. Edison, *J. Am. Chem. Soc.*, **82**, 138 (1960).
[99] J. F. Bunnett, *Angew. Chem. Int. Ed. Engl.*, **1**, 225 (1962).

range in transition states is now widely, though not universally, accepted (see below). Transition states resembling **52**—that is, those toward the E1 end of the spectrum—seem to be least common in protic solvents since α-aryl groups do not greatly accelerate these reactions.[100]

In aprotic media weak proton bases such as thiophenoxide and chloride ions may promote eliminations more effectively than strong bases such as alkoxide ions. For example, t-butyl bromide, which had been thought to undergo only E1 elimination, actually eliminates by a bimolecular mechanism with NaCl in acetone. In this solvent Cl$^-$ is a more effective catalyst than p-nitrophenoxide, although the latter is 10^{10} times stronger as a hydrogen base.[101,102] On the basis of this and related evidence, Winstein and Parker proposed that in aprotic media the spectrum of transition states extends from the E1cB-like transition state (**50**)—called by them E2H—to one in which the base is pushing out the leaving group rather than attacking the proton (**54**). The latter is designated E2C. In the center of the Winstein–Parker spectrum is the E2 transition state (**53**), in which the base pulls off the proton and pushes off the leaving group simultaneously.

E2H	E2	E2C
50	**53**	**54**

As support for their proposal, Winstein, Parker, and co-workers noted that when hard bases are the catalysts, the rate of elimination of a compound depends on the proton basicity of the catalyst, as shown in the Brønsted relationship, Equation 7.47 (where k^E is the rate constant for bimolecular elimination):

$$\log k^E = \log pK_A + \text{constant} \tag{7.47}$$

Conversely, when soft bases are used, elimination rates, as would be expected from transition state **54**, show no such correlation. Instead there is a relationship between the rate of elimination and the rate of S_N2 substitution by the base as shown in Equation 7.48 (where k^S is the rate constant for bimolecular substitution and X is a constant):

$$\log k^E = X \log k^S + \text{constant} \tag{7.48}$$

For example, Figure 7.4 shows a plot of $\log k^E$ vs. $\log k^S$ for cyclohexyl tosylate with a number of soft bases.[103]

[100] E. Baciocchi, P. Perucci, and C. Rol, *J. Chem. Soc., Perkin Trans.* **2**, 329 (1975).
[101] P. Beltrame, G. Biale, D. J. Lloyd, A. J. Parker, M. Ruane, and S. Winstein, *J. Am. Chem. Soc.*, **94**, 2240 (1972), and references therein.
[102] A. J. Parker, M. Ruane, D. A. Palmer, and S. Winstein, *J. Am. Chem. Soc.*, **94**, 2228 (1972).
[103] A. J. Parker, M. Ruane, G. Biale, and S. Winstein, *Tetrahedron Lett.*, 2113 (1968).

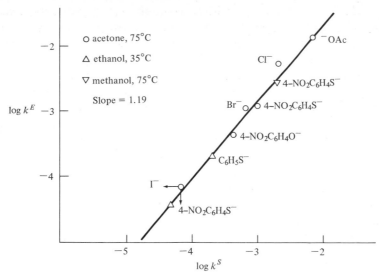

Figure 7.4 Relationship between elimination and substitution rates of cyclohexyl tosylates with soft bases. Reprinted with permission from A. J. Parker, M. Ruane, G. Biale, and S. Winstein, *Tetrahedron Lett.*, 2113 (1968). Copyright 1968 Pergamon Press, Ltd.

The E2C-like transition state (**54**) has been strongly criticized.[104] Correlations such as that in Figure 7.4 are not general for all weak-base-catalyzed reactions. Also, Firestone has pointed out that unless the proton transfer in the transition state is very far advanced, the rate of an E2 reaction in which base attacks proton (transition state **51**) should not even be expected to obey the Brønsted relationship (Equation 7.47) but rather should correlate with the bases' hydrogen-bonding ability.[105] In fact, the correlation between the elimination rates and hydrogen-bonding ability of the bases in weak-base-catalyzed reactions is quite good.

Another argument against the E2C reaction comes from Bunnett,[106] who has pointed out that since the E2C transition state involves partial bonding of the base to C_α, steric effects on weak-base-catalyzed reactions should resemble those on S_N2 reactions—which they do not. Weak-base-catalyzed reactions are much less sensitive to steric bulk on the substrate. Parker has maintained that an E2C transition state is "looser" than an S_N2 transition state—that is, that the C_α– base and C_β– leaving-group bond distances are greater in the E2C transition state.[107] Because electron-releasing substituents at C_α do not much accelerate these eliminations, there cannot be much charge development at C_α. Thus these longer bond distances imply a highly developed π bond at the transition state in the E2C reaction. But problems with this view have also arisen. For example, weak-base-catalyzed elimi-

[104]For a summary of the arguments, see W. T. Ford, *Acc. Chem. Res.*, **6**, 410 (1973).
[105]R. A. Firestone, *J. Org. Chem.*, **36**, 702 (1971).
[106]J. F. Bunnett and D. L. Eck, *J. Am. Chem. Soc.*, **95**, 1897 (1973).
[107]E. C. F. Ko and A. J. Parker, *J. Am. Chem. Soc.*, **90**, 6447 (1968).

nations carried out on **55** give almost entirely **56** rather than the conjugated **57** which would be expected from a productlike transition state.[108]

$$\phi-CH_2-\underset{\underset{OTs}{|}}{CH}-\underset{\underset{CH_3}{|}}{CH}-CH_3 \qquad \phi-CH_2-CH=\underset{\underset{CH_3}{|}}{C}-CH_3 \qquad \phi-CH=CH-\underset{\underset{CH_3}{|}}{\overset{\overset{CH_3}{|}}{C}}-H$$

$$\quad\quad\quad 55 \quad\quad\quad\quad\quad\quad\quad 56 \quad\quad\quad\quad\quad\quad\quad 57$$

Most workers now seem to feel that the E2C mechanism is an unnecessary complication and that weak-base-catalyzed reactions fit into the variable E2 transition state spectrum described by **50–52** and that they lie near the E1 end.[106,109,110] McLennan has suggested that that end of the spectrum resembles **58** more than **52**.[111] In **58** a good leaving group has almost entirely departed from C_α. The weak base B has not gained control of the β proton to any great extent, so π bonding is weak.[112] A nonlinear B—H—C arrangement places B reasonably close to C_α but not so close as to permit significant covalent bonding. The interaction thus is purely electrostatic.

$$\underset{\underset{58}{}}{\overset{\overset{\displaystyle H\quad\cdots B}{\vdots\quad\quad\vdots}}{C_\beta-\underset{\underset{\displaystyle \overset{\cdot}{X}{}^{\delta-}}{\vdots}}{\overset{|}{C}{}_\alpha^{\,\delta+}}-}}$$

In this chapter we will, following the usage of several investigators, call transition states such as **50** E2H transition states and those such as **52** or **58** E2C transition states. We do this because the nomenclature is easier than E1cB-like or E1-like and not because we are subscribing to the Parker–Winstein variable transition state spectrum.

Reaction Coordinate Diagrams for E2 Reactions

The effect of changing the substrate or reaction conditions on the transition state of an E2 reaction can be predicted by using the rules for three-dimensional reaction coordinate diagrams discussed in Section 2.6, (see also Section 4.2).[113] The concerted reaction is broken down into the two stepwise mechanisms of which it is a

[108] D. M. Muir and A. J. Parker, *J. Org. Chem.*, **41**, 3201 (1976); see also I. N. Feit, I. K. Breger, A. M. Copobianco, T. W. Cooke and L. F. Gitlin, *J. Am. Chem. Soc.*, **97**, 2477 (1975).

[109] W. T. Ford and D. J. J. Pietsek, *J. Am. Chem. Soc.*, **97**, 2194 (1975), and references therein.

[110] For an example where weak-base-catalyzed reactions have appreciable control of the β hydrogen as evinced by maximum k_H/k_D values, see D. J. McLennan and R. J. Wong, *J. Chem. Soc., Perkin Trans.* **2**, 1818 (1974).

[111] D. J. McLennan, *Tetrahedron*, **31**, 2999 (1975).

[112] Parker and Bunnett have maintained that since weak-base-catalyzed reactions usually give very high proportions of Sayzteff olefin, these reactions have well-developed π bonds at the transition state. However, elimination from **55** seems to speak against that view. See note 108.

[113] D. A. Winey and E. R. Thornton, *J. Am. Chem. Soc.*, **97**, 3102 (1975); (b) R. A. More O'Ferrall, *J. Chem. Soc. B*, 274 (1970); P. J. Smith, *Isotopes, in Organic Chemistry*, Vol. 2, Buncel and C. C. Lea, Eds., Elsevier, Amsterdam, 1976, p. 231.

composite. The Bunnett E2 reaction is a composite of the E1 and the E1cB mechanisms. Thus Figure 7.5 shows the three-dimensional reaction coordinate diagram for an E2 reaction. The starting material is at the back left-hand corner and the product at the front right. The coordinate from back to front represents increasing C_α—X distance and that from left to right, increasing C_β—H distance. Thus at the front left-hand corner is the E1 ion-pair intermediate and at the back right, the E1cB carbanion intermediate. That this is the reaction coordinate diagram for a concerted and not a stepwise reaction can be seen by the fact that the diagonal pathway has its energy maximum at a lower free energy than either of the pathways along the edges has its energy maximum.

In order to study the effect of variables on the position of the transition state in this reaction we will follow the procedure we used in Section 2.6. We will look down at the diagram in Figure 7.5 and make a two-dimensional projection of this perspective. The coordinates then run along the sides of a square. We use the designation $*_p$ for an energy maximum (transition state) and ∘ for an energy minimum. Figure 7.6 shows the horizontal projection of a three-dimensional reaction coordinate diagram for a fully concerted E2 mechanism—that is, the horizontal projection of Figure 7.5.[114]

In Figure 7.6(a) is shown the effect of changing the leaving group to a poorer one in such a reaction. This change would result in an increase in the energy of the E1 ion-pair intermediate and in the energy of the product and thus in an increase in the energy of the whole lower edge of the diagram. The transition state, then,

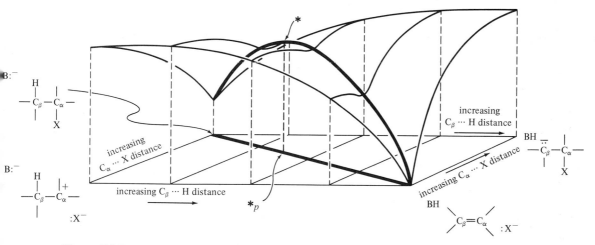

Figure 7.5 Energy surface and reaction coordinate for a hypothetical E2 elimination reaction in which H^+ transfer from carbon to base B$:^-$ and cleavage of carbon–halogen bond occur together, with the transition state at the midpoint of the process. The curved heavy line is the reaction coordinate and the straight heavy line is its projection on the horizontal plane. The transition state is marked by *, and its projection by $*_p$. Vertical dashed lines connect points on the surface with the projections of those points on the horizontal plane.

[114]See problem 10 for the effect of changing substrate and reaction conditions on E2H reactions.

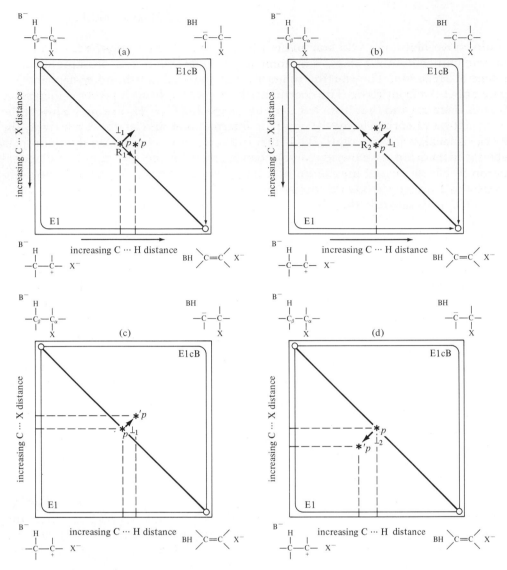

Figure 7.6 Projection in the horizontal plane of a central E2 reaction path. (a) The effect of a poorer leaving group on the position of the transition state. (b) The effect of a stronger base on the position of the transition state. (c) The effect of an electron-withdrawing group at C_β; (d) The effect of an electron-releasing group at C_α.

would move along the reaction path with motion R_1 and perpendicular to the reaction path with motion $\perp_1$. The composite result of the poorer leaving group on a central E2 reaction would be to move the transition state from $*p$ to $*'p$. The extent of C—X bond breaking would not be affected but the C—H bond would be broken more and the carbanion character of the transition state increased.

These predictions are borne out by the ρ values of Table 7.12. More negative charge is localized on C_β when the leaving group is the less reactive $N(CH_3)_3$ than when it is the more reactive I^-. The isotope effects mentioned above fit this explanation if it is assumed that when Br^- is the leaving group, the proton is approximately half transferred at the transition state. The smaller value of k_H/k_D when $^+N(CH_3)_3$ departs is a result of an unsymmetrical transition state in which the proton is more than half transferred.

Figure 7.6(b) shows the effect of increasing base strength on a central E2 mechanism. This change would lower the energy of the E1cB intermediate, the energy of the product, and thus the energy of the whole right-hand edge of the diagram. The transition state would move along R_2 and $\perp_1$. Thus a stronger base would shift the transition state to one in which there would be less C—X bond breaking but about the same amount of C—H bond breaking.[115] Taken together this means more carbanion character at C_β and a decrease in the amount of double-bond formation.

Figures 7.6(c) and 7.6(d) show the effect, on the transition state of a central E2 reaction, of adding an electron-withdrawing group to C_β and an electron-releasing group to C_α, respectively. Note that in each of these last two cases the energy of one of the intermediates is affected but not the energy of the starting material nor of the product. Thus only one corner and not a whole edge is moved up or down.

Table 7.13 summarizes the predictions of Figures 7.6(a)–7.6(d). Below we will see that these predictions concur with experimental observation.

Orientation of double bonds   If the double bond can be oriented toward either of two or three carbons in an E2 reaction, the product depends on where the transition state of the particular reaction lies in the spectrum. In a central E2 reaction the double bond is apparently so highly developed in the transition state that the relative olefin stability is the controlling factor in deciding the product. In an E2H reaction, however, there is so much carbanion character at the transition state that the most stable transition state will be the one with the most stabilized negative charge at C_β. Thus the relative acidity of the hydrogens is of overriding importance to product determination. Since the acidity of the proton, the reactivity of the leaving group, and the strength of the base all help determine where the transition state lies in the spectrum, all these factors affect the ratio of Hofmann to Saytzeff product.

In general, one can use three-dimensional reaction coordinate diagrams to predict the approximate position of the transition state in the variable E2 transition

---

[115] For ranking of leaving-group ability in elimination reactions, see C. J. M. Stirling, *Acc. Chem. Res.*, **12**, 198 (1979).

Table 7.13 Predictions for Change of Transition State Structure for a Central
E2 Reaction Resulting from Change in Reaction Conditions

Condition changed	C—H bond length	C—X bond length	Carbanion character at $C_\beta{}^a$	Double-bond formation
Poorer leaving group	Longer	Same	More	Same
Stronger base	Same	Shorter	More	Less
Electron-withdrawing group at C_β	Longer	Shorter	More	Less
Electron-releasing group at C_α	Shorter	Longer	Less	Less

^{a} The amount of carbocation character is not listed in the table because it seems never to be very high in E2 eliminations [see note 100 and note 113(a)]. However, the amounts of carbanion character at C_β and carbocation character at C_α are related because of the developing double-bond character between these carbons. Thus, for example, an electron-withdrawing group at C_β could result in more carbocation character at C_α rather than less carbanion character at C_β.

state spectrum. From that one can predict the orientation of the double bond. For example, the strong electron-withdrawing ability of fluorine, which renders the β protons acidic, the low reactivity of this halogen as a leaving group, and the strength of the base toward hydrogen assures that Reaction 7.49, when X = F, lies well toward the E2H end of the spectrum (see Figure 7.5). The data in Table 7.14 show that, as expected, the more acidic primary protons are lost preferentially to

$$CH_3(CH_2)_3CHCH_3 \xrightarrow{\ ^-OCH_3\ } CH_3(CH_2)_2CH{=}CHCH_3 + CH_3(CH_2)_3CH{=}CH_2 \qquad (7.49)$$
$$\underset{X}{|}$$

the less acidic secondary ones, giving predominantly Hofmann-type product. In the series of increasing atomic weight, the halogens become simultaneously less elec-

Table 7.14 Orientation of the Double Bond in the Products of Reaction 7.49^a

X	2-Hexene/ 1-Hexene
F	0.43
Cl	2.0
Br	2.6
I	4.2

^{a} Reprinted with permission from R. A. Bartsch and J. F. Bunnett, *J. Am. Chem. Soc.*, **90,** 408 (1968). Copyright by the American Chemical Society.

tron withdrawing and better as leaving groups; therefore as the fluorine is substituted in turn by Cl, Br, and I, Reaction 7.49 moves more toward the E2C end of the spectrum and Saytzeff products become more important.[116]

The importance of the base in determining the nature of the transition state and thereby the product can be seen from Table 7.15. When Reaction 7.50 is carried out with the very strong base potassium t-butoxide, 20 percent of the less substituted Hofmann product (59) is obtained. The weaker base sodium n-propoxide causes a shift toward a more central transition state (Figure 7.5) and less 59 is formed.[117]

$$CH_3CHCH_2CH_3 \xrightarrow[\text{DMSO}]{M^+\bar{O}R} CH_2{=}CHCH_2CH_3 + CH_3CH{=}CHCH_3 \qquad (7.50)$$
$$\overset{|}{I}$$

$$59 \qquad\qquad\qquad 60$$

Brown has suggested[118] that steric factors are of primary and almost sole importance in determining the position of the double bond. Thus Hofmann product becomes important when a large leaving group—large either on its own account or because of heavy solvation—makes it even more difficult for the base to abstract the more hindered protons. That steric effects in the base are not usually the prime cause for the change in orientation of the double bond for the reactions in Table 7.15 was convincingly demonstrated by Bartsch.[117] From the product

Table 7.15 Relative Alkene Proportions from Reactions of 2-Iodobutane with Oxyanion Bases in Dimethyl Sulfoxide at 50.0°C[a]

System	Base	pK_a of Conjugate acid in DMSO	Percent 1-butene in total butenes
1	Potassium p-nitrobenzoate	8.9	5.8 ± 0.1
2	Potassium benzoate	11.0	7.2 ± 0.2
3	Potassium p-nitrophenoxide	11.0	7.5 ± 0.1
4	Potassium o-nitrophenoxide	11.0	7.5 ± 0.2
5	Potassium acetate	11.6	7.4 ± 0.1
6	Potassium p-aminobenzoate	12.7	8.0 ± 0.2
7	Potassium 2,6-di-$tert$-butylphenoxide	15.0	19.2 ± 0.4
8	Potassium phenoxide	16.4	11.4 ± 0.2
9	Sodium 2,2,2-trifluoroethoxide	21.6	14.3 ± 0.2
10	Sodium methoxide	27.0	17.0 ± 0.5
11	Sodium ethoxide	27.4	17.1 ± 0.4
12	Sodium n-propoxide	28.0	18.5 ± 0.3
13	Potassium $tert$-butoxide	29.2	20.7 ± 0.4

[a] Reprinted with permission from R. A. Bartsch, G. M. Pruss, B. A. Bushaw, and K. E. Wiegers, *J. Am. Chem. Soc.*, **95**, 3405 (1973). Copyright by the American Chemical Society.

[116] R. A. Bartsch and J. F. Bunnett, *J. Am. Chem. Soc.*, **90**, 408 (1968).
[117] (a) R. A. Bartsch, G. M. Pruss, B. A. Bushaw, and K. E. Wiegers, *J. Am. Chem. Soc.*, **95**, 3405 (1973); (b) R. A. Bartsch, K. E. Wiegers, and D. R. Guritz, *J. Am. Chem. Soc.*, **96**, 430 (1974).
[118] H. C. Brown and R. L. Klimisch, *J. Am. Chem. Soc.*, **88**, 1425 (1966), and references therein.

composition Bartsch determined, for each reaction system, the difference in the free energies of activation for the formation of 1-butene and *trans*-2-butene. In Figure 7.7 these values are plotted against the pK_a's of the conjugate acids of the bases. A good straight line is obtained for all the bases studied except 2,6-di-*t*-butylphenoxide, for which the difference in energies of activation is smaller than would be expected from the pK_a of 2,6-di-*t*-butylphenol. Thus from this relatively unhindered compound (2-iodobutane), the orientation of the double bond is determined by the base strength, unless really outsized bases are used.

How large the base needs to be before its steric effects are introduced into double-bond orientation does depend on the alkyl substitution in the substrate but not on the size of the leaving group.[119]

Stereochemistry[120] Since all E2 transition states have some double-bond character, E2 eliminations, if they are to go at all well, require that H and X be

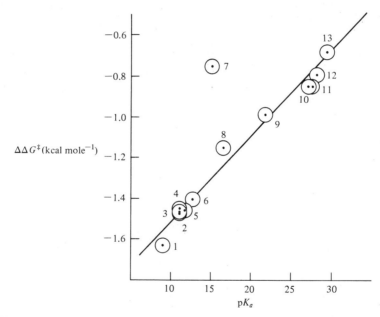

Figure 7.7 Plot of the free-energy difference for formation of 1-butene and *trans*-2-butene vs. the pK_a of the conjugate acid of the base. System numbers refer to Table 7.15. Reprinted with permission from R. A. Bartsch, G. M. Pruss, B. A. Bushaw, and K. E. Wiegers, *J. Am. Chem. Soc.*, **95**, 3405 (1973). Copyright 1973 American Chemical Society.

[119](a) B. A. Bartsch, R. A. Read, D. T. Larsen, D. K. Roberts, K. J. Scott, and B. R. Cho, *J. Am. Chem. Soc.*, **101**, 1176 (1979); (b) M. Charton, *J. Am. Chem. Soc.*, **97**, 6159 (1975).
[120]For reviews, see (a) J. Sicher, *Angew. Chem. Int. Ed. Engl.*, **11**, 200 (1972); (b) S. Wolfe, *Acc. Chem. Res.*, **5**, 102 (1972).

either syn or anti periplanar in the transition state. For an E2C reaction corresponding to **58**, anti elimination would be expected. The two geometries for transition states of the E2H reaction are shown in Figures 7.8(a) and 7.8(b). All other things being equal, steric effects would cause anti elimination to be of lower energy than syn elimination, since the transition state leading to the former [Figure 7.8(b)] is entirely staggered, whereas the transition state leading to the latter [Figure 7.8(a)] is partially eclipsed.[121] Furthermore, Bach has recently pointed out that frontier molecular orbital theory also leads one to predict that anti elimination would be favored. We saw in Chapter 1 that S_N2 displacements can be considered in terms of interaction between the HOMO of the nucleophile and the LUMO of the C—X σ bond. A 1,2-elimination reaction can be considered to be an intramolecular S_N2 reaction in which the leaving group is displaced by the developing electron pair at the β carbon: the electron pair at C_β *is* the nucleophile. A 1,2-elimination will then be favored when the HOMO of the nucleophile (the filled C_β—H σ orbital) can best interact with the LUMO of the σ bond involved in the displacement (the empty C—X (σ^* orbital). The orientation of this HOMO to this LUMO in an anti transition state is shown in **61a** and in a syn transition state in **61b**. The larger lobes of a C—X σ^* orbital are, as shown in **61a** and **61b**, not between C and X but at the back side of C and of X, so that antibonding interactions would be minimized if the orbitals were filled. The transition state for anti elimination, then, allows better HOMO–LUMO overlap (large lobe with large lobe) than the transition state for syn elimination (large lobe with small lobe).[122]

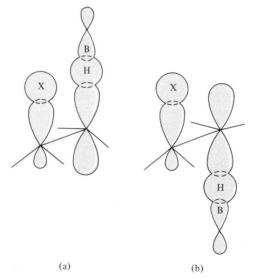

(a)	(b)

Figure 7.8. (a) Transition state for *E*2H syn elimination. (b) Transition state for *E*2H anti elimination.

[121] J. Hine, *J. Am. Chem. Soc.*, **88**, 5525 (1966).
[122] R. D. Bach, R. C. Badger, and T. J. Lang, *J. Am. Chem,. Soc.*, **101**, 2845 (1979).

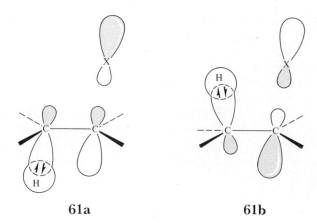

61a   **61b**

Let us now turn to the experimental results to see if the predictions are borne out. It has long been known that E2 reactions normally give preferentially anti elimination. For example, reaction of *meso*-stilbene dibromide with potassium ethoxide gives *cis*-bromostilbene (Reaction 7.51), whereas reaction of the D,L-dibromide gives the trans product (Reaction 7.52).[123] A multitude of other examples exist—see, for example, note 95.

$$\phi \overset{\phi}{\underset{H}{\overset{Br}{\diagdown}}}\overset{Br}{\underset{}{\diagdown}} H \longrightarrow \overset{\phi}{\underset{Br}{\diagup}}C=C\overset{\phi}{\underset{H}{\diagdown}} \tag{7.51}$$

$$Br\overset{\phi}{\underset{H}{\overset{Br}{\diagdown}}}\overset{}{\underset{\phi}{\diagdown}} H \longrightarrow \overset{Br}{\underset{\phi}{\diagdown}}C=C\overset{\phi}{\underset{H}{\diagdown}} \tag{7.52}$$

E2 reactions do, however, give syn elimination when the transition state has considerable anionic character and when (1) an H—X dihedral angle of 0° is achievable but one of 180° is not or, put another way, H and X can become syn periplanar but not anti periplanar; (2) a syn hydrogen is much more reactive than the anti ones; (3) syn elimination is favored for steric reasons; and (4) an anionic base that remains coordinated with its cation, that, in turn, is coordinated with the leaving group, is used as catalyst.[124]

Bach suggests that the overall requirement for syn elimination—that there be considerable anionic character to the transition state—is due to the fact that syn elimination occurs as shown in Equation 7.53. Initial proton abstraction in the syn conformation leaves the HOMO of the C_β—H and the LUMO of the C_α—X bond

[123] P. Pfeiffer, *Z. Phys. Chem. (Leibzig)*, **48**, 40 (1904).
[124] For a review of these effects, see R. A. Bartsch, *Acc. Chem. Res.*, **8**, 239 (1975).

unable to interact effectively as discussed just above. Thus C_β inverts simultaneously with leaving-group departure and the interaction is as in **61a**. The inversion (the second step in Equation 7.53) can only occur after the hydrogen has been substantially pulled from C_β.[122]

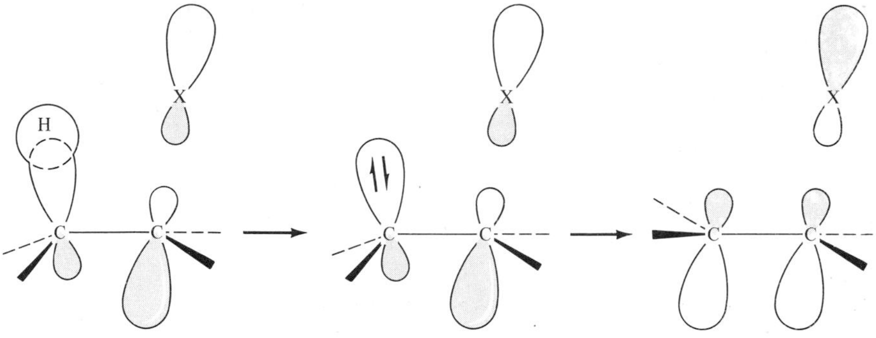

$$(7.53)$$

An example of category 1 is found in the observation by Brown and Liu that eliminations from the rigid ring system **62**, induced by the sodium salt of 2-cyclohexylcyclohexanol in triglyme, produces norbornene (98 percent) but no 2-deuterionorbornene.[125] The dihedral angle between D and tosylate is 0°, but that

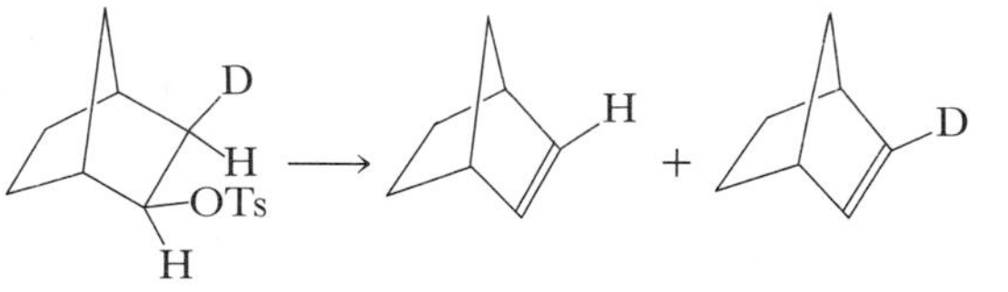

$$(7.54)$$

Crown ether present:			
	No	98%	0%
	Yes	70%	27.2%

between H and tosylate is 120°. However, when the crown ether **63**, which is an excellent complexing agent for sodium ion, is added to the reaction, the amount of syn elimination drops to 70 percent. Apparently coordination of the sodium ion to both the leaving group and the base in the transition state, as in **64**, is responsible for some of the syn elimination from **62** in the absence of crown ether (category 4 above).[126]

Category 2 is exemplified by E2 elimination from **65**, in which the tosylate group can become periplanar with either H_1 or H_2. However, H_1 is activated and H_2 is not; when **65** is treated with potassium *t*-butoxide in *t*-butanol at 50°C, elimination of H_1 is greatly preferred.[127] However, when the crown ether **66** is

[125] H. C. Brown and K.-J. Liu, *J. Am. Chem. Soc.*, **92**, 200 (1970).
[126] R. A. Bartsch and R. H. Kayser, *J. Am. Chem. Soc.*, **96**, 4346 (1974). When the leaving group is positively charged, reduced ion pairing *reduces* the amount of syn elimination [J. K. Borchardt and W. H. Saunders, *J. Am. Chem. Soc.*, **96**, 3912 (1974)].
[127] C. H. DePuy, R. D. Thurn, and G. F. Morris, *J. Am. Chem. Soc.*, **84**, 1314 (1962).

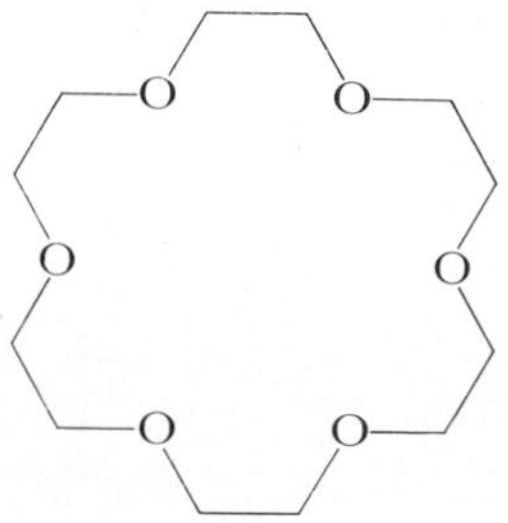

63

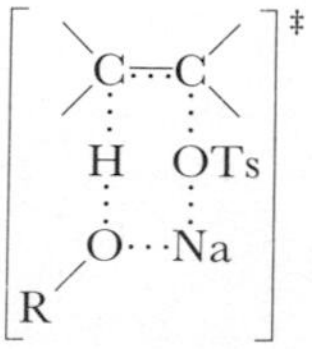

64

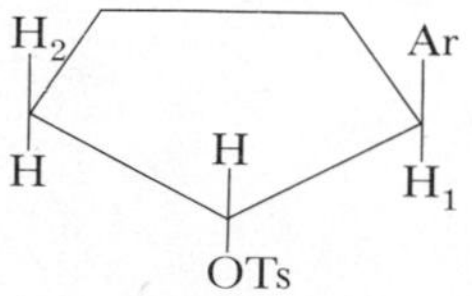

65

added, the amount of syn elimination is reduced. The results shown below are obtained. Again coordination of the cation must be partially responsible for the syn elimination.[128]

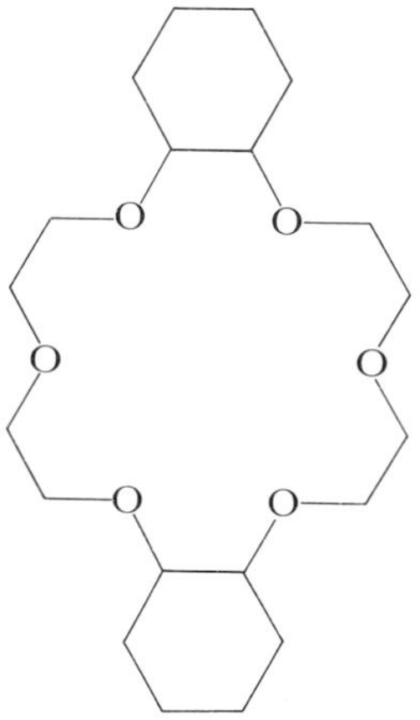

66

The role of steric factors in determining the syn/anti ratio has been investigated by Saunders and co-workers. From experiments with deuterated substrates they calculated that formation of 3-hexene from t-pentoxide-catalyzed decomposition of 3-n-hexyltrimethylammonium iodide (**67**) proceeds 83 percent by syn and 17 percent by anti elimination. They also found that syn elimination gives almost entirely trans olefin, but anti elimination gives cis product, a phenomenon noted previously and called the syn–anti dichotomy. Saunders proposed that the reason for the small amount of anti elimination can be understood if one considers the three staggered rotamers of the C_α—C_β bond (**67a**, **67b**, and **67c**). In all three the bulky trimethylammonium group forces the terminal methyl groups on the n-hexyl moiety as far away from it as possible. Thus in the two rotamers that have an anti hydrogen (**67a** and **67b**), approach to that hydrogen is sterically hindered and anti elimination is difficult. From this model one can also understand why the anti

[128] R. A. Bartsch, E. A. Mintz, and R. M. Parlman, *J. Am. Chem. Soc.*, **96**, 4249 (1974).

[66](M)	Product of elimination from 65[a]	
	 ⬠–φ	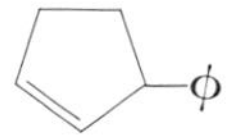 ⬠–φ
0.00	89.2	10.8
0.031	46.5	53.5
0.10	30.1	69.9
0.22	30.8	69.2

[a] Reprinted with permission from R. A. Bartsch, E. A. Mintz, and R. M. Parlman, *J. Am. Chem. Soc.*, **96**, 4249 (1974). Copyright (1974) by the American Chemical Society.

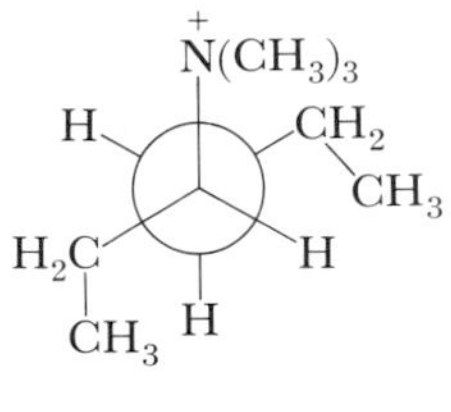

67a 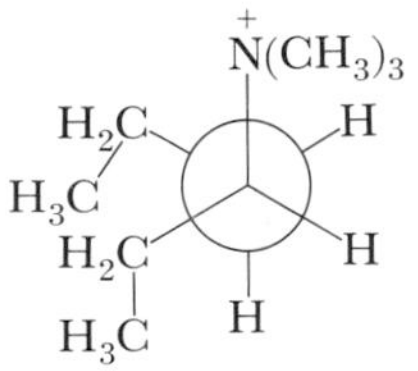   **67b** 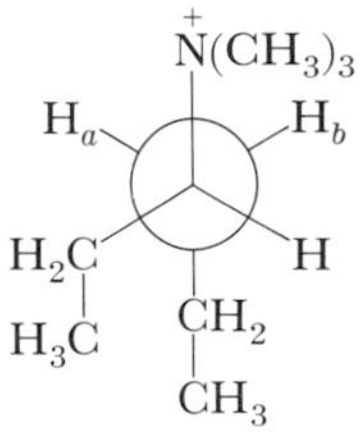   **67c**

addition that does occur gives cis alkene. Steric hindrance to anti attack at **67b** is less than to anti attack at **67a**. But anti attack on **67b** leads to cis alkene. All three staggered rotamers have syn hydrogens but syn elimination from **67b** or loss of H_a from **67c** would cause the alkyl groups in the transition state to be eclipsed. Thus syn elimination would occur from **67a** or, with loss of H_b, from **67c**. Both of these syn pathways would lead to trans alkene.[129]

Saunders's explanation of the syn–anti dichotomy is still somewhat open to question. For example, consistent with it is the fact that 3-hexyl tosylate, with bases of very large effective size, gives substantial amount of syn elimination leading to trans alkene. Inconsistent with it is that alkyl fluorides, with their very small leaving group, also give substantial amounts of syn elimination. Saunders suggests electrical effects—repulsion between hydrocarbonlike ends of the alkyl chains and the leaving group with its partial charge might lead to the same results as the steric effect discussed above.[130] Another experiment, difficult at present to reconcile with Saunders's theory is that 1-phenyl-2-propyl halides (**68**) give greater amounts of trans alkene as the effective size of the base increases.[131] Activated substrates, then, act differently from nonactivated ones.

[129] D. S. Bailey and W. H. Saunders, Jr., *J. Am. Chem. Soc.*, **92**, 6904 (1970).
[130] W.-B. Chiao and W. H. Saunders, Jr., *J. Am. Chem. Soc.*, **99**, 6699 (1977), and references therein.
[131] S. Alumni and E. Baciocchi, *J. Chem. Soc., Perkin Trans.* 2, 877 (1976).

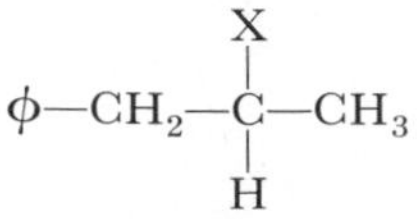

$$\phi-CH_2-\overset{\overset{\displaystyle X}{|}}{\underset{\underset{\displaystyle H}{|}}{C}}-CH_3$$

68

Weak-base-catalyzed reactions give entirely anti elimination—even when the product of anti elimination is the less stable product.[101,132] Thus, for example, **69** with $N(Bu)_4Cl$ gives >99.9 percent **70**, whereas the other diastereomer (**71**) gives >99.9 percent **72**.[133] Of course, **70** is the more stable olefin. This is one of the best pieces of evidence for transition states such as **58**.

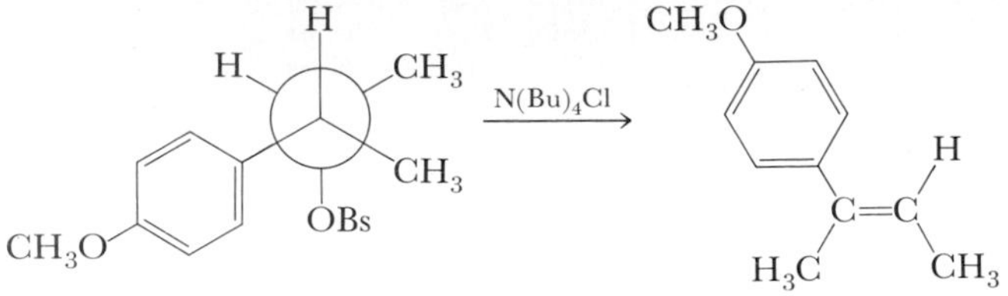

(7.55)

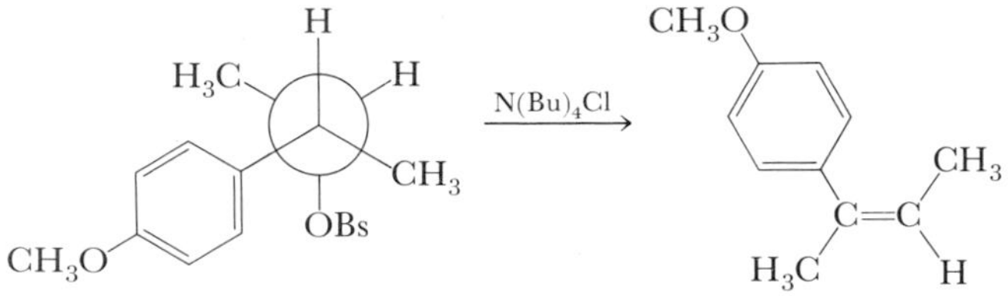

69   70

(7.56)

71   72

Because of the greater acidity of a vinylic than an alkyl proton, vinyl halides, $RHC=CRX$, are more likely than alkyl halides to undergo E1cB elimination. However, when the proton is not rendered even more acidic by a vicinal electron-withdrawing group, and when the basic catalyst is not too strong, E2 reaction obtains. Then anti elimination is much the preferred pathway. Thus, for example, **73** gives entirely **74** when treated with NaOMe in methanol, but under the same conditions **75** gives only the allene **76**.[134]

Substitution vs. elimination   Since in both S_N2 and E2 reactions a Lewis base attacks the substrate and causes another Lewis base to depart from the substrate, these reactions naturally compete with one another. If it is kept in mind that α and β substituents have either little effect on or increase the rate of elimination but greatly retard S_N2 reactions, the predominant product can usually be predicted (see Table 7.16). Thus *t*-alkyl halides give principally elimination products with all

[132](a) G. Biale, A. J. Parker, S. G. Smith, I. D. R. Stevens, and S. Winstein, *J. Am. Chem. Soc.*, **92**, 115 (1970); (b) G. Biale, D. Cook, D. J. Lloyd, A. J. Parker, I. D. R. Stevens, J. Takahashi, and S. Winstein, *J. Am. Chem. Soc.*, **93**, 4735 (1971).

[133]See note 132(b).

[134]S. W. Staley and R. F. Doherty, *J. Chem. Soc. D*, 288 (1969).

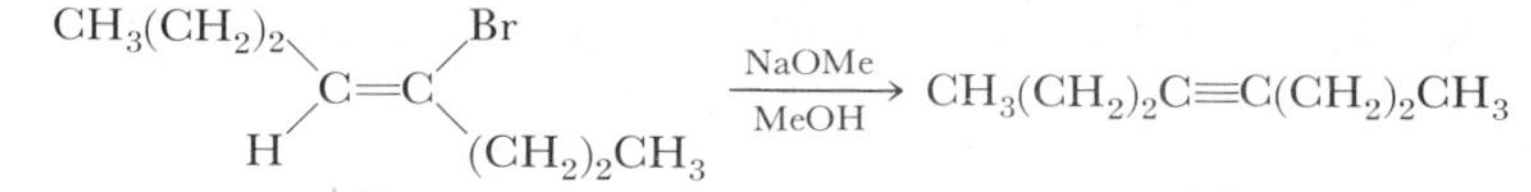

$$CH_3(CH_2)_2 \overset{Br}{\underset{H \quad (CH_2)_2CH_3}{C=C}} \xrightarrow[\text{MeOH}]{\text{NaOMe}} CH_3(CH_2)_2C \equiv C(CH_2)_2CH_3 \qquad (7.57)$$

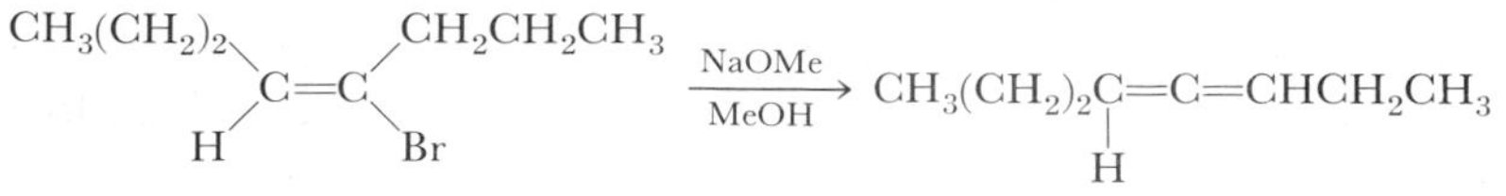

73   74

$$CH_3(CH_2)_2 \overset{CH_2CH_2CH_3}{\underset{H \quad Br}{C=C}} \xrightarrow[\text{MeOH}]{\text{NaOMe}} CH_3(CH_2)_2\underset{H}{C}=C=CHCH_2CH_3 \qquad (7.58)$$

75   76

bases. Secondary substrates are borderline and favor either elimination or substitution depending on the exact reaction conditions. For example, if the attacking reagent is a strong hard base, elimination competes well with substitution. If a soft weak base is used, unhindered secondary substrates give predominantly substitution, but hindered substrates give predominantly elimination.

Table 7.16 THE EFFECT OF SUBSTRATE AND BASE ON THE COMPETITION BETWEEN SUBSTITUTION AND ELIMINATION[a]

Substrate	Carbon	Base	Solvent	Percent elimination
$CH_3-\underset{CH_3}{\overset{CH_3}{C}}-Br$	3°	NBu_4Cl	Acetone	96
$CH_3-\underset{CH_3}{\overset{CH_3}{C}}-Br$	3°	NaOEt	HOEt	100
$CH_3-\underset{Br}{CH}-CH_3$	2°	NBu_4Cl	Acetone	0
$CH_3-\underset{Br}{CH}-CH_3$	2°	NaOEt	HOEt	75
$CH_3-\underset{H_3C \quad Br}{\overset{H}{C}-CH}-CH_3$	2°	NBu_4Cl	Acetone	51.4
$CH_3-\underset{Br \quad CH_3}{\overset{CH_3}{CH}-C}-CH_3$	2°	NBu_4Cl	Acetone	17.6
$CH_3CH_2CH_2-Br$	1°	NBu_4Cl	Acetone	0
$CH_3CH_2CH_2-Br$	1°	NaOEt	HOEt	8.8

[a] Reprinted with permisssion from G. Biale, D. Cook, D. J. Lloyd, A. J. Parker, I. D. R. Stevens, J. Takahashi, and S. Winstein, *J. Am. Chem. Soc.*, **93**, 4735 (1971). Copyright by the American Chemical Society.

Eliminations from primary halides using weak soft bases do not take place at all, but eliminations using hard bases do.[133]

Pyrolysis of Esters[135]

Pyrolyses of esters (**77**) and xanthate esters (**78**), either in the gas phase or in solution, give 1,2-elimination (Equations 7.59 and 7.60). These reactions are syn-

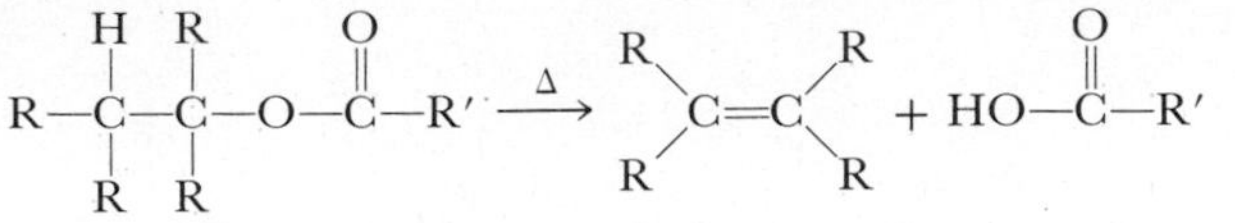

$$
\underset{\substack{|\ \ | \\ R\ \ R}}{\overset{\substack{H\ \ R \\ |\ \ |}}{R-C-C}}-O-\overset{\overset{O}{\|}}{C}-R' \xrightarrow{\Delta} \underset{R}{\overset{R}{>}}C=C\overset{R}{\underset{R}{<}} + HO-\overset{\overset{O}{\|}}{C}-R' \qquad (7.59)
$$

77

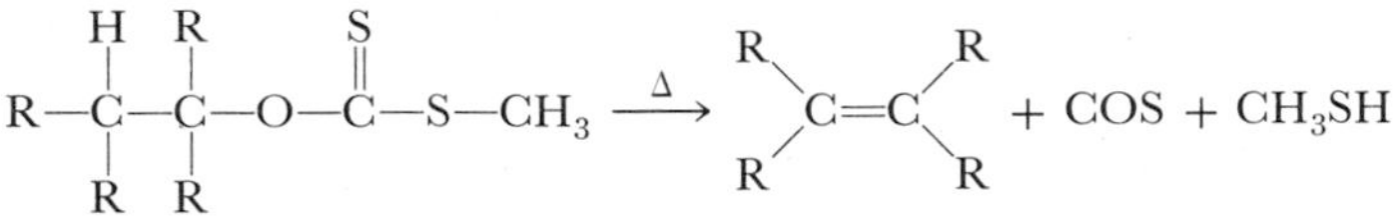

$$
\underset{\substack{|\ \ | \\ R\ \ R}}{\overset{\substack{H\ \ R \\ |\ \ |}}{R-C-C}}-O-\overset{\overset{S}{\|}}{C}-S-CH_3 \xrightarrow{\Delta} \underset{R}{\overset{R}{>}}C=C\overset{R}{\underset{R}{<}} + COS + CH_3SH \qquad (7.60)
$$

78

thetically useful because, unlike E1 and E2 eliminations, there are very few accompanying side reactions such as substitution or rearrangement. Both types of pyrolyses give predominantly Hofmann elimination. Xanthates decompose at considerably lower temperatures than the corresponding esters and therefore often give a higher yield of olefin and a lower yield of tar. They are conveniently prepared in situ by the reactions shown in Equation 7.61.

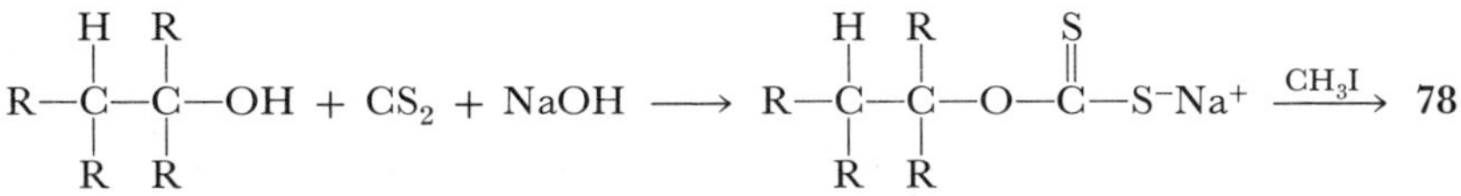

$$
\underset{\substack{|\ \ | \\ R\ \ R}}{\overset{\substack{H\ \ R \\ |\ \ |}}{R-C-C}}-OH + CS_2 + NaOH \longrightarrow \underset{\substack{|\ \ | \\ R\ \ R}}{\overset{\substack{H\ \ R \\ |\ \ |}}{R-C-C}}-O-\overset{\overset{S}{\|}}{C}-S^-Na^+ \xrightarrow{CH_3I} \textbf{78} \qquad (7.61)
$$

Isotope effects and the large negative entropies of activation for the pyrolyses make it appear probable that the transition states for Reactions 7.59 and 7.60 are **79** and **80**, respectively. Substituent effects, however, indicate that the transition states do have some polar character.

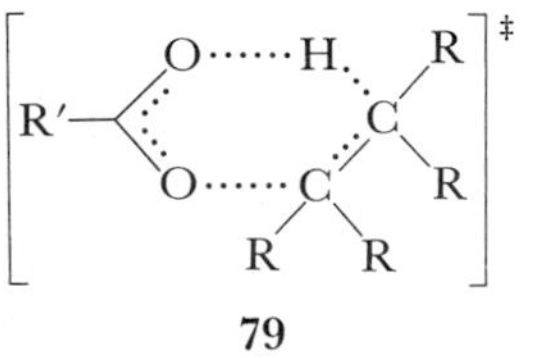

79 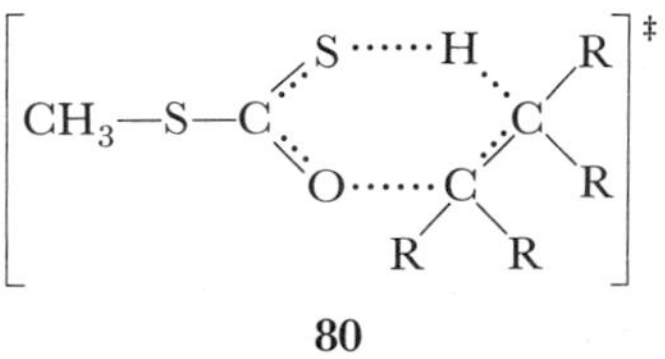   **80**

[135] For general reviews of pyrolytic olefin-forming eliminations, see (a) A. Maccoll, in *The Chemistry of Alkenes*, S. Patai, Ed., Wiley, New York, 1964, p. 203; (b) A. Maccoll, *Chem. Rev.*, **69**, 33 (1969); (c) A. Maccoll, *Adv. Phys. Org. Chem.*, **3**, 91 (1965); (d) G. G. Smith and F. W. Kelly, *Prog. Phys. Org. Chem.*, **8**, 75 (1971); (e) note 77(d); (f) note 77(f); (g) H. B. Amin and R. Taylor, *J. Chem. Soc., Perkin Trans. 2*, 228 (1979), and references therein.

7.3 NUCLEOPHILIC ADDITION TO MULTIPLE BONDS[136]

When the electron density of a carbon–carbon bond is reduced by strongly electron-withdrawing substituents, nucleophilic attack at one of the vinylic or acetylenic carbons may occur. Electron withdrawal may be either by induction or by resonance. Examples of nucleophilic addition are shown in Equations 7.62–7.66.

$$CF_2{=}CF_2 + RSH \xrightarrow{\phi CH_2\overset{+}{N}(CH_3)_3\overset{-}{O}H} \underset{\underset{SR\ \ \ H}{|\quad\ |}}{CF_2{-}CF_2} \qquad (7.62)^{137}$$

$$ArCH{=}C(CN)_2 \xrightarrow[\text{(2) } H^+]{\text{(1) KCN}} \underset{\underset{CN}{|}}{ArCH{-}CH(CN)_2} \qquad (7.63)^{138}$$

$$CH_2{=}CH{-}\overset{\overset{O}{\|}}{C}{-}CH_3 + CH_3\overset{\overset{O}{\|}}{C}{-}\overset{-}{C}H{-}\overset{\overset{O}{\|}}{C}{-}OEt \xrightarrow[\text{(2) protonation}]{\text{(1) addition}} CH_3\overset{\overset{O}{\|}}{C}{-}\underset{\underset{OEt}{\underset{\|}{C{=}O}}}{\overset{\overset{CH_2{-}CH_2{-}\overset{\overset{O}{\|}}{C}{-}CH_3}{|}}{C{-}H}} \qquad (7.64)^{139}$$

$$CH_2{=}CH{-}C{\equiv}N + CH_3\overset{\overset{O}{\|}}{C}{-}\overset{-}{C}H{-}\overset{\overset{O}{\|}}{C}{-}OEt \xrightarrow[\text{(2) protonation}]{\text{(1) addition}} CH_3\overset{\overset{O}{\|}}{C}{-}\underset{\underset{OEt}{\underset{\|}{C{=}O}}}{\overset{\overset{CH_2CH_2{-}C{\equiv}N}{|}}{C{-}H}} \qquad (7.65)^{139}$$

$$CF_3C{\equiv}CCF_3 + {}^-OCH_3 \xrightarrow[\text{(2) protonation}]{\text{(1) addition}} \underset{\underset{CH_3}{O}}{\overset{CF_3}{\diagdown}}C{=}C\underset{CF_3}{\overset{H}{\diagup}} \qquad (7.66)^{140}$$

In general, the mechanisms of nucleophilic additions to carbon–carbon double bonds have not been as much studied or systemized as those of electrophilic addition. Reactions 7.64 and 7.65 are examples of the very useful Michael condensation, in which a carbanion adds to an α,β-unsaturated carbonyl or nitrile compound.

---

[136](a) S. Patai and Z. Rappoport, in *The Chemistry of Alkenes,* S. Patai, Ed., Wiley, New York, 1964, p. 464; (b) E. Winterfeldt, *Angew. Chem. Int. Ed.,* **6**, 423 (1967); (c) Z. Rappoport, *Adv. Phys. Org. Chem.,* **7**, 1 (1969); (d) G. Modena, *Acc. Chem. Res.,* **4**, 73 (1971); (e) J. I. Dickstein and S. I. Miller, in *The Chemistry of the Carbon–Carbon Triple Bond, Part 2,* S. Patai, Ed., Wiley, New York, 1978, p. 813.

[137]W. K. R. Musgrave, *Quart. Rev. (London),* **8**, 331 (1954).

[138]See note 136(a).

[139]J. A. Markisz and J. D. Gettler, *Can. J. Chem.,* **47**, 1965 (1969).

[140]E. K. Raunio and T. G. Frey, *J. Org. Chem.,* **36**, 345 (1971).

The mechanism of the Michael condensation is not actually a 1,2-addition as implied in Equations 7.64 and 7.65, but rather a 1,4-addition as shown in Equation 7.67. Protonation occurs first on the oxygen atom because proton transfer to carbon is slower than to O^-. The stereochemistry of 1,2-addition in the Michael condensation is therefore irrelevant to the mechanism of the condensation.[141] Other nucleophilic additions to alkenes[136] and alkynes[142] go either syn or anti depending on the particular reaction, although alkynes usually add anti.[143]

$$R'^- + R_2C{=}CR{-}\overset{\overset{\textstyle O}{\|}}{C}{-}R \longrightarrow R_2\overset{\underset{\textstyle R'}{|}}{C}{-}\overset{\overset{\textstyle O}{\|}}{\underset{\textstyle R}{C}}{-}C{-}R \longleftrightarrow R_2\overset{\underset{\textstyle R'}{|}}{C}{-}C{=}\overset{\underset{\textstyle R}{|}}{C}{-}R \xrightarrow{\;H^+\;}$$

$$\underset{\textstyle 81a}{} \qquad\qquad \underset{\textstyle 81b}{}$$

$$R_2\overset{\underset{\textstyle R'}{|}}{C}{-}\overset{\overset{\textstyle OH}{|}}{C}{=}\overset{\underset{\textstyle R}{|}}{C}{-}R \longrightarrow R_2\overset{\underset{\textstyle R'}{|}}{C}{-}\overset{\overset{\textstyle H}{|}}{\underset{\textstyle R}{C}}{-}\overset{\overset{\textstyle O}{\|}}{C}{-}R \qquad (7.67)$$

The rate-determining step in nucleophilic additions is usually nucleophilic attack on the multiple bond.[136] For example, the entropy of activation of a Michael condensation is always a large, negative quantity. In the transition state the five atoms, $O{=}C{-}C{-}C{=}O$, of the anion and the four atoms, $C{=}C{-}C{=}O$ (or $C{=}C{-}C{=}N$), of the α,β-unsaturated carbonyl (or nitrile) system are all restricted to a shape that allows maximum π overlap.[139] Nucleophilic additions in which the second step, protonation of the intermediate carbanion, is rate determining are also known.[144]

Alkynes are more reactive to nucleophilic addition than are alkenes. Consideration of the frontier molecular orbital model in Section 7.1 would lead us to the opposite prediction. If an alkene has a higher HOMO than an alkyne, then it should have a lower LUMO. Thus the alkene's LUMO should be better able to interact with the nucleophile's HOMO at the transition state. However, *ab initio* calculations indicate that at the transition state for nucleophilic addition, strong carbon–hydrogen bending takes place that lowers the HOMO of alkynes below that of alkenes so that charge-transfer stabilization of the transition state is greater for alkynes.[145]

[141] R. A. Abramovitch, M. M. Rogić, S. S. Singer, and N. Venkateswaran, *J. Org. Chem.*, **37**, 3577 (1972), and references therein.

[142] For example, see (a) E. Winterfeldt and H. Preuss, *Chem. Ber.*, **99**, 450 (1966); (b) K. Bowden and M. J. Price, *J. Chem. Soc. B*, 1466 (1970); (c) R. W. Strozier, P. Caramella, and K. N. Houk, *J. Am. Chem. Soc.*, **101**, 1340 (1979).

[143] See note 142(c).

[144] L. A. Kaplan and H. B. Pickard, *J. Am. Chem. Soc.*, **93**, 3447 (1971).

[145] R. W. Strozier, P. Caramella, and K. N. Houk, *J. Am. Chem. Soc.*, **101**, 1340 (1979).

7.4 ELECTROPHILIC AROMATIC SUBSTITUTION[146]

The substitution of an electrophile for another group on an aromatic ring is electrophilic aromatic substitution (Equation 7.68). Although the leaving group, E_1, is

$$\text{(structure)} + E_2^+ \longrightarrow \text{(structure)} + E_1^+ \tag{7.68}$$

most often H^+, it may also be another Lewis acid. When it is not H^+—that is, when an electrophile attacks a substituted aromatic ring, not ortho, meta, or para to the substituent but directly at the position bearing the substituent—then attack is at the *ipso* position.[147]

After a brief discussion of the nature of the attacking species in some of the most important types of electrophilic aromatic substitution, we shall examine the mechanism and the effect of substituents on rates and products.

Reagents

Substitution by halogen[148] may be carried out in three ways: (1) by molecular halogenation, in which polarized X_2 itself acts as the electrophile (Equation 7.69); (2) by molecular halogenation with a catalyst, in which the role of the catalyst is to polarize the halogen molecule; and (3) by positive halogenation in which the halogen is the cation of a salt.[149]

$$^{\delta-}X{-}X^{\delta+} + \text{(structure)} \longrightarrow \text{(structure)} + H^+X^- \tag{7.69}$$

Iodination by molecular iodine is slow and operates only when the aromatic substrate is particularly reactive. Iodination can, however, be effected by using ICl, CH_3CO_2I or CF_3CO_2I as reagents. Addition of zinc chloride to an iodination reaction in which ICl is the reagent increases the rate by assisting in breaking the I—Cl bond.[150] Usually positive I^+ is the attacking reagent in these reactions.

Bromination with molecular bromine takes place readily. The reaction is normally carried out in acetic acid. Under these conditions the kinetics are second-

[146]For reviews, see (a) L. Stock, *Aromatic Substitution Reactions,* Prentice-Hall, Englewood Cliffs, N.J., 1968; (b) R. O. C. Norman and R. Taylor, *Electrophilic Substitution in Benzenoid Compounds,* Elsevier, Amsterdam, 1965; (c) E. Berliner, *Prog. Phys. Org. Chem.,* **2,** 253 (1964); (d) L. M. Stock and H. C. Brown, *Adv. Phys. Org. Chem.,* **1,** 35 (1963); (e) R. Taylor, *Aromatic and Heteroaromatic Chem.,* **2,** 217 (1974); **3,** 220 (1975).

[147](a) C. L. Perrin, *J. Org. Chem.,* **36,** 420 (1971); (b) C. L. Perrin and G. A. Skinner, *J. Am. Chem. Soc.,* **93,** 3389 (1971).

[148]For a review, see P. B. D. de la Mare, *Electrophilic Halogenation,* Cambridge University Press, London, 1976.

[149]R. M. Keefer and L. J. Andrews, *J. Am. Chem. Soc.,* **78,** 5623 (1956).

order in bromine; the second molecule of Br_2 polarizes the first, and the overall reaction is that of Equation 7.70.

$$ArH + 2Br_2 \longrightarrow ArBr + H^+ + Br_3^- \tag{7.70}$$

The addition of I_2 to the reaction mixture increases the rate, because I_2Br^- is formed more readily than Br_3^-. HOBr, CH_3CO_2Br, and CF_3CO_2Br can also all be used as sources of electrophilic bromine, the last being particularly reactive.[150] The attacking species is usually the entire molecule, but Br^+ may be formed at times.[151] Lewis acids such as $AlCl_3$ catalyze bromination by forming Br^+ as in Equation 7.71.

$$AlCl_3 + Br_2 \longrightarrow \overset{-}{AlCl_3}Br + Br^+ \tag{7.71}$$

Chlorination with molecular chlorine also occurs readily and is usually first-order in Cl_2. Apparently chlorine is electronegative enough so that an additional Cl_2 is not required to polarize the Cl—Cl bond at the transition state. Stronger Lewis acids such as $FeCl_3$ do, however, catalyze the reaction by assisting in bond polarization. HOCl and CH_3COOCl also act as chlorinating agents, but free Cl^+ is never formed. The reactive species from HOCl are Cl_2O (formed by dehydration of two molecules of acid) and $H_2\overset{+}{O}Cl$, both of which deliver Cl^+ to the aromatic π system.[152]

Direct fluorination of aromatic rings is very exothermic, but the reaction can be carried if the fluorine, diluted with argon or nitrogen, is bubbled into a dilute solution of the aromatic substrate in an inert solvent at $-78\,°C$. The mechanism under these conditions is apparently very similar to other electrophilic halogenations by molecular halogens except that the reaction is heterogeneous, occurring at the gas–liquid interface.[153] Reaction of benzene with the xenon fluorides, XeF_2 or XeF_4, does give fluorobenzene, but the mechanism is probably free radical rather than polar.[154]

Nitration of an aromatic ring[155] to give $ArNO_2$ is most often carried out with nitric acid in sulfuric acid. The attacking species is usually the nitronium ion, NO_2^+.[156] That this ion exists has been abundantly demonstrated. For example, cryoscopic measurements show that each molecule of nitric acid dissolved in sulfuric acid gives rise to four ions. This result is best explained by the equilibria shown

[150] J. R. Barnett, L. J. Andrews, and R. M. Keefer, *J. Am. Chem. Soc.*, **94**, 6129 (1972).

[151] See, however, H. M. Gilow and J. H. Ridd, *J. Chem. Soc., Perkin Trans.* 2, 1321 (1973).

[152] C. G. Swain and D. R. Crist, *J. Am. Chem. Soc.*, **94**, 3195 (1972).

[153] F. Cacace and A. P. Wolf, *J. Am. Chem. Soc.*, **100**, 3639 (1978); F. Cacace, P. Giacomello, and A. P. Wolf, *J. Am. Chem. Soc.*, **102**, 3511 (1980).

[154] (a) M. J. Shaw, H. H. Hyman, and R. Filler, *J. Am. Chem. Soc.*, **91**, 1563 (1969); (b) T. C. Shieh, E. D. Feit, C. L. Chernick, and N. C. Yang, *J. Org. Chem.*, **35**, 4020 (1970).

[155] For reviews see (a) J. G. Hoggett, R. B. Moodie, J. R. Penton, and K. Schofield, *Nitration and Aromatic Reactivity*, Cambridge University Press, London, 1971; (b) L. M. Stock, *Prog. Phys. Org. Chem.*, **12**, 21 (1976); (c) J. H. Ridd, *Adv. Phys. Org. Chem.*, **16**, 1 (1978).

[156] L. M. Stock, *Prog. Phys. Org. Chem.*, **12**, 21 (1976).

in Equations 7.72–7.74.[157] Raman spectra also show that in highly acidic media nitric acid is completely converted to NO_2^+.[158] In fact, nitronium salts such as $NO_2^+BF_4^-$ have actually been isolated and can also be used for aromatic nitrations.[159]

$$HNO_3 + H_2SO_4 \rightleftharpoons HSO_4^- + H_2\overset{+}{O}\text{—}NO_2 \tag{7.72}$$

$$H_2\overset{+}{O}\text{—}NO_2 \rightleftharpoons H_2O + NO_2^+ \tag{7.73}$$

$$H_2O + H_2SO_4 \rightleftharpoons H_3O^+ + HSO_4^- \tag{7.74}$$

It has been demonstrated that the nitronium ion not only exists but also can be the reactive species. For example, the rate of nitration of toluene (and of other aromatics) in solutions of nitric acid in nitromethane were independent of the concentration of toluene.[160,161] Thus the slow step must be the formation of the reactive species *prior to* attack on the toluene ring. This rules out HNO_3 as the nitrating agent. That protonated nitric acid, formed as shown in Equation 7.75, is not the reactive species follows from the fact that the rate does not become first-

$$2HNO_3 \rightleftharpoons H_2\overset{+}{O}\text{—}NO_2 + NO_3^- \tag{7.75}$$

order in toluene when NO_3^- is added to the reaction. A rate first-order in toluene would be expected if $H_2\overset{+}{O}\text{—}NO_2$ were the nitrating agent, because the equilibrium in Equation 7.75 would be driven to the left and toluene would have to compete with NO_3^- for $H_2\overset{+}{O}\text{—}NO_2$ (see the following discussion of the partition effect).[162] Protonated nitric acid has also been ruled out as the reactive species in aqueous sulfuric acid. At various acid strengths the rate of nitration correlates with the

$$AOH + SH^+ \rightleftharpoons A^+ + H_2O + S \tag{7.76}$$

acidity function H_R, which is defined by equilibria of the type shown in Equation 7.76, rather than with the acidity function H_0, defined by equilibria of the type shown in Equation 7.77 (see Section 3.2).[163] The fact that nitronium salts are

$$AOH + SH^+ \rightleftharpoons A\overset{+}{\underset{\underset{H}{|}}{\text{—O}}}H + S \tag{7.77}$$

excellent nitrating agents is direct proof of the ability of NO_2^+ to substitute an aromatic ring.[159]

[157] R. J. Gillespie, J. Graham, E. D. Hughes, C. K. Ingold, and E. R. A. Peeling, *Nature,* **158,** 480 (1946).
[158] C. K. Ingold and D. J. Millen, *J. Chem. Soc.,* 2612 (1950), and references therein.
[159] G. A. Olah, S. Kuhn, and A. Mlinko, *J. Chem. Soc.,* 4257 (1956).
[160] G. Benford and C. K. Ingold, *J. Chem. Soc.,* 929 (1938).
[161] Rates zeroth-order in aromatic substrate occur in some organic solvents and in very concentrated sulfuric acid but not in aqueous H_2SO_4 [J. W. Chapman and A. N. Strachan, *J. Chem. Soc. Chem. Commun.,* 293 (1974).]
[162] E. D. Hughes, C. K. Ingold, and R. I. Reed, *J. Chem. Soc.,* 2400 (1950).
[163] F. H. Westheimer and M. S. Kharasch, *J. Am. Chem. Soc.,* **68,** 1871 (1946).

Sulfonation of an aromatic substrate to produce $ArSO_3H$ is usually brought about by reaction of the aromatic with concentrated sulfuric acid or with sulfur trioxide in organic solvents.[164] When SO_3 is used in fairly dilute solution, the attacking species is SO_3 itself. In concentrated sulfuric acid, however, the mechanism is more complex. Fuming sulfuric acid (in which the mole fraction of $SO_3 > 0.5$) is actually a mixture of SO_3 and ionized and nonionized monomers, dimers, trimers, and tetramers of H_2SO_4 (the three latter formed by dehydration). At higher water content, the tetramer and trimer disappear, and the amount of dimer decreases. The reactive species in sulfuric acid thus depends on the amount of water in the acid and on the reactivity of the substrate. The reactive species in aqueous sulfuric acid are H_2SO_4 and $H_2S_2O_7$, the latter being more important at higher acid concentrations. In fuming sulfuric acid $H_3S_2O_7O_7^+$ and $H_2S_4O_{13}$ are also involved.[165]

Aromatic compounds are usually readily alkylated or acylated by a Friedel–Crafts reaction.[166] The combination of reagents used most commonly for aromatic alkylation is an alkyl halide with a strong Lewis acid (Equation 7.78). How-

$$R—X + A + ArH \longrightarrow AR—R + \bar{A}X + H^+ \qquad (7.78)$$

ever, alkenes, alcohols, mercaptans, and a number of other types of organic compounds also alkylate aromatic rings when a Friedel–Crafts catalyst is present. The order of reactivity of Lewis acids as catalysts varies from reaction to reaction but is most commonly $AlCl_3 > SbCl_5 > FeCl_3 > TiCl_2 > SnCl_4 > TiCl_4 > TeCl_4 > BiCl_3 > ZnCl_2$. The attacking species is sometimes the carbocation itself and sometimes an alkyl halide–Lewis acid complex (e.g., $R—\overset{\delta+}{X}\cdots\overset{\delta-}{AlCl_3}$). For example, benzene reacts with n-propyl chloride at low temperatures to yield predominantly 1-phenylpropane, but at higher temperatures 2-phenylpropane is the major product (Equation 7.79).[167] Isomerization most probably occurs via a

$$(7.79)$$

$-6°C$	60%	40%
$+35°C$	40%	60%

[164] For a review, see H. Cerfontain, *Mechanistic Aspects in Aromatic Sulfonation and Desulfonation,* Wiley, New York, 1968.

[165] A. Koeberg-Telder and H. Cerfontain, *Recl. Trav. Chim.,* **90**, 193 (1971).

[166] For a comprehensive review of all aspects of the Friedel–Crafts reaction, see G. A. Olah, Ed., *Friedel–Crafts and Related Reactions,* Vols. 1–4, Wiley, New York, 1963–1965; see also R. Miethchen and C.-F. Kröger, *Z. Chem.,* **15**, 135 (1975).

[167] V. N. Ipatieff, H. Pines, and L. Schmerling, *J. Org. Chem.,* **5**, 253 (1940).

free carbocation. Free $C^3H_3^+$ ions (that is, free of all counter-ions) formed upon nuclear disintegration of one of the tritium atoms in C^3H_4 have recently been observed to give the products of electrotrophilic substitution when they are generated in the presence of benzene or toluene.[168]

Friedel–Crafts acylations are most often carried out with BF_3 or $AlCl_3$ and an acyl halide, anhydride, ester, or a carboxylic acid (Equation 7.80). Appar-

$$R-\overset{\overset{O}{\|}}{C}-Y + \underset{\text{benzene}}{\bigcirc} \xrightarrow[\text{AlCl}_3]{\text{BF}_3 \text{ or}} \underset{\text{}}{\bigcirc}\overset{\overset{O}{\|}}{\underset{}{C}}-R + HY \qquad (7.80)$$

$$Y = \text{halide}, O_2CR, OR, \text{ or } OH$$

ently the attacking species is most often an acyl cation, $R-\overset{+}{C}=O \leftrightarrow R-C\equiv\overset{+}{O}$.[169]

The action of nitrous acid on aromatic amines produces aromatic diazonium ions (Equation 7.81), which are weak electrophiles. Correlation of the rate of

$$ArNH_2 + HNO_2 \longrightarrow Ar\overset{+}{N}\equiv N \ OH^- + H_2O \qquad (7.81)$$

diazonium coupling, as Reaction 7.82 is called, with pH shows that the reactive species must be the free diazonium ion rather than ArN_2OH.[170]

$$ArN_2^+OH^- + Ar'H \longrightarrow ArN\!=\!NAr' + H_2O \qquad (7.82)$$

Some metals, such as mercury and thallium, that form covalent carbon–metal bonds react in electrophilic aromatic substitutions. Both ionic [e.g., $Hg(ClO_4)_2$] and covalent [e.g., $Hg(OAc)_2$] mercuric compounds react; the attacking species, depending on the reagent and on the reaction conditions, may be Hg^{2+}, HgX^+, or HgX_2.[171] The only reagent that has been found to give high yields of arylthallium compounds is $Tl(OCCF_3)_3$.[172] The nature of the attacking species has not been studied, but presumably it is $Tl(O_2CCF_3)_3$ or $\overset{+}{Tl}(O_2CCF_3)_2$. The products, $ArTl(OC_2CF_3)_2$, are useful in organic synthesis because the thallium group can be introduced into a substituted aromatic ring highly regiospecifically and can then be replaced by another group such as I or CN. An example is shown in Scheme 3. Regiospecific introduction of aromatic substituents by direct means is often difficult to carry out.

[168] F. Cacace and P. Giacomello, *J. Am. Chem. Soc.*, **99**, 5477 (1977).

[169] F. R. Jensen and G. Goldman, in *Friedel–Crafts and Related Reactions*, Vol. 3, G. A. Olah, Ed., Wiley, New York, (1964), p. 1003.

[170] R. Wistar and P. D. Bartlett, *J. Am. Chem. Soc.*, **63**, 413 (1941).

[171] (a) A. J. Kresge, M. Dubeck, and H. C. Brown, *J. Org. Chem.*, **32**, 745 (1967); (b) C. Perrin and F. H. Westheimer, *J. Am. Chem. Soc.*, **85**, 2773 (1963).

[172] (a) E. C. Taylor and A. McKillop, *Acc. Chem. Res.*, **3**, 338 (1970); (b) A. McKillop, J. D. Hunt, M. J. Zelesko, J. S. Fowler, E. C. Taylor, G. McGillivray, and F. Kienzle, *J. Am. Chem. Soc.*, **93**, 4841 (1971); (c) E. C. Taylor, F. Kienzle, R. L. Robey, A. McKillop, and J. D. Hunt, *J. Am. Chem. Soc.*, **93**, 4845 (1971).

SCHEME 3

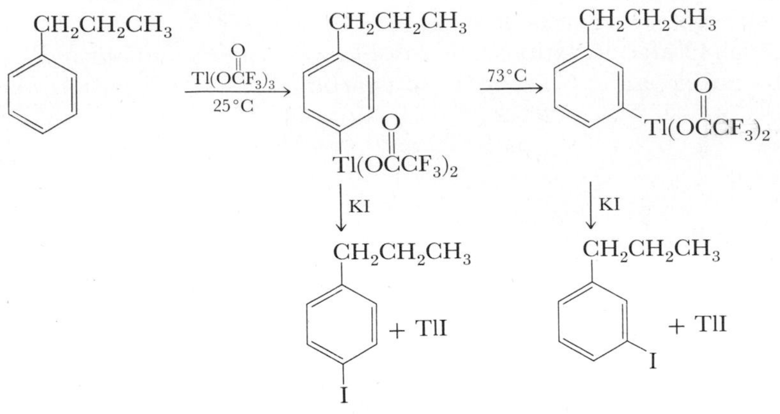

CH$_2$CH$_2$CH$_3$

$\xrightarrow[25\,°C]{Tl(OCF_3)_3}$

CH$_2$CH$_2$CH$_3$

Tl(OCCF$_3$)$_2$

$\xrightarrow{73\,°C}$

CH$_2$CH$_2$CH$_3$

Tl(OCCF$_3$)$_2$

↓ KI

CH$_2$CH$_2$CH$_3$

+ TlI

I

↓ KI

CH$_2$CH$_2$CH$_3$

+ TlI

I

(Note that the replacement of Tl by I is an oxidation–reduction reaction.)

Direct Displacement vs. Multistep Reaction

Until 1950 it seemed possible that electrophilic aromatic substitution proceeded by a direct displacement; the transition state for this mechanism is shown in **82**. The attraction that this mechanism held was that in it the aromatic character of the ring is left relatively undisturbed during the course of the reaction.

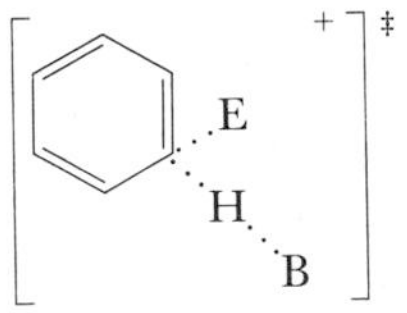

E

H

B

82

However, in 1950 this mechanism was shown to be incorrect. Melander[173] found that in the nitration and bromination of a number of benzene derivatives the tritium isotope effect (k_H/k_T) is not 10–20 as is to be expected if carbon–hydrogen bond breaking occurs in the rate-determining step, but rather is less than 1.3. The direct displacement mechanism was thus ruled out, and a multistep reaction in which the rate-determining step occurs before C—H bond breaking was implicated. The simplest mechanism of this type, in which there is only one intermediate, is shown in Equation 7.83. We shall assume this simplest mechanism and shall see that most of the experimental evidence fits it but that in very fast aromatic substitutions there may be two intermediates on the reaction path.

Examination of the rate equation for the mechanism of Equation 7.83 reveals the probable origin of the small isotope effects observed by Melander. Using

[173] L. Melander, *Ark. Kemi*, **2**, 211 (1950).

$$E^+ + \bigcirc \underset{k_{-1}}{\overset{k_1}{\rightleftarrows}} \underbrace{\bigcirc E^+}_{\substack{\text{complex} \\ \mathbf{83}}} \overset{B}{\underset{k_2}{\longrightarrow}} \bigcirc^{E} + BH \tag{7.83}$$

the steady-state approximation for the concentration of the intermediate complex (**83**), the observed rate is calculated to be

$$\text{rate} = [\text{Ar}][\text{E}^+]\frac{k_1(k_2[\text{B}]/k_{-1})}{1 + (k_2[\text{B}]/k_{-1})} \tag{7.84}$$

When the second step is very fast compared to the reverse of the first step—that is, when $k_2[\text{B}]/k_{-1} \gg 1$—Equation 7.84 can be simplified to

$$\text{rate} = k_1[\text{Ar}][\text{E}^+] \tag{7.85}$$

In this case a primary isotope effect of 1.0 would be expected, since only the rate constant for the first step, in which no bond breaking occurs, is involved in the rate equation. When the reverse of the first step is very fast compared to the second step—that is, $k_2[\text{B}]/k_1 \ll 1$—then the observed rate is linear with k_2 as shown in Equation 7.86.

$$\text{rate} = [\text{Ar}][\text{E}^+]\frac{k_1 k_2[\text{B}]}{k_{-1}} \tag{7.86}$$

In this case a large isotope effect would be expected. If, however, $k_2[\text{B}] \approx k_{-1}$, then Equation 7.84 cannot be simplified, and the rate will depend to a small extent on the magnitude of k_2. Then, even if $k_2 > k_1$, some isotope effect should be observed. The small isotope effect of Melander's experiments make it appear that the first step is slower than the second, but that k_{-1} competes favorably with k_2. When the second step becomes kinetically important in spite of the first step being the slow step, we have an example of the *partitioning effect*—so-called because the kinetic significance of the second step arises from the way in which the intermediate partitions itself. Since 1950 a very large number of electrophilic substitutions have been examined for isotope effects; in the absence of special circumstances (see below), the isotope effects found are usually very small.[174]

Studies of the effect of base concentration on rate also provide strong support for the two-step mechanism. The simple displacement mechanisms with transition state **82** should be first-order in base, as can be seen from the rate equation for this mechanism,

$$\text{rate} = k_3[\text{Ar}][\text{E}^+][\text{B}] \tag{7.87}$$

In the two-step mechanism, if $k_2[\text{B}]/k_{-1} \gg 1$, no base catalysis whatsoever should be observed; if $k_2[\text{B}]/k_{-1} \ll 1$, a linear dependence on base is expected; and if $k_2[\text{B}]/k_{-1} \approx 1$, nonlinear dependence on base should result.

[174]For reviews, see note 146(c) and H. Zollinger, *Adv. Phys. Org. Chem.*, **2**, 163 (1964).

Zollinger observed that Reaction 7.88 is not catalyzed by pyridine and does not show an isotope effect.[175] In this case the two-step mechanism must be operative, and k_2 is so large that $k_2[\text{B}]/k_{-1}$ is always much larger than 1 even at low base

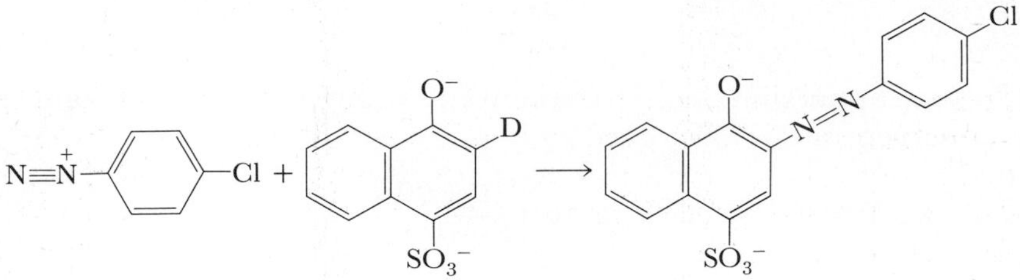

$$(7.88)$$

concentrations. For Reaction 7.89, however, there is a nonlinear correlation between rate and the concentration of pyridine. A deuterium isotope effect (k_H/k_D) of

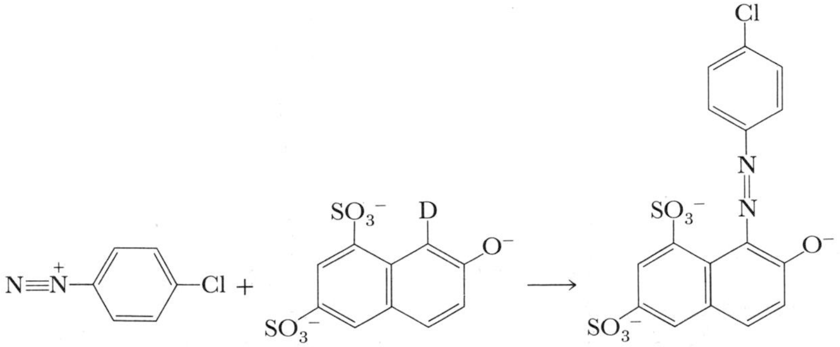

84

$$(7.89)$$

6.55 was found for this reaction in pure water, but at pyridine concentrations of $0.0232M$ and $0.905M$ it decreased to 6.01 and 3.62, respectively.[176] The fact that the rate is not first-order in base rules out both a simple displacement mechanism and a two-step mechanism with proton loss rate determining (see Equation 7.86). We shall return shortly to a consideration of why this reaction is catalyzed by base and has an isotope effect after we have ascertained the usual nature of the intermediate **83** formed in the slow step.

Nature of the Intermediate

Three possibilities for the intermediate complex (**83**) seem most likely: (1) a π complex in which the electrophile is coordinated with the entire π system (**85**); (2) a π complex in which the electrophile is coordinated with a single π bond (**86**);[177] or (3) a σ complex or *benzenium ion* in which the electrophile has formed a σ bond with one carbon of the aromatic ring (**87**).[178]

[175] R. Ernst, O. A. Stamm, and H. Zollinger, *Helv. Chim. Acta*, **41**, 2274 (1958).

[176] H. Zollinger, *Helv. Chim. Acta*, **38**, 1597, 1623 (1955).

[177] (a) G. A. Olah, S. Kobayashi, and M. Tashiro, *J. Am. Chem. Soc.*, **94**, 7448 (1972); (b) D. V. Banthorpe, *Chem. Rev.*, **70**, 295 (1970).

[178] G. A. Olah and Y. K. Mo, *J. Am. Chem. Soc.*, **94**, 9241 (1972).

π complex   π complex   σ complex

85   **86**   **87**

There is an abundance of evidence that both π complexes and σ complexes exist as stable species. For example, nmr studies have shown that the CH_2 protons of the ethyl fluoride–boron trifluoride complex absorb at slightly lower fields in the presence of toluene. Thus a new complex, which includes toluene and in which the CH_2 group bears more positive charge than it does in the absence of toluene, is formed. However, the aromatic protons of toluene absorb at almost the same frequency in the presence of BF_3–FCH_2CH_3 as in its absence;[179] thus the new complex is probably that shown in **88**. Another example is the complex that benzene

88

forms with iodine. The infrared spectrum in a frozen nitrogen matrix shows that in the complex the benzene symmetry in the ring plane is not altered. The π complex (**89**) with the iodine axial, has been proposed as the structure.[180]

89

Sigma complexes of benzene and other aromatics have also been observed in the nmr.[181] For example, when benzene is dissolved in SbF_5–FSO_3H–SO_2ClF–SO_2F_2 at $-140°$ the proton nmr spectrum shown in Figure 7.9 is obtained. The peak at 5.6δ is due to two paraffinic protons, and thus the spectrum fits the Structure **90**. At higher temperatures the spectra of aromatic σ complexes usually change; the lines broaden and eventually coalesce due to intramolecular hydrogen shifts. Protonation of benzenes in the gas phase apparently also lead to σ complexes,

[179] T. Oyama and R. Nakane, *J. Org. Chem.*, **34**, 949 (1969).

[180] L. Fredin and B. Melander, *J. Am. Chem. Soc.*, **96**, 1672 (1974).

[181] (a) G. A. Olah, J. S. Staral, G. Asencio, G. Liang, D. A. Forsythe, and G. D. Mateescu, *J. Am. Chem. Soc.*, **100**, 6299 (1978), and references therein; (b) G. A. Olah, *Angew. Chem., Int. Ed. Engl.*, **12**, 173 (1973); (c) D. M. Brouwer, E. L. Mackor, and C. MacLean, in *Carbonium Ions*, Vol. 2, G. A. Olah and P. v. R. Schleyer, Eds., Wiley, New York, 1970, Chap. 20.

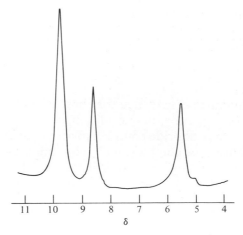

Figure 7.9 Proton nmr spectrum of the benzenium ion (**90**). Reprinted with permission from G. A. Olah, J. S. Staral, G. Asencio, G. Liang, D. A. Forsythe, and G. D. Mateescu, *J. Am. Chem. Soc.*, **100**, 6299 (1978). Copyright 1978 American Chemical Society.

as evidenced by the high level of agreement between the experimentally and theoretically determined alkyl substituent effects.[182] A few stable σ complexes such as **91** and **92** have been prepared and isolated in the form of salts.[183]

The fact that π and σ complexes do form is not proof that either or both are intermediates in electrophilic aromatic substitution. However, Table 7.17 gives strong evidence that usually the formation of σ complexes is rate determining. Electron-donating groups greatly stabilize σ complexes of benzene derivatives but only slightly stabilize π complexes. Thus 1,2,3,5-tetramethylbenzene forms σ complexes that are 2 billion times more stable than those of toluene, but its π complexes are only more stable than those of toluene by a factor of 3. The rate of bromination of benzene derivatives also increases drastically with methyl substitution; and the relative rates are very similar to the relative stabilities of the σ complexes. Apparently the highest-energy transition state resembles the σ complex. Figure 2.6 is a similar piece of evidence for the intervention of σ-complexes. The rate of aromatic bromination (and of a large number of other aromatic substitutions[184]) correlate with the σ+ constants of substituents on the benzene ring but not with the σ con-

[182] W. J. Hehre, R. T. McIver, Jr., J. A. Pople, and P. v. R. Schleyer, *J. Am. Chem. Soc.*, **96**, 7162 (1974).
[183] G. A. Olah and S. J. Kuhn, *J. Am. Chem. Soc.*, **80**, 6535, 6541 (1958).
[184] See note 146(d).

Table 7.17 Relative Rates of Aromatic Substitutions and Relative σ and π Complex Stabilities of Methylbenzenes[a]

Benzene derivative	Relative σ complex stability (ArH + HF—BF$_3$)[b]	Relative π complex stability with HCl[b]	Relative rate of bromination, Br$_2$ in 85% HOAc	Relative rate of chlorination, Cl$_2$ in HOAc
H	1	1.0	1	1
Methyl	790	1.5	605	340
1,2-Dimethyl	7,900	1.8	5,300	2,030
1,3-Dimethyl	1,000,000	2.0	514,000	180,000
1,4-Dimethyl	3,200	1.6	2,500	2,000
1,2,3-Trimethyl	2,000,000	2.4	1,670,000	—
1,2,4-Trimethyl	2,000,000	2.2	1,520,000	—
1,3,5-Trimethyl	630,000,000	2.6	189,000,000	30,000,000
1,2,3,4-Tetramethyl	20,000,000	2.6	11,000,000	—
1,2,3,5-Tetramethyl	2,000,000,000	2.7	420,000,000	—
1,2,4,5-Tetramethyl	10,000,000	2.8	2,830,000	1,580,000
Pentamethyl	2,000,000,000	—	810,000,000	134,000,000

[a] Reprinted with permission from G. A. Olah, *Acc. Chem. Res.*, **4**, 240 (1971). Copyright by the American Chemical Society.
[b] From equilibrium constant measurements.

stants. Thus the substituents are conjugated with almost a full positive charge at the transition state.

Now that we have determined that a σ complex is usually formed in the slow step in electrophilic aromatic substitution, let us return to a consideration of Reaction 7.89. Two factors probably combine to cause the observed isotope effect and base catalysis. First, the strong electron-donating groups stabilize the intermediate **93** (Equation 7.90) and make departure of the proton more difficult than proton loss in many other electrophilic substitutions. [Remember from above, however, that $k_1 < k_2$.] Second, steric interactions between the large diazonium group and the nearby substituents increase the rate of decomposition (k_{-1}) of **93** back to

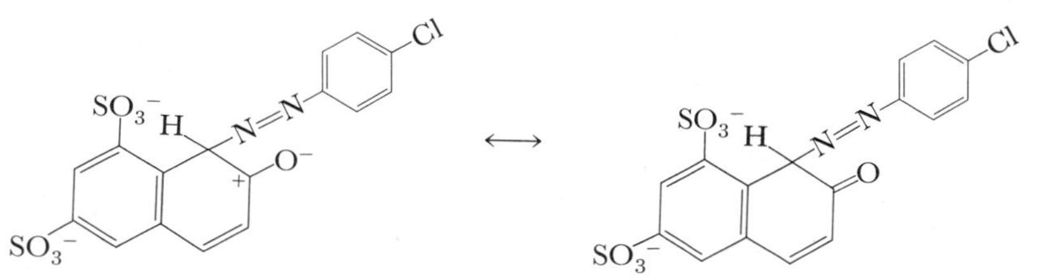

$$(7.90)$$

93

starting material. Both factors, then work together to cause $k_2[B]/k_{-1}$ in water to be small, and a large isotope effect is observed. As [B] is increased, the ratio necessarily becomes larger and the isotope effect decreases.[176] The effect of steric factors on the size of the isotope effect is neatly demonstrated by the fact that the isotope effect for bromination of **94, 95, 96,** and **97** increases in the order **94** $<$ **95** $<$ **96** $<$ **97**.[185] A few examples are known in which destruction of the σ complex is rate determining.

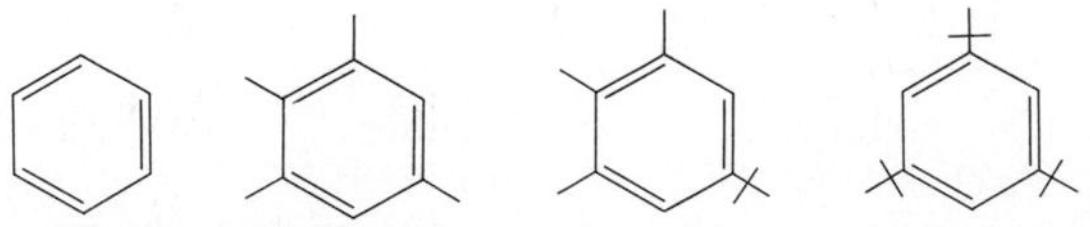

For example, **84** is brominated by Br$_2$ and BrOH at approximately the same rate, even though the latter is usually much the more reactive reagent. Moreover, the rate of reaction is first-order in base.[186]

Effect of Substituents on Rate and Orientation of Substitution

As has already been mentioned, electrophilic aromatic substitutions are accelerated by electron-donating groups on the aromatic ring. Electron-withdrawing groups decelerate them. The overall rate enhancement (or dimunition) arises from a sum of the group's inductive (I) and resonance (R) effects. Table 7.18 gives the relative rates of mononitration of a number of benzene derivatives.

Aromatic substituents that increase the rate relative to hydrogen direct the electrophile predominantly to the ortho and para positions. Substituents that decrease the rate (except for the halogens, see Problem 7) direct the electrophile predominantly to the meta position. To understand why this is so, we must consider the nature of the transition state, but since the transition state is often similar to the σ complex, we shall use the σ complex as a model for the transition state.

If the electrophile attacks the benzene ring at a position ortho or para to a $+I$ substituent (i.e., to one electron-donating by inductive effect), the activated complex will be similar to **98** or **99**, respectively. Resonance structures **98c** and **99c** are of particularly low energy because in these the positive charge is localized on the carbon that bears the $+I$ group. Attack at the meta position does not allow such a resonance structure to be drawn.

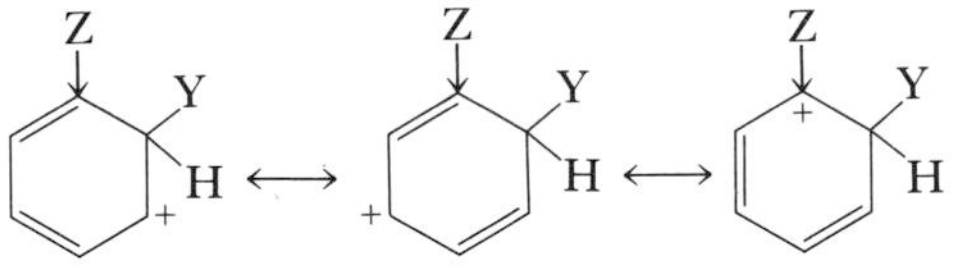

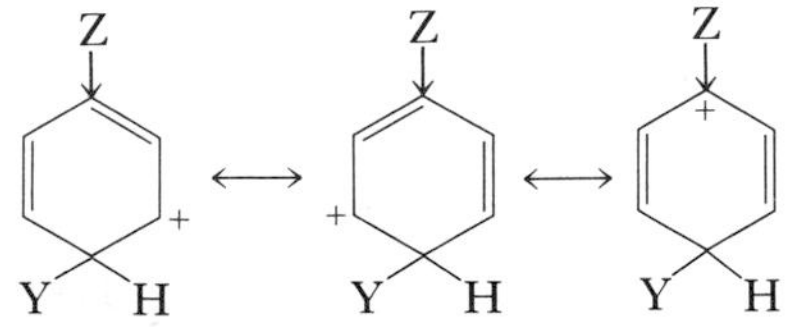

[185] E. Baciocchi, G. I. Illuminati, G. Sleiter, and F. Stegel, *J. Am. Chem. Soc.*, **89**, 125 (1967).
[186] M. Christen and H. Zollinger, *Helv. Chim. Acta*, **45**, 2057, 2066 (1962).

Table 7.18 Relative Rates of Nitration of Benzene Derivatives[a]

R in C_6H_5R	Relative rate
—OH	1000
—CH_3	25
—$CH_2CO_2CH_2CH_3$	3.8
—H	1
—CH_2Cl	0.71
—CH_2CN	0.35
—I	0.18
—Cl	0.033
—$CO_2CH_2CH_3$	0.0037
—$CH_2\overset{+}{N}(CH_3)_3$	2.6×10^{-5}
—NO_2	6×10^{-8}
—$\overset{+}{N}(CH_3)_3$	1.2×10^{-8}

[a] Reprinted from C. K. Ingold, *Structure and Mechanism in Organic Chemistry*, 2d ed. Copyright 1953, Copyright © 1969 by Cornell University. Used by permission of Cornell University Press.

The reaction conditions were not the same for all the nitrations listed, and therefore the relative rates are only approximate. However, apparently none of the nitrations were carried out using conditions under which formation of $^+NO_2$ is rate determining or using the very reactive, unselective $^+NO_2X^-$ salts.

If attack is ortho or para to a group that is electron-donating by resonance (—O—R, —NR$_2$ are, for example, $+R$ groups), an additional resonance structure for the transition state can be drawn (**98d** or **99d**, respectively). There is no such stabilization for meta substitution.

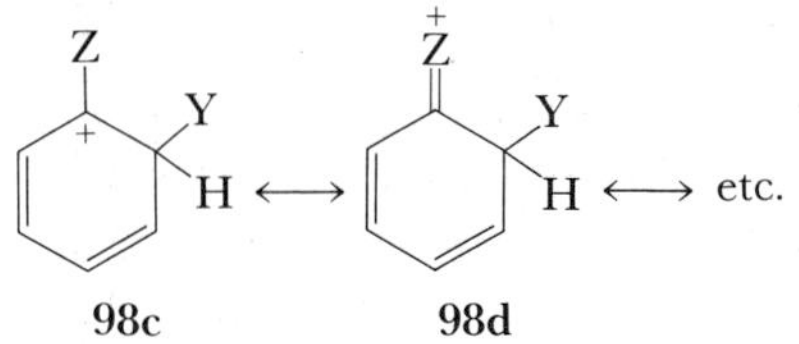

98c   98d

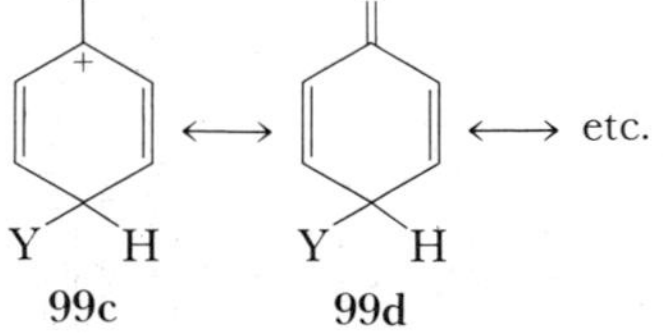

99c   99d

Now we can also understand why meta attack **100** is preferred in a deactivated ring. Only if attack is at that position do none of the resonance structures of the transition state have a positive charge on that carbon that bears the electron-withdrawing group.

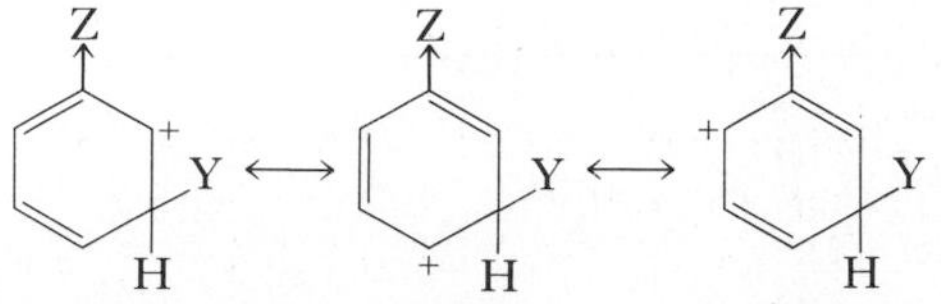

$$100a \qquad\qquad 100b \qquad\qquad 100c$$

Because the rate of substitution varies with position, in a benzene derivative it is more informative and frequently more useful to talk about *partial rate factors* than about relative rates. A partial rate factor is defined as the rate *at one particular position* in the benzene derivative relative to the rate of substitution at one position in benzene. Let us, for example, calculate the para and meta partial rate factors (p_f and m_f, respectively) for bromination of toluene with bromine in aqueous acetic acid. Toluene brominates 605 times faster than benzene under these conditions. The product is 66.8 percent *p*-, 0.3 percent *m*-, and 32.9 percent *o*-bromotoluene. Attack at the para position of toluene occurs 0.668×605 times as fast as attack at all six positions of benzene but ($0.668 \times 605 \times 6 = 2420$) times as fast as at one position of benzene. Therefore $p_f^{CH_3}$ for bromination of toluene under these conditions is 2420. There are only three times as many total carbons in benzene as meta carbons in toluene. Therefore $m_f^{CH_3} = 0.003 \times 605 \times 3 = 5.5$. The definitions of the partial rate factors for monosubstituted benzenes (ϕ—R) are given in Equations 7.91–7.93.

$$p_f^R = \frac{6k_{\phi-R}}{k_{\phi-H}} \times \frac{\%\ \mathrm{para}}{100} \tag{7.91}$$

$$m_f^R = \frac{3k_{\phi-R}}{k_{\phi-H}} \times \frac{\%\ \mathrm{meta}}{100} \tag{7.92}$$

$$o_f^R = \frac{3k_{\phi-R}}{k_{\phi-H}} \times \frac{\%\ \mathrm{ortho}}{100} \tag{7.93}$$

The rates of electrophilic substitutions at the para and meta positions of benzene derivatives can be correlated by the linear free-energy relationships shown in Equations 7.94 and 7.95 (see also Figure 2.6).[187]

$$\log p_f^R = \sigma_p^+ \rho \tag{7.94}$$
$$\log m_f^E = \sigma_m^+ \rho \tag{7.95}$$

The substituents in a benzene derivative may affect the rate of electrophilic attack at the ortho position by steric interaction and secondary bonding (e.g., hydrogen bonding or charge–transfer complexing) as well as by electrical influence. Therefore σ_o^+ is not necessarily constant but depends on the size and nature of the

[187] L. M. Stock and H. C. Brown, *Adv. Phys. Org. Chem.*, **1**, 35 (1963).

electrophile, and a correlation of rates of ortho substitution is less satisfactory.[188] (See Section 2.2.)

Reactivity–Selectivity;[189] the Mechanism of Very Fast Electrophilic Aromatic Substitutions

In general, the less reactive a reagent is, the more selective it is in attacking an activated rather than a deactivated site. In 1953 H. C. Brown observed that the selectivity of an electrophile in choosing between the para and meta positions of toluene is linearly related to its selectivity in choosing between toluene and benzene. If the intramolecular *selectivity factor*, S_f, is defined by Equation 7.96, then the

$$S_f \equiv \log \frac{p_f^{CH_3}}{m_f^{CH_3}} \tag{7.96}$$

inter- and intramolecular selectivities are correlated by Equations 7.97 and 7.98 in which b and b' are empirical constants.

$$\log p_f^{CH_3} = bS_f \tag{7.97}$$
$$\log m_f^{CH_3} = b'S_f \tag{7.98}$$

Figure 7.10 shows the straight line obtained when S_f, for a wide variety of nucleophiles, is plotted against the $p_f^{CH_3}$ factor for these reagents. One point deviates sharply from the line. That point corresponds to nitration with $NO_2^+BF_4^-$.[187] Olah[190] found similar deviations for other very reactive nitrating agents, as shown in Table 7.19. The various nitrating agents listed vary greatly, according to their reactivity, in their selectivity in choosing between toluene or benzene. Their selectivity in choosing the ortho and para positions of toluene as opposed to the meta is, however, invariant and high. The intermolecular selectivites do not increase with increasing substitution. Table 7.20 gives the intermolecular selectivity of $NO_2^+BF_4^-$ toward a number of methylated benzenes. Note that the reagent reacts only three times more rapidly with 1,3,5-trimethylbenzene than with benzene. Compare this to a factor of 2×10^6 when the electrophile is Br_2 in acetic acid (Table 7.17). Olah has attributed the low intermolecular selectivity to an early transition state. He has suggested that when the electrophile is very reactive, the transition state resembles the starting material and is similar to a π complex. Since methylation does not greatly increase the stability of a π complex (Table 7.16), it would thus not much increase the rate of nitration.

According to Olah, the high intramolecular sensitivity in toluene arises

---

[188] For a discussion of the *ortho effect*, see (a) J. Shorter, *Advanc. in Linear Free-Energy Relat.*, N. B. Chapman and J. Shorter, Eds., Plenum Press, London, 1972, p. 71; (b) T. Fujita and T. Nishioka, *Prog. Phys. Org. Chem.*, **12,** 49 (1976).

[189] For a discussion of the reactivity–selectivity principle, see A. Pross, *Adv. Phys. Org. Chem.*, **14,** 69 (1977); for a recent exception in electrophilic aromatic substitution, see E. M. Arnett and R. Reich, *J. Am. Chem. Soc.*, **100,** 2930 (1978).

[190] (a) G. A. Olah, *Acc. Chem. Res.*, **4,** 240 (1971); (b) G. A. Olah and S. Kobayashi, *J. Am. Chem. Soc.*, **93,** 6964 (1971); G. A. Olah, S. C. Narang, J. A. Olah, R. L. Pearson, and C. A. Cupas, *J. Am. Chem. Soc.*, **102,** 3507 (1980).

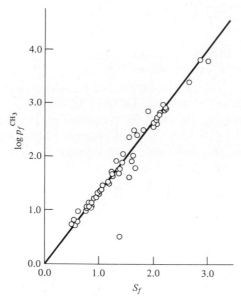

Figure 7.10 The relationship between S_f and $\log p_f^{CH_3}$ for a number of electrophiles. From L. M. Stock and H. C. Brown, *Adv. Phys. Org. Chem.*, **1**, 35 (1963). Reprinted by permission of Academic Press (London).

from the orbital symmetry requirements of this transition state. The electrophile can only interact with two p orbitals that have the same sign in the highest occupied molecular orbital (**101**). Thus transition state **102** and **103** are possible, but **104** is not. The activated complex **102** can open only to the ortho σ complex; **103** opens to the para and (less often) to the meta σ complex.[191]

Table 7.19 Nitration of Toluene and Benzene[a]

Nitrating agent	Solvent	k_t/k_B	% Ortho	% Meta	% Para
$NO_2^+PF_6^-$	CH_3NO_2	1.6	68.2	2.0	29.8
$NO_2^+BF_4^-$	Sulfolane	1.7	65.4	2.8	31.8
$NO_2^+BF_4^-$	CH_3CN	2.3	69	2	29
HNO_3	80% H_2SO_4	4.8			
	77% H_2SO_4	5.0			
	75.3% H_2SO_4	7.2			
	68.3% H_2SO_4	17.2	60	3	37
HNO_3	CH_3NO_2	21	58.5	4.4	37.1
	$(CH_3CO)_2O$	23	58.4	4.4	37.2
CH_3COONO_2	CH_3CN	44	63	2	35
HNO_3	Sulfolane, H_2SO_4	37	61.6	2.9	35.5

[a] Reprinted with permission from G. A. Olah, *Acc. Chem. Res.*, **4**, 240 (1971). Copyright (1971) by the American Chemical Society.

[191] See note 177(a).

Table 7.20 RELATIVE RATES OF AROMATIC
NITRATION OF BENZENE
DERIVATIVES WITH $NO_2^+BF_4^-$
AND WITH $CH_3ONO_2\text{-}BF_3{}^a$

Benzene derivative	$NO_2^+BF_4^-$ in Sulfolane
H	1.0
Methyl	1.6
1,2-Dimethyl	1.7
1,3-Dimethyl	1.6
1,4-Dimethyl	1.9
1,3,5-Trimethyl	2.7

[a] Reprinted with permission from G. A. Olah, *Acc. Chem. Res.*, **4**, 240 (1971). Copyright (1971) by the American Chemical Society.

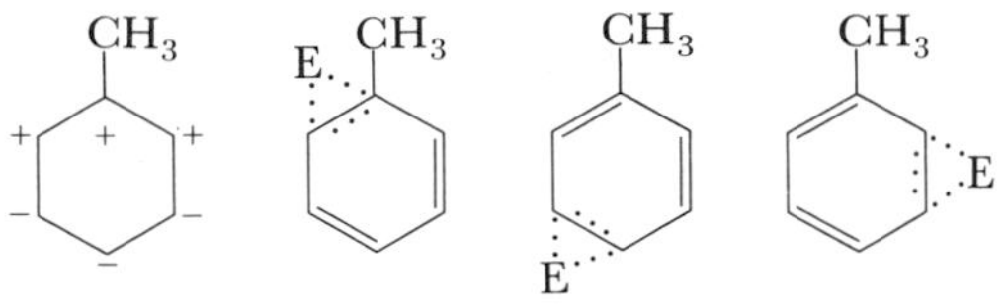

101	102	103	104

Olah's original experiments, in which the intermolecular selectivities were determined by direct competition for the electrophile by toluene and benzene, gave rise to controversy and criticism.[192] Schofield and Moodie first suggested that the reactions in question do have transition states similar to the σ complexes but are so fast that they are essentially over before the reactants are mixed; the ratio of products then depends on the local concentrations of the two aromatic substrates. However, further experiments convinced them that formation of σ complexes is not rate determining in very reactive nitrations.[193] Mononitration of **105**, **106**, **107**, and **108** in HNO_3/H_2SO_4 all occur at the same rate—diffusion-controlled encounter between NO_2^+ and the aromatic compound. If the simple two-step mechanism of

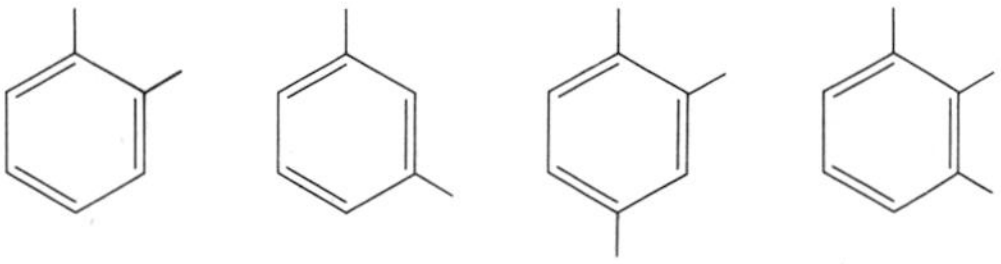

105	106	107	108

Equation 7.83 were operative, the NO_2^+ should nitrate the first position it hits and positional selectivity should be small. However, positional selectivity is not small: **107** gives the products shown in Equation 7.99.

[192] For a summary, see J. H. Ridd, *Acc. Chem. Res.*, **4**, 248 (1971) and also C. D. Johnson and K. Schofield, *J. Am. Chem. Soc.*, **95**, 270 (1973).

[193] J. W. Barnett, R. B. Moodie, K. Schofield, and J. B. Weston, *J. Chem. Soc., Perkin Trans.* **2**, 648 (1975).

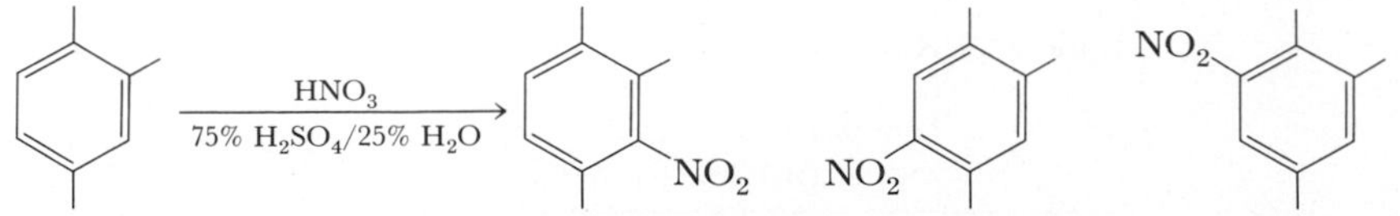

$$(7.99)$$

107　　　　　　　　　　　　**109**　　**110**

12%　　52%　　36%

Moodie, Schofield, and co-workers propose that the rate-determining step is the formation of an encounter complex (**111**), which in a product-determining step rearranges to a σ complex. This proposed mechanism is shown for the formation of **109** from **107** in Scheme 4.

SCHEME 4

$$HNO_3 + H^+ \overset{1}{\rightleftharpoons} H_2O + NO_2^+$$

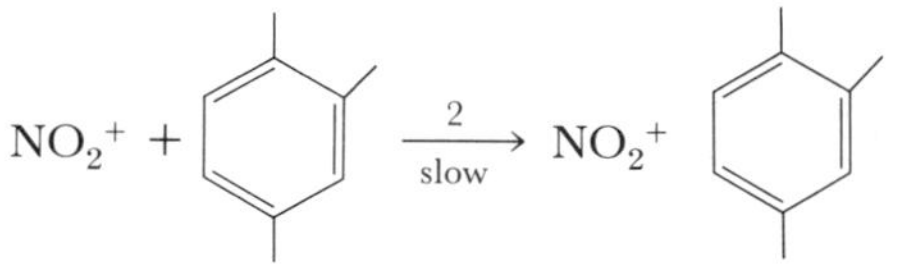

107　　　　　　**111**

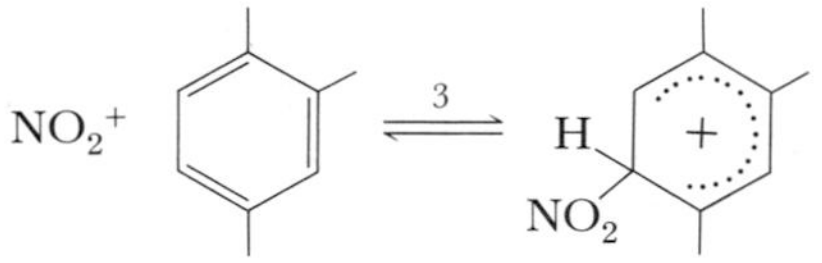

111

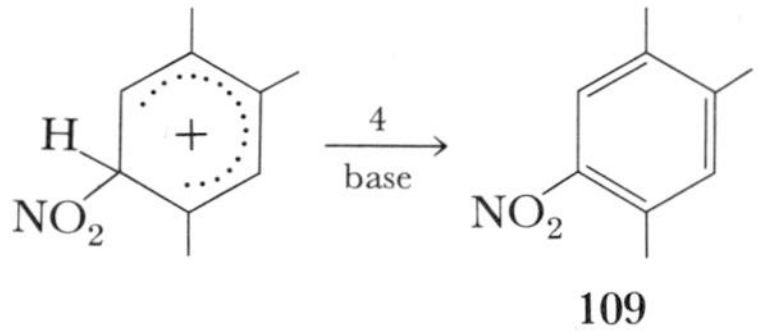

109

According to these investigators the encounter complex **111** may be neither the π complex **85** nor the π complex **86** but a complex held together by solvent with no attractive interactions between NO_2^+ and the aromatic ring. The reason for proposing this structure is that Rys and co-workers reexamined Olah's data correlating the π complex stability of methylated benzenes with the rate of $NO_2^+BF_4^-$ nitration. Instead of making only qualitative correlations as Olah had done, Rys used Olah's data and plotted the product ratios of polymethylbenzenes in the $NO_2^+BF_4^-$ competition experiments as function of π complex stabilities. The correlation he got [Figure 7.11(a)] was visually no better than when he plotted the same product ratios as a function of relative σ complex stability [Figure 7.11(b)].

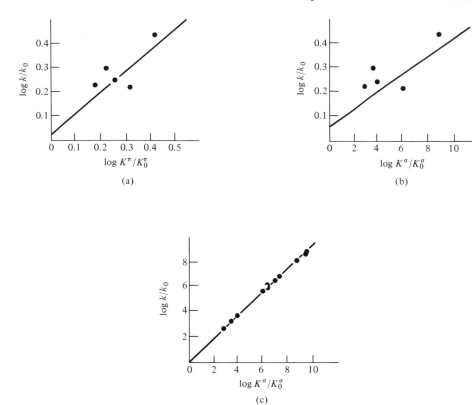

Figure 7.11 (a) Nitration product ratios of polymethylbenzenes with $NO_2^+BF_4^-$ as a function of the relative σ-complex stabilities of the polymethylated benzenes (see Tables 7.17 and 7.20). (b) Nitration product ratios of polymethylbenzenes with $NO_2^+BF_4^-$ as a function of the relative σ-complex stabilities of the polymethylated benzenes (see Table 7.17 and 7.20). (c) Relative rate constants for the bromination of polymethylbenzenes as a function of the relative σ-complex stabilities of the polymethylbenzenes (see Table 7.17). Reprinted with permission from P. Rys, P. Skrabal, and H. Zollinger, *Angew. Chem. Int. Ed. Engl.*, **11,** 874 (1972). Copyright 1972 Verlag Chemie.

Rys contrasted these graphs to the excellent correlation he obtained when he again used Olah's data and plotted the relative rate constants for the bromination of polymethylbenzenes as function of the relative σ complex stabilities [Figure 7.11(c)]. Rys concluded that Olah's correlations of relative rate of $NO_2^+BF_4^-$ nitrations with relative π complex stability were not justified.[194]

The mechamism for aromatic nitration shown in Scheme 4 has been rather widely accepted as the complete mechanism for aromatic nitration.[195] For very reactive nitrating agents the mechanism as drawn is considered correct. For less

[194]P. Rys, P. Skrabal, and H. Zollinger, *Angew. Chem. Int. Ed. Engl.*, **11,** 874 (1972).
[195]L. M. Stock, *Prog. Phys. Org. Chem.*, **12,** 21 (1976), and references therein; J. H. Ridd, *Adv. Phys. Org. Chem.*, **16,** 1, (1978), and references therein.

reactive ones the third step, formation of the σ complex, becomes the slow step and formation of the encounter complex is rapid and reversible. (Because it consists of the same species as the σ complex, the encounter complex is impossible to detect kinetically when σ complex formation is rate determining.)

Perrin has objected to the intermediacy of an encounter pair as an explanation for the high-intramolecular but low-intermolecular selectivity of reactive nitrations. He has suggested that in the very short lifetime of an encounter pair (10^{-10} sec) there is not time for high-intramolecular selectivity to occur. Furthermore, he sees no explanation for the fact that the very reactive NO_2^+, which does not differentiate intermolecularly, should become so selective intramolecularly once it is in the encounter complex.

Instead, Perrin has proposed the mechanism in Equation 7.100.[196] The aromatic compound, in a diffusion-controlled step, transfers an electron to NO_2^+ so

$$NO_2^+ + ArH \xrightarrow[\text{controlled}]{\text{encounter}} NO_2^{\cdot} + ArH^{\cdot +} \longrightarrow HArNO_2^+ \qquad (7.100)$$

that the encounter pair is composed of a radical and a radical cation. In a second step the radical pair collapses to a σ complex intermediate.

Perrin suggests that the "electron transfer mechanism immediately resolves the paradox of intramolecular selectivity without intermolecular selectivity, since $NO_2^{\cdot}$, the attacking species exhibiting the intramolecular selectivity, is different from NO_2^+, the one exhibiting no intermolecular selectivity." Perrin nitrated napthalene under both acid-induced and electrically induced conditions (Equations 7.101 and 7.102) and obtained, within experimental error, the same product

$$(7.101)$$

9.2   1

$$(7.102)$$

10.9   1

ratio in both cases. This, he maintains, is strong evidence for the electron transfer mechanism for nitration.[197]

This mechanism, however, has also been criticized on the grounds that, in the collapse of the encounter pair (**112**), no product is formed in which $NO_2^{\cdot}$ forms a bond to position 4, a position of very high spin density.[198]

[196]C. L. Perrin, *J. Am. Chem. Soc.*, **99**, 5516 (1977).
[197]Perrin does not suggest that this mechanism is general for electrophilic aromatic substitutions since for NO_2^+ electron transfer is exothermic for all aromatics more reactive than toluene but for most electrophiles it is endothermic.
[198]C. E. Barnes and P. C. Myhre, *J. Am. Chem. Soc.*, **100**, 975 (1978).

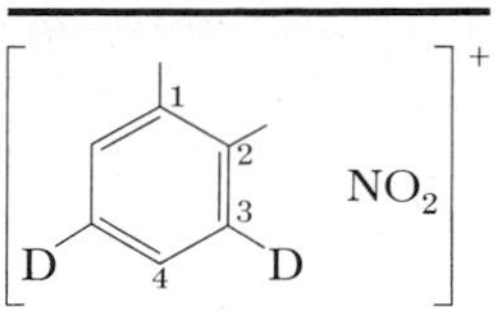

112

It seems that the final sentence to the mechanism of highly reactive aromatic nitrations is still not written.

A number of other very reactive electrophilic aromatic substitutions (for example, Friedel–Crafts acylations), although originally suggested by Olah to be further examples of early, π complexlike transition states, apparently obey the relationship between S_f and $\log p_f^{CH_3}$ (Figure 7.10) quite well and thus most likely have as a rate-determining step the formation of a σ complex.[199]

Ipso Substitution

Perrin has found that in ipso substitutions the order of leaving-group abilities is $H^+ \gg I^+ > Br^+ > NO_2^+ > Cl^+$. Electrophilic attack at alkyl-substituted positions also occurs but when it does, the alkyl group, a very poor leaving group, usually does not depart and the reaction follows one of the pathways shown in Scheme 5.[200]

SCHEME 5

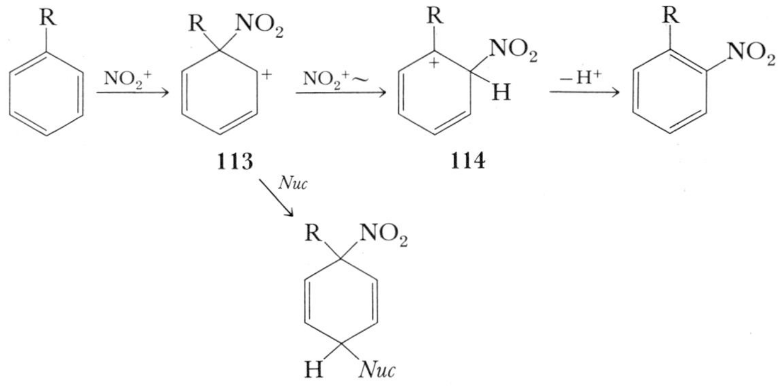

The partial rate factor for attack of a nitronium ion ipso to a methyl group is quite large—three times that for substitution at a meta position.[201] The fate of the nitrated intermediate (**113**) depends on the solvent. In concentrated sulfuric acid it rearranges almost entirely to the ortho-substituted benzenium ion (**114**), which in

[199](a) F. D. DeHaan, W. D. Covey, G. L. Delker, N. J. Baker, J. F. Feigon, K. D. Miller, and E. D. Stelter, *J. Am. Chem. Soc.*, **101**, 1336 (1979); (b) C. Santiago, K. N. Houk, and C. L. Perrin, *J. Am. Chem. Soc.*, **101**, 1337 (1979).
[200]R. B. Moodie and K. Schofield, *Acc. Chem. Res.*, **9**, 287 (1976).
[201]A. Fischer and C. J. Wright. *Aust. J. Chem.*, **27**, 217 (1974).

turn can then lose an H^+.[202] In aqueous sulfuric acid[193,202] or nucleophilic organic solvents,[203] **113** can be captured by a nucleophile to give **115**. Compounds **115**, some of which have been isolated and thoroughly characterized, in turn have a variety of reaction pathways available to them. Which pathways they choose depends on the reaction conditions.[200,203]

7.5 NUCLEOPHILIC AROMATIC SUBSTITUTION[204]

Because H^- is not a good leaving group, nucleophilic displacements on unsubstituted aromatics do not occur. However, if there is a suitable leaving group on the ring, nucleophilic aromatic substitution may take place by one of four mechanisms.

S_NAr Substitutions

If the ring bears strongly electron-withdrawing substituents as well as a good leaving group, nucleophilic displacements (called S_NAr or activated aromatic nucleophilic substitutions) take place under mild conditions. The kinetics are second-order, first-order each in aromatic substrate and in nucleophile. An example is shown in Equation 7.103.[205] A body of evidence has been accumulated that points

$$(7.103)$$

to an addition–elimination mechanism (Equation 7.104) for these reactions that is analogous to the mechanism for electrophilic aromatic substitution.

$$(7.104)$$

116

Meisenheimer first showed that σ complexes (**116**) can exist by isolating **117** in the form of a salt.[206] Since then a host of other *Meisenheimer complexes* have been isolated or identified by spectroscopy or other physical methods.[207]

[202] P. C. Myhre, *J. Am. Chem. Soc.*, **94**, 7921 (1972).

[203] (a) T. Banwell, C. S. Morse, P. C. Myhre, and A. Vollmar, *J. Am. Chem. Soc.*, **99**, 3042 (1977); (b) C. E. Barnes and P. C. Myhre, *J. Am. Chem. Soc.*, **100**, 973, 975 (1978); (c) K. S. Feldman, A. McDermott, and P. C. Myhre, *J. Am. Chem. Soc.*, **101**, 505 (1979).

[204] (a) J. Miller, *Aromatic Nucleophilic Substitution*, Elsevier, Amsterdam, 1968; (b) G. B. Barlin, *Aromat. and Heteroaromat. Chem.*, **2**, 271 (1974). (c) C. F. Bernasconi, *MTP Int. Rev. Sci:, Org. Chem.*, Ser. One, **3**, 33 (1973); (d) F. Pietra, *Quart. Rev.*, **23**, 504 (1969).

[205] P. Carniti, P. Beltrame, and Z. Cabiddu, *J. Chem. Soc., Perkin Trans. 2*, 1430 (1973).

[206] J. Meisenheimer, *Ann.*, **323**, 205 (1902).

[207] (a) M. R. Crampton, *Adv. Phys. Org. Chem.*, **7**, 211 (1969); (b) M. J. Strauss, *Acc. Chem. Res.*, **7**, 181 (1974).

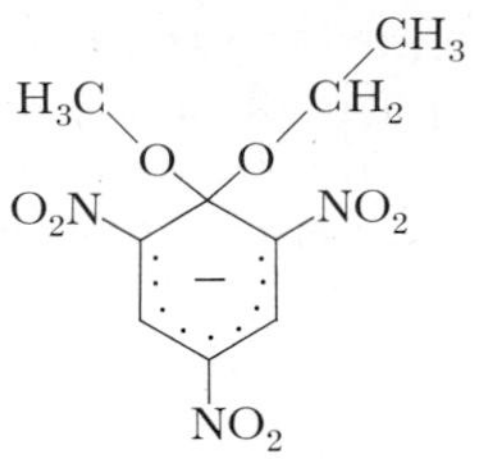

117

Studies of base catalysis have shown that which step of the addition–elimination reaction is rate determining depends on the nucleophile, the leaving group, and the solvent. For example, in dimethylsulfoxide the reaction between 1-ethoxy-2,4-dinitronaphthalene and *n*-butyl amine is specific base–general acid catalyzed. This observation requires that a proton be transferred in the rate-determining step. But simple proton transfers from one base to another are very rapid—much more rapid than the rate of this reaction. Thus some other molecular reorientation must also be occurring in the rate-determining step. The mechanism in Scheme 6 is implicated.[208]

SCHEME 6

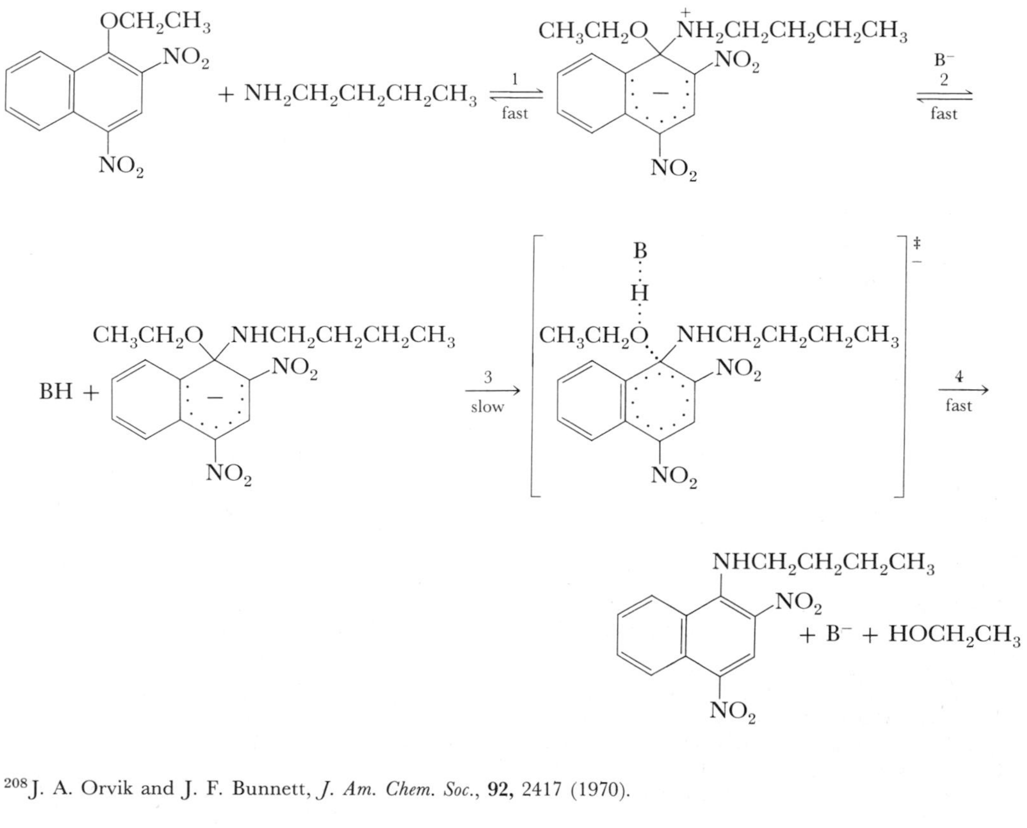

[208] J. A. Orvik and J. F. Bunnett, *J. Am. Chem. Soc.*, **92**, 2417 (1970).

Bernasconi believes that rate-determining loss of the leaving group is general in aprotic solvents.[209] In protic solvents either attack of the nucleophile or departure of the leaving group can be rate determining. For example, Bunnett has demonstrated the effect of the leaving group on the mechanism of Reaction 7.105

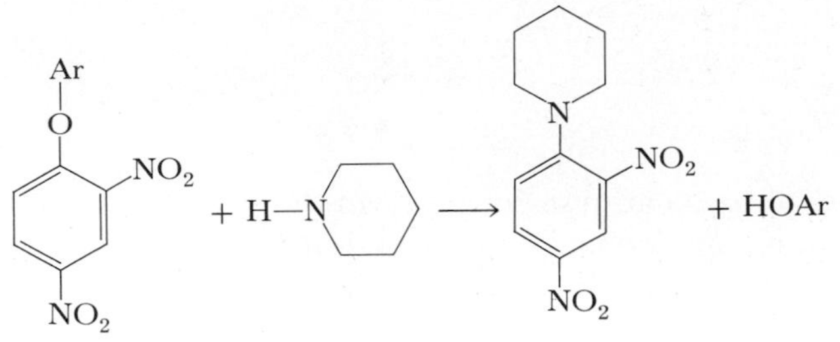

$$\text{(7.105)}$$

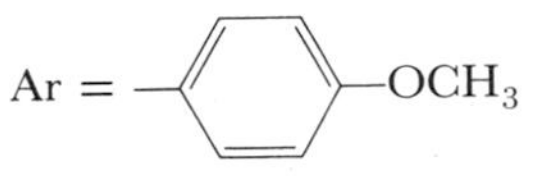

$$\text{Ar} = \text{—} \langle \rangle \text{—OCH}_3$$

118a

$$\text{Ar} = \text{—} \langle \rangle \text{—NO}_2$$

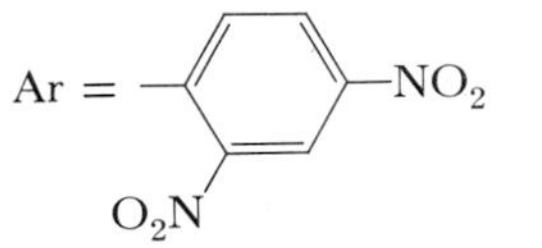

118b

in aqueous dioxane. Expulsion of ArO⁻ is much easier from the intermediate **120** than from **119**. Thus if loss of ArO⁻ is rate limiting, Reaction 7.105 should be base-catalyzed. When **118a** is used as substrate, strong acceleration by base is observed; but when **118b** is the substrate, added base has almost no effect on the rate.[210]

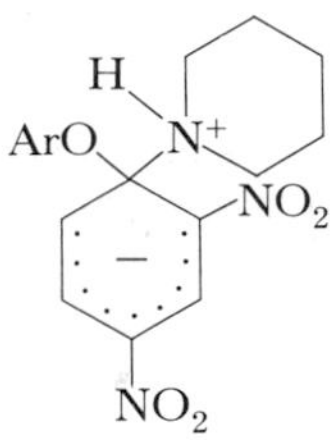

119 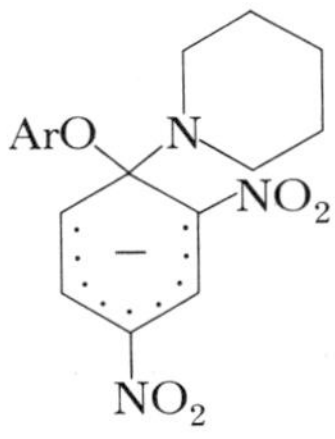   **120**

Benzyne Mechanism

Treatment of iodo-, bromo-, or chlorobenzene with potassium amide yields aniline. In 1953 Roberts observed that when chlorobenzene-1-^{14}C is the substrate, approxi-

[209] C. F. Bernasconi, *Acc. Chem. Res.*, **11**, 147 (1978).

[210] J. F. Bunnett and C. F. Bernasconi, *J. Org. Chem.*, **35**, 70 (1970). See also (b) D. M. Brewis, N. B. Chapman, J. S. Paine, J. Shorter and D. J. Wright, *J. Chem. Soc., Perkin Trans. 2*, 1787 (1974); (c) D. M. Brewis, N. B. Chapman, J. S. Paine, and J. Shorter, *J. Chem. Soc., Perkin Trans. 2*, 1802 (1974).

mately 50 percent of the ^{14}C in the product is found in the 1 and approximately 50 percent in the 2-position. The overall substitution then must go by an elimination–addition mechanism in which the highly strained intermediate, benzyne, is formed as shown in Equation 7.106.[211]

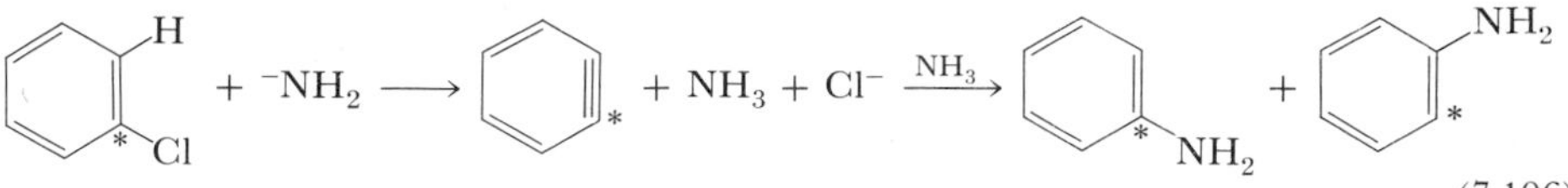

$$(7.106)$$

Comparison of the rates of formation of aniline from bromobenzene and bromobenzene-2-d gives $k_H/k_D = 5.5$; thus the proton is removed in the rate-determining step.[212] Fluorobenzene-2-d, however, exchanges its deuterium with solvent a million times faster than does deuterobenzene, but no aniline is formed.[213] Apparently, when the halogen is weakly electron withdrawing but a good leaving group hydrogen abstraction is the slow step; when the halogen is strongly electron withdrawing but unreactive as a leaving group, its expulsion is the slow step.

The presence of benzyne has been more directly demonstrated by trapping experiments. For example, the generation of benzyne in the presence of anthracene gives the Diels–Alder adduct, triptycene (Equation 7.107).[214]

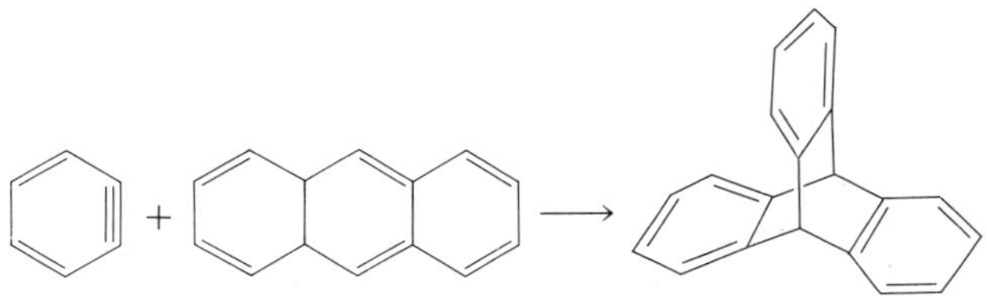

$$(7.107)$$

Nucleophilic Substitutions on Aromatic Diazonium Compounds

A third mechanism of nucleophilic aromatic substitution, specific for substitution on aromatic diazonium salts,[215] is shown in Equation 7.108. Apparently the leaving group must be as reactive as $—N_2{}^+$ in order that the intermediate **122** be formed.

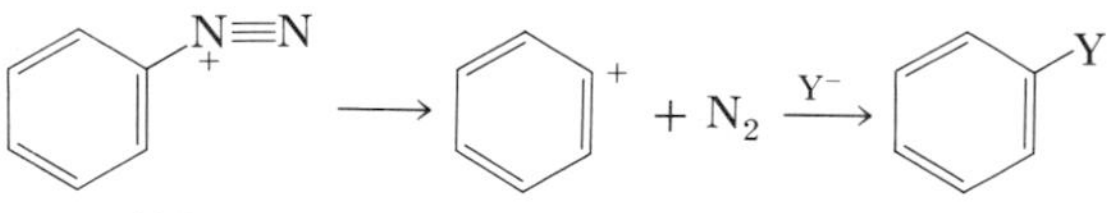

$$(7.108)$$

121   **122**

[211](a) J. D. Roberts, H. E. Simmons, Jr., L. A. Carlsmith, and C. W. Vaughan, *J. Am. Chem. Soc.*, **75**, 3290 (1953); (b) J. D. Roberts, D. A. Semenow, H. E. Simmons, Jr., and L. A. Carlsmith, *J. Am. Chem. Soc.*, **78**, 601 (1956); (c) J. D. Roberts, C. W. Vaughan, L. A. Carlsmith, and D. A. Semenow, *J. Am. Chem. Soc.*, **78**, 611 (1956).
[212]See note 211(b).
[213]G. E. Hall, R. Piccolini, and J. D. Roberts, *J. Am. Chem. Soc.*, **77**, 4540 (1955).
[214](a) G. Wittig and K. Niethammer, *Chem. Ber.*, **93**, 944 (1960); (b) G. Wittig, H. Härle, E. Knauss, and K. Niethammer, *Chem. Ber.*, **93**, 951 (1960).
[215]Substitutions on aromatic diazonium compounds may also take place by a free-radical mechanism. See G.-J. Meyer, K. Rössler, and G. Stocklin, *J. Am. Chem. Soc.*, **101**, 3121 (1979).

The evidence for the intermediacy of **122** is excellent.[216] Ion **121** reacts with D_2O to give only **123**. No **124** is formed. Thus the benzyne intermediate is ex-

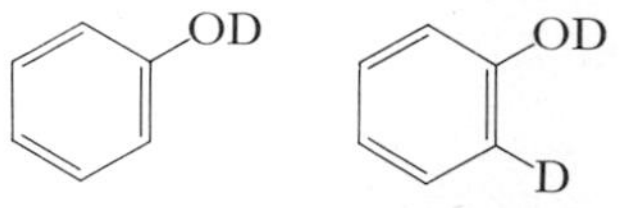

cluded. D_2O and H_2O react at almost the same rate so there can be no nucleophilic involvement of water in the rate-determining step.[217] When the starting material is

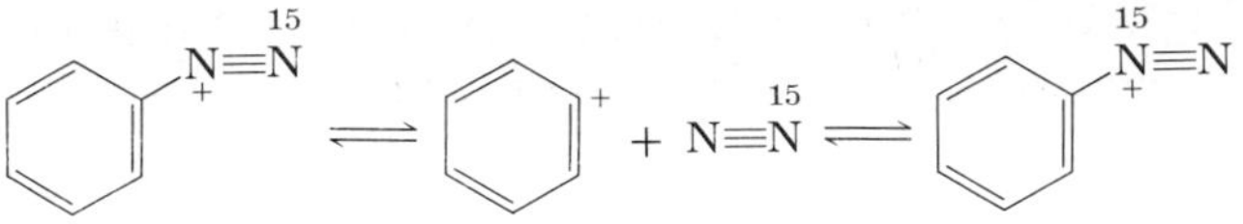

$$\tag{7.109}$$

selectively labeled with ^{15}N, the rearrangement shown in Equation 7.109 takes place. That some of this occurs through dissociation, reassociation, and not through transition state **126** is shown by the fact that when scrambling of nitrogen from **125**

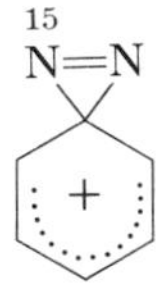

occurs under 300 atm of unlabeled nitrogen, 2.5 percent of unlabeled nitrogen is incorporated into the diazonium cation![218]

The $S_{RN}1$ Mechanism[219]

5-Chloro and 6-chloro-1,2,4-trimethylbenzene **127** and **128**, when treated with potassium in liquid ammonia, give the same ratio of 5-amino- to 6-amino-1,2,4-trimethylbenzene (**133** and **134**, respectively)—a result to be expected from an aryne mechanism. The bromo-1,2,4-trimethylbenzenes **129** and **130**) behave similarly. 5-Iodo- and 6-iodo-1,2,4-trimethylbenzene **131** and **132**), however, each react to give predominantly unrearranged product (**133** and **134**, respectively). Addition of a radical scavenger causes the ratio of products from **131** and **132** to be almost the same as that from **127** to **130** (see Table 7.21).

[216](a) C. G. Swain, J. E. Sheats, and K. G. Harbison, *J. Am. Chem. Soc.*, **97**, 783 (1975); (b) C. G. Swain, J. E. Sheats, D. G. Gorenstein, and K. G. Harbison, *J. Am. Chem. Soc.*, **97**, 791 (1975); (c) C. G. Swain, J. E. Sheats, and K. G. Harbison, *J. Am. Chem. Soc.*, **97**, 796 (1975); (d) R. G. Bergstrom, R. G. M. Landells, G. W. Wahl, Jr., and H. Zollinger, *J. Am. Chem. Soc.*, **98**, 3301 (1976).

[217] See note 216(a).

[218] See note 216(d)

[219] J. F. Bunnett, *Acc. Chem. Res.*, **11**, 413 (1978).

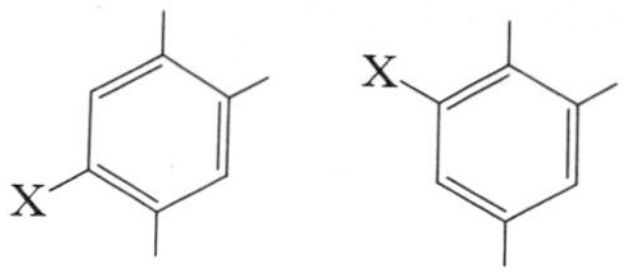

X		
Cl	127	128
Br	129	130
I	131	132
NH_2	133	134

Bunnett proposed the mechanism shown for 5-iodo-1,2,4-trimethylbenzene in Scheme 7. Summation of the three chain propagation steps gives Equation 7.110, an overall nucleophilic aromatic substitution.

SCHEME 7

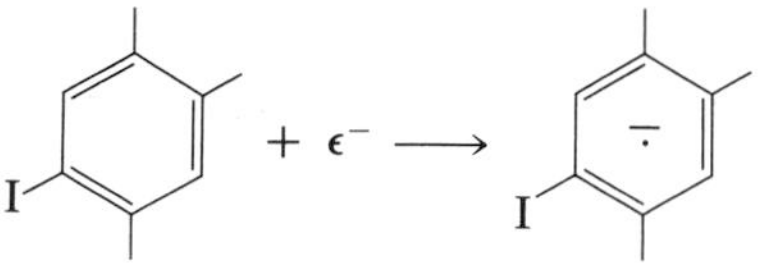

Chain initiation

131

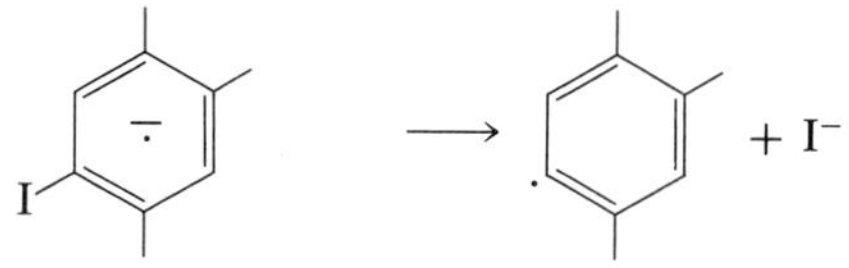

Chain propagation 1

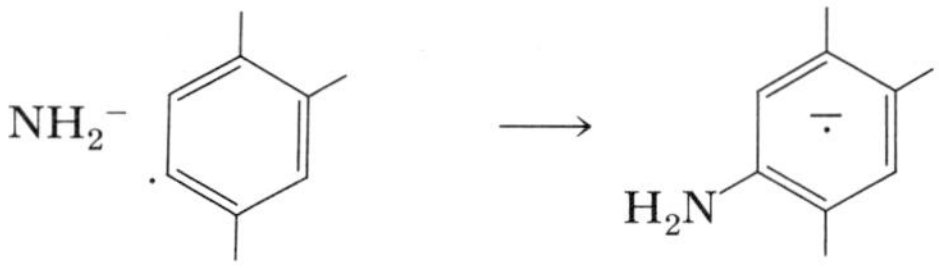

Chain propagation 2

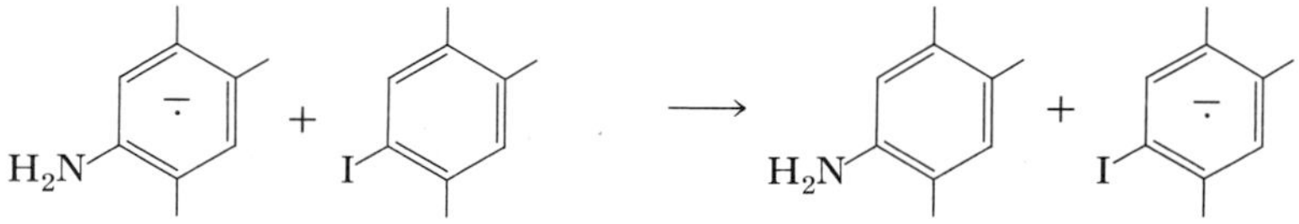

Chain propagation 3

131 133

Chain termination steps

The first two chain propagation steps in Scheme 7 are analogous to the S_N1 reaction, so Bunnett calls this mechanism $S_{RN}1$.

$S_{RN}1$ reactions can be initiated in a variety of ways—electrochemically,[220] photochemically, or with solvated electrons in liquid ammonia (e.g., $K^0 \xrightarrow{NH_3}$

[220] J. Pinson and J.-M. Saveant, *J. Am. Chem. Soc.*, **100**, 1506 (1978).

Table 7.21 REACTIONS OF 5- AND 6-HALOPSEUDOCUMENES
WITH POTASSIUM AMIDE IN LIQUID AMMONIA[a]

Substrate	Conc. of KNH$_2$ (M)	Conc. of radical scavenger[b]	Ratio **133**:**134**
127	0.30	—	1.46
128	0.30	—	1.45
129	0.25	—	1.55
130	0.13	—	1.45
131	0.29	—	0.63
131	0.30	0.008	1.41
132	0.29	—	5.9
132	0.45	0.012	2.0

[a] Reprinted with permission from J. K. Kim and J. F. Bunnett, *J. Am. Chem. Soc.,* **92,** 7463 (1970). Copyright (1970) by the American Chemical Society.
[b] Tetraphenylhydrazine.

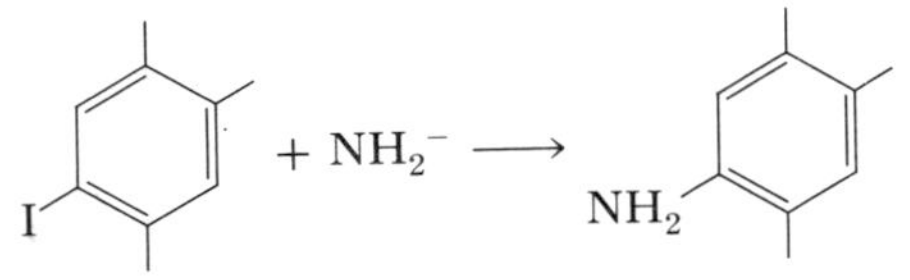

$$+ NH_2^- \longrightarrow \qquad\qquad (7.110)$$

$$131 \qquad\qquad 133$$

$K^+ + \epsilon^-$).[221] In accordance with expectation, photochemical initiation gives the high ($\sim$50) quantum yields indicative of a chain reaction. (See Section 12.4).

$S_{RN}1$ reactions have been found to be quite general. No special substituents are required on the aromatic ring; in fact, simple aryl halides undergo reaction, but alkyl, aryl, and alkoxy substituents do not interfere. All the halogens have been used as leaving groups, as have $-S\phi$, $-\overset{+}{N}(CH_3)_3$, and $-OP(OCH_2CH_3)_2$. A wide variety of groups have been employed as nucleophiles—for example, ketone and

ester enolate ions ($\overset{-}{C}H_2-\overset{\overset{\displaystyle O}{\|}}{C}-X$), thiolate anions (RS$^-$), and NH$_2^-$. Harder bases such as alkoxides and aryloxides, however, are not effective, nor are highly stabilized carbanions such as **135**.

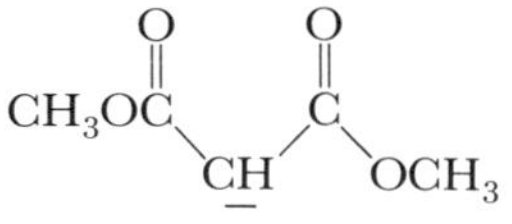

$$135$$

PROBLEMS

1. Explain each of the following observations.

(a) Hydrochlorination of 1,2-dimethylcyclohexene gives a mixture of the two products shown in Equation 1. The ratio of **1** to **2** depends on the solvent. In methanol the product is predominantly **1** and in acetyl chloride it is predominantly **2**.

[221] These conditions may, of course, also lead to an aryne mechanism.

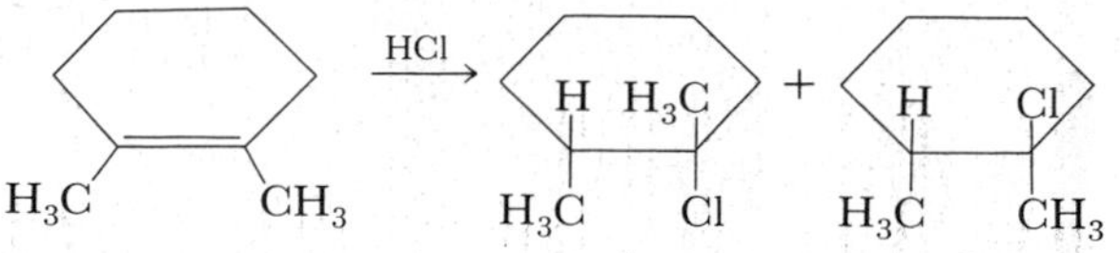

$$(1)$$

(b) In the reaction of Equation 2, if the base is potassium *t*-butoxide, the rate of reaction is independent of the concentration of base if more than an equimolar amount of base is present. If, however, the base is sodium methoxide, the rate is dependent on added base up to high base concentrations.

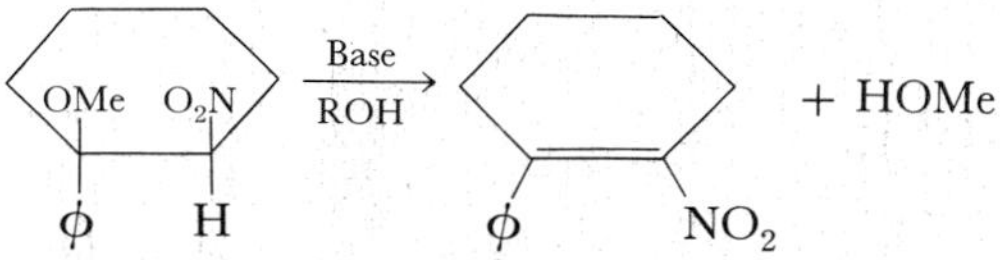

$$(2)$$

(c) Bromination of substituted benzenes with Br_2 has a ρ value of -12.1, whereas bromination with HOBr has a ρ value of only -6.2.

(d) The phenyl halides undergo electrophilic aromatic substitution more slowly than benzene but give predominantly ortho and para substitution.

(e) Treatment of 1-chloronaphthalene with ethoxide ion gives no reaction. Treatment of **3** with ethoxide, however, gives the nucleophilic substitution product shown in Equation 3.

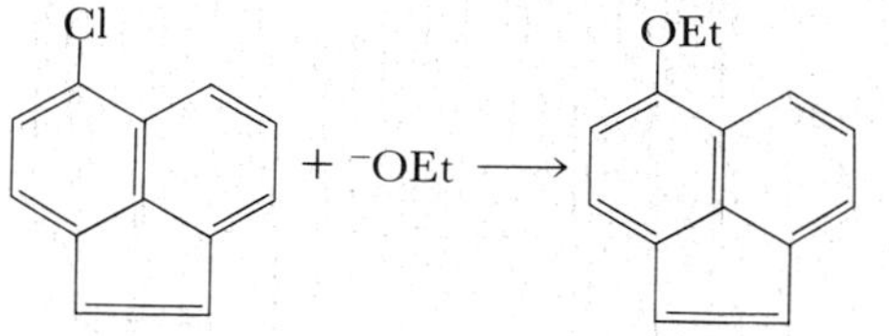

$$(3)$$

(f) Nitration of excess **4** with nitric acid in acetic acid gives very little dinitration. However, adding $NO_2{}^+BF_4{}^-$ to excess **4** in sulfolane gives a high preponderance of products in which both rings are nitrated.

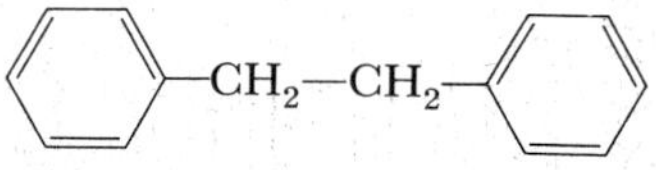

4

(g) The ratio of the specific rate constants of **5** to **6** in hydration is 3.8, whereas in bromination it is 51.9.

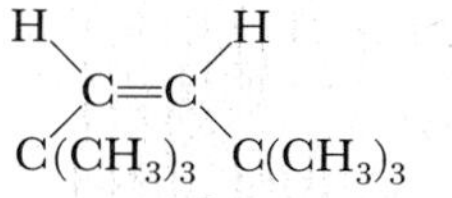

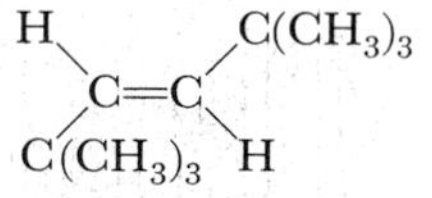

(h) **7** forms a cyclic compound, C_5H_9BrO, on bromination in methanol.

$$HOCH_2CH_2CH_2CH{=}CH_2$$

7

$$\text{(i)} \quad + \text{Br}_2 \longrightarrow \qquad \qquad + \qquad \qquad \text{(4)}$$

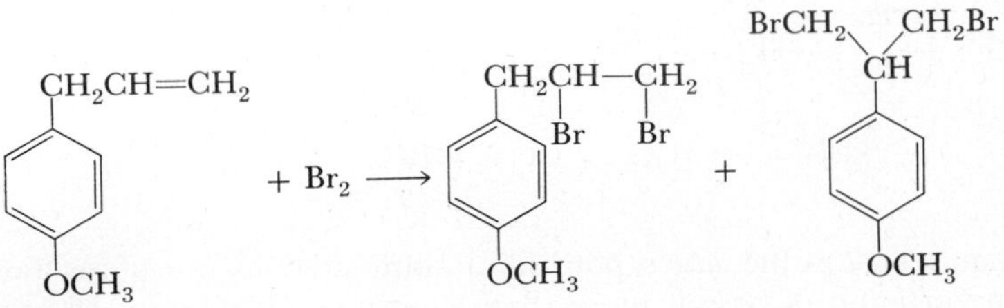

$$\text{(j)} \quad + \quad \xrightarrow[\text{SbF}_5]{\text{HF}} \qquad + \qquad \text{(5)}$$

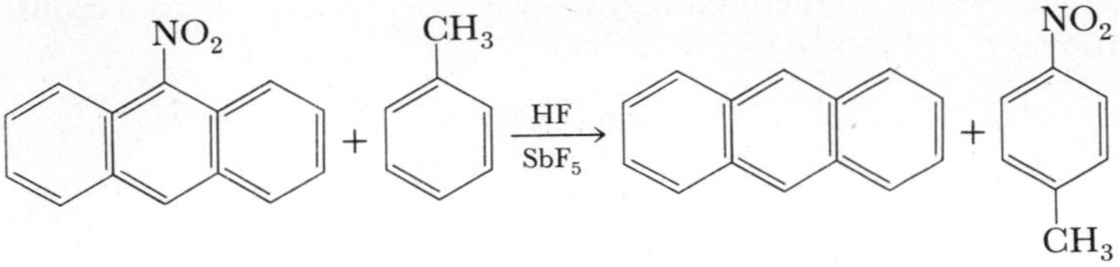

Toluene is nitrated 95 percent para in Reaction 5. Toluene gives 60–65 percent ortho nitration with $NO_2^+BF_4^-$.

$$\text{(k)} \quad \xrightarrow[\substack{\text{acetic acid/} \\ \text{acetic anhydride}}]{\text{HNO}_3}$$

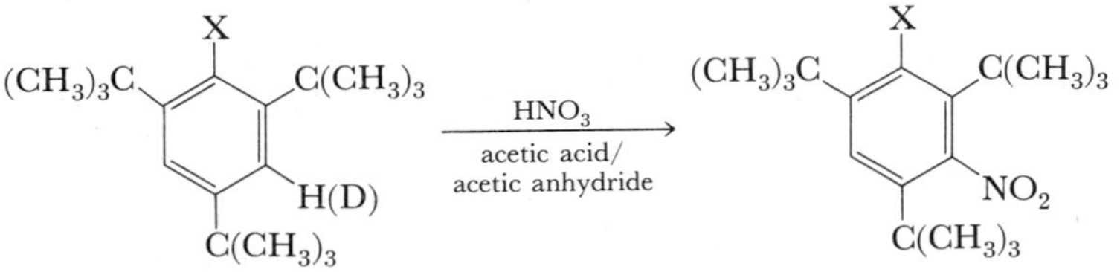

k_H/k_D increases from 1.0 to 3.8 as the size of X increases.

(l) Elimination of HBr from **8** in dimethyl sulfoxide gives one and a half times more cis isomer when one racemic diastereomeric pair of **9** is used as base then when the other racemic diastereomeric pair is used.

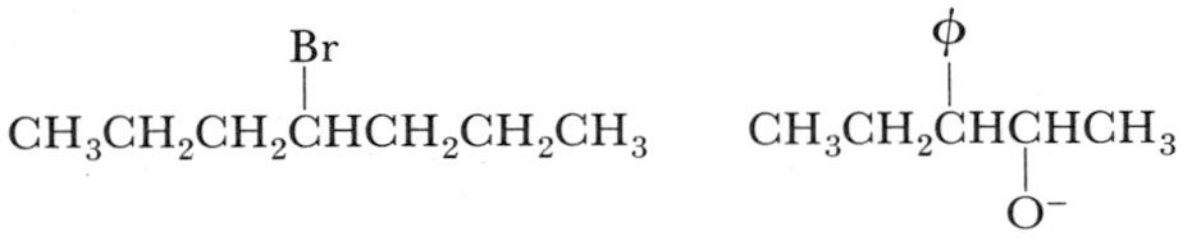

$$\underset{\textbf{8}}{\text{CH}_3\text{CH}_2\text{CH}_2\overset{\overset{\text{Br}}{|}}{\text{C}}\text{HCH}_2\text{CH}_2\text{CH}_3} \qquad\qquad \underset{\textbf{9}}{\text{CH}_3\text{CH}_2\overset{\overset{\phi}{|}}{\text{C}}\text{H}\overset{}{\text{C}}\text{HCH}_3}$$

(m) When the $S_{RN}1$ reaction of *m*-bromoiodobenzene, **10**, with diethylphosphite anion, $(C_2H_5O)_2PO^-$, was interrupted before completion, there was present 28 percent unreacted **10**, 60 percent disubstitution product **11**, and only 7 percent of monosubstitution product **12**. Compound **12**, if subjected to the reaction conditions, is transformed to **11** more slowly than is **10**. Explain.

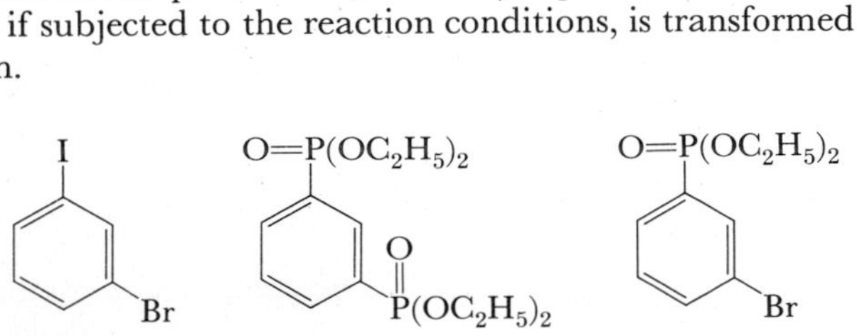

$$\textbf{10} \qquad\qquad\qquad \textbf{11} \qquad\qquad\qquad \textbf{12}$$

2. Predict the following.

(a) To which pair will nitronium-salt nitration show greater intermolecular selectivity?

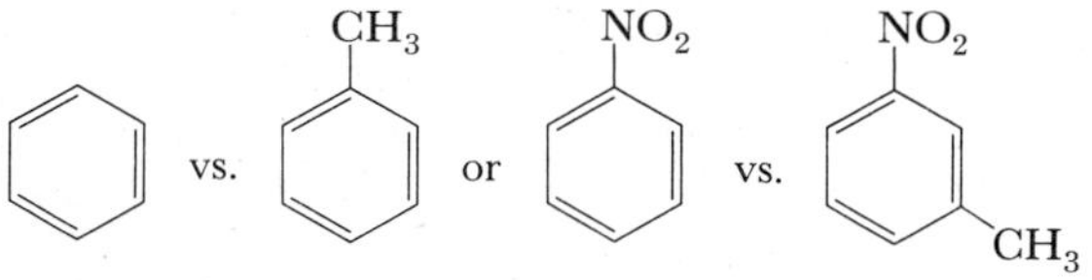

vs.   or   vs.

(b) In nucleophilic substitution of Cl⁻ the relative reactivities of (1) **13a** and **14a**, **13b** and **14b**, **13c** and **14c**; (2) **13a**, **13b**, and **13c**; (3) **13d** and **13e**.

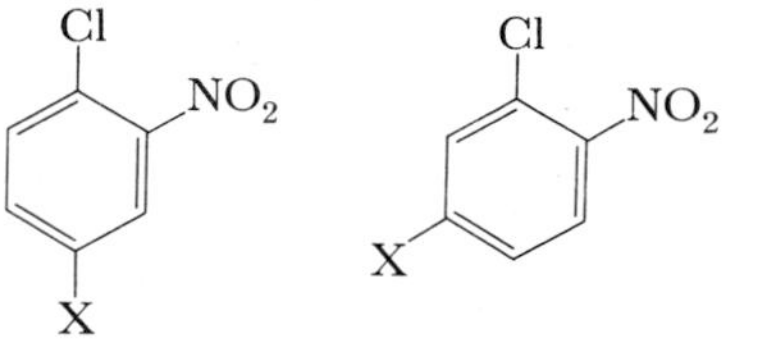

13   **14**

a, X = CH_3
b, X = CN
c, X = NH_2
d, X = $\overset{+}{N}(CH_3)_3$
e, X = SO_2CH_3

(c) Which will give more terminal alkene with potassium tertiary butoxide, **15** or **16**?

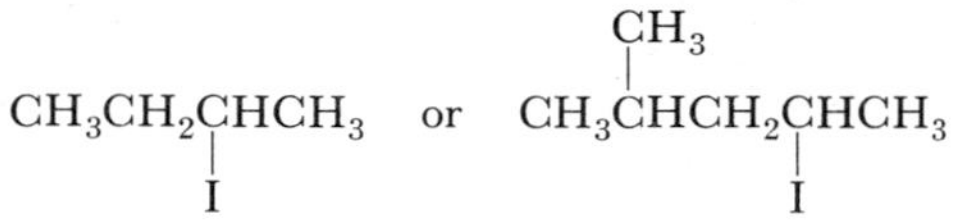

$$CH_3CH_2\underset{\underset{I}{|}}{CH}CH_3 \quad \text{or} \quad CH_3\underset{\underset{CH_3}{|}}{CH}CH_2\underset{\underset{I}{|}}{CH}CH_3$$

15   **16**

3. Propose a mechanism for each of the following reactions.

(a)

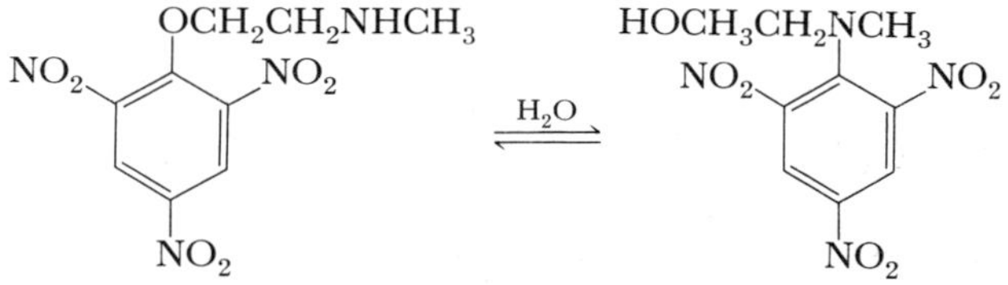

(b)

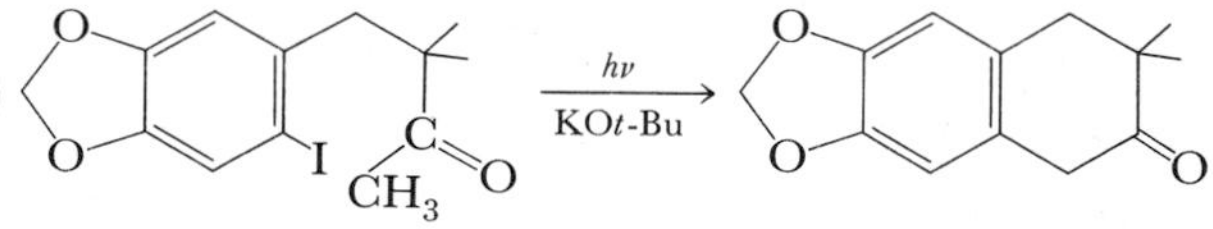

99%

(c) The addition of 2,4-dinitrobenzenesulfenyl chloride to *cis*- and to *trans*-2-phenyl-2-butene.

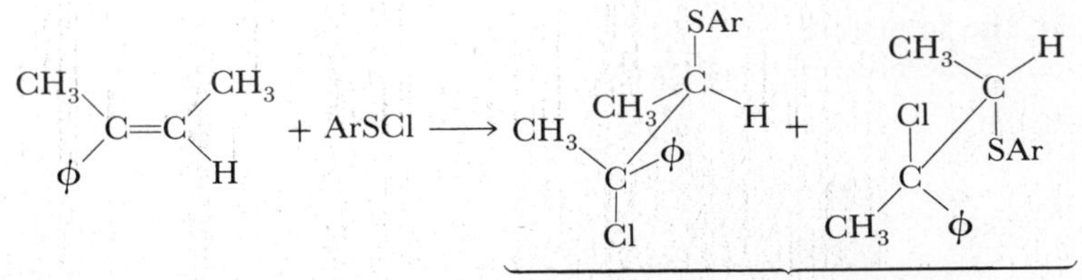

$$CH_3 \underset{\phi}{\overset{}{C}}=C\underset{H}{\overset{CH_3}{}} + ArSCl \longrightarrow CH_3 \cdots$$

Racemate of diastereomer A

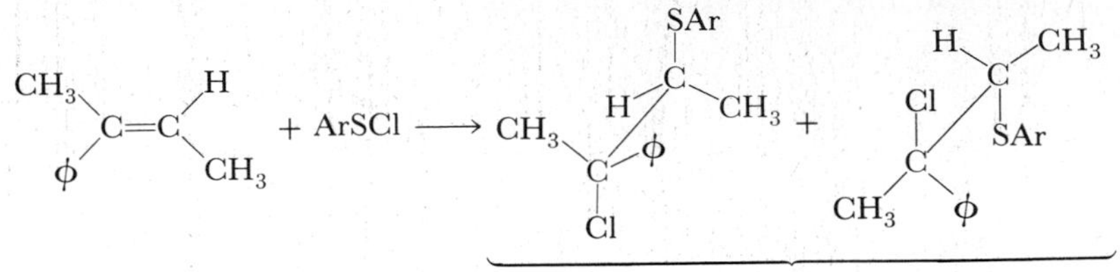

$$CH_3 \underset{\phi}{\overset{}{C}}=C\underset{CH_3}{\overset{H}{}} + ArSCl \longrightarrow CH_3 \cdots$$

Racemate of diastereomer B

(d) The Sommelet–Hauser rearrangement, three examples of which are shown below.

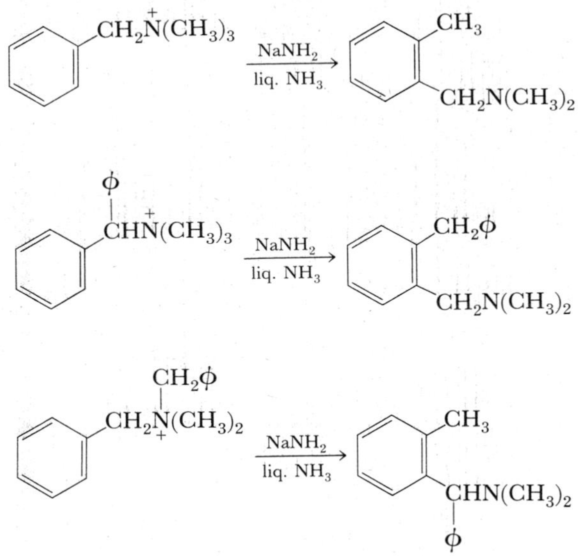

$$\phi\text{---}CH_2\overset{+}{N}(CH_3)_3 \xrightarrow[\text{liq. NH}_3]{\text{NaNH}_2}$$

with products CH_3 and $CH_2N(CH_3)_2$ on the ring

$$\phi\text{---}\overset{\phi}{\underset{}{C}}H\overset{+}{N}(CH_3)_3 \xrightarrow[\text{liq. NH}_3]{\text{NaNH}_2}$$

with products $CH_2\phi$ and $CH_2N(CH_3)_2$ on the ring

$$\phi\text{---}CH_2\overset{\overset{CH_2\phi}{|}}{\underset{+}{N}}(CH_3)_2 \xrightarrow[\text{liq. NH}_3]{\text{NaNH}_2}$$

with products CH_3 and $CHN(CH_3)_2$ with ϕ on the ring

(e)

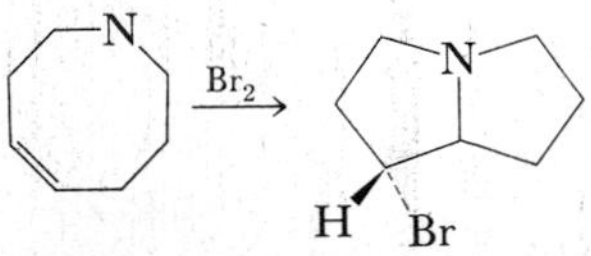

(f) The Smiles rearrangement.

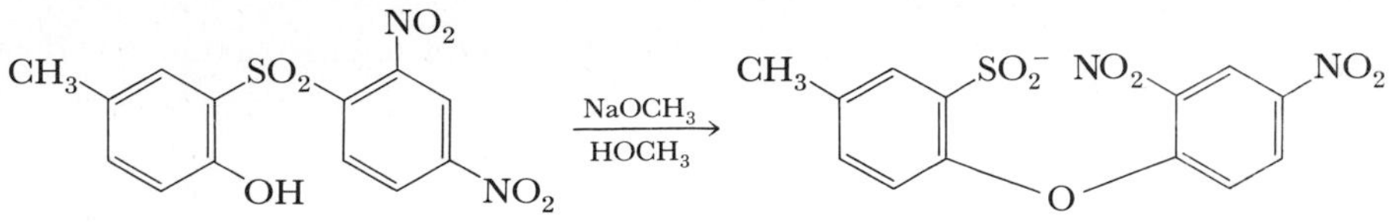

(g) The Wallach rearrangement.

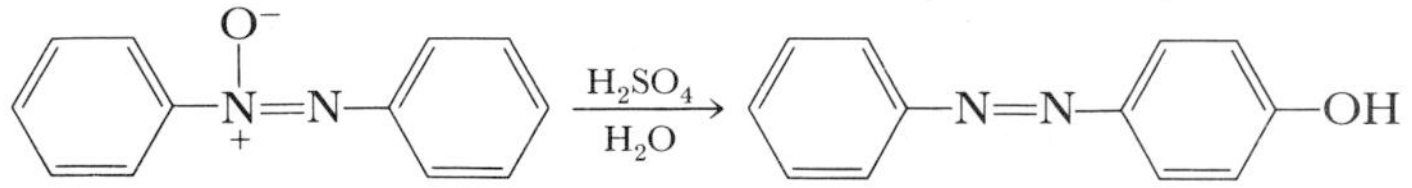

1. A ^{14}N-labeled nitrogen in the starting material is totally scrambled in the product.

2. ^{18}O tracer studies show that the OH in the product comes from the solvent.

3. Kinetic studies show that the transition state is diprotonated.

(h)

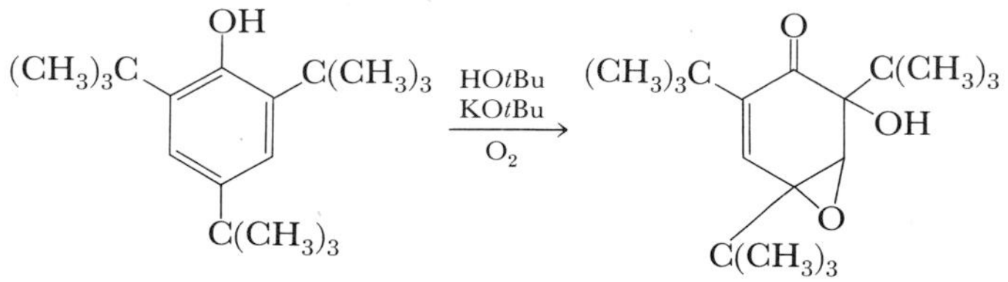

(i)

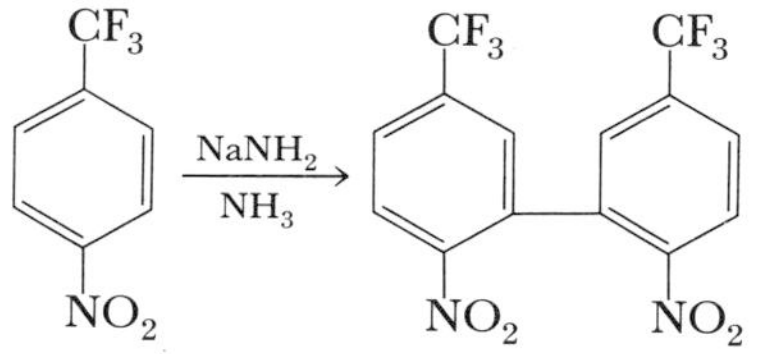

(j)

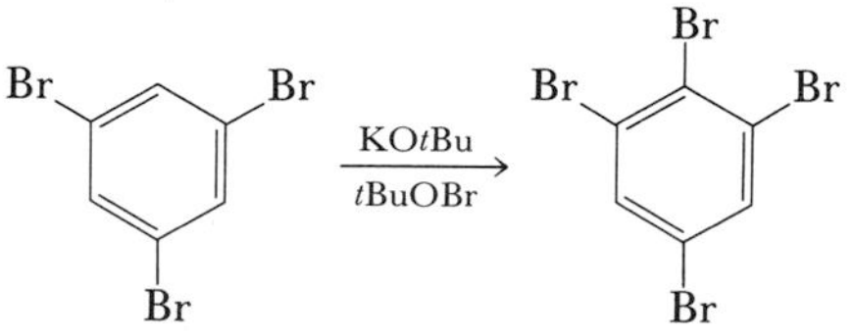

There is no reaction in the absense of KOtBu.

(k)

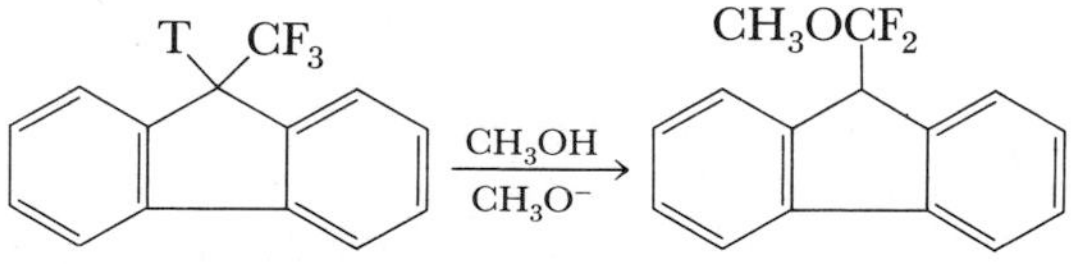

There is rapid tritium exchange with the solvent.

4. On the basis of your mechanism for Problem 3(c), answer the following questions.
(a) Additions of Br$_2$ and ArSCl to [2.2.1]-bicycloheptene give the products shown in Equa-

tions 6 and 7, respectively. Give a mechanism for the products of Reaction 6 and explain why only the bromination gives rearrangement products.

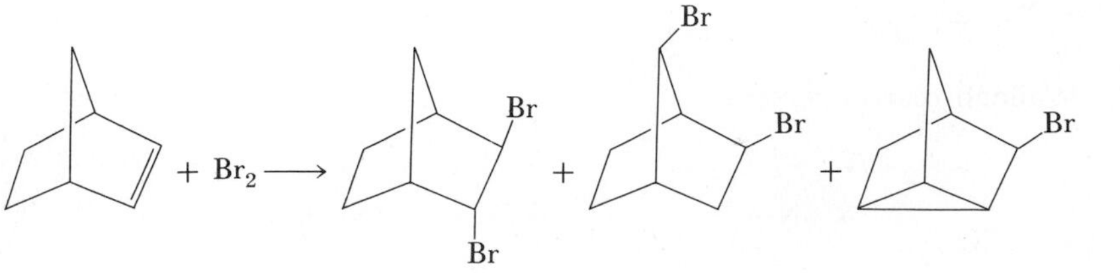

$$+ Br_2 \longrightarrow \qquad + \qquad + \qquad \tag{6}$$

$$+ ArSCl \longrightarrow \qquad \tag{7}$$

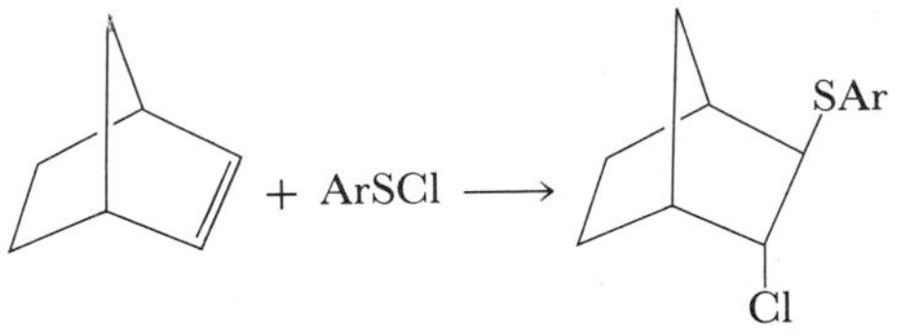

(b) 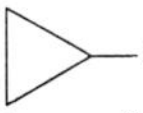 ▷—CH=CH$_2$ reacts 1000 times faster than ϕ—CH=CH$_2$ with Br$_2$-HOAc but only 3.8 times faster with pCl-C$_6$H$_5$SCl in HOAc at 25°C. Explain.

(c) Which would you expect to show a better linear correlation with the rate constants for bromination of a number of alkenes—the rate constants for hydration or those for addition of ArSCl to the same alkenes?

    5. Addition of hypochlorous acid to **17** gives the three products shown in Equation 8.

$$CH_2=CH-\underset{\underset{36}{\overset{|}{Cl}}}{CH_2} + HOCl \longrightarrow$$

17

$$CH_2-CH-CH_2 + CH_2-CH-CH_2 + CH_2-CH-CH_2 \tag{8}$$
$$\overset{|}{Cl}\ \overset{|}{HO}\ \overset{|}{^{36}Cl} \qquad \overset{|}{OH}\ \overset{|}{Cl}\ \overset{|}{^{36}Cl} \qquad \overset{|}{Cl}\ \overset{|}{^{36}Cl}\ \overset{|}{OH}$$

18

(a) Give a mechanism for the formation of **18**.

(b) Much less rearrangement of ^{36}Cl occurs if **17** is treated with HOBr instead of HOCl. Explain.

    6. With what alkene would you begin and what synthetic method would you use to produce pure *threo*-3-*p*-anisyl-2-butanol? Pure *erythro*?

    7. Decide for each of the compounds in Table 7.18 whether the substituent is $+I$ or $-I$, $+R$ or $-R$.

    8. By drawing resonance structures for the respective σ complexes, decide whether attack is more likely at the 1 or the 2 position in electrophilic substitution on naphthalene.

    9. In 80 percent aqueous ethanol, 3-β-tropanyl chloride gives the product shown in Equation 9 in 100 percent yield. Under the same conditions, 3-α-tropanyl chloride reacts at one-twentieth the rate to give only the addition and elimination products shown in Equation 10. Formulate a mechanism for Reaction 9 and explain the difference between the mode of reaction of Reactions 9 and 10.

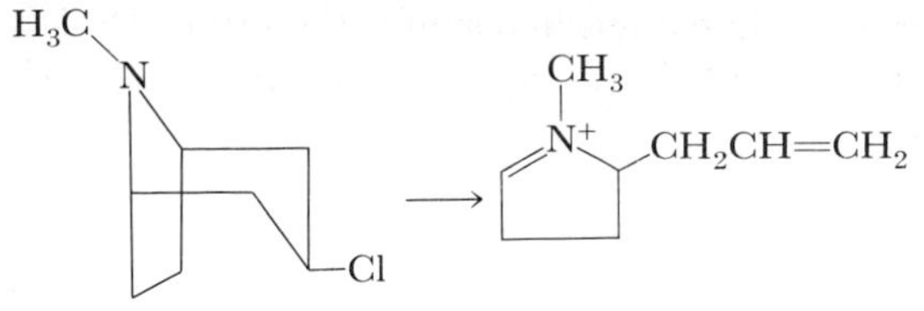

$$(9)$$

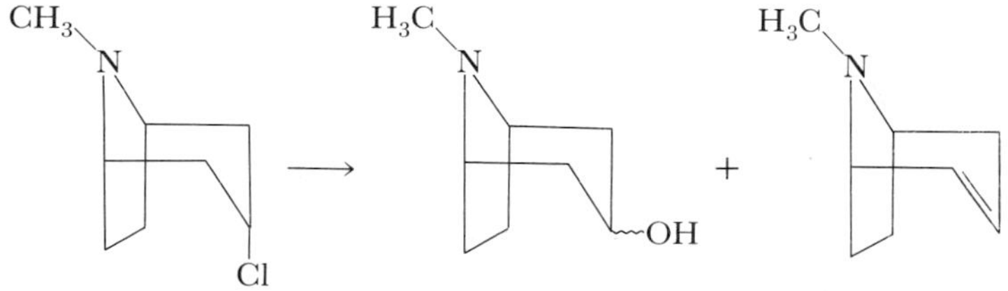

$$(10)$$

10. The data for E2H reactions indicates that at the transition state the reaction coordinate motion involves mainly proton transfer—that is, the carbon–leaving-group bond is already partially broken but it is not being further broken near the transition state. Further C_α—X bond breaking occurs only after the transition state has been traversed.
(a) Construct a reaction coordinate diagram, which corresponds to the description above, analogous to Figure 7.6 but for the E2H (rather than a central E2) reaction.
(b) Using the diagram you have constructed in part (a), predict the effect of: (1) a poorer leaving group, (2) a stronger base, (3) an electron-withdrawing group at C_β, and (4) an electron-releasing group at C_α on the transition state.
(c) Using the diagram you constructed in part (a), predict whether electron-withdrawing groups in the α-aryl group would be more effective in accelerating sodium ethoxide-promoted elimination of HX in ethanol from **19** or from **20**.

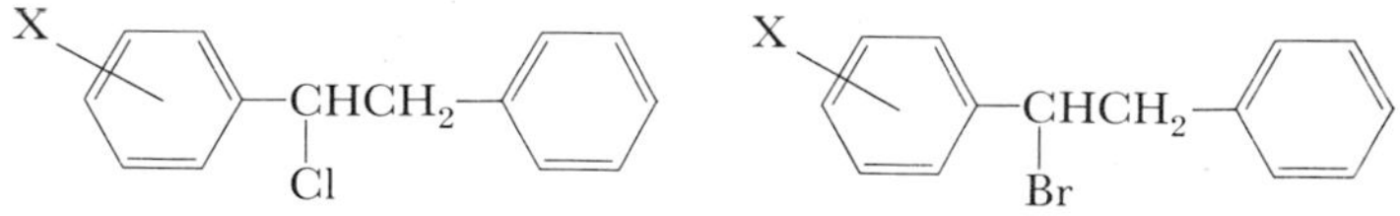

19

20

REFERENCES FOR PROBLEMS

1. (a) R. C. Fahey and C. A. McPherson, *J. Am. Chem. Soc.*, **93**, 2445 (1971); (b) F. G. Bordwell, M. M. Vestling, and K. C. Yee, *J. Am. Chem. Soc.*, **92**, 5950 (1970); (e) M. J. Perkins, *Chem. Commun.*, 231 (1971); (f) A. Gastaminza and J. H. Ridd, *J. Chem. Soc., Perkin Trans.* **2**, 813 (1972); (g) W. K. Chwang and T. T. Tidwell, *J. Org. Chem.*, **43**, 1904 (1978); (h) E. Bienvenüe-Goetz, J.-E. Dubois, D. W. Pearson, and D. L. H. Williamson, *J. Chem. Soc. B*, 1275 (1970); (i) J.-E. Dubois, J. Toullec, and D. Fain, *Tetrahedron Lett.*, 4859 (1973); (j) G. A. Olah, S. C. Narang, R. Malhotra, and J. A. Olah, *J. Am. Chem. Soc.*, **101**, 1805 (1979); (k) P. C. Myhre, M. Beug, and L. L. James, *J. Am. Chem. Soc.*, **90**, 2105 (1968); (l) I. Ishizawa, *Bull. Chem. Soc. Jpn.*, **48**, 1572 (1975); (m) J. F. Bunnett, *Acc. Chem. Res.*, **11**, 413 (1978).

2. (a) G. A. Olah and H. C. Lin, *J. Am. Chem. Soc.*, **96**, 549 (1974); (b) (1, 2) W. Greizerstein, R. A. Bonelli, and J. A. Brieux, *J. Am. Chem. Soc.*, **84**, 1026 (1962); (3) J. F. Bunnett, F. Draper, Jr., R. R. Ryason, P. Noble, Jr., R. G. Tonkyn, and R. E. Zahler, *J. Am. Chem. Soc.*, **75**, 642 (1953); (c) B. A. Bartsch, R. A. Read, D. T. Larsen, D. K. Roberts, K. J. Scott, and B. R. Cho, *J. Am. Chem. Soc.*, **101**, 1176 (1979).

3. (a) C. F. Bernasconi, C. L. Gehriger, and R. H. deRossi, *J. Am. Chem. Soc.*, **98**, 8451 (1976); (b) M. F. Semmelhack and T. M. Bargar, *J. Org. Chem.* **42**, 1481 (1977); (c) D. J. Cram, *J. Am. Chem. Soc.*, **71**, 3883 (1949); (d) S. W. Kantor and C. R. Hauser, *J. Am. Chem. Soc.*, **73**, 4122 (1951); (e) S. R. Wilson and R. A. Sawick, *J. Org. Chem.*, **44**, 287 (1979); (f) F. Galbraith and S. Smiles, *J. Chem. Soc.*, 1234 (1935); (g) E. Buncel, *Acc. Chem. Res.*, **8**, 132 (1975); (h) A. Nishinaga, T. Itahara, T. Shimizu, and T. Matsuura, *J. Am. Chem. Soc.*, **100**, 1820 (1978); (i) Y. Kobayashi and I. Kumadaki, *Acc. Chem. Res.*, **11**, 197 (1978); (j) M. J. Mach and J. F. Bunnett, *J. Am. Chem. Soc.*, **96**, 936 (1974); (k) A. Streitweiser, Jr., A. P. Marchand, and A. H. Pudjaatmaka, *J. Am. Chem. Soc.*, **89**, 693 (1967).

4. (a) H. Kwart and R. K. Miller, *J. Am. Chem. Soc.*, **78**, 5678 (1956); (b) D. G. Garratt, A. Modro, K. Oyama, G. H. Schmid, T. T. Tidwell, and K. Yates, *J. Am. Chem. Soc.*, **96**, 5295 (1974); (c) G. H. Schmid and T. T. Tidwell, *J. Org. Chem.*, 43, 460 (1978).

5. C. A. Clarke and D. L. H. Williams, *J. Chem. Soc, B*, 1126 (1966).

6. E. L. Allred, J. Sonnenberg, and S. Winstein, *J. Org. Chem.*, **25**, 26 (1960).

9. G. A. Grob, *Theoretical Organic Chemistry—Kekulé Symposium*, London: Butterworths, 1959, p. 114.

10. (a) and (b) D. A. Winney and E. R. Thornton, *J. Am. Chem. Soc.*, **97**, 3102 (1975); (c) E. Bacciocchi, P. Perucci, and C. Rol, *J. Chem. Soc., Perkin Trans.* **2**, 329 (1975).

Chapter 8
REACTIONS
OF CARBONYL
COMPOUNDS

Carbonyl compounds comprise a large and important class of organic substances; the chemistry of this functional group is essential to the understanding of many chemical and biochemical processes.[1] In this chapter we use a few fundamental ideas of mechanism to correlate reactions of various carbonyl functional groups.

Carbonyl reactions may be understood in terms of two basic processes: addition of a nucleophile to the carbonyl carbon (Equation 8.1) and removal of a proton from the carbon adjacent to the carbonyl group (Equation 8.2). In the first

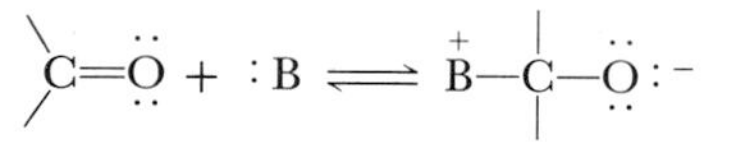

$$\underset{/}{\overset{\backslash}{C}}{=}\ddot{O} + :B \;\rightleftharpoons\; \overset{+}{B}{-}\overset{|}{\underset{|}{C}}{-}\ddot{O}:^- \tag{8.1}$$

$$\underset{\underset{H}{\overset{|}{C}}}{\overset{\backslash}{C}}{=}\ddot{O} + :B \;\rightleftharpoons\; \underset{\overset{|}{C}:^-}{\overset{\backslash}{C}}{=}\ddot{O} \;\longleftrightarrow\; \underset{\overset{|}{C}}{\overset{\backslash}{C}}{-}\ddot{O}:^- + BH^+ \tag{8.2}$$

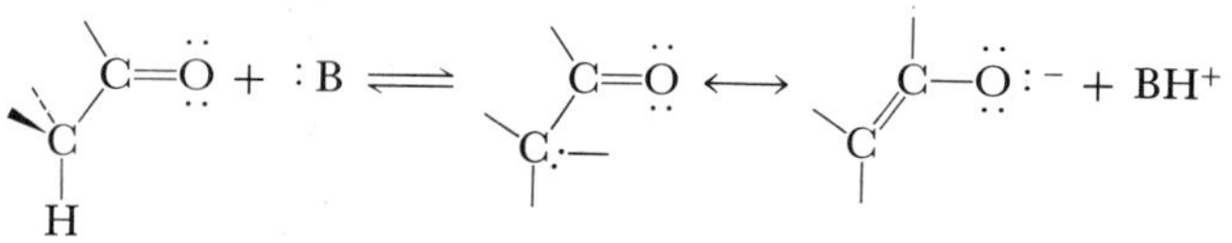

process the carbonyl molecule is acting as a Lewis acid, and in the second (which is, of course, possible only if the molecule bears an α hydrogen) as a Brønsted acid.

---

[1] Reviews of various aspects of carbonyl chemistry may be found in the following sources: (a) W. P. Jencks, *Catalysis in Chemistry and Enzymology*, McGraw-Hill, New York, 1969; (b) M. L. Bender, *Mechanisms of Homogeneous Catalysis from Protons to Proteins*, Wiley, New York, 1971; (c) S. Patai, Ed., *The Chemistry of the Carbonyl Group*, Vol. 1, and J. Zabicky, Ed., *The Chemistry of the Carbonyl Group*, Vol. 2, Wiley, London, 1966 and 1970; (d) S. Patai, Ed., *The Chemistry of Acyl Halides*, Wiley, London, 1972; (e) W. P. Jencks, *Prog. Phys. Org. Chem.*, **2**, 63 (1964); (f) R. P. Bell, *Adv. Phys. Org. Chem.*, **4**, 1 (1966); (g) T. H. Fife, *Adv. Phys. Org. Chem.*, **11**, 1 (1975); (h) E. H. Cordes and H. G. Bull, *Chem. Rev.*, **74**, 581 (1974).

Both processes depend on the electron deficiency of the carbonyl carbon, which is in turn caused by the electronegativity of the oxygen and its ability to accept a negative charge. The second reaction is readily reversible, and the first under most circumstances is also. Coordination of the carbonyl oxygen with a proton or some other Lewis acid will make the oxygen more electrophilic and may be expected to facilitate both addition of a nucleophile to the carbonyl carbon and removal of a proton from the α position. Catalysis by acids and bases is thus a central theme of carbonyl reactions.

We have already considered in Chapter 6 the process of Equation 8.2 in the context of carbanion chemistry. This chapter will concentrate primarily on reactions of the type 8.1, with a brief discussion of some further aspects of Equation 8.2.

8.1 HYDRATION AND ACID–BASE CATALYSIS

We consider first the simple addition of a nucleophile to a carbonyl carbon, preceded, accompanied, or followed by addition of a proton to the oxygen, and the reverse. The overall process (Equation 8.3) amounts to addition of H—X to C=O.

$$\text{\Large$>$}C{=}O + H{-}X \rightleftharpoons \text{\Large$>$}C\begin{smallmatrix}\nearrow OH \\ \searrow X\end{smallmatrix} \tag{8.3}$$

The reaction differs from the additions to C=C discussed in Chapter 7 in two important respects. First, the nucleophile always becomes bonded to the carbon and the proton to the oxygen, so there is no ambiguity concerning direction of addition; and second, the C=O group is much more susceptible to attack by a nucleophile than is C=C.

Hydration

Water adds to the carbonyl group of aldehydes and ketones to yield hydrates (Equation 8.4). For ketones and aryl aldehydes, equilibrium constants of the reac-

$$\begin{matrix} R_1 \\ \diagdown \\ \diagup \\ R_2 \end{matrix} C{=}O + H_2O \rightleftharpoons \begin{matrix} R_1 \\ \diagdown \\ \diagup \\ R_2 \end{matrix} C\begin{matrix} \diagup OH \\ \diagdown OH \end{matrix} \tag{8.4}$$

$$R_1, R_2 = H, \text{ alkyl, or aryl}$$

tion as written are much less than unity, but aliphatic aldehydes are appreciably hydrated in water solution. The equilibrium constant is larger for the lower aldehydes and is largest for formaldehyde. Some representative values are given in Table 8.1. Bulky groups and groups that donate electron density to the electron-deficient carbonyl carbon stabilize the carbonyl form, whereas substituting electron-withdrawing groups or incorporating the carbonyl carbon in a strained ring

Table 8.1 APPROXIMATE EQUILIBRIUM CONSTANTS AT
25°C FOR THE REACTION

$$R_1R_2C{=}O + H_2O \rightleftharpoons R_1R_2C(OH)_2$$

Carbonyl compound	$K[H_2O] = \dfrac{[R_1R_2C(OH)_2]^a}{[R_1R_2C{=}O]}$
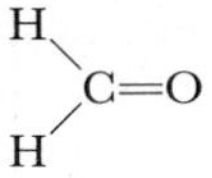 $\begin{matrix} H \\ \diagdown \\ C{=}O \\ \diagup \\ H \end{matrix}$	2×10^3
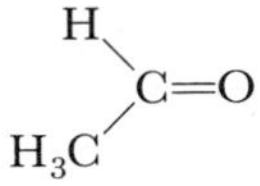 $\begin{matrix} H \\ \diagdown \\ C{=}O \\ \diagup \\ H_3C \end{matrix}$	1.3
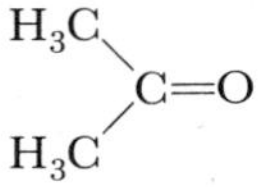 $\begin{matrix} H_3C \\ \diagdown \\ C{=}O \\ \diagup \\ H_3C \end{matrix}$	2×10^{-3}
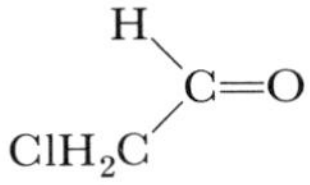 $\begin{matrix} H \\ \diagdown \\ C{=}O \\ \diagup \\ ClH_2C \end{matrix}$	37
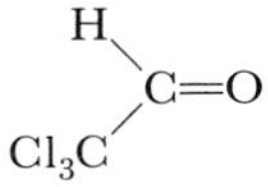 $\begin{matrix} H \\ \diagdown \\ C{=}O \\ \diagup \\ Cl_3C \end{matrix}$	2.8×10^4
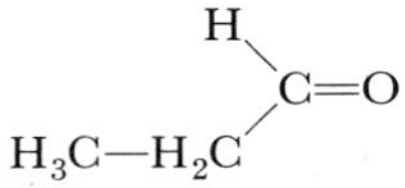 $\begin{matrix} H \\ \diagdown \\ C{=}O \\ \diagup \\ H_3C{-}H_2C \end{matrix}$	0.71
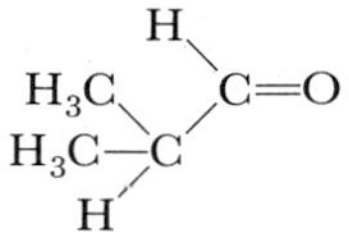 $\begin{matrix} H \\ H_3C \diagdown\diagup \\ C{=}O \\ H_3C{-}C \\ H \end{matrix}$	0.43^b
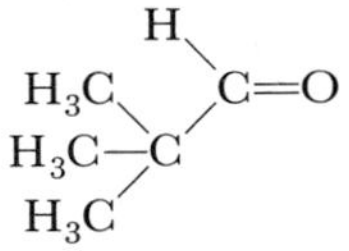 $\begin{matrix} H \\ H_3C \diagdown\diagup \\ C{=}O \\ H_3C{-}C \\ H_3C \end{matrix}$	0.24^b
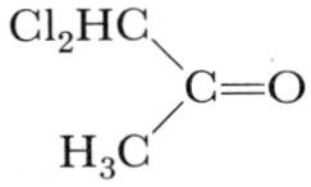 $\begin{matrix} Cl_2HC \\ \diagdown \\ C{=}O \\ \diagup \\ H_3C \end{matrix}$	2.9
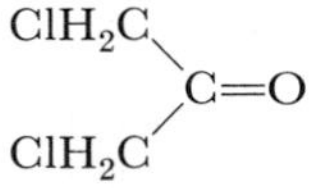 $\begin{matrix} ClH_2C \\ \diagdown \\ C{=}O \\ \diagup \\ ClH_2C \end{matrix}$	10

[a] Except as noted, values are calculated from data given by R. P.
Bell, *Adv. Phys. Org. Chem.*, **4**, 1 (1966).
[b] P. Greenzaid, Z. Luz, and D. Samuel, *J. Am. Chem. Soc.*, **89**,
749 (1967).

favors the hydrate.[2] Equilibrium constants correlate with Taft inductive and steric parameters.[3]

Mechanistic questions in the hydration–dehydration equilibrium center around the acid-base relationships and the precise sequence of events in the addition or elimination of the water molecule. Investigations have relied primarily on kinetics of aldehyde hydration to elucidate the mechanistic details; rates of reaction in both directions have been measured by spectroscopic methods,[4] isotope exchange experiments,[5] heat of reaction,[6] volume change measurements,[7] and by scavenging liberated aldehyde.[8] The reaction is subject to general acid and general base catalysis.[9] Catalytic rate constants have been measured for a number of acids and bases in aldehyde hydration–dehydration, notably by Bell and co-workers.[8] For formaldehyde, $\alpha = 0.24$, $\beta = 0.40$; earlier work[10] gave for acetaldehyde $\alpha = 0.54$, $\beta = 0.45$, and for symmetrical dichloroacetone $\alpha = 0.27$, $\beta = 0.50$.

As we have seen in our earlier discussion of general catalysis in Chapter 7, the observation of general catalysis means that proton transfer must occur in the rate-determining step or, if there is no single rate-determining step, in the group of steps that together constitute the rate-determining process. Scheme 1 shows a possible reaction sequence for addition of a nucleophile to a carbonyl group. ($:Nuc^-$ represents a generalized nucleophile; it need not necessarily be negatively charged.) HA may be the solvent or some other acid. If the nucleophile is strong, the intermediate **1** should be relatively stable, and its rate of decomposition back to reactants,

SCHEME 1

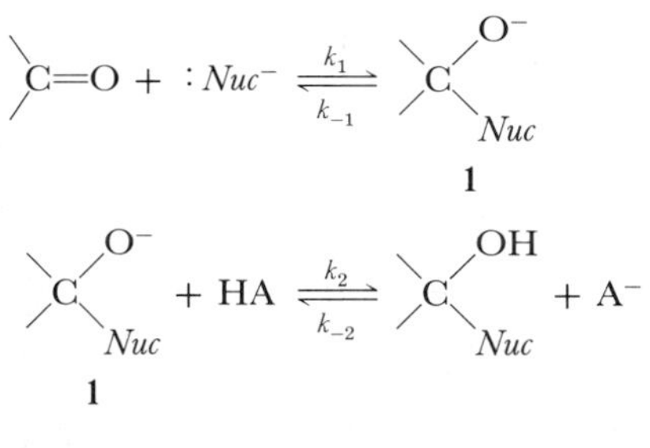

[2](a) H. C. Brown, R. S. Fletcher, and R. B. Johannesen, *J. Am. Chem. Soc.*, **73**, 212 (1951); (b) J. F. Pazos, J. G. Pacifici, G. O. Pierson, D. R. Sclove, and F. D. Greene, *J. Org. Chem.*, **39**, 1990 (1974); see also (c) N. J. Turro and W. B. Hammond, *J. Am. Chem. Soc.*, **88**, 3672 (1966).

[3](a) Y. Ogata and A. Kawasaki, in *The Chemistry of the Carbonyl Group*, J. Zabicky, Ed., Vol. 2, Wiley, London, 1970, p. 1; (b) P. Greenzaid, Z. Luz, and D. Samuel, *J. Am. Chem. Soc.*, **89**, 749 (1967).

[4](a) P. Greenzaid, Z. Luz, and D. Samuel, *J. Am. Chem. Soc.*, **89**, 756 (1967); (b) M.-L. Ahrens and H. Strehlow, *Disc. Faraday Soc.*, **39**, 112 (1965); (c) R. P. Bell and M. B. Jensen, *Proc. R. Soc. London, Ser. A*, **261**, 38 (1961).

[5]M. Cohn and H. C. Urey, *J. Am. Chem. Soc.*, **60**, 679 (1938).

[6]R. P. Bell, M. H. Rand, and K. M. A. Wynne-Jones, *Trans. Faraday Soc.*, **52**, 1093 (1956).

[7]R. P. Bell and B. de B. Darwent, *Trans. Faraday Soc.*, **46**, 34 (1950).

[8](a) R. P. Bell and P. G. Evans, *Proc. R. Soc. London, Ser. A*, **291**, 297 (1966); (b) L. H. Funderburk, L. Aldwin, and W. P. Jencks, *J. Am. Chem. Soc.*, **100**, 5444 (1978).

[9]Discussions of the acid–base catalysis may be found in the reviews cited in note 1.

[10]See note 6 and note 4(c).

k_{-1}, will be small compared with the rate of its protonation to product, $k_2[HA]$. Figure 8.1(a) shows the reaction coordinate for this process. The first step is rate determining. No acid–base catalysis is expected for such a reaction. (Specific base catalysis would occur if a preliminary equilibrium is required to produce $:Nuc^-$ from H—Nuc.) For a less strong nucleophile, intermediate **1** will break down more

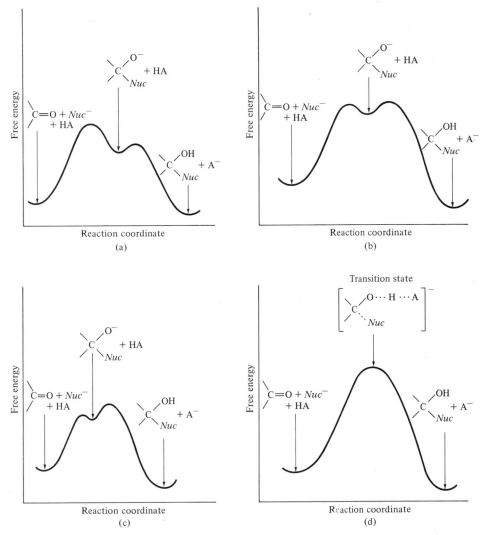

Figure 8.1 Two-dimensional reaction coordinate diagram for addition of a nucleophile $:Nuc^-$ to a carbonyl group. (a) The first step is rate determining; no acid catalysis occurs. (b) Neither step is rate determining; the overall rate shows general acid catalysis because the acid traps the intermediate by protonation. (c) The proton transfer step is rate determining; there is general acid catalysis. (d) The intermediate lifetime has become so short it ceases to exist; nucleophile attack and proton transfer are concerted. There is general acid catalysis.

rapidly to reactants. If k_{-1} becomes comparable with $k_2[HA]$, there is no longer a true rate-determining step. (See Section 2.5.) Figure 8.1(b) shows the reaction coordinate for this process. The proton transfer rate constant k_2 now enters the overall kinetics and the reaction shows general catalysis. Another way of thinking of the situation is to say that the intermediate **1** will revert rapidly to reactants unless protonation by the acid HA traps it; hence HA catalyzes product formation. Jencks has called this type of catalysis *enforced*, because it is required by the short lifetime of the intermediate.[11] The same conclusions regarding catalysis apply if k_{-1} becomes faster than $k_2[HA]$ so that the second step is rate determining [Figure 8.1(c)].

If the nucleophile becomes still weaker, **1** has such a short lifetime that it ceases to exist as a discrete intermediate. Then the proton transfer from HA to the carbonyl oxygen must occur simultaneously with nucleophile attack. The process of Scheme 1 is reduced to a single step (Equation 8.5) and the mechanism is termed a concerted proton transfer. General acid catalysis will occur. Figure 8.1(d) shows the reaction coordinate under these circumstances.[12]

$$\overset{\diagdown}{\underset{\diagup}{C}}=O + :Nuc^- + HA \rightleftharpoons \overset{\diagdown}{\underset{\diagup}{C}}\overset{OH}{\underset{Nuc}{\diagup}} + A^- \tag{8.5}$$

Simultaneous Proton Transfer and Attack of Nucleophile

The alternative that appears to offer a consistent explanation for hydration is that proton transfer occurs simultaneously with addition of the nucleophile.

In Scheme 2 the attack of the water molecule at the carbonyl carbon and transfer of a proton from the acid catalyst to the carbonyl oxygen (to which the acid proton may in some cases already be associated by hydrogen bonding[13]) occur together as the rate-determining step through transition state **2**. The rate of this

---

[11] W. P. Jencks, *Acc. Chem. Res.*, **9**, 425 (1976).

[12] For further discussion, see (a) R. E. Barnett, *Acc. Chem. Res.*, **6**, 41 (1973); (b) note 11.

[13] A possible modification of the mechanism of Scheme 2 is a preassociation of the substrate, nucleophile, and catalyst:

$$H_2O + \overset{\diagdown}{\underset{\diagup}{C}}=O + HA \underset{fast}{\overset{K_{assoc}}{\rightleftharpoons}} H_2O\cdots\overset{\diagdown}{\underset{\diagup}{C}}=O\cdots HA$$

$$H_2O\cdots\overset{\diagdown}{\underset{\diagup}{C}}=O\cdots HA \underset{k_{-1}}{\overset{k_1}{\rightleftharpoons}} \overset{\diagdown}{\underset{\underset{H}{\overset{+}{O}-H\cdots A^-}}{C}}\overset{OH}{\diagup}$$

a

The preassociation mechanism applies when the reverse of the addition step, k_{-1}, is found or estimated to be faster than the diffusion-controlled rate. The ion pair **a** must then exist in the associated state as shown, because there is not time enough for A^- to diffuse away as is implied in Scheme 2. If the most favorable pathway for the breakdown of **a** involves no diffusion of the partners, the formation of **a**, which must follow the reverse of that path, also involves no diffusion of reaction partners. See W. P. Jencks, *Acc. Chem. Res.*, **9**, 425 (1976).

SCHEME 2

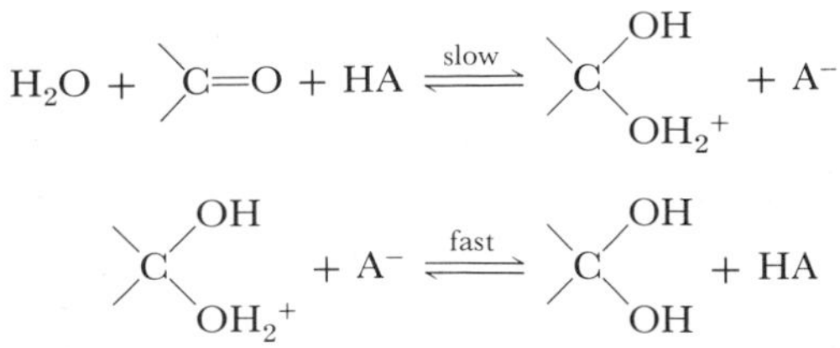

$$H_2O + \ \diagdown C{=}O + HA \ \underset{slow}{\rightleftharpoons} \ \diagup C \diagdown {\overset{OH}{\underset{OH_2^+}{}}} \ + \ A^-$$

$$\diagup C {\overset{OH}{\underset{OH_2^+}{}}} \ + \ A^- \ \underset{fast}{\rightleftharpoons} \ \diagup C {\overset{OH}{\underset{OH}{}}} \ + \ HA$$

step will depend on the nature and concentration of **HA**, and the mechanism is consistent with the observed general catalysis. It should be noted that the reverse

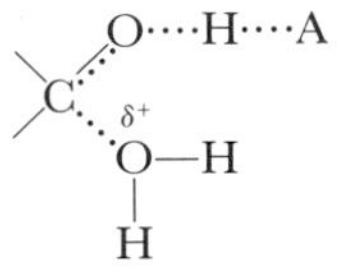

$$\diagdown C {\overset{O \cdots H \cdots A^{\delta^-}}{\underset{\underset{\underset{H}{|}}{\overset{\delta^+}{O}-H}}{}}}$$

2

process consists of a specific acid plus a general base catalysis. A possible general base catalysis mechanism is shown in Scheme 3. The reverse is a specific base plus a general acid catalysis.

SCHEME 3

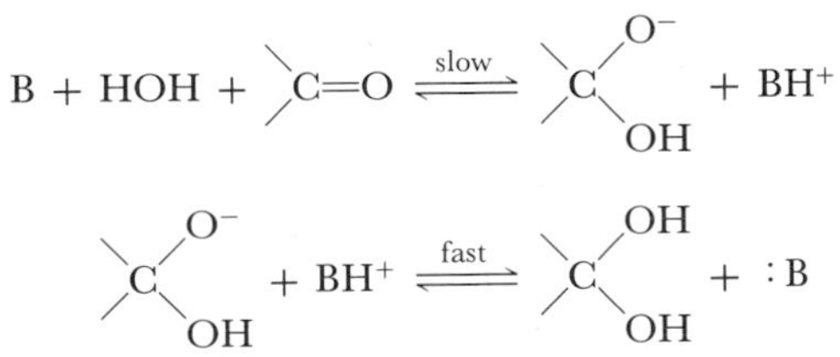

$$B + HOH + \ \diagdown C{=}O \ \underset{slow}{\rightleftharpoons} \ \diagup C {\overset{O^-}{\underset{OH}{}}} \ + \ BH^+$$

$$\diagup C {\overset{O^-}{\underset{OH}{}}} \ + \ BH^+ \ \underset{fast}{\rightleftharpoons} \ \diagup C {\overset{OH}{\underset{OH}{}}} \ + \ :B$$

The mechanism with simultaneous proton transfer and nucleophile attack helps account for the observed behavior of Brønsted catalysis law coefficient α. Eigen found that, for simple proton transfers between oxygen atoms or between oxygen and nitrogen atoms, the proton transfer rate responds as shown in Figure 8.2 to changes in relative acidity of the two acids.[14] In Reaction 8.6, suppose that the structure of acid **HA** is varied so as to change its strength, but **HA** is kept

$$HA + B^- \ \underset{k_{-1}}{\overset{k_1}{\rightleftharpoons}} \ A^- + HB \tag{8.6}$$

$$\log k_1 - \log k_{-1} = \log K = \log \frac{K_{a_{HA}}}{K_{a_{HB}}} = \Delta pK_a \tag{8.7}$$

---

[14] M. Eigen, *Angew. Chem. Int. Ed. Engl.,* **3,** 1 (1964). If the acids are of different charge types, the curve is modified somewhat.

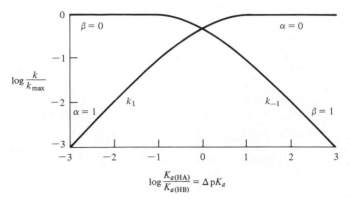

Figure 8.2 Logarithm of the rate of a simple proton transfer of the type

$$HA + B^- \underset{k_{-1}}{\overset{k_1}{\rightleftharpoons}} A^- + HB$$

as a function of the relative strength of the acids HA and HB. When equilibrium lies toward the right (log $[K_{a(HA)}/K_{a(HB)}]$ positive), k_1 is diffusion-controlled and $\alpha = 0$; when equilibrium lies toward the left (log $[K_{a(HA)}/K_{a(HB)}]$ negative), k_{-1} is diffusion-controlled and $\beta = 0$. From M. Eigen, *Angew. Chem. Int. Ed. Engl.*, **3**, 1 (1964). Reprinted by permission of Verlag Chemie, gmbh. See also J. N. Brønsted and K. Pederson, *Z. Phys. Chem.*, **108**, 185 (1924).

substantially stronger than HB. Then the rate (k_1) of proton transfer from HA to B$^-$ will be diffusion-controlled ($k_1 = k_{max} \approx 10^{10} M^{-1}$ sec^{-1}) and will not change as HA changes. Interpreted from the point of view of the Brønsted catalysis law, a rate that is independent of acid strength means that $\alpha = 0$ (k_1 curve, right-hand side of Figure 8.2). The reverse reaction in this same region of relative acid strength would be considered a base catalysis by bases A$^-$ of varying strength. Equation 8.7 shows that, with $k_1 = k_{max}$, log k_{-1} will be linearly related, with a slope of unity, to ΔpK_a when the latter is varied by changing A$^-$. Hence $\beta = 1$, as shown by the k_{-1} curve on the right side of Figure 8.2. The same reasoning, this time with HB as the stronger acid, generates the left side of the figure. Each curve changes slope from unity to zero over a relatively narrow range. These results are in accord with the interpretation based on the Hammond postulate in Section 3.3. Eigen's data suggest that in an acid-catalyzed reaction in which the rate-determining step is a simple proton transfer between oxygens, the α relating rate (k_1) to strength of catalyst ($K_{a_{HA}}$) should be either zero or unity if the catalyst, HA, is a much stronger or a much weaker acid than the protonated substrate, HB. Intermediate values of α should be found when the two are of similar strength.

Brønsted anticipated this kind of behavior when he originally proposed the catalysis law.[15] Subsequent investigators nevertheless found many examples of general acid- or base-catalyzed reactions in which α remains constant at some value between zero and unity over a range of catalysts of widely different acid strength.

[15] J. N. Brønsted and K. Pedersen, *Z. Phys. Chem.*, **108**, 185 (1924).

Carbonyl hydration is an example of such a reaction. The reason for this behavior is that in these reactions some other process is occurring simultaneously with proton transfer. Figure 8.3 gives a schematic representation of the energy surface for a concerted addition and proton transfer; Figure 8.4 is the projection of the reaction coordinate in the horizontal plane.[16] In the figures the reaction begins with carbonyl compound, nucleophile, and acid catalyst (point R) and proceeds directly to point P by simultaneous addition of nucleophile and transfer of proton, thereby avoiding the higher-energy stepwise alternatives through points Q and S.

In order to relate the Brønsted coefficient α to the reaction coordinate diagrams, we interpret α as a measure of the position of the proton at the transition state, α near zero (HA very strong) indicating an earlier transition state with little proton transfer and α near unity (HA very weak) indicating a late transition state with proton nearly completely transferred. We have already given a partial justification for this interpretation of α in the discussion in Section 3.3; we shall return to this point again later. It is nevertheless important to emphasize here that in a process involving simultaneous proton transfer and nucleophilic addition, α measures only the degree of proton transfer at the transition state (location along the back-to-front coordinate in Figure 8.3 and along the top-to-bottom coordinate in Figure 8.4) and not the degree of bonding of the nucleophile. Because the proton is partly transferred at the transition state $*_p$ in Figure 8.4, α will have a value between zero and one. Replacing the catalyzing acid HA by a stronger one is equivalent to increasing the energy along the back edge in Figure 8.3 or along the top edge in Figure 8.4. This change is shown in Figure 8.5. As we have seen in earlier analyses of diagrams of this type in Chapters 1, 4, and 7, the transition state will move along the reaction coordinate toward the configuration of increased energy and will move perpendicular to the reaction coordinate away from the configuration of increased energy. The result of these two motions is to change the reaction coordinate projection to the dashed line in Figure 8.4, with the new transition state location at $*'_p$. Note that although there is a substantial change in the structure of the transition state with respect to the nucleophile–carbonyl group distance, there is little change in the extent of proton transfer and hence little change in α. The coupling of proton transfer with nucleophile attack therefore accounts for the observed slower-than-expected change of α with change of catalyst acid strength.

The Bronsted α and β as Measures of Transition State Location

We have made use above of the idea that the magnitude of α (or β) measures the extent of proton transfer at the transition state or, equivalently, of the position of the transition state along the proton transfer reaction coordinate. Figures 8.6 and 8.7 show, respectively, the reaction coordinate diagrams for Reaction 8.8 in the

$$\mathrm{HX} + \mathrm{Y}^- \rightleftharpoons \mathrm{X}^- + \mathrm{HY} \qquad (8.8)$$

[16](a) W. P. Jencks, *Chem. Rev.*, **72**, 705 (1972); (b) D. A. Jencks and W. P. Jencks, *J. Am. Chem. Soc.*, **99**, 7948 (1977).

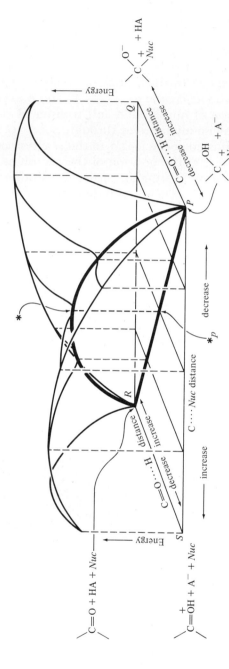

Figure 8.3 Energy surface for addition of nucleophile *Nuc* to a carbonyl group with concerted proton transfer from an acid HA. The lowest-energy path is indicated by the heavy curved line from point R to point P. Points Q and S are the high-energy intermediates of the two possible stepwise paths. The point * is the transition state. The heavy straight line is the projection of the reaction coordinate onto the horizontal plane, and *$*_p$* marks the projection of the transition state.

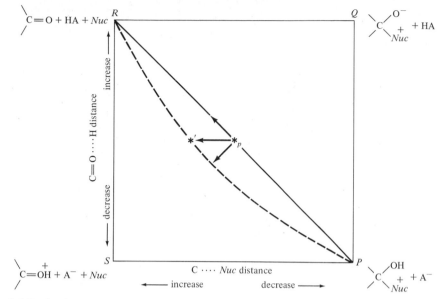

Figure 8.4 Projection in the horizontal plane of the reaction path shown in Figure 8.3 (heavy straight line). Increasing the strength of HA is equivalent to raising the energy along the top edge of the diagram; the energy surface changes as shown in Figure 8.5. The new reaction coordinate projection is the dashed line. The transition state moves along the reaction coordinate toward upper left and perpendicular to the reaction coordinate toward lower left. The net change is to the new transition state location $*'_p$.

extreme cases where HX is a much weaker or a much stronger acid than HY. In the former case (Figure 8.6), $x_{\ddagger}$ close to unity, small structural changes that alter the free energy of products, G_p°, will cause a similar change in transition state free energy, $G^{\ddagger}$, whereas changes in reactant free energy, G_r°, will have little effect. This case corresponds to $\alpha = 1$, $\beta = 0$ in Figure 8.2, that is, reverse reaction diffusion-controlled. Figure 8.7 depicts the opposite extreme, $x_{\ddagger}$ close to zero, where $G^{\ddagger}$ depends on G_r° and not G_p°; this case corresponds to forward reaction diffusion-controlled, $\alpha = 0$ and $\beta = 1$ in Figure 8.2. In the intermediate region (center part of Figure 8.2), α varies from 1 to 0 as the transition state moves from being near products (Figure 8.6) to near reactants (Figure 8.7). If we let the symbol δ be an operator designating change in a thermodynamic quantity caused by structural change in the molecules involved,[17] the predictions of the diagrams may be roughly quantified by using α and β as the parameters relating $G^{\ddagger}$ to G_r° and G_p° as shown in Equation 8.9, where $0 < \alpha < 1$ and $0 < \beta < 1$.[18] If we also assume that $\beta = 1 - \alpha$ (see Figure 8.2) and remember that $\Delta G^{\ddagger} = G^{\ddagger} - G_r^{\circ}$ and $\Delta G^{\circ} = G_p^{\circ} - G_r^{\circ}$, we obtain Equation 8.10 by subtracting δG_r° from both sides of Equation

[17]J. E. Leffler and E. Grunwald, *Rates and Equilibria of Organic Reactions*, Wiley, New York, 1963, p. 26.
[18](a) J. E. Leffler and E. Grunwald, *Rates and Equilibria of Organic Reactions*, Wiley, New York, 1963, p. 156; (b) M. Bender, *Mechanisms of Homogeneous Catalysis from Protons to Proteins*, Wiley, New York, 1971, p. 85.

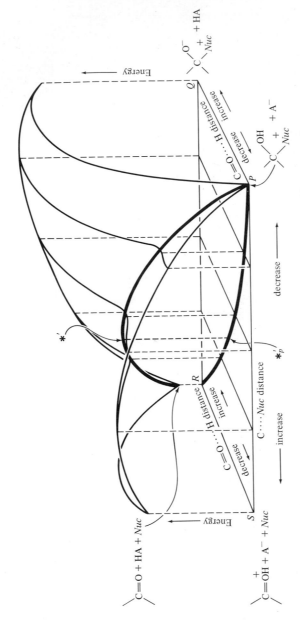

Figure 8.5 Energy surface for addition of nucleophile *Nuc* to a carbonyl group with concerted proton transfer from an acid HA that is stronger than that depicted in Figure 8.3. The heavy dashed line is the projection in the horizontal plane of the new reaction coordinate. This line corresponds to the dashed line in Figure 8.4.

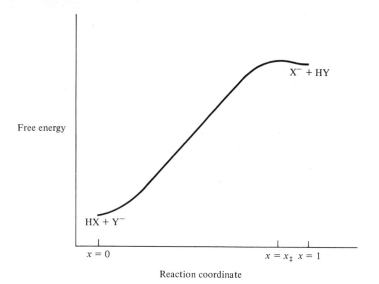

Figure 8.6 Reaction coordinate diagram for the proton transfer $HX + Y^- \rightleftharpoons X^- + HY$, where HX is a weaker acid than HY.

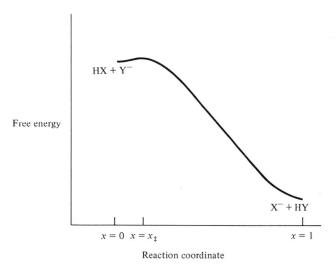

Figure 8.7 Reaction coordinate diagram for the proton transfer $HX + Y^- \rightleftharpoons X^- + HY$, where HX is a stronger acid than HY.

$$\delta G^{\ddagger} = \alpha \ \delta G_p^{\circ} + \beta \ \delta G_r^{\circ} \tag{8.9}$$

8.9 and find relation 8.12 between activation free-energy, $\Delta G^{\ddagger}$, and standard free-energy change, ΔG°. This equation is equivalent to the Brønsted catalysis law as was shown in Section 3.3 (Equations 3.69 and 3.73).

$$\delta G^{\ddagger} - \delta G_r^{\circ} = \alpha \ \delta G_p^{\circ} + (1 - \alpha) \delta G_r^{\circ} - \delta G_r^{\circ} \tag{8.10}$$

$$\delta G^{\ddagger} - \delta G_r^{\circ} = \alpha \{ \delta G_p^{\circ} - \delta G_r^{\circ} \} \qquad (8.11)$$

$$\delta \Delta G^{\ddagger} = \alpha \ \delta \Delta G^{\circ} \qquad (8.12)$$

When interpreted in this way as a rough quantitative measure of degree of proton transfer, the Brønsted coefficient α can be applied directly as a parameter for the proton transfer coordinate of reaction coordinate diagrams (the vertical axis in Figure 8.4). Jencks has proposed that a parameter for the nucleophile attack axis (horizontal axis in Figure 8.5) is β_{Nuc}, derived by analogy with the Brønsted catalysis law as the slope of a plot of reaction rate with various nucleophiles against the pK_a of the nucleophile conjugate acids, Equation 8.13.[19] This relationship cannot

$$\log k_{Nuc} = \beta_{Nuc} \log K_{a_{NucH^+}} + \log C'' \qquad (8.13)$$

be expected to hold over a broad range of nucleophiles because nucleophilicity toward carbon does not follow basicity toward protons; however, for a series of nucleophiles of similar structural type, and particularly with the same nucleophilic atom, it should be satisfactory. With a semiquantitative parameter available for each coordinate, it is possible to use experimental data to locate transition states roughly on the surface, assess their structures, and determine how those structures change with changing substitution patterns and conditions.[19]

Although the interpretation of the Brønsted α given above is a reasonable one, it may not apply to all proton transfers. There are instances, though not of the carbonyl addition type, in which α falls outside the range zero to unity. An example is deprotonation and protonation of nitroalkanes.[20] The explanation for the abnormal behavior of these reactions may lie in the particular features of charge and solvent reorganization that accompany them.[21]

The Mechanistic Ambiguity in General Catalysis

We have assumed that the observation of general acid catalysis implies proton transfer from acid catalyst to substrate in the rate-determining step (Mechanism I, Scheme 4). Mechanism II in Scheme 4 shows that a preliminary fast equilibrium yielding the conjugate acid of the substrate, followed by a rate-determining step in which a proton is transferred from the protonated substrate to the conjugate base of the catalyst, predicts the same kinetic dependence on substrate, catalyst, and nucleophile concentrations. Furthermore, the catalysis law shows that when HA is changed, the observed rate constant of Mechanism I, k_1, is proportional to $K_{a_{HA}}^{\alpha}$. In Mechanism II, k_1' is proportional to $K_{a_{HA}}^{-\beta}$ whereas K is equal to $K_{a_{HA}}/K_{a_{SH^+}}$, and the observed rate constant, $k_1'K$, is therefore proportional to $K_{a_{HA}}^{1-\beta} = K_{a_{HA}}^{\alpha}$. The mechanisms therefore also predict the same dependence of rate on strength of the acid catalyst. The fundamental reason for this kinetic equivalence is that the stoichiometric composition of the transition states in the rate-determining steps are the

[19] D. A. Jencks and W. P. Jencks, *J. Am. Chem. Soc.*, **99**, 7948 (1977).
[20] F. G. Bordwell, W. J. Boyle, Jr., and K. C. Yee, *J. Am. Chem. Soc.*, **92**, 5926 (1970).
[21] J. R. Keefe, J. Morey, C. A. Palmer, and J. C. Lee, *J. Am. Chem. Soc.*, **101**, 1295 (1979).

SCHEME 4[22]

Mechanism I (type e)

$$\text{C=O} + \text{HA} + \text{H}_2\text{O} \xrightarrow[\text{slow}]{k_1} \text{C} \begin{smallmatrix}\text{OH} \\ \text{OH}_2^+\end{smallmatrix} + \text{A}^-$$

(PH⁺)

$$\text{C} \begin{smallmatrix}\text{OH} \\ \text{OH}_2^+\end{smallmatrix} + \text{A}^- \underset{\text{fast}}{\rightleftharpoons} \text{C} \begin{smallmatrix}\text{OH} \\ \text{OH}\end{smallmatrix} + \text{HA}$$

(PH⁺) → (P)

rate $= k_1[\text{S}][\text{HA}][\text{H}_2\text{O}]$

Mechanism II (type n)

$$\text{C=O} + \text{HA} \underset{\text{fast}}{\overset{K}{\rightleftharpoons}} \text{C=OH}^+ + \text{A}^-$$

(S) → (SH⁺)

$$\text{C=OH}^+ + \text{A}^- + \text{H}_2\text{O} \xrightarrow[\text{slow}]{k_1'} \text{C} \begin{smallmatrix}\text{OH} \\ \text{OH}\end{smallmatrix} + \text{HA}$$

(SH⁺) → (P)

rate $= k_1'[\text{SH}^+][\text{A}^-][\text{H}_2\text{O}]$

$$\text{SH}^+ = K\frac{[\text{S}][\text{HA}]}{[\text{A}^-]}$$

rate $= k_1'K[\text{S}][\text{HA}][\text{H}_2\text{O}]$

same in both mechanisms. We may summarize this conclusion by stating that general acid catalysis is not distinguishable by kinetic measurements alone from specific acid plus general base catalysis. The reader may show by similar reasoning (Problem 13) that general base catalysis cannot be distinguished from specific base plus general acid catalysis. Jencks has designated Mechanism I, in which the catalyst donates a proton to the carbonyl compound (electrophile) in the rate-determining step, as type e, and Mechanism II, in which the catalyst accepts a proton from the nucleophile in the rate-determining step, as type n.[23]

One possible method of deciding between Mechanisms I and II is to look at the trend of α in acid-catalyzed additions of various nucleophiles to a carbonyl group.[24] In true general acid catalysis (Mechanism I, type e), the sensitivity of the rate to acidity of the catalyst, and therefore also α, should decrease as the species adding is made more nucleophilic. The reason is that this variation will cause the change in reaction coordinate shown in Figure 8.8, so that in the transition state the proton will be transferred from the catalyst to a smaller extent and the acidity of the catalyst will not be so strongly felt. If, on the other hand, the reaction is actually specific acid- plus general base-catalyzed, (Mechanism II, type n), then analysis of Figure 8.9 shows that the sensitivity of rate to basicity of A^-, and therefore also β, should decrease as the adding species is made more nucleophilic. But if this latter alternative were the correct mechanism and the reaction were erroneously regarded as a true general acid catalysis, one would find experimentally that $\alpha (= 1 - \beta)$ would increase with more nucleophilic adding reagents.

Although making large variations in the nucleophile necessarily introduces uncertainties, the experimental evidence favors the conclusion that the addition of

[22] The mechanisms are simplified by neglecting reverse reactions in the slow steps.
[23] W. P. Jencks, *Acc. Chem. Res.*, **9**, 425 (1976).
[24] W. P. Jencks, *Catalysis in Chemistry and Enzymology*, McGraw-Hill, New York, 1969, pp. 195–197.

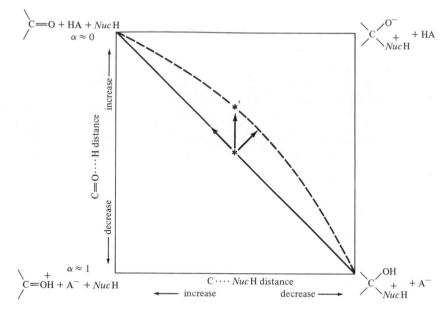

Figure 8.8 Projection in the horizontal plane of the three-dimensional reaction coordinate for the rate-determining step in Mechanism I, Scheme 4. Increasing the nucleophilicity of *Nuc* (equivalent to raising the energy of the left edge) will shift the transition state to *′ and decrease α.

nucleophiles to carbonyl groups follows the true general acid catalysis mechanism, type e. Table 8.2 presents selected data; α decreases with increasing nucleophilicity of the addend.

Another method of removing the mechanistic ambiguity, applicable to specific reactions for which sufficient rate and equilibrium data are available, is to show that one or the other of the alternative mechanisms could produce product at the observed rate only if the rate of one of its steps were to exceed the diffusion-controlled limit. For example, in Scheme 4, the rate of the forward reaction is characterized by an experimentally determined rate constant that is equal to k_1 if Mechanism I (type e) holds and to $k_1'K$ if Mechanism II holds. Jencks and co-workers have shown that for acetic acid catalysis of formaldehyde hydration this constant is $57 M^{-1} \sec^{-1}$, whereas K is less than about 2.9×10^{-8}; therefore if Mechanism II (type n) were correct, k_1' would have to be greater than $57/2.9 \times 10^{-8} = 2 \times 10^9 M^{-1} \sec^{-1}$, and the corresponding value for catalysis by boric acid would have to be greater than $10^{11} M^{-1} \sec^{-1}$. The latter value (and probably the former also, when allowance is made for the fact that it is only a lower limit) exceeds the diffusion-controlled rate of about $10^{10} M^{-1} \sec^{-1}$ and this argument thus provides strong support for Mechanism I.[25]

A third method of resolving ambiguity in general acid- or base-catalyzed reactions is to compare the process of interest with an analog in which one protona-

[25]L. H. Funderburk, L. Aldwin, and W. P. Jencks, *J. Am. Chem. Soc.*, **100**, 5444 (1978).

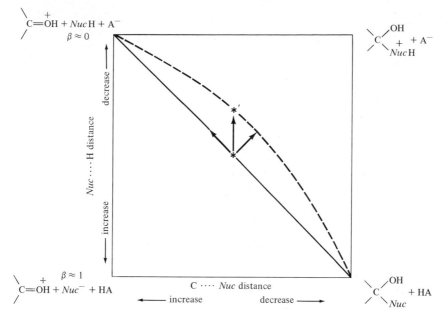

Figure 8.9 Projection in the horizontal plane of the three-dimensional reaction coordinate for the rate-determining step in Mechanism II, Scheme 4. Increasing the nucleophilicity, again equivalent to raising the energy of the left edge, will shift the transition state to $*'$ and decrease β. Because $\alpha = 1 - \beta$, this change is equivalent to increasing α.

Table 8.2 DEPENDENCE OF BRØNSTED COEFFICIENT α ON BASICITY OF THE NUCLEOPHILE IN ADDITIONS TO C=O[a]

Nucleophile	$pK_{a(NucH+)}$	Addition to	α
HOO⁻	11.6	$ClC_6H_4\overset{\overset{O}{\|\|}}{C}H$	0
RS⁻	10	$CH_3\overset{\overset{O}{\|\|}}{C}H$	0
$\phi NHNH_2$	5.2	$\phi\overset{\overset{O}{\|\|}}{C}H$	0.2
$H_2N\overset{\overset{O}{\|\|}}{C}NH_2$	0.2	$CH_3\overset{\overset{O}{\|\|}}{C}H$	0.45
RSH	−7	$CH_3\overset{\overset{O}{\|\|}}{C}H$	0.7
HOOH	−7	$ClC_6H_4\overset{\overset{O}{\|\|}}{C}H$	1.0

[a] Data from W. P. Jencks, *Catalysis in Chemistry and Enzymology*, McGraw-Hill, New York, 1969, p. 198, where a more extensive table may be found. Reproduced by permission of McGraw-Hill.

tion site is blocked by a methyl group so that only one of the alternatives is possible. This method is illustrated in the discussion of sulfite addition in the next section and of enolization in Section 8.6.

8.2 OTHER SIMPLE ADDITIONS

The previous section considered the simple single-step addition of water to the carbonyl group. Certain other nucleophiles undergo similar simple additions.

Addition of Cyanide and Sulfite

Aldehydes and unhindered aliphatic ketones or arylalkyl ketones add hydrogen cyanide to form cyanohydrins (Equation 8.14). As with hydration, the equilibrium

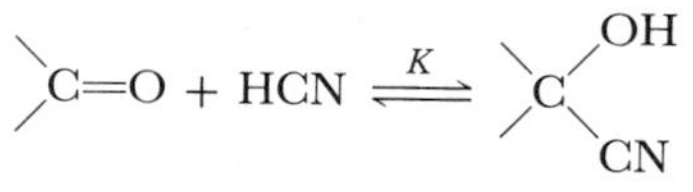

$$\text{\Large$\diagup$}\!\!\diagdown\!\!C{=}O + HCN \; \underset{}{\overset{K}{\rightleftharpoons}} \; C\!\!\diagdown_{CN}^{\,OH} \tag{8.14}$$

lies farther to the right for aldehydes than for ketones. The equilibrium constant K is decreased by electron-donating groups, which stabilize the electron-deficient carbonyl carbon, and by the presence of bulky groups, which will be pushed closer together by the change to tetrahedral cyanohydrin. Table 8.3 gives selected equilibrium constants. The reaction is of considerable synthetic utility, since the cyano group is readily hydrolyzed to yield an α-hydroxy acid.[26]

Cyanide addition was one of the first organic reactions to be elucidated mechanistically. In 1903 Lapworth found that whereas a high concentration of undissociated HCN is desirable to assure a high yield of cyanohydrin, the reaction rate depends on concentration of cyanide ion according to Equation 8.15[27] His mechanism was essentially the same as the one accepted today (Scheme 5).[28]

$$\text{rate} = k \left[\text{\Large$\diagup$}\!\!\diagdown\!\!C{=}O \right] [^{-}CN] \tag{8.15}$$

SCHEME 5

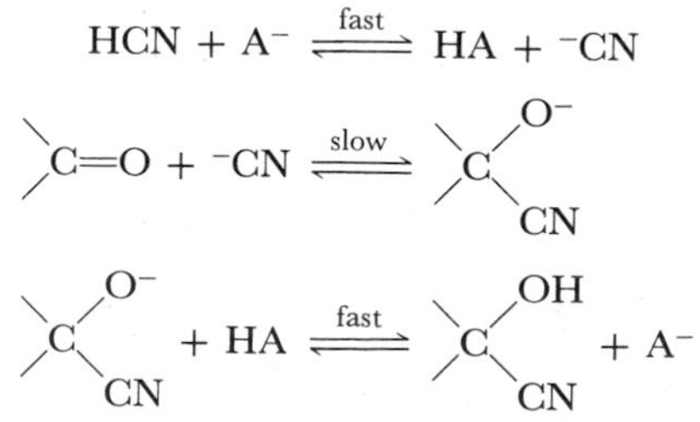

$$HCN + A^{-} \underset{}{\overset{\text{fast}}{\rightleftharpoons}} HA + {}^{-}CN$$

$$\text{\Large$\diagup$}\!\!\diagdown\!\!C{=}O + {}^{-}CN \underset{}{\overset{\text{slow}}{\rightleftharpoons}} C\!\!\diagdown_{CN}^{\,O^{-}}$$

$$C\!\!\diagdown_{CN}^{\,O^{-}} + HA \underset{}{\overset{\text{fast}}{\rightleftharpoons}} C\!\!\diagdown_{CN}^{\,OH} + A^{-}$$

---

[26]HCN adds 1,4 to $\alpha\beta$-unsaturated aldehydes and ketones. See W. Nagata and M. Yoshioka, *Org. Reactions,* **25,** 255 (1977).

[27]A. Lapworth, *J. Chem. Soc.,* **83,** 995 (1903).

[28]The observation of a low yield of chiral cyanohydrin when certain optically active amines are present requires a minor modification of the mechanism to allow for coordination of the carbonyl oxygen with a cation. See (a) V. Prelog and M. Wilhelm, *Helv. Chim. Acta,* **37,** 1634 (1954); (b) H. Hustedt and E. Pfeil, *Justus Liebigs Ann. Chem.,* **640,** 15 (1961).

Table 8.3 EQUILIBRIUM CONSTANTS AT 20°C IN 96 PERCENT ETHANOL FOR THE REACTION

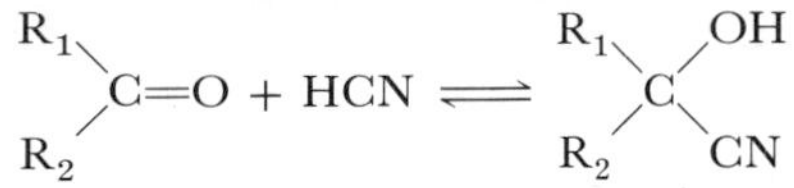

$$\underset{R_2}{\overset{R_1}{\diagdown}}C=O + HCN \rightleftharpoons \underset{R_2}{\overset{R_1}{\diagdown}}\underset{CN}{\overset{OH}{\diagup}}C$$

R_1	R_2	K	Reference
C_6H_5	H	220, 236	a, b
$p\text{-}CH_3\text{—}C_6H_5$	H	110, 114	a, b
$p\text{-}CH_3O\text{—}C_6H_5$	H	32.7	b
$p\text{-}NO_2\text{—}C_6H_5$	H	1820	b
C_6H_5	CH_3	0.77	c
CH_3	CH_3	33	c
CH_3	C_2H_5	38	c
CH_3	$i\text{-}C_3H_7$	65	c
CH_3	$t\text{-}C_4H_9$	32	c
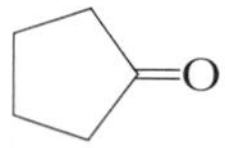 (cyclopentanone)		48	d
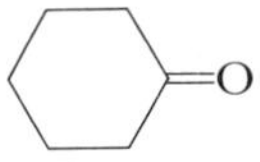 (cyclohexanone)		1000	d
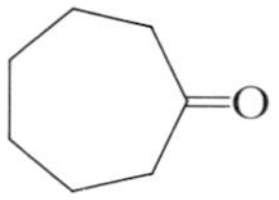 (cycloheptanone)		7.8	d
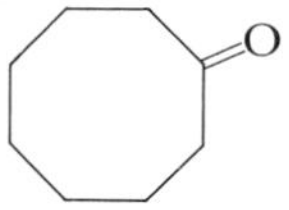 (cyclooctanone)		1.17	d

[a] J. W. Baker, *Tetrahedron*, **5**, 135 (1959).
[b] W.-M. Ching and R. G. Kallen, *J. Am. Chem. Soc.*, **100**, 6119 (1978) (Aqueous solution at 25°C.)
[c] A. Lapworth and R. H. F. Manske, *J. Chem. Soc.*, 1976 (1930).
[d] V. Prelog and M. Kobelt, *Helv. Chim. Acta*, **32**, 1187 (1949).

Cyanide is a strong nucleophile. The reverse of the addition step is slow compared with the final protonation, so no catalysis of the addition step is required, and no general catalysis is observed.[29] (The second and third steps in Scheme 5 correspond to the reaction coordinate shown in Figure 8.1(a).) There is, however, specific base catalysis that arises because of the preliminary equilibrium that produces ⁻CN.

Sulfite ions also undergo simple addition to aldehydes and unhindered ketones. The equilibrium constants (Table 8.4) are again larger for aldehydes than for ketones; equilibrium constants correlate with the Taft σ^* parameters.[30]

At low pH the reaction is subject to general acid catalysis in the direction of breakdown of the adduct to bisulfite and carbonyl compound. The mechanism of

[29] (a) W. J. Svirbely and J. F. Roth, *J. Am. Chem. Soc.*, **75**, 3106 (1953); (b) W. J. Svirbely and F. H. Brock, *J. Am. Chem. Soc.*, **77**, 5789 (1955); (c) W.-M. Ching and R. G. Kallen, *J. Am. Chem. Soc.*, **100**, 6119 (1978).
[30] K. Arai, *Nippon Kagaku Zasshi*, **82**, 955 (1961) [*Chem. Abstr.*, **56**, 5623g (1962)].

Table 8.4 EQUILIBRIUM CONSTANTS AT
0°C FOR THE REACTION

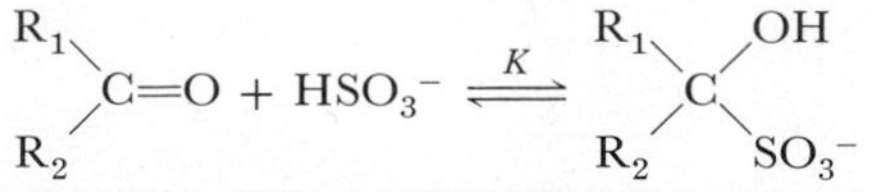

$$\underset{R_2}{\overset{R_1}{>}}C=O + HSO_3^- \overset{K}{\rightleftharpoons} \underset{R_2}{\overset{R_1}{>}}C\underset{SO_3^-}{\overset{OH}{<}}$$

R_1	R_2	K^a
H	CH_3	800
H	C_6H_5	1200^b
CH_3	CH_3	200
CH_3	C_2H_5	40
CH_3	$i\text{-}C_3H_7$	8
CH_3	$t\text{-}C_4H_9$	1.6

^{a} Calculated from data of K. Arai, *Nippon Kagaku Zasshi*, **82**, 955 (1961) [*Chem. Abstr.*, **56**, 5623g (1962)].
^{b} At pH 7.5. F. C. Kokesh and R. E. Hall, *J. Org. Chem.*, **40**, 1632 (1975).

the reaction in this direction is true general acid catalysis, type n (Scheme 6) rather than the kinetically equivalent alternative, type e. (See Problem 3.) This conclusion is reached on the basis of the close similarity of Brønsted α, kinetic isotope effects, and substituent effects between breakdown of bisulfite addition compound and the reaction of the methoxy analog, Scheme 7.[31]

SCHEME 6

Type n

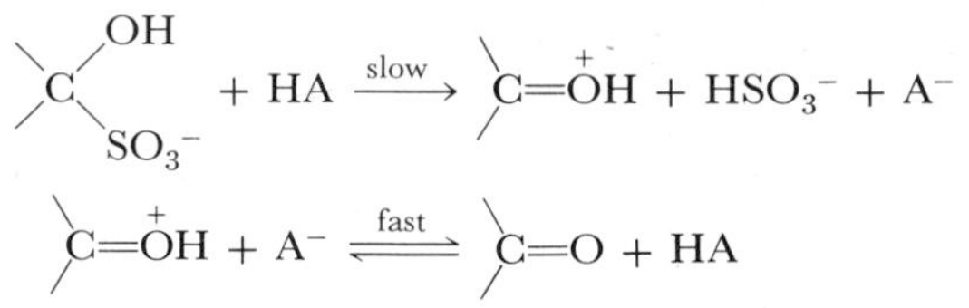

$$\underset{SO_3^-}{\overset{OH}{>}}C{<} + HA \overset{slow}{\longrightarrow} {>}C{=}\overset{+}{O}H + HSO_3^- + A^-$$

$${>}C{=}\overset{+}{O}H + A^- \overset{fast}{\rightleftharpoons} {>}C{=}O + HA$$

SCHEME 7

$$\underset{SO_3^-}{\overset{OCH_3}{>}}C{<} + HA \overset{slow}{\longrightarrow} {>}C{=}\overset{+}{O}CH_3 + HSO_3^- + A^-$$

$${>}C{=}\overset{+}{O}CH_3 \overset{H_2O}{\underset{fast}{\longrightarrow}} {>}C{=}O + CH_3OH$$

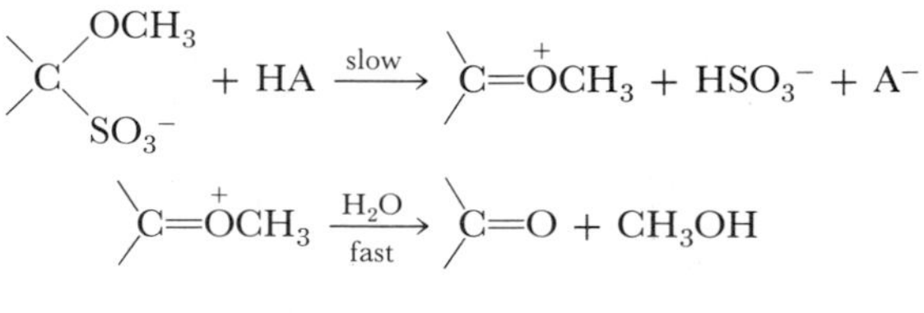

Above pH 3 the mechanism shown in Scheme 8 applies. The first step is a specific base catalysis in which the hydroxyl proton is removed. There is a small

[31] (a) P. R. Young and W. P. Jencks, *J. Am. Chem. Soc.*, **100**, 1228 (1978). See also (b) P. R. Young and W. P. Jencks, *J. Am. Chem. Soc.*, **101**, 3288 (1979).

degree of catalysis of the second (rate-determining) step by hydrogen bonding to BH^+, manifest in a Brønsted β of 0.94.[32]

SCHEME 8

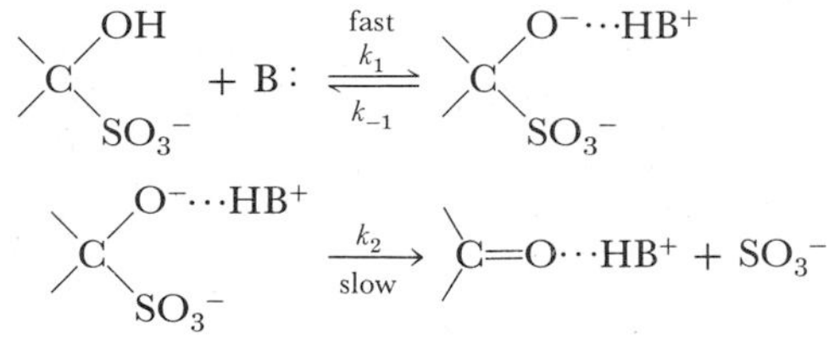

The sulfite adducts have been demonstrated by synthesis,[33] Raman spectroscopy,[34] and isotope effect measurements[35] to have the hydroxy sulfonic acid structure (**3**) rather than the alternative hydroxy sulfite ester structure (**4**). Note that the softer sulfur center rather than the harder oxygen of the ambident sulfite prefers to be bonded to carbon.

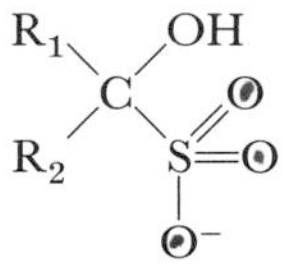

3 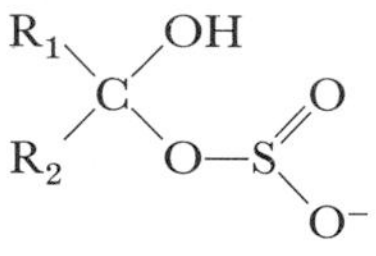   **4**

Addition of Organometallics and Hydrides. Reductions and Oxidations

The compositions of Grignard solutions and of organolithium compounds have been covered briefly in Section 6.1. Although the detailed structure of the reagents, in particular, the nature and degree of association, will vary from one compound to another and from one solvent to another,[36] the organomagnesium and organo-lithium compounds consist essentially of a strong, soft Lewis base coordinated to a hard metal ion Lewis acid.[37] Combination with a carbonyl compound (Equation 8.16) yields a bonding situation of so much lower energy that equilibrium constants are usually very large and the additions are for practical purposes irreversible. Acid–base catalysis of the type we have been considering is clearly out of the

---

[32] P. R. Young and W. P. Jencks, *J. Am. Chem. Soc.*, **99**, 1206 (1977). The known rate constants $k_{-1} \approx 4.6 \times 10^6 \ \text{sec}^{-1}$, $k_2 \approx 4.4 \times 10^5 \ \text{sec}^{-1}$, and $k_1 \approx 6 \times 10^9 M^{-1} \ \text{sec}^{-1}$ (B: = OH^-) show that the first step is not rate determining. Hence catalysis must come in the second step, where no actual proton transfer occurs. Stabilization of the transition state by weak hydrogen bonding to BH^+ explains the observations.

[33] (a) W. M. Lauer and C. M. Langkammerer, *J. Am. Chem. Soc.*, **57**, 2360 (1935); (b) R. L. Shriner and A. H. Land, *J. Org. Chem.*, **6**, 888 (1941).

[34] C. N. Caughlan and H. V. Tarter, *J. Am. Chem. Soc.*, **63**, 1265 (1941).

[35] W. A. Sheppard and A. N. Bourns, *Can. J. Chem.*, **32**, 4 (1954).

[36] (a) E. C. Ashby, *Q. Rev. Chem. Soc.*, **21**, 259 (1967); (b) E. C. Ashby, J. Laemmle, and H. M. Neumann, *Acc. Chem. Res.*, **7**, 272 (1974). See Section 6.1 for further discussion.

[37] In some other organometallics, for example, the soft-acid–soft-base organomercury compounds, the carbon–metal bond has a high degree of covalent character and is sufficiently strong that these substances are considerably less reactive than substances of the organolithium or organomagnesium type.

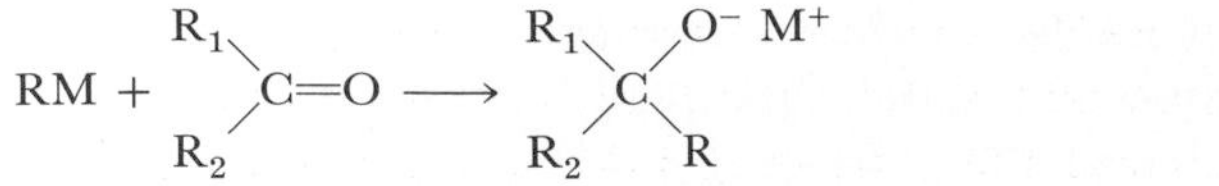

$$RM + \begin{array}{c} R_1 \\ \diagdown \\ R_2 \diagup \end{array}C{=}O \longrightarrow \begin{array}{c} R_1 \\ \diagdown \\ R_2 \diagup \end{array}\underset{R}{\overset{O^- \ M^+}{C}} \qquad (8.16)$$

question here, as the reactions must be conducted under rigorously aprotic conditions if the organometallic reagent is not to be destroyed.

The addition of Grignard reagents to ketones has been studied in some detail, notably by Ashby and co-workers. For some combinations of Grignard and ketone, the mechanism one would expect by analogy with other ketone–nucleophile additions holds. The alkyl group of the RMgX reagent attacks the carbonyl carbon in a polar mechanism to yield directly a magnesium alkoxide (Equation 8.17). This process occurs, for example, when CH_3MgBr reacts with acetone.[38]

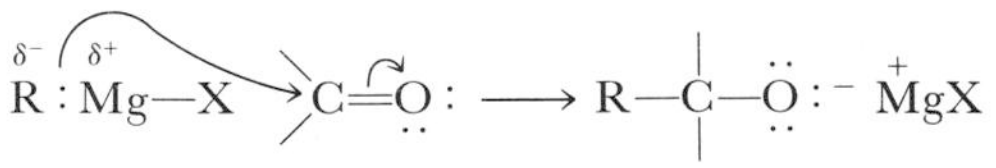

$$\overset{\delta^-}{R}{:}\overset{\delta^+}{Mg}{-}X \underset{}{\overset{}{\diagdown}} C{\overset{\frown}{=}}\ddot{O}: \longrightarrow R{-}\overset{|}{\underset{|}{C}}{-}\ddot{O}{:}^- \ \overset{+}{Mg}X \qquad (8.17)$$

There is an alternative mechanism (Scheme 9), that operates, for example, in the reaction of $(CH_3)_3CMgBr$ with benzophenone and with other conjugated ketones.[39] This process, termed the single-electron transfer mechanism, is a radical reaction in which the Grignard transfers only one electron to the carbonyl group.

SCHEME 9

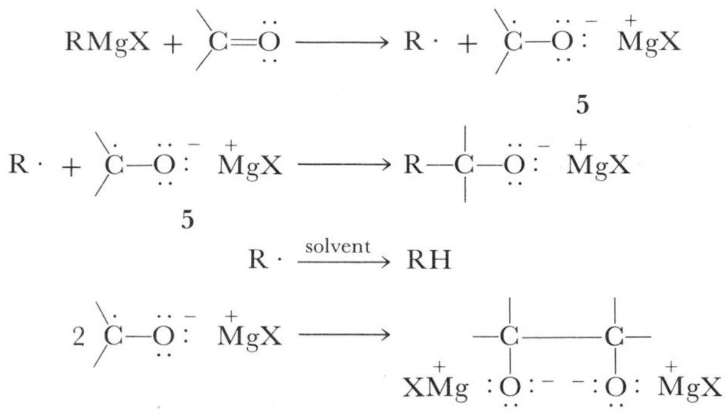

$$RMgX + \begin{array}{c} \diagdown \\ \diagup \end{array}C{=}\ddot{O} \longrightarrow R\cdot + \begin{array}{c} \diagdown \\ \diagup \end{array}\overset{\cdot}{C}{-}\ddot{O}{:}^- \ \overset{+}{Mg}X$$

$$\mathbf{5}$$

$$R\cdot + \begin{array}{c} \diagdown \\ \diagup \end{array}\overset{\cdot}{C}{-}\ddot{O}{:}^- \ \overset{+}{Mg}X \longrightarrow R{-}\overset{|}{\underset{|}{C}}{-}\ddot{O}{:}^- \ \overset{+}{Mg}X$$

$$\mathbf{5}$$

$$R\cdot \xrightarrow{\text{solvent}} RH$$

$$2\begin{array}{c} \diagdown \\ \diagup \end{array}\overset{\cdot}{C}{-}\ddot{O}{:}^- \ \overset{+}{Mg}X \longrightarrow \begin{array}{cc} | & | \\ {-}C{-}\!\!-\!\!-C{-} \\ | & | \end{array}$$
$$\overset{+}{X}Mg \ {:}\ddot{O}{:}^- \ \ ^-{:}\ddot{O}{:} \ \overset{+}{Mg}X$$

The radical $R\cdot$ and the ketyl radical anion (**5**) with its coordinate magnesium cation are left paired in a solvent cage.[40] If the radicals simply combine with each other, the end product is the same alkoxide as produced by the polar mechanism; but the radicals may also escape and react in other ways to produce products from hydrogen abstraction and from coupling. The single-electron transfer process thus accounts for some of the typical side products found in Grignard reactions. Transition metal ions, particularly the iron present as an impurity in the magnesium used to prepare the Grignard reagents, catalyze the single-electron transfer.[41]

[38] E. C. Ashby and J. S. Bowers, Jr., *J. Am. Chem. Soc.*, **99**, 8504 (1977).
[39] (a) I. G. Lopp, J. D. Buhler, and E. C. Ashby, *J. Am. Chem. Soc.*, **97**, 4966 (1975); (b) E. C. Ashby and T. L. Wiesemann, *J. Am. Chem. Soc.*, **100**, 3101 (1978).
[40] Radical cage reactions are discussed in Section 9.2.
[41] E. C. Ashby, J. D. Buhler, I. G. Lopp, T. L. Wiesemann, J. S. Bowers, Jr., and J. T. Laemmle, *J. Am. Chem. Soc.*, **98**, 6561 (1976).

Other side reactions occur in Grignard additions. For example, if the carbonyl compound bears α hydrogens, an enolate ion may form (Equation 8.18); the negative charge then prevents addition. Another possibility is reduction of the carbonyl group by transfer of a hydrogen from the Grignard reagent (Equation 8.19). Reduction also occurs by the action of a small amount of magnesium hydride produced during the formation of the Grignard reagent.[42]

$$RM + \overset{H}{\underset{|}{\underset{\diagup}{\overset{\diagup}{C}}}}{-}C{\overset{\diagup O}{\diagdown}} \longrightarrow \left[\overset{\diagup}{\diagdown}C{=}C{\overset{\diagup O^-}{\diagdown}} \longleftrightarrow \overset{\diagup}{\diagdown}\overset{..}{C}{-}C{\overset{\diagup O}{\diagdown}} \right] M^+ + RH \qquad (8.18)$$

$$\overset{H}{\underset{|}{\underset{\diagup}{\overset{\diagup}{C}}}}{-}\overset{|}{\underset{|}{C}}{-}M + \overset{\diagdown}{\diagup}C{=}O \longrightarrow \overset{\diagdown}{\diagup}C{=}C{\overset{\diagup}{\diagdown}} + \overset{\diagup}{\diagdown}C{\overset{O^- M^+}{\underset{H}{\diagdown}}} \qquad (8.19)$$

The reactions of carbonyl compounds with complex metal hydrides, particularly lithium aluminum hydride, sodium borohydride, and their alkoxy derivatives, have received a great deal of attention because of their importance in synthesis.[43] Although the mechanisms are not yet understood in detail, it seems likely that, in analogy to other reactions of carbonyl groups with nucleophiles, the basic process consists of transfer of hydride to the carbonyl carbon. The reactions are complicated by aggregation and ion pairing of the hydride reagents in ether solvents, by the role of the cation in the addition reaction, and by the occurrence of a sequence of steps as the four available hydrides on each aluminohydride or borohydride ion react with four moles of carbonyl compound.

In diethyl ether $LiAlH_4$ is present largely as a dimer at concentrations above about $0.3M$ and as contact ion pairs at low concentration. In tetrahydrofuran (THF) it is in the form of ion triplets ($Li^+AlH_4^-Li^+$ and $AlH_4^-Li^+AlH_4^-$) at high concentration and solvent-separated ion pairs at low concentration.[44] The kinetics of ketone reduction by $LiAlH_4$ in diethylether[45] and in THF[46] are consistent with a bimolecular reaction between hydride ion pairs and ketone.[47] Comparison of rates between $LiAlH_4$ and $LiAlD_4$ yields isotope effects k_H/k_D on the order of 1.3; the interpretation of this result is complicated because it is an aggregate of a primary effect [from the H (D) being transferred] and secondary effect [from the H's (D's) remaining on the hydride], but it appears consistent with hydride transfer in a rate-determining step with an unsymmetrical (probably early) transition state.

[42] E. C. Ashby, T. L. Wiesemann, J. S. Bowers, Jr., and J. T. Laemmle, *Tetrahedron Lett.*, 21 (1976). See also note 41.

[43] (a) Oxidations and reductions are considered in detail by H. O. House, *Modern Synthetic Reactions*, 2d ed., W. A. Benjamin, Menlo Park, Calif., 1972; (b) reductions with borohydrides and aluminohydrides are discussed by H. C. Brown, *Boranes in Organic Chemistry*, Cornell University Press, Ithaca, N.Y., 1972; (c) O. H. Wheeler, in *The Chemistry of the Carbonyl Group*, Vol. 1, S. Patai, Ed., Wiley, London, 1966, Chap. 11 (reductions); (d) N. G. Gaylord, *Reduction with Complex Metal Hydrides*, Wiley, New York, 1956; (e) E. R. H. Walker, *Chem. Soc. Rev.*, **5**, 23 (1976) (selective complex metal hydrides); (f) C. F. Lane, *Chem. Rev.*, **76**, 773 (1976) (reductions with boranes); (g) E. C. Ashby, J.-J. Lin, and A. B. Goel, *J. Org. Chem.*, **43**, 183 (1978) (copper hydrides).

[44] E. C. Ashby, F. R. Dobbs, and H. P. Hopkins, Jr., *J. Am. Chem. Soc.*, **97**, 3158 (1975).

[45] K. E. Wiegers and S. G. Smith, *J. Am. Chem. Soc.*, **99**, 1480 (1977).

[46] E. C. Ashby and J. R. Boone, *J. Am. Chem. Soc.*, **98**, 5524 (1976).

[47] In diethylether the dimeric nature of the hydride leads to reaction kinetics that are order 0.5 in $LiAlH_4$. See Problem 11.

The cation clearly plays an important role. Rates are different for $LiAlH_4$ and $NaAlH_4$, and the activation parameters (Table 8.5) indicate that the transition state is considerably more ordered with lithium cations than with sodium. Furthermore, no reaction occurs when $LiAlH_4$ reductions are attempted in the presence of one mole of the lithium ion specific cryptand complexing agent **6** for

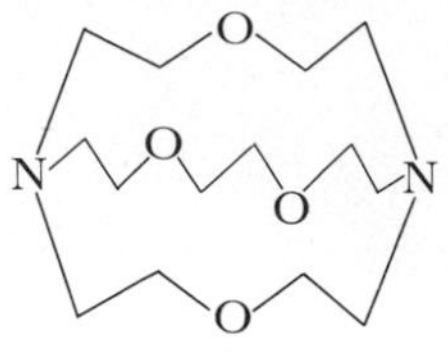

6

each mole of $LiAlH_4$. If more Li^+ ions are added from another source, reactivity is restored.[48] These results point to a transition state in which the oxygen is complexed to a lithium cation (or to some other electrophile)[49] perhaps through a preliminary equilibrium, while the AlH_4^- ion transfers the hydride, as indicated in **7**.

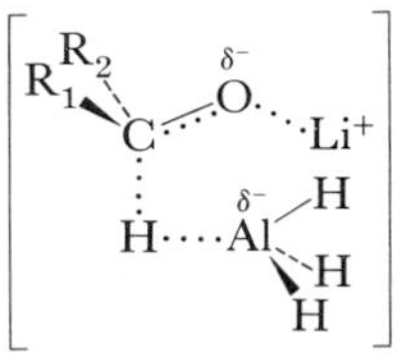

7

A reduction that also consists of hydride addition but in which the hydride donor is an alkoxide ion has found some synthetic use.[50] Equation 8.20 shows this process, referred to as the Meerwein–Ponndorf–Verley reduction. The alkoxide

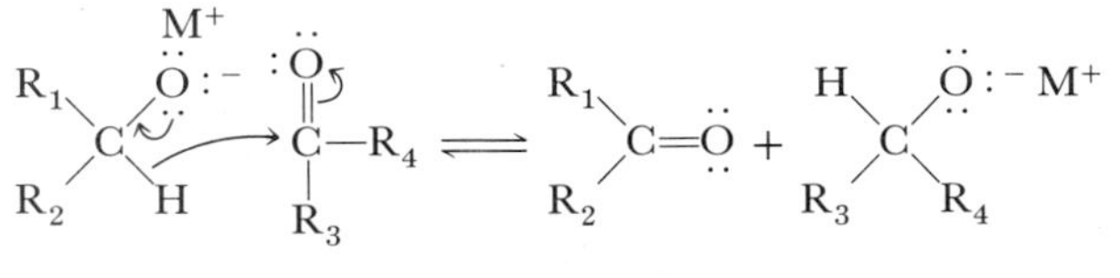

$$(8.20)$$

undergoes a reverse carbonyl addition and hence is oxidized; when looked at from this point of view, the transformation is called the Oppenauer oxidation. The equilibrium may be shifted in the desired direction by using an excess of one of the reagents.

[48](a) J.-L. Pierre and H. Handel, *Tetrahedron Lett.*, 2317 (1974); (b) J. L. Pierre, H. Handel, and R. Perraud, *Tetrahedron*, **31**, 2795 (1975).

[49] D. C. Wigfield and F. W. Gowland, *J. Org. Chem.*, **42**, 1108 (1977).

[50](a) Note 43(a); (b) A. L. Wilds, *Org. Reactions*, **2**, 178 (1944); (c) C. Djerassi, *Org. Reactions*, **6**, 207 (1951); (d) C. F. Cullis and A. Fish, in *The Chemistry of the Carbonyl Group*, Vol. 1, S. Patai, Ed., Wiley, London, 1966, Chap. 2 (oxidations).

Table 8.5 Entropy and Enthalpy of Activation for Reduction of Mesityl Phenyl Ketone by $LiAlH_4$ and $NaAlH_4$ in THF at 25°C[a]

Reagent	$\Delta H^{\ddagger}$ (kcal mole^{-1})	$\Delta S^{\ddagger}$ (cal mole^{-1} K^{-1})
$LiAlH_4$	9.9	-26.2
$NaAlH_4$	17.5	-5.4

[a] Reprinted in part with permission from E. C. Ashby and J. R. Boone, *J. Am. Chem. Soc.*, **98**, 5524 (1976). Copyright 1976 American Chemical Society.

A similar oxidation–reduction is the Cannizzaro reaction, a base-catalyzed disproportionation of aldehydes that have no α hydrogen. The process is initiated by addition of hydroxide ion to the carbonyl carbon in a preliminary equilibrium (Equation 8.21). The intermediate (**8**) then transfers a hydride to a second molecule of the aldehyde in the rate-determining step (Equation 8.22). The reaction is third-order overall, first-order in hydroxide and second-order in aldehyde.[51] (See Problem 17 for further details of the kinetics.)

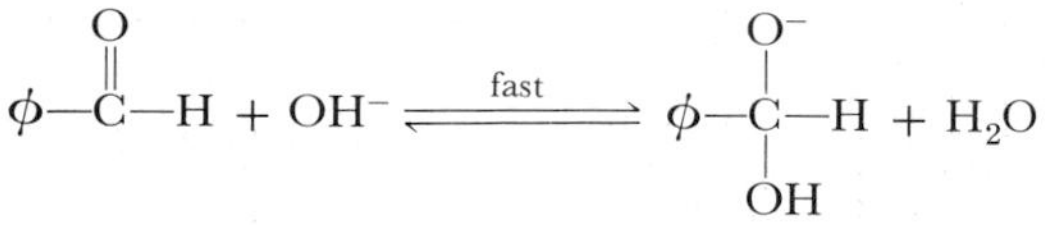

$$\phi-\overset{\overset{O}{\|}}{C}-H + OH^- \underset{\text{fast}}{\rightleftharpoons} \phi-\overset{\overset{O^-}{|}}{\underset{\underset{OH}{|}}{C}}-H + H_2O \qquad (8.21)$$

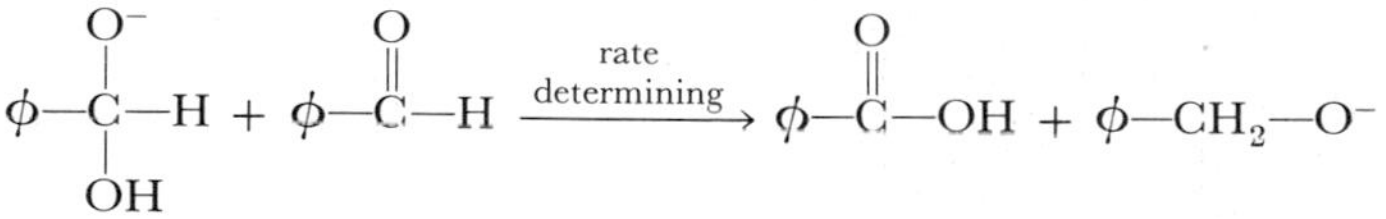

$$\mathbf{8}$$

$$\phi-\overset{\overset{O^-}{|}}{\underset{\underset{OH}{|}}{C}}-H + \phi-\overset{\overset{O}{\|}}{C}-H \xrightarrow{\underset{\text{determining}}{\text{rate}}} \phi-\overset{\overset{O}{\|}}{C}-OH + \phi-CH_2-O^- \qquad (8.22)$$

$$\mathbf{8}$$

$$\phi-\overset{\overset{O}{\|}}{C}-OH + \phi CH_2O^- \xrightarrow{\text{fast}} \phi-\overset{\overset{O}{\|}}{C}-O^- + \phi-CH_2-OH \qquad (8.23)$$

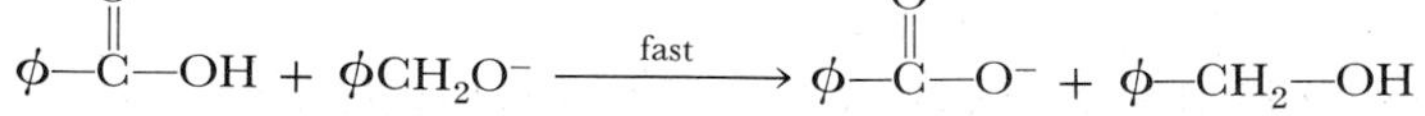

An alternative and more generally used oxidation method employs chromic acid. This process is an exception to our general theme because here the alcohol is transformed into a carbonyl group by removal of electron density from oxygen rather than from carbon. The first step has been shown to be a rapid equilibrium between the alcohol and its chromate ester, followed by rate-determining decom-

[51] C. G. Swain, A. L. Powell, W. A. Sheppard, and C. R. Morgan, *J. Am. Chem. Soc.*, **101**, 3576 (1979). For some aldehydes (formaldehyde is an example) there is a pathway in which the —OH proton is removed from **8** to form a dinegative ion that acts as hydride transfer agent. See Problem 17.

position of the ester in the manner shown in Scheme 10.[52] It will be noted that the species eliminated from the carbon that becomes the carbonyl carbon is a Lewis acid, not a Lewis base.

SCHEME 10

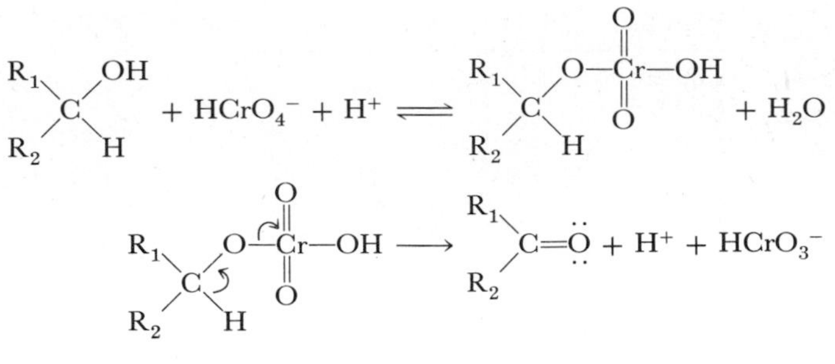

$$
\underset{R_2}{\overset{R_1}{>}}\!C\!\underset{H}{\overset{OH}{<}} + HCrO_4^- + H^+ \;\rightleftharpoons\; \underset{R_2}{\overset{R_1}{>}}\!C\!\underset{H}{\overset{O-\overset{\overset{\displaystyle O}{\|}}{Cr}-OH}{<}} + H_2O
$$

$$
\underset{R_2}{\overset{R_1}{>}}\!C\!\underset{H}{\overset{O-\overset{\overset{\displaystyle O}{\|}}{Cr}-OH}{<}} \;\longrightarrow\; \underset{R_2}{\overset{R_1}{>}}\!C\!=\!\ddot{O} + H^+ + HCrO_3^-
$$

Stereochemistry of Addition

When an effectively irreversible adddition, such as that of an organometallic or hydride, occurs in a cyclic structure or in a molecule that contains a chiral center, two isomers may form. When the chiral center is adjacent to the carbonyl, the additions are stereospecific, and one of the isomers is formed in greater amount than the other. Cram and his collaborators provided the first rationalization for the observed isomer distribution based on conformational analysis;[53] a more recent version of the theory has been given by Karabatsos.[54] Nuclear magnetic resonance investigation of aldehyde and ketone oximes showed **9** to be the lowest-energy conformation. In the transition state for addition of an organometallic or metal

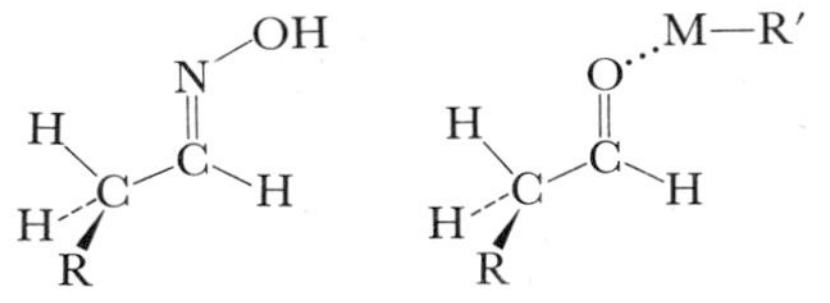

  9   **10**

hydride M—R' to a carbonyl compound, the carbonyl oxygen should be complexed to the electrophile and the analogous conformation **10** should be the preferred one. If the α carbon has three different substituents, **L**, **M**, and **S**, where **L** is the largest, **M**, medium, and **S**, the smallest, the lowest-energy conformations for

[52](a) A. Leo and F. H. Westheimer, *J. Am. Chem. Soc.*, **74**, 4383 (1952); (b) M. Cohen and F. H. Westheimer, *J. Am. Chem. Soc.*, **74**, 4387 (1952); (c) K. B. Wiberg, Ed., *Oxidation in Organic Chemistry*, Academic Press, New York, 1965; (d) R. Stewart, *Oxidation Mechanisms*, W. A. Benjamin, Menlo Park, Calif., 1964, Chap. 4.

[53](a) D. J. Cram and F. A. Abd Elhafez, *J. Am. Chem. Soc.*, **74**, 5828 (1952); (b) D. J. Cram and F. D. Greene, *J. Am. Chem. Soc.*, **75**, 6005 (1953). Cram has also developed models for 1,3-asymmetric induction. See (c) T. J. Leitereg and D. J. Cram, *J. Am. Chem. Soc.*, **90**, 4011, 4019 (1968).

[54](a) G. J. Karabatsos, *J. Am. Chem. Soc.*, **89**, 1367 (1967); (b) G. J. Karabatsos, C. Zioudrou, and I. Moustakali, *Tetrahedron Lett.*, 5289 (1972); (c) C. Zioudrou, P. Chrysochou, G. J. Karabatsos, D. Herlem, and R. N. Nipe, *Tetrahedron Lett.*, 5293 (1972).

transition states leading to the two alternative diastereomers **A** and **B** should be **11** and **12**, the two conformations in which carbonyl substituent R is gauche to the

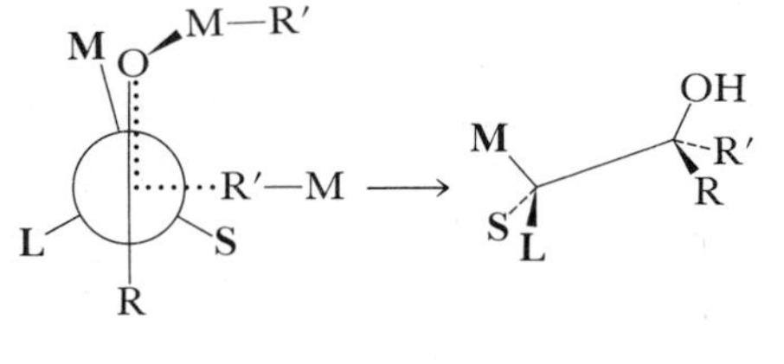

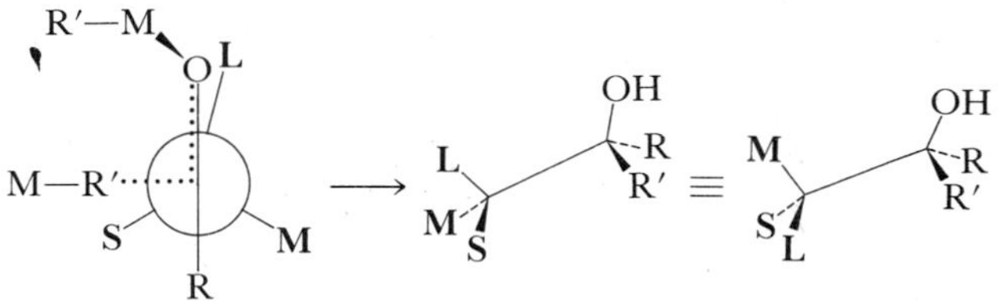

smallest group **S**. Because **11** has an $M \leftrightarrow O$ interaction while **12** has an $L \leftrightarrow O$ interaction, **11** should be lower-energy and the **A/B** ratio should be greater than one. The predictions of this approach are usually in agreement with the earlier model of Cram, which assumed that conformation **13** would be attacked preferentially from the side nearer the small group. The Karabatsos model, however, provides a fairly successful semiquantitative estimate of the experimentally observed

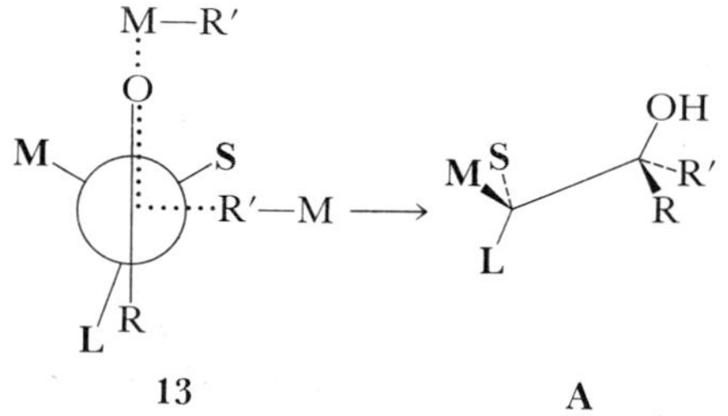

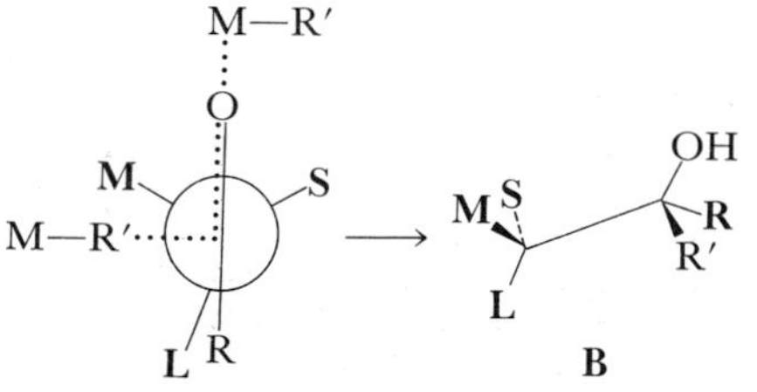

A/B ratio and explains more successfully than Cram's theory trends of the ratio with changes of bulk of substituents.[55]

The stereochemistry of addition to cyclic ketones, particularly addition of hydride from complex metal hydride reducing agents, has presented puzzling aspects. Reduction of 4-*t*-butylcyclohexanone by $LiAlH_4$ yields about 90 percent *trans*-4-*t*-butylcyclohexanol (Equation 8.24). This result is surprising because it

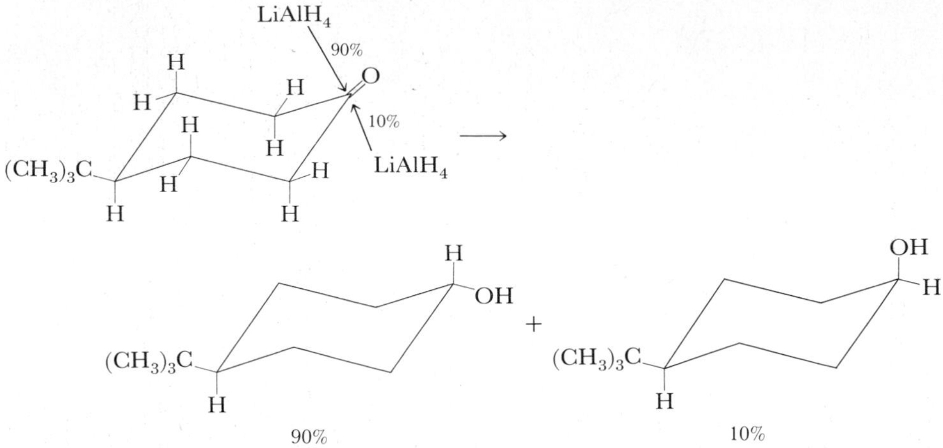

(8.24)

implies predominant attack by the hydride from the more hindered axial side, where there would be steric interference from the axial hydrogens in the 3-positions. Reduction of 3,3,5-trimethylcyclohexanone, on the other hand, yields 80 percent of the product of attack from the less hindered equatorial direction (Equation 8.25),[56] a result in agreement with predictions from steric effects on approach of the reagent. The experimental trends are summarized in Figure 8.10.

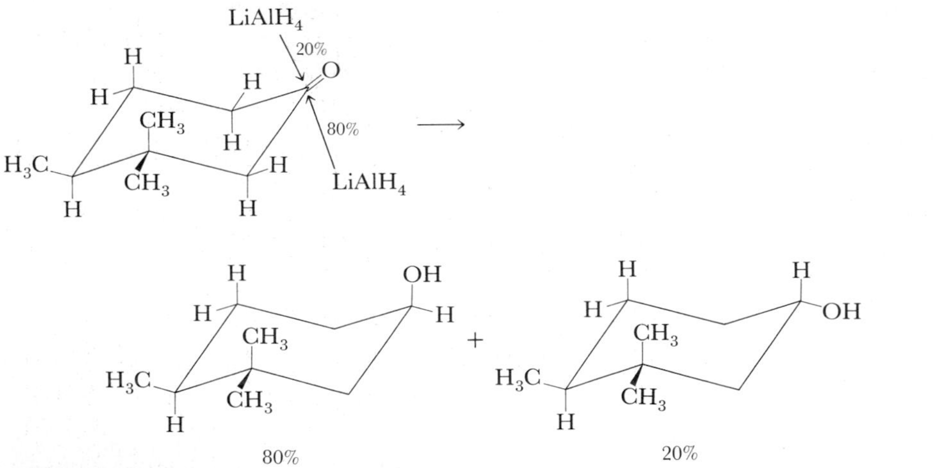

(8.25)

---

[55] Special situations requiring slightly different analyses arise when one of the substituent groups has unshared pairs of electrons or when R has large steric requirements.

[56] E. C. Ashby and J. R. Boone, *J. Org. Chem.*, **41**, 2890 (1976).

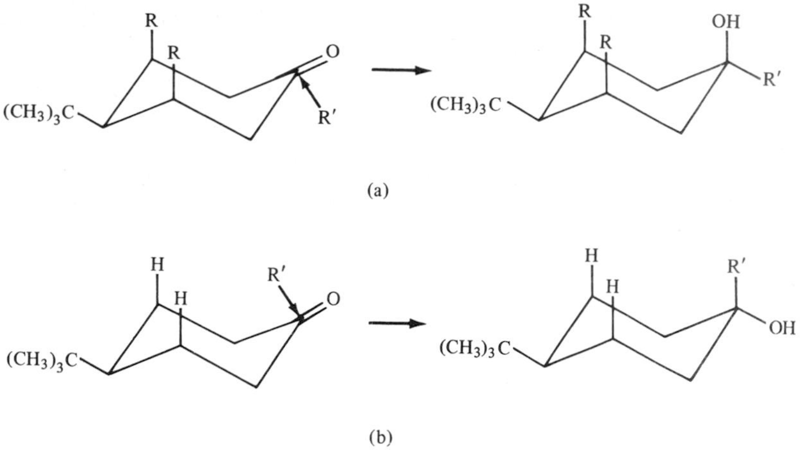

(a)

(b)

Figure 8.10 Stereochemistry of addition to cyclohexanones. (a) R′ enters from the less hindered direction, giving the axial alcohol. This behavior is observed for bulky R′, such as Grignard reagents, $LiAlH(OCH_3)_3$, or when axial groups R in the 3 position are large. (b) R′ enters from the axial direction, giving equatorial alcohol. Observed with small R′, such as $LiAlH_4$, and small axial groups in the 3 position.

In order to explain these results, Dauben and co-workers[57] proposed that, depending upon the bulk of the entering reagent and the presence or absence of axial substituents on the ring two carbons removed from the site of attack, addition to a cyclohexanone could be subject to either *steric approach control* or *product development control*. Steric approach control would direct the entering group to the less hindered equatorial position (predominant pathway in Equation 8.25), whereas product development control would direct the entering group to the more hindered axial position (predominent pathway in Equation 8.24) so that the developing hydroxyl group would be in the less hindered equatorial position.

Early criticisms of this proposal pointed out that it assumed an early transition state in steric approach control and a late one in product development control, whereas it seemed unlikely that the location of the transition state on the reaction coordinate would undergo so drastic a change with a relatively small change of structure.[58] Subsequently, evidence that discredits the product development control idea has appeared. For example, 7-norbornanone (**14**) is an unhindered ketone that should, according to Dauben's hypothesis, be subject to product development control. Hence if an exo methyl group is introduced (**16**) the rate of attack by hydride from the side remote from the methyl to yield the syn alcohol **17** should be slower than the rate of attack on one side of **14**. Because attack on each side of **14** gives the same product (**15**), rate of formation of **17** should be less than half the rate of formation of **15**. If, on the other hand, no product development control is operat-

[57](a) W. G. Dauben, G. J. Fonken, and D. S. Noyce, *J. Am. Chem. Soc.*, **78**, 2579 (1956). For more recent arguments supporting the product development control idea, see (b) M.-H. Rei, *J. Org. Chem.*, **44**, 2760 (1979).
[58](a) D. M. S. Wheeler and M. M. Wheeler, *J. Org. Chem.*, **27**, 3796 (1962); (b) E. L. Eliel and Y. Senda, *Tetrahedron*, **26**, 2411 (1970).

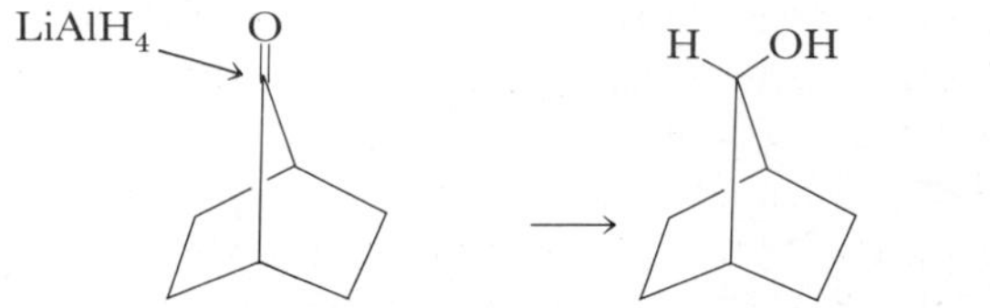

$$(8.26)$$

$$(8.27)$$

ing, rate of attack on one side of **14** should be equal to rate of attack on the side of **16** remote from the methyl group because the entering reagent encounters essentially the same structure in each. In this case rate of formation of **17** should be exactly half the rate of formation of **15**. This latter result is the one found experimentally.[59]

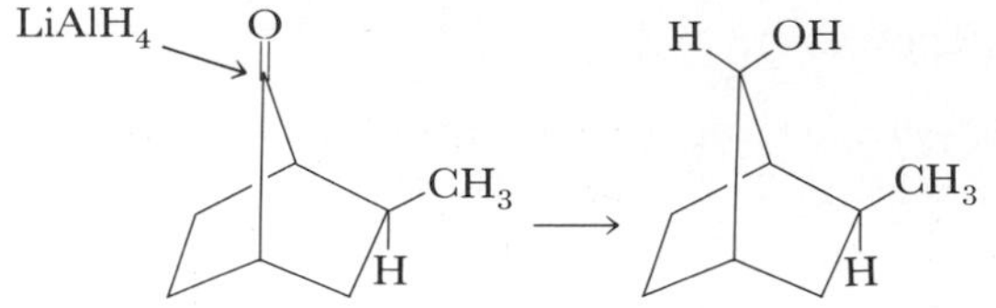

Although the steric approach control idea is a reasonable application of ideas about steric hindrance to the approach of the reagent in a process with an early transition state, an alternative is needed to explain stereochemistry of attack on unhindered cyclic ketones. The idea that appears to be most reasonable, proposed by Richer, and by Chérest and Felkin, is that in the absence of large steric interference to approach of the reagent, the developing eclipsing and resulting torsional strains that occur when the new bond is formed on the equatorial side (**18**) are more serious than the steric interference by the axial hydrogens in the 3 positions that occurs during attack on the axial side (**18a**).[60]

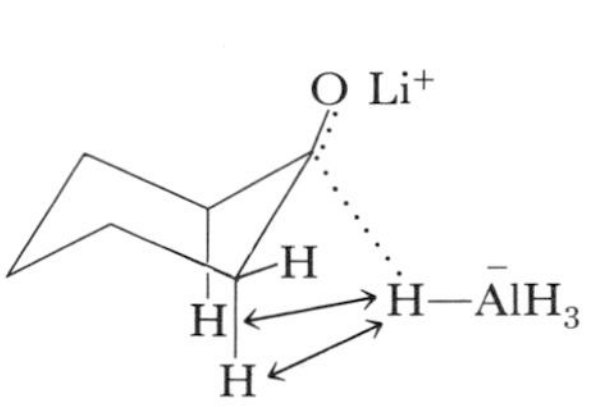

[59](a) E. C. Ashby and S. A. Noding, *J. Am. Chem. Soc.*, **98**, 2010 (1976); (b) E. C. Ashby and S. A. Noding, *J. Org. Chem.*, **42**, 264 (1977). See also note 58(b). Theoretical models of the transition state have also been proposed. See (c) W. T. Wipke and P. Gund, *J. Am. Chem. Soc.*, **98**, 8107 (1976); (d) J.-C. Perlberger and P. Müller, *J. Am. Chem. Soc.*, **99**, 6316 (1977).

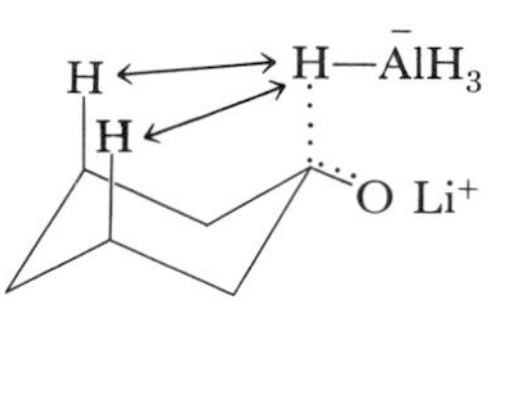

[60](a) J.-C. Richer, *J. Org. Chem.*, **30**, 324 (1965); (b) M. Chérest, H. Felkin, and N. Prudent, *Tetrahedron Lett.*, 2199 (1968); (c) M. Chérest and H. Felkin, *Tetrahedron Lett.*, 2205 (1968); (d) M. Chérest, H. Felkin, and C. Frajerman, *Tetrahedron Lett.*, 379 (1971); (e) M. Chérest and H. Felkin, *Tetrahedron Lett.*, 383 (1971); see also (f) R. A. Auerbach and C. A. Kingsbury, *Tetrahedron*, **29**, 1457 (1973).

While it would appear that the combination of steric interference to reagent approach and torsional strain arising from eclipsing neighboring bonds will account for the main features of stereochemistry, it is also clear that the nature of the cation, the type of ion pairing (contact or solvent-separated) and degree of solvation of the reducing agent, and the possible influence of the cation on the conformation of the ketone all have an effect on the stereoselectivity in these reactions.[61]

8.3 ADDITION FOLLOWED BY ELIMINATION

Many carbonyl additions yield intermediates that undergo further transformations restoring the original carbonyl carbon to a doubly bonded state. These changes are fundamentally the same as the reverse steps of the additions considered in the previous sections, and differ only in that departure of some group other than the original nucleophile is possible. Equations 8.28 and 8.29, where *Nuc* and *Nuc'* are generalized nucleophiles, illustrate two possibilities.

$$\begin{matrix} R \\ \diagdown \\ \diagup \\ R \end{matrix} C=O + \mathit{Nuc}-H \rightleftharpoons \begin{matrix} R & OH \\ \diagdown \diagup \\ C \\ \diagup \diagdown \\ R & \mathit{Nuc} \end{matrix} \overset{H^+}{\rightleftharpoons} \begin{matrix} R \\ \diagdown \\ \diagup \\ R \end{matrix} C=\mathit{Nuc}^+ + H_2O \tag{8.28}$$

$$\begin{matrix} \mathit{Nuc} \\ \diagdown \\ \diagup \\ R \end{matrix} C=O + \mathit{Nuc'}^- \rightleftharpoons \begin{matrix} \mathit{Nuc} & O^- \\ \diagdown \diagup \\ C \\ \diagup \diagdown \\ R & \mathit{Nuc'} \end{matrix} \rightleftharpoons \begin{matrix} O \\ \parallel \\ C \\ \diagup \diagdown \\ R & \mathit{Nuc'} \end{matrix} + \mathit{Nuc}^- \tag{8.29}$$

The product of the elimination step may still contain a highly electrophilic doubly bonded carbon, as in Equation 8.28. In that case, a third step may follow in which a second molecule of the nucleophile adds to yield a final product in which the original carbonyl carbon is tetrahedrally bonded (Equation 8.30). In this section we consider reactions of this kind, and in Sections 8.4 and 8.5 we take up reactions that stop at the stage indicated by Equation 8.28 or 8.29.

$$\begin{matrix} R \\ \diagdown \\ \diagup \\ R \end{matrix} C=\mathit{Nuc}^+ + \mathit{Nuc}-H \rightleftharpoons \begin{matrix} R & \mathit{Nuc} \\ \diagdown \diagup \\ C \\ \diagup \diagdown \\ R & \mathit{Nuc} \end{matrix} + H^+ \tag{8.30}$$

Acetals and Ketals

Addition of an alcohol to a carbonyl group is a straightforward extension of the hydration process. The first product formed will be a hemiacetal or hemiketal (**19**). In the presence of an acid catalyst this intermediate may eliminate the OH group to return to a structure with trigonal carbon, stabilized carbocation (**20**). This ion will then react with a second molecule of the alcohol to yield the acetal or ketal. Hemiacetals and hemiketals, with a few exceptions, are not sufficiently stable to isolate in pure form; their presence in solution has been demonstrated by various

[61] E. C. Ashby and J. R. Boone, *J. Org. Chem.*, **41**, 2890 (1976).

$$
\begin{array}{c}
R_1 \\
\diagdown \\
C{=}O + R'OH \rightleftharpoons \\
\diagup \\
R_2
\end{array}
\qquad
\begin{array}{c}
R_1 \quad OH \\
\diagdown\ \diagup \\
C \\
\diagup\ \diagdown \\
R_2 \quad OR' \\
\mathbf{19}
\end{array}
\qquad (8.31)
$$

$$
\begin{array}{c}
R_1 \quad OH \\
\diagdown\ \diagup \\
C \\
\diagup\ \diagdown \\
R_2 \quad OR' \\
\mathbf{19}
\end{array}
+ H^+ \rightleftharpoons
\begin{array}{c}
R_1 \\
\diagdown \\
C{=}\overset{+}{\underset{\cdot\cdot}{O}}{-}R' \longleftrightarrow \\
\diagup \\
R_2
\end{array}
\begin{array}{c}
R_1 \\
\diagdown\ \overset{+}{} \\
C{-}\overset{\cdot\cdot}{O}{-}R' + H_2O \\
\diagup \\
R_2 \\
\mathbf{20}
\end{array}
\qquad (8.32)
$$

$$
\begin{array}{c}
R_1 \\
\diagdown\ \overset{+}{} \\
C{=}\overset{\cdot\cdot}{O}{-}R' + R'OH \rightleftharpoons \\
\diagup \\
R_2 \\
\mathbf{20}
\end{array}
\begin{array}{c}
R_1 \quad OR' \\
\diagdown\ \diagup \\
C \\
\diagup\ \diagdown \\
R_2 \quad OR'
\end{array}
+ H^+
\qquad (8.33)
$$

physical measurements.[62] Acetals and ketals are stable under neutral or basic conditions but undergo reaction back to alcohol and aldehyde or ketone in the presence of aqueous acid.

Because rapid conversion of hemiacetal to acetal requires more acidic conditions than does formation of the hemiacetal, it is possible to measure the rate of hemiacetal production without complication from the second stage of the reaction. As might be expected, the hemiacetal formation displays characteristics similar to those of hydration; general acid and general base catalysis are observed.[63]

Cyclic hemiacetals and hemiketals with five- and six-membered rings formed by hydroxy aldehydes or hydroxy ketones are considerably more stable than their acyclic counterparts. The most important examples are the sugars.[64] Glucose exists largely in the pyranose form, of which there are two possible structures (**21** and **22**), called, respectively, α-glucose and β-glucose. The β form, having

α-glucose
21

β-glucose
22

23

all hydroxyl groups equatorial, is thermodynamically slightly more stable. In neutral solution the open-chain free aldehyde (**23**) accounts for only about 0.003 per-

[62] See, for example, G. W. Meadows and B. de B. Darwent, *Can. J. Chem.*, **30**, 501 (1952).

[63] (a) G. W. Meadows and B. de B. Darwent, *Trans. Faraday Soc.*, **48**, 1015 (1952); (b) L. H. Funderburk, L. Aldwin, and W. P. Jencks, *J. Am. Chem. Soc.*, **100**, 5444 (1978). See also (c) Y. Ogata and A. Kawasaki, in *The Chemistry of the Carbonyl Group*, Vol. 2, J. Zabicky, Ed., Wiley, London, 1970, p. 1.

[64] See B. Capon, *Chem. Rev.*, **69**, 407 (1969), for a comprehensive discussion of mechanism in carbohydrate chemistry.

cent of the total at equilibrium.[65] The rate of interconversion of the α and β modifications is readily measured by following the change in optical rotation (mutarotation). The reaction proceeds through the open-chain hydroxy aldehyde, and so serves as a conveniently studied example of hemiacetal formation. Mutarotation of glucose was one of the early reactions to be investigated using modern ideas of acid–base catalysis; it is subject to general acid catalysis with $\alpha = 0.27$ and to general base catalysis with $\beta = 0.36$.[66,67] The fact that α is not equal to $1 - \beta$ indicates that the acid- and base-catalyzed mechanisms differ by more than just a proton.[68]

A possibility that was proposed quite early for the glucose mutarotation, and that could conceivably be of importance for other reactions, is simultaneous catalysis by an acid and a base. It will be recalled from Section 8.1 that hydration requires addition of a proton at one site and removal of a proton from another. If both these processes were to occur in one step, either by means of separate acid and base molecules acting together or by action of a single molecule containing both an acidic and a basic center, the reaction would be subject to *simultaneous acid and base* catalysis (Equation 8.34).[69] Swain found that the rate of mutarotation of tetra-

$$
\begin{array}{c}
\diagdown \\
C{=}\ddot{O}\cdots H{-}A \\
\diagup \\
\vdots\ddot{O}{-}H\cdots :B \\
\mid \\
R
\end{array}
\longrightarrow
\begin{array}{c}
\diagdown \quad OH \\
C \\
\diagup \quad OR
\end{array}
+ A^- + BH^+
\tag{8.34}
$$

methylglucose in benzene containing pyridine and phenol is third-order overall (Equation 8.35); he interpreted this result as showing the importance of the simul-

$$
\text{rate} = k[\text{Me}_4\text{glucose}][\text{pyridine}][\text{phenol}] \tag{8.35}
$$

taneous mechanism 8.34 in aprotic solvents.[70] Subsequent work has cast some doubt on this interpretation;[71] and Bell and co-workers have shown that the proposal of simultaneous acid and base catalysis does not apply as generally as Swain had expected.[72] On the other hand, simultaneous catalysis does occur with substances that have an acidic and a basic site in the same molecule and in which the sites have a tautomeric relationship to each other.[73] An example of such a catalyst

[65] J. M. Los and K. Wiesner, *J. Am. Chem. Soc.*, **75**, 6346 (1953).

[66] J. N. Brønsted and E. A. Guggenheim, *J. Am. Chem. Soc.*, **49**, 2554 (1927).

[67] T. M. Lowry, *J. Chem. Soc.*, 2554 (1927).

[68] W. P. Jencks, *Prog. Phys. Org. Chem.*, **2**, 63 (1964).

[69] T. M. Lowry and I. J. Faulkner, *J. Chem. Soc.*, **127**, 2883 (1925).

[70] (a) C. G. Swain, *J. Am. Chem. Soc.*, **72**, 4578 (1950); (b) C. G. Swain and J. F. Brown, Jr., *J. Am. Chem. Soc.*, **74**, 2534 (1952).

[71] H. Anderson, C-W. Su, and J. W. Watson, *J. Am. Chem. Soc.*, **91**, 482 (1969).

[72] R. P. Bell and J. C. Clunie, *Proc. R. Soc. London, Ser. A*, **212**, 33 (1952).

[73] (a) C. G. Swain and J. F. Brown, Jr., *J. Am. Chem. Soc.*, **74**, 2538 (1952); (b) P. R. Rony, *J. Am. Chem. Soc.*, **91**, 6090 (1969); (c) P. R. Rony and R. O. Neff, *J. Am. Chem. Soc.*, **95**, 2896 (1973).

is 2-pyridone, which can catalyze a nucleophile addition in the manner shown in Equation 8.36.[74] Carboxylic acids can also function in this way.[75] The evidence for such a process is the acceleration of reaction rate compared with what one would

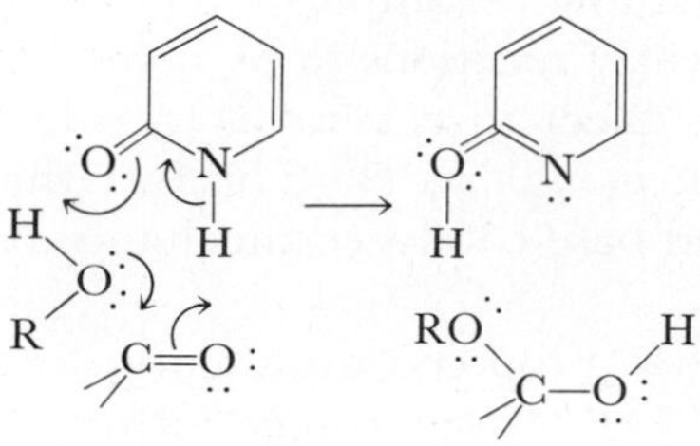

$$(8.36)$$

expect on the basis of the catalyst pK_a and the Brønsted α of the reaction being catalyzed. These examples of simultaneous acid and base catalysis have been found in nonaqueous solvents; although there is little evidence for such a mechanism in aqueous reactions, it remains a possibility for catalysis by enzymes.[76]

The second stage of acetal and ketal formation, the acid-catalyzed elimination of the hydroxyl group as a water molecule and addition of a second alcohol molecule to the resulting carbocation (Equations 8.32 and 8.33), is most conveniently investigated in the reverse direction starting from the acetal or ketal.[77] As Structures **24** and **25** indicate, it is conceivable that either of two bonds could be

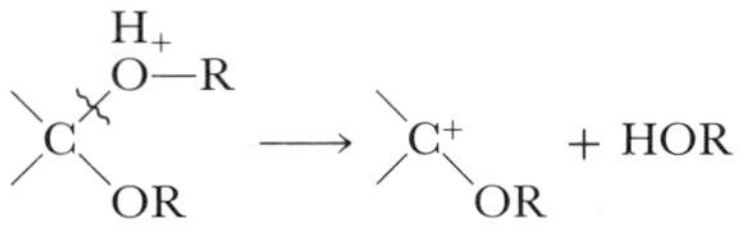

$$(8.37)$$

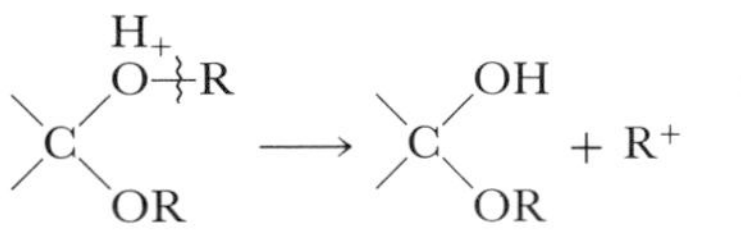

$$(8.38)$$

broken in the hydrolysis. One method of settling the ambiguity is to hydrolyze acetals in which R is bonded to oxygen at a chiral center; such experiments consistently show retained configuration, demonstrating cleavage of oxygen–carbonyl carbon bond (**24**) rather than oxygen–alkyl bond.[78] This type of cleavage is preferred

---

[74] See note 73(a).

[75] See note 73(b), (c).

[76] (a) W. P. Jencks, *Catalysis in Chemistry and Enzymology,* McGraw-Hill, New York, 1969, pp. 211–217; (b) T. H. Fife, *Adv. Phys. Org. Chem.,* **11,** 1 (1975); (c) a report of simultaneous acid and base catalysis of an esterification has appeared: S. Milstien and L. A. Cohen, *J. Am. Chem. Soc.,* **91,** 4585 (1969). See also (d) J. P. Fox and W. P. Jencks, *J. Am. Chem. Soc.,* **96,** 1436 (1974), who have reported negative results in a case that should be favorably disposed toward simultaneous catalysis.

[77] (a) E. H. Cordes and H. G. Bull, *Chem. Rev.,* **74,** 581 (1974), have reviewed hydrolysis of acetals, ketals, and ortho esters. See also (b) E. H. Cordes, *Prog. Phys. Org. Chem.,* **4,** 1 (1967).

[78] (a) J. M. O'Gorman and H. J. Lucas, *J. Am. Chem. Soc.,* **72,** 5489 (1950); (b) H. K. Garner and H. J. Lucas, *J. Am. Chem. Soc.,* **72,** 5497 (1950); (c) E. R. Alexander, H. M. Busch, and G. L. Webster, *J. Am. Chem. Soc.,* **74,** 3173 (1962).

SCHEME 11

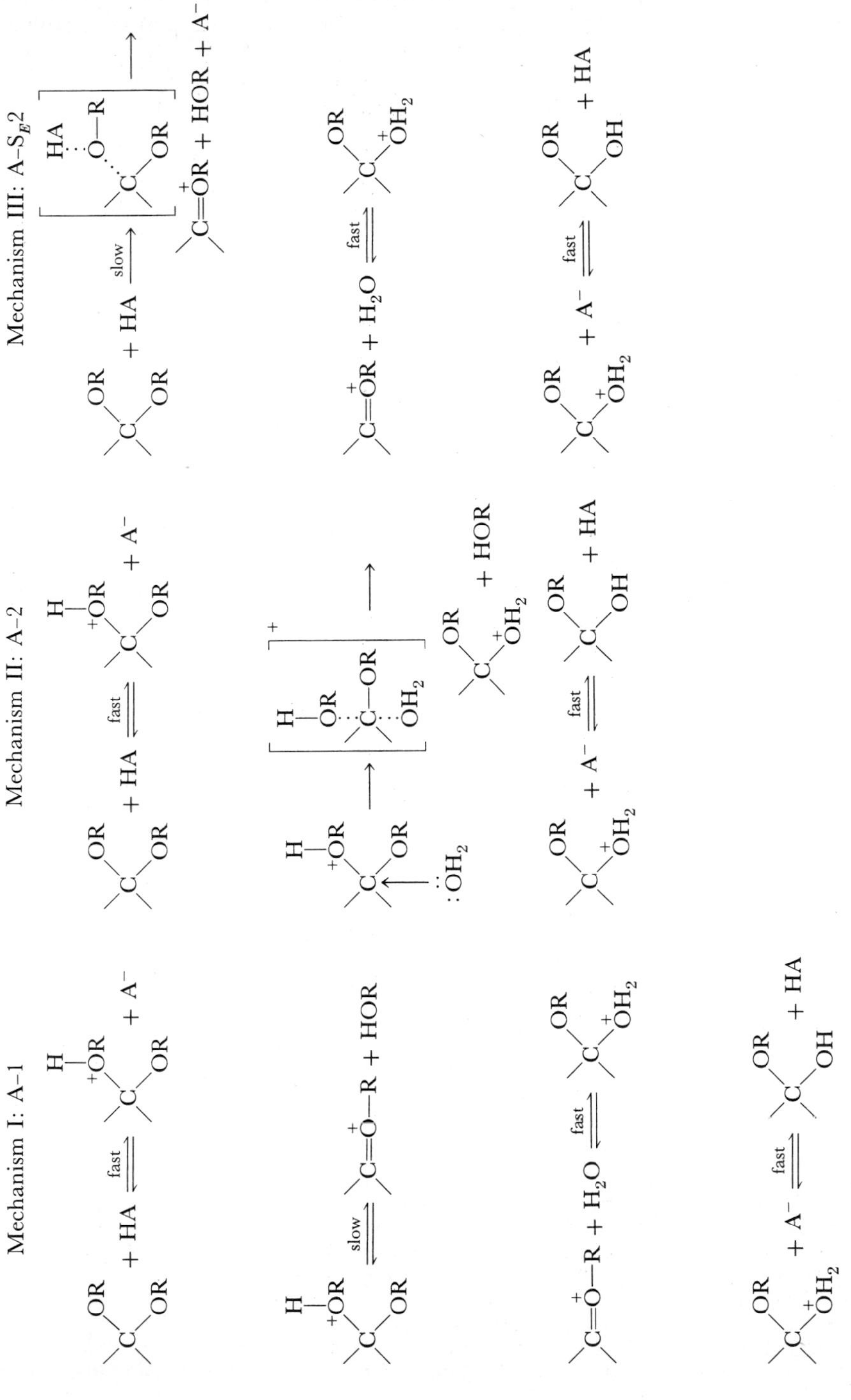
Mechanism I: A-1
fast
slow
fast
fast
Mechanism II: A-2
fast
fast
Mechanism III: A-S_E2
slow
fast
fast

even when R is chosen deliberately so as to make R^+ a good cation.[79] A second type of experiment is to use ^{18}O as a tracer; these investigations lead to the same conclusion.[80]

A somewhat more difficult question is that of the precise mechanism by which the carbonyl carbon–oxygen bond is cleaved. Scheme 11 illustrates three reasonable possibilities; these mechanisms are further clarified by schematic reaction coordinate diagrams in Figure 8.11. Mechanism I is a unimolecular S_N1 ionization of the acetal conjugate acid. It is designated the A–1 (acid-catalyzed unimolecular) mechanism. Mechanism II is an S_N2 displacement of ROH by H_2O, designated A–2, and Mechanism III, A–S_E2, is essentially a bimolecular electrophilic substitution by proton on the oxygen. Mechanisms I and II both predict specific acid catalysis, whereas Mechanism III leads to general acid catalysis.

Specific acid catalysis, but not general catalysis, is found for acetal and ketal hydrolysis.[81] (Exceptions to this statement for particular structures will be discussed below.) Mechanisms I and II therefore remain as possibilities. Of these, the A–1 process, Mechanism I, appears on the basis of a number of criteria to be the correct one for most acetals and ketals. Strong acceleration by electron donation in the carbonyl portion of the molecule has been demonstrated by Hammett σ–ρ and Taft σ^*–ρ^* correlations. For example, in hydrolysis of **26** ρ is in the neighborhood of -3.3, and in hydrolysis of **27** ρ^* is near -3.6.[82] Entropies and volumes of activa-

tion,[83] though less reliable criteria, are in the range usually found for unimolecular reactions and do not agree with values expected for the A–2 process. Solvent isotope effects also are in agreement with the A–1 mechanism.[84] It is also possible, by carrying out the hydrolyses in aqueous sulfuric acid, perchloric acid, or in some other solvent mixture of reduced nucleophilicity, to slow the H_2O addition step of Mechanism I (or III) sufficiently that it becomes rate determining. The cation intermediate can then be observed directly by nuclear magnetic resonance or ultraviolet spectroscopy.[85]

[79] J. D. Drumheller and L. J. Andrews, *J. Am. Chem. Soc.*, **77**, 3290 (1955).

[80] F. Stasiuk, W. A. Sheppard, and A. N. Bourns, *Can. J. Chem.*, **34**, 123 (1956).

[81] (a) K. Koehler and E. H. Cordes, reported on p. 32 of reference 77(b); (b) J. N. Brønsted and W. F. K. Wynne-Jones, *Trans. Faraday Soc.*, **25**, 59 (1929); (c) M. M. Kreevoy and R. W. Taft, Jr., *J. Am. Chem. Soc.*, **77**, 3146 (1955); (d) T. H. Fife and L. K. Jao, *J. Org. Chem.*, **30**, 1492 (1965).

[82] (a) Note 81(d); (b) M. M. Kreevoy and R. W. Taft, Jr., *J. Am. Chem. Soc.*, **77**, 5590 (1955).

[83] E. H. Cordes, *Prog. Phys. Org. Chem.*, **4**, 1 (1967), pp. 13, 14.

[84] (a) C. A. Bunton and V. J. Shiner, Jr., *J. Am. Chem. Soc.*, **83**, 3207 (1961); (b) M. Kilpatrick, *J. Am. Chem. Soc.*, **85**, 1036 (1963).

[85] (a) R. A. McClelland and M. Ahmad, *J. Am. Chem. Soc.*, **100**, 7027; 7031 (1978); (b) R. A. McClelland, M. Ahmad, and G. Mandrapilias, *J. Am. Chem. Soc.*, **101**, 970 (1979).

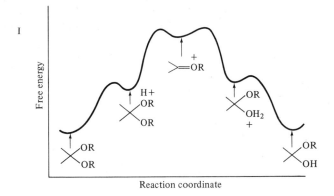

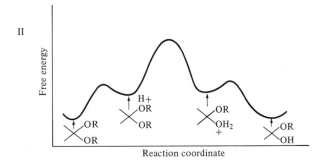

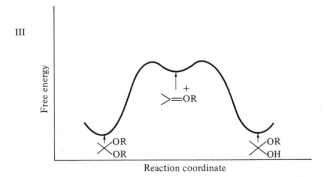

Figure 8.11 Reaction coordinates for the alternative mechanisms for cleavage of acetals and ketals shown in Scheme 11. I: A–1, Mechanism I; II: A–2, Mechanism II; III: A–S$_E$2, Mechanism III.

We have discussed in Section 3.2 the use of correlations of reaction rates with acidity functions to distinguish between A–1 and A–2 mechanisms for acid-catalyzed hydrolyses. Recall the Bunnett criterion (Equation 8.39) relating rate of the acid-catalyzed reaction to the acidity function H_0 and to the water activity

$$\log k_{\text{obs}} + H_0 = w \log a_{\text{H}_2\text{O}} + C \tag{8.39}$$

a_{H_2O}.[86] Negative values of w indicate no involvement of water in the transition state of the rate-determining step, whereas reactions in which water enters as a nucleophile usually show values of w between $\sim +1.2$ and $+3.3$. Table 8.6 gives w values for hydrolyses of a number of acetals. The negative values found in nearly all cases support the conclusion that Mechanism I is correct.

Table 8.6 BUNNETT w PARAMETERS
FOR HYDROLYSIS OF
SOME ACETALS IN
AQUEOUS HCl AT 25°C[a]

Acetal	w
$H_2C(OCH_3)_2$	-5.26
$H_2C(OC_2H_5)_2$	-7.18[b]
$ClCH_2CH(OC_2H_5)_2$	-6.06
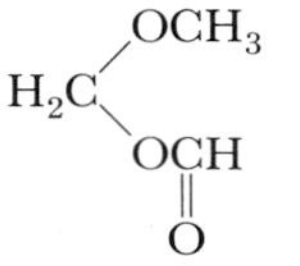 $H_2C(OCH_3)(OCHO)$	$+1.24$
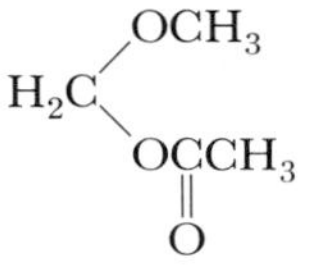 $H_2C(OCH_3)(OCCH_3{=}O)$	-2.39
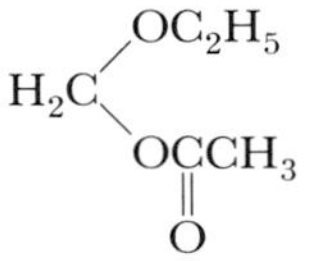 $H_2C(OC_2H_5)(OCCH_3{=}O)$	-3.57
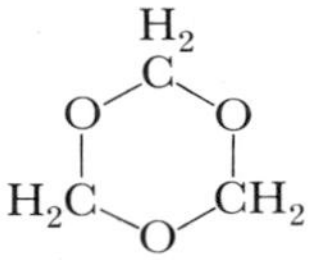 1,3,5-trioxane (H_2C, O, CH_2 ring)	-2.10
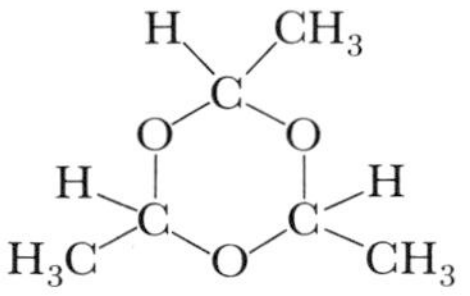 paraldehyde (trimethyl-1,3,5-trioxane)	-4.32

[a] Reprinted in part with permission from J. F. Bunnett, *J. Am. Chem. Soc.*, **83**, 4956 (1961). Copyright 1961 American Chemical Society.
[b] At 10°C.

[86] J. F. Bunnett, *J. Am. Chem. Soc.*, **83**, 4956 (1961), and following papers.

General Catalysis in Acetal Hydrolysis

Increasing knowledge of enzyme mechanisms has spurred a renewed search for general acid catalysis in acetal hydrolysis. The active site in the enzyme lysozyme, which catalyzes hydrolysis of the acetal link in glycosides, has a carboxylic acid and a carboxylate anion in favorable positions to interact with the acetal group, one presumably by general acid catalysis and the other by electrostatic stabilization of the intermediate carbocation.[87] The proposed general catalysis in the enzyme would be more convincing if nonenzymatic analogies could be found.

A guide in searching for general acid catalysis in acetal hydrolysis is found in the close relationship of Mechanisms I and III, Scheme 11. In Mechanism I the carbocation is sufficiently difficult to form that transfer of a proton to the oxygen of the leaving group must be completed in a preliminary step. Specific acid catalysis is predicted. If the carbocation were further stabilized, or if there were a better leaving group, perhaps only partial proton transfer would be required. Then Mechanism III, with general acid catalysis, would occur. Structural changes in the direction suggested by this argument do produce general catalysis; for example, 2-p-nitrophenoxytetrahydropyran (28) hydrolyzes with general acid catalysis and Brønsted $\alpha = 0.5$,[88] and general acid catalysis is commonly found in hydrolyses of ortho esters (29).[89] Hydrolyses of these compounds are best described by Mechanism III.[90]

28   29

Thioacetals and Thioketals

Thiols have a markedly greater tendency to add to carbonyl groups than do water and alcohols. The equilibrium constant for the exchange reaction (8.40) is esti-

$$\begin{array}{c}\diagdown\diagup OH \\ C \\ \diagup\diagdown OH \end{array} + RSH \rightleftharpoons \begin{array}{c}\diagdown\diagup OH \\ C \\ \diagup\diagdown SR \end{array} + H_2O \qquad (8.40)$$

mated to be about 2.5×10^4.[91] Addition of hydrogen sulfide and thiols is qualitatively similar to reaction with alcohols in that there are two stages, formation of

[87](a) T. H. Fife, *Acc. Chem. Res.*, **5**, 264 (1972); (b) B. M. Dunn and T. C. Bruice, *J. Am. Chem. Soc.*, **92**, 2410, 6589 (1970). (c) See, however, G. M. Louden, C. K. Smith, and S. E. Zimmerman, *J. Am. Chem. Soc.*, **96**, 465 (1974), who find that electrostatic stabilization contributes only a small rate acceleration in a model system.
[88] See note 87(a).
[89](a) See note 77(a). For another structure that shows similar behavior, see (b) R. A. McClelland, *J. Am. Chem. Soc.*, **100**, 1844 (1978). See also (c) T. H. Fife and T. J. Przystas, *J. Am. Chem. Soc.*, **102**, 292 (1980).
[90] R. Eliason and M. M. Kreevoy, *J. Am. Chem. Soc.*, **100**, 7037 (1978).
[91] W. P. Jencks, *Prog. Phys. Org. Chem.*, **2**, 63 (1964), p. 104.

hemithioacetal (or hemithioketal) followed by acid-catalyzed elimination of the hydroxy group and substitution of a second —SR (Equations 8.41 and 8.42).

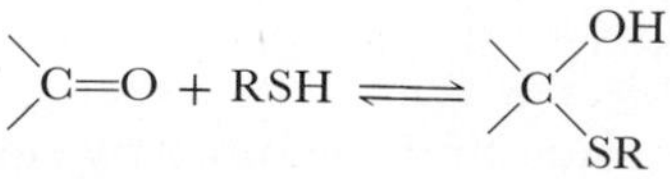

$$\ce{>C=O + RSH <=> >C(OH)(SR)} \tag{8.41}$$

$$\ce{>C(OH)(SR) + RSH <=>[H+] >C(SR)(SR) + H2O} \tag{8.42}$$

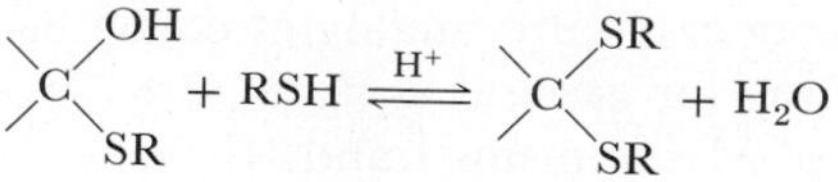

The initial addition is specific base-catalyzed, an observation that implies that RS$^-$ is the reactive species (Equations 8.43 and 8.44). Compared with alcohols,

$$\ce{RSH + A^- <=>[K] RS^- + HA} \tag{8.43}$$

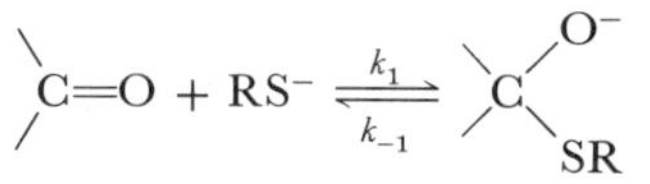

$$\ce{>C=O + RS^- <=>[k_1][k_{-1}] >C(O^-)(SR)} \tag{8.44}$$

$$\underset{\textbf{30}}{\qquad} \qquad \underset{\textbf{31}}{\qquad}$$

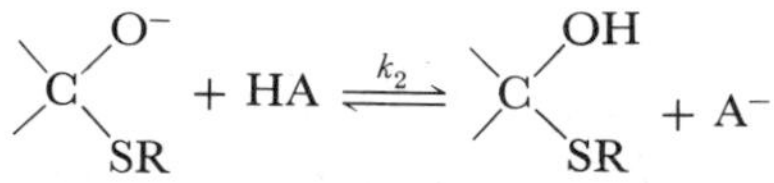

$$\ce{>C(O^-)(SR) + HA <=>[k_2] >C(OH)(SR) + A^-} \tag{8.45}$$

thiols of similar structure are more acidic. Hence thiolate anions are less basic than analogous alkoxides, and the reverse of the initial addition reaction (8.44) is faster than the comparable reaction with an alkoxide leaving group. For the less basic thiolate anions, for example, **30**, R = CH_3—O—CO—CH_2—, k_{-1} is so large that the reverse of 8.44 is faster than the forward step of 8.45. The reaction coordinate diagram for this process is that shown in Figure 8.1(c). The addition step (8.44) rather than being rate determining as it is for most additions becomes another preliminary equilibrium and Reaction 8.45 is rate determining. This new rate-determining step is a simple proton transfer. The reaction therefore shows general acid catalysis of the type that Jencks has termed enforced. The Brønsted α changes from zero to unity rapidly with change of catalyst pK_a. This behavior is expected from the Eigen diagram (Figure 8.2) for a proton transfer that is not coupled with another process. More basic thiolate ions, such as **30**, R = alkyl, have a smaller k_{-1} and for these substances Step 8.44 remains rate determining. These compounds add with no general acid catalysis; their addition mechanism is similar to that of cyanide.[92]

[92] (a) H. F. Gilbert and W. P. Jencks, *J. Am. Chem. Soc.*, **99**, 7931 (1977); (b) W. P. Jencks, *Acc. Chem. Res.*, **9**, 425 (1976).

8.4 ADDITION OF NITROGEN NUCLEOPHILES

A number of important chemical and biochemical processes are initiated by addition of a nitrogen nucleophile to a carbonyl group.[93] These processes have been the subject of extensive study, and we shall not attempt to do more than outline the main features.

Addition of Nitrogen Nucleophiles to C=O

Addition of primary amines to carbonyl groups follows the pattern we have established for other nucleophiles with formation of a carbinolamine (Equation 8.46).

$$\text{C=O} + H_2NR \rightleftharpoons \text{C}\underset{NHR}{\overset{OH}{<}} \tag{8.46}$$

$$\text{C}\underset{NHR}{\overset{OH}{<}} \rightleftharpoons \text{C=N}\underset{R}{\overset{..}{<}} + H_2O \tag{8.47}$$
$$\mathbf{32}$$

These compounds are sufficiently stable to be isolated in some cases,[94] but they usually undergo an elimination to an imine (Equation 8.47). Note that this reaction is analogous to the elimination of H_2O in the second stage of acetal formation; the product (**32**) is isoelectronic with the oxygen-stabilized carbocation (**20**) of Equation 8.32. The imines **32** are usually too reactive to isolate if the substituents on carbon and nitrogen are all alkyl or hydrogen. Imines with hydrogen attached to nitrogen have been demonstrated spectrophotometrically,[95] but they undergo further condensations and cannot ordinarily be isolated.[96] Imines can be stabilized by one or more aryl groups attached to carbon or nitrogen, in which case the compounds are easily isolated and are called Schiff's bases. Stabilization is also achieved if a hydroxyl group or a second nitrogen is attached to the nitrogen. The most common of these structures are the oximes (**33**), semicarbazones (**34**), and hydrazones (**35**).

$$\text{C=N}\overset{}{\underset{OH}{\diagdown}} \qquad \text{C=N}\overset{}{\underset{\underset{H}{\overset{O}{\underset{N-C-NH_2}{|}}}}{\diagdown}} \qquad \text{C=N}\overset{}{\underset{\underset{H}{\overset{N-R}{|}}}{\diagdown}}$$
$$\mathbf{33} \qquad\qquad \mathbf{34} \qquad\qquad \mathbf{35}$$

---

[93] Discussions of mechanisms of addition of nitrogen nucleophiles may be found in the following sources: (a) W. P. Jencks, *Catalysis in Chemistry and Enzymology*, McGraw-Hill, New York, 1969, pp. 490*ff*; (b) W. P. Jencks, *Chem. Rev.*, **72**, 705 (1972); (c) W. P. Jencks, *Prog. Phys. Org. Chem.*, **2**, 63 (1964); (d) R. L. Reeves, *The Chemistry of the Carbonyl Group*, Vol. 1, S. Patai, Ed., Wiley, London, 1966, p. 567; (e) M. L. Bender, *Mechanisms of Homogeneous Catalysis from Protons to Proteins*, Wiley, New York, 1971; (f) L. P. Hammett, *Physical Organic Chemistry*, 2d ed., McGraw-Hill, New York, 1970, p. 336.

[94] (a) P. K. Chang and T. L. V. Ulbricht, *J. Am. Chem. Soc.*, **80**, 976 (1958); (b) E. J. Poziomek, D. N. Kramer, B. W. Fromm, and W. A. Mosher, *J. Org. Chem.*, **26**, 423 (1961).

[95] R. K. McLeod and T. I. Crowell, *J. Org. Chem.*, **26**, 1094 (1961).

[96] F. Sachs and P. Steinert, *Chem. Ber.*, **37**, 1733 (1904).

Another possible reaction of the carbinolamine, observed in the addition of amide or urea nitrogen, is substitution of the hydroxyl by a second molecule of the nucleophile (Equation 8.48).

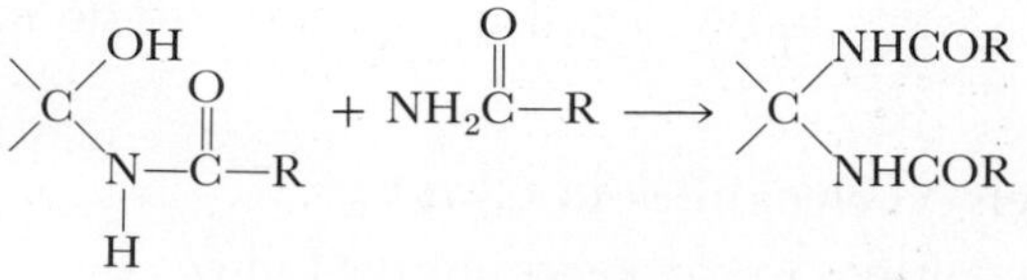

$$
\begin{array}{c}
\text{OH} \\
\diagup \text{C} \diagdown \\[2pt]
\text{N---C---R} \\
\mid \\
\text{H}
\end{array}
+ \text{NH}_2\text{C---R} \longrightarrow
\begin{array}{c}
\text{NHCOR} \\
\diagup \text{C} \diagdown \\
\text{NHCOR}
\end{array}
\qquad (8.48)
$$

Addition of a secondary amine to a carbonyl group leads to a carbinolamine that cannot attain a neutral structure with a carbon–nitrogen double bond; if there is a hydrogen on the α carbon, elimination of water can occur in this direction to yield a product with a carbon–carbon double bond (Equation 8.49), a process similar to that which occurs in aldol-type condensations. The vinyl amines (**36**) are

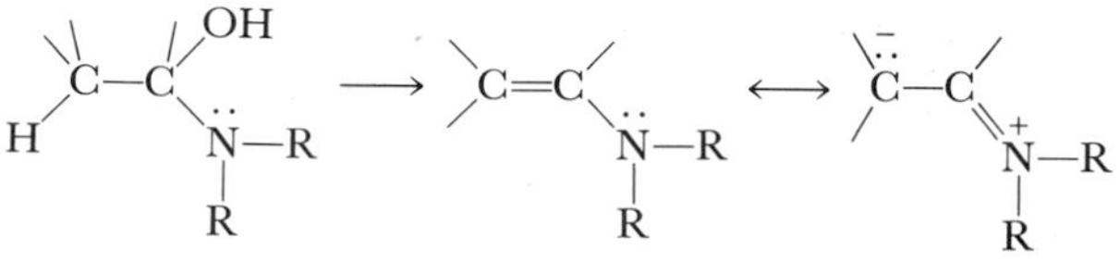

$$
\begin{array}{c}
\text{OH} \\
\diagup \text{C---C} \diagdown \\
\text{H} \qquad \text{N---R} \\
\mid \\
\text{R}
\end{array}
\longrightarrow
\begin{array}{c}
\diagup \text{C}=\text{C} \diagdown \\
\text{N---R} \\
\mid \\
\text{R}
\end{array}
\longleftrightarrow
\begin{array}{c}
\diagup \text{C---C} \diagdown \\
\overset{+}{\text{N}}\text{---R} \\
\mid \\
\text{R}
\end{array}
\qquad (8.49)
$$

36

referred to as *enamines;* they find application in synthesis[97] and are important biochemical intermediates. Enamines serve as carbanion equivalents by virtue of the electron distribution indicated by the resonance structures **36**.

If a tertiary amine adds to a carbonyl group, the carbinolamine is ionic. There is then no possibility of forming a neutral addition compound; the carbinolamine reverts to reactants.

Addition of primary amines to carbonyl groups has been the subject of extensive study, notably by Jencks and co-workers. The most striking feature of these reactions is the characteristic maximum in the graph of reaction rate as a function of pH.[98] Figure 8.12 illustrates the observations for the reaction of hydroxylamine with acetone. It is also found that the sensitivity of rate to acid catalysis, and to substituent effects, is different on either side of the maximum in the pH–rate curve.[99] These phenomena may be understood in terms of the two-step nature of the reaction. Either the addition step or the elimination step may be rate determining depending on the pH.

It is convenient for further analysis to divide the nitrogen nucleophiles into two categories: the strongly basic hydroxylamine and aliphatic amines, $pK_{a_{\text{BH}^+}}$

[97] See, for example, (a) J. Szmuszkovicz, *Advances in Organic Chemistry: Methods and Results,* Vol. 4, Wiley, New York, 1963, p. 1; (b) S. F. Dyke, *The Chemistry of Enamines,* Cambridge University Press, London, 1973.

[98] (a) W. P. Jencks, *J. Am. Chem. Soc.,* **81**, 475 (1959); (b) E. Barrett and A. Lapworth, *J. Chem. Soc.,* **93**, 85 (1908); (c) J. B. Conant and P. D. Bartlett, *J. Am. Chem. Soc.,* **54**, 2881 (1932); (d) J. C. Powers and F. H. Westheimer, *J. Am. Chem. Soc.,* **82**, 5431 (1960). See also note 93(a), (c).

[99] (a) E. H. Cordes and W. P. Jencks, *J. Am. Chem. Soc.,* **84**, 832 (1962); (b) B. M. Anderson and W. P. Jencks, *J. Am. Chem. Soc.,* **82**, 1773 (1960). See also note 98(d).

roughly 6 to 10, and the weakly basic semicarbazide and aryl amines, $pK_{a_{BH^+}}$ about 4. Considering the former class first, we find good evidence that in neutral solution, which is on the basic side of the maximum in the pH–rate curve, the second step (dehydration) is rate determining. For example, it is observed that when the nucleophile and carbonyl compound are mixed in relatively concentrated solution, the characteristic absorption spectrum of the carbonyl group disappears rapidly, and the spectrum of the final product, the imine, appears much more slowly.

SCHEME 12

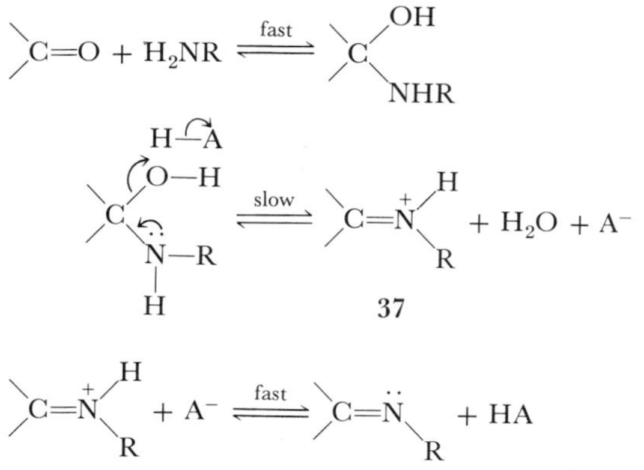

$$\text{\bigg\rangle}C{=}O + H_2NR \overset{fast}{\rightleftharpoons} \text{\bigg\rangle}C\overset{OH}{\underset{NHR}{<}}$$

$$\underset{\underset{H}{|}}{\overset{H{-}A}{\overset{O{-}H}{\underset{N{-}R}{|}}}}C\text{\bigg\langle} \overset{slow}{\rightleftharpoons} \text{\bigg\rangle}C{=}\overset{+}{N}\overset{H}{\underset{R}{<}} + H_2O + A^-$$

37

$$\text{\bigg\rangle}C{=}\overset{+}{N}\overset{H}{\underset{R}{<}} + A^- \overset{fast}{\rightleftharpoons} \text{\bigg\rangle}C{=}\overset{..}{N}\underset{R}{<} + HA$$

R = alkyl or OH

SCHEME 13

$$\text{\bigg\rangle}C\overset{OH}{\underset{NHR}{<}} + HA \overset{fast}{\rightleftharpoons} \text{\bigg\rangle}C\overset{\overset{+}{O}H_2}{\underset{NHR}{<}} + A^-$$

$$\underset{\underset{R}{|}}{\overset{\overset{+}{O}H_2}{\underset{N{-}H}{|}}}C\text{\bigg\langle} + A^- \overset{slow}{\rightleftharpoons} H_2O + \text{\bigg\rangle}C{=}\overset{..}{N}\underset{R}{<} + HA$$

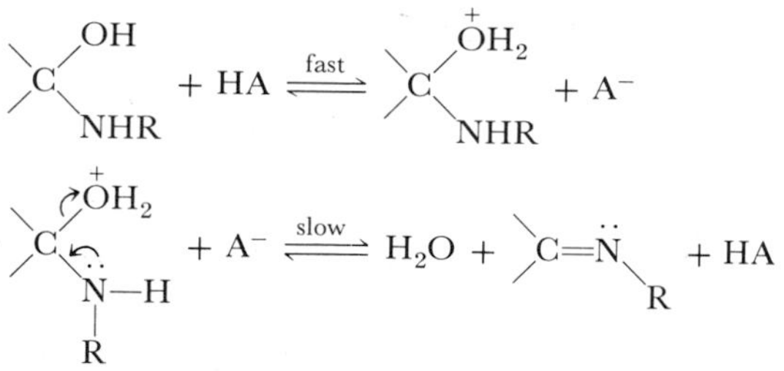

In the pH region where dehydration is rate determining, the overall reaction is subject to general acid catalysis. That the mechanism is probably true general acid catalysis (Scheme 12) and not specific acid plus general base catalysis (Scheme 13) is indicated by the close similarity of characteristics of Reactions 8.50 and 8.51. Both show the same general catalysis, with Brønsted $\alpha = 0.77$. Nitrone **39** corresponds to the intermediate **37** that appears only in Scheme 12. Therefore its formation and hydrolysis must follow Scheme 12 rather than Scheme 13, and the same mechanism is therefore postulated for formation and hydrolysis of oxime **38**.[100]

[100](a) J. E. Reimann and W. P. Jencks, *J. Am. Chem. Soc.*, **88**, 3973 (1966); (b) K. Koehler, W. Sandstrom, and E. H. Cordes, *J. Am. Chem. Soc.*, **86**, 2413 (1964).

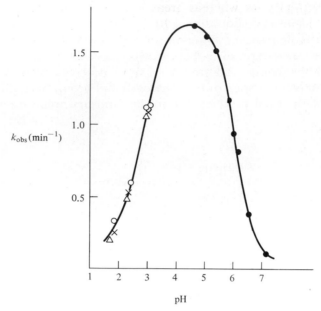

Figure 8.12 Dependence of rate on pH for the reaction

$$\underset{H_3C}{\overset{H_3C}{\diagdown}}C=O + H_2NOH \longrightarrow \underset{H_3C}{\overset{H_3C}{\diagdown}}C=N\underset{OH}{\diagdown} + H_2O$$

Reprinted with permission from W. P. Jencks, *J. Am. Chem. Soc.*, **81**, 475 (1959). Copyright 1959 American Chemical Society.

$$\underset{H}{\overset{p\text{-}Cl-\phi}{\diagdown}}C=O + H_2NOH \rightleftharpoons \underset{H}{\overset{p\text{-}Cl-\phi}{\diagdown}}C=NOH + H_2O \qquad (8.50)$$

38

$$\underset{H}{\overset{p\text{-}Cl-\phi}{\diagdown}}C=O + \underset{H_3C}{\overset{H}{\diagdown}}N-OH \rightleftharpoons \underset{H}{\overset{p\text{-}Cl-\phi}{\diagdown}}C=\overset{+}{N}\underset{CH_3}{\overset{O^-}{\diagup}} + H_2O \qquad (8.51)$$

39

This evidence also establishes the mechanism for the reverse reaction, addition of water to C=N; note that it differs from addition to C=O in that an equilibrium protonation precedes the addition.

In more acidic solutions the rate increases as a result of the acid catalysis of the dehydration step, until the maximum rate is attained, usually at pH between 2 and 5.[101] The decrease of rate that occurs on further decrease of pH can be ex-

---

[101] E. H. Cordes and W. P. Jencks, *J. Am. Chem. Soc.*, **85**, 2843 (1963).

plained by assuming that whereas greater acidity facilitates the dehydration, it inhibits the addition step (Equation 8.46) because only the unprotonated amine is reactive. The first step then becomes rate determining. In this pH region there is no general catalysis with the more strongly basic amines.[102] These bases, like cyanide ion, are strong enough not to require assistance by proton transfer in the rate-determining step; the mechanism is as shown in Equation 8.52. The intramolecular

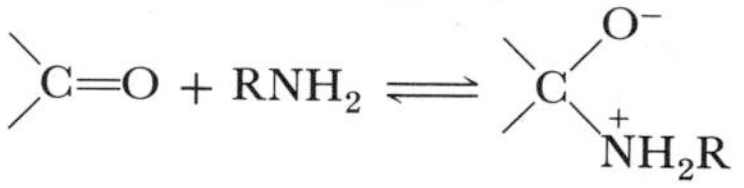

$$\text{>C=O} + RNH_2 \;\rightleftharpoons\; \text{>C} \begin{smallmatrix} O^- \\ \\ \overset{+}{N}H_2R \end{smallmatrix} \tag{8.52}$$

40

proton transfer from nitrogen to oxygen in intermediate **40** is presumed to be fast compared to the addition in these circumstances, although, as we shall see, it can become rate determining under certain conditions. The entire mechanism is shown in Scheme 14.

SCHEME 14

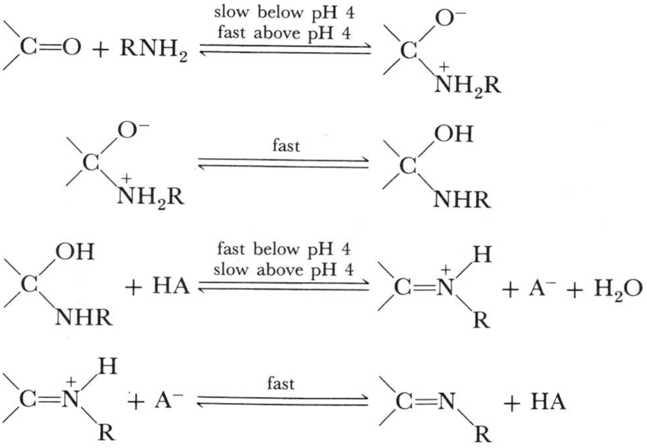

The general pattern we have outlined holds true also for the more weakly basic nitrogen nucleophiles such as semicarbazide or aryl amines. The electron-withdrawing groups retard the addition step, in which the nitrogen unshared pair attacks carbon, but they also retard the dehydration, where again the unshared pair is acting in a nucleophilic manner to expel the hydroxyl group. The change from rate-determining addition to rate-determining dehydration is observed at roughly the same pH as for the more basic amines. The weaker bases, however, require more help from acid catalysts. In the addition step, general acid catalysis is

[102] (a) W. P. Jencks, *Prog. Phys. Org. Chem.*, **2**, 63 (1964); (b) M. Cocivera and K. W. Woo, *J. Am. Chem. Soc.*, **98**, 7366 (1976); (c) M. Cocivera and A. Effio, *J. Am. Chem. Soc.*, **98**, 7371 (1976).

found with α about 0.25.[103] The dehydration step shows general acid catalysis similar to that found with the more basic amines.[104] The detailed mechanism is given in Scheme 15.

SCHEME 15

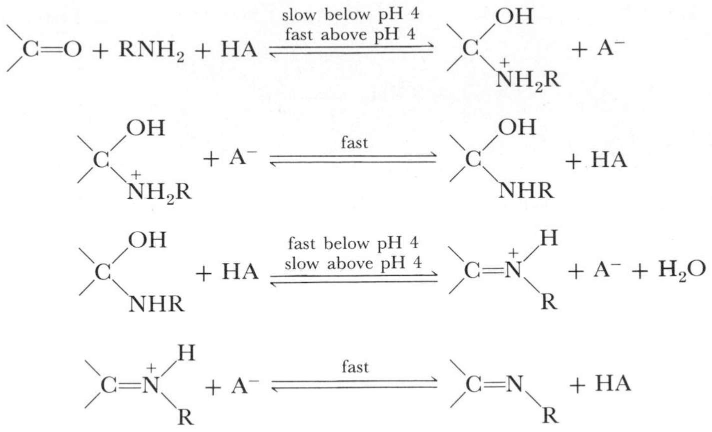

One might inquire at this point about the addition of a secondary amine that cannot yield a stable neutral product by dehydration as the primary amine can and that also cannot form an enamine. Diebler and Thorneley measured rate constants of the addition step for reaction of piperazine (**41**) with pyridine-4-aldehyde in the pH range 5.8–10.8, a range in which the addition step is very fast so that, for primary amines, the kinetics would be determined by the rate-limiting dehydration.[105] They were able to show, by use of fast-reaction measurement techniques, that the general base catalysis observed is a result of a simple rate-determining proton transfer in Step 2 of Scheme 16. The rate constant for the actual addi-

---

[103](a) E. H. Cordes and W. P. Jencks, *J. Am. Chem. Soc.*, **84**, 4319 (1962). Although the general acid catalysis is in some instances due to the concerted addition and proton transfer indicated as the first step of Scheme 13, there are examples in which it results from the proton-trapping (enforced catalysis) mechanism discussed earlier in connection with thiol additions:

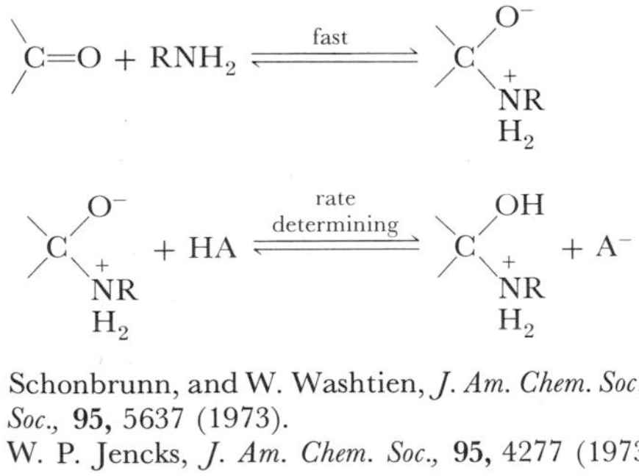

See (b) J. M. Sayer, B. Pinsky, A. Schonbrunn, and W. Washtien, *J. Am. Chem. Soc.*, **96**, 7998 (1974); (c) J. M. Sayer and W. P. Jencks, *J. Am. Chem. Soc.*, **95**, 5637 (1973).
[104]J. M. Sayer, M. Peskin, and W. P. Jencks, *J. Am. Chem. Soc.*, **95**, 4277 (1973). There is also a general base-catalyzed dehydration.
[105]H. Diebler and R. N. F. Thorneley, *J. Am. Chem. Soc.*, **95**, 896 (1973).

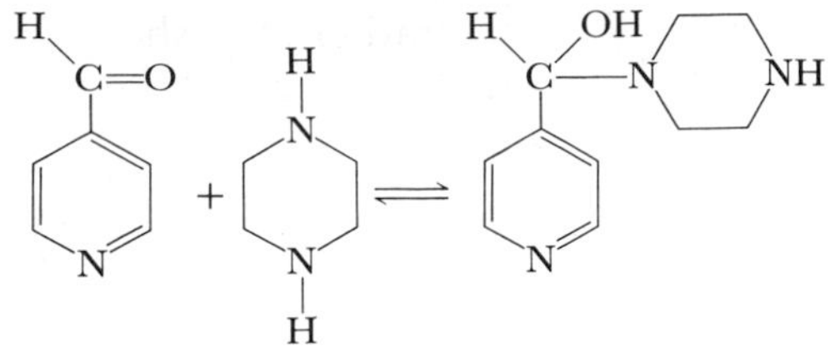

$$(8.53)$$

41

tion (Step 1) is on the order of $10^7 M^{-1} \sec^{-1}$. Note that this reaction is another example of enforced catalysis, in which an intermediate of very short lifetime is trapped by a rate-determining proton transfer. When no base catalyst is present, Step 2 is still rate determining, but it now consists of an intramolecular proton transfer from nitrogen to oxygen.

SCHEME 16

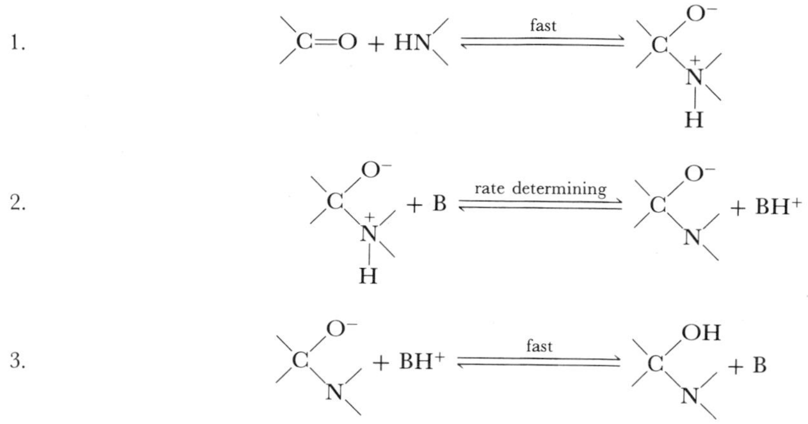

1.

2.

3.

Nucleophilic Catalysis

It has been found that amines frequently are effective catalysts for addition of other nucleophiles to carbonyl groups.[106] The reason for this catalysis is that amines can add rapidly to the carbonyl compound to form an imine; the imine in turn is subject to the same kinds of addition reactions as are carbonyl compounds, but reacts faster because it is more easily protonated. Scheme 17 illustrates this process, which is referred to as *nucleophilic catalysis*. In order for nucleophilic catalysis of addition of Q to occur, it is necessary that (1) the rate of addition of the catalyst to the carbonyl group be greater than the rate of addition of Q; (2) the protonated imine be more reactive than the carbonyl compound toward Q; and (3) the equilibrium favor the Q addition more than amine addition.

[106] M. L. Bender, *Mechanisms of Homogeneous Catalysis from Protons to Proteins,* Wiley, New York, 1971, p. 165.

Scheme 17

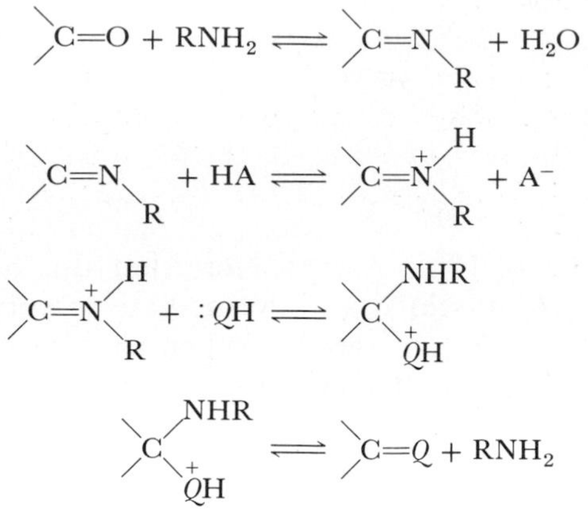

$$\rangle C{=}O + RNH_2 \rightleftharpoons \rangle C{=}N{\diagdown}_R + H_2O$$

$$\rangle C{=}N{\diagdown}_R + HA \rightleftharpoons \rangle C{=}\overset{+}{N}{\diagdown}_R^H + A^-$$

$$\rangle C{=}\overset{+}{N}{\diagdown}_R^H + :QH \rightleftharpoons \rangle C{\diagdown}_{\overset{+}{Q}H}^{NHR}$$

$$\rangle C{\diagdown}_{\overset{+}{Q}H}^{NHR} \rightleftharpoons \rangle C{=}Q + RNH_2$$

8.5 CARBOXYLIC ACID DERIVATIVES

When the carbonyl group bears as one substituent a potential Lewis base leaving group, a second nucleophile can substitute for the first, as illustrated in Equation 8.54.

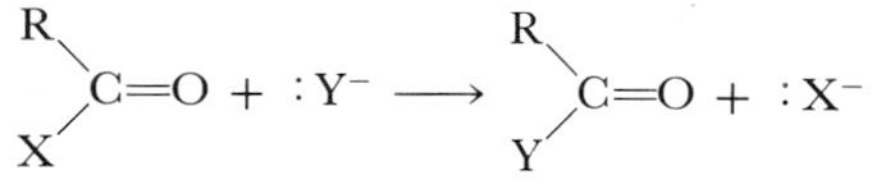

$$\overset{R}{\underset{X}{\diagdown}}C{=}O + :Y^- \longrightarrow \overset{R}{\underset{Y}{\diagdown}}C{=}O + :X^- \tag{8.54}$$

42

Structures **42** are conveniently thought of as derivatives of carboxylic acids, and include acids, esters, anhydrides, acyl halides, and amides. These structures (and others less commonly encountered) can be readily interconverted, either directly or indirectly; the number of different reactions is therefore large.[107] Because these processes occupy an important place in organic chemistry and because carboxylic acid derivatives are of central importance in biochemical systems and therefore of considerable interest in the study of enzyme action, they have been the subject of intensive investigation.[108]

---

[107] Surveys outlining the general features of many of the reactions may be found in (a) J. March, *Advanced Organic Chemistry: Reactions, Mechanisms, and Structure,* McGraw-Hill, New York, 1968; (b) D. P. N. Satchell, *Q. Rev. Chem. Soc.,* **17**, 160 (1963).

[108] Comprehensive discussions are to be found in: (a) M. L. Bender, *Mechanisms of Homogeneous Catalysis from Protons to Proteins,* Wiley, New York, 1971; (b) W. P. Jencks, *Catalysis in Chemistry and Enzymology,* McGraw-Hill, New York, 1969; (c) M. L. Bender, *Chem. Rev.,* **60**, 53 (1960). For more specialized treatments of particular aspects, see (d) W. P. Jencks, *Chem. Rev.,* **72**, 705 (1972), general acid–base catalysis; (e) S. L. Johnson, *Adv. Phys. Org. Chem.,* **5**, 237 (1967), ester hydrolysis; (f) L. P. Hammett, *Physical Organic Chemistry,* 2d ed., McGraw-Hill, New York, 1970, Chap. 10, acid–base catalysis.

Mechanistic Alternatives

The mechanism for Reaction 8.54 that we would expect by analogy with the chemistry of other types of carbonyl compounds is the *addition mechanism,* Scheme 18. Addition of the nucleophile $:Y^-$ to the carbonyl group generates tetrahedral addition intermediate **43**, which either reverts to starting materials or goes on by elimination of the leaving group $:X^-$ to yield a new carboxylic acid derivative. As we shall see in more detail below, this is the pathway most often followed in these reactions.[109]

SCHEME 18

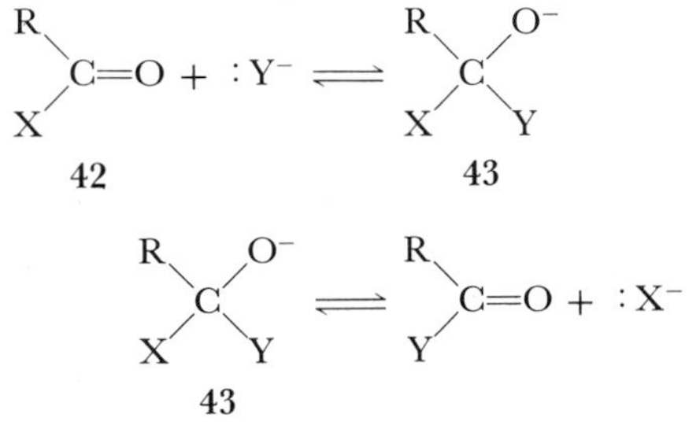

The presence of a leaving group provides the possibility of two other routes, Schemes 19 and 20, the first of which is an acid-catalyzed S_N1 substitution, designated A–1, and the second is an S_N2 substitution. Although these processes are observed much less commonly than that shown in Scheme 18, they do occur under

SCHEME 19

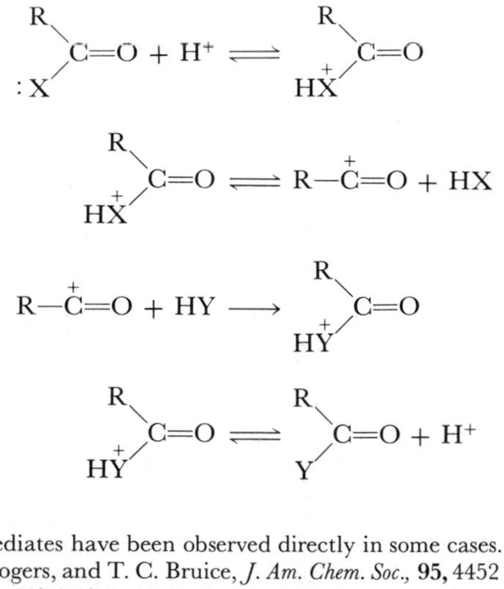

[109]Tetrahedral addition intermediates have been observed directly in some cases. (a) M. L. Bender, *J. Am. Chem. Soc.,* **75**, 5986 (1953); (b) G. A. Rogers, and T. C. Bruice, *J. Am. Chem. Soc.,* **95**, 4452 (1973); (c) N. Gravitz and W. P. Jencks, *J. Am. Chem. Soc.,* **96**, 489, 507 (1974); (d) B. Capon, J. H. Gall, and D. McL. A. Grieve, *J. Chem. Soc., Chem. Commun.,* 1034 (1976); (e) gas phase: O. I. Asubiojo, L. K. Blair, and J. I. Brauman, *J. Am. Chem. Soc.,* **97**, 6685 (1975).

some circumstances, for example, in acid-catalyzed reactions of acyl halides[110] and hydrolysis of esters and amides in water–acid mixtures of high acid content.[111]

SCHEME 20

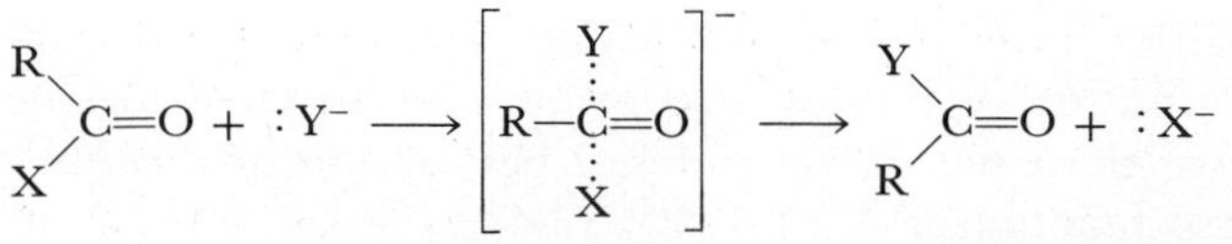

$$
\underset{X}{\overset{R}{>}}C=O + :Y^- \longrightarrow \left[R-\underset{\underset{X}{\cdot\cdot}}{\overset{Y}{\underset{|}{C}}}=O \right]^- \longrightarrow \underset{R}{\overset{Y}{>}}C=O + :X^-
$$

Oxygen exchange Bender introduced an isotopic-labeling method to demonstrate the presence of tetrahedral addition intermediate **43** in hydrolysis reactions.[112] Addition of water to a carbonyl oxygen-labeled substrate (Scheme 21) would yield an intermediate with two oxygens bonded to carbon; a simple proton transfer, which should be very rapid, would make these oxygens equivalent. Reversal of the addition step would then exchange the oxygen, so that unreacted starting material isolated after the reaction was partly complete would have a reduced proportion of labeled oxygen. A similar process, with one less proton, can be written for addition of hydroxide ion. Bender found significant exchange in ester hydrolysis; exchange has also been found in hydrolysis of amides, acyl chlorides, and anhydrides.[113]

SCHEME 21

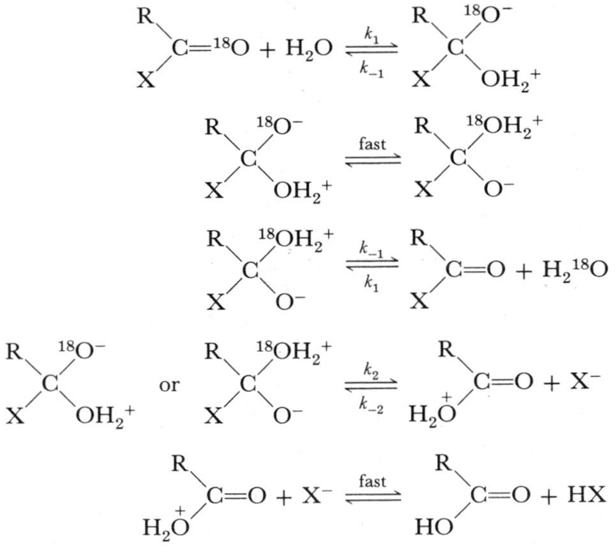

---

[110](a) D. P. N. Satchell, *J. Chem. Soc.*, 558, 564 (1963). See also (b) M. L. Bender and M. C. Chen, *J. Am. Chem. Soc.*, **85**, 30, 37 (1963).

[111]Amides: (a) J. T. Edward, G. D. Derdall, and S. C. Wong, *J. Am. Chem. Soc.*, **100**, 7023 (1978); (b) C. R. Smith and K. Yates, *J. Am. Chem. Soc.*, **94**, 8811 (1972); esters: (c) K. Yates, *Acc. Chem. Res.*, **4**, 136 (1971). Review: (d) A. Williams and K. T. Douglas, *Chem. Rev.*, **75**, 627 (1975).

[112]M. L. Bender, *J. Am. Chem. Soc.*, **73**, 1626 (1951).

[113]D. Samuel and B. L. Silver, *Adv. Phys. Org. Chem.*, **3**, 123 (1965).

Although observation of oxygen exchange confirms the existence of the tetrahedral intermediate,[114] failure to find exchange does not demonstrate its absence. For exchange to occur in Scheme 21, k_{-1} must be comparable to or greater than k_2; that is, either the k_2 step is rate determining ($k_{-1} \gg k_2$) or the two steps are of about the same rate ($k_{-1} \approx k_2$). If the addition step is rate determining ($k_{-1} \ll k_2$), very little of the tetrahedral intermediate will return to starting material and there will be no observable exchange. Failure to find exchange is therefore compatible with Scheme 18 (addition step rate determining), Scheme 19, or Scheme 20.

Catalysis Catalysis by acids and bases is a characteristic feature of carboxylic acid derivative chemistry. Base catalysis is possible, as in other carbonyl group reactions, through deprotonation of the nucleophile. Acid catalysis can operate either by protonation on carbonyl oxygen, as in reactions considered earlier, or by protonation of the leaving group X, which usually has unshared pairs of electrons. Elucidation of mechanism therefore requires determination of both state and site of protonation of the reactive form of the substrate as well as identification of the rate-determining step.

In most instances of acid catalysis, the protonation appears to be on the carbonyl oxygen.[115] This protonation site favors the addition mechanism (Scheme 18), whereas the S_N1 and S_N2 mechanisms (Schemes 19 and 20) would benefit from protonation of the leaving group X. This latter type of catalysis is therefore associated with these less common reaction types. Acid catalysis is expected to be effective for the less reactive carbonyl groups, as in esters and amides; for acid chlorides the electron-withdrawing halogen makes the unprotonated carbonyl so reactive that acid catalysis is not usually observed, except again in those cases where an S_N1 mechanism is being followed.

Nucleophilic catalysis is a process of particular significance in reactions of carboxylic acid derivatives. As an example we may cite hydrolysis catalyzed by a tertiary amine (Scheme 22). The catalysis is effective because initial attack of the amine will be faster than attack by the less nucleophilic water; the amine addition yields the intermediate **44**, which, because of the positive charge, has an extremely reactive carbonyl group and is attacked by water much faster than the original compound. The fact that a given base is acting by nucleophilic catalysis rather than by general base catalysis (removal of a proton from attacking H_2O) can be established by noting deviations from the Brønsted catalysis correlation that demonstrate that a given substance is markedly more effective as a catalyst than its proton basicity would indicate, or by structural variations that show the effectiveness as a catalyst to be more sensitive than proton basicity to steric effects.[116]

Electrophilic catalysis by Lewis acids is also observed. An interesting exam-

---

[114] Strictly speaking, the observation of exchange shows only that tetrahedral intermediate is present, not that it necessarily lies on the reaction path leading to product. But the possibility that the intermediate would be readily and reversibly formed yet not go on to product seems unreasonable.

[115] M. L. Bender, *Chem. Rev.*, **60**, 53 (1960).

[116] See, for example, M. L. Bender, *Mechanisms of Homogeneous Catalysis from Protons to Proteins*, Wiley, New York, 1971, Chap. 6.

SCHEME 22

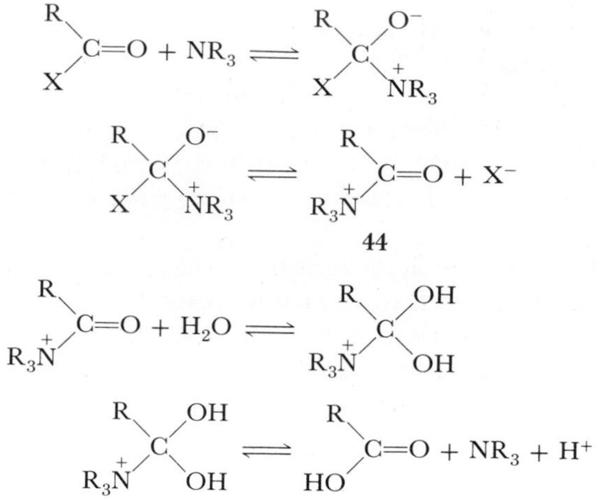

$$\underset{X}{\overset{R}{\diagdown}}C{=}O + NR_3 \;\rightleftharpoons\; \underset{X}{\overset{R}{\diagdown}}\underset{\overset{+}{NR_3}}{\overset{O^-}{\diagup}}C$$

$$\underset{X}{\overset{R}{\diagdown}}\underset{\overset{+}{NR_3}}{\overset{O^-}{\diagup}}C \;\rightleftharpoons\; \underset{\overset{+}{R_3N}}{\overset{R}{\diagdown}}C{=}O + X^-$$

44

$$\underset{\overset{+}{R_3N}}{\overset{R}{\diagdown}}C{=}O + H_2O \;\rightleftharpoons\; \underset{\overset{+}{R_3N}}{\overset{R}{\diagdown}}\underset{OH}{\overset{OH}{\diagup}}C$$

$$\underset{\overset{+}{R_3N}}{\overset{R}{\diagdown}}\underset{OH}{\overset{OH}{\diagup}}C \;\rightleftharpoons\; \underset{HO}{\overset{R}{\diagdown}}C{=}O + NR_3 + H^+$$

ple is the strong catalysis of thiolester hydrolysis by mercuric and silver ions. These soft acids presumably coordinate with the sulfur and, by virtue of the consequent electron withdrawal, make the carbonyl group much more susceptible to attack in the addition mechanism, or, in favorable cases, promote unimolecular S_N1 cleavage of the sulfur–carbon bond.[117]

Amide Hydrolysis

Hydrolysis of amides can occur with either acid or base catalysis. The base catalysis mechanism is shown in Scheme 23. Because H_2N^- is a poorer leaving group than HO^-, k_{-1} is greater than k_2, and the k_2 step is rate determining.[118] As a result, oxygen exchange in unreacted amide is rapid, at least for primary and secondary amides.[119]

Another consequence of the relative magnitudes of k_{-1} and k_2 is that base-catalyzed hydrolysis of amides in aqueous solution is not a very efficient process, and acid catalysis is ordinarily the method of choice for synthesis. It is, however, possible to hydrolyze amides efficiently under basic conditions by using only a

[117] D. P. N. Satchell and I. I. Secemski, *Tetrahedron Lett.*, 1991 (1969).

[118] C. D. Ritchie, *J. Am. Chem. Soc.*, **97**, 1170 (1975), has determined relative leaving-group ability from a tetrahedral addition intermediate for a number of nucleophiles. However, quantitative data for negatively charged nucleophiles are not given.

[119] C. A. Bunton, B. Nayak, and C. O'Connor, *J. Org. Chem.*, **33**, 572 (1968). Tertiary amides, R—CONR$_2'$, show no oxygen exchange. The reason may be that the oxygen equilibration (second step, Scheme 21) is mediated by a proton on nitrogen:

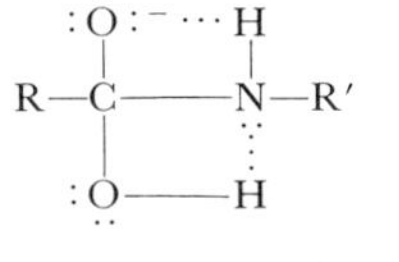

$$\underset{\substack{\,\\ :O:{\overset{\cdot\cdot}{}}}}{\overset{:\ddot{O}:^-\cdots H}{\underset{|}{R{-}C}}}\!\!\underset{\substack{|\\ H}}{\overset{|}{N{-}R'}}$$

SCHEME 23

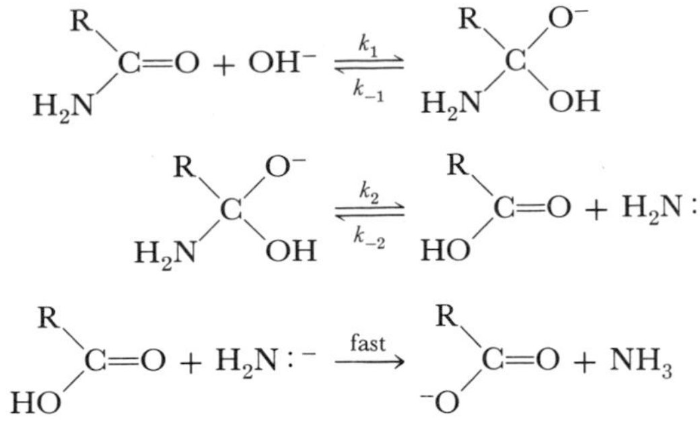

$$R_2N\!\!-\!\!C\!\!=\!\!O + OH^- \underset{k_{-1}}{\overset{k_1}{\rightleftharpoons}} \text{(intermediate)}$$

small excess of water, excess potassium *tert*-butoxide as base, and ether as solvent. Under these conditions the intermediate is presumably deprotonated to dianion **45**, which can then expel the negatively charged nitrogen.[120]

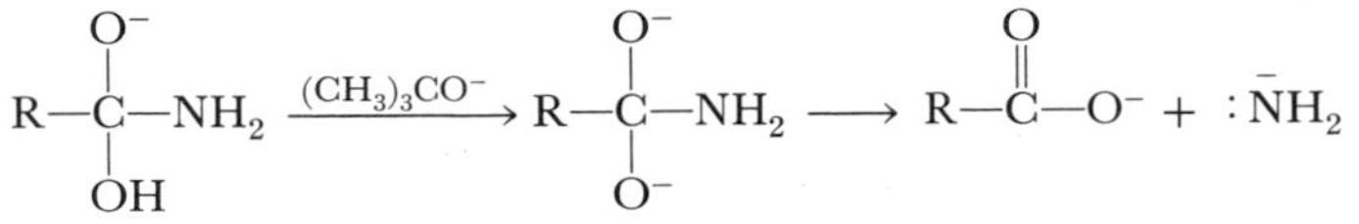

$$R\!-\!\overset{O^-}{\underset{OH}{\underset{|}{\overset{|}{C}}}}\!-\!NH_2 \xrightarrow{(CH_3)_3CO^-} R\!-\!\overset{O^-}{\underset{O^-}{\underset{|}{\overset{|}{C}}}}\!-\!NH_2 \longrightarrow R\!-\!\overset{O}{\underset{}{\overset{\parallel}{C}}}\!-\!O^- + {}^-\!:NH_2 \qquad (8.55)$$

$$\mathbf{45}$$

Acid catalysis of amide hydrolysis is mechanistically more complex than base catalysis. As we have noted above, there are two potential sites of protonation: the amide nitrogen and the carbonyl oxygen. Oxygen protonation leads logically to the addition mechanism, whereas nitrogen protonation lends itself to an S_N1 type of process.

The general tendency for protonation to occur primarily on carbonyl oxygen in carboxylic acid derivatives[121] apparently holds for amides. Gas-phase measurements of nitrogen $1s$ ionization potentials of protonated amides by X-ray electron spectroscopy (ESCA) show marked deviations from values found for protonated amines, indicating O protonation in the amides. Spectroscopic measurements in solution, particularly nuclear magnetic resonance spectroscopy, demonstrate predominant oxygen protonation.[122] As one would expect on the basis of these findings, acid-catalyzed hydrolysis occurs by the addition mechanism (Scheme 24) in most instances. (The observed lack of oxygen exchange in acid-catalyzed amide hydrolysis is explained by $k_2 > k_{-1}$ in Scheme 21.) Evidence includes the behavior of the transition state activity coefficient ($\gamma^{\ddagger}$) compared with that for ester hydrolysis, and compared with activity coefficients of stable ions, in

[120](a) P. G. Gassman, P. K. G. Hodgson, and R. L. Balchunis, *J. Am. Chem. Soc.*, **98**, 1275 (1976). For other examples of amide reactions catalyzed by bases, see (b) J. W. Henderson and P. Haake, *J. Org. Chem.*, **42**, 3989 (1977); (c) R. Kluger and C.-H. Lam, *J. Am. Chem. Soc.*, **100**, 2191 (1978); (d) T. H. Fife and V. L. Squillacote, *J. Am. Chem. Soc.*, **100**, 4787 (1978).

[121]M. L. Bender, *Chem. Rev.*, **60**, 53 (1960).

[122](a) A. R. Katritzky and R. A. Y. Jones, *Chem. Ind.* (*London*), 722 (1961); (b) R. J. Gillespie and T. Birchall, *Can. J. Chem.*, **41**, 148 (1963).

SCHEME 24

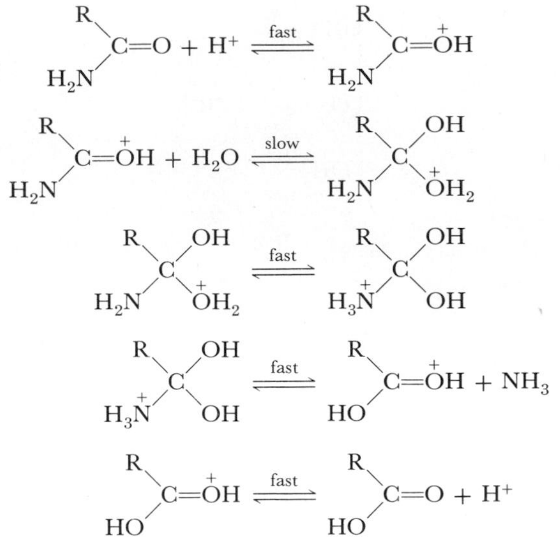

$$\begin{array}{c}R\\ \diagdown\\ \diagup\\ H_2N\end{array}C{=}O + H^+ \underset{}{\overset{fast}{\rightleftharpoons}} \begin{array}{c}R\\ \diagdown\\ \diagup\\ H_2N\end{array}C{=}\overset{+}{O}H$$

$$\begin{array}{c}R\\ \diagdown\\ \diagup\\ H_2N\end{array}\overset{+}{C}{=}OH + H_2O \underset{}{\overset{slow}{\rightleftharpoons}} \begin{array}{cc}R & OH\\ \diagdown & \diagup\\ & C\\ \diagup & \diagdown\\ H_2N & \overset{+}{O}H_2\end{array}$$

$$\begin{array}{cc}R & OH\\ \diagdown & \diagup\\ & C\\ \diagup & \diagdown\\ H_2N & \overset{+}{O}H_2\end{array} \underset{}{\overset{fast}{\rightleftharpoons}} \begin{array}{cc}R & OH\\ \diagdown & \diagup\\ & C\\ \diagup & \diagdown\\ H_3\overset{+}{N} & OH\end{array}$$

$$\begin{array}{cc}R & OH\\ \diagdown & \diagup\\ & C\\ \diagup & \diagdown\\ H_3\overset{+}{N} & OH\end{array} \underset{}{\overset{fast}{\rightleftharpoons}} \begin{array}{c}R\\ \diagdown\\ \diagup\\ HO\end{array}C{=}\overset{+}{O}H + NH_3$$

$$\begin{array}{c}R\\ \diagdown\\ \diagup\\ HO\end{array}\overset{+}{C}{=}OH \underset{}{\overset{fast}{\rightleftharpoons}} \begin{array}{c}R\\ \diagdown\\ \diagup\\ HO\end{array}C{=}O + H^+$$

acid media of varying acid concentration. The transition state for the addition mechanism (**46**) has greater solvation demands than that for the S_N1 process (**47**) because it has more hydrogen-bonding sites; hence $\gamma^{\ddagger}$ for **46** increases more rapidly with increasing acid concentration than does that for **47**. In ester hydrolysis (see

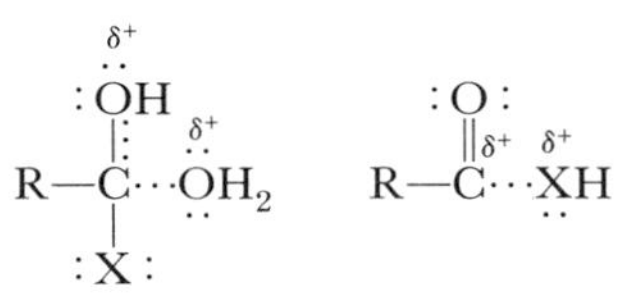

$$\begin{array}{ccc}& \overset{\delta^+}{:\ddot{O}H} & & :\ddot{O}:\\ & |\,\vdots\,\overset{\delta^+}{} & & \|\overset{\delta^+}{}\overset{\delta^+}{}\\ R{-}\overset{|}{\underset{|}{C}}\cdots\ddot{O}H_2 & & R{-}C\cdots XH\\ & :\ddot{X}: & & \\ & \mathbf{46} & & \mathbf{47}\end{array}$$

below) one sees both kinds of behavior depending on the ester structure. Amides ordinarily follow the behavior corresponding to the addition mechanism in water–strong acid mixtures containing up to about 80 percent acid.[123] Above this concentration the S_N1-type A–1 mechanism (Scheme 25) appears to take over.[124]

Ester Hydrolysis[125]

In ester hydrolysis an extra complication is introduced because the cleavage of either of two bonds (Equations 8.56 and 8.57) leads to the same product. Both of

[123] K. Yates and T. A. Modro, *Acc. Chem. Res.,* **11**, 190 (1978). See Section 3.2, for discussion of activity coefficients in strong acids.

[124] (a) J. T. Edward, G. D. Derdall, and S. C. Wong, *J. Am. Chem. Soc.,* **100**, 7023 (1978); (b) J. T. Edward and S. C. R. Meacook, *J. Chem. Soc.,* 2000 (1957). There is some evidence for a wider applicability of the A–1 mechanism: (c) C. R. Smith and K. Yates, *J. Am. Chem. Soc.,* **94**, 8811 (1972).

[125] Reviews: (a) C. H. Bamford and C. F. H. Tipper, Eds., *Ester Formation and Hydrolysis* (Comprehensive Chemical Kinetics, Vol. 10), Elsevier, Amsterdam, 1972, Chaps. 2 (A. J. Kirby) and 3 (R. E. J. Talbot); (b) C. K. Ingold, *Structure and Mechanism in Organic Chemistry,* 2d ed., Cornell University Press, Ithaca, N.Y., 1969, p. 1131.

SCHEME 25

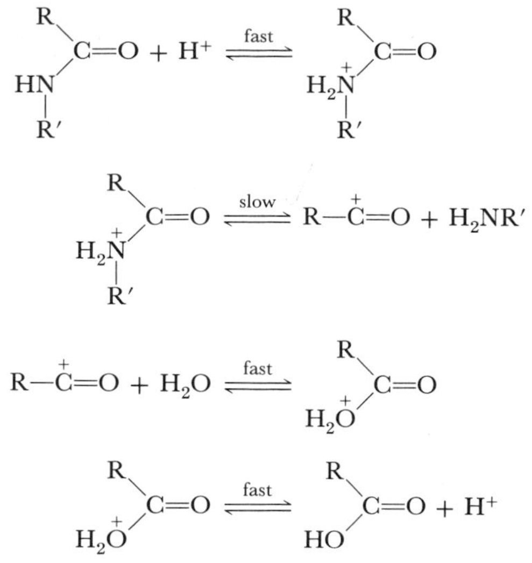

$$\underset{HN}{\overset{R}{\diagdown}}C=O + H^+ \underset{\longleftarrow}{\overset{fast}{\longrightarrow}} \underset{H_2\overset{+}{N}}{\overset{R}{\diagdown}}C=O$$
$$\underset{R'}{|} \qquad\qquad \underset{R'}{|}$$

$$\underset{H_2\overset{+}{N}}{\overset{R}{\diagdown}}C=O \underset{\longleftarrow}{\overset{slow}{\longrightarrow}} R\!-\!\overset{+}{C}\!=\!O + H_2NR'$$
$$\underset{R'}{|}$$

$$R\!-\!\overset{+}{C}\!=\!O + H_2O \underset{\longleftarrow}{\overset{fast}{\longrightarrow}} \underset{H_2\overset{+}{O}}{\overset{R}{\diagdown}}C=O$$

$$\underset{H_2\overset{+}{O}}{\overset{R}{\diagdown}}C=O \underset{\longleftarrow}{\overset{fast}{\longrightarrow}} \underset{HO}{\overset{R}{\diagdown}}C=O + H^+$$

these types of cleavage are observed, as are catalysis by both acids and bases. The result is a rich array of possible mechanisms, of which we will discuss only the most common.

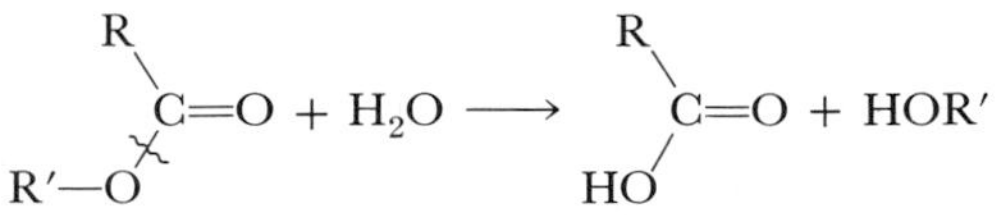

$$\underset{R'-O}{\overset{R}{\diagdown}}C=O + H_2O \longrightarrow \underset{HO}{\overset{R}{\diagdown}}C=O + HOR' \qquad (8.56)$$

acyl–oxygen cleavage

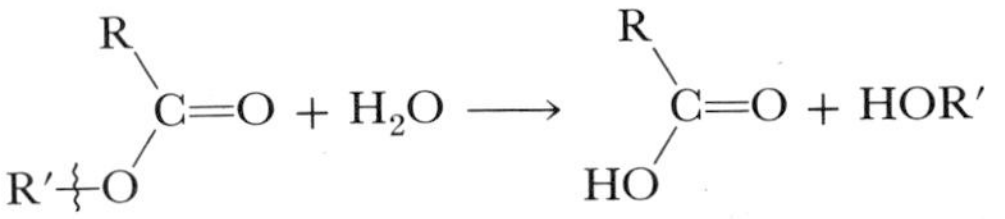

$$\underset{R'+O}{\overset{R}{\diagdown}}C=O + H_2O \longrightarrow \underset{HO}{\overset{R}{\diagdown}}C=O + HOR' \qquad (8.57)$$

alkyl–oxygen cleavage

Acyl vs. alkyl cleavage The occurrence of acyl–oxygen or of alkyl–oxygen cleavage may be determined by isotopic labeling[126] and by stereochemical methods.[127] Schemes 26 and 27 show how these techniques are used.

Acyl–oxygen cleavage is more commonly observed; it corresponds to the familiar addition mechanism and appears as the usual pathway both with base and with acid catalysis. Alkyl–oxygen cleavage occurs for special structures and under particularly favorable circumstances as discussed further below.

[126] D. Samuel and B. L. Silver, *Adv. Phys. Org. Chem.*, **3**, 123 (1965).
[127] (a) B. Holmberg, *Chem. Ber.*, **45**, 2997 (1912). For other methods see (b) E. H. Ingold and C. K. Ingold, *J. Chem. Soc.*, 756 (1932); (c) O. R. Quayle and H. M. Norton, *J. Am. Chem. Soc.*, **62**, 1170 (1940).

Scheme 26

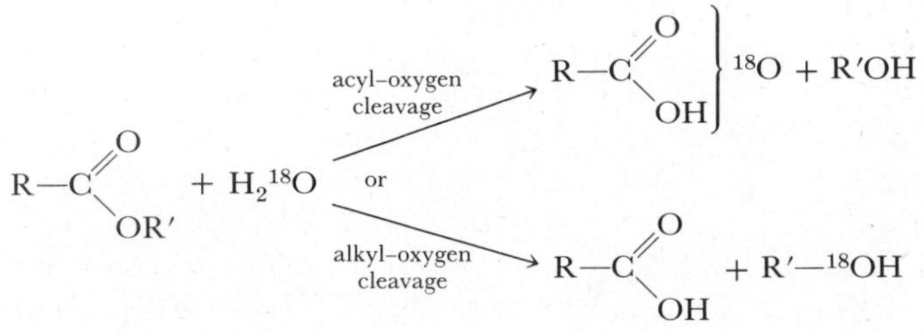

Scheme 27

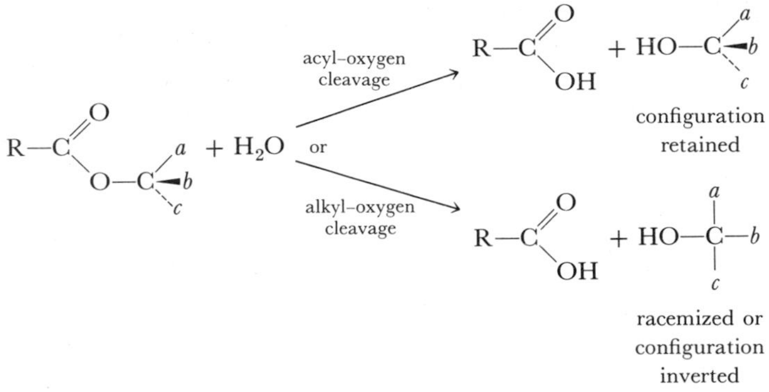

Base catalysis In synthesis, esters are ordinarily prepared with acid catalysis and hydrolyzed with base catalysis. The reason is evident from consideration of the equilibria involved in each case (Equations 8.58 and 8.59). Under basic condi-

$$\left.\begin{aligned} R(R'O)C{=}O + HO^- &\rightleftharpoons R(HO)C{=}O + R'O^- \\ R(HO)C{=}O + R'O^- &\rightleftharpoons R(^-O)C{=}O + R'OH \end{aligned}\right\} \tag{8.58}$$

$$R(R'O)C{=}O + H_2O \overset{H^+}{\rightleftharpoons} R(HO)C{=}O + R'OH \tag{8.59}$$

tions (Equations 8.58) deprotonation of the carboxylic acid shifts the equilibrium toward the hydrolyzed side; hence good hydrolysis yields are obtainable in basic solution, but practical esterification is not possible. The acid-catalyzed reaction, however, is energetically approximately evenly balanced; the yield of either acid or ester may be enhanced by use of excess of one of the reagents. The acid-catalyzed procedure with excess alcohol is practical for synthesis of esters of the lower alcohols, and with excess water is, of course, also useful for hydrolysis if sensitivity to base elsewhere in the molecule rules out the use of the more efficient base-catalyzed method.

Base-catalyzed hydrolysis of most common types of esters proceeds according to the addition mechanism (Scheme 28). This route is designated $B_{AC}2$ (base-catalyzed, acyl–oxygen cleavage, bimolecular).[128] The tetrahedral addition intermediate (48) can lose either OH^- (the k_{-1} step) or $R'O^-$ (the k_2 step); in contrast to the situation with amides, the two steps here are evenly balanced with respect to rate. Which step is faster will be determined by the nature of R'. If $R'O^-$ is a better leaving group than HO^- (as, for example, with phenyl esters), the k_2 step is fast, k_1 is rate determining, and little or no oxygen exchange is expected. If, on the other hand, loss of HO^- is faster, k_2 is rate determining and there will be oxygen exchange.[129] For the base-catalyzed reaction of ethyl benzoate in dioxane–water solution, exchange occurs at about one-tenth the rate of hydrolysis.[130]

SCHEME 28

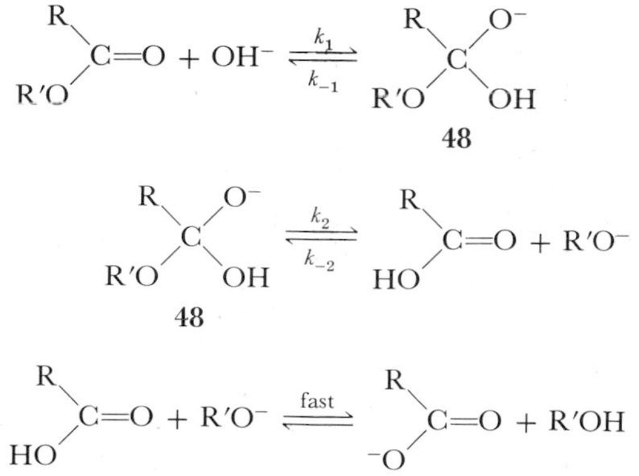

A far less common mode of ester hydrolysis is $B_{AL}2$ (Equation 8.60). This reaction leads to the same products as the $B_{AC}2$ process but is more logically

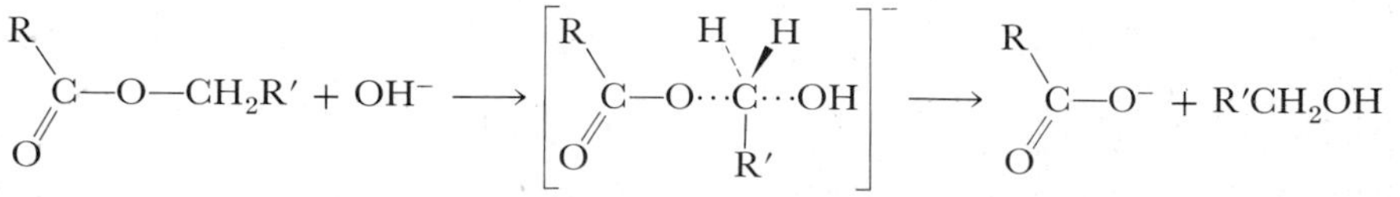

(8.60)

[128] General base catalysis is also observed. See Problem 15. (a) W. P. Jencks, and J. Carriuolo, *J. Am. Chem. Soc.*, **83**, 1743 (1961); (b) S. L. Johnson, *Adv. Phys. Org. Chem.*, **5**, 237 (1967); (c) S. L. Johnson, *J. Am. Chem. Soc.*, **86**, 3819 (1964).
[129] (a) W. P. Jencks, *Catalysis in Chemistry and Enzymology*, McGraw-Hill, New York, 1969, p. 508; (b) M. L. Bender, *Chem. Rev.*, **60**, 53 (1960); (c) C. D. Ritchie, *J. Am. Chem. Soc.*, **97**, 1170 (1975).
[130] M. L. Bender, R. D. Ginger, and J. P. Unik, *J. Am. Chem. Soc.*, **80**, 1044 (1958).

considered as an S_N2 substitution with carboxylate leaving group. It would occur only with an unhindered alkyl group in the ester (Chapter 4).

Acid catalysis The most commonly observed mode of acid-catalyzed hydrolysis is also the addition pathway. Scheme 29 shows the mechanism designated $A_{AC}2$ (acid-catalyzed, acyl–oxygen cleavage, bimolecular). Oxygen exchange is observed; for ethyl benzoate in dioxane–water solution with a dilute acid catalyst, exchange occurs at about one-fifth the rate of hydrolysis.[130]

SCHEME 29

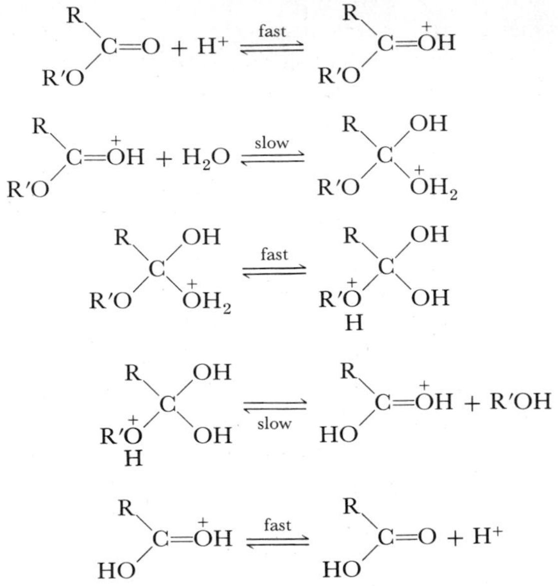

Recall from Section 8.3 and from Section 3.2 the Bunnett criterion for distinguishing A–1 and A–2 hydrolysis mechanisms. Table 8.7 shows results of application of the Bunnett equation (8.61) to ester hydrolysis. Note that most of the

$$\log k_{obs} + H_0 = w \log a_{H_2O} + C \tag{8.61}$$

esters listed have positive values of w, a result consistent with the addition mechanism, in which the water molecule is present in the rate-determining step. Bunnett has proposed that w values of about $+2$ to $+3$ should be characteristic of an addition mechanism with the addition step rate determining, and values of $+6$ should indicate that the decomposition of the tetrahedral intermediate is rate determining. The ester hydrolyses typically have values between these limits; as we noted above, the two steps are likely to have similar rates.[131]

[131] J. F. Bunnett, *J. Am. Chem. Soc.*, **83**, 4978 (1961).

The two esters that show negative values of w in Table 8.7 hydrolyze by different mechanisms. In each a special feature of structure favors one of the less common pathways.

In hydrolysis of *tert*-butyl acetate, the alkyl group can form a well-stabilized cation; there is alkyl–oxygen cleavage in an S_N1-type reaction. The mechanism is designated $A_{AL}1$ (Scheme 30).

SCHEME 30

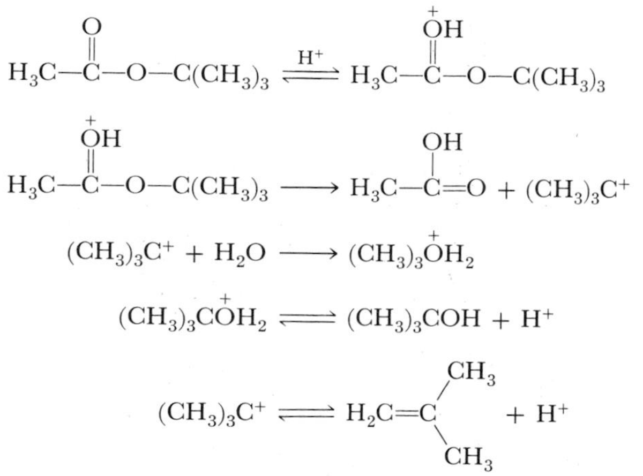

$$
H_3C-\overset{\overset{\displaystyle O}{\|}}{C}-O-C(CH_3)_3 \underset{}{\overset{H^+}{\rightleftharpoons}} H_3C-\overset{\overset{\displaystyle \overset{+}{O}H}{\|}}{C}-O-C(CH_3)_3
$$

$$
H_3C-\overset{\overset{\displaystyle \overset{+}{O}H}{\|}}{C}-O-C(CH_3)_3 \longrightarrow H_3C-\overset{\overset{\displaystyle OH}{|}}{C}=O + (CH_3)_3C^+
$$

$$
(CH_3)_3C^+ + H_2O \longrightarrow (CH_3)_3\overset{+}{O}H_2
$$

$$
(CH_3)_3C\overset{+}{O}H_2 \rightleftharpoons (CH_3)_3COH + H^+
$$

$$
(CH_3)_3C^+ \rightleftharpoons H_2C=C\overset{\displaystyle CH_3}{\underset{\displaystyle CH_3}{\Big\langle}} + H^+
$$

In the case of methyl 2,4,6-trimethylbenzoate, the ortho methyl groups on the ring interfere with nucleophilic attack at the carbonyl carbon, and a pathway involving an acylium ion (Scheme 31) is followed.[132] This mechanism is designated $A_{AC}1$. It probably involves protonation on the alkyl oxygen.

SCHEME 31

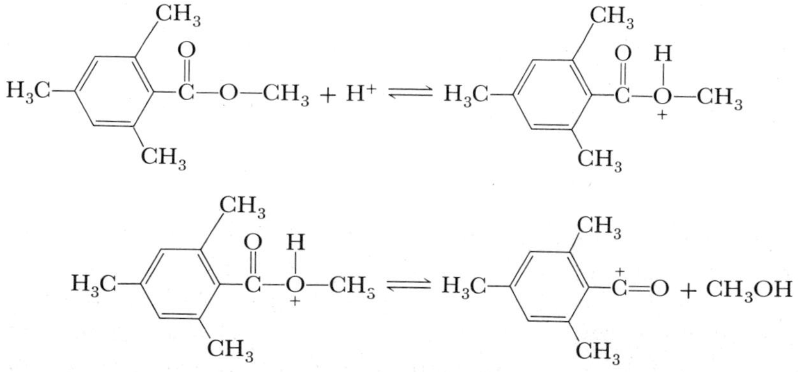

[132](a) M. L. Bender, H. Ladenheim, and M. C. Chen, *J. Am. Chem. Soc.*, **83**, 123 (1961). Other mesitoyl derivatives follow similar reaction paths. See (b) H. Ladenheim and M. L. Bender, *J. Am. Chem. Soc.*, **82**, 1895 (1960); (c) M. L. Bender and M. C. Chen, *J. Am. Chem. Soc.*, **85**, 30 (1963).

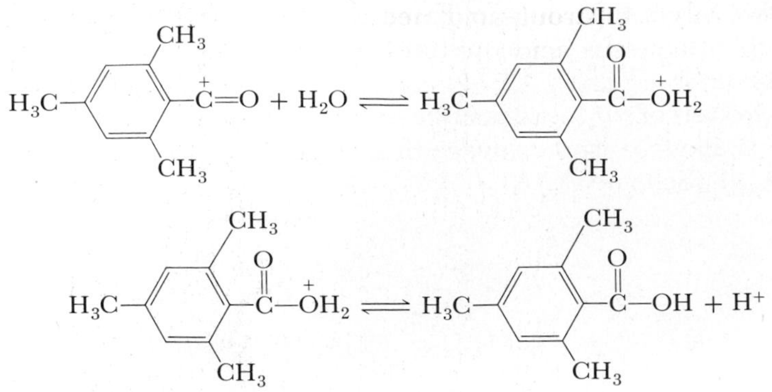

Table 8.7 BUNNETT w PARAMETERS FOR HYDROLYSIS OF SOME ESTERS IN AQUEOUS ACIDS[a]

Ester	Conditions	w
$HCOCH_3$ (O)	HCl, 25°	+4.21
H_3CCOCH_3 (O)	HCl, 25°	+5.83
$H_3CCOC_2H_5$ (O)	HCl, 25°	+4.15
$H_3CCOCH(CH_3)_2$ (O)	HCl, 25°	+4.62
$H_3CCOC(CH_3)_3$ (O)	HCl, 25°	−1.17
$\phi COCH_3$ (O)	HClO$_4$, 90°	+7.02
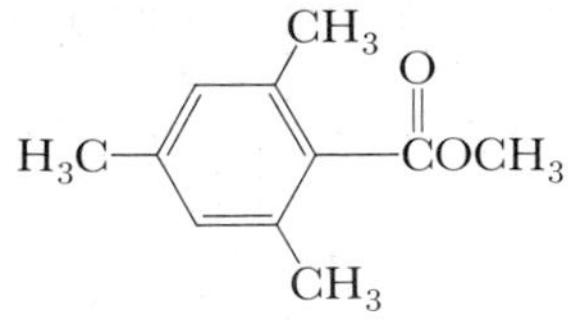	HClO$_4$, 90°	−2.47
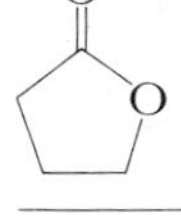	HCl, 0°	+6.11

[a] Reprinted in part with permission from J. F. Bunnett, *J. Am. Chem. Soc.*, **83**, 4956 (1961). Copyright 1961 American Chemical Society.

Ester hydrolysis in strong-acid media Yates has analyzed reaction rates for ester hydrolysis in aqueous acid mixtures in terms of Equation 8.62, a modified form of the Bunnett equation. The symbol r replaces Bunnett's w, and a new parameter, m, which has the value 0.62 for the reactions considered here, is added to correct the H_0 scale so that it is appropriate for ester protonation. Equation 8.63 is an alternative form for use at high acid concentrations where correction is necessary because a significant proportion of the ester is protonated.[133]

$$\log k_{obs} + mH_0 = r \log a_{H_2O} + C \tag{8.62}$$

$$\log k_{obs} - \log \frac{h_0{}^m}{K_{SH^+}^m + h_0{}^m} = r \log a_{H_2O} + C \tag{8.63}$$

The behavior of the parameter r with increasing acid concentration shows a change in mechanism away from the ordinary $A_{AC}2$ mechanism to one of the other acid-catalyzed mechanisms. Figure 8.13 shows plots of rate as a function of percent sulfuric acid in the medium and plots of Equation 8.63 for four different types of esters (**49**). In Types I and II, the initial increase of rate followed by a drop reflects

$$\begin{array}{c} CH_3 \\ \diagdown \\ C{=}O \\ \diagup \\ RO \end{array}$$

49

Type I	Type II	Type III	Type IV
R = primary alkyl	R = secondary alkyl, benzyl, allyl	R = vinyl, phenyl	R = *tert*-butyl, *p*-methoxyphenyl

the ordinary $A_{AC}2$ mechanism catalyzed by acid but requiring as nucleophile free H_2O, which begins to become less available as the proportion of sulfuric acid increases. The final rise, which occurs for Type I and Type II esters, corresponds to a change to a mechanism not requiring nucleophilic participation of water, as shown by the break in the acidity function graph, for Type I presumably to $A_{AC}1$ and for Type II esters probably to $A_{AL}1$. For the Type III esters the acidity function plots demonstrate clearly that in these cases also a change in mechanism occurs; this fact is not evident from the graphs of rate vs. percent sulfuric acid, because the change occurs before the acid concentration becomes high enough to cause the decrease found with Types I and II. Since vinyl and phenyl cations are unlikely, Yates proposed a change to the $A_{AC}1$ mechanism, or, for the vinyl case, to a mechanism involving protonation of the carbon–carbon double bond. *t*-Butyl acetate, Type IV, hydrolyzes by the $A_{AL}1$ route over the whole range, a fact confirmed by isotope tracers.[134] Uncertainty remains about the *p*-methoxyphenyl acetate because, although it behaves like the *t*-butyl ester, most of the reaction proceeds with acyl–oxygen cleavage.

---

[133]K. Yates, *Acc. Chem. Res.*, **4**, 136 (1971).
[134]C. A. Bunton and J. L. Wood, *J. Chem. Soc.*, 1522 (1955).

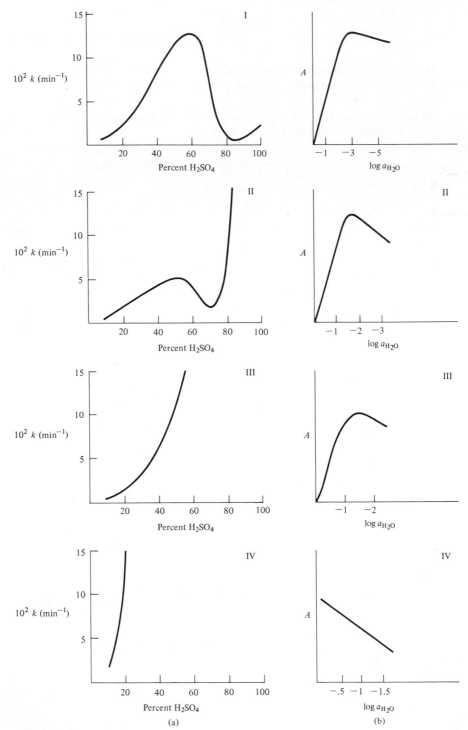

Figure 8.13 (a) Hydrolysis rates of esters of Types I through IV as a function of percent H_2SO_4. (b) Graphs of Equation 8.63 for esters of Types I through IV.

$$A = \log k_{obs} - \log \frac{h_0{}^m}{K_{SH^+}^m + h_0{}^m}$$

Reprinted with permission from K. Yates, *Acc. Chem. Res.*, **4**, 136 (1971). Copyright 1971 American Chemical Society.

   An alternative approach, the use of transition state activity coefficients, also shows the change of mechanism that occurs at high acid concentration.[135]

   Summary of ester hydrolysis mechanisms   Table 8.8 summarizes the ester hydrolysis mechanisms as identified by Ingold.

8.6 ENOLS, ENOLATES, AND ADDITION OF CARBON NUCLEOPHILES TO C=O

The second main aspect of reactions of carbonyl compounds is one we have already touched upon in Chapters 3 and 6. The carbonyl group increases the acidity of C—H bonds on a carbon directly attached to it by many powers of ten over an unactivated carbon–hydrogen bond. Removal of such a proton leaves the conjugated ambident enolate ion (**50**), which can be reprotonated either at the carbon, to give back the original keto tautomer, or at oxygen to give the enol (Equation 8.64).[136] Acid also promotes interconversion between enol and keto forms (Equation 8.65). In some cases it has proved possible by careful exclusion of acids and bases to isolate enol and keto forms,[137] but under ordinary circumstances there are sufficient acids or bases present to cause rapid interconversion.

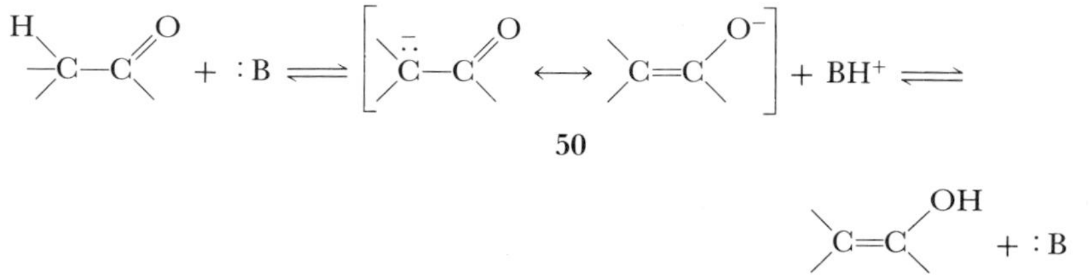

$$\tag{8.64}$$

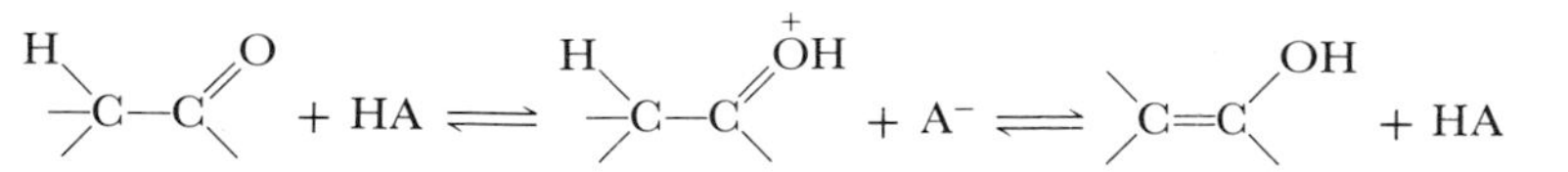

$$\tag{8.65}$$

Enols and Enolate Ions

Table 8.9 shows that the equilibrium mixture consists of almost entirely keto form in the case of simple aliphatic and aromatic ketones. For acetone, the enol tautomer

[135](a) K. Yates and T. A. Modro, *Acc. Chem. Res.*, **11**, 190 (1978); (b) J. T. Edward and S. C. Wong, *J. Am. Chem. Soc.*, **99**, 7224 (1977).

[136](a) For a discussion of enolization, see S. Forsén and M. Nilsson, in *The Chemistry of the Carbonyl Group*, Vol. 2, J. Zabicky, Ed., Wiley, London, 1970, p. 157; (b) E. Buncel, *Carbanions: Mechanistic and Isotopic Aspects*, Elsevier, Amsterdam, 1975, p. 70. (c) Preparative aspects of enols and particularly enolate ions are discussed in detail with numerous references to the literature in H. O. House, *Modern Synthetic Reactions*, 2d ed., W. A. Benjamin, Menlo Park, Calif., 1972.

[137](a) L. Knorr, O. Rothe, and H. Averbeck, *Chem. Ber.*, **44**, 1138 (1911); (b) K. H. Meyer and V. Schoeller, *Chem. Ber.*, **53**, 1410 (1920); (c) K. H. Meyer and H. Hopff, *Chem. Ber.*, **54**, 579 (1921); (d) for further examples, see C. K. Ingold, *Structure and Mechanism in Organic Chemistry*, 2d ed., Cornell University Press, Ithaca, N. Y., 1969, pp. 794*ff*.

Table 8.8 MECHANISMS OF ESTER HYDROLYSIS

Cleavage	Mechanism	Conditions	Abbreviation[a,b]	Mechanism[c]
Acyl-oxygen cleavage	Addition	Acid	*$A_{AC}2$	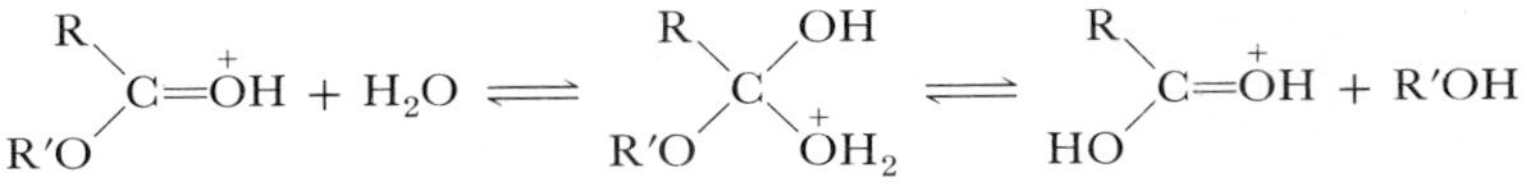 $R-C(O^+H)(OR')=O + H_2O \rightleftharpoons R-C(OH)(OR')(O^+H_2) \rightleftharpoons R-C(OH)(O^+H)=O + HO-R'OH$
		Base	*$B_{AC}2$	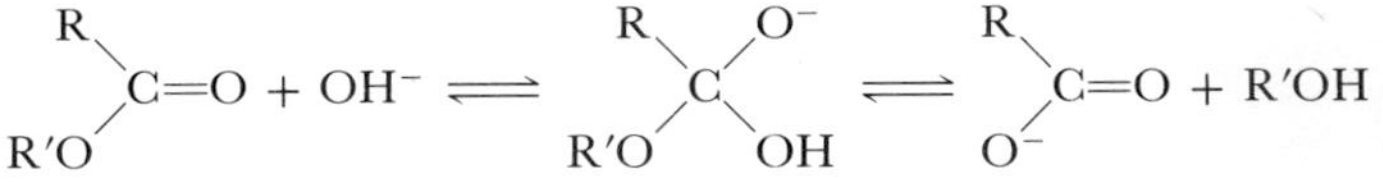 $R-C(OR')=O + OH^- \rightleftharpoons R-C(O^-)(OR')(OH) \rightleftharpoons R-C(O^-)=O + R'OH$
	Unimolecular	Acid	$A_{AC}1$	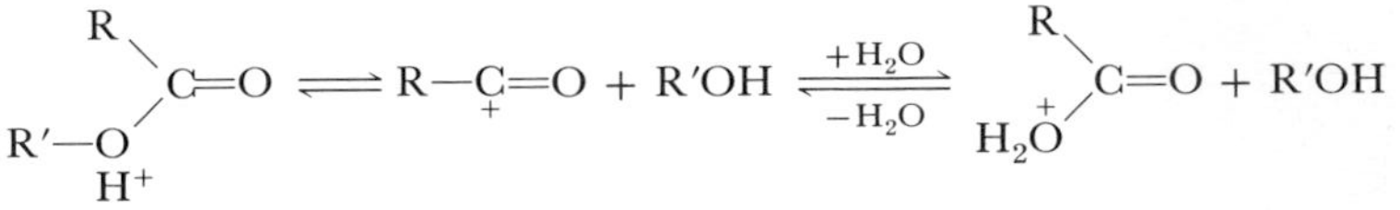 $R-C(OR')=O \rightleftharpoons R-C^+=O + R'OH \;(via\; R'-O^+H) \xrightleftharpoons[-H_2O]{+H_2O} R-C(OH)=O + R'O^-H_2O$
		Base	$B_{AC}1$	Not observed
		Base	$E1cB^d$	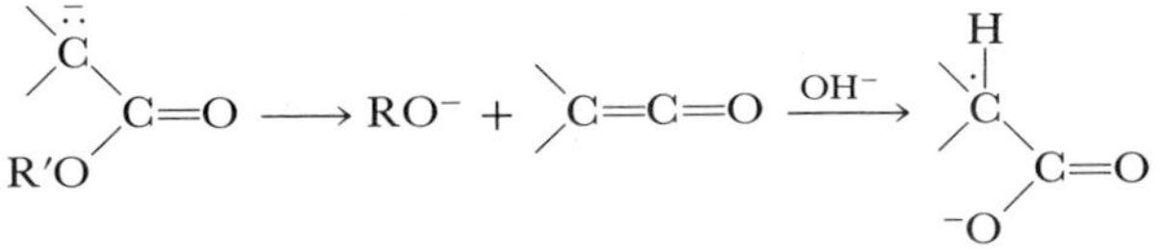 $:\overset{\cdot\cdot}{C}(OR')-C=O \longrightarrow RO^- + \,C=C=O \xrightarrow{OH^-} H-C-C(=O)(O^-)$
Alkyl-oxygen cleavage	Bimolecular (nucleophilic substitution)	Acid	$A_{AL}2$	Not observed
		Base	$B_{AL}2$	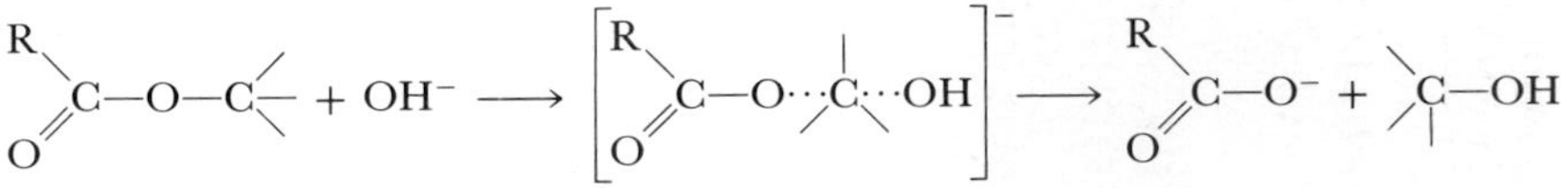 $R-C(=O)-O-C\, + OH^- \longrightarrow \left[R-C(=O)-O\cdots C\cdots OH \right]^- \longrightarrow R-C(=O)-O^- + \,C-OH$
	Unimolecular (nucleophilic substitution)	Acid	$A_{AL}1$	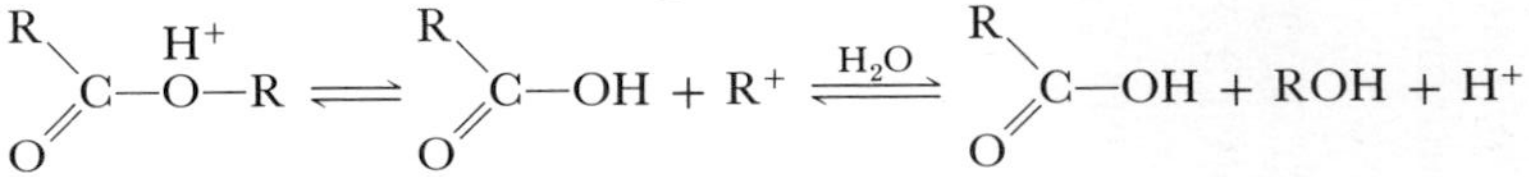 $R-C(=O)-O-R \overset{H^+}{\rightleftharpoons} R-C(=O)-O^+-R \rightleftharpoons R-C(=O)-OH + R^+ \xrightarrow{H_2O} R-C(=O)-OH + ROH + H^+$
		Base	$B_{AL}1$	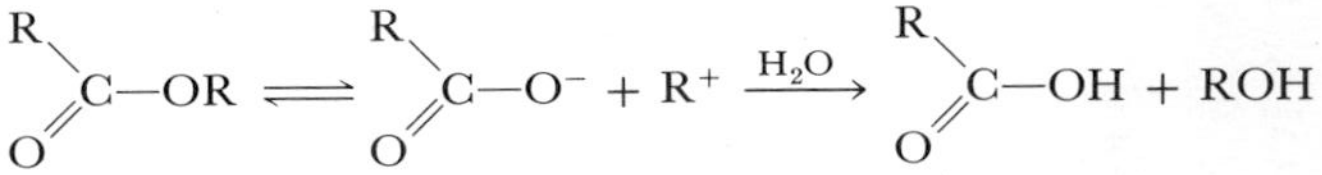 $R-C(=O)-OR \rightleftharpoons R-C(=O)-O^- + R^+ \xrightarrow{H_2O} R-C(=O)-OH + ROH$

*Designates the two most common mechanisms.

[a] C. K. Ingold, *Structure and Mechanism in Organic Chemistry*, 2d ed., Cornell University Press, Ithaca, N.Y., 1969, p. 1131.

[b] A or B: acid or base catalysis; AC or AL: acyl-oxygen or alkyl-oxygen fission; 1 or 2: unimolecular or bimolecular. ElcB designates unimolecular elimination through the conjugate base.

[c] Mechanisms are schematic and do not show initial and final proton transfers that may occur between acid and base sites within an intermediate.

[d] R. F. Pratt and T. C. Bruice, *J. Am. Chem. Soc.*, **92**, 5956 (1970).

Table 8.9 APPROXIMATE VALUES OF KETO–ENOL EQUILIBRIUM CONSTANTS

Compound	$K = $ [enol]/[keto]	Reference
$CH_3\overset{O}{\overset{\|}{C}}CH_3$	$\leqslant 10^{-6}$	a
CH_3CHO	$< 10^{-7}$	b
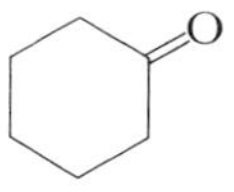 (cyclohexanone)	4.1×10^{-6}	a
$\phi-\overset{O}{\overset{\|}{C}}-CH_3$	3.5×10^{-4}	c
$CH_3-\overset{O}{\overset{\|}{C}}-CH_2-\overset{O}{\overset{\|}{C}}-CH_3$	3.2	c
$CH_3-\overset{O}{\overset{\|}{C}}-CH_2-\overset{O}{\overset{\|}{C}}-OR$	0.09	c
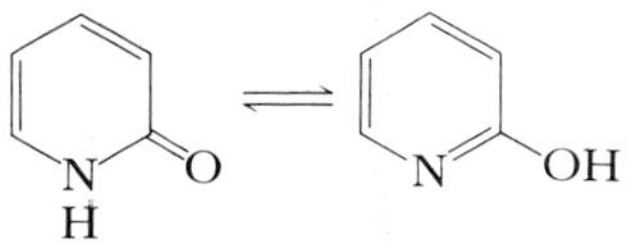 (2-pyridone ⇌ 2-hydroxypyridine)	1.1×10^{-3}	d
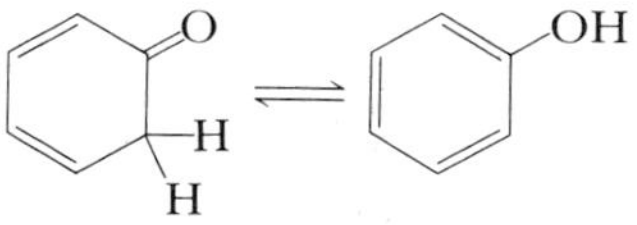 (cyclohexadienone ⇌ phenol)	4×10^{13}	e

^{a} R. P. Bell and P. W. Smith, *J. Chem. Soc. B*, 241 (1966).
^{b} A. Gero, *J. Org. Chem.*, **19**, 469 (1954).
^{c} A. Gero, *J. Org. Chem.*, **19**, 1960 (1954).
^{d} S. F. Mason, *J. Chem. Soc.*, 674 (1958).
^{e} Calculated from $\Delta G°$ estimated by J. B. Conant and G. B. Kistiakowsky, *Chem. Rev.*, **20**, 181 (1937).

is at least 8.2 kcal mole^{-1} higher in energy than the keto;[138] the difference in the gas phase is 13.9 ± 2 kcal mole^{-1}.[139] In β-diketones and β-ketoesters, on the other hand, significant amounts of the enol tautomer are present. In these latter cases, the enol contains a conjugated π electron system and an intramolecular hydrogen bond (**51**).[140] Phenol exists entirely in the enol form, because the alternative keto structure would sacrifice the stability associated with the six π electron aromatic system. In polycyclic phenols, however, it has been possible in a number of cases to obtain

[138] R. P. Bell and P. W. Smith, *J. Chem. Soc. B*, 241 (1966).
[139] S. K. Pollack and W. J. Hehre, *J. Am. Chem. Soc.*, **99**, 4845 (1977).
[140] X-ray photoelectron spectroscopy shows two types of oxygen, in accord with the formulation **51**. R. S. Brown, A. Tse, T. Nakashima, and R. C. Haddon, *J. Am. Chem. Soc.*, **101**, 3157 (1979).

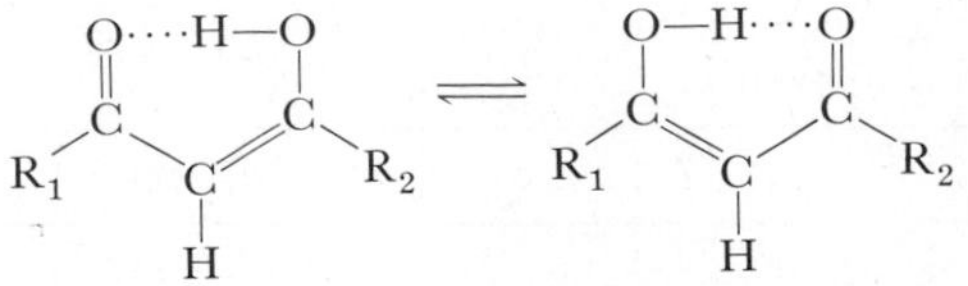

51

keto forms,[141] and the keto form predominates for heterocyclic compounds such as 52 and 53.[142]

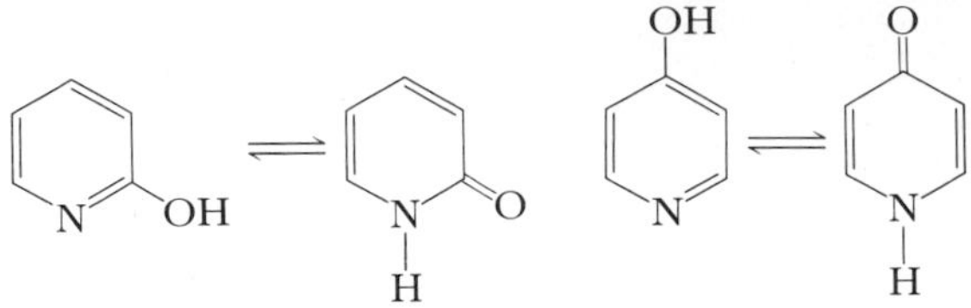

52   **53**

Several reactions of carbonyl compounds that have one or more α hydrogens proceed through the enol form. Reaction of ketones with chlorine, bromine, and iodine result in substitution of halogen for α hydrogen; rates are typically first-order in ketone and independent of halogen concentration and even of which halogen is used. Racemization of ketones with asymmetric centers adjacent to the carbonyl group (54), and hydrogen–deuterium exchange at the α position, proceed at the

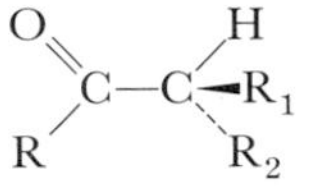

54

same rate as halogenation.[143] Primary hydrogen isotope effects are found; for example, for bromination of methyl cyclohexyl ketone (55), $k_H/k_D = 6.1$ in methanol.[144] It seems reasonable that all these processes involve the enol, with enol formation being the rate-determining step in each case.

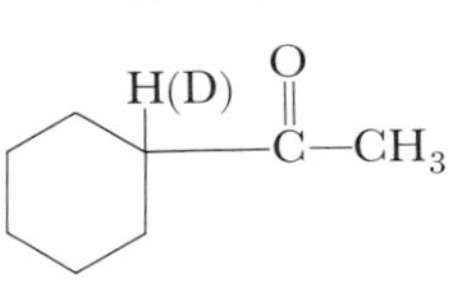

55

[141] S. Forsén and M. Nilsson, in *The Chemistry of the Carbonyl Group*, Vol. 2, J. Zabicky, Ed., Wiley, London, 1970, p. 168.

[142] S. F. Mason, *J. Chem. Soc.*, 674 (1958).

[143] (a) R. P. Bell and K. Yates, *J. Chem. Soc.*, 1927 (1962); (b) C. K. Ingold and C. L. Wilson, *J. Chem. Soc.*, 773 (1934); (c) P. D. Bartlett and C. H. Stauffer, *J. Am. Chem. Soc.*, **57**, 2580 (1935); (d) P. D. Bartlett, *J. Am. Chem. Soc.*, **56**, 967 (1934); (e) S. K. Hsü and C. L. Wilson, *J. Chem. Soc.*, 623 (1936); (f) S. K. Hsü, C. K. Ingold, and C. L. Wilson, *J. Chem. Soc.*, 78 (1938).

[144] Y. Jasor, M. Gaudry, and A. Marquet, *Tetrahedron Lett.*, 53 (1976).

The mechanism for halogenation is shown in Scheme 32. If the carbon α to the carbonyl bears more than one hydrogen, each is replaced by halogen in turn, but the first enolization is rate determining overall.

SCHEME 32

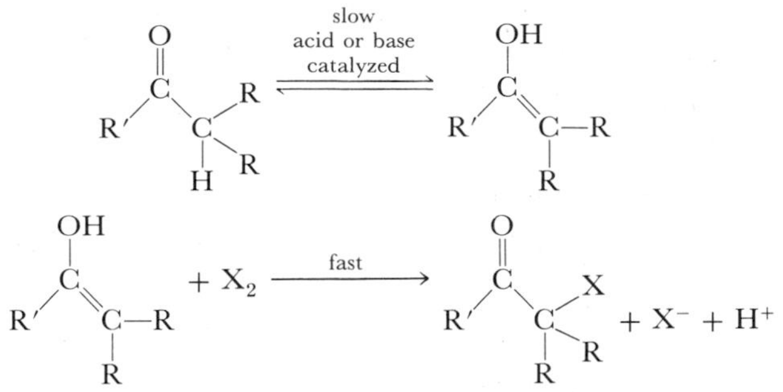

The enolization rate is subject to both general acid and general base catalysis.[145] Base catalysis presumably involves removal of a proton from carbon by the general base to yield the enolate, which will be in equilibrium with the enol. This equilibrium is expected to be rapid, as it involves proton transfer to and from oxygen; halogen can react either with enol, as shown in Scheme 32, or with enolate.

Acid catalysis could be either true general acid (Mechanism I, Scheme 33) or specific acid–general base (Mechanism II). Mechanism II, rate-determining removal of proton by A^- from protonated carbonyl compound, appears to be correct. One expects proton removal from carbon to be the slow process; this intuitive conclusion is substantiated by comparison of rate of hydrolysis of 1-methoxycyclohexene (Equation 8.66) with rate of ketonization of cyclohexanone enol (Equation 8.67). In Equation 8.66 the first step, addition of a proton to the C=C double bond, is known to be rate determining, and is analogous to the reverse of the second step of Mechanism II. Reaction 8.67 proceeds at the same rate as Reaction 8.66; it therefore seems likely that the addition of H^+ to C=C, and not deprotonation of the oxygen, is also the rate-determining step in Reaction 8.67.[146] One may

[145](a) R. P. Bell, *The Proton in Chemistry,* 2d ed., Cornell University Press, Ithaca, N.Y., 1973, pp. 141; 171; (b) R. P. Bell and H. F. F. Ridgewell, *Proc. R. Soc. London, Ser. A,* **298,** 178 (1967); (c) R. P. Bell, G. R. Hillier, J. W. Mansfield, and D. G. Street, *J. Chem. Soc. B,* 827 (1967); (d) R. P. Bell and O. M. Lidwell, *Proc. R. Soc. London, Ser. A,* **176,** 88 (1940).

[146]G. E. Lienhard and T.-C. Wang, *J. Am. Chem. Soc.,* **91,** 1146 (1969). The rate constant for ketonization of cyclohexanone enol (Equation 8.67) was calculated by dividing the rate constant for enolization by the enolization equilibrium constant:

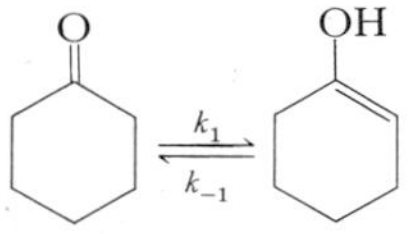

$$K_{eq} = \frac{k_1}{k_{-1}}$$

$$k_{-1} = \frac{k_1}{K_{eq}}$$

The enolization rate was determined by measuring the rate of iodination of cyclohexanone, and K_{eq} was taken from the data of Bell and Smith (Table 8.9).

SCHEME 33

Mechanism I
True general acid catalysis

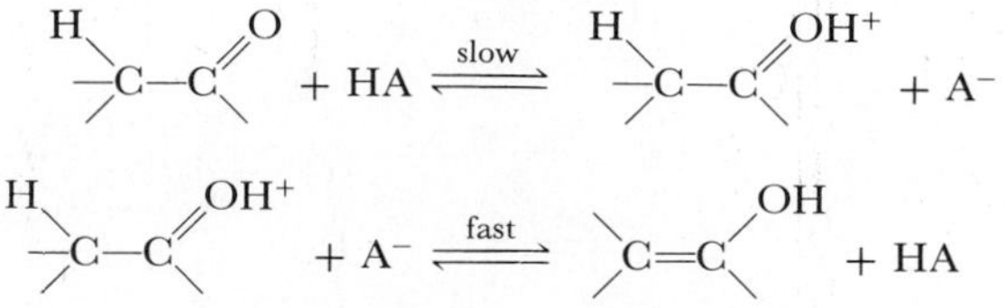

Mechanism II
Specific acid–general base catalysis

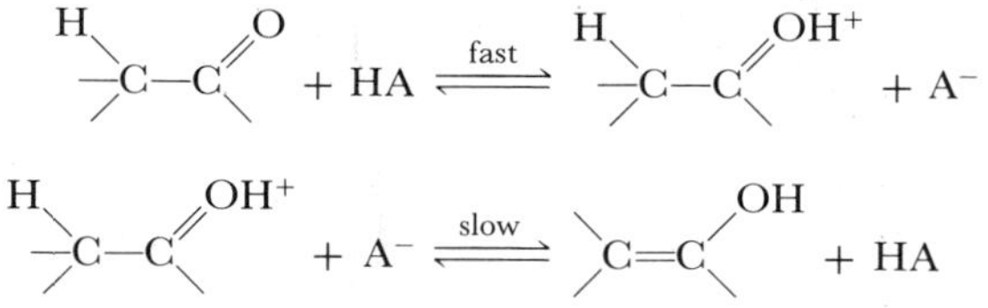

conclude, then, that in acid catalysis of enolization, Mechanism II is correct; that is, addition of a proton to C=C is rate determining in the reverse direction, and proton removal from carbon is rate determining in the forward direction.

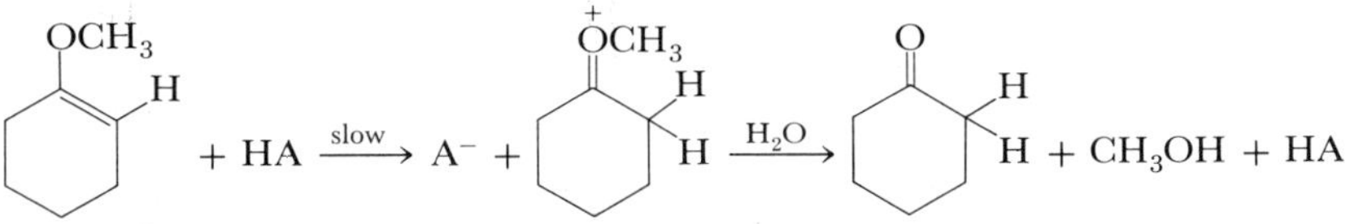

$$\text{(8.66)}$$

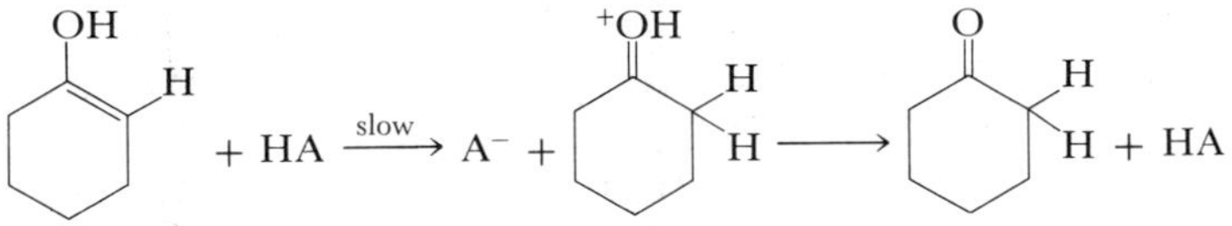

$$\text{(8.67)}$$

As with the additions to carbonyl carbon, there is the possibility here of simultaneous catalysis through a transition state **56** with an acid and a base simultaneously attacking. Although evidence in favor of such a mechanism, primarily

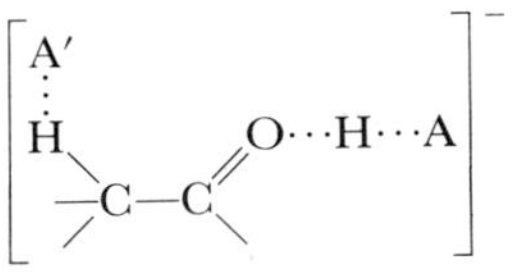

56

Table 8.10 Condensation Reactions Involving Active Methylene Compounds*

Active methylene compound	Product of addition to		
	R'X[b]	R'—C(=O)—R" or R'—C(=O)—H	R'—C(=O)—Z (Z = hal, OR, O—C—R)

(The body of the table consists of chemical structure diagrams for the active methylene compounds — the X,Y-substituted methylene[a], the malonate-type (COOR$_2$)[b], the β-ketoester/CR$_2$-type, and the CH-type — together with the corresponding addition products. Footnote superscript markers appearing in the table: R'X[b]; product column [a],[c],[d],[e]; [h,i]; [e,f,g]; [f,j]; [f,o]; [l]; [r',m,n]; [r,k].)

*No attempt is made to include complete references. Discussions in *Organic Reactions* and in H. O. House, *Modern Synthetic Reactions*, 2d ed., W. A. Benjamin, Menlo Park, Calif., 1972, will provide the reader with access to the literature.

[a] X, Y are most commonly —COOR or —CN, but may be other anion-stabilizing groups.

[b] A. C. Cope, H. L. Holmes, and H. O. House, *Org. Reactions*, 9, 107 (1957); House, p. 510.

[c] Commonly catalyzed by ammonia or an amine, the Knovenagel reaction. Amine may in some cases act as proton acceptor, in other cases as nucleophilic catalyst. Dehydration usually occurs and shifts equilibrium in favor of product.

[d] G. Jones, *Org. Reactions*, 15, 204 (1967); House, p. 646.

[e] F. Freeman, *Chem. Rev.*, 69, 591 (1969).

[f] The equilibrium is shifted to the right by formation of the stabilized enolate, which is not readily attacked by Z⁻. Acid catalysis is also possible for Z = hal., —OR. See House, p. 772.

[g] House, p. 756.

[h] Generally carried out under basic conditions, often starting from the organozinc or organomagnesium derivative of an α-bromoester, the Reformatsky reaction. M. W. Rathke, *Org. Reactions*, 22, 423 (1975). Equilibrium is shifted toward product by formation of a chelate structure with the metal ion.

[i] R. L. Shriner, *Org. Reactions*, 1, 1 (1942); House, p. 671.

[j] C. R. Hauser and B. E. Hudson, Jr., *Org. Reactions*, 1, 266 (1942); House, pp. 734, 762.

[k] House, p. 546.

[l] Aldol condensation. With aldehydes, successful with either acid or base catalysis.

[m] With ketones, conditions (strong-acid or strong-base catalysis) under which dehydration occurs are usually used to shift equilibrium toward the product.

[n] A. J. Nielsen and W. J. Houlihan, *Org. Reactions*, 16, 1 (1968); House, pp. 629–645.

[o] C. R. Hauser, F. W. Swamer, and J. T. Adams. *Org. Reactions*, 8, 59 (1954); House, pp. 747, 762.

Table 8.11 OTHER CONDENSATION REACTIONS OF CARBONYL COMPOUNDS *

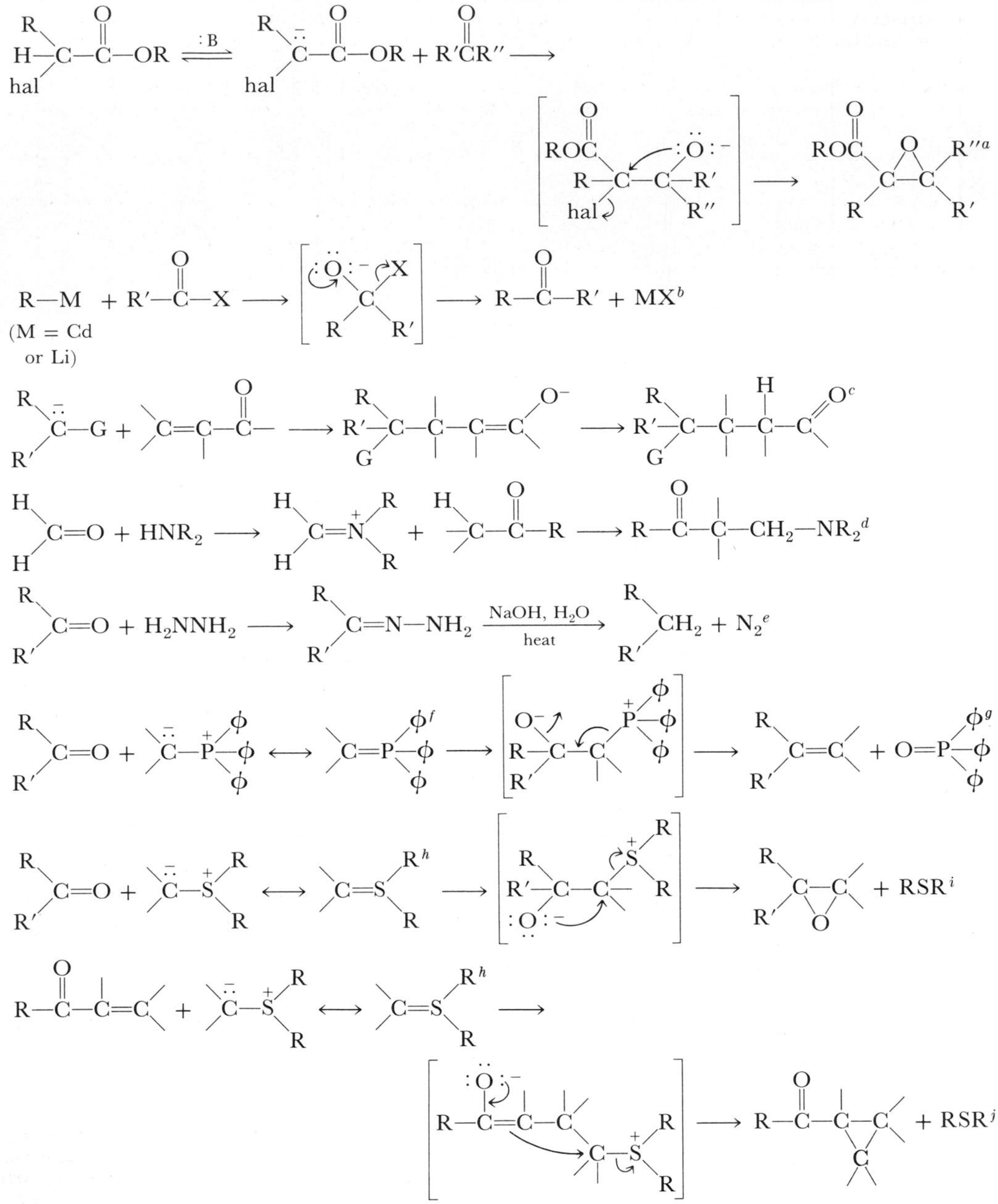

Table 8.11 (*Continued*)

---

*No attempt is made to include complete references. Discussions in *Organic Reactions* and in H. O. House, *Modern Synthetic Reactions,* 2d ed., W. A. Benjamin, Menlo Park, Calif., 1972, will provide the reader with access to the literature.

[a] Darzens condensation: M. S. Newman and B. J. Magerlein, *Org. Reactions,* **5,** 413 (1949); House, p. 666.

[b] M = Cd, X = hal: D. A. Shirley, *Org. Reactions,* **8,** 28 (1954); M = Li, X = OH, M. J. Jorgenson, *Org. Reactions,* **18,** 1 (1970).

[c] Michael addition. Discussed in Section 7.3. See also E. D. Bergmann, D. Ginsberg, and R. Pappo, *Org. Reactions,* **10,** 179 (1959); House, p. 595.

[d] Mannich reaction. The β-amino group of the product is readily eliminated to yield an α,β-unsaturated carbonyl structure. Frequently used in synthesis in conjunction with the Michael addition. F. F. Blicke, *Org. Reactions,* **1,** 303 (1942); J. H. Brewster and E. L. Eliel, *Org. Reactions,* **7,** 99 (1953); House, p. 654.

[e] Wolff–Kishner reduction: D. Todd, *Org. Reactions,* **4,** 378 (1948); House, p. 228.

[f] Reagent may be of either $\overset{\ldots}{\underset{}{C}}-\overset{+}{P}\phi_3$ or $\overset{\ldots}{\underset{}{C}}-\overset{\overset{O}{\|}}{P}(OR)_2$ type, formed by action of base on $\overset{H}{\underset{}{C}}-\overset{+}{P}\phi_3$ or $\overset{H}{\underset{}{C}}-\overset{\overset{O}{\|}}{P}(OR)_2$.

[g] Wittig reaction: A. Maercker, *Org. Reactions,* **14,** 270 (1965); House, p. 682.

[h] The sulfur ylides may be of type $\overset{\ldots}{\underset{}{C}}-\overset{+}{S}R_2$ or $\overset{\ldots}{\underset{}{C}}-\overset{\overset{O}{\|}}{\underset{+}{S}}R_2$, formed by action of base on $\overset{H}{\underset{}{C}}-\overset{+}{S}R_2$ or $\overset{H}{\underset{}{C}}-\overset{\overset{O}{\|}}{\underset{+}{S}}R_2$.

[i] House, p. 709.

[j] House, p. 719.

the observation of a kinetic term of the form k[ketone][acid][base] has been presented,[147] reinvestigations have not always confirmed the findings.[148]

Under strongly basic conditions carbonyl compounds with α hydrogens are converted to enolate ions, either partially or completely, depending on the conditions of basicity. We have discussed these species in Section 6.1.

Carbonyl compounds bearing α hydrogens, and other compounds in which a hydrogen is activated toward bases by one or more electron-withdrawing groups (see Table 6.1) are called *active methylene* compounds. These substances are particularly important in synthesis because their anions, easily produced under base catalysis, act as carbon nucleophiles in a variety of substitution and condensation reactions that constitute some of the most useful methods of introducing new carbon–carbon bonds. Rather than attempt to discuss all these reactions in detail, we summarize in Tables 8.10 and 8.11 the main features of a number of them, with references to more detailed discussions. The tables illustrate the close connections among these reactions and show how the ideas discussed in this chapter can be used to understand them.

[147] (a) H. M. Dawson and E. Spivey, *J. Chem. Soc.,* 2180 (1930); (b) R. P. Bell and P. Jones, *J. Chem. Soc.,* 88 (1953); (c) C. G. Swain, *J. Am. Chem. Soc.,* **72,** 4578 (1950); (d) B. E. C. Banks, *J. Chem. Soc.,* 63 (1962); (e) E. S. Hand and W. P. Jencks, *J. Am. Chem. Soc.,* **97,** 6221 (1975); (f) A. F. Hegarty and W. P. Jencks, *J. Am. Chem. Soc.,* **97,** 7188 (1975).

[148] (a) J. K. Coward and T. C. Bruice, *J. Am. Chem. Soc.,* **91,** 5339 (1969); (b) P. Y. Bruice and T. C. Bruice, *J. Am. Chem. Soc.,* **100,** 4793, 4802 (1978). A bifunctional catalysis of α hydrogen exchange that involves initial formation of an imine with one amino group of a diamine, followed by deprotonation by the other amino group, has been studied by J. Hine, *Acc. Chem. Res.,* **11,** 1 (1978).

PROBLEMS

1. Propose a mechanism for the following transformation:

(1) $H_2C=O$, $(C_2H_5)_2NH$
(2) H_2O, acetate buffer

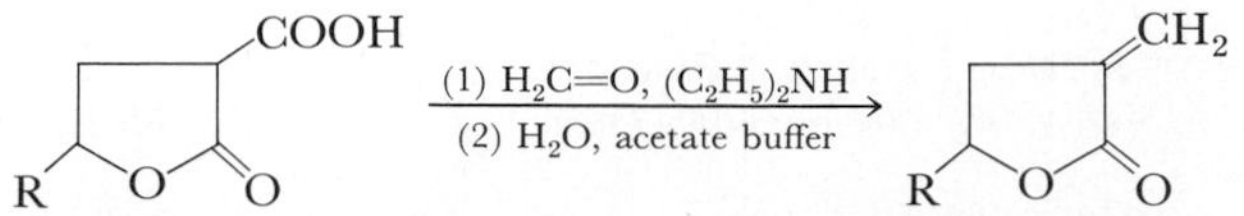

2. Propose a mechanism that will account for the product of the Stobbe condensation:

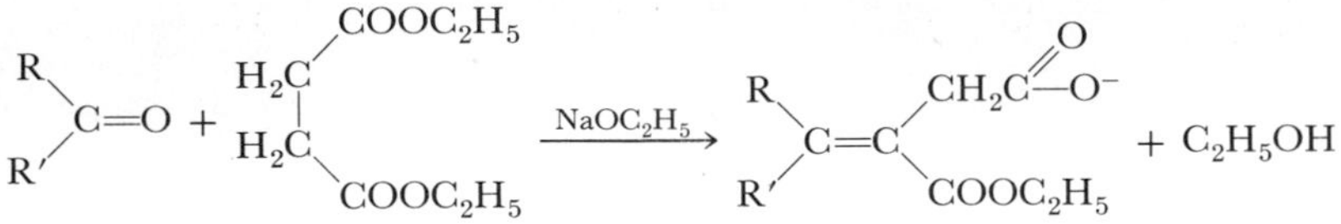

$NaOC_2H_5$

$+ C_2H_5OH$

3. Write the alternate kinetically equivalent mechanism to the bisulfite–carbonyl addition (Scheme 6, p. 614). Identify the type of catalysis in forward and reverse directions, and explain why comparison with the reaction shown in Scheme 7 helps to distinguish between the alternatives.

4. Anilinium ion, ϕNH_3^+, catalyzes semicarbazone formation from benzaldehyde much more effectively than would be expected on the basis of its strength as an acid. Explain.

5. Propose a mechanism to account for the following transformation:

aqueous propanol
KOH, reflux

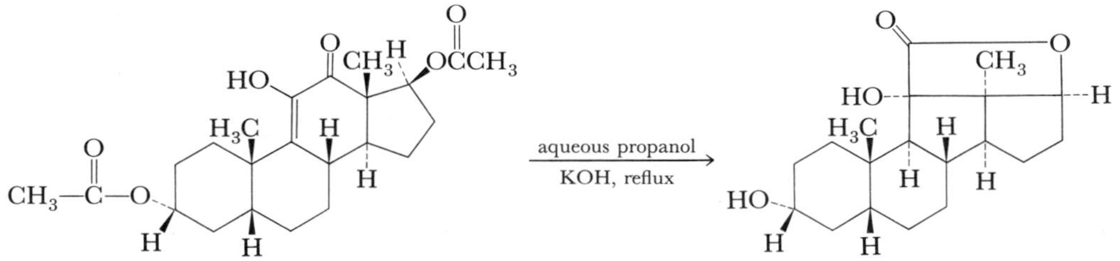

6. Explain why general acid catalysis is found in the hydrolysis of tropone diethylketal (**1**) and of 2-methoxy-3,3-dimethyloxetane (**2**) despite the fact that hydrolysis of dialkyl ketals ordinarily shows only specific hydronium ion catalysis.

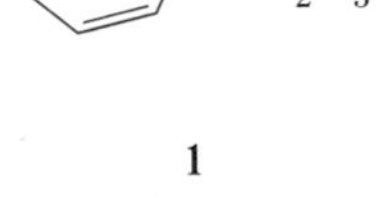

1

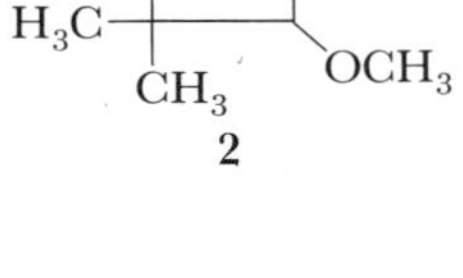

2

7. Explain why compound **3** hydrolyzes at a rate that is independent of pH in the range pH 1.5 to 0.1M NaOH, despite the fact that as a general rule acetals and monothioacetals show strong acid catalysis.

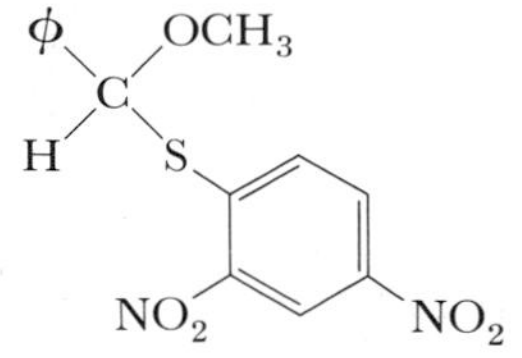

3

8. Propose mechanisms to account for the following transformations:

(a)

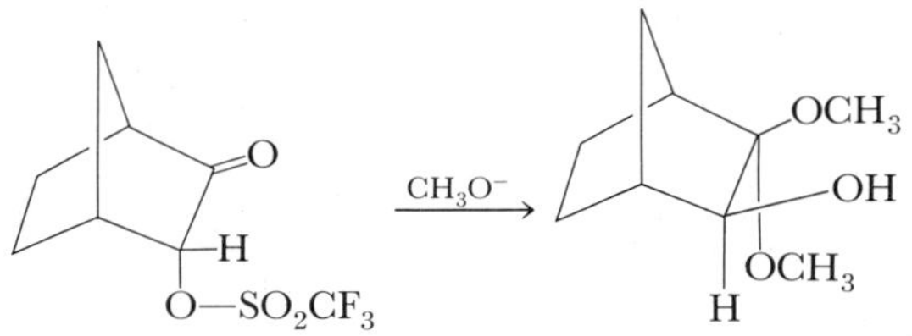

(b)

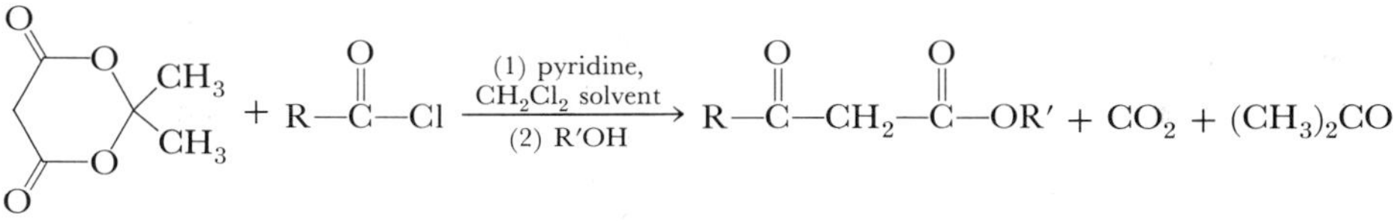

(c)

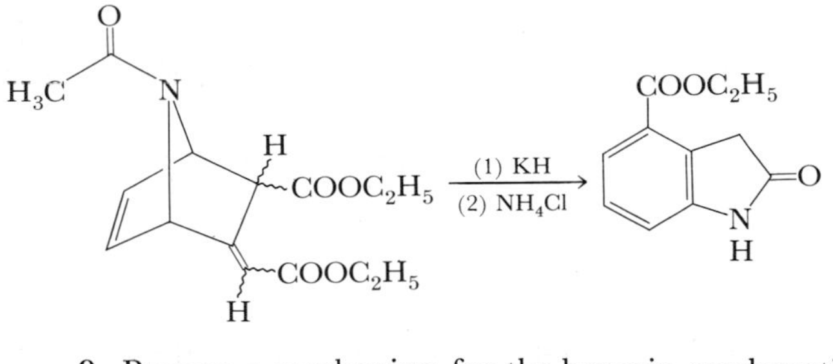

9. Propose a mechanism for the benzoin condensation, which is specifically catalyzed by cyanide:

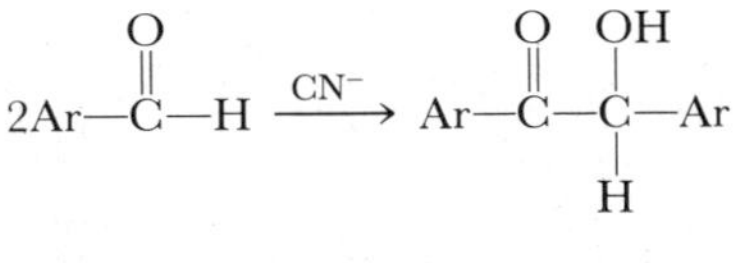

10. Propose a mechanism for the following transformation:

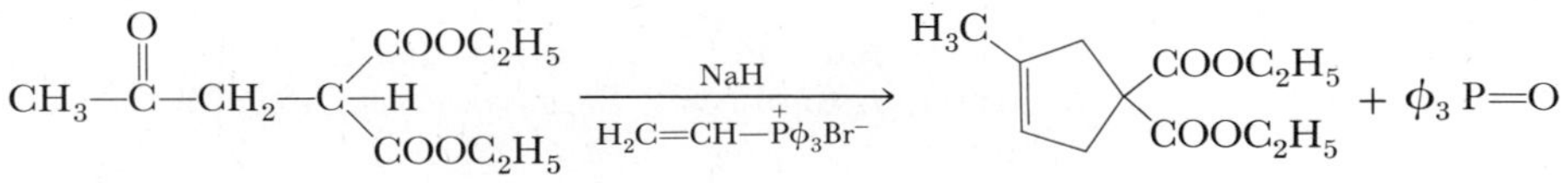

11. If lithium aluminum hydride exists in diethylether solution nearly entirely as a dimer, in equilibrium with a small concentration of monomer, and if the monomer is the reactive species in reducing ketones, show that the rate of reduction is given by

$$k_{obs} = [dimer]^{1/2}[ketone]$$

What is the relation between k_{obs} and the rate constant k_2 for the reduction step, in terms of the equilibrium constant K for dimer dissociation?

12. Show that a mechanism involving general base catalysis is indistinguishable kinetically from one involving a specific base-catalyzed step followed by a general acid-catalyzed step.

13. Draw three-dimensional reaction coordinate diagrams for alternative acetal hydrolysis Mechanisms I and III, p. 629.

14. Compare the mechanisms of the following reactions: Cannizzaro (p. 619); benzilic acid rearrangement (p. 437); Meerwein–Ponndorf–Verley reduction (p. 618).

15. Propose two alternative kinetically equivalent mechanisms for general base catalysis of ester hydrolysis with acyl–oxygen cleavage.

16. Why do high concentrations of acid in acid–water mixtures tend to favor the unimolecular mechanisms for hydrolysis of amides and esters?

17. Find the rate equation for the Cannizzaro reaction mechanism shown in Equations 8.21–8.23, p. 619. What is the relation between k_{obs} and the constants for the individual steps? How would the expression change if the hydride transfer agent were a dinegative ion formed from **8** (p. 619) by removal of the —OH proton by OH⁻ in a second rapid equilibrium?

18. Rates of acid-catalyzed hydrolysis of methyl and ethyl benzoates in 99.9 percent sulfuric acid produce the σ–ρ plots shown below. Suggest an explanation.

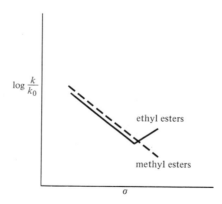

From J. A. Hirsch, *Concepts in Theoretical Organic Chemistry*, Allyn and Bacon, Inc, Boston, 1974, p. 139. Reprinted by permission of Allyn and Bacon Inc.

REFERENCES FOR PROBLEMS

1. (a) J. Martin, P. C. Watts, and F. Johnson, *J. Chem. Soc., Chem. Commun.*, 27 (1970); (b) E. S. Behare and R. B. Miller, *J. Chem. Soc., Chem. Commun.*, 402 (1970).
2. W. S. Johnson and G. H. Daub, *Org. Reactions*, **6,** 1 (1951).

3. P. R. Young and W. P. Jencks, *J. Am.Chem. Soc.*, **100,** 1228 (1978).

4. E. H. Cordes and W. P. Jencks, *J. Am. Chem. Soc.*, **84,** 826 (1962).

5. P. Kurath, *J. Org. Chem.*, **32,** 3626 (1967).

6. (a) T. H. Fife, *Acc. Chem. Res.*, **5,** 264 (1972); (b) R. F. Atkinson and T. C. Bruice, *J. Am. Chem. Soc.*, **96,** 819 (1974).

7. T. H. Fife and E. Anderson, *J. Am. Chem. Soc.*, **92,** 5464 (1970).

8. (a) X. Creary and A. J. Rollin, *J. Org. Chem.*, **42,** 4226 (1977); (b) Y. Oikawa, K. Sugano, and O. Yonemitsu, *J. Org. Chem.*, **43,** 2087 (1978); (c) A. P. Kozikowski and M. P. Kuniak, *J. Org. Chem.*, **43,** 2083 (1978).

9. (a) A. Lapworth, *J. Chem. Soc.*, **83,** 995 (1903); (b) J. P. Kuebrich, R. L. Schowen, M. Wang, and M. E. Lupes, *J. Am. Chem. Soc.*, **93,** 1214 (1971).

10. E. E. Schweizer and G. J. O'Neill, *J. Org. Chem.*, **30,** 2082 (1965).

11. K. E. Wiegers and S. G. Smith, *J. Am. Chem. Soc.*, **99,** 1480 (1977).

15. S. L. Johnson, *J. Am. Chem. Soc.*, **86,** 3819 (1964).

16. K. Yates and T. A. Modro, *Acc. Chem. Res.*, **11,** 190 (1978).

17. (a) Section 2.5; (b) C. G. Swain, A. L. Powell, W. A. Sheppard, and C. R. Morgan, *J. Am. Chem. Soc.*, **101,** 3576 (1979).

18. J. A. Hirsch, *Concepts in Theoretical Organic Chemistry,* Allyn and Bacon, Boston, 1974, p. 139.

Chapter 9
RADICAL REACTIONS

The salient feature of the homolytic process is the presence of the free-radical intermediate. The lack of charge of radicals and the high reactivity of nearly all of those that become involved in a typical organic reaction lead to important differences between homolytic and heterolytic processes.

9.1 CHARACTERISTICS OF ORGANIC FREE RADICALS[1]

The free-radical reaction is ordinarily initiated by a homolytic bond cleavage (Equation 9.1). The radicals may then undergo various processes, for example,

$$A\text{—}B \longrightarrow A\cdot + B\cdot \tag{9.1}$$

substitution, addition, rearrangement, elimination, or fragmentation. Unlike their heterolytic counterparts, these processes seldom produce a stable species. The new radical is likely to be as reactive, or nearly so, as was the original one, and so another step will follow rapidly upon the first, producing still another reactive radical, which enters into a third step, and so on. The series ends when two radicals

---

[1]Comprehensive treatments of free-radical chemistry will be found in (a) C. Walling, *Free Radicals in Solution*, Wiley, New York, 1957; (b) E. S. Huyser, *Free-Radical Chain Reactions*, Wiley, New York, 1970; (c) J. K. Kochi, Ed., *Free Radicals*, Vols. I and II, Wiley, New York, 1973; (d) W. A. Pryor, *Free Radicals*, McGraw-Hill, New York, 1966; (e) D. C. Nonhebel and J. C. Walton, *Free Radical Chemistry*, Cambridge University Press, London, 1974; (f) W. A. Pryor, Ed., *Organic Free Radicals*, American Chemical Society, Washington, D.C., 1978.

encounter each other and either recombine (Equation 9.2) or disproportionate (Equation 9.3). The sequence of radical formation (*initiation*), successive steps gen-

$$A \cdot + B \cdot \longrightarrow A\text{---}B \tag{9.2}$$

$$A \cdot + \cdot \overset{|}{\underset{|}{C}}\text{---}\overset{|}{\underset{|}{C}}\text{---}H \longrightarrow A\text{---}H + \overset{\diagdown}{\diagup}C{=}C\overset{\diagup}{\diagdown} \tag{9.3}$$

erating new radicals (*propagation*), and return to stable molecular species (*termination*) is called a *chain reaction*; although not all radical processes are chain reactions, and nonradical transformations on occasion follow a chain path, chain reactions are so common in homolytic chemistry that we may regard them as typical of free-radical behavior.

The other important difference between heterolytic and free-radical reactions arises from the nature of the termination step. Two radicals, being uncharged and having each an odd number of electrons, can come together to form a stable neutral molecule without encountering any significant barrier; ions, on the other hand, usually do not react with others of their own kind. Hence although it is quite common to find stable solutions containing high concentrations of ions, radicals, with a few important exceptions, can exist only in extremely low concentrations and then only when they are being continuously generated. Activation energies for the termination reactions (Equations 9.2 and 9.3) are very low; those for recombination of small alkyl radicals approach zero.

The rate of a bimolecular reaction with zero activation energy is limited by the rate at which the reaction partners can diffuse through the solution. Diffusion-controlled rate constants depend on solvent viscosity and are on the order of 10^9 to $10^{10} M^{-1}$ sec^{-1} for typical solvents at ambient temperature.[2]

The very low activation energies of the termination reactions mean that the lifetimes of individual radicals are short; hence radical reactions competing with termination are necessarily extremely rapid, and the total time elapsed between the generation of a radical in an initiation step and its destruction or the termination of a chain it starts is typically on the order of one second.[3]

Spectroscopic Investigation of Radical Structure

The rapid self-reaction of most organic radicals makes their direct observation difficult. Exceptions to this generalization are the *persistent* radicals for which reactions are slowed by bulky groups near the radical center. We discuss these systems further in Section 9.2.

With radicals that lack a high degree of stabilization, it is necessary to circumvent the limitation imposed by rapid termination by maintaining low temperature and by using flow techniques, by immobilizing the radicals in an inert solid matrix, or by generating them rapidly, for example, by an intense flash of light (flash photolysis) or with a pulse of energetic particles (pulse radiolysis).

[2]K. U. Ingold, in *Free Radicals*, Vol. I, J. K. Kochi, Ed., Wiley, New York, 1973, p. 39.
[3]C. Walling, *Free Radicals in Solution*, Wiley, New York, 1957, p. 37.

Early structural information about the simple alkyl radicals was obtained by Herzberg, who observed the ultraviolet spectra of $\cdot CH_3$ and $\cdot CD_3$ in the gas phase by flash photolysis.[4] Analysis of the fine structure led to the conclusion that the methyl radical is planar, a deviation from planarity of not more than $10°$ being consistent with the data.[5] The infrared spectrum of $\cdot CH_3$ has also been recorded; it does not yield conclusive structural information.[6] Radicals trapped in a matrix of solid adamantane can be formed by X-irradiation of suitable precursors. In this way spectra can be obtained at room temperature.[7]

Electron spin resonance A more generally applicable technique for spectroscopic observation of radicals is *electron spin resonance* (esr, also known as *electron paramagnetic resonance,* epr).[8] The method is in principle the same as nuclear magnetic resonance but relies on electron spin rather than nuclear spin. The spectrometer detects changes in energy states of the unpaired electron (a spin $\frac{1}{2}$ particle) in the presence of a magnetic field; because only unpaired electrons are detected, it is specific for radicals.

The esr technique has high sensitivity and can detect radicals in concentrations on the order of $5 \times 10^{-7}M$. Radicals can be observed in solution if they are persistent or if they are being continuously generated at a rate fast enough to provide a detectable steady-state concentration. Despite the high sensitivity of esr, the more reactive radicals often must be studied by immobilizing them in a solid matrix, or by spin trapping. In the spin-trapping technique a compound that will react with a transient radical to yield a persistent radical is added to a solution in which radicals are being generated. From esr spectra of the new radicals formed, the structure of the original radical can often be deduced.[9] Phenyl *t*-butyl nitrone (**1**) and 2-methyl-2-nitrosopropane (**2**) are the most commonly used spin traps; their reactions are illustrated in Equations 9.4 and 9.5. The rate of addition of 5-hexenyl radicals at 40°C to **1** is $1.3 \times 10^5 M^{-1}$ sec^{-1}, and to **2** is $9.0 \times 10^6 M^{-1}$ sec^{-1}.[10]

If the unpaired electron is close to a proton or other magnetic nucleus, the

[4]G. Herzberg and J. Shoosmith, *Can. J. Phys.*, **34**, 523 (1956).

[5]G. Herzberg, *Molecular Spectra and Molecular Structure,* Vol. III, Van Nostrand Reinhold, New York, 1966.

[6](a) D. E. Milligan and M. E. Jacox, *J. Chem. Phys.*, **47**, 5146 (1967); (b) L. Y. Tan, A. M. Winer, and G. C. Pimentel, *J. Chem. Phys.*, **57**, 4028 (1972).

[7]J. E. Jordan, D. W. Pratt, and D. E. Wood, *J. Am. Chem. Soc.*, **96**, 5588 (1974).

[8]For further details of techniques and results of esr spectroscopy, see (a) H. Fischer, in *Free Radicals,* Vol. II, J. K. Kochi, Ed., Wiley, New York, 1973, p. 435; (b) H. M. Assenheim, *Introduction to Electron Spin Resonance,* Plenum, New York, 1967; (c) R. Bersohn, in *Determination of Organic Structures by Physical Methods,* Vol. 2, F. C. Nachod and W. D. Phillips, Eds., Academic Press, New York, 1962, p. 563; (d) G. A. Russell, in *Determination of Organic Structures by Physical Methods,* Vol. 3, F. C. Nachod and J. J. Zuckerman, Eds., Academic Press, New York, 1971, p. 293; (e) D. H. Geske, *Prog. Phys. Org. Chem.*, **4**, 125 (1967); (f) J. E. Wertz and J. R. Bolton, *Electron Spin Resonance: Elementary Theory and Practical Applications,* McGraw-Hill, New York, 1972; (g) J. K. Kochi, *Adv. Free Radical Chem.*, **5**, 189 (1975).

[9](a) E. G. Janzen, *Acc. Chem. Res.*, **4**, 31 (1971); (b) C. Lagercrantz, *J. Phys. Chem.*, **75**, 3466 (1971); (c) F. P. Sargent and E. M. Gardy, *Can. J. Chem.*, **54**, 275 (1976); (d) F. P. Sargent and E. M. Gardy, *J. Chem. Phys.*, **67**, 1793 (1977); (e) C. A. Evans, *Aldrichim. Acta*, **12**, (2), 23 (1979).

[10]P. Schmid and K. U. Ingold, *J. Am. Chem. Soc.*, **99**, 6434 (1977).

$$\phi\text{—CH}\overset{\overset{\displaystyle O}{\vert}}{=}\text{N—C(CH}_3)_3 + \text{R}\cdot \longrightarrow \phi\text{—CH—}\overset{\overset{\displaystyle O\cdot}{\vert}}{\text{N}}\text{—C(CH}_3)_3 \qquad (9.4)$$

1

R

$$(\text{CH}_3)_3\text{C—N}\!\!=\!\!\text{O} + \text{R}\cdot \longrightarrow (\text{CH}_3)_3\text{C—}\overset{\overset{\displaystyle }{\vert}}{\text{N}}\text{—O}\cdot \qquad (9.5)$$

2

R

resonance signal in the esr spectrum is split by a coupling similar to that which operates in nuclear resonance spectra. The coupling constant depends on the magnitude of the probability distribution function for the unpaired electron in the vicinity of the nucleus in question and so provides information about structure in organic radicals.

A great deal of the electron spin resonance work has been done with aromatic radical ions produced by adding or removing a π electron from the neutral molecule by the action of a reducing or oxidizing agent.[11] Correlation of data for many systems with molecular orbital calculations of π electron distributions has led to a quantitative expression, the McConnell relation (Equation 9.6), which relates the coupling constant between the odd electron and a hydrogen nucleus attached

$$a_{\text{H}} = Q\rho$$
$$(Q \approx -22\,\text{G}^{12}) \qquad (9.6)$$

to the ring, a_{H}, with unpaired electron density (*spin density*), ρ, at the carbon to which the hydrogen is bonded.[13] Q (in gauss, G) is a constant characteristic of the coupling mechanism in a planar π radical. If the radical or radical ion has been prepared from a substrate enriched in ^{13}C, a spin $\frac{1}{2}$ nucleus, couplings of the unpaired electron to carbon can be observed. The constants are correlated by an expression similar to, but somewhat more complex than, Equation 9.6.[14]

The coupling constant between an unpaired electron and α protons, a few of which are given in Table 9.1, in most nonconjugated alkyl radicals is 22–23 G. This is the value obtained from Equation 9.6 if it is assumed that the unpaired electron is localized at the α carbon ($\rho = 1$) and that the hybridization of the α carbon is trigonal as it is in the aromatic π radical ions on which the equation is based.

Alkyl radicals bearing β protons also show coupling to these positions. The coupling constants are ordinarily on the order of 25–30 G, somewhat larger than those to α protons.[15] Such a large coupling might at first seem inconsistent with the

---

[11] For a review of esr spectroscopy of radical ions, see J. R. Bolton, in *Radical Ions*, E. T. Kaiser and L. Kevan, Eds., Wiley, New York, 1968. Radical ion chemistry is considered further in Section 9.2.

[12] J. A. Pople, D. I. Beveridge, and P. A. Dobosh, *J. Am. Chem. Soc.*, **90**, 4201 (1968).

[13] (a) H. M. McConnell, *J. Chem. Phys.*, **24**, 632, 764, (1956); (b) H. M. McConnell and D. B. Chesnut, *J. Chem. Phys.*, **28**, 107 (1958).

[14] (a) M. Karplus and G. K. Fraenkel, *J. Chem. Phys.*, **35**, 1312 (1961); (b) H. Fischer, in *Free Radicals*, Vol. II, J. K. Kochi, Ed., Wiley, New York, 1973, p. 448.

[15] The algebraic sign is predicted theoretically to be opposite for α and β couplings. Few coupling constant signs have been determined experimentally.

Table 9.1 ELECTRON–PROTON AND ELECTRON–^{13}C COUPLING CONSTANTS OF SOME ALKYL RADICALS (IN GAUSS)

Radical	$\overset{\displaystyle\cdot}{>}$C—H^a	$\underset{\displaystyle}{\overset{\displaystyle H}{>}}$C—$\overset{\displaystyle\cdot}{C}<$ ^{b}	$^{13}\overset{\displaystyle\cdot}{C}$—^{b}
·CH$_3$	23^{c,d}	—	38^e
CH$_3$CH$_2$·	22^{c,d}	27^{c,d}	39^e
CH$_3$CH$_2$CH$_2$·	22^{c,d}	30^{c,d}	—
CH$_3$ >C—H , CH$_3$	22^{c,d}	25^{c,d}	41^h
H C=C , H H (·)	13^c	—	108^e
(CH$_3$)$_3$C·	—	23^{c,f}	45^{h,i}
·CH$_2$F	21^g	—	55^g
·CHF$_2$	22^g	—	149^g
·CF$_3$	—	—	272^g

^{a} a_{H_α} are expected to be negative except for ·CHF$_2$, in which it is probably positive. See reference g, p. 2710, and D. L. Beveridge, P. A. Dobosh, and J. A. Pople, *J. Chem. Phys.*, **48**, 4802 (1968).
^{b} a_{H_β} and $a_{^{13}C}$ are expected to be positive.
^{c} R. W. Fessenden and R. H. Schuler, *J. Chem. Phys.*, **39**, 2147 (1963).
^{d} J. K. Kochi and P. J. Krusic, *J. Am. Chem. Soc.*, **91**, 3940 (1969).
^{e} R. W. Fessenden, *J. Phys. Chem.*, **71**, 74 (1967).
^{f} H. Paul and H. Fischer, *Chem. Commun.*, 1038 (1971).
^{g} R. W. Fessenden and R. H. Schuler, *J. Chem. Phys.*, **43**, 2704 (1965).
^{h} H. Fischer, *Free Radicals*, Vol. II, J. K. Kochi, Ed., Wiley, New York, 1973, p. 443.
^{i} D. E. Wood, L. F. Williams, R. F. Sprecher, and W. A. Latham, *J. Am. Chem. Soc.*, **94**, 6241 (1972).

idea that coupling depends on unpaired electron density, because one would expect to find very little delocalization to the β position in a saturated system. The β protons, however, do not lie in the nodal plane of the radical p orbital as the α protons do; their position (**3**) is much more favorable for interaction, and a small degree of delocalization causes substantial coupling. The delocalization here is hyperconjugation (Section 1.7).

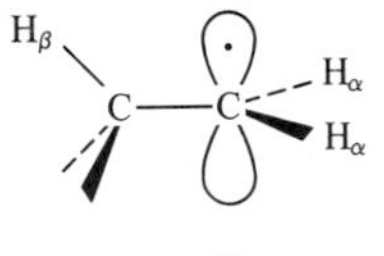

3

The magnitude of the coupling to the β proton will depend on the angle of rotation about the C$_\alpha$—C$_\beta$ bond; the coupling is largest when the C$_\beta$—H$_\beta$ bond is parallel to the axis of the p orbital containing the unpaired electron (**4**), and small-

$$\theta = 0°$$

4

$$\theta = 90°$$

5

est when it is perpendicular (**5**). The relation between the coupling constant and the angle θ is given by Equation 9.7, where A and C are constants with values of 0–5 G and 40–45 G, respectively, and θ is the dihedral angle between the C_β—H_β bond and the C_α-$2p$ axis.

$$a_{H_\beta} = A + C \cos^2 \theta \tag{9.7}$$

This relation yields information about conformations in radicals.[16] When there is free rotation about the C_α—C_β bond, $\cos^2 \theta$ in Equation 9.7 is replaced by $\frac{1}{2}$, its average value. Coupling constants to more remote protons are small, ordinarily less than 1 G.

Carbon-13 couplings provide further useful information about carbon radicals. Table 9.1 includes some representative values. Theoretical considerations suggest that this parameter should be a sensitive function of hybridization at the radical center;[17] the reason in qualitative terms is that deviation from planarity will introduce s character into the orbital carrying the unpaired electron and thus will bring that electron closer to the nucleus where it can couple more effectively with the nuclear spin. Accurate molecular orbital calculations,[18] results from photoelectron spectroscopy,[19] and the ultraviolet spectroscopic data cited earlier all point to a planar structure for $\cdot CH_3$. The large values of $a_{^{13}C}$ shown in Table 9.1 for $\cdot CHF_2$ and $\cdot CF_3$ lead to the conclusion that these radicals are nonplanar, bond angles in $\cdot CF_3$ probably approaching the tetrahedral value.[20]

Results for the t-butyl radical, $(CH_3)_3C\cdot$, have led to controversy. The $a_{^{13}C}$ value of 45 G is not much higher than the $\cdot CH_3$ value; however, the variation of this coupling constant with temperature is not that expected for a planar radical. The temperature variation is thought to arise through the vibration that bends the substituents of a planar radical away from the plane (or of a nonplanar radical toward planarity). The increase in the vibrational amplitude with increasing tem-

[16]See, for example, (a) D. Griller and K. U. Ingold, *J. Am. Chem. Soc.*, **96**, 6715 (1974); (b) J. K. Kochi, *Adv. Free Radical Chem.*, **5**, 189 (1975).

[17]See note 14(a); (b) D. M. Schrader and M. Karplus, *J. Chem. Phys.*, **40**, 1593 (1964).

[18]E. D. Jemmis, V. Buss, P. v. R. Schleyer, and L. C. Allen, *J. Am. Chem. Soc.*, **98**, 6483 (1976).

[19](a) F. A. Houle and J. L. Beauchamp, *J. Am. Chem. Soc.*, **101**, 4067 (1979); (b) G. B. Ellison, P. C. Engelking, and W. C. Lineberger, *J. Am. Chem. Soc.*, **100**, 2556 (1978).

[20](a) R. W. Fessenden, *J. Phys. Chem.*, **71**, 74 (1967); (b) K. Morokuma, L. Pedersen, and M. Karplus, *J. Chem. Phys.*, **48**, 4801 (1968); (c) D. L. Beveridge, P. A. Dobosh, and J. A. Pople, *J. Chem. Phys.*, **48**, 4802 (1968).

perature changes the $a_{13\text{C}}$. Analysis of the temperature behavior of the coupling constant for the *t*-butyl radical trapped in a matrix of solid adamantane led to the conclusion that it has a nonplanar equilibrium geometry with the methyl groups bent about 19° out of the plane and a barrier to inversion of 0.5 to 0.6 kcal mole^{-1}.[21] Substituting π electron donor substituents for hydrogen in $\cdot CH_3$ (for example, in the fluoromethyl radicals cited in Table 9.1) bends the structure away from planarity; a methyl substituent donates electrons by hyperconjugation.[22] Electron acceptor substituents, on the other hand, lead to planar radicals as would be expected for conjugated π systems.[23] There are indications, however, that the *t*-butyl radical may be closer to planarity (about 11°) and have a lower barrier (0.45 kcal mole^{-1}) in solution than it does in the solid matrix.[24] There has also been a claim that this radical is indeed planar, the temperature dependence of the coupling constant being an artifact caused by the matrix.[25] Ingold and co-workers have pointed out that the analysis rests on fitting experimental results to semiempirical molecular orbital calculations and may not be reliable.[24]

Electron spin resonance will also detect molecules with two unpaired electrons. With a few exceptions (O_2 is one), molecules with an even number of electrons are ground-state singlets: they have all electrons paired in the lowest energy state. But in electronically excited states, and in the ground states of certain reactive intermediates (for example, CH_2, Section 6.3), two electrons may be unpaired. As we have noted in Section 6.3, a molecule with two unpaired electrons is said to be in a triplet state. Electrons in a singlet-state species have no net magnetic moment and will not interact with a magnetic field; that is, a singlet molecule has only one magnetic energy state. But when a triplet-state species is placed in a magnetic field, there are three electron spin energy states. (A radical, with one unpaired electron, has just two energy states.) Transitions among these states of the triplet may be observed in the esr spectrum; they occur at different frequencies from those characteristic of radicals and so are readily distinguished.

Chemically induced dynamic polarization[26]   Chemically induced dynamic nuclear and electron polarization spectroscopy are techniques that take advantage of the coupling between electron and nuclear spins to detect radicals and products of radical recombinations by magnetic resonance. The methods are

[21](a) J. B. Lisle, L. F. Williams, and D. E. Wood, *J. Am. Chem. Soc.,* **98,** 227 (1976); (b) P. J. Krusic and P. Meakin, *J. Am. Chem. Soc.,* **98,** 228 (1976).

[22]P. J. Krusic and R. C. Bingham, *J. Am. Chem. Soc.,* **98,** 230 (1976).

[23]R. A. Kaba and K. U. Ingold, *J. Am. Chem. Soc.,* **98,** 523 (1976).

[24]D. Griller, K. U. Ingold, P. J. Krusic, and H. Fischer, *J. Am. Chem. Soc.,* **100,** 6750 (1978).

[25]L. Bonazzola, N. Leray, and J. Roncin, *J. Am. Chem. Soc.,* **99,** 8348 (1977).

[26]For summaries, see (a) H. R. Ward, in *Free Radicals,* Vol. I, J. K. Kochi, Ed., Wiley, New York, 1973, p. 239; (b) A. R. Lepley and G. L. Closs, Eds., *Chemically Induced Magnetic Polarization,* Wiley, New York, 1973; (c) D. Bethell and M. R. Brinkman, *Adv. Phys. Org. Chem.,* **10,** 53 (1973); (d) H. R. Ward, *Acc. Chem. Res.,* **5,** 18 (1972); (e) R. G. Lawler, *Acc. Chem. Res.,* **5,** 18 (1972); (f) R. Kaptein, *Adv. Free Radical Chem.,* **5,** 319 (1975); (g) J. K. S. Wan and A. J. Elliot, *Acc. Chem. Res.,* **10,** 161 (1977); (h) P. W. Atkins and G. T. Evans, *Adv. Chem. Phys.,* **xxxv,** 1 (1976); (i) N. C. Verma and R. W. Fessenden, *J. Chem. Phys.,* **65,** 2139 (1976).

suited to investigation of the dynamics of radical processes, particularly the events just preceding radical recombinations.

Chemically induced dynamic nuclear polarization (CIDNP) was first observed in 1967 by Bargon and Fischer[27] and independently by Ward and Lawler.[28] The phenomenon consists of alterations of the normal intensity patterns of the nuclear magnetic resonance spectrum, with lines occurring sometimes with enhanced absorption intensity and sometimes with negative absorption—that is, emission—when a reaction mixture is observed while a radical process is occurring. Chemically induced dynamic electron polarization (CIDEP) is a similar phenomenon in which abnormal intensities are observed in the esr spectrum. The currently accepted interpretation of the origin of these changes, proposed by Closs[29] and by Kaptein and Oosterhoff,[30] rests on the concept of a *caged* or *geminate* radical pair.

When a pair of radicals is formed in solution by homolytic bond cleavage, it is surrounded by a cage of solvent molecules; there is a competition between recombination and diffusion out of the cage. A caged pair will also arise when two radicals diffuse together from the solution. Once in close association as a pair, interaction of the two electrons will produce either a singlet or a triplet state. The probability of recombination before escape from the cage depends on whether the interacting pair is a singlet, which has spins paired ready for bond formation, or a triplet, which has spins unpaired and so cannot bond. Because the electron and nuclear spins are coupled, the rate of interconversion between singlet and triplet is affected by the interaction of nuclear and electron spin states. The result is that radical pairs that happen to have their nuclei and electrons in certain spin states are likely to yield cage recombination product, whereas radical pairs with other spin states are likely to escape to react outside the cage and give noncage products. Radicals that escape from the cage therefore have abnormal populations of some electron spin states, and various nonradical products have abnormal populations of some nuclear spin states; if the esr or nmr spectrum is observed, one finds abnormal relative line intensities.[31] The Appendix to this chapter gives a more detailed account of the CIDNP mechanism.

Chemical Investigations of Radical Structure

In addition to the spectroscopic investigations, there have been attempts to obtain structural and stereochemical information about radicals by chemical means.[32] The approach generally taken is to generate radicals by one of the methods discussed in the next section at a carbon where stereochemistry can be determined. As an example we may cite the experiment shown in Equation 9.8, in which an

---

[27](a) J. Bargon, H. Fischer, and U. Johnsen, *Z. Naturforsch.,* **22A,** 1551 (1967); (b) J. Bargon and H. Fischer, *Z. Naturforsch.,* **22A,** 1556 (1967); (c) H. Fischer and J. Bargon, *Acc. Chem. Res.,* **2,** 110 (1969).

[28]H. R. Ward and R. G. Lawler, *J. Am. Chem. Soc.,* **89,** 5518 (1967).

[29]G. L. Closs, *J. Am. Chem. Soc.,* **91,** 4552 (1969).

[30]R. Kaptein and L. J. Oosterhoff, *Chem. Phys. Lett.,* **4,** 214 (1969).

[31]Another mechanism for the production of CIDEP occurs in photochemical reactions and depends on rates of crossing from an excited singlet to an excited triplet state. See note 26(g)–(i).

[32]For a summary, see L. Kaplan, in *Free Radicals,* Vol. II, J. K. Kochi, Ed., Wiley, New York, 1973, p. 370.

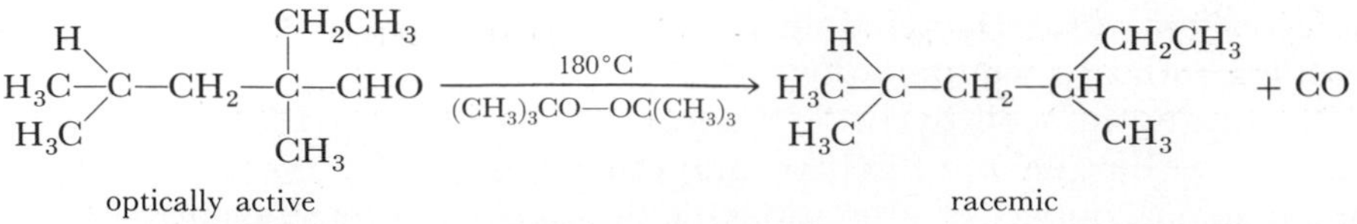

$$
\underset{\text{optically active}}{\overset{\displaystyle\text{H}\diagdown\quad\overset{\text{CH}_2\text{CH}_3}{|}}{\text{H}_3\text{C}\!-\!\text{C}\!-\!\text{CH}_2\!-\!\text{C}\!-\!\text{CHO}}} \;\xrightarrow[\text{(CH}_3\text{)}_3\text{CO}-\text{OC(CH}_3\text{)}_3]{180\,^\circ\text{C}}\; \underset{\text{racemic}}{\overset{\displaystyle\text{H}\diagdown\quad\overset{}{\text{CH}_2\text{CH}_3}}{\text{H}_3\text{C}\!-\!\text{C}\!-\!\text{CH}_2\!-\!\text{CH}}} \;+\; \text{CO} \qquad (9.8)
$$

optically active aldehyde is heated in the presence of a source of radicals.[33] The reaction follows the chain pathway indicated in Scheme 1; the loss of chirality indicates that the radical is either planar or, if pyramidal, undergoes inversion rapidly with respect to the rate (on the order of 10^5 sec^{-1}) at which it abstracts a hydrogen atom from another molecule of aldehyde.

SCHEME 1

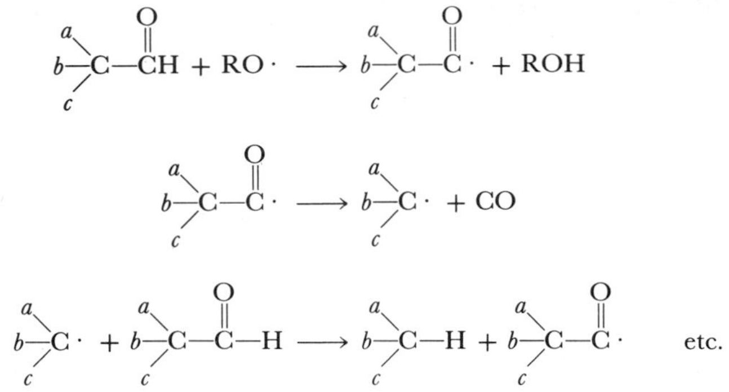

Table 9.2 presents the results of experiments with the 9-decalyl radical, generated as indicated in Scheme 2 from two isomeric precursors.[34] In this reaction the alkyl radicals, which form by fragmentation of the alkoxy radical, convert to stable products by abstracting a chlorine atom from a new molecule of hypochlorite which in turn is left as a new alkoxy radical to continue the chain. The lifetime of the alkyl radicals is thus dependent on concentration of the hypochlorite; the results demonstrate that whereas the radical from trans precursor does not isomerize (product composition approximately independent of concentration), the radical arising from cis precursor isomerizes to the trans radical when the concentration of trapping hypochlorite is low but abstracts chlorine before it has time to do so when hypochlorite concentration is high. The barrier to conversion from cis to trans radical is estimated at 3–6 kcal mole^{-1}. Bartlett and co-workers found a similar result in the trapping of 9-decalyl radicals with varying pressures of oxygen.[35]

One possible interpretation of these results is that the two tertiary decalyl radicals react differently because they are pyramidal (**6** and **7**); on the other hand, the hypothesis that the radicals are planar and differ only in ring conformation (**8**

[33] W. v. E. Doering, M. Farber, M. Sprecher, and K. B. Wiberg, *J. Am. Chem. Soc.*, **74**, 3000 (1952).
[34] F. D. Greene and N. N. Lowry, *J. Org. Chem.*, **32**, 875 (1967).
[35] P. D. Bartlett, R. E. Pincock, J. H. Rolston, W. G. Schindel, and L. A. Singer, *J. Am. Chem. Soc.*, **87**, 2590 (1965).

Table 9.2 PRODUCTS OF DECOMPOSITION OF *cis*- AND *trans*-9-DECALYLCARBINYL HYPOCHLORITE AT −40°C[a]

Starting From	Concentration (M)	Products	
		trans	cis
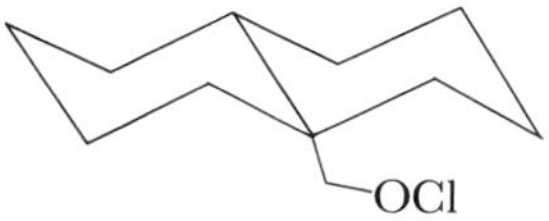 trans	0.009	27	1
	0.8	26	1
	3.3	32	1
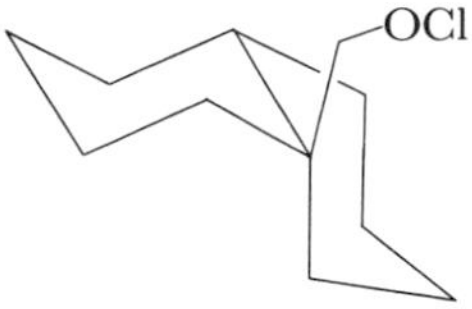 cis	0.008	29	1
	0.1	10	1
	1.0	1.7	1
	3.0	1.1	1

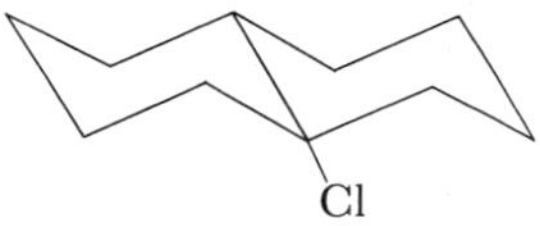

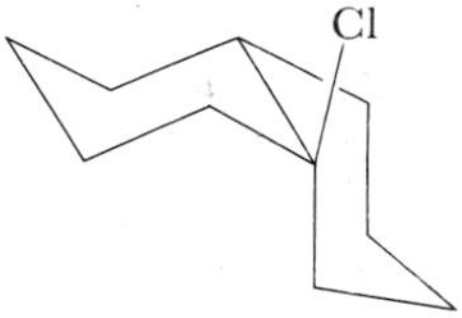

[a] Reprinted with permission from F. D. Greene and N. N. Lowry, *J. Org. Chem.*, **32**, 875 (1967). Copyright by the American Chemical Society. Refer to this paper for results at other temperatures and concentrations.

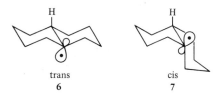

trans
6

cis
7

and **9**) explains the facts equally well, because the barrier to such a conversion might well be expected to be of the correct order of magnitude.[34]

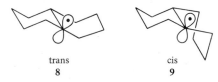

trans
8

cis
9

In cyclopropyl radicals the strain associated with the planar structure might be expected to lead to a higher inversion barrier than other alkyl radicals have. Although the cyclopropyl radical has a nonplanar equilibrium geometry, it inverts

SCHEME 2

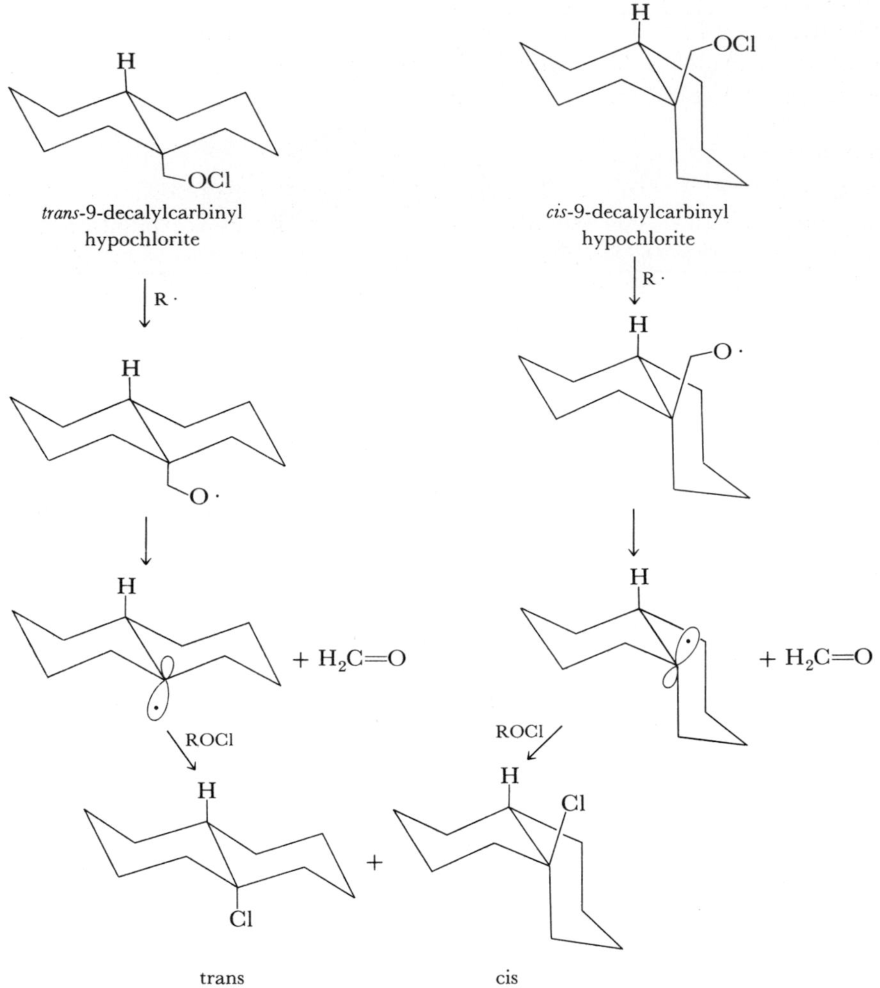
H
OCl
trans-9-decalylcarbinyl hypochlorite
cis-9-decalylcarbinyl hypochlorite
R·
O·
+ H₂C=O
ROCl
Cl
trans
cis

rapidly and cannot be trapped chemically before losing chirality.[36] The α-fluorocyclopropyl radical (**10**), on the other hand, can be generated from a chiral precursor and trapped stereospecifically by hydrogen abstraction from solvent or from an added hydrogen transfer agent.[37]

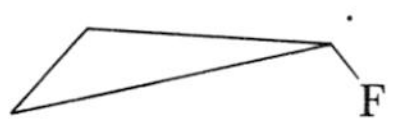
F

10

[36] V. Malatesta, D. Forrest, and K. U. Ingold, *J. Am. Chem. Soc.*, **100**, 7073 (1978).
[37] (a) T. Ando, H. Yamanaka, F. Namigata, and W. Funasaka, *J. Org. Chem.*, **35**, 33 (1970); (b) H. M. Walborsky and P. C. Collins, *J. Org. Chem.*, **41**, 940 (1976).

In summary, the alkyl radicals are either planar or, in the case of tertiary alkyl, nonplanar but very rapidly inverting. Two fluorine substituents or one fluorine substituent and incorporation into a highly strained ring appear to be necessary to produce a nonplanar radical with an inversion barrier high enough to be chemically significant.

Thermochemistry[38]

Of fundamental importance to free-radical chemistry are bond dissociation energies and radical heats and entropies of formation. Bond dissociation energy is defined as the energy required to break a particular bond to form two radicals. More precisely, bond dissociation energy of the R—X bond, $D(R—X)$, is the enthalpy change of Reaction 9.9.[39]

$$R—X \longrightarrow R\cdot + X\cdot \tag{9.9}$$

Radical heats of formation are defined in the usual way, that is, as enthalpy of formation of the radical in question from the elements in their standard states. The heats of formation and the bond dissociation energies are derivable from each other and are based on the same data. Thus in Reaction 9.9 the heat of formation of $R\cdot$ is readily found from the bond dissociation energy by means of the enthalpy cycle shown in Scheme 3 if heats of formation of R—X and $X\cdot$ are known; conversely, $D(R—X)$ may be found once heats of formation of RX, $R\cdot$, and $X\cdot$ are known. The same comments, of course, apply to the corresponding reaction entropies. Because of their lack of charge, radicals interact relatively weakly with solvents, and so thermochemical data obtained from gas-phase reactions can be used reasonably confidently for solutions as well.

SCHEME 3

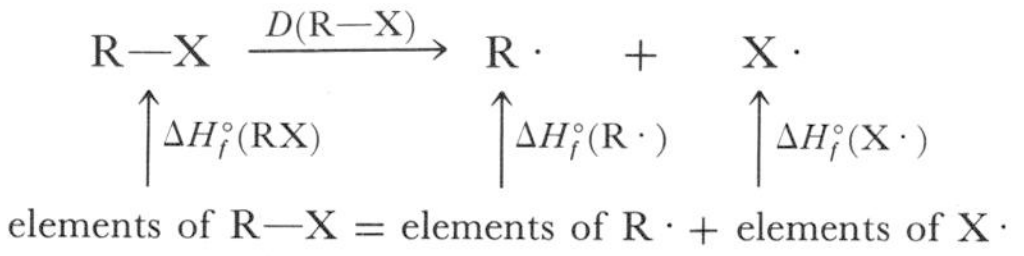

$$R—X \xrightarrow{\;D(R—X)\;} R\cdot \quad + \quad X\cdot$$

$$\uparrow \Delta H_f^\circ(RX) \qquad \uparrow \Delta H_f^\circ(R\cdot) \qquad \uparrow \Delta H_f^\circ(X\cdot)$$

elements of R—X = elements of $R\cdot$ + elements of $X\cdot$

$$\Delta H_f^\circ(RX) + D(R—X) - \Delta H_f^\circ(R\cdot) - \Delta H_f^\circ(X\cdot) = 0 \tag{9.10}$$
$$D(R—X) = \Delta H_f^\circ(R\cdot) + \Delta H_f^\circ(X\cdot) - \Delta H_f^\circ(RX) \tag{9.11}$$

We shall not consider here the methods by which thermodynamic data are obtained for radicals; O'Neal and Benson have reviewed the various techniques and results.[40]

---

[38] For reviews of radical thermochemistry, see (a) S. W. Benson, *J. Chem. Educ.,* **42,** 502 (1965); (b) J. A. Kerr, *Chem. Rev.,* **66,** 465 (1966); (c) D. M. Golden and S. W. Benson, *Chem. Rev.,* **69,** 125 (1969); (d) H. E. O'Neal and S. W. Benson, in *Free Radicals,* Vol. II, J. K. Kochi, Ed., Wiley, New York, 1973, p. 275.

[39] It is important to distinguish *bond dissociation energy* from *average bond strength,* which is averaged over the bonds of a given type in the molecule.

[40] See note 38(d).

Tables 9.3 and 9.4 list selected bond dissociation energies and radical heats of formation. Note particularly that the energy required to remove a hydrogen decreases in the series methane, primary, secondary, tertiary, and that aldehydic, allylic, and benzylic hydrogens have bond dissociation energies substantially lower than do alkyl hydrogens. The bond dissociation energy is a measure of stabilization of the radical, a concept that can be quantified as the stabilization energy, E_s. E_s is defined as the difference between $D(R—H)$ for the radical in question and $D(R—H)$ for the appropriate model alkane (primary, secondary, or tertiary).[41]

Table 9.3 Selected Bond Dissociation Energies and Radical Heats of Formation at $25°C$ (kcal mole^{-1})[a]

R ·	$D_{298}(R—H)$[b]	$\Delta H_f°(R·)$[c]
·CH$_3$	104.9[d]	34.9[d]
·CH$_2$CH$_3$	98	25[e]
·CH$_2$CH$_2$CH$_3$	98	23
(CH$_3$)$_2$ĊH	95	17.6[e]
(CH$_3$)$_3$C ·	92	8.2[e]
$\overset{\overset{O}{\|}}{CH_3C·}$	87	−5.2[e]
ϕ·	112	80.0[e]
H$_2$C=CH ·	≥108	≥68.0[e]
·CF$_3$	106	−111[e,f]
·CCl$_3$	96[g]	
H$_2$C=CH—ĊH$_2$	86.6[h]	39.4[h]
H$_2$C=CH—CH=CH—ĊH$_2$	80	52.9
·CH$_2\phi$	87.9[h]	47.8[h]
·OCH$_2$CH$_3$	110[i]	
·OOR	89[i]	
D(t-BuO-Ot-Bu)	37.5[j]	
D(t-BuO-OH)	42[j]	

[a] Except where otherwise noted, reprinted in part with permission from D. M. Golden and S. W. Benson, *Chem. Rev.,* **69,** 125 (1969). Copyright 1969 American Chemical Society.
[b] Reported uncertainty ⩽1 kcal mole^{-1} in most cases.
[c] Reported uncertainties ⩽2 kcal mole^{-1} in most cases.
[d] M. H. Baghal-Vayjooee, A. J. Colussi, and S. W. Benson, *J. Am. Chem. Soc.,* **100,** 3214 (1978).
[e] Average of determinations by two or more methods.
[f] Erroneously reported as positive in reference *a*. See J. A. Kerr, *Chem. Rev.,* **66,** 465 (1966); H. E. O'Neal and S. W. Benson, in *Free Radicals,* Vol. II, J. K. Kochi, Ed., Wiley, New York, 1973, p. 275.
[g] S. W. Benson, *J. Chem. Educ.,* **42,** 502 (1965).
[h] M. Rossi and D. M. Golden, *J. Am. Chem. Soc.,* **101,** 1230 (1979).
[i] M. Simonyi and F. Tüdõs, *Adv. Phys. Org. Chem.,* **9,** 135 (1971).
[j] S. W. Benson, *J. Chem. Phys.,* **40,** 1007 (1964).

[41] D. M. Golden and S. W. Benson, *Chem. Rev.,* **69,** 125 (1969).

Table 9.4 Bond Dissociation Energies of Halogen Compounds (kcal mole^{-1})^a

$D(X—X)$			$D(H—X)$			$D(CH_3—X)$
F_2	38	HF	135.8	CH_3F		108
Cl_2	58	HCl	103.0	CH_3Cl		83.5
Br_2	46.0	HBr	87.5	CH_3Br		70
I_2	36.1	HI	71.3	CH_3I		56

	$D(CH_3CH_2—X)$	$D\left(\begin{smallmatrix}CH_3\\ \\CH_3\end{smallmatrix}CH—X\right)$	$D((CH_3)_3C—X)$
F	106	105	—
Cl	81.5	81	78.5
Br	69	68	63
I	53.5	53	49.5

^{a}Data from S. W. Benson, *J. Chem. Educ.*, **42**, 502 (1965). Reprinted by permission of the Division of Chemical Education, American Chemical Society. See also J. A. Kerr, *Chem. Rev.*, **66**, 465 (1966), and T. L. Cottrell, *The Strengths of Chemical Bonds,* Butterworths, London, 1954.

Hence E_s for benzyl is $D(CH_3CH_2—H) - D(\phi CH_2—H) = 98 - 88 = 10$ kcal mole^{-1}. Phenyl is destabilized: $E_s = D((CH_3)_3C—H) - D(\phi—H) = 92 - 112 = -20$ kcal mole^{-1}. Stabilization does not necessarily lead to radicals with long lifetimes; as we shall see in Section 9.2, radical lifetime is controlled primarily by steric hindrance.

Just as for molecules, it is possible to estimate radical heats of formation by summing group contributions. Radical entropies and heat capacities are also available by calculation. Table 9.5 lists group contributions for radical heats of formation, ΔH_f°, and entropies, S°. The table is used in the way described in Section 2.3; contributions for groups not involving the radical center are found from the tables in that section, and corrections for strain and gauche interactions are applied as noted there. The additional corrections given in Table 9.5 must be applied to S°.[42] Calculated values for radicals are not as accurate as for molecules; the results should be correct to ± 2 kcal mole^{-1} for ΔH_f° and to ± 2 cal mole^{-1} $^\circ K^{-1}$ for S°. The following example shows how the table is used to calculate thermodynamic properties for the 2-butyl radical (**11**).[43]

$$H_3C—\overset{\bullet}{C}H—CH_2—CH_3$$
11

[42] See H. E. O'Neal and S. W. Benson, in *Free Radicals,* Vol. II, J. K. Kochi, Ed., Wiley, New York, 1973, p. 337, for more detailed discussion of corrections and for data required to correct to other temperatures.
[43] From H. E. O'Neal and S. W. Benson, in *Free Radicals,* Vol. II, J. K. Kochi, Ed., Wiley, New York, 1973, p. 343. Reprinted by permission of John Wiley & Sons, Inc.

Table 9.5 Free Radical Group Additivities[a,b]

Radical	ΔH°_{f298}	S°_{298}	$DH^{\circ}(\mathrm{C-H})^{c}$
$[\cdot\mathrm{C-(C)(H)_2}]$	35.82	31.20	98
$[\cdot\mathrm{C-(C)_2(H)}]$	37.45	10.74	95
$[\cdot\mathrm{C-(C)_3}]$	38.00	-10.77	92
$[\mathrm{C-(C\cdot)(H)_3}]$	-10.08	30.41	—
$[\mathrm{C-(C\cdot)(C)(H)_2}]$	-4.95	9.42	—
$[\mathrm{C-(C\cdot)(C)_2(H)}]$	-1.90	-12.07	—
$[\mathrm{C-(C\cdot)(C)_3}]$	1.50	-35.10	—
$[\mathrm{C-(O\cdot)(C)(H)_2}]$	6.1	36.4	—
$[\mathrm{C-(O\cdot)(C)_2(H)}]$	7.8	14.7	—
$[\mathrm{C-(O\cdot)(C)_3}]$	8.6	-7.5	—
$[\mathrm{C-(S\cdot)(C)(H)_2}]$	32.4	39.0	—
$[\mathrm{C-(S\cdot)(C)_2(H)}]$	35.5	17.8	—
$[\mathrm{C-(S\cdot)(C)_3}]$	37.5	-5.3	—
$[\cdot\mathrm{C-(H_2)(C_d)}]^{d}$	23.2	27.65	—
$[\cdot\mathrm{C-(H)(C)(C_d)}]$	25.5	7.02	—
$[\cdot\mathrm{C-(C)_2(C_d)}]$	24.8	-15.00	—
$[\mathrm{C_d-(C\cdot)(H)}]$	8.59	7.97	—
$[\mathrm{C_d-(C\cdot)(C)}]$	10.34	-12.30	—
$[\cdot\mathrm{C-(C_B)(H)_2}]^{e}$	23.0	26.85	85
$[\cdot\mathrm{C-(C_B)(C)(H)}]$	24.7	6.36	82
$[\cdot\mathrm{C-(C_B)(C)_2}]$	25.5	-15.46	79
$[\mathrm{C_B-C\cdot}]$	5.51	-7.69	—
$[\mathrm{C-(\cdot CO)(H)_3}]$	-5.4	66.6	—
$[\mathrm{C-(\cdot CO)(C)(H)_2}]$	-0.3	45.8	—
$[\mathrm{C-(\cdot CO)(C)_2(H)}]$	2.6	$(23.7)^{f}$	—
$[\cdot\mathrm{N-(H)(C)}]$	$(55.3)^{f}$	30.23	—
$[\cdot\mathrm{N-(C)_2}]$	(58.4)	10.24	—
$[\mathrm{C-(\cdot N)(C)(H)_2}]$	-6.6	9.8	—
$[\mathrm{C-(\cdot N)(C)_2(H)}]$	-5.2	-11.7	—
$[\mathrm{C-(\cdot N)(C)_3}]$	(-3.2)	-34.1	—
$[\cdot\mathrm{C-(H)_2(CN)}]$	$(54.2)^{g}$	58.5	—
$[\cdot\mathrm{C-(H)(C)(CN)}]$	$(52.8)^{g}$	40.0	—
$[\cdot\mathrm{C-(C)_2(CN)}]$	$(52.1)^{g}$	19.6	—
$[\cdot\mathrm{N-(H)(C_B)}]$	38.0	27.3	—
$[\cdot\mathrm{N-(C)(C_B)}]$	42.7	(6.5)	—
$[\mathrm{C_B-N\cdot}]$	-0.5	-9.69	—
$[\mathrm{C-(CO_2\cdot)(H)_3}]$	-49.7	71.4	—
$[\mathrm{C-(CO_2\cdot)(H)_2(C)}]$	-43.9	49.8	—
$[\mathrm{C-(CO_2\cdot)(H)(C)_2}]$	-41.0	-12.1	—
$[\mathrm{C-(N_A)(H)_3}]^{h}$	-10.08	30.41	—
$[\mathrm{C-(N_A)(C)(H)_2}]$	-5.5	9.42	—
$[\mathrm{C-(N_A)(C)_2(H)}]$	-3.3	-12.07	—
$[\mathrm{C-(N_A)(C)_3}]$	-1.9	-35.10	—
$[\mathrm{N_A-C}]$	32.5	8.0	—
$[\mathrm{N_A-(N_A\cdot)(C)}]$	74.2	36.1	—

Table 9.5 (*Continued*)

Mass corrections in conjugated systems

---

1. If the masses on each side of a resonance-stabilized bond in a radical have the same number of carbon atoms (i.e., are roughly equal), add 0.7 cal mole^{-1} K^{-1} to the entropy; for example,

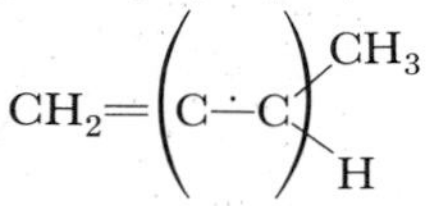

$$CH_2=\left(C\!-\!C\begin{array}{c}CH_3\\ \\H\end{array}\right) \qquad m_1 = CH_2 \qquad m_2 = (CH_3 + H)$$

---

2. If the masses (on each side) differ by one carbon atom, add 0.3 cal mole^{-1} K^{-1} to $S°$; for example,

$$(CH_2=C\!-\!CH_2) \qquad m_1 = CH_2 \qquad m_2 = H_2$$

Internal rotation barrier corrections[i]

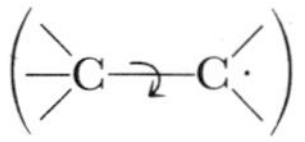

$$\left(\!>\!C\!-\!\!\!\underset{2}{}\!\!\!-C\!<\!\cdot\right) \qquad \begin{array}{c}S°\\+0.5\end{array}$$

---

[a] From H. E. O'Neal and S. W. Benson, in *Free Radicals*, Vol. II, J. K. Kochi, Ed., Wiley, New York, 1973, pp. 338–340. Reprinted by permission of John Wiley & Sons, Inc.
[b] Units are kcal mole^{-1} for $\Delta H_f°$ and cal mole^{-1} K^{-1} for $S°$.
[c] Heat of formation group values are based on these (C—H) bond dissociation energies.
[d] C_d = doubly bonded carbon.
[e] C_B = carbon in a benzene ring.
[f] Values in parentheses are best guesses.
[g] Heats of formation assume resonance stabilizations: 10.8 kcal mole^{-1} in ($\cdot CH_2CN$); 12.6 kcal mole^{-1} in (CH$_3$ĊHCN) and in [(CH$_3$)$_3$ĊCN].
[h] N_A = doubly bonded nitrogen in azo compounds.
[i] This correction assumes that the barrier to rotation in the radical R $\cdot$ is two-thirds the barrier in the corresponding alkane RH. See O'Neal and Benson for further discussion of this point.

Group	Contribution to $\Delta H_f°$ (298°K)	Contribution to $S°$ (298°K)
C—(C)(H)$_3$	-10.08	30.41 (Table 2.6)
C—(C$\cdot$)(H)$_3$	-10.08	30.41
C—(C$\cdot$)(C)(H)$_2$	-4.95	9.42
$\cdot$C—(C)$_2$(H)	$+37.45$	10.74
Barrier corrections:		
H$_3$C—$\underset{2}{}$—Ċ		0.50
Ċ—$\underset{2}{}$—CH$_2$—CH$_3$		0.50
Symmetry number = $3 \times 3 = 9$		
(two internal threefold rotations of —CH$_3$ groups)		
$-R \ln 9$		-4.37
Estimates:	$\Delta H_f° = +12.3$ kcal mole^{-1};	$S° = 77.6$ cal mole^{-1} °K^{-1}.

9.2 RADICAL PRODUCTION AND TERMINATION

As we have noted briefly in the previous section, radical reactions can be broken down into three stages: radical production, reactions yielding new radicals, and radical destruction. In this section we examine the steps that produce and destroy radicals.

Radical Production

The first radical source we shall consider is thermal homolysis. At sufficiently high temperatures, most chemical bonds will break to form radicals, but in the temperature range of ordinary solution chemistry, below 200°C, the bonds that will do so at reasonable rates are limited to a few types, the most common of which are the peroxy bond and the azo linkage.[44] Substances that produce radicals easily in a thermal process are designated *initiators*. Equations 9.12–9.16 illustrate a few typical examples with activation parameters. Radicals can also arise by interaction of closed-shell molecules, as illustrated, for example, in Equation 9.17.

$$C_2H_5O-OC_2H_5 \xrightarrow[\text{(gas phase)}]{134-185°C} 2C_2H_5O\cdot$$

E_a 37.3 kcal mole^{-1}
log A 16.1
$(9.12)^{[45]}$

$$CH_3\overset{O}{\overset{\|}{C}}O-O\overset{O}{\overset{\|}{C}}CH_3 \xrightarrow[\text{(benzene)}]{55-85°C} 2CH_3\overset{O}{\overset{\|}{C}}O\cdot$$

E_a 32.3 kcal mole^{-1}
$(9.13)^{[46]}$

$$C_6H_5\overset{O}{\overset{\|}{C}}O-O\overset{O}{\overset{\|}{C}}C_6H_5 \xrightarrow[\text{(benzene)}]{60-80°C} 2C_6H_5\overset{O}{\overset{\|}{C}}O\cdot$$

E_a 33.3 kcal mole^{-1}
$(9.14)^{[47]}$

$$\underset{H_3C}{\overset{NC}{>}}C-CN=NC-\underset{CH_3}{\overset{CN}{<}}CH_3 \xrightarrow{66-72°C} 2CH_3\underset{H_3C}{\overset{NC}{>}}C\cdot + N_2$$

$\Delta H^{\ddagger}$ 31.2 kcal mole^{-1}
$\Delta S^{\ddagger}$ +12.2 cal mole^{-1} K^{-1}
$(9.15)^{[48]}$

$$C_6H_5-\underset{CH_3}{\overset{CH_3}{|}}C-N=N-\underset{CH_3}{\overset{CH_3}{|}}C-C_6H_5 \xrightarrow[\text{(toluene)}]{40-70°C} 2C_6H_5-\underset{CH_3}{\overset{CH_3}{|}}C\cdot + N_2$$

$\Delta H^{\ddagger}$ 29.0 kcal mole^{-1}
$\Delta S^{\ddagger}$ +11.0 cal mole^{-1}
$(9.16)^{[49]}$

[44] For a review, see T. Koenig, in *Free Radicals*, Vol. I, J. K. Kochi, Ed., Wiley, New York, 1973, p. 113.
[45] C. Leggett and J. C. J. Thynne, *Trans. Faraday Soc.*, **63**, 2504 (1967).
[46] M. Levy, M. Steinberg, and M. Szwarc, *J. Am. Chem. Soc.*, **76**, 5978 (1954).
[47] K. Nozaki and P. D. Bartlett, *J. Am. Chem. Soc.*, **68**, 1686 (1946).
[48] R. C. Petersen, J. H. Markgraf, and S. D. Ross, *J. Am. Chem. Soc.*, **83**, 3819 (1961).
[49] S. F. Nelsen and P. D. Bartlett, *J. Am. Chem. Soc.*, **88**, 137 (1966).

$$(CH_3)_3C-O-O-H + O_3 \xrightarrow{-60°} (CH_3)_3C-O-O\cdot + HO\cdot + O_2$$

$$E_a \approx 7 \text{ kcal mole}^{-1}$$
$$\log A = 7 \tag{9.17}[50]$$

Complications arise in many of these decompositions because other types of processes can occur. There are, for example, heterolytic rearrangements of peroxyesters and diacyl peroxides (Equations 9.18 and 9.19), as well as base-catalyzed

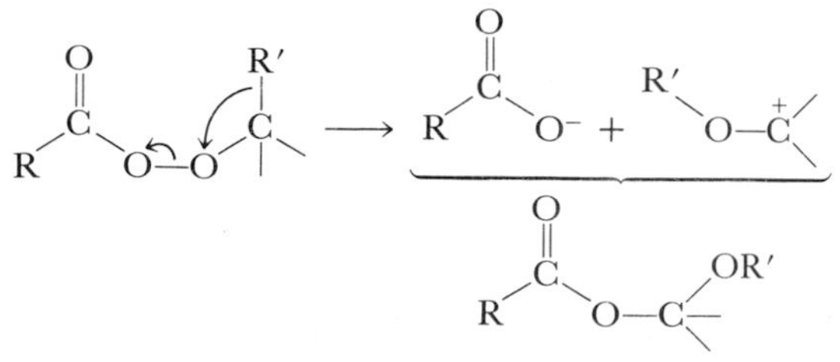

$$\tag{9.18}[51]$$

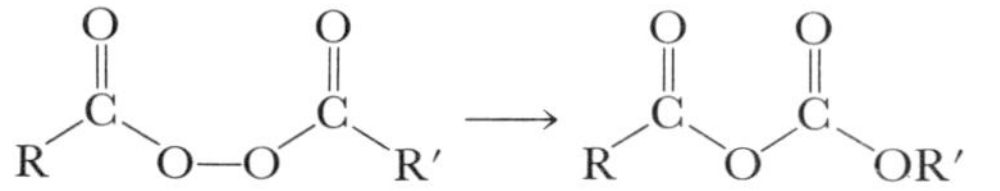

$$\tag{9.19}[52]$$

$$\tag{9.20}[53]$$

processes, for example, Reaction 9.20. These reactions we mention only in passing; they may be understood in terms of the ideas discussed in earlier chapters. Of more immediate concern are the *induced decompositions,* which are chain reactions arising because the radical products of an initial unimolecular decomposition can attack unreacted initiator molecules to yield new radicals that continue the chain. *t*-Butyl hydroperoxide, for example, decomposes when heated in a solvent at 150–180°C.[54] The reaction does not follow first-order kinetics and is a chain reaction initiated by a unimolecular decomposition of the peroxide (Equation 9.21) followed by attack

$$(CH_3)_3CO-OH \xrightarrow{\Delta} (CH_3)_3CO\cdot + HO\cdot \tag{9.21}$$

$$\begin{array}{l} (CH_3)_3O\cdot \\ \text{or} \\ HO\cdot \end{array} + (CH_3)_3CO-OH \longrightarrow (CH_3)_3CO-O\cdot + \begin{array}{l} (CH_3)_3COH \\ \text{or} \\ HOH \end{array} \tag{9.22}$$

$$2(CH_3)_3CO-O\cdot \longrightarrow O_2 + 2(CH_3)_3CO\cdot \tag{9.23}$$

[50]M. E. Kurz and W. A. Pryor, *J. Am. Chem. Soc.,* **100**, 7953 (1978).
[51](a) R. Criegee, *Justus Liebigs Ann. Chem.,* **560**, 127 (1948); (b) E. Hedaya and S. Winstein, *J. Am. Chem. Soc.,* **89**, 1661 (1967).
[52](a) J. E. Leffler, *J. Am. Chem. Soc.,* **72**, 67 (1950); (b) F. D. Greene, H. P. Stein, C. C. Chu, and F. M. Vane, *J. Am. Chem. Soc.,* **86**, 2080 (1964).
[53]N. Kornblum and H. E. DeLaMare, *J. Am. Chem. Soc.,* **73**, 880 (1951).
[54](a) E. R. Bell, J. H. Raley, F. F. Rust, F. H. Seubold, and W. E. Vaughan, *Faraday Discuss. Chem. Soc.,* **10**, 242 (1951); (b) S. W. Benson, *J. Chem. Phys.,* **40**, 1007 (1964).

of the radicals so produced on unreacted hydroperoxide (Equation 9.22). The peroxy radicals are known to react with each other according to Equation 9.23 to yield molecular oxygen and alkoxy radicals, which can return to attack more hydroperoxide.[55] The chain nature of the decomposition is amply confirmed by the kinetics[56] and by the observation that it is accelerated by the addition of an independent source of radicals.[57] In other instances, solvent molecules may become involved in a chain decomposition.[58]

The complexities of the chain-induced decompositions can be minimized by choosing an initiator and solvent without easily abstracted hydrogens; under these conditions peroxy and azo compounds decompose unimolecularly at easily measured rates to provide convenient sources of radicals for use in studying other radical processes.

Peroxy compounds Even the unimolecular decompositions of peroxy compounds can be complicated. Consider, for example, the unimolecular decomposition of benzoyl peroxide in benzene or carbon tetrachloride, where good first-order kinetics indicate that the contribution of induced decomposition is small.[59] Initial homolysis of the O—O bond leads to two benzoyl radicals (Equation 9.24), which can fragment according to Equation 9.25 to yield phenyl radicals and carbon

$$
\underset{\substack{\|\\ \text{O}}}{C_6H_5C}O\text{—}O\underset{\substack{\|\\ \text{O}}}{C}C_6H_5 \longrightarrow 2C_6H_5\underset{\substack{\|\\ \text{O}}}{C}O\cdot \qquad (9.24)
$$

$$
C_6H_5\underset{\substack{\|\\ \text{O}}}{C}O\cdot \longrightarrow C_6H_5\cdot + CO_2 \qquad (9.25)
$$

dioxide. It is conceivable, however, that the homolysis and fragmentation could be concerted so that the phenyl radicals and carbon dioxide are formed in a single step as illustrated by Equation 9.26.

$$
\underset{\substack{\|\\ \text{O}}}{C_6H_5C}O\text{—}O\underset{\substack{\|\\ \text{O}}}{C}C_6H_5 \xrightarrow{?} 2C_6H_5\cdot + 2CO_2 \qquad (9.26)
$$

In the decomposition of benzoyl peroxide, which occurs at convenient rates in the neighborhood of 80°C, addition of iodine[59] or the stable radical galvinoxyl (**12**)[60] to the reaction reduces the yield of carbon dioxide.[61] One interpretation of

[55] We shall return to this reaction later when considering autoxidation in Section 9.4.

[56] See note 54(b).

[57] (a) R. Hiatt, J. Clipsham, and T. Visser, *Can. J. Chem.*, **42**, 2754 (1964); (b) P. D. Bartlett and P. Günther, *J. Am. Chem. Soc.*, **88**, 3288 (1966).

[58] (a) W. E. Cass, *J. Am. Chem. Soc.*, **69**, 500 (1947); (b) P. D. Bartlett and K. Nozaki, *J. Am. Chem. Soc.*, **69**, 2299 (1947).

[59] G. S. Hammond and L. M. Soffer, *J. Am. Chem. Soc.*, **72**, 4711 (1950).

[60] H. J. Shine, J. A. Waters, and D. M. Hoffman, *J. Am. Chem. Soc.*, **85**, 3613 (1963).

[61] Substances of this kind, which react at rates approaching diffusion control with radicals, are referred to as *scavengers*.

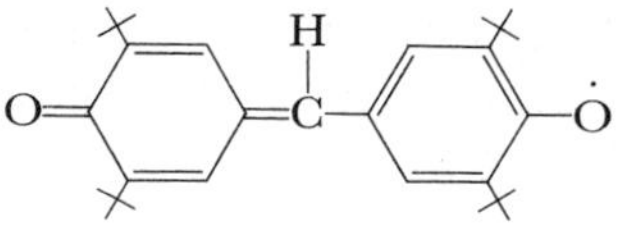

12

this result is that the benzoyl radicals are being trapped before they can decarboxylate; it is nevertheless possible that the reduced carbon dioxide yield results from decomposition induced by the scavenger. An alternative approach to the problem is to look for scrambling of ^{18}O label between carbonyl and peroxy positions in unreacted peroxide. Such scrambling does indeed occur and could be interpreted as in Equation 9.27 to imply nonconcerted reversible homolysis.[62]

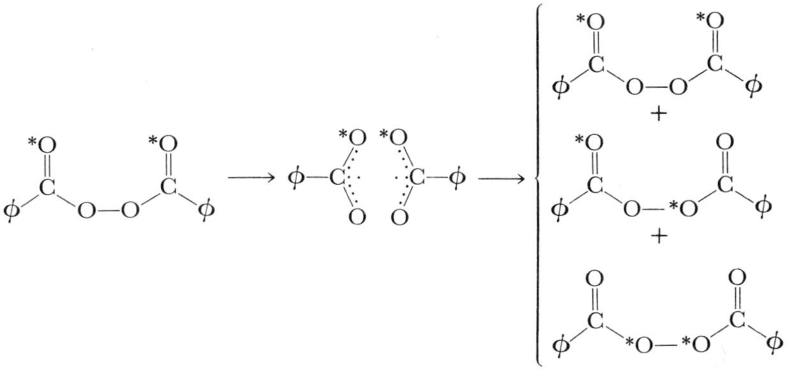

(9.27)

Goldstein and co-workers have pointed out that this oxygen exchange can take place at least partly through nonradical rearrangements 9.28 or 9.29,[63] and isotope effects in acetyl peroxide decomposition are consistent with significant concerted decomposition.[64]

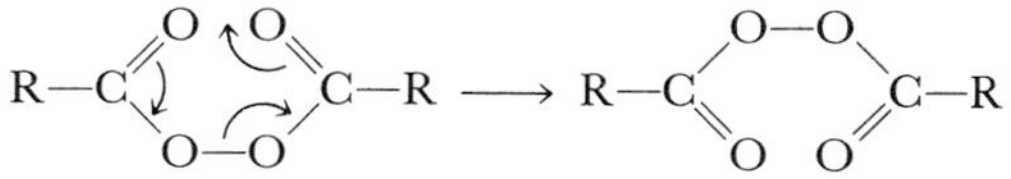

(9.28)

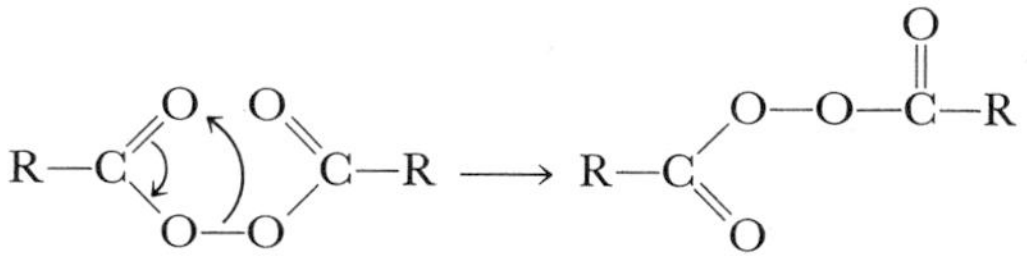

(9.29)

[62](a) J. C. Martin, and J. H. Hargis, *J. Am. Chem. Soc.*, **91**, 5399 (1969); (b) see also W. A. Pryor and K. Smith, *J. Am. Chem. Soc.*, **92**, 5403 (1970), who discuss the influence of solvent viscosity on rate and the use of such data as a criterion for concertedness.

[63](a) M. J. Goldstein and H. A. Judson, *J. Am. Chem. Soc.*, **92**, 4119, 4120 (1970); (b) M. J. Goldstein, M. DeCamp, and W. A. Haiby, *American Chemical Society, 168th Meeting, Atlantic City, N.J., September 8–13, 1974, Abstracts of Papers*, Orgn. 53.

[64](a) M. J. Goldstein, *Tetrahedron Lett.*, 1601 (1964); (b) M. J. Goldstein, H. A. Judson, and M. Yoshida, *J. Am. Chem. Soc.*, **92**, 4122 (1970).

The CIDNP technique gives much useful information about diacyl peroxide decomposition. We shall discuss an example from the work of Ward, the decomposition of propanoyl benzoyl peroxide (**13**), which in the presence of I_2 yields the products in Equation 9.30.[65] Figure 9.1 shows the proton magnetic resonance spec-

$$\phi-\overset{\overset{\displaystyle O}{\|}}{C}-O-O-\overset{\overset{\displaystyle O}{\|}}{C}-CH_2-CH_3 + I_2 \longrightarrow \phi-\overset{\overset{\displaystyle O}{\|}}{C}-O-CH_2-CH_3 +$$

13

$$I-CH_2-CH_3 + \phi-CH_2-CH_3 + CH_3-CH_2-CH_2-CH_3 + CO_2 \qquad (9.30)$$

trum recorded while the reaction is occurring at 100°C. The first step (Scheme 4) will be O—O bond homolysis to yield the caged radical pair **14**, which rapidly loses CO_2 to the new pair **15**. If we follow the CIDNP rules given in the Appendix to this chapter (Table A1.1), we find that the ethyl radical protons should acquire net

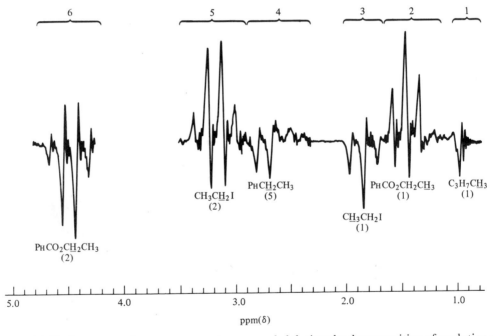

Figure 9.1 Nuclear magnetic resonance spectrum recorded during the decomposition of a solution of propionyl benzoyl peroxide and iodine in *o*-dichlorobenzene at 100°C. The numbers in parentheses below the formulas indicate the relative spectrum amplitudes for the underlined protons. Spectrum groups referred to in the text are indicated at the top of the figure. Reprinted with permission from H. R. Ward, *Acc. Chem. Res.*, **5**, 18 (1972). Copyright by the American Chemical Society.

[65] H. R. Ward, *Acc. Chem. Res.*, **5**, 18 (1972).

polarization such that the CH_2 protons in products of reaction within the cage will show emission and the CH_3 protons in the same products will show enhanced absorption. (The benzoyl radical has the larger g, so Δg is negative. See Table A1.1 in the Appendix to this chapter.) This is indeed the result observed. The CH_2 of ethyl benzoate (**16**), group 6, gives emission, whereas the CH_3 of the same product, group 2, gives absorption. (A small multiplet effect could also be present, but is largely obscured by the net effect.) The result shows that the benzoyl radical is present in the cage, and clearly demonstrates the stepwise nature of the decomposition in this compound.[66]

The caged pair **15** can lose the second CO_2 to yield a new pair **17**; **17** retains the net polarization of **15**, which was emission for the CH_2, but now acquires in addition a multiplet effect in the sense E/A for the CH_2 group. The CH_2 in the

SCHEME 4

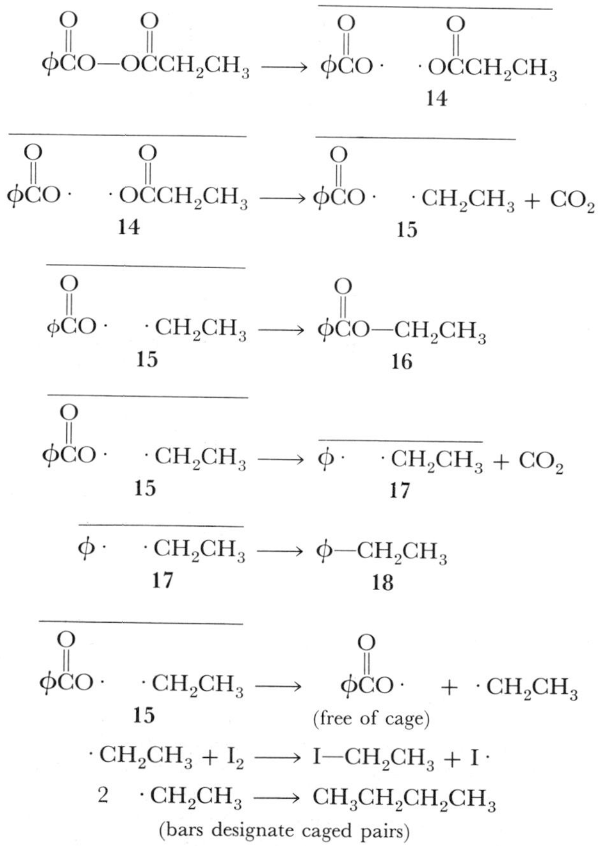

$$\phi\overset{O}{\overset{\|}{C}}O-O\overset{O}{\overset{\|}{C}}CH_2CH_3 \longrightarrow \overline{\phi\overset{O}{\overset{\|}{C}}O\cdot \quad \cdot O\overset{O}{\overset{\|}{C}}CH_2CH_3}$$
$$\textbf{14}$$

$$\overline{\phi\overset{O}{\overset{\|}{C}}O\cdot \quad \cdot O\overset{O}{\overset{\|}{C}}CH_2CH_3} \longrightarrow \overline{\phi\overset{O}{\overset{\|}{C}}O\cdot \quad \cdot CH_2CH_3} + CO_2$$
$$\textbf{14} \qquad\qquad\qquad \textbf{15}$$

$$\overline{\phi\overset{O}{\overset{\|}{C}}O\cdot \quad \cdot CH_2CH_3} \longrightarrow \phi\overset{O}{\overset{\|}{C}}O-CH_2CH_3$$
$$\textbf{15} \qquad\qquad\qquad \textbf{16}$$

$$\overline{\phi\overset{O}{\overset{\|}{C}}O\cdot \quad \cdot CH_2CH_3} \longrightarrow \overline{\phi\cdot \quad \cdot CH_2CH_3} + CO_2$$
$$\textbf{15} \qquad\qquad\qquad \textbf{17}$$

$$\overline{\phi\cdot \quad \cdot CH_2CH_3} \longrightarrow \phi-CH_2CH_3$$
$$\textbf{17} \qquad\qquad\qquad \textbf{18}$$

$$\overline{\phi\overset{O}{\overset{\|}{C}}O\cdot \quad \cdot CH_2CH_3} \longrightarrow \phi\overset{O}{\overset{\|}{C}}O\cdot \quad + \cdot CH_2CH_3$$
$$\textbf{15} \qquad\qquad\qquad \text{(free of cage)}$$

$$\cdot CH_2CH_3 + I_2 \longrightarrow I-CH_2CH_3 + I\cdot$$

$$2 \quad \cdot CH_2CH_3 \longrightarrow CH_3CH_2CH_2CH_3$$

(bars designate caged pairs)

[66] Bartlett and co-workers have found chemical evidence for concerted decomposition of certain peroxy esters. See (a) P. D. Bartlett and R. R. Hiatt, *J. Am. Chem. Soc.*, **80**, 1398 (1958); (b) P. D. Bartlett and D. M. Simons, *J. Am. Chem. Soc.*, **82**, 1753 (1960). See also note 64(b).

product phenylethane (**18**), group 4 in the spectrum, shows superposition of the net emission, E, and a multiplet effect in the predicted sense E/A. (The CH_3 of this product is evidently obscured by the CH_3 of the ethyl benzoate.) Ethyl radicals that escape from the cage either react with iodine to give ethyl iodide, groups 3 (CH_3) and 5 (CH_2), which shows net polarization just opposite to that of the ethyl benzoate, or combine to give butane, of which the CH_3 can be seen in emission (group 1).

Azoalkanes Similar questions of concertedness may be raised for decomposition of azoalkanes. In a concerted reaction (Equation 9.31), two C—N bonds would break simultaneously, whereas a stepwise process would produce first one R · radical (Equation 9.32) and later the second (Equation 9.33). One test is to

$$R—N{=}N—R \longrightarrow R\cdot + N_2 + \cdot R \qquad (9.31)$$

$$R—N{=}N—R \longrightarrow R—N{=}N\cdot + \cdot R \qquad (9.32)$$

$$R—N{=}N\cdot \longrightarrow R\cdot + N_2 \qquad (9.33)$$

measure the effect on decomposition rate of changing one of the R groups. If R—N=N—R and R—N=N—R′, where R · is a more highly stabilized radical than R ·′, are compared, a lower rate should be found for the unsymmetrical compound in a concerted decomposition; in a stepwise one the better radical R · should come off first in either compound, and the rate should be unaffected. The results support the concerted process (Equation 9.31) for the symmetrical azo compounds with R · a well-stabilized radical such as ϕ_2CCN, ϕ_2CH, and so forth, but leave open the possibility that in the unsymmetrical cases the mechanism may change to the stepwise one.[67]

Crawford and Tagaki have examined gas-phase decompositions of several azo compounds with R = CH_3, t-butyl, or allyl; they found that most of these compounds, both symmetric and unsymmetric, decompose by the nonconcerted path (Equation 9.32).[68] Seltzer found by studying secondary deuterium isotope effects that in unsymmetrical azo compounds in which one R group is a much better radical than the other, the bond breaking is stepwise.[69] In a compound such as **19**, substitution of the hydrogen on the α-phenylethyl side gives $k_H/k_D = 1.13$, a

reasonable value for partial breaking of the bond between nitrogen and the carbon at which the deuterium substitution is made. If, on the other hand, deuterium substitution is made in the methyl group (**20**), the k_H/k_D is 0.97 for three deuteriums, or 0.99 per deuterium. These C—H bonds are therefore being altered very little in the rate-determining step, and the conclusion is that the N—CH_3

[67] See T. Koenig, in *Free Radicals*, Vol. I, J. K. Kochi, Ed., Wiley, New York, 1973, p. 143, for a summary of results.
[68] R. J. Crawford and K. Takagi, *J. Am. Chem. Soc.*, **94**, 7406 (1972).
[69] S. Seltzer and F. T. Dunne, *J. Am. Chem. Soc.*, **87**, 2628 (1965).

bond is not breaking. The isotope effect observed when two deuteriums are substituted in **21**, which can produce stabilized benzylic radicals, is $k_H/k_D = 1.27$ (two

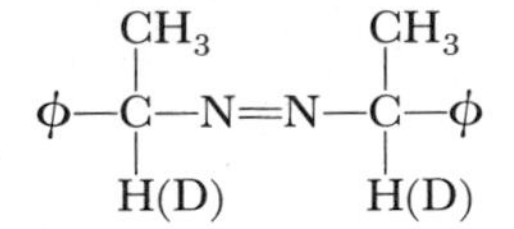

$$\phi-\underset{\underset{\displaystyle H(D)}{|}}{\overset{\overset{\displaystyle CH_3}{|}}{C}}-N=N-\underset{\underset{\displaystyle H(D)}{|}}{\overset{\overset{\displaystyle CH_3}{|}}{C}}-\phi$$

21

deuteriums), or 1.13 per deuterium. Therefore here both C—N bonds are breaking, and force constants of the C—H bonds on both sides are altered at the transition state. Pryor and Smith have found independent evidence favoring the single bond cleavage mechanism for unsymmetrical azo compounds.[70]

Decomposition chemistry of the azo compounds is potentially complicated by the existence of cis and trans isomers (**22** and **23**). The trans form is the more

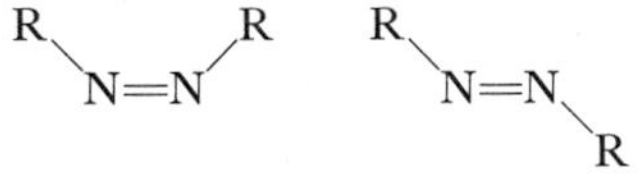

$$\underset{\displaystyle \textbf{22}}{\overset{\displaystyle R}{\underset{\displaystyle R}{N=N}}} \qquad \underset{\displaystyle \textbf{23}}{\overset{\displaystyle R}{N=N}}\overset{}{\underset{\displaystyle R}{}}$$

stable; cis isomers can be prepared by low-temperature photolysis, but they react fairly rapidly at room temperature to yield partly trans isomer and partly radical products. The cis isomers are implicated as intermediates in photochemical decompositions starting from trans, but it is not clear whether they are also involved in thermal decompositions.[71]

Cyclic azo compounds are important as a source of diradicals.[72] Equation 9.34 shows an example from the work of Berson.[73]

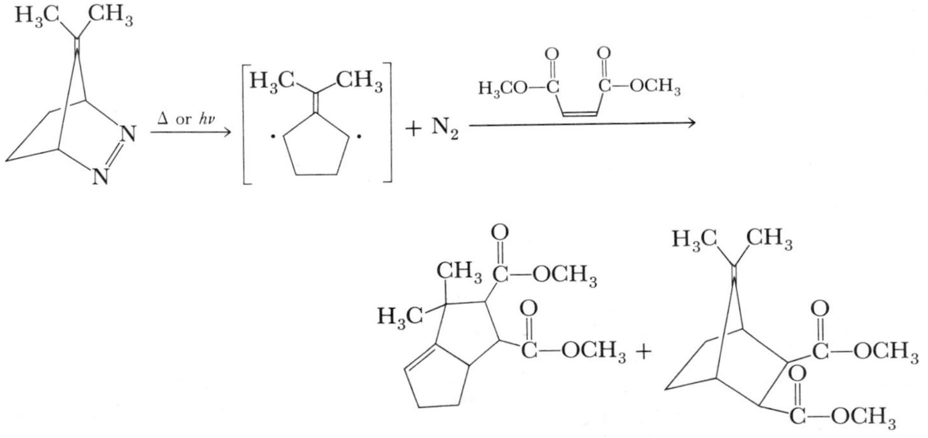

$$(9.34)$$

[70](a) W. A. Pryor and K. Smith, *J. Am. Chem. Soc.,* **92**, 5403 (1970); (b) P. S. Engel, and D. J. Bishop, *J. Am. Chem. Soc.,* **97**, 6754 (1975); (c) J. G. Green, G. R. Dubay, and N. A. Porter, *J. Am. Chem. Soc.,* **99**, 1264 (1977).

[71] N. A. Porter and L. J. Marnett, *J. Am. Chem. Soc.,* **95**, 4361 (1973).

[72](a) R. G. Bergman, in *Free Radicals,* Vol. I, J. K. Kochi, Ed., Wiley, New York, 1973, p. 191; (b) S. Braslavsky and J. Heicklen, *Chem. Rev.,* **77**, 473 (1977).

[73] J. A. Berson, *Acc. Chem. Res.,* **11**, 446 (1978).

Structural effects in bond homolysis The discussion above has pointed out some of the relationships between structure and concertedness in the decompositions of peroxy compounds and azo alkanes. Other structural variations have been investigated also, such as the effect of electron-donating and -withdrawing substituents and of geometry about carbon atoms at which radical character is developing.

Because radicals are uncharged, one would expect that electron donation and withdrawal would be less important than in ionic reactions. Although this expectation is certainly correct, there is nevertheless ample evidence that transition states of radical reactions do have some polar character. In resonance terminology, one might describe the transition state of a homolysis as a resonance hybrid (**24**) of

$$A{-}B \longleftrightarrow A{\cdot}{\cdot}B \longleftrightarrow A^+B^- \longleftrightarrow A^-B^+$$
$$24$$

covalent, radical, and ionic structures, with the importance of the ionic forms subject to influence by substituents in A and B. It should therefore not be surprising to find modest polar effects in radical reactions; indeed, in perester decompositions, for example, of **25**, a Hammett correlation of rate with σ^+ parameters yields ρ in the

$$X{-}\phi{-}CH_2{-}\overset{\overset{\textstyle O}{\|}}{C}{-}O{-}O{-}C(CH_3)_3$$
$$25$$

neighborhood of -1, a result that indicates acceleration by electron donation.[74] The result is that expected if in the transition state of a concerted decomposition the breaking C—CO bond is polarized toward the electron-deficient carbonyl carbon, as shown in **26**, leaving the benzylic position with some carbocation character.

$$X{-}\phi{-}\overset{\delta^+}{C}H_2{\cdots}\overset{\overset{\textstyle O}{\overset{\|}{}}}{C}{-}O{\cdots}O{-}C(CH_3)_3$$
$$26$$

[Note that the peroxyesters of the phenylacetic acids (**25**) undergo a concerted two-bond decomposition, whereas peroxyesters of benzoic acid (Scheme 4) decompose with initial cleavage of the peroxy bond. See Problem 10.] Investigation of alkyl peroxides using the Taft correlation also indicates the polar nature of those transition states.[75]

Azo compound decomposition is much less susceptible to polar substituent effects than is perester decomposition, and so it probably has less charge separation

[74](a) P. D. Bartlett and C. Rüchardt, *J. Am. Chem. Soc.*, **82**, 1756 (1960); (b) J. P. Engstrom and J. C. DuBose, *J. Org. Chem.*, **38**, 3817 (1973).
[75](a) W. H. Richardson, M. B. Yelvington, A. H. Andrist, E. W. Ertley, R. S. Smith, and T. D. Johnson, *J. Org. Chem.*, **38**, 4219 (1973). The idea that polar effects are important in radical reactions has been challenged by (b) A. A. Zavitsas and J. A. Pinto, *J. Am. Chem. Soc.*, **94**, 7390 (1972). This work is, however, controversial. See, for example, (c) W. A. Pryor, W. H. Davis, Jr., and J. P. Stanley, *J. Am. Chem. Soc.*, **95**, 4754 (1973).

in the transition state[76] but is more sensitive to geometrical restrictions. Bridgehead azo compounds decompose at rates lower than expected on the basis of their tertiary nature, a result attributed to nonplanarity of the incipient radical.[77] Results of peroxyester decomposition and of tin hydride reduction of bridgehead halides, although showing a smaller effect than found in the azo compound experiments, were interpreted in the same way.[78] This rationalization assumes that tertiary radicals prefer a planar configuration and thus appears to be in conflict with the esr evidence discussed earlier that indicates a preference on the part of the *t*-butyl radical for a nonplanar geometry.

   Photochemical homolysis   A second general method of obtaining radicals is through irradiation with either light or accelerated particles. The energy transferred to the molecule by the interaction must be of the order of bond dissociation energies or greater to produce homolysis. The energy associated with light quanta in the visible region of the spectrum ranges from around 40 kcal mole^{-1} in the red to 70 kcal mole^{-1} in the blue.[79] These energies are sufficient for only the weaker bonds, but in the readily accessible ultraviolet, down to 200 nm, the energy rises to a little over 140 kcal mole^{-1}, enough to break most bonds. It should not be concluded that all molecules will homolyze readily upon irradiation with ultraviolet light. In the first place, the light must be absorbed, and many substances are transparent down to 200 nm. Second, even when the energy is absorbed, a significant portion of it must be concentrated in vibrational modes leading to dissociation rather than being distributed so as to lead to any of a variety of other pathways open to excited states.

   More detailed consideration of light absorption and consequent chemical changes is left to Chapter 12, but it is appropriate here to summarize briefly the types of compounds that are convenient photochemical radical sources. Many of the substances we have been discussing as thermal radical sources absorb light in the visible or ultraviolet and can be decomposed photochemically. The azoalkanes are particularly versatile; they absorb around 350 nm and decompose cleanly to nitrogen and two radicals just as in the thermal reaction. As we have already noted, a preliminary photochemical isomerization to the cis isomer precedes the homolysis. CIDNP observations confirm a stepwise decomposition pathway and clarify the various reactions of the radicals produced.[80]

   The dialkyl peroxides absorb light below about 300 nm; the quantum energy at this wavelength corresponds to 95 kcal mole^{-1}, and since these compounds

[76](a) J. R. Shelton, C. K. Liang, and P. Kovacic, *J. Am. Chem. Soc.*, **90**, 354 (1968); (b) P. Kovacic, R. R. Flynn, J. F. Gormish, A. H. Kappelman, and J. R. Shelton, *J. Org. Chem.*, **34**, 3312 (1969); (c) B. K. Bandlish, A. W. Garner, M. L. Hodges, and J. W. Timberlake, *J. Am. Chem. Soc.*, **97**, 5856 (1975).

[77]C. Rüchardt, *Angew. Chem. Int. Ed. Engl.*, **9**, 830 (1970).

[78](a) J. P. Lorand, S. D. Chodroff, and R. W. Wallace, *J. Am. Chem. Soc.*, **90**, 5266 (1968); (b) R. C. Fort, Jr., and R. E. Franklin, *J. Am. Chem. Soc.*, **90**, 5267 (1968); (c) R. C. Fort, Jr., and J. Hiti, *J. Org. Chem.*, **42**, 3968 (1977).

[79]The quantum energies are put on a per-mole basis (energy of Avogadro's number of quanta) for convenient comparison with molar bond dissociation energies. Refer to Section 12.1 for further discussion.

[80](a) See note 71; (b) N. A. Porter, L. J. Marnett, C. H. Lochmüller, G. L. Closs, and M. Shobataki, *J. Am. Chem. Soc.*, **94**, 3664 (1972).

have activation energies for thermal homolysis on the order of 35 kcal mole^{-1} for the dialkyl peroxides and 30 kcal mole^{-1} for the diacyl peroxides, there is an excess of at least 60 kcal mole^{-1}. The acyl radicals lose CO_2 rapidly, whereas the alkoxy radicals from dialkyl peroxides, if produced in the gas phase where they cannot readily dissipate their excess energy by collisions, undergo fragmentation.[81]

Ketones constitute another important class that will yield radicals upon ultraviolet irradiation. Nonconjugated ketones absorb weakly at 270 nm and decompose with cleavage at the carbonyl group as indicated in Equation 9.35. The carbonyl radical may then fragment further with loss of carbon monoxide according to Equation 9.36. Although this reaction can serve as a source of the simplest

$$\underset{\substack{\| \\ R-C-R}}{\overset{O}{}} \xrightarrow{h\nu} R\cdot + \underset{\substack{\| \\ \cdot C-R}}{\overset{O}{}} \tag{9.35}$$

$$\underset{\substack{\| \\ R-C\cdot}}{\overset{O}{}} \longrightarrow R\cdot + CO \tag{9.36}$$

alkyl radicals, there are potential complications from other processes.[82] Similar types of dissociations occur in carboxylic acid derivatives. A number of other functional groups, including aliphatic halides and the hypohalites, ROX, dissociate to radicals on absorption of ultraviolet light.

Another important source of radicals in solution is radiolysis, a technique in which gamma rays or high-energy electrons provide the energy needed to break bonds. In aqueous solutions the radiation produces solvated electrons together with radicals $\cdot$H and $\cdot$OH from the breakdown of water. These species react with solute molecules to produce other radicals. The technique is useful for producing radicals for spectroscopic observation and for measuring reaction rates and acidity.[83]

Termination[84]

The most important processes that remove radicals are (1) combination of radicals with each other, either by direct bond formation (recombination) or by hydrogen atom transfer from one to the other (disproportionation), and (2) electron transfer between a radical and an oxidizing or reducing agent.

Radical recombination is the simple bond formation between two radicals, for example, Equation 9.37; disproportionation is the transfer of a β hydrogen from one radical to the other, shown in Equation 9.38.

Rates of recombination and disproportionation of simple alkyl radicals are, as we have indicated earlier, very high, and special techniques are required to measure them. These methods measure total termination rate, recombination plus disproportionation. Direct measurement is limited to those systems where the radi-

[81] See J. G. Calvert and J. N. Pitts, Jr., *Photochemistry,* Wiley, New York, 1966, Chap. 5.
[82] See Section 12.4 for further discussion of ketone photochemistry.
[83] P. Neta, *Adv. Phys. Org. Chem.,* **12,** 223 (1976).
[84] For a review, see M. J. Gibian and R. C. Corley, *Chem. Rev.,* **73,** 441 (1973).

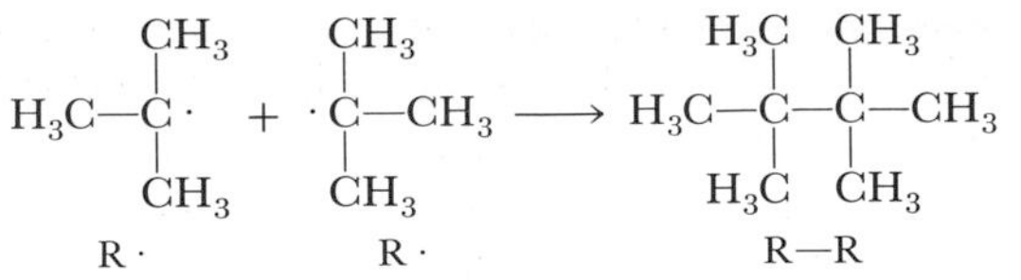

$$\underset{\substack{R\cdot}}{\underset{\substack{\\CH_3\\|\\H_3C-C\cdot\\|\\CH_3}}{}} + \underset{\substack{R\cdot}}{\underset{\substack{\\CH_3\\|\\\cdot C-CH_3\\|\\CH_3}}{}} \longrightarrow \underset{\substack{R-R}}{\underset{\substack{\\H_3C\ \ CH_3\\|\ \ \ |\\H_3C-C-C-CH_3\\|\ \ \ |\\H_3C\ \ CH_3}}{}} \tag{9.37}$$

$$\underset{\substack{R\cdot}}{\underset{\substack{\\H\\|\\H-C+H\\|\\CH_3-C\cdot\\|\\CH_3}}{}} + \underset{\substack{R\cdot}}{\underset{\substack{\\CH_3\\|\\\cdot C-CH_3\\|\\CH_3}}{}} \longrightarrow \underset{\substack{R(-H)}}{\underset{\substack{\\H\ \ \ H\\\diagdown\ \diagup\\C\\||\\C\\\diagup\ \ \diagdown\\H_3C\ \ \ CH_3}}{}} + \underset{\substack{R(+H)}}{\underset{\substack{\\CH_3\\|\\H-C-CH_3\\|\\CH_3}}{}} \tag{9.38}$$

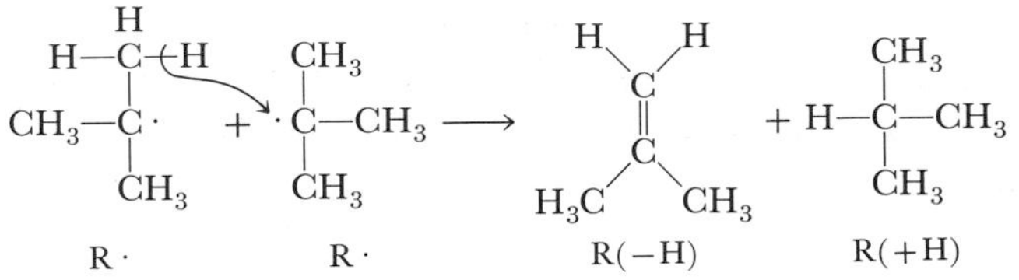

cals themselves can be observed, a requirement that can sometimes be met with flash photolysis or electron resonance. More common are indirect methods, of which one, the *rotating sector* technique, will serve as an example.[85]

One first selects a chain reaction that can be initiated photochemically and that is terminated by the recombination and disproportionation of interest. An example would be the tin hydride reduction of an alkyl bromide, which proceeds according to Scheme 5. Kinetic analysis yields a relation between rate constants of the individual steps and rate of change of some conveniently observed physical property. The initiation rate depends on the square root of light intensity and can be measured directly; the problem is to separate the dependence of overall reaction rate on propagation from its dependence on termination. This task is accomplished by observing the response of the system to a periodic interruption of the light source by a rotating screen with a sector cut from it. As the screen stops the light, initiation ceases and the reaction rate falls rapidly, only to increase again when the light is allowed to pass. By measuring the average rate of reaction and the rate of interruption of the light, it is possible to obtain the propagation and termination rate constants. Equations relating rate constants to the speed of rotation of the sector and to measured reaction rate may be found in the literature.[86]

SCHEME 5

$$
\begin{array}{ll}
\text{Initiation} & \left\{
\begin{array}{l}
\text{Initiator} \xrightarrow{\ h\nu\ } 2\text{In}\cdot \\
\text{In}\cdot + R_3'\text{SnH} \longrightarrow R_3'\text{Sn}\cdot + \text{InH}
\end{array}
\right. \\[2em]
\text{Propagation} & \left\{
\begin{array}{l}
R_3'\text{Sn}\cdot + \text{RBr} \longrightarrow R\cdot + R_3'\text{SnBr} \\
R\cdot + R_3'\text{SnH} \longrightarrow \text{RH} + R_3'\text{Sn}\cdot
\end{array}
\right. \\[1.5em]
\text{Termination} & \quad 2R\cdot \longrightarrow R-R \quad \text{and} \quad R(-H) + R(+H)
\end{array}
$$

[85] (a) For a discussion of this and other similar techniques, see K. U. Ingold, in *Free Radicals,* Vol. I, J. K. Kochi, Ed., Wiley, New York, 1973, p. 40. (b) The rotating sector technique is reviewed by G. M. Burnett and H. W. Melville, in *Technique of Organic Chemistry,* Vol. VIII, Part II, S. L. Friess, E. S. Lewis, and A. Weissberger, Eds., Wiley, New York, 1963, p. 1107.
[86] See note 85(b).

Termination rate constants for alkyl and benzyl radicals in solution range between 10^9 and $10^{10} M^{-1}$ sec^{-1}.[87] These rates correspond quite closely to that calculated for a diffusion-controlled reaction, about $8 \times 10^9 M^{-1}$ sec^{-1} for the common solvents at room temperature.[88] Gas-phase rates are similar: $\cdot CH_3$, $2.4 \times 10^{10} M^{-1}$ sec^{-1}; $\cdot C_2H_5$, $7.8 \times 10^9 M^{-1}$ sec^{-1}; $(CH_3)_2CH \cdot$, $5.0 \times 10^9 M^{-1}$ sec^{-1}; $(CH_3)_3C \cdot$, $2.4 \times 10^9 M^{-1}$ sec^{-1}; $H_2C{=}CH{-}CH_2 \cdot$, $6.5 \times 10^9 M^{-1}$ sec^{-1}.[89] (Gas-phase rates quoted are for recombination only; all are at room temperature, except that for allyl, which is at 625K.)

Not all radicals terminate as rapidly as do the simple alkyl radicals. Placement of several bulky groups near the radical center leads to *persistent* radicals, defined by Griller and Ingold as radicals that under specified conditions have lifetimes significantly greater than do methyl and the simple alkyl radicals, which are termed *transient*.[90] Persistence is conferred on a radical primarily by steric hindrance. The known persistent radicals have bulky substituents, usually *t*-butyl or trimethylsilyl groups, near the radical center. Stabilization has a much smaller effect on persistence; examples of persistent radicals that are stabilized, destabilized, and unstabilized (in the sense defined earlier) are all known, and many stabilized radicals are not persistent. Structures **27** through **36** show some persistent radicals.

Of the two alternative termination reactions, disproportionation in alkyl radicals is ordinarily somewhat slower than recombination but is frequently still fast enough to compete. The total termination rate can be separated into recombination and disproportionation rates by analyzing the products. The number of hydrogens available for transfer at the β position of a radical will contribute a statistical factor; thus isopropyl radicals, with six β hydrogens, would disproportionate twice as rapidly as ethyl, with only three, if other factors were equal. For

$$((CH_3)_3C)_3C \cdot \qquad ((CH_3)_3Si)_3C \cdot$$

$$\textbf{27}^{91} \qquad\qquad \textbf{28}^{91} \qquad\qquad \textbf{29}^{92}$$

[87] (a) Termination rates ordinarily quoted are $2k_t$, where k_t is the bimolecular rate constant for the sum of recombination and disproportionation. The factor of 2 appears because two radicals are removed from the system each time the reaction occurs. See K. U. Ingold, in *Free Radicals,* Vol. I, J. K. Kochi, Ed., Wiley, New York, 1973, p. 43, for references to particular alkyl radical results. See also (b) H. J. Hefter, C. S. Wu, and G. S. Hammond, *J. Am. Chem. Soc.,* **95**, 851 (1973) (allyl); (c) J. E. Bennett and R. Summers, *J. Chem. Soc. Perkin Trans.* 2, 1504 (1977) (*t*-butyl).
[88] See K. U. Ingold, in *Free Radicals,* Vol. I, J. K. Kochi, Ed., Wiley, New York, 1973, p. 39.
[89] (a) D. A. Parkes, D. M. Paul, and C. P. Quinn, *J. Chem. Soc., Faraday Trans. 1,* **72**, 1935 (1976) (methyl); (b) D. A. Parkes and C. P. Quinn, *J. Chem. Soc., Faraday Trans. 1,* **72**, 1952 (1976) (ethyl, isopropyl, *t*-butyl); (c) M. Rossi, K. D. King, and D. M. Golden, *J. Am. Chem. Soc.,* **101**, 1223 (1979) (allyl).
[90] D. Griller and K. U. Ingold, *Acc. Chem. Res.,* **9**, 13 (1976).
[91] G. D. Mendenhall, D. Griller, D. Lindsay, T. T. Tidwell, and K. U. Ingold, *J. Am. Chem. Soc.,* **96**, 2441 (1974).
[92] L. R. C. Barclay, D. Griller, and K. U. Ingold, *J. Am. Chem. Soc.,* **96**, 3011 (1974).

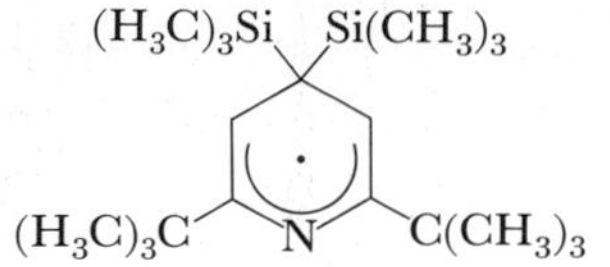

(H₃C)₃Si   Si(CH₃)₃

(H₃C)₃C   N   C(CH₃)₃

30[93]

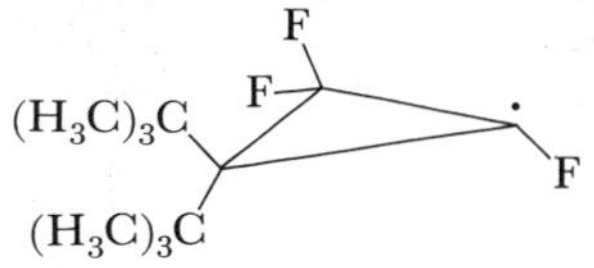

F
F
(H₃C)₃C
(H₃C)₃C   F

31[94]

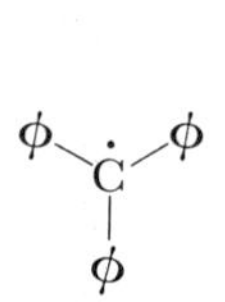

φ   C   φ
φ

32[95]

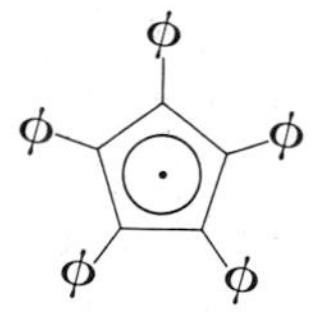

φ
φ   φ
φ   φ

33[96]

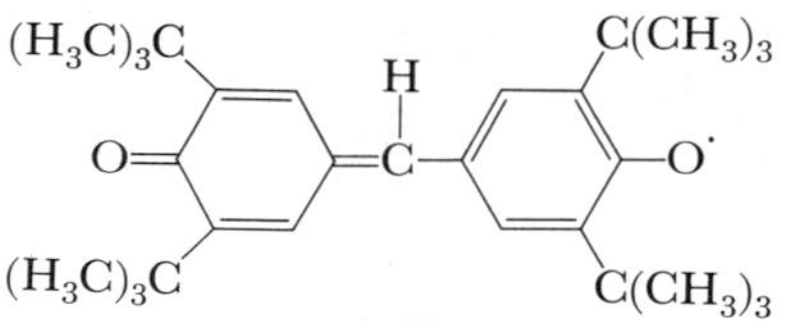

(H₃C)₃C   C(CH₃)₃
O   C   O·
H
(H₃C)₃C   C(CH₃)₃

34[97]

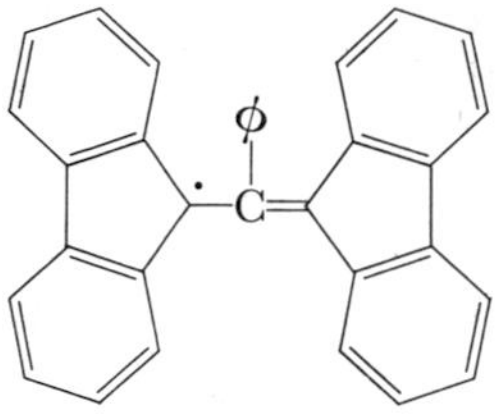

φ
C

35[98]

O·
(H₃C)₃C   C(CH₃)₃

C(CH₃)₃

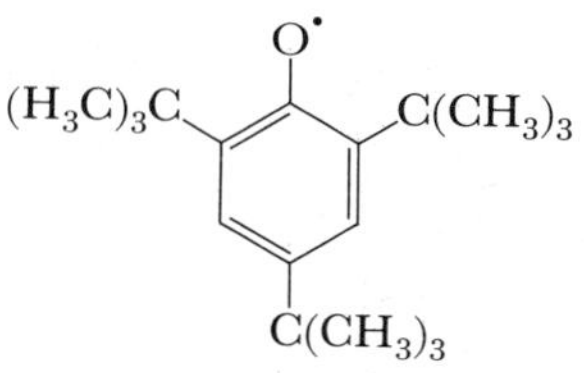

36

this reason comparisons of disproportionation rates among compounds of different types require that the observed rates be corrected by dividing by the number n of β hydrogens. Disproportionation becomes more favorable relative to recombination in the series primary, secondary, tertiary, the ratios k_{dis}/nk_{comb} being roughly

[93] D. Griller, K. Dimroth, T. M. Fyles, and K. U. Ingold, *J. Am. Chem. Soc.,* **97,** 5526 (1975).
[94] V. Malatesta, D. Forrest, and K. U. Ingold, *J. Am. Chem. Soc.,* **100,** 7073 (1978).
[95] (a) H. Lankamp, W. Th. Nauta, and C. MacLean, *Tetrahedron Lett.,* 249 (1968); (b) H. A. Staab, H. Brett-schneider, and H. Brunner, *Chem. Ber.,* **103,** 1101 (1970); (c) J. M. McBride, *Tetrahedron,* **30,** 2009 (1974). The triphenylmethyl radical is in equilibrium with its dimer, which has the structure

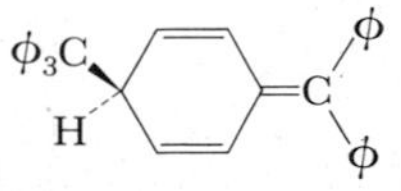

φ₃C   φ
C
H   φ

[96] E. Müller and I. Müller-Rodloff, *Chem. Ber.,* **69,** 665 (1936).
[97] P. D. Bartlett and T. Funahashi, *J. Am. Chem. Soc.,* **84,** 2596 (1962).
[98] C. F. Koelsch, *J. Am. Chem. Soc.,* **79,** 4439 (1957).

0.05, 0.2, and 0.8 in this series.[99] In tertiary radicals stabilized by conjugation, disproportionation is slower relative to combination; values of k_{dis}/nk_{comb} are typically less than 0.01.[100]

Cage Effects

Radicals initially formed in solution by a bond homolysis will be held together briefly in a cage of solvent molecules. Because radical recombinations and disproportionations are so fast, they can compete with diffusion of the radicals through the layer of solvent molecules that surround them. The result is that some of the radicals formed never become available to initiate other processes in the bulk of the solution.[101] These recombinations are termed *geminate recombinations*, and the phenomena that arise from this behavior are *cage effects.*

In the simplest type of bond homolysis, where only one bond breaks, as with a dialkyl peroxide, cage recombination simply regenerates the starting compound and hence is the radical analog of internal return in ionizations.[102] Equation 9.39 illustrates this process; the bar over the radical pair is the symbol for its caged

$$\text{R—O—O—R} \rightleftharpoons \overline{\text{R—O} \cdot \quad \cdot \text{O—R}} \longrightarrow 2\text{R—O} \cdot \tag{9.39}$$

$$\underset{\text{caged pair}}{} \qquad \underset{\substack{\text{radicals} \\ \text{free in} \\ \text{solution}}}{}$$

nature. The effect of the internal return is simply to lower the rate of disappearance of substrate from what it would have been without the return. A key to the experimental detection of this phenomenon is the dependence of cage effect on solvent viscosity. The more viscous the solvent, the more difficult it will be for the radicals to move through the walls of the cage and escape; and the longer they remain trapped, the greater the chance that they will recombine with each other. The theoretical relationship between viscosity and internal return has been applied to the determination of the extent of cage recombination.[103] In other types of structures, such as diacyl peroxides, isotopic labeling may be used as pointed out earlier to detect return; the two methods sometimes agree (acetyl peroxide, for which about 36 percent of radicals formed return in octane[104]) but not always (*t*-butyl perbenzoate[105]).

[99] R. A. Sheldon and J. K. Kochi, *J. Am. Chem. Soc.,* **92**, 4395 (1970). Note that in the *t*-butyl radical, with nine hydrogens, seven times as many radical pairs will disproportionate as combine even though the disproportionation rate per hydrogen is less than k_{comb}.

[100] J. R. Shelton, C. K. Liang, and P. Kovacic, *J. Am. Chem. Soc.,* **90**, 354 (1968).

[101] (a) J. Franck and E. Rabinowitch, *Trans. Faraday Soc.,* **30**, 120 (1934); (b) E. Rabinowitch and W. C. Wood, *Trans. Faraday Soc.,* **32**, 1381 (1936); (c) R. M. Noyes, *J. Am. Chem. Soc.,* **77**, 2042 (1955).

[102] See Section 4.3.

[103] (a) W. A. Pryor and K. Smith, *J. Am. Chem. Soc.,* **89**, 1741 (1967); (b) W. A. Pryor and K. Smith, *J. Am. Chem. Soc.,* **92**, 5403 (1970).

[104] (a) See note 103(b); (b) J. W. Taylor and J. C. Martin, *J. Am. Chem. Soc.,* **89**, 6904 (1967).

[105] T. Koenig, M. Deinzer, and J. A. Hoobler, *J. Am. Chem. Soc.,* **93**, 938 (1971).

When the decomposition in question involves more than one bond in a concerted homolysis, as in an azoalkane (Equation 9.40), the disappearance of substrate is unaffected by recombination, but the number of $R \cdot$ radicals available in the bulk solution to initiate other processes is less than two for each molecule of initiator consumed. Most experimental efforts to determine the amount of cage

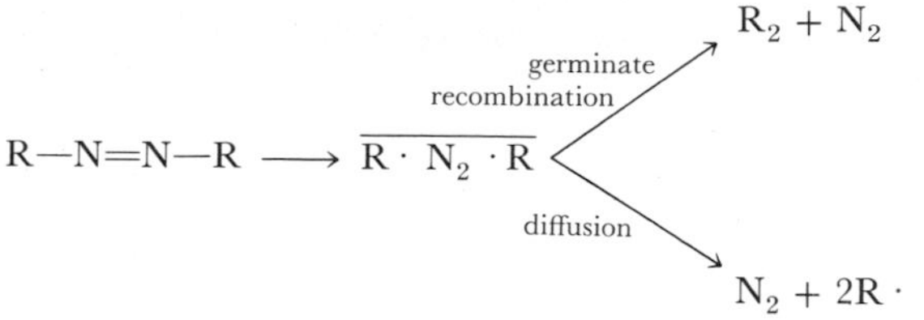

$$R-N=N-R \longrightarrow \overline{R \cdot N_2 \cdot R} \begin{array}{c} \overset{\text{germinate}}{\overset{\text{recombination}}{\nearrow}} \; R_2 + N_2 \\ \\ \underset{\text{diffusion}}{\searrow} \; N_2 + 2R \cdot \end{array} \qquad (9.40)$$

recombination in these instances are of either the crossover or the scavenger type. In a crossover experiment one decomposes a mixture of $R-N=N-R$ and $R'-N=N-R'$; geminate recombination must yield only $R-R$ and $R'-R'$, whereas the separated radicals will recombine randomly to a statistical mixture of $R-R$, $R'-R'$, and $R-R'$.[106]

The scavenging approach is to add a substance that reacts very rapidly with radicals. Some typical scavengers are persistent free radicals such as galvinoxyl (**34**), the Koelch radical (**35**), and diphenylpicryl hydrazyl (**37**).[107] These radicals will not undergo self-recombination but nevertheless react with transient radicals at high rates. Other possible scavengers are substances with particularly easily abstracted hydrogens like thiols. The task of the scavenger is to pick up all the radicals that escape from the cage. Experimentally, one measures the rate of disappearance of a low concentration of a colored scavenger radical in the presence of excess decomposing initiator; rates first-order in initiator and independent of scavenger concentration assure that the scavenger is trapping all available radicals.[108] Results of these experiments show, for example, that azo-isobutyronitrile (**38**) yields only 65 percent of the potential radicals as free scavengable radicals in solution, whereas with di-t-butylperoxyoxalate (**39**), which produces two t-butoxy radicals in

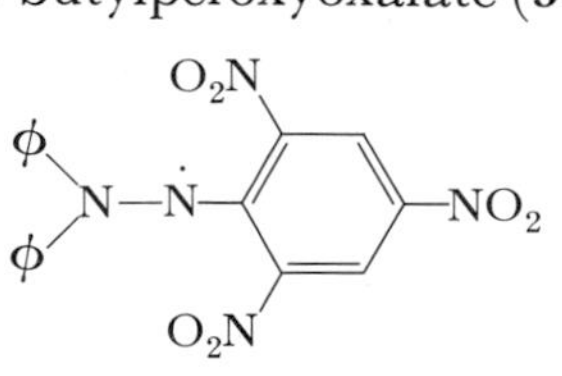

37 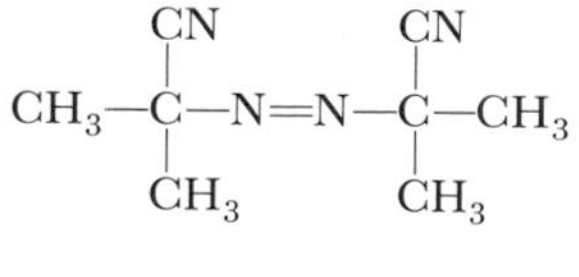 **38**

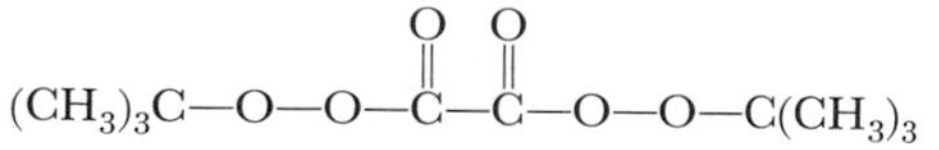

39

[106]The results will require a correction for disproportionation if $k_{\text{dis}}/k_{\text{comb}}$ is different for $R + R$, $R + R'$, and $R' + R'$. See, for example, note 99.
[107]C. E. H. Bawn and S. F. Mellish, *Trans. Faraday Soc.*, **47**, 1216 (1951).

a cage separated by two carbon dioxide molecules, all the radicals produced escape from the cage.[108]

Finally, recall that geminate processes are detected by the spectroscopic technique of chemically induced nuclear polarization. Because the CIDNP spectrum is directly related to properties of the caged radical pair, as explained in the Appendix to this chapter, much useful information about the nature of caged radicals and their fate can be obtained by this method.

Oxidations and Reductions Involving Radicals

An area closely related to ordinary radical chemistry, but on which we shall touch only briefly, is that of oxidations and reductions in which radicals are substrates or products.[109]

Under appropriate circumstances a strong reducing or oxidizing agent, or an electrode, can transfer an electron to or from an organic molecule.[110] The substances most susceptible to reduction have low-energy vacant orbitals and those most susceptible to oxidation have high-energy filled orbitals; the molecules most likely to meet these requirements are those with extended π systems. The product of the electron transfer is a radical anion or radical cation, that is, a molecule with an odd number of electrons and a negative or positive charge. Equations 9.41–9.44 illustrate some typical examples.

The charge of radical ions modifies their properties considerably from those of ordinary radicals. Because dimerization could form a new bond only at the expense of bringing together two like charges in the same molecule, it is possible to prepare stable solutions of many of them. The radical ions are readily studied spectroscopically; electron spin resonance is particularly fruitful.

$$Na + \text{[naphthalene]} \longrightarrow Na^+ \left[\text{naphthalene} \right]^{\overline{\cdot}} \tag{9.41}$$

naphthalene
radical anion

$$Na + \underset{\phi \quad \phi}{\overset{O}{\underset{\|}{C}}} \longrightarrow Na^+ \left[\underset{\phi \quad \phi}{\overset{O}{\underset{\|}{C}}} \right]^{\overline{\cdot}} \tag{9.42}$$

benzophenone
ketyl

---

[108] See, for example: P. D. Bartlett and T. Funahashi, *J. Am. Chem. Soc.,* **84,** 2596 (1962).

[109] I. V. Khudyakov and V. A. Kuz'min, *Russ. Chem. Rev.,* **47,** 22 (1978).

[110] For discussions of the chemistry of radical ions, see (a) E. T. Kaiser and L. Kevan, Eds., *Radical Ions,* Wiley, New York, 1968; (b) J. F. Garst, in *Free Radicals,* Vol. I, J. K. Kochi, Ed., Wiley, New York, 1973, p. 503; (c) N. L. Holy, *Chem. Rev.,* **74,** 243 (1974); (d) M. Szwarc and J. Jagur-Grodzinski, in *Ions and Ion Pairs in Organic Reactions,* Vol. 2, M. Szwarc, Ed., Wiley, New York, 1974, Chap. 1; (e) A. J. Bard, A. Ledwith, and H. J. Shine, *Adv. Phys. Org. Chem.,* **13,** 155 (1976).

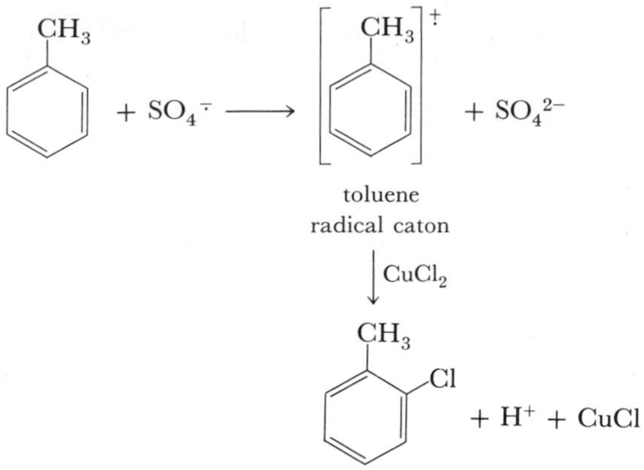

$$\text{toluene radical caton} \quad \xrightarrow{CuCl_2}$$

$$(9.43)^{111}$$

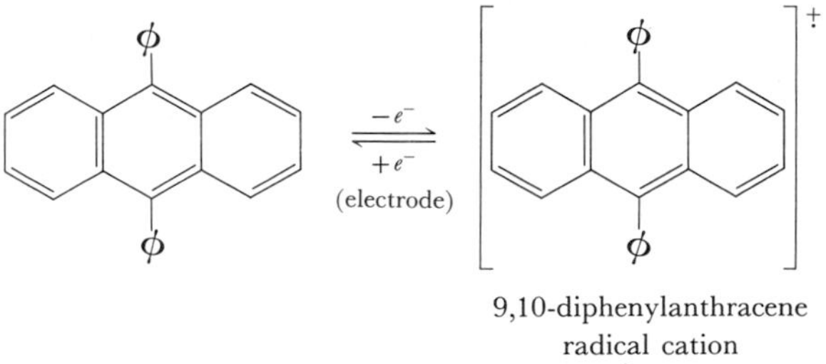

9,10-diphenylanthracene radical cation

$$(9.44)^{112}$$

Radical anions are both bases and reducing agents. Although they do react by proton transfer under appropriate conditions,[113] electron transfer is more common. An example is electron transfer to an alkyl halide. The halide does not remain as a radical ion but dissociates into a halide ion and an ordinary alkyl radical. If that radical encounters another radical anion, it will be reduced further to an alkyl anion.[114] Equations 9.45 and 9.46 show the sequence. Typical ionic reactions follow, such as protonation of $R\!:\!^-$ from the solvent, or nucleophilic addition of $R\!:\!^-$ to the aromatic.

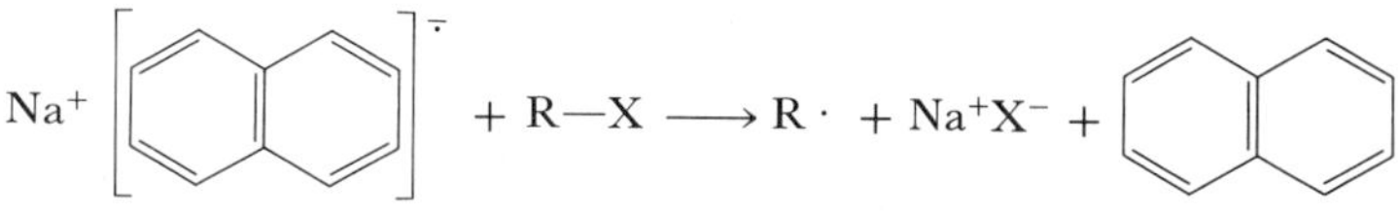

$$(9.45)$$

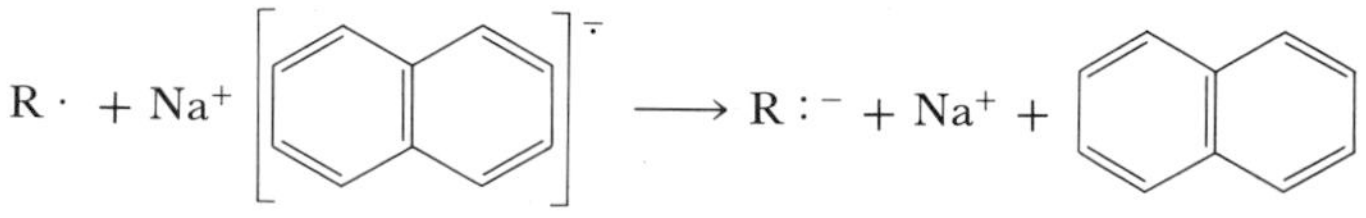

$$(9.46)$$

[111] A. Ledwith and P. J. Russell, *J. Chem. Soc., Chem. Commun.,* 291 (1974).

[112] (a) J. Phelps, K. S. V. Santhanam, and A. J. Bard, *J. Am. Chem. Soc.,* **89,** 1752 (1967); (b) L. S. Marcoux, J. M. Fritsch, and R. N. Adams, *J. Am. Chem. Soc.,* **89,** 5766 (1967).

[113] M. Szwarc, A. Streitwieser, Jr., and P. C. Mowrey, in *Ions and Ion Pairs in Organic Reactions,* Vol. 2, M. Szwarc, Ed., Wiley, New York, 1974, p. 151.

[114] (a) J. F. Garst, *Acc. Chem. Res.,* **4,** 400 (1971); (b) S. Bank and D. A. Juckett, *J. Am. Chem. Soc.,* **98,** 7742 (1976).

Radical ions and radicals are also intermediates in reaction of alkyl halides with metals, as shown by CIDNP experiments. The formation of Grignard reagents, for example, occurs through a one-electron transfer to form a radical ion pair, as indicated in Scheme 6.[115]

Scheme 6

$$\overline{RX + Mg \longrightarrow RX^{\cdot -} Mg^{\cdot +}}$$

$$\overline{RX^{\cdot -} Mg^{\cdot +} \longrightarrow R\cdot + X^- Mg^+}$$

$$R\cdot + X^- Mg^+ \longrightarrow RMgX$$

$$2R\cdot \longrightarrow R{-}R + R(+H) + R(-H)$$

A closely related process is the interaction of organometallics with alkyl halides. Here the electron donor, $RLi \leftrightarrow R^- Li^+$, has an even number of electrons at the start but ends up as a radical (Equation 9.47) after transferring one electron to $R'X$. The radicals can now couple or disproportionate (Equation 9.48); the radical nature of the process has been confirmed by CIDNP, which shows that some of the radicals abstract halogen from unreacted halide (Equation 9.49).[116]

$$RLi + R'X \longrightarrow R\cdot + R'\cdot + LiX \qquad (9.47)$$

$$R\cdot + R'\cdot \longrightarrow R{-}R' + R(+H) + R(-H) \qquad (9.48)$$

$$R\cdot \quad \text{or} \quad R'\cdot + R'X \longrightarrow RX \quad \text{or} \quad R'X \qquad (9.49)$$

Oxidation and reduction can also enter into more conventional radical reactions. Salts of metals with two readily accessible oxidation states can effect chain transfer in radical additions to alkenes, as illustrated in Scheme 7.[117]

Scheme 7

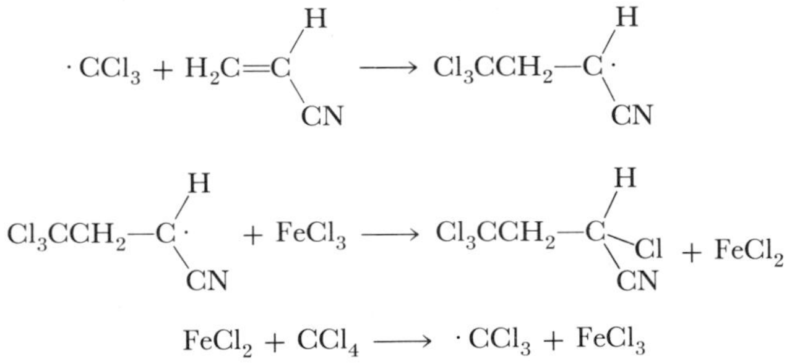

$$FeCl_2 + CCl_4 \longrightarrow \cdot CCl_3 + FeCl_3$$

Electron transfer between two neutral radicals to yield a cation and an anion (Scheme 8) has also been demonstrated by CIDNP spectroscopy.[118]

[115](a) H. W. H. J. Bodewitz, C. Bomberg, and F. Bickelhaupt, *Tetrahedron*, **29**, 719 (1973); (b) E. A. Vogler, R. L. Stein, and J. M. Hayes, *J. Am. Chem. Soc.*, **100**, 3163 (1978).

[116]H. R. Ward, R. G. Lawler, and R. A. Cooper, in *Chemically Induced Magnetic Polarization*, A. R. Lepley and G. L. Closs, Eds., Wiley, New York, 1973, p. 281.

[117]F. Minisci, *Acc. Chem. Res.*, **8**, 165 (1975).

[118]R. G. Lawler, P. F. Barbara, and D. Jacobs, *J. Am. Chem. Soc.*, **100**, 4912 (1978). See also Section 12.5.

SCHEME 8

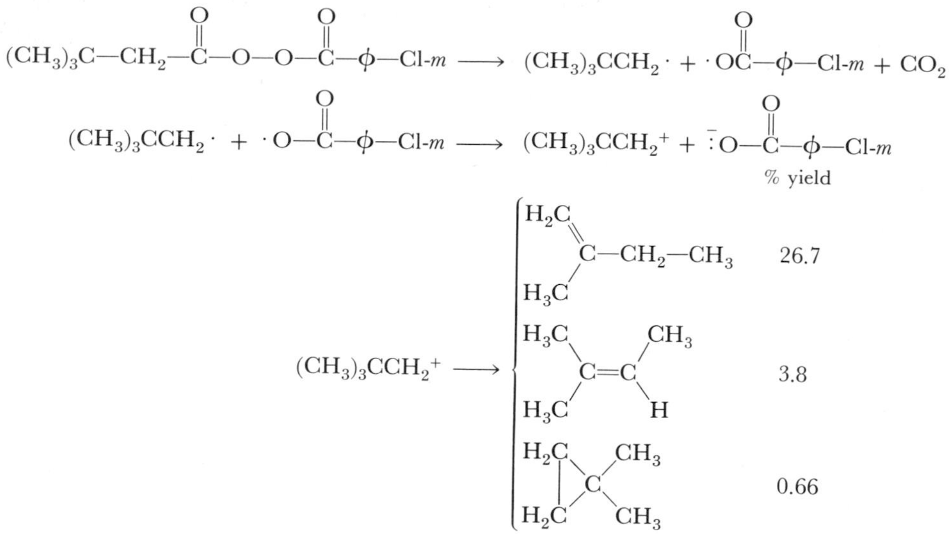

$$(CH_3)_3C-CH_2-\overset{\overset{\displaystyle O}{\|}}{C}-O-O-\overset{\overset{\displaystyle O}{\|}}{C}-\phi-Cl\text{-}m \longrightarrow (CH_3)_3CCH_2\cdot + \cdot O\overset{\overset{\displaystyle O}{\|}}{C}-\phi-Cl\text{-}m + CO_2$$

$$(CH_3)_3CCH_2\cdot + \cdot O-\overset{\overset{\displaystyle O}{\|}}{C}-\phi-Cl\text{-}m \longrightarrow (CH_3)_3CCH_2{}^+ + {}^-{:}O-\overset{\overset{\displaystyle O}{\|}}{C}-\phi-Cl\text{-}m$$

% yield

$$(CH_3)_3CCH_2{}^+ \longrightarrow$$

	% yield
$H_2C{=}\overset{\overset{\displaystyle }{}}{\underset{H_3C}{C}}-CH_2-CH_3$	26.7
$\underset{H_3C}{\overset{H_3C}{>}}C{=}C\overset{CH_3}{\underset{H}{<}}$	3.8
$\underset{H_2C}{\overset{H_2C}{>}}C\overset{CH_3}{\underset{CH_3}{<}}$	0.66

9.3 CHAIN REACTIONS

We turn now to a more detailed consideration of some of the features of chain reactions. As we have noted earlier, these processes consist of three stages: initiation, propagation, and termination. Each of these stages may include one or more individual reactions; reaction schemes can become complex if a number of different radicals are present reacting in a variety of ways, including perhaps with products of the reaction. We shall follow the usual practice of choosing for purposes of illustration systems for which a small number of steps accounts reasonably well for the observations.

Chain reactions can be divided roughly into two types: polymerizations and nonpolymerizations. In polymerizations (Scheme 9), an initiating radical $(R\cdot)$ adds to a substrate alkene (ordinarily termed the *monomer*) to yield a new radical, which adds to another alkene, and so forth. The kinetic chain, that is the sequence of events begun by a given $R\cdot$ radical from the initiator, corresponds in this scheme to the actual growth of the polymer molecule and terminates simultaneously with the growth of the molecular chain as two radicals combine or disproportionate.

A typical nonpolymerization chain reaction (Scheme 10) would be the decomposition (in a solvent without readily abstracted hydrogens) of a tertiary alkyl hypochlorite. Here the kinetic chain may be long without yielding any large molecules, because at each stage a new radical is produced by abstraction rather than by addition. This process is referred to as *chain transfer*, the terminology signifying that the kinetic chain is continued by transfer to a new molecule rather than by addition of a new monomer unit to the radical.

SCHEME 9

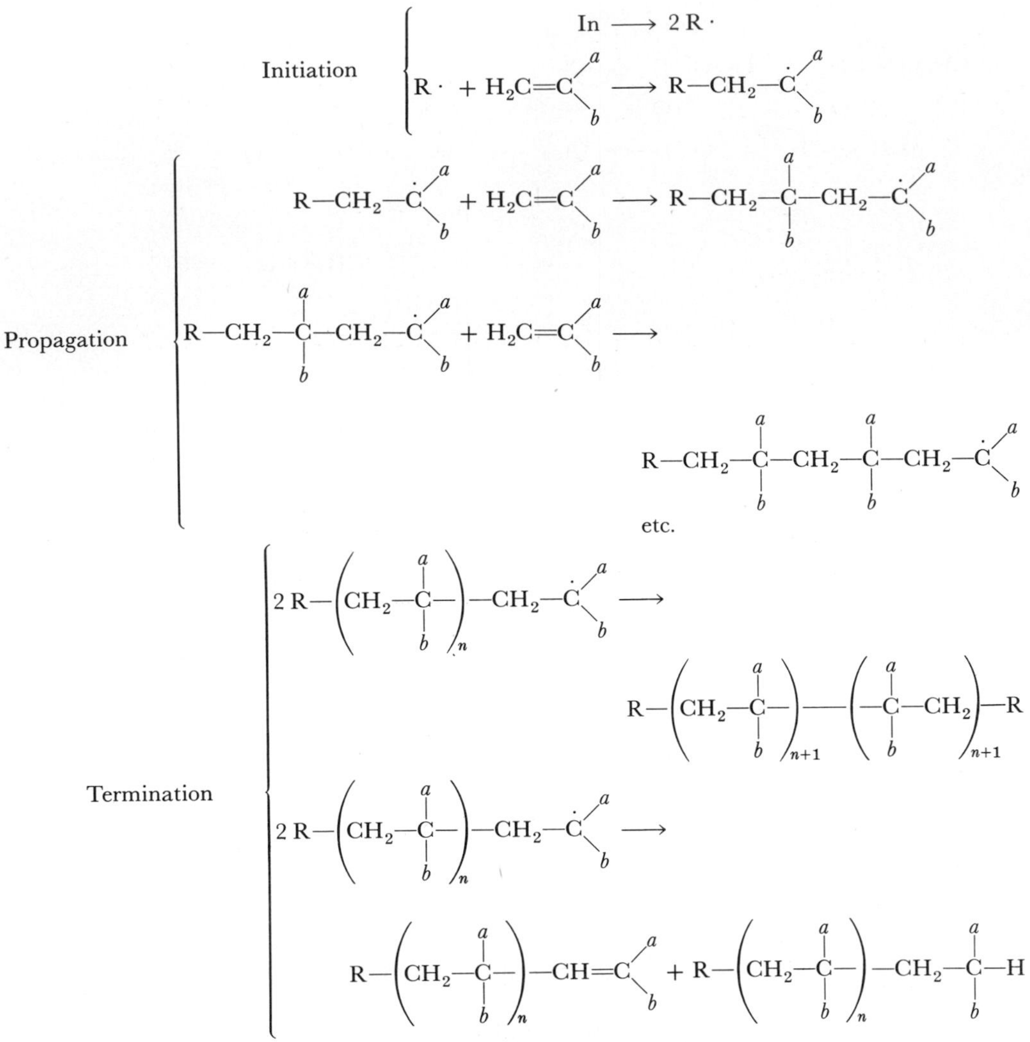

The reactions in Schemes 9 and 10 are only extremes in a continuous grada-tion of chain reactions. Suppose, for example, that there is present in the solution a substance with an abstractable hydrogen or halogen with which the growing mo-lecular chain reacts at a rate comparable to the rate of addition to the next mono-mer unit. Abstraction will lead to chain transfer at an early stage in the molecule's growth, and the product will consist of a distribution of molecular sizes ranging

SCHEME 10

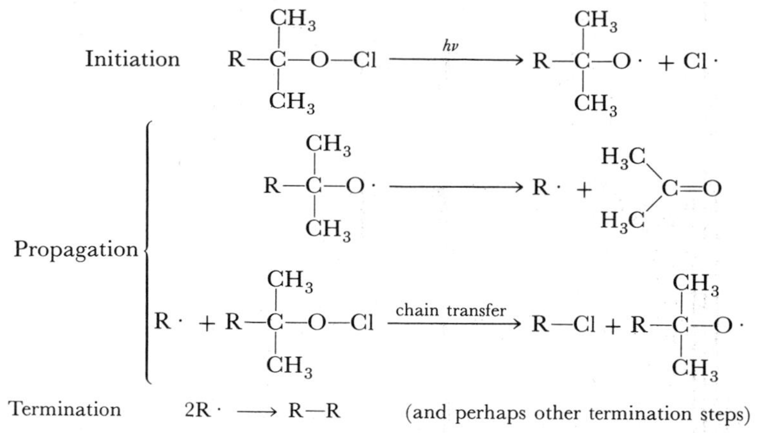

$$\text{Initiation} \quad R-\underset{\underset{CH_3}{|}}{\overset{\overset{CH_3}{|}}{C}}-O-Cl \xrightarrow{h\nu} R-\underset{\underset{CH_3}{|}}{\overset{\overset{CH_3}{|}}{C}}-O\cdot + Cl\cdot$$

Propagation

$$R-\underset{\underset{CH_3}{|}}{\overset{\overset{CH_3}{|}}{C}}-O\cdot \longrightarrow R\cdot + \underset{H_3C}{\overset{H_3C}{>}}C=O$$

$$R\cdot + R-\underset{\underset{CH_3}{|}}{\overset{\overset{CH_3}{|}}{C}}-O-Cl \xrightarrow{\text{chain transfer}} R-Cl + R-\underset{\underset{CH_3}{|}}{\overset{\overset{CH_3}{|}}{C}}-O\cdot$$

Termination $2R\cdot \longrightarrow R-R$ (and perhaps other termination steps)

from one to perhaps the order of ten monomer units. Such a reaction is called *telomerization,* and the chain transfer agent that produces it is a *telogen.* Scheme 11 shows the telomerization of ethylene with CCl_4 as telogen.

SCHEME 11

Initiation
$$In \longrightarrow 2R\cdot$$
$$R\cdot + H_2C{=}CH_2 \longrightarrow RCH_2-CH_2\cdot$$
$$\text{or}$$
$$R\cdot + CCl_4 \longrightarrow RCl + \cdot CCl_3$$
$$\cdot CCl_3 + H_2C{=}CH_2 \longrightarrow Cl_3C-CH_2-CH_2\cdot$$

Chain growth
$$\left\{ Cl_3C(CH_2-CH_2)_n\cdot + H_2C{=}CH_2 \longrightarrow Cl_3C(CH_2-CH_2)_{n+1}\cdot \right.$$

Chain transfer
$$\left\{ Cl_3C(CH_2-CH_2)_n\cdot + CCl_4 \longrightarrow Cl_3C(CH_2-CH_2)_nCl + \cdot CCl_3 \right.$$

Termination
$$2\,Cl_3C(CH_2-CH_2)_n\cdot \longrightarrow (Cl_3C(CH_2-CH_2)_n)_2$$
$$Cl_3C(CH_2-CH_2)_n\cdot + \cdot CCl_3 \longrightarrow Cl_3C(CH_2-CH_2)_nCCl_3$$
$$2\,\cdot CCl_3 \longrightarrow Cl_3C-CCl_3$$

(and other possible termination steps involving $R\cdot$ and $R(CH_2-CH_2)_n\cdot$)

(disproportionations not shown)

The molecular size distribution obtained in telomerization depends on the chain transfer coefficients C_n, defined as (rate constant for chain transfer after addition of monomer unit n)/(rate constant for addition of monomer unit $n + 1$),

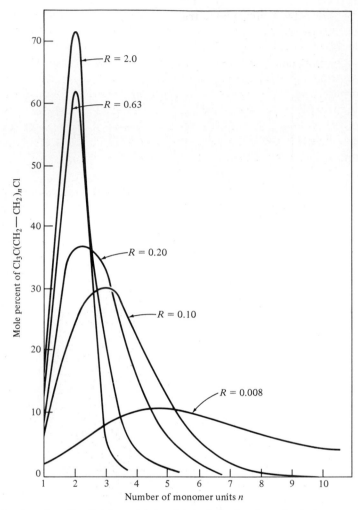

Figure 9.2 Molecular size distribution obtained in telomerization of ethylene and carbon tetra-cholride for various ratios $R = [CCl_4]/[C_2H_4]$. From C. M. Starks, *Free Radical Telomerization*, Academic Press, New York, 1974, p. 9. Reprinted by permission of Academic Press.

and on the relative concentrations of telogen and monomer. Figure 9.2 shows how the molecular size distribution depends on the relative concentrations in the ethylene—CCl_4 system, for which chain transfer coefficients are $C_1 = 0.10$, $C_2 = 3.0$, $C_3 = 7.0$, $C_4 = 10.3$, $C_5 = 13.3$, $C_\infty = 13$.[119]

In chain reactions, if the chains are long,[120] the products are formed mainly

[119] C. M. Starks, *Free Radical Telomerization*, Academic Press, New York, 1974, p. 9.
[120] Kinetic chain length is the number of substrate molecules that react for each chain initiation that occurs. Kinetic chain lengths are typically 10^3 or greater.

in the propagation steps. When chain transfer occurs at each propagation cycle, as in Scheme 10, the products are almost exclusively those of the propagation steps, alkyl chloride R—Cl and acetone in this example, and it makes relatively little difference to the products how the chains initiate or terminate. The same remarks apply in telomerization where most of the product is formed in the chain transfer step. In a polymerization the initiator contributes the end group of each chain, but this group constitutes a relatively insignificant part of a molecule that may contain many thousand monomer units. In this type of chain reaction, however, the relative rates of initiation, propagation, and termination determine molecular weight, a property that may be as important as structure in determining the characteristics of the product.

Another characteristic feature of chain reactions is *inhibition*. Interruption of a single chain will prevent the reaction of a large number of substrate molecules; hence any substance that diverts radicals will dramatically reduce reaction rate. Inhibitors may themselves be persistent radicals or substances (for example, 2,4,6-tri-*t*-butyl phenol) that can react with radicals to yield persistent radicals; the requirement that must be met is that the inhibitor react efficiently with transient radicals and that neither it nor its products be initiators of new chains.

Kinetics of Polymerization

The kinetics of polymerization can be easily worked out with the aid of certain simplifying approximations.[121] We modify Scheme 9 to Scheme 12, where In is an initiator producing radicals R·, M is the monomer, M_n· is the growing polymer chain, and $M_n—M_m$ is the combination product and $M_n(\pm H)$ are disproportionation products. If all radicals R· produced by the initiator were available to start chains, we could write, from the first two reactions, kinetic Equation 9.50 for rate of change of R· concentration. Because of cage recombination, only some fraction f

SCHEME 12

Initiation   In $\xrightarrow{k_d}$ 2R·

Propagation   R· + M $\xrightarrow{k_i}$ M_1·

M_n· + M $\xrightarrow{k_p}$ M_{n+1}·

Termination M_m· + M_n· $\xrightarrow{k_c}$ $M_n—M_m$

M_m· + M_n· $\xrightarrow{k_d}$ $M_m(\pm H)$ + $M_n(\mp H)$

[121](a) C. Walling, *Free Radicals in Solution,* Wiley, New York, 1957, Chap. 3; (b) F. G. R. Gimblett, *Introduction to the Kinetics of Chemical Chain Reactions,* McGraw-Hill, London, 1970.

of the radicals produced will actually be free to start chains, and the equation must be modified accordingly to 9.51. We then assume that when the growing polymer

$$\frac{d[\text{R}\cdot]}{dt} = 2k_d[\text{In}] - k_i[\text{R}\cdot][\text{M}] \tag{9.50}$$

$$\frac{d[\text{R}\cdot]}{dt} = 2fk_d[\text{In}] - k_i[\text{R}\cdot][\text{M}] \tag{9.51}$$

chain adds a new unit, it does so at a rate independent of its length; hence even though there are actually many different structures included under the symbol $\text{M}_n\cdot$, only one rate constant is needed for each process in which they take part. This assumption is clearly necessary if we are not to be overwhelmed with equations; it is justified on the basis of the principle that reactivity of a functional group (in this case $\text{R}\!-\!\text{CR}'\text{R}''$) is to a good approximation unaffected by changes made at remote sites in the molecule (here various lengths of the chain R). To simplify matters further, we may put together the two bimolecular termination processes as indicated in Equation 9.52. We can now proceed to establish Equations 9.53 and

$$\text{M}_m\cdot + \text{M}_n\cdot \xrightarrow{k_t} \text{nonradical products} \tag{9.52}$$
$$k_t = k_c + k_d$$

$$\frac{d[\text{M}_n\cdot]}{dt} = k_i[\text{R}\cdot][\text{M}] - k_p[\text{M}_n\cdot][\text{M}] + k_p[\text{M}_n\cdot][\text{M}] - 2k_t[\text{M}_n\cdot]^2 \tag{9.53}$$

$$= k_i[\text{R}\cdot][\text{M}] - 2k_t[\text{M}_n\cdot]^2 \tag{9.54}$$

9.54 for rates of change of the radical concentrations. Note that the propagation step makes no net contribution to the concentration of $\text{M}_n\cdot$ radicals; one radical is destroyed but a new one takes its place. We now make the stationary-state assumption for the radicals $\text{R}\cdot$ and $\text{M}_n\cdot$; since $d[\text{R}\cdot]/dt = d[\text{M}_n\cdot]/dt = 0$, the initiation rate, $2fk_d[\text{In}]$, and termination rate, $2k_t[\text{M}_n\cdot]^2$, are equal. This point is intuitively obvious once one assumes the stationary state, that is, that the number of radical chains remains constant. Equations 9.55 and 9.56 follow immediately. Equation

$$2fk_d[\text{In}] = 2k_t[\text{M}_n\cdot]^2 \tag{9.55}$$

$$[\text{M}_n\cdot] = \sqrt{\frac{fk_d\,[\text{In}]}{k_t}} \tag{9.56}$$

9.57 gives the predicted rate of monomer consumption. If chains are long, the second term in Equation 9.57 will be much larger than the first,[122] and the approximation 9.58 will be sufficient; Equation 9.59 follows as the predicted rate equation

$$-\frac{d[\text{M}]}{dt} = k_i[\text{R}\cdot][\text{M}] + k_p[\text{M}_n\cdot][\text{M}] \tag{9.57}$$

---

[122] For example, a chain length of 1000 means that for every M that reacts with $\text{R}\cdot$ to start a chain, 1000 react with $\text{M}\cdot$ carrying it on; $k_p[\text{M}\cdot][\text{M}]$ would then be 1000 times $k_1[\text{R}\cdot][\text{M}]$.

$$-\frac{d[\mathrm{M}]}{dt} \approx k_p[\mathrm{M}_n \cdot][\mathrm{M}] \tag{9.58}$$

$$-\frac{d[\mathrm{M}]}{dt} = k_p[\mathrm{M}]\sqrt{\frac{fk_d[\mathrm{In}]}{k_t}} \tag{9.59}$$

for the overall reaction. It is often convenient to substitute for the term $2fk_d[\mathrm{In}]$, which represents the rate of initiation of chains, a more general notation, r_i, rate of initiation. Equation 9.60 then includes also the possibility of photochemical initiation, where r_i will be proportional to light intensity.

$$-\frac{d[\mathrm{M}]}{dt} = k_p[\mathrm{M}]\sqrt{\frac{r_i}{2k_t}} \tag{9.60}$$

The kinetic treatment of a simple polymerization is readily extended to nonpolymerization chain reactions such as that of Scheme 13. Here we again argue that if chains are long, nearly all the R · radicals produced react with tin hydride

SCHEME 13

$$\mathrm{In} \longrightarrow 2\mathrm{R}'' \cdot$$

$$\mathrm{R}'' \cdot + \mathrm{HSnR}_3' \longrightarrow \mathrm{R}''\mathrm{H} + \cdot\mathrm{SnR}_3' \qquad \mathrm{rate} = r_i$$

$$\mathrm{R{-}X} + \cdot\mathrm{SnR}_3' \xrightarrow{k_{p_1}} \mathrm{R} \cdot + \mathrm{X{-}SnR}_3'$$

$$\mathrm{R} \cdot + \mathrm{HSnR}_3' \xrightarrow{k_{p_2}} \mathrm{RH} + \cdot\mathrm{SnR}_3'$$

$$2\mathrm{R} \cdot \xrightarrow{k_t} \text{termination products}$$

to continue the chain; therefore the overall rates of the two propagation steps are approximately the same (Equation 9.61). Termination and initiation proceed at

$$k_{p_1}[\mathrm{R{-}X}][\cdot\mathrm{SnR}_3'] \approx k_{p_2}[\mathrm{HSnR}_3'][\mathrm{R} \cdot] \tag{9.61}$$

$$r_i = 2k_t[\mathrm{R} \cdot]^2 \tag{9.62}$$

the same rate in the stationary state, hence Equation 9.62 is also approximately true, and the rate expression predicted for this scheme is found in Equations 9.63–9.65.

$$-\frac{d[\mathrm{RX}]}{dt} = k_{p_1}[\mathrm{RX}][\cdot\mathrm{SnR}_3'] \tag{9.63}$$

$$= k_{p_2}[\mathrm{HSnR}_3'][\mathrm{R} \cdot] \tag{9.64}$$

$$= k_{p_2}[\mathrm{HSnR}_3']\sqrt{\frac{r_i}{2k_t}} \tag{9.65}$$

Note that both 9.60 and 9.65 show the characteristic dependence of reaction rate on first power of substrate concentration and square root of the ratio of initiation rate to termination rate.

Emulsion Polymerization

In a polymerization the length of the molecular chains produced depends on the balance between propagation rate and termination rate. It may be desirable to change this balance. For example, in butadiene polymerization the propagation rate is too low compared with termination rate to yield high-molecular-weight linear chains needed for synthetic rubber. This problem is circumvented by carrying out the reaction in an aqueous mixture with a water-soluble initiator such as persulfate, $S_2O_8^{2-}$, and the butadiene monomer emulsified with soap into drops about 1 μm in diameter and a very large number of minute droplets called *micelles*, initially about 5 to 10 nm in diameter and containing on the order of 1000 to 2000 molecules. The drops initially contain most of the monomer, but the initiator radicals SO_4^{-} are much more likely to diffuse into the micelles, which, because of their small size and large number, have a much larger surface area. The micelles are so small that most of the time there is no more than one growing chain in a given micelle. Hence termination, which requires a bimolecular encounter of two radicals, is slowed and propagation can continue to high molecular weights. As monomer is consumed, it is replaced by more that diffuses through the solution from the drops and enters the micelles.[123]

9.4 RADICAL SUBSTITUTIONS[124]

The term *substitution* in an unrestricted sense is rather too broad to be useful in classification of radical reactions because most of them result in replacement of one group by another. We have already seen typical examples of bond homolysis, in which a molecule dissociates to yield two radicals which combine with each other or with another molecule. We are primarily concerned in this section with those elementary reaction steps in which a radical attacks directly an atom of another molecule (Equation 9.66), displacing from the site of attack another group, and

$$A \cdot + B—C \longrightarrow [A \cdots B \cdots C] \longrightarrow A—B + C \cdot \qquad (9.66)$$

with the overall reaction schemes in which these elementary reactions occur. The direct substitution steps are analogous to the S_N2 or S_E2 displacements of heterolytic chemistry and are termed S_H2 reactions; radical substitutions that are most reasonably formulated as being initiated by addition of a radical to an unsaturated system (Equation 9.67) (analogous to addition–elimination sequences in heterolytic reactions) are considered in Section 9.5.

$$A \cdot + \underset{D}{\overset{}{B}}{=}C \longrightarrow \underset{D}{\overset{A}{B}}—C \cdot \longrightarrow \overset{A}{B}{=}C + D \cdot \qquad (9.67)$$

---

[123]D. H. Richards, *Chem. Soc. Rev.*, **6**, 235 (1977).
[124]For a review, see M. L. Poutsma, in *Free Radicals*, Vol. II, J. K. Kochi, Ed., Wiley, New York, 1973, p. 113.

Some Examples of S_H2 Processes

S_H2 substitutions frequently occur in chain reaction sequences, some of which we have already encountered in the earlier discussion. Equations 9.68–9.71 illustrate some reactions; the simple stoichiometric equations serve only to emphasize the overall substitution nature of the process and do not reveal the complexities, often considerable, of the actual pathways followed. We shall not attempt to analyze all of these processes in detail, but in order to give a better idea of the mechanisms will describe more fully two of them, halogenation and autoxidation.

$$R-H + X_2 \longrightarrow R-X + HX \tag{9.68}[125]$$
$$R-H + X-Z \longrightarrow R-X + HZ \quad \text{(or other products)} \tag{9.69}[125]$$

$$(X = \text{halogen}; \ Z = -CBr_3, \ -OC(CH_3)_3, \ -SO_2Cl, \ -PCl_4, \ -\overset{+}{N}HR_2')$$

$$R-H + O_2 \longrightarrow R-O-O-H \tag{9.70}[126]$$
$$R-X + HSnR_3' \longrightarrow R-H + XSnR_3' \tag{9.71}[127]$$

Halogenation[128]

Free-radical halogenations by the molecular halogens are adequately described by the chain sequence in Scheme 14. The two propagation steps, 2 and 3, are S_H2 substitutions. Note that the substitutions occur by attack of the radical on a terminal, univalent atom, in one case H, in the other halogen. This feature is characteristic of bimolecular radical substitution steps: attack at multiply bonded sites tends to be by addition (Equation 9.67), and attack at saturated carbon occurs only in highly strained molecules. Because terminal singly bonded centers in organic compounds are nearly always hydrogen or halogen, it is at these atoms that most S_H2 substitutions occur.

S<small>CHEME</small> 14

Initiation	Initiator $\longrightarrow$ X$\cdot$ (1)
Propagation	$\begin{cases} X\cdot + H-R \rightleftharpoons X-H + R\cdot & (2) \\ R\cdot + X-X \rightleftharpoons R-X + X\cdot & (3) \end{cases}$
Termination	$\begin{cases} 2R\cdot \longrightarrow R-R + R(+H) + R(-H) & (4) \\ R\cdot + X\cdot \longrightarrow R-X & (5) \\ 2X\cdot \longrightarrow X_2 & (6) \end{cases}$

Halogenations can be initiated in any of the various ways we have discussed earlier; but since the molecular halogens absorb light and dissociate into the free atoms on doing so, initiation is frequently photochemical, in which case it will yield

[125](a) M. L. Poutsma, in *Free Radicals,* Vol. II, J. K. Kochi, Ed., Wiley, New York, 1973, p. 159; (b) M. L. Poutsma, in *Methods in Free Radical Chemistry,* Vol. 1, E. S. Huyser, Ed., Dekker, New York, 1969, p. 79.

[126](a) J. A. Howard, in *Free Radicals,* Vol. II, J. K. Kochi, Ed., Wiley, New York, 1973, p. 3; (b) F. R. Mayo, *Acc. Chem. Res.,* **1,** 193 (1968).

[127] H. G. Kuivila, *Acc. Chem. Res.,* **1,** 299 (1968).

[128] Poutsma [see note 125(a)] reviews the field and gives references to numerous other reviews as well as to experimental papers.

the $X \cdot$ radical directly. Initiation by some other method produces a radical $R' \cdot$ that will rapidly attack X_2 to produce $X \cdot$. The reaction enthalpies of halogenation (F, -101; Cl, -22; Br, -4; and I, $+16$ kcal mole^{-1})[129] reflect the decreasing bond dissociation energies for H—X and C—X bonds in the series F, Cl, Br, I, and the relatively constant $D(X—X)$ (Table 9.4). The highly exothermic fluorination requires no external initiators and occurs violently and uncontrollably on mixing fluorine with a hydrocarbon either in the gas or liquid phase. Chlorination must be initiated but proceeds readily, whereas bromination frequently requires elevated temperatures. Iodination is rarely successful and indeed is more likely to occur in the reverse direction as the reduction of alkyl iodides by HI.[125]

A prediction of expected kinetic behavior can be made on the basis of the Scheme 14 (Problem 5); experimental observations are in agreement with the radical chain process.[130]

Autoxidation[131]

Autoxidation, oxidation by molecular oxygen, is of great importance and has been the subject of intensive study. Scheme 15 outlines the bare essentials. If the compound being oxidized is an alkene, an addition process involving propagation steps 9.72 and 9.73 can compete and, depending on the structure of the alkene, may

$$\text{ROO} \cdot + \text{C=C} \longrightarrow \text{ROO—C—C} \cdot \qquad (9.72)$$

$$\text{ROO—C—C} \cdot + \text{O}_2 \longrightarrow \text{ROO—C—C—OO} \cdot \text{ etc.} \qquad (9.73)$$

occur to the exclusion of Scheme 15.[132] Autoxidations are a consequence of the special nature of O_2. In its ground state it has two unpaired electrons and so is a triplet-state molecule. Because of this highly unusual feature, oxygen is an efficient trap for radicals.

Termination in autoxidations is rather more complex than in the chain reactions we have considered so far. The peroxy radicals first combine to an unstable tetroxide, ROOOOR.[133] The existence of these compounds when R is tertiary is inferred from isotope tracer studies,[134] and the equilibrium 9.74 is observable by

$$\text{ROOOOR} \rightleftharpoons 2\text{ROO} \cdot \qquad (9.74)$$

$$\text{RO} \cdot + \text{ROO} \cdot \rightleftharpoons \text{ROOOR} \qquad (9.75)$$

40

[129]Values are for substitution of one hydrogen of CH_4: $CH_4 + X_2 \rightarrow CH_3X + HX$. See M. L. Poutsma, in *Free Radicals*, Vol. II, J. K. Kochi, Ed., Wiley, New York, 1973, p. 162.
[130](a) G. Chiltz, P. Goldfinger, G. Huybrechts, G. Martens, and G. Verbeke, *Chem. Rev.*, **63**, 355 (1963); (b) S. W. Benson and J. H. Buss, *J. Chem. Phys.*, **28**, 301 (1958).
[131](a) J. A. Howard, in *Free Radicals*, Vol. II, J. K. Kochi, Ed., Wiley, New York, 1973, p. 3; (b) F. R. Mayo, *Acc. Chem. Res.*, **1**, 193 (1968).
[132]See note 131(b).
[133](a) G. A. Russell, *J. Am. Chem. Soc.*, **79**, 3871 (1957); (b) H. S. Blanchard, *J. Am. Chem. Soc.*, **81**, 4548 (1959).
[134]P. D. Bartlett and T. G. Traylor, *J. Am. Chem. Soc.*, **85**, 2407 (1963).

SCHEME 15

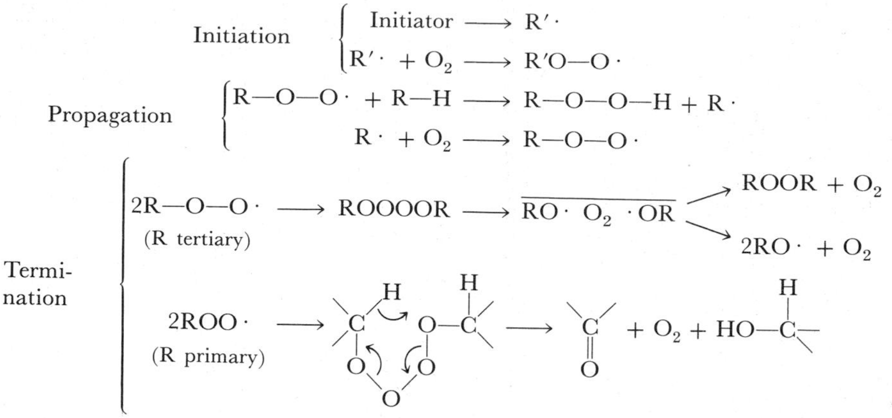

electron resonance near $-80°C.$[135] Another labile compound, the trioxide **40,** also forms at low temperature but dissociates above about $-30°C.$[136] At ambient temperatures and above, combination of peroxy radicals to tetroxide is followed rapidly by its decomposition by the cyclic Russell mechanism, shown in Scheme 15, which yields nonradical products and so terminates the chain.[137] If the original R · was tertiary so that no α hydrogens are available, decomposition occurs by the simple dissociation into oxygen and two alkoxy radicals (Equation 9.76). These

$$ROOOOR \longrightarrow RO \cdot O_2 \cdot OR \qquad (9.76)$$

fragments are produced in a solvent cage; if the two alkoxy radicals combine in the cage, the chain is terminated and peroxide ROOR results. Some of the alkoxy radicals will escape the cage and initiate the chain of events outlined in Scheme 16.[138]

Further complications can be introduced when interaction of some of the R · radicals with oxygen produces hydrogen transfer from the β carbon of the radical to oxygen (Equation 9.77), a reaction that is in essence a disproportionation

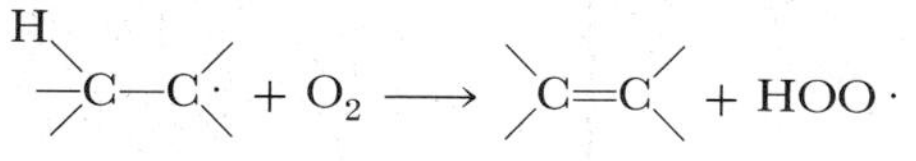

$$(9.77)$$

with oxygen playing the role of the hydrogen-accepting radical. Yet because oxygen is behaving like a diradical, this process is not a termination like an ordinary disproportionation but instead yields HOO · , which can continue the chain. The alkene produced in the disproportionation may itself suffer attack by a radical according to the addition–oxidation scheme (Equations 9.72 and 9.73).

[135] P. D. Bartlett and G. Guaraldi, *J. Am. Chem. Soc.,* **89,** 4799 (1967).
[136] P. D. Bartlett and P. Günther, *J. Am. Chem. Soc.,* **88,** 3288 (1966).
[137] See note 133(a).
[138] F. R. Mayo, *Acc. Chem. Res.,* **1,** 193 (1968).

SCHEME 16

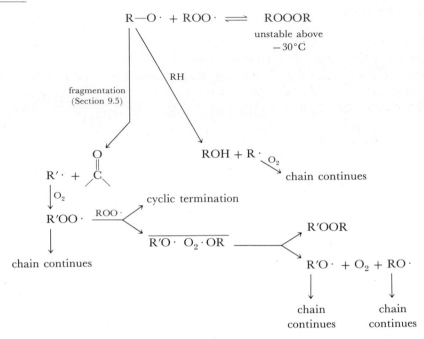

We shall not continue any further into the labyrinth of autoxidation. We shall merely point out that the complexities we have described are multiplied manifold when one considers the situations that will arise in oxidation of an alkane that reacts by a combination of the addition–polymerization and the abstraction routes, or when the temperature is high enough to homolyze the peroxide products and the reaction is thus producing its own initiator, or when there are several nonequivalent hydrogens in the substrate. Furthermore, the products will themselves be subject to oxidation. Clearly the possibilities are almost without limit.

Because any organic compound is subject to destruction by autoxidation (some, of course, being much more susceptible than others), it is a matter of considerable practical importance to prevent this reaction. Inhibition is accomplished either by the addition of substances that will break the kinetic chains or by substances that prevent their initiation.[139] In the former category are compounds such as the hindered phenols, which have in the —OH group a readily abstracted hydrogen but which yield a radical sufficiently unreactive that it will not in turn enter into any processes that will continue the chain. The latter type of inhibitor, particularly useful at higher temperatures where the hindered phenols become ineffective, include substances that destroy peroxide initiators (frequently substances containing unoxidized sulfur), that absorb ultraviolet light without initiat-

[139]Inhibition is discussed by J. A. Howard, in *Free Radicals,* Vol. II, J. K. Kochi, Ed., Wiley, New York, 1973, p. 3.

ing oxidation chains, or that tie up by chelation metal ions that catalyze radical production from peroxides.

Stereochemistry

Since most S_H2 displacements occur at univalent atoms, tests of stereochemistry at the reaction center are ordinarily not possible. Substitutions do nevertheless occur at saturated carbons in highly strained rings.[140] Halogen atoms attack cyclopropane (Equation 9.78) and other strained cyclic compounds with ring opening.

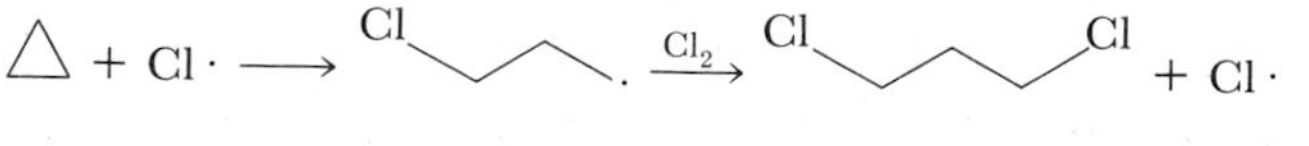

$$\triangle + Cl\cdot \longrightarrow \quad \overset{Cl}{\diagdown\!\!\diagup\!\!\diagdown} \cdot \overset{Cl_2}{\longrightarrow} \quad \overset{Cl}{\diagdown\!\!\diagup\!\!\diagdown\!\!\diagup}\!\!\overset{Cl}{\diagdown} + Cl\cdot \qquad (9.78)$$

Opening of *cis-* and *trans*-1,1-dichloro-2,3-dideuteriocyclopropane has been found to occur with inversion in the experiments summarized in Equations 9.79 and 9.80.[141]

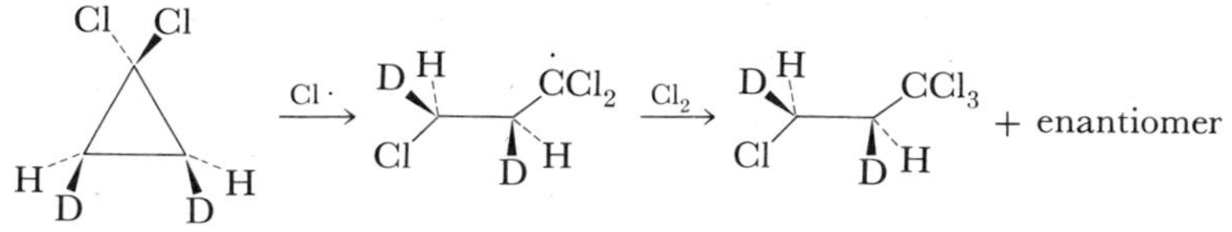

$$(9.79)$$

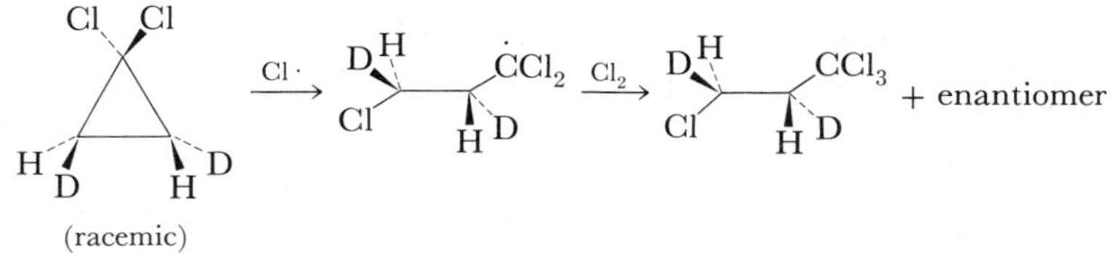

(racemic)

$$(9.80)$$

Relative Reactivities

A large amount of information is available on relative reactivities of different kinds of hydrogens toward free radicals and on variations of behavior among different abstracting radicals.[142] We must note first that in considering relative reactivities of different kinds of hydrogen, a statistical correction is necessary. Thus in the reaction of chlorine atoms with propane, if primary and secondary hydrogens were of equal reactivity, 1-chloropropane and 2-chloropropane would form in a ratio of 3:1 because there are six methyl protons and only two methylene protons. The actual ratio in the gas phase is only 1.41:1 at 25°C, so on a per-atom basis the secondary hydrogens are more reactive; the primary:secondary ratio at this temperature is

[140](a) K. U. Ingold and B. P. Roberts, *Free-Radical Substitution Reactions,* Wiley, New York, 1971, p. 72; (b) M. L. Poutsma, in *Free Radicals,* Vol. II, J. K. Kochi, Ed., Wiley, New York, 1973, p. 142.
[141] J. H. Incremona and C. J. Upton, *J. Am. Chem. Soc.,* **94,** 301 (1972).
[142] See, for example, (a) M. L. Poutsma, in *Free radicals,* Vol. II, J. K. Kochi, Ed., Wiley, New York, 1973, pp. 170, 187; (b) G. A. Russell, in *Free Radicals,* Vol. I, J. K. Kochi, Ed., Wiley, New York, 1973, pp. 283, 299; (c) C. Rüchardt, *Angew. Chem. Int. Ed. Engl.,* **9,** 830 (1970); (d) A. F. Trotman-Dickenson, *Adv. Free Radical Chem.,* **1,** 1 (1965); (e) A. A. Zavitsas and A. A. Melikian, *J. Am. Chem. Soc.,* **97,** 2757 (1975).

1:2.13.[143] Since activation energies for hydrogen abstraction are different at different positions, the ratios depend on temperature.

Tables 9.6–9.8 illustrate some of the trends observed, and Table 9.9 shows activation parameters for abstraction of different types of hydrogen by halogen atoms. Note that in alkanes, hydrogens become more easily abstracted on proceeding along the series primary, secondary, tertiary. The magnitudes of the differences depend on the radical removing the hydrogen; the more reactive ones (F·, Cl·) are less selective and those of lower reactivity (Br·) more selective. Phenyl substitution on the α carbon makes the hydrogens more easily abstracted; again, abstraction by bromine atoms benefits more. The reaction coordinate diagram (Figure 9.3) shows that the consequence of an earlier transition state in abstraction by a more reactive radical should be less sensitivity to stabilization in the product radical. The trend of decreasing selectivity in the series Br· > Cl· > F· parallels the decrease in activation energy illustrated in Table 9.9. Radical selectivity can be modified by the conditions; aromatic solvents, for example, apparently complex with chlorine atoms and increase their selectivity.[144]

A parameter that should be useful in assessing position of the transition state is the primary hydrogen isotope effect. The isotope effect should be largest for a

Table 9.6 Relative Reactivities of Radicals in Abstraction of Primary, Secondary, and Tertiary Hydrogen Atoms[a]

Radical	Temperature (°C) and phase	Hydrogen abstracted			Reference
		$H-CH_2R$	$H-CHR_2$	$H-CR_3$	
F·	25, gas	1	1.2	1.4	b
Cl·	25, gas	1	4	6	b
Br·	40, liquid	1	200	1.9×10^4	c
φ·[f]	60, liquid	1	9	47	d
CH_3·	110, liquid	1	4	46	e
CH_3·	182, gas	1	7	50	e
CF_3·	182, gas	1	8	24	e
CCl_3·	190, gas	1	80	2300	e
CH_3O·	250, gas	1	8	27	e
H·	35, liquid	1	5	40	e

[a] Rates for each radical are relative to its reaction with primary C—H.
[b] M. L. Poutsma, in *Free Radicals,* Vol. II, J. K. Kochi, Ed., Wiley, New York, 1973, p. 172.
[c] G. A. Russell and C. De Boer, *J. Am. Chem. Soc.,* **85,** 3136 (1963).
[d] G. A. Russell, in reference b, Vol. I, p. 299.
[e] J. K. Kochi, in reference b, Vol. II, p. 690. Reprinted by permission of John Wiley & Sons, Inc.
[f] The absolute rate constant for abstraction of hydrogen from mineral oil by phenyl radical is on the order of $3 \times 10^5 M^{-1}sec^{-1}$ at 45° C. R. G. Kryger, J. P. Lorand, N. R. Stevens, and N. R. Herron, *J. Am. Chem. Soc.,* **99,** 7589 (1977).

[143] Data from J. H. Knox and R. L. Nelson, *Trans. Faraday Soc.,* **55,** 937 (1959).
[144] (a) M. L. Poutsma, in *Free Radicals,* Vol. II, J. K. Kochi, Ed., Wiley, New York, 1973, p. 175; (b) J. C. Martin, in *Free Radicals,* Vol. II, J. K. Kochi, Ed., Wiley, New York, 1973, p. 493.

Table 9.7 RELATIVE REACTIVITIES OF RADICALS IN ABSTRACTION OF HYDROGEN FROM PHENYL-SUBSTITUTED CARBON

Radical	Temperature (°C) and phase	Hydrogen abstracted				Reference
		H—CH$_2$CH$_3$	H—CH$_2\phi$	H—CHϕ_2	H—Cϕ_3	
Cl · [a]	40, gas	1	1.3	2.6	9.5	b
			(1 :	2 :	7.3)	
Br · [a]	40, liquid	1	6.4 × 10^4	6.2 × 10^5	1.1 × 10^6	b
			(1 :	10 :	17)	
ϕ · [a]	60, liquid	1	9	69	—	c
			(1 :	7.7	—)	
Cl$_3$C ·	40, liquid	1	50	160		b

[a] Series are given both relative to primary and relative to benzyl.
[b] G. A. Russell and C. De Boer, *J. Am. Chem. Soc.*, **85**, 3136 (1963).
[c] G. A. Russell, in *Free Radicals*, Vol. I, J. K. Kochi, Ed., Wiley, New York, 1973, p. 299.

Table 9.8 SENSITIVITY OF HYDROGEN ABSTRACTION TO POLAR EFFECTS

Radical	Temperature (°C) and phase	Hydrogen abstracted			Reference
		H—CHR$_2$	H—CHR (CH$_2$Cl)	H—CHR (Cl)	
Cl ·	35, gas	1	0.52	0.17	a
Br ·	60, liquid	1	0.47	0.45	a
(CH$_3$)$_3$CO · [d]	40		1	1.2	b
		H—CHR$_2$	H—CHR (CH$_2$F)	H—CHR (F)	
Cl ·	75	1	0.46	0.24	c
Br ·	150	1	0.08	0.1	c
		H—CHR$_2$	H—CHR (CH$_2$CF$_3$)	H—CHR (CF$_3$)	
Cl ·		1	0.28	0.01	c
Br ·		1	0.08	0.01	c

[a] M. L. Poutsma, in *Free Radicals*, Vol. II, J. K. Kochi, Ed., Wiley, New York, 1973, p. 188.
[b] G. A. Russell, in reference a, Vol. I, p. 307.
[c] A. F. Trotman-Dickenson, *Adv. Free Radical Chem.*, **1**, 1 (1965), p. 15.
[d] The absolute rate constants per hydrogen for abstraction of hydrogen by *t*-butoxy radical are in the range 10^4 to 10^5M^{-1} sec^{-1}. S. K. Wong, *J. Am. Chem. Soc.*, **101**, 1235 (1979); R. D. Small, Jr., and J. C. Scaiano, *J. Am. Chem. Soc.*, **100**, 296 (1978).

symmetrical transition state (Section 2.7) and smaller for less symmetrical transition states. Russell has tabulated k_H/k_D for hydrogen abstractions from carbon;[145] although the variety of temperatures makes an evaluation difficult,[146] there appears to be some confirmation of this expectation. For example, k_H/k_D is greater (4.9) for the thermoneutral reaction of Br · with toluene than for the 18 kcal mole^{-1}

[145] G. A. Russell, in *Free Radicals*, Vol. I, J. K. Kochi, Ed., Wiley, New York, 1973, p. 312.
[146] Some isotope effects much in excess of the theoretical maximum for loss of one degree of freedom imply loss of bending modes in addition to loss of C—H stretching.

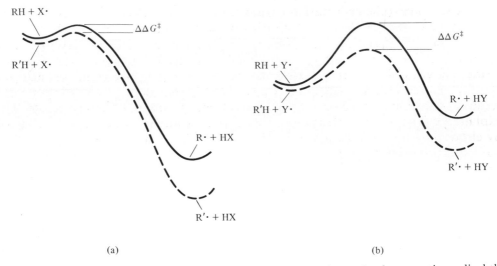

Figure 9.3 Relative reactivities in hydrogen abstraction. (a) In abstraction by a reactive radical the transition state is early and relatively little affected (small $\Delta\Delta_G{}^{\ddagger}$) by changing the structure of R so as to stabilize the radical being formed. (b) In abstraction by a less reactive radical, with a later transition state, structure change in the product is more strongly felt at the transition state (larger $\Delta\Delta G^{\ddagger}$).

Table 9.9 Activation Parameters for the Reaction $R\!-\!H + X\cdot \longrightarrow R\cdot + HX^{a}$

	Hydrogen abstracted					
	$H\!-\!CH_2R$		$H\!-\!CHR_2$		$H\!-\!CR_3$	
$X\cdot$	E_a (kcal mole^{-1})	$\log A$	E_a (kcal mole^{-1})	$\log A$	E_a (kcal mole^{-1})	$\log A$
F·	0	13	0	13	0	13
Cl·	1	13	0.5	13	0	13
Br·	13	13	10	13	7	13

^{a}Approximate values, summarized from data of A. F. Trotman-Dickenson, *Adv. Free Radical Chem.*, **1**, 1 (1965), pp. 9, 11.

exothermic reaction of Cl· with toluene (k_H/k_D 1.3).[147] Pryor and Kneipp have reported isotope effects for hydrogen abstraction from *t*-butyl thiol (Equation 9.81)

$$(CH_3)_3C\!-\!S\!-\!H(D) + R\cdot \longrightarrow (CH_3)_3C\!-\!S\cdot + R\!-\!H(D) \tag{9.81}$$

by various radicals R· chosen so that the enthalpy change of the reaction ranges from -24 to $+18$ kcal mole^{-1}. They found a clearly defined maximum in k_H/k_D when the process is approximately thermoneutral, $\Delta H° \approx 5$ kcal mole^{-1}. For these

[147]M. L. Poutsma, in *Free Radicals*, Vol. II, J. K. Kochi, Ed., Wiley, New York, 1973, p. 963.

reactions, at least, the expectation of maximum isotope effect for a thermoneutral reaction that probably has a symmetrical transition state is confirmed.[148]

Table 9.8 illustrates the second main feature of selectivity in abstraction of hydrogen. Polar substituents affect the product distribution but do so differently for different abstracting radicals.[149] Chlorine atoms attack a hydrogen on a carbon already bearing chlorine less rapidly than on one that does not, whereas phenyl radicals behave in just the opposite way and bromine atoms are intermediate. The explanation is that the chlorine atom is electrophilic and its attack will be hindered by electron withdrawal. In resonance terms, there is an important contribution to the transition state by **41c** and electron withdrawal at the carbon will raise the

$$\dot{C}l \quad H\!-\!C\!\!\stackrel{<}{<} \quad \longleftrightarrow \quad Cl\!-\!H \quad \dot{C}\!\!\stackrel{<}{<} \quad \longleftrightarrow \quad \bar{C}l \quad H \quad \overset{+}{C}\!\!\stackrel{<}{<}$$

| **41a** | **41b** | **41c** |

energy. Bromine atoms are affected similarly and react less rapidly with $H\!-\!CH_2\!-\!CH_2\!-\!Cl$ than with an unchlorinated secondary position, but because of the later transition state they are more susceptible to the radical stabilizing tendency of an α chlorine and so do not suffer a further decline of activity when the Cl substituent is brought to the α position as do chlorine atoms. Hammett σ–ρ correlations also point to the influence of polar effects in the abstraction reactions. The simple alkyl radicals, however, apparently do not show significant polar properties.[150]

9.5   RADICAL ADDITIONS AND ELIMINATIONS[151]

The addition of radicals to carbon–carbon double bonds, ordinarily a chain process, may lead to the formation of either small molecules or polymers depending upon the efficiency of chain transfer. Rates and activation parameters for the addition of various radicals $X\cdot$ to ethylene to form the radical $X\!-\!CH_2\!-\!CH_2\cdot$ are given in Table 9.10.

Addition with Chain Transfer[152]

Chain reactions of radical addition will follow the familiar sequence of initiation, propagation, and termination. We shall be concerned here primarily with the propagation steps, which determine the nature of the products. Suppose that a

[148]W. A. Pryor and K. G. Kneipp, *J. Am. Chem. Soc.,* **93,** 5584 (1971). See also (b) S. Kozuka and E. S. Lewis, *J. Am. Chem. Soc.,* **98,** 2254 (1976), and following papers; (c) E. S. Lewis, *Top. Current Chem.,* **74,** 31 (1978).

[149]G. A. Russell, in *Free Radicals,* Vol. I, J. K. Kochi, Ed., Wiley, New York, 1973, p. 275.

[150](a) D. D. Tanner, P. W. Samal, T. C.-S. Ru;, and R. Henriquez, *J. Am. Chem. Soc.,* **101,** 1168 (1979);(b) T. H. Fisher and A. H. Meierhoefer, *J. Org. Chem.,* **43,** 220, 224 (1978).

[151]Vor a review, see (a) P. I. Abell, in *Free Radicals,* Vol. II, J. K. Kochi, Ed., Wiley, New York, 1973, p. 63. (b) E. S. Huyser, *Free-Radical Chain Reactions,* Wiley, New York, 1970.

[152]Chain transfer is discussed by (a) C. Walling, *Free Radicals in Solution,* Wiley, New York, 1957, Chap. 6; (b) E. S. Huyser, *Free-Radical Chain Reactions,* Wiley, New York, 1970, p. 344, (c) C. M. Starks, *Free Radical Telomerization,* Academic Press, New York, 1974.

Table 9.10 ACTIVATION PARAMETERS FOR THE ADDITION OF ATOMS AND RADICALS TO ETHYLENE[a]

Atom or radical	$\log A$ (l mole^{-1} sec^{-1})	E(kcal mole^{-1})	k(l mole^{-1} sec^{-1} at 437 K)
Cl ·	10.6	0	4.5×10^{10}
S · (^{3}P)	10.0	1.5	1.8×10^9
H ·	11.0	2.8	4.9×10^9
Se · (^{3}P)	10.1	2.8	4.9×10^8
Br ·	9.5	2.9	1.1×10^8
CF$_3$ ·	8.0	2.9	3.5×10^6
CF$_2$Br ·	8.0	3.1	2.8×10^6
CH$_2$F ·	7.6	4.3	2.8×10^5
CCl$_3$ ·	7.8	6.3	4.5×10^4
CH$_3$ ·	8.5	7.7	4.5×10^4
CH$_3$CH$_2$ ·	8.3	7.5	3.5×10^4
(CH$_3$)$_2$CH ·	7.8	6.9	2.2×10^4
(CH$_3$)$_3$C ·	7.5	7.1	8.9×10^3

[a] With permission from J. M. Tedder and J. C. Walton, *Adv. Phys. Org. Chem.*, **16**, 51 (1978), V. Gold and D. Bethell, Eds., p. 57. Copyright by Academic Press Inc. (London) Ltd.

chain is initiated by the addition of some radical X · to an alkene (Equation 9.82). The intermediate alkyl radical produced by the addition has a choice of paths to

$$X \cdot + C{=}C \longrightarrow X{-}C{-}C \cdot \tag{9.82}$$

$$X{-}C{-}C \cdot + X{-}Y \xrightarrow{k_{transfer}} X{-}C{-}C{-}Y + X \cdot \tag{9.83}$$

or

$$X{-}C{-}C \cdot + C{=}C \xrightarrow{k_{polymer}} X{-}C{-}C{-}C{-}C \cdot \tag{9.84}$$

follow: It may attack X—Y, abstracting Y and leaving a new X · to add to another alkene (Equation 9.83), or it may itself add to the next alkene (Equation 9.84). There are also other possibilities, which we shall consider later, such as reversal of the original addition or some other fragmentation. The nature of the products will depend on the relative rates of processes 9.83 and 9.84. If we suppose the concentrations of X—Y and the alkene to be comparable, $k_{transfer} \gg k_{polymer}$ will produce simple addition of X—Y across the double bond, whereas $k_{transfer} \ll k_{polymer}$ will mean that very little of the X—Y will react and the process will produce long polymer chains. If the two rates are about the same, the chances will be good that the first step to follow after Reaction 9.82 will be addition of another alkene, but the chance of adding still another to the same chain will rapidly diminish as the chain grows longer. The product under these circumstances is a telomer, with molecular size controlled by the chain transfer coefficients as described earlier.

The reaction of hydrogen halides with alkenes provides a typical example of radical addition. With HCl the chain transfer step, abstraction of H from HCl by RCH$_2$ ·, is endothermic by about 5 kcal mole^{-1}, and therefore will have an activation energy of at least this value. The competing polymerization step will typically have an activation energy of 6–10 kcal mole^{-1}.[153] For HBr addition, on the other hand, the chain transfer step is exothermic by about 10 kcal mole^{-1}. These data suggest that HBr is more likely than HCl to produce a simple addition of HX to the double bond. Experimentally, simple addition of HCl is rare, although it can be made to occur with excess HCl.[154] In contrast, HBr readily undergoes simple addition to the exclusion of telomerization.[155] Radical addition of hydrogen iodide to alkenes is not successful because now the initial addition of the iodine atom to the double bond is endothermic.

A number of other substances, for example CCl_4, CCl_3Br, and several other alkyl polyhalides, aldehydes, and thiols, add successfully to alkenes.[156] The addition of Cl_2, frequently looked upon as ionic, often occurs as a radical chain reaction, particularly in nonpolar solvents and in the presence of light or peroxides.[157]

The orientation of addition to unsymmetrical alkenes is determined at the addition step, which occurs so as to leave the most stable radical (Equations 9.85 and 9.86). Overall addition is therefore in the sense opposite to that predicted by Markownikoff's rule and to that observed in the polar addition.

$$Br\cdot + H_2C{=}C\underset{CH_3}{\overset{H}{\big<}} \longrightarrow BrCH_2{-}\underset{CH_3}{\overset{H}{\overset{|}{C}}}\cdot \qquad (9.85)$$

$$BrCH_2{-}\underset{CH_3}{\overset{H}{\overset{|}{C}}}\cdot + HBr \longrightarrow BrCH_2{-}\underset{CH_3}{\overset{H}{\overset{|}{C}}}{-}H + Br\cdot \qquad (9.86)$$

Copolymerization and Relative Reactivity in Addition to Alkenes

Having considered earlier the polymerization reaction pathway and kinetics we shall confine the discussion here to a brief consideration of copolymerization. When two different monomers are present in the reaction mixture undergoing polymerization or telomerization, the growing chain may have as its terminal unit either of the two monomers, and when it reacts with the next one, it has the choice of adding either. There are therefore four propagation steps (Scheme 17) rather than just one as in simple polymerization (Scheme 9). In Scheme 17,[158] we assume that the reactivity of a growing chain depends only on which unit added last. This assumption allows us to say that there are, for purposes of kinetics, only two kinds of chains,

[153] K. U. Ingold, in *Free Radicals,* Vol. I, J. K. Kochi, Ed., Wiley, New York, 1973, p. 92.

[154] J. H. Raley, F. F. Rust, and W. E. Vaughan, *J. Am. Chem. Soc.,* **70**, 2767 (1948).

[155] (a) M. S. Kharasch, H. Engelmann, and F. R. Mayo, *J. Org. Chem.,* **2**, 288 (1937); (b) D. H. Hey and W. A. Waters, *Chem. Rev.,* **21**, 169 (1937); (c) C. Walling, *Free Radicals in Solution,* Wiley, New York, 1957, pp. 240, 291.

[156] C. Walling, *Free Radicals in Solution,* Wiley, New York, 1957, Chaps. 6 and 7, and E. S. Huyser, *Free-Radical Chain Reactions,* Wiley, New York, 1970, Chap. 7, give detailed discussions.

[157] M. L. Poutsma, in *Free Radicals,* Vol. II, J. K. Kochi, Ed., Wiley, New York, 1973, p. 185.

[158] (a) C. Walling, *Free Radicals in Solution,* Wiley, New York, 1957, p. 100; (b) T. Alfrey, Jr., and G. Goldfinger, *J. Chem. Phys.,* **12**, 205, 322 (1944); (c) F. T. Wall, *J. Am. Chem. Soc.,* **66**, 2050 (1944); (d) F. R. Mayo and F. M. Lewis, *J. Am. Chem. Soc.,* **66**, 1594 (1944); (e) E. S. Huyser, *Free-Radical Chain Reactions,* Wiley, New York, 1970, p. 352; (f) C. M. Starks, *Free Radical Telomerization,* Academic Press, New York, 1974, p. 218.

SCHEME 17

Monomers: M_1 and M_2

$$M_1 \cdot = \sim M_1 \cdot$$

$$M_2 \cdot = \sim M_2 \cdot$$

Propagation
$$\begin{cases}
M_1 \cdot + M_1 \xrightarrow{k_{11}} M_1 \cdot \\
M_1 \cdot + M_2 \xrightarrow{k_{12}} M_2 \cdot \\
M_2 \cdot + M_1 \xrightarrow{k_{21}} M_1 \cdot \\
M_2 \cdot + M_2 \xrightarrow{k_{22}} M_2 \cdot
\end{cases}$$

those ending in $—M_1 \cdot$ and those ending in $—M_2 \cdot$, and that addition of M_1 to either will leave an $—M_1 \cdot$ chain, whereas addition of M_2 will leave an $—M_2 \cdot$ chain. The predictions of this scheme are correct for most, but not all, systems.[159] The matter of interest from the point of view of radical reactivity concerns the effect of structure on the rate constants k_{mn}. In order to bring the data into a more convenient form, it is customary to define the ratios r_1 and r_2 as in Equations 9.87 and 9.88.

$$r_1 = \frac{k_{11}}{k_{12}} \tag{9.87}$$

$$r_2 = \frac{k_{22}}{k_{21}} \tag{9.88}$$

A value of r_1 greater than unity means the chain $—M_1 \cdot$ prefers to add another M_1; $r_1 = 1$ means that $—M_1 \cdot$ adds at random either M_1 or M_2, and r_1 less than unity means that $—M_1 \cdot$ prefers to react with M_2.[160]

The behavior of copolymerizing systems can be deduced from the r values. For example, when $M_1 =$ styrene and $M_2 =$ vinyl acetate, $r_1 = 55$ and $r_2 = 0.01$; then any chain, whether it ends in $—M_1 \cdot$ or $—M_2 \cdot$, will prefer to add M_1 next. Under these circumstances, at the beginning of the reaction essentially only M_1 will be reacting, and poly-M_1 (polystyrene) will be formed; when nearly all the M_1 is gone, M_2 (vinyl acetate) will polymerize. If r_1 and r_2 are each about unity, as in the case of acrylonitrile and butyl acrylate, chains contain the two monomers in random order. If both r_1 and r_2 are less than unity, as for $M_1 =$ styrene, $M_2 =$ vinylidene cyanide (1,1-dicyanoethylene), $r_1 = 0.005$, $r_2 = 0.001$, there is a regular alternation and the chains have the structure $\cdots—M_1—M_2—M_1—M_2—\cdots$.

[159] See note 158(a).
[160] C. Walling, *Free Radicals in Solution*, Wiley, New York, 1957, p. 116–119, E. S. Huyser, *Free-Radical Chain Reactions*, Wiley, New York, 1970, p. 357, and C. M. Starks, *Free Radical Telomerization*, Academic Press, New York, 1974, p. 220, list values.

From the r values one can obtain relative reactivities of each type of radical $R' \cdot$ toward addition to various alkenes.[161] In Equation 9.89 rates are greatest for

$$R' \cdot \; + \; H_2C{=}\overset{\displaystyle H}{\underset{\displaystyle R}{C}} \quad \longrightarrow \quad R'CH_2{-}\overset{\displaystyle H}{\underset{\displaystyle R}{\overset{\displaystyle \cdot}{C}}} \tag{9.89}$$

conjugating R groups, such as phenyl or vinyl, and decrease roughly in the order $R = -\phi, \; -CH{=}CH_2 > -COR > -C{\equiv}N > -Cl > -COOR > -OR > -CH_3 > -H$. Thus, for example, the r values for the styrene–vinyl acetate pair show that styrene, $R = \phi$, is more reactive by a factor of 55 than vinyl acetate, $R = OAc$, toward styrene radicals, $R' \cdot = \sim HC\phi$. Styrene is more reactive than vinyl acetate by a factor of $1/r_2 = 100$ toward vinyl acetate radicals, $R' \cdot = \sim HCOAc$. These results must be interpreted as a stabilization by $-R$ in the order mentioned of the developing radical which exceeds any stabilization that $-R$ might contribute to the ground-state alkene. Alkenes substituted at one end only are more reactive than those with substituents at both ends, an observation that has been explained, although perhaps not completely, on the basis of steric effects.[162]

Reactivities of various radicals toward a given alkene may be obtained by combining relative rate information, from r values, with absolute propagation rate constants for polymerization of various monomers alone.[163] The results show that, as expected, substituents that make alkenes reactive by stabilizing the developing radicals make the radicals, once formed, less reactive. The effect is greater on the radicals than on the alkenes. Returning again to the styrene–vinyl acetate pair, we find that k_p for styrene at $60°C$ is $145M^{-1} \; sec^{-1}$ and for vinyl acetate is $3700M^{-1} \; sec^{-1}$.[164] From these data we may derive Table 9.11 by noting that the radical bearing a phenyl substituent prefers styrene over vinyl acetate by a factor of $55 = r_1$, and the radical bearing an acetoxy substituent also prefers styrene by a factor of $100 = 1/r_2$. The columns in the table then show that the radical with the acetoxy substituent is more reactive by a factor of over 1000 toward both alkenes. It is the interplay between substituent effects on radical reactivity and alkene susceptibility to attack that leads to the curious fact that styrene alone polymerizes more slowly than does vinyl acetate alone, but when mixed the styrene polymerizes first and the vinyl acetate only after the styrene is nearly gone.

For those monomer pairs that show a strong tendency toward alternation, inconsistencies with the generalizations above appear. With $M_1 =$ styrene, $M_2 =$ maleic anhydride, both r_1 and r_2 are less than 0.05; chains ending in styrene

[161] See also (a) C. M. Starks, *Free Radical Telomerization*, Academic Press, New York, 1974, pp. 64–70; (b) J. M. Tedder and J. C. Walton, *Adv. Phys. Org. Chem.*, **16**, 51 (1978).

[162] P. I. Abell, in *Free Radicals*, Vol. II, J. K. Kochi, Ed., Wiley, New York, 1973, p. 99.

[163] (a) C. Walling, *Free Radicals in Solution*, Wiley, New York, 1957, p. 123; for rate constants see also (b) K. U. Ingold, in *Free Radicals*, Vol. I, J. K. Kochi, Ed., Wiley, New York, 1973, p. 92.

[164] Sue note 163(b).

Table 9.11 RATE CONSTANTS IN THE STYRENE–VINYL ACETATE SYSTEM (M^{-1} SEC^{-1} AT 60°C)^a

	Alkene	
Radical	$=\!\!\!\diagdown_{\phi}$	$=\!\!\!\diagdown_{\text{O}-\overset{\text{O}}{\overset{\|}{\text{C}}}-\text{CH}_3}$
$\cdot\diagdown_{\phi}$	145	2.6
$\cdot\diagdown_{\text{O}-\overset{\text{O}}{\overset{\|}{\text{C}}}-\text{CH}_3}$	3.7×10^5	3700

^{a}Calculated from data given by C. M. Starks, *Free Radical Telomerization*, Academic Press, New York, 1974, p. 220, and by K. U. Ingold, in *Free Radicals*, Vol. I, J. K. Kochi, Ed., Wiley, New York, 1973, p. 92.

units prefer to add to maleic anhydride, but those ending in maleic anhydride prefer to add to styrene. The sequence of stabilization that we have given suggests that styrene should be the more reactive of the two alkenes, a conclusion substantiated by the relatively low reactivity of maleic anhydride compared with styrene toward radicals with carboxyl, cyano, or chlorine α substituents. The unusual affinities found in alternating pairs are attributed to the complementary polar character of the groups. If one of the groups R, R′ is particularly good for stabilization of positive charge and the other for stabilization of negative charge, then in the transition state either **42b** or **42c** will make a significant contribution and the

energy will be lowered. In the styrene–maleic anhydride copolymerization, this criterion is met both for addition of chains ending in a styrene unit to maleic anhydride (R = ϕ, R′ = COO—, **42b** favored) and for addition of chains ending in a maleic anhydride unit to styrene (R = COO—, R′ = ϕ, **42c** favored).

Eliminations

A number of instances are known in which a radical elimination or fragmentation occurs. We have already met some of these, for example, in the decomposition of diacyl peroxides, where the second step (Equation 9.91) is an elimination, or in the fragmentation of alkoxy radicals (Equation 9.92). Perhaps the most obvious elimination is simple reversal of addition of an alkyl radical to a carbon–carbon double bond, the back reaction in Equation 9.93. As we noted earlier, this reaction is exothermic in the forward direction as written and at moderate temperatures proceeds efficiently to the right; the reverse, having the larger activation energy, nevertheless competes more and more effectively as the temperature is raised. At some

$$\phi-\overset{\overset{\displaystyle O}{\|}}{C}-O-O-\overset{\overset{\displaystyle O}{\|}}{C}-\phi \longrightarrow 2\,\phi-\overset{\overset{\displaystyle O}{\|}}{C}-O\cdot \qquad (9.90)$$

$$\phi-\overset{\overset{\displaystyle O}{\|}}{C}-O\cdot \longrightarrow \phi\cdot \;+\;O{=}C{=}O \qquad (9.91)$$

$$R-\overset{\overset{\displaystyle CH_3}{|}}{\underset{\underset{\displaystyle CH_3}{|}}{C}}-O\cdot \longrightarrow R\cdot \;+\; \overset{H_3C}{\underset{H_3C}{>}}C{=}O \qquad (9.92)$$

$$R\cdot \;+\; \overset{\diagdown}{\diagup}C{=}C\overset{\diagup}{\diagdown} \;\rightleftharpoons\; R-\overset{|}{C}-C\overset{\diagup}{\diagdown} \qquad (9.93)$$

critical temperature the rates of polymerization and the reverse (depolymerization) become equal; above the critical temperature polymer chains will come apart to monomer. For styrene the critical temperature is a little over 300°C.[165] Reversibility of addition to alkenes is also manifest in the cis–trans isomerization of double bonds by bromine,[166] iodine,[167] and by thiols[168] under radical conditions. Strain in cyclic compounds can alter the energy relationships so that elimination becomes more favorable relative to addition, as in Equations 9.94 and 9.95.

$$(9.94)^{169}$$

$$(9.95)^{170}$$

Eliminations in alkoxy radicals[171]   Elimination of an alkyl radical from an alkoxy radical with formation of a carbonyl group, usually termed β scission (Equation 9.96), is more favorable thermodynamically than is elimination to form an alkene and often takes place readily in competition with other processes. Using

$$R-\overset{|}{\underset{|}{C}}-O\cdot \longrightarrow R\cdot \;+\; \overset{\diagdown}{\diagup}C{=}O \qquad (9.96)$$

[165] F. S. Dainton and K. J. Ivin, *Nature,* **162,** 705 (1948).
[166] G. A. Oldershaw and R. J. Cvetanović, *J. Chem. Phys.,* **41,** 3639 (1964).
[167] K. W. Egger, *J. Am. Chem. Soc.,* **89,** 504 (1967).
[168] (a) R. H. Pallen and C. Sivertz, *Can. J. Chem.,* **35,** 723 (1957); see also (b) J. T. Heppinstall, Jr., and J. A. Kampmeier, *J. Am. Chem. Soc.,* **95,** 1904 (1973); review: (c) K. Griesbaum, *Angew. Chem. Int. Ed. Engl.,* **9,** 273 (1970).
[169] D. I. Davies, J. N. Done, and D. H. Hey, *Chem. Commun.,* 725 (1966).
[170] (a) J. K. Kochi, P. J. Krusic, and D. R. Eaton, *J. Am. Chem. Soc.,* **91,** 1877 (1969); for further examples, see (b) J. W. Wilt in *Free Radicals,* Vol. I, J. K. Kochi, Ed., Wiley, New York, 1973, p. 398.
[171] Chemistry of alkoxy radicals is reviewed by J. K. Kochi, Ed., *Free Radicals,* Vol. II, Wiley, New York, 1973, p. 665.

the data of Benson and co-workers (Tables 2.9 and 9.5), we can calculate that fragmentation of the *t*-butoxy radical (Equation 9.97) is endothermic by 4 kcal mole^{-1} and elimination of benzyl radical in Equation 9.98 should be exothermic by 12 kcal mole^{-1}. Elimination of hydrogen, on the other hand, is substantially endo-

$$\text{CH}_3\text{—}\overset{\displaystyle \text{CH}_3}{\underset{\displaystyle \text{CH}_3}{\text{C}}}\text{—O}\cdot \longrightarrow \cdot\text{CH}_3 + \overset{\displaystyle \text{CH}_3}{\underset{\displaystyle \text{CH}_3}{\text{C}}}=\text{O} \qquad \Delta H_{\text{calc}} = +4.2 \text{ kcal mole}^{-1} \qquad (9.97)$$

$$\phi\text{—CH}_2\text{—}\overset{\displaystyle \text{CH}_3}{\underset{\displaystyle \text{CH}_3}{\text{C}}}\text{—O}\cdot \longrightarrow \phi\text{CH}_2\cdot + \overset{\displaystyle \text{CH}_3}{\underset{\displaystyle \text{CH}_3}{\text{C}}}=\text{O} \qquad \Delta H_{\text{calc}} = -12.3 \text{ kcal mole}^{-1} \qquad (9.98)$$

thermic and rarely occurs.[172] Rates of fragmentation have been measured relative to rate of abstraction of hydrogen from a hydrocarbon solvent by the alkoxy radical; if one assumes that the abstractions proceed at about the same rate for different alkoxy radicals, the results yield relative rates of fragmentation for different departing R · radicals. Table 9.12 lists relative rate values, which show the increasing rate of fragmentation as the departing radical becomes more highly stabilized.

When there is a choice of more than one group that might cleave, the more stable radical leaves preferentially.[173] A strained bridgehead radical (compare Equations 9.99 and 9.100) is much less susceptible to cleavage than would be

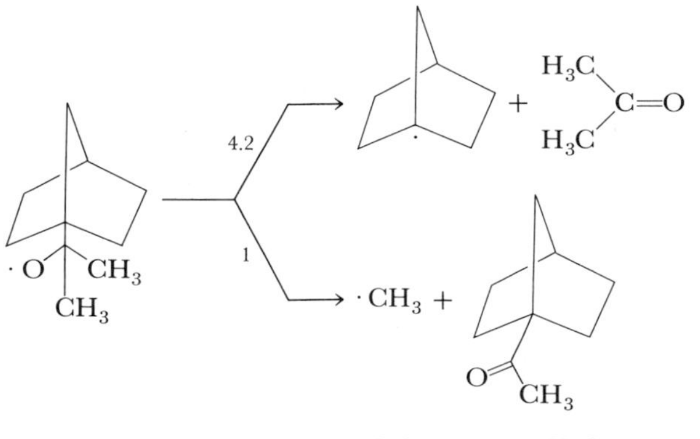

(9.99)

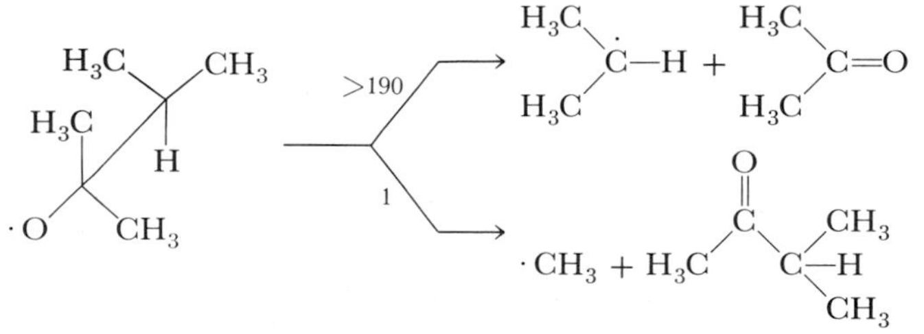

(9.100)

[172] (a) P. Gray and A. Williams, *Chem. Rev.*, **59**, 239 (1959); (b) P. Gray, R. Shaw, and J. C. J. Thynne, *Prog. Reaction Kinetics*, **4**, 63 (1967).

[173] F. D. Greene, M. L. Savitz, F. D. Osterholtz, H. H. Lau, W. N. Smith, and P. M. Zanet, *J. Org. Chem.*, **28**, 55 (1963).

Table 9.12 RELATIVE RATES OF FRAGMENTATION OF ALKOXY RADICALS AT $40°C^a$

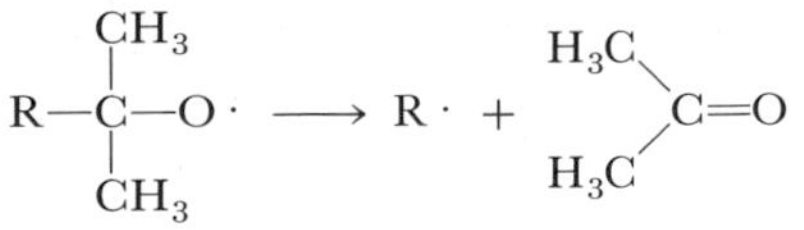

$$R-\underset{\underset{CH_3}{|}}{\overset{\overset{CH_3}{|}}{C}}-O\cdot \longrightarrow R\cdot + \underset{H_3C}{\overset{H_3C}{>}}C{=}O$$

R·	k_{rel}	Reference
·CH$_3$	(1)	b
·CH$_2$Cl	6	b
·C$_2$H$_5$	100	b
·C(CH$_3$)$_2$H	3,600	b
ϕCH$_2$·	11,900	c
·C(CH$_3$)$_2$CH$_3$	>14,000	b
ϕ·	23	b

^{a} Assuming all alkoxy radicals abstract hydrogen from cyclohexane at the same rate, and that there is no interference by chlorine atom chains in the hypochlorite decompositions. See reference c.
^{b} Calculated from data of C. Walling and A. Padwa, *J. Am. Chem. Soc.,* **85**, 1593 (1963).
^{c} Calculated from data of C. Walling and J. A. McGuinness, *J. Am. Chem. Soc.,* **91**, 2053 (1969).

expected for an unstrained tertiary radical, a result that might be interpreted as reflecting difficulty in establishing a radical center at a position that cannot attain planarity. The difficulty of eliminating the norbornyl fragment might, on the other hand, be a consequence of polar contributions to the transition state (**43b**), which would contribute less stabilization if the alkyl fragment is restricted to a nonplanar geometry.[174]

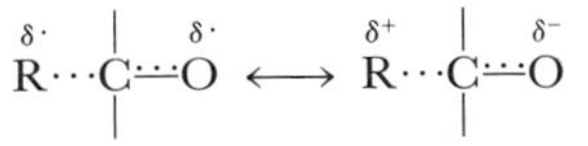

$$\overset{\delta\cdot}{R}\cdots\overset{|}{\underset{|}{C}}\overset{\delta\cdot}{\cdots O} \longleftrightarrow \overset{\delta^+}{R}\cdots\overset{|}{\underset{|}{C}}\overset{\delta^-}{\cdots O}$$

43a   43b

Another important fragmentation occurs in the chain decomposition of aldehydes, the propagation steps of which are shown in Equations 9.101 and 9.102. Applequist has established a relation between bond dissociation energy $D(R{-}H)$ and the ratio k_f/k_a, where k_f is the rate of fragmentation (Equation 9.102) and k_a is

[174] J. K. Kochi, in *Free Radicals,* Vol. II, J. K. Kochi, Ed., Wiley, New York, 1973, p. 685.

$$R\cdot + R\overset{O}{\underset{H}{-C}} \longrightarrow RH + R\overset{\cdot}{-}C{=}O \qquad (9.101)$$

$$R\overset{\cdot}{-}C{=}O \xrightarrow{k_f} R\cdot + CO \qquad (9.102)$$

$$R\overset{\cdot}{-}C{=}O + CCl_4 \xrightarrow{k_a} R\overset{O}{\underset{Cl}{-C}} + \cdot CCl_3 \qquad (9.103)$$

the rate of abstraction of chlorine by the $R\overset{\cdot}{-}C{=}O$ radical from CCl_4 (Equation 9.103).[175]

Aromatic Substitution[176]

Radicals react with aromatic rings by substitution in a manner superficially resembling electrophilic or nucleophilic substitution. The reaction proceeds in steps, as do the heterolytic processes, with initial addition yielding a resonance-stabilized conjugated radical (Equation 9.104).[177] When the radical adding is phenyl, the

$$R\cdot + \longrightarrow \qquad (9.104)$$
$$\underset{\textstyle \mathbf{44}}{H \quad R}$$

addition step is for practical purposes not reversible; subsequent steps are much faster than loss of the phenyl radical. Evidence pointing to this conclusion is found in the lack of an isotope effect with deuterated benzene.[178] Were reversal of Equation 9.104 to compete with the subsequent removal of a hydrogen atom from **44**, the rate of the hydrogen removal step would enter the overall rate expression; the reaction would then show an isotope effect. (See Section 7.4.) In some instances, for example, when the benzoyl radical attacks benzene, the initial addition is apparently reversible,[179] and an isotope effect is found.[180]

These features all find counterparts in the heterolytic aromatic substitutions; the rest of the reaction sequence in the radical additions becomes considera-

[175](a) D. E. Applequist and L. Kaplan, *J. Am. Chem. Soc.*, **87**, 2194 (1965); (b) D. E. Applequist and J. H. Klug, *J. Org. Chem.*, **43**, 1729 (1978).

[176]For reviews see (a) M. J. Perkins, in *Free Radicals*, Vol. II, J. K. Kochi, Ed., Wiley, New York, 1973, p. 231; (b) D. H. Hey, in *Adv. Free Radical Chem.*, **2**, 47 (1967).

[177]The rate constant for addition of phenyl radical to benzene is about $10^5 M^{-1}$ sec^{-1}. E. G. Janzen and C. A. Evans, *J. Am. Chem. Soc.*, **97**, 205 (1975).

[178](a) R. J. Convery and C. C. Price, *J. Am. Chem. Soc.*, **80**, 4101 (1958); (b) E. L. Eliel, S. Meyerson, Z. Welvart, and S. H. Wilen, *J. Am. Chem. Soc.*, **82**, 2936 (1960). These authors point out that there may be isotopic selectivity in product formation because of competition at later stages among C—H bond breaking and other kinds of steps. (c) C. Shih, D. H. Hey, and G. H. Williams, *J. Chem. Soc.*, 1871 (1959).

[179]T. Nakata, K. Tokumaru, and O. Simamura, *Tetrahedron Lett.*, 3303 (1967).

[180]J. Saltiel and H. C. Curtis, *J. Am. Chem. Soc.*, **93**, 2056 (1971).

bly more complex. The first point is that the removal of the hydrogen atom from the carbon that has been attacked in **44** is not unimolecular but requires interaction with some other radical (Equation 9.105) or oxidizing agent (Equation 9.106).

$$\text{44} \quad + \text{Y} \cdot \longrightarrow \quad + \text{HY} \tag{9.105}$$

$$\text{44} \quad + \text{O}_2 \longrightarrow \quad + \text{HOO} \cdot \tag{9.106}$$

If the abstraction is by another radical (Equation 9.105), the process is a disproportionation that terminates the radical chain. Hence the aromatic substitutions are not ordinarily chain reactions, although side reactions that are chain processes may well accompany them.

The necessity for another radical to complete the substitution sequence opens the way for complications. The cyclohexadienyl radical **44** may well react in some other way. It may dimerize, undergo self-disproportionation, or couple with some other radical. Furthermore, since the cyclohexadienyl radical has three positions at which it can react, isomeric products are possible in each case.

Isomer distributions in aromatic substitution Study of isomer distribution in substitution of benzene rings already carrying one substituent presents some potential pitfalls. Inspection of product ratios for ortho, meta, and para substitution, as in the investigation of electrophilic substitution (Section 7.4), might be expected to give misleading results because of the side reactions that occur in radical substitution. The isomeric substituted cyclohexadienyl radicals first formed by radical attack partition between the simple substitution route and other pathways (Scheme 18). In order for the isomer ratios of final simple substitution products to reflect the preference for initial attack at the various positions, the partition between hydrogen abstraction and the other paths must be substantially the same for all three isomers. Careful studies of the effect on product ratios of added oxygen, which diverts more of the intermediate cyclohexadienyl radicals to simple substitution product through Reaction 9.106, suggests that the problem may not be a serious one, but some doubt remains about the reliability of product composition data as a measure of relative reactivity of the different positions.[181]

[181](a) R. T. Morrison, J. Cazes, N. Samkoff, and C. A. Howe, *J. Am. Chem. Soc.*, **84**, 4152 (1962); (b) D. H. Hey, K. S. Y. Liang, and M. J. Perkins, *Tetrahedron Lett.*, 1477 (1967); (c) D. H. Hey, M. J. Perkins, and G. H. Williams, *Chem. Ind.*, 83 (1963); (d) W. A. Pryor, W. H. Davis, Jr., and J. H. Gleaton, *J. Org. Chem.*, **40**, 2099 (1975).

Scheme 18

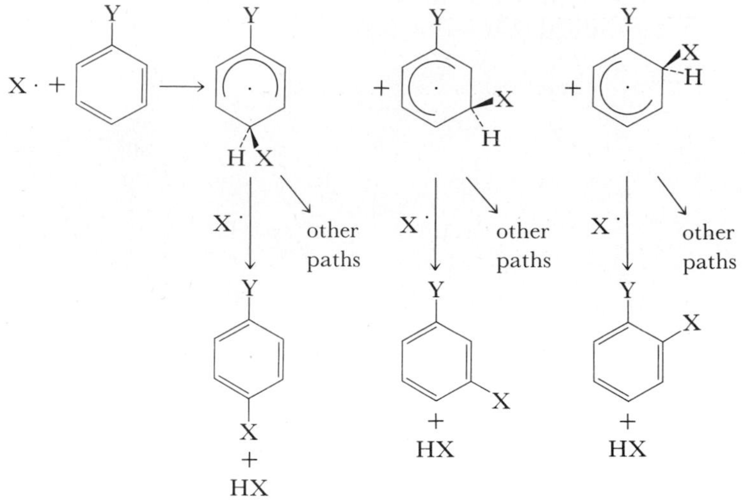

Table 9.13 compares partial rate factors for substitution by phenyl radical with those for electrophilic bromination. Selectivity is clearly much lower for the radical substitution; furthermore, for attacking phenyl radical, nearly all positions in the substituted benzenes are more reactive than in benzene itself, a finding that reflects the tendency of most substituents to stabilize a radical and thus to lower transition state energy for formation of the cyclohexadienyl intermediate when compared with hydrogen. The strong polar effects, which cause the familiar pattern of activation and deactivation in the electrophilic substitutions, are absent. One factor that presumably contributes to the low selectivity in radical attack is an early transition state in the addition step, which is exothermic by roughly 20 kcal mole^{-1}.[182]

Not all radical aromatic substitutions are as immune to polar effects as is attack by phenyl. Some radicals reveal marked electrophilic or nucleophilic character. Oxygen-centered radicals, for example, are electrophilic, as would be expected if there is substantial polar contribution to the transition state. Table 9.14 lists partial rate factors for substitution by benzoyl radicals; note that the orientation and activation trends found in typical electrophilic substitutions have begun to appear but are still modest when compared with the dramatic effects shown in Table 9.13 for a true heterolytic substitution.[183]

---

[182]Estimated from data of Benson, Tables 2.9, 9.3 and 9.5.

[183]Aromatic substitution patterns have been correlated fairly successfully with activation energies calculated by molecular orbital methods. See (a) M. J. S. Dewar, *The Molecular Orbital Theory of Organic Chemistry,* McGraw-Hill, New York, 1969, p. 299; (b) L. Salem, *Molecular Orbital Theory of Conjugated Systems,* W. A. Benjamin, Menlo Park, Calif., 1966, Chap. 6.

Table 9.13 PARTIAL RATE FACTORS[a] FOR RADICAL
AND ELECTROPHILIC SUBSTITUTION

Substrate		Radical		Electrophilic Bromination[d]	
		$\phi \cdot$[b]	$CH_3 \cdot$[c]		
ϕ—CH₃	f_{ortho}	3.30	1.6	600	
	f_{meta}	1.09	0.96	5.5	e
	f_{para}	1.27	0.90	2420	
ϕ—NO₂	f_{ortho}	9.38	9.7	—	
	f_{meta}	1.16	0.97	4.8×10^{-5}	f
	f_{para}	9.05	6.1	—	
ϕ—OCH₃	f_{ortho}	3.56	—	8.7×10^{7}	
	f_{meta}	0.93	—	2.0	e
	f_{para}	1.29	—	1.1×10^{10}	
ϕ—Cl	f_{ortho}	3.09	3.1	—	
	f_{meta}	1.01	0.90	0.00056	g
	f_{para}	1.48	0.90	0.145	

[a] Rate of substitution at position indicated relative to rate at a single position of benzene. See Section 7.4.
[b] R. Itô, T. Migita, N. Morikawa, and O. Simamura, *Tetrahedron*, **21**, 955 (1965).
[c] W. A. Pryor, W. H. Davis, Jr., and J. H. Gleaton, *J. Org. Chem.*, **40**, 2099 (1975).
[d] Data from compilation of L. M. Stock and H. C. Brown, *Adv. Phys. Org. Chem.*, **1**, 35 (1963).
[e] Br₂ in acetic acid–water, 25°C.
[f] HOBr in perchloric acid–water, 25°C.
[g] Br₂ in acetic acid–nitromethane, 30°C.

Table 9.14 PARTIAL RATE FACTORS
FOR SUBSTITUTION BY
BENZOYL RADICALS[a]

Substrate		Attacking reagent $\overset{\overset{O}{\|}}{\phi—C—O \cdot}$
ϕ—CH₃	f_{ortho}	4.1
	f_{meta}	1.4
	f_{para}	3.7
ϕ—OCH₃	f_{ortho}	20.7
	f_{meta}	<0.31
	f_{para}	20.4
ϕ—Cl	f_{ortho}	0.80
	f_{meta}	0.24
	f_{para}	0.98

[a] Calculated from data of M. E. Kurz and M. Pellegrini, *J. Org. Chem.*, **35**, 990 (1970).

9.6 REARRANGEMENTS OF RADICALS

Although rearrangements are less prevalent in radical chemistry than in the chemistry of cations, they do occur under various circumstances.[184]

1,2-Shifts

In radicals alkyl groups and hydrogen do not undergo the 1,2-shifts so common in carbocations. Orbital theory provides a rationalization for the difference in behavior. The three-center transition state for rearrangement has only one bonding level, which can accommodate the two electrons of a rearranging cation. In a rearranging radical, one electron must go into an antibonding level, and the transition state is destabilized.[185] Another point is the relatively small energy differences among primary, secondary, and tertiary radicals; Benson's group additivity method allows an estimate that rearrangement converting primary to secondary and primary to tertiary radicals (Equations 9.107 and 9.108) will be exothermic by only 1.4 and 4.3 kcal mole^{-1}, respectively. The driving forces are thus small; furthermore, in radical systems there are always rapid competing processes that cannot be suppressed.

$$H_3C-\underset{\underset{H}{|}}{\overset{\overset{CH_3}{|}}{C}}-\overset{H}{\underset{H}{C}}\cdot \longrightarrow \underset{\underset{H}{|}}{\overset{\overset{H_3C}{}}{\cdot}}C-\underset{\underset{H}{|}}{\overset{\overset{CH_3}{|}}{C}}-H \qquad \Delta H_{calc} = -1.4 \text{ kcal mole}^{-1} \qquad (9.107)$$

$$H_3C-\underset{\underset{CH_3}{|}}{\overset{\overset{CH_3}{|}}{C}}-\overset{H}{\underset{H}{C}}\cdot \longrightarrow \underset{\underset{H_3C}{}}{\overset{\overset{H_3C}{}}{\cdot}}C-\underset{\underset{H}{|}}{\overset{\overset{CH_3}{|}}{C}}-H \qquad \Delta H_{calc} = -4.3 \text{ kcal mole}^{-1} \qquad (9.108)$$

Rearrangement of phenyl Rearrangement of a phenyl group and of halogen to an adjacent center is more favorable than that of alkyl and hydrogen. 1,2-Shifts of phenyl occur most readily when the rearrangement will yield a more highly stabilized radical, for example, in Equations 9.109 and 9.110. The rear-

$$\phi-\underset{\underset{\phi}{|}}{\overset{\overset{\phi}{|}}{C}}-\overset{H}{\underset{H}{C}}\cdot \longrightarrow \underset{\underset{\phi}{}}{\overset{\overset{\phi}{}}{\cdot}}C-\underset{\underset{H}{|}}{\overset{\overset{\phi}{|}}{C}}-H \qquad (9.109)^{186}$$

45

[184](a) A comprehensive review is J. W. Wilt, in *Free Radicals,* Vol. I, J. K. Kochi, Ed., Wiley, New York, 1973, p. 333; see also (b) R. Kh. Friedlina, in *Adv. Free Radical Chem.,* **1,** 211 (1965). (c) C. Walling, in *Molecular Rearrangements,* P. de Mayo, Ed., Wiley, New York, 1963, Part I, p. 407.
[185](a) H. E. Zimmerman and A. Zweig, *J. Am. Chem. Soc.,* **83,** 1196 (1961); (b) N. F. Phelan, H. H. Jaffé, and M. Orchin, *J. Chem. Educ.,* **44,** 626 (1967); (c) J. W. Wilt, in *Free Radicals,* Vol. I, J. K. Kochi, Ed., Wiley, New York, 1973, p. 335.

$$\text{H}_3\text{C}-\underset{\underset{\text{CH}_3}{|}}{\overset{\overset{\phi}{|}}{\text{C}}}-\overset{\text{H}}{\underset{\text{H}}{\text{C}\cdot}} \longrightarrow \underset{\text{H}_3\text{C}}{\overset{\text{H}_3\text{C}}{}}\text{C}\cdot-\underset{\underset{\text{H}}{|}}{\overset{\overset{\phi}{|}}{\text{C}}}-\text{H} \qquad (9.110)^{187}$$

$$\mathbf{46}$$

rangement does nevertheless occur even when there are no substituents, as demonstrated by label scrambling in the β-phenylethyl radical (Equation 9.111).[188] The

$$\underset{\phi}{\overset{\phi}{\diagdown}}\text{CH}_2-{}^*\dot{\text{C}}\text{H}_2 \rightleftharpoons \dot{\text{C}}\text{H}_2-{}^*\text{CH}_2\overset{\diagup}{\diagup}\phi \qquad (9.111)$$

tendency for **47** to rearrange is less than for **45** or **46**; 2 to 5 percent of radicals **47** rearrange before reacting by hydrogen abstraction (from aldehyde $\phi\text{CH}_2\text{CH}_2\text{CHO}$), whereas **45** rearranges completely and **46** does so to the extent of about **50** percent under similar conditions.[189]

Because raising the concentrations of substances from which the radicals can abstract hydrogen decreases the amount of rearrangement,[190] it is clear that rearrangement follows the formation of the radical rather than being concerted with it; the half-migrated structure **48**, if an intermediate at all, must be a very

$$\text{R}\text{---}\underset{\text{R}}{\text{C}}\text{---}\underset{\text{R}}{\text{C}}\text{---}\text{R}$$

$$\mathbf{48}$$

short-lived one. It has not proved possible to detect it by electron spin resonance spectroscopy at low temperature.[191] Vinyl groups also migrate, probably in a manner similar to phenyl.[192]

Rearrangement of halogen Rearrangement of Cl, Br, or I to an adjacent radical site has been proposed to account for a number of results. Chlorine migra-

---

[186](a) D. Y. Curtin and M. J. Hurwitz, *J. Am. Chem. Soc.*, **74**, 5381 (1952); (b) D. Y. Curtin and J. C. Kauer, *J. Org. Chem.*, **25**, 880 (1960); (c) J. W. Wilt, in *Free Radicals*, Vol. I, J. K. Kochi, Ed., Wiley, New York, 1973, p. 351, estimates the activation energy of Equation 9.109 to be about 9 kcal mole^{-1}.

[187]See note 186(a), (b); (b) E. J. Hamilton, Jr. and H. Fischer, *Helv. Chim. Acta*, **56**, 795 (1973), report $E_a = 10.3 \pm 2$ kcal mole^{-1} for Equation 9.110.

[188]L. H. Slaugh, *J. Am. Chem. Soc.*, **81**, 2262 (1959).

[189]See note 186(a).

[190]See, for example, M. L. Poutsma and P. A. Ibarbia, *Tetrahedron Lett.*, 3309 (1972). Other studies are cited by J. W. Wilt, in *Free Radicals*, Vol. I, J. K. Kochi, Ed., Wiley, New York, 1973, pp. 346–347.

[191]J. K. Kochi and P. J. Krusic, *J. Am. Chem. Soc.*, **91**, 3940 (1969).

[192]See, for example, L. K. Montgomery, J. W. Matt, and J. R. Webster, *J. Am. Chem. Soc.*, **89**, 923 (1967).

tion occurs in reaction sequences such as that shown in Scheme 19.[193] Generation of a radical center adjacent to a carbon bearing a bromine has been postulated to be accompanied by bridging of the bromine to form the radical analog of the bromonium ion (Equation 9.112).[194-197]

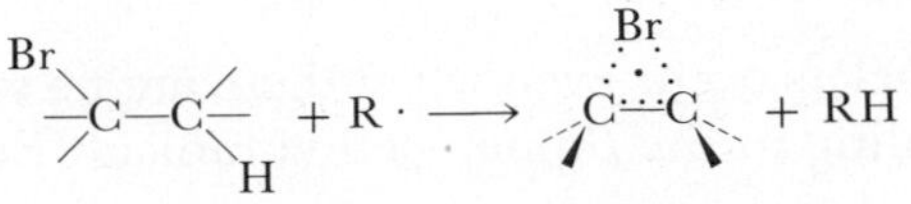

$$\underset{H}{\overset{Br}{\underset{|}{>}}}C-C<} + R\cdot \longrightarrow \underset{49}{\overset{Br}{>}}C\cdots C< + RH \qquad (9.112)$$

Electron spin resonance spectroscopic studies, on the other hand, gave no evidence for a symmetrical halogen-bridged radical of type **49**.[198] Primary radicals with chlorine at the β position do prefer the conformation **50**. The results indicate distortion of the chlorine toward the radical center (arrow in **50**); such a distortion would indicate some interaction.[199]

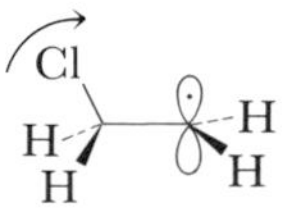

$$\overset{Cl}{\underset{\overset{|}{H}\diagup}{H-}}\overset{}{C}\underset{}{\overset{}{-}}\overset{}{C}\overset{-H}{\underset{H}{\diagdown}}$$

50

SCHEME 19

$$\underset{H}{\overset{Cl_3C}{>}}C=CH_2 + Br\cdot \longrightarrow \underset{H}{\overset{Cl_3C}{>}}C-CH_2Br$$

$$\underset{H}{\overset{Cl_3C}{>}}\overset{\cdot}{C}-CH_2Br \longrightarrow \underset{H\diagup\underset{Cl}{}}{\overset{Cl_2\overset{\cdot}{C}}{>}}C-CH_2Br$$

$$\underset{H\diagup\underset{Cl}{}}{\overset{Cl_2\overset{\cdot}{C}}{>}}C-CH_2Br + HBr \longrightarrow \underset{H\diagup\underset{Cl}{}}{\overset{Cl_2CH}{>}}C-CH_2Br + Br\cdot$$

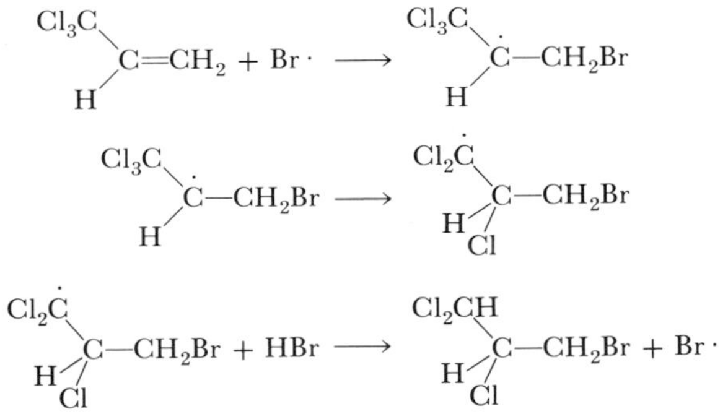

[193] J. W. Wilt, in *Free Radicals,* Vol. I, J. K. Kochi, Ed., Wiley, New York, 1973, p. 362.

[194] (a) P. S. Skell and K. J. Shea, in *Free Radicals,* Vol. II, J. K. Kochi, Ed., Wiley, New York, 1973, p. 809; (b) L. Kaplan, *Bridged Free Radicals,* Dekker, New York, 1972.

[195] W. Thaler, *J. Am. Chem. Soc.,* **85,** 2607 (1963).

[196] P. S. Skell and P. D. Readio, *J. Am. Chem. Soc.,* **86,** 3334 (1964).

[197] D. D. Tanner, H. Yabuuchi, and E. V. Blackburn, *J. Am. Chem. Soc.,* **93,** 4802 (1971); D. D. Tanner, M. W. Mosher, N. C. Das, and E. V. Blackburn, *J. Am. Chem. Soc.,* **93,** 5846 (1971), have claimed that product ratios supporting the bridging hypothesis are not kinetically controlled but result from subsequent reaction of products with HBr. Others have found this work not reproducible [P. S. Skell and K. S. Shea, *J. Am. Chem. Soc.,* **94,** 6550 (1972); J. G. Traynham, E. E. Green, Y. Lee, F. Schweinsberg, and C. Low, *J. Am. Chem. Soc.,* **94,** 6552 (1972)], and have provided further evidence in favor of bridging [P. S. Skell, R. R. Pavlis, D. C. Lewis, and K. J. Shea, *J. Am. Chem. Soc.,* **95,** 6735 (1973)].

[198] (a) J. K. Kochi, *Adv. Free Radical Chem.,* **5,** 189 (1975), p. 274; (b) R. V. Lloyd and D. E. Wood, *J. Am. Chem. Soc.,* **97,** 5986 (1975).

[199] T. Kawamura, D. J. Edge, and J. K. Kochi, *J. Am. Chem. Soc.,* **94,** 1752 (1972).

Other Rearrangements

The various radical reactions we have discussed in the earlier sections can occur intramolecularly. When they do, products have rearranged structures.[200] Some of these processes have been exploited for synthetic purposes, but there remains great potential for development of this field.

Longer-range rearrangements of aryl groups   When a radical is formed in a chain four or five carbon atoms removed from an aryl group, rearrangements that amount to intramolecular aromatic substitutions can occur. The initial radical attack will yield a cyclohexadienyl radical intermediate with an attached five- or six-membered ring, as shown in Equation 9.113. If attack was at the point of

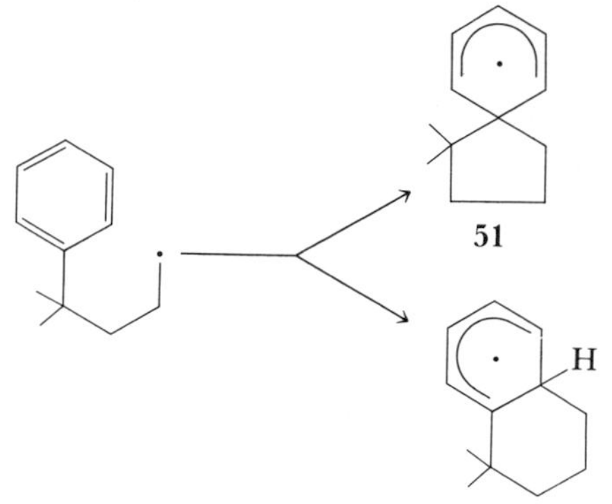

51

52

(9.113)

attachment of the side chain (**51**), the spirocyclic system will open again (Equation 9.114), and a 1,4-migration of phenyl will have occurred. If the attack was at an

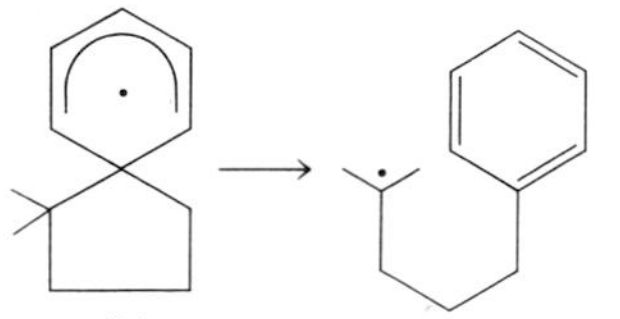

51

(9.114)

ortho position (**52**), the aromatic substitution will proceed in the ordinary manner to yield a fused ring product (Equation 9.115).[201]

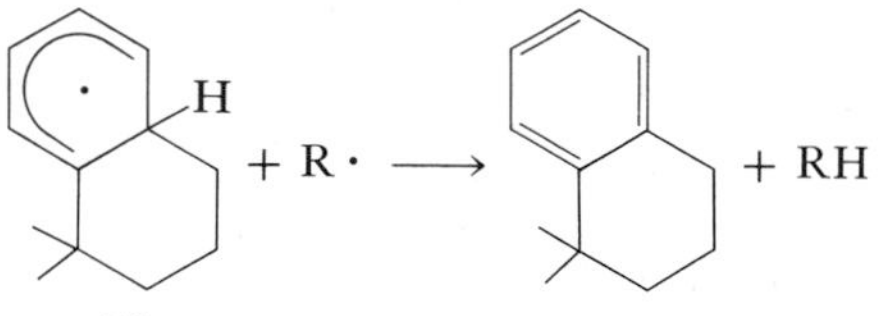

52

(9.115)

---

[200] These processes are extensively reviewed by J. W. Wilt, in *Free Radicals*, Vol. I, J. K. Kochi, Ed., Wiley, New York, 1973, p. 333. Rate constants for a number of rearrangements, and the use of these reactions in determining rates of other radical processes are discussed by D. Griller and K. U. Ingold, *Acc. Chem. Res.*, **13**, 317 (1980).
[201] For an example, see S. Winstein, R. Heck, S. Lapporte, and R. Baird, *Experientia*, **12**, 138 (1956).

1,5-Hydrogen migration An intramolecular S_H2 reaction can occur readily through a six-membered cyclic transition state (**53**). Migration of a hydrogen over

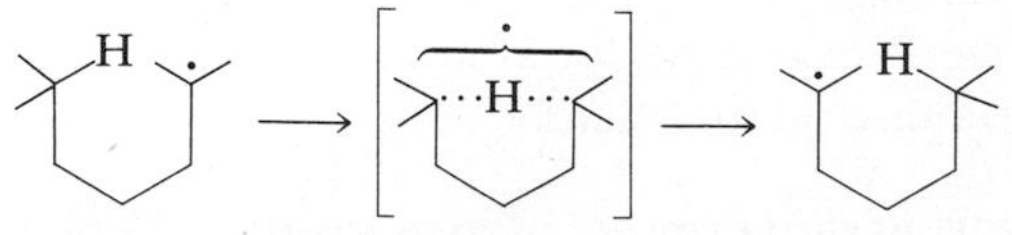

$$(9.116)$$

53

longer distances than 1,5 becomes increasingly difficult as the chain lengthens because of the decreasing probability of the proper conformation being present in appreciable concentration. The migration over shorter distances is unfavorable because of the strain that will be introduced in attaining a colinear arrangement of the hydrogen and the two atoms between which it is moving. The 1,5-transfer is therefore by far the most common of the free-radical hydrogen shifts.[202]

Intramolecular hydrogen abstraction is an important synthetic reaction because it provides a method of introducing a functional group at a saturated carbon bonded to only hydrogen and other carbons, a difficult task to accomplish by other means.[203] The abstracting radical need not be carbon; the Barton reaction (Scheme 20) is a standard method in which the abstracting radical is oxygen.[204] This technique has been used extensively for selective introduction of functionality in steroid synthesis.[203]

SCHEME 20

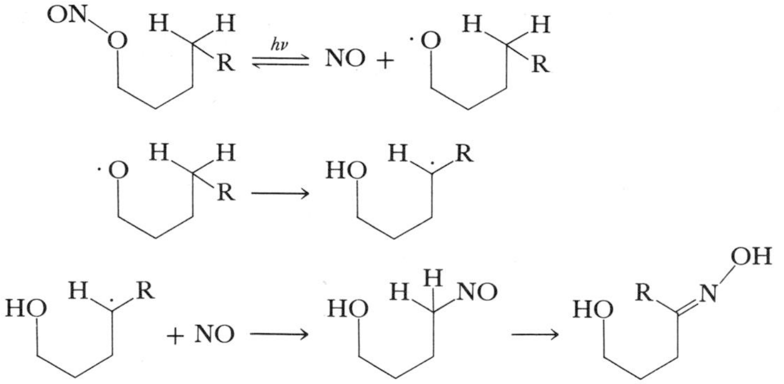

Transannular hydrogen migrations occur readily in medium rings. An example is shown in Scheme 21.[205]

---

[202] Both 1,3- and 1,6-hydrogen transfers are observed in telomerization, but they are much less common than 1,5-transfer. R. Kh. Freidlina and A. B. Terent'ev, *Acc. Chem. Res.*, **10**, 9 (1977).

[203] For a review of applications, see K. Heusler and J. Kalvoda, *Angew. Chem. Int. Ed. Engl.*, **3**, 525 (1964).

[204] (a) D. H. R. Barton, J. M. Beaton, L. E. Geller, and M. M. Pechet, *J. Am. Chem. Soc.*, **83**, 4076 (1961); (b) R. H. Hesse, *Adv. Free Radical Chem.*, **3**, 83 (1969).

[205] J. G. Traynham and T. M. Couvillon, *J. Am. Chem. Soc.*, **89**, 3205 (1967).

Scheme 21

$$R\cdot + CCl_4 \longrightarrow RCl + \cdot CCl_3$$

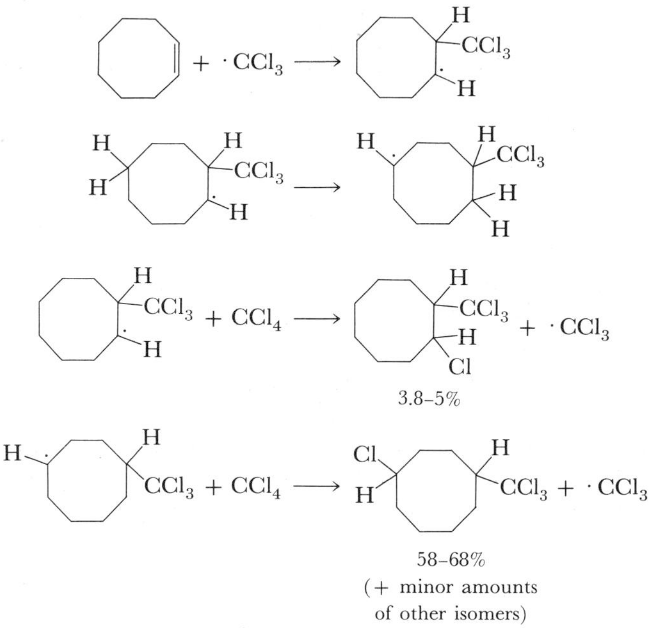
3.8–5%
58–68%
(+ minor amounts
of other isomers)

It is sometimes possible to combine a radical reaction with a heterolytic process to achieve a synthetically useful result, as in the Hofmann–Löffler reaction (Scheme 22).[206]

Intramolecular additions and eliminations Addition of a radical to a carbon–carbon double bond in the same molecule occurs easily if a five- or six-membered ring can form. The most common location of the double bond is at the 5,6 position as shown in Equation 9.117, although cyclizations will also occur by

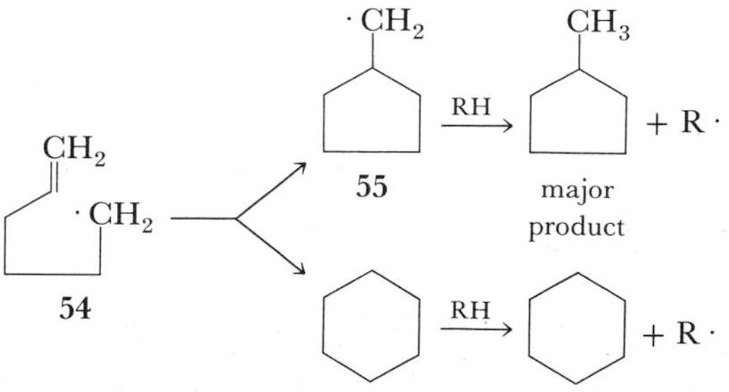
54
55
major
product

(9.117)

[206] For a review, see M. E. Wolff, *Chem. Rev.,* **63,** 55 (1963).

SCHEME 22

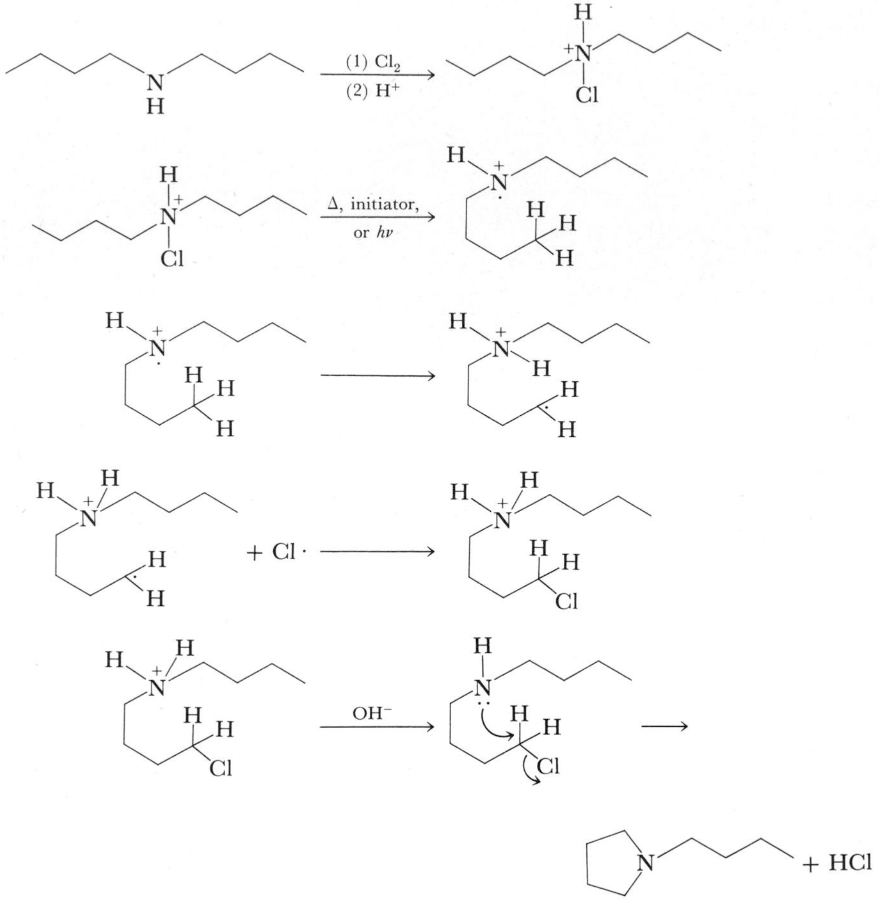

addition to double bonds one carbon closer or one carbon farther from the radical site. Formation of the five-membered ring is faster than formation of the six-membered ring, but the six-membered radical is thermodynamically more stable. In the presence of good hydrogen donor solvents, and if the original radical center is not stabilized, the addition does not reverse and the reaction is kinetically controlled; the more rapidly formed five-membered ring is trapped by hydrogen abstraction. On the other hand, if the original radical is well stabilized and good hydrogen donors are absent, the cyclization is reversible; then the reaction is thermodynamically controlled. (Scheme 23) The equilibrium will favor the six-membered ring radical, and the cyclohexane product will dominate.[207] The rate con-

[207] J. W. Wilt, in *Free Radicals,* Vol. I, J. K. Kochi, Ed., Wiley, New York, 1973, p. 432.

stant for cyclization of 5-hexenyl radical (**54** → **55**) has been determined to be $1 \times 10^5 \text{ sec}^{-1}$ at 25°C, with activation parameters of $\log A = 10.7 \pm 1$ and $E_a = 7.8 \pm 1$ kcal mole^{-1}.[208]

SCHEME 23

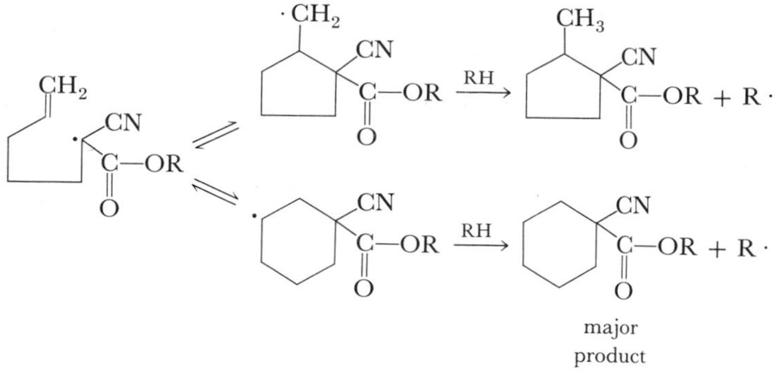

Some remarkable cyclizations have been accomplished with free radicals; Equation 9.118 gives an example.[209]

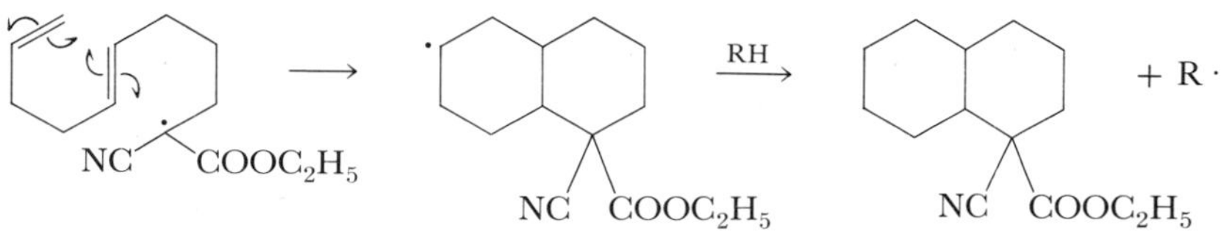

$$\text{(9.118)}$$

From our discussion of intermolecular additions, we would not expect these cyclizations of carbon radicals to occur in the elimination direction, and indeed cyclic radicals in five-membered or larger rings do not ordinarily open according to Equation 9.119. If the ring is small or highly strained, on the other hand, ring

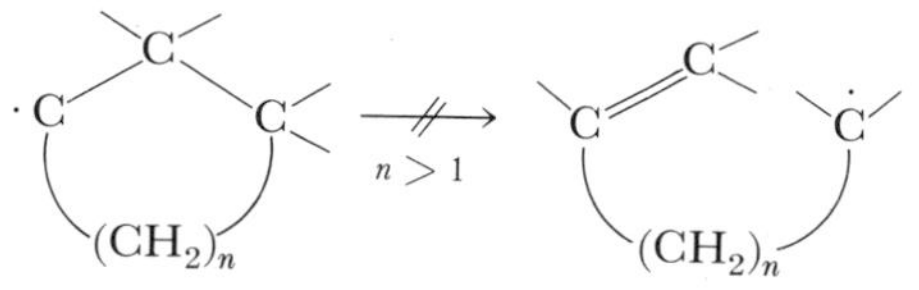

$$\text{(9.119)}$$

opening will occur. An example, Equation 9.120, has been mentioned earlier.[210]

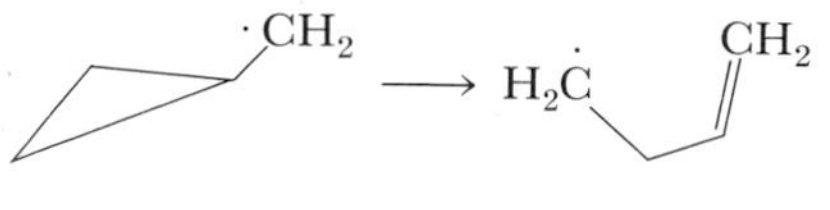

$$\text{(9.120)}$$

Ring opening with formation of a carbonyl group, β scission, is, as we have seen earlier, thermodynamically more favorable. Ring openings occur both when

[208] D. Lal, D. Griller, S. Husband, and K. U. Ingold, *J. Am. Chem. Soc.*, **96**, 6355 (1974).
[209] M. Julia, F. Le Goffic, and L. Katz, *Bull. Soc. Chim. Fr.* 1122 (1964).
[210] See note 170.

the oxygen is originally part of the ring, as in Equation 9.121, and when the oxygen is initially in the form of an alkoxy radical, as in Equation 9.122.

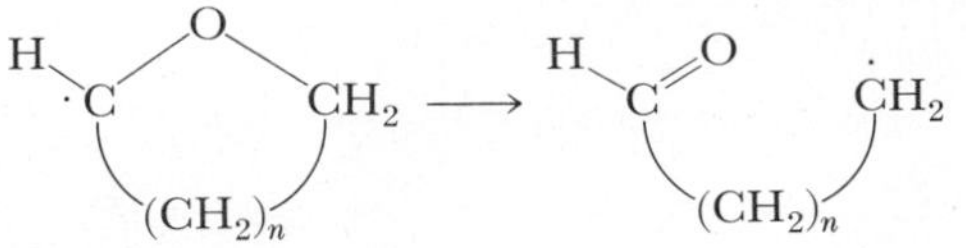

$$(9.121)^{211}$$

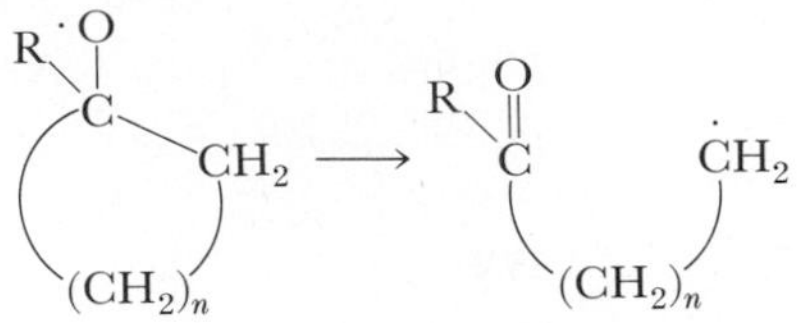

$$(9.122)^{212}$$

Rearrangement of ylides There are a number of rearrangements of the type shown in Equation 9.123, which is the Stevens rearrangement. These processes

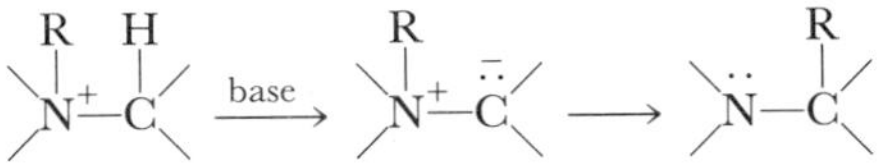

$$(9.123)$$

appear at first sight to be ionic, but some of them give rise to CIDNP spectra and therefore proceed at least partly by a pathway involving radicals. A possible pathway is dissociation to a caged radical pair (**56**) followed by recombination.[213]

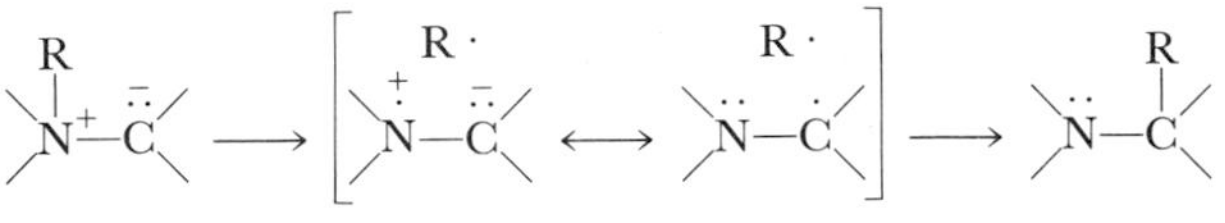

$$(9.124)$$

56

PROBLEMS

1. Propose a pathway to account for the formation of epoxides in autoxidation of alkenes.

2. Explain the following observation:

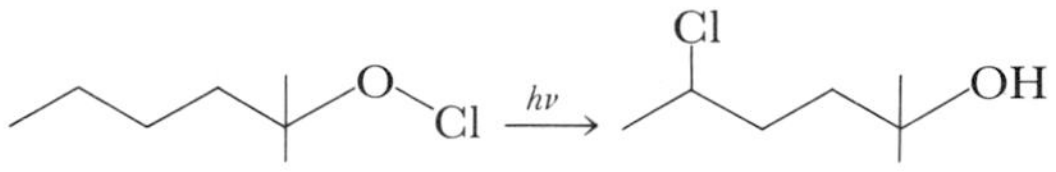

55% yield

(plus other products)

[211] T. J. Wallace and R. J. Gritter, *J. Org. Chem.* **26**, 5256 (1961).

[212] See, for example, (a) J. W. Wilt and J. W. Hill, *J. Org. Chem.*, **26**, 3523 (1961); (b) F. D. Greene, M. L. Savitz, F. D. Osterholtz, H. H. Lau, W. N. Smith, and P. M. Zanet, *J. Org. Chem.*, **28**, 55 (1963).

[213] A. R. Lepley, in *Chemically Induced Magnetic Polarization*, A. R. Lepley and G. L. Closs, Eds., Wiley, New York, 1973, p. 323.

3. Account for the differences in proportions of products in the following:

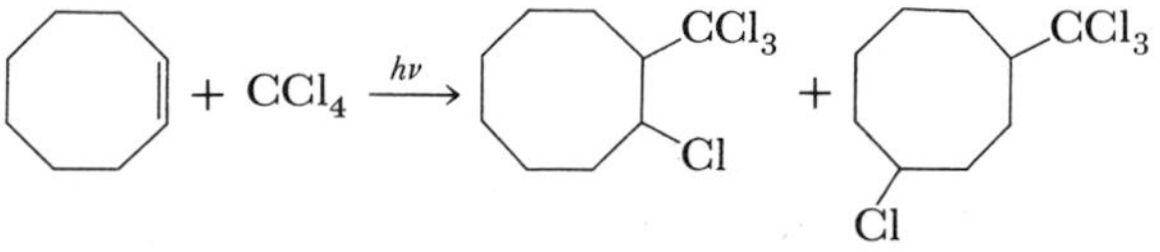

4–5%   56–58%

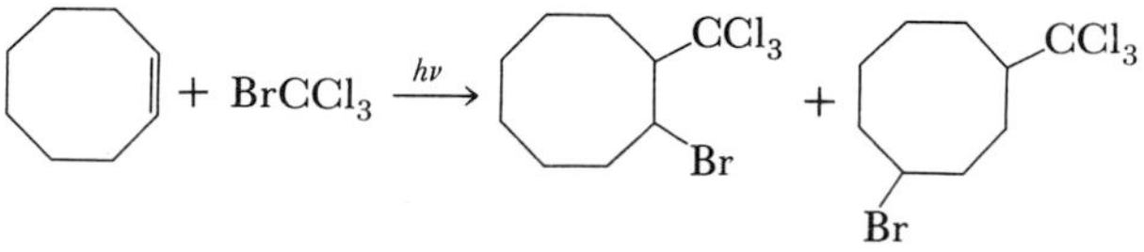

major product   3–10%

4. Propose a pathway to account for the formation of the minor product shown:

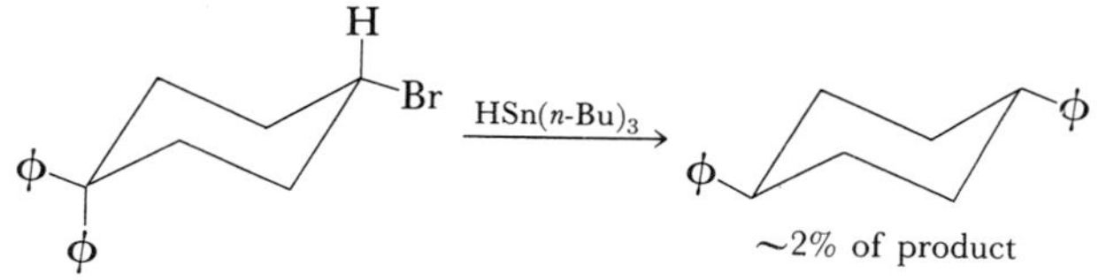

~2% of product

5. Assuming that kinetic chains are long and that radicals are in the stationary state, derive an expression for rate of disappearance of X_2 in halogenation, Scheme 14.

6. Explain the following transformation:

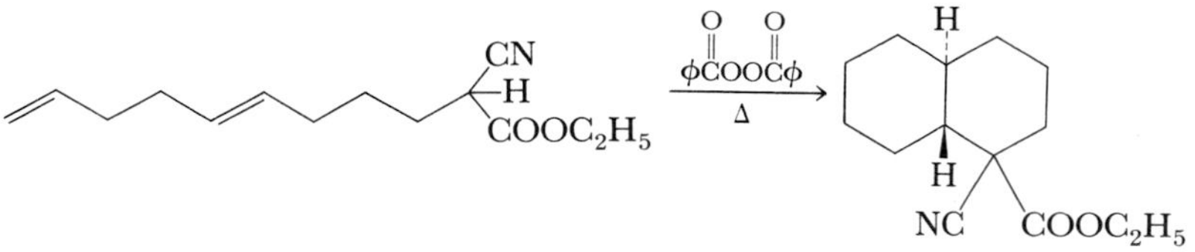

7. Predict the products obtained from thermal decomposition of

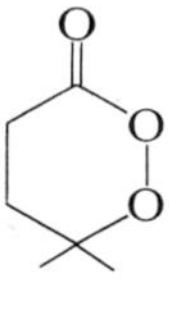

8. Propose a pathway for the following reaction:

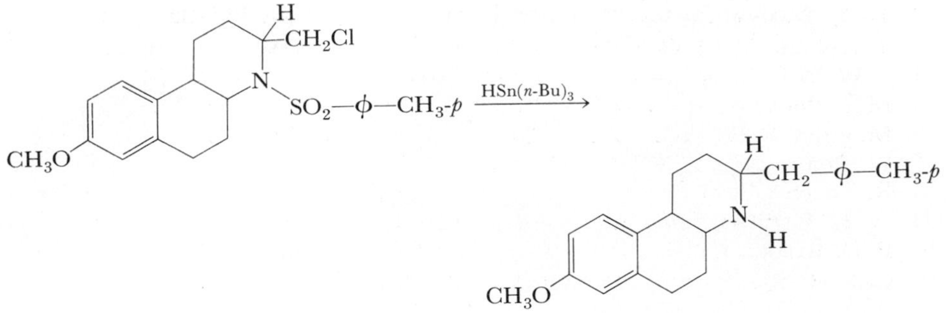

9. Explain the following transformation:

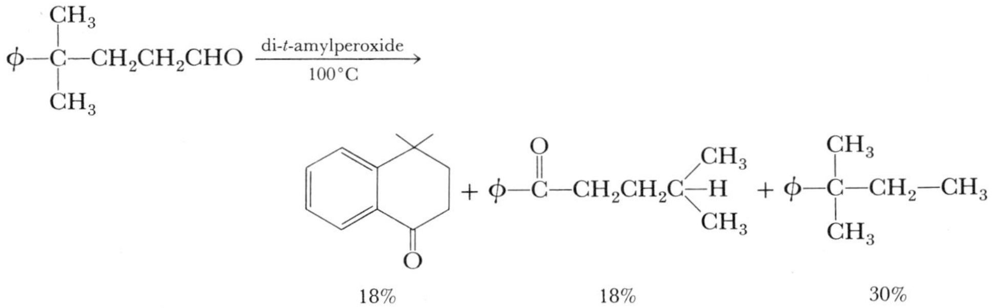

| | 18% | 18% | 30% |

10. Explain why perester **1** decomposes by concerted two-bond homolysis, but **2** decomposes in a stepwise process.

$$\phi—CH_2—\overset{O}{\underset{||}{C}}—O—O—R \xrightarrow{\Delta} \phi—\dot{C}H_2 + CO_2 + RO\cdot$$

$$\mathbf{1}$$

$$\phi—\overset{O}{\underset{||}{C}}—O—O—R \xrightarrow{\Delta} \phi—\overset{O}{\underset{||}{C}}—O\cdot + RO\cdot \longrightarrow \phi\cdot + CO_2 + RO\cdot$$

$$\mathbf{2}$$

11. Propose a mechanism for the Meerwein arylation:

$$ArN_2{}^+Cl^- + H_2C{=}CHZ \xrightarrow{CuCl} ArCH_2—CHZCl + ArCH{=}CZCl + HCl$$
$$Z = CN, COR, Cl, etc.$$

REFERENCES FOR PROBLEMS

1. M. L. Poutsma, in *Free Radicals,* Vol. II, J. K. Kochi, Ed., Wiley, New York, 1973, p. 134.
2. F. D. Greene, M. L. Savitz, H. H. Lau, F. D. Osterholtz, and W. N. Smith, *J. Am. Chem. Soc.,* **83,** 2196 (1961). See also E. J. Corey and W. R. Hertler, *J. Am. Chem. Soc.,* **82,** 1657 (1960).

3. J. G. Traynham and T. M. Couvillon, *J. Am. Chem. Soc.,* **89,** 3205 (1967); J. G. Traynham, T. M. Couvillon, and N. S. Bhacca, *J. Org. Chem.,* **32,** 529 (1967).

4. J. W. Wilt, R. A. Dabek, and K. C. Welzel, *J. Org. Chem.,* **37,** 425 (1972).

5. M. L. Poutsma, in *Free Radicals,* Vol. II, J. K. Kochi, Ed., Wiley, New York, 1973, p. 165.

6. M. Julia, F. Le Goffic, and L. Katz, *Bull. Soc. Chim. Fr.,* 1122 (1964).

7. W. Adam and L. Szendrey, *Chem. Commun.,* 1299 (1971).

8. R. Loven and W. N. Speckamp, *Tetrahedron Lett.,* 1567 (1972).

9. W. H. Urry, D. J. Trecker, and H. D. Hartzler, *J. Org. Chem.,* **29,** 1663 (1964).

10. P. D. Bartlett and R. R. Hiatt, *J. Am. Chem. Soc.,* **80,** 1398 (1958).

11. C. S. Rondestvedt, Jr., *Org. Reactions,* **24,** 225 (1976).

Appendix 1

CHEMICALLY INDUCED

DYNAMIC NUCLEAR

POLARIZATION (CIDNP)[a]

When two radicals are in close association as a pair surrounded by a cage of solvent molecules, the two unpaired electrons will interact with one another just as two electrons do within a molecule. The interaction will yield either a singlet state, if the two electrons have spins paired, or a triplet, if the spins are unpaired. If, for example, the caged pair arose by thermal dissociation of an ordinary ground-state molecule, in which all electrons would have been paired, the state would initially be a singlet, S, whereas if the pair arose in a photochemical reaction from dissociation of an excited molecule in a triplet state, it would be initially a triplet, T.

Let us assume that the system is in the magnetic field of the magnetic resonance spectrometer; then a singlet will have only one energy state, but a triplet will have three. During the small fraction of a second that the radicals are held near each other in the solvent cage, they will move about, approaching and receding from each other. As the distance varies so will the interaction between the electrons. The result is that the energies of the singlet and the three triplet electron spin states will vary, and at certain separations the singlet state and one of the triplet states will come close to each other in energy. Ordinarily the transition between a singlet and a triplet state has a low probability, but if the two states are close in energy, the likelihood of a transition is increased.

A simple model will illustrate how the transitions occur. Figure A1.1 shows a schematic diagram of an electron, with its spin represented by a vector pointing

[a] (a) H. R. Ward, in *Free Radicals*, Vol. I, J. K. Kochi, Ed., Wiley, New York, 1973, p. 241; (b) S. H. Glarum, in *Chemically Induced Magnetic Polarization*, A. R. Lepley and G. L. Closs, Eds., Wiley, New York, 1973, p. 7; (c) S. H. Pine, *J. Chem. Educ.*, **49**, 664 (1972).

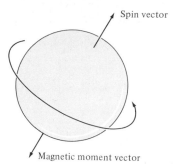

Spin vector

Magnetic moment vector

Figure A1.1 A schematic representation of an electron. The spin axis is designated by a vector. The magnetic moment coincides with the spin vector but is in the opposite direction because of the negative charge.

along the spin axis. The spinning electron also has a magnetic moment, which because of the negative charge points in the opposite direction from the spin vector. (A proton has spin and magnetic moment vectors in the same direction.) If the electron is in a magnetic field, it can be in one of two quantized energy states, represented by the two different orientations of the magnetic moment vector in Figure A1.2. Because the energy and therefore the direction of the magnetic moment vector is quantized, the magnetic moment does not line up with the field direction even though the field is exerting a twisting force that tries to make it do so. Any spinning object subjected to a force trying to twist the direction of its axis of rotation will respond by precessing at a characteristic rate. This precession is represented in Figure A1.2 by the circular lines, which show the path that the tip of the vector traces out during its precession. The precession frequency depends on the magnetic field strength; for an electron in magnetic fields typically found in magnetic resonance spectrometers, it is of the order of 10^{10} Hz. This precession frequency corresponds to the frequency of radiation that will cause transitions of the electron between the two energy states [Figures A1.2(a) and A1.2(b)].

Now suppose that there are two radicals close to each other. The two electron magnetic moments can arrange themselves in any one of the four ways shown in Figure A1.3. At the bottom of the figure is the singlet state. In this arrangement the magnetic moments of the two electrons are pointing in opposite directions. Their vector sum is zero, and there is no magnetic moment. At the top are the three triplet states. On the left, both magnetic moments up, there is a net component of magnetic moment in the direction of the applied magnetic field H_0; on the right, magnetic moments down, there is a component in the direction opposite H_0. In the center one moment is up and one is down, but the situation is different from the singlet because the relative orientations are such that the two vectors do not cancel. These three substates of the triplet state are called, respectively, T_{-1}, T_1, and T_0.

If the two radicals were identical, the precession frequency of the two electrons would be the same, and a particular radical pair would remain in whichever

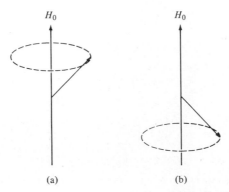

Figure A1.2 The two spin states of an electron in an applied magnetic field H_0. In (a) there is a component of the magnetic moment vector oriented in the same direction as the applied field, and in (b) there is a component oriented in the direction opposite the applied field. The dotted circles represent the precession of the vectors around the field direction.

of the four states it found itself initially.[b] But if the two radicals are different, the two electrons will have slightly different precession frequencies. The precession frequency of an electron is characterized by a quantity called the *g factor*. If one were to observe just one of the radicals of the pair in an electron spin resonance spectrometer, the value of its *g* factor would determine the position of the resonance line in the spectrum; *g* of an electron in a radical is thus analogous to chemical shift of a proton.

The consequence of a difference in *g* for our radical pair is that the two electrons will precess at very slightly different rates, and so over the course of many revolutions the relative phases of the two will change. One spin will gain on the other, and a pair that started out in state *S* will change to T_0, while a pair that started out in T_0 will change to *S*.

Suppose now that the radical pair, initially in state *S*, has Structure **1**, where one radical, let us say the one with larger *g*, has a proton H_A that is coupled to the electron. The electron–nuclear coupling constant *a* will be negative for this structure. The proton itself experiences the magnetic field, and can be in one of two proton magnetic spin states, α or β. The nuclear magnets provide small magnetic fields which add to or subtract from the applied field. The precession frequency of the electron depends on the total magnetic field strength it feels. The frequency of the electron in Radical **1a** will therefore be different depending on whether its

$$\left[\begin{array}{cc} \overset{\displaystyle R_1}{\underset{\displaystyle H_A}{R_2\!-\!\overset{|}{\underset{|}{C}}\!\cdot}} & \overset{\displaystyle R_3}{\underset{\displaystyle R_5}{\cdot\overset{|}{\underset{|}{C}}\!-\!R_4}} \end{array} \right]$$

$\qquad$ **1a** $\qquad$ **1b**

[b] There are other mechanisms for change of spin state in addition to the one discussed here. The radical pair would therefore not remain in the same spin state indefinitely, even if the radicals were identical.

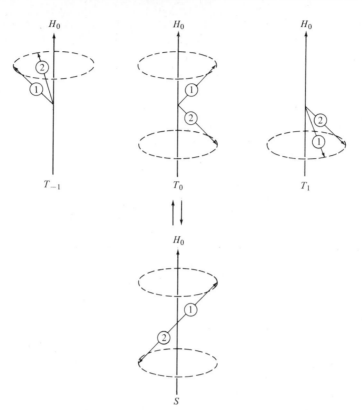

Figure A1.3 Schematic representation of the three triplet states (top) and the singlet state (bottom) for two electrons in an applied magnetic field H_0. The vectors represent magnetic moments. From H. R. Ward, in *Free Radicals,* Vol. I, J. K. Kochi, Ed., Wiley, New York, 1973, p. 242. Reprinted by permission of John Wiley & Sons, Inc.

proton happens to be in state α or state β, and so the rate at which the two electron spins get out of step with each other will depend on the proton spin state. In this example the lower-energy (parallel to H_0) nuclear spin state α, because of the negative coupling constant a, will subtract from the field felt by the electron and will therefore cause the electron spin to precess more slowly. The precession frequency of the electron in Radical **1a** will therefore be brought closer to that of the lower-g Radical **1b**, and the pair will go over to the T_0 state less rapidly than would have been the case had H_A not been present. Conversely, the pairs with H_A in state β will go over to T_0 state more rapidly. Since electron spins must be paired for bond formation to occur, only the singlet pairs can combine with each other to yield **2**. (Disproportionation products can also be formed.) Because those radicals in which the nuclear spin state was β are more likely to have gone over to triplets, which could not combine, cage product **2** will form with an excess population of nuclear spin state α. The radical pairs that were transformed into triplets cannot recombine and so are more likely than the singlet pairs to escape from the cage. Let us suppose that escaped radicals react with a chlorinated solvent to yield chloride **3**; this product will have an excess population of the nuclear spin state β.

Now let us look at the nuclear resonance spectra arising from H_A in products **2** and **3**. Figure A1.4 shows the relative energies of the two nuclear states α and β when H_A is in the magnetic field. Under ordinary circumstances the population in the lower state would be slightly greater than in the higher. This population difference is called *polarization*. As the sweeping magnetic field strength reaches the right value for a given transition, the radio frequency field of the spectrometer causes transfer between levels as indicated by the vertical arrow; since the polarization is normal, with slightly more molecules in the lower state than in the higher, there is a net transfer to the higher state and absorption of energy by the spin system. The resulting absorption spectrum is shown at the bottom of the figure.

Cage product **2** is formed with an excess of protons in spin state α. Figure A1.5 shows the result of this alteration in the normal spin polarization. The proba-

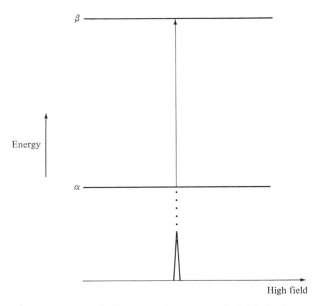

Figure A1.4 Nuclear spin states for a single proton in a magnetic field. At the top the two spin states, α and β, are shown in an energy-level diagram with the transition indicated by an arrow. Below is the single-line spectrum. With normal populations the number of molecules in the lower state is slightly greater than the number in the higher state; normal net absorption occurs.

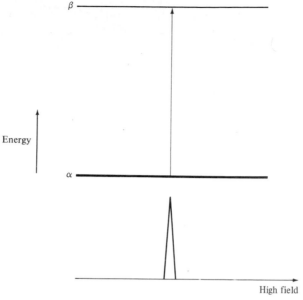

Figure A1.5 Nuclear spin states and spectrum for cage product **2**. The heavy line indicates an enhanced population in state α. The probability of the transition $\alpha \rightarrow \beta$ is increased, and the absorption line is more intense than in Figure A1.4. The spectrum shows enchanced absorption, A. From S. H. Pine, *J. Chem. Educ.*, **49**, 664 (1972). Reproduced by permission of the Division of Chemical Education, American Chemical Society.

bility of upward transition is increased since there is a larger excess of molecules in the lower state to absorb, and the spectrum shows an absorption line of enhanced intensity. This overall change of spectrum intensity is called a *net effect,* and in this instance is in the direction of *enhanced absorption, A.*

The product of escape from the cage (**3**) is formed with an excess population of nuclear spin state β. Figure A1.6 shows the upper state with enhanced population. With this inverted polarization the spin system will emit energy and a negative peak will be observed. There is a net effect in the direction of *emission, E.*

If the two radicals of the pair are identical, there are still mechanisms for polarization. Consider, for example, the radical pair **4**. Now there are four nuclear

$$\begin{bmatrix} R & H_X & H_A & & H_A & H_X & R \\ & | & | & & | & | & \\ & C-C\cdot & & \cdot C-C & & \\ & | & | & & | & | & \\ R & & R & & R & & R \end{bmatrix}$$

4

spin states, $\alpha_A\alpha_X$, $\alpha_A\beta_X$, $\beta_A\alpha_X$, $\beta_A\beta_X$. Figure A1.7 shows the relative energies of these four states, if the nuclear–nuclear coupling constant J_{AX} is positive, and the transitions that can occur between them. With normal populations the spectrum of this spin system would appear as shown at the bottom in Figure A1.7. Since the two

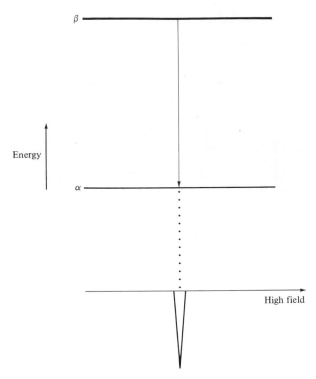

Figure A1.6 Nuclear spin states and spectrum for product **3**. The upper spin state, β, has abnormally high population. The probability of the downward transition, $\beta \rightarrow \alpha$, is greater than the probability of the upward transition, and emission occurs. The spectrum shows an inverted emission peak, *E*. From S. H. Pine, *J. Chem. Educ.*, **49**, 664 (1972). Reproduced by permission of the Division of Chemical Education, American Chemical Society.

radicals are identical, there is no *g* factor difference to cause different precession frequencies in the two, but the various combinations of the nuclear spin states will provide slightly different effective fields for the two electrons. The electron–nuclear coupling constants are about the same magnitude for the H_A protons and for the H_X protons, but negative for H_A and positive for H_X (see Table 9.1). For the proton states $\alpha_A\beta_X$ and $\beta_A\alpha_X$, the magnetic effects of the two nuclear spins as felt by the electron therefore reinforce each other, whereas for $\alpha_A\alpha_X$ and $\beta_A\beta_X$ they tend to cancel. Considering both of the radicals of the pair together, those pairs in which one or both of the radicals is in nuclear spin state $\alpha_A\beta_X$ or $\beta_A\alpha_X$ are the most likely to have the two electrons experience different effective fields. Hence the $\alpha_A\beta_X$ and $\beta_A\alpha_X$ states facilitate the conversion of the pair from singlet to trilet and thus promote escape from the cage. The recombination product (**5**) will therefore form with excess population of states $\alpha_A\alpha_X$ and $\beta_A\beta_X$. The spectrum will appear as shown in Figure A1.8. This result is called the *multiplet effect*. A multiplet in which the lower field members show emission and the higher field members show absorption is designated *E/A*.

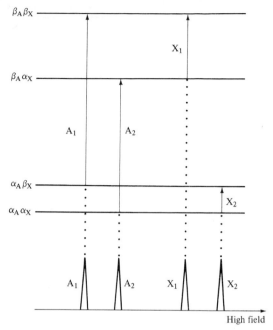

Figure A1.7 Nuclear spin states and normal absorption spectrum for the AX spin system, assuming a positive nuclear–nuclear coupling constant J_{AX}. At the top the four spin states are shown, with the four allowed transitions indicated by arrows. Below is the spectrum, with a line corresponding to each transition. With normal population the number of molecules in the lower state of each pair is slightly greater than the number in the higher; normal net absorption occurs for each peak in the spectrum. From S. H. Pine, *J. Chem. Educ.*, **49**, 664 (1972). Adapted by permission of the Division of Chemical Education, American Chemical Society.

The product of escape from the cage (**6**) shows the opposite effect (Figure A1.9). Here the spin states $\alpha_A\beta_X$ and $\beta_A\alpha_X$ have abnormally high populations, and the lower field members of the multiplets exhibit enhanced absorption while the higher field members show emission, A/E.

The conclusions reached in this example depend on the nuclear–nuclear coupling constant J_{AX} being positive. If J_{AX} is negative, the energy levels and transitions are as shown in Figure A1.10. Note that the normal spectrum looks the same as with positive J_{AX} (Figure A1.7), but the order in which the lines appear has changed. Now excess population in the outer levels, $\alpha_A\alpha_X$ and $\beta_A\beta_X$, will cause an A/E effect, and excess population in the inner levels, $\alpha_A\beta_X$ and $\beta_A\alpha_X$, will cause an E/A effect.

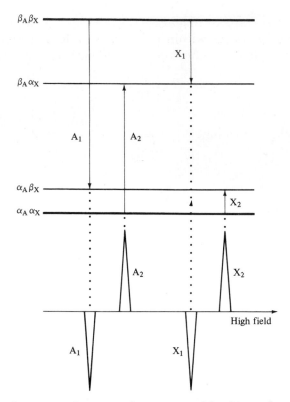

Figure A1.8 Nuclear spin states and spectrum for cage recombination product **5,** for which the inner states $\alpha_A\beta_X$ and $\beta_A\alpha_X$ have been depleted by preferential escape from the cage. At the top the four states are again shown in an energy-level diagram. Heavy lines are the states with enhanced populations. A downward-pointing arrow indicates a net transfer of molecules from an overpopulated higher spin state to a less populated lower one, and corresponds to net emission. The spectrum shows the multiplet effect of type E/A. From S. H. Pine, *J. Chem. Educ.*, **49,** 664 (1972). Adapted by permission of the Division of Chemical Education, American Chemical Society.

The net effect and the multiplet effect may occur together. Figure A1.11 shows how excess population in the lowest level of our AX system, and a smaller excess in the highest level, will yield a superposition of a net and a multiplet effect, $A + E/A$.

Kaptein has worked out simple formulae for deciding the type of spectrum that will be obtained in a given set of circumstances.[c] Two parameters are defined, Γ_n and Γ_m. Γ_n tells whether there will be a net enhanced absorption, A, or a net emission, E, and Γ_m tells whether the multiplet effect will be of the E/A type or of the A/E type. The calculation of Γ_n and Γ_m and their interpretation are given in Table A1.1.

[c]R. Kaptein, *J. Chem. Soc. D*, 732 (1971); (b) R. Kaptein, *Adv. Free Radical Chem.*, **5,** 319 (1975). The rules apply to reactions carried out in the strong magnetic field of the nuclear magnetic resonance spectrometer.

To apply the rule to our first example, product **2** is from a singlet precursor, so μ is negative; and it forms within the cage, so ϵ is positive. We assumed that radical **1a** had the higher g, so Δg is positive. Coupling constant a is negative for $\cdot$C—H. We find

$$\Gamma_n = \overset{\mu}{-} + \overset{\epsilon}{+} + \overset{\Delta g}{+} - \overset{a}{-} = + \tag{A1.1}$$

net absorption. For the product of escape from the cage **3**,

$$\Gamma_n = \overset{\mu}{-} - \overset{\epsilon}{-} + \overset{\Delta g}{+} - \overset{a}{-} = - \tag{A1.2}$$

net emission.

Table A1.1 QUALITATIVE RULES FOR PREDICTING CIDNP EFFECTS[a]

Net Effect:

$$\Gamma_n = \mu\epsilon\ \Delta g\ a_i$$
If Γ_n is $+$, net absorption, A
If Γ_n is $-$, net emission, E

Multiplet Effect:

$$\Gamma_m = \mu\epsilon\ a_i a_j J_{ij} \sigma_{ij}$$
If Γ_m is $+$, E/A
If Γ_m is $-$, A/E

Factors:

μ	$\begin{cases} + \text{ for a radical pair formed from a triplet precursor;} \\ - \text{ for a radical pair formed from a singlet precursor.} \end{cases}$
ϵ	$\begin{cases} + \text{ for products of recombination or disproportionation within the original cage;} \\ - \text{ for products from radicals that escape the cage.} \end{cases}$
σ_{ij}	$\begin{cases} + \text{ if nuclei } i \text{ and } j \text{ are originally in the same radical fragment;} \\ - \text{ if nuclei } i \text{ and } j \text{ are originally in different radical fragments.} \end{cases}$

Δg the sign of the difference in g value ($g_i - g$), where g_i is the g value of the radical containing the nucleus giving the portion of the spectrum under observation, g is the g value of the other radical. (g is larger for radicals of type $\cdot$C—X, X = oxygen, halogen, or carbonyl, than for radicals containing only C and H. See H. Fischer, in *Free Radicals,* Vol. II, J. K. Kochi, Ed., Wiley, New York, 1973, p. 453.)

a The sign of the coupling constant between the electron and the proton giving rise to the portion of the spectrum under observation. (a for $\cdot$C—H is negative; for $\cdot$C—C—H, positive. See Table 9.1.)

J_{ij} The sign of the nuclear–nuclear coupling constant giving rise to the multiplet. (Usually J_{ij} is negative if protons i and j are separated by an even number of bonds (H—C—H; H—C=C—C—H) and positive if protons i and j are separated by an odd number of bonds (H—C—C—H, H—C=C—H). However, o, m, and p coupling constants in a benzene ring are all positive. See J. W. Emsley, J. Feeney, and L. H. Sutcliffe, *High Resolution Nuclear Magnetic Resonance Spectroscopy,* Vol. 2, Pergamon Press, Oxford, 1966, p. 681.

[a] R. Kaptein, *J. Chem. Soc. D.,* 732 (1971). Adapted by permission of the Royal Society of Chemistry.

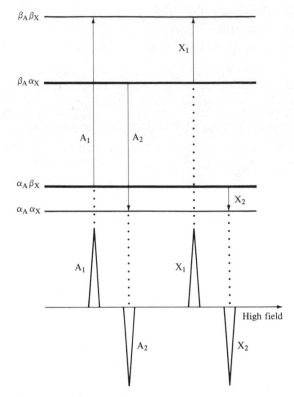

Figure A1.9 Nuclear spin states and spectrum for noncage product **6,** which has excess population of the inner states $\alpha_A\beta_X$ and $\beta_A\alpha_X$. The conventions of the diagram are the same as for Figure A1.7. The spectrum shows the multiplet effect of type A/E.

For the example **4**, the multiplets arise from coupling of nuclei A and X in the fragment **7**. The precursor of the radical pair is a singlet, so μ is negative; σ_{AX} is

$$
\begin{array}{cc}
\text{H}_X & \text{H}_A \\
| & | \\
-\text{C}-\text{C}\cdot \\
| & |
\end{array}
$$

7

positive, since nuclei A and X are in the same fragment; $\Delta g = 0$, since the two radicals of the pair are identical; a is negative for H_A, positive for H_X; J_{AX} is positive. For the cage product, ϵ is positive. We therefore find

$$
\begin{array}{ccccccc}
& \mu & \epsilon & a_A & a_X & J_{AX} & \sigma \\
\Gamma_m = & - & + & - & + & + & + & = +
\end{array}
\tag{A1.3}
$$

$$
\begin{array}{ccccc}
& \mu & \epsilon & \Delta g & & a \\
\Gamma_n = & - & + & 0 & (+\text{ or }-) & = 0
\end{array}
\tag{A1.4}
$$

There should be an E/A multiplet effect with no net effect, as in Figure A1.8. The product of escape from the cage has ϵ negative and an A/E multiplet effect.

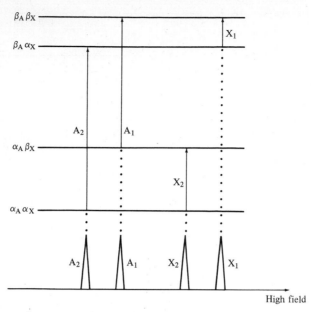

Figure A1.10 Nuclear spin states and normal absorption spectrum for the AX system assuming a negative nuclear–nuclear coupling constant J_{AX}. Excess population of the outer states will cause an A/E multiplet effect, and excess population of the inner states will cause an E/A effect.

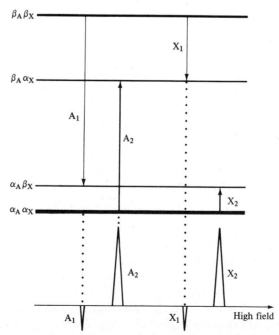

Figure A1.11 The superposition of net enhanced absorption (A) with multiplet emission–absorption (E/A). Positive J_{AX} is assumed. The lowest level, $\alpha_A\alpha_X$, has the largest excess population, and the highest, $\beta_A\beta_X$, a smaller excess. The lines A_2 and X_2, appearing in absorption, are more intense than the lines A_1, X_1, which are in emission. The overall spectrum shows net absorption. From S. H. Pine, *J. Chem. Educ.*, **49**, 664 (1972). Adapted by permission of the Division of Chemical Education, American Chemical Society.

Chapter 10
THE THEORY
OF PERICYCLIC
REACTIONS

In Chapters 10 and 11 we consider pericyclic reactions, processes characterized by bonding changes taking place through reorganization of electron pairs within a closed loop of interacting orbitals. The bonding changes must be concerted in order for a reaction to fit the pericyclic category; that is, all bonds breaking and forming must do so simultaneously rather than in two or more steps. We shall use the term *stepwise* to imply a nonpericyclic pathway in which one or more bonds form (or break) in a first step to yield an intermediate that subsequently reacts by formation (breakage) of other bonds.

The requirement for simultaneous bond formation and cleavage in a pericyclic reaction does not mean that all changes must have taken place to the same extent at all stages along the reaction coordinate. So long as the changes are occurring together, it is possible for bond formation or cleavage at one site to run substantially ahead of bond formation or cleavage at another. A *concerted* reaction is therefore interpreted as simultaneous, but not necessarily synchronous, bond formation and breakage.

Pericyclic reactions represented for many years a difficult mechanistic problem because the apparent absence of intermediates left few concrete features that could be subjected to experimental study. Application of some fundamental principles of orbital theory, initiated in 1965 by Woodward and Hoffmann[1] and since

---

[1] R. B. Woodward and R. Hoffmann, *J. Am. Chem. Soc.,* **87,** 395 (1965).

developed extensively by them[2] and by others,[3] have provided new insight into these reactions and have opened a new field of experimental investigation. This chapter considers the theoretical aspects, and Chapter 11 will take up applications and examples.

10.1 DEFINITIONS

Woodward and Hoffmann[4] have subdivided pericyclic transformations into five categories: cycloaddition, electrocyclic, sigmatropic, cheletropic, and group transfer reactions. Although one can, from a more generalized point of view, regard all these types as cycloadditions, it is convenient to follow this subdivision for purposes of thinking about both theoretical aspects and particular reactions. In this section we shall define and illustrate each class in order to arrive at an overall view of the kinds of questions with which pericyclic reaction theory is concerned.

Cycloaddition

Cycloaddition is a process in which two or more molecules condense to form a ring by transferring electrons from π bonds to new σ bonds.[5] Some typical examples, characterized by the number of π electrons in the reacting components, are illustrated in Equations 10.1–10.3.[6] The bonding changes are sometimes emphasized by the use of curved arrows, as in Equation 10.4. Although this notation serves as a useful bookkeeping device to record the changes in bonding, one must nevertheless be careful when using it for pericyclic reactions not to attach literal significance to the arrows nor to their direction. As we shall see in more detail later, the bonding changes occur through a set of orbitals interacting on the equivalent of a circular path, but a drawing like **1** must not be taken to imply that particular electron pairs are to be found at particular places on the circle or that pairs circulate in a particular direction.

[2](a) R. Hoffmann and R. B. Woodward, *J. Am. Chem. Soc.,* **87**, 2046 (1965); (b) R. B. Woodward and R. Hoffmann, *J. Am. Chem. Soc.,* **87**, 2511 (1965); (c) R. Hoffmann and R. B. Woodward, *Acc. Chem. Res.,* **1**, 17 (1968); (d) R. B. Woodward and R. Hoffmann, *The Conservation of Orbital Symmetry, Angew. Chem. Int. Ed. Engl.,* **8**, 781 (1969); Verlag Chemie, Weinheim/Bergstr., Germany, and Academic Press, New York, 1970.

[3](a) H. C. Longuet–Higgins and E. W. Abrahamson, *J. Am. Chem. Soc.,* **87**, 2045 (1965); (b) H. E. Zimmerman, *J. Am. Chem. Soc.,* **88**, 1564, 1566 (1966); *Acc. Chem. Res.,* **4**, 272 (1971); *Acc. Chem. Res.,* **5**, 393 (1972); (c) M. J. S. Dewar, *Tetrahedron Suppl.,* **8**, 75 (1966); *Angew. Chem. Int. Ed. Engl.,* **10**, 761 (1971); (d) K. Fukui, *Acc. Chem. Res.,* **4**, 57 (1971); (e) J. J. C. Mulder and L. J. Oosterhoff, *Chem. Commun.,* 305, 307 (1970); (f) W. J. van der Hart, J. J. C. Mulder, and L. J. Oosterhoff, *J. Am. Chem. Soc.,* **94**, 5724 (1972); (g) W. A. Goddard III, *J. Am. Chem. Soc.,* **94**, 793 (1972); (h) M. J. S. Dewar, S. Kirschner, and H. W. Kollmar, *J. Am. Chem. Soc.,* **96**, 5240 (1974) and following papers; (i) D. M. Silver, *J. Am. Chem. Soc.,* **96**, 5959 (1974); (j) N. D. Epiotis, *Theory of Organic Reactions,* Springer-Verlag, Berlin, 1978; (k) E. A. Halevi, *Angew. Chem. Int. Ed. Engl.,* **15**, 593 (1976); general: (l) R. G. Pearson, *Symmetry Rules for Reactions,* Wiley, New York, 1976; (m) K. Fukui, *Theory of Orientation and Stereoselection,* Springer-Verlag, Berlin, 1975; (n) K. N. Houk, *Acc. Chem. Res.,* **8**, 361 (1975); (o) T. L. Gilchrist and R. G. Storr, *Organic Reactions and Symmetry,* Cambridge University Press, London, 1972; (p) I. Fleming, *Frontier Orbitals and Organic Chemical Reactions,* Wiley, London, 1976; (q) A. P. Marchand and R. E. Lehr, Eds., *Pericyclic Reactions,* Vols. 1 and 2, Academic Press, New York, 1977.

[4]See note 2(d).

[5]R. Huisgen, *Angew. Chem. Int. Ed. Engl.,* **7**, 321 (1968).

[6]Huisgen (note 5) designates components according to the number of atoms rather than number of electrons.

$$2 + 2 \quad \| + \| \quad \rightleftharpoons \quad \square \tag{10.1}$$

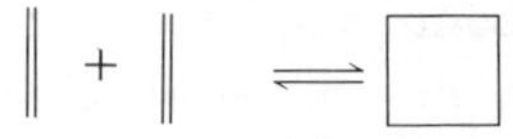

$$2 + 4 \quad \| + \rangle \rightleftharpoons \bigcirc \tag{10.2}$$

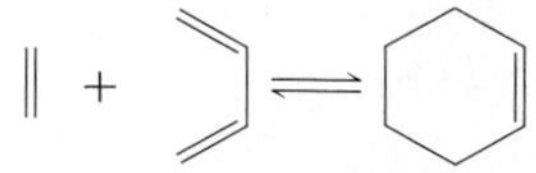

$$4 + 4 \quad \big\langle + \rangle \rightleftharpoons \bigcirc \tag{10.3}$$

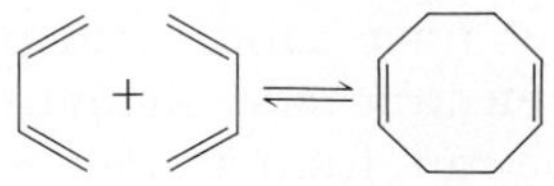

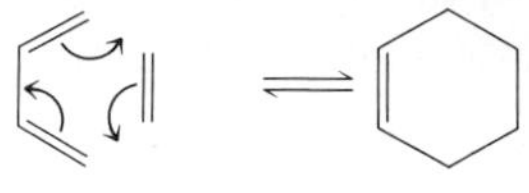

$$\rightleftharpoons \bigcirc \tag{10.4}$$

1

The interesting feature of the cycloadditions is that the ease with which they take place depends on the number of electrons involved in the bonding changes. Whereas the 2 + 4 process (Equation 10.2), known as the Diels–Alder reaction, occurs readily with activation enthalpies of roughly 25–35 kcal mole^{-1}, and has been one of the cornerstones of organic chemistry for many years,[7] the 2 + 2 and 4 + 4 additions are accomplished much less easily. One finds, however, that if one of the reacting molecules is in an electronically excited state, the 2 + 2 and 4 + 4 processes occur more readily than the 2 + 4.[8]

The 1,3-dipolar additions (Equation 10.5), studied extensively by Huisgen and co-workers,[9] also fit in the cycloaddition category. The 1,3-dipole, which can

$$2 + 4 \quad \| + \begin{array}{c} a \\ \searrow b \\ c \end{array} \longrightarrow \begin{array}{c} a \\ \boxed{} b \\ c \end{array} \tag{10.5}$$

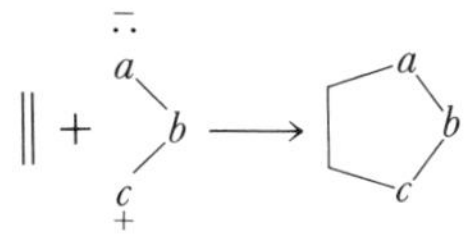

be any of a variety of stable molecules or reactive intermediates—for example, ozone ($O{=}O{-}O \leftrightarrow O{-}O{-}O \leftrightarrow O{-}O{-}O$), nitro ($O{=}N{-}O \leftrightarrow O{-}N{-}O \leftrightarrow O{-}N{-}O$), nitrile oxide ($-C{\equiv}N{-}O \leftrightarrow -C{=}N{-}O$)— typically has four π electrons in an orbital system delocalized over three centers. These reactions, like the Diels–Alder, are easily accomplished.

---

[7](a) M. C. Kloetzel, *Org. Reactions,* **4,** 1 (1948); (b) H. L. Holmes, *Org. Reactions,* **4,** 60 (1948); (c) L. W. Butz and A. W. Rytina, *Org. Reactions,* **5,** 136 (1949); (d) J. G. Martin and R. K. Hill, *Chem. Rev.,* **61,** 537 (1961); (e) R. Huisgen, R. Grashey, and J. Sauer, in *The Chemistry of Alkenes,* S. Patai, Ed., Wiley, London, 1964, p. 739; (f) A. Wassermann, *Diels–Alder Reactions,* Elsevier, Amsterdam, 1965; (g) H. Kwart and K. King, *Chem. Rev.,* **68,** 415 (1968).

[8]See, for example, W. L. Dilling, *Chem. Rev.,* **69,** 845 (1969).

[9]R. Huisgen, *Angew. Chem. Int. Ed. Engl.,* **2,** 565 (1963).

Stereochemistry in Cycloaddition

For a given π electron system there are two stereochemical alternatives, illustrated for the four-electron fragment (a diene) of a 2 + 4 addition by Structures **2** and **3**, in which the arrows show where the two new σ bonds are forming. In **2** both bonds

2

3

form to the same face of the four-electron system; whenever two bonds form on the same face of a unit entering a cycloaddition, that unit is said to enter suprafacially, abbreviated *s*. In **3** addition occurs on the upper face at one end of the diene and on the lower face at the other; in this instance, the diene is entering antarafacially, *a*. In both Structures **2** and **3** the two-electron component (an alkene) is acting in a suprafacial way; in **2** we would say that the combination is (diene *s*) + (alkene *s*), and in **3** it is (diene *a*) + (alkene *s*). Further economy of notation is obtained by indicating only the number of electrons that enter from each unit. The diene is contributing four electrons from its π system and so is designated 4*s* in **2** and 4*a* in **3**. The alkene is contributing two electrons and so is designated 2*s* in each case. Hence the shorthand notation summarizing number of electrons and stereochemistry is 2*s* + 4*s* for **2**, and 2*s* + 4*a* for **3**.

Any of the combinations *s* + *s*, *s* + *a*, *a* + *s*, *a* + *a* is conceivable for a cycloaddition of two components. Comparison of cis–trans stereochemistry of substituents in product to that in reactants establishes which occurred. In additions of relatively short chains the antarafacial interaction is difficult for the molecule to attain, but when systems with appropriate geometry are contrived, it is found that reactions in which one component acts in the antarafacial manner exhibit an inverted preference with respect to ring size: In the electronic ground state the 2 + 2 additions are now favorable and the 4 + 2 are not.

Electrocyclic Reactions

The second category is that of the electrocyclic reaction, illustrated in Equations 10.6–10.8, characterized by opening or closing of a ring within a single molecule by conversion of σ bonds to π bonds or the reverse. Again, a notation with curved arrows (**4**) is helpful if used with appropriate caution. The primary feature of interest here is the stereochemistry of substituents of the σ bond that is broken.

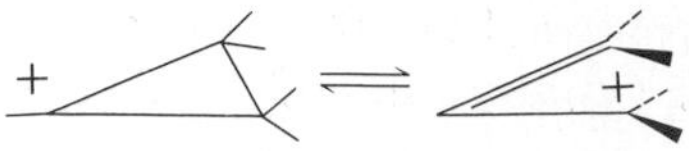

(10.6)

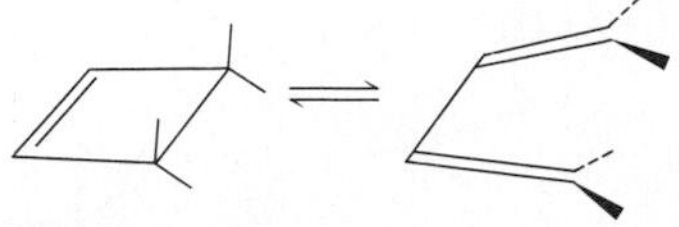

(10.7)

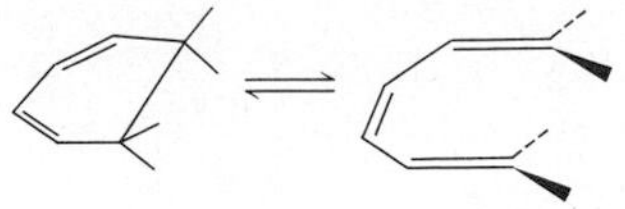

(10.8)

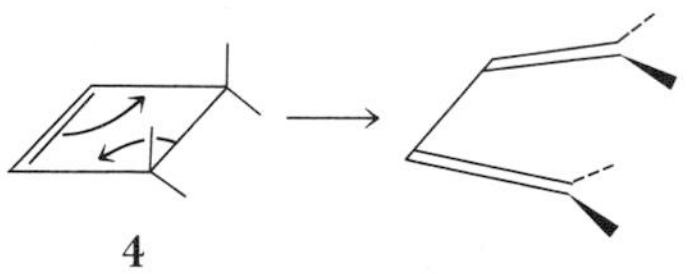

(10.9)

4

These substituents lie above and below the ring plane, and as the bond breaks must move in one of the ways illustrated in Equation 10.10. Woodward and Hoffmann introduced the terms *disrotatory* (rotating in opposite directions) and *conrotatory* (rotating in the same direction) to describe these alternatives.[10] Again the preferred

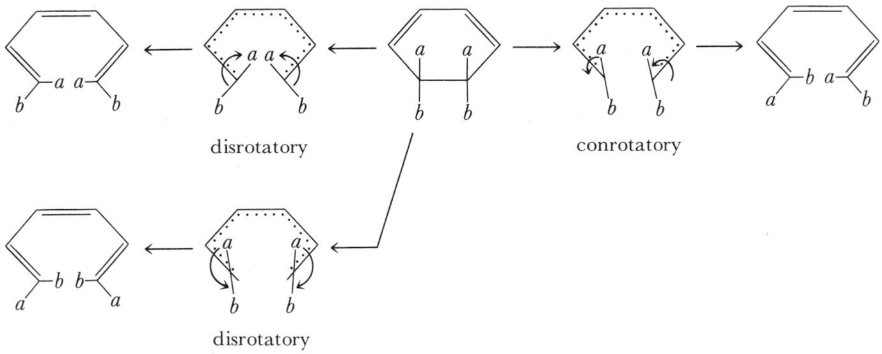

(10.10)

path depends on ring size: Electrocyclic ring opening and closing of a ground-state cyclobutene (Equation 10.7) usually follows the conrotatory route, whereas a ground-state cyclohexadiene (Equation 10.8) prefers the disrotatory. The preferences reverse in the photochemical excited-state reactions.

It will be useful later to be able to apply the suprafacial–antarafacial notation to electrocyclic reactions, and to the other pericyclic reactions described below, as well as to cycloadditions. Structure **5a** represents schematically a π bond interacting suprafacially; the arrows indicate that interactions are occurring from the π bond shown, on the same face of the π bond at each end, to some other unspecified

[10]R. B. Woodward and R. Hoffmann, *J. Am. Chem. Soc.,* **87,** 395 (1965).

bond or bonds. Structure **5b** shows how this concept extends to a diene system. Structures **6a** and **6b** show the π antarafacial interaction, in which the π system

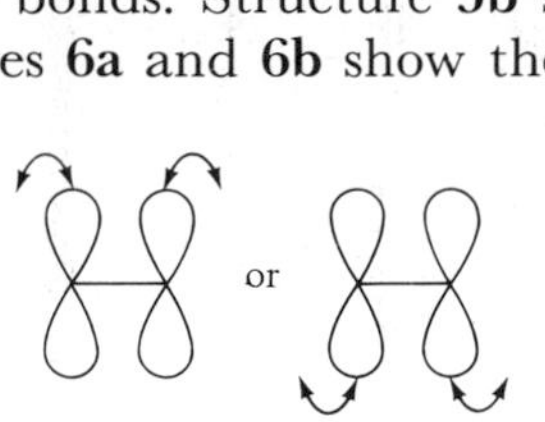

π suprafacial
5a

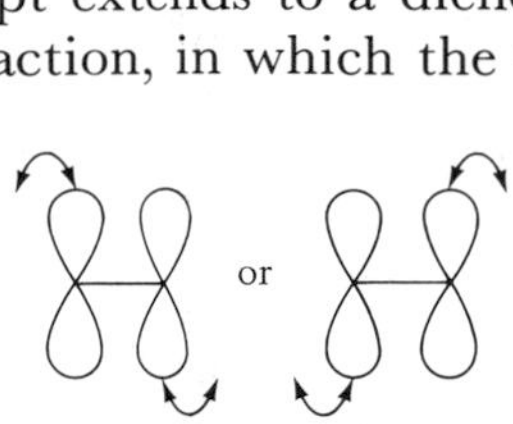

π antarafacial
6a

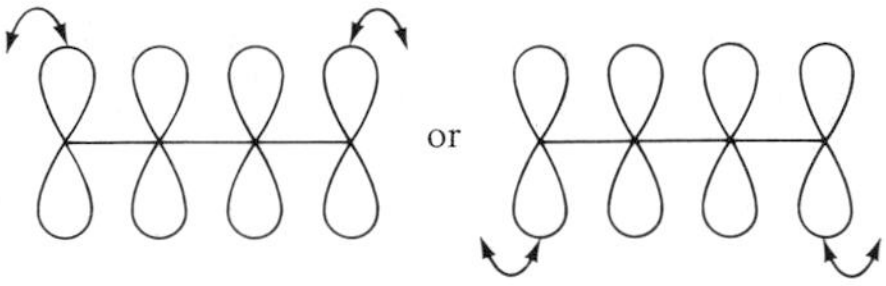

π suprafacial
5b

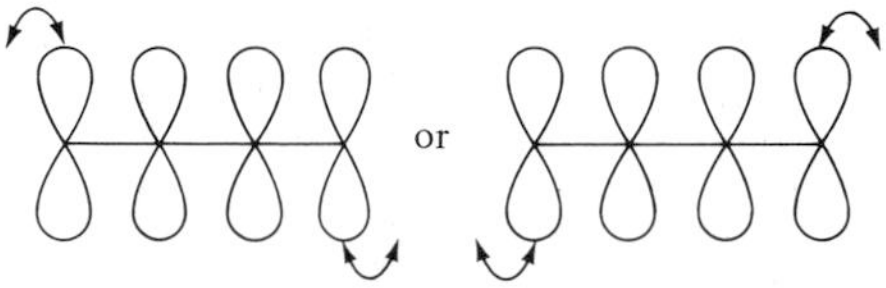

π antarafacial
6b

interacts at the two ends on opposite faces. If we represent a σ orbital in the analogous way as being made up of two hybrids, we can define σ suprafacial interactions by **7** and σ antarafacial interactions by **8**. The analogy between the π and σ systems may be made clear by rotating the σ orbitals in **7** and **8** by 90° and expanding the small lobes to change hybridization from sp^3 to p. We can also define suprafacial and antarafacial interactions for a single orbital as illustrated in **9**.

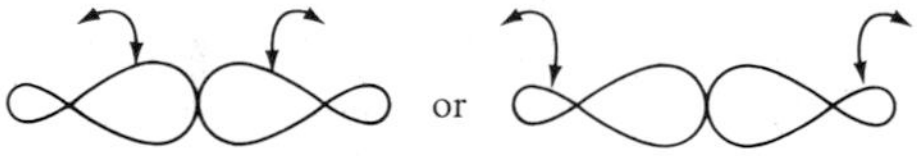

σ suprafacial
7

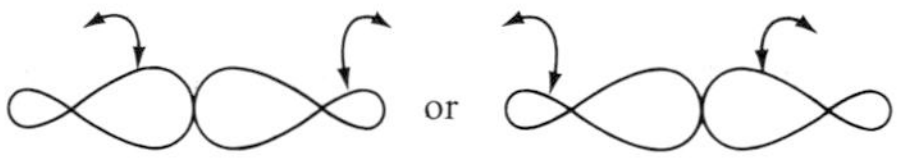

σ antarafacial
8

ω suprafacial  ω antarafacial
9

With the help of the definitions illustrated in Structures **5–9**, we can consider electrocyclic reactions as a category of cycloaddition. For example, in the disrotatory ring opening depicted in Equation 10.11, a σ bond consisting of orbitals 1 and 2 enters suprafacially into a new interaction; the other partner in the reaction is the system of two π bonds, orbitals 3, 4, 5, and 6, which also enter suprafacially. This process can therefore be thought of as a generalized cycloaddition of type $\sigma 2s + \pi 4s$. Equations 10.12 and 10.13 show two equivalent ways of looking at a conrotatory electrocyclic reaction.

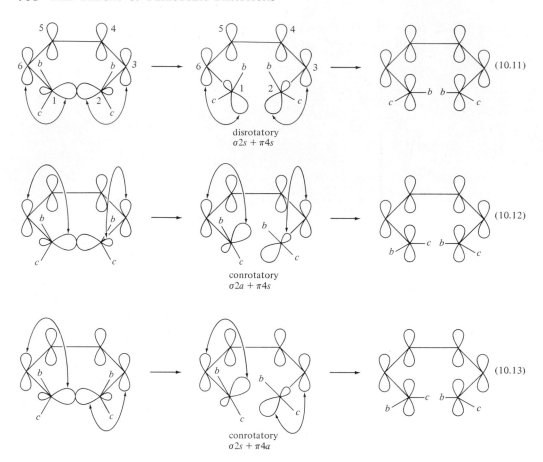

Sigmatropic Reactions

Sigmatropic reactions are those in which a bond migrates over a conjugated system as illustrated in Equations 10.14–10.17, the migrating bond being indicated by a heavy line. The Woodward–Hoffmann nomenclature designates with a pair of numbers in square brackets the change (if any) in the position of attachment of the ends of the bond in question. Thus a [1,2] migration means that one end of the bond remains attached at its original position while the other end moves to an adjacent position; a [3,5] migration would mean that one end moves to a 3 position relative to its original point of attachment while the other end moves to a 5 position.

$$\text{(10.14)}$$

[1,2]

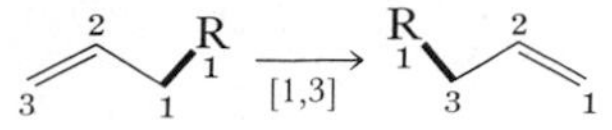

$$\text{(10.15)}$$

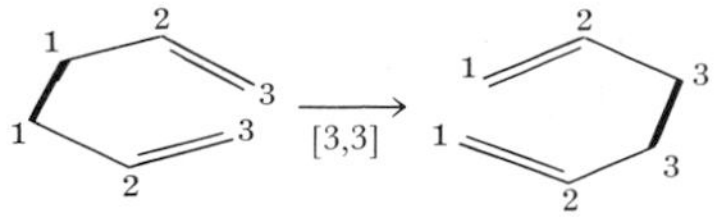

$$\text{(10.16)}$$

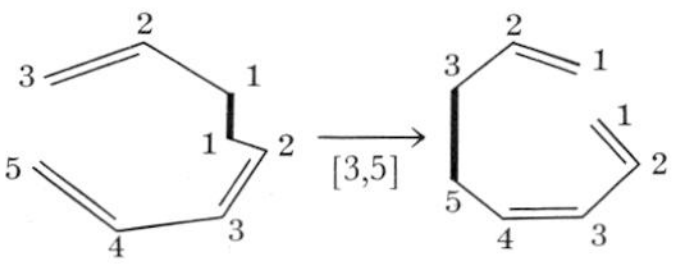

$$\text{(10.17)}$$

Suprafacial and antarafacial possibilities are present here as well; preferred stereo-chemistry is again determined by the number of electrons in the cyclic array of orbitals. Sigmatropic changes can also be regarded as generalized cycloadditions (Equations 10.18–10.21). Note (Equation 10.20) that when a σ bond enters in an antarafacial way in a [1,n] sigmatropic transformation, an inversion occurs at the center to which the migrating bond remains attached.

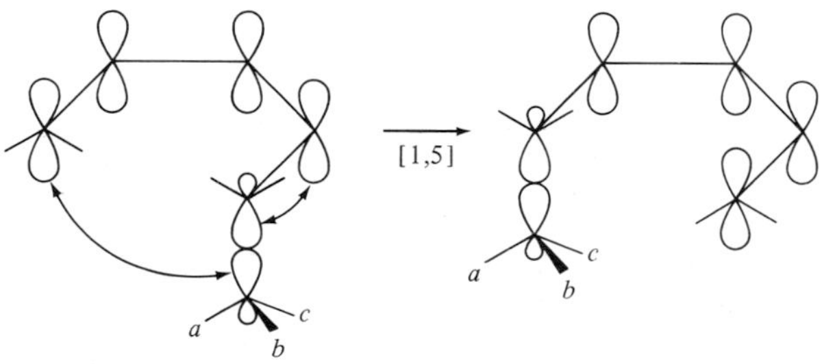

$$\text{(10.18)}$$

$\sigma 2s + \pi 4s$

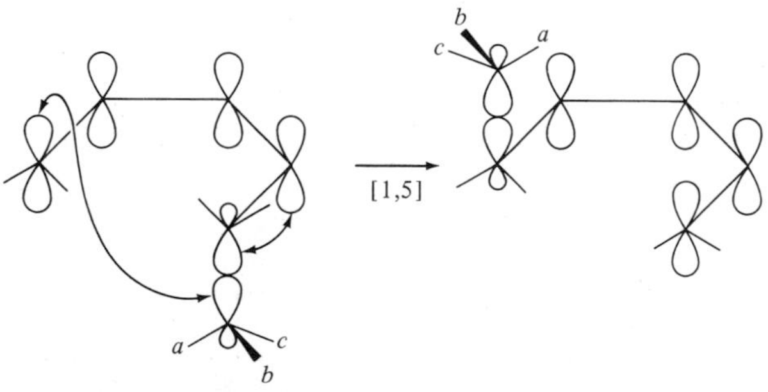

$$\text{(10.19)}$$

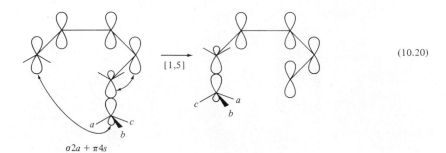

(10.20)

$\sigma 2a + \pi 4s$

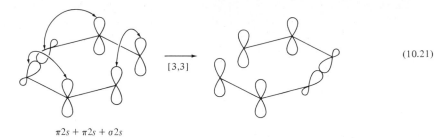

(10.21)

$\pi 2s + \pi 2s + \sigma 2s$

Cheletropic Reactions and Group Transfers

Cheletropic reactions are cycloadditions in which one of the components interacts through a single atom, as in Equation 10.22. Group transfers are characterized by

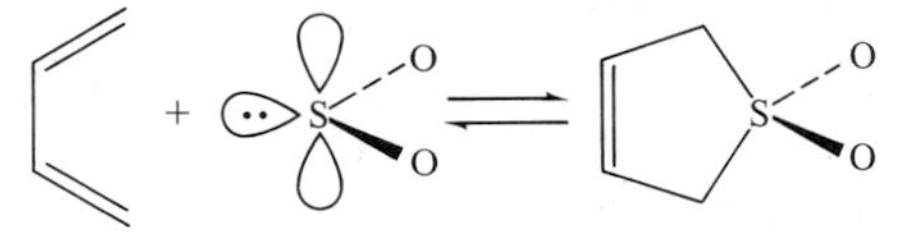

(10.22)

transfer of a group or groups from one molecule to another, as illustrated in Equations 10.23–10.25. Again, one can view these reactions as generalized cycloadditions (Equations 10.26–10.27).

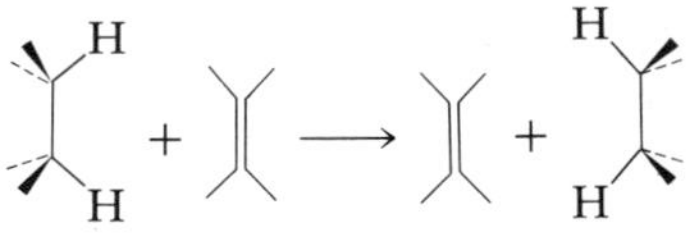

(10.23)

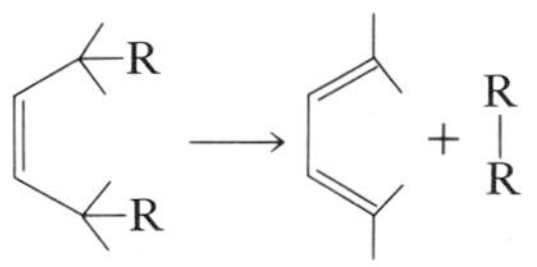

(10.24)

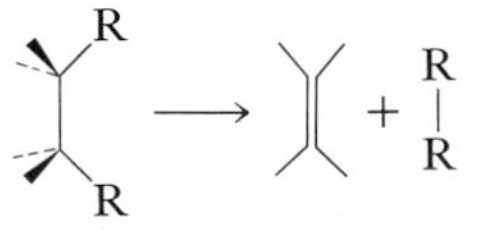

(10.25)

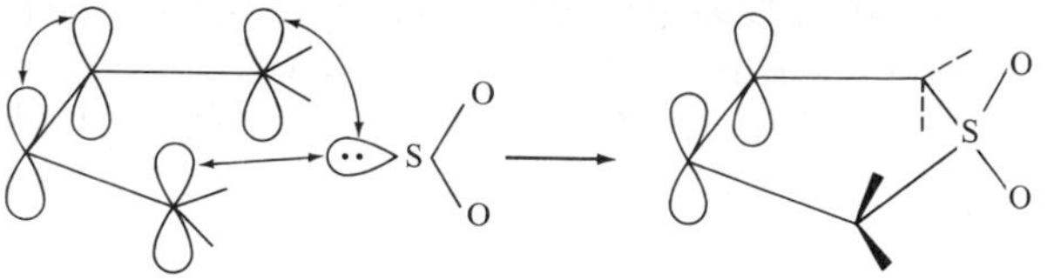

$$\pi 2s + \pi 2s + \omega 2s$$

(10.26)

$$\pi 2s + \sigma 2s + \pi 2s + \sigma 2s$$

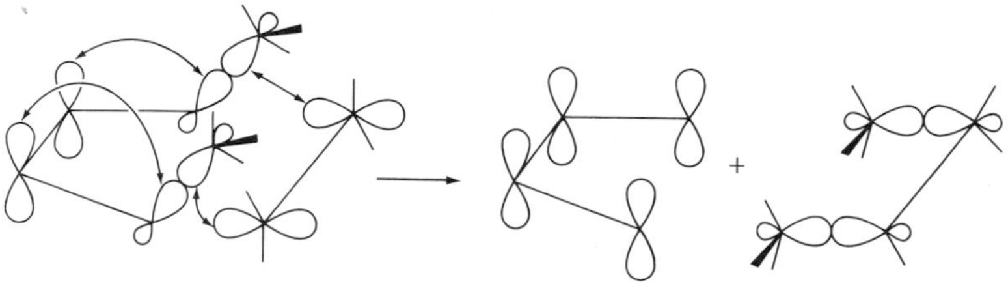

(10.27)

10.2 PERICYCLIC REACTIONS AND TRANSITION STATE AROMATICITY

Before examining pericyclic reaction theory as originally presented by Woodward and Hoffmann, we shall describe an alternative formulation due to Dewar[11] and to Zimmerman[12] that summarizes the predictions of the theory in a particularly convenient way.

Interaction Diagrams

We have illustrated in the previous section how the bonds entering into a pericyclic reaction can be classified as to their type (σ, π, or ω) and their mode of interaction (suprafacial or antarafacial). In that classification we emphasized the interactions occurring between the ends of the two bond systems, as, for example, between the ends of a diene π system and the ends of an alkene π bond in the $\pi 2s + \pi 4s$ cycloaddition pictured in **2**. It is possible also to trace a path of interaction within the bonding system of a molecule; this process defines the fundamental topology of the orbital system and gives rise to the *interaction diagram*.

For any bond or delocalized bonding system we may draw schematically the basis orbitals that would be used to construct molecular orbitals for the system (for the moment without specification of algebraic sign) and then trace with short curved lines a path of interaction among those orbitals. The result is the interaction diagram for the system. We illustrate the process with the π orbitals of butadiene. First draw the basis orbitals, one on each carbon (**10**). Each p orbital interacts with

[11](a) M. J. S. Dewar, *Tetrahedron Suppl.,* **8,** 75 (1966); (b) M. J. S. Dewar, *Angew. Chem. Int. Ed. Engl.,* **10,** 761 (1971).
[12](a) H. E. Zimmerman, *J. Am. Chem. Soc.,* **88,** 1564, 1566 (1966); (b) H. E. Zimmerman, *Acc. Chem. Res.,* **4,** 272 (1971).

its immediate neighbors; the orbitals at the two ends are too far apart to interact with each other. Hence interactions are to be specified from orbital 1 to 2, 2 to 3, and 3 to 4, as shown in **11**. (The lines could just as well have been drawn at the bottom.)

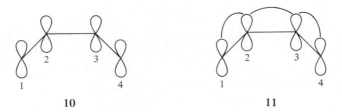

During a reaction a molecule may twist or otherwise distort without breaking the path of interaction among its orbitals. Therefore we want to regard the framework on which the orbitals are placed in an interaction diagram as being flexible; imagine that the interaction lines in **11** are made of thin strands of rubber, so that the molecule can be twisted, bent, and stretched without breaking them. The interaction topology and hence the fundamental properties of the interaction diagram is thus independent of distortion.

In order to make the interaction diagram idea as general as possible, we want to focus on the interacting orbitals and the path of interaction; we shall often omit the framework of the rest of the molecule. Thus Structure **12** shows the

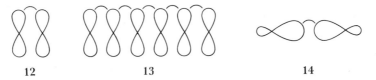

ethylene π bond; **13** would be hexatriene; **14** is a σ bond. Because we are regarding the interaction diagrams as flexible, the precise spatial arrangement and orientation of the orbitals is unimportant as long as the pathway of interaction is unbroken. Thus **15** and **16** are entirely equivalent to **14**; **17** is equivalent to **12**. Further-

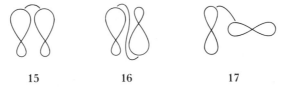

more, for a system of adjacent p orbitals in a π electron system, the interaction at the top lobes is equivalent to that at the bottom; we may therefore choose either for the orbital interaction diagram. For ethylene we could equally well use **18** or **12**; for hexatriene **19** would serve as well as **13**. For convenience we shall ordinarily choose the connections in such a case to be all on one face in the diagram. We note also

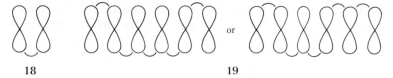

18   19

that aside from the fact that interaction on the two faces is no longer the same, there is no fundamental topological difference between a π bond and a σ bond; hence the representations **15** and **20** for a σ bond are equivalent, and we may if we

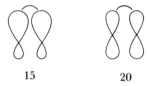

15   20

wish use the p orbital schematic for both hybrids and true p orbitals. Because during a reaction at least some of the reacting orbitals ordinarily change from p to hybrid or vice versa, it will often be convenient to do this.

It is important to remember that the *interaction diagrams depict sets of basis orbitals and not molecular orbitals.*

The examples of interaction diagrams given so far consist of the basis orbitals of single molecules. The utility of the diagrams is in analyzing pericyclic transition states, where two or more molecules, or two or more different sets of orbitals within a molecule, come together to form a continuous cyclic chain of interacting orbitals. Although this ring of interacting orbitals will be distorted from the geometry of a regular plane polygon, the fundamental topology of the orbital interaction path for the distorted ring of the pericyclic transition state will be the same as for π orbitals of a regular planar ring. The transition state of pericyclic reactions will therefore be characterized by aromatic or antiaromatic character as are the cyclic compounds benzene, cyclobutadiene, and so forth. Transition states with aromatic character will be stabilized; those reactions with aromatic transition states are predicted to occur readily in the electronic ground state and are said to be *thermally allowed.* Transition states with antiaromatic character will be destabilized; those reactions with antiaromatic transition states are said to be *thermally forbidden.* (*Allowed* and *forbidden* are only relative terms as used here; a more accurate wording would be *of low activation energy* or *of high activation energy.*) The analysis developed below will show the method of determining whether a transition state is aromatic or antiaromatic.

Before proceeding to the next step, let us look at some examples of interaction diagrams. The $\pi 2s + \pi 4s$ cycloaddition of butadiene with ethylene is shown in Structure **2**. The reaction can be considered to be made up of two components: the $\pi 2s$ ethylene component and the $\pi 4s$ butadiene component. (We shall define later an unambiguous method of determining when a component is s and when a.) Before the reaction, interactions exist between orbitals 1 and 2 within the ethylene and between 3, 4; 4, 5; and 5, 6 within the butadiene, as illustrated in **21**. At the transition state these interactions are still present (although some may be diminished in magnitude) and new interactions 1, 6 and 2, 3 have been established; hence interaction diagram **22**. As we mentioned earlier, the interaction lines within

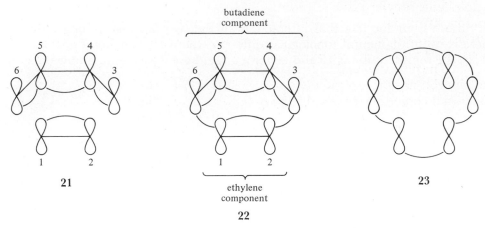

a π component can be drawn on either face; the important point is always to connect the lobes according to the actual topology of the cyclic interaction as it is occurring in the transition state. Thus **23** would be an acceptable alternative formulation for butadiene + ethylene. Consider the same reaction in the reverse direction (Equation 10.28). The components now are one π bond and two σ bonds; interactions 2, 3; 4, 5; 1, 6 exist before the reaction (**24**). New interactions 1, 2; 3,

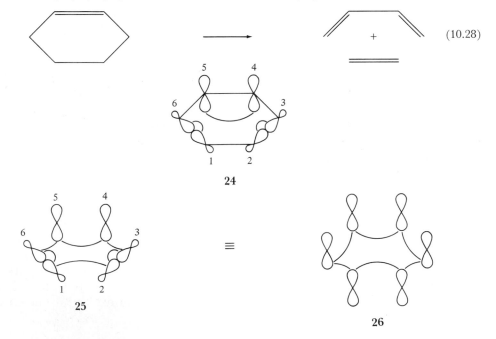

4; 5, 6 are established on going to the transition state; hence interaction diagram **25**. If we distort the orbitals slightly, we have the equivalent **26**, which is the same as **22** or **23**. Equations 10.29–10.33 illustrate some other pericyclic reactions and their transition state interaction diagrams. The equivalences indicated in the examples illustrate how one can, by distorting the orbitals and changing their hybridization, bring out the essential topological features of the cyclic interactions and discern similarities between reactions that superficially look quite different (Equations 10.29 and 10.32; 10.30 and 10.31; 10.33 and 10.28).

The next step in the analysis is to establish rules for assigning algebraic signs to the orbitals of interaction diagrams. Whenever a set of basis orbitals is chosen, the initial choice of sign is arbitrary. In Chapter 1 we chose basis function signs, as

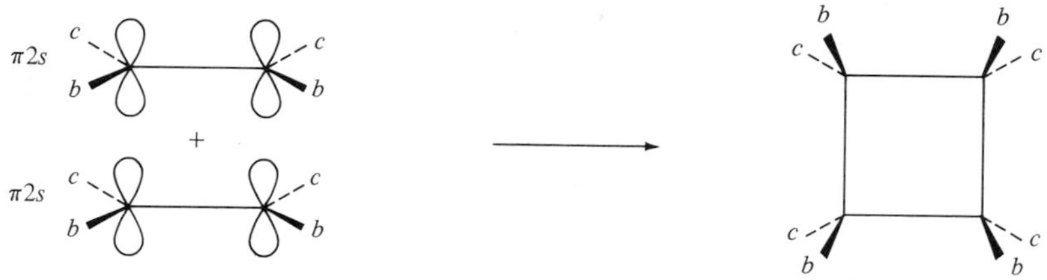

(10.29)

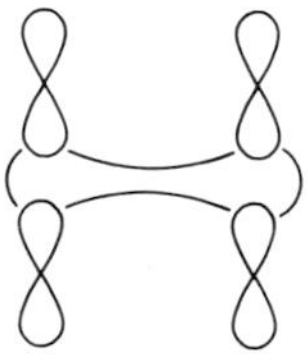

27

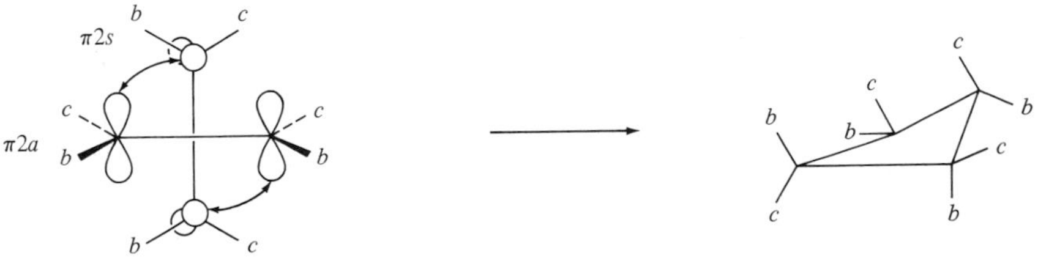

(10.30)

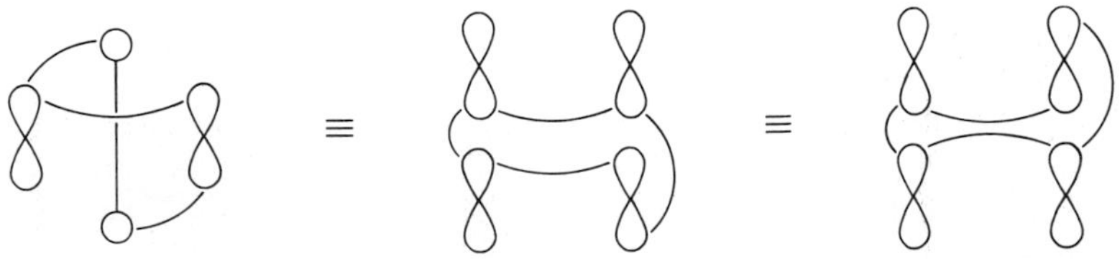

28

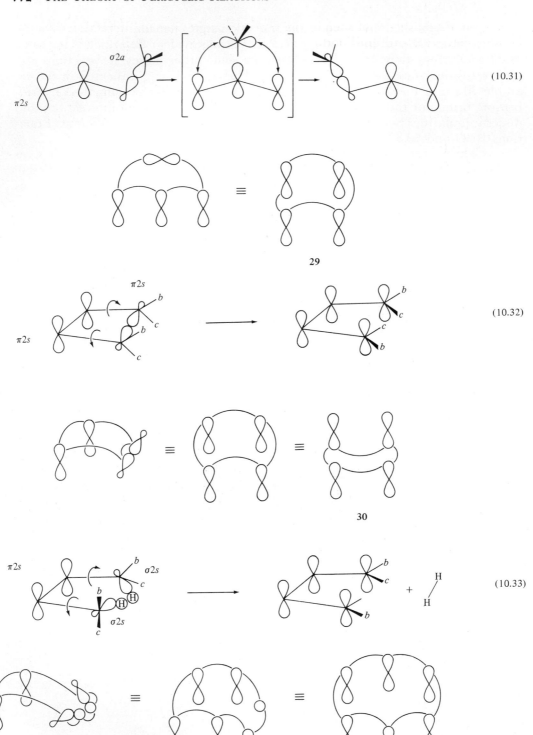

(10.31)

29

(10.32)

30

(10.33)

31

a matter of convenience, in such a way that bonding molecular orbitals would be the algebraic sum and antibonding orbitals the algebraic difference of the basis orbitals, as shown below for H_2:

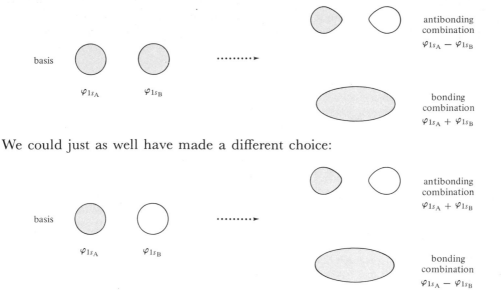

antibonding
combination

$\varphi_{1s_A} - \varphi_{1s_B}$

basis

φ_{1s_A} φ_{1s_B}

bonding
combination

$\varphi_{1s_A} + \varphi_{1s_B}$

We could just as well have made a different choice:

antibonding
combination

$\varphi_{1s_A} + \varphi_{1s_B}$

basis

φ_{1s_A} φ_{1s_B}

bonding
combination

$\varphi_{1s_A} - \varphi_{1s_B}$

The higher-energy molecular orbital still comes out to be the one with a node; the only difference is in its mathematical expression in terms of the basis.

Given that the initial choice of basis orbital signs is arbitrary, we are free to adopt the following convention:

Basis orbital signs of interaction diagrams are to be chosen in such a way that the interaction path connects only lobes of like sign so far as possible.

Take as an example the interaction diagram **31**. Suppose we set the signs of the orbitals at random, say as shown in **32**. This assignment does not conform to the

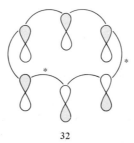

32

convention stated above: there are two points, marked by asterisks, at which inter-action paths connect lobes of unlike sign. A connection between lobes of unlike sign is called a *phase inversion*. Recalling that the interaction path is the fundamental aspect of the diagram, derived from the topology of the transition state under consideration, and thus cannot be changed but that the signs are arbitrary, we simply make a different choice of signs so as to eliminate the phase inversions (**33**).

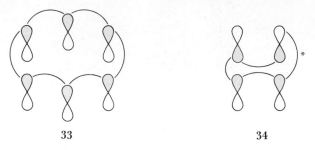

33   34

Now look at diagram **28**. There is no way of choosing the signs so as to avoid a phase inversion; the best we can do is **34**. Other sign choices would only move the phase inversion to another point (**34a**) or create more phase inversions (**34b**). Cyclic interaction diagrams **33** and **34** belong to two fundamentally different classes: diagrams with no phase inversions, and diagrams with one phase inversion.

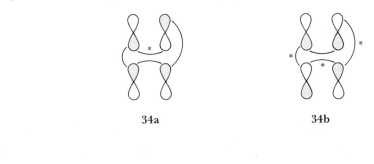

34a   34b

34c   34d   34e

It is easy to see that there are only these two types of monocyclic interaction diagrams. If a particular lobe has a phase inversion on one side but not on the other, changing the sign of that lobe moves the phase inversion from one side to the other but does not change the number of phase inversions in the ring. (The change **34** → **34a** is an example.) If a lobe has phase inversions on both sides, changing the sign of that lobe eliminates both phase inversions. (The change **34b** → **34** is an example.) Therefore if an initial arbitrary assignment of orbital signs in a cyclic interaction diagram produces phase inversions, they can be moved around by changing signs until they meet on either side of a single lobe and can then be eliminated in pairs. (The series **34c** → **34d** → **34e** illustrates this process.) All rings with an even number of phase inversions are thus equivalent to rings with no phase inversions, whereas all rings with an odd number are equivalent to rings with one. Rings with zero (or an even number of) phase inversions are said to be of the *Hückel* type; rings with one (or an odd number of) phase inversion are said to be of the *anti-Hückel* type.

Aromatic and Antiaromatic Transition States

Consider the ground-state benzene molecule in terms of an interaction diagram. Six *p* orbitals are arrayed on a ring, each interacting with its neighbor (35). The

35

transition state interaction diagram of the $\pi 2s + \pi 4s$ cycloaddition (**33**) is identical. The aromaticity theory of pericyclic reactions holds that the $\pi 2s + \pi 4s$ transition state will enjoy a stabilization of the same kind as does benzene ground state. (Stabilization of the same kind, but not of the same magnitude, because distortions are required to make the interaction diagram of the pericyclic transition state conform to a regular ring pattern.) The $\pi 2s + \pi 4s$ process with $(4n + 2)\pi$ electrons ($n = 1$) has an aromatic transition state and is thermally allowed. This analogy between benzene and the Diels–Alder transition state was first pointed out by Evans in 1939;[13] its applicability to a wide variety of reactions, however, was not fully appreciated until after the pioneering work of Woodward and Hoffmann, which we describe in Section 10.4.

The cyclobutadiene molecule has the interaction diagram **36**. The transition state for a $\pi 2s + \pi 2s$ cycloaddition is equivalent; this $4n\ \pi$ electron system ($n = 1$) is antiaromatic, the transition state is destabilized, and the reaction is thermally forbidden.

[13]M. G. Evans, *Trans. Faraday Soc.,* **35**, 824 (1939).

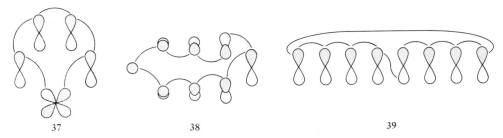

36

Note that the $\pi 2s + \pi 4s$ and $\pi 2s + \pi 2s$ examples both have interaction diagrams of the Hückel type, without phase inversions. Our conclusion is that pericyclic reactions taking place through transition states of Hückel type when thermally induced from the ground state are allowed for $4n + 2$ electrons and forbidden for $4n$ electrons.

We must now consider how the situation changes for an anti-Hückel ring. Craig in 1959 showed that the aromatic character of a ring with a given number of π electrons would change if the conjugated system included d orbitals.[14] He considered systems with an equal number of d and p orbitals, for example, rings composed of alternating phosphorus and nitrogen atoms; the point of importance to the present discussion can be illustrated more effectively with just one d orbital in interaction diagram **37**. Given the interaction pattern specified, there is a single-phase inversion. Heilbronner pointed out that the same situation occurs in a singly twisted ring (**38** or **39**) containing only p orbitals.[15] The topology is that of the

37   38   39

Möbius strip, obtained by joining the ends of a strip containing a single twist. For the Möbius systems, if one ignores the loss of stabilization caused by the twisting, the Hückel $4n + 2$ rule is reversed: The twisted rings are aromatic for $4n$ and antiaromatic for $4n + 2$ electrons. It is the presence of the single phase inversion that causes this reversal. Later, in Section 10.3, we shall present simple arguments, due to Dewar and to Zimmerman, that justify this conclusion.

The $\pi 2s + \pi 2a$ cycloaddition illustrated in Equation 10.30 has a single phase inversion in its interaction diagram (**34**). This transition state is therefore of the anti-Hückel type. Because there are $4n$ electrons the reaction has an aromatic transition state and is thermally allowed.

[14] D. P. Craig, *J. Chem. Soc.*, 997 (1959).
[15] E. Heilbronner, *Tetrahedron Lett.*, 1923 (1964).

We may summarize the procedure for predicting the allowed or forbidden nature of pericyclic reactions by the aromaticity method as follows:

1. Identify the cyclic path of interacting orbitals in the transition state and construct the interaction diagram.

2. Assign algebraic signs to the orbitals so as to minimize the number of phase inversions.

3. Classify the type of ring:

zero phase inversions: Hückel
one phase inversion: anti-Hückel

4. Predict whether the reaction is thermally allowed or forbidden:

Hückel rings: $4n + 2$ electrons; thermally allowed
                  $4n$ electrons, thermally forbidden
anti-Hückel rings: $4n$ electrons, thermally allowed
                  $4n + 2$ electrons, thermally forbidden

Having found a set of pericyclic selection rules, we must add a few cautionary remarks. First, the rules apply to ground-state, thermal reactions only. The Woodward–Hoffmann procedures, which we shall examine in Section 10.5, and Zimmerman's method predict reversal of the rules for photochemical reactions, which occur from excited-state molecules. Second, the rules are not absolute. The analysis considered only the electronic interactions of the pericyclic system and did that in the context of an interaction diagram that in some instances requires distortion to bring it into conformity to the ideal, regular planar polygon to which aromaticity theory applies. Steric effects and the energy costs of nonideal orbital overlap are not taken into account. As we shall see in Chapter 11, the predictions of the theory are in general well substantiated by experiment, but there are many circumstances under which "allowed" reactions have high activation energies for steric reasons, or "forbidden" reactions occur more easily than expected because of relief of ring strain or some similar factor. When these extra effects can be taken into account, better tests of the predictions of pericyclic theory are possible.

10.3 AROMATICITY AND ANTIAROMATICITY IN HÜCKEL AND ANTI-HÜCKEL RINGS

In the Dewar–Zimmerman approach to the stabilization of pericyclic transition states, the basic criterion for deciding whether a process will be allowed or forbidden is whether the transition state is stabilized or destabilized, respectively, compared with an analogous open-chain system.[16] As we have noted above, this criterion is the same as that for aromatic or antiaromatic character of a ground-state system. Dewar has given a simple argument based on perturbation theory that

---

[16](a) M. J. S. Dewar, *Tetrahedron Suppl.,* **8,** 75 (1966); (b) M. J. S. Dewar, *Angew. Chem. Int. Ed. Engl.,* **10,** 761 (1971).

clarifies the origin of aromaticity and also shows why the $4n + 2$ rule is reversed for anti-Hückel rings.

In the appendix to Chapter 1 we defined the Dewar resonance energy, DRE, in terms of a comparison between a cyclic conjugated system and an open-chain system containing the same numbers and types of bonds. Dewar has shown a method of approaching this concept through perturbation theory based on the Hückel orbitals.[17]

When one imagines a conjugated ring containing an even number of p orbitals to be constructed by joining a single carbon with a p orbital to the ends of a chain with an odd number of p orbitals, the perturbation theory gives a particularly simple result for the energy change. We have already seen in Section 1.9 that chains with odd numbers of p orbitals have a nonbonding molecular orbital. This nonbonding MO will be at the same energy as the single orbital being added; the largest interaction will therefore be between these two orbitals of the same energy. The forms of the nonbonding orbitals are simple, and the energy changes for union at one end or at both ends can be found easily in the Hückel molecular orbital approximation. In Dewar's method the other interactions are neglected, since they are between orbitals of different energies and will therefore be smaller.

Figures 10.1–10.4 show schematically the analysis of the four-carbon and six-carbon cases. In Figure 10.1 the nonbonding MO of the allyl radical interacts at one end with a single carbon p orbital. Two new orbitals result, one bonding and one antibonding; these orbitals become the highest bonding and lowest antibonding π MO's of butadiene. For such a union the perturbation theory gives the approximate energy-lowering ΔE as follows:

$$\Delta E = a_{01}a_{02}\beta \tag{10.34}$$

where a_{01} is the coefficient of the nonbonding MO of the first fragment at the point of union, a_{02} is the coefficient of the nonbonding MO of the second fragment at the point of union, and β is the interaction energy between the two p orbitals. The appendix to Chapter 1 gives the method for finding the a coefficients for nonbonding MO's; for allyl the coefficient is $+1/\sqrt{2}$ at one end and $-1/\sqrt{2}$ at the other end, and for a single carbon p orbital the coefficient is unity. Thus the quantity ΔE in Figure 10.1 is $\beta/\sqrt{2}$ and the energy lowering (two electrons each lowered by ΔE) is $2\beta/\sqrt{2} = \sqrt{2}\beta$.

Now suppose we join the single p orbital to both ends of the allyl radical (Figure 10.2). At one end the interaction is $+\beta/\sqrt{2}$ and at the other end it is $-\beta/\sqrt{2}$ because the coefficient of the allyl nonbonding MO is of opposite sign at the two ends. Hence the energy change ΔE is zero. This absence of cyclic interaction is also evident from the mismatch of symmetry with respect to the vertical

[17](a) M. J. S. Dewar, *The Molecular Orbital Theory of Organic Chemistry*, McGraw-Hill, New York, 1969, p. 217; (b) M. J. S. Dewar and R. C. Dougherty, *The PMO Theory of Organic Chemistry*, Plenum, New York, 1975, p. 89.

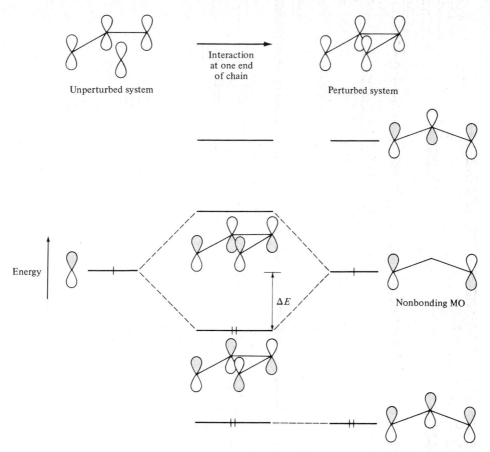

Figure 10.1 Construction of a four-carbon conjugated chain by union of a single carbon with one end of a three-carbon chain. The p orbital being added is at the same energy as the nonbonding orbital. Only the large interaction between orbitals of the same energy is considered; the other, smaller ones are neglected. In the lower-energy combination the interacting orbitals combine in a bonding way and in the higher-energy combination in an antibonding way. The system is stabilized by an amount 2 ΔE (two electrons each decrease in energy by ΔE).

mirror plane passing through the central carbon of the allyl group and the carbon being added. The allyl nonbonding MO is antisymmetric and the single p orbital is symmetric. There is no stabilization, and the ring is less stabilized than the open-chain molecule and hence antiaromatic.

Contrast the four-membered ring with the six-membered ring (Figures 10.3 and 10.4). The pentadienyl nonbonding MO coefficients are $+1/\sqrt{3}$ at both ends of the chain. Union at one end with a single p orbital to yield hexatriene lowers the orbital energy by $\beta/\sqrt{3}$ for a stabilization energy (two electrons) of $2\beta/\sqrt{3}$; union

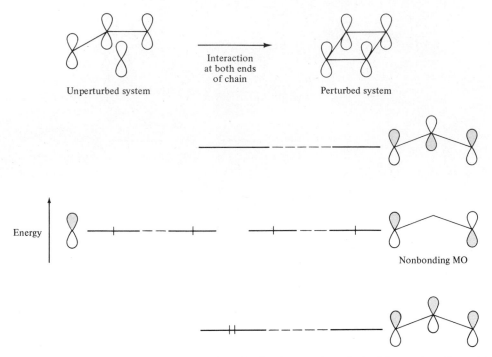

Figure 10.2 Union of a single carbon at both ends of a three-carbon chain. The nonbonding orbital is antisymmetric with respect to the mirror plane bisecting the system and cannot interact with the symmetric p orbital. Alternatively, a bonding interaction at one end is exactly canceled by an antibonding interaction at the other. No stabilization results; to a first approximation the cyclic system is less stable than the open chain (Figure 10.1) by $2\,\Delta E$ and is therefore antiaromatic.

at both ends lowers orbital energy by $2\beta/\sqrt{3}$ and gives stabilization energy $4\beta/\sqrt{3}$. The six-membered ring is more stable than the acyclic model and hence aromatic.

If a ring contains a phase inversion, the effect is to *change the sign of β* between the two orbitals where the phase inversion occurs. We are free to choose the phase inversion to come at the point of the final junction that closes the ring. Then the energy changes for unions to form the linear models are the same as before, but for union to form the four-membered ring, the energy lowering is (two electrons) $2(\beta/\sqrt{2} - (-\beta)/\sqrt{2}) = 4\beta/\sqrt{2}$, and for the six-membered ring it is $2(\beta/\sqrt{3} + (-\beta)/\sqrt{3}) = 0$. Hence the aromaticity rule is reversed: $4n$ rings are stabilized and $4n + 2$ rings are destabilized.

Zimmerman has presented a similar analysis in which an analogy is again made between the pericyclic transition state energy levels and the Hückel π energy levels.[18] The energy-level patterns illustrated in Figure A1.5, Chapter 1 appendix,

[18](a) H. E. Zimmerman, *J. Am. Chem. Soc.*, **88**, 1564, 1566 (1966); (b) H. E. Zimmerman, *Acc. Chem. Res.*, **4**, 272 (1971); (c) H. E. Zimmerman, in *Pericyclic Reactions*, Vol. I, A. P. Marchand and R. E. Lehr, Eds., Academic Press, New York, 1977, p. 53.

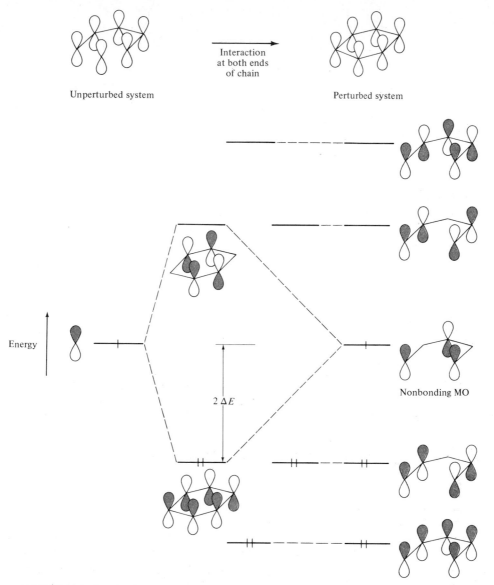

Figure 10.3 Union of a single carbon at one end of a five-carbon chain. Interaction is of the same type as in Figure 10.1; stabilization is $2 \Delta E'$.

show a pair of degenerate nonbonding orbitals for each of the $4n$ antiaromatic Hückel systems. If one starts with two ground-state ethylenes, as shown at the left in Figure 10.5, there will be two pairs of electrons in the two bonding π molecular orbitals. The transition state for their union has the cyclic interaction equivalent to cyclobutadiene; it therefore has one bonding MO and two degenerate nonbonding

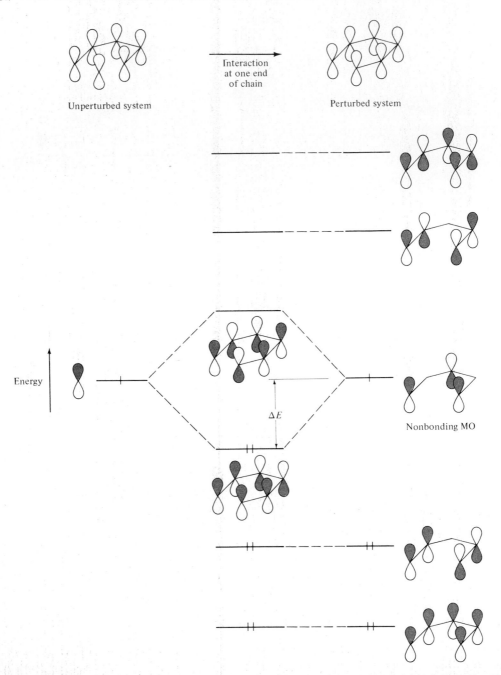

Figure 10.4 Union of a single carbon at both ends of a five-carbon chain. Both orbitals are symmetric, and a bonding interaction occurs at each end. The bonding combination is lowered by $2 \Delta E'$, giving a total stabilization for the two electrons of $4 \Delta E'$. The ring is more stable than the open chain (Figure 10.3) by $2 \Delta E'$ and is therefore aromatic.

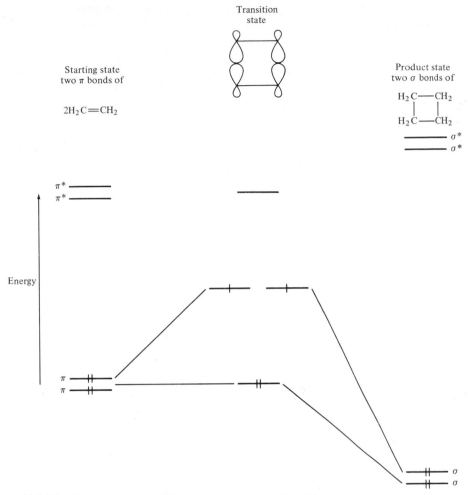

Figure 10.5 The Zimmerman analysis of the $\pi 2s + \pi 2s$ cycloaddition. The diagram traces the energies of the reacting orbitals from starting state (left) through transition state (center) to product (right). At the left, two bonding ethylene π orbitals are occupied by pairs of electrons. In the transition state there is only one bonding orbital; one pair of electrons is forced to climb an energy hill to the nonbonding level. In the product both electron pairs are again in bonding orbitals, this time σ orbitals of cyclobutane.

MO's. Figure 10.5 traces the energies of the relevant orbitals as the two ethylene molecules come together. One pair of electrons can remain in a bonding MO, but the other is forced to climb an energy hill to the nonbonding MO of the transition state before coming back down to a bonding σ orbital in the product cyclobutane. The antiaromatic nature of the transition state thus imposes a substantial energy barrier, and the model explains the high activation energy of the reaction.

As we have noted above, the presence of one phase inversion in the anti-Hückel rings causes the $4n$ rings to be aromatic and the $4n + 2$ rings to be antiaromatic. Zimmerman has pointed out that there is a method of discovering energy-level patterns for anti-Hückel rings similar to that illustrated for Hückel rings in Figure A1.5 in the Appendix to Chapter 1. Figure 10.6 shows regular polygons inscribed with one side downward in circles of radius 2β. Projection horizontally of each vertex gives the energy-level diagram for the anti-Hückel ring.

Several other investigators have also used aromaticity theory to analyze pericyclic reactions.[19]

10.4 FRONTIER ORBITAL INTERACTIONS IN PERICYCLIC REACTIONS

A second approach to pericyclic reactions uses perturbation theory and symmetry to analyze interactions of the frontier orbitals (highest occupied molecular orbital, HOMO, and lowest unoccupied molecular orbital, LUMO).[20] This is the method Woodward and Hoffmann used in their first paper on pericyclic selection rules;[21] it has subsequently been developed by Fukui and others.[22]

Symmetry in Pericyclic Reactions

In Section 1.7 we discussed the formation of symmetry correct molecular orbitals and showed how they are classified as symmetric (S) or antisymmetric (A) with respect to symmetry operations of the molecule.[23] We also illustrated the use of symmetry in assessing interactions among orbitals: Only orbitals of the same symmetry can interact.

Figures 10.7 and 10.8 show examples that review the concepts of orbital symmetry classification. Figure 10.7(a) shows the π bonding orbital of ethylene together with two of the symmetry elements of the molecule, the mirror planes σ and σ'. This orbital is symmetric (S) with respect to reflection in plane σ, but antisymmetric (A) with respect to reflection in the plane σ'. The π^* orbital [Figure 10.7(b)] is of a different symmetry type, antisymmetric (A) with respect to both reflections.

Figure 10.8 shows the π orbitals of butadiene, which are symmetry correct with respect to the two mirror planes and the C_2 axis that are symmetry elements of

[19](a) M. J. Goldstein and R. Hoffmann, *J. Am. Chem. Soc.,* **93**, 6193 (1971); (b) L. Salem, *J. Am. Chem. Soc.,* **90**, 543, 553 (1968); (c) J. J. C. Mulder and L. J. Oosterhoff, *Chem. Commun.,* 305, 307 (1970); (d) W. J. van der Hart, J. J. C. Mulder, and L. J. Oosterhoff, *J. Am. Chem. Soc.,* **94**, 5724 (1972); (e) W. C. Herndon, *Chem. Rev.,* **72**, 157 (1972).

[20]The idea that symmetry properties of orbitals might be important in determining the course of certain reactions apparently originated with L. J. Oosterhoff, quoted by E. Havinga and J. L. M. A. Schlatmann, *Tetrahedron,* **16**, 146 (1961).

[21]R. B. Woodward and R. Hoffmann, *J. Am. Chem. Soc.,* **87**, 395 (1965).

[22](a) K. Fukui, *Theory of Orientation and Stereoselection,* Springer-Verlag, Berlin, 1975; (b) K. N. Houk, *Acc. Chem. Res.,* **8**, 361 (1975); (c) K. N. Houk, in *Pericyclic Reactions,* Vol. II, A. P. Marchand and R. E. Lehr, Eds., Academic Press, New York, 1977, p. 181.

[23]When there are axes of symmetry of threefold or higher, there will be orbitals that transform in more complicated ways. We shall not have occasion to use these higher symmetries in our discussion.

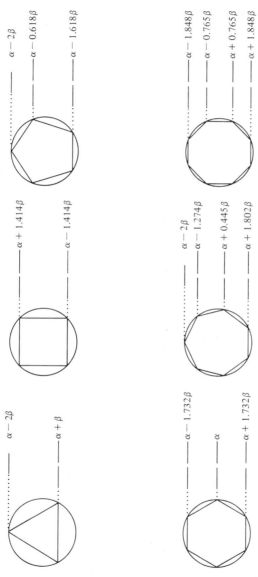

Figure 10.6 Energy levels of anti-Hückel π systems. Polygons are inscribed side down in circles of radius 2β; projection to the side gives the π energy levels. The four-membered ring, for example, has energies $\alpha + \sqrt{2}\beta$; $\alpha + \sqrt{2}\beta$; $\alpha - \sqrt{2}\beta$; $\alpha - \sqrt{2}\beta$. These energy levels are used as models for transition state energies of anti-Hückel pericyclic reactions. From H. E. Zimmerman, in A. P. Marchand and R. E. Lehr, Eds., *Pericyclic Reactions*, vol. I, Academic Press, New York, 1977, p. 64. Adapted by permission of Academic Press, Inc.

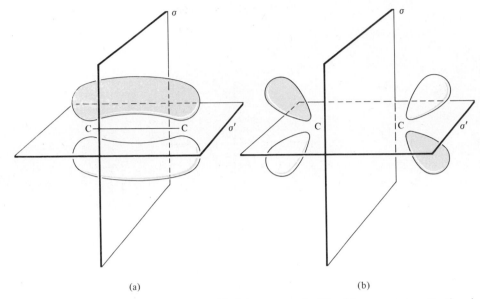

(a)   (b)

Figure 10.7 (a) The ethylene π bonding orbital is symmetric (S) with respect to reflection in the mirror plane σ, and antisymmetric (A) with respect to reflection in the mirror plane σ'. (b) The ethylene π^* antibonding orbital is antisymmetric (A) with respect to reflection in both σ and σ'.

the molecule in the *s*-cis conformation. To avoid confusion we must emphasize that Figure 10.8 depicts *molecular orbitals*. Although the drawings should indicate extended lobes (**40** and **41** for π_1 and π_2), the difficulty of making such diagrams, especially in the more complex structures, dictates a simplified notation in which one merely indicates the relative signs with which the basis functions are combined

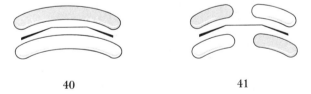

40   **41**

in each MO. Molecular orbitals shown in the manner of Figure 10.8 look very similar to the diagrams we used in Section 10.2 to show sets of basis orbitals. In the remainder of this chapter we shall be discussing molecular orbitals and the diagrams will show molecular orbitals unless otherwise specified.

Figure 10.8 shows that the π orbitals of butadiene all have the same symmetry classification, A, with respect to the mirror plane σ'. Classification with respect to σ' therefore does not contribute to distinguishing among the four orbitals. The purpose of symmetry classification is to separate the orbitals into categories so that interactions can be assessed; symmetry elements that do not assist in this classifica-

Symmetry elements

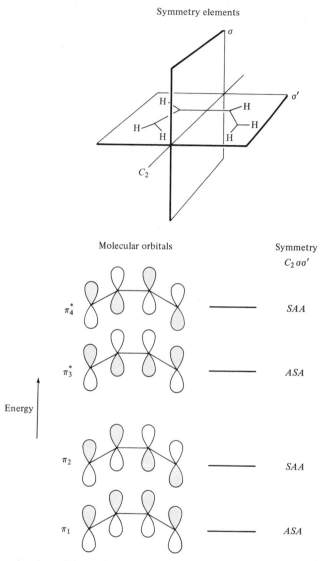

Figure 10.8 The π molecular orbitals of butadiene. At the top are shown the symmetry elements of the molecule, two mirror planes and a C_2 axis. Below are the four molecular orbitals in an energy-level diagram, with their symmetry behavior under each of the symmetry operations listed at the right. Note that the diagrams are of molecular orbitals shown as composites of the constituent basis orbitals, each with its appropriate relative phase.

tion add no useful information and are therefore usually omitted. Such omissions are made in some of the examples we consider below; the extra symmetry elements can always be included, but they will contribute nothing new to the argument. We also need another idea, already introduced in Section 1.9, that in some circumstances it is appropriate to use a symmetry element that is not strictly, but rather only approximately, a correct symmetry element of the molecule. The reason we can do this is that in pericyclic reactions we focus on those orbitals in the molecule that are actually involved in the bonding changes of interest, the reacting orbitals. The system of reacting orbitals may have an intrinsic symmetry that will be appropriate to use, but there may be a substituent in the molecule that, technically speaking, destroys this symmetry. If the substituent does not interact strongly with any of the reacting orbitals, the situation should be the same as if the substituent were not there, and the higher symmetry of the reacting orbitals themselves will be the proper one to use.

Let us suppose, for example, that we wish to know about the interaction of the π orbitals of butadiene with orbitals of some other molecule. If the butadiene is unsubstituted (Figure 10.8), the molecular symmetry and the symmetry needed for the orbital model are the same. If we now consider 1,3-pentadiene Figure 10.9, we find that the introduction of the methyl group has removed two of the symmetry elements. We may nevertheless argue that the methyl group will have only a secondary effect on the π system, the main features of which should be approximately the same as in butadiene itself. For purposes of qualitative arguments, we can therefore ignore the methyl substitution and take the symmetry of the π system itself as the applicable symmetry. The proper symmetry to use in this instance, then, is the same as that of the unsubstituted butadiene. In using approximate symmetries we are doing much the same thing we did in constructing interaction diagrams. We are identifying the intrinsic symmetry properties of the set of reacting orbitals that constitutes the energy-determining part of the interacting system.

We have discussed in Section 1.8 the application of perturbation theory to processes in which two molecules come together. We saw there that the most important interactions will be between the HOMO of one molecule and the LUMO of the other. This method can serve as a useful guide in deciding whether there will be a stabilization as a pericyclic reaction begins to occur. Let us consider ethylene and butadiene approaching each other in the manner shown in **42**. We wish to consider

42

the interactions of the HOMO of ethylene with the LUMO of butadiene, and of the LUMO of ethylene with the HOMO of butadiene. In order to do this, we identify first the symmetry of the aggregate of the two molecules arranged as in **42**; there is one symmetry element, a vertical mirror plane that cuts across the ethylene double bond and the butadiene single bond. In Figure 10.10 we place at the left the

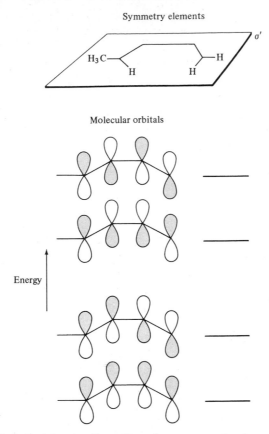

Symmetry elements

Molecular orbitals

Energy

Figure 10.9 The π orbitals for 1,3-pentadiene. Now the correct molecular symmetry consists of only the plane σ', but the π orbitals are still constructed according to their approximate local symmetry, which remains as shown in Figure 10.8.

π orbitals of butadiene, and at the right the π orbitals of ethylene. The symmetry classification of each orbital is with respect to the mirror plane shown. (Note that the other symmetry elements that would be applicable to butadiene alone or to ethylene alone are not correct symmetry elements for the combination **42** and are not included.) The interaction of each HOMO with the other LUMO is permitted by symmetry; as we have seen in Section 1.8, this is just the situation that leads to stabilization, and we may expect this cycloaddition to be facile, as indeed it is.

If we now look at the cycloaddition of two butadiene molecules to each other (Figure 10.11), we find that because of the symmetry mismatch between the HOMO of one molecule and the LUMO of the other, there can be no stabilization. The only interaction is between filled levels, which, as we have seen in Section 1.6, is destabilizing. Hence this process should not occur readily, a conclusion that is again in agreement with experiment.

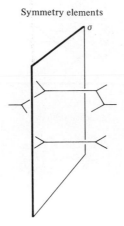

Symmetry elements

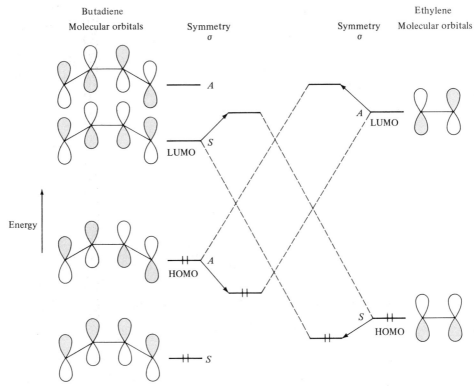

Figure 10.10 HOMO–LUMO interactions in the approach of ethylene to butadiene. The symmetry (σ) is shown at top center. At left are the butadiene π MO's, classified according to their symmetry with respect to σ; at right are the ethylene π MO's, also classified according to σ. The HOMO of each molecule can interact with the LUMO of the other, and a stabilization occurs as they approach one another.

Symmetry elements

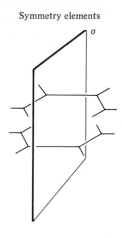

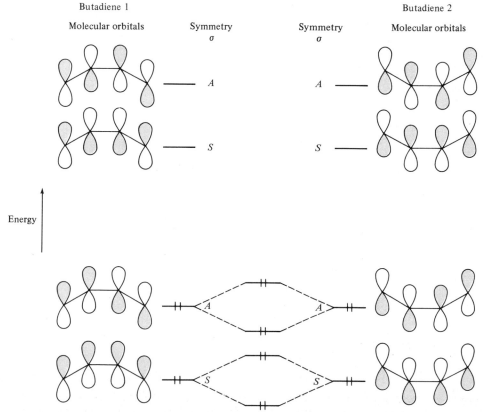

Figure 10.11 The approach of two butadiene molecules. The symmetries do not permit HOMO–LUMO interaction; the interaction between filled levels, permitted by the symmetry, gives no stabilization.

10.5 CORRELATION DIAGRAMS

Although their initial paper looked at pericyclic reactions in terms of frontier orbitals, Woodward and Hoffmann's subsequent discussions have for the most part used orbital correlation.[24] In the orbital correlation method one first identifies in the starting molecules and in the products those molecular orbitals (the reacting orbitals) that correspond to bonds being formed or broken during the pericyclic process. To make the connection with the aromaticity analysis in Section 10.2, we note that these molecular orbitals are just the ones constructed from the basis atomic orbitals that make up the interaction diagram for the reaction. It is, in fact, through the interaction diagram that one may establish the equivalency of the orbital correlation and the aromaticity approaches to pericyclic theory.

Having identified the reacting orbitals, we next find those symmetry elements of the system that are maintained throughout the reaction. The reacting orbitals are then classified according to their symmetries with respect to these elements and are traced (correlated) through the reaction from starting materials to products, always conserving their symmetry properties.

We shall illustrate these principles with several examples. Consider first the electrocyclic interconversion of butadiene and cyclobutene, looked at in the (non-spontaneous) ring-closing direction (Equation 10.35.) It will be recalled that this

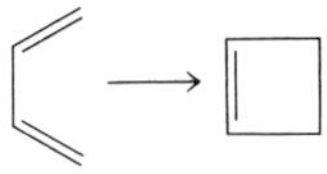

$$(10.35)$$

process can occur in either of two ways: disrotatory (Equation 10.36) or conrotatory (Equation 10.37) and that the thermal reaction exhibits a marked preference for the conrotatory path.

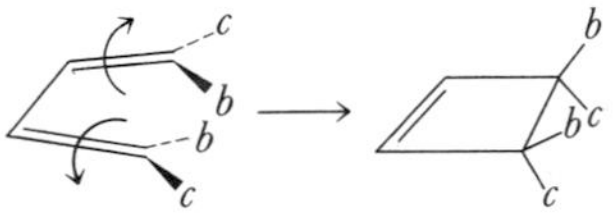

$$(10.36)$$

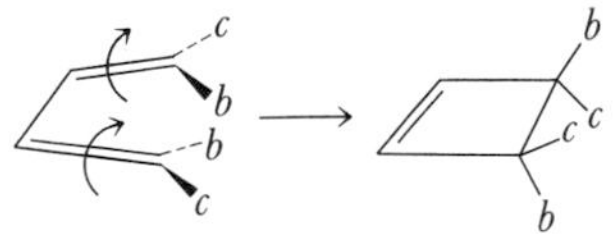

$$(10.37)$$

The reacting orbital system needed to include all the bonds being formed or broken is made up in the reactant from a basis consisting of a p orbital on each of

---

[24] (a) R. Hoffmann and R. B. Woodward, *Acc. Chem. Res.,* **1,** 17 (1968); (b) R. B. Woodward and R. Hoffmann, *The Conservation of Orbital Symmetry,* Verlag Chemie, Weinheim/Bergstr., Germany, and Academic Press, New York, 1970. Although the orbital correlation theory is in some respects more complex than the Dewar–Zimmerman aromaticity theory, it allows better insight into the energy changes of individual orbitals during the pericyclic reaction and gives a better understanding of excited-state reactions.

the four carbon atoms, and in the product of p orbitals on each of the two π bonded carbons and a hybrid orbital, approximately sp^3, on each of the two carbons linked by the newly formed σ bond. Figure 10.12 illustrates the reacting molecular orbitals.

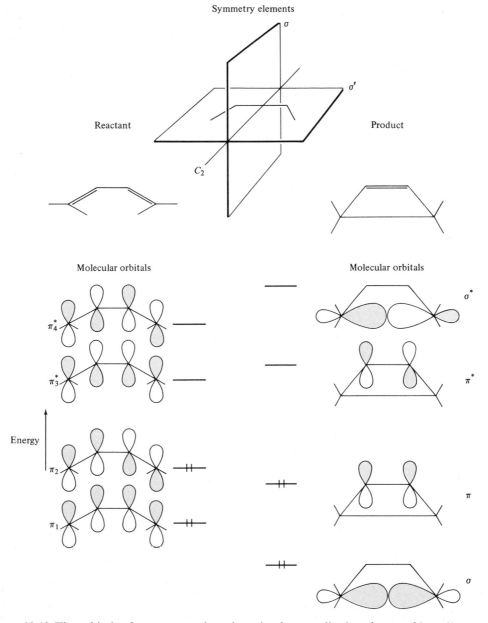

Figure 10.12 The orbitals of reactants and products in electrocyclic ring closure of butadiene.

Consider now the changes that will occur when a conrotatory ring closure takes place. Figure 10.13 shows that as soon as the end carbons of the diene begin their rotation, the symmetry of the reacting orbitals changes. The two mirror planes are no longer present, and only the C_2 axis remains. We now wish to follow the molecular orbitals through the transformation using the principle that a molecular orbital *maintains its symmetry* during a bonding reorganization.

Figure 10.14 reproduces the important molecular orbitals and classifies them according to their symmetry with respect to the C_2 axis, the element that defines the symmetry during the conrotatory process.[25] Orbital π_1, antisymmetric under C_2, must change continuously into an orbital of the product in such a way as to remain at all stages antisymmetric under the C_2 operation. The symmetry conservation principle allows us to reconstruct qualitatively how the orbital π_1 will change. Since it starts out antisymmetric under C_2, it remains so; it can do this only if it ends up as an antisymmetric orbital of the product, say π. As the two end carbons rotate, the contribution of the p orbitals on those end carbons must decrease, finally to disappear altogether. In π_2, symmetry type S, the contribution of the two central p orbitals will decrease, leaving only the end two, which will have rotated onto each other to yield the product σ orbital. The changes in the antibonding orbitals may be visualized in a similar way.

The lines joining reactant and product orbitals in Figure 10.14 are referred to as *correlation* lines, and the entire diagram is an *orbital correlation diagram*. It will be noted that since there are two orbitals of each symmetry type on each side, there is an alternative way the correlation might have been made, namely, π_1 to σ^*, π_2 to

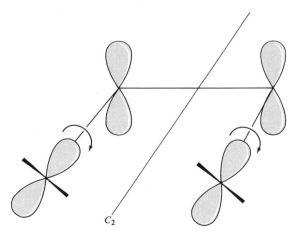

Figure 10.13 Symmetry at an intermediate stage of the conrotatory closure. Only the C_2 axis remains as a symmetry element.

[25] In Figure 10.14, and in subsequent energy-level diagrams we shall construct, the relative energies of reactant and product levels are represented only schematically. One cannot deduce from diagrams of this type the overall thermochemistry of the reaction.

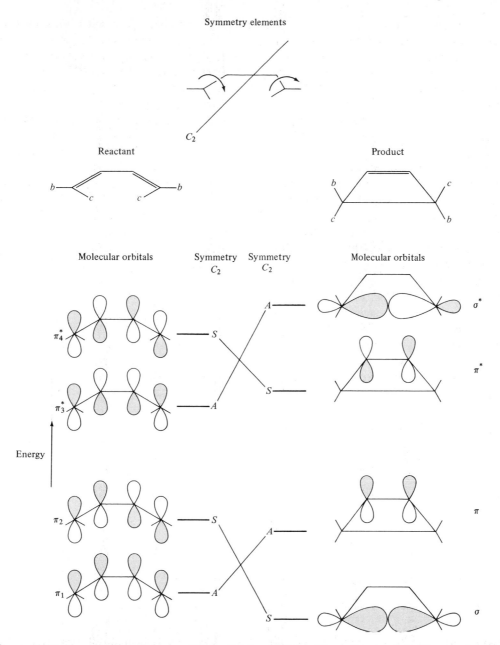

Figure 10.14 Classification of the reacting molecular orbitals of butadiene and cyclobutene for the conrotatory process. Symmetry classifications are with respect to the C_2 axis, S indicating symmetric and A antisymmetric orbitals. The correlation lines are obtained by connecting orbitals of the same symmetry.

π^*, π_3^* to π, π_4^* to σ. This alternative is eliminated by the noncrossing rule: *Orbitals of the same symmetry do not cross.*

Because the noncrossing rule is of fundamental importance to the construction of correlation diagrams, we digress a moment to give a justification for it. We have already seen in Section 1.6 that orbitals of the same symmetry interact in such a way as to push the lower-energy one still lower and the higher-energy one still higher, and that the interaction is stronger the closer the two are in energy. If two orbitals of the same symmetry were to approach each other in energy, the interactions would therefore tend to keep them apart and prevent the lines from crossing. The noncrossing rule does not apply to orbitals of different symmetries, because they do not interact. Hence correlation lines representing orbitals of different symmetry may cross.

Although the noncrossing rule may in many instances be relied upon to determine the correlation pattern where alternatives exist, it is not infallible. In order to avoid difficulties in constructing correlation diagrams, Woodward and Hoffmann cite three precautions that should be observed.[26]

1. Processes that are inherently independent must be considered separately even if they occur in the same molecule.

2. Each reacting system must be reduced, by removing substituents and distortions, to its highest inherent symmetry.

3. Symmetry elements used in the analysis must bisect bonds made or broken in the reaction.

Having obtained the orbital correlation diagram for the butadiene closure, we can now see that the electron pairs that start out in bonding orbitals of the reactant are transferred into bonding orbitals of the product without encountering any symmetry imposed energy barrier. The process is said to be *symmetry allowed.* This principle lies at the heart of the theory of orbital symmetry control of reaction path: *A reaction is allowed in the ground state (thermally allowed) only when all reactant bonding electron pairs are transferred without symmetry imposed barrier into bonding orbitals of the product.*

The Disrotatory Electrocyclic Reaction

We turn now to the disrotatory closure. The orbitals in reactants and products are the same as before, but this time it is the mirror plane σ (Figure 10.15) that is maintained throughout. Figure 10.16 shows that the symmetry dictates correlation of a bonding to an antibonding orbital; the electron pair in orbital π_2, if it remained in its original orbital, would end up in the high-energy orbital π^* of the product, a process that would require a large energy input. A similar situation would arise in the reverse reaction starting from ground-state cyclobutene, levels σ and π occupied. In this case, then, the symmetry imposes a sizable barrier, and the thermal reaction is said to be *symmetry forbidden.*

[26]R. B. Woodward and R. Hoffmann, *The Conservation of Orbital Symmetry,* Verlag Chemie, Weinheim/Bergstr., Germany, and Academic Press, New York, 1970, p. 31.

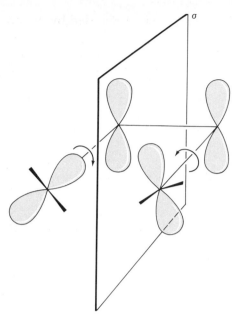

Figure 10.15 Symmetry at an intermediate stage of the disrotatory closure. Only the mirror plane σ remains as a symmetry element.

It is well to recall at this point that the terms allowed and forbidden in this context are not absolute. First, a symmetry allowed reaction will not necessarily take place with low activation energy. Other requirements, such as ease of approach, favorable overlap of orbitals, and steric interactions may well combine to impose a substantial activation barrier where none is predicted on the basis of symmetry alone. Moreover, favorable arrangement of orbital energy levels may permit a forbidden reaction to occur more readily than might have been expected. Second, the symmetry rules are based on the assumption that the reactions are concerted. There are always available nonconcerted pathways going through intermediates; the orbital symmetry predictions are valid only if all the bonding changes occur together.

It is instructive now to turn to the correlation diagrams in Figures 10.17 and 10.18 for conrotatory and disrotatory closure of hexatriene, a six π electron system. The disrotatory mode is now thermally allowed, the conrotatory forbidden. If correlation diagrams for larger systems are constructed, it will be found that with each addition of two carbons and an electron pair the predicted selectivity will reverse.

Photochemical Reactions

The correlation for a thermally forbidden reaction shows a bonding orbital (usually the highest occupied) correlating to an antibonding one (usually the lowest unoccupied). If absorption of a quantum of light of a suitable frequency raises the reactant

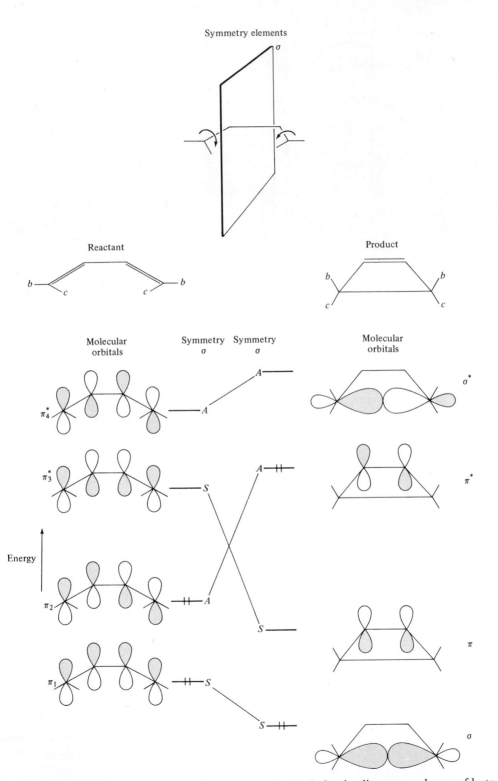

Figure 10.16 Symmetry classification and correlation of orbitals for the disrotatory closure of butadiene. Closure with electron pairs remaining in their original levels would lead to the excited configuration indicated by the orbital occupancy on the right.

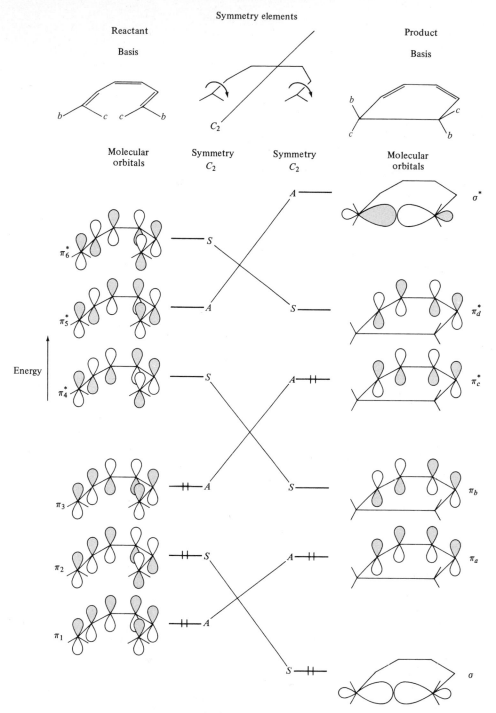

Figure 10.17 Correlation diagram for the thermally forbidden conrotatory closure of hexatriene.

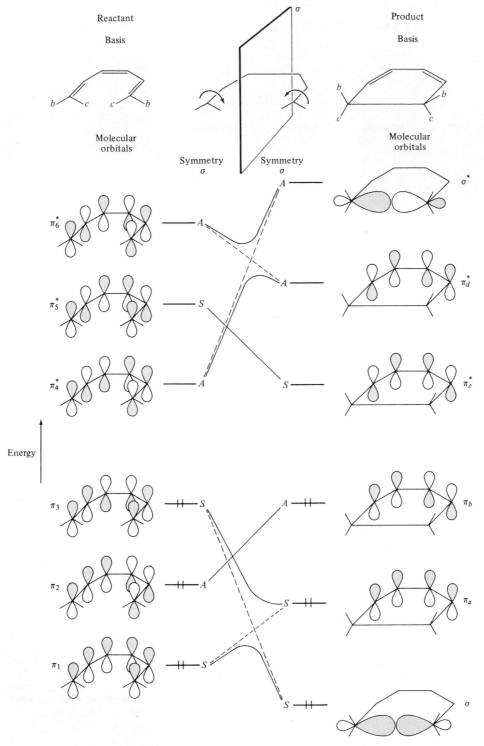

Symmetry elements

Reactant

Basis

Product

Basis

Molecular
orbitals

Symmetry
σ

Symmetry
σ

Molecular
orbitals

Energy

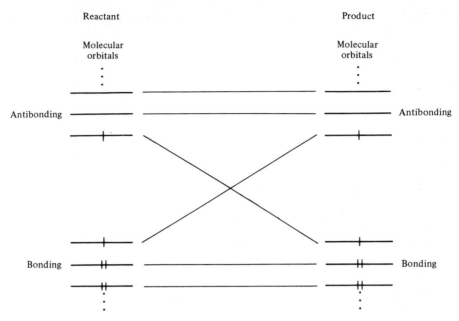

Figure 10.19 Orbital correlation diagram for a photochemically allowed, thermally forbidden pericyclic reaction. Absorption of a quantum of light raises one electron of the starting material from the highest occupied to the lowest unoccupied orbital. The system can pass easily to product because the energy cost of raising one electron is compensated by lowering the other. (Here we are approximating the lowest-energy excited state of reactant and product by the electron configurations shown. See Section 10.6 for further discussion of this point.)

to its first electronic excited configuration, the orbital occupancy will be as indicated in Figure 10.19. Now, during the reaction, as one electron goes up in energy, another comes down; correlation (assuming no switching of orbitals by the electrons) is directly to a product configuration of comparable energy to the starting configuration and the process should be allowed. Inspection of Figures 10.14 or 10.18 indicates, moreover, that an electronic ground-state allowed process should be forbidden photochemically because the lowest-energy excited configuration of reactants correlates to a more highly excited configuration of products. The orbital symmetry rules therefore predict reversal of the allowed or forbidden nature of a

Figure 10.18 Correlation diagram for the thermally allowed disrotatory closure of hexatriene. The forms of the molecular orbitals indicate correlations according to the dashed lines. The dashed-line correlations, however, would cause crossing of orbitals of the same symmetry. As two orbitals of the same symmetry approach each other in energy, they mix and interact strongly. In this mixing process each takes on character of the other. The orbital initially at lower energy remains at lower energy while changing its form; similarly, the orbital initially at higher energy remains higher. In this way the intended crossing is avoided. See Section 10.6 for further discussion of intended crossings.

given process when carried out photochemically. A word of caution is necessary here. Chemistry of electronic excited states is complex and less well understood than that of ground states. The simple correlation diagrams we have been using are inadequate for describing excited states in detail, and the predictions are less reliable than for ground-state reactions.[27] Chemistry of excited states is discussed further in Section 10.6 and in Chapter 12.

Cycloaddition

In Section 10.1 we pointed out that a compound may enter into a cycloaddition in either of two ways, suprafacial or antarafacial. First we shall construct correlation diagrams for the simple $\pi 2s + \pi 2s$ and $\pi 2s + \pi 4s$ cycloadditions.[28]

Figure 10.20 illustrates the symmetry appropriate to the all-suprafacial approach of two ethylenes. We have followed Rule 3 of p. 796 in singling out for attention the symmetry elements that bisect reacting bonds. We have not included others, such as a C_2 axis passing midway between the two molecules perpendicular

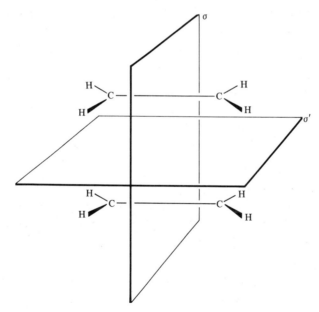

Figure 10.20 Symmetry elements for the suprafacial approach of two ethylene molecules.

---

[27] R. C. Dougherty, *J. Am. Chem. Soc.,* **93,** 7187 (1971) has argued that systems whose ground states are aromatic have antiaromatic excited states and vice versa, and that therefore the universal criterion for allowed pericyclic reactions, both ground and excited-state, is that the transition state be aromatic. The uncertainty of our present knowledge of excited states nevertheless indicates that the more restricted statement given here is to be preferred.
[28] R. Hoffmann and R. B. Woodward, *J. Am. Chem. Soc.,* **87,** 2046 (1965), and Woodward and Hoffmann, *The Conservation of Orbital Symmetry,* Verlag Chemie, Weinheim/Bergstr., Germany, and Academic Press, New York, 1970.

to the page, that do not. (These symmetry elements not bisecting reacting bonds could be included without affecting the outcome of the analysis. There is danger of error only when such elements are used exclusively.) The atomic orbital basis consists of a p function on each of the four carbon atoms; Figure 10.21 illustrates the derived symmetry correct molecular orbitals. As a result of the reflection plane that carries one ethylene into the other, these molecular orbitals are delocalized over both molecules. Figure 10.21 also shows the orbitals of the product; we consider only the two C—C σ bonds formed and ignore the other two, which were present from the beginning and did not undergo any change.

When the two ethylene molecules are far apart, orbitals π_1 and π_2 are essentially the bonding π orbitals, and will therefore each be occupied by an electron pair if both ethylenes are in their ground states. As the molecules approach, the intermolecular interaction of one of these orbitals is antibonding. One electron pair therefore finds itself in an orbital that is increasing in energy, and would, if it remained in that orbital throughout, end up in one of the σ^* orbitals of the product to yield the excited configuration indicated on the right in the diagram. The pattern is the one we associate with a thermally forbidden process.

The situation is reversed for the $\pi 2s + \pi 4s$ addition. Figure 10.22 illustrates this case; now the bonding orbitals all transform directly to bonding orbitals of the product and there is no symmetry imposed barrier. As with the electrocyclic processes, the correlation diagrams illustrate clearly the reason for the striking difference observed experimentally when the number of electrons is increased from four to six. The reader may verify that the $4s + 4s$ reaction will be forbidden. Each change of the total number of electrons by two reverses the selection rule.

Figure 10.23 shows the geometry for a $\pi 2s + \pi 2a$ addition. The correlation diagram presented in Figure 10.24 shows that the change of one component to the antarafacial mode of addition has caused a change from forbidden to allowed. When both ethylene units enter in the antarafacial manner (Figures 10.25 and 10.26), the reaction is again forbidden. It may be seen that a cycloaddition of two molecules will be thermally allowed when the total number of electrons is $4q + 2$, $q = 0, 1, 2, \ldots$, if both components are suprafacial or both antarafacial, and when the total number of electrons is $4q$, $q = 0, 1, 2, \ldots$, if one component is suprafacial and one antarafacial.

The Generalized Woodward–Hoffmann Pericyclic Selection Rules

Comparison of the Woodward–Hoffmann orbital correlation method with the Dewar–Zimmerman aromaticity method shows that the reacting molecular orbitals of a correlation diagram are constructed from exactly those basis orbitals that make up the interaction diagram of the transition state. That interaction diagram can be thought of as being built up from components, where a component is a fragment corresponding to one of the molecular orbital systems used in the Woodward–Hoffmann correlation diagram. A *component,* then, may be defined as *a fragment corresponding to any subdivision of an interaction diagram.* Ordinarily, components are chosen to correspond to a set of orbitals of a particular bond system, but a

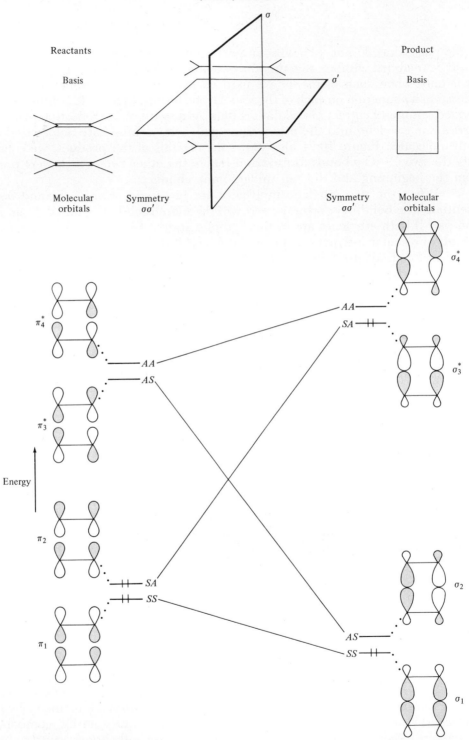

Figure 10.21 Orbitals and orbital correlations for the ground-state $\pi 2s + \pi 2s$ cycloaddition. In this example, at the left side of the diagram, reflection in the mirror plane σ' transforms the p orbitals of one ethylene molecule into those of the other; hence each molecular orbital of the combined system has a contribution from each of the four p orbitals.

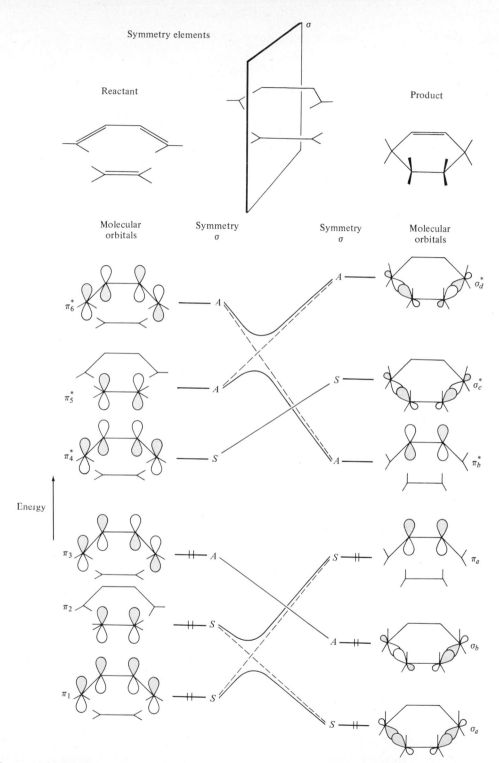

Figure 10.22 Symmetry and correlation diagram for the $\pi 2s + \pi 4s$ cycloaddition. Because no symmetry element transforms orbitals of one molecule into those of the other, ethylene and butadiene orbitals may be considered separately. The intended crossings of orbitals of the same symmetry (dashed lines) are avoided, so that the actual correlations are as indicated by the solid lines.

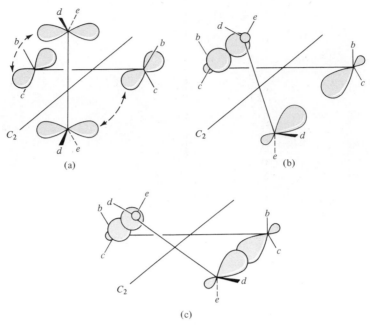

Figure 10.23 The geometry of approach for the $\pi 2s + \pi 2a$ cycloaddition. A C_2 axis of rotation is maintained throughout.

component could be any arbitrarily defined fragment of the interaction diagram. The other requirements we impose for a component are that the basis orbital signs be chosen in such a way that *the interaction path within the component connects only lobes of like sign* and that *the number of electrons a component carries is even.* (This latter restriction is possible because we are considering processes for which the total number of electrons is even.)

With a component defined as described above, we can now state concisely what we mean by suprafacial and antarafacial components:

A suprafacial component is one that interacts with its neighbors through lobes at its ends that have the same algebraic sign.

An antarafacial component is one that interacts with its neighbors through lobes at its ends that have opposite algebraic signs.

We illustrate these ideas by constructing interaction diagrams for some of the reactions for which we have examined correlation diagrams. Look first at the cyclobutene–butadiene electrocyclic reaction. Considered in the ring-opening di-

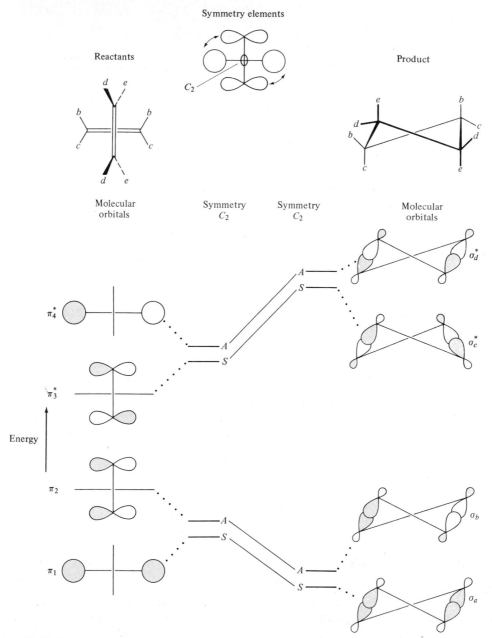

Figure 10.24 Orbital correlation diagram for the $\pi 2s + \pi 2a$ cycloaddition. Reactants are arranged as in Figure 10.23(a), with p orbitals on the more remote ethylene unit seen end-on. The product corresponds to Figure 10.23(c).

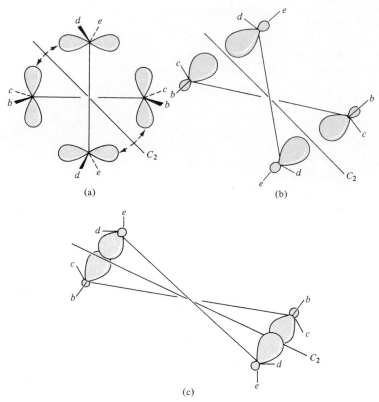

(a)   (b)

(c)

Figure 10.25 The geometry of approach for the $\pi 2a + \pi 2a$ cycloaddition. The C_2 axis is midway between the two molecules.

rection it is a $\pi 2 + \sigma 2$ process. The disrotatory opening (Equation 10.38) is $\pi 2s + \sigma 2s$:

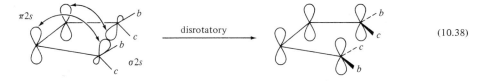

$$\text{(10.38)}$$

Components are

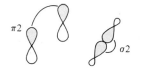

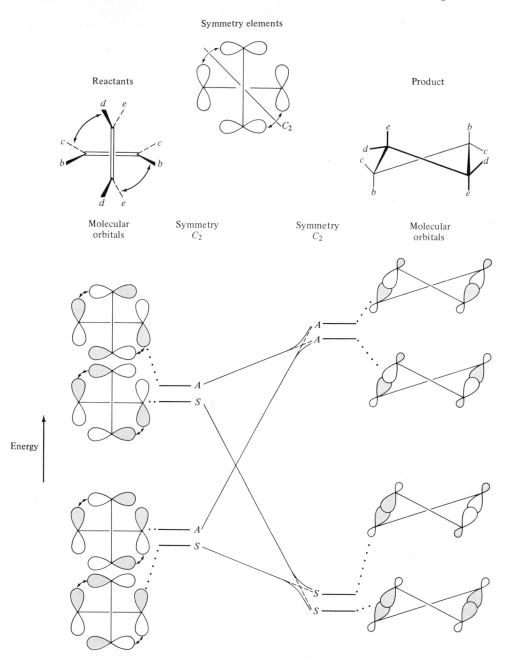

Figure 10.26 Orbital correlation diagram for the $\pi 2a + \pi 2a$ cycloaddition. The forms of the molecular orbitals indicate a correlation according to the dashed lines. The intended crossings of orbitals of the same symmetry are avoided, and the actual correlations are according to the solid lines.

Construct the interaction diagram by joining the components according to the interaction path established by disrotatory opening:

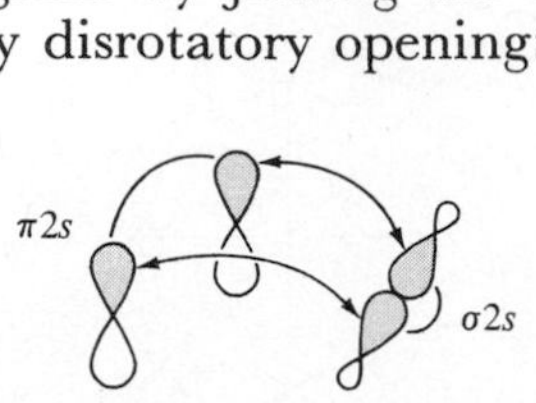

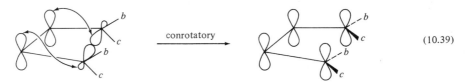

44

Both components are suprafacial. The interaction diagram shows a Hückel ring of four electrons; the process is thermally forbidden.

The conrotatory opening (Equation 10.39) has components as shown in **45**;

$$\text{conrotatory} \tag{10.39}$$

the reaction establishes the interactions shown in **46**. The process is $\pi 2s + \sigma 2a$; the

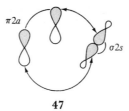

45   **46**

ring is anti-Hückel, and the reaction is thermally allowed for four electrons. We could just as well have called this reaction $\pi 2a + \sigma 2s$, as shown in **47**. The resulting interaction diagram still has one phase inversion and is entirely equivalent to **46**.

47

The $\pi 2s + \pi 4s$ cycloaddition (Figure 10.22) would have components as illustrated in **48**, from which we derive interaction diagram **49**. The interaction diagram shows no phase inversion; a Hückel transition state of six electrons is aromatic and the reaction is thermally allowed. We could have subdivided the orbital systems further by considering each of the double bonds of butadiene as a separate component (**50**) and joined the three components in a $\pi 2s + \pi 2s + \pi 2s$

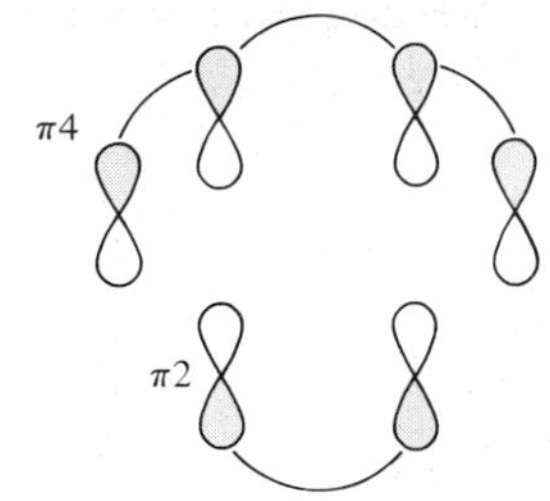

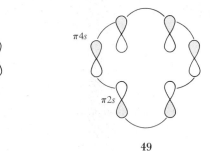

48   49

process (**51**). Our conclusion that the reaction is thermally allowed would have

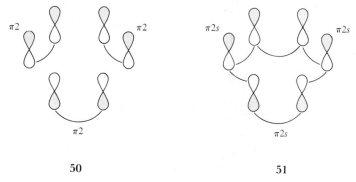

50   51

been unchanged even if we had set up the diagram as shown in **52** or **53**; the two phase inversions introduced by an awkward but permissible choice of signs are equivalent to no phase inversions, and the diagrams still show Hückel rings.

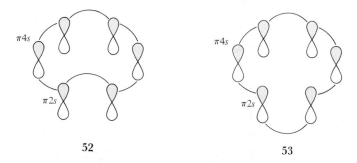

52   53

    The forbidden or allowed nature of the reaction will be determined by the suprafacial–antarafacial properties of the components and the number of electrons carried by each. We have seen that each addition of two electrons to the pericyclic ring reverses the selection rule, as does change of the faciality of a component. A generalized rule can thus be formulated in terms of the classification of the components according to two criteria: whether they are supra- or antarafacial, and whether they contain $4q + 2$ or $4r$ electrons, q and r being any integers including zero. Woodward and Hoffmann established in this way a general pericyclic selection rule:[29]

> **A pericyclic reaction is thermally allowed if the total number of $(4q + 2)s$ and $(4r)a$ components is odd.**

A simplified statement of the rule is possible if the interaction diagram is *always considered to be built from two-electron components,* so that there are no components of the $(4r)$ type:

> **A pericyclic reaction is thermally allowed if the number of suprafacial two-electron components is odd.**

It can be shown that the correspondence between the Woodward–Hoffmann selection rules and the Dewar–Zimmerman aromaticity rules illustrated in the few examples cited above is general. The two statements of the rules for pericyclic reactions are equivalent.[30]

Complex Systems

The reactions we have considered all have interaction diagrams consisting of a single closed loop of orbitals. Such diagrams are said to be *simply connected.* Non-simply connected systems have more than one ring in the interaction diagram. The Woodward–Hoffmann and Dewar–Zimmerman rules apply only to simply connected systems.

Reactions that are not simply connected are sometimes difficult to analyze. Dewar's perturbation theory, which we illustrated in Section 10.3, provides a method for assessing energy changes that occur when extra cross-ring interactions are introduced into a cyclic conjugated system.[31]

Woodward and Hoffmann have shown how to find solutions in complex systems by tracing correlations of orbitals from starting material to products by conserving the nodal structure when there are no symmetry elements bisecting the

[29] R. B. Woodward and R. Hoffmann, *The Conservation of Orbital Symmetry,* Verlag Chemie, Weinheim/Bergstr., Germany, and Academic Press, New York, 1970, p. 169. It is understood that reactions forbidden in the ground state are allowed in the excited state and vice versa.

[30] (a) A. C. Day, *J. Am. Chem. Soc.,* **97,** 2431 (1975); (b) T. H. Lowry and K. S. Richardson, *Mechanism and Theory in Organic Chemistry,* Harper & Row, New York, 1976, p. 611.

[31] M. J. S. Dewar, *The Molecular Orbital Theory of Organic Chemistry,* McGraw-Hill, New York, 1969, Chap. 6.

bonds being broken and formed.[32] They have analyzed the benzene–prismane example in Equations 10.40–10.41. The reaction is formally a $\pi2s + \pi2a + \pi2a$

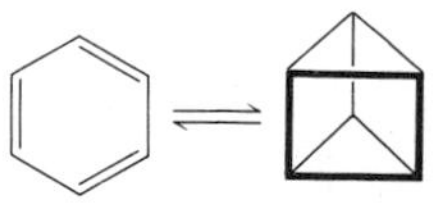

$$(10.40)$$

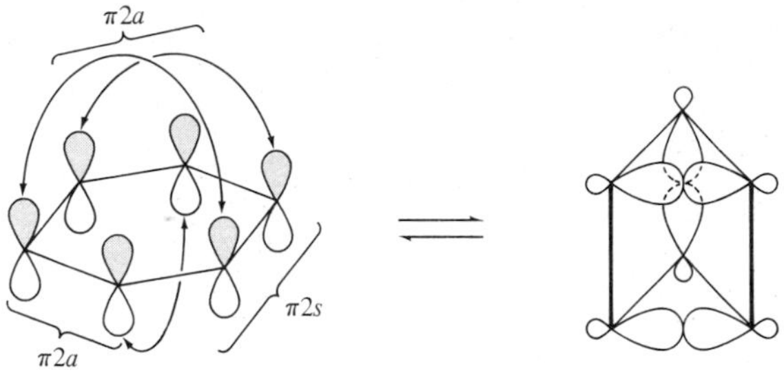

$$(10.41)$$

54

cycloaddition (**54**) and should be thermally allowed. The correlation diagram, however, shows that the symmetry imposes a correlation between bonding and antibonding levels, and so the thermal reaction is actually forbidden. Experimentally, Oth has found that the isomerization of hexamethylprismane to hexamethylbenzene (Equation 10.42) is exothermic by 91.2 kcal mole^{-1}.[33] Despite this large

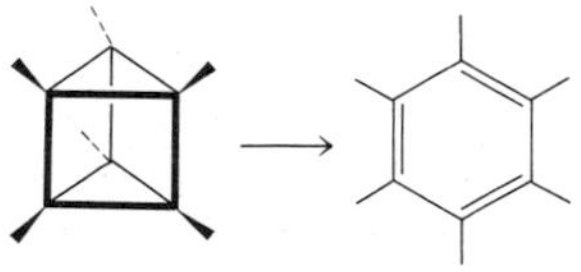

$$(10.42)$$

favorable energy change, the prismane rearranges by way of intermediate structures (Equation 10.43) with the activation enthalpies indicated. Hence the activation enthalpy for the direct conversion (Equation 10.42) must be greater than 33 kcal mole^{-1}. This barrier is remarkably high for a reaction so strongly exothermic; the conclusion that there is a symmetry imposed barrier seems justified.

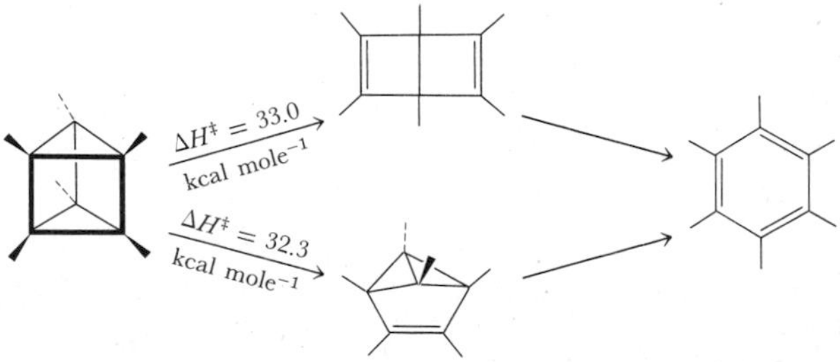

$$(10.43)$$

[32] R. B. Woodward and R. Hoffmann, *The Conservation of Orbital Symmetry,* Verlag Chemie, Weinheim/Bergstr., Germany, and Academic Press, New York, 1970, p. 107.
[33] (a) J. F. M. Oth, *Angew. Chem. Int. Ed. Engl.,* **7,** 646 (1968); (b) J. F. M. Oth, *Recl. Trav. Chim.,* **87,** 1185 (1968).

10.6 CORRELATION OF ELECTRONIC STATES

The Woodward–Hoffmann pericyclic reaction theory has generated substantial interest in the pathways of forbidden reactions and of excited-state processes, beginning with a paper by Longuet–Higgins and Abrahamson[34] that appeared simultaneously with Woodward and Hoffmann's first use of orbital correlation diagrams.[35] We have noted in Section 10.5 that the orbital correlation diagram predicts that if a forbidden process does take place by a concerted pericyclic mechanism,[36] and if electrons were to remain in their original orbitals, an excited configuration should be produced. Similarly, if reactants start out in a singly excited configuration (Figure 10.19), the diagrams imply that one electron will decrease in energy and the other increase, the result being a singly excited product configuration. A number of investigators have pointed out that under these circumstances the electron pairs are ordinarily not expected to remain in their original orbitals.[37]

In order to understand the reasoning supporting these arguments, it is necessary to appreciate the concept of an *electronic state*. Figure 10.27 shows at the top some of the many ways of assigning electrons to the orbitals of a hypothetical molecule in which it is supposed that an adequate model is obtained with six molecular orbitals and which has one symmetry element. Each of these assignments represents a different electronic configuration; the various configurations can be summarized as illustrated at the lower left side of the figure in an energy-level diagram.[38] Each configuration has associated with it a particular symmetry behavior under the symmetry operation of the molecule, found by taking the product of the symmetry ($+1$ for symmetric, -1 for antisymmetric) of each electron.[39]

It is important to realize that in an orbital energy-level diagram (Figure 10.27, top) each horizontal line represents an orbital and several orbitals will be occupied by electrons, whereas in a configuration diagram (Figure 10.27, lower left) each horizontal line represents a separate configuration of the entire molecule.

[34] H. C. Longuet–Higgins and E. W. Abrahamson, *J. Am. Chem. Soc.*, **87**, 2045 (1965).

[35] R. Hoffmann and R. B. Woodward, *J. Am. Chem. Soc.*, **87**, 2046 (1965).

[36] A stepwise pathway, not subject to the same symmetry restrictions, is always possible.

[37] (a) See note 34. See also, for example, (b) M. J. S. Dewar, *Angew. Chem. Int. Ed. Engl.*, **10**, 761 (1971); (c) W. Th. A. M. van der Lugt and L. J. Oosterhoff, *J. Am. Chem. Soc.*, **91**, 6042 (1969); (d) R. C. Dougherty, *J. Am. Chem. Soc.*, **93**, 7187 (1971); (e) J. E. Baldwin, A. H. Andrist, and R. K. Pinschmidt, Jr., *Acc. Chem. Res.*, **5**, 402 (1972); (f) R. Hoffmann, S. Swaminathan, B. G. Odell, and R. Gleiter, *J. Am. Chem. Soc.*, **92**, 7091 (1970); (g) N. D. Epiotis, *J. Am. Chem. Soc.*, **95**, 1191, 1200, 1206, 1214 (1973); (h) N. D. Epiotis and S. Shaik, *J. Am. Chem. Soc.*, **100**, 1, 9, 18, 29 (1978); (i) W. Gerhartz, R. D. Poshusta, and J. Michl, *J. Am. Chem. Soc.*, **98**, 6427 (1976); (j) D. H. Williams, *Acc. Chem. Res.*, **10**, 280 (1977).

[38] Our configuration specification in Figure 10.27 is still oversimplified. We have omitted specification of electron spin and have also shown the orbital energies as though they were independent of occupancy, an approximation that does not hold in the self-consistent-field (SCF) molecular orbital theories that are essential for adequate treatment of excited states. Furthermore, the configurations should be given in the form of Slater determinants, which are products of functions in which individual electrons are permuted among the occupied orbitals in such a way as to include the requirements of the Pauli exclusion principle.

[39] The discussion is restricted to molecules with axes of rotational symmetry of order no higher than twofold. More complex symmetry designations arise when a rotation axis of order three or greater is present.

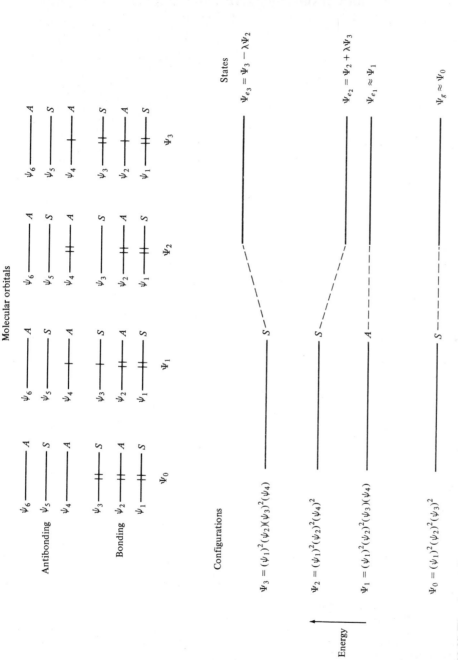

Figure 10.27 The electronic ground state and a few excited states of a hypothetical molecule with one symmetry element. Above, orbital energy-level diagrams specify symmetry properties and illustrate orbital occupancies for various configurations; the molecular orbital set is the same for each configuration in this approximation. Below, left, the configurations are shown on an energy scale. Each configuration function is a product of orbital functions for the electrons, with configuration symmetry determined by multiplying the symmetries of the electrons (+1 for S, −1 for A). The configurations may be considered first approximations to the electronic states. Whenever two configurations of the same symmetry are close in energy, they will interact (just as orbitals do) and mix. At the bottom right, such a configuration interaction is illustrated for the configurations Ψ_2 and Ψ_3. Configurations Ψ_0 and Ψ_1 are farther from others of like symmetry and so interact less. The correct electronic states are the various combinations of configurations obtained by configuration interaction.

Thus the configuration diagram is a summary of the several different configurations in which the molecule might find itself at different times.

The electronic configurations are often taken to be synonymous with the electronic states; however, for an accurate picture one must include another concept known as *configuration interaction*. Any two configurations of the same symmetry can interact in much the same way that orbitals of the same symmetry interact. The result is two electronic states, each made up from a contribution from each of the interacting configurations. This process is illustrated at the bottom right in Figure 10.27. The nearer two configurations are to each other in energy, the greater is the interaction and hence the poorer the individual configurations are as models for the electronic states. Configuration interaction plays a relatively minor role in models of ground-state molecules (except in calculations that aim for high accuracy), because the ground configuration is usually widely separated in energy from excited configurations, interacts little with them, and hence is a good approximation to the ground electronic state. But at higher energies many configurations of similar energy appear, and so individual configurations become poor models for the electronic states.

Ground-State Allowed, Excited-State Forbidden Reactions

Figure 10.28 illllustrates the $\pi 2s + \pi 4s$ cycloaddition from the point of view of electronic states. We are making the approximation here that the three states we are interested in can, in reactant and in product, be represented by single configurations. Note that the ground state, represented by configuration Ψ_g, goes over to product ground state (ξ_g) with no symmetry imposed barrier; hence the reaction is thermally allowed. Reference to the energy behavior of the individual orbitals for this reaction (Figure 10.22) shows, on the other hand, that if one were to start from the excited state represented by configuration Ψ_e, keeping all electrons in their original orbitals, one would end up in the higher energy state represented by configuration $\xi_{e'}$. Thus there is an *intended correlation* between Ψ_e and $\xi_{e'}$. Part way along the path another state of the same symmetry, represented by configuration $\Psi_{e'}$, approaches Ψ_e in energy because it intends to correlate to the lower configuration ξ_e. The resulting *intended crossing* is shown by the dotted lines in Figure 10.28. This intended crossing never actually occurs because when Ψ_e and $\Psi_{e'}$ come close in energy configuration interaction takes over. The two configurations lose their separate identities as the lower energy configuration Ψ_e mixes into itself more and more of $\Psi_{e'}$ and vice versa. The result is that if the molecule starts out in Ψ_e, it follows the solid line and emerges in ξ_e, having climbed the energy hill to a height close to the intended crossing point. Because the orbital symmetry correlations have imposed this energy barrier, the reaction is said to be symmetry forbidden in the first excited state.

Excited-State Allowed Reactions

In Figure 10.21, which shows orbital correlations for the thermally forbidden $\pi 2s + \pi 2s$ cycloaddition, we saw that by raising one electron from orbital π_2 to

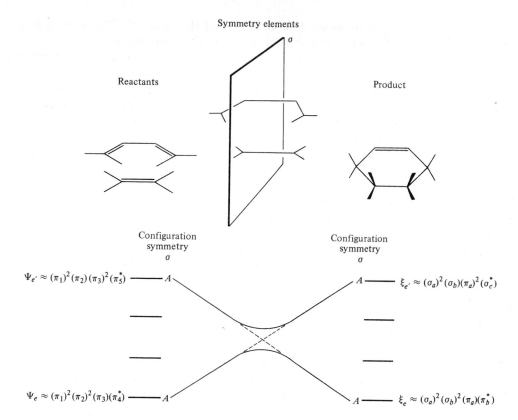

Figure 10.28 Configuration correlation diagram for the allowed $\pi 2s + \pi 4s$ cycloaddition. Refer to Figure 10.22 for the orbital labeling and orbital correlation. Three configurations are shown. We are making the assumption that these configurations are good approximations for electronic states. (This assumption is probably adequate for the configurations Ψ_g and ξ_g, less so for configurations Ψ_e, $\Psi_{e'}$, ξ_e, $\xi_{e'}$.) As the reaction proceeds, there is an intended correlation of the state represented by configuration Ψ_e with that represented by $\xi_{e'}$, and of the state represented by $\Psi_{e'}$ with that represented by ξ_e. The intended crossing is avoided by configuration interaction near the transition state (center).

orbital π_3^* to form a configuration approximating the first excited state, the energy barrier to the ground-state cycloaddition could be overcome. The penalty paid by one electron climbing to higher energy is compensated by the other electron falling to lower energy. Figure 10.29 shows an approximate state correlation diagram for this process. The first excited state, represented by configuration Ψ_e, correlates to the first excited state of the product, represented by ξ_e. There is no symmetry imposed barrier along the path, so the reaction is allowed in the first excited state in agreement with experiment.

Also shown in Figure 10.29 are the correlations of the ground configurations and the doubly excited configurations, Ψ_g, ξ_g, $\Psi_{e'}$, and $\xi_{e'}$. The latter two configurations represent approximately an excited state obtained by promoting two electrons from the highest occupied orbital of reactant or product respectively to the lowest unoccupied orbital. (Refer to Figure 10.21 for the orbital correlations.) Because configuration $\Psi_{e'}$ intends to correlate with ξ_g, it falls rapidly in energy as the reaction proceeds from left to right while Ψ_g climbs. (It is, of course, this rise in energy of Ψ_g that makes the process thermally forbidden.) The intended crossing of the two configurations of the same symmetry is avoided by configuration interaction. At points P, P', however, the doubly excited configuration $\Psi_{e'}$ and the singly excited configuration Ψ_e will cross.[40] This crossing has important consequences for the photochemical reaction, as shown by Oosterhoff and by Devaquet.[41] If the molecule starts out in state Ψ_e, it can cross to $\Psi_{e'}$ at the point of intersection P; once in $\Psi_{e'}$ it continues downhill, finally crossing into Ψ_g in the region of strong configuration interaction mixing near the intended crossing, and emerges to the ground configuration of the product ξ_g. Pathways of this kind, involving crossing of energy surfaces corresponding to different states, explain why photochemical reactions, which start from excited states, usually yield products in ground states. These points are discussed further in Chapter 12.

10.7 STEREOCHEMISTRY AND ENERGY BARRIERS IN ALLOWED AND FORBIDDEN REACTIONS

The pericyclic selection rules predict which reactions are allowed and which forbidden, but experimental verification of the predictions is not always straightforward.[42] The rules apply only to reactions that are concerted: all bonding changes must take place together. But there is always a stepwise nonconcerted pathway available, in addition to the concerted one, to accomplish a given transformation.

[40] N. D. Epiotis and S. Shaik, *J. Am. Chem. Soc.*, **100**, 9 (1978).

[41] (a) W. Th. A. M. van der Lugt and L. J. Oosterhoff, *J. Am. Chem. Soc.*, **91**, 6042 (1969); (b) D. Grimbert, G. Segal, and A. Devaquet, *J. Am. Chem. Soc.*, **97**, 6629 (1975). These authors have carried out *ab initio* calculations that illustrate the state crossings and reaction paths for the disrotatory ring opening of cyclobutene to butadiene. The situation is similar for the $\pi 2s + \pi 2s$ cycloaddition.

[42] Experimental methods of distinguishing concerted from stepwise reactions are reviewed by R. E. Lehr and A. P. Marchand, in *Pericyclic Reactions*, Vol. I, A. P. Marchand and R. E. Lehr, Eds., Academic Press, New York, 1977, p. 1.

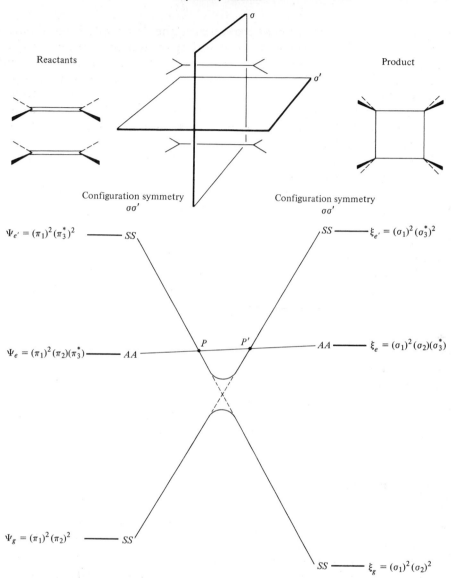

Figure 10.29 Configuration correlation diagram for the thermally forbidden $\pi 2s + \pi 2s$ cycloaddition. The orbital designations are those defined in Figure 10.21. Each configuration is classified according to its symmetry with respect to the two mirror planes σ and σ' shown at the top. Because of the intended correlation of configuration $\Psi_{e'}$ with ξ_g and Ψ_g with $\xi_{e'}$, there is an intended crossing of two configurations of the same symmetry, which is avoided by configuration interaction. The first excited configuration, Ψ_e, crosses the upper SS configuration at points P and P'; these crossing points provide pathways for molecules in the excited state represented by Ψ_e to transfer to the $\Psi_{e'}$–$\xi_{e'}$ SS line where it dips below the AA line and hence through the strong mixing region of the SS configurations to the ground configuration. The figure is approximate; the pattern of configuration energies and crossings follows schematically that found by N. D. Epiotis and S. Shaik, *J. Am. Chem. Soc.*, **100**, 9 (1978).

The difficulties come in trying to tell whether a reaction follows an allowed concerted path, a forbidden concerted path, or a stepwise path, and in trying to define the energy changes that accompany each of these mechanisms.

Stereochemistry is the most reliable criterion and the one most often invoked in attempts to define pericyclic mechanism. A high degree of stereospecificity, as observed, for example, in the 1,3-dipolar additions, Diels–Alder reactions, and electrocyclic ring openings (reactions that we discuss further in Chapter 11) is ordinarily taken as strong evidence in favor of a concerted pathway. When thermal pericyclic reactions are stereospecific, the result nearly always corresponds to that predicted by the selection rules. In Chapter 11 we shall nevertheless encounter a few instances in which steric effects raise the activation energy of the allowed reaction close to that of the forbidden.

Although stereospecificity has usually been considered to be a good criterion for concertedness, stepwise paths need not necessarily lead to mixtures of isomers. A cycloaddition (Equation 10.44) passing through a biradical intermediate (**55**) could lead to stereospecific products if the rate of closure k_2 is fast compared with the rate of bond rotation k_r. This argument has occasionally been invoked in support of claims for stepwise paths.

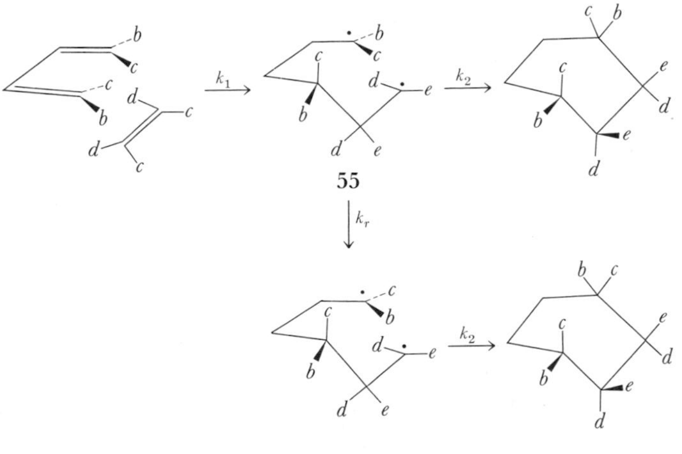

$$(10.44)$$

A postulate of a stepwise reaction is reinforced if it can be shown that the transition state expected for the formation of an intermediate lies close to the energy defined by the observed activation energy of the reaction. An observed activation energy substantially below that estimated for the stepwise path supports an allowed concerted mechanism. For example, ΔH_f° of cyclobutene, obtained from the data in Table 2.9, is $2(-4.76) + 2(8.59) + 29.8 = 37.46$ kcal mole^{-1}. A possible biradical intermediate of the electrocyclic ring-opening reaction is formulated as **56**.

One can estimate ΔH_f° for such a biradical from the data in Tables 2.9 and 9.5. The result is probably no better than a lower limit, however, because a biradical has two spin states, singlet and triplet. The spin state for a hypothetical

$$\Delta H_f^\circ = 37.5 \text{ kcal mole}^{-1} \qquad\qquad \mathbf{56}$$

(10.45)

$$\Delta H_f^\circ \approx 76.9 \text{ kcal mole}^{-1}$$

(lower limit)

biradical constructed according to Benson's group additivity scheme on which the tables are based is undefined. The enthalpy obtained for the biradical is therefore probably between that of the singlet and that of the triplet. Triplet energies are ordinarily lower than singlet, but for a thermal reaction we are interested in the singlet. The actual ΔH_f° of the singlet biradical is therefore likely to be higher than the ΔH_f° estimated from the additivity rules.

In the case of biradical **56**, assuming that one of the radical centers is conjugated with the double bond, data from Tables 2.9 and 9.5 lead to biradical $\Delta H_f^\circ \approx 2(8.59) + 23.2 + 35.82 = 76.2$ kcal mole^{-1} (lower limit). The difference is 38.7 kcal mole^{-1}. The transition state leading to biradical **56** should be higher in energy, perhaps by 4 to 5 kcal mole^{-1}. Hence the expected $\Delta H^\ddagger$ for the biradical path depicted in Equation 10.45 is about 44 kcal mole^{-1} (lower limit). The observed $\Delta H^\ddagger$ for cyclobutene opening is 32 kcal mole^{-1},[43] 12 kcal mole^{-1} lower than the lower limit estimate. A biradical intermediate is therefore unlikely in this reaction and a concerted pathway is plausible on energy grounds. The activation energy difference estimated in this example between the biradical path and the observed thermal reaction is typical;[44] however, the uncertainties in the estimates, and use of different models, have led to controversies, some of which are discussed in Chapter 11.

An even more difficult task is to define the energy requirements for a forbidden pericyclic path. Figure 10.29 provides some insight. As the reaction begins, the ground-state energy surface rises as the unfavorable interactions develop. But somewhere near the center of the reaction coordinate, the ground configuration, which is headed toward an excited configuration of the product, encounters a descending configuration of the same symmetry. The two configurations interact, and the molecule can follow the solid line to product ground state. Hence ground state correlates to ground state, but the intended crossing nevertheless introduces a substantial barrier.

The energy separation between states Ψ_g and Ψ_e for typical unsaturated molecules might range anywhere from 90 to 180 kcal mole^{-1}; if we guess from

[43] See Section 11.2.

[44] (a) J. I. Brauman and W. C. Archie, Jr., *J. Am. Chem. Soc.*, **94**, 4262 (1972); (b) L. M. Stephenson, Jr., and J. I. Brauman, *Acc. Chem. Res.*, **7**, 65 (1974). Wilcox and Carpenter found that Hückel calculations of π energies of reactant molecules and of simple models for pericyclic transition states or biradical intermediates will correlate activation energies for a number of molecules with various conjugating unsaturated hydrocarbon substituents: (c) C. F. Wilcox, Jr., and B. K. Carpenter, *J. Am. Chem. Soc.*, **101**, 3897 (1979).

looking at diagrams like Figure 10.29 that the barrier heights for forbidden reactions might reach the order of $\frac{2}{3}$ the height of Ψ_e, we would conclude that the activation energies for forbidden reactions might range from 60 to 120 kcal mole^{-1}. This estimate is a very crude one, but it is probably of a reasonable order of magnitude. The barriers should certainly be much lower than one would have expected on the basis of a diagram like Figure 10.21, which implies that the product would have to form in a doubly excited state. Furthermore, even allowed reactions often have activation energies on the order of 30–35 kcal mole^{-1}. The conclusion is that whereas the difference between allowed and forbidden reaction activation energies will ordinarily be substantial, there may well be circumstances in which it is rather small.[45]

PROBLEMS

   1. Verify that the selection rules found for the two-component cycloadditions (Section 10.5) agree with the general pericyclic selection rule. What can be said about all-antara $2 + 2 + 2 + \ldots$ cycloadditions? About all-supra $2 + 2 + 2 + \ldots$ cycloadditions?

   2. Classify as sigmatropic reactions and as cycloadditions each of the following, and predict stereochemistry at the migrating center for the ground-state transformation.

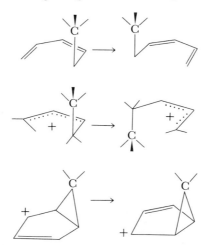

   3. Construct interaction diagrams for conrotatory and disrotatory electrocyclic opening of each of the following:

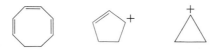

   4. Propose a mechanism for each of the following transformations, and illustrate transition states with interaction diagrams:

[45] See, for example, W. Schmidt, *Helv. Chim. Acta.* **54**, 862 (1971).

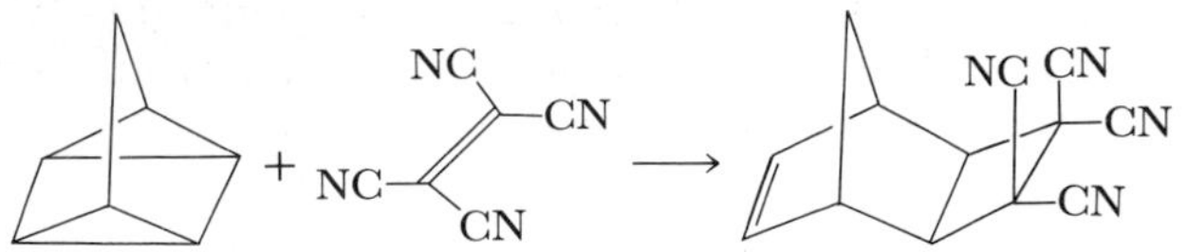

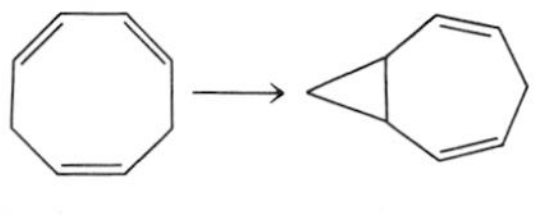

5. Illustrate by orbital drawings at intermediate stages of the reaction the continuous changes that occur in each of the reacting molecular orbitals in the $\pi 2s + \pi 4s$ cycloaddition.

6. Classify each of the following transformations as to type, give the equivalent cycloaddition, and determine whether each will be allowed in ground or excited state.

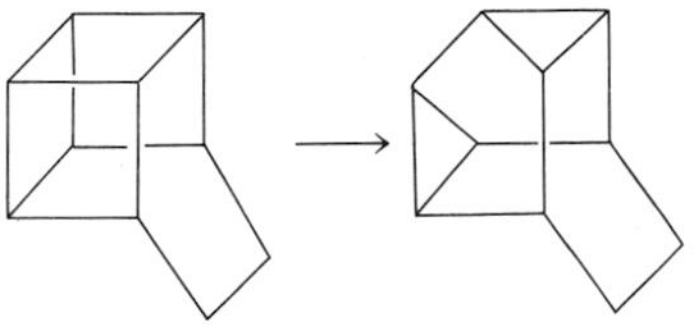

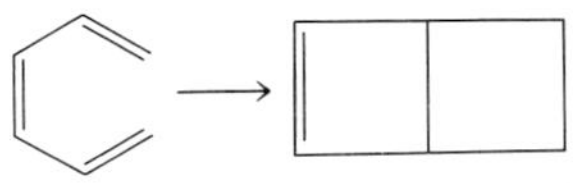

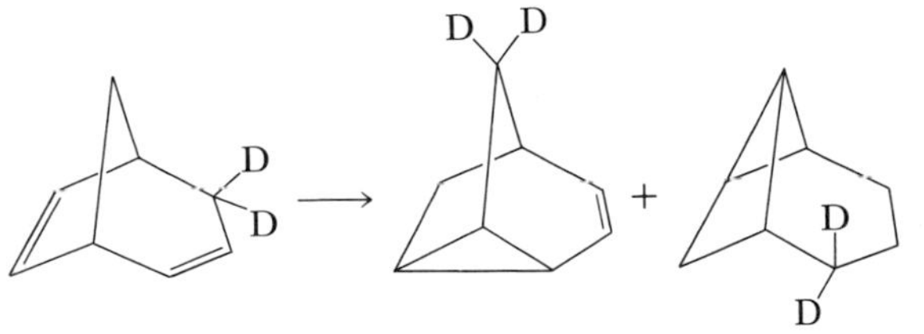

7. Show that the dimerization of sp^2-hybridized CH_2 is forbidden for approach with the geometry shown below but allowed if the π orbital of one methylene impinges on the unshared pair of the other.

8. Show that the frontside displacement reactions discussed in Section 1.8 can be considered as cycloadditions, the nucleophilic forbidden and the electrophilic allowed.

9. Analyze the examples in Figures 10.17, 10.18, and 10.26 by constructing interaction diagrams and applying the Dewar–Zimmerman pericyclic selection rules.

10. Construct a qualitative molecular orbital picture of the O_3 molecule. What would the transition state geometry be for its cycloaddition to ethylene to give the primary ozonide shown below? Construct the interaction diagram and predict whether this addition is thermally allowed or forbidden.

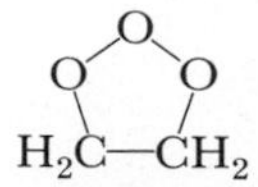

11. In the [1,3]sigmatropic rearrangement shown below, carbon 3 migrates from carbon 2 to carbon 5. This reaction can occur suprafacially (s) or antarafacially (a) with respect to the π bond, and with retention (r) or inversion (i) at carbon 3. Specify the stereochemistry of the product of each of the four possible combinations si, sr, ai, ar, and predict which is allowed and which forbidden.

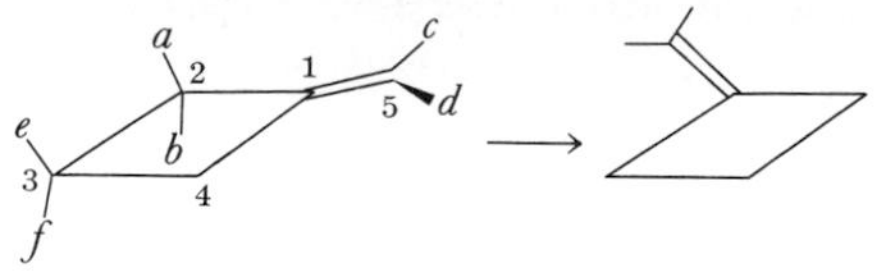

12. In a unimolecular reaction the frontier orbitals that determine the pathway must all be found within one molecule. Find the relevant frontier orbitals for a 1,5-sigmatropic hydrogen migration (example below) and analyze HOMO–LUMO interactions for various stereochemical possibilities.

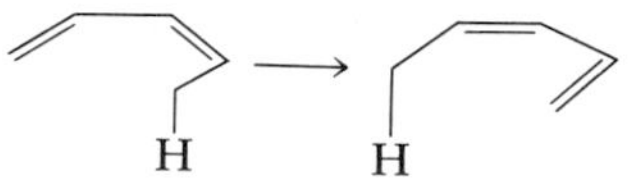

REFERENCES FOR PROBLEMS

1. R. B. Woodward and R. Hoffmann, *The Conservation of Orbital Symmetry,* Verlag Chemie, Weinheim/Bergstr., Germany, and Academic Press, New York, 1970, p. 105.
6. K. Shen, *J. Chem. Educ.,* **50,** 238 (1973); Woodward and Hoffmann, *The Conservation of Orbital Symmetry,* pp. 28; 98.
7. R. Hoffmann, R. Gleiter, and F. B. Mallory, *J. Am. Chem. Soc.,* **92,** 1460 (1970).
11. J. J. Gajewski, *J. Am. Chem. Soc.,* **98,** 5254 (1976).

Chapter 11

APPLICATIONS

OF THE PERICYCLIC

SELECTION RULES

In this chapter we shall illustrate the application of the selection rules to particular examples of pericyclic reactions. The chemical literature provides a wealth of illustrative examples, many collected in review articles.[1]

11.1 CYCLOADDITIONS

Two systems of nomenclature are in use for designating the components of a cycloaddition reaction. One, proposed by Huisgen,[2] specifies the numbers of atoms in each component; the other, introduced with the pericyclic theory[3] and used in Chapter 10, shows the numbers of electrons contributed by each. We shall continue to follow the convention that specifies numbers of electrons. Equations 11.1–11.5 show some examples.

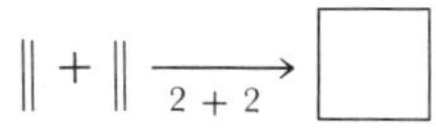

$$\| + \| \xrightarrow[2\,+\,2]{} \square \qquad (11.1)$$

[1] Summaries covering all types of pericyclic reactions may be found in (a) R. B. Woodward and R. Hoffmann, *The Conservation of Orbital Symmetry,* Verlag Chemie, Weinheim/Bergstr., Germany, and Academic Press, New York, 1970; (b) G. B. Gill, *Q. Rev. Chem. Soc.,* **22,** 338 (1968); (c) T. L. Gilchrist and R. C. Storr, *Organic Reactions and Orbital Symmetry,* Cambridge University Press, London, 1972; (d) *Pericyclic Reactions,* Vols. I and II, A. P. Marchand and R. E. Lehr, Eds., Academic Press, New York, 1977, covers both theory and examples. (e) Activation parameters for a variety of pericyclic reactions are given by R. M. Willcott, R. L. Cargill, and A. B. Sears, *Prog. Phys. Org. Chem.,* **9,** 25 (1972). (f) W. C. Herndon *Chem. Rev.,* **72,** 157 (1972), cites numerous more specialized review articles. (g) J. B. Hendrickson, *Angew. Chem. Int. Ed. Engl.,* **13,** 47 (1974), classifies the possible types of pericyclic reactions for six electron systems. We cite reviews covering specific types of pericyclic reactions at appropriate points throughout this chapter.
[2] R. Huisgen, *Angew. Chem. Int. Ed. Engl.,* **7,** 321 (1968).
[3] See note 1(a).

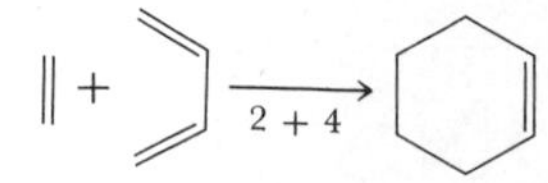

$$\| + \diagup\diagdown \xrightarrow[2+4]{} \bigcirc \qquad (11.2)$$

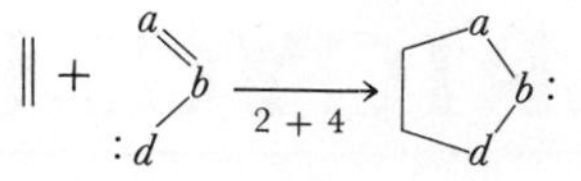

$$\| + \overset{a}{\underset{:d}{\diagdown}}\!b \xrightarrow[2+4]{} \underset{d}{\overset{a}{\bigcirc}}b: \qquad (11.3)$$

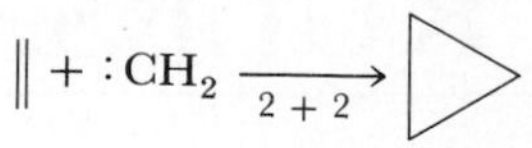

$$\| + :CH_2 \xrightarrow[2+2]{} \triangleright \qquad (11.4)$$

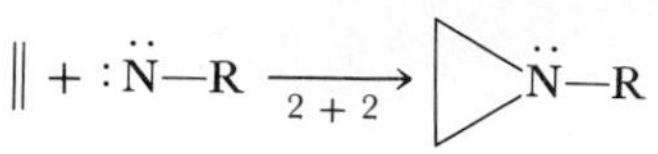

$$\| + :\overset{..}{N}{-}R \xrightarrow[2+2]{} \triangleright\overset{..}{N}{-}R \qquad (11.5)$$

2 + 2 Additions Forming Three-Membered Rings[4]

The simplest example of a 2 + 2 addition about which there is substantial information would appear to be the addition of carbenes and nitrenes to alkenes (Equations 11.4 and 11.5). Woodward and Hoffmann have designated reactions of this kind, in which one component is a single atom, as *cheletropic*.[5]

Our earlier discussion in Section 6.3 has treated the carbene additions; it

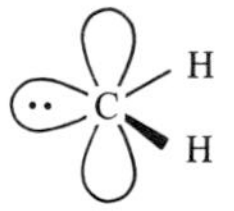

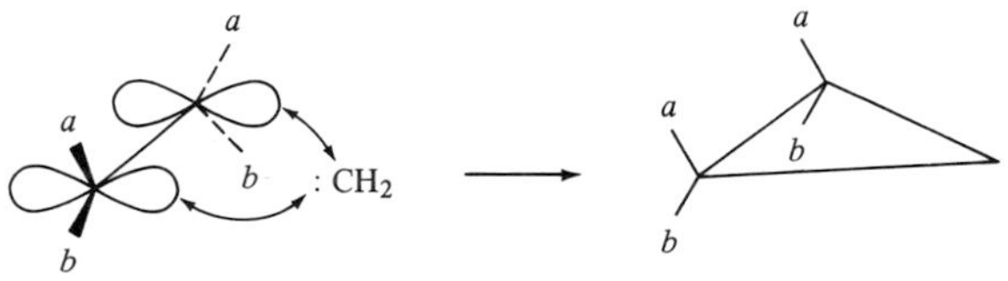

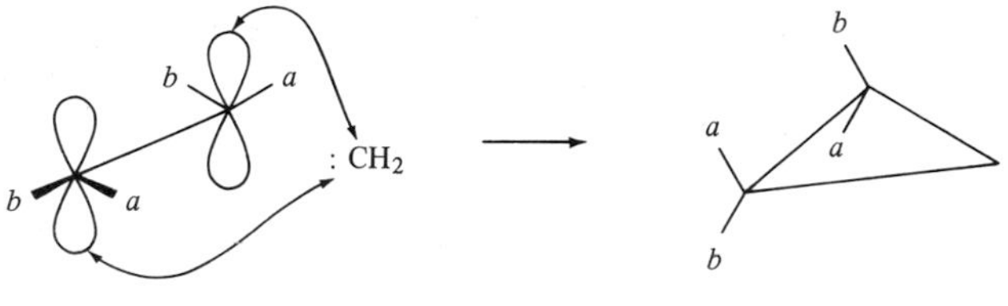

---

[4]For a review of application of 2 + 2 cycloadditions to preparation of three- and four-membered rings, see L. L. Muller and J. Hamer, *1,2 Cycloaddition Reactions*, Wiley, New York, 1967.

[5]R. B. Woodward and R. Hoffmann, *The Conservation of Orbital Symmetry*, Verlag Chemie, Weinheim/Bergstr., Germany, and Academic Press, New York, 1970, p. 152.

remains here to consider the predictions of the pericyclic theory.[6] Recall that the singlet state, with electrons paired in the sp^2 hybrid of the bent structure **1**, adds concertedly in a cis fashion, whereas the triplet gives a mixture of cis and trans addition and is thought to add stepwise. We are concerned here with the concerted reactions of the singlet. The observed cis addition implies that the alkene component enters suprafacially (**2**) rather than antarafacially (**3**). The selection rules then specify that in order for the four-electron addition to be thermally allowed, the carbene component must enter antarafacially, as in Structure **4**, where for purposes

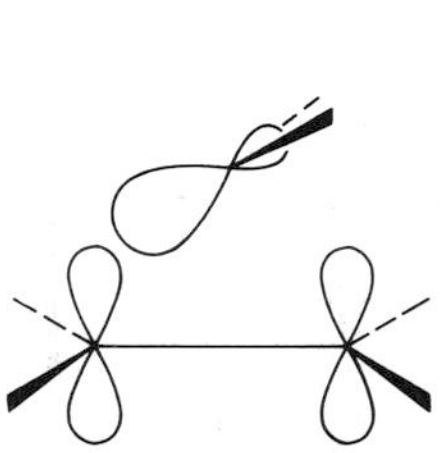

4 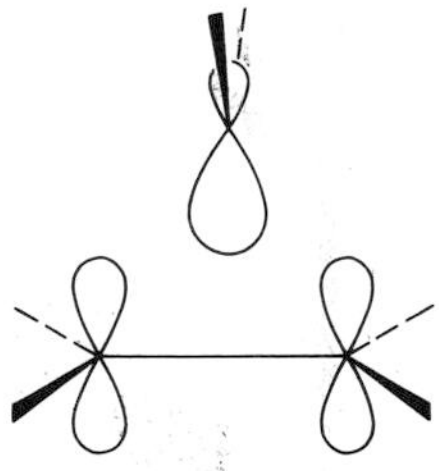   **5**

of clarity we have omitted the vacant p orbital. This geometry is called the *nonlinear,* or π approach to distinguish it from the alternative, more symmetrical *linear,* or σ, approach (**5**), in which the sp^2 orbital enters suprafacially. Although Hoffmann has reported calculations by the extended Hückel method that are in agreement with the mode of addition **4** predicted by the pericyclic theory,[7] and *ab initio* calculations confirm this result,[8] there is little experimental information.

The reverse of the carbene addition (Equation 11.6) occurs photochemically upon ultraviolet irradiation of cyclopropanes.[9] Were the extrusion of the carbene

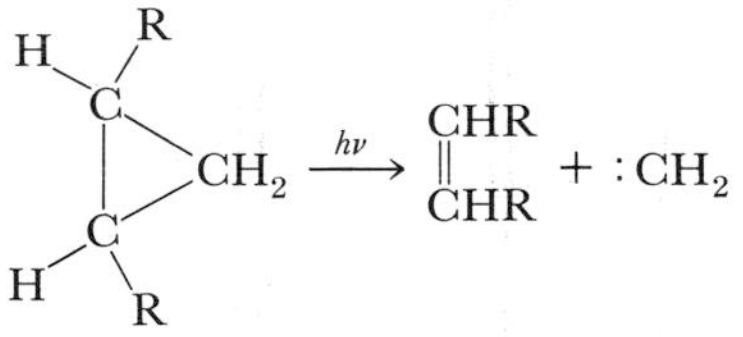

$$ \text{(11.6)} $$

concerted, the pericyclic theory would predict the linear suprafacial–suprafacial path; the observed cis–trans isomerization that accompanies the decomposition to a carbene, as in Equation 11.7,[10] nevertheless implies a stepwise process, with preliminary opening to a diradical (**6**).

[6] W. M. Jones and U. H. Brinker, in *Pericyclic Reactions*, Vol. I, A. P. Marchand and R. E. Lehr, Eds., Academic Press, New York, 1977, p. 109.

[7] (a) R. Hoffmann, *J. Am. Chem. Soc.*, **90**, 1475 (1968); (b) R. Hoffmann, D. M. Hayes, and P. S. Skell, *J. Phys. Chem.*, **76**, 664 (1972).

[8] B. Zurawski and W. Kutzelnigg, *J. Am. Chem. Soc.*, **100**, 2654 (1978).

[9] These reactions are reviewed by G. W. Griffin and N. R. Bertoniere, in *Carbenes*, Vol. I, M. Jones, Jr., and R. A. Moss, Eds., Wiley, New York, 1973, p. 305.

[10] G. W. Griffin, J. Covell, R. C. Petterson, R. M. Dodson, and G. Klose, *J. Am. Chem. Soc.*, **87**, 1410 (1965).

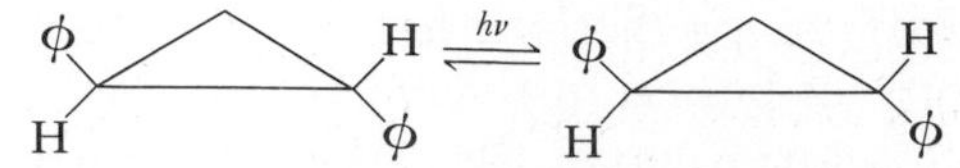

$$\phi\!\!\!\bigwedge\!\!\!H \;\overset{h\nu}{\rightleftharpoons}\; \phi\!\!\!\bigwedge\!\!\!H \tag{11.7}$$

$$\triangle \;\rightleftharpoons\; \cdot\!\!\diagup\!\!\diagdown\!\!\cdot \;\longrightarrow\; /\!\!/ + :CH_2$$

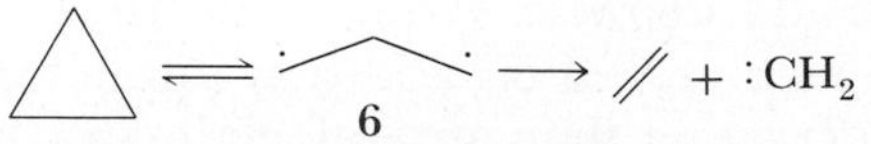

$$\mathbf{6}$$

The examples shown in Equations 11.8–11.11 illustrate a few of the other known reverse $2 + 2$ cheletropic cycloadditions. The first three of these reactions

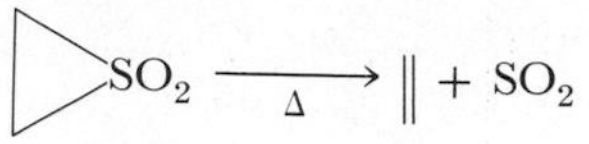

$$\triangleright\!SO_2 \;\xrightarrow{\Delta}\; \| + SO_2 \tag{11.8}[11]$$

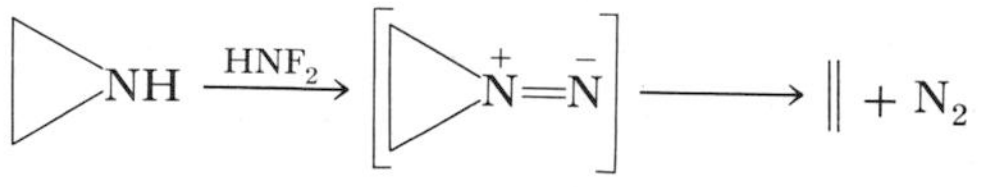

$$\mathbf{7}$$

$$\triangleright\!NH \;\xrightarrow{HNF_2}\; \left[\triangleright\!\overset{+}{N}\!\!=\!\!\overset{-}{N}\right] \;\longrightarrow\; \| + N_2 \tag{11.9}[12]$$

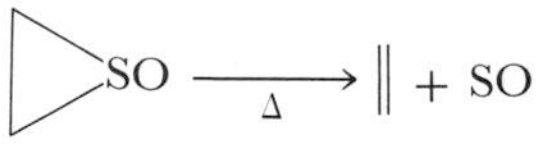

$$\mathbf{8}$$

$$\triangleright\!SO \;\xrightarrow{\Delta}\; \| + SO \tag{11.10}[13]$$

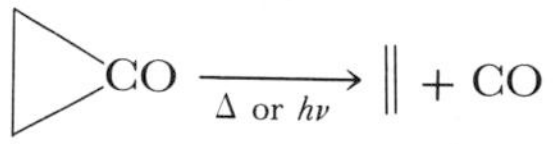

$$\mathbf{9}$$

$$\triangleright\!CO \;\xrightarrow{\Delta \text{ or } h\nu}\; \| + CO \tag{11.11}[14]$$

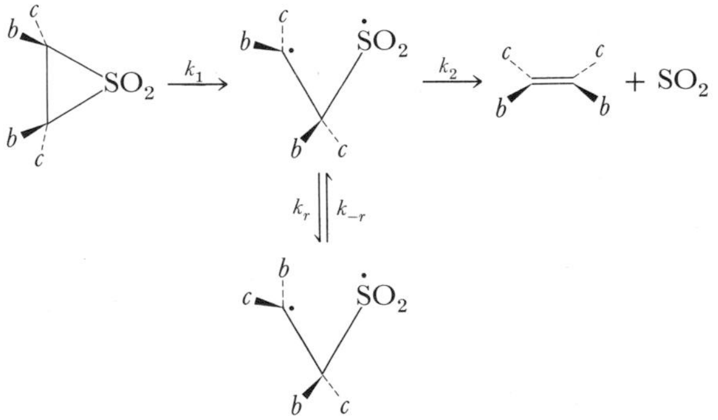

$$\tag{11.12}$$

[11](a) L. A. Paquette, *Acc. Chem. Res.*, **1**, 209 (1968). (b) F. G. Bordwell, J. M. Williams, Jr., E. B. Hoyt, Jr., and B. B. Jarvis, *J. Am. Chem. Soc.*, **90**, 429 (1968), have reported evidence in the case of Equation 11.8 that favors a two-step mechanism with a ring-opened diradical intermediate that loses SO_2 faster than bond rotation can occur.

[12]J. P. Freeman and W. H. Graham, *J. Am. Chem. Soc.*, **89**, 1761 (1967). Note that in this reference the interpretation of the results as a violation of the Woodward–Hoffmann rules presupposes the linear (suprafacial–suprafacial) path; the idea of the nonlinear path, which brings the results and theory into agreement, developed later [see note 1(a)].

[13]W. G. L. Aalbersberg and K. P. C. Vollhardt, *J. Am. Chem. Soc.*, **99**, 2792 (1977).

[14]N. J. Turro, P. A. Leermakers, H. R. Wilson, D. C. Neckers, G. W. Byers, and G. F. Vesley, *J. Am. Chem. Soc.*, **87**, 2613 (1965).

are stereospecific, cis-substituted rings yielding cis alkenes and trans yielding trans. Stereochemistry is not known for Reaction 11.11. Although the stereochemistry of SO and SO_2 elimination from thiirane dioxide and oxide (**7** and **9**) and of nitrogen elimination from diazonium ion **8** are consistent with the pericyclic selection rules for the nonlinear pathway, the observation of stereospecificity does not necessarily demonstrate that this is the path followed. A biradical path (Equation 11.12) with k_2 large compared to k_r is possible.[15]

2 + 2 Additions Forming Four-Membered Rings[16]

The formation of four-membered rings through 2 + 2 cycloaddition is a well-established reaction and the most generally effective synthetic approach to cyclobutanes. Most alkenes cannot be induced to undergo this reaction thermally, a finding that is readily rationalized by the forbidden nature of the $2s + 2s$ addition and the steric difficulties associated with the allowed $2s + 2a$ pathway. From Table 2.9, one finds $\Delta H^\circ_{298} = -19$ kcal mole^{-1} and $\Delta S^\circ_{298} = -44$ cal mole^{-1} K^{-1} for dimerization of ethylene, which gives $\Delta G^\circ_{298} = -5.9$ kcal mole^{-1}. Despite this negative free-energy change, the cycloaddition will not proceed at ambient temperature to any measurable extent because of an activation energy $E_a = 43.8$ kcal mole^{-1},[17] which corresponds to $\Delta H^\ddagger$ of 43.2 kcal mole^{-1}. The ring closure presumably takes place by way of biradical intermediate **10** (Equation 11.13). Using Tables

$$2H_2C{=}CH_2{}^+ \longrightarrow \quad \overset{\cdot CH_2}{\underset{\cdot CH_2}{\Big\langle}} \quad \longrightarrow \quad \square \qquad (11.13)$$

10

2.9 and 9.5, we can make a rough estimate of the enthalpy change for formation of **10** from molecules of ethylene as 37.5 kcal mole^{-1}. Recall from the discussion in Section 10.7 that enthalpies of biradicals are likely to be higher than estimates based on thermochemical additivity rules. Although no firm conclusion can be drawn in this instance, the biradical intermediate is at least plausible.

Although most alkenes do not undergo thermal 2 + 2 cycloadditions, those in which the double bond is substituted by fluorine and chlorine atoms undergo thermal 2 + 2 additions under relatively mild conditions,[18] as do ketenes and allenes.

---

[15] W. L. Mock, in *Pericyclic Reactions*, Vol. II, A. P. Marchand and R. E. Lehr, Eds., Academic Press, New York, 1977, p. 143.

[16] For reviews, see (a) R. Huisgen, R. Grashey, and J. Sauer, in *The Chemistry of Alkenes*, S. Patai, Ed., Wiley, London, 1964, p. 739; (b) J. D. Roberts and C. M. Sharts, *Org. Reactions*, **12**, 1 (1962); (c) L. L. Muller and J. Hamer, *1,2 Cycloaddition Reactions*, Wiley, New York, 1967; (d) R. Huisgen, *Acc. Chem. Res.*, **10**, 117 (1977).

[17] L. M. Quick, D. A. Knecht, and M. H. Back, *Int. J. Chem. Kinet.*, **4**, 61 (1972).

[18] Methylenecyclopropanes—for example,

$$\triangleright{=}\overset{Cl}{\underset{Cl}{}}$$

behave similarly to the *gem*-difluoroethylenes. P. D. Bartlett and R. C. Wheland, *J. Am. Chem. Soc.*, **94**, 2145 (1972).

In the case of the fluorinated ethylenes it is known through work of Bartlett and his collaborators that the addition reaction is stepwise by way of a biradical intermediate.[19] Much of the evidence supporting this conclusion has been obtained from the study of additions of 1,1-difluoro-2,2-dichloroethylene, abbreviated 1122, to dienes. This compound is particularly well suited to addition by Mechanism 11.13 because the fluorines destabilize the double bond and the chlorines stabilize the incipient radical center.

Scheme 1 outlines six possible general routes, each of which has several potential stereochemical variations, that addition of an alkene to a diene could follow. When 1122 reacts with butadiene and simple substituted butadienes, the products are nearly entirely cyclobutanes; consequently, we may restrict attention to the first three mechanisms of Scheme 1.[20] Structure determination showed that

Scheme 1

2 + 2 Addition

Stepwise
(a) Radical

(b) Bipolar

Concerted
(c)

2 + 4 Addition

Stepwise
(d) Radical

(e) Bipolar

Concerted
(f)

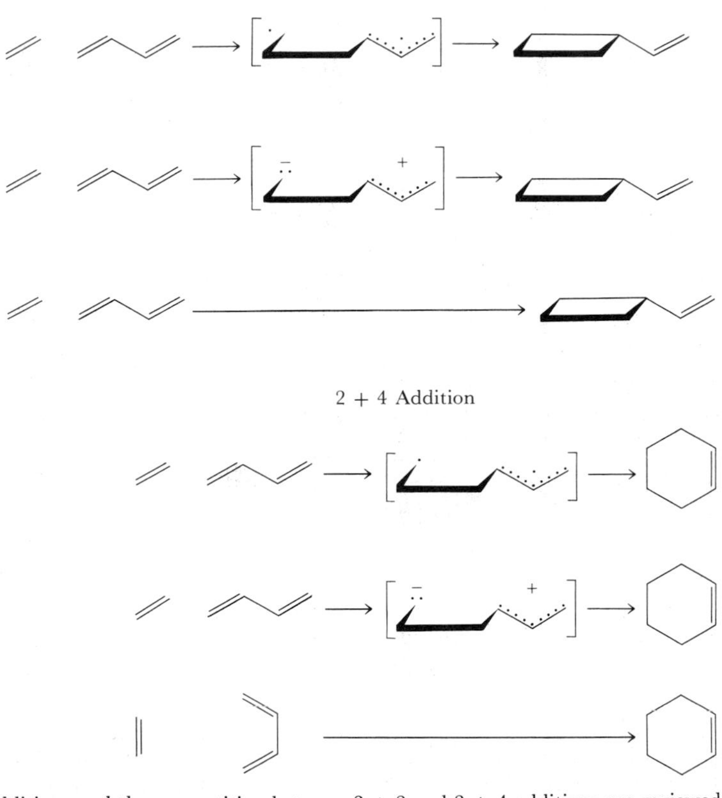

---

[19] Mechanisms of 2 + 2 additions and the competition between 2 + 2 and 2 + 4 additions are reviewed by (a) P. D. Bartlett, *Q. Rev. Chem. Soc.*, **24,** 473 (1970); (b) P. D. Bartlett, *Pure Appl. Chem.*, **27,** 597 (1971); (c) P. D. Bartlett, *Science,* **159,** 833 (1968).

[20] The approximately 1 percent of the reaction that yields cyclohexene products occurs by way of path (d), Scheme 1. P. D. Bartlett and J. J.-B. Mallett, *J. Am. Chem. Soc.,* **98,** 143 (1976).

the products are always of the type **11** rather than the alternative **12**.[21] Because a chlorine substituent is better at stabilizing an adjacent radical center than is fluo-

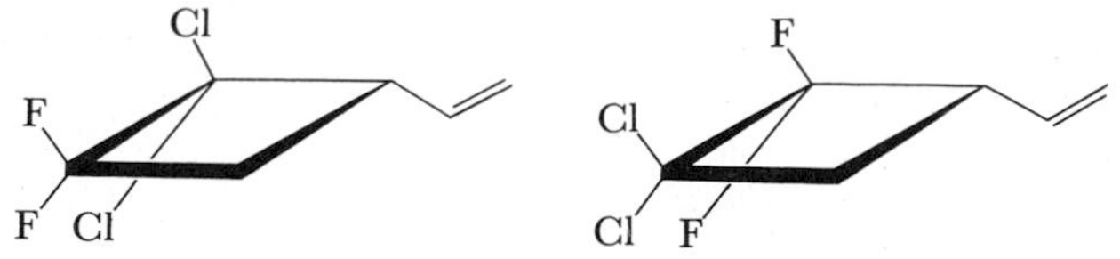

11   **12**

rine,[22] this result suggests path (a) of Scheme 1. The finding that the stereochemistry about the terminal double bond is retained (Equation 11.14) is consistent with the known propensity for allylic radicals to resist rotation about the partial double

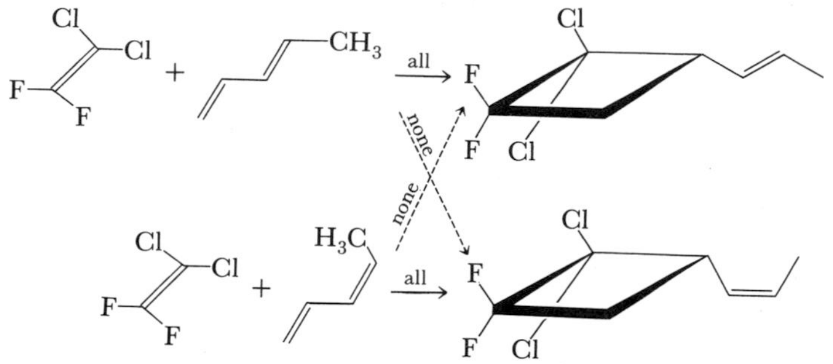

(11.14)

bonds.[23] The observation of a rate change by a factor of only three in addition of 1122 to butadiene upon changing solvent from nonpolar ether or hexane to polar nitromethane or methanol effectively eliminates the possibility of the bipolar path (b) of Scheme 1.[24]

The stereochemistry of the reaction of 1122 with each of the three geometrical isomers of 2,4-hexadiene is shown in Scheme 2.[25] There are only four products, the trans–trans diene yielding two, the cis–cis diene the other two, and the trans–cis diene all four. The numbers over the arrows show the proportions obtained from each. A concerted $\pi 2s + \pi 2s$ mechanism would lead to retention of stereochemistry at the double bond to which the 1122 adds; the trans–trans isomer could then yield only A, the cis–cis isomer only D, and the trans–cis isomer only B and C, predictions clearly inconsistent with the results. The most reasonable way to account for the observations is to postulate an intermediate (Equation 11.15) in which one bond has formed and in which rotation about the former double bond competes with closure of the ring.[26] Scheme 3 shows the pathway leading to each product.

[21]P. D. Bartlett, L. K. Montgomery, and B. Seidel, *J. Am. Chem. Soc.*, **86**, 616 (1964).
[22](a) P. D. Bartlett, L. K. Montgomery, and B. Seidel, *J. Am. Chem. Soc.*, **86**, 616 (1964); (b) P. D. Bartlett and R. R. Hiatt, *J. Am. Chem. Soc.*, **80**, 1398 (1958).
[23]C. Walling and W. Thaler, *J. Am. Chem. Soc.*, **83**, 3877 (1961).
[24]See note 19(a).
[25]L. K. Montgomery, K. Schueller, and P. D. Bartlett, *J. Am. Chem. Soc.*, **86**, 622 (1964).
[26]The estimated ΔH for the addition step is sufficiently small that formation of a biradical is plausible. See note 25.

Scheme 2

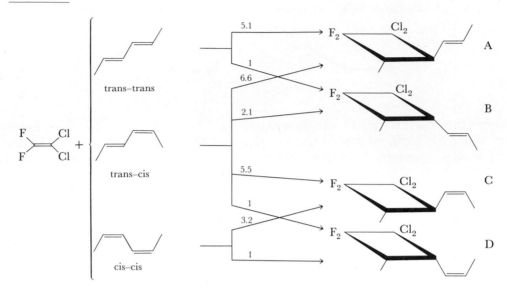

Arrows with loops designate bond rotation preceding ring closure. The proportions reflect the preference for formation of the trans-substituted cyclobutane and the relative rates of bond rotation and of ring closure. Analysis of the data revealed that bond rotation is ten times faster than closure.[27]

(11.15)

Comparison of results with 1122 and tetrafluoroethylene led Bartlett and co-workers to the conclusion that the important barrier to closure in the intermedi-

[27] Another characteristic of the biradical cycloadditions, observed particularly at higher temperatures (above 100°C) is reversal of the initial addition step after bond rotation with resulting isomerization of the double bond in recovered diene. See (a) P. D. Bartlett, C. J. Dempster, L. K. Montgomery, K. E. Schueller, and G. E. H. Wallbillich, *J. Am. Chem. Soc.*, **91**, 405 (1969); (b) P. D. Bartlett and G. E. H. Wallbillich, *J. Am. Chem. Soc.*, **91**, 409 (1969).

SCHEME 3[28]

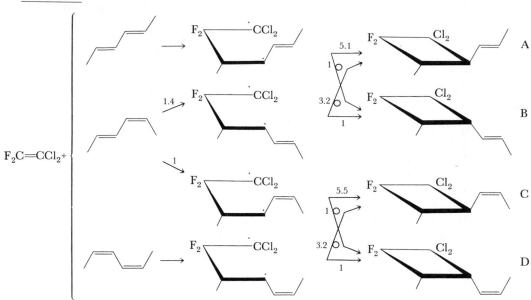

$F_2C=CCl_2+$

ate biradical is a conformational one.[29] The biradical can be formed in any of a number of conformations, with a stretched-out arrangement (**13**) likely to be fa-

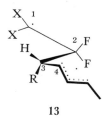

13

vored. Rotations about the 1,2- and 2,3-bonds must occur to bring the two radical centers into the proper relative position and orientation for bonding; meanwhile, there is time for the rotation about 3,4 that determines the stereochemistry of the four-membered ring. Closure occurs with little or no activation energy once the proper conformation is attained. The pattern of loss of stereochemistry is not restricted to addition to dienes but has been demonstrated also in several 2 + 2 additions to alkenes.[30]

[28] Scheme 3 is adapted with permission from Chart I of L. K. Montgomery, K. Schueller, and P. D. Bartlett, *J. Am. Chem. Soc.*, **86**, 622 (1964). Copyright by the American Chemical Society.

[29] See note 27(a).

[30] (a) P. D. Bartlett, K. Hummel, S. P. Elliott, and R. A. Minns, *J. Am. Chem. Soc.*, **94**, 2898 (1972); (b) P. D. Bartlett, G. M. Cohen, S. P. Elliott, K. Hummel, R. A. Minns, C. M. Sharts, and J. Y. Fukunaga, *J. Am. Chem. Soc.*, **94**, 2899 (1972); (c) R. Wheland and P. D. Bartlett, *J. Am. Chem. Soc.*, **95**, 4003 (1973).

Also known are 2 + 2 cycloadditions proceeding by way of a bipolar ion, path (b) of Scheme 1.[31] These reactions occur in situations such as that depicted in Equation 11.16, where the intermediate zwitterion (**14**) is strongly stabilized. Tet-

$$(11.16)$$

racyanoethylene adds by this mechanism to *p*-methoxyphenyl-,[32] alkoxyl-,[33] and cyclopropyl-[34] substituted alkenes. Partial loss of stereochemistry occurs as in the biradical cases, but it is much less pronounced and depends on substituents and solvent. The intermediacy of ionic species is established by such observations as acceleration by a second electron-donating substituent and correlation of rate with solvent polarity parameter E_T.[35]

Equation 11.17 illustrates another variation of the 2 + 2 reaction, the $\pi 2 + \sigma 2$ addition of alkenes activated by electron-withdrawing groups to highly strained single bonds. (The wavy lines indicate a mixture of stereoisomers.) The lack of stereospecificity shows that these are stepwise reactions.[36]

$$(11.17)$$

The evidence we have described shows that in 2 + 2 additions alkenes will avoid the forbidden 2*s* + 2*s* concerted path in favor of a stepwise route. Thermal additions might nevertheless in some instances follow the allowed 2*s* + 2*a* process. Figure 11.1 illustrates the approximate geometry required of a 2*s* + 2*a* transition state; the groups R seriously hinder the approach. In a search for this reaction, it will therefore be advantageous to have the groups R as small as possible. Bartlett and his collaborators have examined the stereochemistry of addition of tetra-fluoroethylene to *cis*- and *trans*-1,2-dideuterioethylene; the results, summarized in Equation 11.18, demonstrate that the reaction is stepwise.[37] Other experiments

[31] Reviews: (a) R. Gompper, *Angew. Chem. Int. Ed. Engl.*, **8**, 312 (1969); (b) R. Huisgen, *Acc. Chem. Res.*, **10**, 117 (1977).

[32] (a) P. D. Bartlett, *Q. Rev. Chem. Soc.*, **24**, 473 (1970); (b) J. K. Williams, D. W. Wiley, and B. C. McKusick, *J. Am. Chem. Soc.*, **84**, 2210 (1962).

[33] (a) R. Huisgen and G. Steiner, *J. Am. Chem. Soc.*, **95**, 5054, 5055 (1973); (b) G. Steiner and R. Huisgen, *J. Am. Chem. Soc.*, **95**, 5056 (1973).

[34] S. Nishida, I. Moritani, and T. Teraji, *J. Org. Chem.*, **38**, 1878 (1973).

[35] See note 31(b).

[36] P. G. Gassman, *Acc. Chem. Res.*, **4**, 128 (1971).

[37] P. D. Bartlett, G. M. Cohen, S. P. Elliott, K. Hummel, R. A. Minns, C. M. Sharts, and J. Y. Fukunaga, *J. Am. Chem. Soc.*, **94**, 2899 (1972).

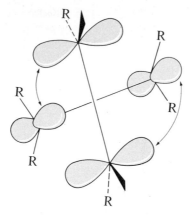

Figure 11.1 Geometry of approach for a $2s + 2a$ cycloaddition.

aimed at detecting the $2s + 2a$ process in alkenes have similarly led to negative results.[38]

$$F_2C=CF_2 +$$

$$F_2C=CF_2 +$$

$$\sim 1:1$$

(11.18)

A suprafacial–antarafacial path is possible when one of the components has an *sp*-hybridized carbon, as in a ketene, so that one of the interfering R groups is absent. Ketenes add to alkenes with retention of the cis–trans geometry about the alkene component and with the orientation and stereochemistry that yields the sterically most hindered product.[39] Thus in additions to cyclopentadiene (Equation 11.19), the larger substituent L assumes the more crowded endo position;[40] addition to cis alkenes (Equation 11.20) yields the syn product, and with 1,1-

[38](a) J. E. Baldwin and P. W. Ford, *J. Am. Chem. Soc.*, **91**, 7192 (1969); (b) A. T. Cocks, H. M. Frey, and I. D. R. Stevens, *Chem. Commun.*, 458 (1969). The lowest singlet state of O_2 (recall that the ground state is a triplet) behaves like a very reactive alkene. It adds to dienes in a $2 + 4$ cycloaddition and to alkenes in a $2 + 2$ cycloaddition. The $2 + 2$ reaction has a nonpolar transition state (small solvent effect on rate) and is stereospecifically cis with respect to alkene. The singlet oxygen molecule might be reacting antarafacially, but there is no way to test this possibility stereochemically because the oxygen atoms lack substituents. See (c) P. D. Bartlett, *Chem. Soc. Rev.*, **5**, 149 (1976); (d) R. W. Denny and A. Nickon, *Org. Reactions*, **20**, 133 (1973).

[39]Reviews: (a) P. D. Bartlett, *Pure Appl. Chem.*, **27**, 597 (1971); (b) R. W. Holder, *J. Chem. Educ.*, **53**, 81 (1976); (c) L. Ghosez and M. J. O'Donnell, in *Pericyclic Reactions*, Vol. II, A. P. Marchand and R. E. Lehr, Eds., Academic Press, New York, 1977, p. 79.

[40](a) M. Rey, S. Roberts, A. Dieffenbacher, and A. S. Dreiding, *Helv. Chim. Acta*, **53**, 417 (1970); (b) P. R. Brook, J. M. Harrison, and A. J. Duke, *Chem. Commun.*, 589 (1970); (c) W. T. Brady and E. F. Hoff, Jr., *J. Org. Chem.*, **35**, 3733 (1970); (d) W. T. Brady and R. Roe, Jr., *J. Am. Chem. Soc.*, **93**, 1662 (1971); (e) A. Hassner, R. M. Cory, and N. Sartoris, *J. Am. Chem. Soc.*, **98**, 7698 (1976).

disubstituted alkenes (Equation 11.21), the product is the vicinal isomer.[41] A negative activation entropy, $\Delta S^{\ddagger} = -20$ cal mole^{-1} K^{-1},[42] small solvent effects on

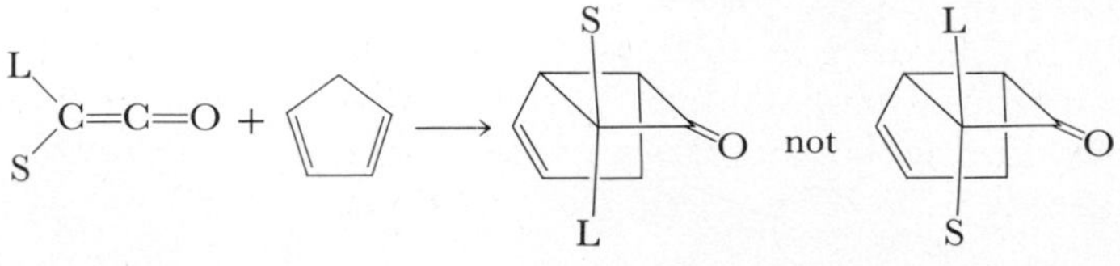

$$
\begin{array}{c}
L \\
\diagdown \\
S
\end{array}
C{=}C{=}O \ + \ \bigcirc \ \longrightarrow \ \text{(bicyclic product with L and L)} \ \text{not} \ \text{(bicyclic product with L and S)} \tag{11.19}
$$

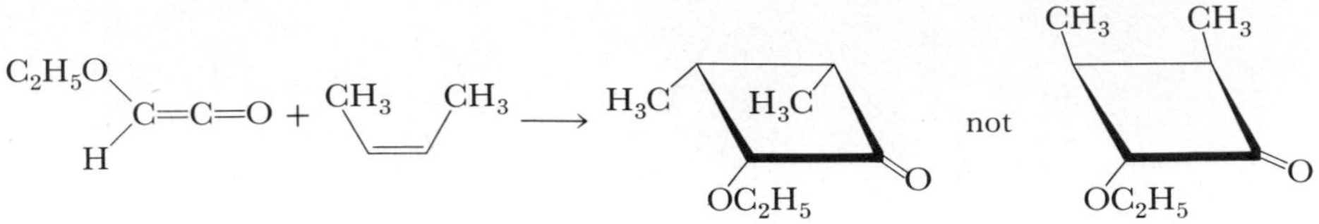

L = larger group; S = smaller group

$$
\begin{array}{c}
C_2H_5O \\
\diagdown \\
H
\end{array}
C{=}C{=}O \ + \ CH_3 \diagup\diagdown CH_3 \ \longrightarrow \ \text{(cyclobutanone product)} \ \text{not} \ \text{(cyclobutanone product)} \tag{11.20}
$$

$$
\begin{array}{c}
C_2H_5O \\
\diagdown \\
H
\end{array}
C{=}C{=}O \ + \
\begin{array}{c}
R \\
\diagdown \\
R'
\end{array}
C{=} \ \longrightarrow \ \text{(cyclobutanone product)} \ \text{not} \ \text{(cyclobutanone product)} \tag{11.21}
$$

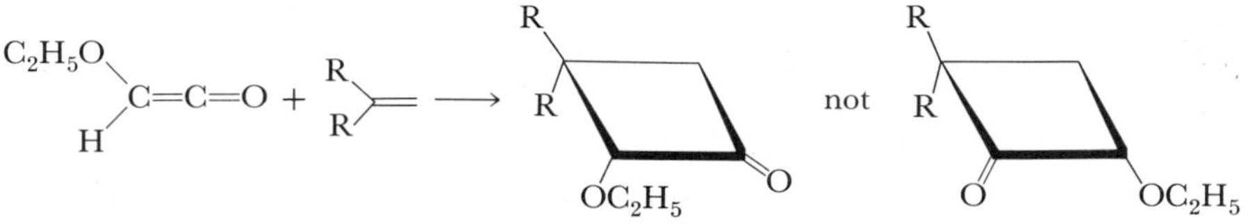

rate,[42] and a Hammett ρ of only -0.73 for addition of diphenyl ketene to substituted styrenes[43] point to a transition state characterized by an ordered structure with little charge separation. A secondary isotope effect of $k_H/k_D = 0.91$ per deuterium for the addition of diphenylketene to β,β-dideuteriostyrene (Equation 11.22)

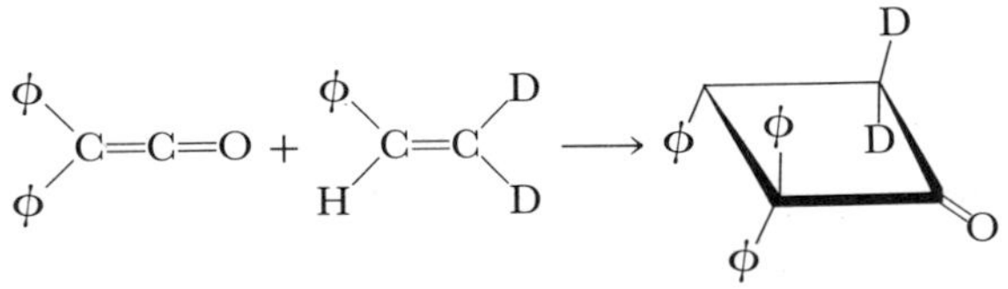

$$
\begin{array}{c}
\phi \\
\diagdown \\
\phi
\end{array}
C{=}C{=}O \ + \
\begin{array}{c}
\phi \\
\diagdown \\
H
\end{array}
C{=}C
\begin{array}{c}
D \\
\diagup \\
D
\end{array}
\ \longrightarrow \ \text{(cyclobutanone product)} \tag{11.22}
$$

$$
\begin{array}{c}
\phi \\
\diagdown \\
\phi
\end{array}
C{=}C{=}O \ + \
\begin{array}{c}
\phi \\
\diagdown \\
D
\end{array}
C{=}CH_2 \ \longrightarrow \ \text{(cyclobutanone product)} \tag{11.23}
$$

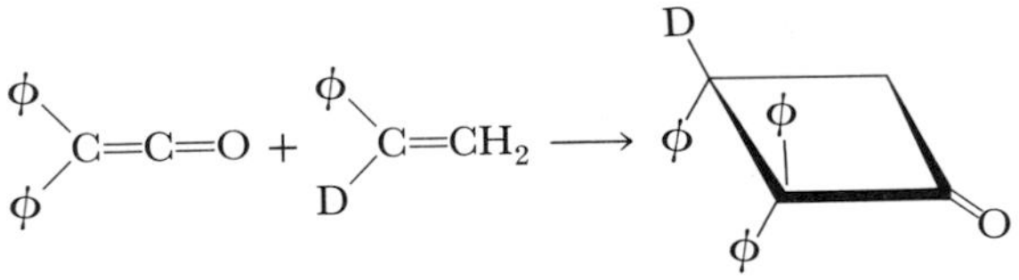

is consistent with a change from sp^2 toward sp^3 hybridization in the transition state;[44] a ^{14}C/^{12}C isotope effect investigation confirmed this conclusion.[45] The isotope effect for deuterium substitution at the α position of styrene (Equation

[41]T. DoMinh and O. P. Strausz, *J. Am. Chem. Soc.*, **92**, 1766 (1970).
[42]N. S. Isaacs and P. Stanbury, *J. Chem. Soc., Perkin Trans.* **2**, 166 (1973).
[43]J. E. Baldwin and J. A. Kapecki, *J. Am. Chem. Soc.*, **92**, 4868 (1970).
[44]J. E. Baldwin and J. A. Kapecki, *J. Am. Chem. Soc.*, **92**, 4874 (1970).
[45]C. J. Collins, B. M. Benjamin, and G. W. Kabalka, *J. Am. Chem. Soc.*, **100**, 2570 (1978).

11.23) gave the unexpectedly large value of $k_H/k_D = 1.23$; this result has not been satisfactorily explained.

The $2s + 2a$ transition state (**15**) accounts for the characteristics of the reaction. The least hindered approach of reactants leads to the most hindered product. The electron-deficient carbon end of the carbonyl π orbital (perpendicular to the page in **15**) impinges directly on the alkene π bond in the $s + a$ approach; this interaction is considered to provide a key stabilizing influence.[46]

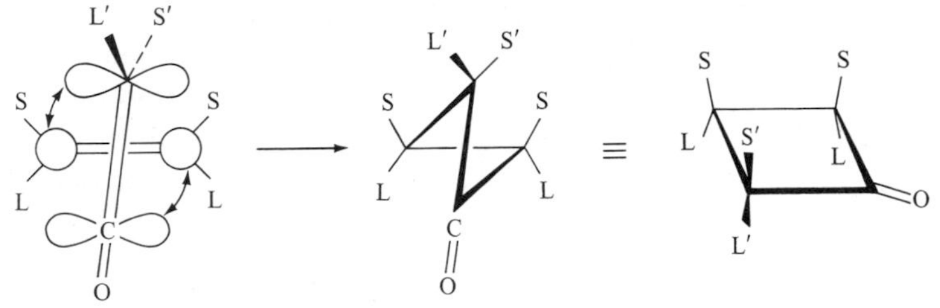

15

An alternative geometry of approach (**16**) has been proposed for the alkene–ketene addition.[47] One end of the alkene double bond interacts with the ketene C=C π bond, but the other end interacts with the carbon end of the perpendicular C=O π bond. Because the two p orbitals on the central sp-hybrid-

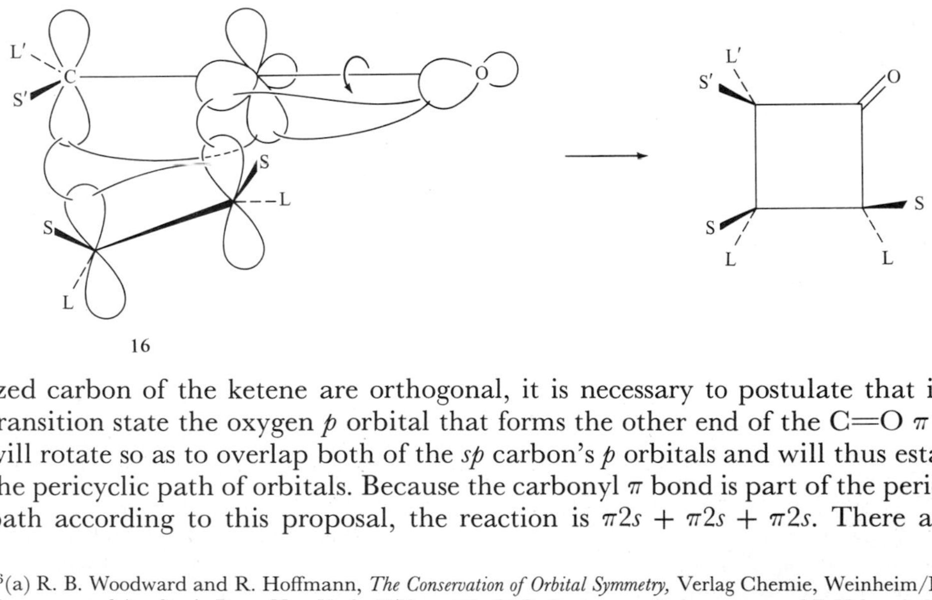

(11.24)

16

ized carbon of the ketene are orthogonal, it is necessary to postulate that in the transition state the oxygen p orbital that forms the other end of the C=O π bond will rotate so as to overlap both of the sp carbon's p orbitals and will thus establish the pericyclic path of orbitals. Because the carbonyl π bond is part of the pericyclic path according to this proposal, the reaction is $\pi 2s + \pi 2s + \pi 2s$. There are six

[46](a) R. B. Woodward and R. Hoffmann, *The Conservation of Orbital Symmetry*, Verlag Chemie, Weinheim/Bergstr., Germany, and Academic Press, New York, 1970, p. 163; (b) R. Sustmann, A. Ansmann, and F. Vahrenholt, *J. Am. Chem. Soc.*, **94**, 8099 (1972).

[47](a) H. E. Zimmerman, in *Pericyclic Reactions*, Vol. I, A. P. Marchand and R. E. Lehr, Eds., Academic Press, New York, 1977, p. 77; (b) L. Ghosez and M. J. O'Donnell, in *Pericyclic Reactions*, Vol. II, p. 87.

electrons and zero phase inversions, so the process is allowed. The predicted stereochemistry is the same as that for the $2s + 2a$ cycloaddition. No experimental distinction has been made between the alternatives **15** and **16**.

Despite the success of the concerted $2s + 2a$ pericyclic pathway in explaining stereochemistry of numerous ketene additions, there are also examples of additions of ketenes that occur by a stepwise path.[48] A number of other heterocumulenes also undergo $2 + 2$ additions,[49] as do allenes. In the case of allene cycloadditions to alkenes, much of the experimental work has been interpreted in terms of a stepwise biradical mechanism,[50] but a concerted $2s + 2s + 2s$ pericyclic transition state of the type shown in **16** has also been proposed.[51]

The most frequently observed $2 + 2$ additions are photochemical. The excited-state additions, allowed by the pericyclic selection rules, occur readily and have proved extremely useful in synthesis. Examples will be considered in Section 12.4.[52]

$2 + 4$ Cycloadditions. 1,3-Dipolar Additions

A 1,3-dipolar cycloaddition is the addition to an alkene double bond of a three-atom, four-electron π system that has dipolar character. These reactions have been studied extensively by Huisgen.[53] A few of the many examples of 1,3-dipoles are shown in Structures **17–21**. The resonance structure that expresses the dipolar

diazoalkanes

17

nitrous oxide

18

[48](a) R. Huisgen and P. Otto, *J. Am. Chem. Soc.*, **91**, 5922 (1969); (b) W. G. Duncan, W. Weyler, Jr., and H. W. Moore, *Tetrahedron Lett.*, 4391 (1973); (c) note 37(b). For a theoretical assessment of conditions under which stepwise additions of ketenes are likely, see (d) K. N. Houk, R. W. Strozier, and J. A. Hall, *Tetrahedron Lett.*, 897 (1974).

[49]See T. L. Gilchrist and R. C. Storr, *Organic Reactions and Orbital Symmetry*, Cambridge University Press, London, 1972, p. 167, for some examples.

[50]See, for example, (a) S.-H. Dai and W. R. Dolbier, Jr., *J. Am. Chem. Soc.*, **94**, 3946 (1972); (b) T. J. Levek and E. F. Kiefer, *J. Am. Chem. Soc.*, **98**, 1875 (1976); (c) J. S. Chickos, *J. Org. Chem.*, **44**, 1515 (1979); (d) L. Ghosez and M. J. O'Donnell, in *Pericyclic Reactions*, Vol. II, A. P. Marchand and R. E. Lehr, Eds., Academic Press, New York, 1977, p. 79.

[51]D. J. Pasto, *J. Am. Chem. Soc.*, **101**, 37 (1979).

[52]Transition metal ions catalyze a number of cycloaddition reactions that are forbidden as pericyclic processes in the ground state. See, for example, (a) L. A. Paquette, *Acc. Chem. Res.*, **4**, 280 (1971); (b) M. J. S. Dewar, *Angew. Chem. Int. Ed. Engl.*, **10**, 761 (1971); (c) P. M. Maitlis, *Acc. Chem. Res.*, **9**, 93 (1976).

[53](a) R. Huisgen, *Angew. Chem. Int. Ed. Engl.*, **2**, 565 (1963); (b) A. Padwa, *Angew. Chem. Int. Ed. Engl.*, **15**, 123 (1976). A number of six-electron cycloadditions in which one component is a cation or anion are also known. See (c) T. S. Sorensen and A. Rauk, in *Pericyclic Reactions*, Vol. II, A. P. Marchand and R. E. Lehr, Eds., Academic Press, New York, 1977, p. 1 (cations); (d) S. W. Staley, in *Pericyclic Reactions*, Vol. I, p. 199 (anions).

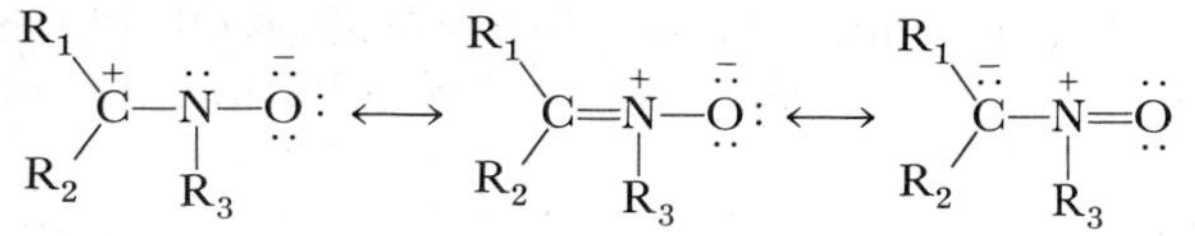

$$\underset{R_2}{\overset{R_1}{>}}\!\!\overset{+}{C}\!\!-\!\!\underset{R_3}{N}\!\!-\!\!\overset{..}{\underset{..}{O}}\!:^{-} \longleftrightarrow \underset{R_2}{\overset{R_1}{>}}\!\!C\!=\!\!\overset{+}{\underset{R_3}{N}}\!\!-\!\!\overset{..}{\underset{..}{O}}\!:^{-} \longleftrightarrow \underset{R_2}{\overset{R_1}{>}}\!\!\overset{-}{C}\!\!-\!\!\overset{+}{\underset{R_3}{N}}\!\!=\!\!\overset{..}{\underset{..}{O}}$$

nitrones
19

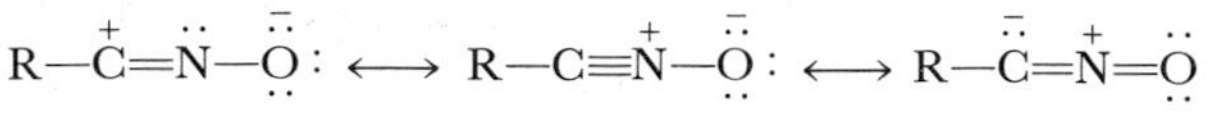

$$R\!\!-\!\!\overset{+}{C}\!\!=\!\!\overset{..}{N}\!\!-\!\!\overset{..}{\underset{..}{O}}\!:^{-} \longleftrightarrow R\!\!-\!\!C\!\!\equiv\!\!\overset{+}{N}\!\!-\!\!\overset{..}{\underset{..}{O}}\!:^{-} \longleftrightarrow R\!\!-\!\!\overset{..}{\underset{-}{C}}\!\!=\!\!\overset{+}{N}\!\!=\!\!\overset{..}{\underset{..}{O}}$$

nitrile oxides
20

$$:\!\overset{+}{\underset{..}{O}}\!\!-\!\!\overset{..}{\underset{..}{O}}\!\!-\!\!\overset{..}{\underset{..}{O}}\!:^{-} \longleftrightarrow \overset{..}{O}\!\!=\!\!\overset{+}{\underset{..}{O}}\!\!-\!\!\overset{..}{\underset{..}{O}}\!:^{-} \longleftrightarrow :\!\overset{..}{\underset{..}{O}}^{-}\!\!-\!\!\overset{+}{\underset{..}{O}}\!\!=\!\!\overset{..}{O}$$

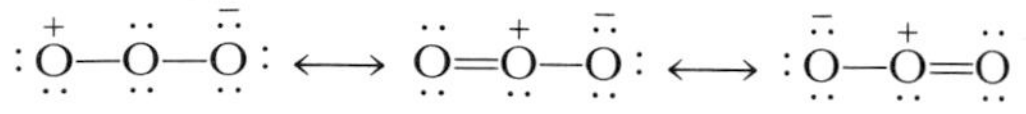

ozone
21

character is shown first in each instance, although it may be of higher energy than the other structures. Equations 11.25–11.27 show a few examples of 1,3-dipolar additions.

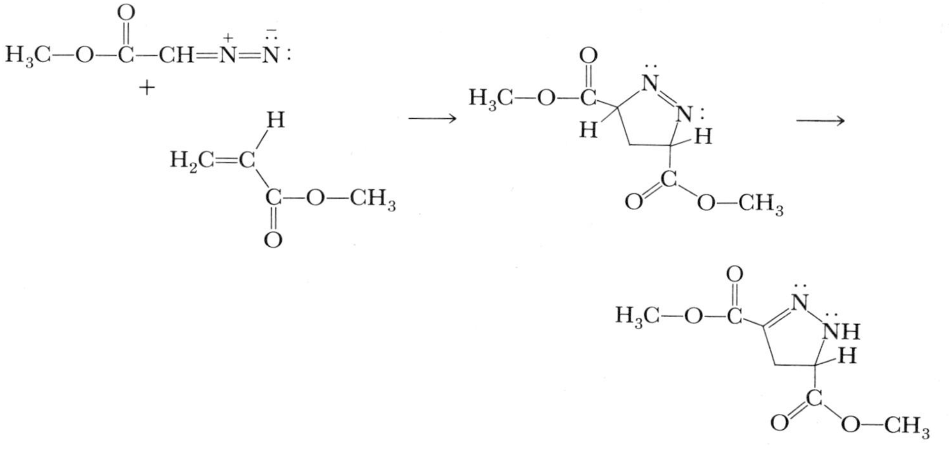

$$(11.25)$$

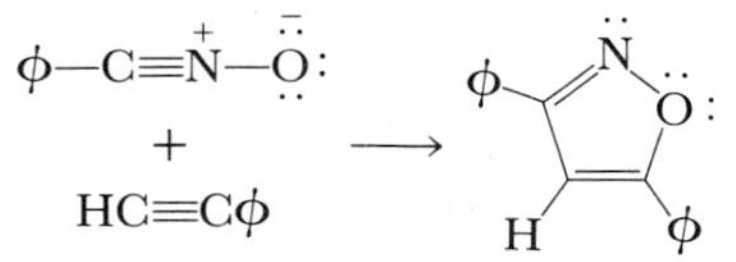

$$(11.26)$$

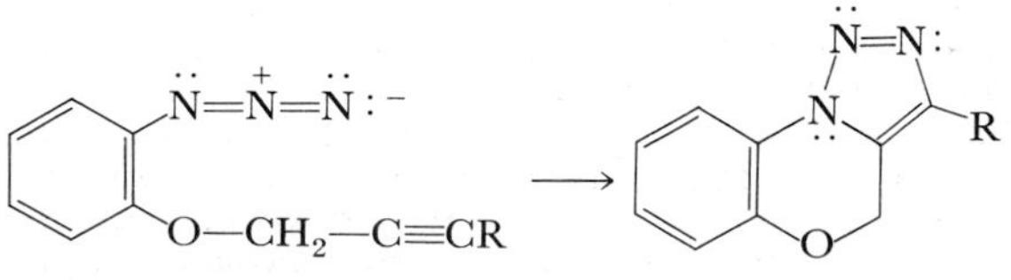

$$(11.27)$$

The ozonolysis of alkenes proceeds through the series of reactions shown in Scheme 4, proposed by Criegee.[54] A 1,3-dipolar addition of O_3 to the double bond leads to the unstable primary ozonide (**22**), which rapidly decomposes through a retro-1,3-dipolar addition to a carbonyl compound and a carbonyl oxide (**23**). These two fragments then recombine in another 1,3-dipolar addition to give the ozonide (**24**), a 1,2,4-trioxolane. The final stage of the ozonolysis is hydrolysis of the ozonide by water to two carbonyl compounds and H_2O_2. (Note that **24** is a cyclic peroxyacetal.)

SCHEME 4

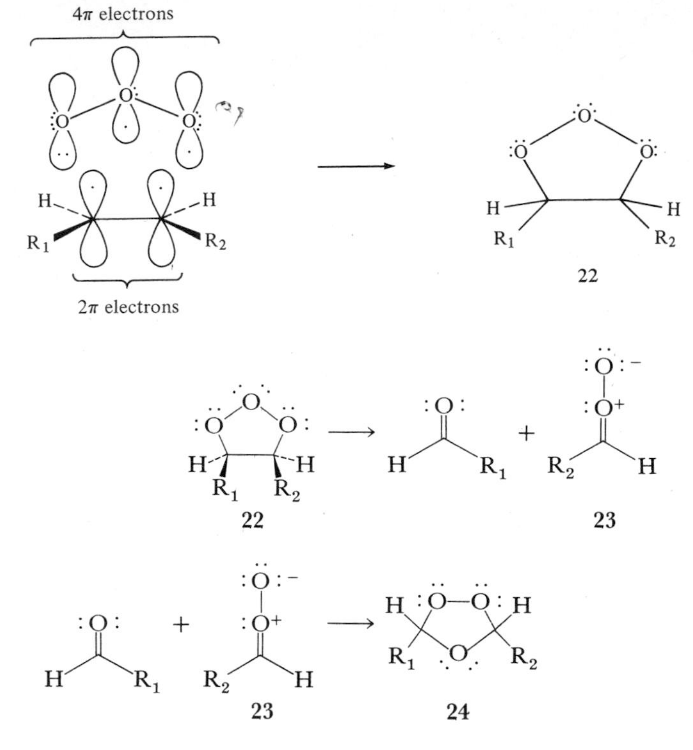

The 1,3-dipolar additions are thermally allowed $2 + 4$ cycloadditions. Experimentally they are stereospecific cis additions with respect to the alkene and they are generally considered to be concerted pericyclic reactions. This conclusion has been challenged by Firestone, who calculated that the activation energies are close to those expected for formation of zwitterionic or biradical intermediates and well above his expectation for the pericyclic path, in which the transition state ought to enjoy full aromatic stabilization. The stereochemistry he explained by proposing ring closure that is rapid compared with bond rotation, $k_2 \gg k_r$ in

[54]R. Criegee, *Angew. Chem. Int. Ed. Engl.*, **14,** 745 (1975).

Scheme 5.[55] He offered no explanation for the supposed failure of the reactions to follow the lower-energy concerted pathway, except to propose the existence of some

SCHEME 5

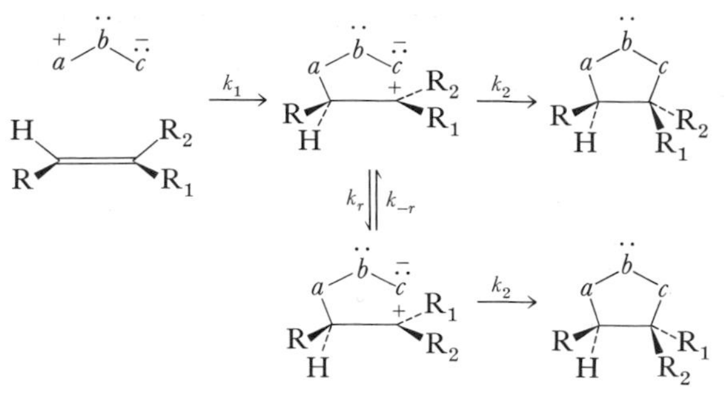

as yet undiscovered selection rule that prevents them from doing so. Huisgen has defended his hypothesis of concerted reactions, pointing out that all known 1,3-dipolar additions are stereospecific, a result unlikely to follow from a stepwise mechanism, and suggesting that Firestone's thermodynamic analysis is incorrect.[56] It has also been suggested by Houk and co-workers that modest energy barriers in thermally allowed reactions are reasonable, because energy must be expended during the approach of reactants to accomplish distortions necessary before favorable HOMO–LUMO interactions start to take effect.[57]

2 + 4 Cycloadditions. The Diels–Alder Reaction

Numerous 2 + 4 cycloadditions have been known since long before the advent of the pericyclic theory; they are among the most powerful of synthetic reactions. The most important of these is the Diels–Alder reaction, of which Equation 11.28 is the prototype. Recognized by Diels and Alder in 1928, it presents a convenient and highly stereospecific route to the ubiquitous six-membered ring.[58] The pericyclic

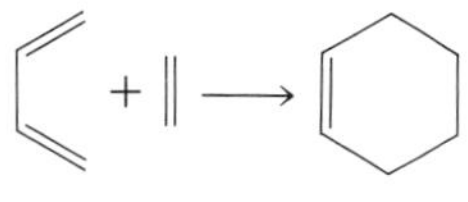

$$\text{(11.28)}$$

[55] R. A. Firestone, *Tetrahedron*, **33**, 3009 (1977).
[56] R. Huisgen, *J. Org. Chem.*, **41**, 403 (1976).
[57] K. N. Houk, R. W. Gandour, R. W. Stozier, N. G. Rondan, and L. A. Paquette, *J. Am. Chem. Soc.*, **101**, 6797 (1979). The analysis is based on semiempirical (MINDO/3) and *ab initio* molecular orbital calculations of trimerization of acetylene to benzene, thermally allowed as a 2s + 2s + 2s cycloaddition.
[58] (a) O. Diels and K. Alder, *Justus Liebigs Ann. Chem.*, **460**, 98 (1928). For references to subsequent work by Diels and Alder and for further discussion the reader is referred to the numerous review articles that have appeared: (b) J. A. Norton, *Chem. Rev.*, **31**, 319 (1942); (c) M. C. Kloetzel, *Org. Reactions*, **4**, 1 (1948); (d) H. L. Holmes, *Org. Reactions*, **4**, 60 (1948); (e) L. W. Butz and A. W. Rytina, *Org. Reactions*, **5**, 136 (1949); (f) J. Sauer, *Angew. Chem. Int. Ed. Engl.*, **5**, 211 (1966); reverse Diels–Alder: (g) H. Kwart and K. King, *Chem. Rev.*, **68**, 415 (1968); components containing hetero atoms: (h) S. B. Needleman and M. C. Chang Kuo, *Chem. Rev.*, **62**, 405 (1962); stereochemistry: (i) J. G. Martin and R. K. Hill, *Chem. Rev.*, **61**, 537 (1961); mechanism: (j) A. Wassermann, *Diels–Alder Reactions*, Elsevier, Amsterdam, 1965; (k) J. Sauer, *Angew. Chem. Int. Ed. Engl.*, **6**, 16 (1967); general: (l) R. Huisgen, R. Grashey, and J. Sauer, in *The Chemistry of Alkenes*, S. Patai, Ed., Wiley, London, 1964, p. 739.

theory, predicting as it does an allowed concerted $2s + 4s$ process, is in agreement with the known facts about the mechanism. We shall review briefly here the salient features.

Although the reaction occurs in the unsubstituted case,[59] it is most successful when the diene and the alkene (referred to in this context as the *dienophile*) bear substituents of complementary electronic influence. Although these are most commonly an electron-donating group on the diene and an electron-withdrawing group on the dienophile, there are also a number of instances that illustrate *inverse electron demand,* that is, electron-withdrawing groups on the diene and donating groups on the dienophile.

The first important point about the reaction is that the diene must be in the s-cis conformation (**25**) in order to react. Thus dienes with one double bond exocyclic, such as **27**, do not give the reaction, whereas endocyclic dienes (**28**) react

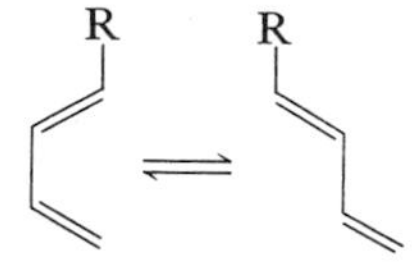

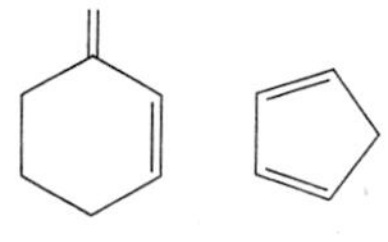

rapidly. Reactivity of open-chain dienes depends on the equilibrium constant for the conformational interconversion. Butadiene itself prefers the s-trans conformation (**26**, R=H) over the s-cis (**25**, R=H) by 2.5 to 3.1 kcal mole^{-1}; the rotational energy barrier separating the two conformations is about 3.9 kcal mole^{-1}.[60] A cis-1-substituent favors the s-trans form (**30**) and retards the addition, whereas a 2-substituent favors the s-cis form (**31**) and enhances the rate. These effects are consistent with a concerted process, where in the transition state bonding occurs simultaneously at both ends of the diene.

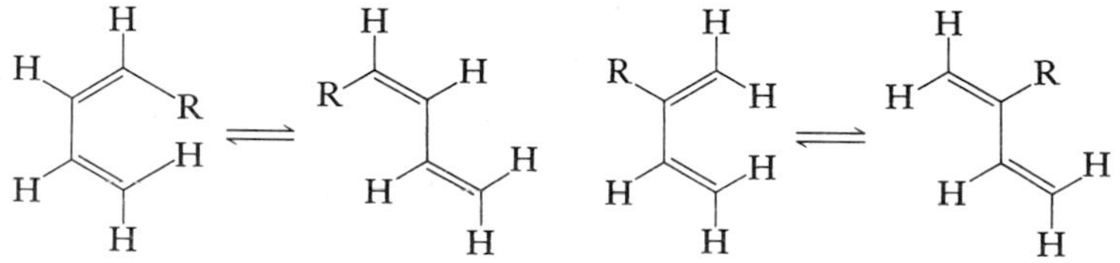

[59]L. M. Joshel and L. W. Butz, *J. Am. Chem. Soc.*, **63**, 3350 (1941).
[60]M. E. Squillacote, R. S. Sheridan, O. L. Chapman, and F. A. L. Anet, *J. Am. Chem. Soc.*, **101**, 3657 (1979).

Preferred orientation in the addition of a monosubstituted diene and a monosubstituted dienophile is shown in Equations 11.29 and 11.30. The regio-

| major product | minor product | (11.29) |

| major product | minor product | (11.30) |

selectivity shown in these equations, and also the propensity for electron-donating groups on the diene and electron-withdrawing groups on the dienophile to enhance reaction rate, may be explained in terms of frontier orbital interactions.[61] Figure 11.2(a) shows the π energy levels of butadiene and ethylene, with HOMO–LUMO interactions indicated. In Figure 11.2(b) an electron-donating group D on the diene will raise the diene HOMO and an electron-accepting group A on the dienophile will lower the dienophile LUMO; the result is the stronger dominant interaction indicated. Figure 11.2(c) shows orbital energies in a case of inverse electron demand.

The explanation of regiospecificity requires a knowledge of the effect of substituents on the coefficients of the HOMO and LUMO orbitals. In a case of normal electron demand, Figure 11.2(b) shows that the important orbitals are the diene HOMO and dienophile LUMO. Donor D in the 1 position leads to HOMO orbital coefficients at the 1 and 4 positions of the diene as shown in **33**, where the view is straight down onto the molecular plane and the circles represent the near lobes of the p orbitals, shaded for positive sign and unshaded for negative. The relative sizes of the circles represent the relative contributions of the respective p orbitals to the HOMO. (Orbitals at positions 2 and 3 are omitted.) It has been shown by Anh and others that the condensation will occur so as to bring together the ends with the largest coefficients.[62] In the normal electron demand example,

[61] I. Fleming, *Frontier Orbitals and Organic Chemical Reactions,* Wiley, London, 1976, Chap. 4.
[62] (a) O. Eisenstein, J. M. Lefour, N. Trong Anh, and R. F. Hudson, *Tetrahedron,* **33**, 523 (1977); (b) N. D. Epiotis, *J. Am. Chem. Soc.,* **95**, 5624 (1973); (c) W. C. Herndon, *Chem. Rev.,* **72**, 157 (1972). It is frequently assumed that although the Diels–Alder condensation is concerted, the bonding at one end will be farther advanced than at the other end in the transition state. See (d) R. B. Woodward and T. J. Katz, *Tetrahedron,* **5**, 70 (1959). The bonding would be expected to advance faster between the positions with highest coefficients in the dominant HOMO and LUMO.

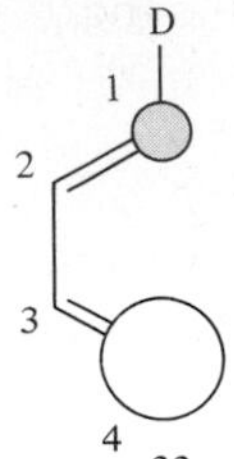

diene HOMO
(contributions at 1,4 only)

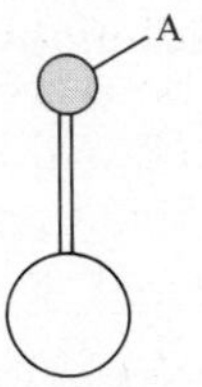

dienophile LUMO

the prediction is as shown in Equation 11.31 and agrees with experiment. If the donor is at position 2, the 1,4-coefficients in the diene HOMO are affected as shown

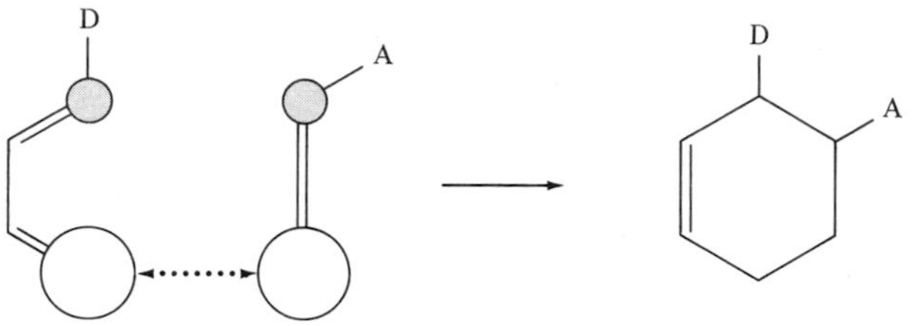

$$(11.31)$$

by **35**. Then the strongest interaction with an acceptor-substituted dienophile is with diene position 1, as shown in Equation 11.32.[63]

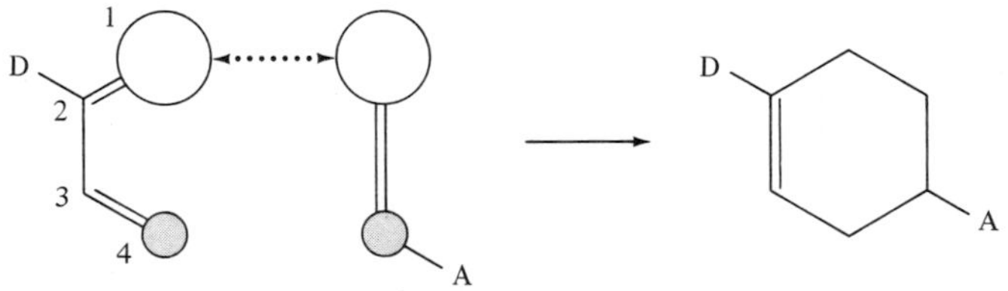

$$(11.32)$$

Stereochemistry provides a strong point in favor of a concerted process. The additions are always exclusively cis, so that the relative configurations of substituents existing in the components is maintained in the product. Thus a cis-substituted dienophile leads to a cis-substituted cyclohexene, and trans to trans (Equation 11.33); trans–trans dienes yield cis substitution, and cis–trans dienes give trans substitution (Equation 11.34).[64] It thus is clear that there is no intermediate of the kind found in 2 + 2 cycloadditions, in which bond rotations leading to mixed stereochemistry can occur. Berson and co-workers have made a careful study of the dimerization of butadiene, where both reactive centers of a potential intermediate biradical would be stabilized by conjugation (36). They concluded that for the Diels–Alder product (37) to be formed through 36 with the observed stereospecificity, bond rotation barriers in 36 would have to be unreasonably high (11 to

---

[63] The forms of the orbitals are found qualitatively as follows. Butadiene HOMO has equal contributions from the p orbitals at positions 1 and 4(a). With a donor substituent at 1, the molecule takes on some of the character of

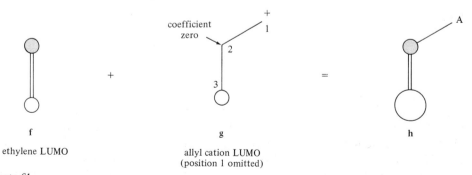

a	**b**	**c**
butadiene HOMO	pentadienyl anion HOMO	
(positions 2, 3 omitted)	(positions 1, 3, 4 omitted)	

a pentadienyl anion (**b**). Superposition of these two orbitals leads to **c**. A donor at the 2 position gives some character of an allyl anion to the double bond to which it is attached:

a	**d**	**e**
butadiene HOMO	allyl anion HOMO	
(positions 2, 3, omitted)	(position 1 omitted)	

An acceptor substituent on ethylene gives it some of the character of an allyl cation:

f	**g**	**h**
ethylene LUMO	allyl cation LUMO	
	(position 1 omitted)	

See note 61.

[64] J. G. Martin and R. K. Hill, *Chem. Rev.*, **61**, 537 (1961).

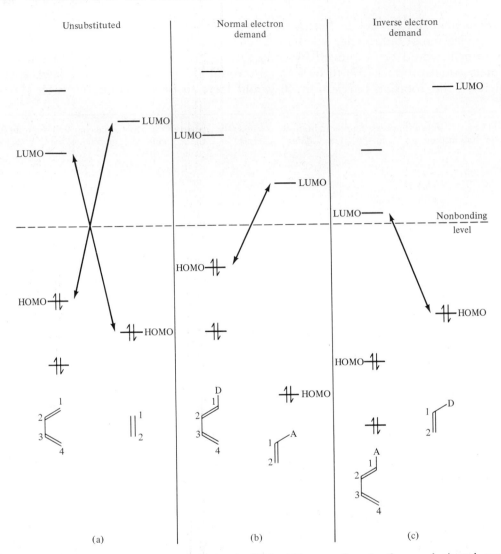

Figure 11.2 Frontier orbital interactions in the Diels–Alder reaction. In the unsubstituted case, butadiene + ethylene (a), both HOMO–LUMO interactions are equally important. Substituting an electron donor D on the diene and an acceptor A on the dienophile leads to the energy-level pattern (b) characteristic of normal electron demand. The interaction diene HOMO–dienophile LUMO now is strong and dominates the reaction. Inverse electron demand occurs for the substitution pattern shown in (c).

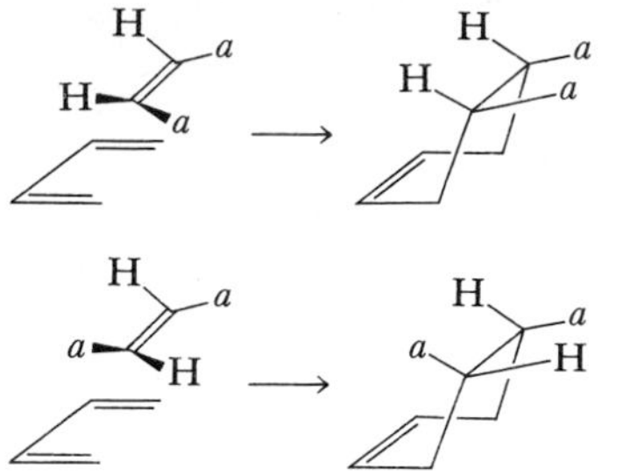

(11.33)

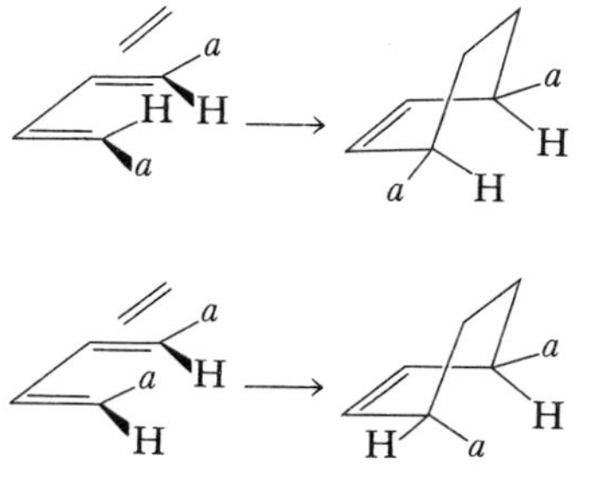

(11.34)

20 kcal mole^{-1}). The biradical **36** may, however, lie as little as 3.6 kcal mole^{-1} higher in energy than the concerted transition state.[65]

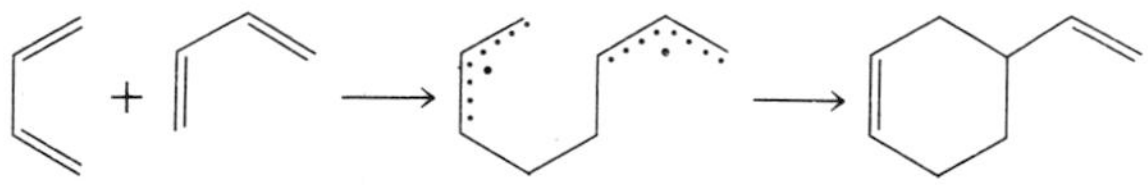

(11.35)

36 **37**

When both diene and dienophile are substituted, the *endo principle* applies. This rule is illustrated in Equation 11.36;[64] the more stable transition state (**38**) is

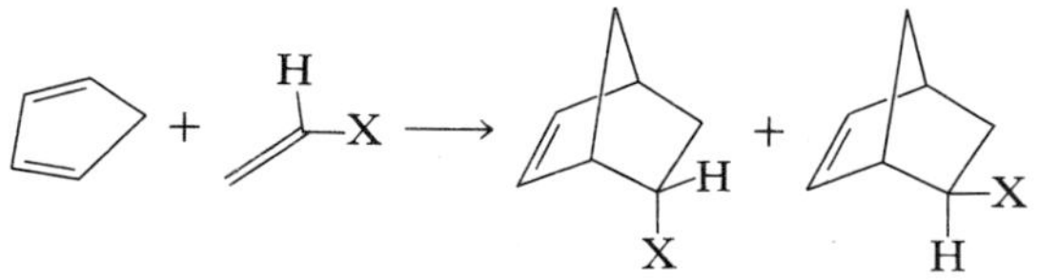

(11.36)

major
product

minor
product

that which has the maximum juxtaposition of unsaturated centers. Frontier orbital theory also explains the endo principle. This time *secondary interactions* are important; these are interactions between portions of the interacting HOMO and LUMO other than the loci of new bond formation. Structures **38** and **39** show the

[65](a) J. A. Berson, P. B. Dervan, R. Malherbe, and J. A. Jenkins, *J. Am. Chem. Soc.*, **98**, 5937 (1976). See also (b) L. M. Stephenson, R. V. Gemmer, and S. Current, *J. Am. Chem. Soc.*, **97**, 5909 (1975).

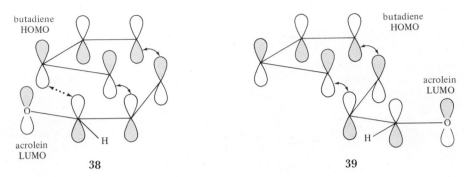

38   39

HOMO of butadiene and the LUMO of acrolein (propenal) when these molecules are oriented in two possible ways. (Note that the acrolein LUMO has the expected nodal pattern for the lowest antibonding orbital of a four-atom π system.) In **38** the primary interactions are indicated by solid arrows; there is also a stabilizing secondary interaction (dotted arrow). In the exo orientation (**39**) the primary interactions are the same but the secondary interaction is absent. Hence the endo orientation is favored.[66] Secondary interactions can also affect regioselectivity. If the butadiene has substituents, the orbital coefficients at positions 2 and 3 may differ; then there is a slight preference for the arrangement that places the dienophile substituent near the position with the larger HOMO coefficient in the diene. These secondary interactions may determine regiochemistry in some instances where the primary interactions discussed earlier do not provide a definitive preference.[67]

In addition to the stereochemistry, there is kinetic evidence favoring a concerted process. The highly negative activation entropies ($\Delta S^{\ddagger}$ typically about -35 cal mole^{-1} K^{-1}) accompanied by low activation enthalpy ($\Delta H^{\ddagger}$ usually less than 25 kcal mole^{-1}) point to an ordered but energetically favorable transition state.[68] Secondary hydrogen isotope effects have been investigated by various workers; although interpretation is difficult, the observation of small inverse effects ($k_{\mathrm{H}}/k_{\mathrm{D}} < 1$) appears consistent with a concerted process with a small degree of change of hybridization at carbon from sp^2 toward sp^3 in the transition state.[69] Although the question of whether the addition is concerted or not might still be considered by some to be open to debate, the weight of evidence favors the conclusion that is is concerted.[70]

[66]R. B. Woodward and R. Hoffmann, *The Conservation of Orbital Symmetry,* Verlag Chemie, Weinheim/Bergstr., Germany, and Academic Press, New York, 1970, p. 145.

[67](a) P. V. Alston, R. M. Ottenbrite, and T. Cohen, *J. Org. Chem.*, **43**, 1864 (1978); (b) T. Cohen, R. J. Ruffner, D. W. Shull, W. M. Daniewski, R. M. Ottenbrite, and P. V. Alston, *J. Org. Chem.*, **43**, 4052 (1978).

[68]Butadiene + ethylene, $\Delta H^{\ddagger} \approx 29$ kcal mole^{-1}, $\Delta S^{\ddagger} \approx -26$ cal mole^{-1} K^{-1}; butadiene + acrolein (propenal), $\Delta H^{\ddagger} \approx 19$ kcal mole^{-1}, $\Delta S^{\ddagger} \approx -32$ cal mole^{-1} K^{-1}. A. Wassermann, *Diels–Alder Reactions,* Elsevier, Amsterdam, 1965, p. 50.

[69]See Section 2.7, and J. Sauer, *Angew. Chem. Int. Ed. Engl.*, **6**, 16 (1967).

[70]Bartlett and co-workers found small amounts of stepwise biradical 2 + 4 cycloaddition accompanying 2 + 2 biradical addition in reactions of dichlorodifluoroethylenes with dienes. In other instances of competing 2 + 2 and 2 + 4 additions, the 2 + 2 mechanism is biradical and the 2 + 4 concerted. (a) P. D. Bartlett and J. J.-B. Mallet, *J. Am. Chem. Soc.*, **98**, 143 (1976); (b) P. D. Bartlett, A. S. Wingrove, and R. Owyang, *J. Am. Chem. Soc.*, **90**, 6067 (1968); (c) R. Wheland and P. D. Bartlett, *J. Am. Chem. Soc.*, **92**, 3822 (1970).

2 + 4 **Cheletropic Reactions**

The pericyclic theory predicts that cheletropic reactions of four-electron with two-electron components may occur by the linear path, that is, suprafacially on the one-atom component, if the diene enters also suprafacially (six-electron interaction diagram **40**). A number of examples are known, most of them in the reverse,

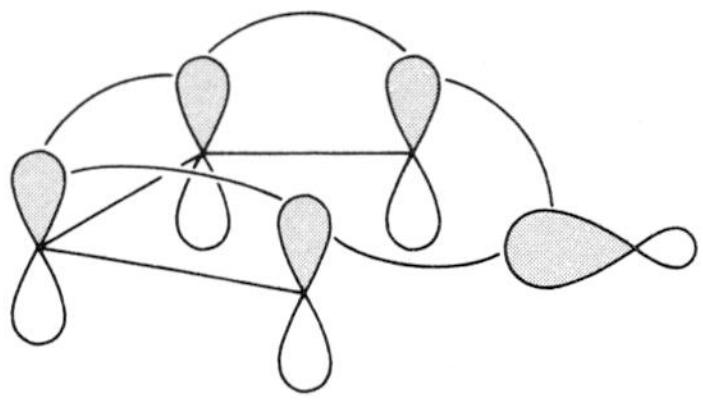

40

fragmentation, direction.[71] The tricyclic dione (**41**) undergoes fragmentation,[72] and attempts to prepare norbornadien-7-one (**42**) yield only benzene and carbon monoxide.[73]

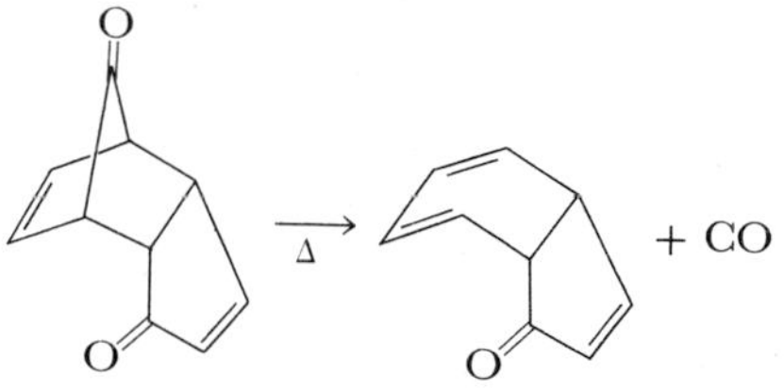

$$+ CO \qquad (11.37)$$

41

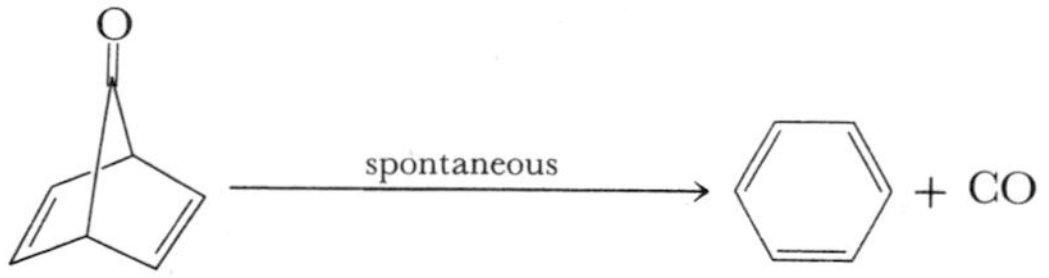

$$+ CO \qquad (11.38)$$

42

Equation 11.39, a reaction that also occurs in the addition direction, is stereospecific. The stereochemistry shows that the diene component enters suprafacially; therefore the pericyclic selection rules specify the linear mode for the SO_2 component, a prediction not readily checked experimentally.[74,75] The isotope ef-

[71](a) R. B. Woodward and R. Hoffmann, *The Conservation of Orbital Symmetry*, Verlag Chemie, Weinheim/Bergstr., Germany, and Academic Press, New York, 1970, p. 156; (b) T. L. Gilchrist and R. C. Storr, *Organic Reactions and Orbital Symmetry*, Cambridge University Press, London, 1972, p. 125; (c) W. L. Mock, in *Pericyclic Reactions*, Vol. II, A. P. Marchand and R. E. Lehr, Eds., Academic Press, New York, 1977, p. 141; (d) S. D. Turk and R. L. Cobb, in *1,4 Cycloaddition Reactions*, J. Hamer, Ed., Academic Press, New York, 1967, p. 13; (e) L. D. Quinn, in *1,4 Cycloaddition Reactions*, p. 47.

[72]J. E. Baldwin, *Can. J. Chem.*, **44**, 2051 (1966).

[73]S. Yankelevich and B. Fuchs, *Tetrahedron Lett.*, 4945 (1967).

[74]W. L. Mock, *J. Am. Chem. Soc.*, **97**, 3666, 3673 (1975).

[75](a) W. L. Mock, *J. Am. Chem. Soc.*, **88**, 2857 (1966); (b) S. D. McGregor and D. M. Lemal, *J. Am. Chem. Soc.*, **88**, 2858 (1966); (c) W. L. Prins and R. M. Kellogg, *Tetrahedron Lett.*, 2833 (1973).

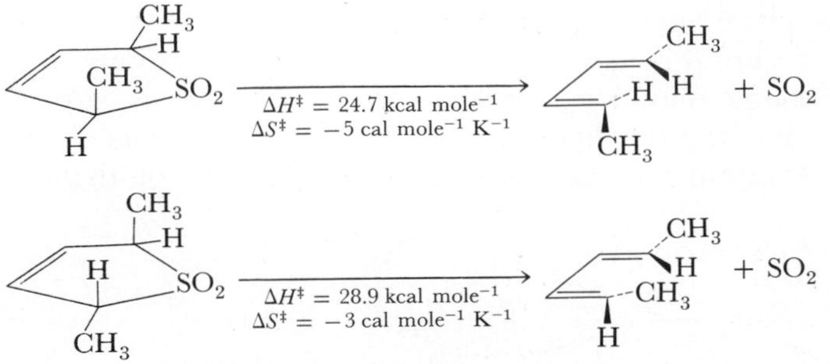

$$\Delta H^{\ddagger} = 24.7 \text{ kcal mole}^{-1}$$
$$\Delta S^{\ddagger} = -5 \text{ cal mole}^{-1}\text{ K}^{-1}$$

$$(11.39)^{74}$$

$$\Delta H^{\ddagger} = 28.9 \text{ kcal mole}^{-1}$$
$$\Delta S^{\ddagger} = -3 \text{ cal mole}^{-1}\text{ K}^{-1}$$

fects for Equations 11.40 and 11.41 provide evidence in favor of a concerted mechanism. The ratio k_H/k_D for the former reaction is equal (within experimental error)

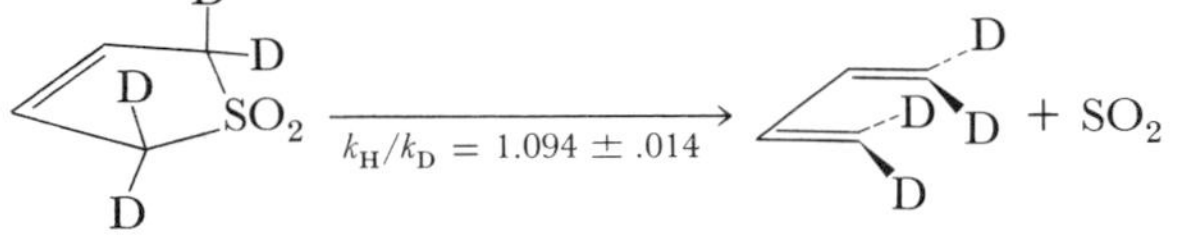

$$k_H/k_D = 1.094 \pm .014 \qquad (11.40)$$

$$k_H/k_D = 1.054 \pm .019 \qquad (11.41)$$

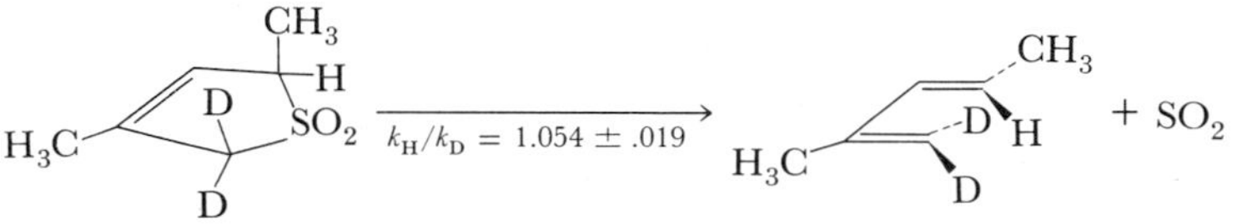

to the square of k_H/k_D for the latter reaction. This result indicates that both C—S bonds are breaking at the transition state; if only one were breaking, it would be the one to the nondeuterated carbon in Equation 11.41 and the isotope effect for this reaction would be smaller.[76]

Sulfur monoxide adds to the isomeric 2,4-hexadienes to yield cyclic sulfoxides in the reaction analogous to the reverse of 11.39, but the addition is not stereospecific. The SO molecule has a triplet ground state and the reaction presumably follows a biradical path.[77]

Photolysis of the sulfolenes shown in Equation 11.39, expected on the basis of the pericyclic theory to yield products with reversed selectivity, produces partial loss of stereochemistry. No definitive interpretation of these results is available.[78]

Higher Cycloadditions

The stereochemistry of the 6 + 2 cheleotrpic sulfur dioxide extrusion shown in Equations 11.42 and 11.43 has been shown to be antarafacial (conrotatory) in the

[76]S. Ašperger, D. Hegedić, D. Pavlović, and S. Borčić, *J. Org. Chem.*, **37**, 1745 (1972).

[77](a) P. Chao and D. M. Lemal, *J. Am. Chem. Soc.*, **95**, 920 (1973); (b) D. M. Lemal and P. Chao, *J. Am. Chem. Soc.*, **95**, 922 (1973).

[78]See note 71(c).

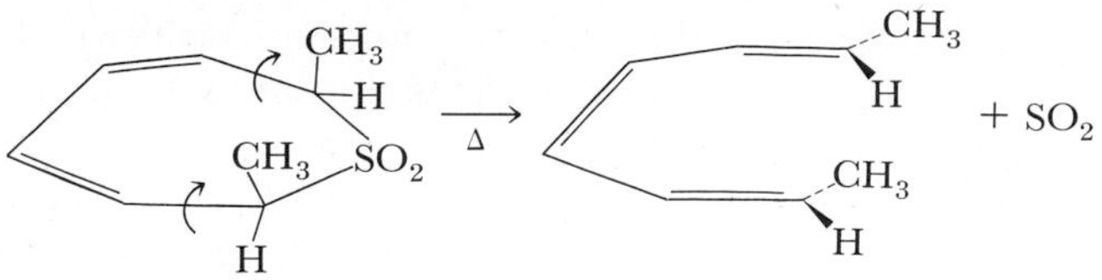

$$+ \; SO_2 \tag{11.42}$$

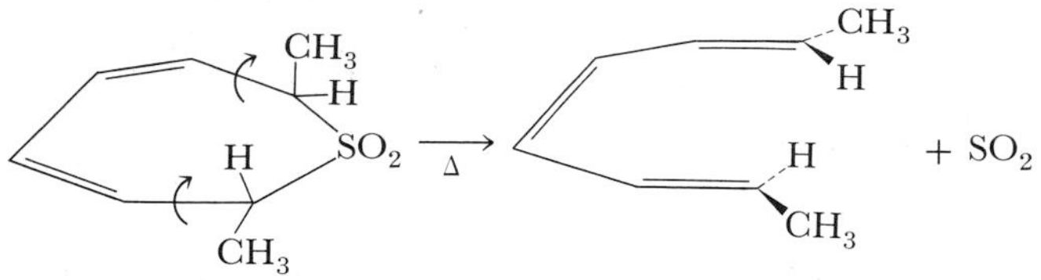

$$+ \; SO_2 \tag{11.43}$$

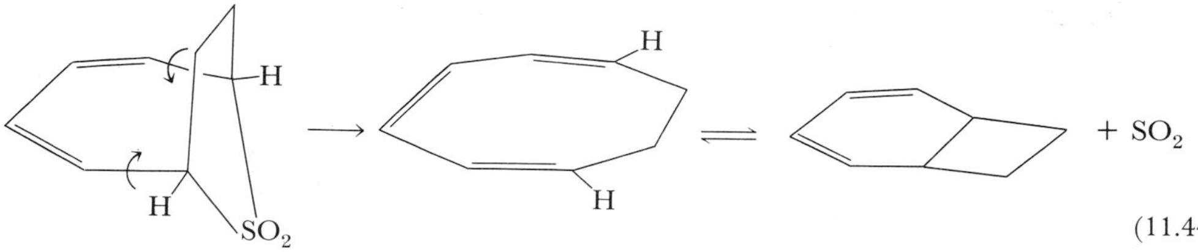

43

$$+ \; SO_2 \tag{11.44}[79]$$

$$\Delta H^{\ddagger} \approx 32 \text{ kcal mole}^{-1}$$
$$\Delta S^{\ddagger} \approx -14 \text{ cal mole}^{-1} \text{ K}^{-1}$$

triene component, in accord with expectations for a linear path.[80] The bicyclic sulfone **43** is constrained by the ring system to fragment suprafacially on the triene and therefore, according to theory, to open antarafacially (nonlinear path)

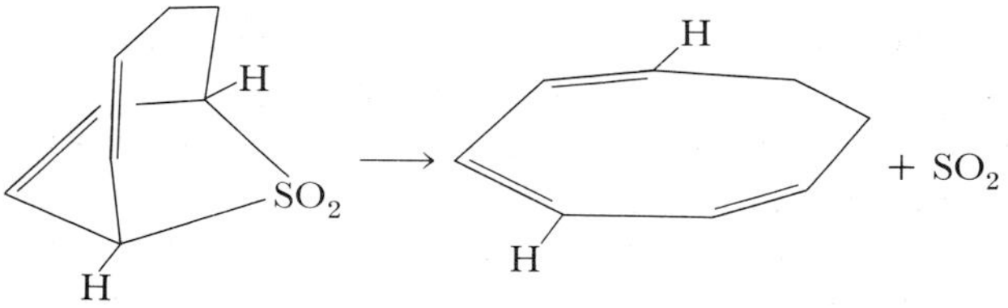

44

$$+ \; SO_2 \tag{11.45}$$

$$\Delta H^{\ddagger} = 29.3 \text{ kcal mole}^{-1}$$
$$\Delta S^{\ddagger} = +2 \text{ cal mole}^{-1} \text{ K}^{-1}$$

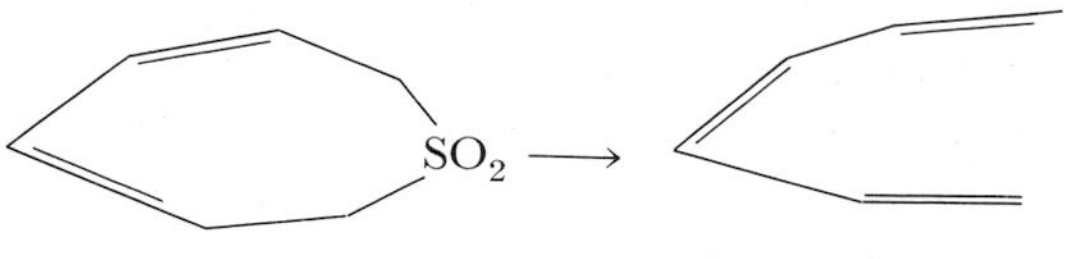

45

$$\tag{11.46}$$

$$\Delta H^{\ddagger} = 29.8 \text{ kcal mole}^{-1}$$
$$\Delta S^{\ddagger} = 0 \text{ cal mole}^{-1} \text{ K}^{-1}$$

with respect to the sulfur dioxide. It reacts slower by a factor of about 6×10^4 at 180°C than the $4 + 2$ counterpart (**44**), and slower by a factor of about 10^4 than

[79] The two C_8H_{10} isomers are in equilibrium at this temperature. See Section 11.2.
[80] W. L. Mock, *J. Am. Chem. Soc.*, **97**, 3666, 3673 (1975).

45.[81] (The activation parameters given for Equation 11.45 are very uncertain.) The selection rules permit the linear path for both **44** and **45**. It is not known whether the decomposition 11.45 is concerted.

There are a few 6 + 2 cycloadditions known; an example is given in Equation 11.47.[82] The selection rules require either that the nitroso group enter antarafacially (the triene must react suprafacially because of the constraint of the ring) or that the reaction be stepwise. The available evidence does not permit a test of this prediction.

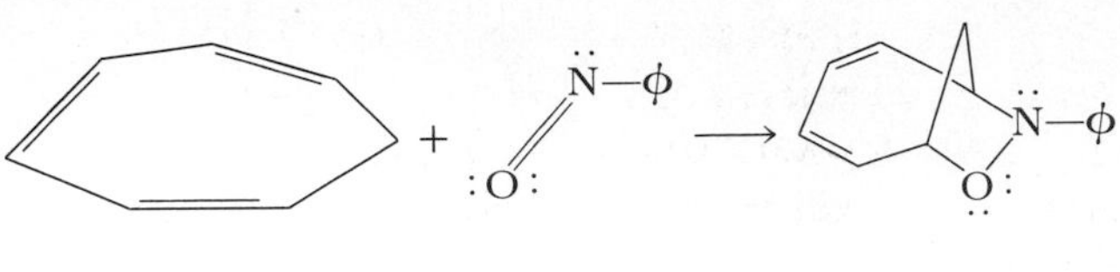

$$(11.47)$$

No concerted thermal 4 + 4 cycloadditions are known; photochemical 4 + 4 additions are observed, but in most cases probably occur through biradicals.[83] It should be noted that in the thermal butadiene dimerization (Equation 11.48), the eight-membered ring arises through the allowed [3,3]-sigmatropic isomerization of *cis*-divinylcyclobutane.[84]

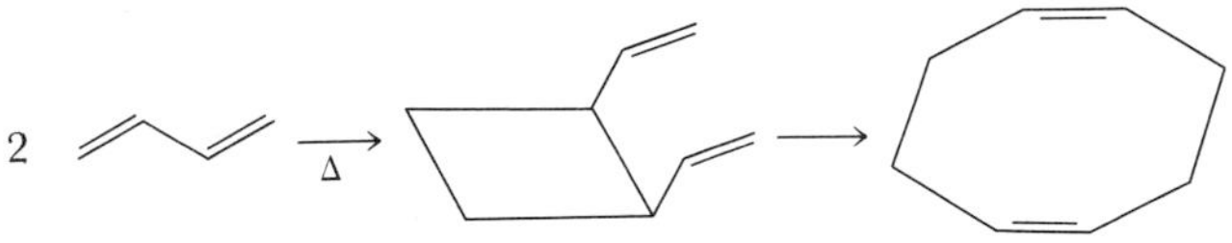

$$(11.48)$$

6 + 4 Cycloadditions, which should be allowed in the suprafacial–suprafacial mode, are known. An example is the addition of tropone to cyclopentadiene (Equation 11.49); a low activation enthalpy and highly negative activation entropy ($\Delta H^{\ddagger} = 15.3$ kcal mole^{-1}, $\Delta S^{\ddagger} = -35$ cal mole^{-1} K^{-1}) suggest a concerted mechanism similar to that of the Diels–Alder reaction.[85]

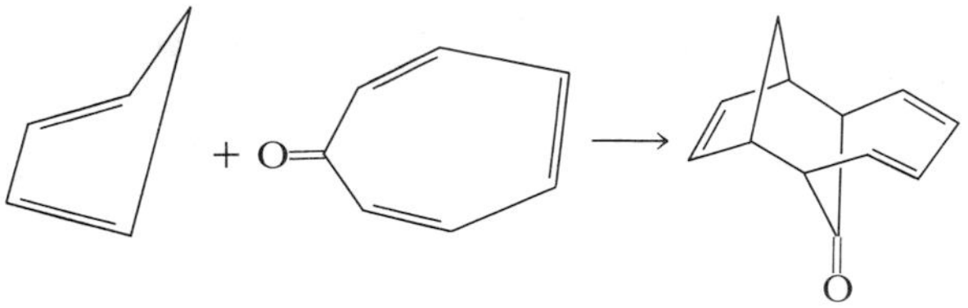

$$(11.49)$$

[81](a) W. L. Mock, *J. Am. Chem. Soc.*, **92**, 3807 (1970); (b) W. L. Mock, *J. Am. Chem. Soc.*, **97**, 3673 (1975).
[82](a) J. Hutton and W. A. Waters, *Chem. Commun.*, 634 (1966); (b) P. Burns and W. A. Waters, *J. Chem. Soc. C*, 27 (1969); (c) for a similar example, see W. S. Murphy and J. P. McCarthy, *Chem. Commun.*, 1155 (1968). (d) W. S. Murphy, K. P. Raman, and B. J. Hathaway, *J. Chem. Soc., Perkin Trans. 1*, 2521 (1977).
[83](a) W. L. Dilling, *Chem. Rev.*, **69**, 845 (1969); (b) G. Kaupp, *Angew. Chem. Int. Ed. Engl.*, **11**, 718 (1972).
[84]See Section 11.3.
[85](a) H. Tanida and H. R. Pfaendler, *Helv. Chim. Acta*, **55**, 3062 (1972). A number of cycloadditions of anions are known, but in most cases the concerted nature of the reaction is not established. See (b) S. W. Staley, in *Pericyclic Reactions*, Vol. I, A. P. Marchand and R. E. Lehr, Eds., Academic Press, New York, 1977, p. 199.

Some interesting examples of additions of tetracyanoethylene to fulvalene systems, as in Equation 11.50,[86] formally a 12 + 2 addition, indicate the potential for development in the field of higher-order cycloadditions.

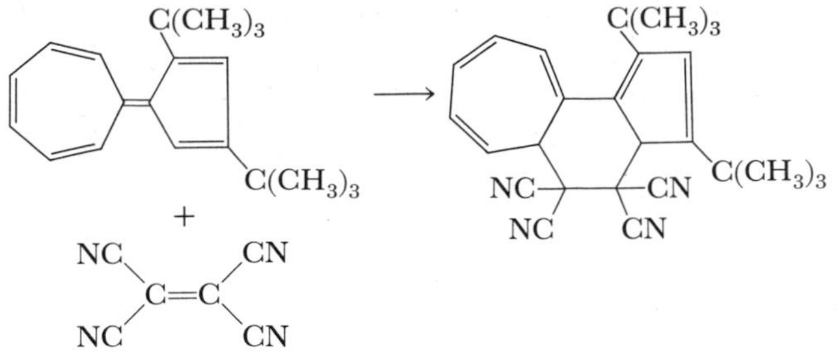

(11.50)

Multicomponent Additions[87]

Reactions of molecularity higher than two are unlikely because of the low probability of a collision of three or more molecules; multicomponent additions are therefore expected only when several of the components are suitably placed within a single molecule. Equations 11.51–11.53 illustrate some examples; for most reactions of this type the question of concertedness remains to be answered.

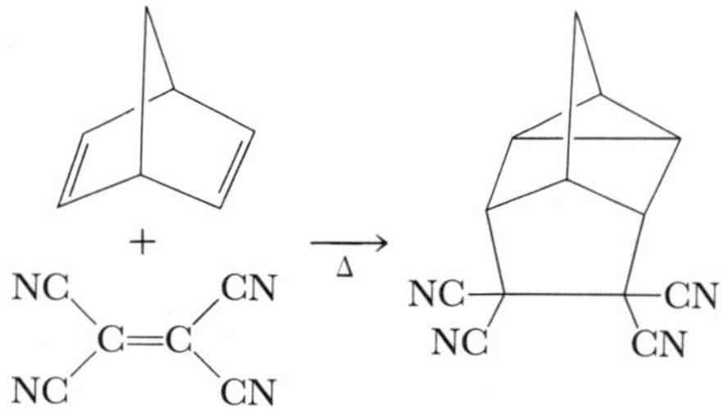

(11.51)[88]

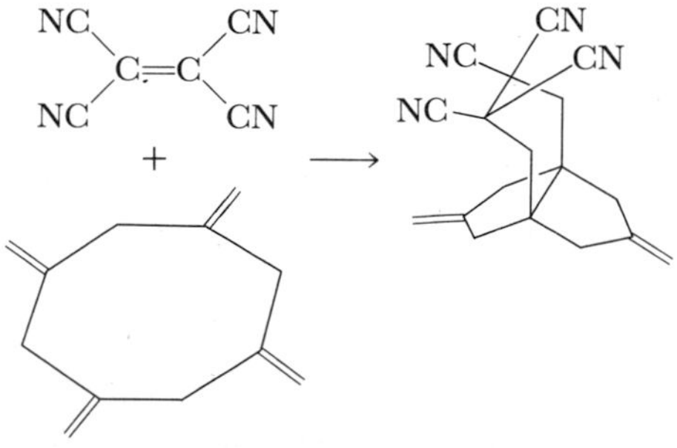

$\pi 2 + \pi 2 + \pi 2$

(11.52)[89]

$\pi 2 + \pi 2 + \pi 2$

[86] H. Prinzbach, *Pure Appl. Chem.*, **28**, 281 (1971).

[87] (a) R. Huisgen, *Angew. Chem. Int. Ed. Engl.*, **7**, 321 (1968); (b) R. B. Woodward and R. Hoffmann, *The Conservation of Orbital Symmetry*, Verlag Chemie, Weinheim/Bergstr., Germany, and Academic Press, New York, 1970, p. 106.

[88] (a) A. T. Blomquist and Y. C. Meinwald, *J. Am. Chem. Soc.*, **81**, 667 (1959); see also (b) H. K. Hall, Jr., *J. Org. Chem.*, **25**, 42 (1960); (c) R. C. Cookson, S. S. H. Gilani, and I. D. R. Stevens, *Tetrahedron Lett.*, 615 (1962).

[89] J. K. Williams and R. E. Benson, *J. Am. Chem. Soc.*, **84**, 1257 (1962).

$$(11.53)^{90}$$

$$H_3C—C\equiv C—CH_3$$

$$\pi 2s + \pi 2s + \pi 2s + \pi 2s$$

11.2 ELECTROCYCLIC REACTIONS[91]

Electrocyclic reactions, considered in the direction of ring opening, occur as shown in Equation 11.54 in rings composed of two saturated centers joined by a single σ

$$(11.54)$$

bond and by a π system. There are two general types, depending on whether the π system has an even or an odd number of atoms; the former are neutral closed-shell molecules, whereas the latter are cations, anions, or radicals.

Recall from the previous chapter that ground-state allowed electrocyclic processes are disrotatory (suprafacial, interaction diagram **46**) for $4n + 2$ and conrotatory (antarafacial, with one phase inversion, interaction diagram **47**) for $4n$ electrons. The ions encountered in odd-membered systems present no difficulty, as they fit readily in the theoretical scheme; radicals are not so easily accommodated.

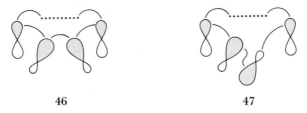

46   **47**

Three-Membered Rings

Figure 11.3 shows the correlation diagrams and the interaction diagrams for opening of a cyclopropyl cation, anion, or radical to an allyl cation, anion, or radical.

[90](a) R. Askani, *Chem. Ber.*, **98**, 3618 (1965); (b) R. B. Woodward and R. Hoffmann, *The Conservation of Orbital Symmetry*, Verlag Chemie, Weinheim/Bergstr., Germany, and Academic Press, New York, 1970, p. 113.
[91](a) R. B. Woodward and R. Hoffmann, *The Conservation of Orbital Symmetry*, Verlag Chemie, Weinheim/Bergstr., Germany, and Academic Press, New York, 1970; (b) G. B. Gill, *Q. Rev. Chem. Soc.*, **22**, 338 (1968); (c) T. L. Gilchrist and R. C. Storr, *Organic Reactions and Orbital Symmetry*, Cambridge University Press, London, 1972.

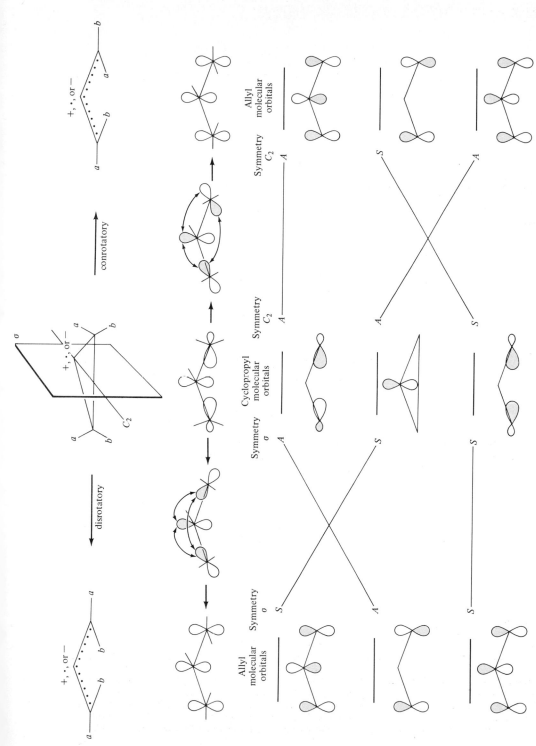

Figure 11.3 Orbital correlation diagram for disrotatory and conrotatory opening of the cyclopropyl system to the allyl system. At the center are the reacting orbitals for the cyclopropyl radical, cation, or anion. Orbitals are the bonding and antibonding σ orbitals and, at the back, the single p orbital that bears the positive or negative charge or unpaired electron. The conrotatory opening is shown at the right, with symmetries given with respect to the C_2 axis; the disrotatory opening is at the left, with symmetries given with respect to the mirror plane. In the cation the lowest level is occupied by two electrons, and disrotatory opening should occur; in the anion the two lower levels are each occupied by two electrons, and the conrotatory opening should occur. In the radical, with two electrons in the lowest level and one in the next, no preference is indicated.

The disrotatory path has a Hückel transition state and is therefore allowed for the cation (two electrons); the conrotatory path is anti-Hückel and so allowed for the anion (four electrons). (The same conclusions follow from the orbital correlations: In the cation only the lowest of the orbitals shown is occupied; in the anion the lower two orbitals are occupied.) In a cyclopropyl radical, with the lowest orbital doubly occupied and the nonbonding orbital singly occupied, the theory at this level allows no choice to be made. Dewar's first-order perturbation molecular orbital analysis also fails to distinguish the two pathways.[92] Attempts to settle the ambiguity theoretically have led to contradictory results: Woodward and Hoffmann concluded, both on the basis of the idea that the highest occupied molecular orbital should be controlling and from extended Hückel calculation, that radicals should behave like the corresponding anions,[93] whereas Dewar and Kirschner, using the MINDO self-consistent field method, decided that the radical should behave like the cation.[92] The latter authors support their conclusion that $4n + 3$ π electron radicals are aromatic by citing experimental evidence of a large stabilization energy in $\cdot C_7H_7$.[94]

Solvolysis of cyclopropyl derivatives leads directly to the allyl cation; the ring opening is disrotatory as predicted. The most direct demonstration is the transformation of the 2,3-dimethyl-1-chlorocyclopropanes at $-100°C$ in strong-acid medium (SbF_5–SO_2ClF) to the isomeric allyl cations (Equations 11.55–11.57), the structures of which were demonstrated by nmr spectroscopy.[95]

$$(11.55)$$

$$(11.56)$$

$$(11.57)$$

[92](a) M. J. S. Dewar and S. Kirschner, *J. Am. Chem. Soc.*, **93**, 4290 (1971); see also (b) G. Boche and G. Szeimies, *Angew. Chem. Int. Ed. Engl.*, **10**, 911 (1971); (c) G. Szeimies and G. Boche, *Angew. Chem. Int. Ed. Engl.*, **10**, 912 (1971).
[93]R. B. Woodward and R. Hoffmann, *J. Am. Chem. Soc.*, **87**, 395 (1965).
[94](a) G. Vincow, H. J. Dauben, Jr., F. R. Hunter, and W. V. Volland, *J. Am. Chem. Soc.*, **91**, 2823 (1969). Other calculations agree with Dewar's conclusion: (b) G. Szeimies and G. Boche, *Angew. Chem. Int. Ed. Engl.*, **10**, 912 (1971); (c) P. Merlet, S. D. Peyerimhoff, R. J. Buenker, and S. Shih, *J. Am. Chem. Soc.*, **96**, 959 (1974).
[95]P. v. R. Schleyer, T. M. Su, M. Saunders, and J. C. Rosenfeld, *J. Am. Chem. Soc.*, **91**, 5174 (1969).

Note that the results shown by Equations 11.55 and 11.57 demonstrate that one of the two possible disrotatory modes of opening is preferred. Woodward and Hoffmann predicted that this specificity should occur;[96] the reason may be readily understood through the following qualitative argument. If the ring C—C bond were to break only after the leaving group had moved far away and left a fully developed cationic center, there would be no difference between the two sides of the ring and the two modes of opening would be equivalent. But because of the relief of ring strain and the gain of charge delocalization that accompanies ring opening (allyl cation is estimated to be about 39 kcal mole^{-1} lower in energy than cyclopropyl cation[97]), the C—C bond breaks as the leaving group departs. The C—C bond cleavage can assist the reaction most effectively if some of the electron density from the C—C bonding orbital can be transferred to the LUMO of the C—X bond. This transfer can occur from the back, as illustrated in Equation 11.58. If, on the other hand, the bond opens downward (Equation 11.59), the electron density comes onto the node of the C—X LUMO, and the transfer is less effective. The two alternatives are essentially equivalent to backside and frontside S_N2 substitution (Section 1.8).

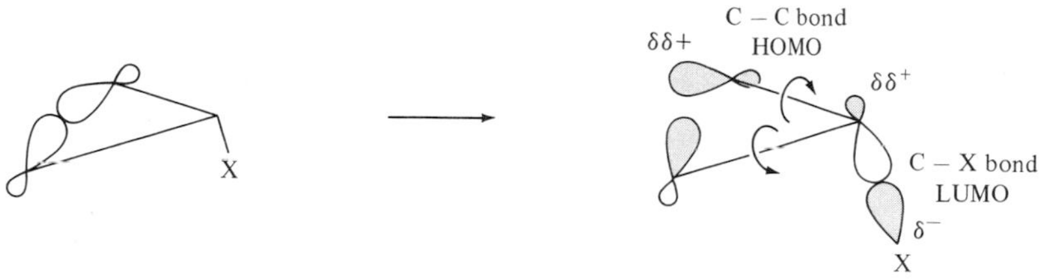

$$(11.58)$$

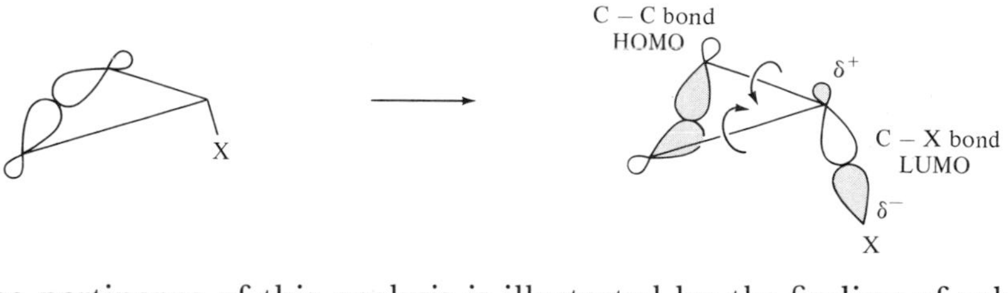

$$(11.59)$$

The pertinence of this analysis is illustrated by the finding of substantial assistance to ionization (estimated rate enhancements of 10^4 to 10^5) in cyclopropyl solvolysis,[98] by the observation of stereospecificity already cited, and by the solvolysis rates of **48** and **49** and of bicyclic compounds **50** and **51**.[99] Compounds **49** and **51** would encounter severe strain in the backside-assisted ionization, which in the case of **49** brings the two methyl groups into contact and in the case of **51** is precluded by the constraint of the second ring.

[96] R. B. Woodward and R. Hoffmann, *J. Am. Chem. Soc.*, **87**, 395 (1965).
[97] L. Radom, P. C. Hariharan, J. A. Pople, and P. v. R. Schleyer, *J. Am. Chem. Soc.*, **95**, 6531 (1973).
[98] P. v. R. Schleyer, W. F. Sliwinski, G. W. Van Dine, U. Schöllkopf, J. Paust, and K. Fellenberger, *J. Am. Chem. Soc.*, **94**, 125 (1972).
[99] P. v. R. Schleyer, G. W. Van Dine, U. Schöllkopf, and J. Paust, *J. Am. Chem. Soc.*, **88**, 2868 (1966).

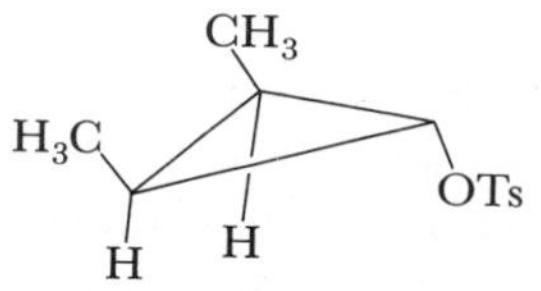

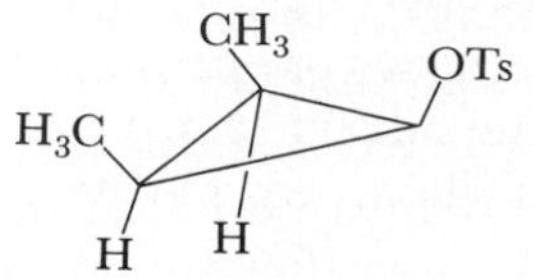

Relative solvolysis rate
(acetic acid, 150°C):   4500   1

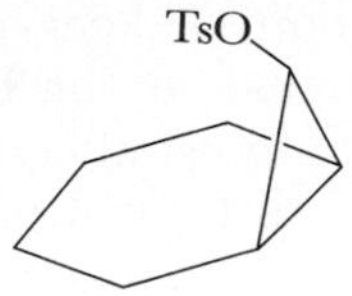

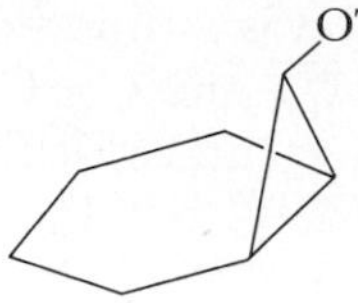

Relative solvolysis rate
(acetic acid, 100°C):   11,000   1

The aziridines **52** and **53**, which are isoelectronic with cyclopropyl carbanions, open in the predicted conrotatory sense. Thus heating the cis and trans isomers in the presence of dimethylacetylenedicarboxylate leads with high stereospecificity, by way of a 1,3-dipolar addition, to the products indicated in Scheme 6. The stereochemistry is reversed, as predicted, for the excited-state process.[100]

SCHEME 6

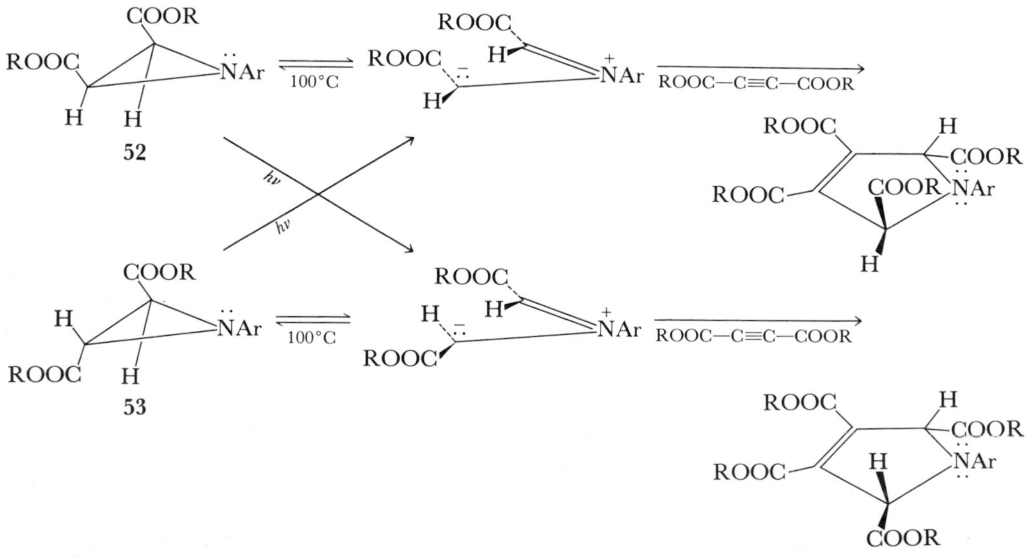

[100] R. Huisgen, W. Scheer, and H. Huber, *J. Am. Chem. Soc.*, **89**, 1753 (1967). For a related study of cyclopropyl carbanions, see (b) N. Newcomb and W. T. Ford, *J. Am. Chem. Soc.*, **96**, 2968 (1974).

Four-Membered Rings

The stereochemistry of opening of cyclobutenes to butadienes was established some time before the advent of the pericyclic theory.[101] It is conrotatory, in accord with the theory, as illustrated by the examples shown in Equations 11.60 and 11.61.[102]

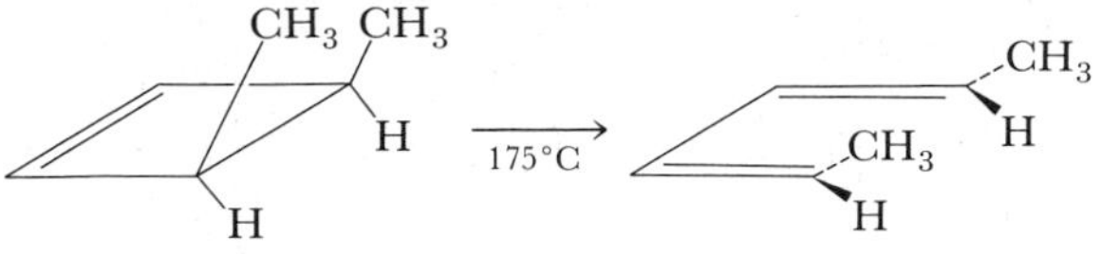

$$(11.60)$$

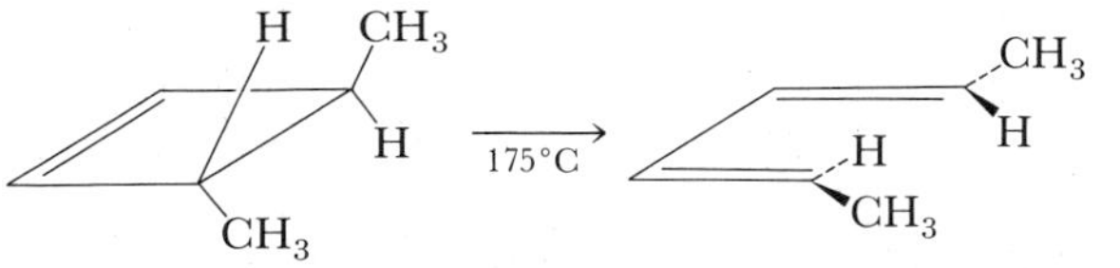

$$(11.61)$$

For the unsubstituted case, $E_a = 32.5$ kcal mole^{-1} ($T \approx 450$ K),[103] and for *cis*-3,4-dimethylcyclobutene, Structure **54**, $E_a = 34$ kcal mole^{-1}, $\log A = 13.88$ ($T = 420 - 450$ K).[104] Because the equilibrium favors the diene (by $\Delta G°$ of about 12 kcal mole^{-1}, as estimated from Table 2.9), the reverse process, closure of an open-chain diene, is not commonly observed. A diene isomerization that presumably takes this route (Equation 11.62) allows an estimate of a lower limit to the

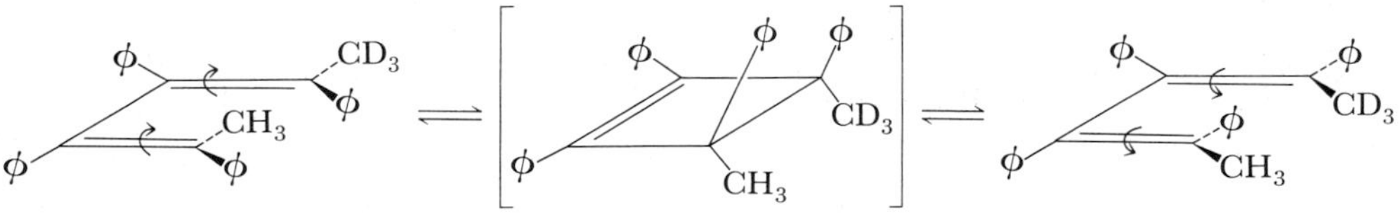

$$(11.62)$$

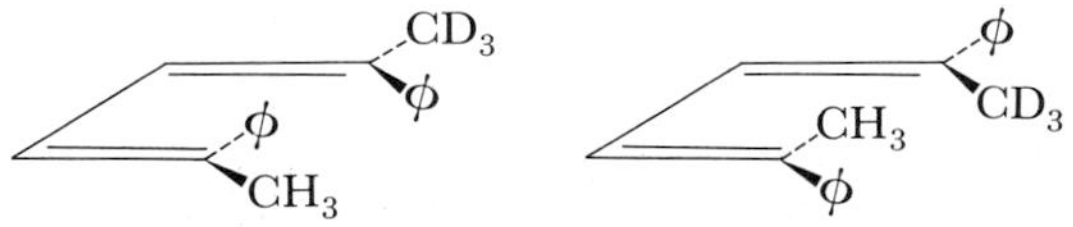

energy difference between the allowed conrotatory and forbidden disrotatory paths. In a time (51 days at 124°C) during which over 2×10^6 ring openings occur, none of the disrotatory products **55** and **56** appear, although 1 percent would have been detected; the conrotatory path is therefore favored by at least 7.3 kcal mole^{-1}.[105] Brauman and Archie found about 0.005 percent of the forbidden

[101](a) E. Vogel, *Justus Liebigs Ann. Chem.*, **615**, 14 (1958); (b) R. Criegee and K. Noll, *Justus Liebigs Ann. Chem.*, **627**, 1 (1959).
[102]R. E. K. Winter, *Tetrahedron Lett.*, 1207 (1965).
[103]R. Criegee, D. Seebach, R. E. Winter, B. Börretzen, and H.-A. Brune, *Chem. Ber.*, **98**, 2339 (1965).
[104]R. Srinivasan, *J. Am. Chem. Soc.*, **91**, 7557 (1969).
[105]G. A. Doorakian and H. H. Freedman, *J. Am. Chem. Soc.*, **90**, 5310, 6896 (1968).

disrotatory product in ring opening of *cis*-3,4-dimethylcyclobutene (Equation 11.63), and estimated, after correcting for steric differences, that the path leading to **57** is on the order of 15 kcal mole^{-1} higher in energy than that leading to **58**.

$$\text{(11.63)}$$

Recalling that E_a for the allowed (unsubstituted) reaction is 32.5 kcal mole^{-1} ($\Delta H^{\ddagger} = 31.6$ kcal mole^{-1}), we can see that $\Delta H^{\ddagger}$ for the minor pathway is about 47 kcal mole^{-1}. A rough estimate of the enthalpy difference between cyclobutene and a hypothetical biradical intermediate **59** in which one radical center is conjugated with the double bond is about 39 kcal mole^{-1}. If we recall from Section 10.7 that the estimate of biradical enthalpy is probably a lower limit, we may conclude that formation of **57** through a stepwise biradical path rather than by the forbidden concerted pericyclic disrotatory path is a possibility.[106]

$$\Delta H_f^{\circ} \approx 23.2 + 35.82 + 2(8.59)$$
$$\approx 76.2 \text{ kcal mole}^{-1}$$
$$\text{(lower limit)}$$

$$\Delta H_f^{\circ} \approx 2(8.59) - 2(4.76) + 29.8$$
$$\approx 37 \text{ kcal mole}^{-1}$$

$$\Delta\Delta H_f^{\circ} \approx 39 \text{ kcal mole}^{-1}$$

When the two saturated carbons of the cyclobutene ring are joined through a second ring, new constraints appear. The conrotatory opening in such a system (Equation 11.64) leads to a trans double bond in the remaining ring. There is no problem if the ring is large enough; thus **60** opens readily at 200°C to *cis-trans*-1,3-cyclodecadiene.[107] When the bridging ring is five- or six-membered, the reaction still apparently leads initially to the highly strained cis-trans diene, which then isomerizes by hydrogen migration in a second, rate-determining step to the cis-cis structure. Temperatures required for ring opening in these compounds are typically on the order of 200° higher than those necessary for monocyclic cyclobutanes.[108] Bicyclo[3.2.0]heptene (**61**) opens to *cis-cis*-cycloheptadiene (Equation

[106] J. I. Brauman and W. C. Archie, Jr., *J. Am. Chem. Soc.*, **94**, 4262 (1972).

[107] P. Radlick and W. Fenical, *Tetrahedron Lett.*, 4901 (1967).

[108] (a) R. Criegee, D. Seebach, R. E. Winter, B. Börretzen, and H. Brune, *Chem. Ber.*, **98**, 2339 (1965); (b) J. S. McConaghy, Jr., and J. J. Bloomfield, *Tetrahedron Lett.*, 3719 (1969); (c) J. J. Bloomfield, J. S. McConaghy, Jr., and A. G. Hortmann, *Tetrahedron Lett.*, 3723 (1969).

11.66) with activation energy 13 kcal mole^{-1} higher than that for opening of cyclobutene itself: it probably does so by way of the cis-trans isomer.[109]

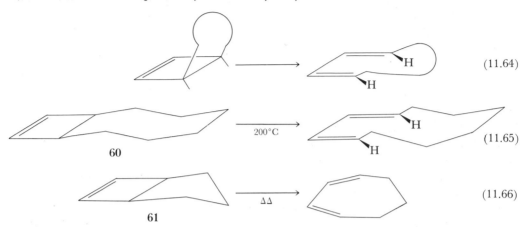

$$\text{(11.64)}$$

$$\xrightarrow{200°C} \quad \text{(11.65)}$$

60

$$\xrightarrow{\Delta\Delta} \quad \text{(11.66)}$$

61

When the bridging ring becomes smaller still, the strain energies are large and ring opening is again facile. The conversion of Dewar benzene (**62**) to benzene (Equation 11.67) has to be disrotatory. A concerted disrotatory opening is ther-

$$\xrightarrow[\text{at 25°C}]{t_{1/2} \approx 2 \text{ days}} \quad \text{(11.67)}$$

62

mally forbidden, yet the $\Delta H^{\ddagger}$ is only 23 kcal mole^{-1} ($\Delta S^{\ddagger} = -5$ cal mole^{-1} K^{-1}).[110] The reaction is exothermic by $\Delta H° \approx 60$ kcal mole^{-1}.[111] The enthalpy of formation of benzene, $\Delta H_f° = 19.8$ kcal mole^{-1}, and this observed $\Delta H°$ for the reaction lead to an estimate of $\Delta H_f° \approx 80$ kcal mole^{-1} for Dewar benzene (**62**). The transition state for Reaction 11.67 therefore has $\Delta H_f \approx 103$ kcal mole^{-1}. The estimated $\Delta H_f°$ for a biradical intermediate (**63**) depends on the degree of conjugative stabilization assumed for the transition state. If the orbitals containing the unpaired electrons were nearly perpendicular to the π orbitals of the double bonds, there would be little conjugation and a lower limit for $\Delta H_f°$ would be on the order of 109 kcal mole^{-1}. In biradical **63** there would probably be some conjugation, which would reduce the lower limit for $\Delta H_f°$ of **63** below this value. Whether this lowering would be sufficient to bring the biradical $\Delta H_f°$ below the observed transition state $\Delta H_f°$ of 103 kcal mole^{-1} is a question that cannot at present be answered with any degree of certainty. An added complication in this reaction is the ob-

[109](a) M. R. Willcott and E. Goerland, *Tetrahedron Lett.*, 6341 (1966); (b) C. F. Wilcox, Jr., and B. K. Carpenter, *J. Am. Chem. Soc.*, **101**, 3897 (1979).
[110]R. Breslow, J. Napierski, and A. H. Schmidt, *J. Am. Chem. Soc.*, **94**, 5906 (1972).
[111]W. Schäfer, *Angew. Chem. Int. Ed. Engl.*, **5**, 669 (1966). The experimental value is 62.4 kcal mole^{-1} for the hexamethyl derivative.

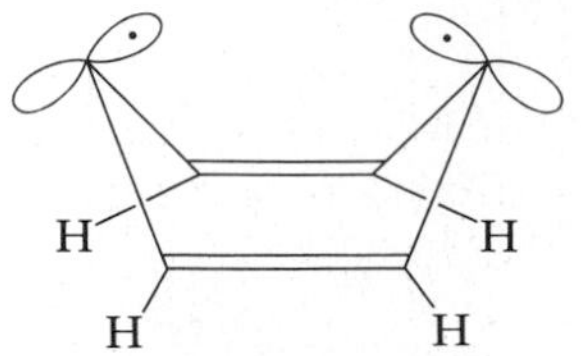

63

served formation of a small amount of triplet state benzene. The singlet and triplet energy surfaces evidently intersect, and some of the reacting molecules cross to the triplet state.[112]

Photochemically, the cyclobutene–diene electrocyclic transformation follows the disrotatory path in agreement with the theory. With the difficulties encountered in bicyclic rings for the thermal conrotatory path thus circumvented, these reactions provide a useful synthetic pathway to various strained ring systems. The cyclic heptadiene **64**, for example, closes as shown in Equation 11.68 in 42 percent yield on irradiation with a mercury arc lamp. Although the process ought

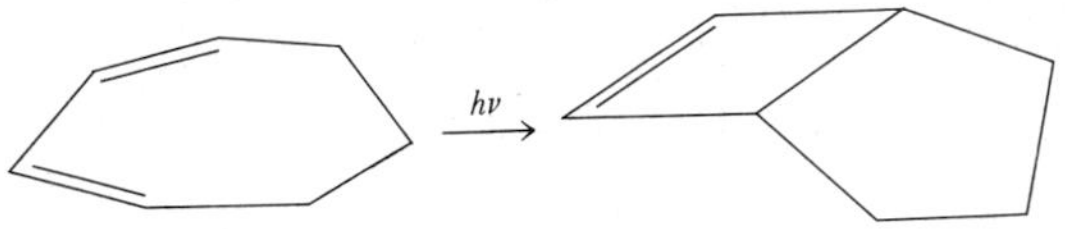

(11.68)

64

to be reversible, the inability of the remaining alkene function to absorb light at the wavelength available, and the orbital symmetry barrier to thermal reversal, effectively prevent reopening. The ring does reopen on vigorous heating; as would be expected from our earlier discussion, the temperature required is 400–420°C.[113]

Five-Membered Rings

Electrocyclic closure of both pentadienyl cation and anion have been observed. Cations generated by protonation of dienones close in the predicted conrotatory manner as shown in Equation 11.69.[114] The isomeric alcohols **65** and **67** generate

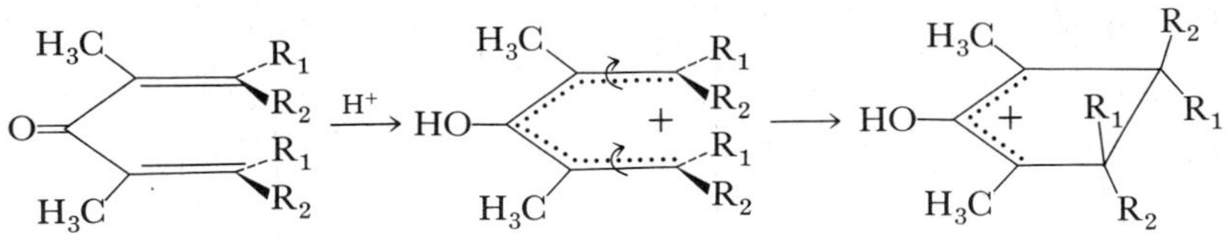

(11.69)

[112] P. Lechtken, R. Breslow, A. H. Schmidt, and N. J. Turro, *J. Am. Chem. Soc.*, **95**, 3025 (1973); (b) M. J. S. Dewar, S. Kirschner, and H. W. Kollmar, *J. Am. Chem. Soc.*, **96**, 7579 (1974). See also (c) W. Grimme and U. Heinze, *Chem. Ber.*, **111**, 2563 (1978).

[113] (a) W. G. Dauben and R. L. Cargill, *Tetrahedron*, **12**, 186 (1961). See also (b) J. I. Brauman, L. E. Ellis, and E. E. van Tamelen, *J. Am. Chem. Soc.*, **88**, 846 (1966).

[114] (a) C. W. Shoppee and B. J. A. Cooke, *J. Chem. Soc., Perkin Trans.* **1**, 2271 (1972); (b) C. W. Shoppee and B. J. A. Cooke, *J. Chem. Soc., Perkin Trans.* **1**, 1026 (1973); (c) see also R. B. Woodward and R. Hoffmann, *The Conservation of Orbital Symmetry*, Verlag Chemie, Weinheim/Bergstr., Germany, and Academic Press, New York, 1970, p. 58.

pentadienyl cations **66** and **68** in strong-acid solution; these ions cyclize as shown in Equations 11.72 and 11.73 at temperatures above about −85°C. Some loss of stereochemistry occurs because bond rotations are required to bring the ion into the right conformation to cyclize, and rotation around a terminal bond of the penta-dienyl ions will interconvert the isomers.[115]

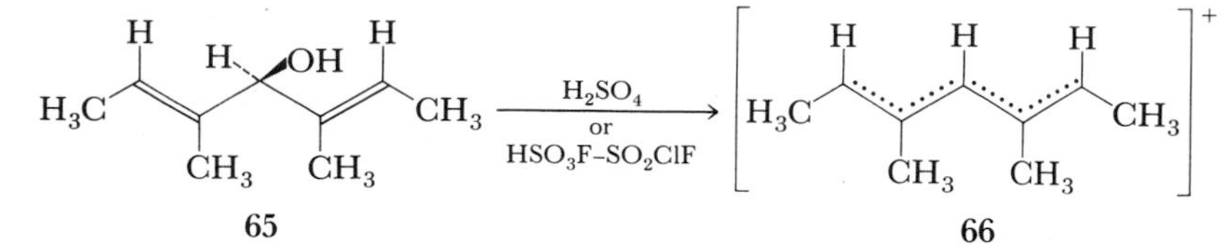

(11.70)

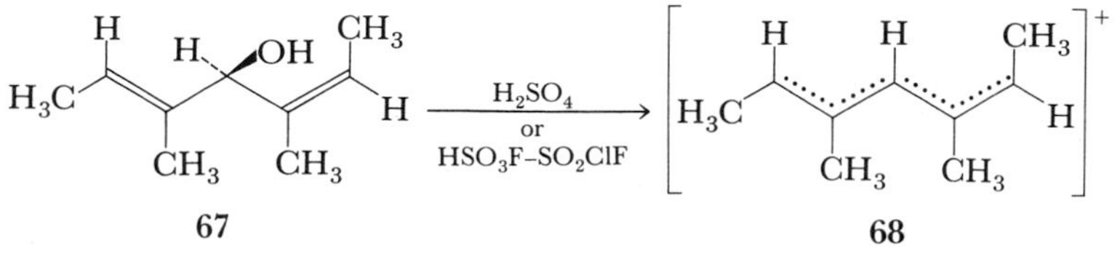

(11.71)

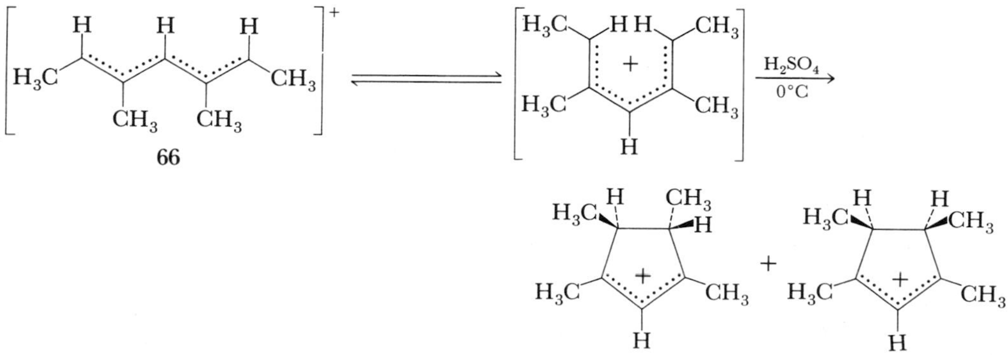

(11.72)

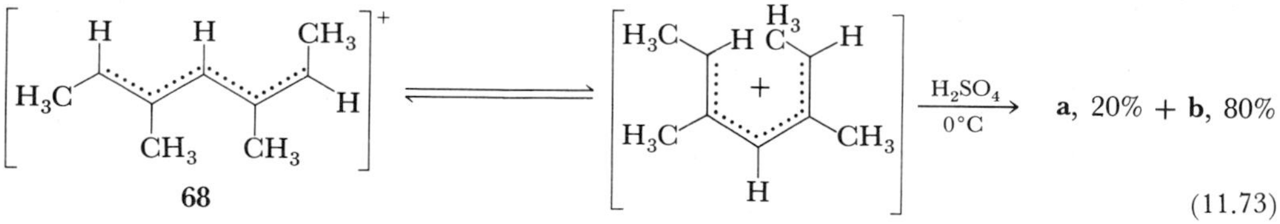

(11.73)

The pentadienyl anion, a six-electron system, should close in the disrotatory sense; a clear example is the rapid isomerization illustrated in Equation 11.74.[116] Photochemical cyclization of pentadienyl cations has been observed; Equation

[115] N. W. K. Chiu and T. S. Sorensen, *Can. J. Chem.*, **51**, 2776 (1973).
[116] R. B. Bates and D. A. McCombs, *Tetrahedron Lett.*, 977 (1969).

11.75 shows an example in a cyclic system.[117] The ready thermal reversion, which should be conrotatory and therefore difficult in the bicyclic system, may possibly occur by a stepwise path.

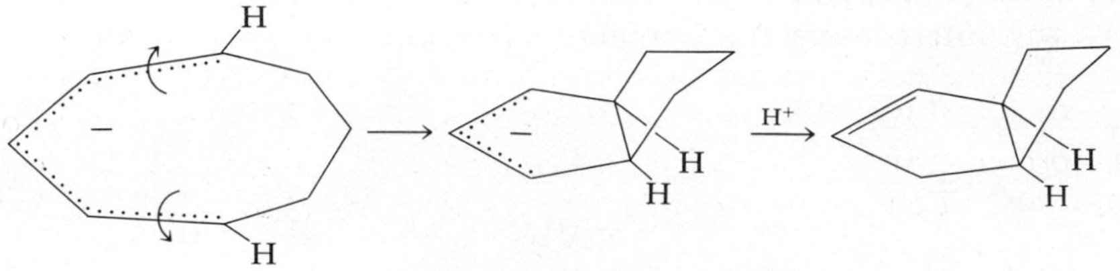

$$(11.74)$$

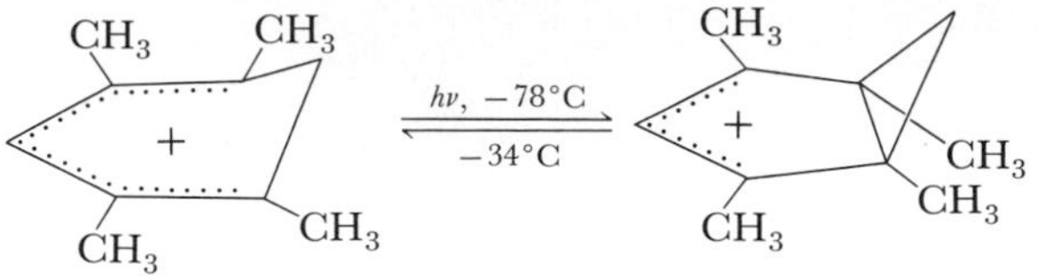

$$(11.75)$$

Six-Membered and Larger Rings

Trans-cis-trans-octatriene (**69**) cyclizes at 130°C in the predicted disrotatory sense to *cis*-5,6-dimethylcyclohexa-1,3-diene.[118] Closure in cyclic trienes also occurs readily according to the generalized structures in Equation 11.77, where the bridging chain indicated by X_n may be made up of carbon or hetero atoms. The simplest example, the cycloheptatriene–norcaradiene interconversion (Equation 11.78), has

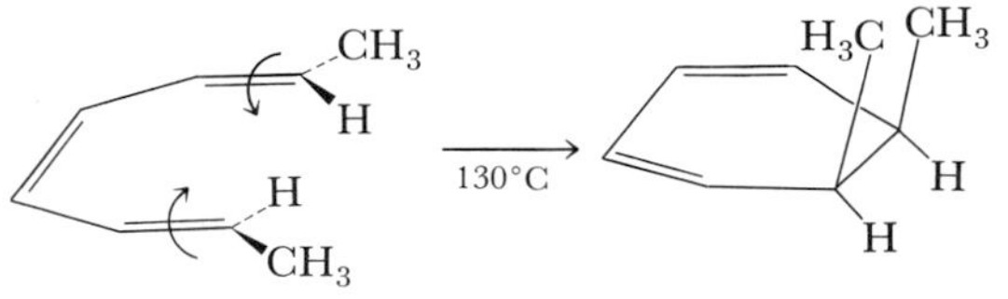

$$(11.76)$$

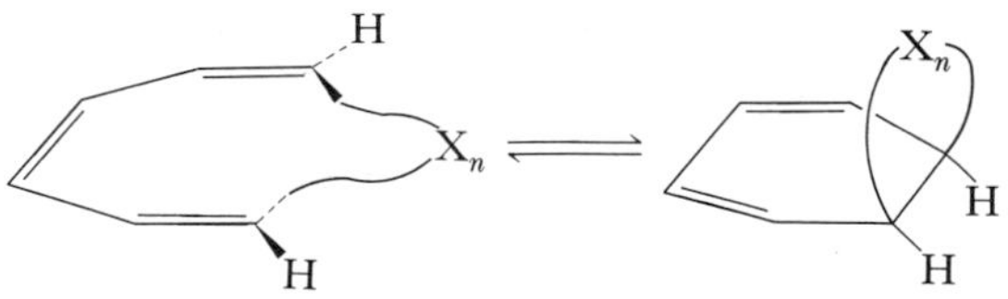

$$(11.77)$$

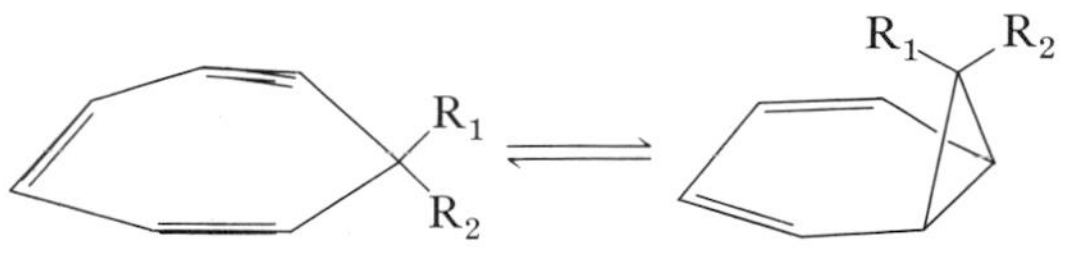

$$(11.78)$$

[117](a) R. F. Childs, M. Sakai, B. D. Parrington, and S. Winstein, *J. Am. Chem. Soc.*, **96**, 6403 (1974); (b) P. W. Cabell-Whiting and H. Hogeveen, *Adv. Phys. Org. Chem.*, **10**, 129 (1973) (photoreactions of carbocations).
[118](a) E. N. Marvell, G. Caple, and B. Schatz, *Tetrahedron Lett.*, 385 (1965); (b) E. Vogel, W. Grimme, and E. Dinné, *Tetrahedron Lett.*, 391 (1965).

received considerable attention.[119] In most instances the equilibrium favors the cycloheptatriene form, but strongly electron-withdrawing substituents R shift the equilibrium toward the bicyclic norcaradiene isomer (**70**).[120] Interconversion rates are so great as to preclude isolation of the separate forms; equilibria can be investigated only by nuclear magnetic resonance. Hoffmann has proposed a simple molecular orbital model to account for the stabilization of the bicyclic form by a π electron acceptor.[121] Briefly stated, the argument is that one of the bonding orbitals of the cyclopropane ring has the symmetry depicted in **71**; the unfilled π^* orbital of a π acceptor such as a cyano group can interact with this orbital (**72**) and withdraw

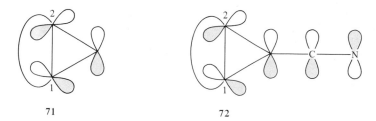

71	**72**

electrons from it. The effect is to reduce electron density in this orbital between carbons 1 and 2. Because the orbital is antibonding between positions 1 and 2, this change strengthens the 1–2 bond. This appealingly simple explanation may not suffice to account completely for the observations. Para-substituted phenyl groups at R_1, R_2 do not appear to affect the equilibrium in a consistent way.[122]

The next homolog in the cyclic series, 1,3,5-cyclooctatriene, also closes in a readily reversible transformation (Equation 11.79) to bicyclo[4.2.0]octadiene.

$$(11.79)$$

Cope and his collaborators reported this valence isomerization in 1952. They were able to separate the isomers, which revert to the equilibrium mixture of 85 percent **73** and 15 percent **74** on heating at 100°C for 1 hr.[123] Huisgen has reported activa-

[119] Reviews: (a) G. Maier, *Angew. Chem. Int. Ed. Engl.*, **6**, 402 (1967); (b) E. Ciganek, *J. Am. Chem. Soc.*, **89**, 1454 (1967), and references cited therein.

[120] (a) See note 119; (b) G. E. Hall and J. D. Roberts, *J. Am. Chem. Soc.*, **93**, 2203 (1971); (c) E. Ciganek, *J. Am. Chem. Soc.*, **93**, 2207 (1971); (d) see also H. Dürr and H. Kober, *Angew. Chem. Int. Ed. Engl.*, **10**, 342 (1971).

[121] (a) R. Hoffmann, *Tetrahedron Lett.*, 2907 (1970); (b) for an example, see M. Simonetta, *Acc. Chem. Res.*, **7**, 345 (1974).

[122] See note 120(b). Hetero atoms in the ring also affect rate and equilibrium. (a) E. Vogel and H. Günther, *Angew. Chem. Int. Ed. Engl.*, **6**, 385 (1967); (b) L. A. Paquette, *Angew. Chem. Int. Ed. Engl.*, **10**, 11 (1971).

[123] A. C. Cope, A. C. Haven, Jr., F. L. Ramp, and E. R. Trumbull, *J. Am. Chem. Soc.*, **74**, 4867 (1952).

tion parameters of $\Delta H^{\ddagger} = 26.6$ kcal mole^{-1}, $\Delta S^{\ddagger} = -1$ cal mole^{-1} K^{-1}, and a free-energy difference between the two tautomers of 1.5 kcal mole^{-1}.[124]

All *cis*-cyclononatriene (**75**) cyclizes in the predicted disrotatory manner at measurable rates at 25–50°C to bicyclo[4.3.0]nonadiene (**76**).[125] The nine-membered ring is large enough to accommodate a trans double bond; *trans-cis-cis*-cyclonoatriene (**77**), obtained by conrotatory photochemical opening of **76**, cyclizes to the trans fused bicyclononadiene **78**.[126]

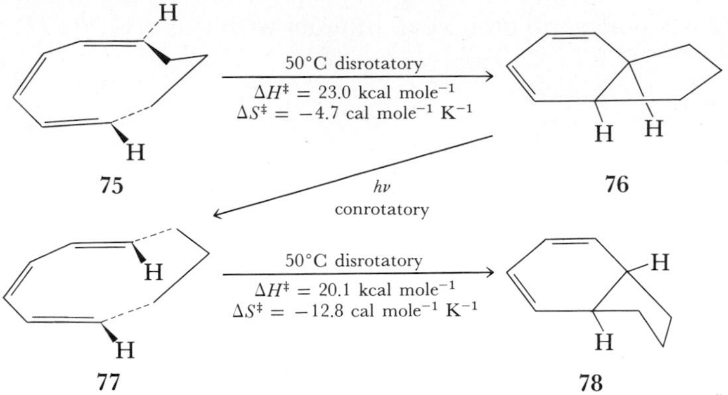

The example just cited provides a verification of the prediction that the excited-state reactions should be conrotatory for six-electron systems. The prototype octatriene–cyclohexadiene interconversion (Equation 11.80) shows the same pattern.[127] The network of photochemical and thermal electrocyclic reactions connected with the formation of vitamin D provide several further examples.[128]

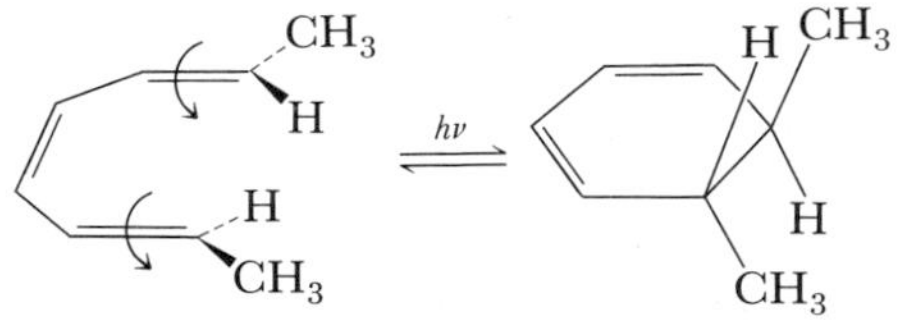

$$(11.80)$$

The examples shown in Equations 11.81–11.83 demonstrate the operation of the selection rules in eight-membered rings. Equation 11.84 shows a twelve-electron electrocyclic ring closure, which occurs in the conrotatory sense as predicted. This reaction, however, is reversed photochemically.

[124] (a) R. Huisgen, F. Mietzsch, G. Boche, and H. Seidl, *Organic Reaction Mechanisms, Special Publ. Chem. Soc. (London)*, **19**, 3 (1965); (b) R. Huisgen, G. Boche, A. Dahmen, and W. Hechtl, *Tetrahedron Lett.*, 5215 (1968).
[125] D. S. Glass, J. W. H. Watthey, and S. Winstein, *Tetrahedron Lett.*, 377 (1965).
[126] E. Vogel, W. Grimme, and E. Dinné, *Tetrahedron Lett.*, 391 (1965).
[127] G. L. Fonken, *Tetrahedron Lett.*, 549 (1962).
[128] (a) R. B. Woodward and R. Hoffmann, *The Conservation of Orbital Symmetry*, Verlag Chemie, Weinheim/Bergstr., Germany, and Academic Press, New York, 1970, p. 52; see also (b) E. Havinga and J. L. M. A. Schlatmann, *Tetrahedron*, **16**, 146 (1961). For further discussion, see Section 12.4.

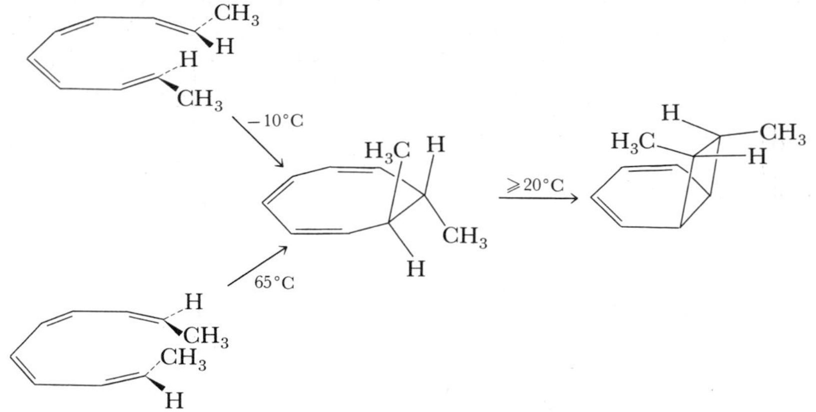

$(11.81)^{129}$

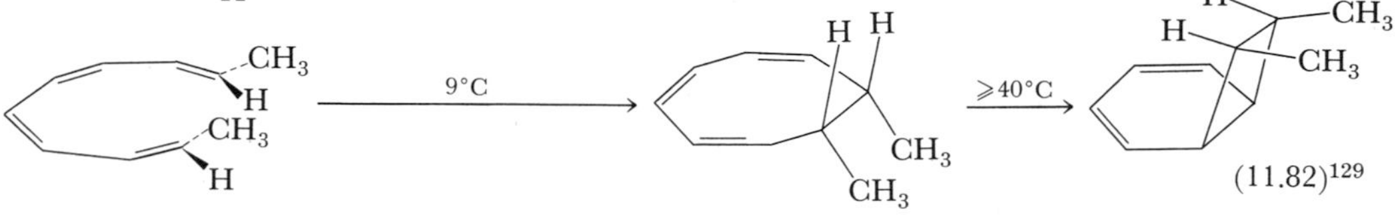

$(11.82)^{129}$

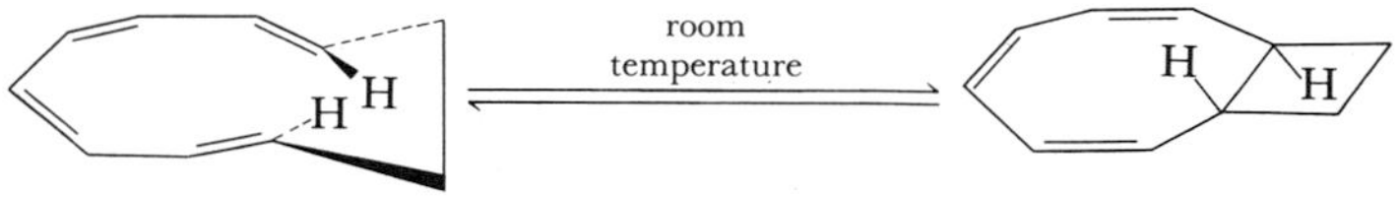

$(11.83)^{130}$

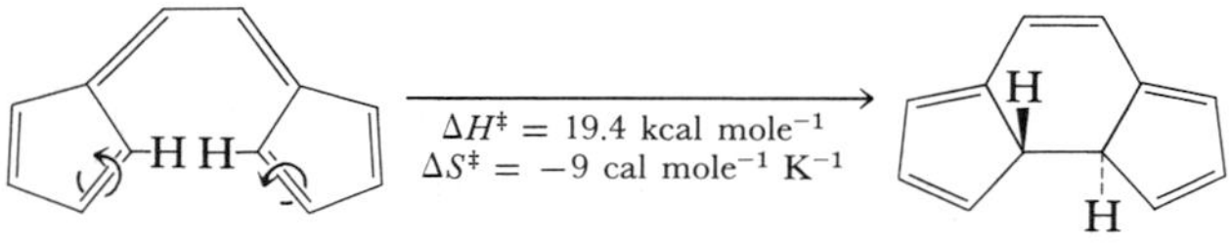

$\Delta H^{\ddagger} = 19.4$ kcal mole^{-1}
$\Delta S^{\ddagger} = -9$ cal mole^{-1} K^{-1}

$(11.84)^{131}$

11.3 SIGMATROPIC REACTIONS[132]

In a sigmatropic reaction a σ bond migrates over a π system to a new location. Either one end or both ends of the bond may move.

[1,n]-Sigmatropic Transformations[133]

In a [1,n] reaction one end of the migrating σ bond remains fixed while the other end moves to the next atom or to a more remote position at the end of an adjacent π system. Equation 11.85 illustrates the generalized process. Recall from our previous

[129] R. Huisgen, A. Dahmen, and H. Huber, *J. Am. Chem. Soc.*, **89**, 7130 (1967).
[130] S. W. Staley and T. J. Henry, *J. Am. Chem. Soc.*, **92**, 7612 (1970).
[131] H. Sauter, B. Gallenkamp, and H. Prinzbach, *Chem. Rev.*, **110**, 1382 (1977).
[132] Sigmatropic processes are discussed in detail by (a) R. B. Woodward and R. Hoffmann, *The Conservation of Orbital Symmetry,* Verlag Chemie, Weinheim/Bergstr., Germany, and Academic Press, New York, 1970. Experimental results are reviewed in (b) G. B. Gill, *Q. Rev. Chem. Soc.*, **22**, 338 (1968); (c) T. L. Gilchrist and R. C. Storr, *Organic Reactions and Orbital Symmetry,* Cambridge University Press, London, 1972; (d) C. W. Spangler, *Chem. Rev.*, **76**, 187 (1976).
[133] See note 132(d).

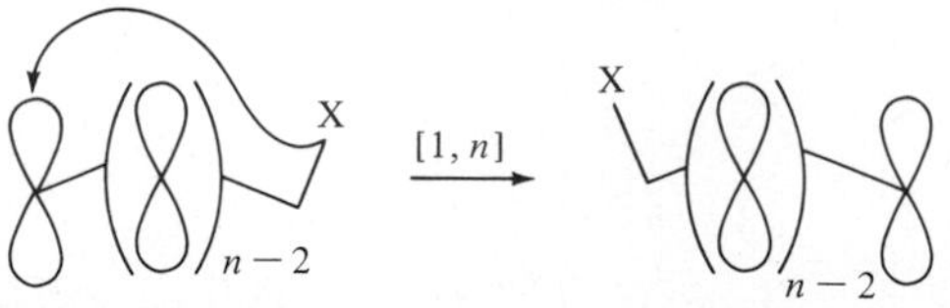

$$[1, n] \qquad (11.85)$$

discussion (Section 10.1) that there is a choice of stereochemistry between suprafacial and antarafacial modes both with respect to the π system and with respect to the migrating group. For short chains it is expected that a migration antarafacial with respect to the π component will be difficult because it requires the migrating group to reach around to the opposite face of the molecule; antarafacial migration will be easier in longer, more flexible chains. Rearrangements suprafacial with respect to the migrating group will yield retained configuration (Equation 11.86), and those antarafacial will produce inversion (Equation 11.87).[134]

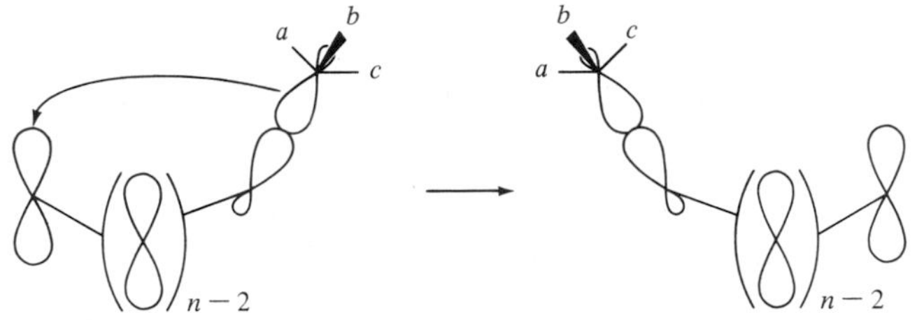

$$(11.86)$$

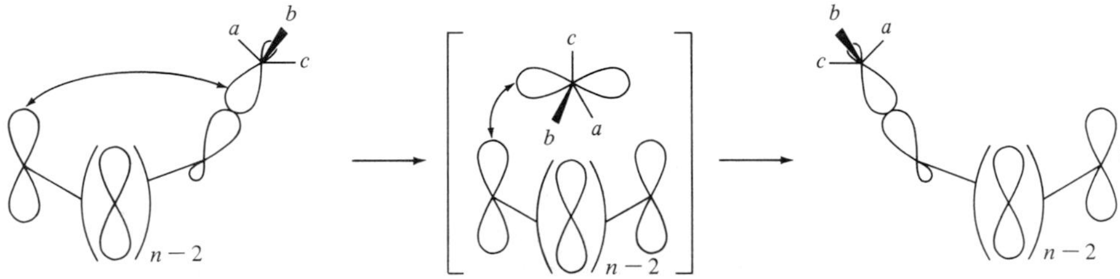

$$(11.87)$$

Two-electron [1,2]-rearrangements Two-electron [1,2]-sigmatropic shifts are the familiar rearrangements observed in carbocations. As these processes have been discussed in detail in Chapter 5, we shall confine our remarks here to noting that the pericyclic rules correctly predict that the rearrangements are suprafacial in both components, with retention of configuration at the migrating group.[135] It is appropriate also to point out the analogy between the transition state for a [1,2]-sigmatropic shift (**79**) and that for a frontside substitution (**80**) (Section 1.8), allowed for the two-electron (electrophilic) case. Note that the bridged carbonium ions (Chapter 5) also correspond to **79**.

---

[134]If the migrating group is hydrogen, presumably only the suprafacial mode is possible, the p orbitals needed for the transition state analogous to that in Equation 11.87 being of too high energy.

[135](a) T. Shono, K. Fujita, and S. Kumai, *Tetrahedron Lett.*, 3123 (1973); (b) see the discussion in Section 5.1.

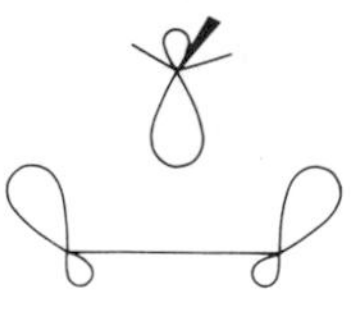

79 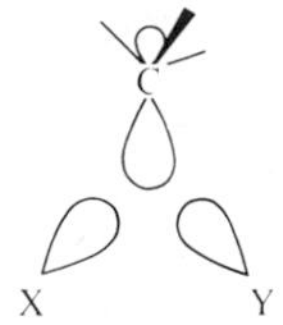   80

Four-electron [1,n]-rearrangements Four-electron processes can take place in the context of a [1,2]-shift (anionic), [1,3]-shift (neutral), or [1,4]-shift (cationic); the selection rules now require the reaction to be antarafacial on one of the components. Thus anions would have to undergo [1,2]-rearrangements by way of **81**, a

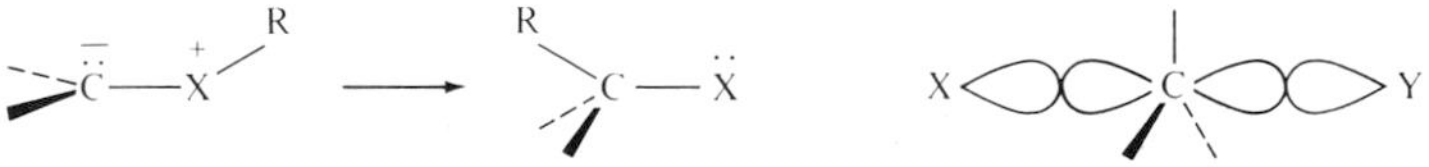

81   82   83   84

geometry that would lead to serious steric difficulties.[136] [1,2]-Rearrangements do occur in ylide structures, **82 → 83** (X = N or S), but CIDNP evidence (Section 9.5), indicates that these reactions take place at least partly through a radical dissociation–recombination mechanism.[137]

The backside S_N2 nucleophilic substitution, transition state **84**, is topologically equivalent to a [1,2]-sigmatropic shift, with the substitution center (antarafacial) playing the role of the migrating group. This process is allowed (Section 1.8), and steric interference is minimized.

In neutral systems [1,3]-shifts of hydrogen should be difficult, since they require transfer to the opposite side of the π system, and indeed these rearrangements are rare. Alkyl groups, on the other hand, can, according to the theory, migrate suprafacially over the π system if they invert configuration (Equation 11.88). Berson has reported extensive investigations of this rearrangement in the

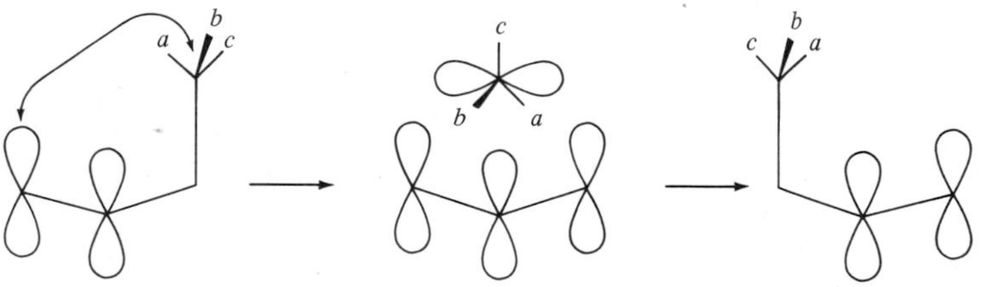

(11.88)

[136]Phenyl groups undergo 1,2-rearrangement in anions by a nonpericyclic addition–elimination pathway similar to the phenonium ion mechanism for cations. E. Grovenstein, Jr., and R. E. Williamson, *J. Am. Chem. Soc.*, **97**, 646 (1975).

[137](a) U. Schöllkopf, *Angew. Chem. Int. Ed. Engl.*, **9**, 763 (1970); (b) J. E. Baldwin, W. F. Erickson, R. E. Hackler, and R. M. Scott, *Chem. Commun.*, 576 (1970); (c) A. R. Lepley, P. M. Cook, and G. F. Willard, *J. Am. Chem. Soc.*, **92**, 1101 (1970); (d) W. D. Ollis, M. Rey, I. O. Sutherland, and G. L. Closs, *J. Chem. Soc., Chem. Commun.*, 543 (1975); (e) U. H. Dolling, G. L. Closs, A. H. Cohen, and W. D. Ollis, *J. Chem. Soc., Chem. Commun.*, 545 (1975).

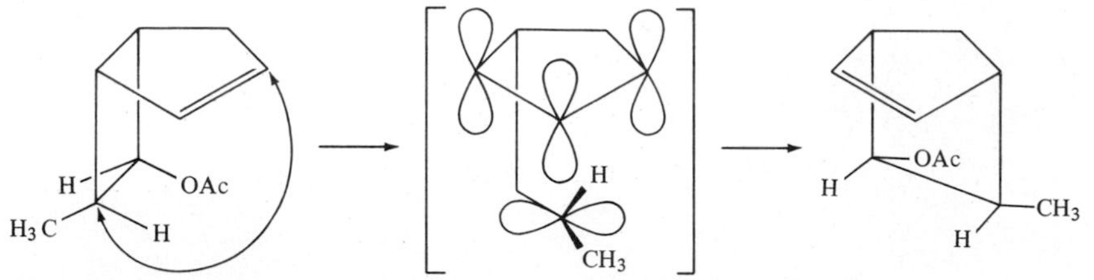

(11.89)

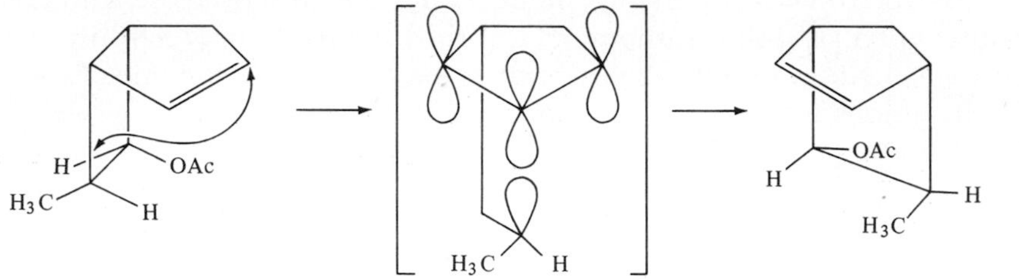

(11.90)

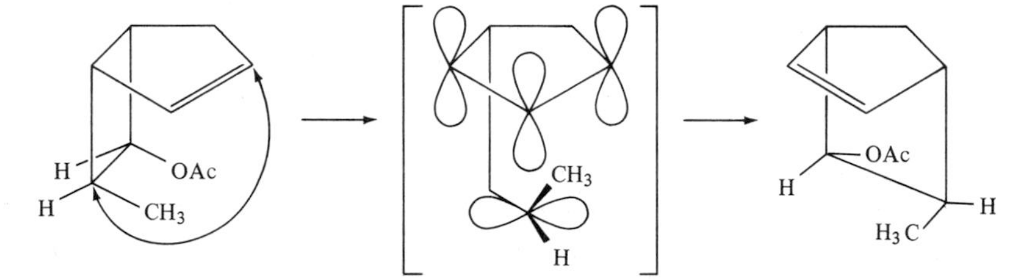

(11.91)

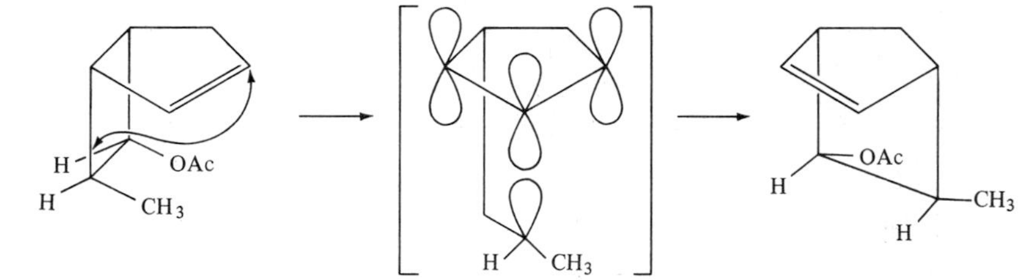

(11.92)

system shown in Equation 11.89.[138] The reaction prefers the allowed path (Equation 11.89) by a factor of ten over the sterically more favorable forbidden one (Equation 11.90). When the steric difficulties of the allowed reaction are increased by requiring the methyl group rather than hydrogen to pass between the migrating carbon and the ring (Equation 11.91), the forbidden path (Equation 11.92) is preferred by a similar factor.

A similar example is provided by the [1,3]-sigmatropic rearrangement of *trans*-1,2-dipropenylcyclobutanes to methylpropenylcyclohexenes (Equation 11.93). Scheme 7 shows the four possible stereochemical pathways for this reaction and the

[138](a) J. A. Berson, *Acc. Chem. Res.*, **5**, 406 (1972); (b) J. A. Berson and R. W. Holder, *J. Am. Chem. Soc.*, **95**, 2037 (1973); (c) J. A. Berson and L. Salem, *J. Am. Chem. Soc.*, **94**, 8917 (1972); (d) W. T. Borden and L. Salem, *J. Am. Chem. Soc.*, **95**, 932 (1973).

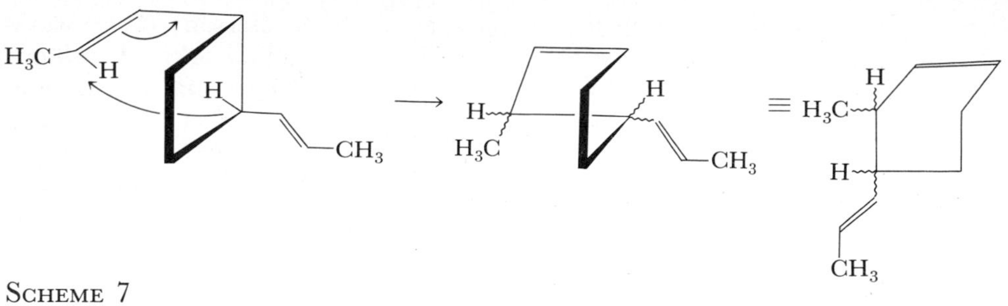
H$_3$C
H
H
CH$_3$
H
H$_3$C
H
CH$_3$
H
H$_3$C
H
CH$_3$

(11.93)

Scheme 7

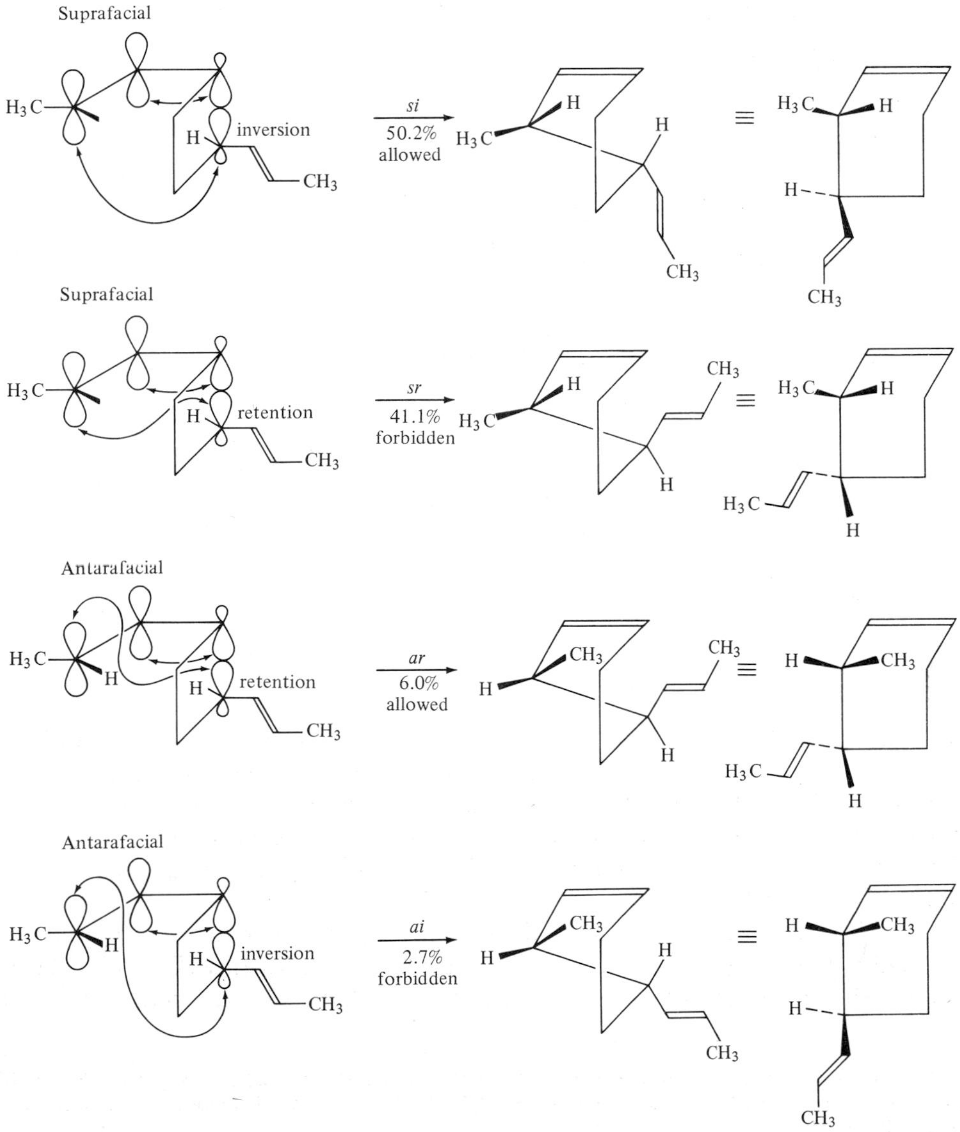
Suprafacial
H$_3$C
H
inversion
CH$_3$
si
50.2%
allowed
H$_3$C
H
H
CH$_3$
H$_3$C
H
H
CH$_3$
Suprafacial
H$_3$C
H
retention
CH$_3$
sr
41.1%
forbidden
H$_3$C
H
CH$_3$
H
H$_3$C
H
H$_3$C
H
Antarafacial
H$_3$C
H
H
retention
CH$_3$
ar
6.0%
allowed
H
CH$_3$
CH$_3$
H
H
CH$_3$
H$_3$C
H
Antarafacial
H$_3$C
H
H
inversion
CH$_3$
ai
2.7%
forbidden
H
CH$_3$
H
CH$_3$
H
CH$_3$
H
CH$_3$

results of a complete stereochemical analysis carried out by Berson and co-workers.[139] The two reactions that are suprafacial on the π system, one allowed and one forbidden, proceed at nearly the same rate, whereas the two antarafacial reactions are slower. A biradical pathway could explain the observed stereochemistry and activation parameters from the point of view of an energy criterion (observed $\Delta H^\ddagger = 34.0$ kcal mole^{-1}; estimated for a biradical path, $\Delta H^\ddagger \approx 30$–$34$ kcal mole^{-1}). However, to explain the predominance of the suprafacial reaction over the antarafacial, it is necessary to propose substantially different rates of rotation about bonds C_3—C_4 and C_5—C_6 in biradical intermediate **85**. This assumption then

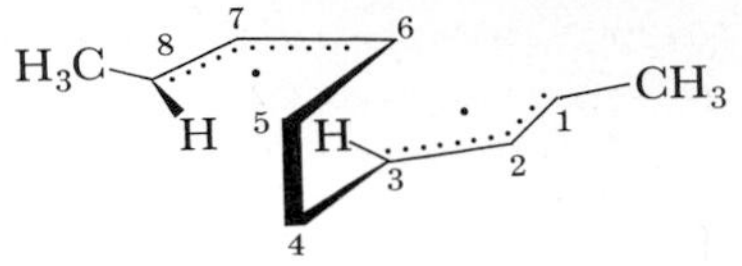

85

makes it difficult to explain why the *sr* reaction, which requires rotation around only bond C_4—C_5, would not be substantially faster than the *si* reaction, which requires rotation about both bonds C_4—C_5 and C_3—C_4. Berson and co-workers proposed on the basis of these arguments that both allowed and forbidden products are formed through concerted pathways. If this proposal is correct, the steric constraint imposed by the inversion requirement in these systems has brought the activation energy of the allowed process up to about the same level as the forbidden.

Another example appears in the ring expansion of vinylcyclopropanes. Baldwin found the product distribution shown in Scheme 8 for the pyrolysis of

SCHEME 8

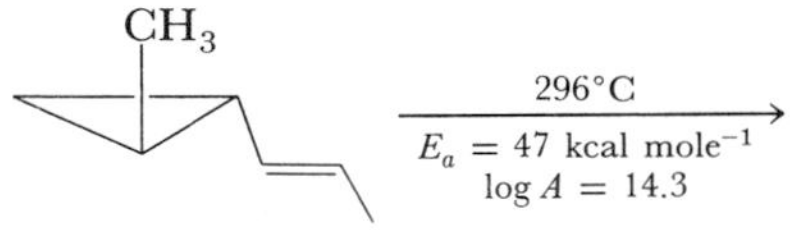

86

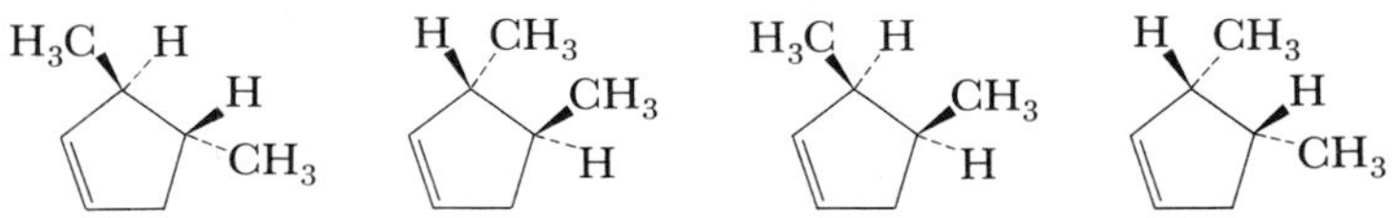

87	**88**	**89**	**90**
65%	8%	22%	5%
si	*ar*	*sr*	*ai*
(allowed)	(allowed)	(forbidden)	(forbidden)

si = suprafacial, inversion
ar = antarafacial, retention } allowed

sr = suprafacial, retention
ai = antarafacial, inversion } forbidden

[139] J. A. Berson, P. D. Dervan, R. Malherbe, and J. A. Jenkins, *J. Am. Chem. Soc.*, **98**, 5937 (1976).

Example. The *si* Path

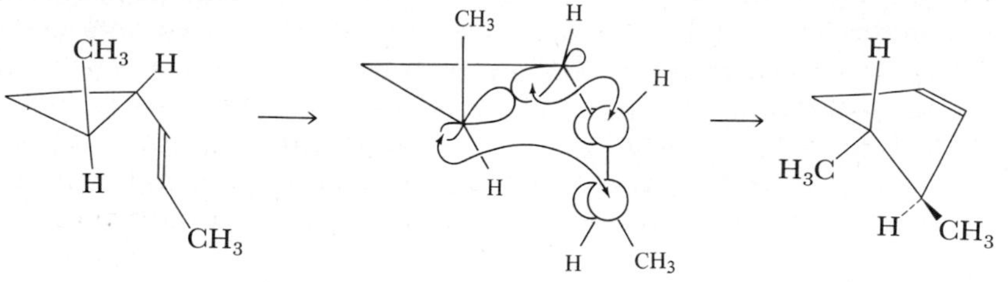

(+)-(1*S*,2*S*)-*trans,trans*-2-methyl-1-propenylcyclopropane (**86**).[140] Allowed reaction is slightly favored over forbidden, but the difference represents only a few tenths of a kilocalorie. The possibility of a stepwise biradical pathway is made implausible by the retention of chirality, that is, by the predominance of **87** over its enantiomer **88** and of **89** over its enantiomer **90**.

The methylenecyclobutane isomerization (Equation 11.94) has been investigated with deuterium labeling that defines the reaction stereochemistry. In this

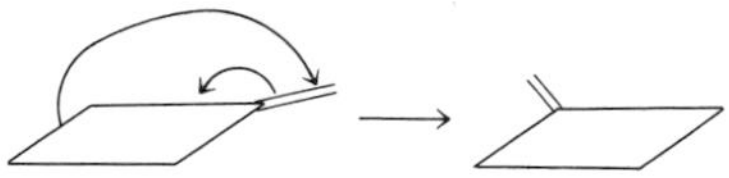

$$\text{(11.94)}$$

reaction, in contrast to those discussed above, the relative rates of formation of products have been interpreted in terms of a competition between the allowed suprafacial–inversion pericyclic reaction (**91**) and a biradical path. The pericyclic

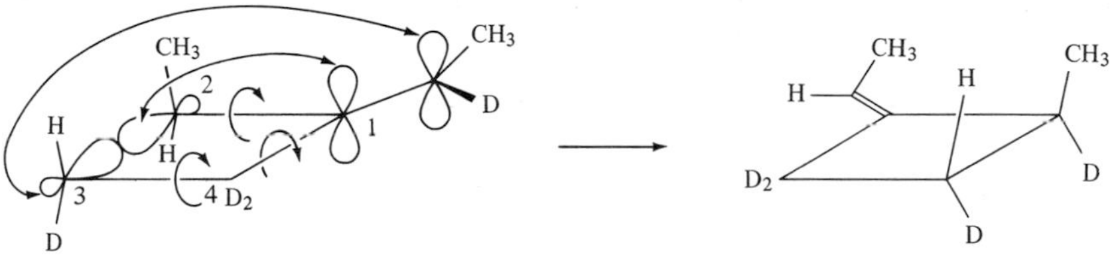

$$\text{(11.95)}$$

91

component of the reaction, which accounts for 36 percent of the product, is stereospecific, whereas the biradical component, which accounts for 64 percent of the product, leads by way of intermediate **92** to loss of stereochemistry at the migrating center. It is proposed that in the formation of **92** the methyl group at C_2 rotates

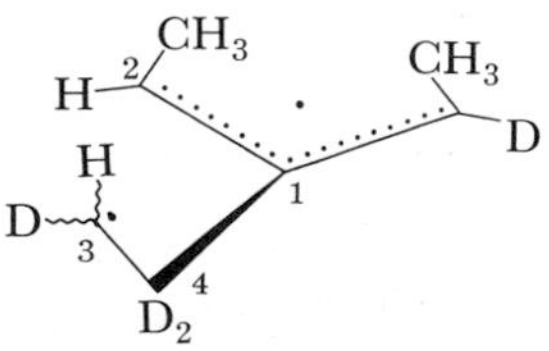

92

[140] G. D. Andrews and J. E. Baldwin, *J. Am. Chem. Soc.*, **98**, 6705 (1976). The product distribution given in Scheme 8 is based on total cyclopentenes. A large amount of acyclic product (largely *cis*-1,4-heptadiene) was also formed.

outward for steric reasons and that in order to maintain maximum overlap C_3 moves upward by rotation about the C_1—C_4 bond. Biradical **92** then closes to rearranged methylenecyclobutane with loss of stereochemistry only at C_3.[141]

Photochemically the [1,3]-rearrangement is allowed in the sterically easily accessible suprafacial–suprafacial mode. A few of the many examples known are shown in Equations 11.96–11.98. Cookson and co-workers have established in the experiments outlined in Equations 11.99 and 11.100 that the stereochemistry is as predicted.[145]

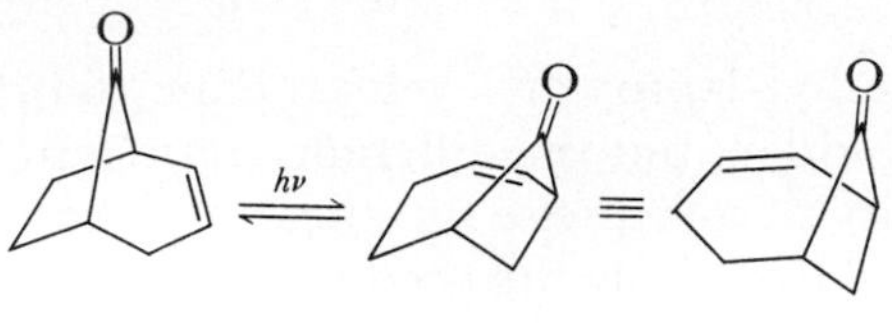

$$(11.96)^{142}$$

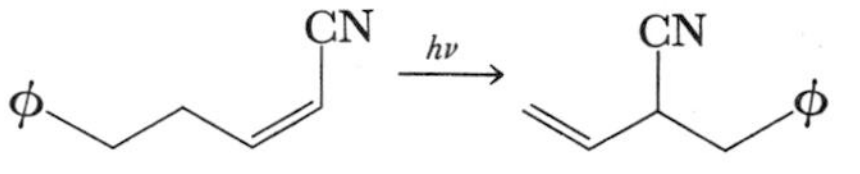

$$(11.97)^{143}$$

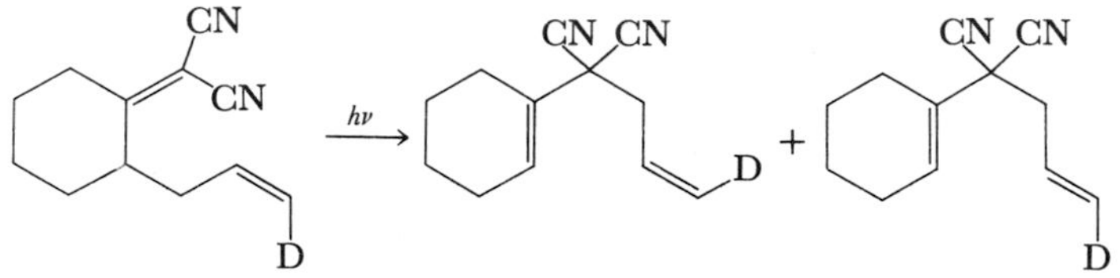

$$(11.98)^{144}$$

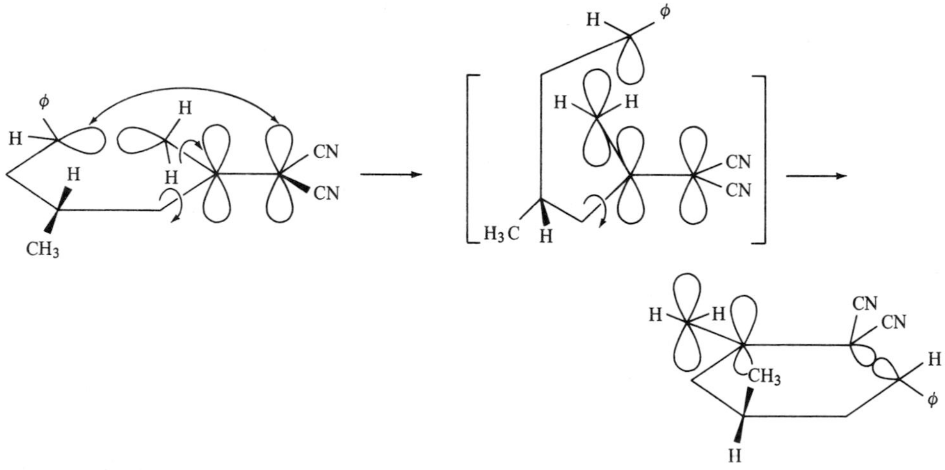

$$(11.99)$$

[141] J. J. Gajewski, *J. Am. Chem. Soc.*, **98**, 5254 (1976).
[142] W. F. Erman and H. C. Kretschmar, *J. Am. Chem. Soc.*, **89**, 3842 (1967).
[143] (a) R. C. Cookson, V. N. Gogte, J. Hudec, and N. A. Mirza, *Tetrahedron Lett.*, 3955 (1965); (b) R. F. C. Brown, R. C. Cookson, and J. Hudec, *Tetrahedron*, **24**, 3955 (1968).
[144] See note 143(b).
[145] R. C. Cookson, J. Hudec, and M. Sharma, *Chem. Commun.*, 107, 108 (1971).

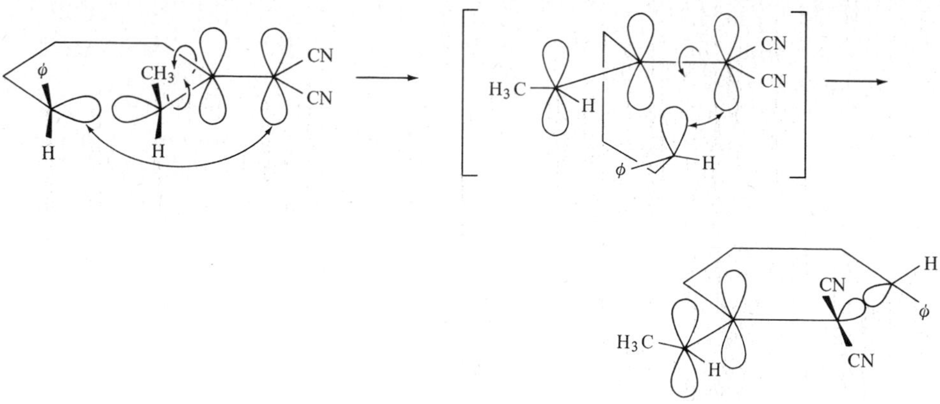

(11.100)

Four-electron [1,4]-rearrangements can occur in allylic cations; as in the neutral four-electron systems, the rules require the suprafacial–antarafacial mode in the ground state. In the bicyclo[3.1.0]hex-3-en-2-yl cation (**93**), the ring fusion

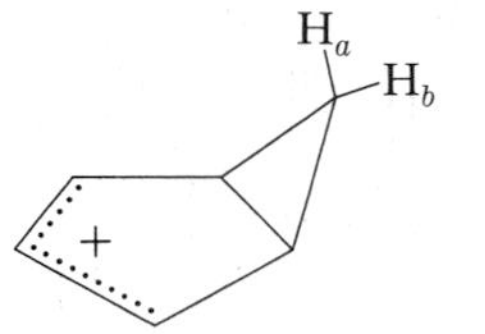

93

prevents access of the migrating —CH$_2$— to the underside of the allylic ion and assures suprafaciality with respect to the π system; the rearrangement can thus take either of the two courses shown in Equations 11.101 and 11.102. In Equation

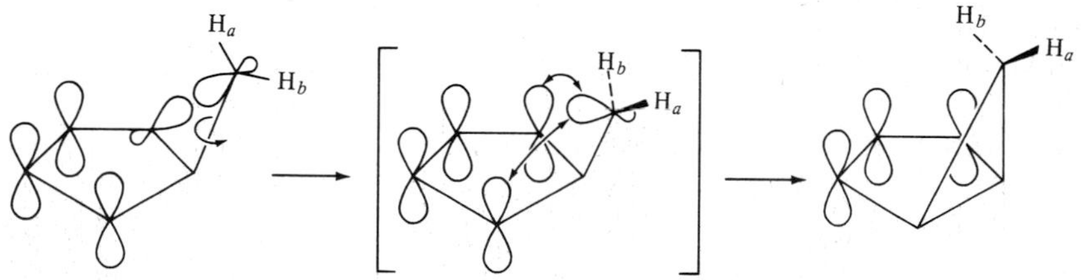

(11.101)

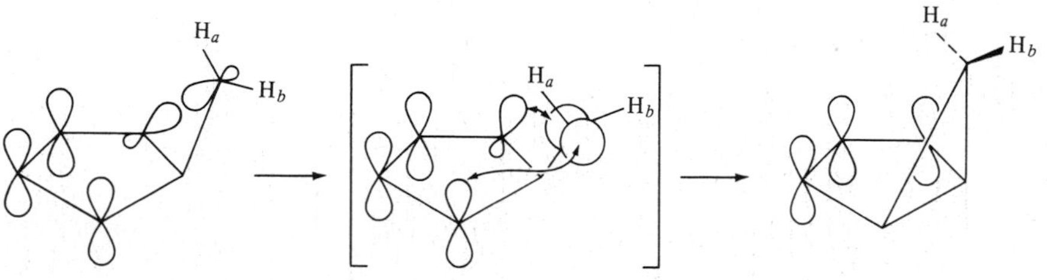

(11.102)

11.101, thermally forbidden, suprafacial on the migrating group, the —CH$_2$— swings around so that the endo (H$_a$) and exo (H$_b$) hydrogens change places, whereas in the allowed reaction (Equation 11.102), the migrating group undergoes inversion and endo and exo positions maintain their integrity. A second rearrangement moves the cyclopropane one position further on the five-membered ring; continuation of the process will walk the three-membered ring all the way around the five. If configuration is retained at the migrating group, the endo and exo substituents interchange at each step; but if configuration is inverted, the endo and exo substituents always remain distinct.[146] It is found experimentally that, in agreement with predictions, the latter process is the one that occurs. The deuterated cation (**94**), prepared by dissolving the chloride **95** in SbF$_5$–SO$_2$ClF at low

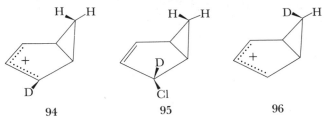

94   95   96

temperature, undergoes a rearrangement, activation free energy $\Delta G^{\ddagger} = 15 \pm 1$ kcal mole^{-1}, which distributes the deuterium equally around the five-membered ring. Deuterium in the *endo*-7-deuterio ion (**96**) remains endo (Equation 11.102). The rate of the rearrangement is at least 10^4 to 10^6 times the rate of loss of endo–exo stereochemistry. The authors point out that because the observed process is much easier sterically than the other alternative, this rearrangement is not a particularly revealing test of the sigmatropic selection rules.[147]

Six-electron [1,n]-rearrangements A variety of sigmatropic rearrangements through six-electron transition states are known. In contrast to the rare [1,3]-migrations, [1,5]-shifts of hydrogen in dienes, suprafacially allowed, occur readily. The experiment outlined in Equation 11.103 confirms the predicted stereochemistry.[148] The authors estimated the alternative antarafacial rearrangement to be at least 8 kcal mole^{-1} higher in activation energy. The lack of dependence of rate on phase and on solvent polarity rules out ionic intermediates,[149] and a large primary isotope effect ($k_H/k_D = 12.2$ at 25°C) indicates a symmetrical transition

[146] R. B. Woodward and R. Hoffmann, *The Conservation of Orbital Symmetry*, Verlag Chemie, Weinheim/Bergstr., Germany, and Academic Press, New York, 1970, p. 132.

[147] (a) P. Vogel, M. Saunders, N. M. Hasty, Jr., and J. A. Berson, *J. Am. Chem. Soc.*, **93**, 1551 (1971). The rearrangement occurs slowly at −90°C. At −20°C, an electrocyclic opening to benzenonium ion takes place; see Equation 11.75. (b) R. F. Childs and S. Winstein, *J. Am. Chem. Soc.*, **96**, 6409 (1974); (c) R. F. Childs and M. Zeya, *J. Am. Chem. Soc.*, **96**, 6418 (1974). A similar rearrangement has been observed in a bicyclic radical: (d) R. Sustmann and F. Lübbe, *J. Am. Chem. Soc.*, **98**, 6037 (1976).

[148] W. R. Roth, J. König, and K. Stein, *Chem. Ber.*, **103**, 426 (1970).

[149] (a) A. P. ter Borg and H. Kloosterziel, *Recl. Trav. Chim. Pays-Bas*, **82**, 741 (1963); (b) D. S. Glass, R. S. Boikess, and S. Winstein, *Tetrahedron Lett.*, 999 (1966); (c) R. W. Roth, *Justus Liebigs Ann. Chem.*, **671**, 25 (1964).

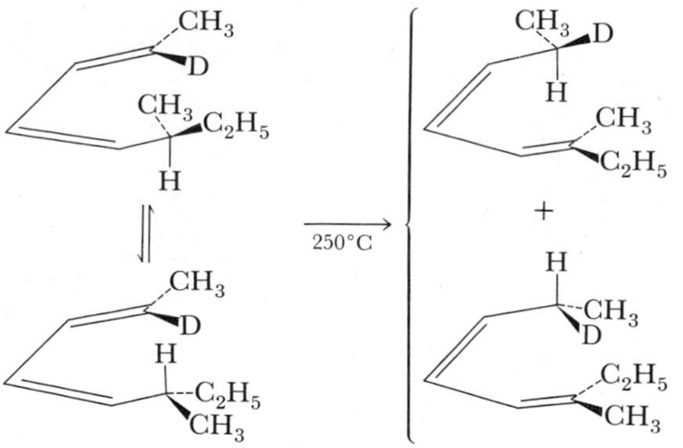

$$(11.103)$$

state.[150] These points together with the stereospecificity, support the conclusion that the reaction is concerted. Enthalpies of activation are in the neighborhood of 30 kcal mole^{-1}, and entropies of activation are -5 to -10 cal mole^{-1} K^{-1}.[149]

In a bicyclo[4.1.0]heptadiene, there should be a [1,5]-walk-around rearrangement analogous to the one we discussed above in the bicyclic allylic cations.[151] For this six-electron system the pericyclic selection rules require the migrating group to move with retention of configuration (Equation 11.104).

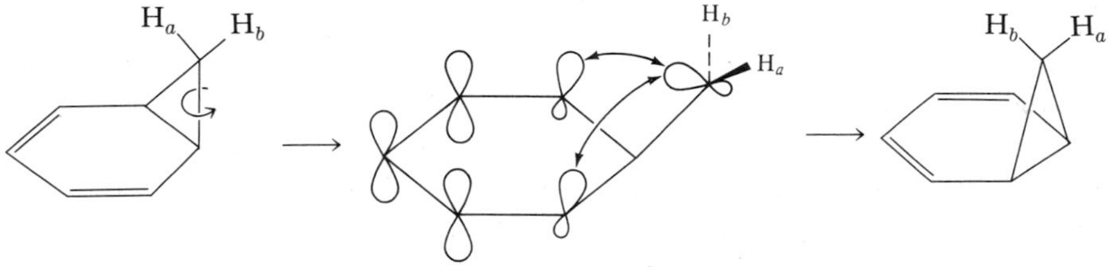

$$(11.104)$$

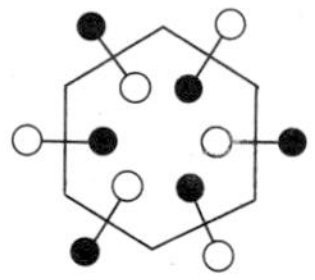

97 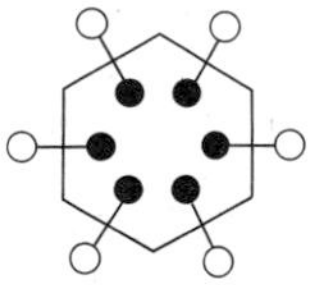   98

Woodward and Hoffmann have introduced the abbreviated notation **97** to summarize this process. The view is directly down onto the plane of the six-membered ring, with the substituents at the 7 position (the one-carbon bridge) indicated by filled and open circles. Structure **97** is a composite picture of all six molecules that would be obtained as the cyclopropane ring moves around the periphery of the cyclohexadiene ring in the manner depicted in Equation 11.104, endo and exo substituents exchanging places with each step. Composite Structure **98** summarizes

[150](a) W. R. Roth and J. König, *Justus Liebigs Ann. Chem.*, **699**, 24 (1966); (b) H. Kloosterziel and A. P. ter Borg, *Recl. Trav. Chem. Pays-Bas,* **84**, 1305 (1965).
[151]Note that these systems also undergo the electrocyclic norcaradiene–cycloheptatriene interconversion (Section 11.2).

the alternative (forbidden) possibility; retention of configuration at the migrating center at each step would leave the substituent initially endo (filled circle) always endo.

The predictions of the pericyclic rules fail for the rearrangement shown in Equation 11.105. The reaction follows the forbidden path with inversion at the migrating carbon; it is proposed that the principle of least motion takes precedence over the pericyclic rules in this instance.[152] The result casts further doubt on the importance of pericyclic control in the rearrangement of cations **94** and **96**.

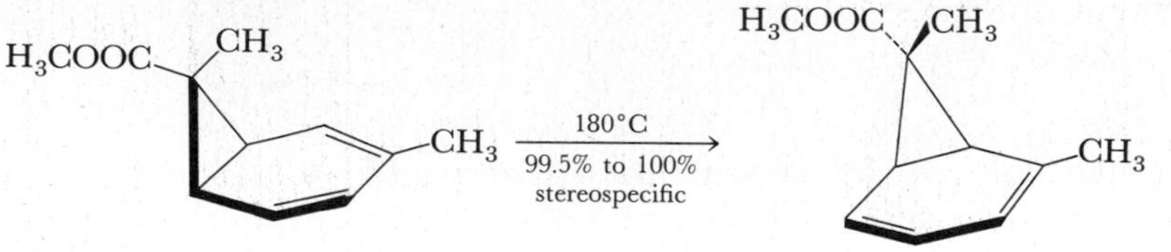

$$180°C \atop \frac{}{99.5\% \text{ to } 100\%}\atop \text{stereospecific}$$ (11.105)

[*m,n*]-Sigmatropic Rearrangements

The majority of known [*m,n*]-rearrangements are six-electron processes, thermally allowed in the suprafacial–suprafacial mode. The most commonly observed examples are the [2,3]- and [3,3]-migrations.[153]

[2,3]-Rearrangements A [2,3]-rearrangement occurs in the anion **99**. The analogous process takes place readily in ylides (Equation 11.107) and in various other isoelectronic situations, for example, Equations 11.108–11.110. In at least

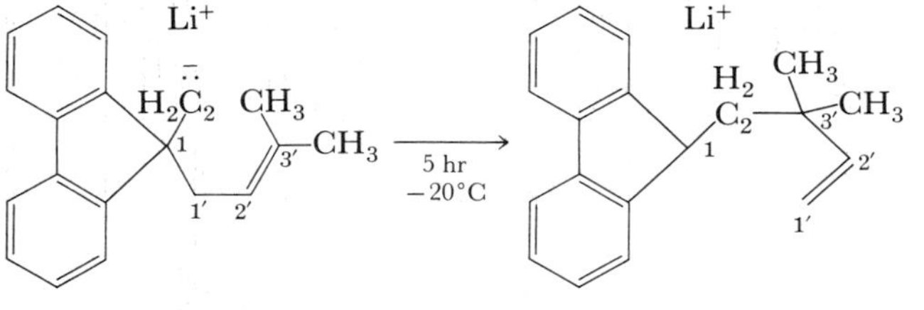

(11.106)[154]

99

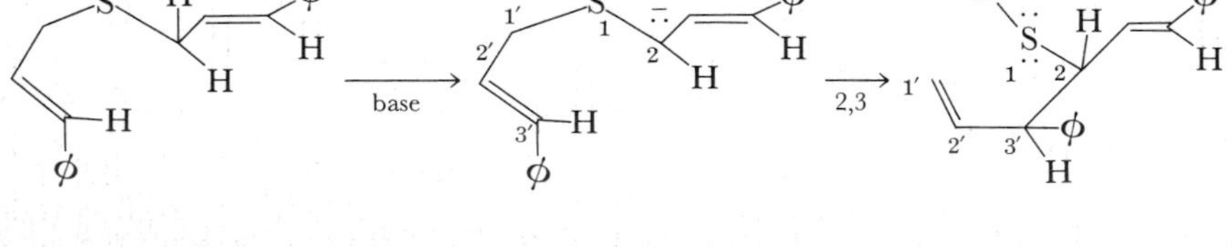

(11.107)[155]

[152](a) F.-G. Klärner, *Angew. Chem. Int. Ed. Engl.*, **13**, 268 (1974); (b) W. W. Schoeller, *J. Am. Chem. Soc.*, **97**, 1978 (1975). This result was challenged by Baldwin but reaffirmed by Klärner. See (c) J. E. Baldwin and B. M. Broline, *J. Am. Chem. Soc.*, **100**, 4599 (1978); (d) F.-G. Klärner and B. Brassel, *J. Am. Chem. Soc.*, **102**, 2469 (1980).

[153]For a more complete discussion of examples, see J. L. Gilchrist and R. C. Storr, *Organic Reactions and Orbital Symmetry*, Cambridge University Press, London, 1972, Chap. 7.

[154]J. E. Baldwin and F. J. Urban, *Chem. Commun.*, 165 (1970).

[155]J. E. Baldwin, R. E. Hackler, and D. P. Kelly, *Chem. Commun.*, 537, 538 (1968).

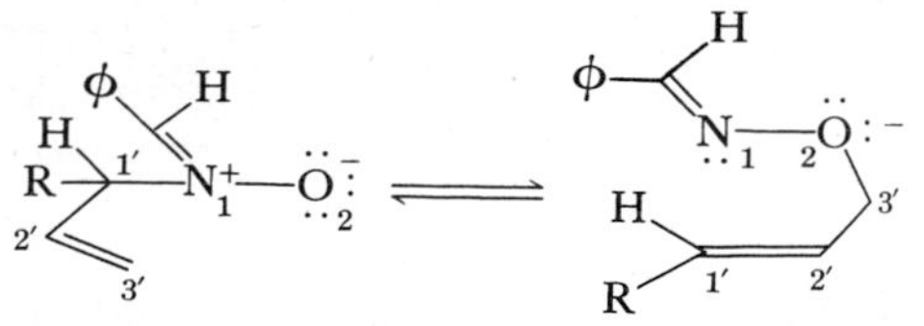

$$(11.108)^{156}$$

some of these rearrangements, product mixtures (for example, Equation 11.111) show that a biradical dissociation–recombination reaction competes with the [2,3]-rearrangement; the biradical reaction has a higher activation energy and hence is favored by raising the temperature.[160]

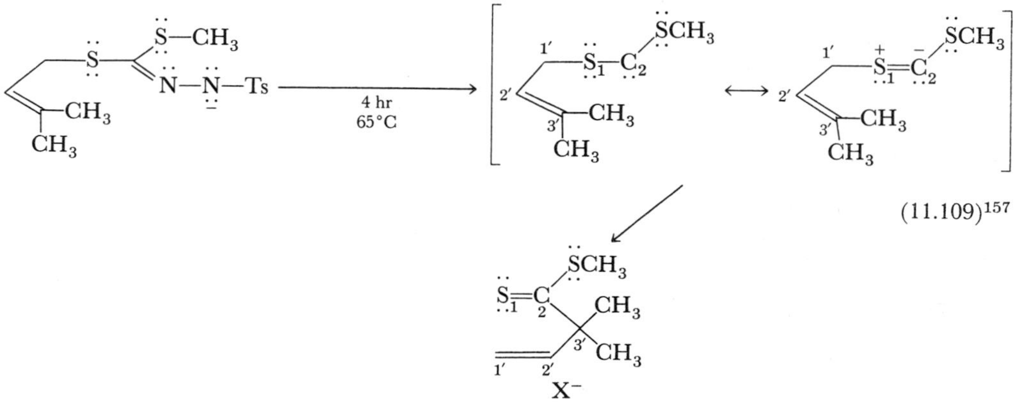

$$(11.109)^{157}$$

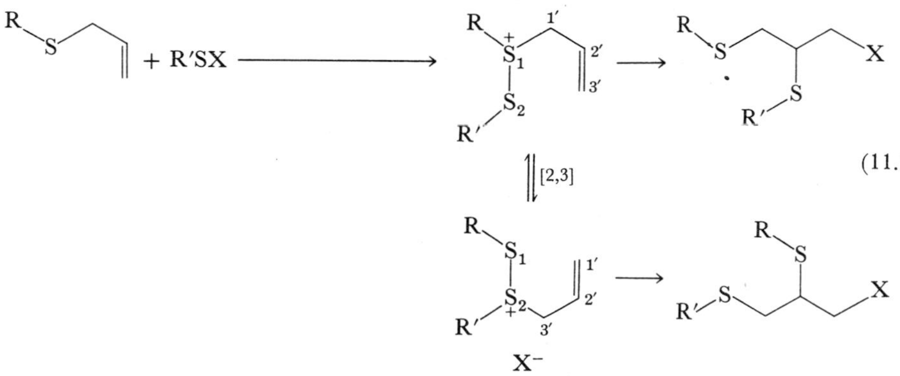

$$(11.110)^{158}$$

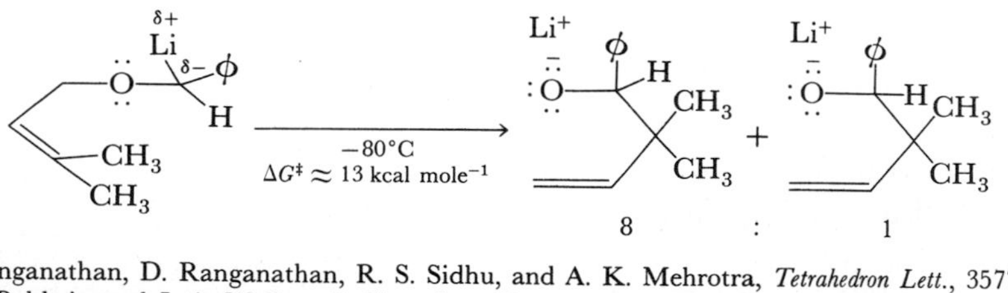

$$(11.111)^{159}$$

8 : 1

[156] S. Ranganathan, D. Ranganathan, R. S. Sidhu, and A. K. Mehrotra, *Tetrahedron Lett.*, 3577 (1973).
[157] J. E. Baldwin and J. A. Walker, *J. Chem. Soc., Chem. Commun.*, 354 (1972).
[158] J. K. Kim, M. L. Kline, and M. C. Caserio, *J. Am. Chem. Soc.*, **100**, 6243 (1978).
[159] V. Rautenstrauch, *Chem. Commun.*, 4 (1970).
[160] See note 159 and J. E. Baldwin, J. E. Brown, and R. W. Cordell, *Chem. Commun.*, 31 (1970).

Baldwin and Patrick have demonstrated by the experiment outlined in Equation 11.112 that the concerted [2,3]-rearrangement takes the suprafacial path over the allyl group.[161] The two products arise from the two conformations **100** and **101**; antarafacial rearrangement would have yielded **102** and **103**, which were not found.

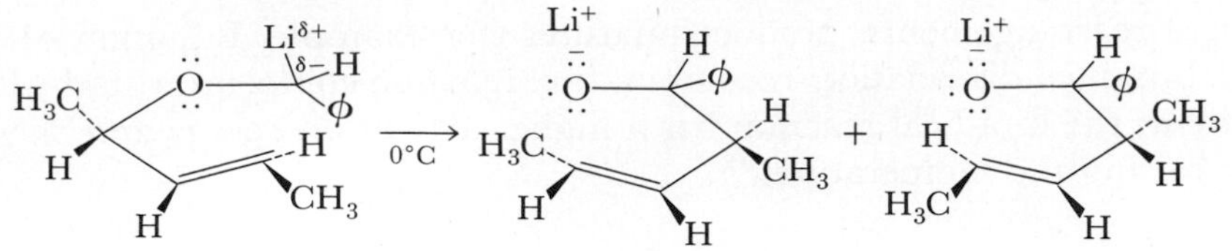

$$(11.112)$$

17%   83%

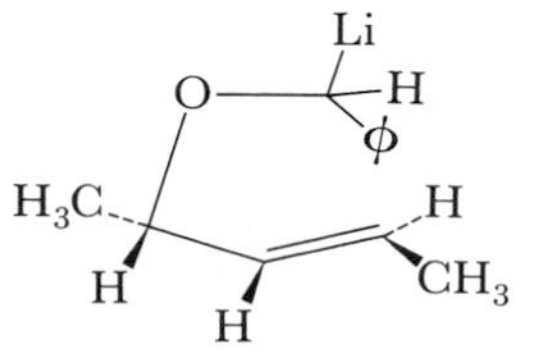

100 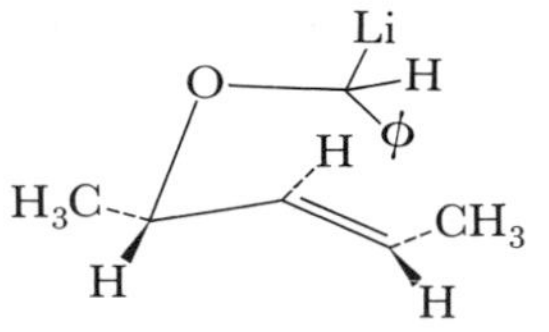   **101**

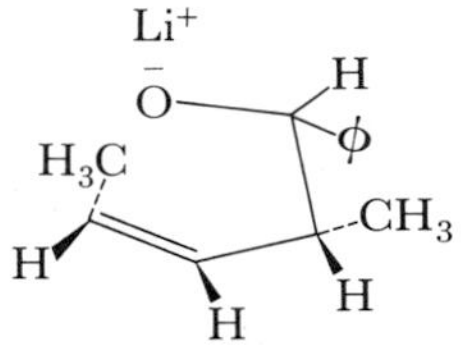

102 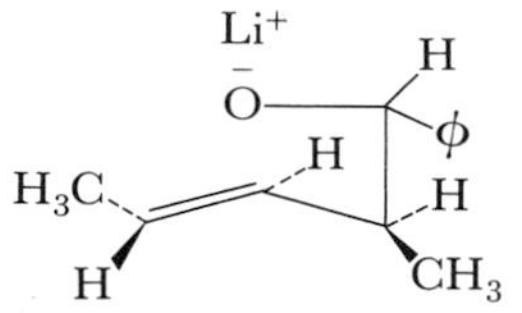   **103**

A transformation closely related to the [2,3]-sigmatropic migration is the elimination illustrated in Equation 11.113. The orbital diagrams **104** for the [2,3]-

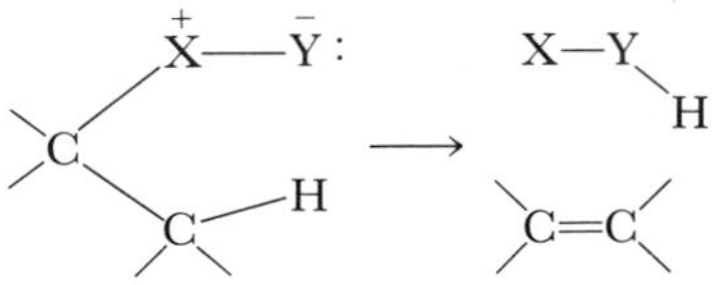

$$(11.113)$$

sigmatropic rearrangement and **105** for the [2,3]-elimination show that they differ only in the substitution of a σ for a π electron pair. An important example of the [2,3]-elimination is the Cope reaction (Equation 11.114), which is a synthetically useful method of introducing unsaturation under mild conditions.[162]

[161] J. E. Baldwin and J. E. Patrick, *J. Am. Chem. Soc.*, **93**, 3556 (1971).
[162] (a) A. C. Cope, T. T. Foster, and P. H. Towle, *J. Am. Chem. Soc.*, **71**, 3929 (1949); the reaction and some similar ones are reviewed by (b) C. H. DePuy and R. W. King, *Chem. Rev.*, **60**, 431 (1960).

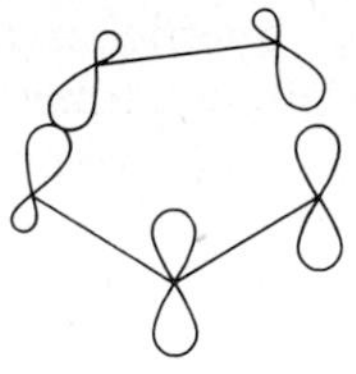

104	105
[2,3]–rearrangement	[2,3]–elimination
$\sigma 2s + \omega 2s + \pi 2s$	$\sigma 2s + \omega 2s + \sigma 2s$

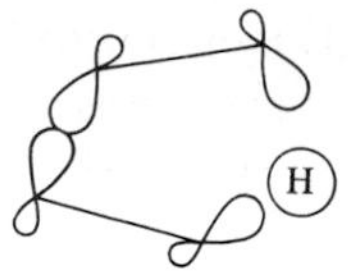

$$\underset{\substack{\sim 100°\text{C}}}{\longrightarrow} \qquad (11.114)$$

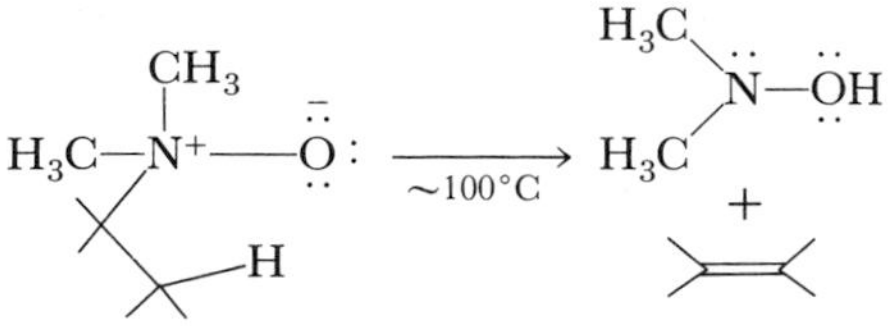

[3,3]-Rearrangements[163] The [3,3]-sigmatropic rearrangements comprise one of the more important classes of pericyclic reactions. The prototype is the degenerate rearrangement of 1,5-hexadiene shown in Equation 11.115.[164] This

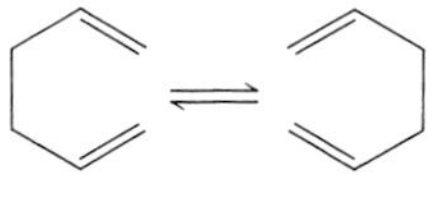

$$\rightleftharpoons \qquad (11.115)$$

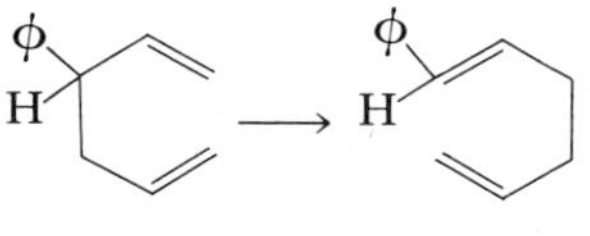

$$\longrightarrow \qquad (11.116)$$

process was discovered and studied by Cope and his collaborators during the 1940s and is now known as the Cope rearrangement.[165] The reaction is common in both acyclic and cyclic 1,5-dienes; in the former, most examples occur in molecules bearing an unsaturated substituent in the 3 position. The rearrangement then brings the group into conjugation (for example, Equation 11.116); the activation energies are in the range of about 25 to 30 kcal mole^{-1} with activation entropies around -10 to -15 cal mole^{-1} K^{-1}, and the reactions proceed readily between 150 and 200°C.[166] In the absence of the unsaturated substituent, somewhat higher temperatures are required: 3-methyl-1,5-hexadiene rearranges at 300°C[167] and 1,1-dideuterio-1,5-hexadiene at 200–250°C, $\Delta H^{\ddagger} = 33.5$ kcal mole^{-1} and $\Delta S^{\ddagger} = -13.8$ cal mole^{-1} K^{-1}.[168]

[163] For reviews, see (a) S. J. Rhoads, in *Molecular Rearrangements,* Part I, P. de Mayo, Ed., Wiley, New York, 1963, p. 655; (b) E. Vogel, *Angew. Chem. Int. Ed. Engl.,* **2,** 1 (1963); (c) W. v. E. Doering and W. R. Roth, *Angew. Chem. Int. Ed. Engl.,* **2,** 115 (1963); (d) S. J. Rhoads and N. R. Raulins, *Org. Reactions,* **22,** 1 (1975).
[164] A degenerate rearrangement is one that changes locations of bonds but leaves the structure unchanged.
[165] (a) A. C. Cope and E. M. Hardy, *J. Am. Chem. Soc.,* **62,** 441 (1940); (b) A. C. Cope, C. M. Hofmann, and E. M. Hardy, *J. Am. Chem. Soc.,* **63,** 1852 (1941).
[166] E. G. Foster, A. C. Cope, and F. Daniels, *J. Am. Chem. Soc.,* **69,** 1893 (1947).
[167] H. Levy and A. C. Cope, *J. Am. Chem. Soc.,* **66,** 1684 (1944).
[168] W. v. E. Doering, V. G. Toscano, and G. H. Beasley, *Tetrahedron,* **27,** 5299 (1971).

The stereochemistry of the Cope rearrangement has aroused considerable interest. Doering and Roth set out to determine whether a boatlike (**106**) or chairlike (**107**) transition state is preferred. Their experiment, the results of which

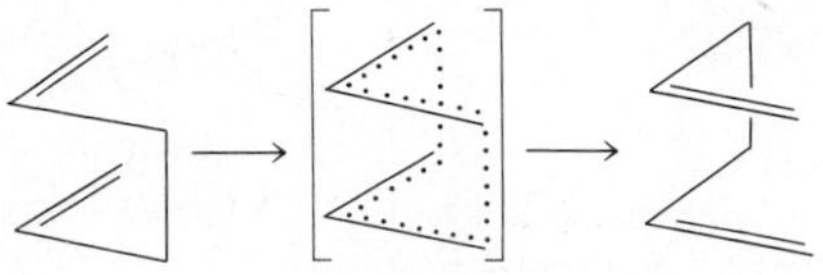

(11.117)

106

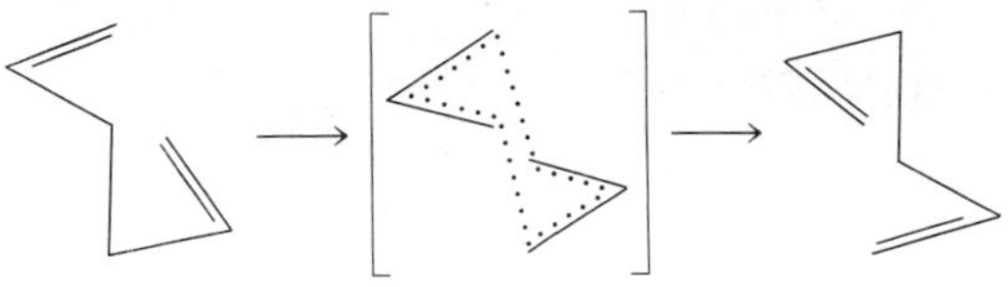

(11.118)

107

are shown in Scheme 9, demonstrate that the chair route (**107**) is the lower-energy one.[169] Goldstein and Benzon showed that the tetradeuterated diene **108** rear-

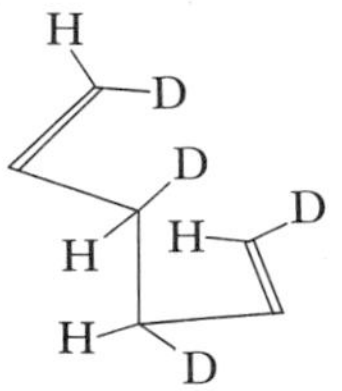

108

ranges through the chair transition state at 230°C, but that the less favorable boat rearrangement begins to compete above 260°C.[170] Their estimate of the difference in free energy of activation, $\Delta\Delta G^{\ddagger} = 5.8$ kcal mole^{-1}, agrees well with Doering and Roth's conclusion of $\Delta\Delta G^{\ddagger} \geqslant 5.7$ kcal mole^{-1}. Woodward and Hoffmann have rationalized the preference in terms of orbital correlation diagrams,[171] and Dewar has done so in terms of interaction diagrams.[172]

The question of stepwise or concerted mechanism for the Cope rearrangement is an interesting one. In the reaction one σ bond breaks and another forms. The concerted process would consist of both these events occurring simultaneously, through a pericyclic transition state (**107**) that has aromatic stabilization. There are two distinct stepwise alternatives: bond breaking could occur first to yield an inter-

[169](a) W. v. E. Doering and W. R. Roth, *Tetrahedron*, **18**, 67 (1962); see also (b) R. K. Hill and N. W. Gilman, *Chem. Commun.*, 619 (1967); (c) R. K. Hill and N. W. Gilman, *Tetrahedron Lett.*, 1421 (1967).

[170](a) M. J. Goldstein and M. S. Benzon, *J. Am. Chem. Soc.*, **94**, 7147, 7149 (1972). A slightly larger difference, $\Delta\Delta G^{\ddagger} = 8.1$ kcal mole^{-1} ($\Delta\Delta H^{\ddagger} = 13.8$ kcal mole^{-1}, $\Delta\Delta S^{\ddagger} = 11.0$ cal mole^{-1} K^{-1}), has also been reported: (b) K. J. Shea and R. B. Phillips, *J. Am. Chem. Soc.*, **100**, 654 (1978). See also (c) J. J. Gajewski, L. K. Hoffman, and C. N. Shih, *J. Am. Chem. Soc.*, **96**, 3705 (1974).

[171]R. B. Woodward and R. Hoffmann, *The Conservation of Orbital Symmetry*, Verlag Chemie, Weinheim/Bergstr. Germany, and Academic Press, New York, 1970, p. 149.

[172]M. J. S. Dewar, *Angew. Chem. Int. Ed. Engl.*, **10**, 761 (1971).

3,4-dimethyl-1,5-hexadiene $\longrightarrow$ 2,6-octadiene

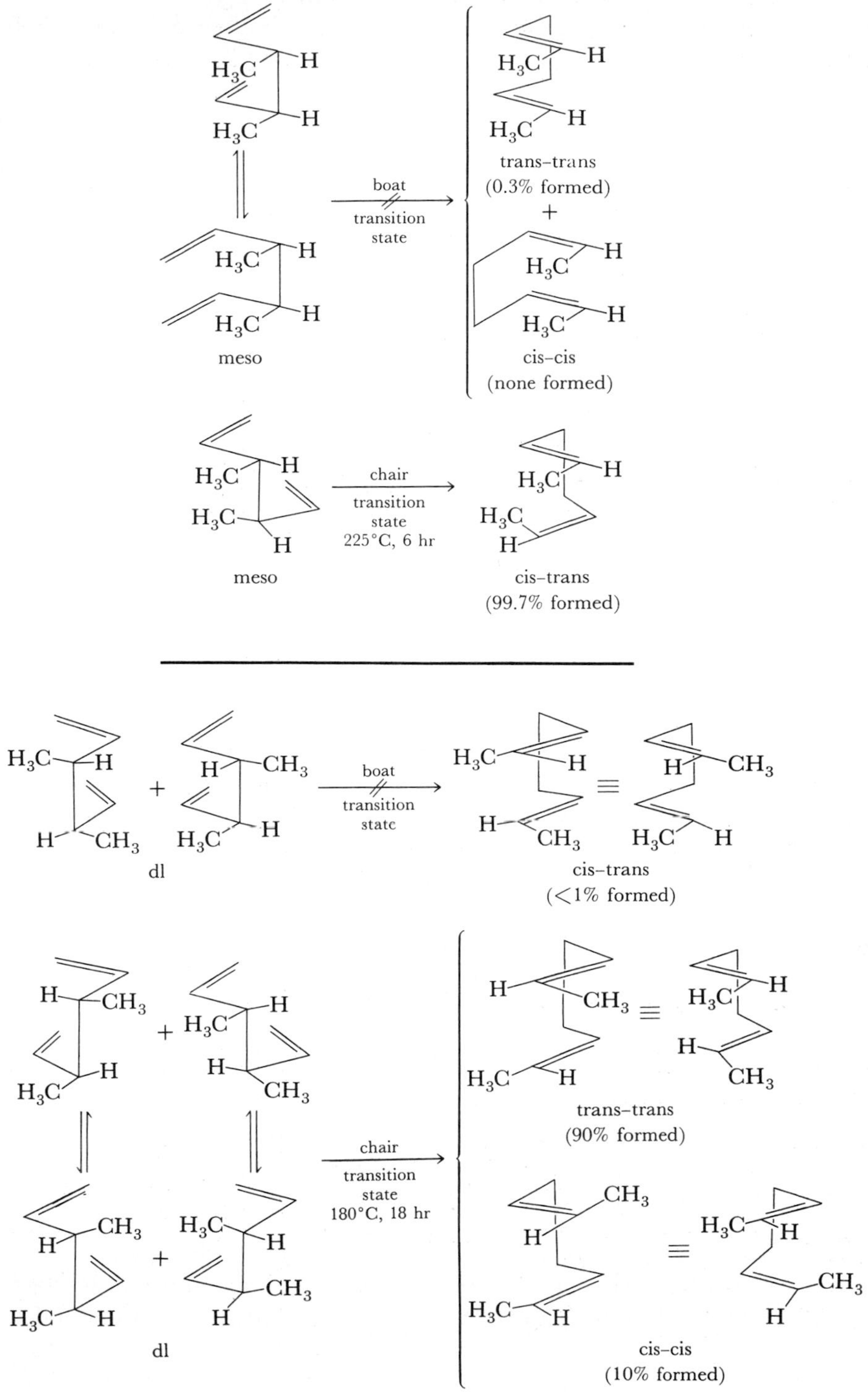

trans–trans
(0.3% formed)
+
cis–cis
(none formed)

meso

meso   chair transition state 225°C, 6 hr   cis–trans (99.7% formed)

dl   boat transition state   cis–trans (<1% formed)

dl   chair transition state 180°C, 18 hr

trans–trans
(90% formed)

cis–cis
(10% formed)

mediate of two allyl radicals (Equation 11.119); or bond making could occur first to yield an intermediate cyclohexane biradical (Equation 11.120).[172] The equations

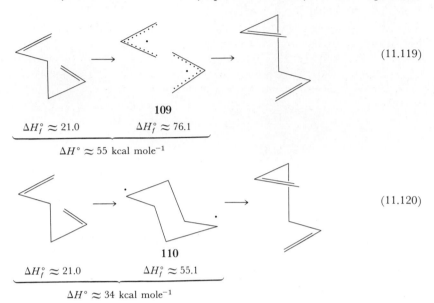

$$\tag{11.119}$$

109

$\Delta H_f^\circ \approx 21.0 \qquad\qquad \Delta H_f^\circ \approx 76.1$

$\Delta H^\circ \approx 55$ kcal mole^{-1}

110

$$\tag{11.120}$$

$\Delta H_f^\circ \approx 21.0 \qquad\qquad \Delta H_f^\circ \approx 55.1$

$\Delta H^\circ \approx 34$ kcal mole^{-1}

show heats of formation of starting material 1,5-hexadiene and rough lower limits for the two possible biradical intermediates, calculated from Tables 2.9 and 9.5 as follows:

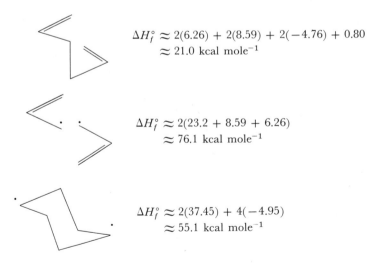

$\Delta H_f^\circ \approx 2(6.26) + 2(8.59) + 2(-4.76) + 0.80$
≈ 21.0 kcal mole^{-1}

$\Delta H_f^\circ \approx 2(23.2 + 8.59 + 6.26)$
≈ 76.1 kcal mole^{-1}

$\Delta H_f^\circ \approx 2(37.45) + 4(-4.95)$
≈ 55.1 kcal mole^{-1}

As we noted earlier, the observed $\Delta H^\ddagger$ is 33.5 kcal mole^{-1}. The bond-breaking path (Equation 11.119) can probably be safely ruled out as having an activation enthalpy at least 20 kcal mole^{-1} above that observed. In the case of the bond-making path (Equation 11.120), however, the energetics do not rule out the biradical route. Biradical **110** would be an intermediate, and so activation energy would be a few kcal mole^{-1} above the calculated $\Delta H°$. Even though the estimate is a lower limit, agreement is close enough to support Equation 11.120 as a possibility that must be confirmed or ruled out experimentally.

Experimental efforts to define the mechanism have relied on substituent effects. Dewar and Wade found that the rate of rearrangement of **112** is four times that of **111**. They reasoned that such an acceleration should occur if the transition state is close to **110**, in which the phenyl group placed as in Equation 11.122 would

$$E_a = 32.5 \text{ kcal mole}^{-1}$$
$$\log A = 10.7$$

111
$$\tag{11.121}$$

$$E_a = 30.2 \text{ kcal mole}^{-1}$$
$$\log A = 10.9$$

112
$$\tag{11.122}$$

stabilize the incipient radical, but not if the transition state is the pericyclic one **107**, in which all positions are essentially equivalent with respect to conjugation.[173] It is likely, however, that the transition state structure is affected by the presence of substituents, shifting more toward the cyclohexyl biradical **110** when radical stabilizing groups are put at the 2 and 5 positions of the diene.

Although Dewar has argued against this view of a variable transition state, others have found evidence for it. Gajewski[174] and also Wherli, Schmidt, and co-workers[175] have constructed three-dimensional reaction coordinate diagrams for the process. In Gajewski's model the free-energy surface is assumed to have the analytical form:

$$\Delta G = ax + by + cxy + d \tag{11.123}$$

This equation represents the simplest possible model for a surface with a saddle point. Boundary conditions are applied to fit it to the known energies at various points. Figure 11.4 shows this surface for the unsubstituted case. The starting material, 1,5-hexadiene, is placed at point 0, 0; all energies are measured relative to it, and so ΔG at this point is zero. Hence parameter $d = 0$. At point 1, 0 is the cyclohexane biradical intermediate **110**; its free energy is 53 kcal mole^{-1},[174] and hence $a = 53$. At point 0, 1 is the double allyl radical intermediate **109**; its free energy is 57 kcal mole^{-1},[174] and hence $b = 57$. The product is placed at point p, p, where p

[173]M. J. S. Dewar and L. E. Wade, Jr., *J. Am. Chem. Soc.*, **99**, 4417 (1977).

[174](a) J. J. Gajewski, *J. Am. Chem. Soc.*, **101**, 4393 (1979); (b) J. J. Gajewski, *Acc. Chem. Res.*, **13**, 142 (1980). See also (c) R. P. Lutz and H. A. J. Berg, *J. Org. Chem.*, **45**, 3915 (1980).

[175]R. Wherli, H. Schmidt, D. Belluš, and H.-J. Hansen, *Helv. Chim. Acta*, **60**, 1325 (1977).

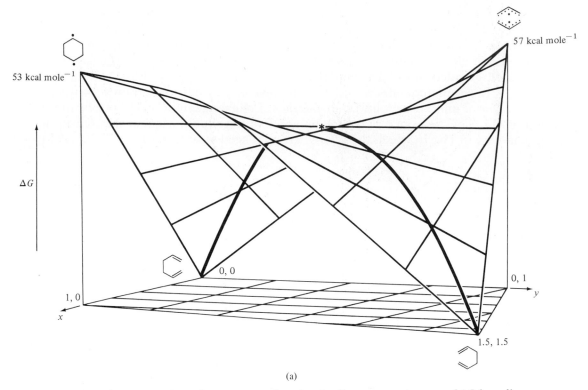

53 kcal mole^{-1}

57 kcal mole^{-1}

ΔG

0, 0

1, 0

0, 1

y

x

1.5, 1.5

(a)

Figure 11.4 Two views of the free-energy surface for the Cope rearrangement of 1,5-hexadiene according to the model proposed by J. J. Gajewski, *J. Am. Chem. Soc.*, **101**, 4393 (1979). View (a) is from below and view (b) is from above. The heavy lines are a grid drawn on the surface with a spacing of 0.25 units in the x and y directions. The lighter lines are the same grid projected onto the x, y plane. The heavy curved line across the surface represents the lowest-energy path, with the transition state, at point $x = 0.77$, $y = 0.72$, marked by *. The segment of the x, y plane covered is not square because the x, y location of the product was adjusted to obtain a fit of energy changes in reactions of a number of substituted 1,5-hexadienes to the functional form assumed for the surface.

is a parameter adjusted so that Equation 11.123 will reproduce $\Delta G^{\ddagger}$ for a variety of substituents. It is found empirically by fitting data for 13 reactions that $p = 1.5$. The final parameter c is then given by

$$c = \frac{\{(\Delta G_{rxn}/p) - a - b\}}{p} \qquad (11.124)$$

where ΔG_{rxn} is the free-energy difference between product and reactant. In the unsubstituted case the product and reactant are the same, $\Delta G_{rxn} = 0$, and $c = (-a - b)/p$. The partial derivatives on the surface can be set equal to zero to find the saddle point, which is the model for the transition state. For the unsubsti-

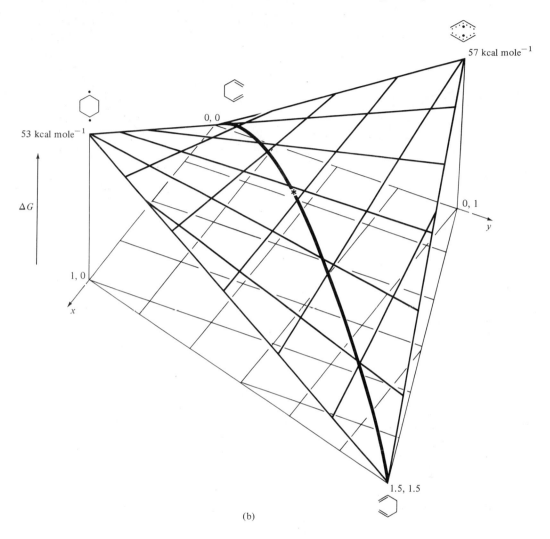

53 kcal mole^{-1}

57 kcal mole^{-1}

0, 0

ΔG

0, 1

y

1, 0

x

1.5, 1.5

(b)

tuted 1,5-hexadiene, the model gives $\Delta G^{\ddagger}_{\text{calc}} = 41.2$ kcal mole^{-1}. The observed $\Delta G^{\ddagger}_{\text{obs}} = 41$ kcal mole^{-1}. (Recall, however, that the parameter p was adjusted to give the best fit to this and 12 other reactions.) The location of the transition state, at point $x = 0.77$, $y = 0.72$, is only very slightly off the center diagonal line representing the symmetrical pericyclic path. The transition state moves farther off the symmetric path, toward the cyclohexyl biradical **110**, as radical stabilizing groups are introduced at the 2 and 5 positions.

Gajewski's model surface suffers from its lack of symmetry. Because of the necessity to adjust product location at point p, p, forward and reverse pathways through various possible intermediates are not the same, as they must be in the unsubstituted reaction by microscopic reversibility. The use of a more complex functional form for the surface, such as that proposed by Jencks,[176] would allow construction of a symmetrical surface, but at the cost of the introduction of another parameter and of a more complicated fitting procedure.

The Cope rearrangement produces a number of interesting phenomena in ring systems. Thermal dimerization of butadiene yields as one product 1,5-cyclooctadiene (Equation 11.125). Formally this dimerization is a 4 + 4 cycloaddition; its actual course (Equation 11.126) is a 2 + 2 addition (presumably biradical) followed by a [3,3]-rearrangement of the *cis*-divinylcyclobutane.[177]

$$2 \quad \diagup\!\!=\!= \quad \longrightarrow \quad \bigcirc \qquad\qquad (11.125)$$

$$+ \quad \longrightarrow \quad \longrightarrow \quad \bigcirc \qquad\qquad (11.126)$$

Cis-1,2-divinylcyclopropane also rearranges rapidly (Equation 11.127). The free energy of activation is 20 kcal mole^{-1}, and $\Delta H^{\ddagger}$ is 19.4 kcal mole^{-1}.[178] The trans isomer, in contrast, rearranges only at 190°C, presumably the temperature required for its isomerization to the cis form.[179]

$$\longrightarrow \qquad\qquad (11.127)$$

Fluxional Molecules[180]

Doering and Roth reported in 1963 the preparation of bicyclo[5.1.0]octa-2,5-diene (**113**), which they found to be surprisingly stable.[181] It withstands heating to 305°C without decomposition; the proton magnetic resonance spectrum nevertheless re-

$$\underset{\textbf{113}}{} \quad \overset{k_1}{\underset{k_{-1}}{\rightleftarrows}} \quad \underset{\textbf{114}}{} \qquad\qquad (11.128)$$

[176]D. A. Jencks and W. P. Jencks, *J. Am. Chem. Soc.*, **99**, 7948 (1977).

[177](a) E. Vogel, *Justus Liebigs Ann. Chem.*, **615**, 1 (1958); (b) C. A. Stewart, Jr., *J. Am. Chem. Soc.*, **94**, 635 (1972); (c) J. A. Berson, P. B. Dervan, R. Malherbe, and J. A. Jenkins, *J. Am. Chem. Soc.*, **98**, 5937 (1976).

[178]J. M. Brown, B. T. Golding, and J. J. Stofko, Jr., *J. Chem. Soc., Chem. Commun.*, 319 (1973).

[179]E. Vogel, *Angew. Chem. Int. Ed. Engl.*, **2**, 1 (1963).

[180]For reviews, see (a) G. Schröder, J. F. M. Oth, and R. Merényi, *Angew. Chem. Int. Ed. Engl.*, **4**, 752 (1965); (b) G. Schröder and J. F. M. Oth, *Angew. Chem. Int. Ed. Engl.*, **6**, 414 (1967); (c) L. T. Scott and M. Jones, Jr., *Chem. Rev.*, **72**, 181 (1972); (d) L. A. Paquette, *Angew. Chem. Int. Ed. Engl.*, **10**, 11 (1971); carbocations: (e) R. E. Leone and P. v. R. Schleyer, *Angew. Chem. Int. Ed. Engl.*, **9**, 860 (1970).

[181]W. v. E. Doering and W. R. Roth, *Angew. Chem. Int. Ed. Engl.*, **2**, 115 (1963).

veals that it is undergoing the degenerate Cope [3,3]-sigmatropic rearrangement indicated in Equation 11.128. The spectrum changes from that expected for **113** alone at $-50°C$ to that of the average of **113** and **114** at $180°C$. Rate constants $k_1 = k_{-1}$ estimated from the spectra are on the order of 10^3 sec^{-1} at $180°C$ and on the order of 1 sec^{-1} at $-50°C$.[182] An accompanying equilibrium ($113 \rightleftharpoons 115$) is avoided in the bridged structure **115**,[183] which rearranges to the identical tautomer rapidly on the proton magnetic resonance time scale down to $-90°C$. The rate constant is 1.2×10^3 sec^{-1} at $-55°C$, and the activation energy E_a is 8.1 kcal $mole^{-1}$.[184]

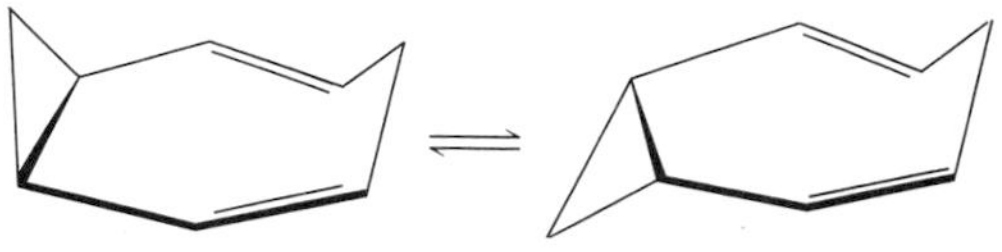

(11.129)

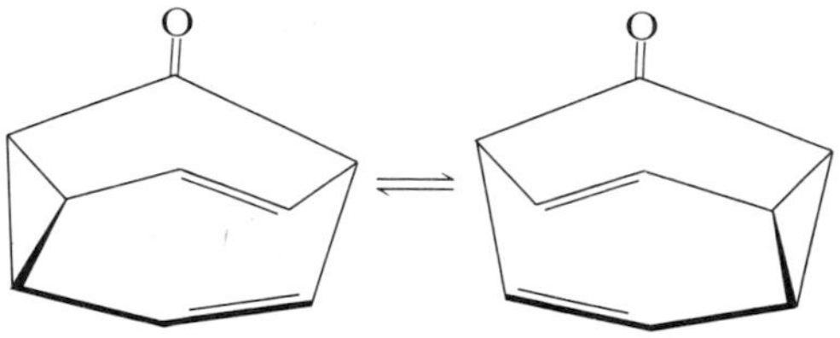

Doering and Roth, at the time of their first investigations of **113** in 1963, proposed that the structure **117**, which they named bullvalene, should undergo degenerate Cope rearrangements that would make each of the ten CH groups equivalent. Equation 11.130 illustrates just a few of these transformations.[185] If

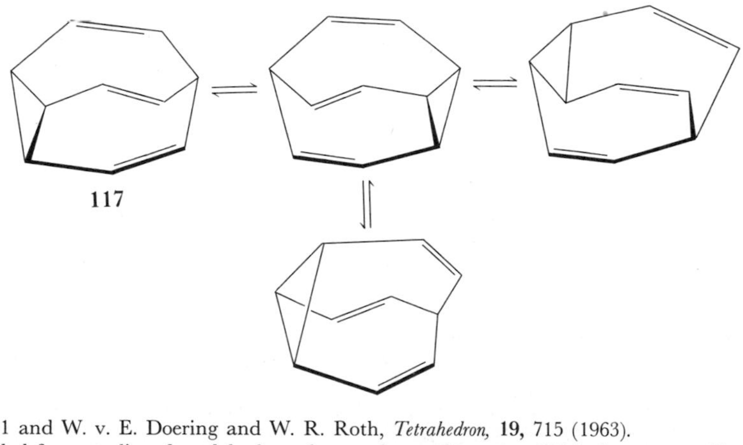

(11.130)

---

[182]See note 181 and W. v. E. Doering and W. R. Roth, *Tetrahedron,* **19,** 715 (1963).

[183]It is concluded from studies of models that whereas the conformation **115** is the more stable one, it does not undergo the [3,3]-rearrangement as readily as does **113**. See notes 181, 182.

[184]J. B. Lambert, *Tetrahedron Lett.,* 1901 (1963).

[185]The structures are drawn so as to make clear which bonds have shifted; close examination will reveal that all are identical.

each of the ten CH groups were individually labeled, there would be 10!/3, or 1,209,600, different ways of arranging them in the Structure **117**.[186] Doering and Roth pointed out that observation of a single line in the proton magnetic resonance spectrum would mean that all these structures were simultaneously present and rapidly interconverting. Later in the same year Schroder announced the synthesis of bullvalene and reported that the proton magnetic resonance spectrum at 100°C is indeed a single sharp line that broadens on cooling and divides into two bands of area ratio (high-field:low-field peak) 4:6 at −25°C.[187] The activation parameters have been determined by Oth, Schroder, and co-workers by ^{13}C nmr spectroscopy: $\Delta H^{\ddagger} = 12.4$ kcal mole^{-1}; $\Delta S^{\ddagger} = 0.8$ cal mole^{-1} K^{-1}.[188] Rearrangement rate is 10^3 at 0°C.[189] When one bridge is saturated (**118**), the rate is faster (about 3×10^5 sec^{-1} at 0°C), largely because of more favorable activation entropy ($\Delta H^{\ddagger} \approx 12$ kcal mole^{-1}, $\Delta S^{\ddagger} \approx +11$ cal mole^{-1} K^{-1}).[190] Decreasing the number of bridging carbons continues to increase the rate, **119** rearranging at $k \approx 3 \times 10^7$ sec^{-1} at 0°C ($E_a \approx 9$ kcal mole^{-1})[191] and **120** even more rapidly, $\Delta H^{\ddagger} = 4.8$ kcal mole^{-1}, $\Delta S^{\ddagger} = -5$ cal mole^{-1} K^{-1}, $\Delta G^{\ddagger}_{-140°C} = 5.5$ kcal mole^{-1}.[192]

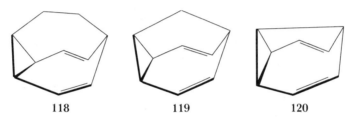

118   **119**   **120**

The Claisen Rearrangement[193]

A close analog of the Cope is the Claisen rearrangement. Historically, the Claisen, having been known and studied since 1912, is much the older of the two. It is a [3,3]-sigmatropic change (Equation 11.131), with oxygen as one atom of the chain.

---

[186]Division by the factor of three is required by the threefold symmetry axis. To each arrangement of CH groups there correspond two others, counted as separate elsewhere among the 10!, which are related by a threefold rotation and so are actually identical.

[187](a) G. Schröder, *Angew. Chem. Int. Ed. Engl.*, **2**, 481 (1963); (b) G. Schröder, *Chem. Ber.*, **97**, 3140 (1964); (c) R. Merényi, J. F. M. Oth, and G. Schröder, *Chem. Ber.*, **97**, 3150 (1964).

[188](a) J. F. M. Oth, K. Müllen, J.-M. Gilles, and G. Schröder, *Helv. Chim. Acta*, **57**, 1415 (1974). See also (b) H. Günther and J. Ulmen, *Tetrahedron*, **30**, 3781 (1974), who report slightly different values using the same method.

[189]M. Saunders, *Tetrahedron Lett.*, 1699 (1963), points out that measured rearrangement rates do not correspond to the rate at which a given structure rearranges to any arbitrarily chosen one of the 10^6 possibilities. Change from one structure to some other particular one will in general require a sequence of steps.

[190]G. Schröder and J. F. M. Oth, *Angew. Chem. Int. Ed. Engl.*, **6**, 414 (1967).

[191]W. v. E. Doering, B. M. Ferrier, E. T. Fossel, J. H. Hartenstein, M. Jones, Jr., G. Klumpp, R. M. Rubin, and M. Saunders, *Tetrahedron*, **23**, 3943 (1967).

[192](a) A. K. Cheng, F. A. L. Anet, J. Mioduski, and J. Meinwald, *J. Am. Chem. Soc.*, **96**, 2887 (1974); (b) R. Hoffmann and W. Stohrer, *J. Am. Chem. Soc.*, **93**, 6941 (1971), have considered theoretically ways to lower the activation energy still further.

[193]Reviews: (a) D. S. Tarbell, *Chem. Rev.*, **27**, 495 (1940); (b) D. S. Tarbell, *Org. Reactions*, **2**, 1 (1944); (c) A. Jefferson and F. Scheinmann, *Q. Rev. Chem. Soc.*, **22**, 391 (1968); (d) S. J. Rhoads and N. R. Raulins, *Org. Reactions*, **22**, 1 (1975); (e) F. E. Ziegler, *Acc. Chem. Res.*, **10**, 227 (1977).

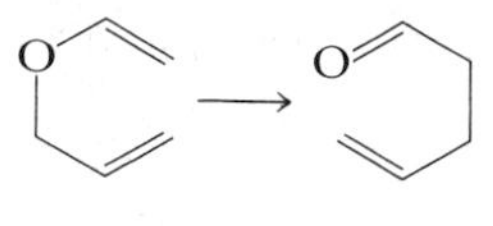

$$(11.131)$$

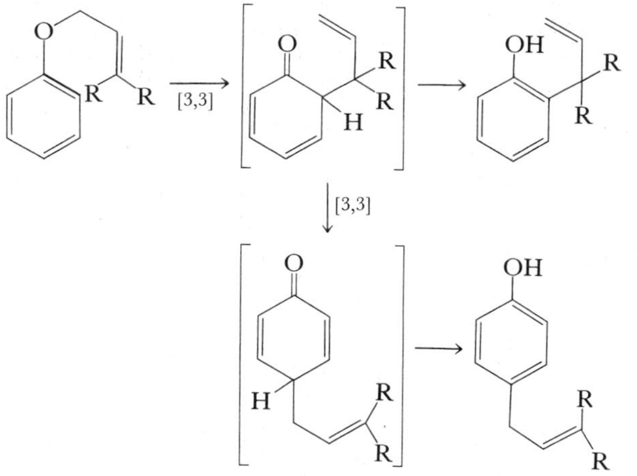

$$(11.132)$$

It is most commonly encountered with unsaturated phenyl ethers of the type illustrated in Equation 11.132, which rearrange to ortho- or parasubstituted phenols. The end of the chain remote from the oxygen becomes attached to the ortho position of the aromatic ring, inverting the allyl group. A second [3,3]-rearrangement, this one of the Cope type, may follow to leave the carbon originally attached to oxygen bonded to the para position. This para rearrangement is particularly likely to occur if the ortho positions are substituted so that enolization is blocked at the first stage. Negative activation entropies, on the order of -10 cal mole^{-1} K^{-1}, moderate activation enthalpies, on the order of 30 kcal mole^{-1},[194] and the allylic inversion accompanying each step indicate a concerted process. The reaction also occurs in aliphatic systems; experiments modeled after those of Doering and Roth on the Cope rearrangement have shown that the chair transition state is preferred in both aromatic and aliphatic Claisen rearrangements.[195]

The Ene Reaction[196]

The process shown in Equation 11.133, in the forward direction termed the *ene reaction* and in the reverse the *retro-ene reaction,* bears the same relationship to the [3,3]-sigmatropic reactions as does the [2,3]-elimination to the [2,3]-sigmatropic shift. It is also closely related to the [1,5]-rearrangement of hydrogen. The reaction

[194]S. J. Rhoads, in *Molecular Rearrangements,* Part I, P. de Mayo, Ed., Wiley, New York, 1963, p. 668.
[195](a) H. J. Hansen and H. Schmid, *Chem. Brit.,* **5,** 111 (1969); (b) G. Fráter, A. Habich, H.-J. Hansen, and H. Schmid, *Helv. Chim. Acta,* **52,** 335, 1156 (1969).
[196]Reviews: (a) H. M. R. Hoffmann, *Angew. Chem. Int. Ed. Engl.,* **8,** 556 (1969); (b) W. Oppolzer and V. Snieckus, *Angew. Chem. Int. Ed. Engl.,* **17,** 476 (1978).

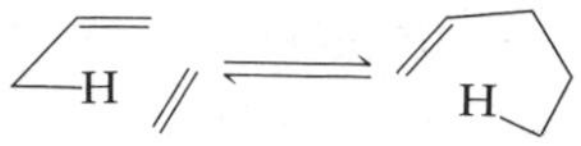

$$(11.133)$$

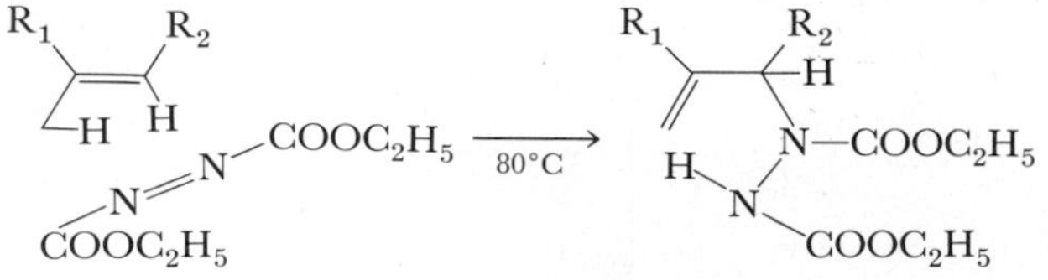

$$(11.134)[197]$$

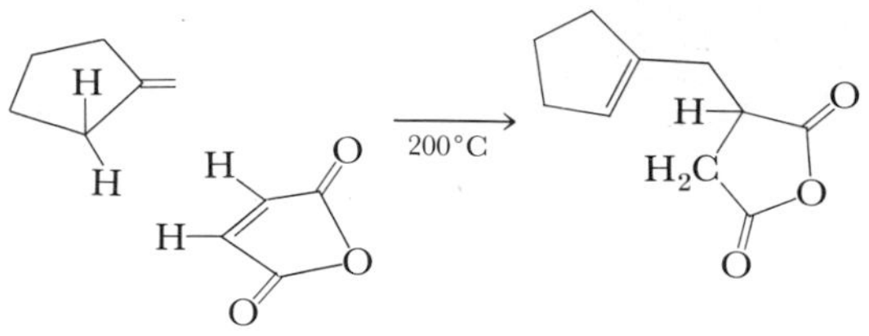

$$(11.135)[198]$$

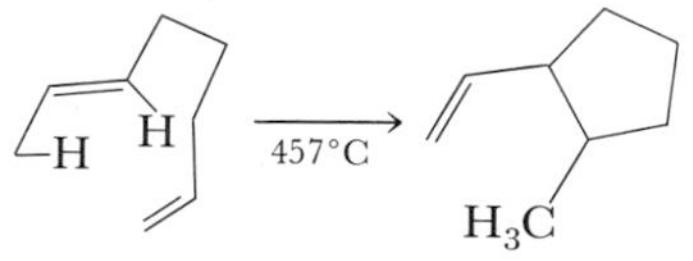

$$(11.136)[199]$$

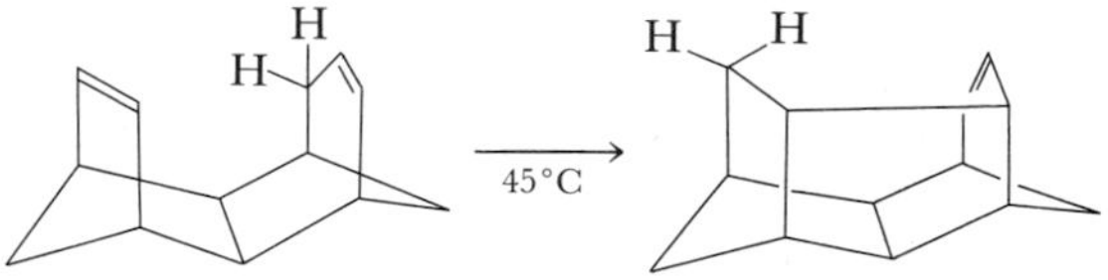

$$(11.137)[200]$$

$$t_{1/2} = 6 \text{ hr}$$
$$\Delta G^{\ddagger} = 26 \text{ kcal mole}^{-1}$$

can occur either intermolecularly, as in, for example, Equations 11.134 and 11.135, or intramolecularly, as in Equations 11.136 and 11.137. The temperature required varies widely depending upon the substitution pattern. The process is favored by electron-withdrawing substituents on the hydrogen acceptor (*enophile*), by geometrical constraints that hold the components in favorable relative positions (Equation 11.137), and by strain in the double bonds.

The ene reactions exhibit the highly negative entropies of activation (on the order of -30 cal mole^{-1} K^{-1} for intermolecular reaction[201]) and cis addition to the enophile expected for a concerted process.[202] There are nevertheless indications that a stepwise biradical path may be followed in some cases.[203]

[197] W. A. Thaler and B. Franzus, *J. Org. Chem.*, **29**, 2226 (1964).
[198] R. T. Arnold, R. W. Amidon, and R. M. Dodson, *J. Am. Chem. Soc.*, **72**, 2871 (1950).
[199] W. D. Huntsman, V. C. Solomon, and D. Eros, *J. Am. Chem. Soc.*, **80**, 5455 (1958).
[200] J. M. Brown, *J. Chem. Soc. B*, 868 (1969).
[201] B. Franzus, *J. Org. Chem.*, **28**, 2954 (1963).
[202] K. Alder and H. von Brachel, *Justus Liebigs Ann. Chem.*, **651** 141 (1962).
[203] H. M. R. Hoffmann, *Angew. Chem. Int. Ed. Engl.*, **8**, 556 (1969).

Higher-Order [*m,n*]-Sigmatropic Reactions

Relatively few sigmatropic reactions of order higher than [3,3] have been observed. Examples are shown in Equations 11.138–11.140.

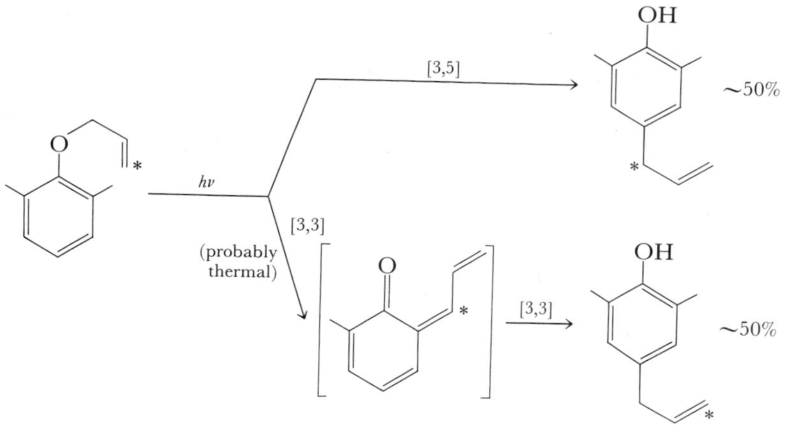

$$(11.138)^{204}$$

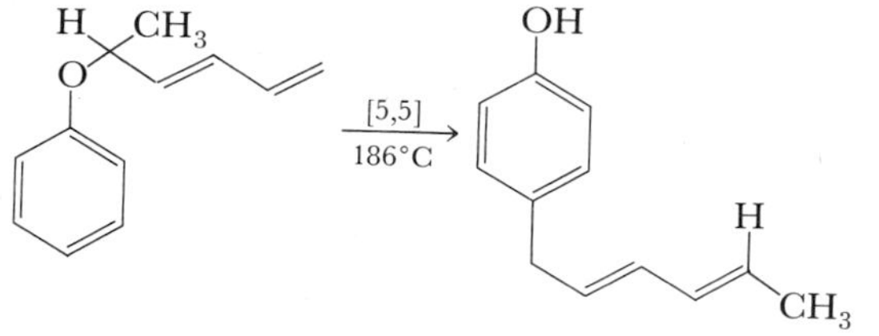

$$(11.139)^{205}$$

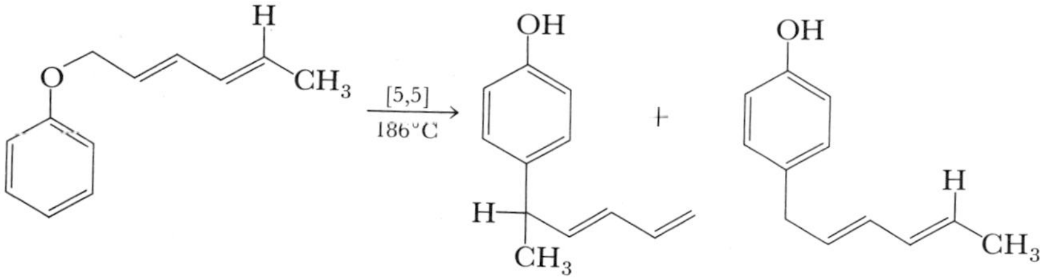

$$(11.140)^{205}$$

PROBLEMS

1. Propose a pathway for each of the following transformations:

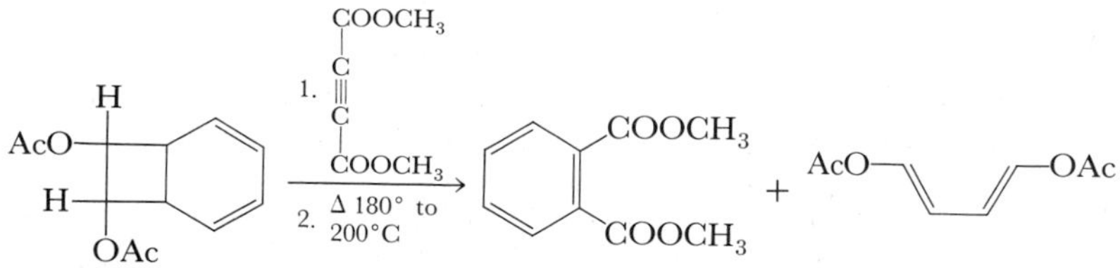

[204] K. Schmid and H. Schmid, *Helv. Chim. Acta*, **36**, 687 (1953).
[205] (a) G. Fráter and H. Schmid, *Helv. Chim. Acta*, **51**, 190 (1968); (b) G. Fráter and H. Schmid, *Helv. Chim. Acta*, **53**, 269 (1970).

(b)

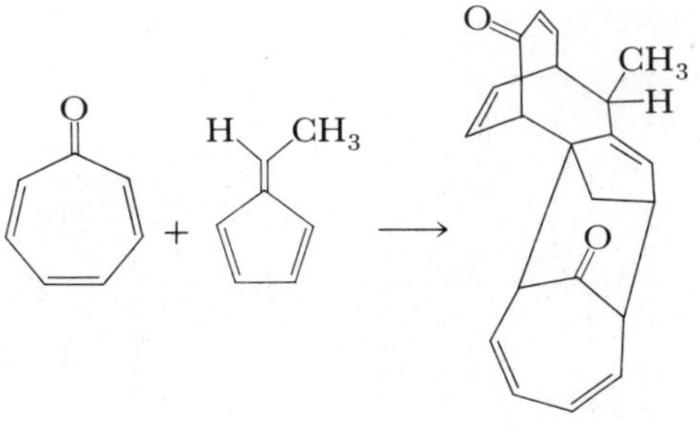

(c) $\xrightarrow{575°C}$ + $\begin{matrix} CH_2 \\ \| \\ CH_2 \end{matrix}$

$\xrightarrow[\text{Lindlar catalyst}]{H_2 \ 0°C}$ 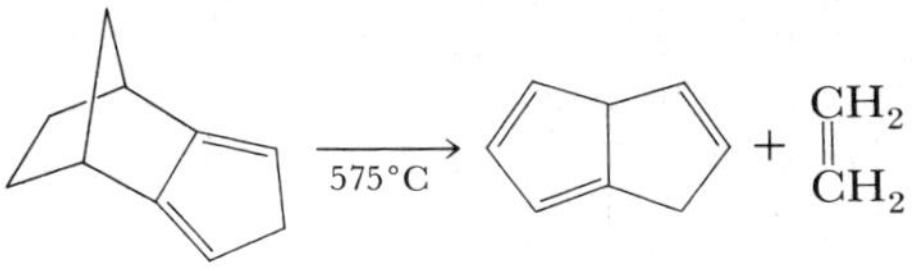 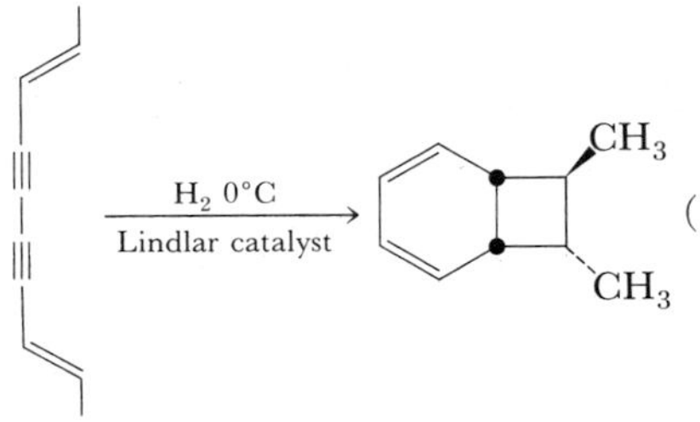 (+ other products)

(d) 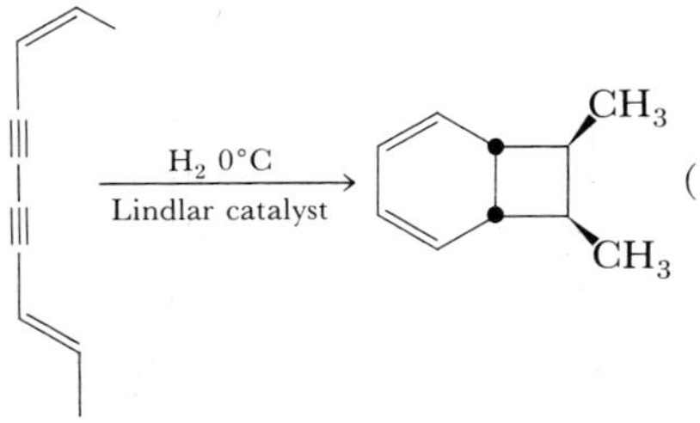 $\xrightarrow[\text{Lindlar catalyst}]{H_2 \ 0°C}$ (+ other products)

(e) 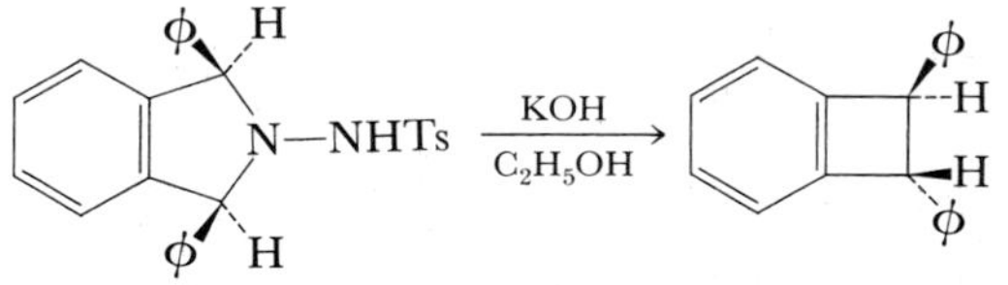 $\xrightarrow[\text{C}_2\text{H}_5\text{OH}]{\text{KOH}}$

(f) 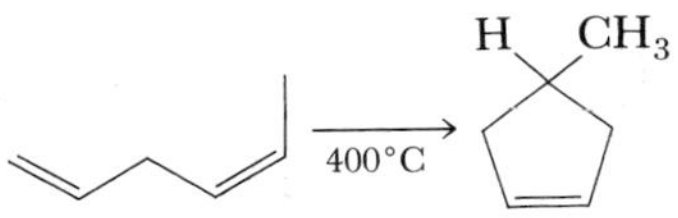 $\xrightarrow{400°C}$

(g) 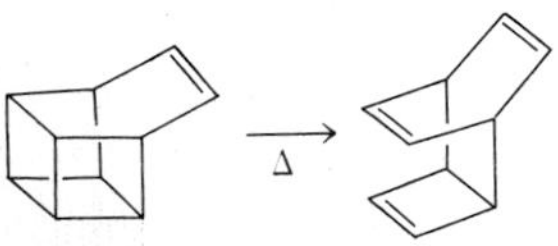 $\xrightarrow{\Delta}$

(h) 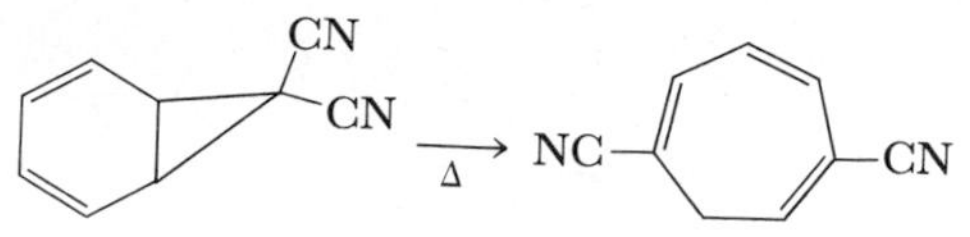 $\xrightarrow{\Delta}$

(i) 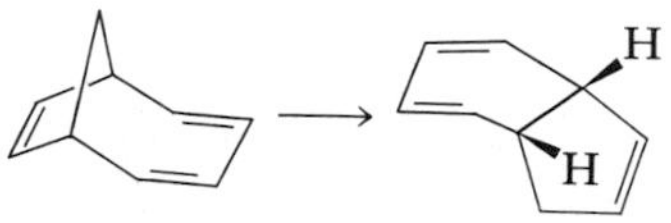 $\longrightarrow$

(j) 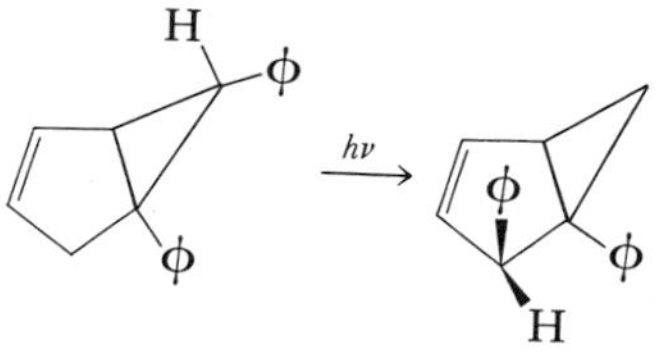 $\xrightarrow{h\nu}$

2. When cyclobutadiene is generated by oxidation of cyclobutadieneiron tricarbonyl, most of the product is a mixture of the dimers 1 and 2. Is this dimerization thermally allowed or forbidden, and which isomer is expected to predominate?

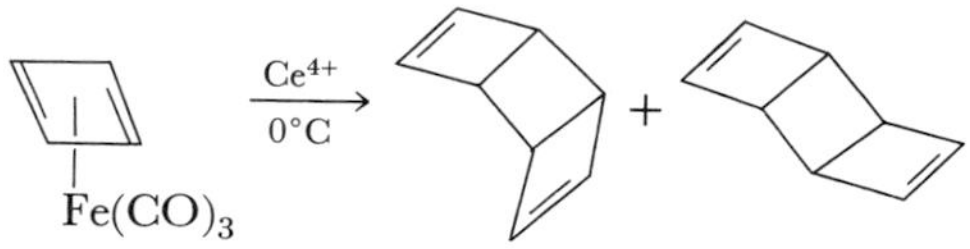

3. Explain the following reaction:

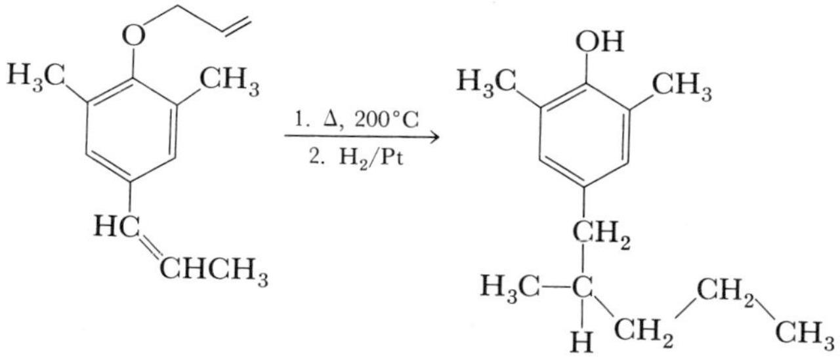

4. Explain why cis alkenes add 1,3-dipoles more slowly than trans alkenes, despite the fact that the cis ground states are higher in energy.

5. Explain the mechanistic significance of the product distributions shown:

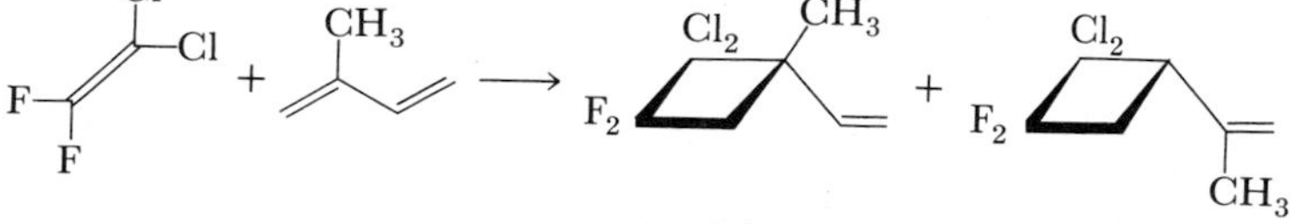

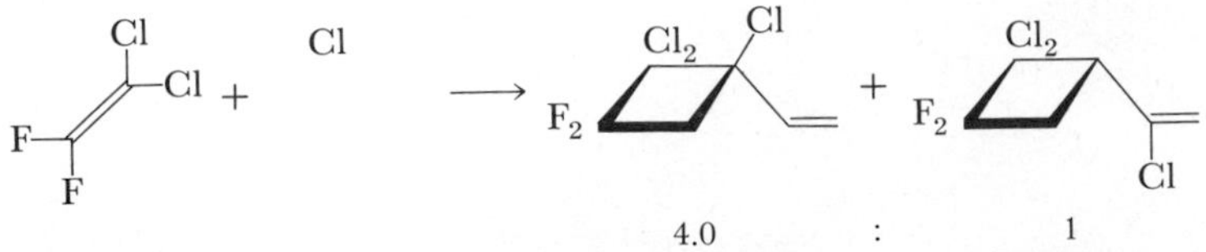

4.0   :   1

6. What would be the stereochemical consequence if the chiral molecule **3** were to undergo a [1,5]-sigmatropic walk-around rearrangement, (a) if it followed the pericyclic allowed path, and (b) if it followed the pericyclic forbidden path?

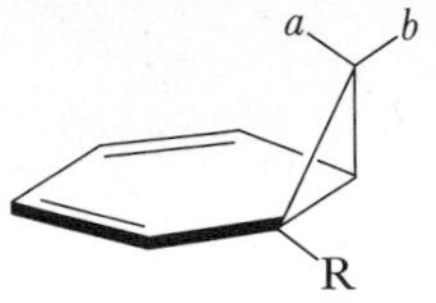

3

7. Propose a mechanism for the following rearrangement (* represents an ^{18}O label).

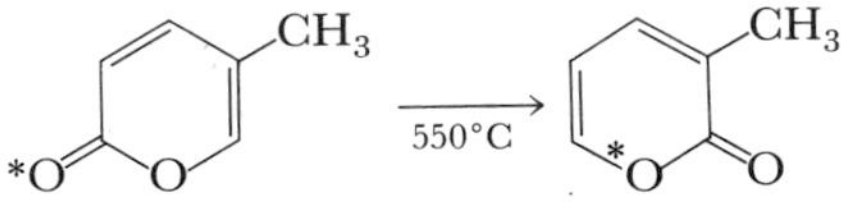

8. A conceivable pathway for conversion of Dewar benzene to benzene is shown below. Classify this process as a pericyclic reaction, predict whether it is thermally allowed, and propose an experiment to distinguish it from a simple disrotatory ring opening or a biradical cleavage of the central bond.

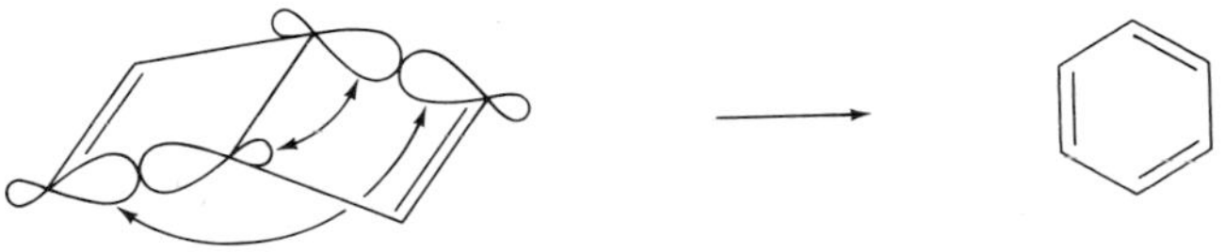

9. Propose a mechanism for the following transformation:

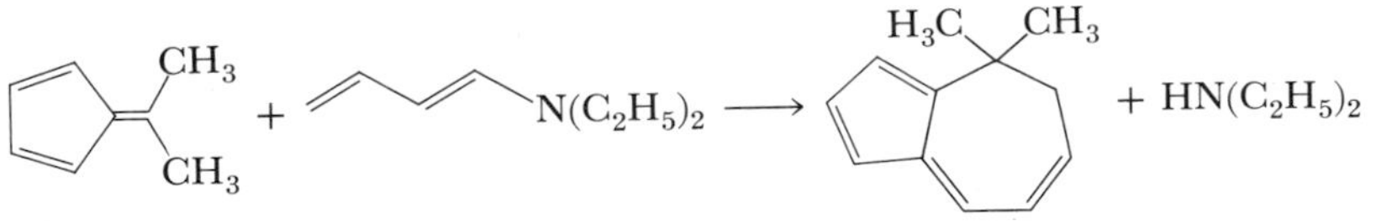

10. Specify a method of accomplishing the following transformation and propose the mechanism.

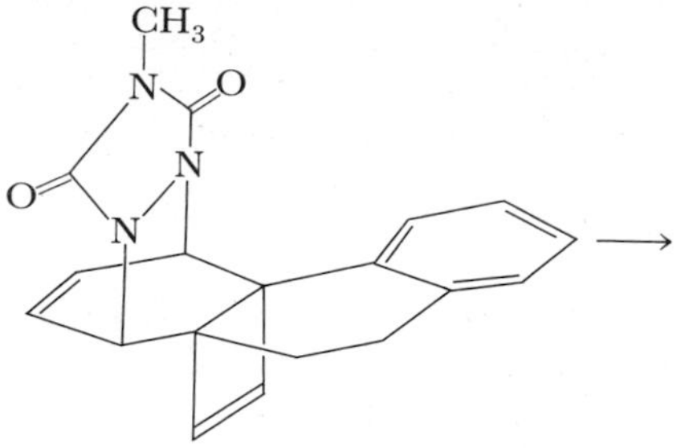

REFERENCES FOR PROBLEMS

1. (a) R. Criegee, W. Hörauf, and W. D. Schellenberg, *Chem. Ber.*, **86,** 126 (1953); (b) H. R. Pfaendler and H. Tanida, *Helv. Chim. Acta*, **56,** 543 (1973); (c) T. J. Katz, M. Rosenberger, and R. K. O'Hara, *J. Am. Chem. Soc.*, **86,** 249 (1964); (d) E. N. Marvell and J. Seubert, *J. Am. Chem. Soc.*, **89,** 3377 (1967); (e) L. A. Carpino, *Chem. Commun.*, 494 (1966); (f) W. R. Roth and J. König, *Justus Liebigs Ann. Chem.*, **688,** 28 (1965); (g) H. H. Westberg, E. N. Cain, and S. Masamune, *J. Am. Chem. Soc.*, **91,** 7512 (1969); **92,** 5291 (1970); (h) E. Ciganek, *J. Am. Chem. Soc.*, **89,** 1458 (1967); (i) J. A. Berson, R. R. Boettcher, and J. J. Vollmer, *J. Am. Chem. Soc.*, **93,** 1540 (1971); (j) H. E. Zimmerman and G. A. Epling, *J. Am. Chem. Soc.*, **94,** 3647 (1972).
2. L. Watts, J. D. Fitzpatrick, and R. Pettit, *J. Am. Chem. Soc.*, **88,** 623 (1966).
3. A. Nickon and B. R. Aaronoff, *J. Org. Chem.*, **29,** 3014 (1964).
4. R. Huisgen, R. Grashey, and J. Sauer, in *The Chemistry of Alkenes*, S. Patai, Ed., Wiley, London, 1964, p. 820.
5. P. D. Bartlett, L. K. Montgomery, and B. Seidel, *J. Am. Chem. Soc.*, **86,** 616 (1964).
6. J. A. Berson, *Acc. Chem. Res.*, **1,** 152 (1968).
7. W. H. Pirkle and W. V. Turner, *J. Org. Chem.*, **40,** 1617 (1975).
8. M. J. Goldsteir and R. S. Leight, *J. Am. Chem. Soc.*, **99,** 8112 (1977).
9. L. C. Dunn, Y.-M. Chang, and K. N. Houk, *J. Am. Chem. Soc.*, **98,** 7095 (1976).
10. L. A. Paquette, T. G. Wallis, K. Hirotsu, and J. Clardy, *J. Am. Chem. Soc.*, **99,** 2815 (1977).

Chapter 12
PHOTOCHEMISTRY[1]

In this chapter we shall present a simplified picture of photophysical processes and a glimpse of mechanistic organic photochemistry. More complete accounts of photochemistry, one of the newest and most intensely studied fields of organic chemistry, are found in the books and articles referred to in note 1.

12.1 LIGHT ABSORPTION

A full explanation of the properties of light requires both the wave theory of electromagnetic radiation and the quantum theory. Most photochemical processes are best understood in terms of the quantum theory, which says that light is made up of discrete particles called quanta or photons. Each quantum carries an amount of energy, $\mathscr{E}$, determined by the wavelength of the light, λ. Equation 12.1, in which

$$\mathscr{E} = \frac{hc}{\lambda} \tag{12.1}$$

[1](a) J. G. Calvert and J. N. Pitts, Jr., *Photochemistry,* Wiley, New York, 1966; (b) R. B. Cundall and A. Gilbert, *Photochemistry,* Thomas Nelson, London, 1970; (c) R. P. Wayne, *Photochemistry,* Butterworths, London, 1970; (d) W. A. Noyes, G. S. Hammond, and J. N. Pitts, Jr., Eds., *Advances in Photochemistry,* Vols. 1-—, Wiley, New York, 1963-—; (e) O. L. Chapman, Ed., *Organic Photochemistry,* Vols. 1-—, Dekker, New York, 1966-—; (f) R. S. Becker, *Theory and Interpretation of Fluorescence and Phosphorescence,* Wiley, New York, 1969; (g) D. R. Arnold, N. C. Baird, J. R. Bolton, J. C. D. Brand, P. W. M. Jacobs, P. de Mayo, and W. R. Ware, *Photochemistry: An Introduction,* Academic Press, New York, 1974; (h) J. P. Simons, *Photochemistry and Spectroscopy,* Wiley, New York, 1971; (i) J. Coxon and B. Halton, *Organic Photochemistry,* Cambridge University Press, New York, 1974; (j) D. O. Cowan and D. L. Drisko, *Elements of Organic Photochemistry,* Phenum, New York, 1976; (k) W. M. Horspool, *Aspects of Organic Photochemistry,* Academic Press, London, 1976; (l) J. A. Barltrop and J. D. Coyle, *Excited States in Organic Chemistry,* Wiley, New York, 1975; (m) N. J. Tiuro, *Modern Molecular Photochemistry,* W. A. Benjamin/Cummings, Menlo Park, Calif., 1978.

h is Planck's constant and c is the speed of light in a vacuum, gives the relationship between energy and wavelength. Equation 12.1 may also be written

$$\mathcal{E} = h\nu \tag{12.2}$$

where ν is the frequency of the light.

A molecule can absorb a quantum if and only if the energy of the quantum is the same as the energy difference between the state the molecule is in and a higher state of the molecule; absorption is accompanied by transition of the molecule to the higher state. According to the law of Stark and Einstein, a molecule absorbs a single quantum to bring about a single transition; thus transition to a first excited state cannot result, for example, from the absorption of two quanta, each of half the required energy. It is important to keep this in mind when considering the effect of light intensity. Increasing it increases the number of molecules that undergo a certain transition but does not alter the nature of the transition. A few exceptions to the Stark–Einstein law have been observed in which a single molecule (or atom) absorbs two or even three quanta simultaneously. These *bi-* or *triphotonic* absorptions require extraordinarily high light intensities and thus are sometimes found in laser photochemistry.

Depending on the wavelength, radiation may bring about a number of different kinds of molecular transitions. Infrared light can excite a molecule to a higher vibrational or rotational level. Very short wavelength ultraviolet light may actually bring about ionization by removal of an electron from the molecule. Light of intermediate energy (ultraviolet or visible) may promote a molecule from its ground state to an electronically excited state (see Chapter 10). Henceforward in this chapter, unless we specify otherwise, we shall designate an electronic state by the electronic configuration of lowest energy in that state.

In line with current practice in the literature, we shall mainly use the nanometer (nm) as the unit of wavelength. One nanometer is equal to 10^{-7} cm. Not long ago the millimicron ($m\mu$) was most frequently used by organic photochemists. One millimicron is equal to one nanometer. Other units often seen are the angstrom (Å), which is equal to 10^{-8} cm, and the wave number, $\bar{\nu}$, which is equal to $1/\lambda$ (the units of $\bar{\nu}$ therefore depend on the unit of λ). The region of the electromagnetic spectrum that brings about controlled electronic excitation and thus is of interest to photochemists is approximately from 200 to 700 nm (2000 to 7000 Å).

The energy, E, of 1 mole of quanta (i.e., of one *einstein*) is given by Equation 12.3, where N_0 is Avogadro's number. If one substitutes the appropriate numerical

$$E = \frac{N_0 hc}{\lambda} \tag{12.3}$$

values for N_0, h, and c, one obtains

$$E = \frac{2.86 \times 10^4}{\lambda} \text{ kcal einstein}^{-1} \tag{12.4}$$

where λ is in nanometers. An einstein of 200-nm light has an energy of 143 kcal, whereas an einstein of 700-nm light has an energy of 40 kcal. Carbon–carbon and carbon–hydrogen bond energies are near 100 kcal mole^{-1}. This is the amount of energy in one einstein of 286-nm light.

The ultraviolet and visible absorption spectra of uncombined atoms in a low-pressure gas consist of a number of sharp lines, which correspond to the promotion of an electron from one orbital to another. Each atomic orbital has a well-defined energy, and therefore only narrow-bandwidth light can be absorbed to effect the transition from one orbital to another. The ultraviolet and visible spectra of molecules are not made up of lines but, instead, of broad bands in which a number of peaks are often discernible. The broader absorption is due to the fact that within the ground state and within each excited state of the molecule there are a number of closely spaced vibrational levels (v), each of different energy and each of which may be populated. The transition of a molecule from the ground state to a certain excited state may therefore be accomplished by light that has a range of energy.

Figure 12.1 shows potential energy vs. internuclear distance curves for the ground state and first excited state of a diatomic molecule A—B. (Note that the energy minimum for the excited state is at greater internuclear distance than the energy minimum for the ground state. This is often the case. Excited states have antibonding electrons that force the nuclei apart.) The horizontal lines are the vibrational levels, and the curve on each horizontal line is the wave function for that vibrational level. The amplitude squared of the function at any internuclear distance corresponds to the probability that a molecule in that vibrational level of that state will have that internuclear distance. Quantum mechanical treatment shows that (1) the wave function of the lowest vibrational level of each state has no nodes; (2) each successive vibrational level has a function with one additional node; and (3) in the higher vibrational levels the amplitude of the function is greatest at the turning points, where the bond reaches its maximum extension or compression.

The excitation of a molecule from one electronic state to another happens much faster than a single molecular vibration (10^{-15} vs. 10^{-12} sec); excitation therefore occurs with almost no change in internuclear distance. The probability that a molecule will undergo a transition from the nth vibrational level of one state to the mth vibrational level of another state is proportional to the overlap of the wave functions of those vibrational levels at the internuclear distance the molecule has at the moment of absorption. For example, as Figure 12.1 shows, in the split second that a molecule of A—B in the lowest (0th) vibrational level of the ground state has internuclear distance x, that molecule cannot absorb a photon to be excited to the 0th vibrational level of the first excited state; the wave function of that excited state vanishes at x, the overlap is zero, and the probability for the transition is also zero. The molecule of A—B can absorb a higher-energy photon and be raised to the 10th vibrational level of the first excited state. Or a split second later, when that molecule has internuclear distance y, there is a small probability that it will be raised to the 0th level of the first excited state. The principle of *vertical transitions* demonstrated in Figure 12.1 was first recognized by Franck and later

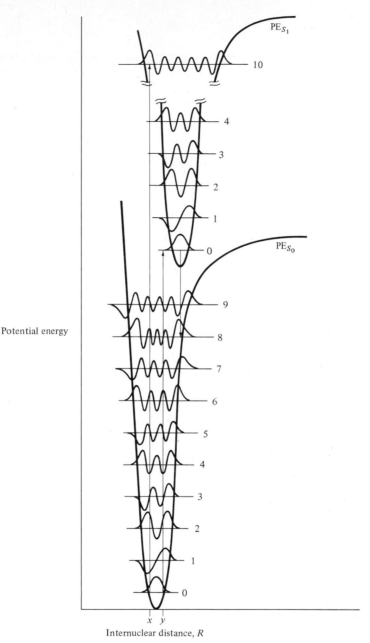

Figure 12.1 Vibrational energy levels and vibrational wave functions for the ground state and first excited state of a diatomic molecule. The two curves labeled PE_{S_0} and PE_{S_1} show the potential energy as a function of internuclear distance R for the two states. The short horizontal lines, numbered 0, 1, 2, . . . , define the vibrational levels of the various possible vibrational states. Superimposed on each of these vibrational energy-level lines is a graph of the vibrational wave function for that vibrational state; the horizontal axis for these wave function graphs is the internuclear distance R, the same as for the potential energy curves, but each graph has its own vertical axis, which corresponds not to energy but to vibrational probability amplitude. The vertical displacements of these graphs from each other, therefore, have no significance and are introduced only to assist in identifying the wave functions with their corresponding energy levels.

elaborated by Condon. It is therefore also known as the *Franck–Condon principle*. Above it has been developed for a diatomic molecule. However, exactly the same argument can be made for a polyatomic molecule. In that case Figure 12.1 represents the internuclear distance vs. potential energy curve for one vibrational mode of the molecule in the ground state and first excited state.

Most organic molecules in the ground state at room temperature are in the 0th vibrational level. To which vibrational level of an excited state they are promoted depends on the relative internuclear distances of the ground and excited states. If the ground and excited states have their energy minima at nearly the same internuclear distance, the probability of transition between the 0th level of the ground state and the 0th level of the excited state is high: The 0–0 absorption band will be strong. If the excited state is displaced to larger internuclear distances as in Figure 12.1, the 0–0 absorption will be weak. Organic chemists usually give the wavelength of maximum absorption when giving the position of an absorption band. It is necessary to keep in mind that this usually does not correspond to 0–0 absorption.

Saturated hydrocarbons do not absorb between 200 and 700 nm. At shorter wavelengths they do absorb to promote an electron from a σ to a σ* orbital. The excited molecule (with one electron still in the original level) is said to be in a σ, σ* state. The photochemistry of σ, σ* states is difficult to study, because all matter absorbs at these wavelengths. Furthermore, it is often not very interesting, since molecules in σ, σ* states fragment indiscriminately through σ bond cleavage. More interesting chemistry is found in the study of molecules that have multiple bonds and/or hetero atoms with nonbonding electrons.

The promotion of an electron from a π to a π* orbital in a simple olefin is also a high-energy process, but increasing conjugation greatly decreases the energy of the lowest transition. For example, ethylene has its lowest-energy λ_{max} at 174 nm, butadiene at 209 nm, and β-carotene (**1**, with 11 conjugated double

1

bonds) at 480 nm. Conjugated polyenes, of course, have several possible π, π* transitions. Carbonyl compounds have two fairly low-energy transitions available to them. The lower, in simple ketones, has λ_{max} at approximately 280 nm and corresponds to the promotion of a nonbonding electron on oxygen to the lowest unoccupied orbital to give an *n*, π* state. The other is promotion of a π electron to give a π, π* state. In unconjugated ketones a π, π* transition requires wavelengths less than 200 nm, but again conjugation shifts the lowest-energy π, π* absorption to longer wavelengths (*red shift*).

The amount of energy required for a certain transition is solvent-dependent. For example, polar solvents cause the n, π^* transition to be shifted to shorter wavelengths. It has been thought that these *blue shifts* are due to solvation of the n electrons in the ground state by the polar solvent, an interaction that would stabilize the ground state. By this hypothesis the excited state can be only partially stabilized, because the solvent molecules do not have time to reorient themselves during excitation. Thus more energy is required to bring about the transition. More recently, the single-mindedness of this view has been challenged by Haberfield. By measuring heats of solution he has calculated, for example, that in going from hexane to ethanol the ground state of acetone is stabilized, but only by 0.97 kcal mole^{-1}. The excited state, however, is destabilized by 2.49 kcal mole^{-1}. Thus most of the blue shift in this solvent change is caused by an increase in the energy of the excited state. In other solvents—for example, chloroform—the blue shift is due to stabilization of the ground state as previously supposed.[2] π, π^* transitions often undergo a red shift in polar solvents.[3]

To specify the occupied orbitals of an excited state does not describe the state fully. We must also specify whether or not all the electron spins in the molecule are paired. In the ground state of most molecules, the bonding orbitals are full and thus all electron spins are paired. After excitation of an electron to an empty orbital, the spins of the two electrons in the half-empty orbitals may be either paired or parallel. If all spins are paired, then the magnetic field generated by one electron is canceled by its partner and the net interaction with an external magnetic field is zero. Because there is only one way in which the electrons of such a species can interact with a magnetic field they are called singlets, and we say they have a *multiplicity* of one. Singlets are diamagnetic.

If the spins of the two electrons are unpaired, their magnetic fields may interact to (1) add together to augment an external magnetic field; (2) add together to diminish an external magnetic field; or (3) cancel each other. Such species are called triplets and are said to have a multiplicity of three. Triplets are of lower energy than the corresponding singlets and are paramagnetic.

The multiplicity, S, of a species is given by Equation 12.5, where s is the sum of the spin quantum numbers of all the electrons in the species. For example, if all spins are paired, $s = 0$ and $S = 1$. If all but two electrons are paired, $s = 1$ and

$$S = 2s + 1 \tag{12.5}$$

$S = 3$. To indicate that the electrons in the n and π^* orbitals of an n, π^* state are paired, we write $^1(n, \pi^*)$. To indicate that they are parallel, we write $^3(n, \pi^*)$. The ground-state singlet of a molecule is abbreviated S_0. The excited singlets are denoted $S_1, S_2, S_3, \ldots, S_n$, where the subscripts $1, 2, 3, \ldots, n$ refer to the first, second,

[2] P. Haberfield, *J. Am. Chem. Soc.*, **96**, 6526 (1974); see also P. Haberfield, D. Rosen, and I. Jasser, *J. Am. Chem. Soc.*, **101**, 3196 (1979), and references therein.

[3] See, for example, R. Rusakowicz, G. W. Byers, and P. A. Leermakers, *J. Am. Chem. Soc.*, **93**, 3263 (1971), and note 1(f).

third, . . . , nth excited singlets, respectively. Similarly, the first, second, third, . . . , nth excited triplets are denoted T_1, T_2, T_3, . . . , T_n. In acetone S_1 is $^1(n, \pi^*)$ and S_2 is $^1(\pi, \pi^*)$. Likewise, T_1 is $^3(n, \pi^*)$.

Not all transitions are equally probable. Improbable transitions are called, dramatically, *forbidden transitions*. Probable ones are called *allowed*. Whether a transition is forbidden or allowed can be predicted on the basis of rules called *selection rules*. The two most important selection rules for the organic photochemist are those that deal with spin-forbidden and space-forbidden transitions.

A transition is spin-forbidden if the multiplicity of the excited state is different from the multiplicity of the ground state. Another way of putting this is to say that during excitation ΔS must equal zero. Thus on absorption the allowed transitions of a ground-state singlet are to excited singlets and of a ground-state triplet (e.g., O_2) are to excited triplets.

A transition is space-forbidden if the orbitals involved occupy different regions in space. An n, π^* transition is forbidden because the greatest probability of finding an electron in an n orbital is around the nucleus—a region of space that corresponds to a node of a π^* orbital. On the other hand, π and π^* orbitals occupy overlapping regions of space and therefore π, π^* transitions are allowed.

Experimentally, we can determine the probability of a certain transition by measuring its *extinction coefficient*, ϵ. When a beam of monochromatic radiation passes through an absorbing system, the intensity of the transmitted beam, I_t, is given by the Beer–Lambert law (Equation 12.6), where I_0 is the intensity of the

$$I_t = I_0 10^{-\epsilon c d} \tag{12.6}$$

incident light, c is the concentration of the absorbing species, and d is the path length of the system. The value of ϵ depends, of course, on wavelength, and also on temperature and solvent, but not on concentration. The often-used logarithmic form of the Beer–Lambert law is shown in Equation 12.7, where A is the *absorbance*.

$$\log \frac{I_0}{I_t} = \epsilon c d = A \tag{12.7}$$

The allowed π, π^* transition in acetone has an extinction coefficient of 900 l mole^{-1} cm^{-1}, whereas the forbidden n, π^* transition has an ϵ of 15 l mole^{-1} cm^{-1}. (Note that forbidden transitions do occur.) The transition to the first excited triplet, $^3(n, \pi^*)$ is rarely observed.

Molecules that contain heavy atoms do not obey the spin-forbidden selection rule. For example, neat 1-chloronaphthalene has an extinction coefficient of $\sim 3 \times 10^{-4}$ l mole^{-1} cm^{-1} for the 0–0 band of its $S_0 \rightarrow T_1$ absorption, but 1-iodonaphthalene has an ϵ of 0.6 l mole^{-1} cm^{-1} for the same transition.[4] The explanation for the *heavy atom effect*[5] is that an actual molecule does not contain pure spin states: A singlet has a certain amount of triplet character and vice versa.

[4]S. P. McGlynn, R. Sunseri, and N. Christodouleas, *J. Chem. Phys.*, **37**, 1818 (1962).
[5]For a review of the heavy atom effect, see J. C. Koziar and D. O. Cowan, *Acc. Chem. Res.*, **11**, 334 (1978).

Mixing comes about through *spin–orbit coupling*. By the theory of relativity, an electron spinning around a nucleus may also be thought of as a nucleus spinning around an electron. The magnetic field produced by the spinning nucleus applies a torque on the electron and may cause it to "flip" (i.e., to change its spin quantum number from $+\frac{1}{2}$ to $-\frac{1}{2}$ or vice versa). The heavier the nucleus is, the greater is the magnetic field generated by it and therefore the more likely it is to cause an electron to flip. Thus for a molecule containing a heavy atom, a value cannot be assigned to S of Equation 12.5. Without a value for S, the stipulation that ΔS must equal zero during a transition has no meaning.

The heavy atom effect does not require that the heavy atom be within the molecule undergoing the transition. For example, when the two colorless liquids 1-chloronaphthalene and ethyl iodide are mixed, a yellow color develops. Spectroscopic examination shows that the color is due to an increase in $S_0 \rightarrow T_1$ transitions in chloronaphthalene.[6] Paramagnetic molecules also increase the probability of $S \rightarrow T$ transitions in neighboring molecules.

12.2 UNIMOLECULAR PHOTOPHYSICAL PROCESSES

Light absorption by a molecule usually produces an excited singlet, S_n, in an upper vibrational level. Normally the following processes follow excitation: Collision with solvent rapidly degrades S_n to its 0th vibrational level, that is, to S_n°; then if n is 2 or higher, this state rapidly undergoes an isoenergetic conversion to the next lowest singlet. The transformation of one excited state into another of the same multiplicity is called *internal conversion* and is shown schematically in the *Jablonski diagram* of Figure 12.2. Jablonski diagrams are one-dimensional diagrams that show the energies of various electronic states of a molecule and transitions between and within them. Processes such as *vibrational degradation* and internal conversion that occur without absorption or emission of light are called *radiationless transitions*[7] and are shown on Jablonski diagrams by wavy lines. Radiative transitions are shown by straight lines. The rapidity of internal conversion between two states depends inversely on the energy separation between them. If it is small, isoenergetic conversion of S_n° populates a low vibrational level of S_{n-1}. This is a transition that allows good overlap of the two wave functions. If ΔE is large, a high vibrational level of S_{n-1} must be populated. The oscillatory nature of the higher vibrational wave functions (see Figure 12.1) precludes good overlap of them with a 0th vibrational-level wave function. The energy gap between upper singlets is usually small, and therefore S_1° is produced very rapidly after excitation. ΔE between S_1 and S_0, however, is almost always large, and internal conversion between these states is slower. Thus it is usually from S_1° that most interesting photochemistry begins.[8]

[6] M. Kasha, *J. Chem. Phys.*, **20**, 71 (1952).

[7] K. F. Freed, *Acc. Chem. Res.*, **11, 74** (1978).

[8] For a review, see (a) N. J. Turro, V. Ramamurthy, W. Cherry, and W. Farneth, *Chem. Rev.*, **78**, 125 (1978); (b) M. A. Souto, J. Kolc, and J. Michl, *J. Am. Chem. Soc.*, **100**, 6692 (1978), and references therein. See also (c) K. Y. Law and P. de Mayo, *J. Am. Chem. Soc.*, **101**, 3251 (1979).

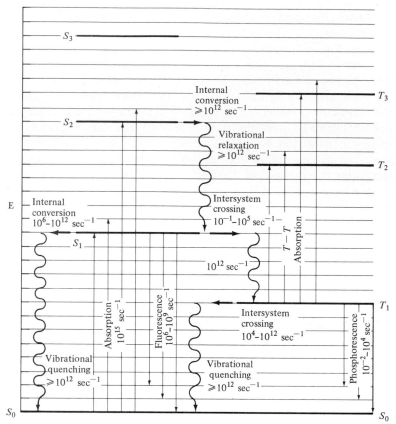

Figure 12.2 Jablonski diagram. Energy levels of excited states of a polyatomic molecule. The lowest vibrational energy levels of a state are indicated by thick horizontal lines; other horizontal lines represent associated vibrational levels. Vertical straight lines represent radiative transitions, wavy lines nonradiative transitions. The orders of magnitude of the first-order rate constants for the various processes are indicated. From R. B. Cundall and A. Gilbert, *Photochemistry,* Thomas Nelson, London, 1970. Reproduced by permission of Thomas Nelson and Sons Limited.

The three unimolecular physical processes that originate from S_1° are internal conversion to S_0, emission of light, and transformation of S_1 into T_1. If internal conversion occurs, the net change resulting from electronic excitation is heat transfer to the solvent. The other two processes are considerably more interesting.

The first excited singlet may return to the ground state by emitting light: $S_1 \rightarrow S_0 + h\nu$. Light emitted from S_1 is called *fluorescence*. The *natural fluorescent lifetime,* τ_f°, is the time required, after the light causing the excitation has been turned off, for the intensity of fluorescence to fall to $1/e$ of its initial value *in the absence of all competing processes*. It depends inversely on the probability of the $S_1 \rightleftharpoons S_0$ radiative transitions (if a transition is probable in one direction, it is also probable in the other); a very rough estimate of τ_f° can be obtained from Equation

12.8, where ϵ_{max} is the extinction coefficient of λ_{max} for the transition $S_0 \rightarrow S_1$. A

$$\tau_f^\circ = \frac{10^{-4}}{\epsilon_{max}} \, \text{sec} \tag{12.8}$$

substance with a highly probable $S_0 \rightarrow S_1$ transition ($\epsilon \approx 10^5 \, \text{l mole}^{-1} \, \text{sec}^{-1}$) will have τ_f° of 10^{-9} sec. A substance whose $S_0 \rightarrow S_1$ transition is space-forbidden ($\epsilon_{max} \approx 10^{-1} \, \text{l mole}^{-1} \, \text{sec}^{-1}$) will have a τ_f° of 10^{-3} sec; such long-lived direct fluorescence is only very rarely observed because competing processes destroy or *quench* S_1 before a quantum can be emitted.

Figure 12.3 shows the absorption and fluorescence spectra of anthracene in ethanol for the $S_0 \rightleftharpoons S_1$ transitions. Since the transitions are allowed, the intensity of light emitted is strong. The quantum yield for fluorescence, Φ_f, is defined by Equation 12.9:

$$\Phi_f = \frac{\text{no. of quanta emitted from } S_1}{\text{no. of quanta absorbed}} \tag{12.9}$$

Note that the absorption and emission spectra of anthracene are not superimposable. Only a single vibrational band overlaps; the rest of the fluorescence spectrum is at longer wavelengths than the absorption spectrum. The reason can be understood by considering Figure 12.1. The initial absorption populates the 0th and higher vibrational levels of S_1. There is a delay, after absorption, before fluorescence occurs, during which all the molecules decay by rapid vibrational relaxation

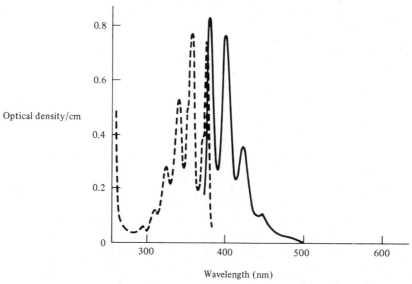

Figure 12.3 Absorption (dashed line; 17.2 µg/ml) and fluorescence (solid line; 1.0 µg/ml) spectra of anthracene in ethanol. From C. A. Parker and W. T. Rees, *The Analyst*, **85**, 587 (1960). Reproduced by permission of *The Analyst* and C. A. Parker.

to the 0th vibrational level of S_1. The actual process of emission is as rapid as absorption; therefore the Franck–Condon principle also applies to emission. The shortest-wavelength light that can be emitted, from the 0th vibrational level of S_1, corresponds to a 0–0 transition. Longer wavelengths are emitted when a molecule falls from the 0th level of S_1 to an upper vibrational level of the ground state. When 0–0 fluorescence has the same wavelength as 0–0 absorption, it is called *resonance fluorescence*. Resonance fluorescence is only sometimes observed. According to the Franck–Condon principle, the initially produced excited molecule must have exactly the same geometry and environment as its ground-state precursor had. If altering either the geometry of the excited molecule or the position of solvent molecules around it reduces the energy of the excited state, and if there is time for these changes to occur before the molecule emits, then 0–0 emission will be of longer wavelength than 0-0 absorption. The shift of emission to longer wavelengths is called a *Stokes shift*. In polar solvents resonance fluorescence of n, π^* transitions is rarely seen. Absorption occurs from a stabilized ground state to an excited state that is not solvated. Before emission can occur, the solvent molecules reorient themselves to stabilize the excited state, but the ground state produced by emission is not stabilized: The emitted light is of longer wavelengths than in nonpolar solvents.

Occasionally absorption occurs from a higher vibrational level of S_0. This leads to *anti-Stokes* lines, in which the fluorescence is at shorter wavelengths than the exciting radiation.

The distance between peaks in a fluorescence spectrum is a measure of the energy differences between vibrational levels in the ground state; likewise, the spacing of the peaks in the lowest-energy band of the absorption spectrum gives the energy differences between vibrational levels in S_1. If the spacings are similar, the absorption and emission spectra are mirror images of one another.

Fluorescence from the S_2 state of a molecule has, in some instances, been observed.[8,9] There appear to be at least two sets of circumstances that favor such emission. The first is when the rate of internal conversion from S_2 to S_1 is very slow. Slow conversion may be due to either very different geometries of, or a large energy gap between, the two states. The best-known example of this type is azulene fluorescence.[10] The absorption band leading to S_2 has a maximum at 340 nm and that leading to S_1 at 585 nm. The fluorescence spectrum has only one band centered at 374 nm. Furthermore the 0–0 bands of the S_2 absorption and of the emission overlap. Clearly the fluorescence originates from S_2. As in all cases when slow internal conversion from S_2 to S_1 is the cause of emission from S_2, the wavelengths and quantum yield of S_2 fluorescence from azulene depends on the wavelength of the absorbed light. S_2 fluorescence can be observed only if S_2 is populated. (This is contrary to the normal situation for fluorescence. When the excited state first

[9]C. F. Easterly and L. G. Christophorou, *J. Chem. Soc., Faraday Trans. 2*, **70**, 267 (1974), and references therein.
[10]G. Binsch, E. Heilbronner, R. Jankow, and D. Schmidt, *Chem. Phys. Lett.*, **1**, 135 (1967).

formed on absorption relaxes to S_1° with high efficiency, the fluorescence spectrum is independent of exciting wavelength.)

The second set of circumstances that favors fluorescence from S_2 is when the energy gap between S_2 and S_1 is very small compared to the S_1–S_0 gap. Then S_2 fluorescence can occur as a result of thermal repopulation of S_2 from S_1. In this case the fluorescence spectrum depends on the temperature but not on the wavelength of the exciting radiation. A number of polynuclear aromatics fluoresce from S_2 by this mechanism.[9]

Conversion of an excited state in the singlet manifold to one in the triplet manifold or vice versa is called *intersystem crossing*.[11] It is a forbidden process, and in the absence of all spin–orbit coupling would occur very slowly. However, when heavy atoms are present within the molecule or in the solvent, the rate is greatly increased. The n, π^* state of ketones usually undergoes intersystem crossing more rapidly than the π, π^* states of symmetric aromatic compounds.[12] This again is due in part to spin–orbit coupling and can be understood as follows. In the $^1(n, \pi^*)$ state **2a**, a ketone has a single electron in a nonbonding orbital on oxygen. This orbital can be assumed to be in the plane of the molecule and to be the p_x orbital. The difference in energy between this orbital and the (half-filled) p_y orbital on oxygen (perpendicular to the plane of the molecule) is very small. Since the probability of a transition is always inversely proportional to the energy gap between the two states, the probability of an electronic transition between p_x and p_y is high. With the transition comes a change in the electron's orbital angular momentum. But the total change in angular momentum of the electron, given by Equation 12.10 (in which $\Delta_{\text{tot ang mom}}$ is the total change of the electron's angular momentum, $\Delta_{\text{orb ang mom}}$ is the change in the orbital angular momentum and $\Delta_{\text{spin ang mom}}$ is the change in the spin angular momentum), must be zero in the transition.

$$\Delta_{\text{tot ang mom}} = \Delta_{\text{orb ang mom}} + \Delta_{\text{spin ang mom}} = 0 \tag{12.10}$$

Thus the change in orbital angular momentum must be accompanied by a change in spin angular momentum and the $^3(\pi, \pi^*)$ state **2b** is formed. From thence internal conversion may produce the $^3(n, \pi^*)$ state. In the $^1(\pi, \pi^*)$ states of aromatic hydrocarbons an electron might move from one p_y orbital to the other as is shown

$$\text{(12.11)}$$

2a **2b**

[11] S. K. Lower and M. A. El-Sayed, *Chem. Rev.*, **66**, 199 (1966).
[12] See T. Azumi, *Chem. Phys. Lett.*, **17**, 211 (1972), and references therein.

in Equation 12.12, but this change produces no change in orbital angular momentum and thus there can be no change in spin angular momentum. In the example of spin-orbit coupling shown above, the angle between the orbitals involved in the coupling is 90°. Much smaller angles may also result in coupling albeit less efficiently.

Another reason for the more probable intersystem crossing between (n, π^*) states than between (π, π^*) states is that the energy gap between S_1 and T_1 (called the *singlet–triplet splitting*) is smaller for (n, π^*) states than for (π, π^*) states.

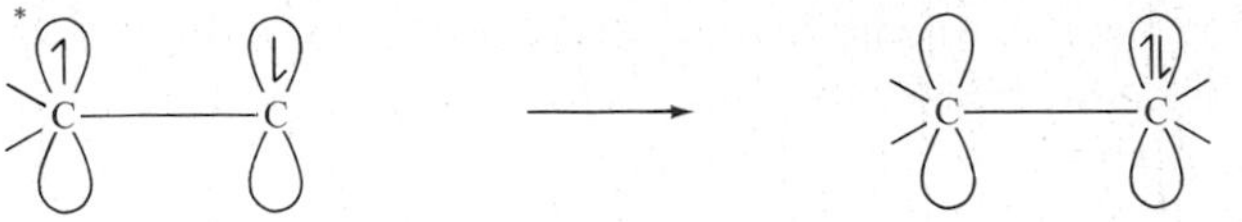

$$ \tag{12.12} $$

Once an upper vibrational level of T_1 has been produced, it rapidly undergoes vibrational degradation to T_1°. This state has three unimolecular physical processes available to it. It may undergo intersystem crossing to S_0; it may undergo intersystem crossing back to S_1; or it may return to S_0 by emitting light called *phosphorescence*. All three processes are spin-forbidden, and therefore at low temperatures the lifetime of T_1° is long. Increasing temperature decreases the lifetime of T_1, mainly by increasing the rate of intersystem crossing to S_0. The reason for the temperature dependence and the exact mechanism of intersystem crossing are not well understood.[13]

Since vibrational degradation, $T_1^v \longrightarrow T_1^\circ$ is very fast ($k \approx 10^{12} \text{ sec}^{-1}$)[10] and since T_1° is always of lower energy than S_1°, intersystem crossing from T_1 back to S_1 requires that T_1° first be reexcited to a higher vibrational level. That this does occur has been shown by emission studies. For example, Saltiel has found that benzophenone emits, along with its long-wavelength phosphorescence, shorter-wavelength light, the energy of which corresponds to the benzophenone $S_1 \rightarrow S_0$ transition. The lifetime of the shorter-wavelength emission is much longer than usual fluorescence lifetimes. Moreover, the relative intensities of the long- and short-wavelength emissions is temperature-dependent, the intensity of the latter increasing with increasing temperature. These observations are consistent with thermally activated repopulation of S_1 from T_1 and the consequent emission of *delayed fluorescence*.[14] (The emission just described is called *E-type* delayed fluorescence. We shall discuss another type of long-lived emission from S_1, called *P-type* delayed fluorescence, below.)

Because of the forbiddenness of the transition, $T_1 \rightarrow S_0 + h\nu$, the natural phosphorescent lifetime, τ_p°, of a triplet state is long—from approximately 10^{-3} sec for an n, π^* triplet to 30 sec for a π, π^* aromatic triplet. At room temperature in

[13] P. J. Wagner and G. S. Hammond, in *Advances in Photochemistry,* Vol. 5, W. A. Noyes, G. S. Hammond, and J. N. Pitts, Eds., Wiley, New York, 1968, p. 21.

[14] J. Saltiel, H. C. Curtis, L. Metts, J. W. Miley, J. Winterle, and M. Wrighton, *J. Am. Chem. Soc.,* **92,** 410 (1970).

solution, phosphorescence is often not observed because intersystem crossing of T_1 to S_0 and quenching of T_1 by impurities and molecular O_2 (see below) competes effectively with phosphorescence. Therefore most phosphorescence studies must be carried out at low temperatures in carefully purified, outgassed, rigid media. Under these conditions the quantum yield of phosphorescence, Φ_p, defined by Equation 12.13, is often high and approaches 1.0 for some aromatic carbonyls.

$$\Phi_p = \frac{\text{no. of quanta emitted from } T_1}{\text{no. of quanta absorbed}} \qquad (12.13)$$

The effect of solvent polarity on the energies of excited states leads to some interesting phenomena. For example, changing the solvent in which ketone 3 is dissolved from isopentane to ethanol increases the phosphorescence lifetime from

3

4.7×10^{-3} to 4.0×10^{-2} sec, but does not much alter the shape of the phosphorescence spectrum. Apparently, in nonpolar media the $^3(n, \pi^*)$ state of phenyl alkyl ketones lies only about 1 kcal below the $^3(\pi, \pi^*)$ state. Change to a polar solvent inverts the order: $^3(\pi, \pi^*)$ becomes T_1 and the triplet lifetime is lengthened. However, the states are so close in energy that even at 77 K they are in thermal equilibrium, with about 10 percent of the triplets occupying the higher, faster-emitting n, π^* state at any moment. This state is then still the chief phosphorescent species. Analogous results are obtained with other phenyl alkyl ketones.[15] Note that the solvent-induced increase in phosphorescent lifetime is a phenomenon similar to E-type delayed fluorescence. In both cases a longer-lived, lower-energy state acts as a reservoir for a higher-energy, faster-emitting state.

Several authors have reported that in polar solvents the overall phosphorescence decay of some phenyl alkyl ketones has a long- and a short-lived component; they attribute this to simultaneous emission from $^3(n, \pi^*)$ and $^3(\pi, \pi^*)$ states that are not in equilibrium with each other. This interpretation assumes that phosphorescence, a spin-forbidden process, occurs more rapidly than internal conversion from T_2 to T_1 and therefore seems improbable. It is more likely that one of the phosphorescent species is a photochemical product of the original ketone.[16,17]

An electronically excited molecule may, under some conditions, absorb another quantum and be raised to a higher excited state. Usually the population of excited species is so low that the probability of this occurrence is very slight. Michl, however, has shown that the T_1 state of **4a**, in a rigid matrix at 77 K, absorbs a

[15] P. J. Wagner, M. J. May, A. Haug, and D. R. Graber, *J. Am. Chem. Soc.*, **92**, 5269 (1970).

[16] N. Y. C. Chu and D. R. Kearns, *J. Am. Chem. Soc.*, **94**, 2619 (1972), and references therein.

[17] For cases where emission from two different triplet states has been suggested, see (a) W. Klöpffer, *Chem. Phys. Lett.*, **11**, 482 (1971); (b) P. de Mayo, *Acc. Chem. Res.*, **4**, 41 (1971), and references therein; (c) E. Migirdicyan, *Chem. Phys. Lett.*, **12**, 473 (1972).

second photon and from the state resulting from that absorption rearranges to **4b**.[18] In recent years the technique of flash photolysis has been developed which allows us

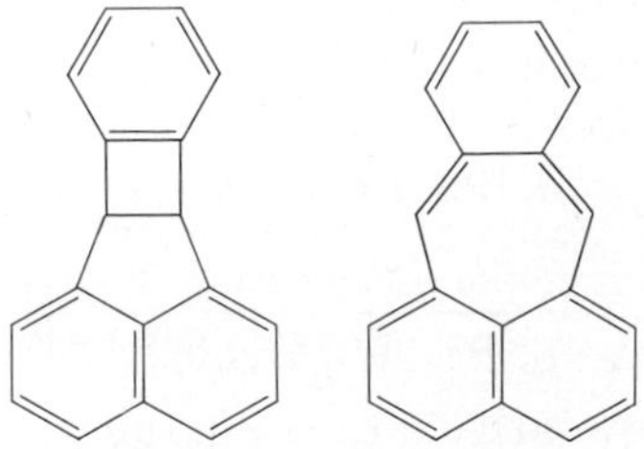

4a   4b

to investigate the absorption properties of excited states. An extremely high intensity laser, which has approximately one million times the power of a conventional spectroscopic lamp, is turned on for a tiny fraction of a second, and a large population of excited species is produced. Immediately after this *photolysis flash* is turned off, a low-power *spectroscopic flash* may be turned on and the absorption spectrum of the already excited system determined. By varying the delay between photolysis and spectroscopic flashes, much can be learned about the absorption and lifetime of singlet and triplet excited states.

12.3 BIMOLECULAR PHOTOPHYSICAL PROCESSES

The intermolecular transfer of electronic excitation energy is a common phenomenon in photochemistry. It is called *photosensitization* and may occur by a number of mechanisms, both radiative and nonradiative. In the radiative process, also called the "trivial" mechanism, the acceptor, A, absorbs a quantum emitted by a donor, D (Equations 12.14 and 12.15).

$$D^* \longrightarrow D_0 + h\nu \qquad (12.14)$$
$$A_0 + h\nu \longrightarrow A^* \qquad (12.15)$$

Nonradiative energy transfer may occur over long or short range. The long-range transfer (≈ 50 Å) occurs by coulombic interactions between donor and acceptor. The transfer must be nearly isoenergetic and usually involves transfer of singlet excitation energy. Short-range transfer occurs only when two molecules collide, and it is this type of transfer that, because of its usefulness, is of enormous interest to photochemists.

Collisional Energy Transfer (Exchange Energy Transfer)
When an electronically excited molecule collides with a ground-state molecule that has an excited state of lower energy, the molecules may come out of the encounter with the electronic excitation energy transferred from the donor to the acceptor.

[18] J. Michl, A. Castellan, M. A. Souto, and J. Kolc, *Jerusalem Symp. Quantum Chem. Biochem,* **10,** 361 (1977).

The spin states (multiplicities) of the donor and acceptor may change during the encounter, but they do so only according to *Wigner's spin conservation rule*. To understand the implications of the rule, it is simplest to think that during the encounter the spin quantum number of each individual electron is fixed but that one electron can move from donor to acceptor if it is replaced by another electron that moves from acceptor to donor.

Thus, for example, an excited triplet donor (D*, ↑↑) could *sensitize* a ground-state singlet acceptor (A$_0$, ↑↓) according to Equation 12.16 or 12.17, depending on

$$D^* + A_0 \longrightarrow D_0 + A^* \qquad (12.16)$$
$$\text{(↑↑) (↑↓)} \qquad \text{(↑↓) (↑↑)}$$

$$D^* + A_0 \longrightarrow D_0 + A^* \qquad (12.17)$$
$$\text{(↑↑) (↑↓)} \qquad \text{(↑↑) (↑↓)}$$

whether the donor is a ground-state singlet or a ground-state triplet. Since most molecules are ground-state singlets, sensitization by a triplet donor usually produces an excited triplet acceptor as in Equation 12.16. This is the process that we spoke of above as being enormously useful.[19] By this method we can produce a vast number of triplets that are not obtainable by intersystem crossing from their singlets. Triplet energy transfer was first observed by Terenin and Ermolaev in rigid media at 77°K.[20] They found, for example, that when mixtures of benzophenone and naphthalene are irradiated at wavelengths where only benzophenone absorbs, the phosphorescence of benzophenone is partially quenched but naphthalene phosphoresces from its T_1 state and does so more strongly than when naphthalene is irradiated directly. The absorption and phosphorescence spectra of the components are shown in Figures 12.4(a) and 12.4(b). As the absorption spectra show, S_1 of naphthalene lies above S_1 of benzophenone; but the T_1 state of naphthalene is lower than the T_1 state of benzophenone. The actual energies are shown in Figure 12.5. Direct irradiation of naphthalene leads to rather inefficient intersystem crossing because of the large S_1–T_1 splitting. When a naphthalene–benzophenone mixture is irradiated with wavelengths too long to excite naphthalene to S_1 but short enough to excite benzophenone to S_1, a large number of benzophenone triplets are produced (benzophenone has a quantum yield for intersystem crossing of approximately 1.0). Once in the T_1 state, benzophenone can transfer its electronic excitation energy to the T_1 state of naphthalene in an exothermic process and does so very efficiently.

Triplet energy transfer also occurs in solution at room temperature. Because most compounds phosphoresce either weakly or not at all under these conditions, energy transfer is often followed by watching the change in reactivity of the acceptor rather than its phosphorescence (see examples in Section 12.4). The rate of

[19] Energy transfer from an excited singlet may also occur [see, for example, P. S. Engel, L. D. Fogel, and C. Steel, *J. Am. Chem. Soc.*, **96**, 327 (1974)], but it is less efficient because of the short lifetime of an excited singlet. It is also less useful, since the product resulting from energy transfer is available by direct excitation.
[20] A. Terenin and V. Ermolaev, *J. Chem. Soc., Faraday Trans. 2*, **52**, 1042 (1956).

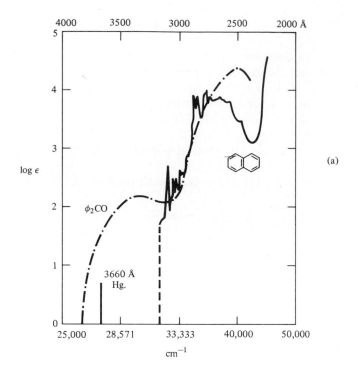

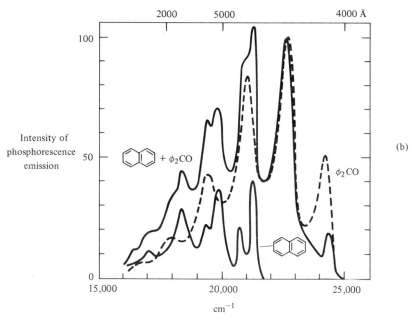

Figure 12.4 (a) Absorption spectra of benzophenone (ethanol, 20°C) and naphthalene (ethanol + methanol, −180°C). (b) Phosphorescence emission spectra at −190°C in ether + ethanol, under steady irradiation at 366 nm. Benzophenone, $2 \times 10^{-2}M$; benzophenone + naphthalene, $2 \times 10^{-2}M$ and $3.2 \times 10^{-1}M$, respectively; concentration of pure naphthalene (solid line) not known. From A. Terenin and V. Ermolaev, *Trans. Faraday Soc.*, **52**, 1042 (1956). Reproduced by permission of the Faraday Society.

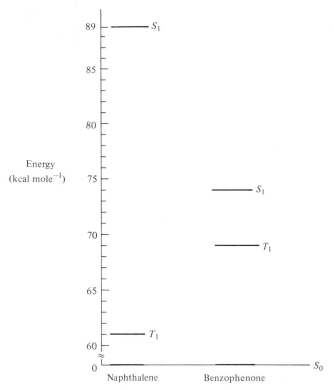

Figure 12.5 Energy levels of the first excited singlets and the first excited triplets of benzophenone and naphthalene.

energy transfer, exothermic by 2 or 3 kcal mole^{-1} or more, is very fast. In viscous solvents the rate is diffusion-controlled, but in very nonviscous solvents such as pentane, the rate is slower than diffusion-controlled.[21] The apparent reason for the dichotomy is that in the more viscous solvents every encounter between donor and acceptor involves a large number of collisions and thus energy transfer, even much less than 100 percent efficient, can take place before the donor and acceptor free themselves from the solvent cage. In nonviscous solvents the donor and acceptor may diffuse apart after only a small number of collisions, before energy transfer has taken place.[22]

Sensitization experiments again provide evidence that occasionally a second excited state may undergo a process other than internal conversion. For example, when dibenzofuran crystals, containing 0.1 percent anthracene and 1 percent naphthalene-d_8 impurities, are irradiated with light that only excites anthracene, naphthalene phosphoresces *with 20 times greater intensity* than it does in the absence

[21]N. J. Turro, N. E. Schore, H.-C. Steinmetzer, and A. Yekta, *J. Am. Chem. Soc.,* **96,** 1936 (1974).
[22]P. J. Wagner and I. Kochevar, *J. Am. Chem. Soc.,* **90,** 2232 (1968).

of dibenzofuran. The relative energies of the relevant excited states are given in Figure 12.6. What apparently happens is that the S_1 state of anthracene undergoes intersystem crossing to its T_2 state. A double energy transfer then occurs, first from T_2 of anthracene to T_1 of dibenzofuran and thence to T_1 of naphthalene.[23]

The fact that one molecule can quench the excited state of another by energy transfer (or by other means—see below) enables us to calculate the lifetime of the excited state in the absence of quencher. As an example we will consider a molecule that is excited to S_1, undergoes intersystem crossing to T_1, and from T_1 either gives products or is quenched. The individual steps and their rates are given in Equations 12.18–12.21.

	Rate forward	Rate reverse	
$h\nu + \text{M}_0 \underset{k_d}{\rightleftharpoons} {}^1\text{M}^*$	I_{abs}	$k_d[{}^1\text{M}^*]$	(12.18)
${}^1\text{M}^* \xrightarrow{k_{ISC}} {}^3\text{M}^*$	$k_{ISC}[{}^1\text{M}^*]$	—	(12.19)
${}^3\text{M}^* \xrightarrow{k_r} \text{product}$	$k_r[{}^3\text{M}^*]$	—	(12.20)
$\text{Q} + {}^3\text{M}^* \xrightarrow{k_q} \text{M}_0 + {}^3\text{Q}^*$	$k_q[\text{Q}_0][{}^3\text{M}^*]$	—	(12.21)

M is the photoreactant and Q is the quencher. Note that the rate of formation of ${}^1\text{M}^*$ is given by the intensity of the absorbed light. Although this may seem strange at first glance, remember that intensity is itself a rate—usually expressed in einsteins $\sec^{-1}$. Each einstein absorbed gives one mole of ${}^1\text{M}^*$.

The rate of formation of product is given by Equation 12.22.

$$\frac{d(\text{product})}{dt} = k_r[{}^3\text{M}^*] \tag{12.22}$$

Using the steady-state approximation for $[{}^3\text{M}^*]$ (and in the process also for $[{}^1\text{M}^*]$), we obtain

$$\frac{d(\text{product})}{dt} = \frac{k_r k_{ISC} I_{abs}}{(k_q[\text{Q}] + k_r)(k_d + k_{ISC})} \tag{12.23}$$

Since

$$\Phi_{\text{product}} = \frac{\text{moles product}}{\text{einsteins absorbed}} \tag{12.24}$$

$$= \frac{\text{moles product } \sec^{-1}}{\text{einsteins absorbed } \sec^{-1}}$$

one can divide Equation 12.23 by the rate of light absorption to obtain

$$\Phi_{\text{product}} = \frac{k_r k_{ISC}}{(k_q[\text{Q}] + k_r)(k_d + k_{ISC})} \tag{12.25}$$

---

[23] R. S. H. Liu and R. E. Kellogg, *J. Am. Chem. Soc.*, **91**, 250 (1969); see also C. C. Ladwig and R. S. H. Liu, *J. Am. Chem. Soc.*, **96**, 6210 (1974), and references therein.

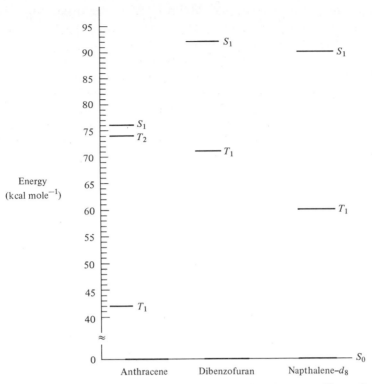

Figure 12.6 Energy levels of the important low-lying states of anthracene, dibenzofuran, and naphthalene-d_8. Reprinted with permission from R. S. H. Liu and R. E. Kellogg, *J. Am. Chem. Soc.*, **91**, 250 (1969). Copyright 1969 American Chemical Society.

In the absence of quencher Equation 12.25 reduces to

$$\Phi^\circ_{product} = \frac{k_{ISC}}{k_d + k_{ISC}} \tag{12.26}$$

where $\Phi^\circ_{product}$ is the quantum yield of photoproduct in the absence of quencher. Dividing Equation 12.26 by Equation 12.25, we obtain

$$\frac{\Phi^\circ_{product}}{\Phi_{product}} = \frac{(k_q[Q] + k_r)}{k_r} = 1 + \left(\frac{k_q}{k_r}\right)[Q] \tag{12.27}$$

If product formation is all that $^3M^*$ does in the absence of quencher, then

$$\tau = \frac{1}{k_r} \tag{12.28}$$

where τ is the actual triplet lifetime of $^3M^*$—the time required for $[^3M^*]$ to decrease to $1/e$ of its initial value. (Note that τ and τ° are not the same; τ° refers to a radiative lifetime.) Substituting 12.28 into 12.27, we obtain the *Stern–Volmer equation:*

$$\frac{\Phi^\circ_{\text{product}}}{\Phi_{\text{product}}} = 1 + k_q\tau[Q] \tag{12.29}$$

A plot of $\Phi^\circ_{\text{product}}/\Phi_{\text{product}}$ versus quencher concentration should give a straight line with slope $k_q\tau$. If k_q is assumed to be diffusion-controlled, then from the Debye equation,

$$k_q = k_{\text{diffusion}} = \frac{8RT}{3000\eta} \tag{12.30}$$

(where η is the solvent viscosity in poise), we can obtain k_q. Now the actual lifetime of $^3M^*$, τ, is readily obtained.

Although the Stern–Volmer equation (Equation 12.29) was specifically derived above for the reaction scheme given in Equations 12.18–12.21, an equation of exactly the same form will be obtained for quenching of product from singlet excited states, for quenching of emission rather than product formation, and for quenching of either product formation or emission when more than one unimolecular process occurs from a single excited state. Under these last conditions the actual lifetime of the state is given by

$$\tau = \frac{1}{\Sigma\, k_u} \tag{12.31}$$

where $\Sigma\, k_u$ is the sum of all the unimolecular decay processes of the particular excited state of M under consideration.

Figure 12.7 shows a Stern–Volmer plot for the quenching of the acetone-sensitized isomerization of 1-chloro-*trans*-2-butene to 1-chloro-*cis*-2-butene by the triplet quencher 1,3-pentadiene. When $\Phi^\circ_{\text{product}}/\Phi_{\text{product}}$ or $\Phi^\circ_{\text{emission}}/\Phi_{\text{emission}}$ is plotted against [Q] and a straight line is not obtained, then reaction or emission, respectively, is occurring from more than one excited state, most often S_1 and T_1.

The lifetimes of excited states, if longer than 10^{-8} sec, may also be measured by laser flash photolysis.

Energy Pooling

Energy transfer may occur when two excited molecules collide. This process, called *energy pooling*, must obey Wigner's rule of spin conservation. Excited singlets are not important in energy pooling, because they usually do not have time to collide with other molecules during their short lifetimes. The most important energy-pooling process is collision of two excited triplets to give an excited singlet and a ground-state singlet. This process is called *triplet–triplet annihilation* and is shown in Equation 12.32. The excited singlet A^*, produced in the S_n state, may then undergo the

same processes it would if formed directly by absorption of light. If it is converted to

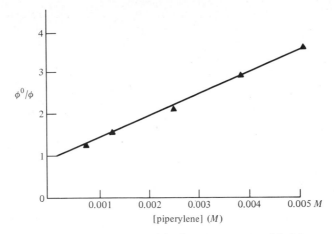

Figure 12.7 Stern–Volmer plot for the acetone-sensitized rearrangement of 1-chloro-*trans*-2-butene to 1-chloro-*cis*-2-butene in the presence of 1,3-pentadiene. Reprinted with permission from S. J. Cristol and R. P. Micheli, *J. Am. Chem. Soc.*, **100**, 850 (1978). Copyright 1978 American Chemical Society.

$$\overset{\uparrow\uparrow}{D^*} + \overset{\uparrow\uparrow}{A^*} \longrightarrow \overset{\uparrow\uparrow}{D_0} + \overset{\uparrow\uparrow}{A^*} \tag{12.32}$$

S_1 and emits, we have what is known as *P-type delayed fluorescence*. It can be distinguished from E-type by the dependence of fluorescence intensity on the intensity of the incident radiation. P-type shows a second-order but E-type a first-order dependence.[24]

Excimers and Exciplexes[25]

For years it has been known that the quantum yield of fluorescence for a number of aromatic hydrocarbons decreases with increasing concentration, but the cause of this concentration quenching was not well understood. In 1955 it was first noted by Förster that increasing concentration not only quenches the normal fluorescence of pyrene (5) but also introduces a new fluorescent component. Figure 12.8 shows the

5

fluorescence spectra of solutions of pyrene at various concentrations in *n*-heptane at 20°C. The high concentration band is not due to emission from a photoproduct:

[24] D. K. K. Liu and L. R. Faulkner, *J. Am. Chem. Soc.*, **100**, 2635 (1978).

[25] (a) Th. Förster, *Angew. Chem. Int. Ed.*, **8**, 333 (1969); (b) B. Stevens, in *Advances in Photochemistry*, Vol. 8, W. A. Noyes, G. S. Hammond, and J. N. Pitts, Jr., Eds., Wiley, New York, 1971, p. 161; (c) M. Gordon and W. R. Ware, Eds., *The Exciplex*, Academic Press, New York, 1975.

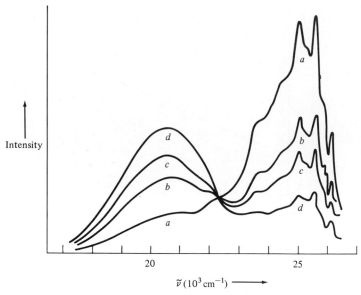

Figure 12.8 Fluorescence spectra of pyrene in *n*-heptane at 20°C; (*a*) 5×10^{-5} mole l^{-1}; (b) 1.8×10^{-4} mole l^{-1}; (c) 3.1×10^{-4} mole l^{-1}; (d) 7.0×10^{-3} mole l^{-1}. From T. Förster, *Angew. Chem. Int. Ed. Engl.*, **8**, 333 (1969). Reproduced by permission of Verlag Chemie, GmbH.

Irradiating pyrene for long periods of time does not change the absorption spectrum. Nor is the new band due to aggregates of pyrene that are formed in the ground state and stay together in the excited state, because, with increasing concentration, no new bands are found in the absorbtion spectrum. The fluorescent component must then be formed after electronic excitation and disappear on emission. Kinetic analysis shows that it is a complex of an electronically excited pyrene molecule (in the S_1 state) and a ground-state pyrene molecule.[26] The term *excimer* is used to described associates of this type.

Examination of the spectra in Figure 12.8 shows that emission from the pyrene monomer has vibrational fine structure but that excimer emission is structureless. Structureless emission (or absorption) is characteristic of a transition to an unstable, dissociative state. Figure 12.9, in which potential energy is plotted against internuclear distance, shows why for a diatomic molecule. The lower state, being dissociative, has no vibrational fine structure and therefore emission to it is not quantized.

Since Förster's original work, a large number of aromatic compounds, including benzene, naphthalene, and anthracene, have been found to have concentration-dependent fluorescence spectra under some conditions. Most of these

[26]Th. Förster and K. Kasper, *Z. Electrochem., Ber. Bunsenges. Phys. Chem.*, **59**, 976 (1955).

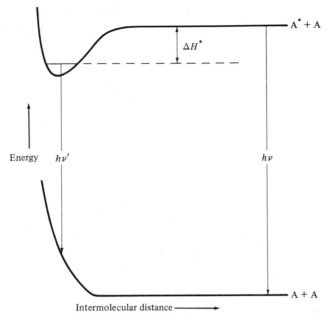

Figure 12.9 Potential energy vs. intermolecular distance for the first excited singlet and the ground state of pyrene excimer. From T. Förster, *Angew. Chem. Int. Ed. Engl.,* **8,** 333 (1969). Reproduced by permission of Verlag Chemie, GMbH.

excimers are not as stable as the pyrene prototype, and require lower temperatures or higher concentrations to be observed.[25] Some crystals also exhibit excimer emission. Crystalline pyrene, for example, has only a single structureless fluorescence band of the same energy as its excimer emission in solution.[27]

The geometrical requirements for excimer emission are stringent. In crystals the molecules must be stacked in columns or arranged in pairs with an interplanar distance of less than 3.5 Å.[28] Chandross and Dempster carried out an elegant investigation of the geometrical requirements for excimer formation in solution. They studied the fluorescence of compounds **6–10** in methylcyclohexane solution and found that only compounds **7** and **9** exhibit strong emission from an intramolecular excimer. These compounds, and these alone of those studied, allow the two naphthalene rings to be parallel, with one lying directly above the other. In compound **6** the two rings must have a dihedral angle of 45° at their closest approach; compound **10** allows the rings to be parallel, but only with approximately 25 percent overlap; and compound **8** requires appreciable deformation of the hydrocarbon chain for the two rings to come within 3 Å of each other.[29]

[27] J. Ferguson, *J. Chem. Phys.,* **28,** 765 (1958).
[28] B. Stevens, *Spectrochim. Acta,* **18,** 439 (1962).
[29] E. A. Chandross and C. J. Dempster, *J. Am. Chem. Soc.,* **92,** 3586 (1970).

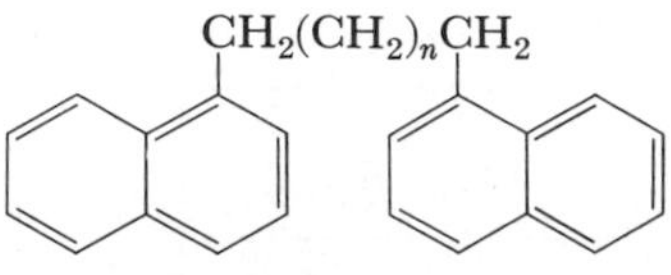

$$CH_2(CH_2)_nCH_2$$

6 $n = 0$
7 $n = 1$
8 $n = 2$

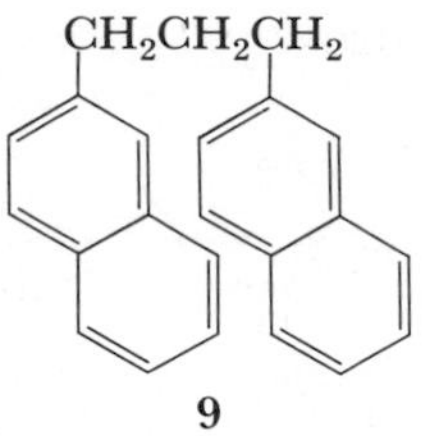

9

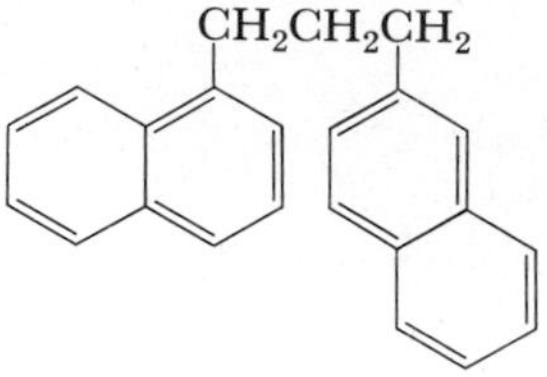

10

What is the nature of the excimer bond? Quantum mechanical calculations of the theoretical position of excimer fluorescence bands give good agreement with experimental positions if excitation resonance and charge resonance are both taken into account—that is, if the structure of the excimer is described by Equation 12.33.[25]

$$A^*A \leftrightarrow AA^* \leftrightarrow A^+A^- \leftrightarrow A^-A^+ \tag{12.33}$$

Excimers may undergo processes other than emission. Nonradiative decay apparently occurs more easily from an excimer than from an excited monomer. Photodimerization may also occur.[29,30] Anthracene, for example, gives **11** through an excimer intermediate. It may be that transient photodimerization is, in fact, responsible for all nonradiative decay of excimers.[25,29]

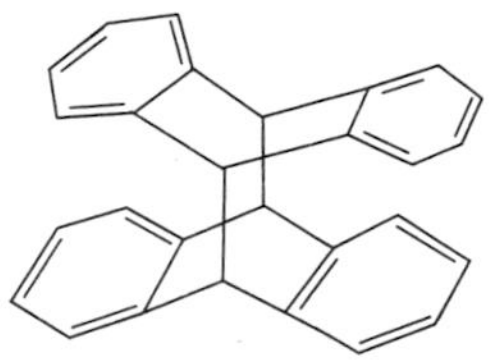

11

Excited-state complexes between two dissimilar entities, called *exciplexes*, are also frequently formed. The most thoroughly studied exciplexes are those between an aromatic compound and either an amine or a conjugated olefin or diene. Compound **12**, for example, has an absorption spectrum identical with α-methylnaphthalene, but its fluorescence is almost entirely structureless and shifted to longer wavelengths than naphthalene fluorescence. An exciplex is apparently formed be-

[30] A. S. Cherkasov and T. M. Vember, *Optics and Spectroscopy*, **6**, 319 (1959).

tween the amine nitrogen and the naphthalene ring involving appreciable charge transfer.[31]

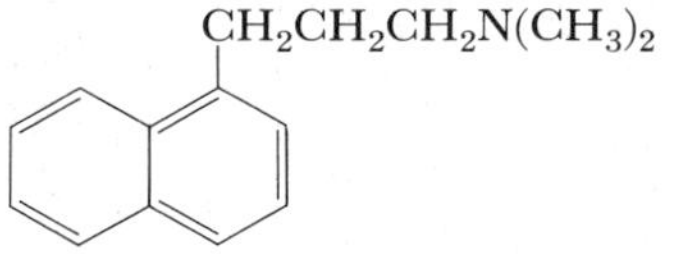

$$CH_2CH_2CH_2N(CH_3)_2$$

12

1,4-Dienes quench naphthalene fluorescence although they have no excited state of lower energy than the S_1 state of naphthalene. Apparently quenching is due to an exciplex that either undergoes radiationless transition to ground-state naphthalene and diene[32] or yields a photoproduct.[33] The photoproducts may often go undetected because of their instability but, for example, when naphthalene is excited in the presence of 2,4-dimethylpiperylene with light of longer wavelength than the diene absorbs, the naphthalene fluorescence is quenched and **13** is iso-

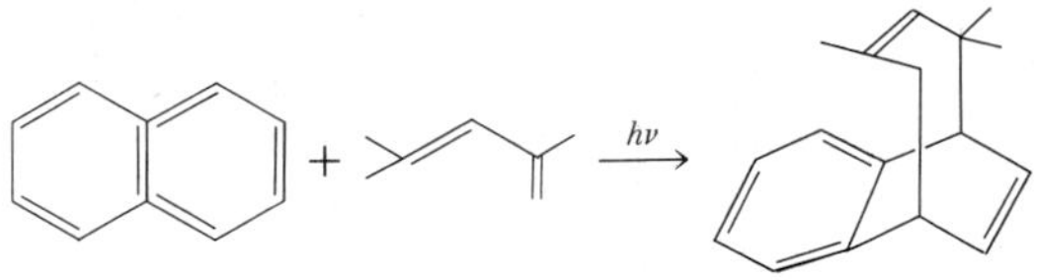

$$\text{(12.34)}$$

13

lated.[34] That cycloaddition definitely can occur via an exciplex and is not a separate quenching pathway has been convincingly demonstrated by Caldwell and Smith. Spectroscopic studies show that **14** and **15** form an exciplex, and product studies show that they form the cycloaddition product **16**. When dimethyl acety-

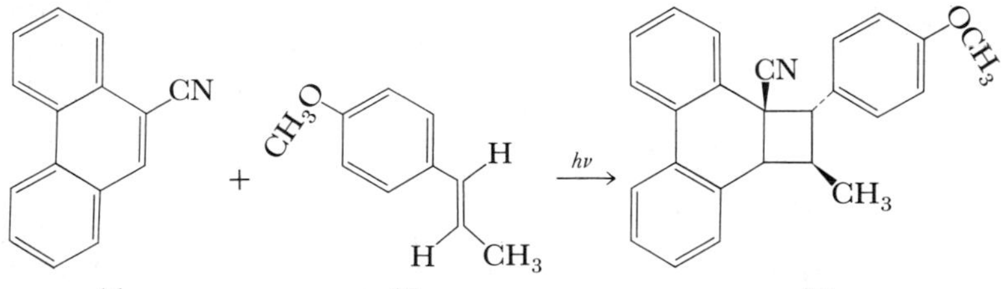

$$\text{(12.35)}$$

14   **15**   **16**

lenedicarboxylate is added to the reaction system, it quenches both exciplex emission and product formation and the Stern–Volmer slope, $k_q\tau$, is identical for both quenching processes.[35]

[31] E. A. Chandross and H. T. Thomas, *Chem. Phys. Lett.* **9**, 393 (1971).
[32] (a) D. A. Labianca, G. N. Taylor, and G. S. Hammond, *J. Am. Chem. Soc.*, **94**, 3679 (1972); (b) G. N. Taylor and G. S. Hammond, *J. Am. Chem. Soc.*, **94**, 3684, 3687 (1972).
[33] (a) N. C. Yang, J. Libman, and M. F. Savitzky, *J. Am. Chem. Soc.*, **94**, 9226 (1972); (b) J. Saltiel and D. E. Townsend, *J. Am. Chem. Soc.*, **95**, 6140 (1973).
[34] See note 32(a).
[35] (a) R. A. Caldwell and L. Smith, *J. Am. Chem. Soc.*, **96**, 2994 (1974); (b) R. A. Caldwell and D. Creed, *J. Am. Chem. Soc.*, **100**, 2905 (1978).

In a systematic study of a number of dienes and substituted naphthalenes, it was found that the quenching efficiency depends on both the ionization potential of the diene and the electron affinity of the excited naphthalene but that the logarithm of the rate of quenching is not a linear function of either parameter. Thus exciplex binding cannot be due to excited-state charge transfer alone.[36] That excitation resonance (see Equation 12.33) accounts for part of exciplex binding has been shown by McDonald and Selinger. They found that exciplexes can be formed by excitation of either component, A or B, and that if the energy levels of the first excited singlets of A and B are not far apart the exciplex can dissociate to produce either excited A or excited B. If the energies are very different, the excited species having the lower energy is produced on dissociation.[37]

The geometrical requirements for exciplex formation are not as stringent as those for excimer formation, especially if charge-transfer resonance is an important part of the exciplex binding. For example, Taylor, Chandross, and Schiebel have studied the exciplexes formed from pyrene on the one hand and N,N-dimethylaniline or 3,5-di-*t*-butyl-N,N-dimethylaniline on the other. In hexane at 23°C these exciplexes have very similar energies (as judged from the wavelength of emission), but that formed from the di-*t*-butyl-substituted aniline is slightly *more* stable. From the small difference in stabilities (0.07 eV), Taylor and co-workers estimate that the excimer formed with N,N-dimethylaniline has the sandwich pair structure **17**, but that formed with the di-*t*-butyl-substituted N,N-dimethylaniline has the structure **18**.[38]

17   18

(If both had the sandwich pair structure **17**, the exciplex formed with N,N-dimethylaniline should be considerably more stable because it would have less steric interference. On the other hand, if both had a structure analogous to **18**, 3,5-

[36](a) See note 32(a). For further evidence for a polar contribution to the exciplex, see (b) R. M. Bowman, T. R. Chamberlain, C.-W. Huang, and J. J. McCullough, *J. Am. Chem. Soc.*, **96**, 692 (1974); (c) R. G. Brown and D. Phillips, *J. Am. Chem. Soc.*, **96**, 4784 (1974); (d) S.-P. Van and G. S. Hammond, *J. Am. Chem. Soc.*, **100**, 3895 (1978).

[37] R. J. McDonald and B. K. Selinger, *Aust. J. Chem.*, **24**, 1797 (1971); see also D. F. Eaton and D. A. Pensak, *J. Am. Chem. Soc.*, **100**, 7428 (1978).

[38] G. N. Taylor, E. A. Chandross, and A. H. Schiebel, *J. Am. Chem. Soc.*, **96**, 2693 (1974).

di-*t*-butyl-N,N-dimethylaniline should form the more stable exciplex because of the electron-donating ability of the *t*-butyl groups.)

Singlet excimers and exciplexes have been proposed as intermediates in a number of chemical reactions; we shall encounter them again in this chapter. Triplet excimers and exciplexes also are formed and bound by both excitation resonance and charge-transfer interactions.[39] They rarely emit[25] but they too have been proposed as photochemical intermediates.

12.4 PHOTOCHEMICAL REACTIONS

A photochemical reaction, a reaction in which one starting material is an electronically excited molecule, may proceed by one of two fundamentally different mechanisms. The starting material may give electronically excited product (Equation 12.36), or it may give ground-state product directly (Equation 12.37). The mechanism of Equation 12.36 has been suggested for a number of reactions. However, in

$$A^* \longrightarrow B^* \tag{12.36}$$
$$A^* \longrightarrow B \tag{12.37}$$

order to show conclusively that it is operative, the product must undergo some detectable photochemical or photophysical process. Clear-cut examples are rare and usually are reactions that also occur in the ground state. For example, Yang and co-workers have found that when **19** is irradiated, anthracene is obtained, which fluoresces from its first excited singlet state.[40]

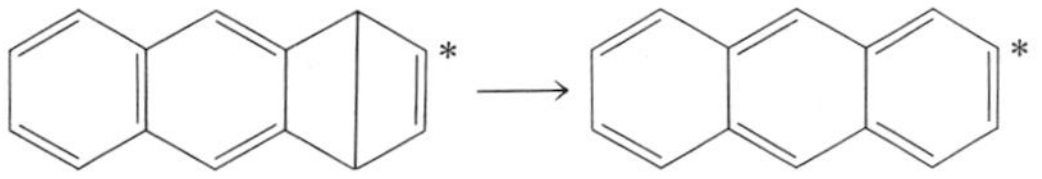

$$\tag{12.38}$$

19

Excited-state proton transfer reactions such as that in Equation 12.39 are also well documented.[41] Hammond has suggested that normally A is converted to B by the

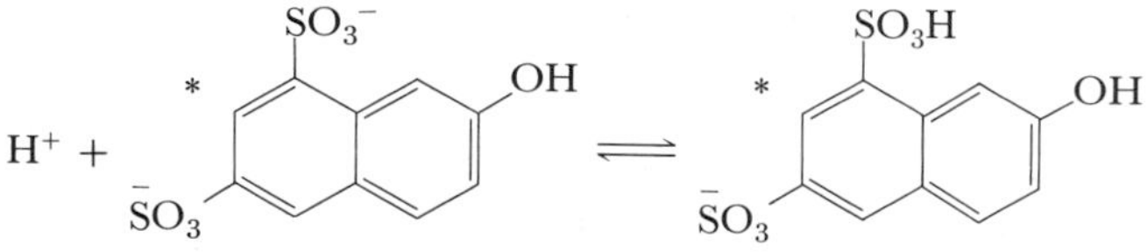

$$\tag{12.39}$$

mechanism of Equation 12.36 at about the same rate as ground-state A is converted into ground-state B. Since the lifetimes of excited states are so short, only very fast reactions, then, can proceed by this mechanism. In the mechanism of Equation

[39] See, for example; (a) S. G. Cohen and G. Parsons, *J. Am. Chem. Soc.*, **92**, 7603 (1970); (b) D. I. Schuster and M. D. Goldstein, *J. Am. Chem. Soc.*, **95**, 986 (1973); (c) N. C. Yang, D. M. Shold, and B. Kim, *J. Am. Chem. Soc.*, **98**, 6587 (1976); (d) R. O. Loutfy and P. de Mayo, *J. Am. Chem. Soc.*, **99**, 3559 (1977).

[40] N. C. Yang, R. V. Carr, E. Li, J. K. McVey, and A. S. Rice, *J. Am. Chem. Soc.*, **96**, 2297 (1974), and references therein.

[41] S. G. Schulman, L. S. Rosenberg, and W. R. Vincent, Jr., *J. Am. Chem. Soc.*, **101**, 139 (1979).

12.37, however, the electronic excitation energy is converted into the driving force of the reaction.[42]

Because of the rapidity with which the first formed excited states are usually converted to the S_1 or T_1 state, most photochemical reactions start from these states. There are, however, a growing number of exceptions to this rule.[8] Some of these follow logically from what has been said above. In gas-phase reactions at low pressures, where energy removal is slow, photochemistry from higher excited states is fairly common. Another obvious exception is when an upper dissociative excited state, in which the molecule immediately fragments, is populated.[43] Furthermore, reaction from second excited states, in the same way as emission from second excited states, sometimes occurs as a result of thermal population of the higher state or of slow internal conversion of S_2 to S_1.[44] There is now, however, some evidence that there are no special requirements for photoproducts to be formed from a second excited state, although the quantum yield will be very low.[45]

Primary photochemical processes are those events that cause a molecule, excited by irradiation, to be converted to another molecule or back to the ground state. They include nonradiative decay to S_0, emission, energy transfer, and quenching through exciplex formation.[46]

Formation of photoproducts may involve only primary processes (Equations 12.36 and 12.37) or may occur in reactions subsequent to the primary processes. Equations 12.40 and 12.41 show the mechanism of photochemical formation of

$$Cl_2 \xrightarrow{h\nu} 2Cl \cdot \qquad (12.40)$$

$$Cl \cdot + H_2 \longrightarrow HCl + Cl \cdot \qquad \text{etc.} \qquad (12.41)$$

HCl from H_2 and Cl_2. The reaction of Equation 12.40 is a primary process, but the product-forming step, Equation 12.41, is a secondary process. According to the law of Stark and Einstein, the sum of the quantum yields of all primary processes is equal to one. Thus if the product is formed entirely in a primary process, the quantum yield of product formation, Φ_{pro}, defined by Equation 12.42, cannot be greater than one. If secondary processes are involved, there is no such limitation. The quantum yield for formation of HCl according to the mechanism of Equations 12.40 and 12.41 may reach 10^6.

$$\Phi_{pro} = \frac{\text{no. of molecules of product formed}}{\text{no. of quanta absorbed}} \qquad (12.42)$$

The rest of this chapter will be devoted to a brief glimpse into the mechanisms of some of the more important organic photoreactions.

[42] (a) G. S. Hammond in *Advances in Photochemistry*, Vol. 7, W. A. Noyes, G. S. Hammond, and J. N. Pitts, Jr., Eds., Wiley, New York, 1969, p. 373; see also (b) L. Salem, *J. Am. Chem. Soc.*, **96**, 3486 (1974).

[43] See, for example, M. Berger, I. L. Goldblatt, and C. Steel, *J. Am. Chem. Soc.*, **95**, 1717 (1973).

[44] See, for example, (a) D. G. Whitten and Y. J. Lee, *J. Am. Chem. Soc.*, **94**, 9142 (1972); (b) P. de Mayo and R. Suau, *J. Am. Chem. Soc.*, **96**, 6807 (1974).

[45] See note 8(b).

[46] W. A. Noyes, Jr., G. B. Porter, and J. E. Jolley, *Chem. Rev.*, **56**, 49 (1956).

Cis–Trans Isomerization[47]

Alkenes, upon electronic excitation, undergo geometric isomerization. To see why this is so, let us consider what happens to *cis*-2-butene when it absorbs a photon. In its ground state *cis*-2-butene is, of course, planar and thus vertical excitation produces a planar excited state. The π^* orbital, however, in agreement with Hund's rule, has an energy minimum when the p orbitals are orthogonal to each other— that is, when the antibonding electrons can be as far apart from each other as possible (see Figure 12.10 for a potential energy vs. torsional angle diagram of ethylene). The relaxation of the excited state produces a twisted geometry as shown in Scheme 1. The twisted geometry produces either ground-state *cis*- or ground-state *trans*-2-butene. Since an energy minimum exists for both S_1 and T_1 at a torsional angle of 90°, isomerization can occur from either state.

SCHEME 1

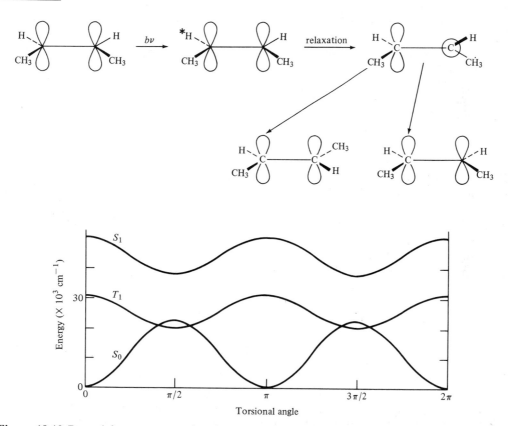

Figure 12.10 Potential energy vs. torsional angle diagram for the S_0, S_1, and T_1 states of ethylene. Adapted from A. J. Merer and R. S. Mulliken, *Chem. Rev.*, **69**, 639 (1969).

[47]J. D. Coyle, *Chem. Soc. Rev.*, **3**, 329 (1974).

Most monoolefins absorb at wavelengths too short for facile study of their behavior on direct irradiation. *Cis-* and *trans-*2-butenes do isomerize on direct irradiation, however, apparently from S_1.[48] *Cis-*cyclooctene, however, isomerizes from a higher π, σ^* state (Rydberg state) since irradiation at wavelengths <200 nm produces *trans-*cyclooctene with a quantum yield of one: Decay of a twisted π, π^* state should produce some cis- as well as some *trans-*cyclooctene and thus $\Phi_{cis\rightarrow trans}$ would be less than one.[49]

Phenyl-substituted alkenes, on direct irradiation, also isomerize from their S_1 states. Azulene, added as quencher, has almost no effect on the cis–trans isomerization of the stilbenes upon direct photolysis, although it has a large effect on their photosensitized isomerization.[50] *p-*Bromostilbene, which undergoes intersystem crossing more readily than stilbene, isomerizes from its S_1 and T_1 states.[51]

If a system reaches the point where further irradiation under constant conditions causes no further change in composition, it is said to have reached its *photostationary state* (pss). The $(cis/trans)_{pss}$ ratio in olefin isomerization depends on the relative extinction coefficients of the cis and trans isomers (i.e., their relative opportunities to isomerize) and on the quantum yields for both directions of the photoisomerization, as shown in Equation 12.43.

$$\left(\frac{cis}{trans}\right)_{pss} = \left(\frac{\epsilon_{trans}}{\epsilon_{cis}}\right)\left(\frac{\Phi_{trans\rightarrow cis}}{\Phi_{cis\rightarrow trans}}\right) \tag{12.43}$$

The absorption spectra of cis and trans isomers usually overlap strongly, so that it is difficult to irradiate one without also irradiating the other. If, however, the absorptions do not overlap, selective irradiation of one converts it completely to the other. An interesting example is the photochemical isomerization of *cis-*cyclooctenone (**20**) to the highly strained, previously unknown isomer (**21**). The cis

$$n, \pi^* \lambda_{max} = 321 \text{ nm} \qquad n, \pi^* \lambda_{max} = 283 \text{ nm}$$
$$\textbf{20} \qquad\qquad\qquad \textbf{21}$$

$$(12.44)$$

isomer, which is able to achieve planarity for its conjugated system, has its n, π^* absorption at longer wavelengths than the trans isomer, which cannot be planar. Irradiation with light of wavelength >300 nm gives 80 percent **21**. (The conversion to **21** is not complete because a tail of the trans isomer absorption extends to wavelengths >300 nm; irradiation of it gives the back reaction, **21** $\rightarrow$ **20**.[52]

[48] H. Yamazaki and R. J. Cvetanović, *J. Am. Chem. Soc.*, **91**, 520 (1969).
[49] R. Srinivasan and K. H. Brown, *J. Am. Chem. Soc.*, **100**, 2589 (1978).
[50] (a) J. Saltiel and E. D. Megarity, *J. Am. Chem. Soc.*, **94**, 2742 (1972); (b) J. Saltiel, A. Marinari, D. W.-L. Chang, J. C. Mitchener, and D. E. Megarity, *J. Am. Chem. Soc.*, **101**, 2982 (1979).
[51] (a) See note 50(b); (b) J. Saltiel, D. W.-L. Chang, and E. D. Megarity, *J. Am. Chem. Soc.*, **96**, 6521 (1974).
[52] P. E. Eaton and K. Lin, *J. Am. Chem. Soc.*, **86**, 2087 (1964).

As already mentioned, photosensitized isomerization of alkenes also occurs. If both cis and trans excited triplets first decay to the twisted (or *phantom*) triplet before they are converted to ground-state alkenes, then

$$\left(\frac{\text{cis}}{\text{trans}}\right)_{\text{pss}} = \left(\frac{k_{t\to p}}{k_{c\to p}}\right)\left(\frac{k_{p\to c}}{k_{p\to t}}\right) \tag{12.45}$$

where $k_{t\to p}$, $k_{c\to p}$, $k_{p\to c}$, and $k_{p\to t}$ are the rates of Reactions 12.46–12.49, respectively.

$$\textit{trans-olefin} + {}^3\text{donor}^* \xrightarrow{k_{t\to p}} {}^3\text{phantom}^* + \text{donor}_0 \tag{12.46}$$

$$\textit{cis-olefin} + {}^3\text{donor}^* \xrightarrow{k_{c\to p}} {}^3\text{phantom}^* + \text{donor}_0 \tag{12.47}$$

$$ {}^3\text{phantom}^* \xrightarrow{k_{p\to c}} \textit{cis-olefin} \tag{12.48}$$

$$ {}^3\text{phantom}^* \xrightarrow{k_{p\to t}} \textit{trans-olefin} \tag{12.49}$$

With sensitizers whose triplet energies lie above the triplet energies of both cis and trans isomers, energy transfer to both isomers occurs at almost the same rate. Then Equation 12.45 simplifies to

$$\left(\frac{\text{cis}}{\text{trans}}\right)_{\text{pss}} \approx \frac{k_{p\to c}}{k_{p\to t}} \tag{12.50}$$

From high-energy sensitizers we can therefore get a direct measure of the relative rates of decay of the phantom triplet to cis and trans isomers.

Figure 12.11 shows a plot of percent cis isomer at the photostationary state vs. the energy of the triplet sensitizer in the cis–trans isomerization of stilbene. The high-energy sensitizers all give approximately the same $(\text{cis}/\text{trans})_{\text{pss}}$ according to Equation 12.50. Then as energy transfer to the cis isomer becomes less efficient (E_T *cis*-stilbene $= 59$ kcal mole^{-1}; E_T *trans*-stilbene $= 49$ kcal mole^{-1}) the amount of cis in the photostationary state increases. Some photosensitized isomerization still occurs when the triplet energy of the sensitizer falls slightly below the triplet energy of *trans*-stilbene. Apparently the higher-energy ground-state isomer, *cis*-stilbene, is able to act as an electronic energy acceptor and be converted directly to the lower-energy twisted triplet. Hammond and Saltiel suggested that this nonvertical energy transfer (which is not a general phenomenon in photosensitized isomerizations) can be explained by a longer interaction time of the donor–acceptor pair in the solvent cage compared with the interaction time of a photon with a molecule.[53,54] A triplet exciplex, then, is a likely intermediate.

Electrocyclic Reactions

In Chapter 10 we saw that the photochemical pericyclic selection rules are reversed from the thermal ones: An allowed photochemical process involving $4n$ π electrons

---

[53](a) W. G. Herkstroeter and G. S. Hammond, *J. Am. Chem. Soc.*, **88**, 4769 (1966); (b) G. S. Hammond and J. Saltiel, *J. Am. Chem. Soc.*, **85**, 2516 (1963); (c) J. Saltiel, J. L. Charlton, and W. B. Mueller, *J. Am. Chem. Soc.*, **101**, 1347 (1979).
[54] For another point of view, see V. Balzani and F. Bolletta, *J. Am. Chem. Soc.*, **100**, 7404 (1978).

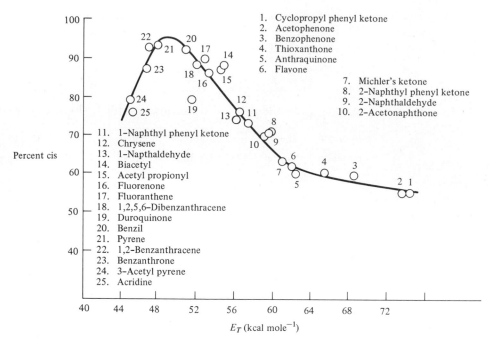

Figure 12.11 Photostationary states for the stilbenes as a function of the sensitizer triplet energy E_T. Reprinted with permission from W. G. Herkstroeter and G. S. Hammond, *J. Am. Chem. Soc.*, **88**, 4769 (1966). Copyright by the American Chemical Society.

must be all suprafacial, whereas one involving $4n + 2$ π electrons must have one antarafacial component. These rules can, at first glance, be easily rationalized in terms of orbital correlation diagrams. For example, Figure 12.12 shows the orbital correlation diagram for the photochemically allowed disrotatory ring closure of butadiene to cyclobutene. (Compare Figure 12.12 with Figure 10.16, the latter being for ground-state disrotatory ring closure of butadiene. The only difference between the two diagrams is the different electronic configurations.) The allowed nature of this photochemical reaction appears to be due to the fact that although a π_2 electron in the starting material must be promoted to a π^* orbital in the product, a π_3^* electron of the starting material correlates with a π electron in the product. Thus the reaction appears to be approximately thermoneutral.

However, as was already mentioned above, excited-state starting material almost never gives excited-state product as Figure 12.12 would predict. And electrocyclic ring closure of butadiene is no exception. Thus this model is not sufficient.

As was already briefly discussed in Section 10.6, the adherence of concerted photochemical reactions to the pericyclic selection rules can often be best understood in terms of state correlation diagrams. Figure 12.13 shows the state correlation diagram for electrocyclic disrotatory ring closure of butadiene to cyclobutene.

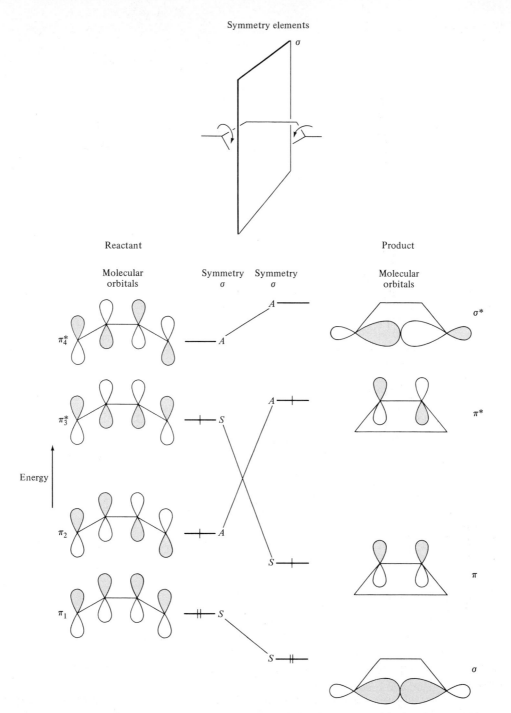

Figure 12.12 Symmetry classification and correlation of orbitals for the photochemical disrotatory closure of butadiene. Closure with electrons remaining in their original levels would lead to the excited state indicated by the orbital occupancy on the right.

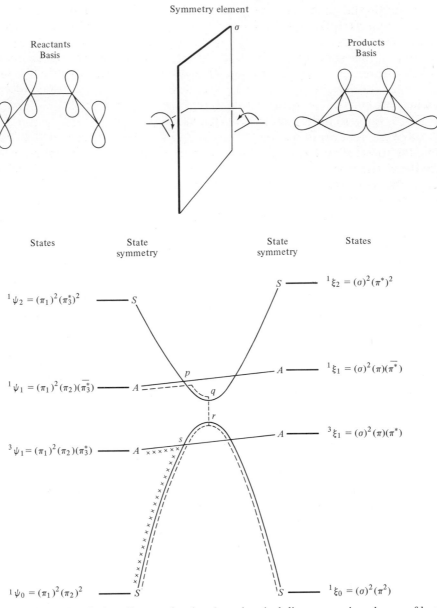

Figure 12.13 State correlation diagram for the photochemical disrotatory ring closure of butadiene. The states are described by the electronic configuration of lowest energy in that state. The solid lines are cross sections of the potential energy surfaces of the various states. The dashed lines show the possible reaction paths from the first excited singlet, and the line of crosses, the reaction path from the first excited triplet.

Cross sections of the potential energy surfaces of the ground state, first excited singlet, first excited triplet, and second excited singlet of both starting material and product are shown. As in Figure 10.29, the states are described by the electronic configuration of lowest energy in that state. In other words, we will be assuming no configuration interaction within a state. (If each of two orbitals in a single electronic configuration is singly occupied, a bar is placed over the symbol for the orbital when the electron is "spin up." No bar is present when the electron has "spin down." This is done to differentiate the electronic configurations of the first excited singlet and of the first excited triplet. The symbol ψ is used for the states of butadiene and ξ for those of cyclobutene. The superscript before the symbol refers to the multiplicity of the state and the subscript after it refers to the level of excitation. Thus $^3\psi_1$ is the first excited triplet of butadiene and $^1\xi_1$ is the first excited singlet of cyclobutene. Although Figure 12.13 is not a quantitative representation, it does show qualitatively the results of calculations of the relative energies of the various states of starting material and product.[55]

The ground states of both reactant and product are symmetrical but they do not intend to correlate with each other because the symmetry of their occupied orbitals is not the same. The intended correlation of the ground state of butadiene is with a second excited state of cyclobutene and that of the ground state of cyclobutene is with the second excited state of butadiene. So far this tells us nothing that Figure 10.16 did not tell us. However, we now see that because the ground states and the second excited states have the same symmetry, their potential energy surfaces have an avoided crossing. Thus ground-state starting material and ground-state product do lie on the same energy surface but with a large symmetry imposed energy barrier between them. The first excited singlet of the starting material correlates with the first excited singlet of the product and the same is true of the first excited triplets.

Now let us examine what happens as the $^1\psi_1$ state of butadiene begins a disrotatory ring closure. At the beginning of the reaction it must climb a small energy hill ($\sim$5 kcal mole^{-1}) since $^1\xi_1$ is of slightly higher energy than $^1\psi_1$. However, it soon, at point p, drops into the "well" of the second excited-state energy surface in an allowed crossing. This well acts as a funnel to the ground state. At point q which is in the region of the avoided crossing, it crosses over onto the ground-state potential energy surface at the ground state's energy maximum, point r, from whence it can be converted to starting material or product. The net result is that first excited singlet butadiene is converted back to ground-state butadiene or to ground-state cyclobutene.

When the first excited triplet of butadiene begins ring closure, it soon encounters the ground-state potential energy surface at point s, and in a forbidden crossing (triplet to singlet) crosses to the ground-state surface. Because a significant activation energy would be required for it to pass over point r to ground-state cyclobutene, it is entirely converted back to ground-state butadiene.

---

[55] D. Grimbert, G. Segal, and A. Devaquet, *J. Am. Chem. Soc.,* **97**, 6629 (1975), and references therein.

This theoretical picture fits experimental experience. Although butadienes, on direct excitation, will close to cyclobutenes in a disrotatory fashion as shown in Equation 12.51[56] photosensitization of butadiene produces starting material.[55]

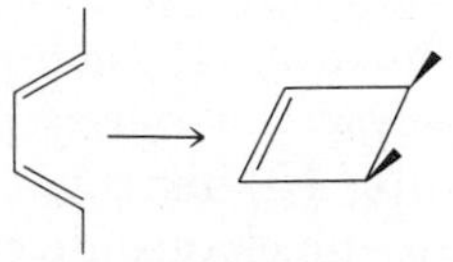

$$(12.51)$$

An important $4n + 2$ photochemical conrotatory ring opening is shown in Scheme 2. Ergosterol, in the presence of light, is converted under the skin to the immediate precursor of vitamin D. Isopyrocalciferol cannot undergo conrotatory ring opening because to do so would introduce a trans double bond in a six-membered ring and thus it does not form previtamin D.[57]

SCHEME 2

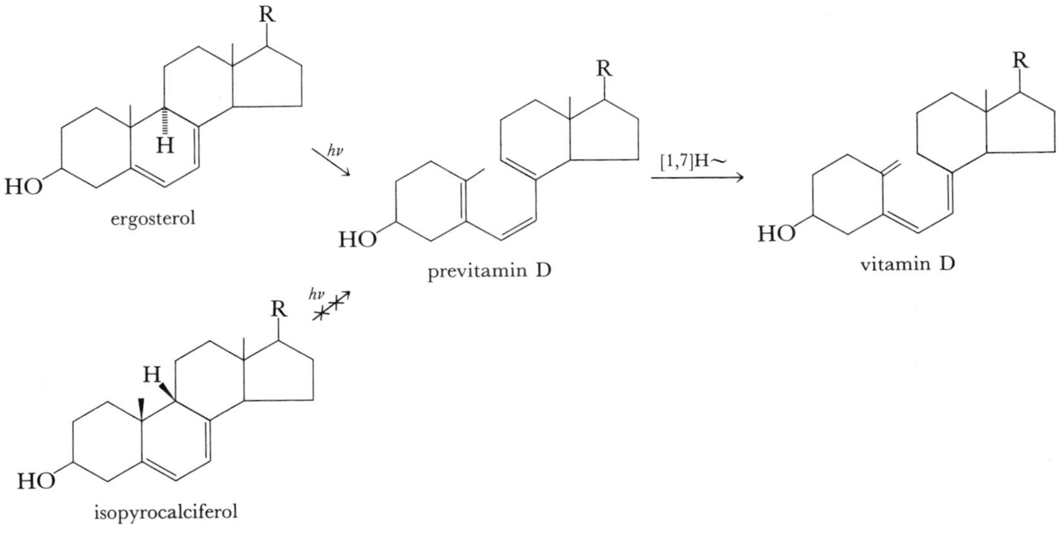

ergosterol

previtamin D

vitamin D

isopyrocalciferol

Photocycloadditions[47,58]

One of the possible reactions of an electronically excited monoolefin is dimerization with a ground-state molecule. If the reaction is concerted, the transition state should be $\pi 2s + \pi 2s$. The state correlation diagram for this process is shown in

[56] J. Saltiel, L. Metts, and M. Wrighton, *J. Am. Chem. Soc.,* **92,** 3227 (1970).

[57] E. Havinga, *Experentia,* **29,** 1181 (1973).

[58] (a) J. S. Swenton, *J. Chem. Educ.,* **46,** 7 (1969); (b) O. L. Chapman and G. Lenz, in *Organic Photochemistry,* Vol. 1, O. L. Chapman, Ed., Dekker, New York, 1966, p. 283; (c) J. D. Coyle and H. A. J. Carless, *Chem. Soc. Rev.,* **1,** 465 (1972).

Figure 10.29.[59] (In that figure ψ_a, $\psi_{e'}$, and $\xi_{e'}$ are not specified as singlets or triplets. That is, however, not important for the overall argument.) The gross features of Figure 10.29 and Figure 12.12 are very similar. As in electrocyclic ring closure of butadiene, in 2 + 2 cycloaddition, excitation produces a first excited state, which crosses to the second excited state and then drops into a well that acts as a funnel for conversion to ground-state reactants or ground-state products.

Although very few simple olefins have been investigated, from the evidence available it does appear that reaction from the singlet state is concerted. The photodimerization of tetramethylethylene (Equation 12.52) occurs upon irradiation without sensitizer; because the sensitizer-produced triplet state of tetramethylethylene does not dimerize, one may conclude that the dimerization observed on direct irradiation occurs from the singlet.[60] Since tetramethylethylene dimerizes from the singlet state, the 2-butenes probably do also; the results, shown in Equations 12.53 and 12.54, demonstrate that each butene molecule maintains its stereochemistry throughout.[61]

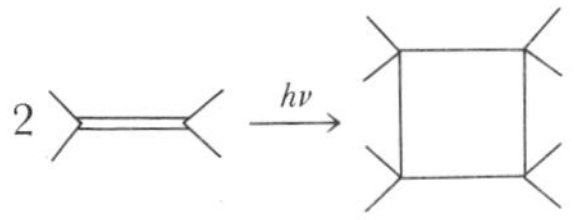

$$(12.52)$$

22

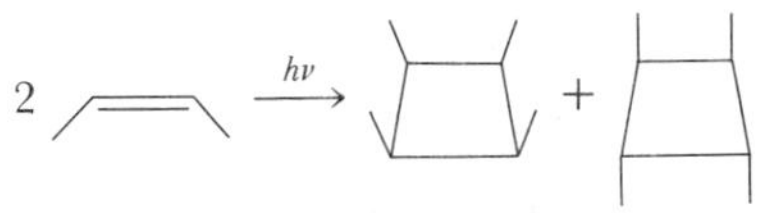

$$(12.53)$$

23   **24**

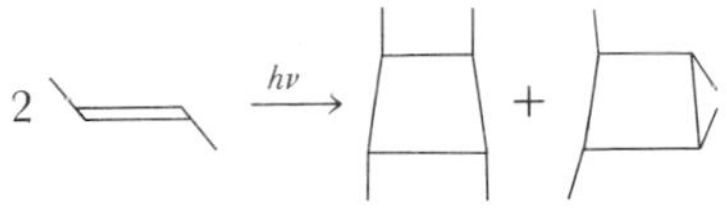

$$(12.54)$$

24   **25**

Photochemically induced intramolecular dimerizations are known. For example, when irradiated with 254-nm light, **26** gives the "housane" products **27** and **28** and butadiene gives **29**. Both these reactions apparently proceed through biradical intermediates.[62]

Some olefins conjugated with electron-withdrawing, unsaturated groups dimerize to cyclobutanes as photoproducts, apparently through their triplet states.

[59] See also (a) J. Michl, *Photochem. Photobiol.,* **25,** 141 (1977); (b) W. G. Dauben, L. Salem, and N. J. Turro, *Acc. Chem. Res.,* **8,** 41 (1975).

[60] D. R. Arnold and V. Y. Abraitys, *J. Chem. Soc. Chem. Commun.,* 1053 (1967).

[61] H. Yamazaki and R. J. Cvetanović, *J. Am. Chem. Soc.,* **91,** 520 (1969).

[62] (a) E. Block and H. W. Orf, *J. Am. Chem. Soc.,* **94,** 8438 (1972); (b) R. Srinivasan, in *Advances in Photochemistry,* Vol. 4, W. A. Noyes, G. S. Hammond, and J. N. Pitts, Jr., Eds., Wiley, New York, 1966, p. 113; (c) R. Srinivasan, *J. Am. Chem. Soc.,* **90,** 4498 (1968).

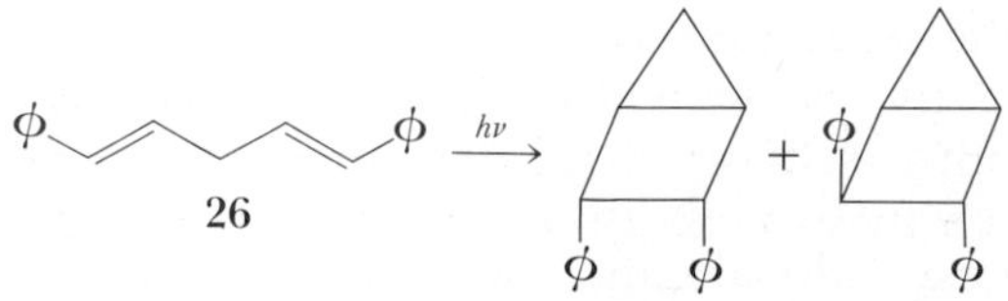

$$\phi \diagdown\diagup\diagdown\diagup\phi \xrightarrow{h\nu} \quad + \quad \phi \qquad (12.55)$$

26   27   28

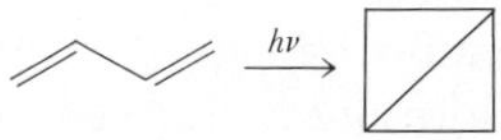

$$\diagup\diagdown\diagup \xrightarrow{h\nu} \qquad (12.56)$$

29

For example, when acrylonitrile in acetone is irradiated under conditions where only acetone absorbs, the dimer **30** is obtained as one of the products. Piperylene,

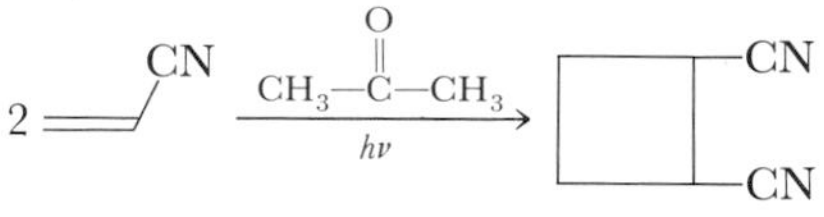

$$2 \overset{CN}{=}\diagup \xrightarrow[h\nu]{CH_3-\overset{O}{\overset{\|}{C}}-CH_3} \qquad \overset{CN}{\underset{CN}{\square}} \qquad (12.57)$$

30

an excellent quencher of triplet energy but a poor singlet quencher, interferes with the formation of **30** when added to the system. It appears that the T_1 state of acetone is populated by intersystem crossing from S_1 and that the triplet energy is then transferred to acrylonitrile.[63] Similarly, cyclopentenone and cyclohexenone dimerize as shown in Equations 12.58 and 12.59.[64] These reactions are also

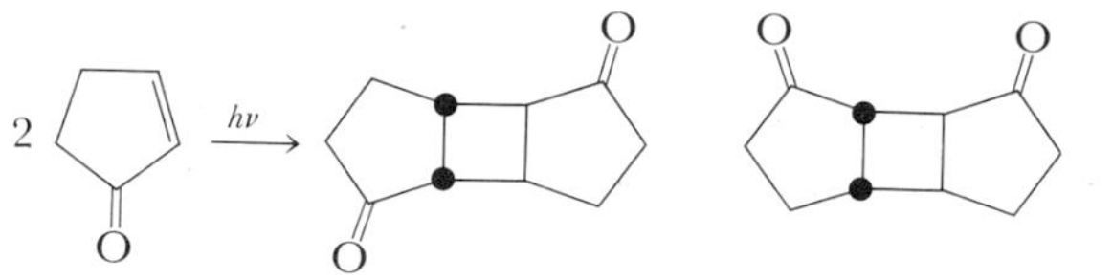

$$(12.58)$$

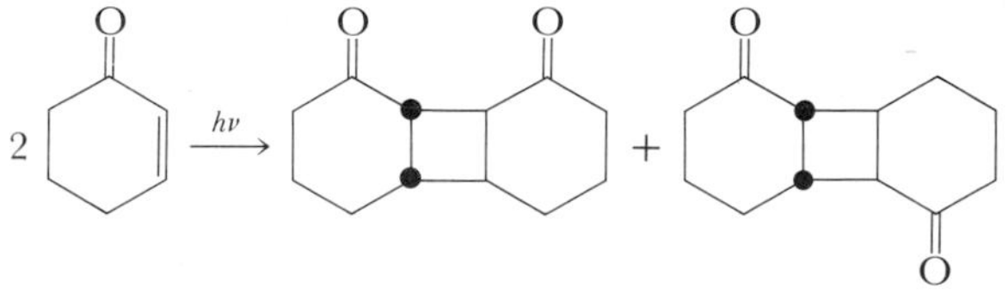

$$(12.59)$$

quenched by the triplet quencher piperylene. The rate of radiationless decay increases as ease of C—C bond twisting increases. Therefore open-chain α,β-unsaturated ketones do not photodimerize.[65]

Exciplexes are often found as intermediates in photochemical addition reactions. An unequivocal example was discussed earlier (Equation 12.35) but there are

[63] J. A. Barltrop and H. A. J. Carless, *J. Am. Chem. Soc.*, **94**, 1951 (1972).
[64] P. E. Eaton, *Acc. Chem. Res.*, **1**, 50 (1968).
[65] P. J. Wagner and D. J. Bucheck, *J. Am. Chem. Soc.*, **91**, 5090 (1969).

a large number of other examples as well.[47,66] For example, on direct irradiation *trans*-stilbene, present in low concentration, and dimethyl fumarate form **31**. Exciplex emission from a stilbene–dimethyl fumarate complex is observed under these conditions. When the concentration of *trans*-stilbene is increased, some **31** is

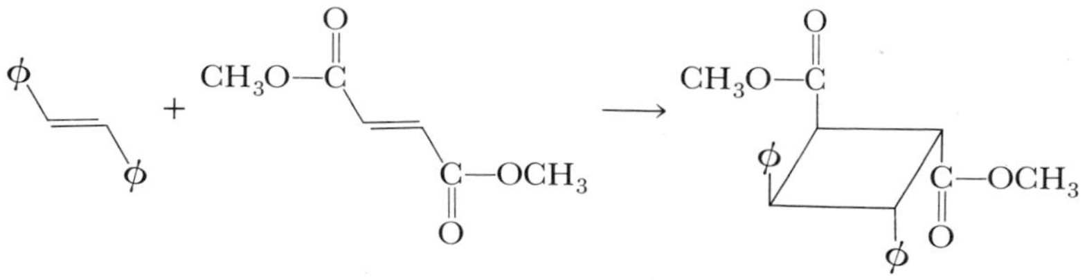

$$\phi\text{—CH=CH—}\phi + \text{CH}_3\text{O—C(=O)—CH=CH—C(=O)—OCH}_3 \longrightarrow \quad (12.60)$$

31

still formed, but **32**, in which stilbene has added to one of the fumarate carbon-oxygen double bonds, is also produced. The quantum yield of **32** increases with increasing concentration of *trans*-stilbene. Both **31** and **32** are formed from singlet

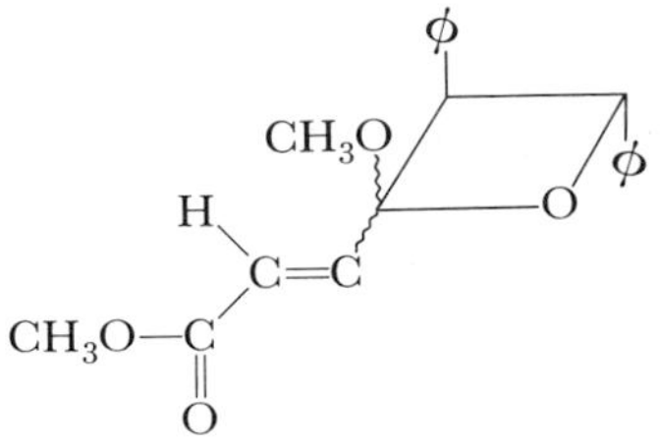

32

trans-stilbene since photosensitizers that cause *trans*-stilbene to isomerize bring about no cycloaddition reaction. Lewis and Johnson suggest that **31** is formed via a stilbene–fumarate exciplex and that **32** is formed by the interception of a stilbene excimer with a fumarate molecule.[67]

   The photochemical addition of a carbonyl compound to an olefin to form an oxetane is called the Paterno–Buchi reaction (Equation 12.61) and is one of the

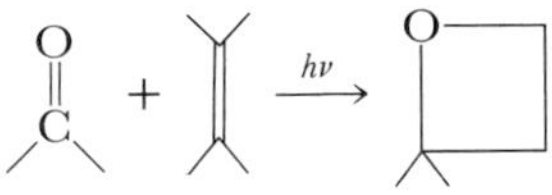

$$\text{C=O} + \text{C=C} \xrightarrow{h\nu} \text{oxetane} \quad (12.61)$$

most common photocyclizations.[47,68] It usually occurs at wavelengths where only the carbonyl compound absorbs and thus involves attack of an excited carbonyl on a ground-state olefin. The stereochemistry of the reaction depends on whether the carbonyl compound is aliphatic or aromatic. Benzaldehyde yields essentially the

---

[66](a) O. L. Chapman and R. D. Lura, *J. Am. Chem. Soc.,* **92**, 6352 (1970); (b) D. Creed and R. A. Caldwell, *J. Am. Chem. Soc.,* **96**, 7369 (1974), and references therein; (c) F. D. Lewis, *Acc. Chem. Res.,* **12**, 152 (1979).
[67]F. D. Lewis and D. E. Johnson, *J. Am. Chem. Soc.,* **100**, 983 (1978).
[68](a) See note 58(c). For a review, see (b) J. N. Pitts, Jr., and J. K. S. Wan in *The Chemistry of the Carbonyl Group,* S. Patai, ed., Wiley, New York, 1966, p. 823.

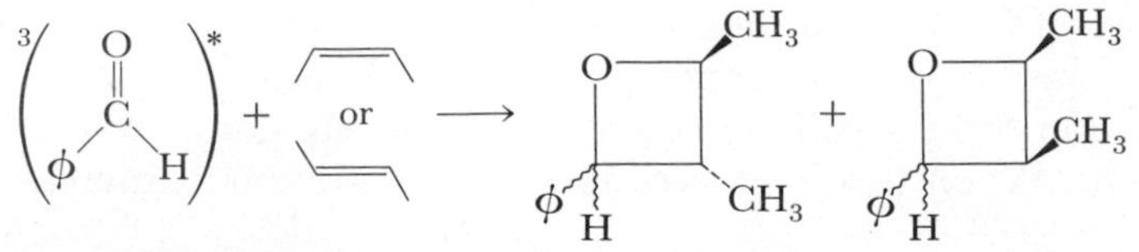

$$(12.62)$$

same mixture of oxetanes with *cis*-2-butene as it does with *trans*-2-butene (Equation 12.62).[69] Acetaldehyde, on the other hand, adds stereospecifically to the isomeric 2-butenes (Equations 12.63 and 12.64).[70]

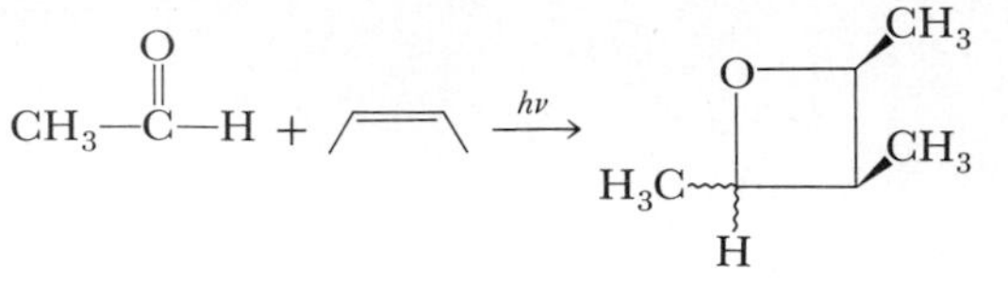

$$(12.63)$$

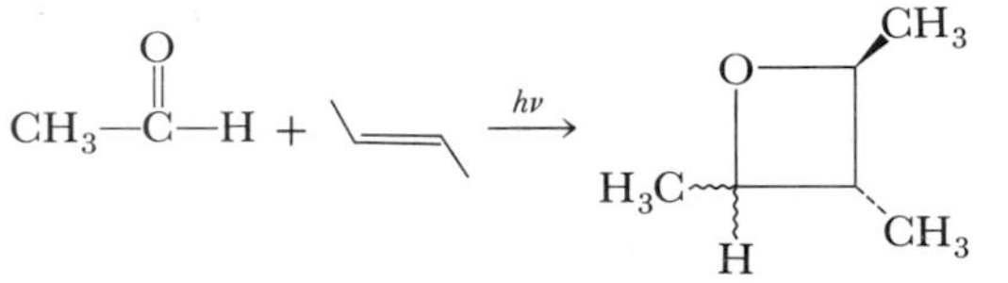

$>90\%$

$$(12.64)$$

$>95\%$

The reason for the difference in stereochemistry is that, as quenching studies show, the reactive state of the aromatic carbonyls (which undergo intersystem crossing rapidly) is T_1 but in aliphatic carbonyls it is S_1. The reaction between acetaldehyde and *cis*-2-butene, then, is either concerted or proceeds by the singlet biradical **33**, whereas the intermediate from benzaldehyde and *cis*-2-butene is the triplet biradical **34**. That the stereochemistry of ring closure of a 1,4-biradical

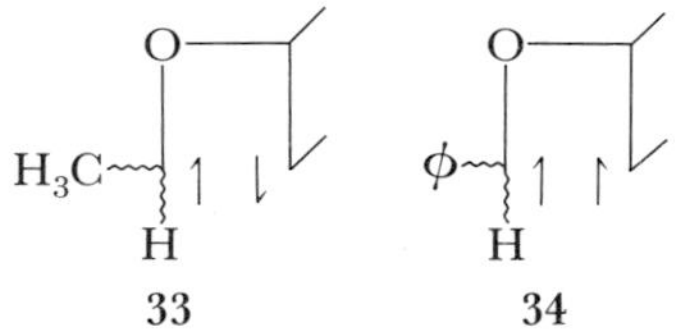

depends on the multiplicity of the biradical was elegantly shown by Bartlett and Porter. When **35** is irradiated, it decomposes through its singlet state to give the singlet 1,4-biradical **36**, which rapidly forms a bond to give predominantly **37** as shown in Equation 12.65. When **35** is photosensitized by a triplet donor, it forms the triplet biradical **39**. Before the slow but necessary process of spin inversion has time to occur, bond rotation in **39** takes place so that a mixture of products rich in

[69] D. R. Arnold, R. L. Hinman, and A. H. Glick, *Tetrahedron Lett.*, 1425 (1964).
[70] N. C. Yang and W. Eisenhardt, *J. Am. Chem. Soc.*, **93**, 1277 (1971).

$$\text{(12.65)}$$

35   **36**   95% **37**   5% **38**

38 is obtained.[71,72] The biradical **33** is analogous to **36**, whereas **34** is analogous to **39**.

$$\text{(12.66)}$$

35   **39**   61% **37**   39% **38**

The exact mechanism of the addition of aliphatic aldehydes and ketones to olefins depends on the electron density of the double bond. Electron-rich olefins react via a short-lived biradical or a concerted mechanism, whereas electron-poor olefins form a preliminary charge-transfer complex.[73]

According to pericyclic selection rules, photocyclic addition of a diene to an olefin in a $(4n + 2)$ π electron process may be concerted if the transition state has one antarafacial component. Such additions are very rare, apparently because the allowed $\pi 2s + \pi 2s$ process is much more accessible (see Section 12.1). Most photo Diels–Alder reactions that have been studied involve attack of a triplet diene on a singlet olefin and thus proceed by an intermediate biradical. Both intramolecular[74] and intermolecular[75] concerted Diels–Alder reactions have, however, been observed. For example, photolysis of **40**, through a $\pi 4s + \pi 2a$ transition state, stereospecifically forms **41**. Similarly, **42** is converted to **43**. Nevertheless, not all intramolecular Diels–Alder reactions brought about by direct irradiation are stereospecific.[76]

Photosensitized dienes dimerize to give $2 + 2$ and $2 + 4$ addition products. For example, triplet butadiene yields the three products in Equation 12.69, but the

[71] P. D. Bartlett and N. A. Porter, *J. Am. Chem. Soc.,* **90**, 5317 (1968).

[72] (a) By the law of spin conservation, the addition of a triplet to a singlet should produce a triplet biradical, which must undergo spin inversion before it can form a second bond. Woodward and Hoffman, however, have suggested that spin inversion may at times be concerted with bond formation so that an intermediate biradical need not always be formed: R. B. Woodward and R. Hoffman, *Angew. Chem. Int. Ed.,* **8**, 781 (1969), Section 6.2. Almost every known example of a triplet adding to a singlet, however, does seem to give an intermediate biradical. (b) Scaiano has recently measured the lifetime of a triplet 1,4-biradical to be ∼100 nanoseconds.: R. D. Small and J. C. Scaiano, *J. Phys. Chem.,* **81**, 828 (1977).

[73] J. A. Barltrop and H. A. J. Carless, *J. Am. Chem. Soc.,* **94**, 8761 (1972).

[74] A. Padwa, L. Brodsky, and S. Clough, *J. Am. Chem. Soc.,* **94**, 6767 (1972).

[75] H. Hart, T. Miyashi, D. N. Buchanan, and S. Sasson, *J. Am. Chem. Soc.,* **96**, 4857 (1974).

[76] See, for example; D. A. Seeley, *J. Am. Chem. Soc.,* **94**, 4378, 8647 (1972).

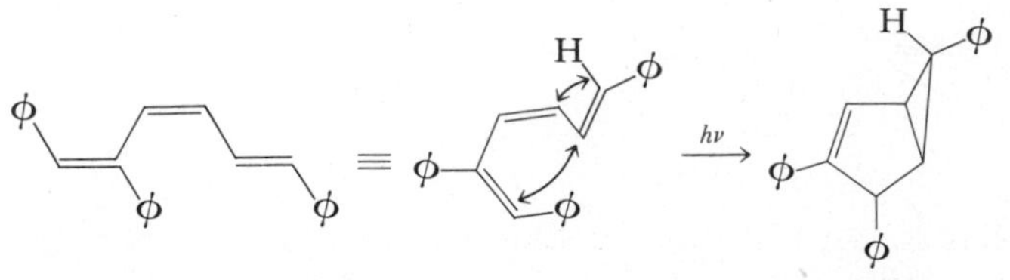

$$(12.67)$$

$$(12.68)$$

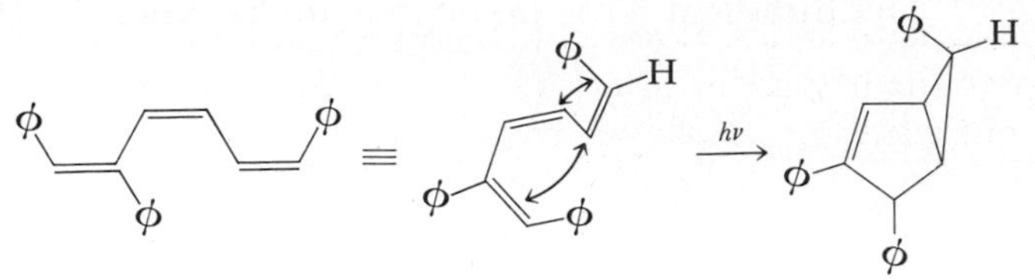

composition of the product mixture depends on the energy of the sensitizer. As shown in Table 12.1, high-energy sensitizers all produce approximately the same ratio of products, but when the sensitizer triplet energy (E_T) falls below 60 kcal mole^{-1}, the percentage of vinylcyclohexene begins to increase. Hammond has explained the product distribution on the basis of two distinct stereoisomeric diene T_1 states. In the ground state the highest filled molecular orbital of butadiene is anti-

Table 12.1 Composition of Products from Photosensitized Dimerization of Butadiene[a]

Sensitizer	Percentage distribution of dimers			E_T (kcal mole^{-1})
	44	**45**	**46**	
Xanthone	78	19	3	74.2
Acetophenone	78	19	3	73.6
Benzaldehyde	80	16	4	71.9
o-Dibenzoylbenzene	76	16	7	68.7
Benzophenone	80	18	2	68.5
2-Acetylfluorenone	78	18	4	62.5
Anthraquinone	77	19	4	62.4
Flavone	75	18	7	62.0
Micheler's ketone	80	17	3	61.0
4-Acetylbiphenyl	77	17	6	60.6
β-Naphthyl phenyl ketone	71	17	12	59.6
β-Naphthaldehyde	71	17	12	59.5
β-Acetonaphthone	76	16	8	59.3
α-Acetonaphthone	63	17	20	56.4
α-Naphthaldehyde	62	15	23	56.3
Biacetyl	52	13	35	54.9
Benzil	44	10	45	53.7
Fluorenone	44	13	43	53.3
Duroquinone	72	16	12	51.0

[a] Reproduced with permission from R. S. H. Liu, N. J. Turro, and G. S. Hammond, *J. Am. Chem. Soc.*, **87**, 3406 (1965). Copyright by the American Chemical Society.

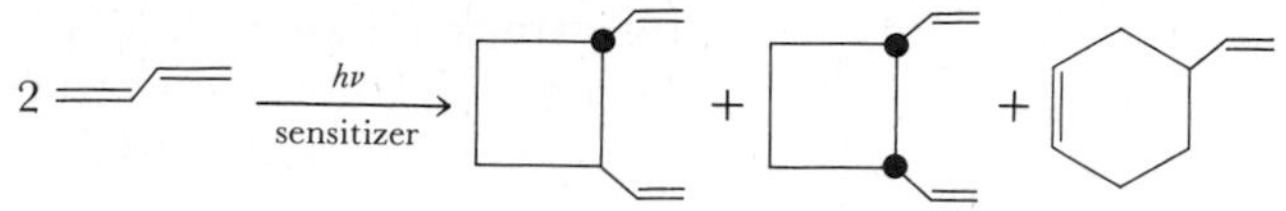

$$2 \quad \xrightarrow[\text{sensitizer}]{h\nu} \quad \boxed{\quad} \quad + \quad \boxed{\quad} \quad + \quad \bigcirc \qquad (12.69)$$

$$\underset{44}{} \qquad \underset{45}{} \qquad \underset{46}{}$$

bonding between C_2 and C_3. In the first excited state an electron has been promoted from this orbital to one that is bonding between C_2 and C_3. There is, therefore, a much greater barrier to interconversion of the s-cis (**47**) and s-trans (**48**) forms in the first excited state than in the ground state. A high-energy sensitizer transfers energy to the first diene molecule it encounters, which is more likely to be in the s-trans configuration ($>95\%$ of butadiene molecules in the ground state at

$$\underset{47}{} \qquad \underset{48}{}$$

room temperature are in that configuration because of the lower nonbonding interactions in it than in the s-cis form). Once produced, an s-trans triplet attacks a ground-state diene molecule, which is again most probably in an s-trans configuration, to form either biradical **49** or **50** (see Scheme 3). These biradicals can easily close to divinylcyclobutanes, but can only close to six-membered rings if the C_2—C_3 bond first rotates so that the alkyl groups on it are close to an s-cis configuration. This bond rotates slowly because it has appreciable double-bond character, and therefore not much vinylcyclohexene is produced.

Sensitizers whose triplet energies are less than 60 kcal mole^{-1} are of lower energy than the butadiene s-trans triplet and they must therefore seek out an s-cis diene (triplet energy of s-cis butadiene is 53 kcal mole^{-1}) before they can transfer

SCHEME 3

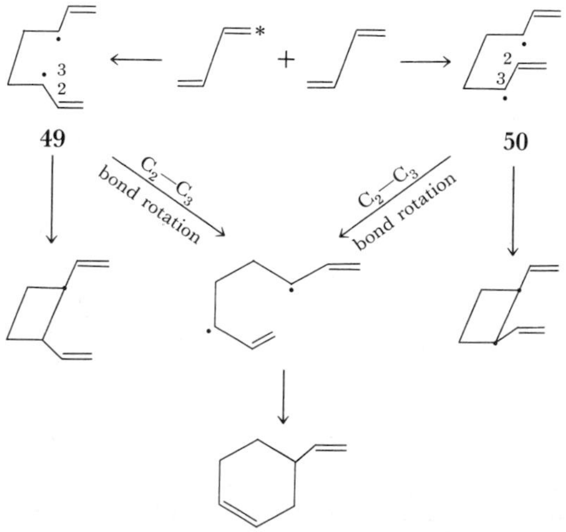

their energy. The biradical produced by attack of an *s*-cis triplet on an *s*-trans ground-state molecule can form a cyclohexene without bond rotation (see Scheme 4), and thus the relative amount of cyclohexene increases. When the energy of the triplet sensitizer falls below 53 kcal, the relationship between product composition and sensitizer energy becomes complicated because of nonvertical energy transfer.[77]

SCHEME 4

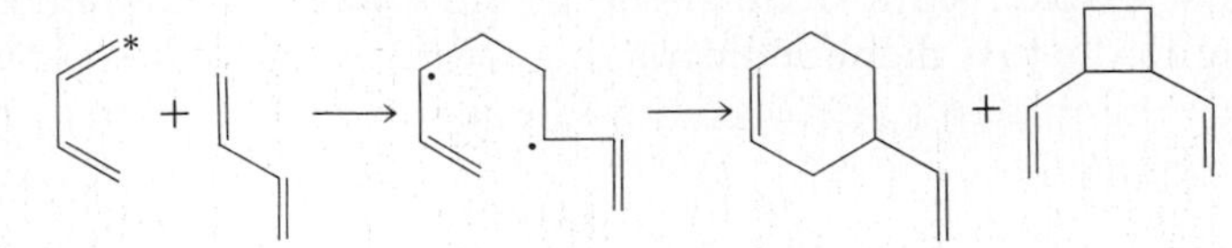

Photosensitized dienes also add to simple olefins; the relative amount of 2 + 2 and 2 + 4 addition product again depends on the energy of the triplet sensitizer. For example, in the photosensitized addition of butadiene to trifluoroethylene (Equation 12.70), the percentage of the 2 + 4 adduct (**51**) increases from 0.5 percent when acetophenone is used as the photosensitizer to 22.5 percent when fluorenone is the photosensitizer.[78]

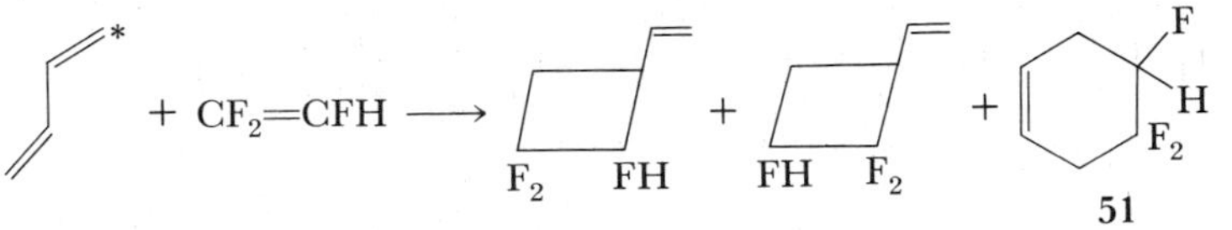

$$\text{(12.70)}$$

Norrish Type I Reaction[68,79]

Most aldehydes and ketones in inert solvents or in the gas phase undergo one or two photoreactions, called Norrish Type I and Norrish Type II processes.

The Norrish Type I reaction, shown in Equation 12.71, may originate from

$$R\!-\!\overset{\overset{\displaystyle O}{\|}}{C}\!-\!R' \xrightarrow{h\nu} R\!-\!\overset{\overset{\displaystyle O}{\|}}{C}\!\cdot\; +\; \cdot R' \longrightarrow \quad \text{products (decarbonylation,} \qquad \text{(12.71)}$$
$$\text{recombination, and disproportionation)}$$

either the S_1 or T_1 state of the carbonyl compound. For example, Type I cleavage of excited cyclohexanone in benzene solution is efficiently quenched by 1,3-dienes, a group of compounds whose S_1 states lie above the carbonyl S_1 states but whose T_1 states are of *lower* energy than the carbonyl T_1 states. A Stern–Volmer plot of $1/\Phi_{pro}$ vs. the concentration of the diene is a good straight line over all diene concentrations. Thus all α cleavage occurs from the T_1 excited state.[80] On the other hand, a

[77] R. S. H. Liu, N. J. Turro, Jr., and G. S. Hammond, *J. Am. Chem. Soc.*, **87**, 3406 (1965).
[78] P. D. Bartlett, B. M. Jacobson, and L. E. Walker, *J. Am. Chem. Soc.*, **95**, 146 (1973).
[79] J. S. Swenton, *J. Chem. Educ.*, **46**, 217 (1969).
[80] P. J. Wagner and R. W. Spoerke, *J. Am. Chem. Soc.*, **91**, 4437 (1969).

Stern–Volmer plot for diene quenching of the Norrish Type I reaction of methyl *t*-butyl ketone shows that in this compound α cleavage is more complicated. At low concentrations of diene there is an inverse relationship between quantum yield of products and concentration of quencher, but at high diene concentrations the quenching becomes markedly less efficient and the curve decreases in slope.[81] Reaction apparently takes place from T_1 and S_1, although much more efficiently from T_1. At high diene concentrations, where most of the triplet reaction has already been quenched, inefficient singlet quenching sets in. The ability of a diene to quench the S_1 state of carbonyl compounds has been shown to depend on its ionization potential, and therefore singlet quenching probably occurs via an exciplex.[82]

Norrish Type I reactions occur from T_1 only if the T_1 state is $^3(n, \pi^*)$. Compound **52** has a phosphorescence lifetime of 10^{-3} sec; furthermore, there is a mirror image relationship between its closely spaced phosphorescence and absorption spectra, indicating that its S_1 and T_1 states are similar in conformation and in energy. Compound **53**, however, has a phosphorescence lifetime of 5.5 sec, and there is a large Stokes shift between its dissimilar absorption and phosphorescence spectra. Apparently both compounds have a lowest (n, π^*) singlet, but whereas (n, π^*) is also the lowest triplet of **52**, $^3(\pi, \pi^*)$ is even lower than $^3(n, \pi^*)$ in **53**. The

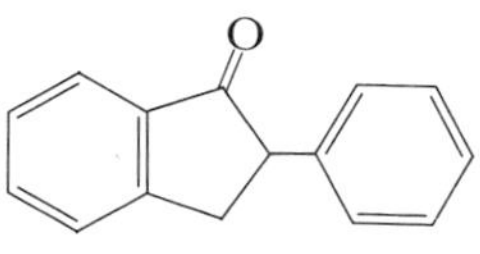

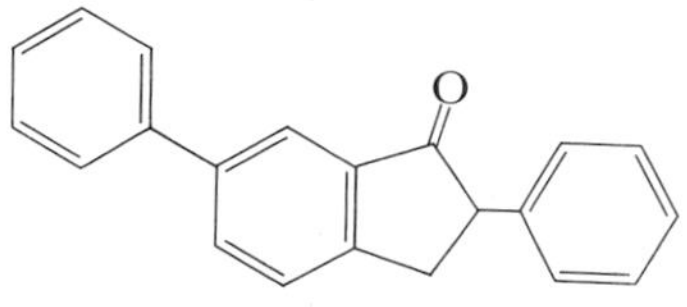

difference in photochemical behavior between the two compounds is striking. Upon irradiation with >320-nm-wavelength light for 9 hr, **52** forms the products

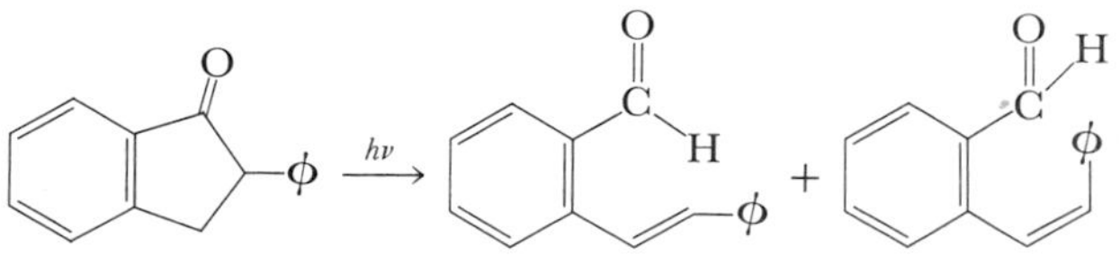

$$\tag{12.72}$$

in Equation 12.72 in 86 percent yield. Under the same conditions **53** gives less than 5 percent aldehydic material and is recovered unchanged in 91 percent yield.[83]

Which of the two R—C bonds cleave in the Norrish Type I reaction of an unsymmetrical ketone depends on the relative stabilities of the two possible R · radicals. Diaryl ketones do not react at all; and **52** cleaves to give a benzyl rather

[81] N. C. Yang and E. D. Feit, *J. Am. Chem. Soc.,* **90**, 504 (1968).
[82] N. C. Yang, M.-H. Hui, and S. A. Bellard, *J. Am. Chem. Soc.,* **93**, 4056 (1971).
[83] A. A. Baum, *J. Am. Chem. Soc.,* **94**, 6866 (1972).

than an aryl radical. From ethyl methyl ketone, an ethyl radical is formed considerably more frequently than a methyl radical.[84]

The fact that the initial cleavage is reversible has been long suspected from the fact that the quantum yield for Norrish Type I products never approaches one, even for very reactive ketones. Positive evidence for the back reaction was provided by Barltrop, who showed that Type I cleavage of *cis*-2,3-dimethylcyclohexanone is accompanied by its isomerization to *trans*-2,3-dimethylcyclohexanone.[85]

Photochemical cleavage of a bond α to a carbonyl group is not unique to aldehydes and ketones. For example, phenyl acetate, from its S_1 state, forms the products in Equation 12.73.[86]

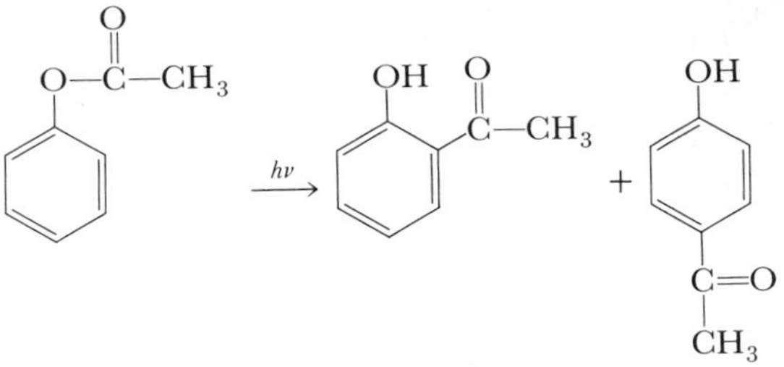

$$ (12.73) $$

Norrish Type II Reaction[68,79,87]

When aldehydes or ketones are irradiated in hydroxylic solvents, they are often reduced to the corresponding alcohol or pinacol. The reduction of benzophenone by isopropanol is shown in Scheme 5. Notably unreactive in such photoreductions

Scheme 5

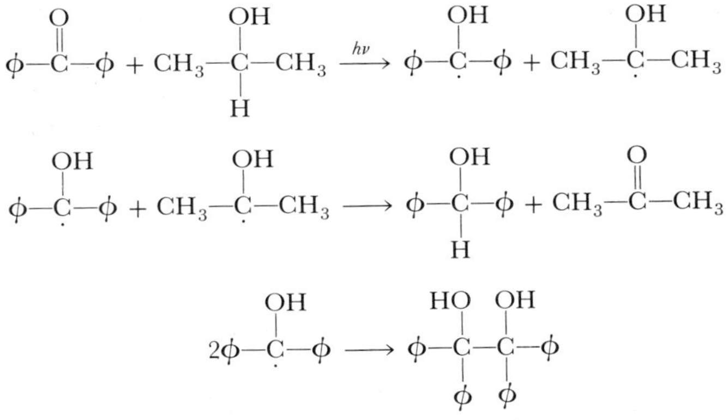

[84] J. N. Pitts and F. E. Blacet, *J. Am. Chem. Soc.*, **72**, 2810 (1950).
[85] J. A. Barltrop and J. D. Coyle, *J. Chem. Soc., Chem. Commun.*, 1081 (1969).
[86] (a) J. W. Meyer and G. S. Hammond, *J. Am. Chem. Soc.*, **92**, 2187 (1970); (b) F. A. Carroll and G. S. Hammond. *J. Am. Chem. Soc.*, **94**, 7151 (1972).
[87] P. J. Wagner, *Acc. Chem. Res.*, **4**, 168 (1971).

are aromatic carbonyl compounds whose lowest T_1 state is mostly π, π^* rather than n, π^* in character. An exception to this rule occurs when the reaction is carried out at a high enough temperature so that reaction can occur by thermal population of T_2—the state with a large amount of n, π^* character.[88]

The intramolecular analog of this reaction is called the *Norrish Type II reaction*. When an excited aldehyde or ketone has a γ hydrogen, intramolecular hydrogen abstraction via a six-membered ring transition state usually occurs. The resulting 1,4-biradical may either cleave or cyclize to give the Norrish Type II products of Scheme 6.

SCHEME 6

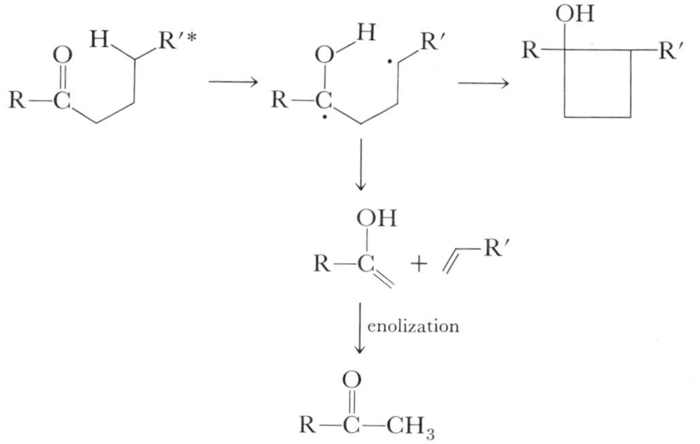

Aliphatic ketones generally undergo Type II reaction from their singlet and triplet states simultaneously, as can be seen from the fact that only part of the reaction can be efficiently quenched by 1,3-dienes. The rates of intersystem crossing for most aliphatic ketones are similar to one another, and therefore the percentage of the singlet reaction depends on how fast it can occur; 5-methyl-2-hexanone (**54**), which has a relatively weak γ-C—H bond, reacts mostly from the singlet state, but 2-pentanone (**55**) reacts mostly from the triplet.[89] Aromatic ketones, which undergo

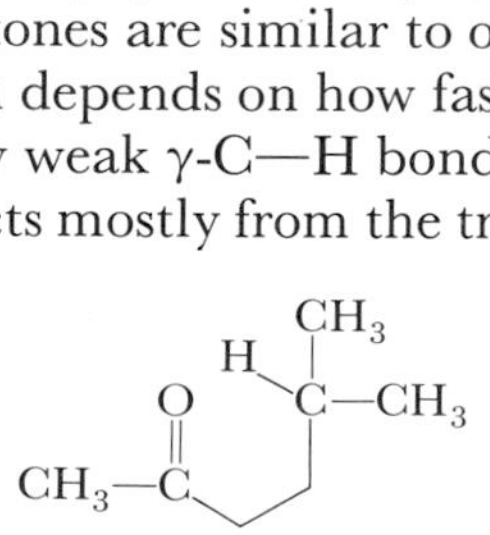

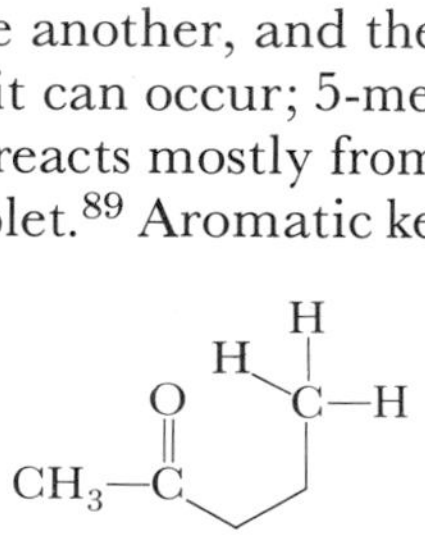

intersystem crossing with great efficiency, give Type II (and Type I) reactions solely from the triplet state. Like the products of intermolecular hydrogen abstraction, Type II products are largely suppressed if the lowest triplet is π, π^*.[90]

[88] M. Berger, E. McAlpine, and C. Steel, *J. Am. Chem. Soc.*, **100**, 5147 (1978).
[89] N. C. Yang, S. P. Elliott, and B. Kim, *J. Am. Chem. Soc.*, **91**, 7551 (1969).
[90] P. J. Wagner and A. E. Kemppainen, *J. Am. Chem. Soc.*, **90**, 5898 (1968).

That the triplet reaction involves a biradical as shown in Scheme 6 has been convincingly demonstrated. For example, (4S)-(+)-4-methyl-1-phenyl-1-hexanone (**56**), on irradiation in benzene, gives the products shown in Scheme 7 with a total quantum yield of only 0.2. The optical activity of recovered "unreacted" ketone indicates that photoracemization has also occurred with a quantum yield of 0.78 ± 0.05. The most reasonable explanation for these observations is that a biradical intermediate is formed with a quantum yield of one, and that the biradical either forms products or returns to racemized starting material.[91]

SCHEME 7

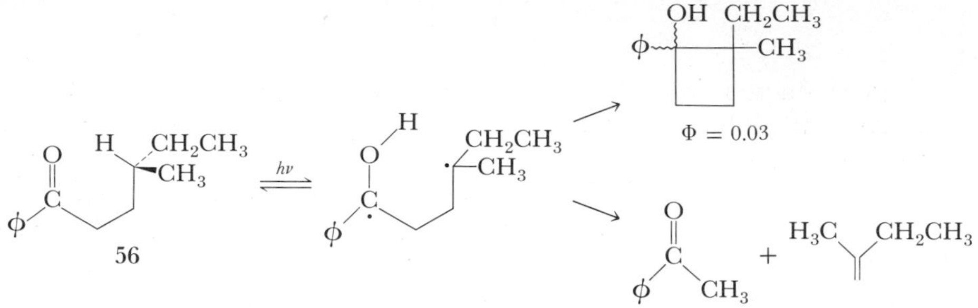

Norrish Type II biradicals have also been trapped. For example, irradiation of a benzene solution of **57**, to which a high concentration of butanethiol-S-d has been added, results in quenching of the Norrish Type II reaction and extensive incorporation of deuterium at the γ carbon of recovered **57**. The thiol, a good hydrogen donor, supplies a deuterium to the radical **58** before **58** can give products or return to starting material as shown in Scheme 8.[92]

SCHEME 8

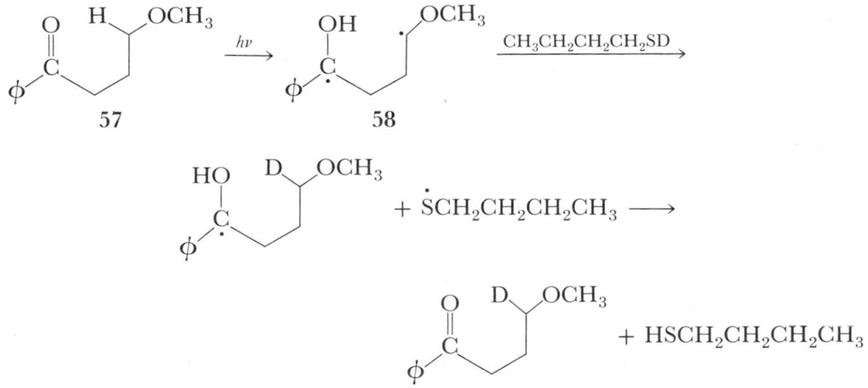

[91] P. J. Wagner, P. A. Kelso, and R. G. Zepp, *J. Am. Chem. Soc.*, **94**, 7480 (1972).
[92] (a) P. J. Wagner and R. G. Zepp, *J. Am. Chem. Soc.*, **94**, 287 (1972). See also (b) P. J. Wagner and K.-C. Liu, *J. Am. Chem. Soc.*, **96**, 5952 (1974).

Norrish Type II processes from the triplet state involve a biradical even when concerted reaction would be exothermic. Photolysis of **59** gives the products in Equation 12.74. Concerted reaction with the formation of triplet stilbene (a process corresponding to Equation 12.36) would be exothermic, but triplet stilbene is not formed: the T_1 state of stilbene decays to a 60:40 cis–trans mixture, but photolysis of **59** gives 98.6 percent *trans*-stilbene.[91]

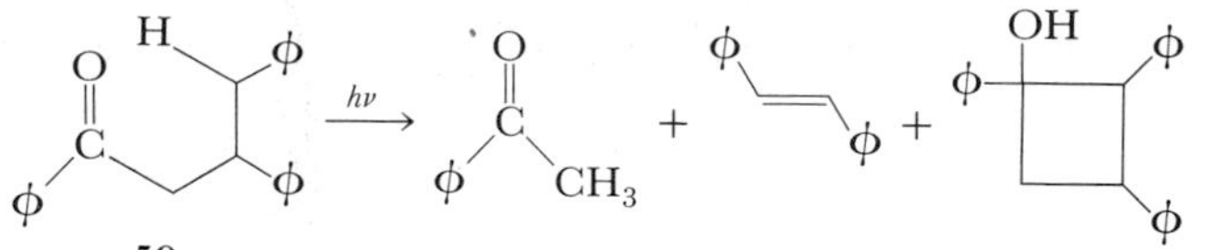

(12.74)

59

Type II reaction from the singlet state is also known to involve a biradical—even though a singlet biradical has been neither trapped nor diverted. Photolysis of *threo*-4-methyl-2-hexanone-5-d_1 (**60**), in the presence of large amounts of piperylene, yields *cis*- and *trans*-2-butene. If reaction were concerted, transfer of H should form cis isomer and transfer of D should form trans isomer. Yet transfer of H gives 10 percent *trans*- and 90 percent *cis*-2-butene. Apparently an intermediate singlet biradical (**61**) is formed, which either cleaves to olefin of retained stereochemistry (cis) or undergoes bond rotation. The rotated biradical can cleave to the trans olefin as shown in Scheme 9. When the photolysis is carried out in the absence of piperylene, triplet-state reaction predominates. Under these conditions the formation of the butenes is much less stereospecific.[93] The shorter lifetime of the singlet 1,4-biradicals is in agreement with the work of Bartlett and Porter.

SCHEME 9

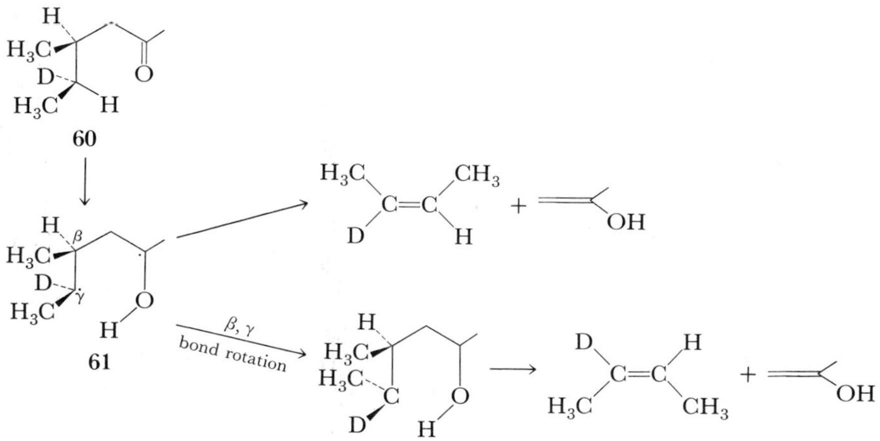

[93](a) C. P. Casey and R. A. Boggs, *J. Am. Chem. Soc.,* **94,** 6457 (1972). See also (b) L. M. Stephenson, P. R. Cavigli, and J. L. Parlett, *J. Am. Chem. Soc.,* **93,** 1984 (1971).

The ratio of Type II fragmentation to Type II cyclization products may depend strongly on the excited state from which reaction occurs. The lowest-energy pathway for fragmentation requires continual orbital overlap between developing p bonds. Cyclobutanol formation, however, has less stringent orbital orientation requirements. When the configuration of the ketone is unfavorable to fragmentation, relatively more fragmentation occurs from the more exothermic, less selective, singlet-state reaction.[94]

If an aldehyde or ketone has an abstractable γ hydrogen, Norrish Type I and Norrish Type II reactions are, of course, competitive. Aliphatic ketones, which have no α side chains (**62**), undergo exclusively Type II reaction in solution;[95] but aliphatic t-butyl ketones (**63**), which can form a stable t-butyl radical on α cleavage, undergo predominantly Type I reaction.[96] Phenyl aliphatic ketones (**64**) form Type I products much more slowly than **63**. The probable reason is that the low triplet energy of **64** makes α cleavage of it almost a thermoneutral reaction, whereas α cleavage of **63** is exothermic by about 5 kcal.[97]

We have now seen that the reactivities of the various low-lying states of the carbonyl compound in Type I and Type II reactions are very similar: Ketones with lowest n, π^* states are very reactive from their $^3(n, \pi^*)$ and less reactive from their $^1(n, \pi^*)$ state; ketones with lowest (π, π^*) states are very unreactive.

The greater reactivities of the n, π^* as opposed to the π, π^* states might simply be due to the greater reactivity of a state in which the oxygen is partially electron-deficient. For the Type II reaction the relative reactivities are also in agreement with the predictions of a diagram correlating the states of the starting material with the states of the intermediate biradical.[98] The problem is that the same sort of diagram for the Type I reaction predicts high reactivity for the π, π^* states.

Let us begin with the Norrish Type II process and let us make the reasonable assumption that during the reaction the original carbonyl group, the α, β, and γ carbons, and the hydrogen being transferred all lie in one plane as shown in **65**. The reaction, then, has as a symmetry element the plane of the ring. Figure 12.14 shows the state correlation diagram for the Norrish Type II process of a ketone that has lowest n, π^* excited states. On the right-hand side are shown the low-lying states of

[94] I. Flemming, A. V. Kemp-Jones, and E. J. Thomas, *J. Chem. Soc., Chem. Commun.*, 1158 (1971).

[95] H. E. O'Neal, R. G. Miller, and E. Gunderson, *J. Am. Chem. Soc.*, **96**, 3351 (1974).

[96] N. C. Yang and E. D. Feit, *J. Am. Chem. Soc.*, **90**, 504 (1968).

[97] (a) P. J. Wagner and J. M. McGrath, *J. Am. Chem. Soc.*, **94**, 3849 (1972); (b) F. D. Lewis and T. A. Hilliard, *J. Am. Chem. Soc.*, **94**, 3852 (1972).

[98] (a) W. G. Dauben, L. Salem, and N. Turro, *Acc. Chem. Res.*, **8**, 41 (1975); (b) A. Devaquet, *Pure Appl. Chem.*, **41**, 455 (1975); (c) A. Devaquet, A. Sevin, and B. Bigot, *J. Am. Chem. Soc.*, **100**, 2009 (1978); see also (d) M. J. S. Dewar and C. Doubleday, *J. Am. Chem. Soc.*, **100**, 4935 (1978).

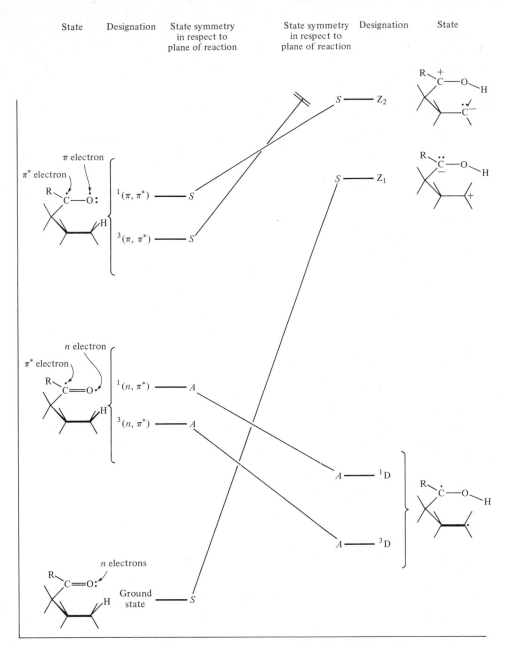

Figure 12.14 State correlation diagram for the first step of the Norrish Type II reaction for a ketone with lowest n, π^* states. The states are described by the electronic configuration of lowest energy in that state. On the left are the states of the starting ketone and on the right, those of the intermediate. D stands for biradical and Z for zwitterion.

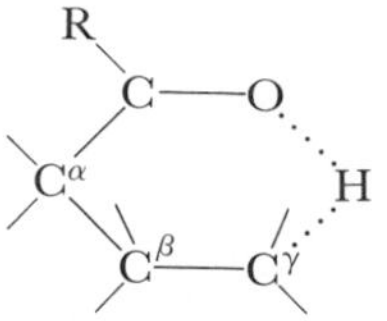

R
C—O
C$^\alpha$
H
C$^\beta$—C$^\gamma$

65

the intermediate together with their overall electronic symmetries in respect to the plane of the reaction. The biradical intermediates (in which two orbitals of approximately equal energy are each singly occupied) are of lower energy than the zwitterions (in which one of the orbitals is doubly occupied and the other is empty) in accordance with Hund's rule. Also in accordance with Hund's rule, the triplet biradical is of lower energy than the singlet. Note that the two biradical states are overall antisymmetric since each has one electron in a p orbital perpendicular to the reaction plane (which contributes an antisymmetric term) and one in a σ orbital in the reaction plane (which contributes a symmetric term). The zwitterionic states are both symmetric since neither has a half-filled orbital.

The low-lying states (again assuming no configuration interaction) of the starting ketone are shown on the left side of Figure 12.14. The (n, π^*) states are antisymmetric since each has one electron in a p orbital perpendicular to the reaction plane and one in a nonbonding orbital in the reaction plane. The (π, π^*) states are symmetric since there are two electrons in p orbitals perpendicular to the reaction plane.

The $^3(n, \pi^*)$ state of the ketone correlates with the lowest energy state of the intermediate, the triplet biradical, and thus reaction from this state is energetically favorable. The $^1(n, \pi^*)$ state of the ketone correlates with a state of the intermediate that is not much higher in energy, the singlet biradical. The (π, π^*) states of the starting material, on the other hand, correlate with states of the intermediate that are considerably more energetic.

What about ketones with lowest π, π^* excited states? The only alteration that would have to be made in Figure 12.14 would be to interchange the two π, π^* states with the two n, π^* states. The state symmetries would, of course, remain unchanged. Thus the π, π^* states would still correlate with those highly energetic states of the intermediate with which they correlate in Figure 12.14 and Norrish Type II reaction from such a ketone would be strongly disfavored energetically. Thus the predictions of state correlation diagrams for the relative reactivities of the (n, π^*) and (π, π^*) states of ketones in the Norrish Type II is in accord with experimental observation.

Now let us turn to the Norrish Type I process. The intermediate in this reaction is the cleavage product (**66**), an acyl and an alkyl or an aryl radical. The

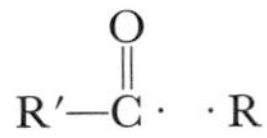

O
‖
R′—C· ·R
66

acyl radical can have its odd electron in a σ or a π orbital, **67σ** or **67π**, respectively. Numerous esr experiments have shown that **67σ** is of lower energy.[99] Figure 12.15,

then, shows the state correlation diagram for the Norrish Type I reaction of a ketone with lowest n, π^* excited states. The symmetry element for the reaction is the plane of the carbonyl carbon and its substituents.

Figure 12.15 implies that the facile Type I reaction from the $^3(n, \pi^*)$ state of ketones involves an internal conversion onto the $^3(\pi, \pi^*)$ potential energy surface which in turn correlates with the ground-state of the biradical cleavage product. The $^1(n, \pi^*)$ state would be of lower reactivity because, although its potential energy surface also crosses the $^3(\pi, \pi^*)$ energy surface, a spin inversion is required at the crossing.

Now let us turn to the Norrish Type I reaction of aryl ketones that have lowest π, π^* excited states. Again, the only alteration that would have to be made in Figure 12.15 would be to interchange the two π, π^* states and the two n, π^* states. Since the $^3(\pi, \pi^*)$ state correlates directly with the ground state of the biradical, we would expect such ketones to be very reactive in Norrish Type I reactions from their T_1 states. But this is exactly contrary to what is found experimentally!

Rearrangements

The scope of photochemical rearrangements is vast. Because an excited molecule is often distorted relative to its ground-state geometry and because the excited molecule is energy-rich, a large number of rearrangements that do not occur in the ground state originate from excited states. Equations 12.75[100] and 12.76[101] show examples of photochemical rearrangements that lead to highly strained products. (The rearrangements in this case are valence isomerizations—two intramolecular 2 + 2 cycloadditions and an electrocyclic ring closure.) In Equation 12.75 Dewar benzene is formed from the S_2 state of benzene and benzvalene from the S_1 state. Often rearrangement products cannot undergo photochemical reversion because

$$\text{(12.75)}$$

Dewar benzene   benzvalene

[99](a) M. B. Yim, O. Kikuchi, and D. E. Wood, *J. Am. Chem. Soc.*, **100**, 1869 (1978), and references therein; (b) P. J. Krusic and T. A. Rettig, *J. Am. Chem. Soc.*, **92**, 722 (1970).
[100]D. Bryce-Smith, A. Gilbert, and D. A. Robinson, *Angew. Chem. Int. Ed. Engl.*, **10**, 745 (1971), and references therein.
[101]G. O. Schenck and R. Steinmetz, *Chem. Ber.*, **96**, 520 (1963).

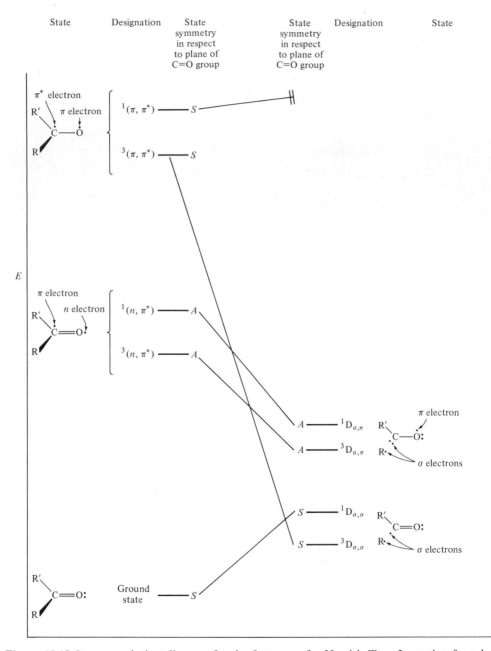

Figure 12.15 State correlation diagram for the first step of a Norrish Type I reaction for a ketone with lowest n, π^* states. The states are described by the electronic configuration of lowest energy in that state. On the left are the states of the starting ketone and on the right, those of the intermediate. D stands for biradical. The first subscript after the D denotes the alkyl fragment, the second, the acyl fragment.

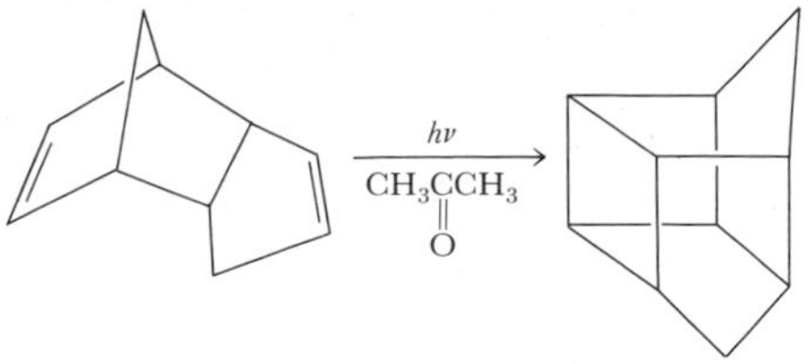

$$\xrightarrow[\substack{CH_3CCH_3 \\ \| \\ O}]{hv}$$

(12.76)

they lack a low-lying excited state and thus are stable if formed and stored at low temperatures.

It is impossible in this chapter even to scratch the surface of photochemical rearrangements. Examples of electrocyclic and sigmatropic rearrangements have already been cited in Sections 11.2, 11.3, and earlier in this chapter. Here we shall limit ourselves to the di-π-methane rearrangement and to cyclohexadiene rearrangements; furthermore, our glimpse of these reactions will be fleeting. The reader who is interested in looking further should consult W. A. Noyes, G. S. Hammond, and J. N. Pitts, Jr., Eds., *Advances in Photochemistry* (Wiley, New York, 1963 ff.) and O. L. Chapman, Ed., *Organic Photochemistry* (Dekker, New York, 1966 ff.), and should also consult the current literature because new rearrangements are discovered every day. It should be stressed that one must use care in applying the pericyclic selection rules to photochemical rearrangements, since many rearrangements occur from the T_1 state and are not concerted.

Compounds that have a structure such as that shown in **68** upon direct irradiation undergo what is called the *di-π-methane rearrangement* to form **69**.[102] Photosensitized **68** rearranges either inefficiently or not at all, indicating that the reactive state of acyclic systems is S_1. Apparently the T_1 state of **68** dissipates its

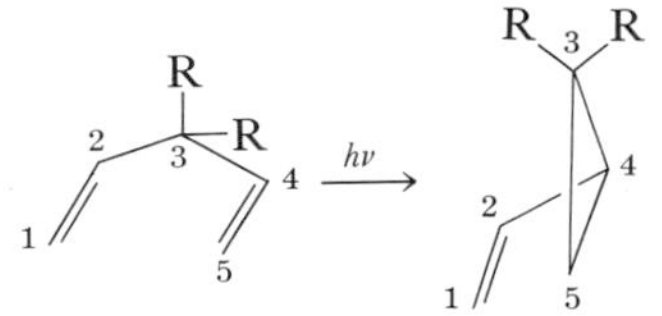

68   **69**

(12.77)

energy by free rotation about the unconstrained π bond. In rigidly constrained systems, however, the di-π-methane rearrangement originates efficiently from T_1 but inefficiently or not at all from S_1. For example, barrelene (**70**), when photosensitized by acetone, rearranges to semibullvalene (**71**), but on direct irradiation yields cyclooctatetraene (**72**).[103]

When reaction is from the singlet state, the rearrangement is stereospecific at carbons 1 and 5. For example, *cis*-1,1-diphenyl-3,3-dimethyl-1,4-hexadiene (**73**) rearranges to **74**, in which the side chain is cis, but *trans*-1,1-diphenyl-3,3-dimethyl-

[102]For a review, see S. S. Hixson, P. S. Mariano, and H. E. Zimmerman, *Chem. Rev.*, **73**, 531 (1973).
[103]H. E. Zimmerman, R. W. Binkley, R. S. Givens, G. L. Grunewald, and M. A. Sherwin, *J. Am. Chem. Soc.*, **91**, 3316 (1969).

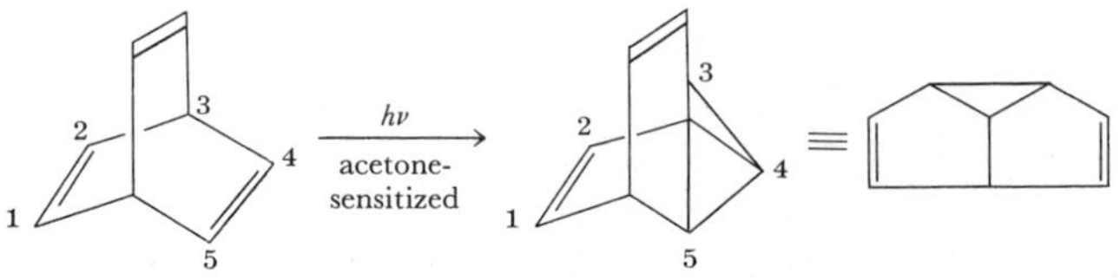

70 71 (12.78)

70 72 (12.79)

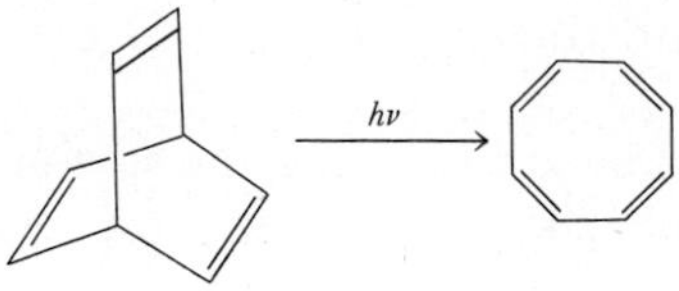

1,4-hexadiene (**75**) rearranges to **76**, in which the side chain is trans.[104] Similarly **77** and **79** give predominantly **78** and **80**, respectively.[105]

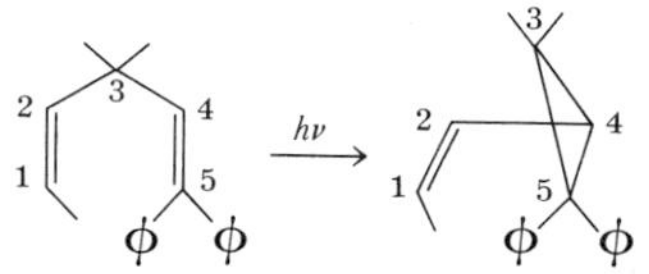

73 74 (12.80)

75 76 (12.81)

77 78 (12.82)

79 80 (12.83)

In order to make it clear what bonding changes are occurring, a stepwise mechanism for the di-π-methane rearrangement is shown in Equation 12.84. However, in any specific rearrangement that takes place from S_1, some or all of the steps

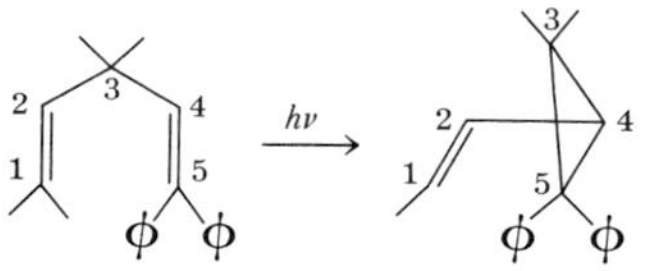

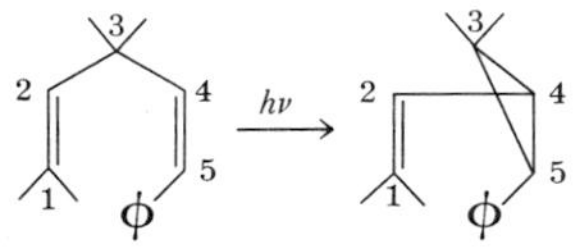

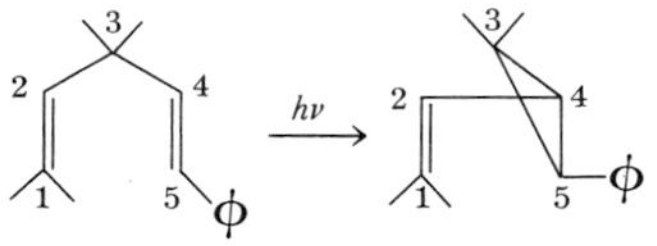

[104] H. E. Zimmerman and A. C. Pratt, *J. Am. Chem. Soc.,* **92,** 6267 (1970).
[105] H. E. Zimmerman, P. Baeckstrom, T. Johnson, and D. W. Kurtz, *J. Am. Chem. Soc.,* **94,** 5504 (1972); **96,** 1459 (1974).

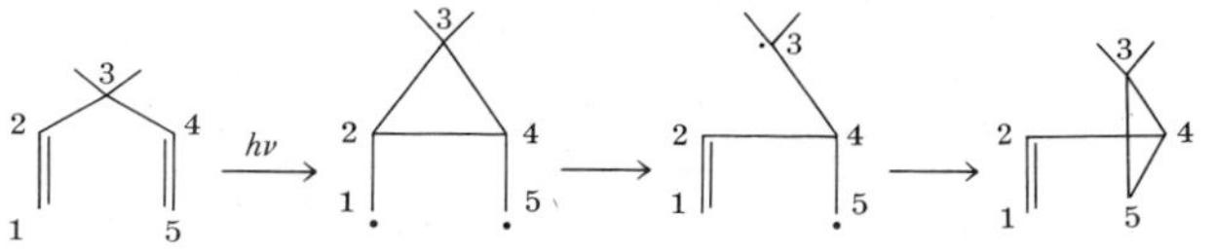

$$(12.84)$$

may merge into one. Zimmerman has proposed the $\pi 2a + \pi 2a + \sigma 2a$ mechanism shown in Figure 12.16 for the singlet-state reaction.[106] This concerted mechanism accounts for the stereospecificity shown in Equations 12.80–12.83 and also for the fact that the stereochemistry at C_3 is found to be inverted by rearrangement from the singlet state.[107] At the transition state, bonding between C_5 and C_3 is minimal. This must be so because of the regiospecificity of the reaction.[108] Note, for example, that in Reactions 12.80–12.83 it is the phenyl-substituted double bond that always becomes part of the cyclopropane ring.[108] This is readily understood if, at the transition state, there is substantial radical character at C_5. Phenyl substituents can stabilize a radical better than alkyl substituents and as a result the carbon bearing the phenyls becomes one of the cyclopropyl ring carbons.

The triplet-state reaction apparently does involve the stepwise mechanism of Equation 12.84. This mechanism is shown for the barralene–semibullvalene conversion (Equation 12.78) in Equation 12.85. One piece of evidence for this

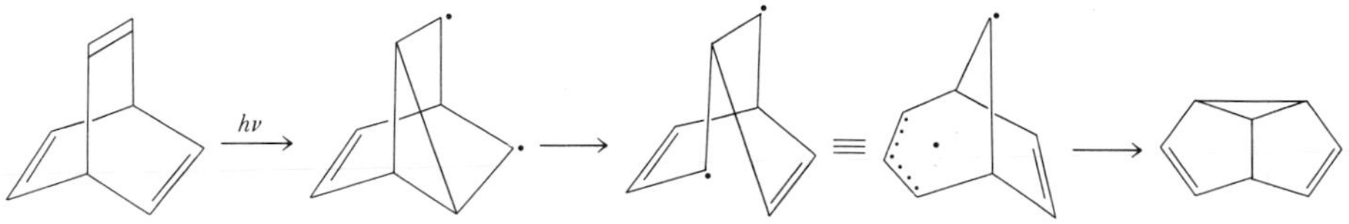

$$(12.85)$$

70   **81**   **71**

mechanism is that independent generation of the triplet diradical **81** by photosensitized decomposition of the barrelene azo compound **82** leads to semibullvalene as the main end product (Equation 12.86). As one would expect from the stepwise

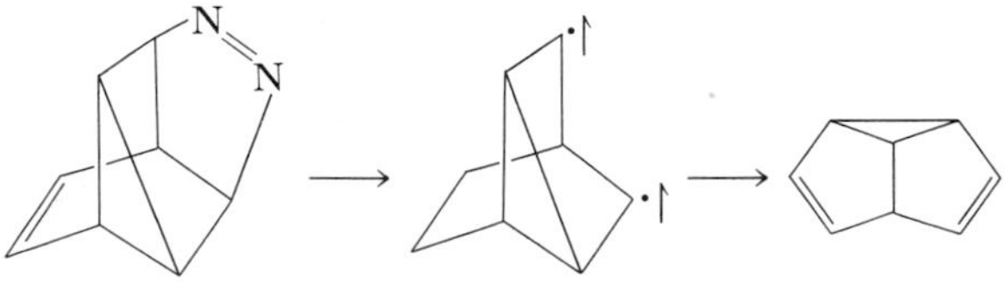

$$(12.86)$$

82   **81**   **71**

mechanism, rearrangement from the triplet state is also highly regiospecific, always giving those products that would be formed from the most stable biradical.[109]

[106] H. E. Zimmerman and R. D. Little, *J. Am. Chem. Soc.,* **94,** 8256 (1972).

[107] H. E. Zimmerman, J. D. Robbins, R. D. McKelvey, C. J. Samuel, and L. R. Sousa, *J. Am. Chem. Soc.,* **96,** 1974, 4630 (1974).

[108] See also (a) H. E. Zimmerman and B. R. Cotter, *J. Am. Chem. Soc.,* **96,** 7445 (1974); (b) H. E. Zimmerman and T. R. Welter, *J. Am. Chem. Soc.,* **100,** 4131 (1978).

[109] (a) L. A. Paquette, A. Y. Ku, C. Santiago, M. D. Rozeboom, and K. N. Houk, *J. Am. Chem. Soc.,* **101,** 5972 (1979); (b) A. Y. Ku, L. A. Paquette, M. D. Rozeboom, and K. N. Houk, *J. Am. Chem. Soc.,* **101,** 5981 (1979).

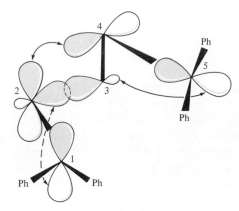

Figure 12.16 $\pi 2a + \pi 2a + \sigma 2a$ mechanism for the di-π-methane rearrangement.

When C_3 is unsubstituted, the di-π-methane rearrangement may still take place but, because of the greater difficulty in breaking the C_2—C_3 bond under these circumstances, a different stepwise mechanism involving hydrogen–radical shifts is followed.[110]

The photochemical rearrangements of cross-conjugated cyclohexadienones in general, and of 4,4-diphenylcyclohexadienone (**83**) in particular, have been intensively studied.[111] When **83** is irradiated in dioxane–water, first 6,6-diphenyl-bicyclo[3.1.0]hex-3-one-2 (**84**) is obtained which, on further irradiation, forms **85**, **86**, and **87**. The primary photorearrangement product (**84**) can also be obtained by photosensitization of **83**, but not by irradiation of **83** in piperylene. Therefore **84** is formed from the lowest triplet of **83**. The subsequent rearrangements of **84** appar-

$$\text{(12.87)}$$

$$\begin{array}{ccccc} \textbf{83} & \textbf{84} & \textbf{85} & \textbf{86} & \textbf{87} \end{array}$$

ently proceed from two different excited states: Formation of the acid (**87**) cannot be entirely quenched by triplet quenchers, whereas formation of the phenols (**85** and **86**) can. Thus **87** is formed partly from S_1, but **85** and **86** are formed from T_1.

Zimmerman proposed the mechanisms in Schemes 10 and 11 to account for the rearrangements. The zwitterionic species **88** and **89** were postulated to resolve an inconsistency. The rearrangements in the final steps of Schemes 10 and 11 are

[110](a) See note 62(a); (b) H. E. Zimmerman and J. A. Pincock, *J. Am. Chem. Soc.*, **94**, 6208 (1972).

[111]For reviews, see (a) P. J. Kropp, in *Organic Photochemistry*, Vol. 1, O. L. Chapman, Ed., Dekker, New York, 1966, p. 1; (b) O. L. Chapman, in *Advances in Photochemistry*, Vol. 1, W. A. Noyes, G. S. Hammond, and J. N. Pitts, Jr., Eds., Wiley, New York, 1963, p. 183; (c) H. E. Zimmerman, in *Advances in Photochemistry*, Vol. 1, 1963, p. 183; (d) D. I. Schuster, *Acc. Chem. Res.*, **11**, 65 (1978).

SCHEME 10

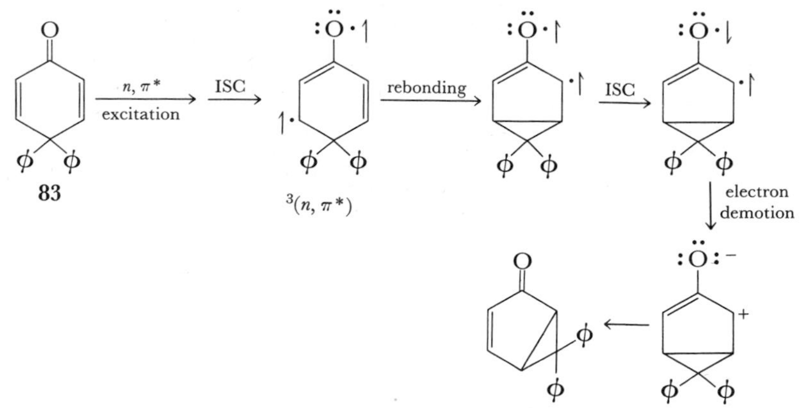

83           $^3(n, \pi^*)$ 84   88

typical of migrations to electron-deficient carbon. However, n, π^* states usually have electron-rich π systems. Migratory aptitudes are also typical of those to carbocations. For example, although p-cyanophenyl is a better migrating group than phenyl in n, π^* excited states, when **90** is irradiated, only phenyl migrates.

SCHEME 11

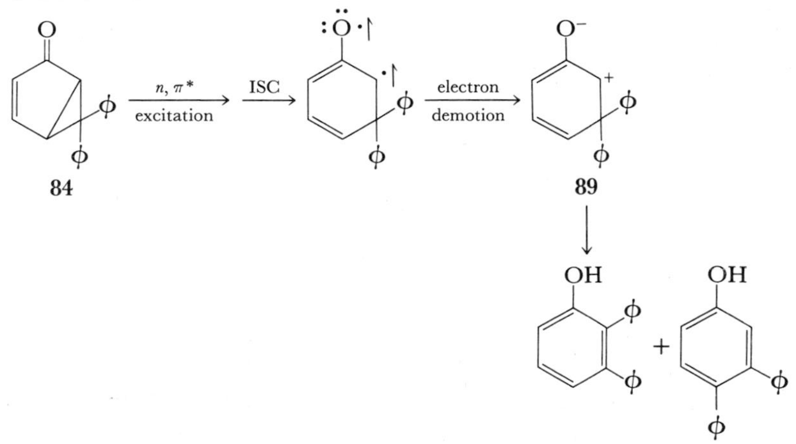

84   89   85   86

     Zimmerman has independently synthesized the proposed zwitterions **88** and **89** by the attack of potassium t-butoxide on **91** and **92**, respectively. **84** is formed from the zwitterion **88** in 74 percent yield.[112] Furthermore, **89** yields **85** and

[112]H. E. Zimmerman and D. S. Crumrine, *J. Am. Chem. Soc.*, **90**, 5612 (1968).

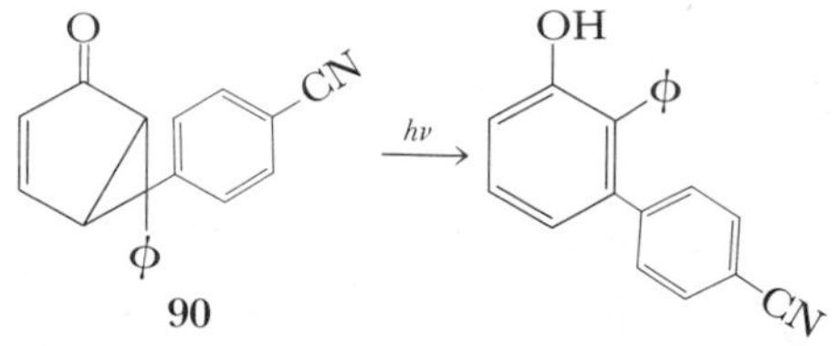

(12.88)

86, and although the ratio of **85** to **86** is solvent-dependent, it varies in precisely the same way as the ratio of **85** to **86** produced by photoreaction in different solvents.[113]

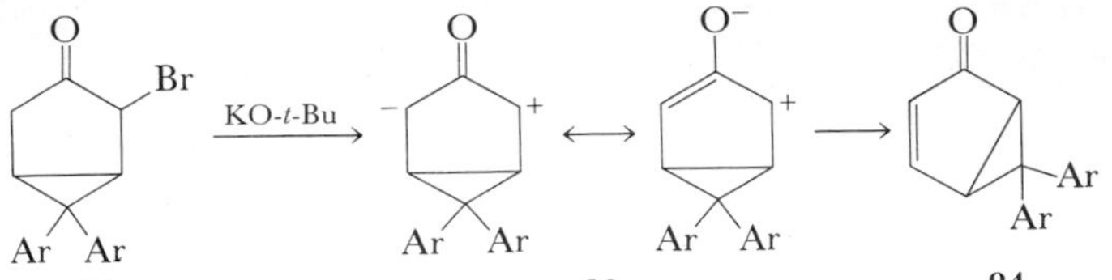

(12.89)

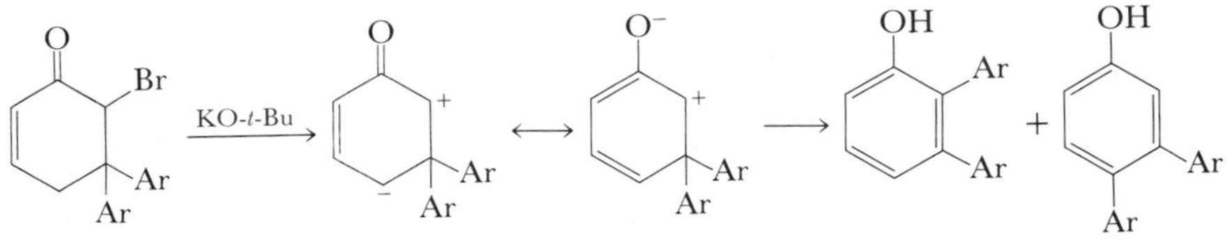

(12.90)

12.5 THERMAL GENERATION OF EXCITED STATES

Under special circumstances a molecule may be promoted to an electronic excited state by thermal rather than electromagnetic energy. The thermal energy cannot however, come from simply heating the compound. Electronic excited states are generally so energetic in comparison to the ground states that decomposition would occur long before electronic excitation. For example, the Boltzmann distribution predicts that a temperature of $\sim 8500°C$ would be required to achieve a 1 percent population of the T_1 state of acetone (78 kcal mole^{-1} above the ground state). The thermal energy required for the excitation process, then, must come from a chemical process. When light is emitted from an excited state that has been generated chemically, the phenomenon is called *chemiluminescence*.

Figure 12.17 shows the energetic requirement for chemical generation of excited states. On the left side of the figure is a reaction coordinate diagram for the ground-state reaction $R_0 \rightarrow P_0$. On the right side of the figure is shown the energy of an excited state of P, P*, relative to P_0. P* may be formed from R_0 if the transition state for the reaction $R_0 \rightarrow P_0$ lies at a higher energy than P*. From Figure 12.17 it is clear that chemical generation of an excited state may occur if the

---

[113]H. E. Zimmerman and G. A. Epling, *J. Am. Chem. Soc.*, **94**, 7806 (1972).

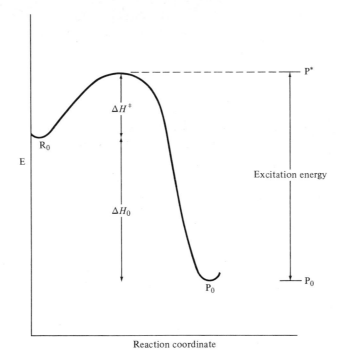

Figure 12.17 Energetic requirements for chemiluminescence. If $\Delta H_0 + \Delta H^\ddagger$ for the reaction $R_0 \leftarrow P_0$ is greater than the energy difference between the ground state and an excited state of P, then the reaction of $R_0 \rightarrow P_0$ may lead to that excited state of P rather than to its ground state.

starting material, R_0, is very energetic or if the activation energy is very high or if both of these conditions obtain.

The well-established mechanisms for thermal excitation usually fall in one of three categories. In the first a peroxide undergoes concerted elimination of singlet oxygen. Oxygen is, of course, one of the few molecules that have triplet ground states. Thus the singlet oxygen is in an excited state. Common precursors for thermal generation of singlet oxygen are *endoperoxides*—compounds formed by the photochemical 1,4-addition of O_2 to cyclic dienes. Singlet oxygen is formed thermally when the endoperoxides undergo symmetry allowed retro-Diels–Alder reaction. An example is shown in Equation 12.91.[114] How does this reaction fit the energetic

$$\xrightarrow{\Delta} {}^1O_2^* + $$

(12.91)

[114]H. H. Wasserman, J. R. Scheffer, and J. L. Cooper, *J. Am. Chem. Soc.*, **94**, 4991 (1972).

criterion for formation of excited states as defined in Figure 12.17? The starting peroxide is not particularly energetic in relation to ground-state products but the activation barrier for peroxide decomposition is quite high—approximately 28 kcal mole^{-1}—and in this case the excited state formed, $^1O_2^*$, is of relatively low energy: The S_1 state of O_2 is only 23 kcal mole^{-1} above the ground state. (Radiation of such low energy has a wavelength greater than that of visible light and thus emission of a visible photon from this state is impossible. Chemiluminescence has, however, been observed after energy transfer to a suitable acceptor molecule from *dimers of singlet oxygen*.[115])

A second category of thermal reactions that sometimes produce electronic excited states is pericyclic-forbidden cleavage reactions of high-energy starting materials. Because symmetry forbidden reactions usually have high activation barriers, such reactions are sometimes able to meet the energetic requirements for electronic excitation (see Figure 12.17). For example, tetramethyl-1,2-dioxetane (**93**) undergoes thermolysis to yield two acetone molecules in 100 percent yield.

$$\text{(Equation 12.92)}$$

Approximately 30 percent of the starting dioxetane molecules produce one electronically excited acetone molecule.[116] The $\Delta H^{\ddagger}$ of reaction is 27 kcal mole^{-1} and the ΔH_0 is -63 kcal mole^{-1}. Thus the transition state of the reaction lies approximately 90 kcal mole^{-1} above the ground state of the products. Since the S_1 and T_1 states of acetone are approximately 85 and 78 kcal mole^{-1}, respectively, above its ground state, either singlet or triplet acetone could be formed in the ring-opening process. It was found by both emission and trapping studies that the ratio of triplet to singlet acetone formed is $\sim 100:1$ and that the triplet is formed directly—that is, not after intersystem crossing from the singlet. For example, when **93** is thermolyzed in the presence of *trans*-dicyanoethene, two products are formed: *cis*-dicyanoethene and oxetane (Equation 12.93). The isomerization reaction and the addition reaction are processes that were shown, in separate experiments, to occur with only triplet and with only singlet acetone molecules, respectively. Thus from the yields of these two products the yield of singlet and triplet acetone in the decomposition reaction could be ascertained. Furthermore, when a deaereated sample of **93** is thermolyzed, chemiluminescence is observed that is almost pure acetone phosphorescence (λ_{max} 430 nm). If oxygen, an excellent triplet quencher, is admitted, the chemiluminescence intensity immediately drops to about 1 percent of its former value and is pure acetone fluorescence (λ_{max} 400 nm).

[115]A. U. Khan and M. Khasha, *J. Am. Chem. Soc.*, **88**, 1574 (1966).
[116](a) N. J. Turro, P. Lechtken, N. E. Schore, G. Schuster, H.-C. Steinmetzer, and A. Yekta, *Acc. Chem. Res.*, **7**, 97 (1974); (b) W. Adam, C.-C. Cheng, O. Cueto, K. Sakanishi, and K. Zinner, *J. Am. Chem. Soc.*, **101**, 1324 (1979).

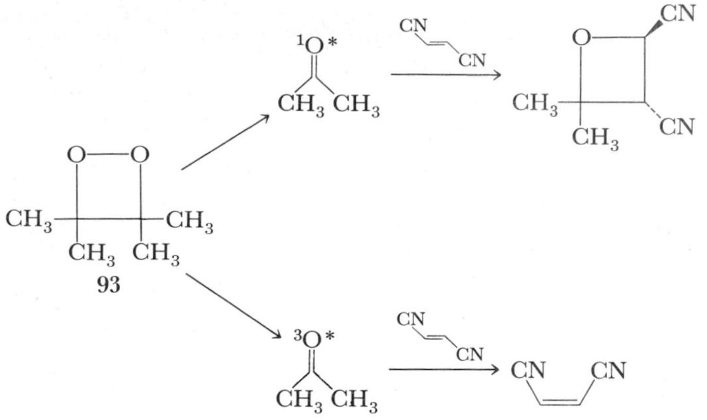

$$(12.93)$$

The excitation mechanism most likely is that shown in Equation 12.94. Rate-determining oxygen–oxygen bond cleavage generates a singlet biradical that easily, by spin–orbit coupling, is converted to the almost isoenergetic triplet biradical. This then cleaves to a ground-state and an excited-state acetone.[117]

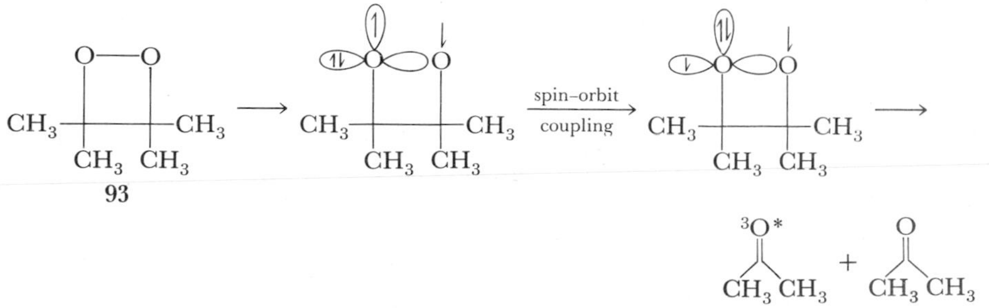

$$(12.94)$$

Another pericyclic-forbidden process that leads to excited-state product is the thermolysis of Dewar benzene (Equation 12.95). Heating Dewar benzene gives ground-state, and in very low yield (0.1 to .01 percent) triplet, benzene.[118] The reaction does not give direct chemiluminescence because triplet benzene does not phosphoresce efficiently. However, indirect chemiluminescence can be observed by energy transfer from triplet benzene to a suitable acceptor.

The third category of thermal electronic excitations is electron-exchange

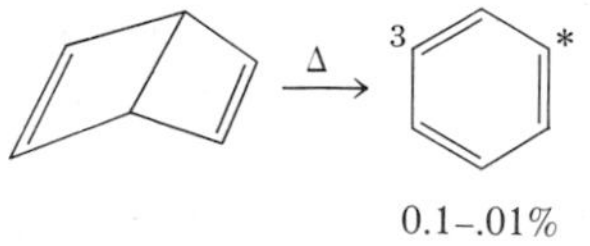

$$(12.95)$$

0.1–.01%

[117](a) W. H. Richardson, F, C. Montgomery, M. B. Yelvington, and H. E. O'Neal, *J. Am. Chem. Soc.,* **96,** 7525 (1974); (b) J.-Y. Koo and G. B. Schuster, *J. Am. Chem. Soc.,* **99,** 5403 (1977); (c) K. A. Horn and G. B. Schuster, *J. Am. Chem. Soc.,* **100,** 6649 (1978); (d) L. B. Harding and W. A. Goddard III, *J. Am. Chem. Soc.,* **99,** 4520 (1977); (e) N. Suzuki, *Angew. Chem. Int. Ed. Engl.,* **18,** 787 (1979).
[118]P. Lechtken, R. Breslow, A. H. Schmidt, and N. J. Turro, *J. Am. Chem. Soc.,* **95,** 3025 (1973).

excitation.[119] In this process a radical cation and a radical anion collide and the radical ions are annihilated as shown in Equation 12.96. If the combined oxidation potential of A and reduction potential of D are greater than the energy of the lowest excited state of D or A, an excited state may be formed in the collision. Figure 12.18 shows the annihilation–excitation step in terms of orbital occupancy.

$$D^{\cdot -} + A^{\cdot +} \longrightarrow D + A \tag{12.96}$$

The radical cation and the radical anion may be generated electrochemically at an anode and cathode, respectively,[120] or they may be generated in a chemical reaction.[119] For example, when diphenoyl peroxide (**94**) is heated, CO_2 splits out of the peranhydride and benzocoumarin (**95**) is formed (Equation 12.97). Although this process is sufficiently exothermic to permit formation of electronically excited **95**, no chemiluminescence could be found. However, when 9,10-diphenylanthracene was added to the reaction system, it was not consumed but it

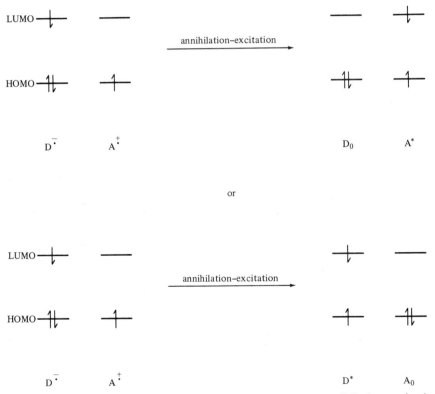

Figure 12.18 The two possible changes in orbital occupancy during an annihilation–excitation state of two oppositely charged radical ions.

[119]G. B. Schuster, *Acc. Chem. Res.*, **12**, 366 (1979); See also E. H. White, M. G. Steinmetz, J. Miamo, P. D. Wildes and R. Morland, *J. Am. Chem. Soc.*, **102**, 3199 (1980).
[120](a) D. M. Hercules, *Science*, **145**, 808 (1964); (b) E. A. Chandross and F. I. Sonntag, *J. Am. Chem. Soc.*, **86**, 3179 (1964); (c) K. S. V. Santhanam and A. J. Bard, *J. Am. Chem. Soc.*, **87**, 139 (1965).

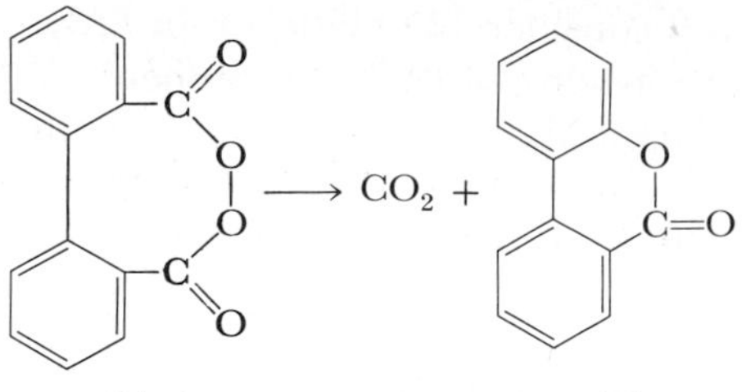

$$\longrightarrow CO_2 + \qquad\qquad (12.97)$$

94   95

(a) catalyzed the decomposition and (b) fluoresced efficiently from its S_1 state! Schuster proposed the mechanism shown in Scheme 12 for the excitation of the diphenyl anthracene.

SCHEME 12

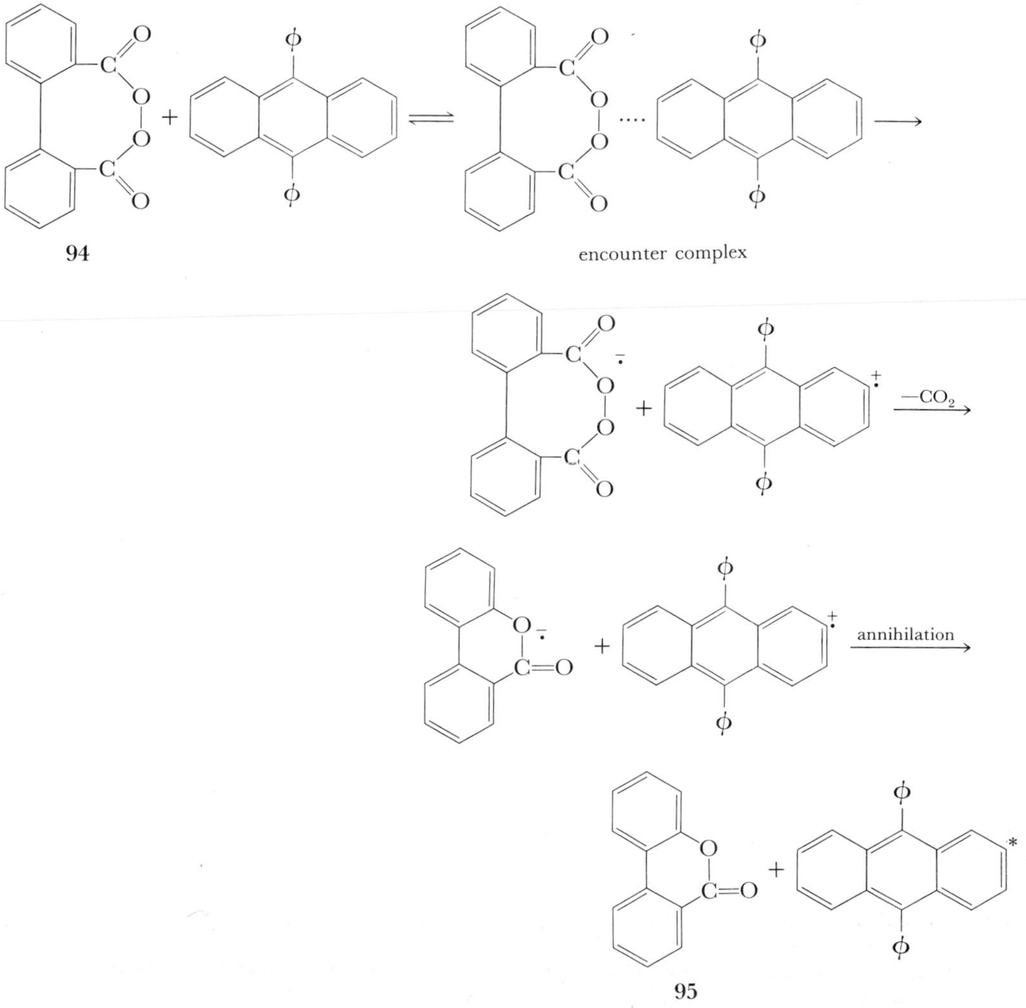

94   encounter complex

$-CO_2$

annihilation

95

Schuster has found considerable evidence in favor of this mechanism. For example, aromatic hydrocarbons other than diphenylanthracene accelerate the rate of decomposition of **94** and also fluoresce. Their efficiency at catalyzing the reaction depends inversely on their oxidation potential. This is exactly what one would predict from Scheme 12 if any step other than the first were rate determining. Figures 12.19 and 12.20 give the pertinent data. Figure 12.19 shows how the

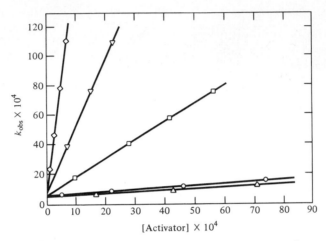

Figure 12.19 Dependence of the observed pseudo first-order rate constants for reaction of diphenoyl peroxide (**94**) on the concentration of various aromatic catalysts: △ = coronene; ○ = diphenylanthracene; □ = perylene; ▽ = naphthacene; ◇ = rubrene. Reprinted with permisssion from G. B. Schuster, *Acc. Chem. Res.*, **12**, 366 (1979). Copyright 1979 American Chemical Society.

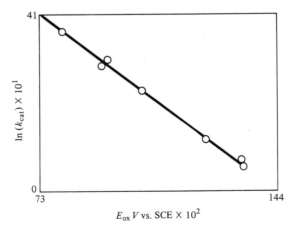

Figure 12.20 Correlation of the magnitude of k_2 for catalyzed decomposition of **94** with the one-electron oxidation potential of the aromatic catalysts. Reprinted with permission from G. B. Schuster, *Acc. Chem. Res.*, **12**, 366 (1979). Copyright 1979 American Chemical Society.

pseudo first-order rate constants at high diphenoyl peroxide concentration depend on the concentration of catalyst for a number of different catalysts. The lines all have a nonzero intercept and thus the observed rate constant can be expressed by Equation 12.98.

$$k_{obs} = k_1 + k_2[\text{catalyst}] \tag{12.98}$$

The intercept, k_1, is the rate constant for unassisted cleavage of the peranhydride. The slopes of the lines, the k_2's, differ by a factor of ~ 150, with rubrene being a much more effective catalyst than diphenylanthracene. Figure 12.20 shows the relationship between the natural logarithm of the k_2's of the various catalysts in Figure 12.19 as well as some other aromatic catalysts and the oxidation potentials of these catalysts. A straight line with negative slope is obtained.

The reaction responsible for firefly bioluminescence is shown in Equation 12.99. Schuster has suggested that this reaction also occurs by a CIEEL (chemi-

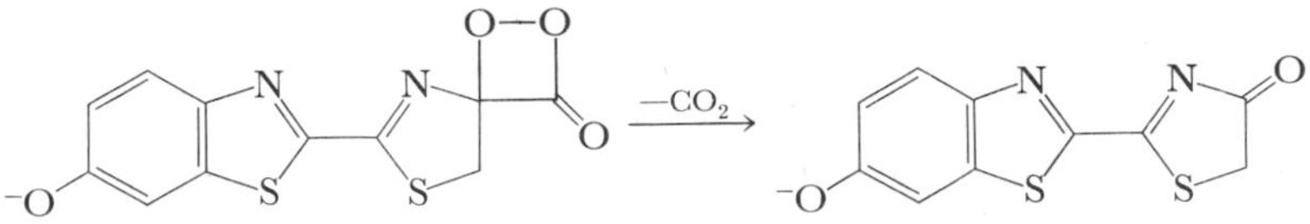

$$\tag{12.99}$$

cally initiated electron-exchange luminescence) mechanism, this time intramolecular.[119]

PROBLEMS

1. What is a doublet?

2. When a mixture of I_2 and O_2 is irradiated with light that only excites I_2, an excited state of O_2 is produced by collisional energy transfer. What is that excited state likely to be?

3. Which will have the longer τ_p°, naphthalene or iodonaphthalene?

4. Free radicals readily quench phosphorescence. This phenomenon may be due to a variety of factors. Explain in terms of short-range collisional energy transfer.

5. Although **1a** undergoes the usual anthracene dimerization, **1b** does not. Explain.

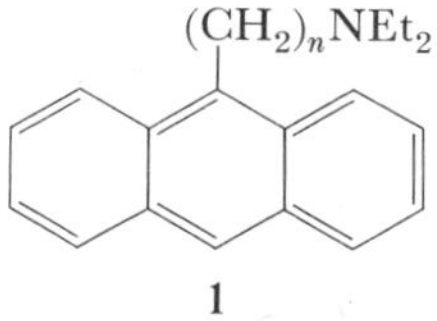

$$(CH_2)_n NEt_2$$

1

a $n = 1$
b $n = 3$

6. (a) Predict the products that will be formed from Reaction 1. (b) Do you expect the product ratio to depend on the sensitizer energy?

$$\tag{1}$$

7. Predict the predominant product in the following reaction:

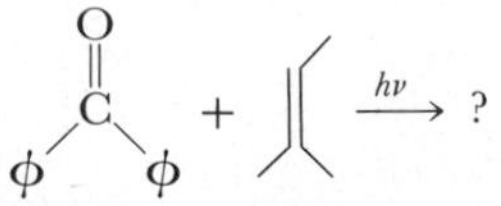

$$\phi-\underset{\underset{\phi}{\|}}{C}-\phi \quad + \quad \underset{}{\big\backslash\!\!\big\backslash} \quad \xrightarrow{h\nu} \quad ? \tag{2}$$

8. The combined quantum yields for Norrish Type I and Norrish Type II processes from **2** are similar to those from **3**. However, **2** forms predominantly Norrish Type II product and **3** forms predominantly Norrish Type I product. Suggest an explanation.

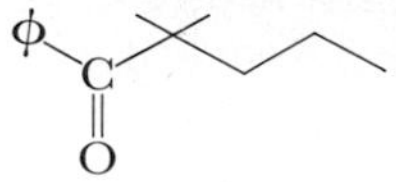

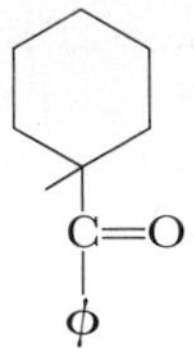

    2   3

9. On irradiation, **4** gives **5** plus the usual Norrish Type II products. Explain.

$$\phi-\underset{\underset{}{\|}}{\overset{O}{C}}-CH_2CH_2CH_2CH_2OCH_3$$

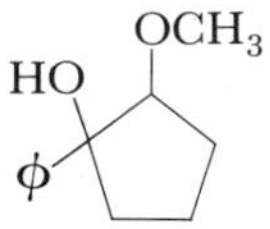

       4   5

10. What would be the product of the di-π-methane rearrangement of **6**?

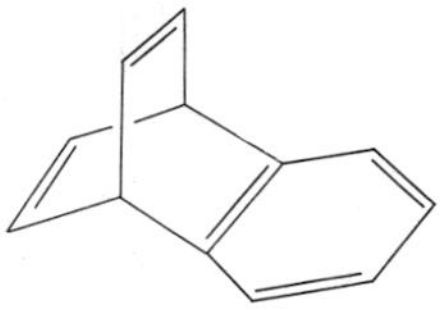

       6

11. The quantum yield for dimerization of acenaphthalene (**7**) increases from 0.01 to 0.18 as the concentration of ethyl iodide is increased from 0 to 10 mole percent. Explain.

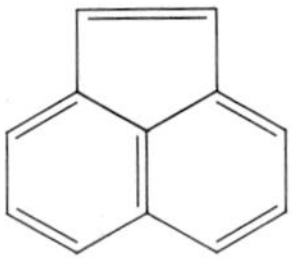

       7

12. The lowest spectroscopic triplet of ketone (**8**) is 68 kcal mole^{-1} above the ground state. 1,3-Cyclohexadiene ($E_T = 53$ kcal mole^{-1}) efficiently quenches the triplet state of **8** but piperylene ($E_T = 59$ kcal mole^{-1}) and naphthalene ($E_T = 61$ kcal mole^{-1}) do not. Explain.

8

13. Compound **9** photodimerizes to give head-to-head and head-to-tail dimers. When the reaction is run in the presence of the triplet quencher cyclooctatetraene and the quantum yields of the dimers are plotted against the concentration of cyclooctatetraene, Figure 12.21 is obtained. Explain the significance of Figure 12.21.

9

14. Propose a mechanism for the following transformation:

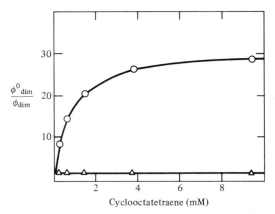

Figure 12.21 Quantum yields of photodimers of **9** in the presence of cyclooctatetraene: ○ = head-to-head dimer; △ = head-to-tail dimer. Reprinted with permission from M. J. Hopkinson, W. W. Schloman, Jr., B. F. Plummer, E. Wenkert, and M. Raju, *J. Am. Chem. Soc.*, **101**, 2157 (1979). Copyright 1979 American Chemical Society.

15. Both anthracene and 9,10-dicyanoanthracene form fluorescent exciplexes with 1,3-cyclohexadiene. The fluorescence for the anthracene exciplex is only slightly red-shifted on change to a more polar solvent, but the 9,10-dicyanoanthracene exciplex is very strongly red-shifted on the same change. Reaction from the anthracene exciplex gives the $4\pi + 4\pi$ cycloaddition product shown in Equation 3, whereas the 9,10-dicyanoanthracene gives the $4\pi + 2\pi$ cycloaddition shown in Equation 4. Explain.

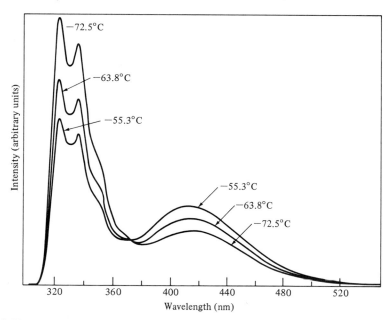

16. When the fluorescence of **10** is studied at a number of different temperatures, the spectra shown in Figure 12.22 are obtained. Explain this figure.

Figure 12.22 Fluorescence spectra of **10** at a number of temperatures. Reprinted with permission from M. Goldenberg, J. Emert, and H. Morawetz, *J. Am. Chem. Soc.*, **100**, 7171 (1978). Copyright 1978 American Chemical Society.

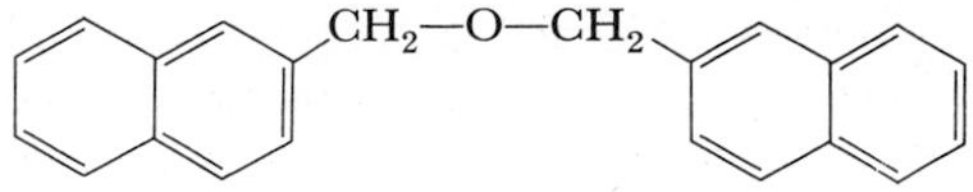

$$\text{CH}_2\text{—O—CH}_2$$

10

REFERENCES FOR PROBLEMS

2. J. Olmsted and G. Karal, *J. Am. Chem. Soc.*, **94**, 3305 (1972).
3. D. S. McClure, *J. Chem. Phys.*, **17**, 905 (1949).
4. R. E. Schwerzel and R. A. Caldwell, *J. Am. Chem. Soc.*, **95**, 1382 (1973).
5. D. R. G. Brimage and R. S. Davidson, *J. Chem. Soc., Chem. Commun.*, 1385 (1971).
6. B. D. Kramer and P. D. Bartlett, *J. Am. Chem. Soc.*, **94**, 3934 (1972).
7. D. R. Arnold, R. L. Hinman, and A. H. Glick, *Tetrahedron Lett.*, 1425 (1964).
8. F. D. Lewis and R. W. Johnson, *J. Am. Chem. Soc.*, **94**, 8914 (1972).
9. P. J. Wagner, P. A. Kelso, A. E. Kemppainen, and R. G. Zepp, *J. Am. Chem. Soc.*, **94**, 7500 (1972).
10. H. E. Zimmerman, R. S. Givens, and R. M. Pagni, *J. Am. Chem. Soc.*, **90**, 4191, 6096 (1968).
11. D. O. Cowan and J. C. Koziar, *J. Am. Chem. Soc.*, **96**, 1229 (1974).
12. D. I. Schuster, *Acc. Chem. Res.*, **11**, 65 (1978).
13. M. J. Hopkinson, W. W. Schloman, Jr., B. F. Plummer, E. Wenkert, and M. Raju, *J. Am. Chem. Soc.*, **101**, 2157 (1979).
14. D. I. Schuster and D. J. Patel, *J. Am. Chem. Soc.*, **90**, 5145 (1968).
15. N. C. Yang, K. Srinivasachar, B. Kim, and J. Libman, *J. Am. Chem. Soc.*, **97**, 5006 (1975).
16. M. Goldenberg, J. Emert, and H. Morawetz, *J. Am. Chem. Soc.*, **100**, 7171 (1978).

INDEX

Senior authors specifically named in the text are indexed, as are compounds and compound types specifically named.

Symbols: *fig.*, figure; *n.*, note; *pr.*, problem, followed by the problem number; *t.*, table.